A Science on the Scales

The Rise of Canadian Atlantic Fisheries Biology, 1898–1939

Jennifer M. Hubbard

UNIVERSITY OF TORONTO PRESS
Toronto Buffalo London

Toronto Buffalo London
Printed in Canada

ISBN 0-8020-8859-7

Printed on acid-free paper

Library and Archives Canada Cataloguing in Publication

Hubbard, Jennifer Mary, 1960–
A science on the scales : the rise of Canadian Atlantic fisheries biology, 1898–1939 / Jennifer M. Hubbard.

Includes bibliographical references and index.
ISBN 0-8020-8859-7

1. Fisheries – Canada – History. 2. Fisheries Research Board of Canada – History. 3. Marine biology – Canada – History. I. Title.

SH229.H83 2006 639.2′0971 C2005-900427-4

University of Toronto Press acknowledges the financial assistance to its publishing program of the Canada Council and the Ontario Arts Council.

University of Toronto Press acknowledges the financial support for its publishing activities of the Government of Canada through the Book Publishing Industry Development Program (BPIDP).

This book has been published with the help of a grant from the Canadian Federation for the Humanities and Social Sciences, through the Aid to Scholarly Publications Programme, using funds provided by the Social Sciences and Humanities Research Council of Canada.

This book is fondly dedicated to Eric Mills, historian and oceanographer, who has done so much to foster the history of oceanography and marine science as a field and as a community.

Contents

Acknowledgments

There are many to whom I wish to offer thanks for their support and scholarly advice. Dr Trevor Levere has my gratitude for his gentle encouragement of both my graduate studies and the completion of this work. I wish especially to thank Dr Eric Mills, a former mentor and a historian of oceanography, for his support and the ideas he has shared with me, and for his enthusiastic endeavours in creating a community of historians of oceanography and marine science, through personal communications and the forum he has created in his *History of Oceanography* newsletter. Dr M.P. Winsor was always a source of exciting and stimulating ideas. Dr Steven Turner, who gave me a wonderful introduction to the history of science in the first place, has always given the best kind of encouragement. Dr Vera Schwach of the Norwegian Institute for Studies in Research and Higher Education kindly offered comments and corrections concerning the work of Johan Hjort in chapter 3. I especially owe a debt of gratitude to Dr Sandra McRae, who pointed me toward the history of Canadian marine science, and had more inspiring ideas for potential research topics than anyone I have ever met. Dr Rudy Stocek also deserves a special note here, as he introduced me, in an undergraduate course on fisheries biology, many years ago, to critical perspectives on this science that have since proved to be of enormous usefulness.

I have been fortunate in receiving the help and friendship of many archivists in the course of doing research. The archivists at University of Toronto Archives, especially the wonderful Ms Marnee Gamble, and the archivists at the National Archives of Canada were a tremendous source of help. Graeme Durkheim at the Department of Fisheries and Oceans Library, Ottawa, was very friendly, helpful and helpful during my doctoral research. At the Pacific Biological Station, DFO, Mr Gordon Miller supplied the materials I needed and his discussions awakened me to several broader historical interpretations. At the St Andrews Biological Station, I would like to thank Ms Darlene Tam, librarian, Brenda Best, the administrative assistant, and Suzanne Taylor, communications director, for their kind and generous help.

This book would have been impossible to have produced without the financial support in the form of a postdoctoral fellowship and a grant in aid of publication from the Social Sciences and Humanities Research Council. My deepest thanks. Dr Carla Cassidy, Dean of Arts at Ryerson University, generously aided the completion of this book with a small grant.

I would never have finished this book had I not received help along the way from many friends. Louise Earl provided many thoughtful suggestions, and her insights will always be remembered with gratitude. My thanks also go Dr Wilfrid Lockett, for his technical advice and sharing of philosophy. I would like to thank my colleagues in the History Department at Ryerson University for making working there such a happy experience. Dr Ron Stagg, former Chair Al Wargo (retired), and Dr David MacKenzie gave me their encouragement and interest during my dark sessional days when both career and writing seemed to be stalled. I wish also to thank Mr Rod Langley, who involved me in his film project. Working on *The Fisherman's Friend: A.G. Huntsman, 1883–1973* rekindled my ambition to finish this book. Trust a Brit to be good with words: my thanks to Gil Parsons for suggesting the title.

While pushing a book through the editorial process is a painful experience, Mr Lennart Husband, my very understanding editor, is wonderfully calming and sometimes makes the whole process even seem fun. I owe Len more than I can say. My readers, Dr Eric Mills and Dr Tim Smith, caught and corrected mistakes and errors and offered important criticisms that proved immeasurably helpful. My copy editor, Mr Matthew Kudelka, caught many mistakes and smoothed out my prose, and my student Marlene Comeau helped put the index together. I am very grateful.

Finally, I wish to thank my family. My father and late mother, Dr Francis and Mrs Ruth Coghlan, were a tremendous source of encouragement, love, and support. My late father-in-law Roy Hubbard, and mother-in-law, Eileen, gave generous help on a number of occasions by looking after their grandchildren, providing much-needed opportunities for my research and writing. My sons Sebastian, Adrian, Nathaniel, and Tristan had much to endure from a busy and distracted mother over all these years. Lastly, and most of all, I would like to thank my husband Danny, for his patience, support, constructive criticism, and affirmation of my work, and for keeping the faith.

A SCIENCE ON THE SCALES

Introduction

The collapse of the Northwest Atlantic cod and other groundfish stocks across the Grand Banks in the early 1990s has drawn much dismayed attention to an industry that many people had taken for granted. Historically, most of the fish harvested on those banks was destined for tables in distant lands, since until the 1980s, few North Americans had much interest in making fish a part of their diet. One of the ironies of this catastrophe is that the gourmet and health potential of the lost stocks was finally becoming appreciated just as their numbers were dwindling. Besides being an economic disaster for Atlantic Canada, the collapse represents a perilous loss of a valuable food source for the world's people. Moreover, the whole situation has called into question the value of the science of fisheries biology, which seeks to measure and predict the size of fish stocks, and to study and explain how the environment, human activities, and other factors affect fish populations.

Before these events, I had embarked on a dissertation on the history of pre–Second World War Atlantic Canadian fisheries, focusing on the purposes, goals, and accomplishments of the pioneering fisheries scientists in Canada and elsewhere. That study forms the core of this book, but I have added some material from the postwar era. Although this material will not fully explain the role of scientists in the stock collapses, it should throw some light on the institutional mindset within the government-science establishment that enabled events to proceed as they did. The sequence of events and political choices that led to the collapse is recounted with bitter clarity by Michael Harris in *Lament for an Ocean.*[1] His book and others make it clear that there was fierce disagreement within the Department of Fisheries: a number of junior scientists questioned how the data were being interpreted, but were not free to examine them fully and to disseminate their contrasting assessments.[2] There is no doubt that the Department of Fisheries and Oceans and its scientists played a role in the eventual collapse of the world's economically most important groundfish stocks.

But it must also be strongly emphasized that without fisheries biology, the world's fisheries would be in dramatically worse shape than they currently are. For example, in all probability, Canada's Pacific salmon fisheries would have vanished. Yet the questions remain: Could fisheries biologists have done more to foresee and forestall not just the Northwest Atlantic groundfish stock collapse, but other drastic declines in major fish stocks? What has fisheries science accomplished?

The answers to these questions necessarily involve understanding the origins and development of the science through the individuals and institutions that shaped it, in the context of the more general scientific preoccupations of the times. Also, since fisheries science is ultimately an applied science, economic, environmental and political circumstances must be incorporated into any interpretation of its evolution and achievements. The Canadian story deserves to be told for a number of reasons. Canadian oceanographers and fisheries scientists held an enviable status in the marine science community in the years leading up to the Second World War, and indeed, have done so ever since. Furthermore, in the interwar years they established many close links with scientists in the international marine science communities. Also, important marine scientists at the Biological Board of Canada (after 1938 the Fisheries Research Board of Canada) showed a remarkable interest in the history of their immediate community: oceanographer Henry B. Hachey and fisheries biologists Archibald G. Huntsman and Wilbert A. Clemens all wrote long and detailed memoirs or 'histories,' which, however, lack any thematic interpretations such as those later rendered, for example, by historian of oceanography Eric Mills. Finally, there are wonderful archival resources at the University of Toronto and the National Archives of Canada. Of particular interest is the extensive correspondence of A.G. Huntsman, a former director of two of the scientific stations, who had a deep and critical interest in the goals, methods, and underlying purposes of his science. His letters drew Canadian, American, British, Scandinavian, and other scientists into discussions not only of immediate concerns, but also of the methods by which fisheries science should proceed.

Fisheries biology is a science that was born out of conservation concerns that began to arise in Northern European countries in the mid-1800s. The field's purpose was to help the fisheries respond to changes in stock densities. The spectre of dwindling stocks first arose in the 1860s, when the new, highly destructive trawling and purse-seining technologies were married to boats propelled by steam. Fishermen could now travel farther than ever before, and catch more fish to feed Europe's burgeoning cosmopolitan populations. Inshore fishermen noticed that catches were declining in the traditional near-shore fishing grounds, and demanded governmental conservation measures.

In Britain, the response was to set up investigatory commissions. The lack of scientific background knowledge relating to fisheries problems led to what can charitably be called inconclusive findings. This highlighted the need for scientific information on commercial fish stocks. Biologists interested in fisheries problems, and in marine life generally – especially as it related to evolutionary and ecological studies following Darwin's 1859 work, *On the Origin of Species* – capitalized on this demand. Starting around 1872, scientists began organizing oceanographic expeditions, establishing marine laboratories, and founding the new disciplines of fisheries biology and oceanography. The story of oceanography and its problems is excellently described and analysed by Margaret Deacon in *Scientists and the Sea* and by Eric Mills in his *Biological Oceanography*.[3]

In the English-speaking world, private enterprise was slow to grasp the research and development possibilities offered by science, and governments were reluctant to squander (from their perspective) taxpayers' money on scientific research that lacked military application, except for geological surveys (to locate mineral wealth) and the increasingly popular agricultural research stations toward the end of the century. The monetary returns from mineral discoveries and higher agricultural yields justified such expenditures, but governments were not certain that funding fisheries science would result in any economic returns, especially in countries where the fisheries were a marginal industry. However, by the early 1900s, governments were beginning to realize that supporting science brought valuable economic and political dividends.

Indeed, government support of fisheries biology and other sciences contributed to Canada's emergence as a scientifically based industrial society, and ultimately ensured that Canada would be able to develop independent fisheries policies and to strengthen its fishing industry. Fisheries science helped establish rational fisheries regulations; it also increases exploitable fish stocks. Also, following the Halifax Award of 1877, the Canadian government eventually realized that scientific information gave it leverage in international fisheries disputes. This was because, due to Canada's ongoing fisheries statistics, and the Americans' neglect of the same, Canadian fishermen successfully claimed damages for American fishing in territorial waters.

As a consequence, in 1898, the Canadian government agreed to sponsor a marine biological station. The first marine biological laboratory, diminutive, resembling an old-fashioned trolley car, was transported from place to place on a scow. The scientists who founded and ran it, most of them university professors, later formed the Biological Board of Canada, which maintained marine stations on the Atlantic and Pacific. In 1938, this board was transformed into the fully professional Fisheries Research Board of Canada. From humble beginnings as a summer 'retreat' for university researchers, it grew into a full-time and permanent enterprise, which has been chronicled in Kenneth Johnstone's rather antiquarian *Aquatic Explorers*[4] (which contains, nevertheless, many revealing anecdotes and some wonderful interviews).

Since they relied on government funding, the scientists emphasized the station's economic potential for solving fisheries problems: this forged the first link in what was to become a broad network of interests, from fishermen to the refrigeration industry. Nevertheless, it is difficult to assess the value of the scientific returns for the government. Certainly, the biologists gathered much information that led to the elimination of reams of useless protective legislation and that pointed the way toward more sensible measures. However, this information resulted in government memos, not scientific publications. More problematical for the historian of science approaching fisheries biology from the outside is trying to figure out just what fisheries biologists actually saw themselves as doing.

At the time, fisheries biology was in its infancy and was just beginning to develop its own methodologies and subdisciplines. The Biological Board of Canada and its stations grew alongside these developments, but far from the mainstream new research traditions beginning to establish themselves in Europe. There is no evidence that the first scientists who worked in Canada's marine stations had any idea that fisheries biology requires an evolving, continuous assessment of conditions, and that fisheries problems are never solved once and for all.

But in Europe, fisheries biologists were already feeling their way toward measures for assessing fish stocks and finding out how to distinguish between natural and fishing-induced stock fluctuations. Indeed, new developments in fisheries biology were occurring throughout the period covered in this book, and for this reason I will offer no introductory chapter which gives the reader a crash course in the history of the science. Rather, I will chronicle new developments in fragments in chapters

1, 2, 3, and 6 as they relate to Canadian scientists' arrival at these stages, to give a more impressionistic understanding of how the Canadians encountered and adapted to the growing science. Although this study is not strictly chronological, it does follow the main areas of research and other activities in the sequence in which they dominated the attention of the Biological Board. Thus, chapters 1 and 2, dealing broadly with the first two decades of the board, incorporate some general descriptions of fisheries biology to 1910, but as the board's work was unfocused and mainly exploratory, there is no detailed discussion of the major advances that occurred between 1900 and 1914. These chapters focus on the Canadian pioneering fisheries biologists' motives, their relations to government, and their approaches to the problem of basic versus applied science. A discussion of the relationship between general marine biology and fisheries science is also included.

One important theme throughout this book is the relation between basic and applied science at marine biological stations. For the Biological Board's fisheries work, this issue poses important questions. With the exception of agricultural research, the development of the applied sciences has received little attention from historians of science.[5] As a result, stereotypes persist that tend to downplay the originality and importance of these sciences. Their scientists are tacitly stigmatized as second-raters unless they escape the mill of rote, applied research and branch into basic studies. The idea persists that applied scientists could only make progress by appropriating developments from basic sciences, and that the 'real' scientists among them chafed at having to do applied research.

This problem in the history of marine science is exacerbated by the fact that the two most-studied marine biological stations, which were among the world's earliest and most famous, were created as centres for advancing the most basic and exciting problems in biology at the time. These stations were the Statione Zoologica di Napoli, founded in 1872, and the Marine Biological Laboratory at Woods Hole, Massachusetts, opened in 1888. Both initially concentrated on developmental and then experimental morphology, and both served as models for later marine laboratories and have extensive bibliographies in the history of science, especially in the writings of Jane Maienschein and Christiane Groeben.[6]

Many later stations, such as that of the Marine Biological Association of the United Kingdom, opened at Plymouth in 1888, and the station founded at St Andrews, Scotland, in 1884 by William Carmichael M'Intosh, undertook both basic research and fisheries science. Thus, they served two main communities of interest: the one centred around basic science, and the one that displayed a real and often intense interest in learning about fish and the marine environment with the goal of devising programs for managing commercial fisheries. It is wrong to draw the simplistic conclusion – as many have been tempted to do – that the founders of these stations, either cynically or in earnest, deliberately used fisheries problems as a means to get funding and support with the intention of only really concentrating on basic research. Although they did indeed use the problems facing the fisheries to arrange funding for basic research, many of the pioneer marine scientists, including E. Ray Lankester in Britain, Edward E. Prince in Canada, and Henry B. Bigelow, had a genuine interest in those problems. Besides, as will be argued in chapter 2, early fisheries science was general enough and had enough new territory to cover that in many respects it differed little from basic, albeit non-experimental, biology.

Major developments in fisheries biology are described in chapter 3, which tells about the Canadian Fisheries Expedition of 1914–15. This expedition was led by the world's leading fisheries biologist, the Norwegian Johan Hjort, at the Biological Board's invitation. This chapter argues that advances in an applied science, and even new research paradigms, can arise internally as a result of developing methodologies or increasing experience. For example, Hjort constructed the important theory that fish population fluctuations are caused by differing survival rates of fish spawned in different years. He relied mainly on recent developments within fisheries biology itself (such as fish scale analysis) for his 'year–class' concept. His work was specific to fisheries biology, and it provided a new and important tool, which was then incorporated into all later analyses of fish stocks. Hjort introduced Canadian biologists to European fisheries biology, which had recently reached something approaching its modern form through his own work and that of scientists in the International Council for the Exploration of the Sea.

While chapter 3 recounts the subsequent influence of this expedition on Canadian practitioners of fisheries science, it also introduces a second major theme of this book: the board's involvement with fish processing research and with educating Canadian Atlantic fishermen, also begun by Hjort. Indeed, Hjort's career refutes the notion that oceanographers shunned applied work. In coming to Canada to lead the Canadian Fisheries Expedition, he hoped to introduce to Canadian fishermen a new, patented fast-freezing technique so as to improve the marketability of their products. Later, Huntsman in Canada and Michael Graham in Great Britain happily devoted a great deal of attention to practical problems confronting the fishing communities in their nations.

Chapters 4 and 5 develop this theme and offer a hiatus from the exploration of emerging fisheries science. Instead, they look at the board's more practical contributions to the Canadian fisheries, from developing new fish-processing techniques to implementing education programs for fishermen relating to everything from conservation to canning. The fact that fisheries biology was often carried out under government auspices did tend to divert fisheries scientists toward developing improved fishery technologies and increasing fishing potential by locating new exploitable stocks. In this respect, its history in Atlantic Canada can serve as a typical and even paradigmatic example. There, biologists had to contend with regional underdevelopment, poorly educated fishermen, and the special problems posed by Canadian geography: long-distance supply and demand, and transporting and preserving fish products for inland markets in the days before mass-refrigeration technology. This work dominated the board until the end of the Great Depression, after it built two fisheries experimental stations in 1924: one in Halifax, the other in Prince Rupert. The scientists involved in fish-processing research, fish population analysis, and other applied work were for the most part motivated by an ideal of service. They were happy to undertake such work, seeing it as being helpful to fishermen. However, when the board began monitoring fishing and fish-processing standards, its relationship with its funding department, the Department of Marine and Fisheries, subtly changed. Scientists found themselves more closely bound to the industry and its leaders, as well as upper-level bureaucrats in the department. This necessitated new efforts to retain the board's independence as much as possible.

The addition of fishing industry leaders to the Biological Board's membership, and their closer work with its scientists in the 1920s, led to a forging of ties between board scientists and fishing 'industrialists.' Industry leaders tended to be educated businessmen heading large firms. By the 1930s fisheries scientists had adopted the position of the industry's more highly capitalized sector. For example, they defended the trawling industry in 1928, against the findings of the 1927–8 Royal Commission Investigating the Fisheries of the Maritime Provinces and the Magdalen Islands, chaired by the Honourable Mr Justice A.K. Maclean. The Maclean Commission favoured long-line and fixed-net fishermen against the steam trawlers. In the 1980s, these links with the more capitalized portion of the industry would work to the detriment of inshore fishermen.

The final three chapters again examine Canadian contributions to fisheries ecology. Chapter 6 argues that Canadians were inordinately influenced by the arguments of Thomas Henry Huxley, the English comparative anatomist, defender of Darwin, and sometime 'expert' on several of Britain's periodic commissions of investigation into her sea fisheries. Huxley, whose essential argument was that overfishing was impossible, was such a towering figure in nineteenth-century English science that his pronouncements were given greater weight by other scientists than they merited. In Canada, this was especially so, because the Biological Board's founding chairman, Edward Ernest Prince, served as a young man with the last of Huxley's fishery commissions. The resulting bias against the idea of serious fish stock depletion, coupled with the immensity of offshore fish stocks in the Northwest Atlantic, delayed Canadian Atlantic fisheries biologists' involvement in science geared toward conserving the fish stocks. As chapter 7 recounts, this included rejecting mathematical modelling as a method for estimating the size and age structure of fish stocks. But other work, geared to preserving fish stocks from more local threats, such as tidal dams, did draw a serious and passionate engagement by the Canadians.

The last chapter tells the story of the increasing professionalization of the Biological Board of Canada as it became the Fisheries Research Board and did away with volunteer university researchers in favour of its own professional staff. It also documents the board's shift toward a more narrowly focused research agenda and the declining fortunes of the Atlantic Biological Station, which slumped from its former status as world-recognized institution to become a scientific backwater.

From the perspective of the early twenty-first century, fisheries biology has not enjoyed unalloyed success. Although it was the first science to centre on the conservation of wild populations and environments – specifically, the conservation of fish stocks for future exploitation – its successes have been limited. Ironically, although fisheries science should be well situated as an ecological, environmental science, the links between environmental ecology and fisheries biology became tenuous. One would instinctively suspect that fisheries biology would have been influenced by early ecological programs seeking to uncover linkages and boundaries for new concepts such as ecosystems, communities, niches and so on, but little of this occurred. Perhaps this was because fisheries biology temporally parallelled ecology's emergence as a branch of biology. Indeed, the first ecological society in the world was formed only in 1913 (by British plant ecologist Alfred Tansley at Cambridge University); by that time, fisheries biology was well under way. Only since the late 1980s have historians of science begun closely investigating the history of ecology,[7] and a cohesive understanding of ecology's development is only recently being formed. The links between fisheries biology and ecological science remain unclear, although this monograph will show that even in the absence of ecological models, the Victorian tradition of natural history studies – excellently described by Suzanne Zeller in its Canadian context in *Inventing Canada*[8] – ensured a measure of ecological understanding. Early Canadian, British, American, and Scandinavian fisheries biologists were determined to discover the biological and the physical factors that shaped the habitats and determined the success of marine populations' reproduction and productivity. Unfortunately, the expense and complexity of studying such factors for various large fish populations overcame the initial interest in these problems. By the 1920s, organizations such as the International Council for the Exploration of the Sea were turning their backs on these approaches in favour of studies of fish population structures and abundance.

As a result, only one recognized branch of ecology, population dynamics, had a lasting effect on fisheries science. Italian population ecologist and physicist Vito Volterra strove to understand predator–prey interactions based on mathematical and statistical studies of commercial fish populations. As Sharon Kingsland's *Modelling Nature* shows, these studies helped promote a problematical and controversial mathematical modelling approach to nature. The link with fisheries science is clear, and one Biological Board scientist, A.G. Huntsman – who will feature prominently in the following pages – even collaborated in Charles Elton's animal population studies.[9] Indeed, according to Tim D. Smith – whose 1994 *Scaling Fisheries* offers a thorough history of the science of estimating fish populations – mathematical population dynamics, which focused on a single species at a time, had such a pervasive influence on fisheries science that fisheries biologists ignored, to the science's detriment, general ecology. Such ecology would have led to a broader understanding of how fish populations interact with other species, the environment, and indeed, the humans who fish them.[10]

Fisheries biology, admittedly, had no choice but to serve two masters: science and government. The closing pages of this book will argue that the weaknesses of Canadian fisheries biology did not emerge from its focus on applied science. Rather, the problems arose due to the conflicting agendas of its funding agency, the federal government, which sought on the one hand to conserve the fish stocks but on the other hand to make its policies appeal to those maritime constituencies where the economy was based on fishing and fish processing. Also, in 1977, when the government had absorbed the quasi-independent Fisheries Research

Board of Canada into the new Department of Fisheries and Oceans for the sake of efficiency, it had at the same time imposed bureaucratic ideals of productivity upon its scientists. Scientists were encouraged to publish articles rather than solve fisheries problems. As I argue in the epilogue, in an admittedly impressionistic analysis of developments following the catastrophic cod stock collapses of the Northwest Atlantic, what has been lost in organizations overseeing fisheries research is the ideal of service that was so powerful in the early decades in Canada and elsewhere. In some cases science has been shackled to a predetermined agenda that has no dedication to the dissemination of dispassionate, non-partisan knowledge and information. A major problem has been data hoarding by senior scientists for the purposes of improved publication records and promotion, in a heavily bureaucratized science organization. And bureaucratization surely followed as the Biological Board's fortunes became more closely integrated within the Department of Marine and Fisheries, especially after its later incarnation, the Fisheries Research Board of Canada, was folded into the Department of Fisheries and Oceanography in 1973. On a positive note, the epilogue also documents an important change taking place in fisheries biology at the turn of the twenty-first century: Canadian fisheries biologists are helping forge new paradigms for a more ecologically and conservation focused science. This offers hope that fisheries management will eventually integrate environmental and human needs. Fish stocks may stage a slow recovery, given enough political will.

The development of Canadian Atlantic marine science and fisheries biology is presented as a case study of the functioning of marine biological stations prior to the Second World War, and of the growing importance of science in the day-to-day workings of government. It takes into account the development of analogous institutions elsewhere, such as the International Council for the Exploration of the Sea, and marine stations in the United States, Great Britain, Scandinavia, France, and Italy. In addition, this history will show how Canadian scientists borrowed ideas from numerous sources, so that many different influences shaped the Biological Board of Canada and its marine stations. The strong relationship between the Biological Board and Canadian universities will be presented as essential to the evolution of the new profession of fisheries biology. The Biological Board and its successor, the Fisheries Research Board, enabled Canada to hold its own in international negotiations over offshore fisheries resources. Thus, the science begun by the Biological Board was vital to Canada's management of its marine and fisheries resources. This study will occasionally glance across the continent to the board's Pacific work; however, the emphasis will be on the board's Atlantic work, which was dominant throughout much of this period. The Pacific agenda became increasingly important, especially in the 1930s, and gets attention in the final chapters; that said, it had such a different emphasis from the Biological Board's concerns on the Atlantic that its full consideration is beyond the scope of this history.

Chapter 1

Scientists at Sea: Expeditions and Seaside Stations

> In fisheries research as in many other things Canada through her intermediate position is powerfully influenced by both Great Britain and the United States, but their differences have permitted her to 'call her soul her own' and to follow a somewhat independent course.[1]

On 16 August 1897, the year of Queen Victoria's Diamond Jubilee, Toronto was abuzz with excitement. The scientists were coming to town. Toronto had scored a coup in enticing the British Association for the Advancement of Science to hold its 1897 annual meeting there. The society would thus be reaffirming its imperial role. The meeting ran from Tuesday, 17 August, to Wednesday, 25 August. Although the Age of Science, in which science was the provenance of the common man, was winding to its close, the event still garnered extensive attention in the local press. The *Toronto Daily Mail and Empire* gushed:

> there cannot but be a feeling of gratification in the minds of those who belong to the British Empire as they remember that the assembly of eminent men that will be here this week is the British Association ...
>
> Britain takes no second place in the world of science, and her savants are justly respected throughout the world. The meeting therefore forms a link with the Mother Land such as appeals to the patriotic sentiment of our entire people.[2]

Local papers summarized the main talks and speeches, and educated readers by running glowing biographies of the more famous visitors.

The British Association meeting, then, had the air of a carnival or twentieth-century World Summit. Famous scientists from the empire's heart and from elsewhere were paraded in the press before a dazzled readership: Lord Kelvin, Prince Kropotkin, Lord Lister, Dr Charles Richet, Michael Foster, and Anton Dohrn were only a few. Diversions for the hard-working participants included picnics, parties, balls, and excursions (including a trip to Niagara Falls), in addition to daily scientific meetings and wives' meetings, all of which were reported faithfully.

The meeting itself wound down after a week; however, Canada's small coterie of university-based scientists were determined to make sure it brought lasting benefits to Canada. Several joint Anglo-Canadian initiatives were launched, and committees were formed to establish or revive major scientific projects. A meteorological observatory was to be built on Montreal's Mount Royal. Two committees for ethnographic studies, one reviving a long-standing ethnological program (sponsoring Franz Boas's research) begun at the 1884 British Association meeting in Montreal.

With more lasting implications, a committee was formed to promote a permanent hydrographic survey within the Canadian government (which was threatening to terminate its sparse tidal surveys, begun in 1890). The British Association's work helped create the Coast Hydrographic Survey.[3] Also, two committees were set up with the complementary ends of having a 'lacustrine' biological station built on Georgian Bay's Go-Home Bay,[4] and a marine biological station on the Gulf of St. Lawrence.

In fact, the formation of a Canadian marine biological station was a key objective of the conference's Toronto organizers. University of Toronto professor A.B. Macallum was president of this committee. He was assisted by James Loudon, president of the University of Toronto and champion of university-based scientific research, by Zoology Department Chairman R. Ramsay Wright, and by zoology professor James Mavor. Of these, only Loudon would have no later involvement with the Canadian marine biological station.

The need for a marine biological station was acutely felt by biology professors at Canadian universities. At Woods Hole, Massachusetts, and Naples, Italy, exciting developments in experimental biology and developmental morphology were taking place during the 1890s. Questions central to biology – relating to evolution and ontological development in organismic growth – were being addressed at marine biological stations using marine organisms. Canadian zoologists chafed at their lack of such facilities. James Playfair McMurrich (1859–1939), professor of biology and horticulture at the Ontario Agricultural College at Guelph, was the first to argue publicly, in 1884, that Canada needed these stations to conduct biological research.[5]

Indeed, the direction biology was taking in the nineteenth century virtually demanded a focus on marine forms. At the beginning of that century, the particular taxonomical challenges presented by marine organisms were beginning to attract a great deal of attention. This was a result of the taxonomical groupings proposed by French anatomist Georges Cuvier (1769–1832), to whom most zoologists turned as their starting point.[6] The real meaning of these classification systems awaited Darwin and Wallace's theory of natural selection. Before them, Cuvier had tried to give the animal kingdom some semblance of order based on gross similarities. He 'divided it into four primary divisions, or *embranchements*: Vertebrata; Mollusca; Articulata; and Radiata. His arrangement elevated the significance of invertebrates ... Jellyfish, starfish and polyps, constituting one of Cuvier's four *embranchements*, stood equal in importance to all the birds, fish and mammals put together. That division, the Radiata, happens to be the only one of his branches that has no modern equivalent.'[7]

The primary members now constitute the separate groupings of the Echinodermata (Starfish) and Coelenterata (Jellyfish). Since most of the *embranchement* of Mollusca is also composed of marine species, and since there are many marine Articulata, Cuvier's scheme focused attention on marine organisms. Not only were new species and species relations to be found, but the complexities posed by Cuvier's Radiata demanded that this classification be unpicked and teased apart, then rewoven into a sounder fabric of new classifications, supported by intensive life history studies, morphology, and embryology.

Darwin's theory of natural selection gave meaning to the natural order; even so, Darwin's opponents were sometimes won over only with great difficulty. Most notable among the hold-outs were the influential Swiss-born American paleontologist, geologist, and ichthyologist Louis Agassiz (1807–73) and the young Alexander Agassiz (1835–1910), who followed in his father's footsteps but later changed his mind while pursuing his own paleontological and embryological studies. Embryology proved to be a powerful explanatory tool for evolutionists. It hinted at earlier forms through which species might have passed in evolving, and as such, it was important in converting sceptics to the evolutionary theory. Embryology itself was best studied through easily collected marine forms, whose eggs tend to be transparent; the contents, the developing embryos, are thus more easily observed than in terrestrial forms. This factor propelled biologists to the seaside.[8] Darwin's *Origin of Species* itself was to provide a powerful enticement for studying marine forms: 'For evolutionists, the study of the oceans had special significance, since the sea was believed to be the ancestral home of all life. Further, compared with terrestrial life, the flora and fauna of the oceans were remarkably diverse in character and abundant in quantity.'[9]

Canadians, however, were so busy surveying and inventorying the land forms on their vast new territory that they turned rather late to marine science. Early marine investigations on Canadian shores were carried out almost exclusively by American natural historians extending their surveys northwards. In 1852, William Stimpson (1832–72), the Massachusetts marine invertebrate specialist, commenced a systematic scientific study of the invertebrata of Grand Manan.[10] In 1877, the peripatetic summer laboratory of the United States Fish Commission was situated in Halifax, Nova Scotia; Spencer Fullerton Baird, its founder and Assistant Secretary of the Smithsonian Institution, and his associates, including Yale's Addison E. Verrill (another invertebrate expert) catalogued characteristic Canadian marine fauna. Only toward the end of the nineteenth century did Canadians such as the University of New Brunswick's Philip Cox, an ichthyologist, and William Francis Ganong (1864–1941) begin to survey the Atlantic Canadian coastal fauna.[11]

But pressure for a Canadian marine station was also to come from a non-academic direction. Perhaps the most vocal proponent of a marine station was Edward Ernest Prince (1858–1936), who had been appointed Canada's Commissioner of Fisheries in 1892. Prince was anxious to place Canadian fisheries on a more scientific footing. This interest was a direct result of his own training. He was born in Leeds, England in 1858 and did undergraduate studies at the University of St Andrews, Scotland. In 1885, after further studies at Edinburgh and Cambridge, he became a naturalist at the marine laboratory in St Andrews, where he had a close, happy, and productive association with the laboratory's founder: William Carmichael McIntosh (1838–1931).

McIntosh (usually written M'Intosh) had been trained in medicine at Edinburgh. His first important paper, 'On the Structure of the British Nemerteans and some New British Annelids, brought him recognition as an expert on marine worms. In 1882 he was appointed to the Chair of Civil and Natural History at the University of St Andrews. In October 1883 he was invited to serve on the Trawling Commission under Lord Dalhousie (1847–87). As the centre of a long-important fishing district, St Andrews was a marvellous base from which to probe interesting and important questions regarding the natural history of food fishes and the problems of oyster cultivation.[12] The commission's trawling gave rise to so much material that a seaside laboratory was needed in order to store and analyse it. As a result of Lord Dalhousie's lobbying with the new Fishery Board for Scotland (formed in 1882), M'Intosh acquired the use of a temporary station. This fulfilled his life's ambition to establish a marine station – an ambition that had been inspired by childhood vacations at St Andrews, 'when ... his bedroom was over-run by specimens dead and alive.' Dalhousie arranged for an unused fever hospital to be rented and for equipment and seawater piping to be installed. This was created 'the first marine biological laboratory in Britain and ... the focus of fisheries research ... for the next twelve years.'[13] At the time, M'Intosh held the St Andrews chair of zoology, so he integrated teaching with the marine station's work and had students investigate marine and tidal organisms under natural and laboratory conditions. In 1896 a permanent building, the Gatty Marine Laboratory, was erected and named after its private benefactor.

As scientific consultant to the Fishery Board for Scotland, M'Intosh's work came to be dominated by the project of establishing a scientific basis for legislative control of the fisheries. The board had sent Prince to serve as M'Intosh's assistant. Prince was the first naturalist to be attached to the laboratory, and became M'Intosh's most loyal collaborator. Prince was thus introduced to fisheries biology and his life's work. By 1888, M'Intosh and Prince had elucidated the life histories of some forty species of fish from fertilized ova to maturity. Their *Life History of the Marine Food Fishes* (1888) provided a solid British foundation for the subject. In the course of their work, Prince coined a useful new scientific term, 'demersal,' to describe fish that live on or near the bottom.[14] In 1890, Prince was appointed professor of zoology and comparative anatomy at St Mungo's College in Glasgow. Meanwhile, in Canada, in 1892, the Department of Marine had merged with the Department of Fisheries under a common deputy minister. The new department resolved to hire a Commissioner of Fisheries to coordinate Canadian fisheries management.[15] The department turned to M'Intosh, regarding him as the world's 'most prominent fishery scientist '[who had] developed fishery investigations of outstanding character.'[16] M'Intosh recommended Prince for the position.

It may seem surprising that the Dominion government felt the need for a university-trained scientist to fill an administrative position at a time when fisheries research was still in its infancy and the government was still largely insensitive to the needs and uses of science. Actually, it hired Prince – an experienced fish embryologist – in response to the popularity of fish culture in North America at that time. The artificial hatching of fish eggs and the planting of fish fry in depleted rivers had begun in France in the 1830s. In the 1850s, fish culture was introduced to North America, where it grew rapidly. Its proponents, and the public at large, believed that fish culture was arresting a fearsome decline of fish populations. The U.S. Commission of Fish and Fisheries was formed in 1871 to improve the operations of existing state fish hatcheries, and to run its own fish hatchery in Woods Hole. On confederation, in 1867, the Dominion government had taken over the responsibility for fish hatcheries from the Ontario and Quebec governments. By 1890 the old expert, Samuel Wilmott, was nearing retirement, and the department needed a new expert to oversee its hatcheries.

In choosing an academic scientist, the department was planting in its midst a strong proponent of research. 'The new appointee immediately advocated an idea already prevalent in academic circles that there should be a marine biological station for fishery research.'[17] Prince wanted to place Canadian marine fisheries resource management on a scientific footing and to promote basic scientific research. However, the government was not at first receptive to his suggestions. So he seized his chance when the British Association came to Toronto, and was joined by the other Canadian participants, who would help found the station.

Why was a station so necessary? At the time, there were many problems associated with studying marine organisms. Their capture requires boats, nets, and other specialized equipment (especially for specimens found any distance from the shore). New techniques were needed to 'fix' the specimens for anatomical study. If they were to be studied alive, tanks filled with well-aerated seawater were necessary. This was obviously beyond the scope of even a handful of individuals, and would require a large funding base and cooperative efforts. Nineteenth-century marine science was in these ways quite similar to present-day 'big sciences' of physics and astronomy. Two different routes were taken by marine scientists. The earlier path involved mounting seagoing expeditions to make extensive collections while recording conditions in the marine environment. The material was then worked up by scientific societies or in university laboratories. The later course, taken by those interested in the systematic study of marine organisms, was to build marine laboratories. Either route required strongly focused concerted activity, expensive and specialized collecting apparatus, storage facilities, and boats.

The first route was taken by the scientists who launched the famous British *Challenger* Expedition of 1872–5. Oceanic expeditions are not a major focus of this discussion. However, they contributed significantly to the growth of nineteenth-century marine biology. Edward Forbes (1815–54), the Manx naturalist, gave these expeditions a biological program by proposing that due to pressure, darkness and coldness, life could not exist below 300 metres in the sea. It was he who discovered characteristic species groupings along different types of shores and at different depths along European coasts[18] and thereby pioneered littoral ecology. In 1839 he encouraged the British Association to form the Dredging Committee to dredge for life forms in coastal waters. His own deep-dredging studies of the barren eastern Aegean in 1842 had led him to believe that marine life could not exist below 300 metres. His 'azoic theory' became dogma until dispelled by 'discoveries of the higher forms, the sponges, rhizopods, echinoderms, crustaceans and molluscs' from below 500 fathoms during the *Lightning* Expedition of 1868,[19] organized by Charles Wyville Thomson (1830–82) and William B. Carpenter (1813–85). These two men specifically set out to prove the existence of life on the deep-sea floor between Scotland and the Faroe Islands. The deepwater cruises of the H.M.S. *Porcupine* in 1869, under the direction of Thomson, Carpenter, and John Gwyn Jeffreys (1809–85), succeeded in dredging marine organisms from a depth of 2,435 fathoms, thus disproving the azoic hypothesis.[20]

Interest in deep-sea organisms continued to grow after the publication of Darwin's *On the Origin of Species* in 1859. Darwin linked evolutionary change with environmental change, which seemed to suggest that more ancient forms might well be found in the dark, unaltering abyss. Indeed, some newly discovered deep-sea forms, notably the living stalked crinoids, did seem to resemble fossils. Thomas Henry Huxley, probably Darwin's most enthusiastic supporter, was thrilled by these, and contended that the abyss was inhabited by living fossils and might represent 'the very cradle of life.' The search for such deep-sea 'missing links' strongly motivated Thompson when he and Carpenter planned a circumnavigatory expedition to increase knowledge of deep sea biology, oceanography, and geology. With the help of the Circumnavigating Committee of the Royal Society, Carpenter and Thomson borrowed the steam corvette H.M.S. *Challenger* for a three-year period beginning in 1872.[21]

The *Challenger's* 68,930-mile voyage around the world took three-and-a-half years. Three naturalists assisted Thompson: Henry Nottidge Moseley (1844–91), John Murray (1841–1914) – a Canadian of Scottish parentage – and Rudolph von Willmöes Suhm. These men analysed the contents of dredges, nets and trawls, preserving specimens for later work. They also determined the specific gravity, carbonic acid content and dissolved gases of water samples, and measured surface and occasionally subsurface currents. Water temperatures were taken at different depths, and bottom muds were sampled.

The compilation and publication of the expedition's work took a further nineteen years. There were enormous collections to be analyzed, including more than 13,000 kinds of animals and plants, as well as rocks, oozes, and water samples. A temporary government department, the *Challenger* Expedition Commission, was given quarters in Edinburgh (where Thompson was Regius Professor of Natural History). Thomson incurred the wrath of nationalists when he farmed out the *Challenger* collection among specialists in Europe and the United States. Louis Agassiz worked up the echinoderms, Ernst Haeckel produced a three-volume report on Radiolarians, and Georg Sars studied several crustaceans. British experts Albert Günther and William C. M'Intosh worked up the fishes and annelids respectively; also involved were Huxley, Carpenter, Gwyn Jeffreys, and E. Ray Lankester. The *Challenger* team also inherited specimens sent in by other expeditions, since only the *Challenger*'s specialists could tell whether recent finds were truly new. The first of the *Challenger* volumes[22] came out in 1880, the fiftieth and last in 1895. In the course of this endeavour, Thomson built up a network of personal relations among scientists from different countries, and drew many of them to Edinburgh, including Agassiz, Haeckel, Anton Dohrn, and Johan Hjort.[23]

Deep-sea expeditions were all very well for making collections, but in temporal terms they had drawbacks. This is why biological stations became popular. While the best of these were too expensive for any one individual or small group to operate, biological stations covered the whole spectrum from tiny, private seaside laboratories to large, government-funded institutions. From a seaside laboratory,[24] collections could be made during boat excursions or by plying dip nets and so on along the shore. Specimens could be captured, kept alive, and studied in aquaria and other facilities, into which fresh seawater was pumped directly. Biologists could now conduct extended life-history studies and experiment with living marine organisms and their eggs.

The British Association and Marine Biological Stations

Marine biological stations thus became centres for leading-edge research by the growing ranks of professional biologists. Yet by the time of the 1897 British Association meeting in Toronto, nothing had been done to answer James Playfair McMurrich's 1884 call for a Canadian station. So Canadian naturalists decided ask the British Association for help.

Canadian researchers had often relied on prestigious bodies to help bureaucrats see the importance of improving Canadian science. For instance, in 1849, the British meteorologists Edward Sabine (1783–1883) and John Henry Lefroy (1817–1890) called on American meteorologists to persuade the Upper Canadian government to maintain the meteorological and magnetic observatory founded in Toronto by the Royal Navy in 1839. The American Association for the Advancement of Science adopted resolutions recognizing that the Toronto Observatory's proximity to the focus of maximum magnetic intensity made its observations of great value to scientific advance, 'especially as connected with similar operations in the U.S.' The American Ambassador to The Court of St James appealed personally to the British prime minister, Lord Palmerston, telling him that 'the objects to be accomplished are strictly international.' Copies of these resolutions were sent to the Canadian government, and Sabine and Lefroy used them to garner support from the president of the Royal Society of London.[25] Their success established the meteorological station on a permanent basis with government funding.

Late-nineteenth-century Canada was notable for its Imperialist sentiment, so Canadian biologists called on the British Association to affirm the importance of a Canadian marine biological station for the British Empire. There was, however, a more compelling reason than this for choosing the British Association as a pressure group: it had an outstanding record in helping found and support marine biological stations. Its 'Committee appointed for the purpose of promoting the Foundation of Zoological Stations in different parts of the World' was instrumental in helping Anton Dohrn (1840–1909) found one of the earliest and greatest of these marine laboratories: the Stazione Zoologica di Napoli. This station opened its doors the same year the *Challenger* began her tremendous voyage around the world.

Dohrn's station was actually inspired by the French ones. France at the time led the rest of the world in founding stations: permanent seaside stations had been established at Concarneau in 1859, at Arcachon in 1863, and at Roscoff in 1872. Concarneau and Arcachon were tiny and offered only limited facilities to visiting researchers, and in this sense they little resembled the station soon to be built at Naples. The Laboratory of Marine Zoology and Physiology at Concarneau was founded by Professor J.J. Coste; spaces were available there for eight investigators, and competent visiting investigators were welcome free of charge. No advanced or elementary instruction was given. In contrast, the Biological Station of Arcachon, opened in 1863 by the local Société Scientifique d'Arcachon, promoted the study, advancement, and popularization of natural science and aquaculture. Unlike Concarneau, it incorporated an aquarium and museum in order to popularize its work. The aquarium was fitted with a research laboratory; separate research facilities were approved in 1867, but owing to financial problems and the society's desire to uphold the facilities' autonomy, they were not completed until 1902. Because the station was formally affiliated with the University of Bordeaux's Faculty of Physiology, physiological investigations tended to predominate.[26]

The Biological Station of Roscoff was opened by Sorbonne professor Henri Lacaze Duthiers in 1872, and so did not actually predate Dohrn's creation. In fact, much of it seemed to be modelled on the Naples laboratory. Open the year round, it grew into a large and important institution, an annex of the University of Paris and the Sorbonne's Faculty of Science. By 1908, it had thirty-eight research rooms, laid out for independent research by its staff and by visiting investigators.[27] It differed from Naples mainly in that it provided elementary and advanced instruction to biology students. Instruction was never part of Dohrn's plan.

Despite these French precursors, the Statione Zoologica was Dohrn's brainchild. Dohrn had studied medicine and zoology at various German universities; however, he had been inspired most deeply by a sojourn at Jena in 1862, where his zoology professor was Ernst Haeckel (1834–1919). Haeckel had filled him with 'piercing' enthusiasm for Darwin's works.[28] He became fascinated by comparative morphology, which employed embryology (the study of embryonic development) and anatomy to reconstruct phylogenetic histories of various organisms. He felt that Darwin had given zoology two chief goals: to explore the basis of natural selection; and to study the evolutionary origins of animal adaptation via embryology. In 1865, Dohrn accompanied Haeckel to Heligoland to study marine organisms. Poor weather, the lack of library and other facilities, and the difficulties inherent in capturing marine organisms led Dohrn to realize how useful a seaside laboratory would be, where biologists could arrive 'and find the "table laid" for their work.'[29]

In 1868, Dohrn had attended the British Association meeting at Dundee. There, he interested Thomas Henry Huxley in his proposed station. At the 1870 meeting at Liverpool, they formed a zoological stations committee. 'During the difficult period (1870–1874) when he was trying to build the Zoological Station, Dohrn received the greatest encouragement and help from his British colleagues.'[30] By 1871, Dohrn had selected a site at Naples for his zoological station. He had also enlisted support from the Academy of Belgium and from the German government, and lined up financial aid from various governments, universities, and other organizations.[31] The station building was finished in September 1873. It contained a Department of Morphology (a large laboratory for about twenty scientists), twelve smaller labs, and a library on the first floor. There were living quarters for custodians and assistants; pumps, machines, store rooms and seawater tanks in the basement; and a public aquarium on the ground floor. The aquarium's entrance fees paid for a permanent assistant and augmented the station's revenues. As the station expanded, extensions were built for Departments of Botany (1876), Physiology (1882), and Bacteriology (1887). With donated biological publications, Dohrn also built up a formidable reference library.

Dohrn ably promoted his station, which soon 'became a "must" for every aspiring biologist around the world and for every visitor to Naples.' It also became a centre for improving and disseminating new histological techniques, thus furthering Dohrn's program, which was to make marine stations the essential foci of biological research. By providing 'perfect working conditions,' the station filled 'the need for organization that Dohrn saw so lacking in zoology at the time. Good organization contributed to saving money, time, and energy for research.'[32] The complete research freedom offered by the station, and its state-of-the-art instruments and procedures, inspired many visiting scientists to promote similar stations in their own countries. These included Charles Otis Whitman, who visited in 1881–2 and who later became first director of the Marine Biological Laboratory at Woods Hole, and E. Ray Lankester who helped found the Marine Biological Laboratory at Plymouth in 1888. Spencer Fullerton Baird (1823–1909), head of the United States Fish Commission, asked Dohrn's advice on the aquarium and seawater supply for his fisheries laboratory at Woods Hole in 1881. Like Wyville Thomson's *Challenger* work, Dohrn's international vision for his station helped transform marine science into a field which transcended national barriers.

After 1874 the British Association also became involved in founding stations through its ubiquitous formation of special committees.[33] It helped found the peripatetic Scottish Zoological Station (1880); the St Andrews Marine Laboratory (1882); the Granton Marine Biological Station, also in Scotland (1885); the Marine Biological Laboratory at Plymouth (1888); and finally, in 1897, the Canadian Biological Station.

The largest and most important of these was the Marine Biological Laboratory in Plymouth, England, opened on 30 June 1888. The forceful Edwin Ray Lankester (1847–1929) – then a professor of zoology at University College, London (1874–91) – is credited as the driving force behind this station. In 1833, as President of Section D (Biology) of the British Association, he lamented in his address: 'There is no such laboratory on the whole of the long line of the British coast' in spite of British naturalists' interest 'in the exploration of the sea and marine organisms' and the dependency of Britain's fisheries on 'the knowledge which a well-organised laboratory of marine biology would help us gain.'[34] Lankester feared that Britain was trailing far behind other nations, especially Germany, in supporting basic marine science, but he was just as upset by the state of the British fisheries.[35] He adamantly rejected the findings of the 1863 Royal Commission on the Sea Fisheries, which had concluded that overexploitation was not occurring. He urged that a society be formed for the scientific and practical study of marine life and its conservation.

The Marine Biological Association of the United Kingdom came into existence on 31 March 1884 in the rooms of the Royal Society in London. It immediately resolved to establish one or more laboratories on the British coast 'where accurate researches may be carried on leading to the improvement of zoological and botanical science, and an increase of our knowledge as regards the food, life, conditions, and habits of British food fishes and molluscs in particular, and the animal and vegetable resources of the sea in general.' When the Plymouth Laboratory opened in 1888, its permanent scientific staff immediately began year-round basic and directed scientific research.[36]

The Plymouth station was similar to the one in Naples, being a large laboratory open to many visiting researchers, with a ground-floor aquarium open to the general public. Unlike Naples, however, Plymouth's naturalist, Walter Garstang (1868–1949), began offering marine biology courses to university students in 1897. These were continued after he left in 1902. The station soon had to expand, when researchers began flooding the facilities during summers and Easter holidays.[37] The Plymouth laboratory had two programs, both open to visiting scientists: basic biological research, and fisheries biology – a new science that looked at the habits and life histories of fish and that explored the effects of fishing on their populations. J.T. Cunningham, the laboratory's first naturalist, carried out lengthy studies of the distribution, reproduction, and development of commercial species. The first visitors and staff naturalists also collected, surveyed, and described local flora and fauna. A later director, F.S. Russell, would remark that this 'was a first requirement, because in all branches of science the systematic observations must come first,'[38] and especially before competent assessments of fisheries problems.

The British Association provided a modicum of financial support; it also formed a committee to promote the establishment of the new laboratory. However, until the First World War, Naples remained its top funding priority in terms of annual grants (£75 until 1884, £100 thereafter). It regularly sponsored the research of British (and occasionally Imperial) biological investigators, maintaining one or two tables at Naples for the purpose. Fewer funds were forthcoming for the Plymouth laboratory: a contribution of £500 in 1888 entitled the British Association to nominate a life governor to the Marine Biological Association, as well as the permanent right to appoint annually one person to occupy one table for one month free of charge. Grants of up to £50 were given to sponsor additional researchers. Section D monitored the work done by committee-sponsored scientists, as well as the finances and research programs at both Naples and Plymouth.

The British Association's involvement with other marine stations was much more limited. It formed no lasting ties with them; rather, it donated funds for equipment (to the St Andrew's laboratory and the Scottish Marine Laboratory) or to support specific researches. The Granton Marine Biological Station was also given a few grants for buildings and equipment.

Section D began its final effort to help found a marine biological station at the 1897 meeting in Toronto. Famous scientists helped raise the profile of the new project. Anton Dohrn himself was there, as a guest of the University of Toronto's zoology chairman, Ramsay Wright, who supported the idea of a Canadian marine station. Also attending was Sir William A. Herdman (1854–1924), honorary director of the Lancashire Sea Fisheries Committee and a veteran of the *Challenger* Expedition and its aftermath of tedious report writing (he himself had been responsible for three volumes of the *Challenger* reports). He had also founded the Liverpool Marine Biological Committee and its stations. Others there included Carl Eigenmann, the American ichthyologist; Walter Garstang, naturalist at the Plymouth laboratory; and James Playfair McMurrich, who had been the first to call for a Canadian station. McMurrich knew firsthand the value of such stations: he had assisted C.O. Whitman, first director of the Marine Biological Laboratory, as Docent of Animal Morphology at Clark University from 1889–92, and as a summer researcher at the Marine Biological Laboratory in the 1890s. A graduate of the University of Toronto's zoology department, McMurrich later became chairman of the Biological Board of Canada. However, at this time he was remote from Canadian affairs: having taught at several American universities, starting with Johns Hopkins in 1884, he was now professor of zoology at the University of Michigan (1895–1907).

Anton Dohrn and Sir William Herdman rated long biographical notices in the *Toronto Daily Mail and Empire's* preconference hype. That paper informed interested readers that Herdman was 'a leading authority on problems of marine biology,' whose 'investigations have always been most carefully done, while his conclusions are invariably sound and reliable.'[39] The article on Dohrn did more to aid the cause, declaring the Naples station one of the most excellent institutions in the world: 'The British Association proposes to create a similar station in Canada for the purpose of biologically examining the inland lakes. The Naples Station has this year celebrated its 25th anniversary, having started in 1872. During this period ... more than thirty similar institutions have sprung into existence. Scientists declare ... that as their forerunner the Naples Station has entirely changed the conditions of modern biological study.'[40] The importance of biological stations was reiterated by the president of Section D, L.C. Miall, who contended that 'the Zoological Stations now maintained by most of the great nations [are] most valuable.'[41]

The Thursday afternoon session of Section D on 17 August focused on these stations. Dohrn read a paper on the 'Naples marine station and its work,' and acknowledged 'his gratitude and indebtedness to the British Association for their support, which has extended from the early critical period of the station's existence up to the present time.'[42] Ramsay Wright read a paper 'On a Proposed Lacustrine Biological Station.' Later, Section D adopted a resolution requesting government support for a marine research station in Canada. It also appointed a committee to present its requests to the ministers of Marine and Fisheries and of Militia and Defence. The Canadian Commissioner of Fisheries, Edward Ernest Prince, was made chairman of the 'Committee to Promote the Establishment of a Biological Station in the Gulf of St. Lawrence.' Other members were Professor D.P. Penhallow of McGill (secretary), John Macoun (the Geological Survey's naturalist), Professor E.W. McBride, Dr A.B. Macallum, and British academic naturalists T. Wesley Mills and W.T. Thiselton-Dyer.

The Dominion department to which the proposed station would necessarily be attached was the Department of Marine and Fisheries. Needless to say, this department did not see the need for a marine research station, and the committee would require a strong argument to win it over. This argument would have to be prepared by those who understood the Canadian system intimately – that is, the Canadian scientists who wanted the station in the first place. Chief among these was Prince, who was already working within the government.

Building a Station and Strengthening a Nation

The British Association helped Prince in three ways. First, it established a Canadian marine station committee to raise the profile of efforts to gain the Dominion government's 'cooperation and financial aid in the establishment of a Biological Station in the Gulf of St. Lawrence.'[43] This committee was short-lived – it would be dissolved at the British Association 1898 meeting in Bristol. No money was forthcoming; instead, the Canadians received a loan of moral support. Second, by forming this committee and including Mills and Thiselton-Dyer, both of whom had served on the Naples and Plymouth laboratory committees, the association gave Canadian scientists the benefits of its experience with the Naples and Plymouth biological stations. Third, it helped instruct Canadians as to how such stations were run; to this end, in 1899–1900, it sponsored Ramsay Wright for a ten-week stay at Naples, where he familiarized himself 'with the methods in use at the station, as well as with the Naples Fauna in a living condition.' He hoped 'to utilise the technical experience gained at the New Canadian Marine Laboratory.'[44]

The committee was coordinated by its secretary, D.P. Penhallow (1854–1910), a professor of botany at McGill University. Canadian universities and scientific societies heartily responded to his request for additional moral support. When it came to considering the laboratory's form, Penhallow liked the idea of a floating station, stating that he did not know otherwise 'quite ... how we are to gain definite knowledge of the most desirable location ... within a ... short time by any other method.'[45] The allusion to a 'desirable location' indicates that the committee had in mind a permanent biological station, which would be built once the best site had been found after several seasons at different locations in a moveable station. This permanence, however, was not emphasized when a committee delegation formally met with the Minister of Marine and Fisheries, Sir Louis H. Davies, on 20 April 1898.

The delegation also included members and representatives of Canadian universities (Toronto, Laval, and McGill) and scientific societies (the Royal Society of Canada, the Nova Scotian Institute, and the Natural History Society of Montreal). The delegates placed a five-year plan before the minister for a station to 'be devoted primarily to investigations on the nature and the sources of the food of fish, oysters and lobsters' – although the facilities might also be used for scientific investigations by 'specially qualified students' from 'the various universities where biological work is carried on.'[46] Presumably, the committee hoped that during this time, the station would produce enough strong results to convince the federal government that it, continued services were required. This is reminiscent of the strategy followed by William Logan (1798–1875) in his successful campaign to place the Geological Survey of Canada on a permanent footing in the 1850s and 1860s.[47]

Six recommendations were placed before the minister. The first two recommended that a floating station, to be moved every year, be established in the Gulf of St Lawrence for a period of five years. The next two asked that the station's researchers be qualified investigators from Canadian universities, who would be offered opportunities for independent research. The final two recommendations proposed that the station be a government institution, but administered through 'a special Board consisting of one or more representatives from the Department of Marine and Fisheries, and one representative from each of the universities represented in support of the petition'; and that it be given an appropriation of $15,000: $5,000 for building and outfitting the station, and $10,000 for running it for five years. The two thousand dollars per year would also cover salaries for 'a janitor or keeper and also ... a Director appointed by the Committee.'[48]

The delegates' arguments were of a kind long used by scientists to get government funding. These 'classical' arguments began with an appeal to national pride: 'Canada is the only civilized country where no Marine Biological Station has been established.'[49] But this argument only works if very tangible benefits are promised. They argued that the fishing industry would generate greater revenues through improved practices. Furthermore, a Canadian station would provide precise scientific observations relating to the condition of fish stocks, their size and locations, how they could be increased, how harvesting methods could be improved, and so on. But this would require 'exact scientific investigations into such questions.'[50]

However, it was not enough to extol the benefits to the fishing industry. Since the federal government was being asked to spend $15,000, the delegates hastened to add that it would also 'benefit ... from co-operation with the different Universities and Scientific Associations in its dealings with the fisheries and in the consideration of methods for their preservation.'[51] University researchers would aid the Dominion government in solving fisheries problems, and the government would have first claim on the information they generated.

Canadian universities would also profit. At that time, to find experimental research facilities and biological material for their investigations, 'many of the qualified men trained in Canadian Universities' had to travel to or obtain specimens from different stations in the United States and other countries.'[52] The British Association had already sponsored Canadians to work at Naples and Plymouth: in 1888, E.W. MacBride had gone to Plymouth to study techniques for raising echinoderm larvae and in 1892 he had gone to Naples. Davies was told that the establishment of a marine station 'would be of incalculable service to the different universities of the Dominion, not only in furnishing them material in Canada but in adding to the material thus obtained an accurate scientific knowledge of the fishes and marine life generally characterizing our northern waters and distinct from the marine Fauna and Flora found in the vicinity of the various United States Stations.'[53]

That Canadians were forced to go abroad was a sting to Canadian national pride, one that Davies himself felt. In fact, the delegates' appeal to nationalism was three-pronged. They were urging the government to raise Canada's international status; they were calling for material assistance to Canadian universities; and they were urging the government to support the kind of inventory science that Suzanne Zeller claims was a component of Canadian nation building.

Zeller contends that by discovering the uniqueness of Canadian geology, meteorology, and flora and fauna, Canada was building its self-image as a distinct nation with a unique identity and thus a reason for existence. Natural history was 'one of the few cultural interests that united inhabitants of all the British North American colonies in a common pursuit during the nineteenth century.'[54] By taking botanic inventories, Canadians were mapping the geographical distribution of two kinds of flora – that which was unique to British North America and that which had apparently migrated from the Old World and had then adapted to Canada's harsh climate. Victorian Canadian botanists such as George Lawson and George Mercer Dawson had been influenced by Joseph Dalton Hooker, son of Sir William J. Hooker, Director of Kew Gardens (and later the director there himself). Hooker noted that 'the northern hemisphere supported two distinct floras, the one American and the other European

[and] expected great discoveries from his North American colleagues.'[55] Lawson's theories about plant migration and evolution helped shape a national Canadian self-image. Plants had migrated to Canada from Northern Europe and had adapted and grown hardy; so, too, would the people who had followed them and the nation they were building. Canadians were using a scientific theory for their own ideological purposes, to create the *idea* of a commonality among Canadians as a transcontinental nation. The model of geographical distribution, with its implications for the evolution of a Canadian nation forged by its northward expansion, offered an attractive interpretation of Canada's role in North America.[56]

In offering to make an inventory of Canadian marine biota, Prince and the other delegates were appealing to the nation-building instincts of the leaders of a fledgling nation. More than this, they were making a subtle appeal to the anti-American feeling then prevalent. An inventory of marine flora and fauna would distinguish Canadian forms characteristic of 'northern waters' from American forms. It was necessary to learn about these forms not only for learning's own sake, but also to discover their Canadian, non-American qualities.

The delegates' submission was appealing directly to the Imperialistic sentiment that permeated the Canadian scene in the late 1880s and 1890s: 'Certain questions relating to the Fisheries of Great Britain ... can only be fully understood by a study of fish fauna on our coast and vice-versa and by the establishment of a Canadian marine Biological Station and its co-operation with similar stations in Great Britain, mutual benefits would certainly result valuable to ... both countries.'[57]

Davies later told the House of Commons that the British Association had 'first represented the necessity of this station.'[58] Obviously, the Imperial connection would be vital to project's success.

The argument that Canada was the only civilized country without a marine biological station, and the 'uncomfortable' information that Canadians had to go to the United States to do research at marine stations, is what most influenced Davies. However, this was not, as has been claimed, 'the only argument Davies felt called upon to put forward in the House of Commons.'[59] In fact, he was also impressed by the delegates' reassurance that the Department of Marine and Fisheries need not 'assume any responsibility for the administration of the Station, but that a committee would be appointed by the various Universities and Scientific Institutions with a representative from the Department of Marine and Fisheries which Committee would be responsible for all arrangements and expenditures and for the administration of this Station.'[60]

Davies told the House of Commons that the government would not be responsible for running the station – only for funding it. In other words, the government would not have to hire or train the expertise.

The 'free' nature of the services being offered by university biologists was brought up by the station's executive committee time and again over the next twenty years in efforts to win concessions from the government. In a sense, the scientists were trying to convince the government that it was getting something for nothing: a very small price tag attached to the marine station, and no work. The truth was otherwise. Enormous fiscal headaches were to drive the marine station's scientists almost to distraction and were to generate much paperwork in the departmental bureaucracy. This was solely because of the department's reluctance to grant the station's management board any fiscal responsibility.

In the short debate over the first appropriation of $7,000 for the new station, approval was also swayed by information relating to Prince's involvement. One MP questioned whether the station should be managed by a committee of university appointees. Having been reassured by Davies that Prince would be a director,[61] he concluded, 'That is all right.'

When the request was finally granted, the triumph belonged mainly to Prince, although Penhallow and Macallum had shared much of the labour. Prince had long agitated for the station, and his personal qualities had been important to the success in obtaining funding. He had established a reputation as a sound and trustworthy administrator and bureaucrat within the Department of Marine and Fisheries, and he would be chairman of fifteen of the department's fishery commissions before he retired in 1924. His official colleagues' respect and trust aided his cause; he was so sincerely dedicated to the welfare of the fisheries that no one could doubt his claim that a marine biological station would aid them. He provided guarantees that scientists would not abuse funds or ignore Canadian fisheries problems. The monies granted would not be 'wasted' on pure and disinterested research, although some of this would go on as well.

The other members of the newly created Board of Management of the Marine Biological Station for Canada were all university professors: D.P. Penhallow (secretary-treasurer); E.W. MacBride from McGill University; Macallum and Ramsay Wright from the University of Toronto;[62] Loring W. Bailey of the University of New Brunswick; A.P. Knight of Queen's University; A.H. MacKay of Dalhousie University; and the Reverend V.A. Huard of Laval University. Among these, too, Prince's qualifications and personality led to trust and respect. M'Intosh described Prince as 'earnest and genial,'[63] and the label 'genial' stuck. Everywhere he was referred to as 'the genial Professor Prince,' and this quality stood him in good stead when he set out to accomplish things through the good offices of others. He was also an experienced scientist with one foot in the academic door. All of this made him the ideal person to mediate between academics and bureaucrats. He understood the desires of the first group and the reassurances required by the second. A large part of the delegation's pitch to Sir L.H. Davies – such as the offer to give the government first claim to any information generated by the new station – reflected Prince's understanding of government. The argument that the government would not have to put any effort into running the station was an ingenious means of gaining the complaisance of both partners in the enterprise: for the government, it created the illusion of efficiency, and for academics, the illusion of research freedom.

It was largely through Prince's influence and the support of the British Association, then, that the floating station – a long, narrow building mounted on a scow – duly opened in the summer of 1899. Prince's early exposure to Scottish marine biology only strengthened the already extraordinary influence that Scottish science was having on nineteenth-century Canadian science. Throughout much of the nineteenth century, English universities emphasized a narrowly classical education; in contrast, Scottish universities offered undergraduates 'practical subjects' such as natural philosophy (science). The influential botanist George Lawson (1827–89) was a Scot trained in science at the University of Edinburgh. Montreal-born William Edmond Logan (1798–1875), director of the Geological Survey of Canada from its inception in 1842 until 1869, had studied medicine in Edinburgh. Robert Graham Dunlop (1789–1841), the man appointed to develop a plan for the geological survey, had studied science in Glasgow and Edinburgh.[64] Nova Scotian Sir William Dawson, Principal of McGill University (1855–93), who had taken a degree in Edinburgh, 'was of special importance in establishing the study of science at McGill and in Canada.'[65] Scottish higher education had done more than anything else to create a scientific tradition in Canada. Men educated in Scotland were leading the Canadian scientific establishment and introducing science into Canadian university curricula. Prince was merely the latest of these.

The floating station was first located at St Andrews, New Brunswick, which would later be its permanent home. Similar stations had already been used in Scotland and Denmark. However, the design for this one was inspired by a floating station in Michigan. Penhallow, during a board meeting in Montreal on 20 January 1899, had 'submitted a plan and specifications of a Biological Station which had been erected in Michigan, and was on a scale and of a design to a large extent applicable to the projected Canadian station.' The board agreed with Prince that 'initial work could be most favourably commenced in the waters of Passamaquoddy Bay' due to its richness of marine life and the conveniences offered by St Andrews. Once it opened, Prince reported happily 'that the station is really a beautiful bright structure admired by the numerous Americans and other visitors who see it. Four able workers are here from Toronto and elsewhere busy at work on important fishery investigations.'[66]

These were very humble beginnings; the marine biological program in Canada had a long way to grow. I have emphasized its English and Scottish roots so far, but Canadian scientists in the following years did not restrict themselves to the Old Country in searching for the best models for marine science operations. Pioneering Canadian marine biologist A.G. Huntsman observed that both Great Britain and the United States strongly influenced the early history of Canadian marine biology. This has been the historic pattern for many scientific and other institutions in Canadian history; English-speaking Canadians have teetered on the fence between their British roots and Imperial ideals and the reality of their powerful southern neighbour. However, this tradition was very much attenuated in the case of the Canadian marine station, due to the equally strong international tradition in marine science developed by the *Challenger* expedition and the Statione Zoologica di Napoli.

By the time the Canadian marine biological station was established, there were thirty-eight marine biological stations already in existence in Europe and another half-dozen in the United States. The Canadians had many examples from which to model their station: some were designed exclusively for basic biological research, others for teaching science students, and still others focused on educating fishermen and skippers. The Zoological Station at Naples was viewed as the ideal owing to its internationalism and complete freedom of research. The most prominent American marine station, the Marine Biological Laboratory at Woods Hole, Massachusetts, also provided an enviable example. So it is surprising that neither was to exert much influence on the Canadian station.

Woods Hole was officially opened on 17 July 1888 and shared with the Naples station an early preoccupation with basic developmental questions 'posed within a solid morphological tradition.'[67] Its precursor was the Anderson School of Natural History, a summer school founded by the Swiss-American paleontologist and ichthyologist Louis Agassiz on Penikese Island in 1873, with financial backing from a wealthy New Yorker. It was meant to offer schoolteachers practical experience in natural history; however, Agassiz died after running it for only one year, and the school foundered the year after that. One of its students, Alpheus Hyatt, then set up a seaside laboratory in Annisquam, Massachusetts. He moved it to Woods Hole in 1887. Woods Hole had pure water and abundant marine life; it was also the site of the U.S. Fish Commission, whose director, Spencer Fullerton Baird (1823–87), was Hyatt's friend. Hyatt became president of the laboratory's Board of Trustees; Dr Charles Otis Whitman (1842–1910) was named its first director.

Wood Hole and Naples were in the front ranks of the revolt against Haeckelian phylogenetic embryology. They promoted instead new methodological and conceptual approaches to the problems of development. According to Haeckel, developing embryonic forms revealed phylogenetic relationships among animals and the history of their evolution. Younger researchers complained that Haeckel's theories were too speculative and were impossible to verify. German embryologist Wilhelm Roux (1850–1925) argued that 'the study of the embryo was interesting in its own right.'[68] He advocated experimentation on embryos to discover what factors that controlled their development – testing the effects on embryonic growth of centrifugation, electric and magnetic fields, and chemical treatments. Naples became a major research centre for the new experimental approaches of Roux's *Entwicklungsmechanik* (developmental mechanics) after Roux's colleague, Hans Driesch (1867–1941), the 'uncompromising and arrogant defender' of that approach, visited the station in 1891. There, he began a whole school of investigation devoted to experimental embryology using easily manipulated sea-urchin eggs.[69] Researchers at Woods Hole, excited by developments at Naples and Germany, also focused on experimental embryology, morphology, physiology, and the problems of regeneration.[70]

It is important to note that Woods Hole and Naples were essentially biological laboratories in marine settings, not marine biological laboratories. Embryology was by no means the only research done, but marine botany suffered at Woods Hole, which lacked a botanist 'with interests parallel to Whitman's.'[71] Naples gave more weight to investigations of the local marine environment than Woods Hole, where fish and environmental studies were already the domain of the U.S. Fisheries Commission's laboratory. At Naples, systematic surveys resulted in a series of monographs: the famous *Fauna und Flora des Golfes von Neapel*. As Dohrn noted in 1873, one of the station's 'foremost duties' was to identify local fauna, to overcome existing 'confusion ... with regard to systematic and zoological questions of the Mediterranean fauna.'[72]

Like Naples, Woods Hole was independently run, funded by donations from educational organizations and the private sector. Thus, it was unfettered by demands for mission-oriented research. The trustees of Woods Hole were usually elected from the corporation, which was in turn usually composed of all workers at the laboratory willing to become members. Thus, those most interested in unfettered research were responsible for running the Marine Biological Laboratory.[73] In contrast, the Canadian station, although almost exclusively staffed by academic biologists, was, in the great Canadian tradition, entirely government funded. This funding was too sparse to support the growth of research traditions analogous to those of Naples and Woods Hole, and was tied to an implicit understanding that fisheries research would be undertaken.

This is why other marine laboratories had a more immediate influence on the first Canadian biological station. Prince's training at the St Andrews station in Scotland, which was dominated by fisheries research, inclined him to model the Canadian station after St Andrews. The Marine Biological Laboratory at Plymouth also emphasized marine and fisheries research, rather than general biology, and served as another attractive model. Soon, Canadians became especially anxious to emulate Northern European fisheries research, and made contact with scientists from Norwegian fisheries research institutions. Professionally speaking, Naples and Plymouth were developing new niches for research-trained biologists, since both required full-time directors, naturalists, curators, and collectors. At Naples, each department – for example, physiology and botany – required its own resident specialist. The Canadian station would hesitantly follow suit; thus, the Biological Board would provide some of the first professional positions for trained biologists outside of Canadian universities.

Clearly, the organization and objectives of biological stations varied greatly between countries. It is hardly surprising that the early Canadian marine biological stations would themselves make independent and unique progress, with a distinctively Canadian twist, highly flavoured by the ubiquitous role of government. It now seems inevitable that conflict would arise between those who expected the new biological station to improve the state of Canadian Atlantic fisheries, and those who wanted the station to provide a setting for academic experimental research – a kind of extension of Canadian university zoology departments. But this conflict would be slow in coming and did not dominate the early years of the new biological station.

Chapter 2

Fishing for Ideas: Approaches to Marine Biology and Fisheries Science to 1914

Alive without breath,
As cold as death;
Never thirsty, ever drinking,
All in mail never clinking.

The Hobbit[1]

In the early years of this century, basic and applied marine biological science were both cultivated so as to produce a harmonious research area for Canadian biologists. That this observation arises at all in this history is a result of the beguiling arguments employed by apologists for both categories as they try to arrange all science into the camps of either pure or applied science. Proponents of basic or pure science argue that it is the only true source of real new knowledge. If the public were ever to stop funding this kind of research, scientific progress would bog down in a welter of special interest studies, and new knowledge would never be made free to the world. For their part, apologists of applied science argue that valuable practical spinoffs from pure research are rare, that it is an inefficient route to follow, and that applied research is vital to developing new and immediately useful knowledge and technologies. What tends to get lost in this argument is the experience of many scientists, that much of science is not so easily clumped into either category. Fisheries biology provides a classic example of this, as illustrated by the early experience of the Biological Board of Canada and fisheries research organizations elsewhere.

As it developed, the Board of Management of the Marine Biological Station charted a unique course. No single station in Europe or the United States could have served as its model. Entirely government funded, it was composed mainly of professors representing universities where biological research was done. No university had power over it, and the Department of Marine and Fisheries had influence only through its chairman, E.E. Prince, the Dominion's Commissioner of Fisheries. Since Prince's background was academic, and since professors made up the Board of Management, only the board's reliance on government funding would lead us to suspect that it was not a harbour for basic science. Applied fisheries biology did in fact, thrive, but it was not because the department (at this stage) was especially insistent that applied work be carried out.

When historians have considered the role of fisheries biology at various marine biological stations, they have perhaps been swayed by a natural but misleading parallel with another great applied science related to food production: agricultural research. The early aquatic biological stations were modelled in part on agricultural research stations, which predated them by about thirty years. The Statione Zoologica di Napoli, opened in 1872, was among the first marine stations, but Jean Baptiste Boussingault (1802–87) established the first agricultural experimental station in 1834 in Alsace, and in 1852 the first government agricultural experimental station was opened at Moeckern, in Saxony, to develop methods for applying science to agriculture as inspired by Liebig's treatise on plant nutrition.[2] Victor Hensen (1835–1924) and Karl Brandt (1854–1931), in their attempts to measure oceanic productivity at the Kiel Commission (the Kommission zur wissenschaftlichen Untersuchung der deutschen Meere, established in 1870), were influenced by Liebigian agricultural chemistry. Brandt turned to Hermann Rodewald, a director at Kiel's Agricultural Institute, for advice in applying techniques from agricultural chemistry to measure planktonic chemical composition and productivity.[3] In the 1920s the Biological Board modelled its fisheries experimental stations on agricultural stations. Since fisheries biology was meant to do for fisheries what experimental farms were doing for agriculture – use science to improve food production – one might expect conflicts in research programming such as those which indeed arose at agricultural stations.

In agricultural research, conflicts between basic and applied science are probably more acute than in any other area except medical science. In the United States in the late nineteenth century, scientists at agricultural stations were especially at the mercy of a largely unsympathetic public, including politicians and farmers hostile to and suspicious of scientific learning. To gain acceptance, experimental farms and research stations had to turn themselves into 'fertilizer control agencies,' and scientists found themselves burdened with numbingly repetitive soil and fertilizer analyses. Farmers wanted individually tailored advice about which fertilizer to buy for their soil; the stations provided it, and also monitored fraudulent claims made by fertilizer companies. Seed testing was another dull and repetitive task. In the United States, these agricultural stations were generally linked to universities or land-grant colleges, which meant that their staffs often carried heavy teaching loads, leaving little time for original research. They had to nurture their contacts with farmers and put out regular bulletins even when they had nothing new to disclose. Some stations became general information bureaux, where scientists were expected to answer every question farmers asked them.[4]

Similarities with early fisheries biology are fleeting, however. The repetitive testing of soils, nutrients, fertilizers, and seeds that was so much a part of agricultural science had no analogy in early fisheries biology. Agricultural biology had little to do with mainstream biology until experimental genetics came along. In contrast, even as late as 1930, much of fisheries biology was barely distinguishable from fundamental marine biology. The main work involved inventorying marine life for different localities (an aspect of basic ecology) and that grab bag of research touching on fish life histories, including developmental and comparative morphology, and studies of predator–prey interactions as well as migratory, reproductive, and other behaviours.

In those early years, marine biology was still largely exploratory, and a fisheries biologist could easily conduct research that he himself would consider both basic and applied. More likely, he would not have used such limiting definitions, although he would have recognized gradations in the applicability of his work. There is in fisheries biology a range between research aimed at longer-term application, and more immediate applied work. Studies of the effects of dynamite fishing offer an example of the applied extreme; esoteric taxonomical studies, on the other hand, would not have any immediate practical applications. Much of fisheries biology occupied a less well-defined region between these extremes. At the time, 'so little was known about the oceans' denizens that any new discovery produced knowledge valuable both as an end in itself and as a basis for fish management – indeed, it could be argued that any information whatsoever would help the fisheries. But even though there was no real conflict between basic and applied science, not all scientists were happy to occupy the middle ground: Thomas Henry Huxley was highly antagonistic to what he regarded as the trivial demands of fisheries research, whereas A.P. Knight of the Biological Board was happiest doing strongly practical work. Most fisheries scientists were not Huxleys, and their attitudes fell between these extremes. Canadian practitioners at least were not particularly prone to distinguishing which kind of science they were doing. In part, this was because they did not really have a vision of what fisheries biology should be, and were content to explore the near-shore waters in the Victorian natural history tradition. The main drawback was that this type of fisheries biology was not highly experimental, and thus not at the fashionable leading edge in biology.

In 1900, fisheries biology was a science without much of a research tradition: it was still searching for a methodology, still seeking the best phrasing for asking questions of nature. In this era, the early post-partum of Canada's first marine station, fisheries science was open to many approaches; it was still undefined, and it offered little guidance to its first practitioners. Its lines of research, such as they were, had not yet ossified into rigid forms that could be labelled 'pure' or 'applied.' And there are many indications that the relation between early applied and basic marine biology was not strictly dichotomous. Fisheries Research Board biologists W.J.K. Harkness and J.W. Leonard, and Berkeley professor Paul R. Needham, wrote in 1954: 'The line between so-called basic and applied research is hard to discern, if it exists at all. It is almost equally difficult to draw a line between research and its application in management.'[5] This leaves questions, however, as to how 'basic' marine biological research is defined, and how it relates to fisheries biology if the latter is seen as more 'applied.'

In one sense, marine biology can be viewed as any study centred around marine organisms, and thus a field as broad as biology itself. Indeed, marine biological laboratories at the time were not necessarily centres for 'marine' biology, and they still are not. For example, the Naples and Woods Hole laboratories were both centres for general biological research. In 1902, the Woods Hole laboratory director C.O. Whitman argued that '"Biological Station" would better express the character and aim' of his laboratory. He even admitted: 'the word "marine" is therefore a somewhat misleading reminiscence of an early stage of development, when sea forms alone occupied attention.' At both Woods Hole and Naples, scientists focused on general morphological and physiological principles rather than on the marine origins of their experimental subjects. The Woods Hole publication *Biological Bulletin* and Naples's *Zoologischer Jahresberich* and *Mittheilungen aus der Zoologischen Station zu*

Neapel were general biological journals; many articles came from research outside the stations.[6]

In its modern sense, marine biology is the study of the ecology of marine organisms and their entire ecosystem. It examines the dependence of living organisms upon abiotic (physical and chemical) conditions, their interdependence, and their influence upon the environment. Marine biology is therefore an ecological science requiring knowledge of both plant and animal systematics and 'physiological and ethological studies ... indispensable for an understanding of biological processes in the sea.'[7]

Fisheries biology, as it was carried out prior to the Second World War, was hard to distinguish from marine biology. It involved the study of all aspects of commercial fish and shellfish, including their physiology, ecology, life histories, migratory and other behaviours and natural population fluctuations. It also involved studying how fishing activities, pollution and other human interventions affect commercial fish populations. Much of this is hard to distinguish from related, non-applied biology, except for its focus on commercial species. However, fisheries biology also entailed and continues to entail investigations of fish preservation and handling, fishing techniques, and fish culture. These incline it to be categorized as an applied science. This has been reinforced by the unfortunate overemphasis on fish population dynamics studies and population modelling that has characterized the discipline since the Second World War, to the detriment of the more complex ecological vision of its earlier practitioners.

Despite the early complexities – and perhaps because of the later 'industrial' emphasis of fisheries science – modern historians have tended to conclude that early marine biologists were staking out territory for basic science in what was more properly an applied field. The work of Spencer Fullerton Baird (1828–1887), Assistant Secretary of the Smithsonian Institution and founder of the U.S. Commission of Fish and Fisheries, has been viewed in this way. Baird's Commission was created by an Act of Congress on 22 January 1871 to organize American fisheries investigations and fish culture.[8] Historian Dean Conrad Allard claims, however, that Baird's 'real' interest was biological research.

Baird's commission got its start because of a heated dispute in 1870 between weir and fixed-net fishermen and line fishermen in New England. The line fishermen complained that coastal traps were capturing most fish before they could spawn and reproduce, causing an alarming decline. Baird believed he could solve the problem of the decline of coastal fishes by conducting a scientific investigation, and he was 'keenly interested in proving the utilitarian value of science.' But Baird's strong support of basic science clashed with Congress's motives for funding the Fish Commission, Allard contends, stating that Baird subordinated practical investigations to abstract research. Baird's Fish Commission summer laboratory was initially peripatetic, ranging from Maine in 1872 and 1873 to Connecticut and Massachussetts in the following years, and finally Halifax, Nova Scotia, in 1877. According to Allard, Baird carried out his top personal priority – primary oceanic research – at each of these coastal summer stations. In 1881, Baird built a permanent laboratory at Woods Hole (the Woods Hole Fisheries Laboratory), the first to locate here. The commission's research vessel *Albatross* was used primarily for scientific exploration, while the Fish Commission launched the first sustained biological study of North American waters, making 'scores of contributions in systematic biology, ecology, and embryology.'[9]

Baird's first report to Congress, Allard notes, remained silent about his 'ambitious' basic scientific studies of coastal and oceanic waters. However, fisheries studies rely on this kind of initial work. After only one summer's research on the question of scup depletion, Baird followed his intuition and recommended restrictions on fishing operations, predicting, if no action were taken, 'the virtual extinction of the scup and other inshore fishes.' Ironically, although no new regulations were applied, after 1871 the yearling scup catch increased sharply as did the total supply. Baird now realized that general studies would be necessary in order to understand the fluctuating abundance of sea life, as a basis for regulating fisheries.[10] Thus the Fish Commission became the first body to take up Norwegian fisheries biologist G.O. Sars's earlier studies of the development of commercial fishes to maturity.

The research of Henry B. Bigelow, founding director of the Woods Hole Oceanographic Institution, has also been described in terms of Bigelow's supposed avoidance of applied problems wherever possible. In 1912, Bigelow convinced the U.S. Bureau of Fisheries to sponsor his intensive area oceanographic studies of the Gulf of Maine, where the fish supply was diminishing due to increasing fishing intensity. Historian Jeffry P. Brosco tells us that Bigelow, as a Bureau of Fisheries advisor, was mostly able to control his direction of work, but occasionally 'Bigelow knew he had to respond to USBF questions regardless of his personal interest in them.' For example, in 1925 his study of currents in Massachussetts Bay indicated that cod eggs would be carried out towards the Nantucket shoals and Georges Bank, so he recommended that fishermen be directed to fish there. Brosco admitted that 'Bigelow's work in physical oceanography and plankton was ... crucial for predicting fish populations,' and that 'much of the information necessary for ... fish supply forecasts came from Bigelow's Gulf of Maine survey,' on the basis of which predictions of mackerel population fluctuations were begun in 1928. Indeed, '"applied" fisheries problems were also "pure" oceanographic problems because of the underlying unity of the sea.'[11] But Bigelow could not have felt quite the aversion to applied problems that Brosco implies. Had this been the case, it is unlikely he would have been president of the North American Council on Fisheries Investigations from 1923 to 1938. This organization, the history of which is told in Chapter 6, was devoted entirely to fisheries science.

Even at places like the Naples station, a certain amount of practical fisheries work was undertaken. An 1886 trawling study was begun in Naples in response to antitrawling agitation, at the instigation of the Italian Ministry of Agriculture and Commerce. Since most food fishes produced floating eggs, 'the supposed injurious effect of trawl-fishing' was 'proved to be an illusion.' The station recommended abandoning laws restricting trawling.[12] This practical work was not far removed from the ontological emphasis of Naples's basic research, as the starting point of developmental morphology was always the fertilized ova. But it also resembled the fisheries science being done in Scotland, Plymouth, and elsewhere at the time.

In 1956, Michael Graham, one of the outstanding twentieth-century fisheries biologists, grappled with the problem of defining fisheries biology. English fisheries biology, he claimed, was launched by the work of E.W.L. Holt, and immediately became distinguishable from 'marine biology' as exemplified by the work of J.T. Cunningham. Holt and Cunningham were young naturalists working for the Marine Biological Association, Holt at Grimsby, Cunningham at Plymouth. Holt in 1892 went to sea in a trawler, collected statistics, and proposed size limits to rest small-fish grounds; he also experimented with and published trawl designs to allow small fish to escape. On the other hand, Cunningham's 'Treatise on the Common Sole' (1890), although a classic, 'has not the flavour of fishery research.' When association funding dried up in the 1890s, Holt became Chief Inspector of Fisheries in Ireland, while Cunningham became a university professor and wrote an influential book on animal sexual dimorphism. However, according to Graham, 'the distinction was not ... and is not, entirely water tight'; indeed, 'the distinction could be made too much of,' since 'Cunningham's intention was fully practical and so are those of the M.B.A. and the Scottish M.B.A.' Graham contended that most fisheries science was 'of intermediate and medium-term application in the fisheries,' while the work of grant-aided laboratories such as the Plymouth laboratory at mid-century was 'of the same nature, but of medium and long-term application.'[13]

Graham had to grapple with these questions for the same reasons that applied science is so frequently shoved aside by historians of science. Many are happier concentrating on histories of basic, fundamental research, preferably of the revolutionary sort (in the Kuhnian sense). The influential nineteenth-century sociologist August Comte elevated the abstract, positive sciences, including physiology; in his view, they were more important because they concerned fundamental laws governing elementary facts of nature. Furthermore, the 'concrete' sciences, including zoology and botany (and by extension, the applied sciences) merely noted present circumstances and concrete facts.[14] By focusing on theoretical science, historians evade the scorn that Comte heaped on 'concrete' sciences. Nathan Reingold noted that American historians of nineteenth-century American science 'complain[ed] about the paucity of great men and great contributions to major scientific developments in the United States ... reflect this awe for the Comtean hierarchy,' as they place emphasis on 'the lack of American contributions to the cell theory, to the germ theory of disease, and to physiology, all areas presumably characterized by a respectable degree of experimentation, quantification and theoretical structure.'[15]

The practice of opposing pure against applied science became more important after the Second World War II, perhaps in response to the odium attached to that most applied scientific exercise of all, the Manhattan Project. Karl Popper denigrated applied science in disagreeing with Thomas Kuhn's theory that scientific revolutions punctuate 'normal' science:

> In my view, the 'normal' scientist, as Kuhn describes him, is a person one ought to be sorry for ... He has been taught in a dogmatic spirit: he is a victim of indoctrination. He has learned a technique which can be applied without asking for the reason why ... He has become what may be called an *applied scientist*, in contradistinction to what I should call a *pure scientist*. He is, as Kuhn puts it, content to solve 'puzzles.'[16]

Applied science indeed has frequently been confused with technological applications and thus with technology, and the relationships between science and technology in turn have often been misconstrued by historians of science. Pure science is portrayed as 'a "search for new law(s) of nature" ... conducted in [a] "free and lofty spirit,"' whereas 'technology is a middle ground between science and arational rules, and it participates in the objectivity of science precisely because it is an application of science.' This view which declares pure science to be the ascendant form of objectively valid knowledge, and which renders technology '"mindless," bereft of its own intellectual method,' was in the past shared by a number of contributors to *Technology and Culture.* Mario Bunge argued that 'the line must be drawn if we want to account for the differences in outlook and motivation between the investigator who searches for a new law of nature and the investigator who applies known laws to the design of a useful gadget.'[17]

I would argue that this kind of assumption is too rarely questioned by historians of science. The implication of identifying normal science with applied science is that 'scientific revolutions' or methodological discoveries do not occur in applied science. Applied science is seen as the passive recipient of advances in the pure or basic sciences. However, such is patently not the case. Edwin Layton's 'Mirror Image Twins' pointed out that 91 per cent of all important developments between 1945 and 1966 arising from U.S. Department of Defense R&D spending were technological, 8.7 per cent were from applied science, and only 0.3 per cent were a result of basic research. The received (and persistent) model of science–technology relationships is misleading, since science and technology 'constitute different communities, each with its own goals and systems of values.' Fisheries biology also refutes commonly held understandings of pure and applied science. It had a strong apologist in the mid-twentieth-century fisheries biologist Michael Graham, who said of it that it 'had pre-eminently the flavour of life. That, I think, comes from real mastery of any problem in applied science.' Fisheries biology had developed its own paradigms, such as the year–class model of population fluctuations in the 1910s, and the theory of fishing, centred on population modelling as developed largely within the discipline in the 1930s and later (see chapters 3, 6 and 7) Also, the Biological Board's involvement with fish processing and fish freezing research (see chapters 4 and 5) drew some of its scientists into highly theoretical research, and to participation in the International Institute of Refrigeration. Commission III of this institute studied the 'relation of refrigeration to enzyme action at low temperatures and to enzyme action in material which has been frozen and thawed,'[18] work of both theoretical and applied interest.

Beginnings of Fisheries Biology

Early twentieth-century fisheries biology was indeed engaged in building its own research community with separate goals. A history of the developments taking place in this new field will be dealt with in some detail in chapters 3 and 6, but at this point, a general introduction is essential. Fisheries biology started with life-history studies of commercial species, and its first real practitioner was Professor G.O. Sars of the University of Christiania (now Oslo), Inspector of Fisheries for Norway. The expanding cod fishery of the 1840s and 1850s had failed in the Vest Fjord, and Sars was appointed in 1864 by Norway's government to find out why. He discovered that the Lofoten Islands' fishing grounds were teeming with floating cod eggs; he collected some of these and hatched them in jars of seawater. From his discovery that cod eggs floated, he realized that hydrographic events such as fluctuating currents could affect larval and juvenile fish. In 1893, Sars showed that good herring catches were linked with warm water off the Norwegian coast.[19]

In Sars's lifetime, sea-fishing underwent deep and lasting changes. With the Industrial Revolution, urban populations had grown rapidly, and so had their demand for more and cheaper food. The new railways were allowing large quantities of fresh fish to be transported to industrial centres, especially in Britain, and this created enormous demand. More and more trawlers were being built that could fish farther offshore than line- and drift-fishing boats. Trawlers, purse seiners, and bag-netters offered gave fishermen highly efficient means of harvesting fish. Then, in the 1880s and 1890s, steam engines were installed in purse seiners, trawlers, drifters, and long-liners. All were more effective than their precursors. The steam trawler, now equipped with a double-barrelled steam winch, 'could shoot and haul the trawl in half an hour as compared with three hours by hand,' while the purse seiner with its power block 'eventually became the most destructive engine for killing fish' because its small mesh captured small fish with large.[20]

North Sea catch rates declined in the 1890s, as did those for Pacific halibut by 1910, and the fish caught were becoming smaller. Overfishing and declining catches drove fishermen to explore farther offshore. By the 1900s, the English steam-trawling fleet, the world's largest, had been drawn farther afield, to Iceland and the Barents Sea. Ice became essential 'because, as stock density fell, vessels stayed at sea longer,' and also because fresh fish required cold storage in the ports. This sequence came later to North America, but in the east, 'railways were in place by the 1870s and 1880s' and factories were being built from the Chesapeake to Nova Scotia.[21]

In Britain, where the steam-trawling fleets were growing fastest, line fishermen began to complain loudly about trawlers. They accused trawlers of depleting fishing grounds, and insisted that heavy beam trawls were tearing up sea bottoms, where fish were believed to spawn and their young to grow. This led in 1883 to the Commission on Trawl Nets and Trawl Fishing. The Secretary for Scotland, Lord Dalhousie, was chairman. Commissioner Thomas Henry Huxley asked W.C. M'Intosh to serve as naturalist. M'Intosh was to trawl 'some piece of ground ... at regular intervals for ... six months or so,' as would a fishing trawler, and was to record what each haul brought in. Fisheries biology was not yet statistical, so the commission made no use of those kinds of data. At the Marine laboratory at St Andrews, M'Intosh and his assistant E.E. Prince concentrated on the life histories of some forty species of fish, from egg to maturity, and published their findings in 1888 as a 270-page paper, the *Life History of the Marine Food Fishes* – 'a milestone in the British treatment of the subject,' according to M'Intosh's biographer, A.E. Gunther. The commission concluded that there was no proof that trawling destroyed fishing grounds; however, if a limited area did become depleted, it could be restocked by 'resting' the area. Trawlers were found not to destroy fish spawn, which floated at the surface. One of the commission's positive results was that detailed fisheries statistics first began to be kept in Britain. Statistical work was first organized by the Fishery Board for Scotland, established by an Act of Parliament in 1882 to investigate British food fishes and provide a scientific basis for legislation.[22]

Many British marine biological stations owed their existence to fisheries biology, a scenario repeated in other maritime nations where fishing was of at least regional importance. Charles Kofoid's 260-page guide, *The Biological Stations of Europe* (1910), surveyed station research activities: in many cases there was a mixture of fisheries and other research. Perhaps predictably, fisheries research was found most often at state-funded stations. Among twenty-three state-funded stations, including those built for the International Council for the Exploration of the Sea (ICES), only in three was fisheries research absent, and in each of these exceptions, funding came through ministries of culture or education rather than fisheries. Many publicly funded stations were formally affiliated with a university or a museum, from which scientific staff were often drawn without extra remuneration. The staff were then answerable to the affiliated biology or zoology departments as well as to some government ministry.[23] Despite their academic affiliations, most publicly funded stations used fisheries biology to justify their existence. For example, when in 1891 the Bergens Museum applied to the Norwegian Storting to fund its biological station, it argued that 'practical scientific fisheries investigations ... would find a biological station a valuable support.'[24]

Kofoid described fifty-one marine biological stations in detail; of these, only twelve were private. They tended to be preserves of basic research. The benefactors of private stations included scientific societies, private patrons, business interests, and sometimes (at arm's length) governments. The most famous, the Naples station, was outstanding for tapping many sources to remain independent. It was supported by Dohrn's 'table system,' by which universities, scientific associations, or individuals could rent a table for the use of one scientist for one year. Income also came from the specimen supply program, which sold collections of preserved marine animals to museums, institutes, and individuals throughout the world. Several governments gave grants, but none had any power to direct the institution.

By contrast, at the Plymouth laboratory, scientists embraced fisheries research, in part because substantial state funding required practical results. When Plymouth offered the Marine Biological Association a site for its laboratory, local businessmen with fisheries interests, other local benefactors, a number of universities, and the Fishmongers' Company offered generous financial assistance. But the greatest single grant to this private organization came from the British government, which gave £5,000 to the building fund and £500 per annum over five years for running costs. The fisheries' increasingly international character meant that the government needed scientific information to draft good legislation; the grant came with provisos that the association be audited yearly, submit annual reports, procure practical results regarding food-fish breeding and management, give aid and space to other agencies' fisheries investigators, and collaborate with quasi-governmental agencies elsewhere, such as the Fishery Board for Scotland.'[25]

The relations between general marine biology and applied fisheries biology at these various stations still require thoughtful elucidation. Histories of Naples and Woods Hole tend to focus on the role these stations played in the birth of experimental biology, and histories of the U.S. Fish Commission have focused on the basic research done by its scientists.[26] Little has been done to describe any of the economic work done at these institutions, and in-house histories of the British institutions do not raise any issues surrounding research motivations.

However, there are now a few studies which indicate that scientists often had mixed motives, and did not particularly recognize any discrepancies between applied and basic research goals. For example, American marine geophysics, a new area of geodesy, emerged as a new research tradition as the direct result of a highly successful submarine expedition in 1928 during which the theoretical research goals of Dutch geodesist F.A. Vening Meinesz were married with the more practical goals of the U.S. Navy. Vening Meinesz wanted to extend land-based geodesy by measuring the acceleration of gravity in the Gulf of Mexico and the Caribbean; by discovering areas of isostatic anomaly, he would be able to interpret the earth's crustal stresses, movements and forms. All of these were fundamental to geologic theory. At the same time, the expedition took depth recordings for topographic corrections, current readings, and other measurements, consistent with the U.S. Navy's requirements for more accurate latitude corrections and other data to solve navigational problems. The presence of geologists on the cruise also coincided with the navy's interest in discovering petroleum reserves to reduce American dependence on imported Mexican oil in case of hostilities. Navy Secretary Curtis Wilbur was highly supportive of the venture. He believed strongly in creative innovation, and in fact, out of the expedition came a new scientific instrument, the torsion gravimeter, patented by geologist–inventor Fred Wright, who accompanied Vening Meinesz and naval scientist Elmer B. Collins on the expedition. Indeed, Vening Meinesz and Wright, as well as the Chief of the Geodesy Division of the U.S. Coast and Geodetic Survey, William Bowie, who helped organize the expedition, were all 'exceedingly productive scientists, and part of the pleasure of their work seems to have been that it *was* both purposeful and significant [to theory].' So writes the historian Naomi Oreskes. In other examples, Nathan Reingold, while denying that he is an apologist for applied research, has highlighted the importance of the ideal of service in American federally funded science. He has also shown how closely related were the functions of 'pure' and 'applied' science in the realms of astronomy, meteorology, and geodetic surveys (the 'humbler' practical results of which contributed mainly to navigation aids, maps, and charts). For example, in the American nineteenth-century research tradition to which Alexander Dallas Bache, supervisor of the U.S. Coast Survey, belonged, 'routine technical work had to have an accompanying research component' and both the research and the routine work 'were ideally related to a theory in mathematical form – the theory giving direction to the research and the routine work, while the resulting data confirmed, extended or refuted the theory.'[27]

Furthermore, at American agricultural research stations, many agricultural scientists were trying to combine their own needs with those of their agricultural constituencies. They reacted in several ways to funding constraints and narrow interpretations of their tasks by politicians and farmers. Although a few did pour scorn on practical work, many came to believe that the conditions most appropriate for 'high-grade' research were also the best ones for economic growth. They strove to create new applied science disciplines with norms more appropriate to the demands of experimental stations than to the older, pure-science fields, although a younger generation of scientists, who enjoyed more funding, power, and professional autonomy following the Adams Act of 1906, rebelled against such compromises. Station directors also used entrepreneurial strategies to tailor their research policies to farmers' needs while maintaining the values and realities of scientific research. For example, W.A. Henry of the University of Wisconsin's agricultural program changed his station's research program to meet influential farmers' needs, so as to improve relations with them. To have a fair chance of solving the farmers' problems, he selected well-trained scientists. Finally, his program was designed to help farmers adapt to an economy that was shifting away from grain production toward other crops. Eugene Davenport, agricultural station director at the University of Illinois, invited producers' associations to lobby for specific appropriations for research in livestock and in certain crops. He claimed repeatedly that 'money devoted to investigation pays, and pays immediately.'[28]

Inevitably, parallels were drawn with the strategies employed by the Biological Board. Its chairman, E.E. Prince, gave the board's stations much-needed publicity. He knew how important it was to popularize the board's work, especially among fishermen. He often resorted to hyperbole when describing the board's progress in magazine articles, and, as he was 'in much demand as a lecturer from coast to coast,' in lectures. When *Canadian Fisherman* first appeared in 1914, Prince immediately grasped the opportunities it offered for disseminating his gentle propaganda; as he well know, fishermen were notoriously suspicious of change and interference. Prince exaggerated what he called the 'considerable economic and practical value' of the work done by emphasizing that it included 'in every case some problem that has been a serious difficulty to the Department.'[29] Nevertheless, many problems were addressed with the needs of fishermen and the Department of Marine and Fisheries in mind.

The board was far from highly professional, although Prince tried to make it sound otherwise. Prince's first-year report claimed that workers at the Atlantic station 'conducted labourious and valuable scientific investigations,' which 'conferred upon it ... a status, which many marine laboratories in other countries have attained only after many years of struggling effort and slow progress.' This station, which opened at St Andrews in 1899, was reminiscent of a pullman car. It had a central, well-lit laboratory, provided with salt and fresh water, that could accommodate twelve investigators. At either end were two smaller rooms for the director's use, for storage of glassware and reagents, and for fresh and salt water tanks. There was a small reference library, including the complete *Challenger* Reports, a gift from the British government. When the station was relocated, it was placed on a large barge and towed to the next site. Queen's University zoology professor A.P. Knight's report of the station's first season rings more true. The researchers 'did little more than make a beginning. The carpenters were not out of the building until August ... No regular servants were employed. The workers were sailors, fishermen, labourers and scientists by turn. We swept floors, washed glassware, delved for worms, waded in mud ... argued, grumbled, swore, but all were determined to return next season.'[30]

The dual imperative of the Canadian Biological Station – that is, basic research and practical fisheries work – was reflected in the sites chosen for the station. Choices were based largely on the richness of local fauna and flora, but political expediency was also a factor. After two seasons at St Andrews, for example, it moved to Canso, Nova Scotia, in 1901 and 1902, after petitions by Canada's largest fish-packing company, A.N. Whitman and Son of Canso, which reminded Canso's MP that Canso 'was the centre of the inshore fisheries of the province as well as the rendezvous of the banking fleet during a large part of the year.'[31] Thus

> from the start, the Board ... attempted to combine thoroughly scientific fundamental investigation with the practical, as instanced in its first two publications ... of the structure of the fins of the mackerel shark, and of the effects of sawdust and explosives on fish life. The important local herring (sardine) and clam fisheries were investigated at St. Andrews in 1899. When the Station was at Malpeque, Prince Edward Island in 1903 and 1904, most of the effort was devoted to study of the famous Malpeque oysters.

The 1905 and 1906 seasons were spent at Gaspé, Quebec, centre of one of the New World's oldest fishing industries, where two firms from the 1790s were still operating.[32] In 1907 the scow transporting the station sprang a leak and did not reach the next site. It had to be beached on the Gulf of St Lawrence's south shore. There, its contents were scavenged by the scientists, a reduced staff of whom worked that year in a fish house attic at Sept-Îles. The mobile station was abandoned.

In 1908 the permanent biological station at St Andrews was built. The board also extended its reach across the country by founding a similar station at Departure Bay near Nanaimo, British Columbia. The Reverend George W. Taylor was first curator of the Pacific Biological Station. Its main building housed a laboratory big enough for eight scientists, as well as a library, a photographic room, a dining room and four bedrooms upstairs. In another building were the kitchen and the caretaker's apartments. Taylor predicted the station would 'be a workshop in which all the various problems connected with marine life – whether economic or purely scientific – will be studied.'[33]

It appears that the professors, students, and volunteer researchers entered into all kinds of research with equal gusto. Their work is difficult to characterize, because there was no theory around which it focused, and the main activities were not one or two, but myriad. Prince himself estimated that the Biological Board had investigated at least two hundred different problems by 1923. He classified these into ten categories, including physical oceanography (of which there was little done), fishery surveys, and fish life-history, pathology, and migration studies. Other efforts included biochemical and bacteriological research into cured, canned, and frozen fish; studies of lobster industry problems and the usefulness of lobster hatcheries; and long-term studies of oysters and other shellfish.

Much of this research dealt with fisheries biology or applied problems. Of some sixty-eight scientific papers from the Atlantic and Pacific stations up to 1912, one-quarter dealt explicitly with fisheries biology. A further half were biological surveys or taxonomies of local species – ongoing work that set the stage for all further marine biological research. The Biological Board also studied fish resources, and began programs to educate fishermen and consumers on the food value of unutilized edible fishes, which in eastern Canada meant most of the flatfishes, which were commercially so important in Europe. Other studies involved fish culture problems and the effects of pollution, dynamite fishing and predation on fish populations. The continuing importance of oyster and other fisheries biology is indicated by Biological Board publications for the period 1912–21: nearly 30 per cent dealt with applied (fisheries) research.[34] However, looking at percentages of different kinds of research tells little about how applied work related to basic marine biological research, and what importance scientists attached to it. The question is, did these studies satisfy the scientists' own research needs?

The answer lies in the way in which the board approached this problem. Up to 1924, the board's main leaders were Prince, D.P. Penhallow, A.B. Macallum, and Ramsay Wright. Each played a different role according to his personality; each strove to balance the needs of the board's researchers with those of the board's chief clients: fishermen and the federal government. First and foremost was Prince; he stepped down as chairman in 1921, but remained involved until his stroke in 1927, and he was influential as the board's secretary-treasurer until his retirement from the civil service in 1924. Prince strove to make the board a centre of scientific excellence and to keep amateurs out of important positions. For example, when G.W. Taylor's stroke in 1911 left him incapable of carrying the full directorship of the Pacific station, Prince begged him to stay on: 'I have two or three applications for the position, two of them I think are ex-Presbyterian ministers, whom I should be sorry to see forced upon the Station by politicians.'[35]

As a consummate diplomat of science, Prince mediated between the Department of Marine and Fisheries and the board. Indeed, he helped create a firm base of support by applying his diplomatic talents to tailor the stations' work to the perceived needs of fishermen and the department. Every year he listed eleven or so research projects, both basic and applied. Thus, when the movable station was located at St Andrews in 1899 and 1900, at his behest Wright studied the foods eaten by commercial fish species; and E.W. MacBride worked on starfish life histories and tested the theory that fishermen helped starfish multiply by throwing chopped-up starfish overboard. A.P. Knight examined the destruction of feeding grounds by dynamite fishing, as well as the effects of pollution on fish – a study that lasted until 1908. In 1901 a Mr Cornish was assigned to test the British beam trawl (not then used in Canada). When the station was situated near Malpeque Bay's famous oyster beds, Prince assigned studies of the life histories, food, species variations, and artificial rearing of oysters.

He had little time to spare for the board after its first several years. As Dominion Commissioner of Fisheries, he was overwhelmed by administrative duties. As chairman of fifteen separate fishery commissions, he arranged for witnesses and evidence, wrote reports and recommendations, and presided over public and executive meetings. This took him from the Atlantic to the Pacific, usually several times a year. He was Canada's agent on the International Fisheries Commission under the Fishery Treaty of 1908; and with the American representative, the famed ichthyologist David Starr Jordan, he surveyed boundary waters along the entire

American–Canadian border. In 1914, as its guest, Prince made the first complete survey of New Zealand's inland and sea fisheries, and issued a report for the government. Unfortunately, the board sometimes had trouble reaching its chairman. Wright, who met some of Prince's friends in England, told him: 'They chiefly think of you as a bad correspondent.' It sometimes took two or three letters to get a reply from Prince. Wright once remarked: 'I shall be glad to know that you are alive.'[36]

The negative aspect of Prince's diplomacy was that he avoided controversy, preferring 'to conduct business through social amenities rather than by pressing issues.' This made him a delightful companion,[37] but perhaps not the best champion of his scientific colleagues. Prince left confrontations with officious bureaucrats to others. Fortunately, Penhallow – and later Macallum and Knight – were willing to risk the confrontations Prince avoided.

David Pierce Penhallow (1854–1910), the first secretary-treasurer, took care of practical arrangements; his abilities complemented Prince's. He became director of the Atlantic station in the crucial years 1907 and 1908, when a permanent site was being sought. 'An outstanding administrator' he held many important positions in learned societies besides being botany professor at McGill University (1883–1910). From 1879 to 1882, he had been professor of botany and chemistry at the Imperial College of Agriculture in Sapporo, Japan. His forthrightness and businesslike approach provided a much-needed antidote to Prince's more happy-go-lucky style.

When the Atlantic Biological Station's permanent site was being chosen, two criteria mattered most: the richness of regional biota, and the importance of local fisheries. Five locations were considered: St Andrews in New Brunswick, and Chester, Yarmouth, Lunenburg, and Digby in Nova Scotia. Scientific considerations included biological diversity, facilities for shallow-water collecting and deep-water dredging, tides and currents, and water depth and purity. Other, more pragmatic factors included whether land was available for erecting a laboratory; whether fresh water, supplies, hotels, and shelter and docking facilities were at hand; accessibility from major urban centres; and various social and other collateral factors.

The choices were narrowed to Mahone Bay, Nova Scotia, and St Andrews, New Brunswick. Board member Dr Joseph Stafford, in charge of evaluating their respective merits, found that prevalent fogs off Mahone Bay would hamper offshore work and that there was less faunal diversity than at St Andrews. Also, although 'more than two hundred fishing vessels find their home ports at Chester and Lunenburg,' these fished off the distant Grand Banks. Their catch usually went directly to Boston or was barrelled and shipped to foreign markets: very few live fish were handled in port. Chester had few boat or household supplies, and worst of all, it was more expensive to get to from central Canada.[38]

While St Andrews was no better for supplies such as groceries, which could 'be obtained only in small lots and hence at unsatisfactory prices,' it had easy access to Saint John and other supply centres. Furthermore, St Andrews was 'the real fishing centre of eastern Canada. The large number of clams, herring, and other fish annually taken in Passamaquoddy Bay and adjacent regions, make it probable that ... it is the immediate centre of a larger live fish industry than any other town on the coast.' This made St Andrews 'ideal' for 'a Station engaged in solving problems associated with the fishing industry.'[39]

St Andrews also had major practical advantages. A town of around one thousand inhabitants, it was an important centre for tourism, a substantial herring fishery, and for the building of rowboats and small pleasure craft. An attractive summer resort, frequented by American magnates and Canadian peers alike, it was a port of call for passenger steamers running between Eastport, Maine, and towns along the St Croix River. As the terminus of a railway line, it was easy to reach from Montreal and Boston. Tourist amenities also proved attractive: 'While scientists engaged in the absorbing work of such a station are not expected to engage largely in social diversions ... their best work cannot be expected in the complete absence of all opportunities for pleasant mental and physical relaxation and recreation.' The board found that St Andrews offered an 'admirable' social life and that its summer amusements would 'add very materially to the successful working of the station.' In fact, it is possible that the Scottish influence on Canadian science extended to the game of golf, as St Andrews mirrors its Scottish counterpart in having a golf course with a lovely, old-fashioned fairway. In 1900, Ramsay Wright told Prince he hoped that the Station 'might be still at St Andrews so that ... I might be able to combine a little golf with the scientific work.'[40]

So Penhallow chose St Andrews – specifically, a place known both as Brandy Cove and as Smuggler's Cove, about two miles from the town. Here, on three acres purchased from the Canadian Pacific Railway, a laboratory some 79 by 31 feet was constructed, with space for nine junior and three senior investigators. It was deemed 'ample for the scientific work of the Station for many years to come.' A dwelling house was also built, with room for the entire junior staff and the curator. Built with 'simplicity but attractiveness of design,' it provided 'the simple comforts of life, since ... men could not be asked to grant free service without a reasonable amount of those surroundings which tend to the maintenance of a proper moral tone.'[41]

As Prince was unable to spend much time at the Atlantic station, in 1901 Ramsay Wright (1852–1933) assumed the role of assistant director. He provided mainly scientific direction, rather than overtly political action, and he continued his own research on plankton. His most important contribution, however, was that he had introduced of laboratory work into Canadian undergraduate biology and emphasized research for advanced students. He transformed the University of Toronto's zoology department, of which he was the first head, by launching a research program comparable to Harvard's and to Henry Newell Martin's at Johns Hopkins. Born in Alloa, Scotland, he had been assistant zoology professor at the University of Edinburgh. Invited to join the *Challenger* expedition, he chose intead, in 1874, to go to Toronto to take the position of professor of natural history (he soon changed his title to 'Professor in Biology'). Wright introduced practical laboratory work and experimental biology for advanced students, while maintaining his own parasitological research. He also helped found a new Faculty of Medicine in 1887; graduates trained by him helped carry forward his ideal for a strong research program in medicine and biology. From 1901 until he retired to Oxford in 1912, he was also Dean of Arts and Vice-President of the University of Toronto, and the university's current zoology building has been named in his honour. Wright's interest in fisheries biology was genuine; in 1892 he published a *Preliminary Report on the Fish and Fisheries of Ontario*, a comprehensive report for the provincial government. His graduates, who included James Playfair McMurrich, A.B. Macallum, and A.G. Huntsman, 'powered and built the Fisheries Research Board of Canada' in its first forty years.[42]

Archibald Byron Macallum (1858–1934), the future founder of Canada's National Research Council, became the board's secretary when in 1910 Penhallow retired (he died soon thereafter), exhausted by his efforts to build up the Atlantic Biological Station. A 'doer,' Macallum took life seriously and was apt to frown on levity.[43] One of Wright's first students, he returned to Toronto in 1883 as a biology lecturer while working toward a medical degree. From Johns Hopkins, he received a PhD in medicine in 1888; there, he was strongly influenced by Cambridge-trained physiologist Henry Newell Martin. In 1890 he took an MB degree at Toronto, and was soon appointed to the new chair of physiology. Macallum helped Wright make the medical school a leader in scientific medicine, aided by two other former pupils of Wright – J.P. McMurrich, who became professor of anatomy, and J.J. Mackenzie, a professor of pathology – and by medical professors Alexander McPhedran and Irving Cameron. In 1908 a new chair in biochemistry was created for Macallum. Yet he continued to pursue his own research, much of which was done at night during term time.[44]

From 1911 until he was forced to resign in 1920, the eminently capable Macallum was the real head of the board.[45] It was he who pushed for the creation of the autonomous Biological Board of Canada. He spearheaded a move among the board's scientists for independence from the Department of Marine and Fisheries. The main reason for this push was the department's meddling, not in the board's research but in its finances. The department was unwilling to advance lump sums for the daily operation of the stations. In 1899, Prince complained to A.W. Owen, the department's accountant: 'As the Department has not seen fit to adopt my suggestion to make an advance to me as Director of the Station ... $872.75 sanctioned by the Board for the *purchase of necessary books of reference* ... and some other items of expenditure *will lapse.*'

The accountant responded coldly that 'this is no funeral of mine' since Prince was aware, no purchases 'chargeable to a fiscal appropriation can be made after the expiry of the year and chargeable to the appropriation of the preceding year.' That is, Prince should have applied months earlier for the books in question.[46] These were the conflicts encountered by the Board of Management, as opposed to complaints that it ignored practical problems. But such conflicts were grave indeed. As Knight observed, until the laboratory was fully equipped with 'apparatus, chemicals and books of reference ... it is manifestly impossible for workers to undertake any research worth speaking of.' But it was hard to equip a station under the department's accounting system, and in 1899 Prince spent some $300 of his own money – almost a year's salary in those days.[47]

The department was also slow in reimbursments. E.W. MacBride complained two months after a 1903 board meeting in Ottawa that his expenses had not been paid. A.H. Mackay decided to miss an upcoming meeting: 'If my presence should be of great value to the country I would gladly make the sacrifice of attendance; for it costs something to go to Ottawa and back again. I have had a good many experiences of costly sacrifices for what has not been worth the price of martyrdom: so that I am convinced that I shall feel better and you equally as well if I just support the action you are to take ... by staying home and praying.

Wright also complained about having to pay bills and salaries out of his own pocket, to the extent that at one point 'he was between $400 and $500 out of funds.'[48] From the accountant's perspective, the problem lay with scientists unwilling to conform with bureaucratic ideals of accounting: their idea was, he commented, 'that the amounts voted are at their service irrespective of dates and they ask for advances when they have no intention of using the same within the period for which they are available, so that each year accounts are dragged into the following year in order that every cent possible may be expended.'

But it was 'a clerk's veto of a request for purchase of scientific literature because the latter was in foreign languages' that caused a final revolt 'and resulted in the creation in 1912 by Act of Parliament of the Biological Board of Canada.' The 'Act to Create the Biological board of Canada' made the Board an independent body under the minister; it was now in charge of all Canadian biological stations and free to spend the money voted to it. Also, it was now able to expand its membership to other Canadian universities with active biology researchers, and to bring in fisheries experts from northern Europe and elsewhere for periodic intensive investigations.[49]

The board's second decade was marked by attempts to develop a clear program – attempts that did not always succeed, as there was no one on the board who could give the work his full-time attention. Even as late as 1916, however, Prince was still supplying researchers with lists of 'problem': a study of diseases in salmon; another of Bay of Fundy shad food (for the Shad Commission); and another of the effects of sewage pollution and chemicals on fish. There were studies of the scales of smelt and attempts to find out when swordfish spawn. On the more practical end there were experiments with haddock smoking, studies to find out how long fish stayed fresh in cold storage, and even an effort to design a fish elevator. Prince also passed along departmental requests. In 1912 the department requested a study on the possibility of rearing lobsters from larvae; and in 1913 it requested a biological inquiry into the halibut fishery to help it negotiate a Pacific halibut fishery protection treaty with the United States.[50]

Among the research projects carried out at the Atlantic station, perhaps the most important involved oyster and clam studies by Joseph Stafford (1864–1925), who was the station's curator from 1900 to 1910 and curator to the Pacific station from 1910 to 1912. Stafford was a student (and protégé) of Ramsay Wright. He graduated in 1890, got his MA and PhD from Leipzig, and returned to lecture at Toronto from 1898 to

1901. He was assistant professor of biology at McGill University from 1901 until his death in 1925. At the biological stations he was in charge of biological surveys, but he also made himself an expert on clams and oysters. By 1908 he had worked out the entire life history of *Ostrea* – something never done before. In 1913 Stafford published a classic monograph on the Canadian oyster's biology and fishery for the Canadian Commission of Conservation: *The Canadian Oyster: Its Development, Environment and Culture.* He 'laid the basis for scientific oyster farming in Canada' and became known 'to oyster biologists the world over.'[51]

There was no attempt to emulate the Marine Biological Laboratory at Woods Hole – morphology did not figure large at the Canadian station. When in 1908 J.P. McMurrich first came to St Andrews, his initial work on actinarian and copepod systematics was not inconsistent with his previous morphological studies at Woods Hole. But by 1911 even he was doing fisheries biology, studying the age of Pacific salmon returning from the sea at the board's Pacific station, commenting that 'a much more complete knowledge of their habits and life cycles' was needed in order to 'determine the proper steps to be taken for their successful conservation.'[52] In 1912 he experimented with a small-mesh drag seine near Vancouver to ascertain whether these nets, used in the smelt fishery, were killing salmon.

In truth, few of the researchers arrived with strong ideas about what they wanted to do. Ramsay Wright, although he defended free research, conceded: 'It is very important that [the] economic aspect of the Laboratory should not be lost sight of and that ... the importance of the Fisheries should be carefully considered.' Many scientists coming to St Andrews asked to be assigned specific problems. Stafford asked Prince to 'put me at some problem that will be likely to interest this country.' In 1900, James Fowler, professor of botany at Queen's University, wanted to study marine algae and mollusca for his own work, but he also asked Prince, '[Is there] any question of interest to your Department, connected with these subjects, to which I might endeavour to find an answer?'[53] It is scarcely surprising that practical fisheries biology turned out to be a major feature of the station's work in its earliest years.

Perhaps this openness to applied problems was a consequence of Canadian biologists' extremely recent exposure to the German academic research ideal. Even at the University of Toronto, the commitment to the research ideal only began after James Loudon became university president in 1892. This commitment was consolidated in 1898 when the University Senate passed a motion proposed by Loudon in 1883 'to introduce the Ph.D. degree "for the purpose of encouraging research."' Loudon, a professor of physics and mathematics, wanted to make Toronto a noted research university like Johns Hopkins; in this, he was supported by the likes of Ramsay Wright and A.B. Macallum.[54]

Yet at the time, many Canadian zoologists and botanists had no set expectations of the new facilities provided by marine biological stations. They reflected the tradition of Victorian amateur naturalists. Most had no prior training in the biology of marine organisms, and several, such as Loring Woart Bailey (1839–1925), were enthusiastic generalists of science. Bailey, the board's oldest founding member, was the son of Jacob W. Bailey, the West Point Military Academy's professor of chemistry, mineralogy, and geology. Jacob pioneered the study of North American diatoms, using samples brought back by the U.S. Coast Survey after 1848, and was an authority on foraminifera and diatoms. Loring became the University of New Brunswick's first professor of natural history (1861-1907), specializing in geology, although he was interested in botany and zoology as well. He studied Canadian diatoms at the Atlantic station. Another Victorian naturalist was Canon V.-A. Huard (1853–1929) of Laval University, editor of *Le Naturaliste canadien.* Huard, who had a doctorate from Laval, served with the board until 1928, and wrote articles on history, travel, zoology, general science, and entomology, as well as books on zoology, hygiene, and entomology. Thus he was truly a generalist.[55]

In sum, the founding members of the Biological Board could not be noted for highly developed professionalism. Sociological studies suggest that conflict over applied science is avoided in contexts where professionalism is not highly developed. These Victorian professors often resembled new PhD graduates in science. Studies of professional scientists have revealed that although new researchers often hope to pursue knowledge for its own sake, many of them find assigned research in an industrial (applied) setting 'equally taxing and fascinating, and fully as satisfying, as ... thesis research had been.' Since thesis topics in science are often suggested by the supervisor, the young graduate 'has probably never initiated a research project of his own, and thus finds industrial research not so different from what he knew at the university.' Solving assigned problems often requires so much self-reliance, ingenuity, and elegance of execution that the new researcher finds them as intriguing as any work done in the academic setting.[56] A study of British industrial scientists and postgraduate students found that 'only a quarter of young industrial scientists wanted to determine their own research goals.'[57] So there is no reason to suspect that the Victorian scientists arrived at the marine station with strongly fixed ideas. They were filled with the basic need to explore – to conduct biological inventories and surveys – and they generally had no specific research problems in mind. Thus, they would have been receptive to 'applied' problems suggested by the Department of Marine and Fisheries.

Another clue to this lack of conflict in the board's early years comes from examining the careers of individual board scientists. These reveal a spectrum of research patterns. Some scientists rarely ventured beyond applied questions: A.P. Knight is a good exemplar. His articles in the board's journal, *Contributions to Canadian Biology*, had titles such as 'A Further Report upon the Effects of Sawdust on Fish Life.' Knight also looked at the effects of dynamite fishing, which was practised by some Bay of Fundy fishermen.[58] Of the eleven articles he published between 1900 and 1921, only one was rather basic: 'Histology of the Flexor tendon in the crushing claw of the lobster,' in *Contributions to Canadian Biology 1918–1920.* Others did nothing but basic science, as exemplified by Biological Board secretary-treasurer A.B. Macallum. His physiological and biochemical researches on inorganic salts in invertebrate and vertebrate body fluids – he established the close correlation between inorganic salt concentrations in animals and seawater[59] – meant that he had a real interest in making Canada's marine station a centre for experimental research. He also devised histological techniques for identifying sodium, potassium, phosphorus and chloride ions in different areas of the cell and for determining their concentrations. Unlike most of the board's scientists, his work at the station was at the cutting edge of physiological and biochemical research.

Many followed a middle path between the two extremes represented by Knight and Macallum. For example, the experienced, self-directed E.W. MacBride (1866–1940), a well-known embryologist, spent several summers conducting mission-oriented oyster culture research. Born in Belfast, Ireland, he graduated from Cambridge in 1891 and worked in 1891 and 1892 under Dohrn at Naples. MacBride came to Canada as the first professor of zoology at McGill University. He was the youngest member of the Board of Management at its inception, but left Canada in 1909 to become assistant professor of zoology at the Imperial College of Science and Technology, retiring in 1934. At Canso in 1900, he worked on his particular specialty, the embryology and developmental morphology of echinoderms. But in 1903 and 1904, when the station was at Malpeque, Prince Edward Island, he began oyster studies with Joseph Stafford. In 1908 he returned to set up an experimental oyster bed for long-term studies; he also investigated complaints that the little-neck clam (*Venus mercernaria)* was damaging oyster beds. After returning to England, MacBride published his *Text-book of Embryology* (1914), which would set the standard in invertebrate embryology for several decades. However, his involvement with fisheries biology did not end. He became a member of the Marine Biological Association of the United Kingdom, and was a vocal supporter of fisheries research and of a new separate Ministry of Fisheries. He served on the Advisory Committee on Fishery Research for the Development Commission from its creation in 1920 until 1939, becoming chairman in 1933.[60]

These dual interests were also reflected in the career of Archibald Gowanlock Huntsman, as will be seen in the rest of this book. By the 1910s, his research interests encompassed in areas as diverse as embryology and fish processing methods. Huntsman was the best example of the second generation of scientists attached to the Biological Board. As director of the Atlantic station over two decades (1914–34), he instilled in the board a greater degree of definition and professional identity. Yet his involvement with the Biological Board in fact grew out of chance. Born on a farm in Tintern, Ontario on 23 November 1883, Huntsman thought he would follow his father as a farmer. At St Catherines Collegiate Institute, however, young Huntsman acquitted himself so well that he won a scholarship to the University of Toronto, which he entered in 1901. Not wishing to become a teacher or a lawyer (he was 'too fond of the truth' for the latter), he decided on medicine. As a child, he had admired a local doctor, and he equally admired his mother's aunt, Jenny Kidd (Gowanlock) Trout, the first woman licensed to practise medicine in Canada.[61] He graduated with a BA in 1905 and earned his Bachelor of Medicine in 1907. The University added the MD degree in 1933, although he never practised. Instead his interest was captured by biology. A reorganization in 1887 closely linked the Faculty of Medicine with the Department of Biology; the two shared equipment, and their courses were even listed in common until the calendar of 1908–9. Robert Bensley (187?–1956), a demonstrator there, became in 1901 the first director of the Georgian Bay Biological Station at Go Home Bay (on Georgian Bay). It was he who drew Huntsman into aquatic research.

The University of Toronto's biology department had very strong affinities with aquatic and marine research, and its biologists were inordinately influential in early Canadian marine science. Huntsman, later a professor there, was strongly influenced by his biology professors, who included Ramsay Wright, Macallum, and the brothers Robert Russell

Bensley and Benjamin Arthur Bensley. All were involved with the Georgian Bay station. This station was run originally by the Madawaska Club, which had been formed in 1898 by various University of Toronto faculty members for the specific purpose of establishing a biological station. Opened in 1901, it came under the supervision of the Board of Management as a precondition a $1,500 per annum government grant. Full management was transferred to the board in 1904. Bensley's brother Benjamin (1875–1934), a lecturer, succeeded his older brother as director of the Go Home Bay laboratory in 1902, and continued at that post until friction with Marine and Fisheries officials over reports and accounting led to his retirement in 1911. The station closed in 1913, At that point, B.A. Bensley organized – as a distinct unit within the zoology department – the Ontario Fisheries Research Laboratory.[62] Bensley succeeded Wright as department head in 1912, and until his retirement in 1934 his fisheries laboratory flourished: many in this laboratory went on to important positions in the Fisheries Research Board of Canada.

The University of Toronto faculty was packed with researchers active in marine and aquatic biology, and this made Toronto a central focus in Canada for these disciplines. So it is not surprising that Huntsman was soon deeply involved in marine biology. As an undergraduate, he went to Georgian Bay in the summer of 1904 at the request of B.A. Bensley. There, at the small lakeside laboratory, with its tank room, boat house, and nearby meteorological station, he worked with Davidson Black (who later discovered 'Peking Man'), surveying the area's wildlife and studying the feeding habits of black bass. Fortunately for Huntsman, a small house for workers had just been erected, which meant no more tents. He returned in the summers of 1905, 1906, and 1907, receiving a small stipend for studying several fish-life histories.

The zoology department noted Huntsman's growing competence in biology and his strong work ethic, and in 1907 hired him as a lecturer. Now drawing a salary, he married Florence Marie Stirling in 1908. In that same year and the following one, the newlyweds went to the newly established Departure Bay marine biological station near Nanaimo, British Columbia. Very likely, they lived there in a tent, as did many researchers and their wives well into the 1920s. He began a general collection of marine animals for his department; he also studied local tunicates, his early research specialty. In 1910 he travelled to the Atlantic station, where he continued to study tunicates. In 1911 he took charge as curator, filling Penhallow's shoes. He continued as curator, then director, of the Atlantic station until 1934, except for a brief period in

1914, which he spent in Germany. He soon became involved in the study of fish population dynamics – a fundamental feature of fisheries biology, which would soon overtake his other interests.

From the examples provided by Knight, Macallum, and McBride, it is clear that scientists responded in a variety of ways to the opportunities offered by the Canadian marine stations. Some happily took on Prince's projects; others carried on in their own preferred areas. On top of this, there is abundant evidence that some board members and employees, including Knight and Huntsman, strongly believed that the board's mission was to improve Canadian fisheries, as will be seen in chapters 4 and 5. They were filled with an ideal of service and a zeal to help fishermen and fishing communities.

This was by no means unique: many early-twentieth-century American agricultural researchers 'did not regard it derogatory to their status to seek solutions to problems put by persons who were not scientists. They often had a close sense of identity with the farmers on whose practical problems they worked.' It did not seem to bother them to be handed specific problems by administrative superiors. Indeed, 'they had strong local attachments to agriculture as a way of life' and to their own working locality.[63] Likewise, British scientists at Plymouth and St Andrews were convinced that fisheries work was worthwhile. J.T. Cunningham, the Plymouth laboratory's naturalist from its opening until 1897, did lengthy studies of commercial species' distribution, reproduction, and development. In the 1890s, Plymouth opened several lines of fisheries research: it began a thirty-year history of North Sea trawling grounds; it collected Dogger Bank fishing statistics; it experimented with beam-trawl nets; it determined the age and size of sexual maturity in commercial fishes; and it produced artificial bait for long-line fishing. It also began hydrological and planktonic studies of the English Channel,[64] – Plymouth's signature area of study for the next half-century. Edgar J. Allen, director from 1894 to 1936, happily declared that 'research having a direct bearing on the fishing industry has been carried on side by side with research in pure science, and the two have mutually helped and supported each other.'[65] This, then, was a fairly solid marriage of perceived economic and scientific goals.

Fisheries research was largely free of basic-vs-applied conflicts because of its catholic nature, its incomplete professionalization, and (in the Biological Board) the fact that individuals wishing to avoid applied investigations could be accommodated owing to the board's structure. Besides this, after 1912 the Biological Board had a great deal of control over its own work. The 'inevitable strain' between conflicting missions was thus delayed until the Department of Marine and Fisheries became interested in exerting more control around 1919 (a story told in chapter 4). Even then, conflict was mitigated, and the fledgling profession of fisheries biology was shielded from the department by the board's powerful substructure and by the mediation of its successive chairmen. As the board matured, these 'scientist-administrators' became increasingly important. Prince was followed by the very capable A.P. Knight, then by McMurrich, and then by the autocratic A.T. Cameron, under whom the Biological Board evolved into the Fisheries Research Board. The growing Fisheries Research Board continued to show 'a lack of distinction in project-selection between applied and basic research [but rather] a criteria of selection of projects based on the extent to which a project will contribute to the solution of practical problems.'[66]

Considering all this, the lack of angst over practical research among the university professors running Canada's marine biological stations should come as no surprise. One would expect such angst, if it existed, to show up in private correspondence, but complaints of government pressure for applied research did not show up until the later 1920s. The potential for conflict did indeed exist, since the board very much needed to prove its usefulness in order to prolong its existence, but also needed to satisfy visiting researchers. In fact, conflict did not begin to build until the 1920s, spurred by the construction in 1924 of the fisheries experimental stations in Halifax, Nova Scotia, and Prince Rupert, British Columbia. By now, the federal government had grasped the importance of fish processing research. For the first time, the Department of Marine and Fisheries had some control over the Biological Board's activities, and it began directing more funds to the newer experimental stations. Modelled on agricultural research stations, these produced tangible and practical results. Biological Board correspondence shows a growing realization that the government had to be made aware of the importance of basic research. But even at this point, many individuals were undertaking both kinds of research willingly, in the hope of improving the social welfare of those whom their more practical science touched.

Chapter 3

The Canadian Fisheries Expedition, 1914–1915

All men are equal before fish.

Herbert Hoover[1]

The central problem of fisheries biology is determining the causes of fluctuations in fish population, with the corollary that its main practical aim is to predict future fish catches. Around 1900, the science of fisheries biology was undergoing birth pangs, and many of the methods now in use had yet to be developed. The crucial breakthrough in understanding natural (as opposed to fishery-induced) population fluctuations came in 1913–14, when following a long study of Norwegian cod and herring, the Norwegian oceanographer Johan Hjort, one of the world's greatest fisheries biologists, introduced the concept of 'year classes.' The development of Hjort's concept will be examined in this chapter. His idea received its first re-enforcement – and early Canadian fisheries biology its real impetus – from the celebrated Canadian Fisheries Expedition of 1914–15. This expedition, carried out to investigate western Atlantic herring, was led by Hjort himself, at the invitation of the Biological Board of Canada.

The Canadian Fisheries Expedition left a permanent mark on the Biological board's work. Hjort himself strongly influenced several of the Board's scientists, most notably Archibald Gowanlock Huntsman, the young curator of the Atlantic Biological Station, who later became one of the board's most prominent scientists. I noted in the previous chapter that the scientists attached to the Biological Board generally did not quibble over whether they were doing basic or applied research. Hjort himself, the greatest fisheries biologist of his time, provides a sterling example of a scientist dedicated with equal passion to the twin goals of extending basic marine science, and of using his ground-breaking research to alleviate as directly as possible the hard lot of Atlantic fishing communities. He worked to modernize the fisheries and to introduce new fish products, such as frozen fish, for inland markets and for export. He brought both goals with him to Canada, where he strongly urged the government to sponsor his fish-processing experiments. Hjort's lasting legacy was his vision that basic fisheries research was without merit unless scientists also attended to the needs of fishermen and their families and communities.

During the expedition, Hjort introduced Canadian scientists to methods developed by himself and by the Conseil International pour l'Exploration de la Mer (in English, the International Council for the Exploration of the Sea, or ICES). This expedition was the first Canadian Atlantic oceanographical survey along the lines of the Intensive Area Studies[2] pioneered by ICES scientists. Intensive Area Studies involve repeated cruises for several seasons and the taking of oceanographic and biological samples, for the purpose of tracking hydrographic and biological changes in a region throughout the year. The expedition gave Canadian biologists their first clues to offshore spawning areas; it also introduced them to ecological life histories, (that is, the movements of fish from feeding to spawning grounds and other migratory points), and enabled them to work out the influences of currents, temperatures, depths, and food concentrations. It also inspired the Biological Board to focus on more directed long-term research. For several years, as a consequence, it carried out similar expeditions on a smaller scale.

In international terms, the Canadian Fisheries Expedition's main scientific legacy – and it was an important one – was that it showed the universality of Hjort's model of fish populations; the population dynamics of the northwest Atlantic herring were similar to those of the northeast Atlantic herring, covered in Hjort's recent and monumental *Fluctuations in the Great Fisheries of Northern Europe*, which was 'one of the most important contributions to fisheries science.'[3]

In the two decades leading up to 1914, Johan Hjort and ICES scientists had developed the fundamental tools, techniques, and agenda of fisheries biology. The evolution of this new science needs to be described here if the reader is to understand the significance of the Canadian Fisheries Expedition. Hjort and the ICES scientists were motivated by the need to place the fisheries on a scientific footing. This need was especially great in countries like Great Britain, Scandinavia, and Germany, where the fisheries were of great economic importance.[4] In the predominantly agrarian Scandinavian countries, cod and herring were the mainstay of coastal peoples. Many coastal farmers relied on fishing as a source of food and income, and violent fluctuations in these fisheries brought severe economic hardship.

The first official and scientific response to these problems came in 1864, when the Norwegian government appointed Georg O. Sars to find out why the previously expanding Norwegian cod fishery had failed. After discovering that cod eggs floated, he realized that fluctuating currents and hydrographic events could affect the food of fish and the larval and juvenile fish themselves. Perhaps this caused them to hatch and grow in new areas (where they would return as mature fish to spawn) and thus elude fishermen. In 1873, Sars also linked good herring catches with the presence of warm water off the Norwegian coast.[5]

Johan Hjort (1869–1948) succeeded Sars in 1893 as research fellow in fisheries at the University of Christiania (now Oslo). He was the son of Johan S.A. Hjort, professor of ophthalmology at the University of Christiania. As an undergraduate he studied zoology under Richard Hertwig in Munich. In 1892 he went to the zoological station at Naples to investigate the budding of ascidians for his dissertation. Returning to Norway, he became 'enthralled with the vast scientific problems underlying the fishing industry.' He made an early practical contribution in 1897 when he discovered stocks of the deep-sea prawn *Pandalus borealis* in Norwegian fjords; this led to a new fishery. His earliest important work, the *Hydrographical Studies of the Norwegian Fisheries*, was published in 1895.[6] His eventual contributions to the fisheries were thus twofold: he was outstanding both as a fisheries biologist and as a hydrographer. His abiding desire to help fishermen led to many successes, especially in locating new fishing grounds and fisheries. He saw himself primarily as a marine biologist, and he always sought to subordinate physical oceanography to fisheries problems.

In the mid-1890s, a poor cod fishery coincided with diminished herring catches, due in part to a cold period, which was also disastrous for Norwegian farmers. Hjort's primary task as the Director of Fisheries of the Fiskeridirektorat (the Norwegian government's new fisheries bureau, located in Christiania, created in 1900), was to discover why these fishing declines occurred.

This was an overwhelming task for one man, at a time when few conceptual tools had been developed to aid such an investigation. But others had also been conducting research, and their findings would provide a basis for understanding the productivity of the seas. Gustaf Ekman, a Swedish physical oceanographer, began in 1877–8 to relate fluctuations in fish stocks to the hydrography of Swedish waters. Otto Pettersson (1848–1941), professor of chemistry at the Stockholm Polytechnic, joined Ekman in 1890. Their work, wonderfully described in Eric Mills's *Biological Oceanography*, departed from that of traditional oceanographic expeditions, which tended to sketch broad and static outlines of oceanic conditions. They explored, rather, a limited sea area, the Skagerrak, but they explored it intensively and at all seasons of the year in order to discover its dynamics. For the first time, the great changes in strong surface water movements were charted.

Pettersson and Ekman made great progress; even so, they recognized that 'only international collaboration could provide enough information to establish connections among sea, climate, fisheries and agriculture.' Different countries could be assigned specific large areas of the sea for detailed surveys; this was done in a limited way in the 1893 'Copenhagen Program': Scandinavian, German and Scottish scientists made quarterly sections in the Skaggerak, the Kattegat, and the North and Baltic seas. In 1895, Petterson proposed a more ambitious scheme, and Hjort was among the pioneers who helped found the Conseil International pour l'Exploration de la Mer. With John Murray and the Norwegian explorer Fridtjof Nansen, he called for a conference to organize an international oceanographic program.[7]

The conference was held between 15 and 23 June 1899, in Stockholm, at the invitation of King Oscar II of Sweden and Norway. Delegates from Germany, Denmark, the United Kingdom, Holland, Russia, Sweden, and Norway laid out a program to found an international commission to explore the sea. Each nation was to be responsible for investigating designated areas of the Baltic or northeast Atlantic over a five-year period. The international body would coordinate the quarterly research cruises and collect statistics on each nation's fisheries. It was hoped that all of this would provide scientific underpinnings for 'rationally' exploiting the sea.[8]

ICES evolved into a permanent organization, and as Mills comments, was 'inordinately significant in the development of marine science after 1900.' It promoted systematic biological and hydrographic ocean research. It standardized methods and developed new instruments and techniques. Its formation also inspired a series of courses for marine scientists from all nations, taught by Hjort, his assistant H.H. Gran, and oceanographers Adolf Appellöf and Bjørn Helland-Hansen between

1903 and 1913 at the Bergens Museum. These courses made it possible for the latest techniques and discoveries to spread rapidly through a large scientific community.[9] ICES itself concentrated on three main areas: the hydrography of European seas; food fish distribution, migrations, feeding, reproduction, and growth, in reference to the 'biological conditions of the animal and plant worlds' of these seas; and ascertaining how commercial fish harvesting was balanced by stock regeneration locally and generally. The British mainly wanted to see if trawling depleted adult spawning of demersal fish; the Scandinavians mainly wanted to develop a program for predicting cod and herring stock fluctuations.[10] The latter program predominated.

Regular quarterly research cruises began on 1 August 1902 and were coordinated to provide a complete picture of the entire northeast Atlantic. Temperature, salinity, and current measurements were made within the same week over the entire area during cruises conducted each August, November, February, and May. This Humboldtian approach allowed detailed analyses 'of seawater chemistry, plankton, and the fisheries from year to year.'[11]

Scientists were hoping to discover an absolute correlation between fish abundance and some physical factor. This would enable a projection from hydrographic data of the next year's fishing success. 'Since it was not known what this factor might be ... every kind of data that could be collected, was, and every one of the measurements was recorded and filed away.'[12] Thus ICES scientists began to assemble the elements of modern fisheries biology – methods and practices that began to move the science beyond descriptive life histories towards a more experimental, quantitative, and mathematical approach. The first significant step involved large-scale fish-tagging experiments. In the 1890s, C.G. Johannes Petersen, founding director of Denmark's small mobile biological station, pioneered studies in which large numbers of plaice were tagged every year. Recaptures of marked fish yielded data on migration patterns, rates of growth, and ages, and thus helped differentiate fish stocks.

In another vein, Dr Friedrich Heincke of Germany pioneered herring race studies. He measured numbers of vertebrae and rays (biometric measurements), lengths, widths, thicknesses, and the scales of hundreds of herring. Although no one trait distinguished herring from any region, Heincke found constant correlations between groups of several measurements, and used these to divide herring from different locales into several races. He believed these races were distinct breeding stocks, living in fairly separate regions of the sea. This contradicted 'the still-popular belief that all herring lived in the Arctic and swam now to Norway, now to Scotland.'[13]

Heincke's and Petersen's work indicated that extensive migrations of single fish stocks did not cause catch fluctuations. Scientists also did not believe overfishing to be responsible for poor catches of open-sea fish; there was, for instance, no statistical correlation between numbers of fishermen working the northern cod grounds and the fluctuations in their catches.[14] Therefore, some other explanation had to be sought.

Returning to the story of Johan Hjort in Norway, as fisheries director from 1900 to 1916, an important part of his job was to find the causes of disastrous declines in cod and herring – a job too big for one man. Therefore he gathered about him brilliant marine scientists, including Einar Lea, Oscar Sund, and Paul Bjerkan; this group was 'so productive that the years 1900–1916 have been called the "Golden Age of Norwegian Fisheries Research."' Hjort had no biological station of his own, only a laboratory at his offices for investigating fish-processing methods. However, Hjort's government built and equipped a splendid research vessel, the *Michael Sars*, which became central to all his work. He and his assistants made a series of cruises in Norwegian waters to study cod and herring spawning, distribution and movements, and stock size-compositions. Hjort initially used fish lengths to estimate their age, thus misestimating 'the composition of the stock in this respect.' Later it was found that this method applied only to immature fish. In full-grown fish, growth rates slowed so that age did not so directly affect size.[15]

At the 1904 ICES annual meeting, Heincke introduced Hjort to methods developed by C. Hoffbauer at the Trachenberg Pond-Culture Station. In his studies of carp culture, Hoffbauer had demonstrated that the concentric rings on fish scales represent annual growth. Since the growth rings are not always so clear in marine fish, Heincke also told Hjort about the work of Johannes Reibisch of the Kiel Laboratory of the Prussian Commission. In 1899, Reibisch had demonstrated that the rings in fish vertebrae and otoliths (ear bones) correspond to fish growth rates.[16] Wide bands correspond to summer growth; narrow bands are formed by slow winter growth. Growth rings can thus be used to estimate fish ages.

Back in Norway, Hjort and his assistants began to test and apply the method. They methodically counted scale rings from thousands of fish and correlated them with fish measurements from around Norway. They found that fish can live more than twenty years. Furthermore, they found a broad range of ages, from two to twenty years, in many catches. And instead of a population bulge of middle-aged fish (three- to five-year-old fish), which fish weighing had seemed to indicate, five- and eight-year-olds might predominate, for instance, while fish from other years would be scarce. Hjort was among the first to use this new technique; for example, scientists at Britain's Board of Agriculture and Fisheries only began to test the method for herring around 1910. Unlike his peers, Hjort attacked fishery problems from many fronts, using statistics, life histories, plankton studies, and hydrography. He analysed Norway's 1868 to 1900 cod-fishing statistics and found no link between the number of fishermen and catch fluctuations. This satisfied him that fluctuations have natural causes.[17]

In 1907, Hjort applied to fisheries research three indicators taken from the science of vital statistics (which studied age compositions, sex ratios, and so on of human societies): these were birth rates, age distribution, and migration. The ICES's extensive collection of herring samples allowed Hjort and his assistants – notably Knut Dahl, Paul Bjerkan, and Einar Lea – to study spring herring from Norway, Iceland, the North Atlantic, the Faroes, the North Sea, and the Baltic. In the 1907 catch, he discovered a preponderance of three-year-old herring. In 1908, more than 34 per cent of the total catch consisted of four-year-olds. In 1909, five-year-old fish made up 43 per cent, and in 1910, six-year-old fish made up 77 per cent. Obviously, fish born in 1904 (what Hjort called the 1904 'year class'), were more numerous than broods of preceding or following years. Even in 1913, the 1904 year class comprised more than 60 per cent of the catch.[18]

In 1914, Hjort published his classic *Fluctuations in the Great Fisheries of Northern Europe*. A turning point in fisheries science, this 228-page report united all the strands of fisheries biology and gave it a new paradigm; in it, Hjort gave the first solid explanation of natural fish-stock fluctuations. These were due not to overfishing or migrations, but to the varying success of year classes. In 1904 the survival rates of herring and cod fry were spectacularly high; that year class provided a rich yield in the fisheries of 1906 and for several years thereafter. Hjort predicted that the fishery would then decline until another productive year occurred.[19]

Hjort's breakthrough, although made possible by new fisheries science techniques (such as Einar Lea's masterly scale analysis, the relating of fishing statistics to indicators of fishing effort, and racial [biometric] studies), was not an inevitable product of these developments. Others had been working toward understanding the same problem, using the same techniques, but making little progress. Hjort was greatly helped by his involvement with ICES, which gave him many resources beyond his or Norway's sole reach. However, the nationalistic Hjort was strongly motivated to put these developments and the work of ICES into practice, to solve the riddle of catch fluctuations and then to work beyond it. Like many Norwegians at that time, he was engaged in the task of validating the concept of a Norway independent from Sweden, and the fisheries were an important aspect of the Norwegian economy. His great urge was to help the fisheries by putting theoretical advances into practice. In a science heavily reliant on statistics, his genius led him to look beyond basic catch statistics and to incorporate elements of vital statistics with the common tools of fisheries science. The new model he forged in Bergen transformed fisheries biology.

Ironically – but given the problem, not surprisingly – although Hjort succeeded in explaining the nature of fish population fluctuations, he was unable to determine what caused good or poor year classes. He suspected that the early stages of a fish's life were critical. His investigations showed that spawning success was not directly linked to the size of the parenting year-class; rather, the survival of fish larvae seemed the important factor. Hjort surmised that if herring eggs hatched during the plankton bloom, a strong year-class would follow. If the eggs hatched too early or too late, or if currents carried them over deep water, where plankton was scarce, few would have enough nourishment to survive. Hjort's hypotheses are 'almost impossible to test at sea, and both continue to be respectfully regarded.'[20] However, Hjort's concept of year classes was enthusiastically accepted by most marine scientists, and would form the basis of subsequent fisheries biology.

Hjort's 'year class' discovery had no precursor; there were no rival claims for his breakthrough. His theory went forward uncontested in terms of priority. It was, in Mertonian terms, one of the rarer 'singletons' among scientific discoveries.[21] However, in its content Hjort's theory was contested by D'arcy Wentworth Thompson, who held the chair of biology at the University of Dundee, and who was sceptical about the correlation between herring scale rings and exact herring ages. Thompson argued that evidence of scale rings corresponding one-to-one with each year of a fish's life was purely circumstantial, even if abundant. He argued that scientists needed to follow the scales of specific fish from year to year in order to empirically demonstrate the accretion of a new ring per year. He observed that Anne Massey's Irish oyster studies showed that oyster shells in the same age group have variations in their ring-layers of between two and seven layers, while the majority show three layers. Thompson believed that a shoal of herring comprised one year-class, which had spawned together and had since migrated, swam, and fed together. Most scientists, including the entire Biological Board, had been fairly completely convinced by Hjort when his paper first appeared, but Thompson and a small band of British fisheries biologists continued to reject Hjort's conclusions well into the 1920s. Thompson finally capitulated and announced his conversion publicly in 1930 at an ICES meeting.[22]

In the meantime, perhaps in response to criticisms, Hjort quickly sought to reinforce his evidence. To this end he accepted an invitation from the Biological Board of Canada to study herring in the northwest Atlantic. Earlier, in 1910, with Sir John Murray, he had taken the *Michael Sars* on an extended oceanographic expedition, one that covered both sides of the North Atlantic. Out of this voyage came a classic of marine ecology, *The Depths of the Ocean* (1912), co-authored by Hjort and Murray.[23] He was anxious to return to the western Atlantic for further hydrographic and fisheries studies and what resulted was the Canadian Fisheries Expedition of 1914–15; an intensive investigation of Canadian Atlantic waters.

Hjort was interested in solidifying his theory; the Biological Board had other motives for inviting him over. Up until 1914, the board's research projects had been carried out from May to September, by volunteer workers – primarily board members, university professors, students, and a few teachers. As the previous chapter makes clear, this work was fairly informal. Studies ranged from morphology and systematics to fishing experiments, faunal surveys, studies on fish and shellfish development, and so on. This work was hard to characterize because it lacked a central theoretical focus; the main activities were not one or two, but myriad. Generally, however, it was ephemeral in nature; there were no long-term projects carried over from year to year (except for Stafford's study of oyster development). Prince hoped that Hjort's visit would help the Biological Board develop a more relevant methodological orientation.

There is no record of the Biological Board's invitation to Hjort early in 1914. It is clear, however, that he quickly accepted, as he wanted to compare the herring fluctuations of the Atlantic's two sides. The initial project only involved a short study beginning in October 1914. The board's secretary, Macallum, felt that Hjort's stature was such that even a few month's work would enable 'the Board to determine what lines of investigation are to be taken in the future.'[24]

In October and November of 1914, Hjort journeyed along the coast from Boston to Newfoundland, conversing with herring fishermen and businessmen. From fishermen's catches, he collected scales and other samples:

> I had no other means at my disposal than such as could be contained in the not very extensive luggage of an ordinary traveller, and it was thus useless to think of anything beyond samples drawn from catches brought in by the fishermen themselves [and ascertaining] what kind of implements were employed for the capture of herring ... In some places, I was able myself to study the fishing in progress.[25]

He returned to Norway early in January 1915. His work so far was pointing to a number of questions. Were there several herring races in Canadian waters? Did growth rates vary in herring from different coastal waters? Did stock fluctuations mirror those of European herring?

If there were different herring races, then it might be possible to establish each race's distribution and migration by studying herring samples from different parts of the coast. But Hjort felt that 'outdated' North American fixed-net fisheries were catching only mature herring, which were not representative of offshore herring stocks. 'Only through an expedition, equipped with gear for the catching of all sizes of herring and with a sea-going vessel ... will a satisfactory scientific study of the herring be possible.'[26]

Hjort had particularly noticed the outdated Canadian technologies. In the spring, prevailing currents drove spawning herring close to shore, where they were caught by stationary trap and gill nets. The mature herring thus caught were inferior to the young 'fat-herring' caught by Norwegians using purse seines and drift nets many miles out to sea. As a result, Canadian herring was disdained by consumers and fetched poor prices. Hjort argued that 'every possible effort' be made to find exploitable Canadian offshore herring stocks: 'This information can be obtained only through the same fishing experiments as are necessary for a thorough scientific investigation of the life-history of the herring. These two sides of the same question ought ... to be attacked at the same time.'[27]

Macallum and Prince petitioned the Department of Marine and Fisheries to subsidize the proposed expedition. Hjort had seen 'ample evidence that enormous supplies of herring occur ... off our Atlantic coasts, where the fishermen are not aware of their existence.' If localized, a great herring industry could be developed. It was an opportune time, 'as the European war has practically stopped the Great North Sea Fisheries ...

and for some years to come ... Canada could find an illimitable market.'[28] They requested $20,000 to allow Hjort to study herring 'from Labrador to the Bay of Fundy.'[29] They asked for a further $30,000 to introduce a cheap quick-freezing technology to fishermen, to build a frozen-fish market (see next section).

The department agreed to the former. Hjort was to survey 'the races of herring in the Gulf of St. Lawrence and off the Cape Breton and Nova Scotian shores'; to ascertain their 'rates of growth and the variations in size'; and to 'decide the constancy of the renewal of the stock or of fluctuations in the stock of Atlantic herring.'[30] A grant of $20,000 was to pay for research assistants, travel expenses, the services of a drift-net steamer, sixty to seventy drift nets, otter trawls, purse and shore seines, plankton nets, thermometers, water bottles, and laboratory equipment. Hjort would get an honorarium of $350 per month, plus expenses.

The Norwegian government quickly granted Hjort a leave of absence, since the First World War was disrupting his work. He was to be in Canada from February to September, accompanied by Thor Iversen, an experienced herring captain. Although Hjort's time in Canada was to be short, Prince's hope was that the Biological Board would continue similar work in later years, and extend it to the Pacific.[31]

Hjort was to be joined by Biological Board scientists James W. Mavor and A.G. Huntsman. The latter had just been appointed permanent curator to the Atlantic Biological Station, and would devote all his time free from teaching that year to assisting Hjort in his herring experiments at the University of Toronto's Zoology Department. Mavor's involvement with the board had begun in 1912, when he came to St Andrews from Harvard University to study fish parasites. He was station curator in 1914 and 1917, by which time he was a professor at Union College, Schenectady, New York. His work at the station involved hydrography and studies of fish growth, until his last year there in 1921. Dr Arthur Willey (1867–1942) of McGill University also became involved, to Hjort's pleasure: he had already been on a three-year South Pacific expedition in search of embryonic forms of *Nautilus macromphalus.* Willey had received his DSc from University College, London, in 1894; the 'star pupil' of E. Ray Lankester, he had published the important *Amphioxus and the Ancestry of Vertebrates* (1894). A tutor in zoology at Columbia University from 1892 to 1894, Balfour Student at Cambridge in 1898-99, and Director of the Columbo Museum in Ceylon from 1903 to 1910, he became head of McGill's Department of Zoology in 1910, (retiring in 1932), and a member of the Biological Board the same year. James Playfair McMurrich (1859–1939) was a late but eager addition to the expedition. McMurrich, now professor of anatomy at Toronto, had joined the Biological Board in 1912. He brought with him experience from earlier marine biological laboratories: he had been at the Chesapeake Zoological Laboratory, Beaufort, N.C., in 1881; and later at Woods Hole (1889–1901), where he had taught invertebrate zoology from 1889 to 1891. Hjort's assistant, Dr Paul Bjerkan, was a last-minute addition: Hjort kindheartedly offered Bjerkan the job of titrating water samples at Souris, to help Bjerkan overcome his grief over the recent sudden deaths of his wife and only child.[32]

Hjort proposed a series of three hydrographic cruises on the *Princess*. The first was to be in May, when herring and other fish spawn and winter conditions still prevail. Another, in June, would collect fish larvae to locate spawning grounds, and a last one in late July or August would study young fish distributions and summer conditions. Each cruise would follow the same course to enable comparisons as the season progressed. Cruises were to run from Point Escuminac, New Brunswick, to Anticosti Island, then to the Bay of Islands in Newfoundland, and finally from Bay St George in Newfoundland to Pictou Island. Sampling stations would be spaced at every twenty miles, the stops lasting one-and-one-half hours. Huntsman was also to collect material from the Bay of Fundy.[33] The steam herring drifter *33* was fitted for herring fishing in the Gulf of St Lawrence, and the steamer *Acadia* was to survey the Scotian Shelf. Souris, Prince Edward Island, was to be the expedition's base. Rooms were set up there for examining material and titrating water samples.

The expedition suffered several setbacks. First, Captain Iversen, the hydrographic equipment, and biological gear were delayed in crossing the Atlantic due to the war. Therefore Hjort, accompanied by his wife, went to Boston to borrow hydrographical instruments from Harvard University, 'through the great kindness of Dr. H.B. Bigelow ... without other cost than sending of the gear.' As Bigelow had 'been doing some good work in the Gulf of Maine and off the coast of Nova Scotia,' Hjort, greatly impressed by Bigelow, tarried there, learning about conditions off Nova Scotia, which, said Hjort, 'could only be [learned of] in this way and I thought it more valuable to us than my staying in Toronto.'[34]

In the first week of May, Hjort arrived in Souris, where he found the harbour blocked by ice. Locals declared that the season was late. Hjort commented: 'Around the Magdalen Islands there is much ice and no fish. There is therefore no fear that we will be late.'[35] Then the *Princess*'s first cruise was interrupted by the grave (and fatal) illness of Commander Wakeham, Inspector of Fisheries for the Gaspé, who had been chosen as ship's commander for his intimate knowledge of local conditions. Wakeham's illness 'grew worse and worse during the cruise and it was at last absolutely necessary ... to bring him straight home to Gaspé.'[36] 'Remarkable' ice conditions and the late arrival of herring made Hjort postpone the next cruise. Now only two Gulf of St Lawrence cruises could be fitted into the season.

Hjort suggested that a cruise be made 'as soon as possible from Halifax ... to allow continuous scientific work.'[37] The *Acadia* therefore left Halifax on 29 May to investigate the seas between Halifax and southern Newfoundland and the entrance to the Gulf of St Lawrence. Hjort wanted to discover whether low temperature (–1.7°C) water layers off Sable Island originated in the Gulf of St Lawrence or the Labrador Current. He found 'free arctic water on the banks south of the Placentia Bay ... proving that a branch of the Labrador current is going Westwards and Southwestwards.' To gain a general understanding of the currents in relation to fisheries, the cruise also measured early summer hydrographical conditions. Plankton, food-fish eggs, and larvae were collected to define spawning areas, but fish eggs were scarce, which Hjort thought was due to low water temperatures. Most of the food fish, Hjort surmised, must spawn 'at the same time of the year as on the Grand Banks, where I have previously found the Cod spawning in July.'[38] The *Acadia*'s cruise, which ended on 4 June, was favoured by excellent weather that allowed continuous operation; thirty-six stations were taken. The *Princess* then surveyed the Gulf of St Lawrence from 9 to 15 June, taking twenty-three stations.

By 23 June, Hjort had established himself, Willey and Huntsman in the small laboratory at Souris, where they made preliminary examinations of water and plankton samples. Bjerkan measured salinities in the water samples; he had brought standard water samples from the Central Laboratory in Copenhagen to be used for calibration.[39] The *Acadia*'s second cruise from 21 to 29 July made fifty-four stations. Right after this, the *Princess*'s second cruise of the gulf, from 3 to 12 August, took twenty-three more stations. Meanwhile, the *33* was conducting experimental fishing as well as collecting material in the Gulf of St Lawrence.

Altogether, the *Acadia* and the *Princess* made 162 stations. Hydrographical material consisted of more than four hundred temperature readings and the same number of water samples taken from different depths. Plankton was collected in one or more vertical hauls and one long surface haul at each station. The Bay of Fundy was also included: in July, McMurrich supervised a ten-day expedition on the Biological Board's boat *Prince*: scientists took samples from nine stations for comparison with Gulf plankton.[40]

Johan Hjort and Fish Processing

Hjort had come to Canada with two objectives, each as important as the other: to lead the Canadian Fisheries Expedition itself, and to introduce a patented method for quick-freezing fish. Hjort had noticed the inferior quality of Canadian 'fresh' and 'frozen' fish, which was hurting fishing communities, as their products were unable to command respectable prices on any market. Even Pacific salmon and halibut were poorly frozen. He felt that these circumstances were 'amongst the most important for scientific fishery work in Canada as in Norway ... In both countries the fishing is mostly carried out from a great number of small harbours, where ... no great and costly cold storage plants could be built. It [is] a great desideratum to find a freezing method, which could be applied for the needs of people having but little capital.'

Thus he insisted that during his expedition it would be necessary 'at the *same time to consider the preservation of the herring for the fishermen, i.e. the question of freezing*. This problem is also of *the greatest importance for the eventual establishment of an enlarged herring fishery*.'

Fisheries expeditions combined with practical work were nothing new to Hjort. During a 1913 Norwegian expedition, he had had on board some fish-freezing apparatus. At every small harbour, the *Sars* was visited by as many fishermen as would fit on board: 'Very little of our work has had so much interest as this particular part of it.'[41]

To place the Canadian Fisheries Expedition 'in the closest possible touch with the industry and give practical results,' Hjort planned to conduct fish-freezing experiments in tandem with the expedition. Prince and Macallum backed him on this, telling the minister that Ottesen's patent method was 'the only method of freezing which is cheap, rapid and perfectly preserves all the best food qualities of fish.' Hjort had demonstrated the method to fishermen and merchants in Halifax and to scientific staff at the University of Toronto, all of whom were impressed. He froze large fish in a barrel of brine and ice, at –15°F, in twenty to forty minutes. *Canadian Fisherman* noted that air-freezing methods, then in vogue, took up to three times as long to freeze fish through.[42]

Hjort was motivated by a genuine desire to improve the lot of Canadian fishing communities. The new Ottesen patented method was cheap as well as accessible to small coastal communities. He hoped to demonstrate it in many small places and to sell the product in inland markets; if the results were satisfactory, he wanted the patent made free for Canadian fishermen. So important was this scheme to Hjort that he felt it pointless to conduct the expedition without it: 'It would [be] very unpleasant for me to make an expedition for the Canadian Government having accepted so much support and confidence and this expedition not ... corresponding to the lines expected by the supporters [members of the fishing community].'[43] However, it would have cost the Canadian government $10,000 to buy the right for Canadian fishermen to use the Ottesen method for one year. Permanent patent rights would have cost another $60,000. Also, a further $12,500 would have been needed to construct Ottesen freezers and pay for demonstrations, and to ship the frozen fish to 'distant Dominion markets.'[44] Canada had just gone to war, so it is hardly surprising that the Canadian government showed no interest.

In late December 1914, Hjort was advised by Macallum that the government would only sponsor oceanic herring investigations. Hjort responded that he 'must' withdraw the whole plan, as he was 'convinced that an effort to solve the great actual practical questions must embrace their most important sides.' Prince tried to change Deputy Minister G.J. Desbarats's mind: he emphasized the technology's small scale, the cost of each community's apparatus being about one hundred dollars.[45] Hjort's hopes in this regard were dashed, but the Biological Board persuaded him to go on with the expedition.

The final arrangements made no provisions for fish-processing research. However, Hjort tried to slip in marketing experiments with herring caught and salted by the expedition. Hjort worked out this lesser scheme with members of the Halifax Board of Trade: the expedition's carefully salted herring would be packed following Norwegian, Scottish, and other methods, and the barrels sent to Souris and Halifax. Leading fish merchants were to use this superior-grade herring for sales experiments. He told Desbarats that this would not interfere with the expedition, or cause it to exceed its appropriation,[46] as income from the herring sold would go to the Department.

Exasperated, Desbarats sent Hjort a strongly worded refusal. The department insisted, he said, that 'any curing experiments or attention to developing the commercial side of the herring industry ... are outside the real plan of work discussed with you when you were in Ottawa.' Hjort backed down, stating that he had not planned to spend much time on curing experiments, and that he had only intended to dispose 'of the herrings, which else might be wasted.'[47] If Hjort did get away with some much-reduced testing of the marketability of cured offshore herring, there are no records of this.

The Expedition's Impact on Canadian Oceanography and Fisheries Science

Hjort's Canadian Fisheries Expedition built on the North Atlantic survey he had made in 1910 with Canadian-born Sir John Murray in the *Michael Sars*. The hydrographic work was reported by J.W. Sandström and Paul Bjerkan. In the final report, temperature readings, salinities, and densities from some 150 stations yielded 'charts and a detailed exposition of ... currents and physical conditions.' Compilation of the report proved difficult, however, as the material had to criss-cross the Atlantic. Also, Hjort and Iversen were distracted by war duties – Iversen now directed fish supplies for Norway and had no free time. In the end, Hjort was unable to work up his own section on fish eggs and larvae, herring measurements and scale readings; however, he handed this section over to competent people such as Einar Lea.[48] Huntsman had charge of all plankton forms except copepods, which Willey covered. Phytoplankton samples, preserved according to a method developed by H.H. Gran, were referred to Gran for study. Iversen's fishing experiment results and Mavor's cod growth studies were not finished in time to be included in the report.

Sandström studied currents, water movements, and hydrography, while Bjerkan reported on salinities and temperatures. These results will not be detailed here. What is interesting is that Sandström devoted some forty-six pages to a textbook-style introduction to physical oceanography, which only goes beyond lay comprehension when he discusses Bjerknes's 'Solenoid' theory of oceanic circulation. This work could well have provided theoretical background for oceanographic work undertaken thereafter by the Biological Board, although such work proved to be very sparse: the board did not hire a full-time oceanographer until 1928. The summary of the expedition's findings comprised only four written pages, but a mass of charts, pictorial representations, and tables compressed a great deal of information about the Gulf of St Lawrence and Scotian Shelf.[49]

Canadian waters had never before been subjected to quantitative planktonic studies, although the importance of plankton as the basis for most marine food chains had been recognized since Victor Hensen's work. Indeed, such quantitative studies were recent even in American waters: a beginning had been made in 1912 and 1913 by H.B. Bigelow working on the American Fish Commission cruiser *Grampus* in the Gulf of Maine and the coastal waters between Nova Scotia and Chesapeake. During the Canadian expedition, plankton was collected using a method developed by H.H. Gran, who had, according to Hjort, 'succeeded in obtaining the first real view of the true plant production in the sea.' Gran himself worked up the expedition's material, and found many similarities between Canadian and Northern European plankton, except that Arctic forms were more common in Canadian waters. However, North American plankton seemed 'both qualitatively and quantitatively poorer than that of the European waters.' Gran believed that perhaps the expedition had missed the greatest plankton maxima of the Arctic spring bloom, owing to its late start.[50]

The fish-egg studies provided many interesting questions. Eggs and larvae of Arctic forms were found in the northern Gulf of St Lawrence, whereas those of southern forms, such as mackerel, were found in the south. Intriguingly, billions of cod and other gadoid eggs were found in the gulf, but very few larvae (young hatched fish); even at the end of July, Alf Dannevig found only a few larvae per thousands of eggs. Hjort noted that fishermen confirmed 'that in the gulf of St. Lawrence, very few young fish of any species are known to occur at all.' Dannevig postulated high egg-death rates due to cold or to unusually intense predation. However, as he was dealing with preserved material, Dannevig had no way of knowing how many of the cod eggs were alive or dead. Hjort, moreover, found it hard to believe that cod would spawn under conditions that would doom billions of eggs to destruction, and he doubted that cold temperatures caused the dearth of cod larvae. He asked the International Bureau for the Study of the Sea (in Copenhagen) to undertake hatching experiments at different temperatures.[51] There August Krogh and Dr A.C. Johansen found that temperature alone was not responsible – cod eggs and larvae could thrive in the warmer surface waters of the Gulf of St Lawrence. Also, cod eggs and young did not survive well in brackish waters.[52] Lacking the hydrographic data, Dannevig could not correlate all these findings when he was writing his report. Hjort believed that the hydrography and Huntsman's plankton work showed that cod eggs were carried en masse by gulf currents far from their spawning grounds.

Hjort and Einar Lea studied thousands of herring scales, length and weight measurements, and observations of sex, state of sexual organs, and fat. They used biometrical observations – numbers of fin rays, vertebrae, and keel scales – to distinguish herring races according to Heincke's methods. The 'Report on "Age and Growth of the Herring in Canadian Waters"' began with a thirty-two-page discussion of Lea's scale-aging methods, and could have served as a beginner's manual. 'Lea's report,' wrote scientist Johan T. Ruud, was 'one of his most important works. It includes a revision and extension of his previous studies of the structure of herring scales, and the formation of winter and summer zones, the basis for age determination and growth analysis.'[53]

At least four distinct groups of herring, characteristic of different regions, were found. Lea remarked: 'The difference between fish from different localities made itself apparent partly in the ... distinct marking of the annual rings on the scales ... partly in the fact that the dominant age-groups (year-classes) differed in samples from different localities, and finally, in the different character of the growth, as indicated by the scales.'[54] This fulfilled one the aims of the expedition – to find out whether there were different racial types. The four types were designated 'Newfoundland,' 'southern Gulf of St. Lawrence,' 'Cape Breton,' and 'southwestern Nova Scotia.' Unfortunately, Hjort and his associates failed to synthesize this material with the hydrographic and plankton results, so the environmental causes of these variations remained unexplained.

Hjort fulfilled his own main aim for the expedition when it was confirmed that like their European brethren, Western Atlantic herring had unequal reproductive success from year to year. The abundant Newfoundland herring had a predominant 1904 year-class; Lea remarked that this bore witness 'to a state of things similar to that noted in the case of the Norwegian stock, i.e., a single year-class taking up a dominant position, which is maintained for several seasons.' The year-class concept thus received important reinforcement and was shown to be universal. In contrast, the other Atlantic Canadian stocks all had a dominant 1903 year-class, with later additions of dominant 1908, 1910, or 1911 year-classes. An unexpected and surprising finding was that each racial type was also characterized by a distinct pattern of year-class abundance. This gave increased support, according to Lea, to the supposition that the regional types of herring were indeed different racial types or tribes.[55]

Johan Hjort's stay in Canada altered the style and direction of Canadian marine research. The Canadian scientists' encounter with cutting-edge fisheries biology and oceanography inspired the more motivated among them to adopt the new forms of these emerging sciences, albeit in a manner limited by available boats, equipment, and funding. Huntsman, Hjort's Canadian assistant, recalled: 'His tremendous vitality and clear thinking made a great impression on us all ... His main contribution was to apply to our Atlantic waters the methods of investigation which had been developed in Europe and for which he had been to a considerable extent responsible ... He ... opened up for us in marvellous fashion the fishery-biological problem throughout these waters.'[56]

In following years, the Biological Board undertook smaller surveys and expeditions, applying the intensive area surveys introduced by Hjort. For the most part, these were combined operations involving both biological and hydrographical work. They included physical oceanography, qualitative and quantitative plankton studies, and studies of commercial fish eggs and larvae.[57]

Hjort also increased the Canadian scientists' awareness of the need for research and development to improve fish-processing procedures. If Hjort did not directly inspire the Halifax and Prince Rupert experimental stations to engage in fish processing research – which work dominated the board for the ten years after these stations opened in 1924 – he certainly influenced the man who inspired many of the Biological Board's fish-processing and other research policies for the two decades following the expedition: Archibald Gowanlock Huntsman. Huntsman's work at the Pacific and Atlantic stations from 1908 to 1913 had revolved around the comparative morphology of ascidians. But after the Canadian Fisheries Expedition, Huntsman immersed himself in fisheries biology. His conversion was important, because when Prince appointed him permanent curator in 1915, he became the Biological Board's first full-time professional employee. He became director of the Atlantic station in 1919, a position he held until 1934. In the meantime, in 1907 he had been appointed lecturer at the University of Toronto; he became associate professor in 1917 and professor of marine biology in 1927, but after 1915 was not paid by the university; rather, he received all his salary as an employee of the Biological Board.

After the Canadian Fisheries Expedition, Huntsman was placed in charge of later surveys, and followed up Hjort's 'broad survey with intensive local investigations in successive seasons.' He regarded the 1915 expedition as 'only a beginning in what is needed to understand and use our fishery resources.' His encounter with scale reading led him to publish a paper on the method's shortcomings.[58] Following his new interests, he also published in 1918 'a mathematically unsophisticated but perceptive paper in which he showed how age distributions should change in response to varying fishing rates.'[59]

The later small expeditions started in the vicinity of the Atlantic Biological Station in 1916 and ended in the region of the Strait of Belle Isle between Newfoundland and Labrador in 1923. In 1916 the department arranged for herring drifter *33* to resume its survey work.[60] Hydrographic and biological surveys were made of St Mary's Bay, the Annapolis Basin, and Yarmouth Harbour in Nova Scotia and of Kennebecasis Bay, New Brunswick, in that year and again in 1919. In 1917 the Biological Board began to map undeveloped fishery resources in the Gulf of St Lawrence, and the *Prince* was used for studies of chemical and physical oceanography. In 1918 the board surveyed the Miramichi river from its head of tide seaward thirty miles into the Gulf of St Lawrence, following methods used by the Canadian Fisheries Expedition. Huntsman described in detail the hydrography of the thirty-mile-long Miramichi estuary, where the mixing of sea and river water affected the temperatures and salinities of surface coastal gulf waters for many miles. This allowed species not normally found that far north to survive in the region.[61]

An expedition to the Strait of Belle Isle in 1923, organized by Huntsman, was the last one conducted by the Biological Board along the ICES model. The work was also stimulated by the North American Council on Fishery Investigations, founded in emulation of the ICES in 1921 by the United States, Canada, and Newfoundland (France joined in 1923). Huntsman wished to discover why the Newfoundland cod fishery extended from south of Newfoundland up along the west coast through the Strait of Belle Isle and then down the east coast. He believed (and the expedition subsequently showed) that prevailing westerly winds heaped up warm Gulf of St Lawrence surface waters at the mouth of the strait. These waters passed through the strait along the Newfoundland shore, around Cape Bauld, then streamed to the east coast. He felt, however, that the Labrador Current partly entered the strait and passed westward along the Labrador shore, making the Gulf of St Lawrence north shore so cold that its waters supported Arctic fauna.[62] William Bell Dawson's hydrographic survey in the 1890s had shown heavier water at the surface along the north shore than at the bottom along the south shore. To Huntsman this could only mean a double-movement through the strait.

Newfoundland was asked to send a scientist for the cruise of the *Arleux*, as the expedition was investigating conditions affecting its staple, the cod. Alan Gardiner of Cambridge University served for Newfoundland, and University of Toronto physics professor L. Gilchrist was the expedition's physical oceanographer. The board's sixty-foot motor launch *Prince* joined the fisheries patrol boat *Arleux* in the work. Sections were made across the Labrador Current and its course was followed by means of drift bottles. The French representative on the North American Council, Dr Eduard LeDanois, placed two series of drift bottles across St. Pierre Bank.[63]

Gardiner collected seventeen new species and found a mixture of 'Arctic (in northern) and temperate forms.' Besides increasing knowledge of currents around the Gulf of St Lawrence, the expedition showed that Labrador Current waters mixing with warmer southern waters became very productive of phytoplankton, contributing to the richness of the eastern fishing banks. Unfortunately, the Belle Isle Expedition reports were never published; the Department of Marine and Fisheries got only a summary of the results. While much of a report was written up, it was abandoned when Huntsman was asked in 1924 to help develop a fish-handling experimental station at Halifax. The results were nevertheless used by later oceanographers. Immersed in other duties, Huntsman was unable to undertake further hydrographical surveys,[64] and ICES-style oceanographic work languished at the Biological Board for a long time thereafter.

Nevertheless, the impact of the Canadian Fisheries Expedition cannot be denied. The final report of the Canadian Fisheries Expedition contains a mass of charts and tables that 'contain exactly what a physical oceanographer would have wanted to see,'[65] and the experience gained helped prepare the ground not only for the shorter-term work under Huntsman, but also for the board's later emphasis on oceanography, which began when, under Huntsman's prodding, the Biological Board hired its first full-time professional oceanographer, Henry B. Hachey, in 1928.

Yet the experience gained by Huntsman and other biologists in the latest methods of fisheries science proved to be of much more immediate value to the Biological Board than the information gained from the expedition report. Canadian Atlantic herring were shown to exhibit 'the same peculiarity ... so markedly apparent in several European ... waters, to wit, the pronounced fluctuation in the increment of young individuals from year to year, whereby the age composition of the stock reveals an enormous predominance of some few extremely rich year-classes, while others hardly contribute.'[66]

Einar Lea's pioneering report on the age and growth of Canadian herring was most important, although it did little to synthesize any relationship between oceanography and herring distribution and natural history. It would have been much stronger had this material been interwoven more than Hjort attempted in his introduction. As it is, the report seems fragmented and lacking in interpretation. Furthermore, as some important work was never completed,[67] cod and other commercial fishes did not receive the attention they merited.

The Editorial Committee on Governmental Publications pronounced the Canadian Fisheries Expedition report to be 'of a most abstruse and technical character,' useful 'to only a very limited number of persons in the Dominion.' The committee went so far as to state that 'from an economic standpoint the value of the report to the fishing industry of Canada will be nil.'[68] Desbarats retorted that the Biological Board would consider the report to have 'great practical value,' and that it would 'probably be regarded as of great importance amongst fishery authorities at home and abroad.'[69]

The report was published in 1919. Ironically, the Editorial Committee had given a better valuation of the report than did Desbarats, since, indeed, the Canadian Fisheries Expedition's *Investigations in the Gulf of St. Lawrence and Atlantic Waters of Canada* had no economic significance. The department had only wanted Hjort to find new herring stocks for exploitation. Fishermen had no interest in these, however – they had as many inshore fish as they could handle, and they lacked expensive offshore fishing equipment. Canadian fishermen were rather in need of a more extensive market, and greater fish handling and preserving skills.

Hjort had realized this. His fish-processing projects might have gone a certain way toward helping Canadian fishermen, by creating new demand for superior fish products. But the Canadian government thwarted these treasured projects. Hjort comfortably married his pure and applied research activities; in contrast, the Canadian government reversed the usual government–science relationship by forcing this world-renowned scientist to abandon applied work, insisting on sponsoring only what turned out to be important basic research, which did not yield the expected short-term economic dividends. The net result of the expedition was to augment basic knowledge.

In spite of its shortcomings, the report of the Canadian Fisheries Expedition was a valuable addition to fisheries biology. Its eleven memoirs by specialists, and Hjort's summary of results, filled five hundred pages and provided the best report ever published in Canada 'dealing with the resources of the Canadian seas.' The expedition also gave 'the first critical assessment of oceanographic conditions in the general area,' as well as directing attention to questions related to the supply of food fish in the region. Much remained ill understood, and Hjort argued that it was too early to think of drawing together 'the various separate investigations ... into a single whole.' However, Lea's and Dannevig's papers gave 'the first foundation' for understanding herring and other important species' distributions and spawning areas, and the great fluctuations in their reproduction in Canadian waters.[70] And Hjort got the important confirmation he had been seeking for the universality of his year-class hypothesis.

The Canadian Fisheries Expedition had important consequences, only some of them foreseen by Prince. Through Hjort, the Biological Board was introduced to sophisticated oceanographic and fisheries biological methodologies, including Intensive Area Studies, and this changed the tenor of Canadian fisheries research. It was now characterized by longer-term studies in areas such as physical oceanography, fish-life histories and pathologies, and the effects of pollution, fishing, and predation on fish populations. Furthermore, the encounter with Hjort sensitized Canadian researchers to the value of applied research. The Biological Board soon began fish processing research and educational programs for fishermen, and the board's work in the 1920s and 1930s was outstanding for its emphasis on service to fishing communities.

Chapter 4

Ottawa, 1919: Bureaucrats versus the Biological Board

The husbandman has rent to pay,
  Blow winds, blow!
And seed to purchase every day,
  Row boys, row!

But he who farms the rolling deeps,
Though never sowing always reaps.
The ocean's fields are fair and free,
There are no rent days on the Sea.

Old Sea Song[1]

From its start, Canada's Department of Marine and Fisheries was responsible for fisheries legislation, licensing, and regulation; but its conservative bureaucracy was reluctant to enter into new endeavours, especially positive interventions to help the fisheries. Even before 1900, fishermen were demanding special education, the introduction of fish inspection regulations, and other interventions. Some of these were taken up in a half-hearted and lacklustre manner by the department, but so meagrely as to bring about no lasting changes. They tended to involve a minimum of departmental expertise; one of the best received was railway subsidies for transporting fresh fish from the Atlantic provinces to Montreal and Toronto. It was the Biological Board that finally altered the department's agenda and understanding of its role, by involving the department in board schemes for fishermen's education, the upgrading and quality control of fisheries products, and scientifically based conservation efforts.

Indeed, the Biological Board did more than any other group to convince the department to take on technical education of Atlantic fishermen and quality control of fish products. These efforts were begun in the 1920s and were directly enabled by board facilities. The board was willing to challenge the department to make reforms so that Canadian fish products would be more attractive to world markets. More importantly, the board itself initiated educational measures for fishermen and departmental fisheries overseers, and introduced other reforms. Once the fisheries experimental stations were in place in Halifax and Prince Rupert, the means existed to upgrade these activities to among the most important undertaken by the board. By the time the Board moved on to other agendas (around the beginning of the Second World War), these reforms were well enough entrenched that the department recognized education and inspection as both being important aspects of its national responsibilities. Admittedly, this recognition was not necessarily matched by legislative fiat.

The Department of Marine and Fisheries itself gradually evolved in its responsibilities both in response to Canada's growth as new provinces joined the Dominion, and in response to an increasingly scientific climate for regulating and understanding fisheries problems. In Canada, the regulation of the fisheries – and later, of fisheries scientific organizations – was a federal responsibility, overseen since Confederation by the Department of Marine and Fisheries. By contrast, in both England and the United States, national fisheries departments appeared only well after regulative roles had become entrenched in local districts or states. British and American national fisheries departments grew out of the need for scientific fisheries and fish processing investigations. They were secondary to local bodies such as American state legislatures or British district councils. The national administrations had little to say about fishing gear and regulations, excepting, in England and Wales, the salmon fisheries. Instead, they undertook scientific investigations to improve fish and shellfish conservation through aquaculture and other measures, and later, to improve fish-processing techniques. A comparison of how the relevant organizations in the United States, England, Scotland, and Ireland operated will indicate the significance of these routes of development: the new government fisheries authorities were based on existing science-oriented organizations, which had already highlighted the importance of new interventions for the fisheries. Thus the new bureaucracies in Britain tended to focus on very effective, practical strategies to aid the British fisheries. The contrasting approaches taken by the Dominion's Department of Marine and Fisheries will be highlighted below, so that the contributions made by the Biological Board will be better appreciated.

In the United States, powerful state governments controlled fisheries legislation. The U.S. Fish Commission, founded by Spencer Fullerton Baird in 1872, evolved into the American federal office responsible for the fisheries. In 1903 this body was renamed the Bureau of Fisheries and placed under the administration of the U.S. Department of Commerce. Its primary task was to prevent overexploitation of fishery resources, while aiding 'the utilization of the resource to the fullest possible extent.'[2] It had little power, however, since power remained vested in the states.

In Great Britain, fisheries administration was fragmented between Scotland, and Wales and England. Before 1882, no government department was responsible for fisheries as such; fishery matters were directed by the Board of Trade. In 1882 an Act of Parliament created the Fishery Board for Scotland, which gave Scotland the most advanced fisheries administration in the British Isles. This board was to 'take cognisance of everything relating to the coast and deep sea fisheries of Scotland,' and was expected to improve fisheries within its fiscal limits 'without interfering with any existing authority or private right.' From its marine station at Aberdeen, the board directed scientific investigations to improve fisheries legislation. Prior to this most fishery legislation had been shaped 'in accordance with local desires or prevailing popular opinion.'[3] The board's mandate was to eliminate useless and costly legislation; to collect fisheries statistics; and to create and enforce fisheries and fishery harbour legislation.[4] It was also to oversee Scottish fish hatcheries, and dredge and clear fishery harbours. In the 1890s, it established programs to educate fishermen to improve fishing and fish-curing techniques, and introduced a standard brand for cured herring. At that time, the Scots were losing markets as a result of careless fish-curing methods. The board spurred production of high-quality, carefully packed barrels for export by paying a bounty for every barrel produced according to its standards. Soon, Scottish herring was the most highly prized in Continental markets.

Ireland possessed a Fishery Board much like that of Scotland; but England and Wales had none. Instead, authority was vested in the Board of Trade, which in 1886 organized a fishery department to administer salmon and freshwater fisheries. But the department had no jurisdiction over the seafisheries. Rather, England and Wales were divided into eleven fisheries districts for administrative purposes; this resulted in regulations that varied from one district to the next. The Board of Trade did not fund scientific work; research came under the Marine Biological

Association and certain local fisheries committees.[5] In 1903 the Board of Agriculture took control of the Department of Fisheries. Most fisheries legislation, however, remained vested with local interests.

In Canada, by contrast, between 1867 and 1899 the federal Department of Marine and Fisheries held all the important legislative reins for the fisheries; the provinces had only secondary powers. Under the British North America Act the Dominion shared legislative power over coastal and inland fisheries with the provinces, but most authority rested with the federal government. Fisheries affairs were only one aspect of the department's work. Founded at the time of Confederation, its main concerns were regulating the merchant marine and maintaining navigational and port facilities.[6] However, it also set fishing seasons, issued fishing licenses, collected fisheries statistics, regulated gear, maintained Canada's fishing boundaries, and enforced regulations through inspectors. Since the department had mainly a regulatory agenda, the Biological Board had to earn its place by proving the value of fisheries science.

The federal government did not have unequivocal power over the fisheries: squabbling over federal and provincial jurisdictions continues to this day. Ontario, desiring fisheries revenues, in the 1890s contested federal authority; as a result, in 1898, the High British Tribunal declared that fisheries 'property rights' were vested in the provinces, but also assigned the Dominion the duty of framing fishery laws, along with limited powers to exact licence fees. In 1899 the Privy Council and Supreme Court allocated to the provinces licensing revenues for their internal fisheries (including the Great Lakes fisheries). But they argued that to properly manage the Canadian sea-fisheries, 'jurisdiction and property rights or licensing rights should be all vested in one authority, as the experience in the United States of the conflicting rights of the various states and their dissimilar laws and their unwillingness to recognize Federal authority, has had a most serious result on the fisheries of that great country.'[7]

In 1900, New Brunswick and Nova Scotia decided to defer authority to the Dominion. Quebec, on the other hand, demanded and obtained control of its coastal fisheries to the three-mile limit, arguing that as sea beds to this limit were contiguous to and part of the province, fish swimming above this bottom also belonged to Quebec. For this reason, the Biological Board had little to do with Quebec until the 1930s, when further legislation changed matters. The federal government withdrew its fisheries officers from Ontario and Quebec, although to protect Canadian fishing interests, a fisheries steamer was maintained to patrol Lake Huron and Lake Erie. The Dominion, however, still retained the 'absolute and uncontrolled right to declare the close season, the seasons when fish may be taken, and also the manner of catching the fish.'[8]

To safeguard the fisheries, protective laws and regulations were enforced by a large staff of fisheries inspectors and by armed cruisers on the Atlantic coast, the Great Lakes, and Hudson Bay. There were 24 permanent fisheries inspectors and special officers prior to 1910; 112 fisheries overseers with magisterial powers *ex officio*; and as many as 440 fisheries guardians hired temporarily as the need arose.[9] Close seasons were enforced to protect spawning fish. The Department of Marine and Fisheries also protected certain spawning grounds, and attempted to prohibit water pollution (with little success) and obstructions to migrating fish such as dams. From time to time, the department also organized fishery commissions to take evidence from fishermen, canners, and packers about conditions in certain fisheries. Fisheries legislation had long been based on evidence from such commissions. In the 1910s the Biological Board began to influence legislation.

Canada began collecting annual fishing statistics long before the United States. Statistics gathered from 1867 onward helped Canada win the Halifax Award in 1877, by offering clear proof that that Americans had been fishing excessively in Canadian waters since the Reciprocity Treaty of 1854.[10] Only later did the United States begin collecting fishery statistics, and only at intervals decided by the different regions. As late as 1930, the Bureau of Fisheries was still not canvassing all marine sections annually.[11]

The department also resorted to positive measures to aid the fisheries. Bounties were paid to Atlantic fishermen from interest on the $4,500,000 Halifax Award of 1877. This interest amounted to about $160,000 annually, and was divided among all vessels, boats, and men in the Canadian Atlantic deep-sea fisheries. A fisheries intelligence bureau was opened in 1889 to telegraph information on coastal movements of important fish and on the state of bait supplies in principal fishing centres. In 1913, this bureau also began daily to report the locations of bait species to point fishermen toward areas of abundance.[12]

After 1900 the department expanded its initiatives to help fishermen. In 1899 it began to provide refrigeration facilities for preserving bait, so that fishing operations would not have to halt owing to lack of bait. It subsidized larger companies, and it gave local fishermen's societies start-up grants for smaller bait freezers – up to 50 per cent to a limit of $1,000. By 1904, twenty-three bait freezers had been built, and more were under construction.[13] But the cooperatives soon fell apart; fishermen lost interest once three-year government subsidies of one dollar per ton of bait stored ended, and in a few years all of the bait freezers ceased to function, in part because frozen bait 'had acquired a bad reputation.' In 1906, scientists at the Atlantic Biological Station conducted fishing trials and found that frozen bait needed to be thawed in cold water to remain firm – a practice not often followed.[14] In 1907 the department began a railway subsidy to lower the costs of shipping fish to central Canada. The department also established four dogfish oil and fertilizer works. Dogfish sharks played havoc with netted and long-lined deep-sea fish, so a small payment was made to fishermen per ton of dogfish brought in; the sharks were ground into manure, processed for oils, and so on. The reduction plants were run at a loss; farmers preferred better-known commercial fertilizers.[15]

Before the Biological Board, science had only a tiny place in the department. The Dominion operated an extensive system of fish hatcheries under Samuel Wilmot, one of Canada's pioneer salmon culturists. When Wilmot retired in 1895, E.E. Prince took on many of his responsibilities. By 1914 there were more than thirty-three government fish hatcheries, twenty of them for Atlantic and Pacific salmon. Eight lobster hatcheries purportedly turned out 700 million lobsters each year. Other facilities hatched whitefish, pike, and black bass, as well as lake, brook, and Pacific trout. Young fry were planted soon after hatching to reduce costs, although it was thought that rearing fish to later stages offered better survival rates. In 1910 a special Pisciculture Branch was created, with its own superintendent and inspectors and, by 1914–15, a budget of $400,000.

Besides Prince, the department also employed Andrew Halkett, curator of the Fisheries Museum in Ottawa. Born in Forfarshire, Scotland, in 1854, he had only a grammar school education, but a great interest in natural history. He came to Canada in 1873 and for six years travelled the country, studying its wildlife. In 1879 he was appointed as departmental naturalist, and in that position worked on Atlantic, Pacific, and Great Lakes fishing boats to gain direct experience. He took part in the *Neptune* expedition to the Arctic and in many minor inland lake and river expeditions.[16] He provided the department with information about fish natural history (later in collaboration with the Biological Board) until around 1930. A.G. Huntsman considered Halkett a valuable observer, although his explanations lacked scientific rigour.[17] Another department expert was Ernest Kemp, hired in 1895 to survey Atlantic oyster beds and to plant new ones. He had been an oyster culture expert at England's Whitstable oyster beds; in Canada, he restored decayed oyster beds, using a specially equipped vessel, the *Ostrea.*

However, because its main concerns were regulatory, the department's positive interventions to help the Atlantic fisheries tended to be ill conceived. For example, much of Kemp's work was in vain. Year after year he levelled, cleaned, and restocked beds depleted by overfishing or deteriorated by silting. He brought many depleted beds back into fine condition, but since there was no leasing system for private oyster culture, these beds were wiped out by local fishermen as soon as the oysters reached harvestable size.[18] Only leasing rights would encourage oyster farming, by guaranteeing farmers the sole right of harvest. Since the Dominion was fighting with the Maritime provinces for jurisdiction over oyster bottoms and oyster lease revenues, the bottoms remained open to all, oyster beds continued to decline, and fishermen remained uneducated about oyster culture's advantages. Although only the federal government was making any effort to encourage oyster farming and to enforce regulations, in 1911 the Maritime provinces obtained leasing rights.[19]

The department's more usual role involved detecting and punishing transgressions. The Member of Parliament for (MP) Guysborough in 1907 observed that fishery officers had 'developed into a kind of detective.' As a result, fishermen rarely consulted a fishery officer or sought 'information from him about his business.' In contrast, 'the officer of the Agricultural Department who runs the travelling dairy or inspects the fruit, is always welcome ... The same relations should exist between the fishery officer' and fisheries workers.[20]

Positive interventions – such as teaching fishermen better fish-curing methods – were not undertaken systematically and were non-existent prior to 1900. Minimal educational work was only begun after repeated negative comparisons with the Department of Agriculture, often in Parliamentary debate. That department's 'persistent method of generally enlightened education' had brought farmers 'up by large strides,' yet fishermen were receiving no similar help.[21]

Later federal attempts to educate fishermen worked better in theory than in practice, being both poorly thought out and inadequately funded. One of the earliest involved demonstrating better fish-curing methods. Maritime fishermen turned out a gruesome, barely edible product; Scotland, in contrast, had a formidable reputation for excellent products. So in 1905 the department brought J.J. Cowie over from Scotland to experiment with herring drift-net fishing and the Scottish cure. Born in 1869 in Moray, Scotland, Cowie entered his father's extensive fish-curing business (which had branches in the Outer Hebrides, the Orkneys, and the Moray Firth, as well as in Yarmouth, Lowestoft, and Cornwall) rather than Aberdeen University. Working at each of his father's plants, he learned all aspects of fish processing. Not surprisingly, his Canadian cured herring proved excellent. The department therefore offered Cowie a permanent position[22] as Chief Inspector of Fish Curing.

In 1905 the department also opened an experimental fish-curing establishment at Souris, Prince Edward Island. In 1907, however, Parliamentarians complained that it was being used 'simply as a business establishment by the government.' Fish bought from fishermen were 'cured for the government and sold by the government ... in New York, Havanna, Porto Rico [*sic*] and other ports.'[23] The Souris plant was supposed to show how properly cured fish could command higher prices. Unfortunately, fishermen were not given hands-on instruction, nor did they get the higher profits – the government did. Thus there was no inducement for them to adopt improved methods. Likewise, Cowie had shown Canadian fish to be equal in quality to European fish, but all that reached fishermen was a report and a few demonstrations of the Scottish method.

Needless to say, these half-hearted efforts failed, although in 1907 the Acting Minister of Marine and Fisheries William Templeman told Parliament that Cowie's fish-curing experiments had been highly successful and had 'fully justified the expenditure incurred.' 'It is remarkable,' opposition critic Mr Ganong responded laconically, 'how frequently the experiments of the government are remarkably successful.'[24] Instead of concluding that more effort was needed, the department shrugged its shoulders; later, it actually discouraged Hjort's improvement schemes.

In general, the department resisted suggestions that it devote itself to improving fisheries through education and quality control. In 1909, Minister of Marine and Fisheries L.P. Brodeur was 'not ready to accept the idea of forming a board [like that] in Scotland. I do not think the time has come when such a board can be formed here to advantage.'[25] In contrast, the U.S. Bureau of Fisheries' Division of Fishery Industries attempted to increase fish consumption by improving fish handling, canning and preserving, and merchandising methods.[26] In Britain, this kind of work was begun by the Fishery Board for Scotland, and later continued by the Board of Agriculture's Fishery Department; by the 1920s the department of Scientific and Industrial Research's expert, Sir William Hardy, had made a speciality of refrigeration science at the Aberdeen Station. There was in 1909 no hint that the Biological Board had ever been intended to help educate fishermen or improve processing techniques. But the scientific linkages involved in its work naturally – though not inevitably – suggested Biological Board involvement at some point.

In fact, it was the work of the Biological Board – in particular, its lobster investigations – that changed the Department of Marine and Fisheries' view of its own role vis-à-vis Canadian fisherman. First, however, the board, being a latecomer in the department, would have to prove its worth. This it did.

The Biological Board supplied information for legislating open and close seasons, fishing gear, and areas to be used as reserves for spawning fish. For example, the department lifted a fishing ban in a long-established spawning herring reserve area off Grand Manan, New Brunswick, following investigations by Huntsman in 1917. Huntsman found that spawning areas vary from season to season and that Grand Manan's protected herring stocks had declined, whereas unprotected Magdalen Islands stocks had increased. The Commissioner of Fisheries added the argument that since the United States had no such reserves, fish protected from Canadians so near the American border were supplying Maine's sardine industry.[27] On the Pacific, the board helped regulate salmon fisheries and was integral to the International Halibut Commission. It did much to provide scientific as opposed to merely rational-intuitive bases for fisheries legislation.

But what most impressed the department was the lobster conservation work of Archibald Patterson Knight (1849–1935). Ontario-born Knight graduated from Queen's University in 1872, and in 1875 was awarded an MA for his thesis 'The Place of Science and Classics in a University Education.'[28] After taking an MD at the University of Toronto in 1886, he was appointed professor of biology and physiology at Queen's in 1892. There, he forged a lasting link between biology and medicine (as Ramsay Wright did in Toronto). A founder of the Board of Management, his early marine studies related to the effects of pollution and dynamite explosions on fish. The adversarial Knight was an articulate champion of applied science.[29] Huntsman said of him that 'he did not pretend so much to discover ... basic things ... as to try to get the results of scientific investigation practically applied to the fisheries.'[30] In this he succeeded remarkably, and nowhere better than in his work on lobster hatcheries, conservation, and canning.

Fishermen thought lobster hatcheries a great success. In 1901 the Minister of Marine and Fisheries had told Parliament: 'A large number of the small lobsters, innumerable almost ... are the result of the Pictou lobster hatcheries'; their presence in large numbers where hatchlings were planted 'led 'to the reasonable conclusion that they are the product of the fish hatchery.'[31] In fact, the lobster industry was in decline: more and more fishermen were entering this lucrative fishery, but overfishing and disregard for lobster conservation meant that fewer lobsters were being caught each year. In 1899 the catch in four Nova Scotia fishing counties using 178,055 traps had been 122,188 cwt. By 1916–17 the catch had fallen to 82,412 cwt, even though 366,238 traps were by then in use and motorboats were allowing a greater trapping range.[32]

In 1915 the department began to give the decline serious study. Noting a shortage of scientific information on lobster life habits and reproduction, the department turned to the Biological Board. The board turned the work over to Knight, who spent many of the following years studying lobsters in various regions as the Department's scientific consultant. On his own initiative, Knight immediately began an 'epoch-making' study of lobster hatcheries:

> For more than fifty years ... authorities ... in Canada, the United States, and other countries, had been engaged in hatching fish eggs and placing the fry in the lakes and streams of their respective countries. In all this time there had *not been one single attempt* made to ascertain what became of these fry after they were 'planted.' What percentage ... lived did not seem to concern the so-called 'fish culturists' at all ... Dr Knight was the first to question the efficiency of fish culture.[33]

He found lobster hatcheries to be a waste of money. Hatched lobsters had high mortality rates, and there was no evidence that they increased lobster populations. He even convinced the department that hatcheries 'had an unfavourable effect' on lobster reproduction,[34] in that they prevented captured berried (egg-bearing) lobsters from being returned to hatch young under natural conditions.

Lobsters usually extrude fertilized eggs, which are attached to and carried by the mother through eleven months' incubation. The mother, using her swimmerets, keeps these eggs oxygenated. Knight discovered that no hatchery could be as efficient even in merely hatching the eggs. Conversely, hatchery larvae suffered from '(1) *cannibalism,* that is the larvae eating one another if not properly fed; (2) *diatoms,* and (3) *a mycelial fungus,* the spores of which attack and penetrate the larvae and eventually kill them.'[35] Cannibalistic young lobsters had to be safeguarded from one another, and this required exceptional treatment. Even then there was no proof that the numbers successfully reared justified the expense. Over considerable opposition from the Fish Culture Branch, Knight succeeded in having the lobster hatcheries closed.

An exchange with M.H. Nickerson, a brash Nova Scotian who now managed the Boston Lobster Company, shows how the lobster industry preferred hatcheries to restrictive conservation measures. Nickerson, who had served on two federal lobster commissions, and who considered *himself* an expert, commented: 'The Government has given a lot of college dons the ultimate say on lobsters! Who is Prof. Knight? ... Then Dr Klugh, another of the faculty now in Nova Scotia gravely informs us 'the "lobster industry has declined 300 per cent!" It is a mathematical impossibility ... no quantity can be reduced to less than nothing.'[36]

West of Halifax, an area of alleged depletion, there were no lobster hatcheries: 'If the localities favoured with breeding stations are holding up so well, the plain inference is, the persistence is due to the hatcheries. The logical conclusion would be to ... establish new ones in the neglected western district.'[37]

By this time, most lobsters had become so small that the industry would collapse if size limits were strictly enforced,[38] and fishermen were ignoring many regulations, such as those to protect berried lobsters. Lobster fishermen and canners desperately needed educating about conservation, and this became a major thrust of Knight's work. Once fishermen knew what protective regulations were meant to achieve they would surely respect them.

Rising Department of Marine and Fisheries administrator W.A. Found, about whom we shall hear a great deal later in this chapter, agreed that educating lobster fishermen would help law enforcement. He and Knight made a beginning at a Conference of Lobster Fishers and Packers, held at Halifax on 7 August 1918. Many delegates denied that there had been any significant decline. Tellingly, one remarked with confidence: 'There is no fear of the government putting a size limit in, because they would put the whole gulf of St. Lawrence out of business.'[39]

The Deputy Minister of Fisheries occupied the chair, and Knight explained to all the rationale for size limits, using an agricultural analogy. If a rancher were forced to sell more of his herd than he desired, said Knight, his best policy would be sell off the immature cows, as 'he can best increase his stock by saving his breeding cows ... if he must choose between the two alternatives.' Applying this principle to lobsters, it became apparent that 'whatever we do ... we must conserve our large breeding lobsters,' especially since it took longer for immature lobsters to become breeders than for calves to become adults. Traps should have 'hoops or rings so small, say 3½ inches in diameter, that the large lobsters, male and female – cannot enter. They would therefore be left in the sea to mate and produce eggs. All the mature breeding lobsters, say above 11″, and all the immature ones under say 7″, should be saved.'[40] He also wanted to shorten the fishing season.

While balking at standardized traps, participants wanted the department to follow Knight's advice to enforce a two-month open season for lobster fishing in each district. If the department properly enforced berried lobster and close season regulations, 'the industry would be self-maintaining.'[41]

The department now began a campaign, organized by Knight, to teach lobster fishermen about lobster preservation. Knight met small groups at lobster canneries throughout the Maritimes. In the spring and summer of 1918 and 1919 he and five others addressed many public meetings of lobster fishermen and canners. Andrew Halkett would continue this for two months each winter for many years.[42] These instructors also distributed illustrated pamphlets explaining lobster conservation in non-technical language.

Knight's educational campaign was a great success. Changing attitudes in the industry were relayed in 1921 by W.H. Tidmarsh, a Prince Edward Island lobster packer and exporter. He not only commended to a convention of lobster canners the department's 'biological research work' on lobster 'perpetuation,' but also the fact that 'wise laws have been enacted and ... at least during the last few years, have been fairly well enforced.' Furthermore, 'Fishermen have been educated to liberate the spawn-bearing female lobster, and there is general sentiment ... that this is the right thing to do ... There is even at the present time indications that [lobsters] are increasing ... The best form of fishery protection is education.'[43]

The department also followed Knight's recommendations for closed seasons in some areas. Prior to 1919, Northumberland Strait lobster had been fished in spring and summer. As a result of Knight's work, the season was pushed to the end of August through to mid-October, to protect egg-bearing lobsters. Again, this measure was supported by an educational campaign. But Knight's lobster investigations bore other fruits. Fishermen were now more sensitive to lobster conservation, and the Department of Marine and Fisheries now recognized the value of science.

Knight's revelations about the wastefulness of lobster culture were seen by the department as a great cost saver, and increased its estimation of the Biological Board's potential. In the short term this proved to be a boost for the Biological Board; in the long term, its bane. With the increased esteem came a desire to wrest control of the board from its annoyingly independent scientist–administrators, who were in charge by virtue of the Biological Board Act of 1912.

When the Biological Board Act of 1912 was considered, A.B. Macallum, the capable secretary-treasurer, told Prince that 'it will be a great relief to be so free and independent as the Bill will make us as a Board. We can then develop a policy which will stretch forward into the years.'[44] Over the following decade, the board did indeed enjoy greater independence from departmental interference. However, the act did not free the board from scientific duties. Rather, section 5 stated: 'The Board shall have charge of all biological stations in Canada, and shall have the conduct and control of investigations of practical and economic problems connected with marine and fresh water fisheries, flora and fauna, and such other work as may be assigned to it by the Minister.'

Rather than moving away from applied fisheries research, if anything the board became even more committed to it. The 1912 act was in fact framed mostly to enable fiscal independence. Instead of being reimbursed following specific expenses, the board drew up annual budgets and administered its annual appropriations. Many of the earlier fiscal problems disappeared, except for one overriding one: there never was enough money in the first place, so results were often 'far short of what the Board desired.'[45]

In practice, the board did not refuse any work assigned to it by the Minister of Marine and Fisheries unless funds were short. Between 1912 and 1919, Prince as chairman passed along many requests for investigations and information from the minister, and there is no record that any scientist demurred. Nearly one-quarter of the publications arising from Atlantic research for this period dealt specifically with fisheries biology, as scientists grappled with understanding the feeding habits, diseases, life histories, and spawning seasons of commercial fishes, to help the department create effective fishing and closed season regulations. The board also tackled problems in fish processing, in oyster culture, in bait effectiveness, and so on.

In spite of this, the Biological Board's history was marked by periodic episodes of conflict with the department. A cursory look at these episodes might lead one to suppose that the scientists' research independence was being pitted against the department's applied agendas. Indeed, Kenneth Johnstone, in *Aquatic Explorers*, limits his explanation of the 1919 conflict (about to be described) to old platitudes about pure versus applied science.[46] Such views are inaccurate: the relationship between department and board was in fact more one of cooperation than of conflict. When conflict did occur, it tended to reflect power politics or administrative intricacies rather than philosophical differences between pure and applied science.

Some politicians in the department apparently neither knew nor cared what the scientists were doing. Raymond Prefontaine, who became Minister of Marine and Fisheries in 1902, 'repeatedly disclaimed both the paternity and any understanding of the biological research programme, and even indulged in a bit of humorous play-acting on the subject to the amusement of the House of Commons.'[47] Departmental bureaucrats were not so indifferent, however. Individuals such as William Ambrose Found (1873–1940), whose departmental career parallelled the Biological Board's, came to value its work and to want tighter control over its activities.

Born in Prince Edward Island in 1873, Found graduated from the Prince of Wales College in Charlottetown and became a schoolteacher. In 1898, the year the Board of Management of the Marine Biological Station was created, the Department of Marine and Fisheries hired him as secretary to the head of the Fisheries Branch. There, Found discovered his life's calling. In 1911 he was promoted to the newly created position of Superintendent of Fisheries – the chief administrative officer of the fisheries service under the deputy minister. On being promoted to Assistant Deputy Minister of Fisheries in 1920, he became the most powerful individual dealing with the Biological Board until he retired in 1938.

Found ardently desired to change the Biological Board. He chafed at the way the board wiggled out of certain requests, claiming lack of funds or insufficient volunteer personnel – although, in fact, the board rarely ignored or refused questions referred by the department. Furthermore, he was irked by criticisms of his department, which he knew was not doing all it might for the fisheries. Lobster and other fishermen were repeatedly asking that fisheries administration be placed under local boards similar to the Scottish Fishery Board, so that problems could be dealt with by those familiar with local conditions. Politically motivated appointments of fisheries inspectors had often resulted in failure to properly enforce the laws. In a rather emotional memorandum to Deputy Minister G.J. Desbarats, Found admitted: 'There is no question that our outside Service in the Maritime Provinces is deplorably weak.' But he then continued: 'The undersigned strongly feels that the departmental administration has never had an opportunity to vindicate itself, owing to our weak organization, particularly ... on account of political influence being permitted to exercise so much control ... But when the reorganization the Civil Service is engaged in is completed ... the Department will for the first time be in a position to serve the public as it should do.'[48]

One way to improve the department's record would be to incorporate a scientific unit. This would be possible only once scientists could be hired as civil servants under the reformed Civil Services Act, which was to come into effect in 1919. In October 1918, Found told the deputy minister that although 'the Biological Board, as at present organised, is doing good work,' it was 'in the everyday political work of the Department that need for certain lines of investigation become evident'; for efficient coordination, 'scientific work should form a division of the Department's activities.'[49]

Anyone who has watched the British comedy *Yes, Minister,* and laughed as Sir Humphrey adroitly manages his minister, can easily imagine subsequent events. Found played this role to the deputy minister, George Joseph Desbarats who, as events unfolded, also acted as 'Sir Humphrey' to the minister, convincing him that his lack of control over the Biological Board was a serious problem. The Minister of Fisheries, C.C. Ballantyne, was in fact almost totally ignorant about scientific research in his department. Following the initiation of legislation to change the board, he sent a memo requesting information about it. He thought the academic members of the Biological Board were appointed by an Order-in-Council and that a 'Scientific Branch' in the fisheries department had existed prior to the board's creation.'[50]

Found convinced the deputy minister, (who *was* familiar with the board) that 'while Section 3 of the [Biological Board] Act provides that the Board shall be under the control of the Minister, in practice this is hardly possible.'[51] He pointed out that the department could not build up a scientific division as long as the Biological Board Act gave 'the Board control of *all* marine and aquatic biological research in Canada.' He interpreted section 5 as giving the board 'exclusive conduct and control of investigations ... connected with the Fisheries' – an opinion upheld by the Department of Justice.[52] The act would have to be altered before a departmental scientific division could be established. However, the board, with its experts, might prove beneficial to retain for a while, as volunteer consultants,[53] so Found advised only that section 5 of the Biological Board Act be rescinded – the part that gave the board control over its stations and research.

In 1919, with the minister's approval, the deputy minister initiated Parliamentary Amending Bill No. 106 to make the Biological Board a division of the department. Specifically, it would make the biological stations and board infrastructure a division of the department while doing away with the board's executive power. Found's planned departmental scientific division would be ready to take over by 1920.[54]

The board was told about this amendment to the Biological Board Act, but was misled as to its intent. It was told only that the minister wanted to change section 5, which read: 'the Board shall have charge of the Biological Stations of Canada.' This clause, according to Desbarats, took 'away from the Minister and the Department the liberty of having a Biological Station, or of conducting Biological work, if they wished. There is no reason why the Minister should abdicate these powers ... While the Department may not wish to undertake work of this kind, circumstances may well occur under which it would be advisable.'[55]

Board secretary A.B. Macallum concurred, and suggested that the act's wording be changed to indicate that 'the Board shall have charge of the Biological Stations of Canada at St. Andrews and Nanaimo.'[56] This advice met with no response. At this stage, the Biological Board was not especially worried about the proposed legislation, not understanding that it spelled the board's demise and that the minister was acting in bad faith. The bill was introduced into Parliament without its contents being divulged to the board, which was only told that the bill would enable the department to engage and appoint biological experts. As this object, though of interest to the board, did not affect its functions, no opposition was contemplated.

The deception was not discovered until a few hours before Bill No. 106 was read a second time in the House of Commons. Macallum obtained a copy and was aghast at its contents. He immediately went to the minister and was told by him that Desbarats wished to have control over the board; but Desbarats had told Macallum repeatedly that it was the minister who wanted this.[57]

Macallum, the board's most fiercely independent member, who had acted as its real head in his capacity as its secretary-treasurer since 1911,[58] spearheaded the Biological Board's resistance to the bill. He scolded Desbarats:

> In this Bill, absolutely contrary to your statement of three months ago and maintained by you on every occasion until I saw a copy of the Bill, there is nothing whatever concerning the power of the Department to engage biological experts for the scientific work of the Department. On the other

> hand, the Bill takes away from the Board the control of its own two Stations and the power to initiate such work as it has been doing for over twenty years. You say that this control is not taken away. The Deputy Minister of Justice gives quite a contrary opinion.

Desbarats had stated 'again and again that, if the Bill passes, the Board will continue in control.' Why, asked Macallum, had not Desbarats advised the minister to provide 'for that Control explicitly in the Bill?' Moreover 'the Board never did monopolize, or wish to monopolize, one of the scientific branches of the Department.' Indeed, it had supported the department's hiring of Hjort: 'Where, then,' he asked, 'is any monopoly by the Board.?'[59]

In the meantime, Macallum began to rally Senate support to defeat the bill. An angry Desbarats told him: 'The Minister wishes to have complete freedom in the administration of his own department.' He was livid that Macallum was lobbying the Senate to prevent him 'from gaining the full control of the activities of this department.'[60] Macallum was just as furious, demanding to know whether the Biological Board had 'no right to say a word in protest when an attempt is made to impair its functions.'[61]

Macallum's lobbying succeeded. Although the bill passed in Parliament, it was strongly blocked by the Senate, whose members stated that 'the preamble to this Bill has not been proved to their satisfaction.'[62] They found no evidence that the Biological Board claimed a monopoly on all Canadian marine research. Perhaps Macallum had pointed out to them that a departmental scientific division would be more or less a farce because Found and Desbarats would not know what to investigate. Their scientists 'would get, at the best, but meagre salaries, such as are allowed in the new classification of the Civil Service. The result, ultimately, would be researchers who could get positions nowhere else, and the result of whose efforts would be largely those shown in padded Government Reports.'[63]

Macallum had saved the board, but the department demanded his resignation.[64] The same day (7 July 1919), he told his minister that he wished 'to resign my position as representative of your Department of the Biological Board,' which position he had held for twenty-two years. He was resigning because of 'the attitude of your Department and some of its leading officials ... The Board['s] eminent biologists ... have carried on their work ... for over twenty years, without any remuneration whatever, in the belief that what they did was appreciated ... [recent] events ... have shown what an illusion this was.[65]

Macallum's private opinion was that the minister 'of course, does not care a rap for anything except personal victory and in this he was balked.' He assured A.G. Huntsman that he would not lose interest in the board, and would 'endeavour to keep it independent of any petty control that might be attempted by Mr Desbarats and Mr Found, and, in the last resort, by the Minister. The last-named is, of course, wholly ignorant of what is involved. He has taken his cue from Mr Desbarats.'[66] In spite of his intention to keep in touch, however, Macallum quickly dropped out of the picture.

Interestingly, Prince did nothing to champion Macallum. This was somewhat unfortunate, for Macallum's efforts had saved the board. Instead, and revealingly, Prince told Ballantyne that although Macallum would be missed, 'yet, as Chairman of the Board, I often felt he had too much influence, and that occasionally the policy adopted was not fully in accordance with my own views, both as Chairman and as Dominion Commissioner of Fisheries.'[67] Prince and other board members were uncomfortable with Macallum's strident insistence on pure research. Macallum had 'imbibed the dogma of "pure science" as a graduate student at Johns Hopkins,'[68] and although he had helped found Canada's National Research Council in 1917, becoming its first president, he had no knowledge of industry, and promoted pure science in that body. He had told the Royal Society of Canada in his 1917 presidential address: 'The advancement of pure science ... has sanctions deeper and more sacred than those derived from its utilitarian ends, valuable as these are in serving our physical life. Every agency that can promote this advancement ought to be engaged as in the performance of a high duty, a duty with a religious significance.'[69]

However, his plan of action for the NRC also did not receive approval. Macallum, 'by seeing himself too literally as Canada's "czar of research" ... antagonized personnel in government agencies who would otherwise have been his natural allies. After he resigned in 1920 as chairman, he stayed on and tried to interfere, especially when H.M. Tory proposed something different in 1927, which became the true beginning of the NRC.'[70] It appears that a similar situation prevailed at the Biological Board, and few regretted Macallum's departure; they had different ideals for the board.

The 1919 episode illuminates two things. First of all, Found's actions were not motivated by dissatisfaction with the results of the Biological Board's research. Rather, they reflected raw power politics: Found's and Desbarats's attempts to appropriate the Biological Board reflected their desire to speed up research, but were mostly a push for self-aggrandizement and increased personal power. Second, this episode showed the extent to which the board had impressed an initially dismissive department. The board's science had helped rationalize fishing legislation for lobster and other commercial fish species, and Knight's work especially had disclosed how science could release the department from costly programs that did not work, such as the lobster hatcheries. Thus in no way can the conflict with the entire board be categorized as one over pure versus applied research. Throughout the 1920s, with the building of fisheries experimental stations at Halifax and Prince Rupert, the Biological Board continued to show more sympathy than disdain for the department's desire to apply research to fisheries problems.

The board did not survive the skirmish of 1919 unchanged. The Minister of Marine and Fisheries and his deputy both insisted that the Biological Board adopt amendments requiring that the minister approve all board policies; all annual estimates; and all scientific and administrative appointments in the Board.[71] Revised bylaws were adopted at the board's annual meeting in 1920.

The bureaucrats did not want to stop here. Found asked the Board of Agriculture and Fisheries in England how its fisheries research was administered. Henry G. Maurice, Assistant Secretary to the Board of Agriculture, wrote back that although '*marine investigations* are largely influenced by ... the *International Council* for the Exploration of the Sea,' the 'local institutions may be said to "act on their own initiative."' Proposed research schemes were submitted to the board beforehand, and reports and accounts of the year's work 'furnished in due course. The *institutions* also occasionally *take up investigations* at the *request* of the *Board.*' [72] Obviously, there was little here to aid Found in his quest to subjugate the board.

However, Found did begin to make small, discrete adjustments to the board. First, he had Desbarats ask A.P. Knight to fill the Biological Board vacancy created by Macallum's departure: 'The Minister has had in mind the extremely good work you have accomplished in the last few years on the lobster question. He understands that you will shortly be leaving Queen's University and would therefore cease to represent that University on the Board.'[73] Found intended to see Knight elevated to board chairman, certain that Knight would play along with his schemes for reshaping the board. At the board's annual meeting on 18 May 1920, Prince ended his long chairmanship of the board, and Knight took over. Prince became secretary-treasurer until his retirement in 1924, and remained an *ex officio* member until his death in 1936. Knight was far more congenial to the minister than Macallum had been. He deliberately moved the board toward more practical studies, and 'his most significant accomplishment as chairman was to engineer a mutually agreeable rapprochement between the Board and the Department, which permitted a rapid increase and diversification of fishery research activities.'[74]

In 1923, Knight agreed with Found that a proposed new science division in the Fisheries Branch would probably 'duplicate' the Biological Board, and that this same end could be 'most efficiently and economically brought about' by reorganizing the board 'to bring it into closer relation with the Department.'[75]

A new bill to reorganize the board was, according to Found, 'not only fully concurred in by the Biological Board, but it [was] asked for by it.'[76] An acceptance of the need to change had permeated the board. This was highlighted in 1922, when the Atlantic Biological Station, following up the oceanographic program inaugurated by Hjort in 1914–15, made an expedition to the outer coast of Nova Scotia. For publicity, Huntsman arranged consultations with the fisheries committees of the Halifax, Lunenburg, and Liverpool boards of trade. But 'with only one exception, the members of the Halifax committee had not a good word to say about ... the Board,' and the Halifax committee chair, A. Handfield Whitman, 'remarked that he couldn't say that the Board had done any harm, but the money it spent might as well have been pitched into the sea.'[77] The board realized then that it had to become more relevant to the fishing industry. Thus, it eagerly acceded to the deputy minister's proposal that board membership now include representatives from the industry.

The 1923 Bill to Amend the Biological Board Act integrated the board more closely with the department. Under it, the department appointed to the board an additional departmental officer and two industry representatives. The Atlantic representative turned out to be A. Handfield Whitman, its former severe critic, who quickly became one of the board's greatest advocates. To facilitate practical research, two substations were established, one at Halifax, the other at Prince Rupert. These stations were to investigate and demonstrate new fish-handling techniques and to educate fishermen, fish curers, and manufacturers. Found anticipated that the experimental stations would 'do for the fisheries all that a well equipped experimental farm can do for the farmers.'[78]

These changes were quite benign, even positive; but the department was still not satisfied. In 1925, Deputy Minister Alexander Johnston, who had taken over from Desbarats in 1920, wanted to legislate the board into eastern and western sections because 'Pacific and Prairie members of the Board are apparently not much interested in Atlantic problems and *vice-versa.*' He argued that this division would improve it 'from the standpoint of economy.'[79] But the board suspected the department of attempting another power grab.

In fact, the idea originated with the Biological Board's Pacific industry representative, John Dybhaven. He wanted to find some mechanism to make the board pay more attention to Pacific fisheries. A.H. Whitman submitted an alternative proposal, to set up advisory committees for each coast. Knight received Whitman's proposal enthusiastically, but the deputy minister preferred Dybhaven's idea,[80] seizing on it as a way to bring the board's activities under greater ministerial authority:

> Johnston's proposed Bill amended Section 3 to read that the Board would 'consist of an eastern and western section.' The eastern section would deal with problems arising from Atlantic coast inclusive to and including Ontario, and the western section with those arising in the remainder of Canada. The Treasurer of the Board shall be the Chief Accountant of Fisheries. The Board shall meet at Ottawa when called upon to do so by the Minister.

Not surprisingly, this proposal unleashed a storm of wrath from the Biological Board. Once all accounts had to be certified by the Chief Accountant of Fisheries for payment, serious delays in everyday board transactions would ensue. The board suspected the department of trying to reassert intolerable pre-1912 accounting conditions. Board member R.F. Ruttan told Toronto member J.P. McMurrich: 'I feel that every move made by the Department is made with a view to getting more and more control over the activities of the Board.' To McMurrich the Bill portended, 'so far as I can see, the final obliteration of the Board.'[81]

There was some thought that the bill had been precipitated by grumbling Pacific station scientists, who were aggrieved by the Atlantic station's favoured status. Director Wilbert Clemens protested that he was 'not the author of this movement and ha[d] had no part in it whatever.' Even so, he seized the chance to complain that 'no member of the Executive with the exception of Dr Knight' had visited his station for years and thus no one 'really understands circumstances here.' Owing to the tremendous distances involved, communications with Knight and Ottawa were suffering. Nevertheless, he had 'sufficient faith in the Board to feel that it will endeavour to meet our needs. I hope no members will place local desires or needs above those of the whole.'[82]

Knight informed Johnston that 'every scientific member will deny that the essential functions in fish life on the two coasts, are different.' Superficial differences were not enough to justify dividing the 'Biological Board of *Canada*.' Such division would destroy coordination. 'As well divide Parliament and expect unity in legislation, as to divide the Biological Board and expect unity in planning for the scientific development of the Fisheries.' He pointed out that Huntsman was investigating different methods of smoking fish. 'Manifestly, this same investigation should not be started at Prince Rupert. Otherwise effort and expense will be doubled.'[83]

Knight urged members to criticize the bill to Johnston. Industry representative Whitman said it was 'very badly drawn up' and 'would only lead to confusion.' In calling for 'two Biological Boards' it failed to 'lay down any programme as to how these two Boards are to be constituted.' Clemens and Huntsman advised Johnston that 'separatist movements are not advisable and should be resorted to only when no other solution is possible.' A better solution was a Pacific Coast committee that could 'deal with special problems and details as they arise.' C.J. Connolly of St Francis Xavier University condemned the bill as a 'retrograde step' economically, scientifically, and nationally, to be opposed 'on national grounds' as one should oppose 'a separation of any Government department into an Eastern and a Western section.'[84]

Johnston had hoped to play on the separatist forces always present in the Canadian political and social context. He did not (at first) understand the strength and unifying force of the culture of science. However, the board shouted back the message that it must control policy and expenditures to ensure 'that permanence and freedom so essential to scientific work.'[85] This episode provides perhaps the clearest example of attempted political interference that had everything to do with power mongering and little to do with the content of the board's work. The deputy minister conceded defeat, and the project was abandoned; instead, as the board wished, eastern and western committees were formed to deal with the special needs of both coasts. However, Found never gave up his idea of making the Biological Board in constitution the department's scientific division, and it was to this end that it was reconstituted as the Fisheries Research Board in 1937 (see chapter 8).

Influencing the Department of Marine and Fisheries

Even while the department was meddling, the Biological Board was working, perhaps much harder, to change the department. For example, Knight's lobster fishery work showed him that Maritime lobster canners had a compelling need for sanitation education. The department, although aware of the problem, never attempted this. The only people qualified to give instruction were scientists within the board itself, or trained Departmental employees. But the board did not have enough people, time, or money. It really was up to the department to supply personnel to instruct canners – and who better than the fishery officers and inspectors already in its employ?

A comment made in 1902 by the Minister of Marine and Fisheries reveals much about the qualifications of fishery inspectors and officers early in the last century. Manitoba's fishery inspector had been a hardware merchant with no fisheries experience; the minister expostulated: 'I am not aware that any expert knowledge is required of the occupancy of this position ... Any sensible man could discharge the duties of the office.' Since fisheries officers and inspectors were required to disseminate information and gather fisheries statistics, it is not surprising that in 1908, Opposition Leader Robert L. Borden complained that they were insufficiently trained and that fisheries affairs had drifted along too long without any helpful intervention.[86]

Yet it was only around 1920, once the Biological Board became deeply involved in the fishing industry's affairs, that the department began comprehensive efforts to help the Atlantic Canadian fisheries. That these measures were begun at all was due to the Biological Board's constant efforts. First, it pushed for science education for fishery officers and overseers – the department's main contacts with fishermen. Younger fishing industry representatives were given courses at the experimental stations, in order to make them 'proficient Inspectors of fish ... competent to demonstrate ... how fish should be prepared.'[87] But even earlier than this, A.P. Knight had campaigned to eradicate lobster canners' extreme ignorance, and the equal ignorance with which the Department dealt with sanitation problems.

Complaints about blackening in cans of Maritime lobster led the Biological Board in 1919 to ask the department to sponsor an inquiry by Knight. To Knight this was as important as lobster conservation, for many canneries were filthy. He inspected fifty-three lobster canneries in Prince Edward Island and New Brunswick in the summers of 1920 and 1921 and produced two important reports. He found that although the shellfish canning regulations of 1915 and 1918 had been intended to effect radical improvements in lobster factory sanitation, those regulations had not been enforced; furthermore, improvements 'in the past quarter of a century have been almost nil, if I may judge from the report of Sir Andrew McPhail (1896).' McPhail had shown that deterioration in canned lobster was 'putrefaction ... due to the action of microorganisms'; he had seen factories that were 'mere hovels with inadequate appliances for ordinary cleanliness,' and he had warned that even in the best circumstances the offal generated would promote bacterial growth.[88]

Like McPhail, Knight found that the Maritime lobster canneries were unregulated. The canner chose how he washed the meat, how long the total sterilizing and canning operation lasted, the strength of the pickle, how sealed cans were bathed and whether or not they were 'brogued,' or 'blown' – that is, punctured to remove trapped air, to prevent later distortion of the tins in market shelves. (The holes were then soldered and the cans bathed in boiling water.) Most canneries did not have lavatories and did not promote employee hygiene. Knight warned that if such conditions continued, 'the bad [will] reach Europe and the United States along with the good, and help to depress prices.'[89] Many a factory was really 'a mere shack, with incomplete flooring, walls that are unplastered; with no ceilings; no proper drainage ... More than half those which I saw were unsanitary and should be closed.' The large packing firms, when asked why they did not make improvements, answered, according to Knight: 'Why should we spend an additional $2,000 or $3,000 in perfecting our establishments and still come into competition with the "shanty" canneries? ... Our out-put is so small in comparison with the whole output of Canada, that we could not command any higher price.'

The remedy was to standardize cannery equipment and procedures by scientific means. For example, Knight found many variations in pickle strength, from 2 to 12 per cent: 'All ... cannot be equally good.' Macallum had determined inorganic salt concentrations in lobster blood serum to be between 1.25 and 2.85 per cent. The Portland Packing Company tested this pickle strength for Knight, with 'excellent results.'[90] With standardization, producers would no longer be penalized by poor prices owing to lack of consumer confidence.

Working for Knight, Jennie McFarlane, BA, of the University of Toronto, studied lobster 'smut' – or blackening – at St Andrews in 1919 and 1920, and found that it arose from bacteria from lobster intestines. In poorly sterilized cans, these anaerobically produced hydrogen sulphide gas, which eroded tins around the seams and formed black metallic sulphides of iron, tin, and lead on the meat. Although harmless in small amounts, smut looked disagreeable. Brownish, grey, or pink patches were bacterial: 'As time passes the colour, the taste, and the smell, become so altered that no one can eat the meat.' A major cause of bad cans was frequent delays in canning operations: 'When catches were small, lobsters were often allowed to lie uncovered on cooling tables for up to two days, even in warm weather, until a sufficient number of lobsters had accumulated to make it worth-while to complete the canning operation.'[91] Knight said that boiled lobsters should never lie for even half an hour on any table. Also, packed cans should have lids soldered on immediately; the practice was often to solder lids on at day's end.

Cleanliness was the most important thing, however. Knight saw canneries varying from filthy hovels to one as clean as a good housekeeper's kitchen. Few owners provided privies, and in only five of fifty-three canneries did Knight see soap, washbasins, and hand towels. Workers' clothing and hands were often dirty. He saw sterilized white aprons and caps in only one: 'The wonder is that so many cans turn out good.'[92] Tables tended to be surfaced with galvanized steel or zinc, which once rusted was impossible to keep free of bacteria. Often, cannery sites had poor drainage, and offal was merely flushed into adjacent sands. This offal attracted flies, which then abounded within the factories. Most canneries drew water from wells sunk under cannery floors, and these were liable to drainage contamination. In one case, water was drawn from a marsh. Knight saw water he likened to that of mud puddles used for washing lobster meat; often the same water was used repeatedly.

Knight recommended porcelain, glass, or marble surfaces. He urged that regulations for licensing in 1921 require cannery owners to have a separate room 'exclusively for extracting the meat, washing, packing, and weighing the same.' All handling of lobsters would have to be done in this room, which would be designed so that it could easily be kept scrupulously clean. He wanted canneries to install force pumps to provide an ample cold and hot, pure, fresh water for 'flushing, washing, and sterilizing utensils, tables, walls, floors, and gutters' as a precondition for licensing by 1922: 'If the expense ... is a hardship for the poor canner, the obvious answer is that the interests of the public come first.' Those who could not comply 'must cease to can lobsters.' Knight also recommended sterilization of cans in a hot, dry oven for half an hour; similarly, packing brine should be boiled for twenty minutes. Sealed cans should be boiled for three hours or sterilized in an autoclave or retort to destroy bacterial spores. By 1923, in order to be licensed, cannery owners would have to have installed an autoclave and separate cloak rooms and lavatories for male and female operatives.[93]

Government inspection, where it existed at all, was as slipshod as many of the canneries. Knight's reports did not flatter the department and placed Found on the defensive. He found Knight's *Official Report on Lobster Canning* 'admirable' but 'in detail ... really a bit unfair to our officers and to our Branch.'[94] To Ward Fisher, Chief Inspector of Fisheries for the Maritimes, Knight's 'rather depressing' report was justified, but newer fishery officers could not be blamed, 'as they doubtless were of the opinion that as the canneries had been running for years under [existing] conditions ... The canneries therefore must have been satisfactory to the Department.'[95]

Education was essential, according to Knight. It was fishermen who suffered most when prices were depressed by poor-quality goods. Ultimately the problem lay with shoddily trained factory managers. Twenty-five years earlier, dairy schools at Guelph and Kingston had revolutionized Ontario's dairying industry, thereby securing higher prices and bigger markets. Knight firmly believed that the same could be done for the canned lobster industry. He suggested that the department rent, renovate and equip a cannery and run it during autumn and winter as a school for lobster canners.[96] Also, newly hired fishery officers should learn how to inspect canneries: 'With a short course ... they would become efficient inspectors.'[97] In the meantime, a mobile laboratory transported by motor truck could be used 'to demonstrate the bacteriological basis upon which scientific canning depends.'[98]

Not surprisingly, given the department's record on education, the task of teaching canners fell primarily to Knight, with three assistants. Over 1921 and 1922 these four toured canneries, demonstrating the presence and effects of bacteria. In 1921 they visited eighty-seven factories. Operating independently, each spent two days at each group of factories, making bacterial cultures from filth, from lobster meat allowed to stand after boiling, and from dirty cans, utensils, hands, clothing, and dust. These cultures were shown to factory managers and operators, and where possible also to factory owners and fishermen. The instructors suggested improvements in equipment and processes. During these seasons and in 1923, Knight, Dr Louis Paré, Professor G.B. Reed, and Clarence J. Tidmarsh also carried out inspections; inspection reports were published and distributed. During conferences for fisheries officers, Prince, Reed, Huntsman, and Knight gave addresses on fish life, bacteriology, sanitation, and the principles of canning.[99]

Some packers were suspicious or openly hostile; few admitted to ever packing bad cans, and some 'declared discoloration and their defects to be a myth of a declining market.'[100] Their 'chief interest ... lay in getting lobsters into the can and then selling them – and "damn" the quality.' Knight felt that this indicated that standards needed to be enforced by the department.[101] Many packers, however, reacted more favourably. Older packers admitted to seeing thousands of ruined cans buried in sand. Prince Edward Island lobster packer and exporter W.H. Tidmarsh greeted news of the demonstrations with joy. He told Found that 'it is quite impossible to sell P.E. Island lobsters ... 80% of the lobsters I handled last year ultimately turned out unsatisfactory.' He claimed that 'most of the packers will welcome the information and instruction.'[102]

In spite of this praise, the department had to be pressured to continue in 1922. What nearly stopped the program was the slowness of fishery officers to understand the work's importance. On Prince Edward Island, 'neither the Inspector of Fisheries nor the Overseers devoted any time to the study of the demonstrations or improved methods.' They were so poorly versed in the underlying principles that they were unable 'to advise each canner on his own particular problems'; and they were so inexperienced in 'the practical application of the new methods' that they could not tell canners how to make improvements. The department short-sightedly concluded that if fishery officers could not learn what was required, there was no hope for change among the canners.[103]

Knight asked that a canning school be established to train district inspectors and factory superintendents.[104] Found, however, could not approve the expense. J.J. Cowie, the department's Chief Inspector of Fish Curing, also thought that a school of canning would be impractical. Knight's reaction was that 'ignorance is still more expensive, as Mr. Cowie would probably realize if he would count up the annual losses of the canners for the past few years.'[105] However, Cowie urged Found to continue and to expand the program for educating cannery workers, even extending it to the Gaspé. But as for Knight's treasured scheme of grading canneries for licensing, Cowie revealed appalling backwardness. He advised Found that 'the proposed plan of grading canneries by ... assigned values ... for equipment and from which marks are to be deducted for lack of certain things, and to which marks are to be added if certain appliances are present, is quite unnecessary at this stage.'[106]

In the end, Found continued the educational campaign and also allowed Knight to implement grading inspections. Perhaps he was influenced by a *Toronto Globe* article in which the Minister of Agriculture was quoted as stating that to compete in the world's great markets, Canada must employ a uniform system of produce grading to allow consumers everywhere to develop confidence in the quality of Canadian goods.[107]

Nevertheless, he wanted to mitigate Knight's requirements – for example, wanted to endorse wooden tubs for washing lobster. Knight retorted that 'they smell like hell,' since small particles of meat stuck to the wood and putrefied. The department reduced his penalty for dirty privies to a loss of one mark, and there was no insistence on a steam retort, which was Knight's most cherished recommendation. Knight insisted that since steam retorts cost only twenty-five dollars, they were no impossible hardship for even poor canners. 'What is the use,' he inveighed, 'of the Government spending $8000 or $10,000 on scientific investigations and advice during the past three years, if the advice is not to be followed?'[108] Many of the department's retrograde modifications were then dropped. Cannery inspection and grading in 1923 was under the direction of Knight, an inspector, and several overseers. Most canneries received a passing score; some were only a few marks off, and there were signs of much improvement.[109]

Sadly, Found could not continue this mandatory grading scheme. The department did not have the power to revoke licenses or otherwise penalize delinquent canners; by then, Found had 'lost a key case that challenged Ottawa's power to intervene in the canning industry in any fashion.' In British Columbia in 1924, a canner named Francis Millerd had decided to start a floating salmon cannery, which according to the Fisheries Act was supposed to remain in one location. For several seasons he shifted locations in defiance of regulations. At first, the department turned a blind eye, hoping he would find a final ideal site and stay put. But when, in 1927, he also began to can clams without a licence, Found finally took action and pressed charges. But both a local magistrate and then, on 23 September 1927, a B.C. Supreme Court justice acquitted Millerd on the grounds that the Dominion government 'had no constitutional right to regulate canneries.'[110] This decision affected canneries on both coasts of Canada.

Fortunately, by 1925 Knight was reporting that except for some canners still lacking steam retorts, many of his recommendations had been implemented.[111] The response to Knight's efforts was entirely favourable. When the 1928 Royal Commission on Atlantic Fisheries asked fisheries inspectors to comment on his grading scheme, one inspector replied that it was 'without a doubt ... a good one and the canneries in this district have improved one hundred per cent.' Inspector A.G. McLeod stated that Knight's grading scheme in Sydney had given impetus to great advancements, 'with the result that the quality of the pack has shown a marked improvement.'[112] Ward Fisher told the commission that inspectors unanimously agreed that Knight's grading scheme had been 'very beneficial': 'Although the actual grading ... stopped in 1924, the effect of the one year's grading is still being felt ... It would therefore seem that a good object would be achieved if Dr. Knight's grading scheme was again put in force after being properly supported by legislation.'[113]

Fishery officers also required technical training to disseminate information among canners as they went about their inspection duties. This training was offered at the Halifax Experimental Station from 1928 onward. Also, in 1927 a short course for lobster cannery foremen was given; however, this would not be repeated until 1933, after requests by local unions of the United Maritime Fishermen.[114]

In 1926, Found told Knight that 'it will not be necessary for you to continue this particular line of work.' By now, Found felt that the department's inspectors, having had two courses in lobster cannery sanitation taught by Knight, and three years' experience in cannery inspection, could competently inspect the canneries according to Knight's formula. In the Department of Agriculture, he pointed out, all inspectors were not specialists, but were able to fulfil their requirements through similar special courses.[115] However, the department was not as advanced as Found thought: in 1930, Knight's former aid, Dr Guilford B. Reed of Queen's University, was asked by Cowie to take on this inspection work once more.[116] Cowie had found that fishery officers had been unable to achieve reasonably similar standards. Therefore he wanted two or three of the fisheries experimental station staff to grade canners for a year or two, to establish uniform grading standards, while fishery officers observed how the marking was done. Knight's grading plan was to be used, and graders were to discuss with the canners the ways in which their equipment and processes fell short of requirements.[117]

Reed and his co-worker Ernest Hess undertook grading work in 1930 and 1931,[118] and special instruction was 'given to all Inspectors in Cannery Grading at the Course of Instruction.'[119] Knight's inspection reforms were thus finally instituted within the Department of Fisheries, except that the department had no power to revoke the licences of noncompliant canneries.

The enmity between departmental officials such as Found and members of the Biological Board seems to have abated in the late 1920s, due to the board's demonstrated eagerness to help promote Canadian fisheries and a growing respect for each other's integrity. But perhaps this is too positive an assessment: the relationship might better be described as one of grudging tolerance. Knight wrote to J. Playfair McMurrich in December 1926:

> Our Prime Minister has announced that the Gov't intends to create a Department of Fisheries. Good and well. That means a Minister and a Deputy Minister. Now there is only one man in Canada today, who at any rate in my judgement, is fit for the latter position – Huntsman. If a politician goes in, worse still, if one of the present civil servants goes in, it means years of delay and waste of money in aquaculture or fish culture.[120]

By 1928, however, Found had been appointed Deputy Minister of Fisheries. By now he knew that too obvious attempts to strengthen departmental control over the board would generate strong resistance. Eager for results, he told the board chairman, J.P. McMurrich, in 1929 that the board had satisfied the department that 'unless the best methods are used in every phase of the industry, [fisheries] development will not be nearly as rapid, or as great as it otherwise would be.' Therefore, his department would support requests for the 'larger expenditures ... necessary to assure the employment and maintenance of an adequate staff, as well as to carry out the work in hand.'[121] The board was willing to cooperate in this and to see itself become the department's scientific division, so long as this would ensure the growth in breadth and importance of its own activities.

Chapter 5

Rescuing Canada's Sinking Atlantic Fishing Industry, 1924–1939

'Oh fancy,' said Jane, as she gazed at the dish
containing our breakfast, 'They say
That a casual glance at the scales of a fish
will tell you his age to a day.'

'A subject like that I can freely dismiss,'
With a sniff of the nose I replied:
'The question that seems more important is this:
What time has elapsed since he died.'

'Science and Sense'[1]

The Department of Marine and Fisheries, once it recognized that positive interventions helped the fishing industry, placed the Biological Board on its front lines. Even with good intentions, however, its efforts were hampered by federal indifference to the Atlantic fisheries; there was never enough funding for the fisheries' real needs. And the Atlantic fishing industry's needs were enormous: little modernization had occurred in this impoverished sector, relative to the revolutionary changes in production methods in the richer agricultural and other food industries.[2] The fisheries, unlike agriculture, had barely been touched by modern mass production between the wars. Few Atlantic Canadian fisheries products had changed in either form or quality. By contrast, North American agriculture had introduced careful grading to ensure freshness and quality, improved packaging to attract buyers, and used more advertising to build up consumer demand. New approaches to production, processing, and marketing were being taken to hold down costs and keep prices competitive. Mechanized farms now used tractors, combines, balers, and milking machines. Meat-packing plants were being mechanized, and dairies were installing bottle washers. All of these things were moving agriculture toward large-scale development.[3]

Competing against all this were fish products. Fishermen and fish packers caught and landed, processed and packed fish almost entirely by manual labour, even within the largest plants. Salted fish, the Atlantic Canadian traditional staple product, was losing out; North American housewives increasingly preferred fresh or canned goods. The Maritime fisheries adjusted to this change only slowly and painfully. By 1939 'some small changes were made ... away from salted fish to the fresh and frozen types.' These posed no threat to the dynamic and competitive agricultural food industries, however.[4] And in fact, most changes that were made were instigated largely by the Biological Board.

The Biological Board had helped change the government's view of its own role in the fishing industry. Unfortunately, the magnitude of the Maritime fisheries' problems quite overwhelmed the Department of Marine and Fisheries. The Maritime industry declined from the 1890s until the Second World War, and the Canadian government's response compared poorly with those of other important fishing nations. By the late 1920s, reinforced by the 1928 Royal Commission Investigating the Fisheries of the Maritime Provinces and the Magdalen Islands (the Maclean Commission), the Biological Board had became the crutch extended by the department to help the Atlantic Canadian fisheries. This was the result of the indifference and underfunding experienced by the ministry as a whole, as this chapter will show. The scientists responded generously, so that fish-processing research and educational work were probably the greatest contributions made by the Biological Board between the wars.

It is time to describe the Canadian Atlantic fisheries and the changes this sector was facing in this period, so that the contributions made by Biological Board scientists can be appreciated properly. The traditional staple of the Canadian Atlantic fisheries was dry-salted fish, particularly cod. Salted fish was also an important export for Norway, Iceland, Greenland, the United Kingdom, and Newfoundland. The Canadian dried-fish industry had its golden age from 1869 to 1886, with accessible markets in the British and Spanish West Indies and the United States, and the National Policy and the new Intercolonial Railway opening up markets in the west. However, after 1886, the Canadian trade entered a long decline. Dried fish was consumed mainly by Mediterranean and Latin American countries; once tariffs were introduced in 1886 to protect sugarbeet producers in France, Germany, Russia and Austria-Hungary, West Indian cane sugar prices dropped from five cents per pound to two cents by 1900. The West Indies – the most important consumers of Canadian salt fish – lost their purchasing power, and their imports of dried Canadian cod declined 40 per cent by 1909.[5]

Indeed, although national production increased in value by $8 million between 1900 and 1910, the increases came solely from British Columbian and inland fisheries. The eastern fisheries' aggregate value had almost stood still since 1886, fluctuating around $15 million. Meanwhile, British Columbian fisheries grew in value from $2.5 million to nearly $14 million. To J.J. Cowie, the problem lay with Maritimers' adherence to salt fish, when refrigeration had accustomed people to fresh meats. As another writer put it, 'salt junk, so-called, no longer appeals to the palate.'[6] But remote fishing outports, lacking means to transport perishable goods, could not produce fresh or smoked fish.

A brief respite came during the First World War. Norwegian exports fell by three-quarters; Canada and Newfoundland expanded their markets and enjoyed a period of great prosperity. But after the war, Norway and Iceland improved their cures, reduced their costs and export prices, and began a period of intensive international competition.[7] Advertising and convenient packaging made Norwegian fish products attractive, and on top of this the Norwegian and Icelandic governments approved huge industry subsidies. Norwegian fish was consigned at extremely low prices to Europe, Brazil, and Cuba; and Iceland undermined Newfoundland's and the Gaspé's exports by offering a cheap, high-grade cure. Canada and Newfoundland receded not only from strong new markets in Havana and Brazil, but also from previously secure Caribbean markets.[8]

Much worse was to come. In the mid-1920s, a permanent depression in this trade set in, through world overproduction, and affected all salted cures, including green-salted and pickled cures. This affected all of eastern Canada, and neither the fresh nor the smoked trades could provide substitute outlets. 'Prices fell to levels that called for relief – provided by the Salt Fish Board as "deficiency payments" in 1939 and 1940, after a long period of acute distress in many fishing communities.'[9]

The decline after 1920 was caused by world overproduction and by the poverty of the main buying countries. By 1929, Norway had nearly reached pre-war production levels, Iceland had doubled its production, and Newfoundland was supplying greater amounts than ever. But since Latin American and West Indian incomes were falling, prices sagged. Brazil, the world's largest consumer of salt cod, suffered a disastrous crisis in the coffee industry in 1929 and two revolutions: for part of 1932, Brazil burned a million bags of coffee a month to reduce supplies and used heavy export taxes to control prices. It subsequently restricted foreign trade and drastically cut codfish imports. Brazil's inability to purchase salt cod in quantity after 1929 led to other markets being flooded, which further lowered the world price for salt fish. Thus, owing to external conditions, Canadian dry-salted and pickled fish export prices fell between 60 and 80 per cent between 1919 and 1939; fresh cod prices also dropped by half. Only fresh and canned lobster prices had returned to near-1919 price levels by 1939.[10] The Great Depression really hit the Maritimes in the 1920s, in all aspects of this region's economy: the Depression's onset in the rest of the world really did nothing but deepen the Maritime's sufferings, with the main 'resolution' being outmigration.[11] The fishermen who stayed increasingly turned to the more remunerative lobster trade, but this could not compensate for the loss of other fishery revenue because of restrictive lobster quotas.[12]

Some fish-supplying countries, but not Canada, protected exports through subsidies and bilateral trade agreements. Also, most countries controlled exports through some kind of national board; but in Canada, independent exporters competed with one another as well as with foreign, centralized national selling agencies. On top of this, Canadian products were not guaranteed by national grading. With no quality control or centralized marketing, Canada lost ground. Superior low-priced Norwegian-cured fish, sold in the West Indies, forced Canadians to drive down their own, unsubsidized dried fish prices to unprofitable levels; eventually, they lost the market altogether. The eastern Canadian trade gradually veered toward supplying the United States with green-salted and boneless fish. Between 1920 and 1939, salted fish declined from 223 to 126 million pounds, but green-salted fish (fish cured in a pickle brine and used to make fish cakes) and filleted cured fish maintained their volumes. Exports to the United States grew from 32 per cent of the total value in 1920 to 62 per cent in 1939.[13]

Even Americans were overproducing, especially in the canned-fish trade. Some Maine sardine canners were processing offal into farm fertilizer. Also, through a process called 'reduction,' oil and water were pressed out of fish. The dried residue was ground into meal, which, mixed with the oil, provided livestock feed rich in vitamins A and D. In 1922 and 1923, American canned sardine prices fell below production costs; they would not recover until the 1940s. Nearly every sardine cannery in California had a reduction plant, and one-third of the state's massive sardine harvest was processed into by-products, especially poultry feed. Reduction became 'the real business of the sardine packers, and the canning ... a by-product affair.'[14] Despite poor quality, American canned sardines were so cheap that they forced down Canadian import prices.

Unfortunately, the American market for Canadian boneless-salted cod and pickled fish was limited. Reliance on one main foreign market left the Canadian trade vulnerable to American tariff changes. The Fordney-McCumber Tariff Act of 1922 imposed a 25 per cent tax on all fish entering the United States; this tariff rose to 35 per cent in the 1930s. Canadian fresh and frozen fish was competing with the growing New England fishing industry, and to overcome the trade barrier, had to be sold at price sacrifices. Losses fell on the backs of Maritime fishermen, who were facing a high-priced North American consumer goods market yet were producing a Third World commodity. Canadian exporters treated merchandising costs as 'overheads in the trade as a whole, and if for any reason export prices had to fall, these overheads tended to remain fixed.' As a result, Canadian fishermen took the hit.[15] Finally, in 1939 the Dominion government began supporting fish prices by making deficiency payments through the Salt Fish Board.

The domestic market could not make up for lost foreign markets. Canada enjoyed the world's greatest fishery wealth yet it had the lowest home consumption of fish – around 20 pounds per capita, compared with 40 pounds per capita in the United Kingdom and 70 pounds per capita in Norway. In 1915, J.D. Hazen, Minister of the Naval Service (as the Department of Marine and Fisheries was known during the First World War), noted that moderate meat prices meant Canadians had grown up 'without a taste for fish,' and worse, with 'a widespread opinion that fish is not a nourishing food.'[16] Fish also had to contend with Canada's traditional internal trade barriers, its immense size, and its small population. Unlike most other maritime nations, Canada's major markets were distant from sea fishing ports. Fish could not be landed in Montreal, Toronto, or Winnipeg, as they were in Boston, New York, Seattle, London, Amsterdam and other large cities close to or on seacoasts.

Futhermore, the American ports of Portland, Gloucester, and Boston were considerably closer to Montreal and Toronto than were Halifax and Mulgrave. In the 1900s, American shippers had better express and freight rates, which enabled them to overcome the half-cent duty imposed per pound of fish. Central Canadian fish dealers thus preferred to buy from American ports (where the fish landed often came from the banks off Nova Scotia).[17] The situation became so bad that in 1907 the Department of Marine and Fisheries began subsidizing fresh fish shipments to Montreal and Toronto on the Intercolonial and Halifax and Southwestern Railways. The slow freight service suffered frequent delays, so in 1908 the department began subsidizing express shipments, paying one-third of the costs of carload lots as far as the eastern border of Manitoba. As a result, sales of Canadian fresh sea fish rose sharply in Ontario and Quebec. Between 1906 and 1910, American imports declined from nearly two million pounds a year to about three-quarters of a million pounds.[18] However, this service did not provide refrigerated cars – disastrous in the summer, and nearly as bad in the winter, when the cars were heated. In 1914 a limited, three-day-a-week refrigerator express service began. Although the fresh-fish trade benefited, this cost the department $37,818 annually. By 1916, Hazen was hinting this was too expensive.[19] Railway subsidies were phased out by 1919.

Meanwhile, Canadian Atlantic fishermen were doing little to help themselves. In the absence of compulsory inspection, they were slow to make improvements; in particular, they did not bleed their fresh-caught fish to give the flesh a white, wholesome appearance. *Canadian Fisherman* remarked: 'Careless handling of fish by the fishermen and fish packers results in heavy annual losses to the Industry ... Sticking forks in fish and walking over them does not improve them. Yet that is what is done at present.'[20] Robert Gray, Inspector of Pickled Fish for the Maritime Provinces, and a former employee of the Fishery Board for Scotland, was shocked by Canadian packed fish. The barrels looked 'more like the job of a hammer and saw man than of a cooper.' These were carelessly packed with split herring, which were buried in salt and covered with a pickle made from 'water drawn not many yards from the outlet of a sewer.' He was especially surprised to see 'fish handled with a fork in the same way a farmer would handle fertilizer.'[21] Norwegian salt mackerel, exported to the United States since the 1910s, was in far greater demand than the Canadian product, long sold there, because it was 'split, washed, and put in salt a few minutes after being taken from the water ... and the fish, when cured, are white.'[22]

To counteract this carelessness, in 1914 the department brought in the Pickled Fish Inspection Act, which authorized the inspection of pickled herring, salmon, alewives, and mackerel, and of packaging containers. To obtain a government brand, fish had to be graded, cured, packed, and marked according to the act's regulations, in standard barrels and containers.[23] Around the Maritimes, J.J. Cowie and other fishery inspectors demonstrated required methods. But to succeed, the Fish Inspection Act needed enthusiastic cooperation, since inspection was not mandatory. Because the government was offering no fiscal incentives, unfortunately fishermen shunned the new standard barrels, which cost much more than traditional ones. The act had minimal impact. Canada's best market for pickled fish, the United States, still favoured British, Dutch, and Scandinavian products; even in Canada, foreign fish products were preferred.

Surveying the situation in 1944, Stewart Bates (Deputy Minister of Fisheries in 1946) commented that 'for a country bordering two oceans ... the maritime interests of Canada were small.' Even in the Maritimes themselves, the fishery, the shipbuilding industry, and the merchant marine drove only a small part of the economy. These provinces were 'a continental littoral, rather than a maritime, economy.' The Maritime governments regarded the sea fishery as a federal responsibility – but since the fisheries were relatively insignificant in Dominion affairs, Atlantic fishermen were barely heard. Further fragmenting the East Coast industry was its division among four provinces. And again, British Columbia's fisheries were so unlike the Atlantic fisheries that East Coast and West Coast MPs seldom found a 'unity of endeavour.' Quebec's jurisdiction over its own sea fisheries further reduced Maritime fishermen's political influence in the federal house.[24]

Even Maritime provincial governments budgeted practically nothing for the fisheries. They had perhaps some justification: with international fleets fishing just a few miles offshore, these provinces felt that the fisheries were best left to federal care. In fact, only in 1943 did Nova Scotia, with the region's most important fisheries, establish a departmental fisheries division, 'and did not upgrade it to departmental status until 1964.' For Canada's policymakers, the fishing industry 'was marginal'; its fragmentation prevented any 'united, interprovincial demand for government action.' In the Maritimes, Newfoundland, and the Gaspé, the fisheries were too often seen as the employer of last resort. In contrast, independent Prairie wheat producers were resorting to collective action, through marketing boards, to buttress a single-staple economy, and the Canadian Wheat Board enjoyed federal legislative support.[25] A sure sign of the eastern industry's predicament was that its landings of over half a billion pounds of fish annually – half of Canada's total – only brought in one-third of Canada's fisheries revenues.

Fish are not necessarily a bad investment, provided that the fishing industry is surrounded with all the necessary forward, backward, and final demand linkages required to make the industry and its region thrive. But for many staple industries, this requires government intervention, and the Atlantic fisheries had the lowest government priority of all leading industrial sectors.[26]

Forward linkages are investments in industries that use the output of one (export) sector as their own raw material. For example, cleaned and lightly preserved fish is processed into a wider range of products. Such forward linkages may augment the industry's value considerably. Final demand linkages are investments in firms that make clothing, foodstuffs, appliances, and many other consumer goods. These linkages form when the primary industry produces enough capital to support such demand. Backward linkages, the most difficult to create, and usually the last to occur, are essential for industrial development. They make the goods needed to run the export sector, including equipment for harvesting, processing, and exporting the staple; as well as railways, energy systems, and wharf facilities, which are not directly productive. Ideally, each of these will develop new backward and forward linkages of its own: for example rail, locomotive, and rolling stock companies for railways; and fish processing equipment companies for fish processors. The lack of such linkages is a clear sign of economic underdevelopment or dependency,[27] and this lack is what characterized the Maritime fishing industry.

The staple is not necessarily to blame for the absence of such linkages. Often the poor performance of the fisheries in Newfoundland and the Maritimes was blamed on the fish. But the problem lay more in how that staple was handled by merchants and in government neglect. Iceland provides an alternative scenario. Iceland was 'staggeringly impoverished ... with no significant commercial fishery before 1890 and with fewer alternative resources than Newfoundland.' Small catches of fish were exported through Danish merchants in return for supplies. But Iceland suffered from restrictive agricultural practices that tied people to the land through a 'farm bond.' Iceland differed from Newfoundland only in that it had an agricultural, not a fishery, export sector.[28]

Icelandic fishing wages were higher than farm wages, since labour was scarce as a result of the farm bond. After the bond was abolished in 1893, the fishing population grew. In the past, fishermen-crofters had sold fresh fish for goods; later they made cash sales, which served to create final demand, and 'by 1870 Iceland had a network of small, local savings banks – unthinkable in either Gaspé or Newfoundland at that date.' By 1886, these banks were making loans to fishermen who wanted to buy fishing smacks. The Harbour Acts of 1911 shifted loans so as to assist the industry at the national level. Iceland now treated the fishery as the national economy's 'engine of growth ... supported by businessmen and financiers.'[29]

Forward linkages came with the introduction of saltfish, then iced fish, and finally frozen fish, which helped Iceland capture markets. Consumer industries grew, especially in Reykjavik, where consumer-goods manufacturing comprised 40 per cent of all production by 1940; the same year, fishing and some farming made up 43 per cent. Also, the vital producer goods sector (engineering, shipbuilding, and ship and gear repairs) thrived, and the Harbour Acts encouraged improvements in all these technologies, thus providing backward linkages for self-sustained growth. By the 1930s, Iceland had a mature *fishery-based* economy, built up through local and state capital and entrepreneurship.[30]

Newfoundland, in contrast, turned its back on its obvious source of wealth. Its government was trying repeatedly, and futilely, to build up other, more prestigious industries. The extensive truck system, whereby fishermen exchanged fish for credit with local merchants, reduced wages, kept income levels low, and thus limited markets for products of local manufacture, since such goods had quality and price disadvantages compared to imported goods. But the fisheries themselves were not to blame. 'The Canadian view of the fish staple has been parochial.' It has been characterized by a 'debilitating attitude towards the industry and the region.'[31]

In another example, New England's fisheries suffered from the same attitudes after the Second World War. The prosperous offshore industry, built up around government support in the form of tariff barriers, slumped considerably when those barriers were removed after the war. In the 1950s and 1960s, fishermen, boat owners, and processors operated without tariff relief. Legislators did little to help; they viewed the industry as dying, but did not much care. Compared with other industries, especially agriculture, the fisheries received little federal assistance. Merchant vessels were given construction cost subsidies, whereas fishing vessels were not. Domestically constructed boats cost twice as much as foreign craft, but laws forbade fishermen from using foreign-built boats. Also, as part of war reconstruction programs, American aid helped finance other nations that were rebuilding their own fisheries. Some of these countries then imported fish to the United States, causing job and income losses in the New England fisheries. However, Congress feared that tax payers would balk at subsidizing yet another industry, and President Eisenhower in 1956 rejected recommendations for higher tariffs, favouring international projects over protecting domestic fisheries.[32] Thus, the industry declined through the 1960s.

These cases show that government attitudes can make or break an economic sector. The Canadian government adopted a hands-off approach to its Atlantic fisheries. During the interwar years, the Canadian Atlantic fishing industry was impoverished and largely unsubsidized; its problems seemed to grow every year. And what did the government do to help Atlantic fishermen?

It wrote fish cookbooks. The Department of Marine and Fisheries, short of funds, chose to strengthen Canadian demand for fish. Its pamphlets taught the Canadian housewife 'the great food value of fish, and [show] her how she may serve it up in many tasty and appetizing ways.' The department was not without precedent in this, however. The British National Fisheries Protection Association's Fish for Food campaign had distributed free recipe books through fish retailers throughout Britain. The Canadians, in emulation, had W.A. Found and Andrew Halkett organize a Fisheries Exhibit for the 1913 Canadian National Exhibition. They did this annually thereafter, and fish dealers especially liked the free fish-cookery booklets; the department printed and gave away about 250,000 of them. In 1915 the department opened a restaurant at the CNE that sold substantial fish dinners for twenty-five cents. It sold around 1,600 meals per day.[33]

Another area the department tried to improve was fish retailing. In England, fish was sold at hygienic fishmongers; Canadian retailers, by contrast, were neither equipped to handle fresh fish nor motivated to increase fish sales.[34] Under a Montreal bylaw, butchers could retail fresh fish, whereas grocers could sell only prepared, smoked, or cured fish. At the butcher's, 'fish will be brought from the wholesale house and thrown in a heap in some corner of the shop, to wait a couple of days, sometimes, for the exposition in the front of the store on Friday morning ... In the summer time, after two or three hours on a piece of wood or in a box without ice, the appearance of the fish is not only non-inviting to the purchaser, but in some cases, it is really shocking.'[35]

In winter, smoked fish was often left outside to freeze by day, and at night was brought back into the heated store. After a week the fish were inedible. Huntsman, who spent part of each year in Toronto and part in St Andrews, said of the fish sold in Toronto that his family would be 'tempted to try some, eat it in part, then say "Not again!"' On the coast, his family ate fish nearly every day.[36] Of course, some well-equipped stores provided modern display conveniences that kept fish in good condition. But poor practices were common. Canada's first national fisheries trade journal, *Canadian Fisherman*, devoutly supported the department's educational campaign to increase fish consumption: 'Fish in a box or barrel is unattractive, but the same fish, cleaned and laid out upon a marble slab with running water or chopped ice upon it, and tastefully decorated with parsley, red peppers and lemons, takes the eye right away and acts as a silent salesman.'[37]

After 1936, the fisheries experimental stations surveyed the quality of fresh fish retailed in Toronto, Montreal, and Halifax, and found that approximately 90 per cent showed some spoilage, and that this spoilage was advanced in two-thirds of the fish. Wholesalers held fish in storage too long; yet that fish shipped fresh on ice had begun as the best arriving in port.[38] Still, the department would not go beyond propaganda and trying to educate retailers through *Canadian Fisherman* and other trade journals; it had no jurisdiction over the retail acts that enforced sanitation.

The federal government was especially negligent in not inaugurating sufficient technical education for fishermen. In tandem with mandatory inspection this would have improved product quality and competitiveness. It could not plead ignorance, either. The Royal Commission on Technical Training of 1913 recommended that travelling instructors give short courses to fishermen, along with winter courses of up to two weeks – similar to those given in Britain at Peel and Aberdeen – to teach fish processing and marketing, as well as navigation and engine maintenance.[39] The example of agricultural colleges and experimental farms was often invoked by an envious fishing industry, which felt that a fisheries college, set in a port handy to prolific fishing grounds, was a fair request. After all, 'Agricultural Colleges and Experimental Farms for the better education of the farmer have been located all over the Dominion ... trained lecturers tour the farming districts and deliver valuable lessons ... Special trains with cars fitted up with agricultural exhibits tour the country from coast to coast: handbooks ... are distributed free ... Nowadays, if a farmer don't know his business, it is not the fault of the Government.'[40] Canada was far behind most world governments in this. Japan had already established fisheries technical schools (even in 1914 Japan was considered the world's foremost fishing nation). In Britain, Ireland, France, and Norway, both government and private interests had established fisheries schools. In 1913, 891 men attended the fisheries technical school in Grimsby, England, 'evidence that the fishermen feel the need of instruction.' Technical education was offered 'in practically every other important fishing country except the United States.'[41]

The Biological Board also advocated technical training. Prince argued that a brief fishing course would teach fishermen to abandon 'injurious and foolish [fishing] methods such as the use of explosives, etc.' Also, if fishermen 'were taught to think and to appreciate what I might call the rights of the fishes themselves,' they would learn to preserve immature fish. The minister, J.D. Hazen, agreed in 1916 that with agricultural education, 'a great many of the farmers are better off today ... We in the Department are trying to accomplish a similar result in a similar way, amongst the fishermen.' But what Hazen meant was the useless Pickled Fish Inspection Act, with its voluntary system of quality control. The department, however well intentioned, could not help fishermen through this. Fisheries officers also distributed information among fishermen, but their informational pamphlets were mostly cast away unread. These could not take the place of personal advice, demonstration, and guidance.[42]

The department's officials suffered serious impediments, however, 'not the least of which has been the evident disinterest of the government in the fishing industry.'[43] In 1918, because fisheries funding was so short, Found suggested that the Halifax Award's annual $160,000 bounty be diverted to fund fishermen's technical education and to establish 'local experimental stations along the coasts,' to act for fishermen as the experiment farms did for the farmers.[44] Not surprisingly, considering their poverty, fishermen did not wish to give up their share of the bounty.

As these problems deepened, calls for federal aid increased. The industry received the traditional Canadian response. In 1927, a Royal Commission was set up to inquire into the eastern Canadian fisheries.

This was brought about largely through the work of one man, the Reverend Dr J.J. Tompkins. The former vice-president of St Francis Xavier University was so concerned 'to bring the university to the people' that his bishop sent him as pastor to the fishing community of Canso in 1923. In some Nova Scotia and Cape Breton outports, fishermen and their families were barely surviving, living in earth-floored huts, their children often barefoot. He began to question their circumstances, arousing them to the point that on 1 July 1927 they 'signed a petition to the federal government asking for somebody to come and see the conditions of their fishing industry.' The *Halifax Chronicle*, prompted by Tompkins, published 'a series of seven sympathetic articles under the heading "Save the Fishermen."' Also, forty priests signed a resolution requesting a thorough study of fishermen's problems, and sent telegrams about this to Halifax and Ottawa. At a later meeting in Canso, attended by

John A. Walker, Minister of Natural Resources, fishermen passed a resolution requesting 'an investigation, by Royal Commission, touching all phases of the fishing industry and to make recommendations as to means of improving conditions so that fishermen will be enabled to continue their calling.'[45]

On 7 October 1927 the Governor General of Canada authorized the 1927–8 Royal Commission Investigating the Fisheries of the Maritime Provinces and the Magdalen Islands. It was chaired by the Honourable Mr Justice A.K. Maclean.

Travelling throughout the Maritimes, the Maclean Commission discerned a deep need for technical education, for cooperatives, and for better inspection practices. Nearly everybody who spoke before it wanted all fish to be inspected and expressed their willingness to cooperate with inspection. They knew that Norwegian and Icelandic mandatory government inspection had helped these countries capture the former markets of Nova Scotia and Newfoundland.[46] The commission listened, and recommended 'grading and inspection of dried fish for export be made obligatory for the year 1929,' and that inspection be extended to canned, frozen, and smoked fish. In spite of this, inspection remained voluntary; compulsory inspection would have required a new administrative infrastructure and many more fishery inspectors. But since fisheries jurisdiction also rested with several provinces (and these measures would have cost money), the federal government pleaded that compulsory inspection was hampered by the British North America Act. The only positive result of the commission's recommendations was that the Biological Board established size and quality standards in green salted fish for the department's *voluntary* Fish Inspection Act. In 1934, economic analyst Ruth Fulton Grant compared this with the Department of Agriculture's inspection and branding of meat in packing factories, and was scandalized that 'the purchaser of fish is assured of no such protection, yet fish is a more perishable commodity than meat.' She added: 'The question arises whether the government ... can conscientiously recommend the consumption of fish as a healthy food.' Throughout the 1960s and 1970s, recommendations for compulsory government inspections would continue to be made by the Fisheries Research Board. This was never to occur in the long, sad history of the Canadian Atlantic fisheries, even though such a need was reiterated in the 1990s, after the stock collapses. The Department of Fisheries always hedged, claiming that it required provincial agreement and more trained personnel.[47]

By contrast, in Norway, where inspection was compulsory, the government appointed thirty-six inspectors and six travelling inspectors, whose wages were paid by the exporters. A large firm usually required the continuous services of one inspector; a smaller firm might need one two days a week. Iceland's inspection laws were similarly rigid. These inspections made their fish products highly competitive on foreign markets. But other factors also helped: inspected fish products were graded and branded; scientific and technological research improved cures; foreign markets were studied by fishery agents abroad; and finally, domestic banks offered ample financial assistance to fishermen.[48]

These measures were costly for these governments. In contrast, the Canadian Atlantic fisheries received little federal aid. Indeed, the Canadian government was delinquent. In the 1920s it intervened to help the iron-and-steel and forest industries. Also, although there were federal and provincial agricultural credit systems, there were no fishery credit systems. Fishery credit systems were in place in Germany, France, Denmark, Norway, and Iceland. For example, France helped fishermen's cooperatives form credit associations. Interest-free loans were proportional to the capital put up by these associations, to help with boat and gear purchases, with transporting fish to inland markets, and with establishing cold-storage and fish-curing plants.[49]

Larger Canadian fishing corporations probably had no more difficulty in getting financing than any large local business. However, no loan boards were established to capitalize smaller fish businesses. Fishermen were less able to obtain capital than farmers; they lacked savings. Small merchants and outfitters could only make short-term advances; fishermen relied on a merchant credit system for supplies, and although some merchants helped carry fishermen through bad times, others worked through a 'truck' system, whereby fishermen handed over their catch in return for all their supplies of food, clothing, and gear, 'often at valuations which kept them constantly in debt.'[50]

Thus Maritime fishermen were confined to backward fishing methods and small boats. Even the expensive Lunenburg schooners were not the actual fishing instruments; small dories, with two men to a boat actually netted the fish. There were only three Canadian trawlers left following the Maclean Commission, which wanted them abolished because they pre-empted sales from small fishermen. There were also 45 schooners over 40 tons, and more than 10,600 boats under 10 tons, most 'little better than row boats.'[51]

Obviously, the Atlantic Canadian fisheries were beset with problems from one end of the industry to the other. These were the problems that the Fisheries Experimental Station at Halifax was set up to address. In 1923 the board and the department decided to establish experimental stations in Halifax and Prince Rupert. At these, scientists would investigate and demonstrate fish-processing methods, research economic fisheries problems, and educate fishermen and fisheries officers. The new stations were to be centrally located so that their scientists and staff would be readily available to the fishing industry; they were to be everything for fishermen and the fishing industry that agricultural stations and experimental farms were for farmers and the agricultural industry.[52]

Halifax's experimental station, situated as it was in a great commercial fishing port, was to include a chemical laboratory for researching fish preserving and curing, for analyzing fish oils and salts, and for finding ways to process fish wastes into fertilizer, oils, and glues. A model fish-drying plant and smokehouse would be used to test and demonstrate variations in fish curing and smoking methods. Similarly, the best methods for pickling fish were to be investigated and then demonstrated. A museum or lecture room was to be filled with photographs and models of boats, nets, and other fishing gear used by Canada's major competitors; and of foreign curing establishments and the utensils used therein (the Pacific Experimental Station had similar facilities).

On 4 April 1925 the Atlantic Experimental Station opened. A.G. Huntsman became its first director, and from 1924 until 1928 directed the stations at both St Andrews and Halifax. He held both positions because the board wanted St Andrews to cover the more fundamental research and the problems from one end of the industry to the other, from production to marketing. The Halifax station was to help the Atlantic fishing trade diversify its products and adapt to changing markets, and to educate fishermen in ways to improve fishing and fish-processing techniques. The Halifax station also became involved in setting industry standards. In sum, the new experimental station was intended to make up for many of the Dominion government's shortcomings in its dealings with Atlantic Canadian fishermen – all the problems, that is, that could be dealt with without a huge outlay of capital.

Educating Fishermen: The Fisheries Experimental Station

After the 1928 Maclean Commission, the Biological Board became the central pillar of the Department of Fisheries' new, two-pronged scheme for educating fishermen and fisheries officers. Incidentally, the same commission had also called for a new and separate Department of Fisheries, which was finally established. (Fishermen had been calling for a separate, dedicated department since before the First World War.) The educational measures that were taken occurred in tandem with the other major initiative that followed the commission: the formation of fishermen's organizations for economic and educational improvements.

The Maclean Commission had made note of an unfortunate lack of cooperation, both in standardizing product quality and in marketing. It recommended that fishermen apply cooperative methods to sell and 'jealously' safeguard the reputation of Maritime products. In Canada, cooperatives already existed for egg producers, fruit, potato, and wool growers and lifestock breeders. But fishermen still had no bargaining power, and little to say about selling prices; often they were forced to accept whatever price they could get, which was sometimes below the cost of production. 'If the shore fishing industry is to succeed,' stated the commission, 'co-operation among fishermen is absolutely and immediately essential.' It therefore recommended that the department help fishermen organize, stating such an undertaking was not 'beyond the scope of governmental responsibility or Government aid. The fisheries are a basic industry and are reasonably entitled to assistance and encouragement. A similar venture had already been aided for agriculture.' With cooperatives, fishermen could form savings-and-loan societies to help them purchase boats, equipment, and insurance. Cooperative marketing would bring them pricing according to product quality, along with a new bargaining power.[53]

Accordingly, the department in 1929 appointed the Reverend Dr M.M. Coady, of St Francis Xavier University, Antigonish, to promote fishermen's organizations in the Maritimes and the Magdalen Islands. Many other university professors and parish priests were also involved in the 'Antigonish Movement.' 'Despite extensive lethargy of fishermen in many places and despite the opposition of big fish dealers, on June 25 and 26, 1930, two hundred delegates gathered in Halifax' and established the 'United Maritime Fishermen.'[54]

The UMF helped fishermen educate themselves and promoted buying and selling along cooperative principles. Started with a $5,000 grant from the Department of Fisheries, its finances were based on its two-dollar membership dues. By 1931, nearly 150 cooperatives had been formed in the Maritimes and the Magdalen Islands. These cooperatives, many of them locals of the UMF, worked together to purchase gasoline, oil, rope, paint and other necessities; they achieved savings for fishermen by buying supplies in quantities beyond the needs of local units. The UMF successfully developed cooperative marketing for lobster, fresh salmon, and smelt, and later fresh or frozen groundfish, forcing buyers to pay more. The organization also built cold-storage plants at central points; they could do so because they had access to more capital than could be raised by most fishing communities and individual local units. By 1938 there were 2,275 study groups in the Maritimes and 125 in the Magdalens, with a total membership of 20,600. There were some 350 credit unions, 47 cooperative stores, 36 cooperative lobster fisheries, 10 cooperative fish plants, 8 community industries, and 12 smelt-marketing organizations.[55]

The cooperative movement also petitioned the department for more travelling instructors to provide technical training. The Maclean Commission had already asked the Atlantic Experimental Station to offer courses that would qualify young fisheries officers as special instructors in modern methods of navigation, in the 'salting, curing, smoking, pickling, canning, packing and marketing of the various fish products,' and in the 'building of simple ... bait freezers, smoking houses and driers, the care and repair of engines and other accessories, nets and other fishing gear.'[56] Travelling fishery officers were to be specially trained to become instructors. To some extent been done before, but the fishermen's cooperatives wanted to see more being done. Indeed, educating fishermen was to become the Department of Fisheries' main strategy for improving the Maritime fisheries in response to the Maclean Commission.

In 1926 the experimental station was already offering courses expressly designed to train these officers. Furthermore, the station was using these instructors to disseminate information about advances in fish processing methods. Over the next few years, and especially after the Maclean Commission submitted its report, this work intensified. After 1927, overseers were required to complete a six-week course at the Halifax Experimental Station in order to qualify as fish inspectors. In this way more trained officers were made available, and they achieved better results in helping fishermen. Some officers were trained specifically as instructors, and the department hired seven of these – of whom George R. Earl was the chief – to teach improved curing methods. The department selected the 'Gaspé' cure, since products from the Gaspé and Caraquet-Shippegan still enjoyed premium prices in Italy, despite the world slump in dried fish.[57]

The station trained Philip Mercier and William Mercier, experienced cod curers from the Gaspé, as instructors. These two men then travelled around northern New Brunswick teaching the Gaspé cure. Their work brought about marked improvements. Wholesaler J.G. Robichaud, of

Shippigan, wrote to Found that he was now receiving fish 'second to none in quality ... It is gratifying to note the excellent quality of fish produced.' In 1930, at the Islands' urgent request, Philip Mercier was sent to the Magdalens, where he stayed until 1933. Fish wholesaler L.P. Gaudet saw a 'wonderfull change ... It will be a Godblessing for these Islands, the day that they will realize that they are getting double money ... on account of the curing alone.'[58]

George R. Earl also garnered glowing reviews. He taught fish-curing methods in the Atlantic station's Course for Fishermen from its inception in 1928. In 1929 he and two fisheries officers toured Prince Edward Island codfish-packing plants, teaching the locals how to improve split boneless cod. Earl offered a higher price for fresh fish that had been bled as the catch was pulled in. His new product, 'as white as snow,' quickly 'met with the hearty approval of wholesalers.' By 1931, island producers had stopped importing Nova Scotia salt cod and were exporting their own product to Canadian and American markets.[59]

In 1930, R.A. Merchant of the Frank C. Pearce Company, fish wholesalers of Gloucester, Massachusetts, told Cowie that although his company been rejecting Prince Edward Island fish as a result of past experience, in 1930 it had chanced purchasing 'several cars of Pickled Cured Codfish, and since then we have purchased a number of cars more.' The company's changed attitude was due 'to the work which Mr Earl has been doing there. Because of this we felt that we should write you personally.' Merchant added that 'providing the fishermen and producers at Prince Edward Island continue along the[se] lines ... they will find ... a ready sale for all the fish they can produce and will feel extremely well paid for the extra attention ... given to their product.'[60] In 1930 his company purchased more than half a million pounds of the island's boneless cured cod. Production and prices rose markedly, for which, Earl claimed, 'our department work here is absolutely responsible.'[61] The degree to which these educational schemes succeeded was largely due to the Biological Board.

Earl and his staff worked in tandem with the Biological Board, and their work was coordinated by the experimental station. But the board also wanted to use the Halifax station to educate fishermen themselves. Real educational work began after the Maclean Commission asked the board to found a centralized training school and, with the newly formed Department of Fisheries, help fishermen start educational cooperatives.

Even before 1924 the board had been educating lobster fishermen, canners, and inspectors, as discussed in the last chapter. However, extensive technical education of fishermen had still not been carried out. In 1927, fishermen had demanded technical training, and the Maclean Commission backed their demands. The Rural Conference of the Roman Catholic Church in Nova Scotia's Guysborough and Antigonish counties promised twenty-five scholarships.[62] The Department of Fisheries did not want to place the administrative burden on the staff of the Atlantic station,[63] but, as in the past, the department had nowhere else to turn; thus, the Biological Board was saddled with establishing the Course for Fishermen, albeit with help from Dalhousie University and the Truro Agricultural College. Perhaps this was a natural choice: after all, the board was staffed mainly by members of the didactic profession *par excellence*: university professors. At any rate, the board had already planned such courses.[64]

It designed a course for the winter months, so as not to interfere with the fishing season, and it agreed to pay return railway fares plus $45 to each of the twenty-five fishermen who completed the six-week course. Students were required to have grade six arithmetic, reading, and writing skills.

The first Course for Fishermen, offered in 1928, received only twenty applications.[65] Practical courses on motor engines and navigation were offered by Captain H.M. O'Hara of the Halifax School of Navigation. George Earl and Robert Gray, the Supervisor of Fisheries for Halifax, taught a hands-on course on preparing pickled, dried and boneless fish. A 'Science' course offered by board employees had sections on chemistry and physics, 'Biology and Conditions in the Sea,' bacteriology, refrigeration, fish oils, and food chemistry. The 'Natural Resources' course, taught by professors from King's College, the Truro Agricultural College, St Francis Xavier University, and Dr M. Cumming of the Nova Scotia Department of Natural Resources, covered economics, gardening, cooperative marketing, and the use of natural resources. In 1929, instruction in processing fresh fish was added.[66] Students were examined on all subjects.

This course was a great success. After 1928 it was advertised annually in thirteen Maritime newspapers and in 211 coastal post offices. It was limited to twenty-five fishermen, owing to restricted accommodation and facilities. No one older than thirty-five was to attend. Found received applications from several men too old to attend, one of them wishing to qualify as a fishery inspector. Another had qualified as an inspector but had not been hired. He had hoped the course would improve his chances, since he had only a small income and a wife and seven dependent children.[67] Found was moved and asked that the age limit be abolished, but the board refused, pointing to its limited facilities and its expectation that young men would be more likely to use and pass along their new knowledge.

Found wanted inspectors to be required to take the Course for Fishermen. However, the education committee feared that this would interfere with the course's original intent. It suggested separate courses. And so a course for fishery officers was offered during March and the first half of April.[68]

The fishermen sang the praises of the course.[69] One Prince Edward Island fisherman said that the new fish handling methods gave him 'the highest price ... and I got the name for puting up the best fish on the Island and it was the school that don it for me.' J.F. Sutherland had improved his fish cures and 'in the care of a motor engine I think I can say that I learned more and made more use of it than any other subject taught at the station.'[70] Ernest Arsenault testified: 'We could never keep fish in good condition after a certain time but this year I salted the fish myself and its in good condition yet, I gave some to different persons and they said they had never ate such good fish they also said that next year they would have their fish salted by me ... If I ever have a chance to attend another course I would be very glad.' George A. LeClair stated that during that summer he had helped friends 'bothered with little red specks on their fish ... I advised them to wash place and fish containers out thoroughly and lime same which they did and found good results.'[71]

Unfortunately, just as the Department of Fisheries was beginning to see results from its educational programs, the stock markets crashed, the Great Depression set in, and buyers in the United States were no longer able to support the new trade. In 1932, Gorton Pew Fisheries of Gloucester, Massachusetts, did not come to Prince Edward Island due to financial instability. But the department's assurance of cooperation in monitoring fish quality made the company president relent. The Frank C. Pearce Company told Earl that it was only interested in fish over which the 'Department particularly supervises in the curing etc.'[72]

As market shares dropped, and as markets for Maritime fish dried up, so too did the department's financial resources. No changes were made to the course in 1932, but in 1933 it was cut to four weeks. In 1934, reflecting the Depression's depth, the number of applicants swelled to 95. In 1935, 110 men applied but only 20 were accepted, and Cowie felt that three weeks would suffice. Other cost-cutting measures were initiated: separate courses for fishermen and officers were beyond the department's means, so a combined course was given.[73]

Although the Depression dampened its success, the educational campaign led many fishermen to cooperate with the department in adopting standards for the American market. The results of organized fishermen's cooperatives were 'startling. Prices for their catch increased substantially and prices of commodities ... were reduced more than even the optimists believed to be possible.' Private buyers were forced to pay more for fish and to reduce prices for supplies. Even fishermen too poor to join cooperatives 'benefitted because the new prices were substantially higher than before.'[74]

In 1936, as a result of their improved products and some market upturn, fishing communities began to anticipate a sunnier future. The larders of fishing families were now stocked with 'flour, lard, tea, even vegetables and some of those necessities of the city family which sound like luxuries to ... those who live in the shore hamlets.'[75] As a reflection of slightly better times, the Course for Fishermen was lengthened back to four weeks and opened to twenty-five men.

The Halifax station's educational and outreach programs, and the fishermen's cooperative movement, had aided many. The *Halifax Herald* reported that many fishermen had no more need of government assistance: 'United States buyers, impressed by the satisfactory quality of the fish produced ... in districts where the department instructors had been at work, have placed substantial orders ... where, otherwise, operations would have been negligible or nearly so.'[76]

Maritime fish products were much improved by these measures, but in terms of saving the interwar fishing industry, they proved to be too little, too late. The Dominion government simply had other priorities, and the department was too severely underfunded to extend even a portion of the aid that the industry so badly needed. The real exigency of the fishing industry, which continues to this day, is mandatory government inspection of fish handling and products. The lack of such oversight had its most significant and scandalous impact in the late 1980s, when 'Tunagate' saw Canadians fall ill and some die after eating tuna processed at the last Atlantic Canadian tuna cannery at St Stephen, New Brunswick. The cannery was closed, with a resulting loss of jobs. Since that time, all tuna consumed in Canada has been canned in the Far East. Putting that aside, in the 1930s, Canadian fishermen were fighting for shares in a shrinking market, and product improvements could offer only a modicum of relief. True relief for Maritime fishermen awaited the Second World War.

Perfectly Delicious: Inventing Frozen Fish for Inland Markets

In the 1920s and 1930s, technical education and the promotion of fishermen's cooperatives were two of the Department of Fisheries' most important approaches for showing up the troubled Atlantic fisheries. As has been seen, both activities involved the Biological Board, which took the lead in training instructors as well as fishermen. The department's third major prop for the industry was research and development, and here the Biological Board stood alone.

Research involved not just fish-processing investigations, but searches for new fish stocks and efforts to boost scallop fishing and oyster culture on Prince Edward Island. In the 1920s, however, fish processing research tended to dominate. The trade's first request to the Atlantic Experimental Station had been for a study of fish smoking: 'Current methods vary greatly and are for the most part comparatively crude.'[77] The station's scientists worked out how to make a good uniform product at the least possible expense, and their ideas were soon taken up by local firms. Indeed, the Maclean Commission recommended that their research on curing and packing mackerel and herring be expanded. What the station's director, A.G. Huntsman, pushed, however, was research into fast-freezing to help build a Canadian frozen-fish retail trade.

The Canadian experimental stations had adopted 'the best features' of 'similar stations and schools in other countries.'[78] The Biological Board sent a delegation to Washington, D.C., to examine the U.S. Fisheries Products Laboratory. In a two-storey annex of the Bureau of Fisheries Building, they found an upstairs kitchen, clerical spaces, a museum filled with sample fish products and transparencies illustrating fisheries methods, and a main floor with cold-storage equipment, a smoke room, autoclaves, can-capping machines, and a chemical laboratory. They were most struck by an apparatus for rapidly and continuously freezing fish, designed by the laboratory's former director, Mr Taylor, consisting of

> a long, strongly built, tunnel-like box with insulated walls and provided with a system of pipes containing sodium chloride brine ... cooled to the desired temperature. The fish ... are suspended in the box and pass gradually from one end to the other, [sprayed by] ... cooled sodium chloride brine by which they are frozen. On entering the box they are sprayed to

> clean the surface and ... [another] spray of water ... washes off the brine and at the same time glazes the fish.

The laboratory was excellent except for its location, which was too far from any important deep sea fishing centre. Haddock, shipped from Boston, was too poor 'to give the best results.'[79] Also, Dr Taylor had left for a higher-paid position directing fish-freezing research and development for National Seafoods Corporation. In the United States, the fish food industry was sufficiently capitalized that it was willing to underwrite its own research, to capture a competitive edge; in Canada, this role was played by the Biological Board.

Brine freezing was first suggested in 1889, when two different English patents were granted. Not until 1919 did Americans see its possibilities. In the late 1920s, New England expanded its fish market into the American Midwest and south by developing fillets and the quick freezing method. Together, these resulted in a standardized and attractively packaged product that was easy to prepare.[80]

In Canada, the First World War highlighted the importance of fish-handling research. Canadian and British troops in France were complaining bitterly about the Canadian frozen fish that formed part of their rations. Board scientists investigated, and found it 'tough, papery and tasteless. It looked nice when frozen, but when it was cut and thawed, the juice ran from it like water from a sponge.' Canadians in peacetime also shunned frozen fish: 'Canadian housewives' had learned that fish was often sold frozen because it had decayed 'beyond the point where it could be sold in any other form.'[81] Yet Huntsman and the Biological Board had been deeply impressed by Johan Hjort's demonstration of the Ottesen quick-freezing process.

In 1924 the St Andrews scientists constructed for the Halifax station a simpler 'automatic Brine Freezing System.' In the course of his research into frozen fish, Huntsman invented what he called 'jacketed cold storage.' One defect in refrigeration at that time was that it dried out stored products; any humidity in stored food, or in the air, was condensed and frozen on the refrigerating pipes. To reduce this moisture loss, fish were glazed with water to produce a coating of ice. Yet frosted-up pipes still made refrigeration less efficient. Huntsman now realized the obvious:

> The refrigerating pipes ... should be in a jacket in the wall of the room. A jacketed cold storage room was constructed in the Demonstration Building of the Halifax Station in 1927, and it was found to reduce the drying of the fish to one-fifth. An improved type was constructed at the St. Andrews Sta-

> tion in 1929 ... This decided advance in refrigerating practice, although brought to general notice by publication, has evoked no interest whatever throughout the industries that might benefit by it ... Only through the personal efforts of Mr. Otto C. Young, the Mechanical Engineer at the Board's Pacific Fisheries Experimental Station, has the device been slowly brought into use.[82]

Huntsman's innovation took about twenty-five years to catch on, but is now standard in refrigeration devices around the world. The first company to use it was the Canadian Fishing Company of Vancouver, B.C., at two of its plants. 'Board engineers, led by Otto Young, conducted a sustained and finally successful campaign to have railway cars and trucks fitted with individual refrigeration units that would get fresh fish to inland markets in edible condition.' 'Jacketed' cold storage was later adapted to frozen storage on fishing vessels, and is now used widely for long-term storage.[83]

In 1927, Huntsman went to Boston and Gloucester to observe General Sea-Foods Company's production of two frozen products in one-pound waxed-paper cartons, called sea loaf and frosted fillets. These consisted of haddock – minced in the sea loaf and filleted in the frosted fillets. The idea was a good one, stated Huntsman, 'and we had been planning something of the sort,' but he believed that the American company's overhead was 'extremely high and the scale of expenditure altogether too elaborate.'[84]

At the experimental station, Huntsman wanted to develop a less expensive way of making quality frozen fish. The Biological Board assented, and in 1928 set up a Research Committee on Fish Handling. Huntsman decided to make a demonstration in which the station 'would not only freeze the fish quickly, but would take only really fresh fish and would control the handling of the frozen fish until delivered to the consumer.'[85] For this he selected haddock fillets, to be sold in Toronto.

Research showed that the thickness of fish flesh was more important than freezing temperatures. Trials at the experimental station showed that a piece of fish one inch thick, immersed in ordinary salt brine at 0° F, would freeze solid within fifteen minutes, whereas a two-inch thick slab would take an hour to freeze, even in a special brine held at –40° F, Huntsman decided that the more economical course would be to reduce fillet thickness and to bring freezing brine into contact with both surfaces in order to speed freezing. Fish were placed on horizontal wet metal plates on a conveyer; lowered slowly into the tanks, the fish froze fast to the plates, and thus were held flat in the brine, which circulated freely about them. Scientists designed a tank for freezing half a ton of fish per hour, using ammonia refrigeration. Two endless conveyer chains carried the plates in the circulating brine at such a rate that the fish were fast frozen by the time they had passed through the tank. An ejector at the far end automatically removed the fillets and left them ready for packing.

Huntsman had noted an increasing demand by consumers for food as fully prepared as possible. In response, the Biological Board developed a process for cutting fillets and forming them into half-pound blocks, five inches long, three inches wide, and nearly an inch thick.[86] Wrapped in waxed paper, these were frozen rapidly in appropriately shaped forms on conveyer belts, and packed, two in each one-pound carton. Named 'Ice Fillets,' they were to be kept frozen until cooked, which went against the usual practice of defrosting fish before cooking.

For a year, the station produced haddock Ice Fillets and sold these in Toronto. For quality control, the Biological Board selected a single wholesaler, the F.T. Jones Co., and a single retailer, the Robert Simpson Co., both firms having agreed to follow instructions. Sales began in January 1929, with much publicity: samples were cooked for customers in the basement of Simpson's. The board provided one thousand pounds of Ice Fillets per week, which each week quickly sold out, even though priced 50 per cent higher than 'fresh' fillets of haddock kept on ice. 'The quality of those fish is still remembered in Toronto,' Huntsman claimed in 1945. 'There was immediate proof that the Toronto public would buy frozen fish of high quality and continue to come back for more. Thirty-five tons were sold in a year at that high price.'[87]

After a year, the committee urged that the new, established market be turned over to the private sector. The Board insisted that it be allowed to inspect the product to maintain high quality. But the department demurred, stating that it had no authority to inspect fish for domestic sale. The board was left with the vain hope that the industry would follow instructions and not retail fish of poor quality.

Only two companies on the Atlantic coast were both willing and able to produce Ice Fillets, having satisfactory facilities for holding and transporting them in good condition. These were the Lockeport Company and the Lunenburg Sea Products. The Biological Board had to foot the expense of any plant inspections, but soon discovered that frozen fish retailers needed as much monitoring as the producers. It had to insist that Ice Fillet retailers be equipped to hold fish solidly frozen.

This product was ahead of its time. The firms producing Ice Fillets soon let standards fall. They could not resist using second-rate fish held longer than six days, and they failed to keep the product stored at constant low temperatures. Their product became as tough and tasteless as the fish rejected by Canadian troops during the First World War and by the American market. Ice Fillets lost their reputation, and the attempt failed. It took the industry nearly a quarter of a century to begin marketing frozen fish of Ice Fillet quality.[88]

Canada lagged seriously behind the United States in refrigeration, frozen storage, and freezing facilities. In those days, ordinary home refrigerators were not equipped for storing brine-frozen fillets, and fish had to be consumed soon after purchase. Another big problem was getting enough stores to provide proper cold-storage display cases, and here the United States' size and investment capital gave it an advantage. When George Earl went to Gloucester, Massachusetts, to investigate General Foods' fast-frozen fish production, 'Mr Birdseye very frankly explained' that marketing was the big problem. By 1935, 845 retail stores were selling General Foods' Birds Eye Frosted Foods. However, expansion was handicapped by the high price of cold-storage sales cabinets, which cost up to $1,800 apiece. In 1934, General Foods was trying to develop a $300 display cabinet, and in 1934–5 its Birds Eye Frozen Foods Division was hoping to turn a profit for the first time.[89]

In Canada, fish consumption was uneven; it seemed to be limited to one day a week (Friday). Also, the population was small. As a consequence, specialized fish stores remained few and far between. The quick-freezing method could not on its own solve the problem of low demand for fish. Later, the standards at Birds Eye's Gloucester plant also began fall. The product imported into Toronto in the late 1930s looked 'rather like Neopolitan ice-cream,' according to Huntsman, owing to the different stages of decomposition of the fillets in one carton. Although frozen food sales expanded considerably throughout the 1930s, it was not until the Second World War that this product established a firm foothold. Wartime conditions advanced public acceptance of frozen foods by years: 'Frozen fruit and vegetables were on the ration-free list in the U.S. at a time when the general public had more money than earlier with which to buy them.'[90]

The shortcomings and failures that dogged aspects of the Biological Board's fishermen's education program and fish processing schemes were, as can be seen, due to factors that were beyond the control of the scientists. Their work in itself maintained high levels of execution and delivery, and because of their competence, in the 1920s and 1930s, the Biological Board was central to the Department of Fisheries' work for the fishing industry. The Atlantic Experimental Station provided a vital link, by serving as the technical training centre for fisheries officers and by directly educating the fishermen themselves. The fish-processing research, beyond a doubt, did a great deal to improve Maritime product quality in the short term. Indeed, the Maclean Commission made a point of commending the station's work, 'under the direction of Dr. A.G. Huntsman ... for its persistent and patient efforts and its successful results, all of which have been of incalculable benefit to the fishing industry ... We cannot emphasize too strongly the invaluable results of their work.'[91] But these results were, sadly, only effective in the short term, since the legislative measures to maintain them were never enacted.

The department, in making the board central to its improvement schemes, had taken an approach fairly typical of other low-priority, underfunded government departments, as can be seen in the case of the similarly neglected American fisheries. After the Second World War, the cash-strapped New England fishing industry could only get government help in the form of research and development funding to increase yields and fish-product demand. The Bureau of Commercial Fisheries did research into marketing fishery products and 'offered information to food specialists and consumers about new types of fish and ways to prepare fish.' In 1954 the New England Fisheries Committee's agenda included better statistics and biological and oceanographic research to sustain fish stocks; exploratory fishing to find new fish concentrations; gear development; and research into better fish processing and uses for underutilized species.[92] Such research and development, it was (vainly) hoped, would jump start the fisheries economy, thus circumventing the need for large-scale government policies.

The Biological Board's new central role in the Department of Fisheries gave grounds for the first dissatisfaction expressed over the board's direction in applied work. D.B. Finn, director of the Prince Rupert Experimental Station, complained: 'Conditions make it necessary for us to ... do work which should properly be done by private companies.' The board, he said, should be supplying 'plant laboratories with formulae' developed by its own 'pure science.' J.P. McMurrich, the Biological Board's chairman after Knight, was swayed, and told Found in 1931 that although he favoured close relations to keep the board aware of industrial problems, and industry informed of the board's latest innovations, the board, in manufacturing and marketing fish products in commercial quantities, had stepped 'entirely outside' its purposes. Huntsman's 'brilliant demonstration' of Iced Fillets 'was a proper undertaking for the Board, but the further commercial exploitation of the product was not.'[93]

By the late 1920s and early 1930s, younger scientists were assuming positions of responsibility within the Biological Board, and they shared a different vision of the board. In 1934, A.T. Cameron took over the chairmanship. Although McMurrich had been happy with the board's applied and educational work, Cameron was not. He believed that field educational work was better carried out by competent non-scientists than by junior scientists, and that 'any further demands ... for educational work from the senior members of the staffs at Halifax and St. Andrews must militate against the successful accomplishment of the research work.'[94] The Second World War ended the Course for Fishermen.

This disengagement was hinted at by Cameron in 1938, when he told Found that many excellent research scientists 'cannot give good elementary teaching ... although we are at present fortunate in having a number of man at Halifax and St Andrews who [can].' Found countered with his own conviction (mirroring that of earlier scientists such as Prince and Huntsman) that contact with industry made scientists better investigators: 'Discussions held with the industry ... have better equipped me for administrative work. I cannot see that it would be different with an investigator.' However, Cameron believed that the board should no longer rely on research staff for transmitting knowledge to fishermen. Thus it is not surprising that the board took advantage of the disruption offered by the Second World War to move away from educational activities. In fact, technical education for fishermen suffered a hiatus until the late 1950s, by which time it had become a provincial responsibility, and the Fisheries Research Board was not involved.[95]

Nevertheless, this educational work and its industrial research in the 1920s and 1930s both fulfilled the Biological Board's initial promise to aid the fisheries, and laid the foundations for later important applied research by the Fisheries Research Board. But although it improved Maritime fishery products, and helped Maritime fishermen weather the Depression, it could not provide a panacea.

What the fishing industry desperately needed was investment capital. This came with and after the war. Wartime demands for increased fisheries production focused attention on the Maritime industry's deficiencies: there was in fact an unprecedented labour shortage, as men had been abandoning this low-prestige, low-paying occupation. With greater funding, the Department of Fisheries was at last able to take effective measures to modernize and expand the Atlantic fisheries. In 1941, New Brunswick diverted three-quarters of a $100,000 fisheries grant to have Gorton-Pew Fisheries of Gloucester, Massachusetts, construct a cold-storage plant at Caraquet. This was the beginning of state interventions to speed up development. When Ernest Bertrand (Laurier) was appointed Minister of Fisheries in 1942, 'the modernization policy mushroomed with subsidies for dragger construction and schooner conversion to trawling gear' to upgrade efficiency among independent and corporate producers.'[96]

In 1946, Canadian fishermen got unwitting help from New England fishermen when the latter went on strike for higher returns for their catches. Wholesalers who had bought New England frozen groundfish turned to Canadian suppliers, who offered lower prices. Ex-vessel prices of Canadian groundfish were one-third to one-half the prices at the Boston Fish Pier. St Louis wholesalers, for example, 'turned permanently to Canada ... The ability to process fish was transferred to Nova Scotia and Newfoundland during the 21 weeks of strike.'[97] Also, the new market for fish sticks further aided the Maritimes' groundfishery. By 1955 at least 55 per cent of households in the Northeast were eating fish sticks, and producers preferred imported fish. They even moved their processing factories from Boston 'to new or expanded plants in Canada, Maine and Gloucester.' In 1957, General Foods' Birds Eye division sold its fishing operations to National Sea Products, the largest fish-processing firm in Nova Scotia, because Canadian processing workers' wages were lower even though both countries used the same filleting technology. Birds Eye wanted to limit its activities to processing and distribution.[98]

In the 1950s and 1960s both federal and provincial governments took active measures to modernize the fishing industry. The federal government, through subsidies and tax concessions, and the province of Nova Scotia, through loans, began to aid fishing firms and some fishermen. These two decades in fact constituted the golden age of Atlantic Canadian fishery modernization. Policy centred around the reorganization of the Fisheries Loan Board, 'large capital contributions for freezing plants, and financial assistance for small and intermediate port facilities.'[99] Maritime fishermen at last had the competitive advantage their fisheries needed, and a period of growth ensued until other problems – those relating to collapsing fish stocks – came to haunt the northwest Atlantic.

Chapter 6

Huxley's Red Herring

'O Oysters,' said the Carpenter,
'You've had a pleasant run!
Shall we be trotting home again?'
But answer there came none –
And this was scarcely odd, because
They'd eaten every one.

Lewis Carroll, 'The Walrus and the Carpenter'

Thomas Henry Huxley's name hardly conjures up images of the fisheries, but in truth, Huxley exerted an enormous and pathological influence over fisheries science, especially of the Canadian variety, and one that lasted well into the middle of the last century. This influence was out of proportion to his late involvement with and limited personal dedication to fisheries biology issues. Huxley's declarations that the open-sea fisheries were inexhaustible, and that the efforts of man could in no way compromise the extreme fecundity of the oceans, were enshrined by certain younger scientists. He would not have been surprised, since, although allowing himself the luxury of extreme scepticism, he reacted unfavourably wherever others doubted his scientific opinions.[1]

Fisheries biology was essentially a branch of science born out of the growing recognition that human activities were altering the environment. As evidence of overfishing became increasingly obvious in the late nineteenth century, questions arose that only scientific study could answer. As seen in chapter 3, by the 1890s the characteristics of fisheries biology were emerging, 'blended ... of older ingredients – zoology and statistics principally,' plus elements of the emerging science of oceanography. The result was both a new form and a function: 'The form was knowledge like that of a fisherman, but deeper; and the function was guidance to better use of the stocks of fish.'[2]

Fisheries biology can be seen as essentially a form of population ecology, although it has not always recognized itself as such. Like ecology, its history is one of changing goals 'for imposing order upon nature,' as well as changing criteria for achieving these goals. In ecology 'there is no final end toward which this all tends, no fixed markers of progress: what is sought is a better understanding of the connectedness of nature's parts.'[3] Similarly, in fisheries biology, 'as each step in management reveals new problems, the scientific advice changes and new steps in management are taken. This changing pattern of science and management is likely to persist because there are no pristine rules. But the manager's job remains the same, to obtain the greatest gain in value or in yield that she or he can, with the least loss of jobs.'[4]

Huxley's views impeded fisheries biologists' recognition of the need for fish stock conservation, especially in Canada.

Thomas Henry Huxley (1825–95), who is most famous today for championing Darwin's theory of evolution, had an odd relationship with fish. Beginning in the 1860s, he was involved in three of the British government's infrequent and dilettantish assessments of the fisheries; yet he was neither sympathetic to nor perceptive about fisheries conservation.

By mid-century the British were beginning to realize that their fisheries were imperilled by rampant industrialization. In 1862, James Fenwick, Liberal MP for Sutherland, single-handedly gathered information from around the country and found that 'the almost universal cry is that our fisheries are falling off year by year.' He asked for a Royal Commission to investigate the matter. The House of Commons agreed. Late in 1863 a Royal Commission was created to investigate the sea fisheries 'with the view of increasing the supply of a favourite and nutritious article of food for the people.'[5]

Huxley served as the commission's scientific expert. One of England's first professional scientists, he was a vocal apologist for disciplined (and paid) practitioners of science. Already well respected and famous, he had made his name as a brilliant comparative anatomist and embryologist. He gained notoriety as the leading champion of Charles Darwin's theory of evolution through natural selection. The son of a school teacher, he began his scientific career as ship's surgeon on the *Rattlesnake*'s 1847–50 expedition to Australia's Torres Straits. During that voyage, on his own initiative, he studied the organisms then classified as Cuvier's radiata. He sent several papers to the Linnean Society and a major paper to the Royal

Canada's first biological station, which was mounted on a scow and towed from place to place. Its yearly location was decided based on petitions for investigations by different influential fisheries sectors. It first opened at St Andrews, New Brunswick, in 1899 and remained there the following year. It then went to Canso, Nova Scotia, in 1901 and 1902; Malpeque, Prince Edward Island, in 1903 and 1904 for oyster studies; and the Gaspé, Quebec, in 1905 and 1906. In 1907 the scow sprang a leak and was beached on the Gulf of St Lawrence's south shore. Its contents were scavenged by the scientists.

The original permanent laboratory at St Andrews, built in 1908.

The original laboratory, with the 1921 addition on the waterfront side set at right angles to the first laboratory, which added facilities for bacteriology and biochemical studies. The entire building was destroyed by fire in 1932. The station's water tower is located on the left, and behind it is the three-storey residence, still standing today.

Another view of the Atlantic Biological Station, this time from the waterfront, showing the wharf in the foreground. Also visible are the fish-handling and carpentry shop to the right, the water tower to the left, and the top of the three-story residence.

The steamer CGS *Acadia*, one of three ships used to survey the Gulf of St Lawrence and the Scotian shelf during the Canadian Fisheries Expedition of 1914–15. The work included hydrographical surveys, plankton tows, and fish and fish-egg sampling.

The great Norwegian oceanographer and fisheries biologist Johan Hjort, left, on board the CGS *Acadia*. Beside him is Captain W.A. Robson.

Dr Arthur Willey, leaning over the rail, is watching as a plankton net is hauled in by two deckhands during a Canadian Fisheries Expedition cruise.

Scientists and volunteers who worked at the Atlantic Biological Station in the summer of 1913. They are, left to right standing: Dr E.E. Prince; Prof. J.D. Detweiler; Eleanor Huntsman, in her father's arms; Dr A.G. Huntsman; Prof. H.E. Perry, of Acadia University; Dr Philip Cox, of the University of New Brunswick; and G.A. Miller, of Queen's. Sitting are, left to right: Dr A.P. Knight; J.W. Mavor, N.A. Wallace, and A.R. Cooper, students at the University of Toronto; and Prof. W.H. Martin of the University of Toronto.

Dr A.P. Knight (1849–1946) on the wharf at the Atlantic Biological Station (no date). Knight was largely responsible for pushing the Biological Board's fish-processing research into the forefront of its efforts in the 1920s. His work on lobster canning and educating lobster processors showed how the board could help fishermen materially.

Dr Arthur Gowanlock Huntsman (1883–1973), the most dynamic and influential individual associated with the Biological Board, became permanent curator of the Atlantic Biological Station in 1915 and director in 1919. He was forced out of this position in 1934, and was 'kicked upstairs' into the (powerless) position of consulting director of the board. He trained many marine scientists of the next generation, expanded the board's research horizons to include physical oceanography and fish-processing research, and continued to influence research as the editor of what became the *Journal of the Fisheries Research Board* from 1934 to 1949. His personal links to the most important marine scientists of his era, including Norway's Johan Hjort, Henry B. Bigelow, first director of the Woods Hole Oceanographic Institution, and Michael Graham, first director of the permanent Lowestoft Fisheries Laboratory, helped put the Atlantic Biological Station in the leading ranks of the world's marine stations in the 1920s and early 1930s.

The interior of main laboratory at St Andrews, 1908.

Dining room in the residence in 1923, with tables set. The forgotten side of scientific research is laboratory administration and seeing to scientists' material needs. A.G. Huntsman, who was so instrumental in enlarging the Atlantic Biological Station, was ably assisted by his secretary, Margaret Rigby, and his wife, Florence Marie (Mary) Huntsman (née Stirling). Indeed, Miss Rigby was an equal partner in administering the station, and was given great credit for her work by Huntsman. Mary Huntsman, who had wanted to become a professional interior designer (and was thwarted as a younger, unmarried woman, by family considerations) was in charge of running the dining hall, helped run the residence, and acted as hostess to important scientific guests and chaperone to spirited young science students. She handled the logistics of bringing in food supplies, including ordering dry goods from Toronto, brought in by train. She redesigned the residence interior to enlarge the dining room, creating a cantilevered ceiling to replace a support wall. Mary Huntsman also designed the chairman's and director's cottages, which are still standing (personal family communication).

Henry Byrant Bigelow (1879–1967), pioneering biological oceanographer, first director of Woods Hole Oceanographic Institution, and permanent chairman of the North American Council on Fishery Investigations (1923–38). It was through this organization that he met A.G. Huntsman, who became NACFI's permanent secretary. The two became firm friends, which helped establish tighter U.S.–Canadian research links and drew American scientists into the work of the International Passamaquoddy Fisheries Commission (1931–3).

FNACFI meeting, 13–14 September 1933. Members present are, left to right: Mr Elmer Higgins, chief of the U.S. Bureau of Fisheries' Division of Scientific Inquiry; A.G. Huntsman; Dr James Playfair McMurrich (Biological Board chairman, 1926–34, who had been the first to call for a marine biological station for Canada in an 1884 article); Dr Harold Thompson, the British fisheries scientist who represented Newfoundland; Henry B. Bigelow; and William A. Found, the Deputy Minister of Fisheries.

The fire of 8 March 1932 that destroyed the original laboratory at St Andrews, along with much of the data from the International Passamaquoddy Fisheries Investigation. The aftermath of this fire, which occurred during the height of the Depression, left the future of the Atlantic Biological Station in doubt.

The new fireproof yellow brick laboratory that Huntsman built without Biological Board authorization. He rightly feared that the cash-strapped board would permanently close the station and transfer all Atlantic research to the Halifax Fisheries Laboratory. Biological Board chairman A.T. Cameron used Huntsman's insubordination as an excuse to remove him as station director.

The staff at one of the Biological Board's substations, the Ellerslie Biological Station on Prince Edward Island, founded in 1930 to study Malpeque oysters and shellfish problems. In this 1936 photograph are shown, left to right: W.A. Found; F.R. Hayes (later chairman of the Fisheries Research Board from 1964 to 1969); Miss Underhill; oyster expert J.C. Medcof; and substation director A.W.H. Needler, who later became director of the Atlantic Biological Station (1941–54).

Society, which elected him a Fellow. In 1854 he succeeded Edward Forbes as lecturer in natural history at the Royal School of Mines. Soon after, he was also appointed naturalist to the Geological Survey. He would hold both appointments for over thirty years.[6] Lyon Playfair, his superior at the London School of Mines, and a crony of the Prince Consort, was responsible for having Huxley appointed to the 1863 Sea Fisheries Commission. Huxley had already served with him on the 1862 Royal Commission on the Herring Fishery in Scotland,[7] which reached conclusions similar to those of the more important 1863 Sea Fisheries Commission. Huxley, seeking to supplement his low income, would devote more and more of his time to similar public inquiries, to the detriment of his scientific work.

The Sea Fisheries Commission sat for two years and gathered evidence in every important coastal community. It learned that trawling, a fishing method destructive of immature fish, had mostly displaced older fishing methods. Hundreds of witnesses complained that there were no measures for protecting immature fish, and testified that although total catches had greatly increased, the fish taken were becoming smaller and selling for less. Despite this evidence, the Royal Commission reported in 1865 that the supply of fish 'obtained upon the coasts of the United Kingdom has not diminished of late years, but has increased; and it admits of further augmentation.'[8] Indeed, the commission went so far as to claim that 'beam trawling in the open sea is not a wastefully destructive mode of fishing, but is one of the most copious and regular sources of the supply of eminently wholesome and nutritious fish. Any restriction upon this mode of fishing would be equivalent to a diminution of the supply of food to the people.'[9]

The commissioners believed the oceans to be so vast that commercial fisheries could cause no real damage to fish stocks. They even suggested that laws dating from the time of James I dictating mesh size, fishing seasons, and so on, be abolished, claiming these would hurt fishermen if enforced.

However, the commission had overlooked several important factors. First, although total catches had indeed increased, there were in 1865 many more trawlers operating than ever before, each boat was harvesting fewer and smaller fish. To improve profits, many owners borrowed heavily to finance bigger fleets, further overstressing the fisheries.[10] In ignoring this, the commission dealt a severe blow to sea fisheries conservation; based on its findings, Parliament refused to regulate trawling.

Later, in 1881, Huxley became one of the two Inspectors of Salmon Fisheries for England and Wales. This sinecure was offered to him through the machinations of Prime Minister Gladstone's home secretary,

Sir William Harcourt. Harcourt raised the pay to £700 per annum[11] and reduced the duties in the job description in order to attract Huxley, who accepted. Thus he was finally relieved of 'the necessity of bread-making' – that is, holding down several jobs in order to earn a decent wage as a scientist at a time when it was not yet widely recognized that British professional scientists needed money to live.[12] The position he filled had been created following the 1860 Royal Commission of Inquiry into the Salmon Fisheries of England and Wales, which found that salmon were disappearing due to poaching, raw sewage, and industrial waste. The 1862 Salmon Act created a central inspectorate of salmon fisheries, which operated under the Home Office. William Ffennell and Frederick Eden, Irishmen experienced in managing salmon rivers, were the first inspectors of the salmon fishery. They were empowered to issue edicts concerning fishing practices and the industrial pollution of rivers.[13]

When both Eden and Ffennell were taken ill and resigned in 1866, Spencer Walpole, the new home secretary, replaced them with his own son, Spencer Walpole, Jr (a lawyer who ably correlated local by-laws with general fishery acts passed by Parliament), and Frank Buckland. Buckland and Walpole enthusiastically collected statistics and scientific information on salmon and their fisheries.[14] Buckland's approach to this position was diametrically opposed to that of his successor, Huxley.

Frank Buckland (1826–80) was the son of geologist William Buckland, who had been appointed Reader of Geology at Oxford in 1819. As a student, Frank Buckland began studying the habits of trout and perch in his spare time. In 1853 he was hired as assistant surgeon in the Life Guards. His light duties allowed ample time for him to pursue his interest in natural history. In 1857 his first book, *Curiosities of Natural History*, was a huge success. Around 1860, he developed an interest in fish culture, and in 1862 he was asked to establish a model fish hatchery in the office windows of *The Field* magazine, located on The Strand. This hatchery was later placed on permanent display at the Science Museum in South Kensington. In 1863, he expanded a lecture on fish culture at the Royal Institute into his second well-received book, *Fish Hatching*. Meanwhile, Buckland deepened his salmon and trout studies and became interested in oyster culture.[15]

As a result of his reputation as a fish expert, Buckland was appointed to the salmon inspectorate in February 1867. He was well qualified for the job: he understood the salmon fisheries' technical problems, and he enjoyed working outdoors. He got along well with Spencer Walpole, Jr, and he liked to meet people and explain his ideas. His work involved, as he said, 'advising, wheedling, coaxing, exhorting and instructing fishing boards, fishermen and mill owners in all aspects of the proper management and conservation of the salmon rivers.'[16] Nor did the salmon inspectors confine their activities to salmon inspection. As the only official experts in Britain's central administration empowered to deal with the protection of any fisheries, they also investigated marine dynamite fishing (prohibited following their recommendations) and the lobster and crab fisheries. With the Home Office's complicity, they sought to gain permanent status as inspectors of the sea fisheries. However, there arose a jurisdictional conflict with the Board of Trade, which had overarching responsibilities for Britain's sea fisheries, although it was never inclined toward conservation policies. After Walpole and Buckland published an indictment of this lack of policy in 1879, the Board of Trade issued a memorandum that ended the interference of the Home Office Inspectors.

'In 1879, Buckland became ill; he died on December 23, 1880. He had been an extremely adept administrator and his success in arousing the interest of the nation in conservation of the fisheries was unequalled by any other individual in the nineteenth century. The driving force behind his actions was a Christian ethical and moral utilitarianism.'[17]

This was in tremendous contrast with the character of his successor, Huxley. Huxley preferred the laboratory to the field, and he did not share Buckland's joy in dealing with fish and fishermen in their natural habitats. He cared little for the opinion of any fishermen – in accepting the post as Buckland's replacement he predicted he would be 'jamming common-sense down the throat of fools'[18] – and he was thankful that the position required 'no serious labour.'[19] The post was, as far as Huxley was concerned, only part-time, and he retained his other positions. Although he immersed himself in studies of fish anatomy,[20] he found tedious the politics and 'drivel' of fisheries conservation. To a friend he complained: 'The mere thought of having to occupy myself with the squabbles of these idiots of country squireens and poachers makes me sick – and is, I believe, the chief cause of the morbid state of my mucous membranes ... All of this week I shall be occupied in hearing one jackass contradict another jackass about questions which are of no importance.'[21] He disliked 'sitting all day in a crowded court, hearing a disputed case of fishing rights, or examining witnesses who stick firmly to views about fish which had long been exploded by scientific observation.'[22]

Good relations between local communities and the inspectors had been essential to the success of fisheries conservation measures. However, Huxley, swamped by correspondence on the fisheries and other issues, had no desire to maintain contact with local groups; nor was he interested in explaining his directives. With this attitude he managed to alienate many local boards.[23]

Although Huxley upheld the importance of Buckland's restrictions on lobster and crab fisheries, the quality of his salmon fishery reports declined after Walpole resigned in 1882. In 1883 he told Walpole: 'The Office would be quite perfect, it they did not want an annual report. I can't go in for a disputation on river basins after the manner of Buckland, and you have exhausted the other topics. I polished off the Salmon Disease pretty fully last year, so what the Deuce am I to write about.'[24]

Fortunately, grave ill health found Huxley to resign his position in May 1885. However, Huxley's damaging influence on the fisheries did not stop here. To be fair, and in what follows, it must be remembered that this great scientist and spokesman for science took on this fisheries work later in life, when he was already in declining health. The work did not have much intellectual attraction for him, and he took it on mainly for financial reasons. Quite possibly, his ill health had a great deal to do with his querulous attitude toward the public relations side of his duties as a fisheries inspector. Also, as Huxley himself had proclaimed when he was thirty, few scientists did valuable work after the age of sixty, and it would (his younger self believed) be better if they were throttled.[25] It was as an aging scientist that he did his worst damage.

In 1886, a new central department for fisheries was being devised within the Board of Trade; experts were to be hired who would help base policies on scientific findings. Huxley was against such a department: dedicated to *laissez-faire* ideals, he felt that a permanent body of government scientists would extend the arm of the state into areas in which it had no proper business.[26] He wrote in the *Times*:

> I am of opinion that the less the Government interferes with any branch of the industry the better ... The cost and trouble of obtaining such scientific information ... for ... a branch of industry ought to fall upon those who profit by it, and not upon the general body of the tax-payers ... I have had something to do both with science and with administration, and it is in the interest of both that I express my strong conviction that they be kept separate.[27]

Huxley recognized that English fisheries did need protection from foreign encroachment, and that different classes of fishermen had conflicts of interest requiring government mediation; nevertheless, he felt that government should only consult outside scientists on specific questions. He wished to

> repudiate as strongly as possible, in the name and interests of science, no less than in that of the working fishermen ... the proposal ... to appoint a body of scientific men to 'manage' the fisheries ... Men of science are no more the right people to be trusted with managing fishery affairs than a landsman who happens to be master of the theory of navigation is the right man to be trusted with steering an ironclad.[28]

By the time the new fishery department of the Board of Trade was constituted in 1886, 'Huxley's pratings,' as historian Norman Jester called them, had left their mark. No scientists were appointed.[29] The Board kept personal contact between fishermen and the inspectors to a minimum, and the inspectors were responsible only for collecting statistics. On-the-spot inspections decreased, and more information was gathered by correspondence. Fisheries regulation and law enforcement languished, as local fisheries councils and boards of trade found it difficult to coordinate policies.

Finally, in 1902, a Royal Commission advised that all fishery affairs be codified and transferred to the Board of Agriculture, which would be able to address fisheries conservation and management through scientific investigations. This proposal proved popular, and was put through in 1903. Under the new Board of Agriculture and Fisheries, scientists finally began to create fisheries conservation policies – a development that Huxley had helped delay by twenty years.[30]

Yet even this was not Huxley's most damaging legacy. Appropriately, his most lasting negative influence was intellectual. At the London International Fisheries Exhibition of 1883, which he helped organize, Huxley gave the opening address. He told his large audience that there was no likelihood of fish scarcity through overfishing, and that with existing fishing methods, it was inconceivable that the great sea fisheries could ever be exhausted.

Huxley admitted that salmon and all river fish could be endangered by river pollution, and by nets stretched from bank to bank to capture entire migrating fish populations. He also realized that oyster beds could be dredged out of existence, although 'it ceases to be worth while to dredge long before this limit is reached.' Regarding the sea fisheries, however, he stated:

> I believe that it may be affirmed with confidence that, in relation to our present modes of fishing, a number of the most important sea fisheries, such as the cod fishery, the herring fishery, and the mackerel fishery, are inexhaustible. And I base this conviction on two grounds – first, that the

> multitude of these fishes is so inconceivably great that the number we catch is relatively insignificant; and, secondly, that the magnitude of the destructive agencies at work upon them is so prodigious that the destruction effected by the fishermen cannot sensibly increase the death-rate.[31]

Huxley noted that shoals of cod in the Vestfjord, Norway, were up to 55 metres thick; if the fish were a metre apart, there might be 55 million cod below one square kilometre. Lofoten fishermen took around 30 million cod in an exceptionally good annual catch. If each cod ate one herring each day, that shoal took 840 million herring each week; yet all Norwegian fisheries took less than 3,400 million herring.[32] Therefore, argued Huxley, 'nothing we do seriously affects the numbers of the fish' in comparison with natural predation, 'and any attempt to regulate these fisheries seems consequently, from the nature of the case to be useless.'[33]

During the International Fisheries Exhibition, E. Ray Lankester appealed for the study of British marine life for scientific and commercial ends. A group of eminent scientists resolved to form a society and build a laboratory on the British coast. Huxley did not sign this resolution. He disagreed vehemently with Lankester,[34] who argued that intensive fishing *did* pose a real threat to fish stocks: 'It is a mistake to suppose that ... the grand total of fish [is] so numerous in comparison with man's depredations as to make his operations in this respect insignificant.'[35] But by the time the Marine Biological Association was formed on 31 March 1884, Huxley had been persuaded to give his support – presumably because the society would also pursue general marine biological science. He was even elected its president. He joked to his former protégé, Lankester, who had pushed him into this position, 'that he would do as little as possible to deserve the honour.' Nevertheless, he worked hard at collecting funds to build a marine station at Plymouth.[36]

The Marine Biological Association's laboratory at Plymouth received its biggest funding allocation from the government, which awarded it a capital grant of £5,000 for the laboratory and £500 per annum for running expenses.[37] In 1889 the Marine Biological Association asked for an increased annual grant to investigate the effects of trawling on immature fish. This reignited the dispute about overfishing, and Huxley resigned the association's presidency when most of its members supported Lankester's views. His ill health was a factor, but there is no doubt that having his views crossed is what caused the breach.[38]

Most of the members of the Marine Biological Association agreed that overfishing could occur. As scientific evidence of overfishing increased in the 1890s, fisheries science 'began to be studied continuously.' The Dane C.G. Johannes Petersen, who developed the method of attaching numbered tags to sea fish, made a vital contribution: he 'successfully thought out a workable definition of "overfishing," in terms of growth, mortality, and rate of fishing.'[39] Working with North Sea plaice fishery statistics, he developed a new model of overfishing. What later became called 'recruitment' overfishing occurred when fishermen left too few adult fish to replace the population. What later was termed 'growth' overfishing was fishing down the stock, that is, harvesting progressively younger fish before they reach the reproductive age at which they can produce the maximum number of eggs. Such young fish are valuable as they are in fact growing, and adding to the stock biomass, faster than they are lost to the stock through natural mortality. Older fish tend to have higher natural mortality rates, and their loss from fishing does not have the same kind of impact on the overall health of the stock. This then lowers the total yield from the fishery. Petersen noted that tagged plaice did not leave the Kattegat: the stock was stationary, had few natural predators, and was not diseased. This meant that the marked loss of larger, older fish had to be due to fishing. Petersen hoped that fishermen themselves would put off taking smaller fish once they realized 'the economic advantage of ... allowing the smaller ones to survive and grow.'[40]

Similar work was taken up in Britain in 1902, when, after a dry spell owing to lack of funds, the Marine Biological Association resumed its fishery work under Walter Garstang. He was provided with a trawler, the *Huxley*, and a new laboratory at Lowestoft. He introduced two important procedures. First, he came up with the idea of using the average annual catch per unit of effort as an index of stock. Second, he devised 'a rough index of fishing power in that one steamer was considered equivalent to four smacks, and he derived an estimate of fishing effort, smack units.' He demonstrated that the English North Sea trawl fishery's catch per unit effort had declined by nearly half between 1889 and 1898, the years during which steam trawlers had multiplied.[41]

Garstang left to take a university position in 1908, and his work was taken over by two Board of Agriculture scientists, A.T. Masterman and E.S. Russell. After the first World War, Russell was appointed director,[42] and under his leadership, the Lowestoft laboratory became the main school of research which propagated the theory of fishing – finding means of detecting, predicting mathematically, then regulating the effects of fishing – that would dominate fisheries biology from the 1930s onward. However, interest in the problems of overfishing was not universal.

Indeed, well into the 1960s, many fisheries biologists also believed in increasing fish catches through positive measures. There were four main routes of positive action: first, increase the fish population through fish culture, an idea largely passé by the 1960s but still pursued in places, and later superseded by fish farming; second, find new, unexploited stocks – a strategy favoured by the great Norwegian fisheries biologist Johan Hjort and still favoured by the Department of Fisheries and Oceans in Canada; third, give fishermen new methods to better locate and exploit the stocks; and fourth (and perhaps the most theoretical), discover indicators by which to predict the likely abundance of fish stocks at least a year in advance. Predicting the success of future fishing seasons was perhaps the single most important project in fisheries biology; the hope was that fishing effort could be directed according to predicted abundance.

The U.S. Bureau of Fisheries preferred to increase commercial fish populations through fish hatcheries. Over 30 per cent of its yearly budget was allotted to fish culture, 5 per cent or less to scientific research. In 1920 one observer wrote: 'The policy of the Bureau as regards fish culture has been ... that it is better to expend a small amount of public money in making fish so abundant that they can be caught without restriction, thus serving as cheap fish for the people at large, than to expend a much larger sum in preventing the people from catching the few fish that remain after generations of improvidence.'[43] Unfortunately, although this solution to fisheries problems was politically expedient, later critical evaluation revealed it to be ineffective.

Standard Canadian fish-hatchery practices were largely abandoned in the late 1920s, or replaced by 'rearing the fish to larger sizes before release, or giving assistance to natural propagation.' The Biological Board 'played a major role in this change of emphasis'[44] following critical work on lobster hatcheries by A.P. Knight around 1920. In a classic study, between 1925 to 1938, board employee R.E. Foerster investigated the efficiency of sockeye salmon hatcheries at the biological station at Cultus Lake, British Columbia; he 'showed that hatcheries were contributing very little to the total sockeye production.'[45] On this basis, the salmon hatcheries were closed.

Like Hjort in Norway, Canada's Biological Board favoured finding new stocks and developing better fishing tools and methods. The purpose of the Canadian Fisheries Expedition of 1914–15, directed by Hjort, had been to locate new herring stocks. Grand Bank and St Pierre Bank haddock concentrations were discovered through research explorations; commercial exploitation began about 1946.[46] However, Canadian fishermen needed improved technology more than anything else. Thus, at the Prince Rupert Experimental Station, for example, 'mid-water trawls were developed that have provided a basis for present commercial models.'[47] Norwegian, Canadian, and Newfoundland scientists encouraged fishermen to locate cod using a thermometer, since cod were known to favour certain temperatures. Similarly, Alister C. Hardy, professor of zoology at the University College of Hull's Fishery Research Department, taught fishermen to recognize good herring fishing conditions by monitoring concentrations of tiny, shrimplike *Calanus,* using a small plankton indicator he had developed.[48]

Clearly, fisheries science, especially before the Second World War, was not uniformly devoted to fisheries regulation. It is true that 'negative' measures were always important: since 1903, Britain's Board of Agriculture and Fisheries had been collecting fisheries statistics to find 'the abundance and distribution of food fish ... with a view to determining whether the stock is decreasing, and to form a basis for ... determining what protective measures are required and where.'[49] Even those who were inclined to take as gospel Huxley's dictum that the great sea fisheries were inexhaustible recognized that this was not universally applicable, and supported measures to help conserve especially smaller fish stocks found in near-shore areas and in semi-enclosed bays and seas. But Huxley's notion that the great offshore fisheries were inexhaustible struck them as true and affected their practices. Their fears that the great fish stocks could be depleted were perhaps as remote as our current fears about human activities rendering the mosquito extinct.

Nowhere is this clearer than in the Canadian tradition fostered by the Commissioner of Fisheries, Edward Ernest Prince. As has been noted elsewhere, Prince had been chief assistant at William Carmichael M'Intosh's marine laboratory at St Andrew's, Scotland. M'Intosh was selected by Huxley as scientific expert for yet another British fisheries commission on which Huxley served.[50] This 1883 Royal Commission on Trawl Nets and Trawl Fishing was appointed to investigate complaints by line fishermen that steam trawlers were destroying their fisheries. From January to August 1884, M'Intosh made a series of fortnightly trawling cruises for this investigation. Unfortunately for the line fishermen, he did not find that trawlers damaged their fisheries.

In 1898, M'Intosh published *The Resources of the Sea,* in which he promoted his view that the ocean's resources were inexhaustible. He claimed that he had arrived at these views independently of Huxley. His 'assurance that the periodical scares of the depletion of the sea of fish were without foundation and that science, associated with the name of the great Huxley, was showing that the nation's food supply was secure in the inexhaustible productivity of the oceans, was much to the taste of the Victorian middle class.'[51]

Arguing against M'Intosh was Walter Garstang, who in his article 'The Impoverishment of the Sea' pointed out that ten years of English Channel investigations showed that 'the bottom fisheries were not only not inexhaustible, but in rapid and continuous process of exhaustion; that the rate at which sea fishes multiply and grow, even in favourable seasons, is exceeded by the rate of capture.'[52]

E. Ray Lankester and Sir William Herdman also rejected M'Intosh's claims, yet in spite of the weight of scientific opinion against M'Intosh, we find Prince in Canada stating in 1913: 'It is practically impossible to exterminate sea fish on account of the abundance of their eggs.'[53] Prince, as M'Intosh's closest protégé, was in a sense a direct heir of Huxley. Like M'Intosh he totally ignored the 'immense advances in the engines of capture by a new generation of powerful steam trawlers during the 1890s,'[54] perhaps because trawling had not reached any intensity in Canada. The deference shown by Canadian scientists to leading scientists from the Old Country meant that there was little opposition to Huxley's views in Canada in Prince's time and after. Pro-British, imperialistic sentiments meant that contrary views were rarer here than in the United States, and even than in Britain itself.

A.G. Huntsman, director of the Atlantic Biological Station from 1919 to 1934, and perhaps the most influential individual on the Biological Board during that time, espoused the views of Prince, M'Intosh, and Huxley. He argued that because of high reproductive rates, fish quickly outstripped available food supplies, so there must be concomitant high death rates. Therefore, 'fishing will never be so intensive as to decrease the take of fish in the long run, except where practically the whole stock can be easily removed,'[55] as in salmon river fisheries.

Huntsman declared that the idea of fish depletion was 'little more than a bogie to frighten the credulous ... Those who raise this bogie ... conveniently ignore incompatible facts,' such as in 1939 a higher yield of whitefish than ever before in intensively fished Lake Erie.[56] Instead of worrying about fish depletion, 'fishermen in order to improve their economic position need to choose between (1) restricting their fishery so as to increase their take per unit of effort and (2) making their efforts more effective.'[57] He felt that much of fisheries research ought to be directed toward reducing fishing costs. As a station director, and as a university professor who directed the theses of many young biologists and future fisheries biologists, his opinions carried enormous weight during the Biological Board's formative years. Huxley was casting a long shadow over Canadian fisheries science.

Yet the Canadians and McIntosh were not the only 'Huxleyists.' On the American Pacific was the influential Wilbert McCleod Chapman (1910–70). Unlike Huxley he knew from experience that fish stocks could collapse, but like Huxley, he retained an ultimate faith in the oceans' supreme fecundity. During the Second World War he surveyed the tropical Pacific for fish for American troops. He later directed the College of Fisheries in Seattle. He was the U.S. State Department's main fisheries advisor between 1948 and 1951. After this he advised the fishing industry from his base in San Diego, directed research for the American Tunaboat Association, and worked for the Van Camp Sea Food Company.[58] Arthur McEvoy, author of *The Fisherman's Problem,* stated that Chapman 'firmly believed that the ocean had vast, untouched reserves of food and that U.S. entrepreneurs had a mission to develop those resources for the benefit of humankind ... The ocean's potential was "practically speaking, unlimited." Overfishing did not greatly concern him.'[59] Chapman knew about the great California sardine fishery collapse in the 1950s, and the later mackerel stock collapse, but 'the depletion of individual fisheries' did not concern him 'because he believed that there were always more to be had, with scientific help.'[60]

The influence of such individuals lay only in part in their colleagues taking them seriously – and some highly influential Canadian fisheries scientists did. For example, Dr D.B. Finn, the first director of the Prince Rupert Technological Laboratory in 1929, later Deputy Minister of Fisheries (1940–3), and then the first Director for Fisheries in the UN's Food and Agricultural Organization (FAO), created in 1943, held fast to Huxley's teachings that most fish stocks are inexhaustible, commenting that 'you don't have to be afraid of extermination except in very few cases,' including, obviously, the salmon.[61] The larger problem lay in their stating what government and industry wanted to hear. Thus, J.A. Paulus, a fisheries firm president who had frequent contact with Huntsman, told the Canadian Fisheries Association in 1916: 'To my mind the fish industry is worth infinitely more ... than any of our other resources, because the sources of supplies in this case are practically inexhaustible ... It is not so with our mines, our forests, and even in our agriculture ... while with our fisheries ... the more we ask of them the more they will yield.'[62]

Democratic governments postpone intervening in an industry, even for its own good, to avoid inflicting hardship on voters. They have to consider not only the resource, but the legal and social problems arising from infringements on the individual 'right' to exploit resources. Any expert who opposes restrictions will be listened to most eagerly. Huxley's pronouncement served those with a *laissez-faire*, pro-industry bias.

Indeed, Huxley was of such stature that even opponents of his views tried to reconstruct his statements. In 'The Impoverishment of the Sea,' Garstang defended Huxley as being misunderstood. Since Huxley recognized that theoretically an oyster bed could be dredged clean, and since trawl fisheries effectively operated like oyster dredging, with sluggish, bottom-dwelling flatfishes as their target, Garstang felt that 'it is clear ... that Professor Huxley limits his conviction as to the inexhaustibility of sea fisheries to the drift-net fisheries.' Garstang stated that 'Professor Huxley's view on this important question has been ... widely misunderstood'; and felt justified in going 'a step further than Professor Huxley's words authorise' to suggest that Huxley's comment that 'those who take up the subjects of trawling and of the shell fisheries will discuss the question in relation to those fisheries' recognized that bottom-dwelling flatfishes 'nearly approximate ... the oyster, as regards the conditions of their inexhaustibility.' It may be that Huxley grudgingly accepted that some fisheries were volunerable to overfishing; nevertheless he remained opposed to fisheries restrictions unless overwhelming evidence of depletion existed, and even then he did not see how these could be enforced.[63]

In 1948, the leading Lowestoft fisheries biologist, Michael Graham (1898–1972), told the British Association that Huxley was to be thanked for pointing out the relationship between natural fish mortality and fishing mortality. Furthermore,

> Huxley ... left English fishery science a legacy that may be even more valuable. This is the principle that men should not interfere with the activities of other men without exceedingly good reason, because he who creates a regulation thereby creates a new [possibility] of legal offense. This is abhorrent to English thinking ... The onus for showing that any regulation of fisheries is necessary lies upon those who advocate it ... I would assert that ... the idea of regulation to prevent over-fishing before it occurs is not acceptable.

Too often, regulations had been made 'for the care of fish stocks, which on proper examination have been found to be able to stand further exploitation, in fact to be in need of it if they are to be rightly used.'[64]

I do not advocate the view that Huxley was the reason why some of the more extreme 'inexhaustibilists' held their views. Rather, Huxley's stature is what inspired opponents of overfishing theory to persist despite growing negative evidence. Fisheries biology was a young science seeking status, and the Huxley connection lent it respectability. As a consequence, even those who vehemently disagreed with Huxley's contentions, such as Graham and Garstang, sought to rehabilitate his contributions and to extract for their own use the few nuggets of common sense on this issue that Huxley left for later generations. That said, although fisheries biology involves more than detecting overfishing and protecting against it, conservation has arguably been its underlying justification, and Huxley's utterances made this work seem less urgent than it was.

The North American Council on Fishery Investigations

The years following the First World War were ones of relative optimism in fisheries research. This was the time when fisheries biology started to become the mainstay of the Biological Board of Canada. According A.W.H. Needler, who directed both the Atlantic Biological Station and the Pacific one at different points in his career, this research 'took somewhat different directions on the two coasts because of the very different conditions in the sea and in the fishing industry.' Although there was 'increasing attention to oceanography, important to all fisheries,' what came to be recognized as classic fisheries biology – concerned with preventing overfishing while maintaining the highest possible economic yield – did not develop on the Atlantic. Rather, it was advanced by the Fisheries Research Board on the Pacific. Needler attributed this to the two coasts' very different fisheries: 'On the Pacific coast, salmon contribute close to three-quarters of the fishery's value and ... were so easily caught as they entered their rivers that the fishery could survive only if very strictly regulated ... On the Atlantic the fisheries depended mainly on species much harder to overfish and there was then little public recognition of the need for regulation.'[65]

Also, research on the Atlantic had to be apportioned among lobsters, groundfish, herring, salmon, trout, scallops, oysters, clams, and seaweeds. More importantly, fisheries development research had top priority. However, Needler's theory alone does not sufficiently explain the non-appearance of classical fisheries biology on Canada's east coast. Another contributing factor was the Canadian government's implementation of the 1928 Maclean Commission's majority recommendation to restrict steam trawlers on the Canadian Atlantic. Fishermen's protests against trawlers had comprised the most important evidence given before the commission,[66] although there were only eleven steam trawlers operating from Canadian ports. The government opted to impose a licence fee of $500 per trawler;[67] only three were still in operation by 1933. Scientists thus could even argue that Canadian Atlantic fisheries resources were being underexploited. But in fact, it was Huntsman's and some of his colleagues' intellectual commitment to Huxley's views, outlined above, that blocked work on the theory of fishing. Huntsman was an exceptional figure in Canadian Atlantic fisheries research: many board scientists hired in the late 1920s and the 1930s had been his graduate students at the University of Toronto, where they absorbed his views to at least some extent.[68] Predicting the success of future fishing seasons at least a year ahead was one of their primary research objectives, but one that required no acknowledgment of fisheries depletion.

In its Atlantic fisheries research, the Biological Board worked within an umbrella organization. Just as British and European fisheries research was coordinated by the International Council for the Exploration of the Sea (ICES), an international forum for coordinating northwest Atlantic research was developed by Canada, the United States, and Newfoundland.

The North American Council on Fisheries Investigations (NACFI) was formed after a meeting held in Ottawa on 23 September 1920. It was pushed into existence through the lobbying of the Canadian Fisheries Association, an organization that had been formed in 1915 by fishing industry leaders to help boost the sagging fortunes of the Canadian Atlantic fisheries. The CFA was disgusted by the apparent lack of any practical applications from the Biological Board's fisheries research, and pushed for closer ties between the scientists and the fishing industry. The Canadian Fisheries Association was convinced that to get results, Canadian scientists would have to band together with American and Newfoundland scientists. In August 1918 they called for a joint organization that would mirror ICES in Europe. It fell to Huntsman, a member of the CFA, to push the idea with his American and Newfoundland counterparts and the Department of Marine and Fisheries. At the first meeting of NACFI, the CFA was represented by Captain F. William Wallace. Although thereafter NACFI meetings were attended only by scientists and government officials, the CFA continued to report on NACFI's activities. Also among those present at the first meeting were E.E. Prince, William Found, Canada's Deputy Minister of Fisheries Alexander Johnston, the U.S. Commissioner of Fisheries, Dr H.M. Smith, and Newfoundland's Minister of Marine and Fisheries, S.P. Whiteway.[69] They agreed that only informal cooperation was desirable for the participating countries, none of which belonged to ICES, although the United States had been a member from 1912 to 1916.[70] It was felt that although ICES's results were 'of interest ... information pertaining specifically to the northern Atlantic [is] of infinitely greater importance to Canada, the United States and Newfoundland.' U.S. Bureau of Fisheries advisor Henry Bigelow emphasized the need to find methods for foretelling the course of a fishery on 'a given section of the coast.' Whiteway pointed out that 'time is being lost and money is being lost because the migration of the different fish ... is unknown to Newfoundlanders ... If more detailed information was available ... fishermen [could] locate the fish, and not await their coming.'[71]

There were no notions of conservation in this shared research emphasis; indeed, for the Americans, the North American Council was in part, to begin with, a vehicle promoted by the scientists involved to establish a demand for fisheries scientists; so far, few were involved in this research, due to minimal government support.[72] With this new international organization, the U.S. government would have to add more scientists to the staff of the Bureau of Fisheries in order to meet its new international obligations. But, as will be seen in the next chapter, there were few fisheries scientists of a high enough professional stature to truly meet the needs of the organization in the United States: the 60 per cent increases in pay that followed U.S. Commissioner of the Bureau Henry O'Malley's successful argument in 1928, that fisheries biology was emerging as an important natural science,[73] could not immediately supply the trained fisheries scientists that would be required in order to accomplish all that the North American Council envisioned.

The International Committee on Marine Fishery Investigations (after 1930, the North American Council on Fishery Investigations) held its first meeting in Montreal on 23 June 1921. A program was sketched out for cod and haddock tagging and for studies of fish life histories, catch statistics, and plankton using 'European' methods.[74] American efforts were organized by the U.S. Bureau of Fisheries' Division of Scientific Inquiry, which had by 1930 a staff of sixty-five biologists (forty-three during the Depression). This division's chief, Elmer Higgins, was a council member throughout its history, together with Henry Bigelow and the U.S. Commissioner of Fisheries (H.M. Smith in 1921, Henry O'Malley from 1922 to 1933, and Frank T. Bell from 1934 to 1938). Oscar E. Sette, Director of the

Fisheries Laboratory in Woods Hole, was in charge of North Atlantic investigations. For Canada, Huntsman coordinated the Atlantic Biological Station's work with the council until replaced as director in 1934 by A.H. Leim.[75]

By 1922, each country had appointed up to three representatives – the Canadians were Found, Huntsman, and J.P. McMurrich – and twice-yearly meetings were being held (once a year after 1930) at different cities in Canada, the United States, and (once) Newfoundland. Much of the responsibility for the direction of the NACFI fell to Bigelow and Huntsman. In 1922, Huntsman was appointed permanent secretary, to ensure continuity in the council's records and documents. In 1923, Bigelow was elected permanent chairman.

Henry Bryant Bigelow (1879–1967) was then a professor of zoology at the Museum of Comparative Zoology of Harvard University. A former student of Alexander Agassiz, he had taken part in several of the Agassiz expeditions. He was an authority on *medusae* systematics and distribution as well as one of the U.S. Bureau of Fisheries' top scientific advisors. This enabled him, starting in 1908, to undertake oceanographic investigations ranging from Halifax to the Delaware. His Gulf of Maine studies in the 1910s and 1920s resulted in three classic monographs on these waters' fishes (1925), plankton (1926), and hydrography (1927). When the Rockefeller Foundation wanted to establish an American oceanographic institution, Bigelow was asked to report on the proposal. Later he would be named the first director of the Woods Hole Oceanographic Institution (WHOI).[76] The NACFI thus carried out a certain amount of research under the auspices of the WHOI.[77] Bigelow retired in 1938 from both the council and the directorship of the WHOI.

Although NACFI had been created in emulation of ICES, it must be noted that in fact, ICES had done little yet to advance fishing theory. Its scientists were hampered by their inability to use catch records from the research trawlers' highly variable samples, which were considered too small to be representative until R.A. Fisher introduced small sampling statistics in 1925.[78] Furthermore, H.M. Kyle, ICES's statistician, in 1905

> attacked the idea that 'statistics of the average catch per boat were alone sufficient to prove the existence of overfishing and coming exhaustion of the fisheries' ... Kyle thought that the catch per unit of effort might be reduced merely because there were more vessels in the area, independently of any change in stock density ... He also disputed ... that the decrease in size of fish was a sign of overfishing.[79]

The problems of small-sampling statistics, and Kyle's attack, meant that Garstang's work would not be followed up for two decades.[80] However, ICES did help standardize the highly varied catch statistics of member countries, which seldom recorded data relating to 'fishing effort.'[81] The Scottish Fishery Board's system, which did record water depths, the location and amount of fishing, and proportions of catch by size of fish, was adopted and modified by D'Arcy Thompson in 1909, when he reformed ICES's fisheries statistics.[82]

On the other side of the Atlantic, NACFI saw the importance of fishing effort as an indicator for overfishing. However, in 1920 there was not even any consistency in how each country's basic catch was reported: the French recorded the weight of green-salted cod; Newfoundlanders, the weight of dry-salted cod; Canadians listed fish 'caught and landed' (often gutted and beheaded); and Americans, 'fresh' and 'salt' fish. Accuracy was hampered by fishermens' reluctance to divulge their secrets. In Britain, G.C.L. Howell recorded one as stating: 'We do not give [statistics collectors] any information; ... We are afraid of his giving information to our neighbours.'[83] In 1921 the North American Council recommended a form, with a section for vessel captains to record the vessel's daily position and amount of gear used. This met with considerable resistance in Canada and had to be abandoned.[84]

In spite of this, NACFI brought about many improvements in member countries' catch statistics. In 1925 it recommended including information on the number of men and days spent fishing,[85] as a step toward approximating fishing effort. In 1931 it asked that the catch be recorded by separate species, as caught fresh before dressing. Also, since the council had, from its inception, approved 'informal contacts between it and the International Council for the Exploration of the Sea ... for the purpose of exchanging information,'[86] it readily complied with a 1930 request from ICES to adopt the European system of dividing coastal fishery waters into statistical areas.[87] By 1932, it had a map of the Northwest Atlantic indicating such statistical regions. Newfoundland soon adopted the council's statistical recommendations.[88] However, Canada lagged behind the United States in relating catches to gear, fishing craft, and fishermen,[89] and New England fisheries statistics were collected only every two years.

NACFI's approach was informally to coordinate investigations on problems of international interest, and to make the results available to all members. Research to find a method to predict stock abundance focused on cod, haddock, and mackerel. Cod was commercially the most important, and received most attention, as 'unexpected fluctuations in its abundance frequently entailed very considerable hardship' for fishermen.[90] The Americans perceived the Atlantic mackerel fishery to be a fishery in trouble, and in 1924 a cooperative study of the mackerel was also proposed.

In investigating a commercial fish stock, biologists must first establish a species' distribution and general life history. Constant environmental appraisals should be done to reveal how fish respond to changes in temperature, salinity, currents, and so on. Then estimates must be made of recruitment and mortality. The council encouraged a number of Canadian and American studies on feeding and growth responses of fish and fish eggs to different temperature and salinity regimes in laboratory tanks.

The council's first major effort involved tagging cod to study their migration patterns. Thousands were tagged: the Americans alone tagged 47,187 cod from 1923 to 1932. After 1925, mackerel studies also became important. Sette had field technicians sample mackerel catches in trap nets; this investigation intensified in 1926 to include records of catches in terms of abundance, fish lengths, 'scale counts,' and so on. The American studies revealed that most of the fish had spawned in 1923 and had reached the length of 37 inches by 1926. As no strong year-classes followed the 1923 year class, and the 1923 group declined in numbers, Sette used the data gathered to predict the immediate future success of the mackerel fishery. Unfortunately for Sette, his predictions for a poor 1929 catch were wrong, because the 1928 year-class proved to be exceptionally strong and fast growing, and the catches increased.[91] This highlights one of the problems of fisheries science: it is exceptionally difficult to predict all of the factors involved in population growth and reproduction.

Nevertheless, Sette is to be credited for attempting one of the most intensive fishery studies ever conducted in North America. In 1926 and 1927 he began a series of cruises to sample the abundance of mackerel fry and eggs. Later he designed further cruises, from the mouth of Chesapeake Bay to Cape Cod, to test Hjort's suggestion that the number of eggs spawned by a fish population did not determine the strength of the ensuing year-class. As outlined in chapter 3, Hjort thought the that abundance of food, and strong and unpredictable current fluctuations, might play a role. Sette located the mackerel spawning grounds, and by 1931 'had perfected his sampling procedures' so as to set fairly uniform parameters at different places and times. He wanted to undertake annual experiments on the 'survival rates of mackerel from the egg through their first three months of life.' Unfortunately for Sette, just after he had made nine of eleven planned cruises in the spring and summer of 1932, the Research Vessel *Albatross II* was taken out of service, due to austerity measures imposed by the American government following the onset of the Depression. These austerity measures were not lifted for many years. Although WHOI helped him carry out the remaining two cruises that year, the long-term program was lost. Intriguingly, his results indicated that there was no correlation between poor year-classes and high mortality rates in the larval stages of mackerel.[92] The question remains unresolved to this day.

For Canada and Newfoundland, investigations of the cod stocks and of the effects of the fisheries were of greater interest. To relate the cod fisheries to oceanographic features, Huntsman led the Belle Isle expedition off Newfoundland in 1923, under the auspices of NACFI (see chapter 3).

A summary of northwestern Atlantic cod catches by Newfoundland, France, Canada, Portugal, and the United States, 'running, respectively, back to ... 1804, 1874, 1869, 1896, and 1880,' was prepared by Sette, for whom statistical studies were a priority. This afforded 'no evidence of any marked alteration in the productivity of the fishes.'[93] Newfoundland scientists found that since about six annual broods composed the Grand Banks cod stocks, fluctuations were 'very much smoothed down ... There is little reason to anticipate striking changes.'[94]

Newfoundland, in fact, offered little help until 1930, when it opened its Bay Bulls Laboratory, following a survey of its fisheries by its government and the British Empire Marketing Board. Station director Harold Thompson then joined the Minister of Marine and Fisheries as council representative for Newfoundland. The cod-fishing research of the *Cape Agulhas*, a trawler equipped for hydrographic and plankton-collecting work and trawl and other fishing, 'definitely' showed 'that the occurrence and size of cod depended upon the temperature.'[95] Thompson hoped to establish a cod fishery forecasting service, based on weekly temperature reports from stations around the coast.

This work was aided by French scientists. In 1922, because of its important Grand Banks fisheries, France asked to join the council. Edouard Le Danois (1887–1967), then assistant director of the Scientific and Technical Marine Fisheries Office, wrote that 'the French adhesion' would profit 'the committee of Canado-American Researches' by introducing 'scientifical methods already tested' and 'the French oceanography being conducted on the Banks of Newfoundland by a staff already specialized.'[96] This request was promptly approved, and Le Danois became France's representative until the council was informally dissolved in 1939.

Portugal was invited to join in 1925, because of its important Grand Banks fisheries, but declined the offer.

Le Danois, an expert in the service of ICES, had a great deal to impart. Consider, for instance, the sophistication of his hydrographic work relative to North American endeavours. In 1921, Bigelow and Huntsman had recommended studies of currents with drift bottles. Navigationally important currents were well known, but not their long-term fluctuations, which affect 'floating fish eggs, fish larvae, planktonic food organisms ... and life conditions in general.'[97] The Biological Board, with Newfoundland, France, and the International Ice Patrol, undertook drift bottle experiments from the Bay of Fundy northward; the U.S. Bureau of Fisheries took charge of those to the south. By 1930, more than 12,000 bottles had been released and had been swept throughout most of the North Atlantic, even as far as Africa's northwest coast. The council also began keeping records of surface water temperatures: 'No physical factor,' it stated, 'is of greater importance in determining the time, nature, and extent of the fisheries.'[98] By 1928, fifty lighthouses and lightships were involved, from Maine to the Gulf of Mexico. Canada was taking morning and evening water temperatures at twelve stations, including the two biological stations; the Quebec Department of Fisheries followed suit in the Gulf of St Lawrence. In 1926 the Biological Board installed thermographs on two Canadian National steamships on the Bermuda and West Indies runs, to gain information on seasonal shifts in the Gulf Stream.

By contrast with this important but rather simplistic work, after 1922 Le Danois began investigating ties between hydrographical conditions and the Grand Banks's cod fisheries. Happy to devote himself to practicalities, he had the French fisheries department take comprehensive fisheries information 'and print ... it in clear and simple language ... with maps which make his meaning unmistakable.'[99] In 1927, Commander Beaugé took over and also became a member of the North American Council. In 1929 and between 1931 and 1935, French hydrographic studies on the Grand Banks and around southern Greenland verified Le Danois's prediction that when warm conditions depressed Newfoundland's fisheries, Greenland's banks tended to be flooded with waters suitable for cod. He characterized warm years as disastrous: waters warmer than 10°C would flood the banks, driving cod away to colder waters. Also, cod liked water with a salinity of 33 to 35 parts per thousand and temperatures of 3° to 5°C. Moderate years would leave the bottoms suitable for cod, while warming the upper layers, where squid (the bait species used by fishermen) would then abound. In cold years there would be plenty of cod but an absence of bait. Beaugé and Le Danois succeeded in forecasting the success of southern banks fisheries from hydrological studies conducted earlier in the year. Le Danois recommended that fishermen learn about the bottoms of the grounds, and how to locate 'cod water.'[100]

It is noteworthy that this research involved environmental factors affecting the cod fishery, not overfishing. Tellingly, overfishing was not the focus of the single most important project undertaken for Canada by the North American Council on Fishery Investigations: the International Passamaquoddy Fisheries Commission of 1931 to 1933, the subject of the next chapter. To be sure, Oscar Sette and some of the American fisheries biologists were not followers of Huxley's ideas, having witnessed severe fluctuations in the mackerel and other catches in American waters, which were less prolific than the Canadian banks; they were also influenced by the prominent Pacific fisheries biologist William F. Thompson, a graduate of Stanford, who developed fisheries investigations at the California Fish and Game Commission, and who later headed the International North Pacific Halibut Commission. Thompson was a vociferous opponent of Huxley's ideas – the record of the Pacific halibut fisheries and the fate of the Pacific salmon, chased to near extinction in the American salmon rivers, gave the lie to any theory of inexhaustibility. Tellingly, it was the American researchers with the U.S. Bureau of Fisheries who carried out NACFI's experiments on large-mesh trawl nets, to reduce the catch of immature, undersize haddock, an outgrowth of American concerns that the haddock stocks were under stress. The Americans were not blinded by any sentimental bonds to the British Empire and its leaders and intellectuals; in contrast it is highly probable that the still-strong ties with the British Empire and culture felt in Canada reinforced the respect for Huxley's ideas among Canadian scientists.

But the Americans were not alone in setting the agenda of NACFI. The Huxleyist Huntsman, with his enormous influence on Canadian fisheries science in that period, had no interest in the idea of overfishing, and to the end of his life was sceptical of the idea. Nor, it must be reiterated, was the spectre of stock collapse yet even a remote possibility for the Grand Banks fisheries, given the fishing intensity and technologies of that era. Here the fish stocks were in a healthy state, and all but three Canadian trawlers had been banned. In any case, with control of the fisheries only out to a limit of three miles, concerns about the state of the international offshore fisheries required international attention, so the Canadians' efforts were necessarily directed more strongly toward concerns closer to shore.

But other forces now appeared to threaten an important Bay of Fundy fishery, and Huntsman involved the entire North American Council on Fishery Investigations when it appeared that a proposed tidal power project would threaten the environment that was supporting some of the inshore fish stocks. As will be seen in the next chapter, which discusses the history of the International Passamaquoddy Fisheries Commission, the Canadian rejection of overfishing did not stem from indifference to the health and survival of the fish stocks. On the contrary, there was deep concern, closely linked to the desire to preserve and aid the fisheries themselves.

Chapter 7

An Environmental Assessment: The International Passamaquoddy Fisheries Commission, 1931–1933

The most concerted activity undertaken by the North American Council on Fishery Investigations was the International Passamaquoddy Fisheries Commission of 1931–3. This was overseen primarily by Henry Bigelow and A.G. Huntsman, and involved an in-depth investigation of the circumstances surrounding the abundantly productive herring fishery of the lower Bay of Fundy, in particular Passamaquoddy Bay, which was being targeted for a tidal power project. The ensuing investigation opened great opportunities for Canadian researchers to be in the vanguard of further development of fishing theory, and of new mathematical, statistical approaches to analysing the effects of fishing on fish stocks. As will be seen, this commission brought to Canada one of the emerging leaders in the theoretical development of fishing theory, the British Ministry of Agriculture and Fisheries's Michael Graham, the principal naturalist at the Lowescroft marine station. The opportunities offered by this contact were ignored, however – which reveals a great deal about the preoccupations of the Biological Board and its scientists in the 1930s. Since fishing theory focuses on conservation, and on managing fish stocks that are in danger of being overfished, the Canadian scientists did not become intellectually engaged by the new research program, which was to increasingly dominate European fisheries science.

Their lack of engagement was also due to several features of the board's Atlantic Canadian enterprise: first of all, the one scientist who might have pursued this new area, even though he was a staunch Huxleyist, was A.G. Huntsman. But Huntsman was removed from his position as station director shortly after the commission was wrapped up; without the full resources of the station behind him, he retreated into intensive studies of the Atlantic salmon. Also, monitoring fish stocks, with careful investigation of egg densities, the survival of fry and young fish, and the age profile of the catches, is – as NACFI's scientists discovered – extremely expensive. Given that the most important fish stocks were found on Newfoundland's Grand Banks, and that Newfoundland was still a foreign country, there was little motivation for Canadian scientists to become involved as yet in the new fishing theory. In addition, as shall be seen in the next chapter, there was a shortage of fisheries biologists in North America in general and this surely made it less likely that the few overworked individuals already in the field would have the energy and enthusiasm to follow the new developments closely. Also, few of the older biologists were willing to deal with the mathematical challenges posed by the new fishing theory. Finally, at the time, few Biological Board scientists on Canada's Atlantic coast were ready to admit that overfishing was possible; as it turned out, Huxleyist beliefs were to persist long after Huntsman.

The International Passamaquoddy Fisheries Commission of 1931–3 focused on a very specific problem.[1] Would hydroelectric tidal power dams between Passamaquoddy and Cobscook bays and the Bay of Fundy adversely affect the fisheries? The Bay of Fundy has the highest tides in the world, and various schemes have been put forward to harness these tides for energy. The earliest one involved two bays adjoining the Bay of Fundy between Maine and New Brunswick. Passamaquoddy Bay, lying within Canadian waters, and the adjoining smaller Cobscook Bay, entirely within Maine, are ringed with islands, and in 1919 Dexter P. Cooper became interested in building hydroelectric dams between these islands. Passamaquoddy Bay was to be filled during high tide and Cobscook Bay emptied at low tide; the upper pool would be discharged into the lower pool and also directly into the Bay of Fundy. On 2 January 1924, Cooper applied to the U.S. Federal Power Commission for a permit to build the dams. Maine granted him an act of incorporation in April 1925.[2] On 15 June 1926, he was also granted a charter by the Canadian federal government, which, however, stipulated that construction in Canada would have to commence within three years and be completed within the following three years.[3]

As early as 1927, Huntsman was theorizing that damming the Passamaquoddy system would make the Bay of Fundy a more perfect funnel, and remove the slowing friction created by water flowing around the islands ringing these bays. He also predicted a one-foot increase in the high tide at the head of Fundy, which 'might make a difference for the dyked land' there. He engaged Henry B. Hachey, then a lecturer in physics at the University of New Brunswick and later the 'dean of Canadian oceanographers,' to study the problem at St Andrews. Hachey's research revealed that indeed, 'the system of channels at the mouth of Passamaquoddy Bay offer a definite drag on the oscillation of the Bay of Fundy.'[4]

Meanwhile, opposition to the project mounted. Fishermen and fish processors turned to Huntsman for support. In February 1928, Huntsman testified before the Maclean Commission, predicting considerable damage to the Passamaquoddy fisheries from the dams, especially to the herring, clam, and pollock fisheries.[5] He also predicted that the oceanography, ecology, and climate of the entire Fundy region would be altered.[6] In spite of this, New Brunswick's legislature approved the project.

Huntsman next published his testimony in serial form in Saint John's *Telegraph Journal* (later, it would be issued as a pamphlet). His main concern was that the dams would destroy the exceptional herring and pollock fisheries of Passamaquoddy Bay. He hypothesized that the bay had unusually high fertility, owing to the mixing of surface waters with nutrient-rich deep seawaters brought up by the 'Passamaquoddy mechanism.' Also, he figured that these conditions were circulated along the Maine coast and across to Digby, Nova Scotia; thus, herring fisheries in these other regions might also suffer. But even if these did not, the Passamaquoddy fishery, as he passionately argued, was well worth saving in itself: its 1919 total catch was over twice the catch of the next richest Fundy area, that of Knox, Maine, and six times that of Digby, Nova Scotia. He argued that Cooper's dams would also destroy Passamaquoddy Bay's clam beds, which steadily provided about 75 per cent of the clams from the Bay of Fundy. Cod and haddock would also be reduced.[7]

After New Brunswick's legislature approved Cooper's project, Huntsman drew international attention to the proposed dams. He sent his pamphlets to the Bureau of Fisheries in Washington, D.C., which brought the matter before Congress.[8] Huntsman then referred the question to the North American Council on Fishery Investigations.

Thus began a series of headaches for Bigelow. At the North American Council's October 1928 meeting at the University of Toronto, Found, on behalf of Canada, asked the council to make a 'pronouncement ... as to what would happen to the fisheries, if the proposed dams were constructed.' Bigelow, absent due to illness, was named to a committee with Huntsman, Found, and O'Malley, to 'study all available information bearing upon the predictable effects of the project on the fisheries of the region, and report findings to the Governments of the United States and Canada.' Bigelow complained: 'It was unlucky for me that I couldn't come ... If I had I should certainly have managed to wriggle out of ... the jobs you-all have wished onto me!' He felt that 'to attempt to predict quantitatively the amount of damage ... is a very different thing from foreseeing the direction of change,'[9] and beyond their mandate.

A meeting of this committee took place at Eastport, Maine, on 12 December 1928, attended by Canadian and American hydraulic engineers as well as by Dexter P. Cooper and the Maine Commissioner of Sea and Shore Fisheries. Neither Huntsman nor Bigelow was pleased with the outcome; it was declared that although the inshore weir fisheries would most likely be 'wholly eliminated,' 'the committee as a whole is not prepared to forecast whether these results will or will not follow, believing that a fuller investigation is needed.'[10] Huntsman was 'fearful that a hasty decision might be reached of the nature of "I don't know," which ... would quite fail to ... get a reasonable prediction or opinion.' Bigelow felt that the affair was distorting the council's purpose: 'The Committee was organized for the purpose of coordinating and encouraging the scientific investigations of the various Governments. As long as we stick to that we do useful work, and get along famously. If we spread our activities beyond that ... it is hard to see where this might land.'

Bigelow wanted to avoid giving either advice or predictions in the matter. As the discussion dragged on, Bigelow, annoyed, and fearing the council would be drawn into advocacy issues, complained, 'I am beginning to think that there is far too much Eastport in my cosmos!' Huntsman tried to sympathize but could not really: 'I can at times reciprocate with your views regarding ennui in this Passamaquoddy matter, but when I recover I find it again extremely interesting.'[11] In truth, Huntsman recognized that this was his chance to tie together diverse interests and make the Atlantic Biological Station the centre of an international scientific research project.[12] If he could interest enough scientists and fish processors, and the American and Canadian governments, and thereby attract enough money to support the station's work, perhaps he could make the Bay of Fundy an important locus for ongoing research into the problem of how the environment 'creates' important fisheries. For Huntsman, this was perhaps the central question of fisheries biology.

In February 1929, Bigelow told Huntsman: 'The Passamaquoddy Pot begins to boil again.' Found's department was being asked about 'its attitude in the matter.' On 24 March 1929, NACFI issued a statement that the herring fishery inside the dams would be wiped out. As to Huntsman's predictions of the effects outside the dams, the council was 'not prepared to forecast whether these results will or will not follow, believing that a further investigation is needed.'[13] Huntsman, in discussing the proposed investigation's purpose with Bigelow, let slip some very revealing remarks: 'You mentioned at Eastport the matter of prejudice, and I have been trying to get that clear in my mind. We are expected, I presume, to be prejudiced in favour of the fisheries, to be speaking for them only ... without ulterior influence.' He protested that Bigelow's concerns about his bias were misplaced: 'Certainly there has been no political influence, no influence from power interests, and no personal influence unless from you.'[14] A review of Huntsman's extensive correspondence clearly reveals that he single-handedly raised the entire concern and agitated for an environmental assessment, driven by an immensely strong passion to save the fisheries around the St Andrews Biological Station. Huntsman could not possibly remain neutral when the fisheries might be at risk.

In the meantime, Cooper had failed to begin construction within the three years stipulated by the Canadian government, and in 1929 he applied for an extension. The Private Bills Committee turned him down. On 15 May 1929 the United States Legation in Ottawa asked Canada to reconsider the extension, suggesting a fishery commission to investigate the project. Thus, in July 1929, NACFI 'recommended a two-year study [and] proposed that the two governments share the costs ... estimated at $45,000 per annum.' The two governments agreed and appointed an international commission 'for a two years investigation of (1) the herring, (2) the zooplankton, (3) the phytoplankton, and (4) the hydrography.'[15]

Four scientific specialists were required. The first choice for the zooplankton specialist was Dr Charles J. Fish of the Buffalo Museum of Science, who had worked for the Bureau of Fisheries for many summers. As negotiations for his appointment went forward, Bigelow gleefully remarked: 'Fish can probably be landed.'[16] Indeed, a two-year leave of absence was readily obtained by him. As hydrographer, the commission chose Dr E.E. Watson, a professor of physics at Queen's University in Kingston, Ontario. There was a drawback, however. Watson would only be available when the university was not in session. The station's resident oceanographer, H.B. Hachey, was not considered; station staff were not to be engaged directly by the commission.

The other positions proved harder to fill. Limnologist and ecologist Dr Chauncey Juday (1871–1944) was asked to become the phytoplankton specialist, and accepted. Juday was a good choice. He had many years' experience assisting University of Wisconsin zoology professor and limnologist Edward Birge in studying Lake Mendotta and other Wisconsin lakes. Together they changed the scientific understanding of lake structures. They determined that water at different levels had strikingly different temperatures, and that these warmer and cooler waters stratified lakes into clearly defined layers. Juday later tried to come up with a picture of how an entire lake functions: he wanted to determine the 'energy budget' (or flow of energy) through an entire lake ecosystem, to see what proportion of sunlight energy was absorbed by photosynthesis, and how much energy was lost at each level of transfer to herbivores and carnivores.[17] He was eminently suited to investigate and analyze the productivity profile of the Bay of Fundy and Passamaquoddy Bay. Unfortunately for the commission, however, soon after signing on he was offered a permanent position at the University of Wisconsin, and backed out of the project. Bigelow and Huntsman settled on the well-published Haaken Hasberg Gran (1870–1955), professor of botany at the University of Oslo, Norway. Gran had gathered planktonic material for his doctoral work during cruises of the *Michael Sars*, which led to his lifelong collaboration with Johan Hjort. He was appointed to the chair of botany in Oslo in 1905, and became director of the Norske Videnskaps-Akademi in Oslo. He had also worked up the phytoplankton results of the Canadian Fisheries Expedition of 1914–15. His interest lay in correlating changes in phytoplankton populations with changes in the local environment,[18] and in discovering how light intensity affected productivity.[19] The promise of a salary and the nature of the problem convinced him to come. However, he was unable to arrange a two full years' absence,[20] and would have to travel back and forth between Norway and New Brunswick.

The most important position to fill was that of fishery investigator. This turned out to be a challenge. A.E. Parr, curator of the Bingham Oceanographic Foundation of Yale University, was tentatively hired, but proved to be too concerned about controlling all aspects of the commission's work. By the time the Canadian appropriation came through on 20 April 1931, a specialist had not been lined up. Huntsman, meanwhile, made the best of things. To provide background information, he set 'all available personnel' at the Atlantic station to make 'an intensive study of the Bay of Fundy fish and fisheries.' With a new research vessel, *Zoarces*, Hachey and J.M. Morton studied oceanography. He himself investigated herring. R.A. McKenzie looked at haddock and alewives; C.L. Newcombe, clams; V.D. Vladykov, pollock, halibut, and flounders; W. Templeman, lobsters; and Dr Helen I. Battle of the University of

Western Ontario, hake, cusk, and mackerel. Others studied Passamaquoddy's scallops, cod, smelt, salmon, and seaweeds.[21]

It was probably Huntsman who suggested that Dr Michael Graham (1898–1972) be offered the position of chief herring investigator. Graham, although he had not yet published his most important work, was among the world's foremost fisheries biologists – as Bigelow put it, he was 'on the map.'[22] A graduate of Kings College, Cambridge, he was assistant naturalist at the fisheries laboratory at Lowestoft. His appraisal of the North Sea cod fishery, published in 1933, 'established the cod's life cycle, its spawning grounds, its age composition in the fishery and the identity of its stocks.' In this, and through his important 1935 and 1939 contributions in the *Journal of the International Council for the Exploration of the Sea*, in his Buckland lectures of 1939, and in *The Fish Gate* of 1943, 'he proceeded to develop ... the theory of fishing.'[23]

Graham responded enthusiastically to Huntsman, setting out a research program, and concluding: 'Good luck to it; and may I have the pleasure of being one of the team.' However, Henry G. Maurice, head of the Board of Agriculture and Fisheries, felt unable to spare him. Huntsman told Maurice that since 'Graham is [already] engaged in a study of herring,' he and Bigelow believed 'that there might be very considerable advantage for both investigations if he took charge of them concurrently, spending part time at each.' The chance to have 'a good man compare simultaneously the very different conditions for the herring existing on the two sides of the Atlantic is too good ... to be missed.'[24] A few days later, Graham's help was secured.

By June 1931, Watson had assembled equipment at the commission's headquarters at St Andrews and had begun observations. Graham arrived in September with his wife and two young children. Fish set himself up at Woods Hole, where as executive secretary he could coordinate the commission's efforts with Bigelow's Gulf of Maine work. Gran also arrived in September. Dr Martin W. Johnson was hired as Fish's assistant, while Gran brought over Trygve Braarud of the University of Oslo. Charles Hughes of Fredericton was hired as field assistant hydrographer, and Charles Bates assisted Graham.

The commission was beset by disasters, not the least of which was the onset of the Great Depression. The commission's budget for the final year was reduced from $45,000 to $37,000, which shortened data collecting. Gran wrote a pathetic letter from Oslo on 29 February 1932, to tell Huntsman that his wife had just died. Huntsman's reply of condolence broke stunning news. On 8 March 1932 'we had a very serious loss in the burning down of the main laboratory at St. Andrews.' Graham, Hughes, and Bates had stored their things there over winter, with resulting serious losses.[25]

The commission lost not only equipment, but also samples and records: 'a part of Graham's data and collections and all routine zooplankton collections from February 1st to the date of the fire.' Fortunately, Graham had already examined his plankton and had records aboard the research vessel *Nova.* Fish's seasonal sequence was broken, but this 'will not prove a serious matter,' wrote Huntsman, 'as I have records of several former years which will serve as a check.' Gran's winter collections were gone, but it was Watson who 'suffered most ... All records of Hugh's hydrographic work were lost together with a large part of Watson's own records ... All of his work of last summer was a complete loss.'[26]

Huntsman was determined that the work continue. Laboratory work would be 'carried on in the Fish handling building, which is heated with a stove.' St Andrews' station staff were 'using the Residence kitchen as a laboratory where Braarud and Bates seem quite at home,' and he hoped to 'get a stove for the hatchery so that we can all use it immediately.' He gave Graham and Fish copies of Hachey's hydrographic papers to make up for Watson's lost reports. Bigelow invited Gran to the Woods Hole Oceanographic Institution as guest investigator. He accepted: the arrangement would be good for the commission and 'for ... cooperation in future between scientists on both sides of the Atlantic.' At St Andrews, Braarud continued collecting and making chemical determinations of plankton nutrients; Gran also spent part of July and August there 'to get unity in the work.'[27]

By October 1933 the International Passamaquoddy Fisheries Commission's report had been submitted to both governments. In the joint statement, the dams' physical effects were projected to be minimal: 'The physical effects of the present mixing mechanism appear to be local and although ... the dams would influence the hydrographic conditions in the passages, it is not expected that their influence would extend far beyond the outer Quoddy region.' Watson found that most of Fundy's waters were mixed at the head of the bay, where the tidal streams are especially violent. Water within the Quoddy Passages was highly homogenous, but it was stratified only four miles away at the heart of the Outer Quoddy region.[28] Similarly, the dams would have a minimal effect on phytoplankton. The mixing of water layers brings nutrients from the depths to surface waters, where plants use them. Gran found that in the Gulf of Maine, increasing water stratification from spring to late summer 'leads to a depletion of nutrient salts in the upper layers [to 400 metres] ... Where light is sufficient for plant growth ... The production comes to an early standstill ... the greatest production of plant plankton was found in ... mixing areas.'[29]

However, the Bay of Fundy had a high degree of turbidity caused by violent tidal movements and large amounts of silt; this resulted in only a narrow band of surface water, 20 metres at most, into which enough light could penetrate for plant growth. Gran concluded that a moderate reduction in mixing would improve productivity, as would occur in outer Quoddy waters if the dams were built.

As for the fisheries, Graham discovered that 'the rich fishery in the Quoddy Region is not due to a localized abundance of zooplankton.' Indeed, the zooplankton supply was found to be 'mainly produced in areas beyond the influence of the Quoddy mixing mechanism and transported passively by ocean circulation into the region,' and 'any influence of the proposed dams upon this supply would probably be insignificant.' Since the herring also originated elsewhere, the dams would have no deleterious effect on Bay of Fundy herring, or on fisheries 'along the coast of Maine or even seriously at Grand Manan.' However, 'the herring fishery inside of Passamaquoddy Bay would almost certainly be reduced to negligible proportions.'[30] The commission supported Huntsman's predictions of events inside the dams.

The summer zooplankton of Passamaquoddy Bay was found to be mostly of local origin and restricted to the bay itself. In the winter, zooplankton was carried in from outside waters. Fish concluded that the effect of the dams 'on the outer waters would not be noticeable,'[31] or would even be good for zooplankton, because of reduced turbulence.

Graham could not explain why Passamaquoddy Bay had an especially rich sardine fishery, but he supported Huntsman's contention that the Passamaquoddy sardine fishery was exceptional. However, this was not due to waters outstandingly rich in food for the fish. Rather, 'the disproportion in landings of sardine herring ... as compared with the rest of the Gulf of Maine, Bay of Fundy and Nova Scotian waters is partly to be explained on the grounds of especial ease of capture in the sardine area.' He predicted that 'the exceptionally rich fishery ... inside the dams would be reduced to almost negligible proportions, since it seems dependent on immigration.' On the other hand, outside the dams, 'there is little possibility of a wide-spread effect.'[32]

Four official papers resulted from the commission. Gran's was 'highly important ... to our knowledge of conditions of plankton production in northern waters,' in that it described 'how production would be limited if phytoplankton cells were constantly mixed below the depth where their respiration was just equal to the photosynthetic rate.'[33] Other papers resulted from parallel investigations by Biological Board scientists.

The International Passamaquoddy Fisheries Commission was undoubtedly the most impressive work done under the aegis of NACFI. However, as pointed out at the beginning of this chapter, it had nothing to do with the problem of overfishing. Rather, it amounted to a fairly thorough environmental assessment, designed, at the instigation of Huntsman, to thwart what he and other scientists at that time thought was a serious threat facing an important fishery. He regarded himself as an advocate for the fisheries and fishermen, and his passionate defence of this economic sector and the natural resources that enabled it to flourish shows clearly that the position taken by Canadian scientists in support of Huxley's views did not mean they were indifferent to the health of fish stocks or the fisheries. They just thought that in most cases, excluding the migratory salmon, overfishing was extremely unlikely. This position led to a high degree of indifference to the new fishing theories that began to emerge in the 1930s elsewhere, as described below.

Fishing Theory

While research into overfishing languished in Europe in the 1920s, for lack of proper statistical tools, and got little attention on the Canadian and American Atlantic, on the Pacific it was being pushed forward by William F. Thompson, a Stanford-trained zoologist, who viewed the purpose of fisheries research as being 'that which is necessary to the perpetuation and prosperity of the fishery.' Stanford had established close ties to the California Fish and Game Commission, created by the California legislature in 1870. This link, and the commission's interest in the scientific study of the fisheries, resulted in graduates like Thompson, who had the ability to focus on the key problems of the fisheries. Thompson kept abreast of European developments and incorporated the findings of Hjort, C.G.J. Petersen, and others into his work.[34]

Beginning in 1915, when he was hired by the California Fish and Game Commission, and conjointly employed by the government of British Columbia, Thompson investigated the decline in Pacific halibut, which he was certain was due to overfishing. Like Garstang, he recognized that higher fishing effort was directly linked to lower stock density,[35] but he also understood the importance of studying natural stock fluctuations. Following Thompson's work, an international joint commission met in 1918. As a result of this, in 1923, Canada and the United States signed a treaty that established the International North Pacific Halibut Commission in 1924, with Thompson as its first director.[36] His principles for regulating the halibut fisheries were based on fish statistics and tagging experiments.

In the meantime, NACFI was trying to establish basic patterns of fish population fluctuations on the Atlantic coast. For cod and mackerel the ultimate aim was to predict the success of a season's fishing. For example, Oscar Elton Sette's mackerel year-class studies, begun in 1925, showed that successful year-classes were uncommon, and that the cause of catch fluctuations was the different sizes of different year-classes. Studies of the relative abundance of year-classes, their distributions, and their effects on commercial fishery success gave the basis for 'reasonably reliable' annual predictions of mackerel abundance, which appeared in U.S. Bureau of Fisheries circulars from 1931 onward.[37] Sette's efforts to uncover links between the number of eggs spawned and year-class success began in 1932, but came to a quick end (as recounted in this book's last chapter) owing to government research cutbacks and the withdrawal of the research vessel *Albatross II* as an economy measure. Even so, his analysis of egg and larval mortality rates indicated that fewer than one fish would survive from one million eggs laid, although later fishing records would show that his results were based on an exceptionally poor year resulting in a weak cohort.[38]

It was through its haddock investigations that NACFI become involved in overfishing research. In 1929, A.W.H. Needler prepared a synopsis of western Atlantic haddock fishery statistics; the haddock catch had been increasing since 1903, due chiefly to otter trawling. There was 'no evidence of depletion,' but as there were also 'no data on the fishing effort, this was inconclusive.'[39] In fact, there were increasing numbers of trawlers using more efficient fishing gear. In 1930 the Americans began analyzing statistically the haddock fisheries and haddock year-class compositions. They were concerned by the haddock decline on the Georges Bank, and they suggested that this problem would be 'of growing importance to Canada [as] a continued and intensive fishery can be expected on the Nova Scotia banks.' The capture and destruction of immature haddock, which at times reached 75 per cent of the total catch, constituted 'a complete economic and biological loss.' Gear studies during 1931 and 1932 showed that 'the use of the proper size of mesh in the otter trawls will reduce this destruction by about 4/5.'[40]

In 1934 NACFI recommended a five-year 'joint programme of intensified research into the biology of the haddock,' and further experimental work, to determine the precise mesh size required to allow undersized haddock to escape.[41] However, Canadian haddock investigations never reflected this management-oriented program, although in 1931 and 1932 R.A. McKenzie and V.D. Vladykov did some haddock life-history studies.

The tardiness of the Biological Board's Atlantic biologists in recognizing the new theory of fishing was not due to lack of exposure to it. It is ironic that it was A.G. Huntsman himself who started this new approach to fisheries science. In 1917 he analyzed catches of plaice – a species for which a new fishery was just developing – from one net haul each from the Gulf of St Lawrence, Passamaquoddy Bay, and Chedabucto Bay. He noted that the Passamaquoddy fish big enough to be caught by the net mesh size were two years and older in the Passamaquoddy Bay, but three years and older in the Gulf of St Lawrence. Of the fish retained by the net, the most abundant were the youngest; the numbers quickly declined with increasing age. He noted that in a natural, unfished population, one should expect regular declines with each age group. But the declines he witnessed were not really that regular – fish declined precipitously in the Passamaquoddy sample at age four; but for the Gulf of St Lawrence this decline was most marked in age nine fish. Huntsman 'took the important step of ignoring the irregularities, and asked what the age distributions might look like under "constant conditions."' He illustrated his answer with diagrams that look like concave pyramids, with large bases to indicate the abundance of younger fish, and dwindling spires to indicate how the numbers progressively decreased among older year-classes. This gave a quick visual understanding of the age structures of the plaice stocks (each age group was represented by a horizontal line dissecting the pyramid, and the actual catch numbers were represented by a thicker line, which sometimes stuck out beyond the edges, so that readers could see how he used the raw data.) He used these diagrams to work out the death rates. From what he saw, he argued that when a new fishery starts, the catch is large at first, but declines as the population is fished up, until fishing effort stabilizes with the rate of input of new fish through reproduction and growth. The average fish will be smaller and younger after the fishery is established, and if overfishing occurs, there is no remedy in the fished area – rather, 'fishermen can easily increase their catches in such an event by seeking new grounds.'[42] According to Tim D. Smith, who published an illuminating history of the development of the science of measuring the effects of fishing: 'Huntsman's paper did not receive wide circulation, and did not receive the attention it deserved.' Part of the problem was that Huntsman had directed this paper, published in the first of the Biological Board's *Bulletin* series, at fishermen rather than scientists. Nevertheless, Smith contends that Huntsman's ideas influenced W.F. Thompson, whom he visited more than once at the Halibut Commission's headquarters in Seattle in the 1920s. Although Thompson's approach to 'analyzing the direct effects of fishing on the halibut population was closely related to Huntsman's analysis,' unfortunately it is difficult to establish this, as 'Thompson did not often directly reference the work of others in his publications, a habit for which others [including Michael Graham] would later chastise him.'[43]

Thompson used more standard x-y coordinate graphical representations rather than Huntsman's pyramids, and made the 'important conceptual leap' of thinking of 'the year-by-year fate of any one group of fish as a representation of the entire population in any one specific year.' He simplified by assuming, unrealistically, 'that the number of young fish entering the fishery was constant each year, and that fishing and natural mortality rates did not change.' He then used these assumptions to determine the effects of different fishing rates, using scale studies and tagging experiments to determine the effects of fishing on each year-class. Based on a priori assumptions, he determined that as fishing intensity increased, 'the numbers and weight of fish in the harvestable population' would become lower, and 'the catch rate in both numbers of fish and in weight of fish per unit of fishing effort would be lower. Thus, while the immediate effect of increasing fishing effort would be to increase catches, the ultimate effect would be to decrease the total catch. Simultaneously the amount of fish caught per unit of fishing effort would decline.'[44]

His mathematical models achieved a fair match with the decline in Pacific halibut catches following an increase in fishing intensity in 1910 and again in the years 1918–26. Since his calculations supported 'his contention that the recent increases in the halibut fishery intensity had been to the detriment of both the halibut population and the fishermen themselves,' the halibut commissioners 'agreed, and based on his analysis steps were taken to reduce the amount of fishing effort by seasonal restrictions.' Thompson's work and the effects of the Halibut Commission's measures were 'closely watched in Europe as an experiment to see if in fact the effects of too much fishing could be controlled effectively.' Thompson reduced fishing effort by lowering catch quotas to below the previous levels of yield, and through licensing. A 1930 treaty closed two nursery areas, enforced a minimum landed size of twenty-six inches, and restricted the number of boats through licensing.[45] Within a few years, catches had increased even with a greatly reduced effort. However, the catch per unit of effort jumped so spectacularly in 1931 that some scientists, including Huntsman, felt that a good year-class must have coincidentally emerged just as the regulations were put into place.[46]

The development of more complex mathematical analyses of fish populations seems not to have attracted much interest from Canadian fisheries scientists, including Huntsman himself, in the 1920s and 1930s, perhaps because they were so busy grappling with the economic problems faced by Maritime fishermen. But there is no doubt that they knew what Thompson was up to, since his work was aided by Canadian volunteers and employees of the Biological Board. Scientists at the Atlantic Biological Station were also exposed to the mathematical treatment of the fisheries, through O.E. Sette, who had trained at the California Fish and Game Commission's State Fisheries Laboratory, and who had worked under Thompson.[47] Sette was one of NACFI's most prolific contributors (see previous chapter). Other exposure to the nascent fishing theory certainly came from Michael Graham, who was on the verge of transforming it when he came to Canada for the Passamaquoddy investigations in 1931.

Graham and E.S. Russell, Lowestoft's director, were among the first to place the theory of fishing on a mathematical footing. Russell formulated the overfishing problem as the sum of annual gains in weight by recruitment and growth minus annual losses by death due to fishing and to natural causes. In 1931 he stated the axioms of fisheries science clearly: $S2 = S1 + (A + G) - (C + M)$ – where S1 is the weight of stock at the beginning of the year and S2 that at the end; A is the weight of recruits entering the fishable stock; G is the growth in weight of fish; C is the weight caught; and M is the weight of fish lost through natural causes. Russell observed that since recruitment fluctuates, the 'ideal of a stabilized fishery yielding a constant maximum value is impracticable.' However, if population fluctuations could be foretold a year or so in advance, it would be possible to adjust the optimal fishing effort each year.[48]

After Russell visited population biologist Raymond Pearl, Graham learned about the logistic equation. Graham's 'considerable achievement was to adapt the logistic equation [in 1934] to the age distribution of fishes and to the data collected on stock density and fishing effort.'[49] He started with two propositions: first, with reduced death rates the average age of a stock should increase; and second, there is a 'profitable age at which to harvest.' The maximum yield was not always the most profitable, as a further economy could still be made by reducing the number of fishermen; fewer fishermen competing for the limited resource could each take a larger share.

While Russell conceived of equilibrium as occurring when a stock's growth plus recruitment is equal to natural and fishing mortalities, Graham noted that an equilibrium can be established at more than one stock level, and that in intensively exploited stocks, the level achieved will depend largely on the rate of fishing mortality. There was, however, only one level of fishing mortality that would give the 'maximum yield.'[50] If the age at first capture was raised by one year, natural mortality was unlikely to increase in compensation, 'so yield would be no less.'[51]

Following the Second World War Graham directed the Lowestoft Fisheries Laboratory. By 1959, its staff had increased from a handful to thirty-five scientists. He devoted himself equally to the theory of fishing and service to mankind, and developed in Lowestoft the foremost school of fishing theory. Two young scientists he brought there – Sidney Holt and R.J.H. Beverton – raised the theory to new heights of sophistication. By the time Graham retired in 1958, his contributions to fisheries biology could only be described as 'massive.'[52]

Yet not all scientists agreed with Graham. Huntsman thought that regulations to preserve small fish would not necessarily increase long-term yields. He argued that these had no sound scientific basis, since older fish had higher death rates and slower individual growth rates. Thus total poundage would decline from year to year.[53] Garstang had also opposed the 'very large consensus of scientific and business opinion' that catching large numbers of immature fish reduced the stock. In fact, he believed that catching fewer small plaice would result only in overcrowding and food shortages.[54]

Huntsman was fond of pointing out that overfishing is not easy to diagnose. Natural population fluctuations, and temporary or long-term environmental changes, may also be reflected in dropping catches per unit of effort. He castigated the tendency of fishing theorists to treat every situation as simple overfishing and to impose new restrictions on that basis. Like Huxley, he reasoned: 'Capture of fish in lake or sea is really so difficult that it becomes unprofitable to fish long before the fish population has been reduced below the number sufficient to provide all the young that the region can support.'[55] This was written in the days before (un)employment insurance made it possible for fishermen to make a living even without catching many fish.

It must be pointed out that even Graham and Russell considered recruitment overfishing (causing declining stock numbers) unlikely as 'it was "never actually met with so far as fish are concerned."'[56] But at the same time they were convinced of the economic benefits of exploiting fish stocks at a maximum sustainable yield – that is, a fishing level calibrated to bring in the greatest possible catch annually without causing stock levels to go into a decline (beyond an initial mild decline due to the adjustment of the fish stocks to this level of fishing intensity).

NACFI was not a factor in post war fisheries biology. By 1935 the council's dynamics were changing. In 1934, Huntsman had been replaced as director of the Atlantic Biological Station; his successor, A.H. Leim, had less interest in the council. U.S. Commissioner of Fisheries Henry O'Malley had been an active partner in the council; Bigelow began to complain that O'Malley's successor, Frank T. Bell, showed a 'lack of interest in the Council.'[57] In 1936, Thompson resigned to become Director of Fishery Investigations in Australia. In 1937, the Bay Bulls Laboratory burned down, and few records were saved.[58] The same year, Sette went to Stanford to organize South Pacific fishery investigations.[59] The last meeting was held on 6 October 1938. There was no intention that it be the last meeting, but the enterprise had lost its vigour. Bigelow stated firmly that he wished to retire from it, just as he was retiring from WHOI, and Huntsman was made chairman.[60] A meeting was planned for September 1939, but the war broke out and Huntsman had to cancel it.[61] After the war, Huntsman and Elmer Higgins decided that 'the international situation is too confused for any formal action at present.' Huntsman advised Deputy Minister of Fisheries Stewart Bates: 'The North American Council on Fishery Investigations should be written off.' He added: 'So far as Canada is concerned, the problem ... is one of markets, not of supply of fish. We could take from the waters off our coast many times the quantity of fish we now do without endangering future production of fish in those waters. It is almost purely a question of economics – how to make fishing pay.'[62]

Fisheries biologists of the Atlantic Biological Station had no commitment to a theory of overfishing, although A.W.H. Needler, who became director of the station in 1940, was beginning to talk about Graham's overfishing models before the war.[63] On the Pacific coast, W.E. Ricker was improving the theory of fishing even before 1945, but work in this area was not addressed on the Atlantic coast until the 1950s. It was only then that international exploitation of eastern groundfish became a cause for concern. In practice, the paradigm of maximum sustainable yield has involved waiting until fish stocks are obviously overfished, then trying to maintain the fishing yield at exploitation rates near a crisis point.

In Europe, the Second World War prevented development of Graham's fishing theory by ICES until the Overfishing Convention of 1946: 'The Convention, held in London, proposed minimum landing sizes for demersal fishes and minimum sizes for trawls, except those used for herring and for shrimp, but there was no agreement to limit entry to a fishery ... The political solution to the problem of growth overfishing was to agree [on] minimum mesh sizes ... This was the major achievement of Russell and Graham.'[64]

The outcome of the Overfishing Convention was the founding of the Permanent Commission in 1953 (later the Northeast Atlantic Fisheries Commission), and the International Commission for the Northwest Atlantic Fisheries (ICNAF) in 1949. These two commissions would dominate the next era of fisheries management, which ended roughly with the establishment of the two-hundred mile limit. International regulations were dominated by the theory of maximum sustainable yield; scientists opted to increase mesh sizes rather than to limit entry to the fisheries.[65] 'The initial delight of the Beverton and Holt solution to growth overfishing was that overfishing could be cured without putting fishermen out of work.' However, this pleasure was 'premature.' In the world's predominantly mixed trawl and purse-seine fisheries, mesh regulations only work for the smallest species: the mix of species caught in a trawl 'encourages a least regulation rather than an optimal one.'[66] It must be pointed out, also, that the large mesh size helps only so long as the trawl net is nearly empty: once a layer of fish is captured against the net walls, smaller fish get trapped among the larger ones anyway. It is, at best, a flawed solution.

Canada played a prominent role in the ICNAF, the first headquarters of which was at St Andrews. A.W.H. Needler and D.B. Finn had attended the 1946 Overfishing Convention in London, at which the small Canadian delegation had pushed for the development of a North Atlantic fisheries organization to monitor the fish stocks. Although the Americans initially resisted the formation of a larger international organization for the Northwest Atlantic that would not be under American control, the Canadians remained adamant, and American capitulation on this issue allowed the nucleus of the new organization to emerge. The first executive secretary of the commission was W.R. (Bob) Martin, who had gained his doctorate in fisheries biology at the University of Toronto. After a stint in the Canadian Air Force during the Second World War, he joined the staff at St Andrews as the head of a new groundfish program. Incidentally, his research focused on developing new fishing techniques to allow the harvest of unused groundfish species and to improve fishing efficiency for 'under-utilized' fish stocks.[67] The focus of the ICNAF was on developing an enlightened exploitation of the northwest Atlantic groundfish stocks, to replace the ongoing free-for-all that had long prevailed. This would require standardized statistics. The ones kept by several of the member countries were initially quite haphazard; therefore, the ICNAF brought together statistics and research results from all nations fishing in the region. It also increased net mesh sizes to release small fish, and established quotas. It divided the northwest Atlantic fishing grounds into a number of regulatory and management areas. Finally, in the late 1960s, once it became apparent that larger mesh openings in the trawls were not an adequate solution, the ICNAF began imposing national quotas (total allowable catches, or TACs). Even this solution presented major problems in terms both of assessing safe quotas that would satisfy the participating nations, and of policing those quotas. These problems were exacerbated by the different nations' different fishing technologies, which ranged from small vessels up to and including the increasingly efficient factory trawlers of the Soviets and Japanese, introduced in the 1950s.

The work done by the Canadians and the Newfoundlanders for NACFI laid the foundations upon which later science would be built: 'the age structure, growth, and movement of a number of stocks were determined at that time.'[68] However, the formerly important role of the scientists at the Atlantic Biological Station in developing this science was ended when the headquarters for the ICNAF was moved from St Andrews to Halifax in the late 1950s.

In light of later developments, it can be argued that the ICNAF, despite the non-cooperation of certain member governments and other difficulties under which it laboured toward the end, was making progress in stabilizing the northwest Atlantic groundfish stocks; it was taking drastic measures to cut quotas and conserve the fishery. Unfortunately, little was accomplished even under the later ICNAF auspices regarding how to conserve fish stocks in terms of more basic science. For many years, fishery managers and biologists failed to heed general ecological and ecosystem factors. These came to be recognized only in the 1980s, and even then they seem to have been ignored by many fisheries managers. Traditional fisheries management ignored ecosystem disruptions unless these became intense. Rather, the ratios of 'benefits and costs of management options' provided the basis for optimizing the fisheries.[69] In fact, fishing activities at levels as high as those allowed by maximum sustainable yield can seriously undermine fish stock stability; it has been demonstrated that the quality of eggs produced by young or first-time spawners is inferior to that of repeat spawners, and 'with the reduction in the number of spawning age classes, a failure in egg or larval survival for any reason is potentially far more catastrophic in its effect on long-term abundance.'[70] In the 1950s, the nightmare of stock collapses made its dreadful appearance in the California sardine fisheries; this was followed by similar collapses in the Peruvian anchovy and European herring fisheries. Other factors lying outside the fishery, the immediate fish population, and its physical environment may also play a role. By the late 1980s, fisheries biologists were beginning to understand the importance of 'ecosystem integrity,'[71] and at least some had begun to question the idea of 'surplus production' of fish within a stock. The earlier paradigm of the theory of fishing was that humans could remove the 'surplus production' of a fish stock with no harm to the stock. Current ecosystem theorists argue that very little biomass of any population is truly surplus, as it is recycled in various forms within the ecosystem.[72] But this understanding was developed by ecosystem biologists from outside the community of fisheries biologists; only very recently has it been imported into fisheries science. Long before these developments, in 1977, the two-hundred mile offshore limits had been declared, the Canadian government had taken charge of managing the northwest Atlantic groundfish stocks, and Atlantic fisheries research was no longer integrated under the ICNAF. By this time, the Atlantic Biological Station's leadership in Canadian fisheries research had become a distant memory, for reasons that will be outlined in the next chapter.

Chapter 8

Ebb Tide at the Atlantic Biological Station

In the years 1932 to 1934, four circumstances converged, and their combined weight pulled the Biological Board from its original moorings. As a consequence, the board evolved into the more narrowly focused Fisheries Research Board of Canada. The overwhelming circumstance was the deepening Great Depression, which affected the other three: the retirement of J.P. McMurrich as the Biological Board's chairman and his replacement by A.T. Cameron; the fire that razed the Atlantic Biological Station laboratory; and Cameron's subsequent sweep of personnel, which saw A.G. Huntsman removed as director of the Atlantic Biological Station. Huntsman's unilateral decision to rebuild the laboratory without board approval gave Cameron an excuse to brush him aside; Cameron resented Huntsman, whose enormous power within the board as director of its flagship station enabled him to dominate its program decisions. Cameron also eliminated the volunteer researchers who had helped build the board's scientific base and reputation, and replaced them with a small but growing corps of professional marine biologists.

Out of these changes emerged the Fisheries Research Board, which would be dominated by professional fisheries biologists and oceanographers, and in which few volunteer researchers would be welcome. Due to the Depression, the board and the Department of Fisheries chose to focus on immediately practical problems, and to reduce their commitment to unrelated basic science. The Fisheries Research Board emerged from the Depression, and subsequently the Second World War, stronger in many respects, since it had a clear professional agenda; but also with an impoverished vision relative to earlier times. The deliberate cutting of direct links with university researchers was a great loss, and many of the board's scientists recognized this.

This period also marked the beginning of the Atlantic Biological Station's decline in tandem with the rising status of the Pacific Biological Station. A number of factors led to this, not least of which was the removal of Huntsman. Without Huntsman, the Atlantic station was cast adrift among various players who lacked Huntsman's vision and commitment. Paradoxically, Huntsman was at least in part to blame: by moving unilaterally to contract for the stations rebuilding after the fire, he alienated board members and employees (for various reasons) and pre-empted a vitally important debate about the future of Canadian Atlantic marine science. Subsequent history strongly suggests that the station might have continued as a leading scientific institution had it relocated from idyllic St Andrews; but as the local fisheries faded in importance relative to the still-burgeoning Grand Banks and Nova Scotia fisheries, and as Halifax became the centre of Atlantic Canadian oceanography and marine science education, the Atlantic Biological Station lost its relevance. The Pacific Biological Station gained in stature, because it was not hampered by these considerations, because it was well located with regard to the most economically important British Columbia fisheries, and because it enjoyed strong links and easy access to the University of British Columbia, which was training many of its future scientists.

The fully professional Fisheries Research Board of the late 1930s emerged as very distinct from its original, amateur roots. When the board of management of the Marine Biological Station was formed in 1898, its members had no idea they were laying the foundations for a new professional discipline in Canada. As botany and zoology professors, they saw the biological station as way to extend their researches into Canada's marine environment and as a welcome summer diversion; as a bonus, they would also be helping the fisheries. This was clearly recognized: A.G. Huntsman commented that the Biological Board's 'fertilizing agent' was 'the desire of University Biologists for means to study aquatic and particularly marine organisms,' which 'fused with a popular belief that science could aid the fisheries.'[1]

The first Canadian marine biological stations were staffed by university professors; thus, although 'professionals,' they were not trained in marine and fisheries biology per se. Yet the strong relationship between the stations and Canadian universities was crucial to the board's professionalization, and allowed Canada to develop its own research program independently. But as the Department of Marine and Fisheries asked the Biological Board to shoulder increasing educational and research burdens, the marine stations became centres of slow but steady growth in professional biology outside of universities. Most of the experts hired once the fisheries biological profession emerged in Canada in the 1930s and 1940s were 'home grown' – educated and trained in Canada. Fisheries biology became a recognized profession in the 1930s, the years in which the Biological Board changed its name to the Fisheries Research Board of Canada.

Long before this, however, the board's summer research agenda made it ripe for such professionalization: it was doing essential work for the government. To succeed, a new profession must be perceived to be filling new and important needs. The perception that fisheries problems could be remedied by scientific intervention only dawned late in the nineteenth century, as international fishing efforts intensified and it became evident that the fisheries were a finite resource that required management. As a consequence, governments came to rely more and more on an emerging cadre of marine scientists to discover new solutions.

Often, new professions 'arise from the development of some scientific or technological discovery which may be applied to the affairs of others.'[2] This is readily illustrated by the emergence of computer engineering and a host of specialized computer-related professions in the late twentieth century. Out of new technologies or techniques arise new problems. In the fisheries, technological changes such as steam trawling and steam purse-seining greatly intensified the fish harvest. Concurrently, oceanographers' new specialized equipment allowed improved sampling techniques and oceanographical analysis,[3] but expertise was required in order to use and interpret the results. Because of the expense, marine scientists had to be independently wealthy, such as Prince Albert of Monaco, and Alexander Agassiz, the son of Louis Agassiz, who built his wealth managing Michigan copper mines and later personally funded deep-sea faunal dredging and oceanographic expeditions in the Caribbean and the mid-Pacific. That, or they had to be professionals with government or institutional support.

The Biological Board, under the auspices of Canada's Department of Marine and Fisheries, had only a short step to professionalize (most science professions in Canada emerged 'by the gradual crystallization of career paths in the federal scientific services').[4] After Knight's lobster work, government officials saw the board's usefulness. Almost immediately, they wanted to hire fisheries scientists as government-retained professionals in the civil service.

Even outside the sciences, professionalization in its modern, Western sense only began to appear in the nineteenth century, as a result of industrialization.[5] Within science, professionalization enabled scientists with diverse interests and expertise to organize and undertake massive research and development projects. Through such interventions, 'science has come to exert a major influence upon economic growth.'[6] All of this followed a related change in social perceptions of science: by 1800, upper-class practitioners of science were being supplanted by middle- and working-class individuals, who organized local natural-history societies. These societies in the provincial cities of England and Europe 'gave a decidedly industrial and utilitarian flavor to much of the scientific discourse in the early decades of the nineteenth century.'[7] This was in contrast to the unfettered science of earlier independent practitioners.

In Britain in the early Victorian era, there was little professional science: 'Apart from the British Museum and a handful of minor institutions ... the only full-time posts at first were the chairs in natural history subjects, under varied names, at the six or seven English, Scots and Irish universities.'[8] In fact, most posts available were designed for men of independent means free to pursue science as a full-time avocation. Since salaries tended to fall below subsistence levels (being worse than the pay for secretaries and clerks), brilliant scientists who had to work for a living, such as Thomas Henry Huxley, had to hold several different posts at the same time in order to live. 'As late as 1874 ... Huxley was still able to complain that no amount of proficiency in the biological sciences "will surely be convertible into bread and cheese."'[9]

In France and Germany, the utility of science was recognized by the state earlier than in England and North America; scientists found places within the civil services, and there was wide-scale, especially 'industrially oriented,' science training. Germany by the 1860s had a large force of professional scientists in industrial laboratories in the new chemical and other scientifically oriented industries.[10] Moreover, in Germany and France, status derived 'less from a given occupation and more from attendance at a state-controlled, élite institution of higher education';[11] this 'assured ... élite positions in the civil service or other technical-managerial positions.'[12]

By contrast, before 1900, most English and North American professional scientists taught in universities, technical institutes, or secondary schools. The few professional scientists had government positions 'in arsenals, mines, observatories, public health services, geodetic surveys or civil engineering projects,' or in rare cases the dyestuffs and electrical industries. However, in the early 1900s, as science proved itself useful for product testing for improving production efficiencies, and for 'conserving raw materials and finding various uses for by-products,' entrepreneurs began to hire scientists in order to strengthen their competitiveness.[13] As the demand for them grew, scientists began to demand professional recognition and salaries. In 1852, scientific salaries had been so low that T.H. Huxley lamented, 'Science in England does everything – but *pay*. You may earn praise but not pudding.' By the end of the century, 'pudding' was more widely available.'[14]

Professionals possess skills 'based on systematic, theoretical, and esoteric knowledge,' which they acquire through specialized training. They are then accredited as specialists in their discipline. For example, in German universities the PhD was a kind of 'public certification of scientific competence' based explicitly on demonstrated research competence in the laboratory. Professions also tend to create autonomous organizations to enforce standards, maintain a sense of collegiality, and uphold the ideal of 'altruistic, though remunerated, service to clients and to society at large.'[15] Professional associations usually form soon after a profession emerges, to promote recognition and also 'to protect their members from undue governmental influences, encroachment of other occupational groups, and interference of the public at large.'[16]

However, in Canada, professional fisheries biology societies did not form immediately, perhaps because the original board members were drawn from universities. Those in the humanities and sciences held the term *professional* in some disdain, rejecting 'a utilitarian definition of the university. They are interested in the pursuit of truth, artistic or scientific, for its own sake.'[17] Humanities and science professors tended to be imbued with the German ideal of *Wissenschaft*. This term, as John Theodore Merz noted, had broader implications than the English and French terms for 'science.' *Wissenschaft* was 'an idea specially evolved out of the German university system, where theology, jurisprudence, medicine, and the special philosophical studies were all held to be treated "scientifically," and to form together the universal, all-embracing edifice of human knowledge.' *Wissenschaft* ('the true principle of research') became central to the 'higher and general education of Germany,' which so strongly influenced higher learning in America and elsewhere by the end of the nineteenth century. Having been linked with the liberal ideal of education, it stigmatized 'trade professions,' and as a consequence, professors of science and arts eschewed the professional label.[18]

This explains why the original members of the Biological Board and the university volunteer scientists had no notion of creating a new professional organization with a very different character. But 'it is not difficult to appreciate how this close relationship between academic scientists and the Board contributed over the years to the standing of the Board as a scientific organization.'[19] From the start, most Canadian universities were represented on the board and had an interest in its research and operations. Thus it was easy to recruit professors and students as volunteers, from whose ranks the future professional, permanent researchers would ultimately be drawn. The close links between the Biological Board and Canadian universities ensured that the board saw itself as coequal in professional status with the universities: this could hardly be otherwise when its members and early staff were mostly either professors themselves or in line to become such.

One ideal shared by the board's professors and professionals was the 'service' ideal. Nathan Reingold observed that 'the common definitions of a profession assume an applied component requiring a service ideal.'[20] The board indeed reflected this ideal in its research goals, which included helping fishermen and government and to training young researchers. Indeed, fisheries biology was always the *raison* d'etre of the board.

But in the early years, other interests were also encouraged. A look at marine biological studies from the amateur period yields an chaotic list of volunteers' diverse and sometimes ephemeral interests. Clearly, however, marine biological investigations of almost every variety flourished under the volunteer system. Even after full-time fisheries biologists were hired, many visiting scientists continued to follow their own interests. Pacific polychaete worms were studied over forty years by volunteer investigators Edith and Cyril Berkeley, who provided a standard guide to that group. (Cyril Berkeley would serve as assistant curator of the Pacific Biological Station in 1920 and 1921.) The hydroids of both coasts and of the Arctic were 'described exhaustively in three monographs and many papers by C. McLean Fraser.'[21] In contrast, studies of specific fisheries problems were conducted mainly by board members.

Initially, the stations were administered by seasonal resident directors assisted by a curator, who received an honorarium.[22] The Biological Board took a faltering step toward professionalizing its services when in 1912 it hired its first scientific employee, C. McLean Fraser, as curator of the Pacific station to replace Reverend Taylor, who had just died. Fraser stayed until 1924, although from 1912 to 1920 the station had few visitors (in 1916, just Fraser and J.B. Collip were there). The Atlantic station was far more important.

In 1909 the Atlantic station's first resident director, D.P. Penhallow, argued that a full-time scientific curator was needed. In 1911, a year after Penhallow died, A.G. Huntsman, then a lecturer at the University of Toronto's zoology department, became the curator. At first this was a summer job. However, in 1916 he became the station's first permanent curator. E.E. Prince had argued that 'we certainly cannot go on much longer without some permanent scientific worker,' and told Huntsman, 'I hope that there is a profitable and pleasurable career opening up for you.' Huntsman looked forward to not having his energies 'so scattered as they have been in the past doing both University and Station work.'[23]

Huntsman became director of the Atlantic station in 1919. After that year the Biological Board's work expanded rapidly, to the point that it needed more staff to deal with the influx of researchers, who were no doubt attracted by the living and travel expenses that enhanced the research experience. During 1920 and 1921, Huntsman and seven support staff at St Andrews were joined by around twenty-five scientists (university professors, assistants, and advanced students). Their work ranged from studying oyster culture on Prince Edward Island, to making marine collections, to conducting inshore and deep-sea researches on the vessel *Prince*.[24] They also staged several hydrographic and plankton expeditions and faunal surveys in the Bay of Fundy and Minas Basin.[25]

Meanwhile, the Department of Marine and Fisheries was manoeuvring to absorb the Biological Board as its own scientific branch (see chapter 4). This move, which would later strip the fisheries biologists of their capacity to render independent and politically dangerous assessments of fisheries policies, was averted by the board until 1973. In 1923, Prince was arguing against W.A. Found's continuing agenda: 'Men of great eminence and distinction are today working under our Biological Board whom the government could not induce to become members of a branch of a department.'[26] But the increasing need for practical solutions to fisheries problems required more permanent scientists on staff, since volunteers did not have enough time to give to broader, more involved inquiries.[27]

A first step toward professionalization was made in 1921, when it was decided that prospective summer researchers would have to meet stringent new requirements: they would have to formally apply to the board's secretary and be approved by its executive committee.[28] Volunteers not tied to universities or other research organizations were no longer welcome. In a sense, the board was now acting as a professional association monitoring the quality of would-be researchers. Senior workers (members of the Biological Board and Canadian university professors) were to do research without an honorarium, but if they agreed to work on economic problems, they were given board and travelling expenses. If the board appointed qualified investigators for special investigations, these might also receive an honorarium; investigation doing other than economic work were given laboratory space only. Biology faculty from any Canadian university could use the station to make collections, but could expect no financial aid, and newcomers and junior workers required a recommendation from their department's senior university professor.[29]

The laboratories and residences were opened from 1 June until 15 September. Scientists had to provide their own microscopes; however, the board loaned various hydrographic and other scientific instruments to competent workers. All investigators were expected to spend at least two months at the Station, and had to submit a research report before the board's next annual meeting, which the board could publish if deemed important. The board filed unpublished reports in a station library.[30]

As long as the volunteer system persisted, general marine biology continued to flourish. Marine biology includes any aspect of the biology, physiology, or ecology of marine organisms, so its progress would be extraordinarily difficult to describe, yet it was essential to further advances. In 1922, several researchers tried to extract insulin – newly discovered by Canadian scientists Frederick Banting, Charles Best, and J.B. Collip – from fish pancreases; other physiological studies looked at fish digestive physiology and fish muscle protein chemistry. Systematic studies predominated. Huntsman observed: 'Perhaps the most important undertaking ... has been to work out the conditions of life generally in the waters along our Atlantic coast ... much pioneer work remains to be done.' Arthur Willey studied copepods; Professor C.J. Connolly of St Francis Xavier University examined the life histories and distributions of larger crustacea. A.B. Klugh surveyed Passamaquoddy Bay and Miramichi algae, while Loring Bailey with Miss C.W. Fritz of the University of Toronto did exhaustive studies of the diatoms. Klugh, along with W.R. Sawyer of Queen's University and Miss E.C. Odell of the University of New Brunswick, observed how light affects marine animal movements. Other researchers explored the relationship between light intensity and the vertical distributions of plankton, and how light affected seaweed growth.[31] Volunteer investigators thus provided a valuable background – albeit in the tradition of Victorian natural history studies – for more intensive fisheries research.[32]

Fisheries research itself also involved grappling with how various physical factors affect fish abundance. Researchers studied the ways that 'combinations of temperature, salinity, and oxygen content of the water' affected hatching and fry survival in several groundfishes. Helen I. Battle and her assistants also studied 'maturation, spawning, egg and larval development of herring, goldeye, salmon, lobsters, crabs, various molluscs, and other animals.'[33] Huntsman, aided by Miss M.E. Reid of the University of Toronto, worked out the life history of *Sagitta*, a floating chaetognath (arrow worm), which herring and other fish gobble up with gusto. J.P. McMurrich studied how plankton quantities fluctuated with seasonal changes. Professor J.W. Mavor studied cod, pollock, and flounder life histories; Huntsman did similar work on herring, smelt, and gaspereau. A.H. Leim for several years researched the life history of shad to determine how to improve that then declining fishery. H.C. White of Queen's experimented with the best conditions for brook trout fry development. A.W.H. Needler, a Toronto student, did vital work on the long-term distributions of good and poor haddock catches for the North American Council on Fisheries Investigations.[34]

In the meantime, more professional, full-time scientific employees were being hired. In 1922, between twenty and thirty volunteer workers spent about three months at the stations, yet Huntsman was the only full-time scientist; Dr C. McLean Fraser at the Pacific station was the only other scientific employee (and he only gave 'one-third of his time to the work'). By this time, the department was looking for permanent qualified employees who could work full-time in fisheries biology without being distracted by university duties. Found commented: 'While there are obvious advantages in the Board keeping in close contact with our Universities ... depending upon volunteers has its limitations.'[35] A.P. Knight, when he took over as chairman – in one of the shortest tenures in board history (1921–5) – responded by stating that he would 'orient the Board towards more practical studies'[36] and began to build a permanent board staff. In 1923, with the board's cooperation, the Department of Marine and Fisheries amended the Biological Board Act to encourage such professionalization.

A.H. Leim, an expert on shad, fisheries technology, and fish taxonomy, in 1924 became the second full-time scientist at St Andrews. Likewise, the Nanaimo station gained the permanent services of R.E. Foerster, who pioneered 'intensive studies on the life history, propagation, and ecology of Pacific salmon – studies that continued for 40 years and made him a foremost international authority on this important group of commercial fishes.'[37]

Knight was succeeded as chairman in 1926 by J.P. McMurrich. McMurrich in his 1884 article 'Science in Canada' had been the first to call for Canada to establish coastal biological stations. However, he did not help found the Biological Board, as he was then teaching at the University of Michigan, where he developed an international reputation as an anatomist.[38] He became head of the University of Toronto's anatomy department in 1907. His deep interest in marine embryology dated from his Ph.D. studies at Johns Hopkins and summer work from 1881 onwards at the Chesapeake Zoological Laboratory, run by Johns Hopkins University, and his later involvement with the Marine Biological Laboratory at Woods Hole. He joined the Biological Board in 1912. Retiring as Professor Emeritus at Toronto in 1930, he continued as chair of the Biological Board until failing health forced his retirement in 1934. However, he was involved in the 1935 formation of an organization for professional biologists in Toronto, and remained active until his death in 1939.

McMurrich's chairmanship (1926–34) 'marked a clear shift from university dominance of Board research to that of government-employed scientists.'[39] By the time he retired, the Biological Board comprised eighteen members, controlled four stations and several substations, and employed 34 scientists and 150 employees of other ranks, up from the mere four scientists who held positions when he was appointed chairman. The board was becoming heavily professionalized, although volunteer investigators continued to make important, even vital, contributions.[40]

Before any growth was possible, however, the board first needed to expand its facilities, which were inadequate even for the volunteers. Huntsman in 1920 complained that 'makeshift' laboratory accommodation was so poor that 'investigators have refused to continue work under such conditions.' The Atlantic station's museum, which both educated visitors and advertised the station, created havoc as tours passing through the laboratory disturbed investigators. The library was now so big that it had to be housed some distance away, in the residence. Added to this, as more workers came, accommodation and even kitchen and dining facilities had 'been so strained in recent years that it ha[d] not been possible to provide for all who would come.'[41] Throughout the 1920s, St Andrews gradually expanded its laboratory facilities, although space remained tight. A car shuttle service between the town and the station was established to avoid the need to build new residence facilities.

One barrier to professionalization was the lamentable lack of facilities at the station for year-round research. Until 1928, scientists did not remain at the Atlantic station throughout the year. Although year-round research was supposedly possible on the Pacific coast, researchers seldom stayed at the Pacific station beyond summer, since they led to return to their respective universities. Until 1929, Huntsman and Leim spent winters at the University of Toronto's biology department; the Atlantic station was officially open only from 1 June to 15 September, although some workers arrived earlier and stayed later. The station lacked proper heating, so that even in the summer it was often too cold 'for successful work except for such as involves definite and continuous physical exercise.' Yet fisheries biology demanded year-round work, 'in following up conditions in the sea, in making varied collections of material, and in carrying on experimental work in fish handling.'[42]

Finally, oceanographer Harry B. Hachey stayed at the station through the winter of 1928–9. In 1929 he was hired permanently as a hydrographer, together with a pathologist, a scientific assistant in aquaculture, and a scientific assistant in limnobiology, following the recommendations of the 1927–8 Maclean Commission. Although makeshift arrangements enabled a longer season, work remained 'out of the question during the rigors of winter,' so the University of Toronto provided the staff with excellent facilities during the winter.[43] A new heating system installed in late 1931 enabled the station to open year-round in 1932; staff and even a volunteer were there in February, all 'commuting' from St Andrews on a specially provided bus or by car.[44]

Professionalization also presumes a 'critical mass' of specialists in a given field. The Biological Board could not professionalize prior to 1930 because there were too few adequately trained biologists. For example, in the 1920s Canada had too few systematists to conduct inventory biology without outside help. In 1923, A.H. Leim asked Charles B. Wilson, of the State Normal School of Westfield, Massachusetts, to identify specimens of argulids (fish lice). Biological Board worker Fritz Johansen sent Hudson Bay mollusks collected in 1930 to the Smithsonian for identification.[45] This is reminiscent of Canadian biology in the Victorian era, when Dominion naturalist John Macoun (1831–1920) relied on specialists at Kew Gardens and at the Smithsonian, among other foreign natural history museums, to identify the botanical specimens he collected. Likewise, the Abbé Leon Provancher (1820–93) relied heavily on Americans such as Asa Gray of Harvard and Ezra T. Cresson of Philadelphia's Academy of Natural Sciences to accurately identify the Quebec flora and insects that he collected.[46]

In 1923, Huntsman inaugurated work on a new collaborative publication, *Canadian Atlantic Fauna,* which was to be a series of handbooks, 'issued in parts as ready ... by the foremost systematists in their respective fields.' Of the twenty-one experts, eighteen were American, one was British, and only two were Canadian. Thus he asked Henry B. Bigelow to do the fishes, Professor W.F.K. Fisher of Pacific Grove, California, to do the starfish, Dr Mary Rathburn of the U.S. National Museum to do the crabs, shimps, and lobsters, and Dr H.L. Clark of the Museum of Comparative Zoology to do the sea cucumbers. The two Canadians were Percy A. Taverner, the Canadian bird expert at Ottawa's Victoria Memorial Museum, and Dr Arthur Willey of McGill University, for the copepods.[47] A.R. Cooper, of the University of Illinois, did not have enough time: 'I feel sure, however, that you must have some young man there ... quite competent to look after the tapeworms.' Huntsman replied: 'I should desire that in the preparation of such a Canadian publication, Canadians be well represented among the authors, [but] we have so few qualified for the task.'[48]

The international Passamaquoddy fisheries investigations (see chapter 7) also highlighted the shortage of professional marine scientists. Bigelow and Huntsman had to lure suitable scientists away from permanent positions on two-year leaves of absence. Charles Fish, the American zooplankton specialist, was the only scientist they managed to hire full-time. Hydrographer E.E. Watson, physics professor at Queen's University, Kingston, had to teach during the first winter. When Dr Chauncey Juday backed down from the position of phytoplankton specialist, Bigelow wrote to Huntsman: 'The reason I haven't proposed someone sooner to take Juday's place is that I was rather stuck. There is no one else in the United States, so far as I know, whom we could turn loose on such a complex problem.'[49] That was why H.H. Gran of the University of Oslo, Norway, was eventually chosen; but Gran was required to be in Norway for much of the term. The hardest position to fill was biologist in charge of fishery investigations. When A.E. Parr, curator of the Bingham Oceanographic Foundation of Yale University, proved to be too concerned about his prerogatives, Huntsman and Bigelow could find no other qualified investigator in North America. They settled, as seen in the last chapter, on Michael Graham of England's Lowestoft laboratory, who was enthusiastic, but whose initial request for a leave of absence from the Board of Agriculture and Fisheries met with a rebuff.

Huntsman's pleas to H.G. Maurice (1874–1950), fisheries secretary to the Board of Agriculture and Fisheries,[50] led Maurice to wire back: 'Would it be possible you send a trained naturalist to substitute Graham. Stop. Experience probably useful your man. Stop. Will examine possibility here on hearing from you.' A covering letter explained that Britain was short of adequately trained men to replace Graham, and there was a

> risk that ... if we can lend one of our principal Naturalists for a long spell of work abroad, we appear to be over-staffed, and that the staff might advisedly be reduced; in point of fact, we could do with more staff, but that is another question ... We might get over most of our difficulties – not all of them – if you could arrange to send us a competent Naturalist to work with our staff while Graham is with you.[51]

Huntsman was amazed that even Britain was short of fisheries biologists: 'We had thought that our position was perhaps peculiar, owing to the development of the work here having been so long delayed.' He regretted that he did not have a 'trained man to send as a substitute for Graham; as it would be such an excellent opportunity for one of our men to get very valuable experience.' Only he himself and A.W.H. Needler were qualified, and Needler was starting up the program 'at the station established in Prince Edward Island last year,'[52] while Huntsman himself could not go. The solution, as recounted earlier, was that Graham took charge of the Passamaquoddy fishery investigation while continuing his study of herring concentrations in European waters.

Just as problematic for the Biological Board was the lack of funds for attracting and keeping qualified scientists. In 1927, McMurrich complained that the board was having trouble building a full-time staff 'especially at the Experimental Stations' because even a well-grounded young scientist 'requires a considerable amount of training ... before he can successfully carry out the investigations desired.' But 'men who have received this additional training and experience become at once attractive to the Universities, and receive from them offers which the Board has, in some cases, been unable to meet.' The board thus lost its newly trained scientists just as they 'had become valuable in the work of the Board.'[53]

Industry was also predatory. In 1929 a 'large and influential' commercial firm offered D.B. Finn, director of the Prince Rupert station, a position with a much higher salary, to establish and direct 'a research department within their organization.' Finn's conditions for staying with the board included matching the firm's offered salary and the board adopting 'a policy of expansion of its work on the Pacific.' Finn was strongly tempted to go even though some 'conditions in the industry ... would not be as desirable,' since other conditions 'might be more desirable, especially when one considers the unavoidable delay which sometimes attends governmental procedures.'[54]

Finn was persuaded to stay. He would direct the Prince Rupert station for a few more years before being transferred to the Atlantic coast in 1933 to direct the Halifax Experimental Station.[55] However, the problem of keeping underpaid staff from the clutches of marauding universities and expanding fish-processing concerns became particularly acute in the late 1930s.

Since the Department of Marine and Fisheries' demand for trained fisheries scientists could not be met, the Biological Board had to offer specialized training to provide potential recruits. The provision of qualifying training is an important step toward establishing a new profession. Before the 1920s there were no programs for training marine and fisheries biologists in Canada, and the scientists who devised the new programs were pioneers. Huntsman, who played such a prominent role on the Biological Board, only possessed a medical degree, just as did earlier American biologists such as Asa Gray and F.V. Hayden.[56] To build up the ranks of competent professional fisheries researchers, two things were needed: first, qualified institutions had to offer appropriate courses; and second, to attract students, there had to exist some impression that those who acquired accreditation would have a chance of employment in the field.[57]

In training graduate students, the Biological Board experienced marked success, and this was part of an important trend. H.J. Deason's 1941 'Survey of Academic Qualifications for Fishery Biologists and of Institutional Facilities for Training Fishery Biologists' recorded that the number of North American academic institutions offering specialized courses related to fishery biology had 'increased considerably during the last 15 years due to the growth of fishery science as a recognized profession and the ever-increasing number of opportunities for employment in the field.' In the United States and Canada, forty-three of the eighty institutions he surveyed had facilities to train fisheries biologists.[58]

The University of Toronto led the way, because Huntsman and other biologists there had strong connections with the Biological Board. Although the stations offered no undergraduate courses, they did train future researchers. Board members sponsored students as research assistants; honours, Master's, and PhD students based their dissertation on their work at the stations. In 1933, Huntsman himself supervised Miss Viola Davidson (diatoms); C.L. Newcombe (growth in clams [*Mya arenaria*]); Misses M.H. Campbell and V.Z. Lucas (life histories of barnacles and other species); M.W. Smith (fertilization of ponds and lakes); Wilfrid Templeman (ecology of the lobster); and A.A. Blair (Atlantic salmon of the Miramichi River).[59] All board biologists hired from the

1930s onward were trained in marine or aquatic research – an obvious sign of the increasing professionalism in fisheries science.

Many students were attracted both by the research opportunities and by the pay offered to summer students helping board staff or members with their researches. The usual rate was fifty dollars per month plus expenses. For example, Paul F. Elson, a fresh BA in 1934, came recommended by Helen I. Battle and helped Huntsman with his salmon investigations; he went on to do a PhD with Huntsman. C.J. Kerswill, sponsored by A.W.H. Needler in 1938, studied the growth and distribution of the quahog (*Venus mercenaria*)[60] and became a graduate student of Huntsman. He joined the scientific staff at St Andrews, and in 1950 became one of the senior staff at the newly formed advisory headquarters in Ottawa. He later directed the Arctic Biological Station at Ste-Anne-de-Bellevue, Quebec.[61]

The Biological Board tended to hire its own trained, proved researchers. In 1941, Huntsman told applicant Dr Hilary B. Moore, who was working at the Woods Hole Oceanographic Institution and the Bermuda Biological Station, that there was no position for him: 'Our trend ... has been towards confining our work to attack on specific fishery problems, and graduate students [working on these] almost inevitably are taken on at first temporarily and then permanently.'[62] Many of Huntsman's students, such as J.C. Medcof, A.W.H. Needler, Wilfred Templeman, and A.A. Blair, became senior members of the Fisheries Research Board.

Some found positions elsewhere. In 1938 the director of the Chesapeake Biological Laboratory, R.V. Truitt, told Huntsman that a former student, Dr C.L. Newcombe, was 'carrying a heavy part of our load at Solomon's Island and is doing very good work for us' and expressed interest in hiring A.A. Blair, Huntsman's former student, if financial strictures eased. Blair was hired. Another graduate, Dr William S. Hoar, in 1944 became a professor of zoology and fisheries biology at the University of British Columbia and turned that department into a centre for education in fisheries science. Hoar's fisheries-related courses explored the conservation, exploitation, and economics of fish stocks and 'develop[ed] the theories of Baranov, Russell, Graham, Thompson & so forth.'[63]

As new, fully qualified scientists joined the staff, those hired earler sought to upgrade their qualifications. In 1929, D.B. Finn, director of the Prince Rupert station since 1924, requested two years' leave of absence to complete a PhD at Cambridge: 'I feel that I must take this step ... The Director should possess not only more research experience but a higher academic standing than I have at present.' McMurrich agreed that PhD training 'would, of course, strengthen your position and your usefulness to the Board.'[64]

The board had succeeded in creating a demand for its services. It attracted motivated graduate students to enter marine and fisheries biology, and it retained many as trained researchers. This fulfilled E.E. Prince's educational expectations of the first biological stations, 'to give an unequalled opportunity to young biologists in the various universities of Canada to carry on original scientific researches.'[65] Many summer volunteer students eventually attained senior rank in the Fisheries Research Board, the Department of Fisheries, the National Research Council, and Canadian universities. Several were appointed to senior positions in other countries or in international fisheries organizations. But the Department of Marine and Fisheries and even some board members were still not satisfied; they wanted to further integrate the Biological Board with both the department and the fishing industry.

By the 1930s the Biological Board had survived a number of (unsuccessful) attempts to annex it to the Department of Fisheries and had experienced growth in its activities, its membership, and its professional staff, with no essential change in its mandate and character. But the Great Depression and the early war years saw key changes in the board's nature and in its activities as well. After McMurrich retired as chairman in 1934, he lived to see his successor's changes. Under A.T. Cameron the Biological Board was transformed into the Fisheries Research Board of Canada, and its character was dramatically altered, not always for the better. Cameron loosened the board's ties with the universities and did away with the volunteer system, closing the Nanaimo and St Andrews residences to volunteer researchers, a move that 'Dr. McMurrich regretted very keenly.'

Alexander Thomas Cameron (1882–1947) was entirely different from the gentlemanly and unassuming McMurrich. Born in England, he was the first non-biologist to head the Biological Board. He specialized in chemistry. After receiving a BSc and MSc from the University of Edinburgh (1904 and 1906), he took two more years of training at University College, London, under Sir William Ramsay, and in Germany under Fritz Haber (who later won a Nobel Prize). In 1909 he was appointed lecturer in physiology at the University of Manitoba where he became a pioneering endocrinologist. The rest of his career was spent in Canada, except for three years during the First World War, during which, as a captain in the Royal Army Medical Corps, he worked as chemistry officer for water purification with the British Expeditionary Force in France. He published more than a hundred articles and four textbooks in biochemistry. Cameron was a man of great energy, 'strong convictions [and] indomitable will,' and was not always easy to work with.[66]

Cameron became chairman of the Biological Board at the height of the Depression, at a time when the board was facing deep funding cuts. He soon eliminated the earlier policy of training young scientists in summer research work. He noted: 'Good results were undoubtedly attained, and some of the brighter men were ultimately recruited to the permanent service of the Board.' But with the Depression, in order to meet the many demands being made by the department and the fishing industry, 'we decided that the function of the Board did not include such training of young scientists in research, and that the time of the Directors was too valuable to be largely occupied in summer in such a way. The residences [built largely for these and other volunteers] were closed.' They would remain closed throughout the Depression. As well, facilities for senior trained scientists 'had, regretfully, to be temporarily withheld.' Graduate and advanced students were no longer admitted unless they were known independent workers, since 'the Directors cannot be expected to give post-graduate training.'[67] If the work was not applicable enough, directors could ask the students to leave.

As the Depression worsened and Biological Board funding was slashed, few opportunities remained for volunteer workers unless they paid all their own expenses. H.B. Hachey, who knew Cameron and who chronicled the Fisheries Research Board's history in 1963, commented: 'It is quite probable that Dr. Cameron never had a very high opinion of voluntary workers, particularly the younger ones. A photograph found in his effects after his death, records his impressions of those who "toil not neither do they spin."'[68]

Volunteer workers had laid the emerging Fisheries Research Board's foundations through their able contributions to the development of scientific fisheries research. Cameron's decision to eliminate volunteers was lamented by most board scientists and was perhaps his least popular measure. In the 1950s, Dr J.R. Dymond argued that 'as soon as adequate space and facilities can be provided University scientists should again be encouraged to engage in research at the Stations.' He pointed out that the board had 'profited in many far-reaching ways' from a large number of visiting Canadian biologists: 'The work of the Board will lose much if closer contact between the Universities and its work is not re-established.'[69] W.E. Clemens, as director of the Pacific station, recalled 'with a few exceptions the investigators applied themselves steadily and faithfully to their problems not only during the summer months but after their return to the universities ... with a relatively small monetary outlay on the part of the Board.' He lamented: 'That this or some similar system was not re-instated [after the Depression] seems to me to be unfortunate, because the association of university and station personnel was mutually stimulating and beneficial.'[70]

Huntsman commented in 1953 that despite the board's phenomenal growth and success, 'I still believe ... that its action twenty years ago in doing away with volunteer research was unwise and that organized research is decidedly expensive and unproductive scientifically in comparison with volunteer research.'[71] Indeed, given the financial situation, it was ironic that Cameron curtailed support of volunteer workers and scientists, who could be highly cost-effective. A few 'visiting scientists ... worked gratis on special projects for short periods in later years,' but the volunteer investigator system was never restored.[72]

Almost without exception, scientists and board members who remembered the volunteer era (and even some who did not) regretted its passing. Through its openness, it had encouraged collegiality, as well as shared ideas and experience. The painstaking fundamental studies carried out by voluntary workers 'from Canadian universities [and] from the United States and Europe ... was mutually stimulating and valuable as all who experienced it can testify.' Indeed, the work of Mr and Mrs C.J. Berkeley, carried on totally unsupported at the Pacific station from 1921 until Mrs Berkeley's death in 1963, had added prestige to the Fisheries Research Board of Canada. They had became 'internationally recognized experts' on the polychaetes, and were sent 'collections for examination, identification, and recording ... from all quarters of the world.'[73] They were symbolic of the breadth of interests that could be supported by the Biological Board through its volunteers.

The volunteer system was not the only victim of the Depression: stringent cutbacks led to other problems as well. The Atlantic stations' appropriations in 1930–1 amounted to $202,748.14; they had fallen to $79,671 by 1934–5. On the Pacific coast, funding fell from a high of $156,662 in 1930–1 to $77,082 in 1934–5. Allocations for the entire board had grown from $70,000 in 1925–6 to $360,000 in 1930–1, the interwar peak funding year. In 1932, funding dropped to under $200,000, and remained at or below this through the Depression, with much of the budget being directed to the experimental stations. One austerity measure for the entire civil service, including Biological Board staff, was to cut wages by 10 per cent – a measure that lasted many years. Even during the Second World War, funding only rose to about $250,000, until 1945 when a budget of $368,000 amounted to a 56 per cent increase over the previous year.

Yet the number of problems requiring solutions had by no means decreased along with the funding. For this reason, Cameron concentrated staff on solving 'practical problems of the Fishing Industry and other fisheries' interests.' Research expansion was impossible, hirings were halted, and 'the skeleton year-round scientific staffs were denied adequate assistance.' Some personnel were discharged, and some were retained at reduced salaries, and field work suffered as a consequence.[74]

It was during this already trying time that the Atlantic station burned down on 8 March 1932. The main laboratory and most of the library and equipment were lost. 'To say that we have no equipment left would be a better way of stating the case,' fish pathologist R.H. M'Gonigle ruefully admitted. Huntsman acted with iron determination to save his station: 'Our greatest asset consists in our investigators, and the fine corps which we have has been developed during the last 20 or 25 years. It is important that they should be kept together.' Huntsman requested $20,000 to begin building a fireproof station. J.J. Cowie demurred: 'I am afraid there will be considerable difficulty in finding that kind of money.'[75] Huntsman then acted without authorization: he decided that temporary cheap buildings would waste money and contracted a permanent fireproof, yellow-brick structure, the basement of which would house a temporary laboratory. To raise the needed money, he cut staff wages – a further reduction for these poor sufferers already reeling from a 10 per cent cut. Huntsman reasoned that he was saving their jobs. There is some debate over whether also he cut his own wages; there is no question that his staff resented his actions and withdrew some of their support for him.

He had been panicked into making these moves when some board members said they did not want the station rebuilt. Industry member A. Handfield Whitman argued that 'most careful consideration should be given as to whether the work at St. Andrews should be materially curtailed and greater efforts be put into the practical end of the Fisheries Department for which the Halifax Station was put into operation.'[76] By forging ahead without permission, Huntsman saved the station from being shut down permanently. But the differing visions of those who agreed with Whitman, and those who agreed with Huntsman, were to persist for the remaining decades of the Fisheries Research Board and were to mark the rebuilt Atlantic station for a much diminished future.

Moreover, rebuilding the station was an enormous unforeseen expense for the cash-strapped Biological Board. Cameron complained that

'Directors ... by proposing Budgets which, during the period of depression, are merely ridiculous' were biasing the executive 'even towards their legitimate requests.'[77] This shot was directed specifically at Huntsman. In 1934, Huntsman was removed as director of the Atlantic station, and was, as he observed, 'kicked upstairs' to become editor of the board's publications.

His actions after the fire provided a good excuse, but removing Huntsman had been Cameron's objective on becoming chairman. To him Huntsman was the competition. Indeed, Huntsman had been the board's de facto director from the time he became director of the Atlantic station. Anyone who has served on a committee or board of directors will appreciate why. Most Biological Board members met only a few times a year to discuss and determine the next course of action, and only focused on board-related issues when the agenda was placed before them just before meetings. Huntsman, by contrast, devoted himself to the board's affairs full time. A perusal of his enormous correspondence makes it abundantly clear that board policies from 1920 to 1934 overwhelmingly originated from Huntsman's suggestions and advice. Indeed, even the Maclean Commission's recommendations regarding the board were first drafted by Huntsman. Building the technological experimental stations was partly his idea, and while still directing the Atlantic station he had also directed the fisheries experimental station in Halifax from 1924 to 1928, which further increased his influence.

While holding these positions, Huntsman operated from a position of strength, because he continued to advance at the University of Toronto, becoming an associate professor in 1917 and professor of marine biology in 1927. He occupied a strange position in the annals of institutional history, since, although as a professor he worked and taught graduate students at Toronto most of the year (spending summers at St Andrews), he drew no salary from the university after 1916: rather, he was on the federal payroll through the Biological Board of Canada. He was pivotal in linking the two institutions and in drawing graduate students into the marine sciences in the 1920s and 1930s. He later strongly promoted the hiring of full-time professional scientists. These scientists provided 'the leadership, and dedication' that would shape the board over the next half-century. But Huntsman

> towered above all the others in this period, and indeed his brilliant mind was a major force throughout most of the Board's history ... Huntsman investigated virtually every area of aquatic science, had practical ideas that

> were decades ahead of his time ... and influenced countless students during his 50 years' association with the University of Toronto ... he was the significant link in this period of the Board's transition to a permanent staff.[78]

That the Biological Board had done so well is a testament to Huntsman's vision and command of the essential problems.

However, Huntsman, like all great men, had his detractors and critics, and plenty of faults to fuel their resentments and valid complaints. He was, in late twentieth-century parlance, a micro-manager who wasted too much energy minding the pennies. A later chairman of the board, Ronald Hayes, recalled that while Huntsman was recuperating from ulcer problems in the later 1920s (exacerbated by his being director at both St Andrews and Halifax) in Bermuda (where he was a long-standing trustee of the Bermuda Biological Station, involved in reorganizing it in the mid-1920s), he 'left word that for every item to be taken from the stock room' of the technological station, 'you had to send the chit down to Bermuda by boat, to be okayed by Huntsman and then sent back to Halifax. Then you could get the test tube or whatever it was ... He was unable to delegate authority and was incapable of authorizing expenditure of funds.'[79]

And yet, as a scientist, a director of two stations, a trustee of the Bermuda Biological station, a founding trustee of the Woods Hole Oceanographic Institution and a friend of Henry Bigelow, Michael Graham, and other oceanographers and marine scientists in the United States, Britain, and Scandinavia, Huntsman was one of the towering marine scientists of that era. Indeed, he was considered for the position of director of the Scripps Institution of Oceanography, as recounted by its first director, Wayland Vaughan. Vaughan told a mutual acquaintance: 'Huntsman! He is really a big man and one of the best – When this place was being proposed I was consulted as to the choice of a Director and I recommended Huntsman but they finally gave me the job. I don't suppose Huntsman knows that, but now that I am leaving you may tell him.'[80]

He was so eminent that he could not be dismissed outright, and was thus retained as editor and consulting director of the Biological Board, and permitted to carry on whatever scientific investigations he desired. Huntsman later reminisced: 'The privilege granted me was remarkable because the Board was at the time doing away with the system of volunteer investigators at its biological stations.'[81] Now free of pressing responsibilities, he immersed himself in scientific work, especially all aspects of salmon studies.

Huntsman's successors were competent scientists, but they lacked his flair, his scope, and his connections (not least his university connections), so it is not surprising that the Atlantic station's image on the international radar screen faded and blurred. It remained important in the immediate postwar era, but more by virtue of its relative proximity to the great banks fisheries off Canada's coast than because of its scientific agenda.

Rise of the Pacific Biological Station

Having broken Huntsman's stranglehold on the Biological Board's policymaking, Cameron turned his attention to the board's character and priorities. One of the biggest changes under Cameron – perhaps attributable in part to Huntsman's loss of power – was that the Pacific station saw its fortunes rise.

For the board's first three decades, the Pacific station had always been the poor, distant cousin. It was hard to get to from the central and eastern provinces, and remote from the University of Toronto, which was the board's key university connection. It was smaller, and had fewer scientists and visiting researchers, and received much less funding. For example, in 1923, the St Andrews station was allocated $18,600 but the Pacific station received only $6,175. In 1926 the Atlantic station received $25,750, the Pacific station only $16,250.[82] Only in years of extraordinary expenses, when new buildings were erected, did this change – in 1928, the Nanaimo station received $53,500, which was $20,000 more than the Atlantic station – but this was to build a residence.

The first laboratory, built in 1908, could only accommodate about eight scientists. Under the Pacific station's first two directors, the Reverend Mr Taylor (1908–12) and Dr C. McLean Fraser (1912–24), volunteers' scientific studies built up 'a great store of scientific knowledge of the coastal waters.' Studies of the life histories, reproduction, and growth of fish (including salmon, halibut, herring, and shellfish), of physical and chemical conditions in the sea, of marine food sources, of wood-boring pests, and of seaweeds provided essential background for later fisheries research.[83]

When the next director, Wilbert Clemens, took over in 1924, he found that the station had meagre equipment and supplies; his first priority was to build up these stocks. The station itself was a two-storey house that also served as the director's residence. It included a long, one-room laboratory, a small library, and an entrance hall extending from one side. In 1923 a basement had been dug under the building for a museum, storage, and a director's office-laboratory. Eight volunteer researchers lived in tents on the grounds in 1924. The station boat, the *Ordonnez*, was rotting. Clemens's pre-emptive efforts to contract a new boat forced the Department of Fisheries to build a new boat for the station, the 50-foot *A.P. Knight.* Owing to Clemen's rebuilding agenda, in 1928 the Pacific station's allocation exceeded that of its Atlantic counterpart; with it, Clemens built a new residence with a large kitchen and two dining rooms, which could accommodate a large staff, including newlywed scientists and spouses. It opened in 1929. Volunteer scientists, many of them eminent in their fields, did most of the research with little or no remuneration.

In 1926, Dr R.E. Foerster began an intensive study of the relative efficiency of natural and artificial propagation of sockeye salmon at Cultus Lake. He was the second full-time scientist (after the director) at the Pacific station. He started the Cultus Lake studies in 1922 as a graduate student, and continued under the board to compare the numbers of natural versus fish-culture hatched fry that survived to the seaward migration stage. 'Over a period of "eight broods"' this research, 'classic in the field of salmon research,' revealed the high mortality rates of salmon fry. It also showed that the popular fish-culture programs did not produce enough fish to be worth supporting, and led to salmon hatcheries in British Columbia being closed. This vitally important science helped change the status of Canadian Pacific fisheries research. By 1933, important new studies had been begun: critical examinations of hatchery procedures; factors limiting the production of young salmon; and studies on the usefulness of transplanting sockeye salmon from one area to another. Extensive tagging operations uncovered migration routes of various salmon species and considerable information on their life histories. The province's fisheries department collaborated in annual statistical analyses of sockeye salmon runs in the four major river systems.[84]

In addition, pilchard and Pacific herring, crabs, clams, scallops, and oysters were studied. The pilchard once fuelled a reduction industry for oil and meal; herring were also used for this and were exported to the Far East. Scientists wanted to determine 'the relation of the fishing drain to the stock' through 'statistical analysis and sampling of the catches, supplemented by biological studies.'[85]

In 1930 and 1931 the board appointed its first full-time staff members in oceanography. The new oceanography program concentrated on the coastal waters, but a sporadic and 'modest' offshore program was begun when occasionally the Royal Canadian Navy provided a suitable ship. Work on pulp-mill effluent dispersion patterns at the Alberni Inlet determined 'measures needed to prevent damage to fisheries. This was the forerunner of many future problems relative to pollution and fisheries.'[86] By 1933, eleven full-time scientists were on the staff of the Pacific station.

Since the Pacific station was now producing economically significant science, and the elegant Cultus Lake experiments were offering clear guidelines about fisheries practices, it naturally grew in lustre for the Biological Board and for the government as well. So far, the St Andrews station had been drawing a larger proportion of the total funding ($78,190 in 1931 compared to Nanaimo's $45,970, partly due to the Passamaquoddy investigations). This situation now changed. The Depression levelled the funding. Relative parity was achieved in 1935. In 1936, St Andrews received $44,310 while Nanaimo was allocated $48,909; the following year, the allocations were $53,852 to $61,209 respectively. Glancing well forward, in 1953 the estimates were $430,814 for St Andrews compared to $460,092 for Nanaimo. Clearly, the Atlantic station had declined in significance.

That said, the Depression also hurt the Pacific station's research program, by forcing it to cut back on volunteer researchers and by narrowing the scope of investigations. R.E. Foerster, who became director in 1940, in 1942 observed recent 'very definite and significant change in the character of the research carried on at the Pacific Biological Station,' with more emphasis 'on economic fisheries problems.' The objects of the 1942 program – 'life history studies of the Pacific salmon, herring, pilchards, ling cod, clams, and oysters, and statistical investigations of the commercial fisheries thereon' – were, first, to give 'biological information upon which to base adequate regulatory measures,' and second, to evaluate 'the condition of the fishery and its evident trend.' Oceanographic field work, he noted, 'has been considerably reduced.'[87]

The general programs slowed during the Depression, and 'progress was reduced further during the early years of World War II owing to loss of scientific staff on war service.' Nevertheless, Pacific fisheries biology was essential in order to closely monitor the Pacific salmon fishery. In 1937 the International Pacific Fisheries Commission was formed, and Fisheries Research Board scientists were responsible for the Canadian portion of its intensive examination of Fraser River sockeye salmon.

The Fisheries Research Board

Cameron's chairmanship consolidated the trend toward more industry-related research, although more fundamental research was not completely abandoned. This trend was soon to be reflected in the change in name of the Biological Board to the Fisheries Research Board of Canada. A.W.H. Needler, director of the Atlantic station from 1940 to 1954, commented during The Second World War that assisting fisheries 'by guidance of government policy' and improving fish culture required 'discovering the fundamental principles involved,' but not as much of this was done as formerly. He argued that much more 'remains to be learned especially of the effects of the fisheries on the stocks, and perhaps more important, of the maximum yields which can be maintained.'[88] But even with this end, directed toward fisheries management, this fundamental research was undercut by underfunding and by growing competition between the biological stations and the experimental stations' fish-processing research.

After the fisheries experimental stations opened in the mid-1920s, the board's funding came under two separate votes. To the scientists' consternation, the experimental stations received much larger allocations. McMurrich warned the deputy minister that the 'fundamental' work of the biological stations 'directly or indirectly supplies the data which may be applied at the Halifax and Prince Rupert Stations.' Cutting the basic work would 'react unfavourably, sooner or later, on the other Stations.' He contended that applied science would 'grow sterile' without the input of pure science, as recognized by Eastman Kodak and General Electric, which encouraged 'investigations in pure science [of] no direct application to the industry.'[89]

It is a testament to Cameron's commitment to fundamental science that he was able to convince the government to favour the original stations. In 1932 the Halifax station was given an estimate of $62,500 compared with the St Andrews station's $43,350, by 1936 the figures had shifted so that St Andrews got $44,310 while Halifax got $39,645.

Nevertheless, pressure was being applied for two very different research agendas. University-inclined scientists wanted broader basic research; the government, in the person of W.A. Found, wanted instant practical results. Therefore Found again stepped up efforts to place the Biological Board under civil service control. Found had never given up on the idea of making the board in fact and in constitution the Department of Fisheries' scientific division.

In 1931, Found inquired into the structures of the U.S. Bureau of Fisheries and the British Ministry of Agriculture's fisheries research institutions. In the former, the Bureau of Fisheries' sixty-five or so scientists had been selected by the commissioner and deputy commissioner of fisheries and were civil servants, whose salaries were governed by civil service regulations. Applicants were required to have a BSc at minimum, to take civil service entrance examinations, and to pass a physical examination. Senior researchers required at least two years of field experience. In the eyes of civil service bosses, the typical hard fieldwork, in remote regions under exposure to the elements, ruled out hiring women scientists: 'While women are by no means excluded from the bureau's service, the opportunities for ... work free from the exposure incident to field investigations are naturally limited in number.'[90]

In Scotland, the fishery board's superintendent directly controlled Aberdeen's marine station. The twenty-two staff, civil servants all (with pension benefits), included two senior naturalists, five junior naturalists, and fifteen technical and laboratory assistants and fish measurers, as well as crews and officers for the two research vessels.[91]

In England and Wales, the Ministry of Agriculture and Fisheries directly controlled the Lowestoft laboratory and the thirteen scientific staff as well as a large number of laboratory assistants. All were civil servants under the Civil Service Commission. A marine station at Conway, North Wales, where three scientists on staff were engaged in cleaning mussels and oysters of sewage bacteria, was also controlled by the ministry.[92]

Armed with these examples, Found set out again to reform Canadian fisheries biology. In February 1931 he informed the deputy justice minister, W.S. Edwards, that the fisheries minister wanted the fisheries experimental and biological stations to be 'administered directly by the Department.' McMurrich and Cameron argued back that 'any action that savored of the annihilation of the Board would result in the loss of the advice and assistance of scientific men from the universities.'[93]

Found proposed a bill to reduce membership in the board, which had grown unwieldy with new representatives from western universities (it now numbered eighteen members). He wanted only six university members, and two men each from the industry on either coast, to act as an advisory board, supervised by a permanent executive officer under the minister. Found also proposed a new name: the 'Advisory Fisheries Research Board of Canada.' The new board was to consider and recommend research agendas, 'leaving the actual control and executive work to an officer to be appointed directly for that purpose.' This was to be secretary-treasurer J.J. Cowie, who already 'acts in that capacity from year to year in a voluntary way.' These changes would link the board's work 'more closely with that of the Department.'[94]

The 'Act to repeal the Biological Board Act' was approved by the Privy

Council on 18 March 1931. However, the Depression halted action until 1936. Fortunately, the board had been busy, and saved its executive powers.[95] The board now consisted of nine scientists and two industry representatives from each coast, plus two representatives of the Department of Fisheries. The other proposed changes came into being. In 1937 the Biological Board of Canada became the Fisheries Research Board of Canada. The new name signalled an official change in research emphasis,[96] which had already occurred in fact if not on paper. The scientific members were determined to safeguard the board's independence in hiring scientists and that choice of research priorities. But the Second World War struck before the reorganization was complete.

The war marked a real changing of the old guard. Found retired in 1938 and died in 1940. Cowie, secretary to the board since 1923, retired in 1940 and died in 1943; A. Handfield Whitman, a board member since 1923, retired in 1941, and his Pacific counterpart, John Dybhavn, died in 1943. Huntsman's influence was much attenuated, and many of his fellow former station directors left for university or other positions. W.A. Clemens, director of the Pacific station since 1924, resigned in 1940 to take up the zoology chair at the University of British Columbia. Thus, in more ways than one, the war marked a watershed in the board's history. The board shook off much of its history and traditions, and with new staff outnumbering and replacing many of the older and familiar scientists, it set off in new directions.

Cameron oversaw the stations throughout these changes and remained chairman through the lean years, when seven experienced scientists fled the board in 1937 after their salaries were frozen. One went to the National Research Council, two to the Salmon Commission, two to universities, and two to private industry (the Atlantic Coast Fisheries Corporation and British Columbia Packers). The board was like 'a happy hunting ground for recruits to universities and private industry.' In 1939 the permanent scientific staff numbered forty-one, and Cameron had to fight the Department of Fisheries for better conditions. He succeeded. By the time of his death in 1947, the board had grown to a staff of seventy-eight scientists and eighty non-scientists, as well as fifty-seven scientists employed temporarily for summer fieldwork.[97]

The Pacific station now emerged as the board's flagship station, and to this day is much better known than its Atlantic counterpart. The reasons for this are not far to seek. In a new board defining its purpose in terms of solving fisheries problems, the Pacific coast offered problems both immediate and solvable. Salmon dominated the fisheries, and their capture was far different from practices in the Atlantic fisheries, where trawlers predominated. To complete their life cycles, salmon must travel upriver to spawn; in doing so they expose themselves to easy capture. This also means their populations are much easier for scientists to monitor. Since the 1920s, scientists have learned a great deal about salmon life histories and physiology. Measures to save the populations are easy to monitor and test, and unsuccessful strategies can quickly be corrected.

In contrast, on the Atlantic coast, up until the 1960s scientists keeping track of the fisheries and the size of groundfish stocks did not foresee any serious problems with the then enormous fisheries. There was also no chance for Canada unilaterally to police these fisheries, the bulk of which were conducted in international waters. Canada joined the International Commission for Northwest Atlantic Fisheries, formed in 1949 to manage fishery resources and exploitation, but the Grand Banks and other fishing grounds were being fished by more than ten nations, over which Canada had no control. Signatories to the Northwest Atlantic Fisheries Treaty on 8 February 1949 included representatives from Canada, the United States, Newfoundland, Britain, Norway, Denmark, Iceland, Portugal, Spain, France, and Italy. The treaty, ratified in 1950, established the International Commission for Northwest Atlantic Fisheries (ICNAF), which provided 'the machinery for international cooperation in the scientific investigation and development of the fishery resources.' It had no regulatory powers; instead it made recommendations (many of them ignored) 'to the respective governments for necessary measures to maintain the stocks of fish in the convention area.'[98]

Canada's proximity to this important fishery was recognized when the commission's first meeting, an organizational one, was held at St Andrews. The Atlantic station's program was now closely allied to a cooperative program organized through the ICNAF, which worked to develop regulations to safeguard and sustain the northwest Atlantic fisheries. The irony here was that although Canadian territories abutted the great northwest Atlantic banks fisheries, St Andrews was very distant from them. With the research program increasingly focused on offshore fisheries, the St Andrews station's location was a horrible disadvantage.

After the fire, the St Andrews station showed its weakness both as a centre for fisheries science and as a centre for Woods Hole's kind of general biological science. It suffered a slow loss of prestige after Huntsman's departure. His successors, A.H. Leim and A.W.H. Needler, were competent but quiet scientists who did not call attention to themselves and their agendas. Thanks to the board's rejection of its former educational role, the St Andrews station was unable to compete with Woods Hole as a centre of excellence. Had the station been relocated to Halifax, the story might have turned out differently. In Halifax it would have become the central laboratory around which the oceanographic facilities of the later Bedford Institute of Oceanography would have been established. The fish-handling work would have been in close proximity, and the postwar marine biology programs at Dalhousie University also would likely have overcome the board's policy of distancing its work from the universities.

Earlier, the importance of education had been much more clearly understood. Indeed, in 1927 the Biological Board became involved in a premature attempt to establish undergraduate fisheries biology in Canada. A.P. Knight wanted Dr A. Stanley Mackenzie, president of Dalhousie University, to institute a degree in fisheries, for which the board 'might furnish the special instruction in fisheries subjects.'[99] Zoology professor James N. Gowanloch at Dalhousie had developed courses in marine biology 'with considerable success.' But Knight wanted undergraduate instruction 'such as is given at the laboratory at Woods Hole,' and which board policy avoided.[100]

Mackenzie agreed both that the fisheries would benefit greatly from 'men with thorough training in science'[101] and that Dalhousie should institute a four-year B.Sc. in fisheries, with specialization in marine zoology, histology, and physiology in the third year; and applied fisheries sciences in the fourth year. He felt, however, that university staff were not qualified to teach the latter.

The board was willing to provide qualified scientific instructors. But Gowanloch requested a biological station with boats, equipment, and a library for Dalhousie for teaching ecology and marine biology. Not surprisingly, the university could not afford the $20,000-plus annual funding.[102] (This would not have even been an issue had the Atlantic station been in Halifax.) With the minister's support, the Department of Marine and Fisheries agreed to provide $25,000.[103] But the money could not go directly to Dalhousie. Also, McMurrich thought it 'highly inadvisable that the Board should embark upon University instruction work, except indirectly as we do it at present.'[104]

At a meeting between Cowie, the board executive, and Dr Mackenzie in Ottawa on 26 November 1927 Huntsman argued that the St Andrews station was hopelessly overcrowded and that the university station would be a helpful board expansion. McMurrich pointed out, however, that the board could not direct this laboratory, even as 'a feeding station for investigations in fisheries,' if it was only to benefit Dalhousie. Any laboratory under the board's jurisdiction had to be open to all Canadian universities. Mackenzie responded: 'All we ever wished for ... is what you are proposing.' When the project was approved, Mackenzie exulted. 'You have almost started here today a small sized Woods Hole in Canada.'[105]

The Biological Board was responsible for the 'Eastern Passage Laboratory' from its inception in 1927 until 1936. Unfortunately, there was no resemblance to Woods Hole. Ruth Fulton Grant noted in 1932 that 'few students enrol in the course,' that the laboratory 'needed considerable equipment,' and that it 'could hardly be said to fulfil efficiently "the purpose of giving an opportunity to universities for training men in fisheries science."' It was another victim of the Depression,[106] forgotten as the board struggled to maintain its primary stations. Also, the driving force behind the program had originally been Gowaloch; unfortunately, he had lost his position at Dalhousie at 1930 owing to a complicated sexual scandal of which he may have been the innocent victim.[107] In 1932 the fisheries course had only four students. In 1933, two Icelanders on Iceland government scholarships were the only graduates.[108] Few graduated before the course was curtailed for lack of interest.

Had the Atlantic station been located in Halifax, interest in the project would have revived following the war, and collaboration with the universities would have benefited the entire marine science agenda on Canada's eastern seaboard; if not another Woods Hole, still a very reputable enterprise would have emerged.

The political geography of Canada also figured in the slow decline of the Atlantic station, which became only one of a constellation of lesser stations. Despite the Depression, the board had continued to grow. New biological stations had been added earlier, including the temporary one at Cultus Lake in 1924, and a permanent substation, recommended by the Maclean Commission, at Ellerslie, Prince Edward Island in 1929. The mandate of the latter was to improve oyster culture, instruct oysterers, and establish experimental oyster 'farms.' Here, during the 1930s director A.W.H. Needler convinced local oysterers that they needed to lease grounds to control the harvest. He then helped establish a stable oyster culture industry.[109] In 1936 a small fisheries experimental station was established at Grande-Rivière in the Gaspé.[110] Later the board added the Newfoundland Biological Station and a technological unit in St John's after the Confederation of Newfoundland in 1949; and the Arctic Biological Station at Ste-Anne-de-Bellevue, Quebec, in 1955. This list does not mention several lesser substations.[111] Each of the five provinces on the Atlantic, therefore, had at least one station. As a result, funding was diverted away from the Atlantic station.

Contrast this with the situation on the Pacific coast, where British Columbia is the only province. The Pacific Biological Station at Nanaimo and the Fisheries Experimental Station in Prince Rupert (shifted to Vancouver in 1943) were the only two stations. Temporary substations at Cultus Lake (1924–36) and elsewhere were always closed once an investigation ended (the Cultus Lake station was taken over by the International Pacific Salmon Commission in 1938).

The focus of this book has been marine biology in Canada and elsewhere, but the Biological Board and its successor also had important freshwater programs. These became more important after the war. Thus, further dividing the Fisheries Research Board's energies and funding was the Central Fisheries Research Station in Winnipeg, Manitoba, founded in 1944. This was relocated to London, Ontario, in 1957 and simply called the Biological Station. Also in London was a technological unit for fish-processing research, founded in 1955. In Burlington, Ontario, the FRB Great Lakes Biolimnology Laboratory was established in 1967.

Ambiguity regarding St Andrews dramatically weakened support for the former proud pilot station of Canadian marine science. When A. Handfield Whitman came out against rebuilding the distroyed station in favour of reinforcing the board's Halifax station, his prophetic ideas opened a protracted debate. When Dr J.L. Kask became chairman of the Fisheries Research Board (1953–63) he made a concerted effort to 'close out St Andrews.' Kask had the support of both Stewart Bates, the deputy minister, and the fisheries minister himself, James Sinclair (Prime Minister Pierre E. Trudeau's future father-in-law). Even with all this support, Kask did not pull this off: the Fisheries Research Board's chief oceanographer, H.B. Hachey, 'got the support of C.D. Howe, who had more umph than Sinclair and stopped it.'[112]

In many respects it is a shame that Kask failed. One industry representative on the board, O.F. MacKenzie, 'went over to St. Andrews one time and found that the ships were going to leave St Andrews, sail around to Halifax, provision, fix themselves up, and go off to a cruise, come back to Halifax and drop off their gear, and sail around to St Andrews again.' This was their regular routine for Atlantic coast research. MacKenzie, 'running a fleet of vessels for profit,' was outraged, and 'was determined that the St. Andrews station should be washed out.' But even after Kask's failure, the station could have been revitalized. Instead it was handled very badly. There was poor morale 'in St Andrews because they didn't know then and they have never known until very recently [1972] whether they were going to be moved to Halifax or what was going to happen.'[113] First, the board's oceanographers were moved to Halifax. Dr John L. Hart, director at St Andrews from 1954 to 1968, said that St Andrews was 'a station that essentially in 1954, and every two years thereafter, was condemned to death. I did manage to keep the body breathing, and all I can say is that John Anderson had something to work on when he came ... This is literally true.' He commented bitterly that 'the senior staff was bled off for Ottawa and other stations and great chunks of the station were moved to other areas. Everything that became important was moved someplace else. They moved the salmon investigation to St. John's. They moved oceanography and bottom studies to Dartmouth.' The best scientists were also moved away.[114] Eminent scientist and later temporary chairman W.E. Ricker commented that his predecessor, Dr F.R. Hayes, who became chairman of the Fisheries Research Board in 1964, viewed St Andrews as a 'scientific backpool,' and that Hart retired early because of Hayes.[115] Indeed, Hayes later observed: 'I don't think anybody would put a laboratory in St. Andrews, if they were building it now. It was put there because it was historically a good place.'

Hart believed that 'rather than moving it limb by limb,' it would have been better had the station been moved wholesale. He also complained: 'I can't see the rationale of this multiplicity of stations all over the place. I think that it is extravagant in equipment. I think that it is extravagant in effort ... Stations on the Atlantic coast ... are competing.'[116]

The truth of this is evident in the Pacific station's uncontested growth and increasing success; it never had to compete with other federal Pacific stations. In truth, the St Andrews station, without a strong advocate, almost faded out of sight, where once in its glory it had been in the same league as Woods Hole.[117] By the 1960s, St Andrews was competing with many other Fisheries Research Board establishments on the Atlantic coast; furthermore, it was 'really in competition with a very rapid development of marine science in the Halifax–Dartmouth area, which was really building up to the level of marine sciences competitive with Woods Hole and La Jolla and all the big marine communities.'[118]

A small resurgence for St Andrews came in the late 1960s, when Dr John Anderson became its director. In 1968 he opened up the station to university scientists, providing a large double-trailer for their work, as well as making boat and laboratory facilities available. Because of the interest of university researchers, a permanent establishment was opened in 1970, run by a board of directors from member universities across the Maritimes and Ontario, as well as representatives from the Woods Hole Oceanographic Institution, the Fisheries Research Board, the New Brunswick Department of Fisheries and the Environment, and the State of Maine. This moderate-sized marine laboratory was endowed with a magnificent older house above the station, which now serves as a residence. Here university biologists follow their own research programs, and a vigorous program of undergraduate instruction in marine biology through intensive week-long courses is carried out throughout the summer, with students and their instructors coming for scheduled courses offered by their own universities, including those in Ontario and Maine. Anderson, in the words of one Fisheries Research Board scientist, 'should have credit for a lot of imagination and leadership in recognizing that the only way St. Andrews was going to survive [Halifax's growing marine science community] was to ... recognize the base on which St. Andrews was established in the first place, that is the need of all universities interested in aquatic science in the eastern half of Canada to have a place where they could go and work.'[119]

An aging A.G. Huntsman was on hand to open the facilities, which were named in his honour. With the Huntsman Marine Laboratory, the story comes full circle, since the marine laboratory exists to fulfil the same functions for which the first Atlantic and Pacific biological stations were originally established – research by the sea, instruction of students, and work that may have some importance for helping the fisheries.[120] The success of this parallel facility indicates that the Biological Board in its original incarnation filled a very real need that the Fisheries Research Board chose to abandon, ultimately to the detriment of both the science followed by the board and its independence from government interference.

Epilogue: Balancing the Scales

> The public demands that fishery management decisions be scientifically credible ... When controversy does arise it is usually because there is high scientific uncertainty and low predictability of the future ... In these circumstances, science has relatively little to offer and the minister may decide to go with his/her political instincts. Departmental scientists may find this situation frustrating but in the final analysis it is the minister, not the scientist, who is accountable to the public.[1]

> A [Cabinet] minister does not run his department – he has neither the time nor the freedom to do so ... I acted as the spokesman for my department ... My experience as a minister tells me, no matter how much I tried, I never had the control and power over my departments that would have given me the ability to answer for all that went on within them.[2]

Canadian fisheries science, begun at the turn of the twentieth century by a jauntily confident coterie of university biologists, optimistic that their studies would result in material gains for science and fishermen alike, ended the century in a very sombre condition. The catastrophic northwest Atlantic cod and groundfish stock collapses of the early 1990s shook the confidence of fishermen (who were already sceptics about the science) and lay observers, and led to recriminations and some much-needed introspection within the fisheries science community. Questions were asked about the competence and integrity of the biologists involved and even the value of fisheries science itself. It was a dark period, and there is still no sign of stock recovery.

However, this ecological and economic disaster did have one very positive outcome: it showed fisheries biologists that their operating scientific paradigm, that of maximum sustainable yield, was seriously flawed, and forced them to reassess their fundamental premises. Even if, as some argued, temperature fluctuations were more important than fishing intensity in the disappearance of the cod, the fact remained that the cod were missing. And fishing had to be a factor in this, no matter what other elements were involved.

By the beginning of the new millennium, new models were emerging worldwide, rooted in the precautionary principle, such as ecosystem-based fishery management. The precautionary principle proposes that an ecosystem should not be pushed to its limits in the name of efficiency; rather the fishery ecosystem's complexities and uncertainties should be respected, and risk-prone management decisions should be replaced by systems that allow a wide margin of error to favour the survival of fish stocks.[3] Like all sciences, fisheries biology is progressing by reassessing and discarding failed models and by finding new theories that better serve humanity's increasing understanding of the world. Unfortunately, the crises that exposed the old paradigm as a failure were more painful for non-scientists than the crises encountered in most sciences. The failure of the northwest Atlantic groundfish in the 1990s, and the earlier collapses of the Californian sardine fishery in the 1950s and of the Peruvian anchovy fishery in the 1970s, threw many out of work and compromised important food and ecological resources of world importance. Yet it seems to have taken the destruction of what was once the world's most prolific fishery to drive home the failure of the old model, even though Canadian fisheries biologist Philip Larkin penned 'An Epitaph for the Concept of Maximum Sustained Yield' as far back as 1977.[4] But now that scientists are developing new approaches, there can be some hope that more conservative harvest projections will aid the cause of saving the fisheries for future generations.

The cod stock collapse has many dimensions, and its political and social agents, such as rapacious greed, international scrabbling over a limited resource, the overcapacity of factory trawlers, and the Canadian use of the fisheries as a welfare extension device, have already been chronicled in many other books. More pertinent to the themes within this study are the structures that allowed an impending crisis to go relatively undetected in the 1980s, and the responses of the scientists themselves to the emerging crisis. Although the following examination of the issues that have riven recent Canadian fisheries biology may seem out of place in a book that has focused on a science's emergence up until the Second World War, it is important to discuss how the later critical developments may have been exacerbated by structures and conventions already in place earlier in the century, as well as by the rejection of other very important approaches that had at one time served well the scientific and the fishing communities.

The collapse of the northwest Atlantic cod stocks heralded not just difficulties within the Canadian practice of fisheries science but problems endemic within the entire fisheries biology community. As has been the practice in the rest of this book, I propose to look at the general problem first, and then examine the Canadian context in somewhat greater depth.

Endemic Problems in Fisheries Biology, 1950–2000

Fisheries science organizations, until the chastening failures of the late 1980s and 1990s, reflected the beliefs and concerns of the conservationists and wildlife managers of the early 1900s – the 'Gospel of Efficiency,' highlighted famously in Samuel P. Hays's book.[5] Managers of national parks, forests, and other resources, including the fisheries, were wont to describe their conservation goals as a 'business proposition.'[6] In the early years, the new conservation and management programs were tempered by disagreements between preservationists and 'wise use' advocates. During the middle decades of the twentieth century the wise-use advocates dominated, which allowed, for example, moderate logging and other harvesting practices to persist in national parks. In fisheries science, the gospel of efficiency shared by most if not all scientists was, until the Depression, moderated by the continuing Victorian naturalist tradition, so that many scientists were engaged not so much in trying to figure out how to manage the fisheries, as in trying to understand fish and their environment. The Depression, however, brought economic concerns to the fore, and thereafter the primary characteristic of fisheries science was a business-oriented and technocratic approach. Scientists saw their science as 'user friendly,' especially to fishermen who were willing to invest in more efficient harvesting technologies (which scientists helped develop) and to fish 'underexploited' or unexploited fish stocks (which the scientists helped locate). Government and business needs were paramount.[7]

This business-oriented approach led scientists in effect to turn their backs on less efficient forms of technology, and by extension, their users, who were primarily the poorer near-shore and fixed-net fishermen. This trend first became evident in Canada in 1928, when the Maclean Commission recommended restricting Canada's steam trawler fleet to three, in favour of traditional fishing boats – a move strongly opposed by the Biological Board and its scientists, who argued that trawlers offered a better opportunity for a steady supply of fresh fish in good condition that would be more palatable to inland markets. Motivating the scientists were not biological factors, but business considerations and the desire for efficiency. Such efficiency, board scientists argued, would mean that fewer fish would be wasted – and thus fish populations ultimately would be better conserved.[8]

Fisheries management was now viewed as a business with more or less fixed parameters, and this entailed taking a highly mathematical approach to assessing stock abundance. Granted, estimating fish abundance is of paramount importance in any management scheme, but the chosen approach, as will be discussed in depth later, was to follow the quantitative assessment theories developed by Graham, Thompson, Schaefer, Ricker, and Beverton and Holt, with the goal of setting 'specific catch recommendations for individual species, using biological reference points.' By moving the frame of reference for fisheries science toward complex and specialized series of formulae, scientists 'began to assume the responsibility for framing management objectives for the industry,'[9] at a time when economic forces, new technologies that allowed high harvesting efficiency, and large business interests were placing fish stocks and small-scale fishermen alike under increasing pressure.

This in turn changed the industry's power structure. Poorer fishermen – and indeed, even the larger corporations – lost any control over how the fishery was conducted. Management was carried on in a top-down fashion, as the power for deciding and allocating the annual fish harvest quotas was placed in the hands of government bureaucrats being advised by the Northwest Atlantic Fisheries Organization (NAFO) or the earlier International Commission for the Northwest Atlantic Fisheries (ICNAF). Later, after the extension of the two-hundred mile limits in 1977, management decisions were made in the centralized bureaucracies of the European Union or in Canada's Department of Fisheries and Oceans, where the fisheries minister was given the ultimate responsibility. There was little decision-making input from small-scale fishermen in any country. This shutting out of fishermen is illustrated by the events surrounding the northern cod stock collapse: fishermen's warnings, and pleas for a fishing moratorium through the late 1980s, were ignored by the Department of Fisheries and Oceans and most of its scientists.

Having been disenfranchised, fishermen in Canada and elsewhere grew increasingly cynical. With no stake in management, many developed a truant's attitude, underreporting catches, ignoring quotas, and failing to report the dumping of prohibited fish species, undesirable species, and undersized fish caught. These are supposed to be reported as part of the catch quota, since the dumped fish are dead, but this would have left the fishermen with a large portion of their catch being counterprofitable. By not reporting the dumped by-catch, they could carry on fishing, claiming that they had not yet met their quotas. When fishermen are not stakeholders in conserving a stock, it actually makes sense for them to fish a stock with low productivity to extinction, and then to direct their capital investment elsewhere.

Various nations, competing for a limited resource on international waters, have ignored the quotas recommended by NAFO or ICNAF. Now that their own territorial limits have been extended, first to fifty miles and then to two hundred miles, they have gone into other nations' territorial waters for fish stocks protected under national conservation measures. British trawlers did this during the Cod War of 1972–6. Iceland defended its fisheries using military vessels, and live shot was fired on several occasions, striking a British trawler and tug boats. Seven British war frigates were despatched to protect British fishing vessels, but Iceland prevailed in the EEC, and Britain was forced to withdraw.[10] Later, in spite of a fishing moratorium, Portuguese and Spanish vessels strayed within the Canadian two hundred mile limit to carry on fishing after the cod stock collapse in 1991. Despite the importance of the fishing sectors in their own economies, none of the transgressor nations took any direct interest in the health of the fish stocks themselves.

Public officials have no real stake in the effects of their decisions, since they will not lose any income or livelihood if the stock fails. For example, bureaucrats ignored scientists' warnings in the decade leading up to the 1977 North Sea herring fishery collapse, in part because of the uncertainty involved in the quantitative assessment theory (about which more will be said shortly); again, the officials had no personal stake in the results of their choices.[11] Indeed, official attitudes were of no help to those concerned about stock conservation. For example, the first fisheries director of the FAO (established in 1943 and made into a special service of the UN in 1945) was a former Fisheries Research Board scientist, Dr D.B. Finn, who accepted T.H. Huxley's contention that most fish stocks are inexhaustible. Finn, interviewed at the FAO headquarters in 1972, remarked: 'Marine fish. You don't have to be afraid of extermination except in very few cases, the whale possibly, the salmon possibly.' He pointed out that when the North Sea fishery was closed for four years, due to overfishing, 'the damn fish started to go up the rivers.'[12] It is worth remarking that Finn's background was not the fisheries science of Johan Hjort, Michael Graham, and the rest, but biochemistry, which he studied together with physiology under A.T. Cameron. He was appointed the first director of the Prince Rupert Technological Laboratory in 1929, progressed through the government, and became deputy fisheries minister in 1940. Thus he never wrestled with the tangled problems of marine ecology and fish population dynamics in his personal scientific work. Given their adherence to the idea that fish are inexhaustible, such officials were not likely to lose sleep wondering whether overfishing might cause lasting problems. Finn is typical of what evolved by the late twentieth century, during which a system developed in which no one who had any say over the direction of fisheries management – no scientist or politician – had any real personal stake in the system, whereas those who bore the greatest risks had no power.

Adding to these managerial weaknesses were the difficulties inherent in the quantitative assessment theory, which left a high degree of uncertainty as to the actual status of fish stocks. The general problems here can be illustrated through Canadian examples.

Quantitative assessment theory was developed by scientists working closely with fisheries under pressure. For example, as was argued in chapter 7, very little use was made of mathematical modelling by Canadian fisheries biologists working on the Atlantic coast until the 1950s. Overfishing was not a serious issue there until after the Second World War, when the sails and small steam trawlers of Maritime fishermen began to disappear, as they began to lose the Darwinian struggle with highly capitalized foreign fishing fleets. In 1953, to meet growing demand for fresh fish, the Russians introduced the factory trawler. Other nations followed suit. Factory ships trawled, processed, and froze entire schools of fish aboard ship. The stern-Otter trawl swept up all fish in its path, while the rollers on the bottom beam scoured the bottom and ravaged its ecology. Canadian and other scientists began to focus on the status of the northwest Atlantic fish stocks; but they were actually beginning these studies long after the populations had started showing signs of being under intense fishing pressure, as later historical analysis has shown. Yet the reigning approach was still to further maximize fishing efficiency, tailoring the fishing catches to the perceived status of the populations, arrived at through quantitative studies and fishing equations.

The range of error that creeps into predicting future stock size can be seen from scientists' long and valuable experience in assessing Pacific salmon stocks. After the 1920s the Pacific Biological Station focused on two fisheries in trouble: the Pacific salmon and halibut fisheries. The halibut fishery, begun in the 1880s, was in serious decline by 1920. The salmon fisheries began experiencing severe declines after 1892, when the nets of American canneries captured enormous numbers of Fraser River salmon migrating through American territory.[13] The Hell's Gate landslide on the Fraser River in 1913 wiped out the major sockeye salmon run of one of the most important salmon rivers. It is notable that the Atlantic groundfish stocks are widely dispersed; in contrast, an entire run of Pacific salmon can be wiped out by nets stretched across a stream. Overfishing, river habitat erosion by dams and factories, deforestation, and pollution all threaten wild Pacific salmon.[14] These problems opened the vexed question of how to guarantee the future Pacific salmon and halibut fisheries.

When Dr Wilbert Clemens took over as director of the Pacific station in 1924, his student R.E. Foerster's famous Cultus Lake spawning salmon studies marked the beginning of an intensive investigation into the five Pacific salmon species. These studies included physiology, life histories, and tagging programs to uncover migration patterns. Canadian biologists also participated in the International North Pacific Halibut Commission, begun in 1924 under American biologist W.F. Thompson. Thompson, like Huntsman before him, used a priori methods to understand the effects of fishing on the age structure of fish populations. However, Thompson checked his ideas against fishing data from 1918 to 1926. He was also more cautious than mathematician Vito Volterra, who had introduced fish population modelling. Thompson's more sophisticated models took into account his concern that fisheries scientists should not assume constant rates of fishing effort, fish growth, and recruitment. He worried that elaborate mathematical models might 'give the illusion of complicated scientific theory' yet fail to address reality.[15] The Halibut Commission's work resulted in new fishing restrictions that kept the fishery from collapsing in the 1920s and 1930s, and that served as a model for later fish management schemes.

Later scientists such as William (Bill) E. Ricker developed mathematical models to analyze the effects of fishing and other factors. An aquatic biologist, he worked in 1931 at Cultus Lake and served on the International Pacific Salmon Commission. Later, as professor of aquatic science at the University of Indiana, he worked on fish population dynamics. Marked or tagged lake fish allowed a very accurate assessment of the total number of tagged fish caught, compared with the number of untagged fish caught. In this way, fish population could be estimated. Ricker returned to Canada in 1950, and at the Pacific station further developed fish population dynamics studies. He developed the *spawner and recruit theory* and designed elaborate fishing equations for commercial fish population modelling.[16] Ricker was among the pioneers of this kind of science, and it became very popular on both coasts thereafter.

The brilliant thing about the efforts of Dr Ricker and later salmon biologists to refine mathematical models of fish population dynamics, was that the models could be corrected, year after year, by carefully counting the salmon returning to the rivers. The very vulnerability of Pacific salmon in the river fishery is paradoxically, the source of their future survival (given enough political support from Ottawa), since population sizes are relatively easy to monitor. Yet even for Pacific salmon, the errors in mathematical modelling remain huge. In 1987, Dr H.D. Smith admitted that even with more than twenty-five years of data on river stocks, their best models gave 'forecasts [which] commonly differ from the actual [salmon] return by more than a factor of 3.'[17] Smith suggested that scientists did not know enough about actual sea conditions. When the Fraser River sockeye salmon stock failed in 1994, the total allowable catch (TAC) had been set for 80 per cent of the returning salmon. Scientists were shocked when the expected three million salmon failed to appear. High catch rates certainly did not help, but there were other unforeseen stresses. Warm river conditions and events at sea may have been factors, but biologists do not know for certain.[18]

Given such huge errors in stock size assessments for a species for which relatively instant confirmation or rejection of assessments is possible, the problems inherent in modelling populations of oceanic species are apparent. Oceanic stocks may be scattered, migrating, or shoaling irregularly, and in any case they are mostly invisible.

Worse still, scientists here and elsewhere have failed to reflect certain factors in their equations. In retrospect, these omissions seem surprising. For example, scientists have never been able to discern a relationship between the number of adult spawners in a population and the recruitment of young fish. While 'logic would suggest an affirmative answer ... given natural fluctuations in the environment and in the corresponding data available to scientists, there is often no scientific "proof."' Since no sense could be made of this problem, it was ignored. No algorithm for the estimated number of spawners and subsequent probable year-class size was ever incorporated into fishing equations used for Atlantic groundfish stock assessments. As Canadian fisheries biologists Ransom Myers and Jeffrey Hutchings noted in their 1994 analysis of the cod stock collapse, the effort dedicated to collecting biological data such as 'age- and size-specific schedules of survival and fecundity,' and 'the biological interactions that influence a species' abundance,' was limited. These data are hard to collect and interpret, so scientists instead used straightforward indices such as abundance, population structure (numbers in each year class), harvesting mortality, and fishing effort. Also, 'the fundamental idea of maintaining a critical spawning stock biomass (to preserve the reproductive process) was not put into practice.'[19]

On top of these uncertainties, fisheries biologists have conceded that overfishing effects can sharply increase the stresses on fish populations caused by natural environmental events, which can include competition with and predation by other species. A population may decline suddenly due to unusually warm water; if overfishing is occurring at the same time, that decline may turn into a calamitous collapse. This is what happened to the Peruvian anchovy fishery in the early 1970s, when El Niño warmed Peruvian coastal waters.

The problem is that scientists still do not know enough about the dynamics of the entire ecosystem to be able to predict the effects of all these interactions. Indeed, 'first and foremost, the [quantitative assessment] theory takes little account of the tendency towards instability within the oceanic environment.' Food supply, numbers of predators, parasites and diseases, water temperatures, currents, and other factors may vary greatly from year to year, affecting the numbers of fish that survive and spawn. These are difficult to monitor annually, so fisheries biologists have chosen to write into their equations and computer models a 'natural' mortality factor of 20 per cent. These and other 'weaknesses ... erode confidence in the accuracy of the results.'[20] Scientific models also oversimplified 'behavioral characteristics of different fish stocks' and ignored complex species interactions *'through its insistence to date on reference to single species.'* They also shrugged off the effects of scarce and fluctuating resources, technological development and human behaviour.[21] As one commentator, J.A. Percy, put it, the typical models considered only the target species and made no 'allowance that every species relies on others as food, and in turn serves as prey for yet others. Thus separate models [were] used to calculate Optimum Sustainable Yields for capelin and for cod.' Yet capelin are the preferred food of cod, and this links the population cycles of the two species. Also, the models ignored 'fluctuations in the numbers of the many other species that the cod intimately shares the ocean with, from shrimp to seabirds to seals. Each commercial species is assessed, modelled and managed as though it lived all by itself in the ocean.'[22] Yet the reigning program in fisheries management was to maximize the efficiency of the fisheries, not allowing any 'excess' fish to be 'wasted' by being lost to natural mortality. This demands that we ask: What exactly is an excess fish in a natural ecosystem?

Adding to these problems has been ambivalence regarding the role of scientists in fish management. To avoid the perception that they, and not politicians, were framing management policy once two-hundred-mile limits were imposed, scientists in ICES and in the Department of Fisheries and Oceans after 1977 began to offer governments a range of 'safe' levels for TACs, given different scenarios. The managerial levels of government, then, could supposedly set quotas to pre-empt problems arising from the fishing effort and constantly varying physical and biological environments. In theory, quotas would be set to preserve a larger spawning biomass if other conditions were unfavourable. The problem was, of course, that these other factors could not be predicted at the beginning of the fishing season. To appease fishing interests, the higher, supposedly 'safe,' TAC levels were always chosen under these circumstances.[23] Thus, yearly quotas fished stocks to the edge (and beyond) of commercial extinction, all the while ignoring the 'claims' or needs of other species within the ecosystem for the fished species.

Yet another weakness was that fisheries managers used convenient geographical management boundaries rather than natural biological boundaries. For example, more than a century of local knowledge acquired by fishermen, and in some cases confirmed by scientific studies as far back as the mid-twentieth century, showed that northwest Atlantic cod substocks have widely differing behaviours. These varying behaviours relate to migration routes and times, age of migration, spawning areas, spawning ages and times, feeding areas, and so on. These variations affect the subpopulations' susceptibility to fishing, and thus create specific conservation problems. Yet the natural boundaries that had been identified for these subpopulations, and championed in 1965 by researchers such as Wilfrid Templeman, director of the Newfoundland Biological Station near St John's, were ignored by the ICNAF in favour of boundaries set by latitude and geographical convenience.[24] Canadian managers kept using these boundaries even after the two-hundred mile limits were established.

Scientifically verified traditional knowledge was rejected for reasons of convenience. Furthermore, the trend from the 1970s onwards in Europe, Canada, and elsewhere was toward scientists ignoring fishermen's input completely. Scientific ecological knowledge has been characterized as objective, analytical, and rational, whereas the ecological understanding generated by actual working fishermen has been seen as subjective, intuitive, sketchy, local,[25] and self-interested, and thus of little practical value. In the Canadian context, this trend was noted by sociologists Barbara Neis and Paul Ripley and by fisheries scientist Jeffrey Hutchings: 'Another feature of fisheries science and management in the 1980s was a substantial reduction in reliance upon input from and careful consideration of the local knowledge of fisheries workers, particularly those working in the inshore fishery [which] marked a significant shift from the past. Fishermen's knowledge came to be seen as "anecdotal" and of relatively little use in stock assessments and fisheries management.'[26]

Scientists preferred the 'hard' statistical science of computerized population modelling, based on sporadic scientific research surveys and 'reliable' (i.e., seemingly standardized, steady, and easily accessed) catch data from the industrialized offshore fishing fleet. The statistical methodology used by the Canadian scientists was virtual population analysis, which tracks each year-class and estimates its mortality, assuming a natural mortality of 20 per cent. But this method has a major delay factor – it takes about five years before scientists know how many fish there are. This situation led fishermen, buoyed by their 'confidence in their knowledge about populations' structures and migrations,' to challenge the fisheries managers and scientists' reliance on large management areas and on statistical methods. This criticism only intensified once the northwest Atlantic cod, haddock, and pollock groundfish stocks collapsed – an event that finally led, as outlined above, to a general agreement, not only among fishermen but also among many fisheries scientists and even bureaucrats, that 'the old ways of managing the fishing industry have failed badly and ... new approaches are urgently needed.'[27]

Problems in Canadian Fisheries Science and Management

The problems described above were seen in fisheries science around the world as well as in Canada. But within the Canadian context the science and management of the fisheries faced its own set of challenges and problems, some of them historical in origin. Perhaps the best place to begin is with the end of the Fisheries Research Board of Canada.

The Fisheries Research Board was relieved of responsibility for its research facilities and programs in 1973. Jack Davis, the acting Minister of the Department of the Environment (the 1971 reincarnation of the newly created Department of Fisheries and Forestry), wanted the Board's 'research and development' functions to line up better with the ministry's objectives, which included stronger efforts to address fisheries problems and to strengthen Canada's international bargaining position. In the past, the eighteen-member board had set objectives and policies for its ten research establishments, and enjoyed 'considerable support and good relations with its major clients: the universities, the fisheries industry and the Ministry of Fisheries ... providing research support for the fisheries' and advancing basic research.[28] However, it was unable to survive the Trudeau government's efficiency policies, whereby agencies with similar mandates were to be integrated, including all agencies involved with renewable resources. The final blow came in 1979, when the Fisheries Research Board was dissolved by an Act of Parliament and its work and scientists were absorbed into the newly created Department of Fisheries and Oceans.

Even before these events, however, the board had lost ground in its resource and environmental mandate. The Atlantic groundfish stocks' real decline in the 1960s came too late before the board's end to shape the attitudes of Canadian Atlantic fisheries biologists, with the possible exception of the St John's station's director, Wilfrid Templeman. Canada's Atlantic scientists generally had concentrated on economic problems from the 1900s onwards – on improving fish-processing methods and gear efficiency, and on finding new exploitable resources – in an effort to help strengthen the economy of the Atlantic fishing industry. Unfortunately, the board was slow to respond to the new political and environmental priorities emerging in the late 1960s, such as, for example, pollution research. Only in 1968, after the fishing industry began to express concerns that pollution was seriously threatening the fisheries, did the board gear up for pollution research as a major new initiative. The government, however, under the 1970 Clean Water Act, handed this research agenda to the new Marine Service Branch of the Department of Energy, Mines and Resources.[29] This new branch was then transferred to the newly formed Department of Fisheries and Forestry, as was the Fisheries Research Board. Ironically, the board in the early 1960s had given up its oceanographic work to the precursor of these departments, the Department of Mines and Technical Services. The board thus found itself competing for 'resource management' research territory with an organization ultimately descended from its own earlier Atlantic and Pacific oceanographic groups.

The last board chairman, J.R.Weir, tried to shift the Fisheries Research

Board toward environmental science, as his reports to the board in 1972 reveal. But the Marine Sciences Branch won this race, and as a result the Fisheries Research Board lost an important chance to expand its research activities,[30] which might have given it more clout in combating its ultimate absorption into the Department of Fisheries and Oceans.

I contend that during the 1960s, the Fisheries Research Board helped weaken its own case, and even sowed the seeds of its later weaknesses in the years leading up to the catastrophic collapses of the northwest Atlantic groundfish stocks. Forces within the board made it impossible for fisheries biologists to fulfil their mandate. For example, F.R. Hayes, Chairman from 1964 until 1969, had an active disdain for fisheries biology, and pushed the organization toward more exact sciences. He favoured biochemistry and physiology – cutting-edge, experimental sciences – over old-fashioned environmental and life-history work. During the 1960s, as fisheries biologist Philip Larkin later observed, studies in physiology dominated FRB research.[31] Indeed, Hayes did not think his initial research under Huntsman, which involved salmon work, was 'a suitable exercise for such amount of brains as I have to be fooling around in rivers with salmon.' When he served as board chairman, he scaled back basic life-history and ecological fisheries research. He complained about systematics and general population studies: 'The case history method of repetitive observation is of very little value in science beyond a limited number of years ... Charting the Pacific salmon runs, for example, in every little river and recording these salmon ... is a waste of time and money since I don't think you make progress by recording this type of repetitive observation.'

Yet this is the essence of fisheries science geared toward conservation and management! He pushed the board in the direction of supporting experimental work: 'I think in all other branches of biology the experimentalist has made the progress and the classical man, the case history man, has had very little progress.' Kask, his predecessor as chairman, had completely centralized FRB policymaking in Ottawa – an arrangement that Hayes liked. As Hayes later recalled with relish, when Philip Larkin, director of the Pacific station, disagreed with Hayes's decision to curtail a costly program of high-seas fisheries research, 'Larkin went ahead and we came to a showdown and Larkin unfortunately broke mentally.' Larkin ended up resigning.[32]

This episode, incidentally, undermines an argument that has recently been made: that an independent science organization, funded by the Canadian government, would have better served the fisheries industry and conservation goals.[33] It is strongly apparent, nevertheless, that the culture of science developed within the Department of Fisheries and Oceans up until the mid-1990s was also severely flawed.

First and foremost was the loss of the ideal of service that marked the work of the independent professors and early professionals associated with the Biological Board, as described in chapter 2. This is evident in the valuable interviews that form a large part of A.C. Finlayson's *Fishing for Truth,* a sociological analysis of scientists and others involved in Atlantic Canadian fish stock assessments in the late 1980s and early 1990s. For a variety of reasons, the race for professional prestige overtask interest in directly helping fishermen and conserving the fisheries. Not only that, but improving one's professional status involved a focus that was actually counterproductive to the stated mandate of the department, which was to make the highest standard of scientific information available to the government, the public, and the fishing industry, in aid of helping plan, use, and regulate of the fisheries. This dissonance was in fact enshrined in the department's reward structure, in which scientists were held to a standard that has unfortunately long infected academia – the publication imperative. Promotions and job security depended upon one's publication record, not on actual service in aid of the fisheries. As Finlayson concluded from his many interviews: 'The reward and advancement of a DFO scientist is determined exclusively by his or her performance as measured against traditional scientific [and] academic standards; number of publications in peer-reviewed journals and relative reputation within the international community of fisheries scientists. It is largely irrelevant whether or not a scientist's work follows from the institutional mandate.'

Scientists, then, got neither promotions nor credit for maintaining contacts and improving information sharing with fishermen and fisheries-related organizations. For the most part, therefore, they selected research problems of interest to the scientific community over mission-related work. Those few who took the high road, who continued to work with fishermen – such as Henry Lear – were punished by being passed over for promotion, even if they worked overtime collecting valuable data and doing important work with fishermen. Lear commented:

> You're so tied up in doing your job that you just don't have time to publish ... If there's a brush fire, you get called out. You're the one who's got the experience and you've always been there and it's so easy, right? And you hire someone ... They're brilliant and they come in and you've got this

> wealth of data you haven't published and they say 'Well, this is not right, This demands publication.' So you hand it over and they get half a dozen papers and next thing you know they're two levels ahead of you.'[34]

Under such circumstances, it is hardly surprising that the ideal of service that so strongly marked the first four decades of fisheries research in Canada was largely lost. Those who held to this ideal were in effect punished through a form of administrative blindness.

In universities the demand for publication in quantity may merely serve to make academics' lives more harried and unproductive of work of real value. In the Department of Fisheries and Oceans, such a culture had far more invidious effects. Some scientists, mindful that only publication was rewarded, and less naive than service-oriented scientists, simply hoarded their data and shared it with nobody. Nor did they necessarily get around to publishing all their work, as administrative duties kept some from focusing on collation and analysis. As Finlayson's book indicates, and as a more recent interview with leading fisheries scientist Ransom A. Myers reiterates, the failure to share masses of data, which turned out to be critical to accurate stock assessments, meant that some of the brilliant young scientists working for the department in the late 1980s and early 1990s were unable to follow up their suspicions regarding the fish stocks' true status.[35] As Myers put it, people 'just trying to do their job' were prevented from doing so. Although the stocks were already on the verge of crashing in the late 1980s, high catch rates continued to be recommended to the Department of Fisheries because of a lack of openness about fundamental data and critical analyses, and because of the prevailing 'group think' of the senior managers.[36]

This 'group think' was that the stocks were recovering since the extension of the two-hundred-mile fishing limits, and that overfishing was not occurring. Senior managers and scientists argued that declines in the inshore cod catch were due to colder than normal conditions, not overfishing. Junior scientists were not made privy to the information used by the Canadian Atlantic Fisheries Scientific Advisory Committee (CAFSAC) in making its recommendations about fishing effort and resource management to the Department of Fisheries and Oceans. Formed in 1977, CAFSAC was made up of the department's regional science directors, as well as the chairs of seven subcommittees, a departmental economist, and four outside experts. These senior science personnel conducted their meetings behind closed doors, and junior scientists were not given access to the data on which they based their decisions. Such a sequestering of information, and the resort to what can only be called a managerial style, was antithetical to the free exchange of ideas and information that is the lifeblood of good science.

Nor were CAFSAC members open to criticism. For example, they rejected the 1986 Keats Report. This study was conducted, at the behest of the Newfoundland Inshore Fisheries Association, by Memorial University biologists Derek Keats, Don Steele, and John Green. They concluded that the department's 1986 assessment 'had severely overestimated the size of the northern cod stock.'[37] The Keats Report pointed out that commercial deep-sea trawlers' catch rates, which formed the major basis for stock assessments, were misleading. These vessels were using high-tech locator devices and were able to find and fish up remaining cod stock concentrations, and to maintain catch rates despite a massive decline in overall stock levels. Although the Keats Report was based on the department's own published data, and although evidence from the falling inshore catch was mounting, the report was dismissed by CAFSAC and senior department officials as being the work of scientists lacking the experience and credentials to deal properly with the information.[38]

The rather patronizing attitude toward scientists outside the inner circle was also extended toward fishermen. Scientists were not asked to lie about certain situations, but according to one of them, scientists 'certainly have, at various times, been discouraged from telling the whole truth. Every government has to do that with its civil servants. You can't have everything that's going on in the halls of government ending up in the newspaper the next day. You have to allow the people whose job it is to make policy ... talk about what the advice is, what it means, come to the conclusions, and make the policy.' They also told inshore fishermen – who were witnessing the effects of the stocks' destruction, and who were begging for a fishing moratorium – that they had no idea what they were talking about.[39]

The Department of Fisheries and Oceans had other priorities. In the Canadian context, the Atlantic fisheries had become part of the welfare support system, and the goal was to maximize employment. Fishermen's unemployment insurance had been introduced in 1957; by 1964, with no increase in Newfoundland's general population, there was a 33 per cent increase in the number of inshore fishermen. The catch per fisherman declined by 50 per cent, but the federal goal was for the fishermen of this impoverished province to work long enough to gain annual insurance coverage.[40] Catch quality and other fisheries-oriented goals were secondary. For example, Newfoundland fishermen until the end continued to pitchfork their fish when handling them; the puncture holes lowered prices through increased spoilage and decay. Yet this practice was never regulated against, as it had been since the early 1900s in Scandinavian countries. Fisheries conservation was subsidiary to the department's political aims, and in light of this agenda, whistleblowers distressed by fish stock declines were not well received by the department. Even after acknowledging the precipitous drop in the cod stocks in 1991 and declaring a moratorium in 1992, the department allowed limited fishing the following season. This was stopped when it became apparent that the 1993 biomass was only 3 per cent of the 1990 cod stock, and that the stock's decline was continuing even though very little fishing was being done.

The official cause of the stock declines, declared the Department of Fisheries and Oceans, was probably unusual oceanographic conditions – colder water temperatures than usual, which had to be influencing other factors, which in turn had an impact on the cod stocks. But several junior scientists were not buying this explanation. Ransom A. Myers, and his then graduate student Jeffrey Hutchings, went back over the scientific assessment data and in 1994 published a 'now famous article' in *Canadian Journal of Fisheries and Aquatic Sciences* (successor to *Journal of the Fisheries Research Board of Canada)*. In 'What Can Be Learned from the Collapse of a Renewable Resource?' they made a number of telling criticisms. They noted that the success of science-based management strategies is hard to assess, given, for example, the absence of a control, unfished population. As long as the fish are found, fish managers can argue that their methods are effective; it is only when commercial extinction occurs, and a harvest ceases to be economically viable, that they can discern their failure (or as they put it, 'that the reliability of a given strategy can potentially be evaluated'). Myers and Hutchings reviewed one location's depth-stratified temperature data going back to 1946, and also analyzed NAFO fish statistics going back to 1979. They sought general evidence for cod displacement by colder water, and evidence for unnaturally high mortality rates, lower fecundity, survival of juvenile fish, and changed relations between adult and juvenile abundance. They also critiqued the flaws in traditional stock analysis, especially the reliance on commercial trawler data.

The historical evidence pointed unequivocally toward overfishing as the culprit. Large harvests had occurred in the late nineteenth and early twentieth centuries when conditions had been much colder, so the official explanation that unusually cold conditions had caused the collapse had to be wrong. Rather, since overfishing takes out maturer, larger cod, which produce more eggs and young than younger, smaller fish, reproduction rates had declined. The analysis by Myers and Hutchings also indicated that long-distance trawlers had been overfishing the stocks since at least 1962; that is, high fishing rates had prevented the stocks from sustaining themselves, and the stocks had neared commercial extinction in 1977. They concluded, caustically, that 'the cod collapse and previous commercial extinctions in the history of fisheries management (eg. California sardine, Peruvian anchovy, African/Namibian pilchard) should provide ample justification for politicians, policy makers, industry, and management to limit the urge to attribute resource collapses to vaguely understood or even imagined environmental causes. The ecological and socioeconomic consequences of so responding to repeated failure are too great.'[41] In sum, high exploitation rates may make fish stocks more vulnerable to environmental stresses, but these stresses do not directly cause stock collapse

This all leads to another question: Why had senior managers strongly argued that the fish stocks had been increasing, given their near commercial extinction in 1977? It appears that CAFSAC was influenced by the 'the exceptionally strong year-classes of northern cod between 1978 and 1981, the highest recruitment since the late 1960s,' which 'coincided with the lowest level of trawling effort observed since the late 1950s.' This apparent strength, reinforced by their knowledge that the TACs were nowhere near historical highs set in the 1960s, led them to the erroneous (as it turned out) conviction that fish stocks were recovering. The record historical catch was 810,000 tons of cod in 1968 (reported catch – the actual catch was probably much higher) whereas the TACs of the 1980s for Canadian trawling fleets were a mere 266,000 to 318,000 metric tonnes. But the actual catches ranged from 140,000 tonnes to 270,000 tonnes in 1988, the peak year under Canadian management. The catch in 1992 had fallen to 97,000 tonnes,[42] by which time the problem was apparent.

In response to the disaster, in 1994 CAFSAC was terminated, and a new fisheries management organization, the Fisheries Resource Conservation Council (FRCC), was created to oversee fishing options and conservation measures. This organization has more recently shown itself to be more sensitive to the dangers of overfishing and to conservation requirements, although until 1997 it apparently responded to political pressure by recommending a limited food fishery in the still declining cod stocks.

The Department of Fisheries and Oceans remained adamant that environmental factors and not overfishing caused the stock collapse. As recounted in Michael Harris's *Lament for an Ocean*, Ottawa was not pleased when Myers and Hutchings's analysis contradicted the official position, and exerted pressure on those involved in the publication process. Larry Coady, head of science in the Newfoundland region, 'stood his ground when asked by Ottawa if their papers had been put through internal review'; he had meticulously 'followed government guidelines of how scientific papers were to be published.' Hutchings recounted: 'For that he was criticized by Ottawa, because what was being published was inconsistent with what the departmental positions were. He actually stood up to Ottawa admirably.' Only in 1997, at the Summit of the Sea conference in St Johns, did the Department of Fisheries and Oceans admit officially that overfishing had caused the collapse.[43]

The department's attempts to suppress conflicting interpretations led Hutchings, by now a biology professor at Dalhousie University, to question the wisdom of embedding scientific research within the bureaucratic and political infrastructure of a government department. He and two other professors, Carl Walters and Richard Haedrich, published in the May 1997 *Canadian Journal of Fisheries and Aquatic Sciences* a paper that asked, 'Is Scientific Inquiry Incompatible with Government Information Control?' They contended that conflating fisheries science with fisheries management led to a political–bureaucratic structure in which a 'suppression of scientific uncertainty' and a failure to admit 'comprehensively legitimate differences in scientific opinion' was antithetical to conservation measures. They put it bluntly: 'Political and bureaucratic interference in government fisheries science compromises the DFO's efforts to sustain fish stocks, and thereby, the socioeconomic well-being of fishing people and fishing communities.' But they were careful to state that their criticism was directed at systems, not individuals. They wanted, ideally, 'a politically independent organization of fisheries scientists,' or at the very least, some 'reorganization of the link between scientific research and the management of natural resources.'

They showed how in the past the Department of Fisheries and Oceans had suppressed important scientific information. For example, the department had been silent concerning valid 'differing opinions on the health of a fish stock,' and had failed to quantify 'sources of variability,' which weakened the reliability of its abundance models. Thus scientific information given 'to fishery managers [bore] little resemblance to the means by which scientific information is communicated, debated, and accepted by scientists in general.'[44] For example, in 1986, George Winters of the St John's station had argued that CAFSAC's data, used to argue that low water temperatures caused low catch rates for 1985, could also be interpreted to show that overfishing was occurring. CAFSAC responded that uncertainty levels were 'not unusual in comparison with the assessment of other cod stocks in the Northwest Atlantic,' and set the problem aside. Unfortunately, since fisheries management decisions were mostly based on CAFSAC's collectively issued scientific documents, and the public service's rule is that 'only one set of advice goes forward to the minister,' the suppression of data on uncertainty had terrible results. Even after the cod stock collapse, 'fishing people and the Canadian public' were 'ill served by the bias that can exist in Stock Status Reports.' Also, the 1995 stock status reports and 'key reviews' were 'biased against research that has identified overfishing as the primary cause of the present stock collapses.'[45]

Worse still, 'a DFO research scientist in Newfoundland' received an official reprimand for telling the *Globe and Mail* that (25 August 1995) that 'what happened to the [East Coast] fish stocks had nothing to do with the environment, nothing to do with seals. It is simply overfishing.' Hutchings and colleagues recount what that scientist was told:

> Your comments, as presented by the media, did not give a balanced perspective on the issue of the status of the cod stocks and were inconsistent with the June 1995 Newfoundland Stock Status Report. [We] have cautioned you regarding statements which do not take into account peer reviewed scientific information. Your ... disregard for both departmental policy on communication with the media and the professional opinions of your colleagues warrant the disciplinary action of a written reprimand. In the future, you are expected to respect both the system of primary spokespersons and peer conclusions on matters within your area of expertise.

Yet, as Hutchings and his colleagues pointed out, the scientist (Ransom Myers) had based his comments on his own, anonymously peer-reviewed scientific publications, and his conclusions were not inconsistent with those of other scientists. The resulting stress and feelings of job insecurity from such treatment are debilitating to scientists. Myers later quit the DFO and decamped to Dalhousie University, in protest over the censorship of a related, coauthored article, one which argued that there was no link detected between the harp seal populations and the cod fishery collapse (another pet government theory). 'Bureaucratic action of this kind cannot help but have a stifling effect on proper conduct of science in the organization. Freedom to raise scientific debate should be integral to the conservation and management of Canada's natural resources.'[46]

This article garnered an immediate defensive response. William G. Doubleday, the department's director of fisheries and oceans science, led an attempt to put Hutchings in the wrong. Doubleday argued in a 1997 opinion paper that public service guidelines bound scientists to providing only factual information, within their areas of responsibility, to the public and media, regarding only programs and policies that had been announced or implemented by the government. Scientists were not to go 'beyond this discussion of factual information.' Doubleday accused Myers of having done just that in the *Globe and Mail* interview, and worse, of attributing his overfishing theory 'to Myers et al. (1995). In fact, there is no such statement in this 1995 paper.' He thus implied that Myers had gone beyond discussing factual information, and he sniped at how 'illogical' Hutchings and his colleagues were to demand that departmental scientists fully acknowledge scientific uncertainty, and yet 'condemn the Department for pointing out to a scientist who makes a categorical and extreme statement. [that he] did not give a balanced perspective.' Doubleday argued that Myers's opinions ignored the regional stock status reports' scientific uncertainties. 'Hutchings et al ... criticize statements that ignore uncertainty and variability in scientific opinion yet criticize the Department for asking a scientist to present a balanced view.'[47] Doubleday condemned Hutchings for backing up Myers's overfishing theory by using, not the 1995 article mentioned (which did not refer to the issue, and may have been mentioned by Myers in his interview by mistake), but fourteen other published papers, many dating *after* the newspaper article. Notwithstanding Doubleday's nitpicking, however, Hutchings and Myers's 1994 peer-reviewed paper had amply supported their argument that overfishing, not natural causes, had caused the northern cod stock collapse. So did several other papers by Myers and other scientists published or completed by 1995. Since later scientific investigations have confirmed Myers's (and many other scientists') conclusions about overfishing, it is hard to take seriously Doubleday's charges, and his department's, that Myers's comments were scientifically unbalanced and that they failed peer review standards.

The 1997 articles generated some much-needed debate. Doubleday blasted Hutchings and his colleagues for ignoring reforms that had already been made. Hutchings, Haedrich, and Walters replied that the reforms had not been nearly as complete as Doubleday was implying. For example, fishermen and outsiders were now allowed into assessment meetings, but reports were 'not written in a form readily accessible to non-scientists' and were not always available for review before the meetings. Often, outsiders were merely allowed to comment after hearing evidence at the meetings; this was far less effective than well-considered responses based on a familiarity with the data. Hutchings and his co-writers strongly recommended that people who were not department employees be involved in actual stock assessment data analysis. This still was not occurring, which left 'the production of, and the ability to comprehensively review, stock assessment documents ... a DFO "in-house" affair in many important respects.'[48]

Michael Harris's *Lament for an Ocean* gives other examples of official harassment and attempts to influence scientific publications and public statements, including two episodes recounted by the outgoing editor of *Canadian Journal of Fisheries and Aquatic Sciences*, who oversaw the publication of Hutchings, Walters, and Haedrich's contentious paper. In each case the officials were concerned that the information about to be related would embarrass the department.[49]

Some of the charges put forward by Hutchings and his colleagues, and the problems related by Harris, are troubling indeed. They also lead one to ask whether Canadian fisheries biology was unique in suffering from these antiscientific practices, given the troubled status of many of the world's major commercial fish stocks. Indeed, given the fundamental nature of bureaucracies, it is not surprising to learn from recent articles that similar problems have been identified in other government-run fisheries management organizations. In the February 2001 issue of Britain's *Fishing News*, a 'former fishery scientist with many years of [experience in] stock assessment' published anonymously a severe critique of fish assessment science. He pointed out the fragility of data acquired by traditional methods, such as fishermen's self-reporting of catch and discard data. He complained that scientific sampling was not thorough enough to rectify this situation. Official organizations were failing to publish the error factor in their population calculations, he observed, and, in addition, 'the system of peer review is not used for fish stock assessments, thereby casting doubt on their reliability.' These criticisms echo those made by critics of the Department of Fisheries and Oceans. Similarly, ICES fisheries scientist Ad Corten, who worked at the Netherlands Institute of Fisheries Research (RIVO) 'argued that there was a strong element of collective solidarity at the working group meetings, closing ranks against contentiousness: "Openly questioning the reliability of someone else's data would spoil the spirit of comradeship on which the functioning of the working group depends."' The anonymous ex-fisheries scientist also 'claimed that scientists from EU member states are not always independent,' but are 'often briefed by their governments and have a political agenda.' As a result, 'stock assessment is not science, it's simply a very flawed statistical exercise ... The whole thing is political.'[50]

These are clearly troubling charges. When the Department of Fisheries censored 'undesirable' but scientifically peer-reviewed interpretations of scientific data, it was, apparently, acting like government–science organizations elsewhere. The Department's censorship of scientists, its denunciation of the 1986 Keats Report and other independent reports, and its suppresson or misrepresentation of alternative hypotheses in public reports and government statements, led Hutchings and his colleagues to conclude that 'non-scientific influences' had rendered government fisheries research 'incompatible with normal scientific inquiry': 'We would argue that bureaucratic intervention has deleteriously influenced the ability of scientists to contribute effectively to fisheries management. Viability of fish stocks, sustainability of employment in fisheries, and persistence of coastal fishing communities would appear to be poorly served by the present institution in which fisheries science is inextricably linked to, and affected by, a political bureaucracy.'

They recommended more openness about data uncertainty, variability of model parameters, and scientific disagreements about stock status. All scientific information should be made available to the public, not just the minister, so that the public could evaluate the minister's management decisions for themselves – and fishermen and fishing interests would indeed do just that. They also suggested that 'the most preferable change in the status quo [would be] the formation of a publicly funded body of scientific inquiry, completely independent of political influence.'[51] They gave as an example the independent Fisheries Research Board.

The call for a publicly funded scientific body 'completely independent of political influence' is an attractive proposal. Unfortunately, even within the Biological Board, scientists did not enjoy complete freedom to air, in public, ideas or facts that contradicted the official line. This climate developed under A.T. Cameron's directorship. At an executive meeting of the Biological Board held in Ottawa on 3 November 1934, the board passed the following resolution, put forward by industry representative A.H. Whitman:

> Whereas, on several occasions during the past year the Board and the Department have been caused embarrassment through unwarranted and misleading speeches and letters by scientists in the Board's employment drawing conclusions, *whether right or wrong*, from the investigations which are still in progress and incomplete.
>
> Therefore if such procedures cause embarrassment in future, disciplinary action may become necessary.
>
> Consequently, Directors of Stations must instruct all the employees of the Board that material concerning the results of the Board's investigations into problems which are in any way controversial must not be communicated to the press or to the public until such results have been considered by the Executive of the Board.
>
> The Executive wishes the Directors to understand that they are held primarily responsible for the carrying out of these instructions by the officers under their direction. The instructions naturally apply also to the Directors themselves.

The Executive then sent a reprimand to R.E. Foerster, director of the Pacific station, and informed the minister of this. Foerster's 'crime' was disclosing preliminary results from the Cultus Lake fish hatchery studies, before the studies were completed, in a speech to the Duncan Rotary Club.[52]

In the same period, one season of scientific studies conducted by the International Pacific Salmon Commission had allowed the scientists involved, including Foerster and the young W.E. Ricker, to conclude that to allow the salmon stocks on the Fraser River to be rebuilt, fishing of the early part of the run would have to be stopped. The Salmon Commission's terms, however, meant they would have to keep silent until the eight-year study was completed before putting forward these recommendations – a situation that considerably frustrated Ricker.[53] Clearly, the Biological Board's relative independence did not prevent its executives and their political bosses from gagging scientists whenever they wanted. Furthermore, under certain strong-willed individuals with their own agendas, such organizations can be derailed from their original mandate, as was the case under FRB Chairman F.R. Hayes, who effectively sabotaged P.A. Larkin's worthwhile fisheries programs, as discussed earlier.

It is apparent that government and civil service norms have for a very long time posed challenges both for scientists employed by the government and for scientists within government-funded organizations like the Biological Board. The solution, therefore, is not necessarily an independent and publicly funded scientific body. The creation of such a body might reduce bureaucratic and political meddling in scientific work, but it would not guarantee that officials would desist from obstructing scientists in their normative publication and communications activities, especially where scientists would sometimes dispute or contradict the official message. It should be beyond argument, however, that only by protecting such scientific freedom will government-funded research generate useful knowledge.

New Paradigms and Practices in Fisheries Science and Management

The human-generated catastrophe that overtook the northern cod and groundfish stocks in the late 1980s and early 1990s has had worldwide ramifications. The commercial extinction of the world's former largest fishery was a monumental tragedy for Newfoundland fishermen and their communities. Unfortunately, there is still no sign of recovery. Scientists who have studied the long-term collapses of various Great Lakes fish stocks, and who see many parallels with the northern cod fisheries, say that such a recovery will be a long time coming if it ever comes at all – the groundfish stocks may never recover their former abundance, and the genetic diversity of the substocks that is so essential to their former strength may be lost forever.[54] Yet at the same time, the very magnitude of this fisheries disaster has borne some highly beneficial results. Around the world, complacent governments, fisheries managers, and fisheries scientists have been considerably humbled by this failure, which has finally forced them to confront the fatal flaws in their accepted management practices. Ruling fisheries paradigms and past standards are being reexamined and rejected. Fisheries biologists are repudiating their former fixation on maximizing harvest and efficiency while keeping a theoretical minimum of the spawning stock intact (at levels that leave no room for error). More comprehensive and varied conservation measures are being sought, to buffer fish stocks from over exploitation; these would offer some hope for the future of fish populations and ecosystem management.

In Canada the sea change began with the Harris Report, released on 30 March 1990. Dr Leslie Harris, a historian and president of Memorial University in Newfoundland, was appointed in 1989 to chair the Northern Cod Review Panel. Harris chose fisheries scientists Lee Alverson and John Pope to aid him. The Harris Report gave Canadians their first official notice that the cod stocks were in trouble. It called for drastic catch reductions, and it criticized the work of the Department of Fisheries and

Oceans' science branch, within which the focus on mathematical data and models had supplanted the study of the species. Over three-quarters of the department's science budget had been allocated to statistical stock analysis. The Harris panel pointed out that proper management of cod was unlikely in the absence of an understanding of cod behaviour and life histories, of the effects of fishing, predation, and the environment, or even of the effects of trawling on the spawning grounds. Indeed, in 1990, then fisheries minister Tom Siddons defended the latter practice, because there was no recorded evidence in the scientific literature that 'fishing on the spawning grounds does measurable damage to the cod stocks.' What he either failed to point out, or was not told, was that the extent of the damage was unknown because no one had ever done the research on this problem.[55]

The federal government responded by establishing the Northern Cod Science Program of 1990–5, with funding of $33 million. Fisheries biologists were now allowed to do true science, and not just the fish counting and number crunching that had for so long dominated their work. By the program's end, this new research had generated 230 scientific papers, which substantially increased our understanding of cod behaviour, feeding, diet, cannibalism, migrations, spawning behaviours, and responses to changing water temperatures, and of the behaviours of various cod stocks. One study showed that trawling does disturb shoals of spawning cod in ways that probably reduce their ability to spawn.[56] Scientists also investigated phytoplankton and zooplankton production, and the food species of cod, such as capelin. These studies are like the work of earlier scientists, such as Hjort, and Huntsman during his expedition years. In a sense, fisheries biologists were returning to their beginnings.

Although cod research funding was sharply reduced once the program ended in 1995, it continued to get much more support than it had received before 1990. On an encouraging note, the Department of Fisheries and Oceans has continued to extend grants to outside scientists, including some who had been its fiercest critics. Ransom Myers has received 'various grants working with the DFO' and has also been 'working with them on a regular basis.' He told me: 'You know, I went to a seminar today at the DFO. I have ... a willingness to criticize ... but also. [am] still being able to get money and grants from the DFO ... even though I am very critical of policies if necessary.'[57] A new maturity has apparently enabled the department to recognize the value of criticisms from fiercely dedicated scientists like Myers.

The crisis in the world's fisheries led to several important international conferences focused on improving the management of fisheries and ocean resources. One such was the Summit of the Sea conference, held in 1997 in St John's, attended by delegates from thirty-seven countries. Through this and other conferences, such as the international environmental conferences for the protection of the North Sea held in London (1987), the Hague (1990), and Esbjerg (1995), the idea of the 'precautionary principle' has gained momentum within the scientific and management communities. The term originated in Germany in the mid-1970s as *Vorsorgeprinzip*, and was initially applied to specific human behaviours that cause outright dangers as opposed to risks. It has since broadened in application to encompass approaches to any human behaviours that can damage the environment in a potentially irreversible fashion.[58] In the marine sciences, the precautionary principle is being applied particularly to problems with marine pollution and to overfishing. The implication of the precautionary principle is that fishery target catches should be set at conservative levels, 'well below the limits and critical thresholds that compromise the productive potential and stability of the ecosystem,'[59] and taking into account the requirements of other species within the ecosystem. The 1995 North Sea Ministerial Meeting led to an agreement that the North Sea's fish stocks should be managed in accordance with the precautionary principle; international treaties under the 1995 'FAO Code of Conduct for Responsible Fisheries' have also endorsed the precautionary approach to managing fish stocks.[60]

In the United States, Congress assigned the National Marine Fisheries Service (descended from Spencer Fullerton Baird's U.S. Commission of Fish and Fisheries) the mandate to establish an Ecosystem Principles Advisory Panel to recommend improved principles for managing American fisheries and fish ecosystems. The panel, which met three times in 1997 and 1998, included members of industry, academia, conservation organizations, and fishery management agencies. Echoing earlier European recommendations, this panel, in its report to Congress, advocated applying the precautionary principle and an ecosystem-based management system for all fisheries. Fisheries management would have to be based on the following premises: ecosystem behaviour is not well understood and is unpredictable; ecosystems have limits of flexibility which if exceeded may cause major and potentially irreversible ecosystem restructuring; and all components within ecosystems are linked although they act on different scales. Ecosystems are open, and they change over time, and this adds to the complexity of dealing with them. Given these parameters, the goal of fisheries management is to maintain, not theoretical maximum sustainable yields per se, but ecosystem health and sustainability, including that of the fisheries. All of this meant that ecosystem managers would now have to 'change the burden of proof' and 'apply the precautionary approach' by using methods that would build in '"insurance" against unforeseen adverse ecosystem impacts,' all the while 'learning from experience.' The panel also recognized the importance of local management and the participation of fishermen and other interested parties in an equitable management system.[61]

At the heart of the Ecosystem Principles Advisory Panel's report was the repeated emphasis on acquiring a scientific understanding of the ecosystem, through hydrographic, and bathymetric studies of physical, chemical, and climatic oceanography, as well as biological studies of food webs, predator–prey interactions of commercial and non-commercial species, trophic structures, and dynamics. They recommended that experts devise indices to determine the health of a given ecosystem, to identify pre-catastrophic indicators of overfishing. As the members observed, you cannot take an ecosystem approach to fisheries management without understanding the ecosystem. On top of this, the fishermen and the fishing mortality that their fishing and by-catch activities caused would have to be incorporated into our understanding of the food web. Long-term monitoring of the fished ecosystems was essential, as well as close inspection of fishing vessels and catches.[62]

Wherever the precautionary principle is being considered as the new basis for fisheries management, there is an emphasis on the need to change the 'burden of proof' of overfishing. Historically, scientists sided with the fishermen. From the 'inexhaustibilist' T.H. Huxley to the conservationist and saviour of the Pacific halibut, W.F. Thompson of the Pacific Halibut Commission, the underlying principle was that conservation measures should inconvenience fishermen as little as possible. In 1919, Thompson – who was deeply concerned about overfishing – wrote that 'proof that seeks to change the way of commerce and sport must be overwhelming.' But recently, scientists have argued that fishing should not be allowed unless regulations are in place that take risk into account, 'protect all elements of the ecosystem,' and that allow the fishery's participants to share responsibility for managing the resource. 'Changing the burden of proof will mean that, when the effects of fishing on either the target population, associated species, or the ecosystem are poorly known ... fishery managers should not expand existing fisheries ... and should not promote or develop new fisheries for so-called "underutilized" species.'[63]

Putting forward recommendations is all very well, but there have to be practical ways of implementing them. Scientists and fisheries managers have come up with a number of ideas, some new and some actually very old, ideas that for a long time were considered obsolete. Among the new ideas, the Americans' Ecosystem Principles Advisory Panel recommended developing applied mathematical models of ecosystems, such as ECOPATH, which can provide insight into some fundamental ecosystem questions. 'ECOPATH provides a framework for summarizing natural rates of growth and consumption of marine populations. This allows small scale studies or models ... to be viewed in a common currency in the context of the ecosystem as a whole.' Such models, using improved computer software programs and more powerful computers, will help us understand a fisheries ecosystem's dynamics; this will improve scientists' and managers' management of fisheries.[64]

Among the older ideas, the practice of establishing marine protected areas is being revisited. Marine protected areas, created in the nineteenth century, were established around important spawning areas for commercial species. The idea was that prohibiting fishing within their boundaries during the spawning season, or even year round, would maintain a reserve of spawners to help replenish stocks. In the early twentieth century, scientists like Huntsman argued that since fish are generally highly mobile, and since their spawning areas were not necessarily the reserve areas, and since in any case the fish stocks were doing well, there was no need for such reserve areas. But scientists are now promoting the idea of marine protected areas, off limits to all fishing, that would mirror national parks on land, and that would form a kind of insurance against overfishing. 'Protecting parts of the ecosystem from exploitation can insure future sustainability. Reserves also serve as baseline areas to evaluate natural variation in animal and plant populations that are free from fishing impacts.' If a reserve area were large enough, it might preserve a substantial proportion of a fished stock from exploitation; this in turn might reduce the need for costly monitoring elsewhere, since it would reduce the problem of the uncertainty in the resource's status.[65]

Reserves can also be created by severely limiting the fishing season. As Carl Walters points out, the success in conserving the Pacific salmon, which could easily be wiped out through intensive fishing of migrating spawning fish, is owed not to accurate stock analysis, but rather to policies that close 'most of the ocean to fishing for most of the time.' He notes that 'salmon fishing areas near rivers are characteristically very small, and open for only a few hours or days at a time,' and he argues that such severe restrictions on other oceanic stocks, including the Atlantic northern cod, may be required as well. The Pacific herring roe fishery has long been managed – very successfully – under the same principles, instead of by adjusting the catch quota up and down to match scientists perceptions of stock abundance. He argues that the precautionary principle might end up acting like the traditional TAC quota method, because 'cautious management can still be utterly destructive if it is based on assumptions and analyses that are not even in the right general ball park in the first place.'[66] Severely limiting fishing times or areas is a far less costly measure than the intensive stock assessment studies required in order to set TAC quotas; the latter can end up costing more than the profits brought in by some of the smaller fisheries.

Another approach being examined and attempted is for fishermen themselves (the greatest stakeholders in the fishery economy) take some responsibility for conservation. In Norway this practice goes back more than a century, to the 1890s, when the Lofoten Act assigned responsibility for regulating the Lofoten Islands cod fishery to the fishermen. The fishermen elected inspectors from among themselves, and formed a public agency to oversee the enforcement of regulations. This system, with a few changes, has continued in operation ever since, with the Norwegian government content to take a back seat in this matter.[67]

In the meantime, in Newfoundland the fishery no longer underpins the economy. So many fishermen have been forced out of their traditional livelihoods that sociologists fear that traditional fishing skills will be lost as a generation of fishermen find other work. Tragic though the circumstances are, perhaps one can clutch at a silver lining: some of those 'skills' – such as using pitchforks to handle the catch – were better jettisoned. Twenty years from now, or however long it takes for the stocks to rebuild, a new generation of fishermen can be trained in superior fish-handling methods and new, more ecologically sensitive fishing techniques. On the other hand, the traditional ecological knowledge of these fishermen, potentially extremely valuable to science, may be mostly forgotten. Elsewhere in Canada, community-based co-management enterprises began organizing themselves in Atlantic Canada in the 1990s to take charge of local fishing levels and conservation.

In Atlantic Canada, the Coastal Communities Network was created in 1992 to take back some control of the fishing industry from the aloof, centralized structures in Ottawa. The new approach was described as 'co-management' to make the scheme 'attractive ... to government,' as the Fisheries Council of Canada insisted that not only local fishermen but also government and corporations be involved. The latter raised problems for most fishermen. The government agreed to support geographically based community management boards (CMBs), which all fishermen were encouraged to join, under conditions that made refusal to join unappealing. This program began in 1995 with the Halifax West group of fishermen and soon spread to other communities. The federal government allocated a quota to each CMB based on the catch history of each licensed fisherman in the previous ten years and on principles of individual transferable quotas (ITQs). 'This process utilized numerous input sources, including DFO, for data analysis and a mediator to resolve differences in opinion with respect to community sharing.' The quota calculations also depended on the historical landings from the fish processors. The CMBs could trade quotas and exchange or trade members, and deal with the fishery in a businesslike way within boundaries demanded by the precautionary approach.[68]

This organization was far superior to the old, entirely top-down administration of the fisheries because CMBs were now responsible for monitoring landings, fishing effort (size and type of vessel used, equipment, etc.), and real-time collection of catch data through the dockside monitoring program. In addition, the fishing industry members were required to subsidize the costs of monitoring the fishing effort through official observers on board fishing boats (although they were not expected to help fund air surveillance). But the real value of these organizations is that they can now employ peer pressure to deter illegal fishing. Now that member fishermen have a stake in the total outcome, and are concerned about more than their own short-term gains, there is considerable pressure on members to conform to measures designed to ensure conservation or industry harvest plans. The fishermen themselves determine the penalties, which have tended to be more severe than traditional, court-ordered penalties; they include reducing the offender's quota or time that can be spent out at sea.[69]

The community-based co-management activists still found themselves locked in a power struggle, however, as the government was slow to share its authority, and it remained fixated on the 'efficiency' and 'viability' of the fisheries. To these ends, it gave the big corporations quota rights that were exclusive, freely traded, and durable, at the expense of small fishermen. But in 1999, the Department of Fisheries and Oceans put forward a draft 'Framework and Guidelines for Implementing the Co-Management Approach,' which declared, as one of five guiding principles, that 'the right to own fishing licenses and quotas must be restricted to professional independent owner-operators ... to keep licenses from falling into the hands of corporations or absentee investors with no attachment to the fishing industry or coastal communities.' The department also finally recognized that it must promote a smaller fishery, in which 'participants are able to earn a livelihood without government subsidy ... a fishery made up of a core group of professional, full-time fish harvesters.'[70]

Fishermen have become more heavily involved in initiatives like the 'Local Knowledge and Local Stocks' project, in which they share their local knowledge of spawning areas, spawning times, and other specialized information. The goal here is to develop conservation schemes and to increase understanding of stock structures. The database, published by Saint Frances Xavier University as *Local Knowledge and Local Stocks: An Atlas of Groundfish Spawning in the Bay of Fundy*, has been sent to scientists, fishermen's associations, and managers, and can be seen as part of a larger effort on the part of fishermen and scientists to work more closely in fisheries conservation. Also, fishermen have been tagging fish in the Bay of Fundy tagging programs for the Department of Fisheries and Oceans, and not leaving this activity to the scientists. 'Fishermen have expressed appreciation for these efforts to include them more fully in the research process.'[71] Certainly, anything that would improve the often adversarial relations between fishermen and some government scientists and fisheries managers is an excellent idea, and one that many fisheries biologists would also welcome.

Fisheries scientists are divided on the merits of some other approaches to solving the problem of fish shortages, such as fish farming and fish culture. Critics argue these may pollute the environment, or encourage the introduction of exotic species, or in the case of native and genetically engineered native species, create genetic drift that could damage wild populations if farmed fish escape. Cod farming is nevertheless being investigated in Norway, Scotland, and Canada. Another approach being experimented with involves increasing wild populations by capturing prey species in distant waters and releasing the (frozen and then defrosted) captured fish in a controlled fashion to enhance the diet of desirable local stocks. This has been tried in Iceland, for example, 'resulting in a large increase in the growth rate of free-ranging cod in the feeding area.' One author of these studies points out that when fish are farmed, prey species have had to be fished in any case, since farmed fish are fed a fish meal derived from processed fish catches.[72] The notion of catching the fish and simply freezing them until they are required to feed a wild population is not as outrageous as it might appear.

The solutions described above are technological fixes and are highly controversial. But most scientists now agree that the old methods of fisheries science have failed, and that the science needs new methodologies and tools, and new goals, some of which have been described above. Most also agree that the future of the fisheries will be bleak indeed unless both society and government embrace the newly identified goal of conserving marine and aquatic ecosystems, and oppose 'business is usual' through recognizing that the entire public, not just fishermen and large fishing corporations, are 'owners' of fisheries resources. Taking the management of these resources out of the hands of 'irresponsible resource users who operate at the public's tolerance' is essential to the recovery of these ecosystems.[73] Although fishermen's co-management schemes offer an important improvement over the old, irresponsible system, by giving fishermen a direct stake in the future of the fish stocks, a question remains: Have governments, in Canada and elsewhere, really absorbed the highly unpalatable truth that the fisheries – and even the survival of larger predatory fish species – have a precarious future if the status quo is preserved? In May 2004 the Canadian government heavily publicized its pursuit of Portuguese fishing vessels using illegal nets off the nose and tail portions of the Grand Banks outside the two-hundred-mile limit, where cod are supposed to be under some protection. But Canada refused to impound these vessels until the action was approved by the European Union and the Northwest Atlantic Fisheries Organization (NAFO) – and under NAFO rules Canadian inspectors gave advance warnings of the inspections, giving the vessel captains the opportunity to jettison their illegal nets and cargoes. This does not indicate a high level of resolve to conserve the fisheries on the part of the Canadian government. More troubling is the information imparted by critics that the government had been tailing these ships for months, and only became more 'proactive' once it became apparent that the governing Liberals would probably soon call an election.

In the United States, the Sustainable Fisheries Act of 1996 has helped many fish populations stage partial recoveries by restricting quota increases and giving priority to the recovery of the fish stocks. However, the Gilchrest–Farr 'Fisheries Recovery Act,' proposed in 2000 to amend the Magnuson–Steven Fisheries Conservation and Management Act, would have gone further in embodying the precautionary approach to managing the federally controlled fisheries. It would have made law most of the recommendations of the Ecosystem Principles Advisory Panel, including the requirement for fisheries ecosystem plans for federal fish management schemes. Unfortunately, the Gilchrest–Farr Act was defeated in Congress.

The North Atlantic fisheries present a far grimmer picture. In the 1980s the cod populations of the North Sea and the Barents Sea were approaching collapse, so in 1990 the Norwegian government imposed, despite massive political opposition from fishermen and trawling companies, a moratorium on the fishery, and subsidized trawlers to travel to other fishing grounds. Following this crisis, the capacity in the fishing fleet was reduced, and there was a move to smaller vessels. By 1992 the spawning biomass of the Barents Sea cod stock was larger than it had been for twenty-five years. The fishery reopened the following year with much-reduced quotas. The cod stock seemed to be recovering, but in 1998 it plummeted again, until by 2002 it had reached historical lows. In October 2003, a fisheries moratorium was again recommended, this time by the International Council for the Exploration of the Sea, for cod stocks all around Europe, including those of the Irish Sea. ICES also recommended fishing moratoria on North Sea plaice, southern hake, Barents Sea capelin, and Irish Sea whiting.[74] All too typically, the European Council of Fisheries Ministers did not deliver; instead of imposing moratoria, or reducing quotas, it chose to reduce the fishing season and to increase surveillance on board fishing vessels. The hard lessons of the 1990s, although understood by scientists and many fishermen, have yet to be learned by politicians; even some fishermen's groups oppose the moratoria.[75]

The future of the fisheries may indeed be bleak, and the preferred food fishes of the past may actually become extinct. They could be supplanted as predator species by, among other things, jellyfish (horrors!) if these lessons continue unlearned. Reforms in fishery management and in government attitudes would go a long way toward improving the health of the fisheries. The problem is that the solutions are very long-term, and require not-so-short-term hardships, especially for fishermen and gourmets. Long after the California mackerel and sardine fisheries collapsed in 1950s, for example, a scientific analysis in 1971 showed that even a minimal fishery involving a few fishing ships, recovering a mere 250 tons per year, would prevent the fish stocks from recovering.[76]

And indeed, their strong recovery did not come about until the late 1990s, following a complete fishing moratorium imposed in 1973 and a slow partial reopening in 1986 with a thousand-ton quota. In 1999 the State of California issued a press release with the good news that the sardine resource had 'surpassed a million tons and is now considered fully recovered for the first time since the heyday of Cannery Row in the mid-1940's.' The fishing quota was set at 132,800 tons, still considerably lower than the unsustainable 800,000 tons landed in 1936–7, and the biomass was estimated to be around one-third of the 1937 biomass. The new fishery will have to be carefully managed to consolidate this success story, but the outcome vindicates the efforts of marine scientists and fishery managers.[77]

Fisheries biology has undergone some dramatic changes since 1990. In *Scaling Fisheries: The Science of Measuring the Effects of Fishing 1855–1955*, an insightful and vital contribution to the history of fisheries science, Tim D. Smith made two perceptive and telling criticisms of his science. The first was that it had failed to embrace multispecies population ecology – or indeed, any kind of ecology – despite the overlap of interests between fishery biologists and population ecologists. The second and perhaps more valuable criticism was that fisheries biology suffered a severe 'historical myopia' that allowed 'periodic rediscovery by fishery biologists and managers of the importance of ecological interactions in the dynamics of fishery resources, and of the ever shifting scapegoats (currently climatic changes and pollution) used to explain the ongoing problems of our fisheries.'[78]

Although the first claim contains important facets of truth, it is not entirely accurate, as early fisheries biology, moulded by the Victorian natural history tradition, had many ecological elements present, and practitioners such as Huntsman were self-consciously ecological in their interests – Huntsman had direct contact with population ecologist Charles Elton, for example. But Hawaiian fisheries biologist R.W. Gauldie agrees with Smith about the ecology content of more recent fisheries science: 'There simply isn't any.' In a historically focused criticism of the mathematical modelling approach, he argues that 'it is at the level of an explanatory theory of ecology where I think that the problem of fisheries management lies.' He concluded that 'the problem of a theory of abundance is still the biologists' to solve. Unless we achieve this first objective then fisheries management science will soon join all of the other withering fig leaves that we have used to hide our shameful ignorance of the ecology of fish.'[79] With the precautionary principle now becoming entrenched as the new paradigm for a more ecologically focused fisheries biology, there are indications that this failure is being addressed, at least to some extent.

As for the second, telling criticism of fisheries science – that it has ignored its own history – there are signs that this, too, is changing. The ongoing fisheries crises have forced scientists to become more self-reflective about their science and their priorities, which makes the perusal of recent articles and books very rewarding. What is most striking about many of the research articles and opinion pieces in scientific publications are the references made to earlier scientific work, and not just from the post–Second World War period: scientists have been harking back to the work of early pioneers such as Hjort, Thompson, Holt, and Petersen. One finds the ghost of Huxley's 'inexhaustible fisheries' theory being resurrected by a Canadian scientist in 1997, hopefully for the last time, and being demolished and excoriated by an exasperated Ransom Myers.[80] But this does not mean that scientists are just looking backward. They are also looking forward, using history as one of their new tools. The tool of history is being promoted in several forms. Scientists are revaluing past contributions and approaches, and are also beginning to take notice of the wealth of information archived in past studies – especially where data collections have survived and can be reanalysed.[81]

Computers offer an important tool for dealing with increasingly complex systems, and their power will have to be harnessed if fisheries biologists and managers are to achieve the goal of maintaining and restoring fisheries ecosystems. It can be argued that the initial move toward using computers for fish population modelling was premature and gave scientists a false sense of mastery. Indeed, the mathematical basis of fisheries science led to problems because the perception became 'that fisheries science ought to be as precise and as predictable as ... physics and engineering'; as a result when fisheries scientists admitted to their 'uncertainty of knowledge [this was] often reinterpreted as a lack of knowledge owing to the public misperception of science as a set of known facts, rather than a process.'[82] More recent tools such as ECOPATH require powerful computers and improved software to summarize natural growth and consumption rates of marine populations to improve our understanding of a fish population's ecosystem dynamics.

Fisheries biologists such as Daniel Pauly at the University of British Columbia's Fisheries Centre are establishing computer-based fish and fishery databases (e.g., Pauly and R. Froes's *FishBase*), which ideally incorporate interdisciplinary knowledge, long-time series of data, and even historical data collections. Computers and large databases can bring together global collections of data from around the world, including results buried in 'grey literature and protected data files' and archived data collections previously 'unavailable to the larger scientific community.' Such databases allow scientists to compare current fish population structures with the status of these populations historically.

The Internet provides another computer-based tool for fisheries science, since the collected databases can potentially be made available through the Internet, as is the case with *FishBase.* Laura J. Richards and Jon T. Schnute compare the potential of such data collections with the great accomplishments in physics and astronomy resulting from the data on planetary motions collected by the Danish astronomer Tycho Brahe (1546–1601), which made it possible for Johannes Kepler (1571–1630) to come up with his theories of planetary motion, and for Isaac Newton (1642–1727) to come up with his general laws of dynamics. They acknowledge that fisheries science is not based on replicable laboratory experiments, but argue that nevertheless, fisheries data 'must be shared to enable objective, scientific development.'

Improved software programs, and new statistical and analytical tools, are needed to deal with the complexities of adjusting for the differing scales and scopes of different studies. Such programs and tools will enable a reliable meta-analysis of the data. This work has recently occupied Ransom Myers, who has applied meta-analysis based on synthesizing data from hundreds of fish populations to make generalizations about the effects of density, temperatures, and reproductive rates on fish stocks.[83] Canadian scientists are at the forefront of the latest research models and approaches.

During the early shock of the cod crisis, I asked W.E. Ricker, the Pacific Biological Station's best known fisheries biologist, 'What use is fisheries biology?' Somewhat acidly, he reminded me that it had ensured the continuity of the Pacific salmon.[84] Indeed, it must always be remembered that, despite serious past failings, fisheries biologists in Canada and elsewhere have accomplished great things. The legacy of the Biological Board of Canada and its successors, and the British, Scandinavian, American and other fisheries research organizations described in this book, has been the building up of a body of knowledge that offers our only hope for understanding and respecting marine ecosystems and resources, and being able to conserve them for the benefit of future generations. Scientists in Canada and the United States have already saved several fisheries from extinction.

A century after the whole enterprise was begun by E.E. Prince in Canada, the scientific work he and his colleagues – Ramsay Wright, A.P. Knight, A.B. Macallum, A.G. Huntsman, R.E. Foerster, and the rest – helped launch is still thriving, although the work has become far more urgent. There is no doubt that had this discipline not evolved from the early marine scientists' work, fishermen, industrialization, and indifferent governments would have long since eliminated the Pacific salmon, halibut, mackerel, and sardines, and many North Atlantic stocks. Recent fisheries publications are full of new approaches, new perspectives, and a new vitality. Despite its spotty past, fisheries biology, with its emerging new goals and methods, offers hope that this science will be able to deliver even better outcomes in the future.

Notes

Introduction

1 See Harris, *Lament for an Ocean*. The cod stock collapse led to a spate of books by individuals striving to understand what went wrong, including Berill, *The Plundered Seas* and Kurlansky, *Cod*.

2 This is brought out in Harris's book and very strongly by Finlayson in *Fishing for Truth*. Ransom Myers, one of the leading fisheries biologists, a young scientist at the Department of Fisheries and Oceans at the time, says that senior scientists during the years leading up to and including the crisis tended to stockpile and safeguard the government fisheries data for their own use, blocking other scientists who questioned their conclusions from examining and interpreting the raw data (personal communication, 27 June 2003).

3 See Mills, *Biological Oceanography*; Deacon, *Scientists and the Sea*; and Schlee, *On Almost Any Wind: The Saga of the Oceanographic Vessel Atlantis* (Ithaca, NY: Cornell University Press, 1978). More recent studies include Benson and Rehbock, eds., *Oceanographic History* and Rozwadowski, *The Sea Knows No Boundaries*.

4 Johnstone, *Aquatic Explorers*.

5 An exception is a valuable study on the Bergen school of meteorology: Friedman, *Appropriating the Weather*. Weir's *An Ocean in Common* does a superb job of showing how oceanographers benefited by sharing goals with the U.S. Navy and pushing forward applied science.

6 For a history of the work focus at Woods Hole, see Maienschein, *One Hundred Years Exploring Life* and *Defining Biology* as well as articles by Maienschein and Groeben in *Biological Bulletin* 168 (1985).

7 See, for example, Bowler, *The Fontana History of the Environmental Sciences*; Golley, *History of the Ecosystem Concept in Ecology*; Shortland, ed., *Science and Nature*; Worster, ed., *The Ends of the Earth*.

8 See Zeller, *Inventing Canada*. Also, for a refreshing look at early conservationism in Canada, see Foster, *Working for Wildlife*.
9 For Elton's work, see Crowcroft, *Elton's Ecologists*.
10 See Smith, *Scaling Fisheries*. Other historical descriptions of the development of fishery science are given by Cushing, *The Provident Sea*, and Lee, *The Directorate of Fisheries Research*.

1. Scientists at Sea

1 Huntsman, 'Fisheries Research in Canada,' 117.
2 *Daily Mail and Empire*, 16 August, 1897: 4.
3 de Vecchi, 'Science and Government in Nineteenth-Century Canada,' 286–91. For a discussion of the BAAS's role in helping to found Canadian scientific institutions, see de Vecchi, 'The Dawning of a National Scientific Community in Canada, 1878–1896,' *Scientia Canadensis* 8 (1984): 32–58.
4 A £75 grant was given to help the University of Toronto's Madawaska Club (formed in 1896 to study lake biology) build a station. The committee included Professor L.C. Maill, chairman; R. Ramsay Wright, secretary; Dr G.M. Dawson; Professor W.H. Ellis; E.E. Prince, and Professor John Macoun. In 1898 the club received land at Go-Home Bay from the Province of Ontario.
5 He called for a marine station in his article, 'Science in Canada' published in *The Week*. See Huntsman, 'James Playfair McMurrich.'
6 Winsor, *Starfish, Jellyfish and the Order of Life*, 5.
7 Ibid.
8 Lillie, *The Woods Hole Marine Biological Laboratory*, 12.
9 Allard, *Spencer Fullerton Baird*, 65–6.
10 Jackson, 'The Canadian Marine Biological Station,' 308.
11 Johnstone, *The Aquatic Explorers*, 22. Son of the founder of New Brunswick's Ganong confectionary, Ganong took his PhD in Munich in 1894 and taught botany at Smith College, Northampton, Massachusetts, from 1895 to 1932. He published over 150 papers and often visited the St Andrews Biological Station. See Margaret E. Macallum, 'Ganong, William Francis,' in *The Canadian Encyclopedia, Year 2000 Edition* (Toronto: McClelland and Stewart, 2000), 948–9.
12 Gunther, *Life of William Carmichael M'Intosh*, 60, 75, 79.
13 Ibid., 85.
14 'Dr E.E. Prince Retires after 30 Years,' *Canadian Fisherman* 11 (October 1924): 286–7. The word is still used, although 'benthic' is more common; Gunther, *Life of William Carmichael M'Intosh*, 93.
15 Gough, *Fisheries Management in Canada*, 12.
16 A.G. Huntsman, 'Edward Ernest Prince,' 1–2.

17 Huntsman, 'Fisheries Research in Canada,' 117.
18 Deacon, *Scientists and the Sea*, 281.
19 Ibid., 309.
20 Ibid., 313.
21 Schlee, *Edge of an Unfamiliar World*, 92, 98; Deacon, *Scientists and the Sea*, 333.
22 Schlee, *Edge of an Unfamiliar World*, 125–6.
23 Deacon, *Scientists and the Sea*, 388.
24 Certain floating laboratories – such as those in Denmark, Scotland, Michigan, and Canada – functioned essentially like those built permanently on shore, since the main purpose for floating them was to transport them from one locale to another. Work was undertaken at different sites only while the 'arks' or stations were fixed in position.
25 Zeller, *Inventing Canada*, 143.
26 Kofoid, *Biological Stations of Europe*, 81–4, 89–90.
27 Ibid., 95.
28 Ibid.
29 Ibid., 13–14, 8.
30 Groeben, 'Anton Dohrn,' 5–6.
31 Dohrn, 'Report of the Committee,' 192.
32 Groeben, 'Anton Dohrn,' 11, 15.
33 Among the prominent scientists who belonged to these committees were Thomas Henry Huxley, E. Ray Lankester, Michael Foster, Adam Sedgwick, Henry Moseley, William A. Herdman, J. Gwyn Jeffreys, Sir C. Wyville Thompson, John Murray, and William C. McIntosh.
34 Lankester, 'Address to Section D.,' 525.
35 Mills, *Biological Oceanography*; 191–2.
36 Allen and Harvey, 'Laboratory of the Marine Biological Association,' 735, 735–6.
37 Ibid., 744.
38 F.S. Russell, 'The Plymouth Laboratory,' 763.
39 *Daily Mail and Empire*, 14 August 1897.
40 Ibid., 17 August 1897.
41 Ibid., 20 August 1897.
42 Ibid.
43 'The Canadian Biological Station – First Report of the Committee consisting of Professor E.E. Prince, (Chairman), Dr. T. Wesley Mills, Dr. A.B. Macallum, Professor John Macoun, Professor E.W. McBride, Mr. W.T. Thistleton-Dyer, and Professor D.P. Penhallow (Secretary), on the Establishment of a Biological Station in the Gulf of St Lawrence,' *Report of the British Association for the Advancement of Science* (1898): 582.
44 Wright, 'Report on the Occupation of a Table at Naples,' 388.

45 Letter from D.P. Penhallow to E.E. Prince, 28 March 1898; and letter from Penhallow to Prince, 12 February 1898, National Archives of Canada (NAC), RG 23, Vol. 312, File 2548 Part 1 (Reel T-3997).
46 'The Canadian Biological Station,' 582.
47 Zeller, *Inventing Canada*, 51–77.
48 'Canadian Biological Station,' 582; and 'Extract from a Report of the Committee of the Honourable the Privy Council, approved by His Excellency on the 9th May, 1898,' NAC, RG 23, Vol. 312, File 2548, Part 2 (Reel T-3997).
49 Letter from D.P. Penhallow (Secretary) to His Excellence the Governor General of Canada, 30 April 1898, NAC, RG 23, Vol. 312, File 2548, Part 1 (Reel T-3997).
50 'Extract from a Report of the Committee of the Honourable the Privy Council, approved by His Excellency on the 9th May, 1898,' NAC, RG 23, vol. 312.
51 Ibid.
52 Ibid.
53 Ibid.
54 Zeller, *Inventing Canada*, 272.
55 Ibid., 205.
56 Ibid., 267.
57 'Extract from a Report.' For a discussion of Canadian imperialism and anti-Americanism, see Berger, *Sense of Power*; de Vecchi, 'Science and Government' especially chap. 3, 77–130; and de Vecchi, 'The Dawning of a National Scientific Community,' 32–58.
58 Canada, House of Commons, *Debates*, 10 June 1898, 7733.
59 For this argument, see de Vecchi, 'Science and Government,' 299.
60 'Extract from a Report.'
61 Canada, House of Commons, *Debates*, 10 June 1898, 7733. This was by virtue of his position in the Department of Marine and Fisheries. See letter from D.P. Penhallow to E.E. Prince, 23 June 1898, NAC, RG 23, Vol. 312, File 2548 Part 1 (Reel T-3997).
62 Wright, although not originally a board member, turned up persistently at its meetings, in spite of bylaws calling for only one trustee from each university. Macallum, a strong personality, had worked hard to found the station and did not give way to Wright. Things were worked out amicably when it was resolved that the federal government should have two members on the board: 'the second appointee was Macallum, who thus liberated the Toronto place for Ramsay Wright' (de Vecchi, 'Science and Government,' 307–8).
63 W.C. M'Intosh, *The Life Histories of the British Marine Food Fishes*, 7.
64 Zeller, *Inventing Canada*, 27.
65 Masters, 'The Scottish Tradition in Higher Education,' 259. Carl Berger

noted the ties between Canadian chairs of natural history with 'the Athens of the North'; see Berger, *Science, God, and Nature,* 1983, 7.

66 'Marine Biological Station for Canada: Report of Progress'; letter from E.E. Prince to the Deputy Minister of Marine and Fisheries, Major Gourdeau, 28 July 1899, NAC, RG 23, Vol. 312, File 2548, Part 1, Reel T-3997.

67 Jane Maienschein, 'Agassiz, Hyatt, Whitman,' 26.

68 Monroy and Groeben, 'The "New" Embryology,' 36.

69 Ibid., 37–8.

70 Maienschein, 'Agassiz, Hyatt, Whitman,' 31–3.

71 Maienschein, 'Introduction,' *Defining Biology,* 32.

72 Dohrn, 'Report of the Committee,' 409–10.

73 D. Ebert, 'Evolving Institutional Patterns for Excellence,' 183.

2. Fishing for Ideas

1 Gollum's riddle in J.R.R. Tolkien, *The Hobbit: or There and Back Again* (London: George Allen and Unwin, 1937), 87.

2 Arnon, *Organisation and Administration of Agricultural Research,* 2–4.

3 Mills, *Biological Oceanography,* 43–74.

4 Rosenberg, 'Science, Technology and Economic Growth,' 184–7.

5 Harkness, Leonard, and Needham, 'Fishery Research at Mid-Century,' 214.

6 Ebert, 'Carnegie Institution of Washington and Marine Biology,' 178; and Russell-Hunter, 'From Woods Hole to the World,' 201.

7 Friedrich, *Marine Biology,* 1.

8 Allard, *Spencer Fullerton Baird,* 68. See also Allard, 'The Fish Commission Laboratory.'

9 Allard, *Spencer Fullerton Baird,* 76–7; 334, 354.

10 Ibid., 101, 317–18, 93, 98.

11 Brosco, 'Henry Bryant Bigelow,' 255–6.

12 'Report of the Committee, consisting of Professor Ray Lankester,' 151.

13 Graham, 'Science and the British Fisheries,' 1n, 2–3.

14 See John Stuart Mill, *August Comte and Positivism,* 33–41.

15 Reingold, 'American Indifference to Basic Research,' 47.

16 Popper, 'Normal Science and Its Dangers,' 52–3. Italics in original.

17 Staudenmaier, *Technology's Storytellers* 1985, 96–71, 102; Bunge, 'Towards a Philosophy of Technology,' 29–30.

18 Layton, 'Mirror Image Twins,' 565–6; Graham, *The Fish Gate,* 115; and letter from S.C. Prescott to Commission III of the Institut International du Froid, 20 April 1938. Huntsman Collection, University of Toronto Archives, Accession No. B78-0010 Box 56.

19 Mills, *Biological Oceanography*, 77–8.

20 Gunther, *William Carmichael M'Intosh* 81; Cushing, *The Provident Sea*, 295.

21 Cushing, *The Provident Sea*, 130–1.

22 Gunther, *William Carmichael M'Intosh*, 81–2, 86, 91, 93.

23 Kofoid, *The Biological Stations of Europe*. Stations funded and run by one university are not included here; these were characterized by basic research and an emphasis on undergraduate teaching.

24 Christensen, 'A Century of Fisheries Research and Management,' 5–6; Brattström, 'The Biological Stations of the Bergens Museum,' 26.

25 Mills, *Biological Oceanography*, 191.

26 See Maienschein, *One Hundred Years Exploring Life*; Maienschein, *Defining Biology*; Groeben, 'Anton Dohrn'; Maienschein, 'Agassiz, Hyatt, Whitman.' Volume 168 of the *Biological Bulletin* is a special issue devoted to histories of the Marine Biological Laboratory and the Statione Zoologica di Napoli. See also Allard, *Spencer Fullerton Baird*, and de Vecchi, 'Science and Government,' 296–7.

27 Oreskes, 'Weighing the Earth from a Submarine,' emphasis in original. See also Reingold, 'Alexander Dallas Bache,' 120, and 'Cleveland Abbe at Pulkovo.'

28 Rosenberg, 'Science, Technology and Economic Growth,' 189, 190–1, 195, 199–202.

29 Huntsman, 'Edward Ernest Prince,' 2; Prince, 'Memorial Re. Increased Appropriation for the Marine Biological Station,' to the Honourable Raymond Prefontaine, n.d., NAC, RG 23, Vol. 312, File 2548, Part 2, Reel T-3998.

30 E.E. Prince, 'Marine Biological Station of Canada,' NAC, RG 23, Vol. 312, File 2548, Part 2, Reel T-3997; Rigby and Huntsman, 'Materials Relating to the History of the Fisheries Research Board,' 31.

31 Letter from E.C. Whitman to the Honourable D.C. Fraser, 25 January 1901, NAC, RG 23, Vol. 312, File 2548 Part 1 – Reel T-3997.

32 Huntsman, 'The Fisheries Research Board of Canada,' n.d., Huntsman Collection, University of Toronto Archives, Accession No. B79-0048, Box 6; Rigby and Huntsman, 'Materials Relating to the History of the Fisheries Research Board,' 41.

33 'The New Biological Station Built at Departure Bay,' *Nanaimo Free Press*, 10 June 1908.

34 These figures concern Atlantic research only. See Huntsman and C.M. Fraser, 'List of Publications'; Rigby and Huntsman, 'Materials Relating to the History of the Fisheries Research Board,' 51–5, 94–6; Hart, 'Fisheries Research Board of Canada,' 1133.

35 Letter from E.E. Prince to George W. Taylor, 11 March 1911, NAC, RG 23, Vol. 378, File 3371, Part 1, Reel T-3369.

36 Letters from R. Ramsay Wright to E.E. Prince, 12 August 1900 and 11 March 1902, NAC, RG 23, Vol. 312, File 2548, Part 1, Reel T-3997.

37 Huntsman, 'Edward Ernest Prince,' 2.

38 D.P. Penhallow, 'Report on the Atlantic Coast Station, St. Andrews, N.B., for 1908,' NAC, RG 23, Vol. 381, File 3407, Part 1, Reel T-3371.

39 Ibid.

40 D.P. Penhallow, 'Report of a Special Committee on an Examination of Localities for the Permanent Location of the Marine Biological Station,' 22 November 1906, University of Toronto Archives, Huntsman Collection, Accession No. B78-0010, Box 11; Letter from R.R. Wright to E.E. Prince, 21 August 1900. NAC, RG 23, Vol. 312, File 2548 Part 1 – Reel T-3997.

41 Penhallow, 'Report on the Atlantic Coast Station.'

42 Johnstone, *The Aquatic Explorers*, 44.

43 Seaborn, *The March of Medicine*, 374–6.

44 J.B.L., 'Archibald Byron Macallum,' *Obituary Notices.*

45 Johnstone, *Aquatic Explorers*, 74.

46 Letter from E.E. Prince to A.W. Owen, 5 August 1899, and letter from A.W. Owen to E.E. Prince, 11 August 1899 (emphasis in original), NAC, RG 23, Vol. 312, File 2548, Part 1, Reel T-3997.

47 Letter from A.P. Knight to E.E. Prince, 23 August 1899, and letter from E.E. Prince to A.W. Owen, 8 August 1899, NAC, RG 23, Vol. 312, File 2548, Part 1, Reel T-3997.

48 Letter from E.W. MacBride to E.E. Prince, 26 March 1901, and letter from A.H. Mackay to E.E. Prince, 18 March 1903, NAC, RG 23, Vol. 312, File 2548 Part 2, Reel T-3997; letter from A.P. Knight to J.P. McMurrich, 23 March 1925, Huntsman Collection, University of Toronto Archives, Accession No. B78-0010, Box 13.

49 A.W. Owen, 'Memo for Deputy Minister,' 9 March 1907, RG 23, Vol. 313, File 2548 Part 3, Reel T-3998; Huntsman, 'Fisheries Research in Canada,' 117; 'Meeting of the Biological Board, April 19th, 1912,' NAC, RG 23, Vol. 381 File 3407, Part 1, Reel T-3371.

50 'Biological Problems at the St Andrews Station 1916,' and letter from E.E. Prince to A.G. Huntsman, 18 July 1912, Huntsman Collection, University of Toronto Archives, Accession No. B78-0010, Box 54; W.A. Found, 'Memorandum for Professor Prince re protection of the Halibut Fishery, Pacific Coast,' Ottawa, 4 July 1913, NAC, RG 23, Vol. 404, File 4232 Part 1, Reel T-3392.

51 Dr J.C. Medcof, Fisheries Research Board, 1961, quoted in Johnstone, *The Aquatic Explorers*, 45; Penhallow, 'Report on the Atlantic Coast Station.' Stafford overstepped his bounds as Pacific Biological Station curator, but the

records are silent about his misdemeanour. He refused to apologize, and was ejected from the board in 1913.

52 Letter from J.P. McMurrich to R. Ramsay Wright, 28 September 1911, NAC, RG 23, Volume 393, File 3712 Part 1, Reel T-3383.

53 Letters from R. Ramsay Wright to E.E. Prince, 21 August 1900; from Joseph Stafford to E.E. Prince, 10 May 1899; from James Fowler to E.E. Prince, 21 April 1900, all NAC, RG 23, Vol. 312, File 2548, Part 1, Reel T-3997.

54 McKillop, 'The Research Ideal and the University of Toronto,' 93.

55 See an entertaining biography, *Loring Woart*, by his son, J.W. Bailey; and Johnstone, *The Aquatic Explorers*, 41–2.

56 Beer and Lewis, 'Aspects of the Professionalization of Science,' 119–20.

57 Workers with more experience desired greater autonomy, which many tended to achieve. See Cotgrove and Box, *Science, Industry and Society*, 124–6.

58 'This was ... wasteful because up to one half of the fish were not recovered, although the range of effectiveness ... of dynamite proved to be much less than anticipated.' Ricker, *The Fisheries Research Board*, 13–14.

59 J.B.L., 'Archibald Byron Macallum,' *Proceedings*, 288.

60 D.M. Simpkins, 'MacBride, Ernest William,' in *Dictionary of Scientific Biography* (New York: Charles Scribner's Sons, 1981), 586; and Lee, *The Directorate of Fisheries Research*. 83, 102.

61 Interview of 3 May 1972 with Huntsman (interviewer: Elizabeth Wilson), Part of the Oral History Program at the University of Toronto, University of Toronto Archives, Reading Room Finding Aid No. B74-0021, transcript p. 3.

62 Craigie, *A History of the Department of Zoology*, 62, 40; Letter from A.G. Huntsman to W.A. Clemens, 26 December 1951, University of Toronto Archives, Accession No. B78-0010, Box 15.

63 Nicholson, 'Autonomy and Accountability of Basic Research,' 59.

64 Calderwood, 'On Recent Investigations of the Marine Biological Association'; and Cunningham, 'On the Growth of Food-fishes.'

65 Allen and Harvey, 'The Laboratory of the Marine Biological Association, 735. However, things changed in 1919–20, when the Development Commission became the source of funding for most British marine research. Because of mounting costs, the Marine Biological Station was only able to survive because the Development Commission provided most of its support. (Graham, 'Science and the British Fisheries,' 3). After this, fisheries laboratories at Lowestoft and Aberdeen, entirely under the Ministry of Agriculture and Fisheries, became centres of fisheries research, while the university-affiliated Plymouth laboratory veered toward basic research in biological oceanography. In spite of this bifurcation, some fisheries research continued to be carried on at Plymouth. (Mills, *Biological Oceanography*, 203, 189–257.)

66 Anderson, 'Policy Determination of Government Scientific Organizations,' 312.

3. The Canadian Fisheries Expedition, 1914–15

1 Herbert Hoover, quoted in Hugh Sidney, 'The Presidency: A Force, Fame, and Fishing,' *Time*, 11 March 1991, 57.
2 The expression 'Intensive Area Study' is taken from Brosco, 'Henry Bryant Bigelow,' 239–64.
3 Went, 'Seventy Years Agrowing,' 48.
4 Mills, *Biological Oceanography*, 75.
5 Ibid., 77–8.
6 D. Merriman, 'Hjort, Johan,' in *Dictionary of Scientific Biography* (New York: Charles Scribner's Sons, 1981), 442.
7 Mills, *Biological Oceanography*, 81–2; Hjort, 'The Formative Years of European Marine Science,' 4; Went, 'Seventy Years Agrowing,' 170.
8 Went, 'Seventy Years Agrowing,' 10.
9 Mills, *Biological Oceanography*, 86.
10 Herwig, Pettersson, and Hoek, 'The Aims of the International Council,' 11; Mills, *Biological Oceanography*, 85.
11 Mills, *Biological Oceanography*, 86.
12 Schlee, *The Edge of an Unfamiliar World*, 216.
13 Ibid., 220.
14 Ibid., 223.
15 Kofoid, *The Biological Stations of Europe*, 297; Went, 'Seventy Years Agrowing,' 23–4; Hjort, 'Fluctuations in the Great Fisheries of Northern Europe,' 58.
16 Tim D. Smith discusses the genesis of these methods in *Scaling Fisheries*, 80–1.
17 Hjort, 'Fluctuations in the Great Fisheries,' 59; Lee, *The Directorate of Fisheries Research*, 79; Schlee, 222–3.
18 Hjort, 'Fluctuations in the Great Fisheries,' 59–60; Schlee, *The Edge of an Unfamiliar World*, 225–6.
19 Thomassen, introduction to Hjort, 'Fluctuations in the Great Fisheries,' 53.
20 Schlee, *The Edge of an Unfamiliar World*, 228.
21 Robert Merton's studies revealed the rather surprising finding that multiple discoveries are very common in the history of science, and perhaps more common than uncontested (single) discoveries. See Merton, 'Singletons and Multiples in Science.'
22 See Tim D. Smith, *Scaling Fisheries*, 134–8.
23 Thomasson, introduction to Hjort, 'The Formative Years of European Marine Science,' 1.

24 The Biological Board actually wanted to hire Hjort permanently. Hjort, at forty-six, 'the greatest living authority on fish and the fishing industry' was willing to come to Canada – he wanted to return to 'scientific and economic investigations' from administrative work. The board, however, was unable to offer him enough money. See letters A.B. Macallum to J.D. Hazen, 25 April, 1914 and from Johan Hjort to A.B. Macallum, 25 November 1914, NAC, RG 23, Volume 409, File 4743, Part 1; and Vol. 1204, File 726-2-4, Part 1.

25 Johan Hjort, 'Introduction to the Canadian Fisheries Expedition, xv.

26 Letter from Hjort to Macallum, 25 November 1914.

27 Ibid. see also J.J. Cowie, 'Drift Net Fishing for Herring,' 42–3.

28 E.E. Prince and A.B. Macallum, 'Memo re Herring Fishery Development Investigation,' 10 December 1914, NAC, RG 23, Vol. 1204, File 726-2-4 Part 1.

29 E.E. Prince, 'Memo: Re Dr. Hjort's Fishery Investigations Season 1915,' NAC, RG 23, Vol. 1204, File 726-2-4, Part 1.

30 Letter from G.J. Desbarats to Johan Hjort, 6 May 1915, NAC, RG 23, Vol. 1204, File 726-2-4, Part 1.

31 Prince and Macallum, 'Memo re Herring Fishery'; Prince, 'Memo: Re Dr. Hjort's Fishery Investigations Season 1915.'

32 Letter from A.G. Huntsman to E.E. Prince, 21 November 1914, Huntsman Collection, University of Toronto Archives, Accession No. B78-0010, Box 9; Letter from Johan Hjort to J.G. Desbarats, 24 April 1915, NAC, RG 23, Vol. 1204, File 726-2-4, Part 1; Letter from Johan Hjort to H.B. Bigelow, 13 March 1915, Harvard University Archives, H.U.G. 4212.5., Bigelow General Correspondence, Box 2, Folder H-I-J; Huntsman, 'Arthur Willey,' 95–8.

33 Johan Hjort, 'Memorandum regarding hydrographical and biological investigations in the Gulf of St. Lawrence and adjacent waters May to August 1915,' 16 March 1915, NAC, RG 23, Vol. 1204, File 726-2-4, Part 1; Letter from Johan Hjort to A.G. Huntsman, 4 May 1915, Huntsman Collection, University of Toronto Archives, Accession No. B78-0010, Box 9.

34 Letter from Johan Hjort to J.G. Desbarats, 24 April 1915, NAC, RG 23, Vol. 1204, File 726-2-4, Part 1; and letters from Johan Hjort to A.G. Huntsman, 12 March 1915, and n.d., Huntsman Collection, University of Toronto Archives, Accession No. B78-0010, Box 9.

35 Letter from Johan Hjort to E.E. Prince, 4 May 1915, NAC, RG 23, Vol. 1204, File 726-2-4, Part 1.

36 Letter from Johan Hjort to G.J. Desbarats, 12 May 1915, NAC, RG 23, Vol. 1204, File 726-2-4, Part 1.

37 Ibid.

38 Letter from Johan Hjort to G.J. Desbarats, 4 June 1915, NAC, RG 23, Vol. 1204, File 726-2-4, Part 1.

39 Paul Bjerkan, 'Results of the Hydrographic Observations,' 352–3.
40 Letter from J.P. McMurrich to A.G. Huntsman, 3 August 1915, Huntsman Collection, University of Toronto Archives, Accession No. B78-0010, Box 9.
41 Letters from Johan Hjort to E.E. Prince, 24 December 1914, and from Hjort to Macallum, 25 November 1914 (emphasis in original), NAC, RG 23, Vol. 1204, File 726-2-3, Part 1.
42 Prince and Macallum, 'Memo re Herring Fishery,' 10 December 1914; and 'Freezing Fish in Brine as Soon as Caught,' 5.
43 Letter from Hjort to Prince, 24 December 1914.
44 Ibid., and Prince and Macallum, 'Memo re Herring Fishery.'
45 Letters from Hjort to Prince, 24 December 1914; E.E. Prince, 'Memo: Re Dr. Hjort's Services in Canada,' 30 December 1914, NAC, RG 23, Vol. 1204, File 726-2-3, Part 1.
46 Letter from Johan Hjort to G.J. Desbarats, 29 April, 1915, NAC, RG 23, Vol. 1204, File 726-2-4, Part 1.
47 Letters from G.J. Desbarats to Johan Hjort, 6 May 1915, and from Johan Hjort to G.J. Desbarats, 7 May 1915, NAC, RG 23, Vol. 1204, File 726-2-4, Part 1.
48 Letters from Johan Hjort to G.J. Desbarats, 22 August 1915, 21 February 1916, and 22 May 1917, NAC, RG 23, Vol. 1204, File 726-2-4, Parts 2 and 3; and Hjort, 'Introduction to the Canadian Fisheries Expedition,' xx.
49 Sandström, 'The Hydrodynamics of Canadian Atlantic Waters.'
50 Hjort, 'Introduction to the Canadian Fisheries Expedition,' xxv. In Gran's method for estimating phytoplankton biomass (as an indicator of productivity), water samples preserved in 'Flemming's liquid' were centrifuged; the plankton deposit depth at the bottom of the test-tube was then quantified. See H.H. Gran, 'Quantitative Investigations as to Phytoplankton and Pelagic Protozoa,' 490, 491–2.
51 This is how Hjort in his 'Introduction to the Canadian Fisheries Expedition' referred to what was originally called the 'Central Bureau of the International Commission for the Investigation of the Sea' (Kofoid, *The Biological Stations of Europe*, 316). This was the ICES headquarters, linked with the Central Laboratory, and founded in 1902, for coordinating its work and arranging for its publications. (See Rozwadowski, *The Sea Knows no Boundaries*, 38–40.)
52 Dannevig, 'Canadian Fish Eggs and Larvae,' 36, 44–8; Hjort, 'Introduction to the Canadian Fisheries Expedition,' xxvii; and letter from Johan Hjort to G.J. Desbarats, 21 February 1916, NAC, RG 23, Vol. 1204, File 726-2-4, Part 2.
53 Ruud, 'The Herring Scale Dispute,' 64.
54 Lea, 'Report on "Age and Growth of the Herring,"' 115.

55 Ibid., 123–4, 130.

56 Letter from A.G. Huntsman to Johan T. Ruud, 12 April 1949. Huntsman Collection, University of Toronto Archives, Accession No. B78-0010, Box 9.

57 Hart, 'Fisheries Research Board of Canada,' 1133–4.

58 See A.G. Huntsman, 'Miramichi Fisheries Investigation,' Huntsman Collection, University of Toronto Archives, Accession No. 879-0048, Box 7. Huntsman, 'The Scale Method of Calculating the Rate of Growth in Fishes,' 48.

59 Huntsman, 'The Canadian Plaice.' The quotation is from an expert on fish population dynamics: Gulland, *Fish Population Dynamics*, 14. The same year, F.I. Baranov devised a sophisticated model describing the effects of fishing pressure on fish age distributions, but his paper was almost unknown outside Russia until translated by W.E. Ricker in the 1940s.

60 Letter from A.G. Huntsman to Johan T. Ruud, 12 April 1949, Huntsman Collection, University of Toronto Archives, Accession No. B78-0010, Box 9; letter from G.J. Desbarats to Johan Hjort, 13 March 1916, NAC, RG 23, Vol.1204, File 726-2-4 Part 2.

61 Hart, 'Fifty Years of Research in Aquatic Biology,' 1134–5.

62 Letter from A.G. Huntsman to W.A. Munn, 25 January 1923, Huntsman Collection, University of Toronto Archives, Accession No. B78-0010, Box 92.

63 Letter from W.A. Found to the Deputy Minister, Department of Marine and Fisheries, Newfoundland, 5 June 1923; and letters from A.G. Huntsman to H.B. Bigelow, 31 July 1923; to Alan Goodridge, 17 July 1923, and to W.A. Found, 22 August 1923. Huntsman Collection, University of Toronto Archives, Accession No. B78-0010, Box 92.

64 Hart, 'Fifty Years of Research in Aquatic Biology,' 1135–6; letter from Alan Gardiner to A.G. Huntsman, 1 November 1923; and letters from A.G. Huntsman to W.B. Bailey, 22 December 1950; to Johan Ruud, 12 April 1949; and to David C. Nutt, 30 December 1948, University of Toronto Archives, Accession No. B78-0010 Boxes 91, 92 and 94.

65 Personal communication from E.L. Mills.

66 Hjort, 'Introduction to the Canadian Fisheries Expedition,' xxviii.

67 Ibid.

68 Letter from Fred Cook to G.J. Desbarats, 12 December 1917, NAC, RG 23, Vol. 1204, File 726-2-4, Part 3.

69 Letter from G.J. Desbarats to Fred Cook, 21 December 1917, NAC, RG 23, Vol. 1204, File 726-2-4, Part 3.

70 E.E. Prince, 'Department of Marine and Fisheries: The Biological Board of Canada,' address delivered at Research Conference Ottawa, 20–22 February 1923, Huntsman Collection, University of Toronto Archives, Accession No. B78-0010, Box 72; Hart, 'Fifty Years of Research in Aquatic Biology,' 1133–4;

Hjort, 'Introduction to the Canadian Fisheries Expedition,' xxi; and letter from Johan Hjort to G.J. Desbarats, 7 June 1916, NAC, RG 23, Vol. 1204, File 726-2-4, Part 3.

4. Ottawa, 1919

1 *Canadian Fisherman* 3 (1916): 325.

2 'Department of Commerce: Bureau of Fisheries, Washington: Conservation Work of the Bureau of Fisheries.' Information sheets sent to W.A. Found, 16 May 1930, NAC, RG 23, Vol. 1467, File 769-1-2, Part 6.

3 Fishery Board for Scotland, *Tenth Annual Report* viii–x.

4 Cunningham, *The Natural History of the Marketable Marine Fishes*, 19.

5 Ibid., 20; and Jenkins, *The Sea Fisheries*, 1920, 245.

6 Chambers, *The Canadian Marine*, 33.

7 Prince, 'The Fishing Industries of Canada,' 17.

8 Canada, House of Commons, *Debates*, 23 May 1900, vol. 52, no.2, 5928–30.

9 Ibid., 5 April 1907, vol. 80, 5829.

10 An excellent account of the role of statistics' in the Halifax Award is given by Allard, *Spencer Fullerton Baird*. 1978.

11 'Department of Commerce: Bureau of Fisheries, Washington.'

12 'Government Daily Bait Reports.'

13 Chambers, *The Canadian Marine*, 114.

14 Letter from E. Hawken to M.G. Schrader, 21 February 1921, NAC, RG 23, Vol. 1318, File 729-3-13, Part 1; and letter to A.K. McLean, 14 April 1928, Huntsman Collection, University of Toronto Archives, Accession No. B78-0010, Box 26.

15 Canada, House of Commons, Debates, 5 April 1907, vol. 80, 5921–23.

16 'Who's Who in the Fishing World,' 110. Halkett's important work, *A Check List of the Fishes of the Dominion of Canada* (1914), included 566 different species and their geographical distributions.

17 Letter from A.G. Huntsman to W.A. Found, 22 January 1925, NAC, RG 23, Vol. 1158, File 7234-2, Part 4.

18 Prince, 'The Fishing Industries of Canada,' 19.

19 Hubbard, 'The Commission of Conservation,' 45. However, this dispute remains ongoing. The federal government regained control of the oyster grounds in 1928, following the Maclean Commission, but in the 1980s British Columbia attempted to wrest control back from the Dominion.

20 Canada, House of Commons, *Debates*, 22 January 1908, vol. 82, 1700–1.

21 Ibid., 3 February 1909, vol. 97, 513–14.

22 'Who's Who in the Fishing World,' 165.

23 Canada, House of Commons, *Debates*, 5 April 1907, vol. 80, 5924.
24 Ibid., 5917.
25 Ibid., 3 February 1909, vol. 87, 517–18.
26 'Department of Commerce: Bureau of Fisheries, Washington.'
27 Prince, 'Memorandum for the Deputy Minister re Dr. Huntsman's Grand Manan Report.' NAC, RG 23, Vol. 1205, File 726-2-7, Part 1.
28 Klugh, 'The Makers of Queen's,' 97.
29 Smallman, Good, and West, *Queen's Biology*, 54.
30 See Johnstone, *The Aquatic Explorers*, 75.
31 Canada, House of Commons, *Debates*, 22 February 1901, vol. 54, 311.
32 Letter from G.J. Desbarats to M.H. Nickerson, 8 July 1918, NAC, RG 23, Vol. 1218 File 726-6-5, Part 1.
33 Klugh, 'The Makers of Queen's,' 98.
34 Letter from G.J. Desbarats to J.H. Rainville, 10 January 1919, NAC, RG 23, Vol. 1218, File 726-6-5, Part 1.
35 Letter from A.B. Macallum to G.J. Desbarats, 8 July 1915, NAC, RG 23, Vol. 1217, File 726-6-2, Part 1.
36 Letter from M.H. Nickerson to J.H. Sinclair, 3 June 1918, NAC, RG 23, Vol. 1218, File 726-6-5, Part 1.
37 Letter from M.H. Nickerson to A.P. Knight, 7 June 1918, NAC, RG 23, Vol. 1218, File 726-6-5, Part 1.
38 Circular letter from G.J. Desbarats, 5 June 1918, NAC, RG 23, Vol. 1218, File 726-6-8, Part 1.
39 'Conference Regarding Lobster Fishery,' 8 August 1918, NAC, RG 23, Vol. 1218, File 726-6-8, Part 4.
40 Ibid.
41 'Resolution adopted at the Conference of Lobster Fishers and Packers, held at Halifax on August 7th, 1918,' NAC, RG 23, Vol. 1218, File 726-6-8, Part 4.
42 A.P. Knight, letter to the Education Committee, Biological Board of Canada, 1 March 1924, Huntsman Collection, University of Toronto Archives, Accession No. B78-0010, Box 12.
43 'Able Address on the Lobster Industry by Mr. F.W. Tidmarsh,' *Island Patriot* (Charlottetown, PEI), 21 September 1921, NAC, RG 23, Vol. 1216, File 726-6-15, Part 1.
44 Letter from A.B. Macallum to E.E. Prince, 6 March 1912, NAC, RG 23, Vol. 1466, File 769-1-2, Part 1.
45 Letter from A.B. Macallum to G.J. Desbarats, 17 December 1918, NAC, RG 23, Vol. 1477, File 769-2-1, Part 1.
46 See especially 99–103.
47 De Vecchi, 'Science and Government,' 299–300.

48 W.A. Found, 'Memorandum: Re Resolutions adopted at lobster fishery Conference at Halifax, August 8 1918,' 18 September 1918, NAC, RG 23, Vol. 1218, File 726-6-8, Part 4.
49 Letter from W.A. Found to G.J. Desbarats, 31 October 1918, NAC, RG 23, Vol. 1466, File 769-1-2, Part 1.
50 Memo to C.C. Ballantyne, 7 July 1919, NAC, RG 23, Vol. 1466, File 769-1-2, Part 1.
51 Letter from W.A. Found to G.J. Desbarats, 27 February 1919, NAC, RG 23, Vol. 1466, File 769-1-2, Part 1.
52 Ibid.
53 Letter from Found to Desbarats, 31 October 1918, NAC, RG 23, Vol. 1466, File 769-1-2, Part 1.
54 Ibid.
55 Letter from G.J. Desbarats to A.B. Macallum, 30 May 1919, NAC, RG 23, Vol. 1466, File 769-1-2, Part 1.
56 Letter from A.B. Macallum to G.J. Desbarats, 16 June 1919, NAC, RG 23, Vol. 1466, File 769-1-2, Part 1.
57 Ibid.
58 Johnstone, *Aquatic Explorers*, 74.
59 Letter from Macallum to Desbarats, 16 June 1919, NAC, RG 23, Vol. 1466, File 769-1-2, Part 1.
60 Letter from G.J. Desbarats to A.B. Macallum, 11 June 1919, NAC, RG 23, Vol. 1466, File 769-1-2, Part 1.
61 Letter from Macallum to Desbarats, 16 June 1919, NAC, 1466, File 769-1-2, Part 1.
62 Rigby and Huntsman, 119.
63 Letter from A.B. Macallum to A.G. Huntsman, 5 August 1919, Huntsman Collection, University of Toronto Archives, Accession No. B78-0010 Box 8.
64 Letter from C.C. Ballantyne to A.B. Macallum, 7 July 1919, NAC, RG 23, Vol. 1466, File 769-1-2, Part 1.
65 Letter from A.B. Macallum to C.C. Ballantyne, 7 July 1919, NAC, RG 23, Vol. 1466, File 769-1-2, Part 1.
66 Letter from A.B. Macallum to A.G. Huntsman, 5 August 1919.
67 Letter from E.E. Prince to C.C. Ballantyne, 9 July 1919, NAC, RG 23, Vol. 1466, File 769-1-2, Part 1.
68 Creet, 'Watch Hollis Godfrey.'
69 Macallum, 'President's Addreses.'
70 Creet, 'Watch Hollis Godfrey.'
71 E.E. Prince, 'Revised By-Laws' (draft), NAC, RG 23, Vol. 1466, File 769-1-2, Part 1.

72 Letter from Henry G. Maurice to W.A. Found, 10 September 1919, NAC, RG 23, Vol. 1466, File 769-1-2, Part 2.

73 Letter from G.J. Desbarats to A.P. Knight, 5 December 1919, NAC, RG 23, Vol. 1466, File 769-1-2, Part 2.

74 Johnstone, *The Aquatic Explorers*, 104.

75 Letter from W.A. Found to A.P. Knight, 15 May 1923, NAC, RG 23, Vol. 1266, File 769-1-2, Part 2.

76 Memorandum from W.A. Found, 28 May 1923, NAC, RG 23, Vol.1466, File 769-1-2, Part 2.

77 Huntsman, 'Investigations of Fish Handling' (manuscript), Huntsman Collection, University of Toronto Archives, Accession No. B79-0048, Box 1.

78 W.A. Found, 'Memorandum Re Bill to Amend the Biological Board Act,' 9 June 1923, NAC, RG 23, Vol. 1466, File 769-1-2, Part 2.

79 Letter from A. Johnston to A.P. Knight, 3 March 1925, Huntsman Collection, University of Toronto Archives, Accession No. B78-0010, Box 13.

80 Letter from A.P. Knight to A.H. Whitman, 6 December 1924, Huntsman Collection, University of Toronto Archives, Accession No. B78-0010, Box 8.

81 'Bill: An Act to Create the Biological Board of Canada' [1925]; letter from R.F. Ruttan to J.P. McMurrich, 23 March 1925; and letter from J.P. McMurrich to A.P. Knight, 20 March 1925, Huntsman Collection, University of Toronto Archives, Accession No. B78-0010, Boxes 8 and 13.

82 Letter from W.A. Clemens to A.G. Huntsman, 14 April 1925, Huntsman Collection, University of Toronto Archives, Accession No. B78-0010, Box 8.

83 Letter from A.P. Knight to A. Johnston, 30 March 1925, Huntsman Collection, University of Toronto Archives, Accession No. B78-0010, Box 13.

84 Letter from A.H. Whitman to A. Johnston, 3 April 1925; and from A.G. Clemens to A. Johnston, 9 April 1925, Huntsman Collection, University of Toronto Archives, Accession No. B78-0010, Box 8. Letter from C.G. Connolly to W.A. Found, 3 April 1925. NAC, RG 23, Vol. 1467, File 769-1-2, Part 4.

85 Letter from A.G. Huntsman to A. Johnston, 9 April 1925, Huntsman Collection, University of Toronto Archives, Accession No. B78-0010, Box 8.

86 Canada, House of Commons, *Debates*, 28 February 1902, vol. 56, 547; and 22 January 1908, vol. 82, 1714–15.

87 W.A. Found, 'The Canadian Fisheries and the Work of the Department of Fisheries,' broadcast Wednesday, 24 February 1932, CNRO Ottawa, PAC, RG 23, Vol. 17, File 1.

88 Knight, *Report upon the Conditions*, 2–4.

89 Knight, *Official Report on Standardization in Lobster Canning*, 3.

90 Knight, *Report upon the Conditions*, 5, 7–8, 24.

91 Ibid., 9, 13.

92 Ibid., 5.

93 Ibid., 2, 15–18.

94 Letter from W.A. Found to A.P. Knight, 8 October 1920, NAC, RG 23, Vol. 1219, File 726-6-14, Part 1.

95 Letter from Ward Fisher to W.A. Found, 27 June 1921, NAC, RG 23, Vol. 1219, File 726-6-14, Part 2.

96 Knight, *Official Report*, 11.

97 Letter from W.A. Found to A.P. Knight, 8 October 1920, NAC, RG 23, Vol. 1219, File 726-6-14, Part 1.

98 Knight, *Report upon the Conditions*, 20.

99 A.P. Knight, letter to the Education Committee, Biological Board of Canada, 1 March 1924, Huntsman Collection, University of Toronto Archives, Accession No. B78-0010, Box 12.

100 Reed and Tidmarsh, 'Demonstrations,' n.d., NAC, RG 23, Vol. 1219, File 726-6-5, Part 1.

101 Letter from A.P. Knight to W.A. Found, 8 June 1921, NAC, RG 23, Vol. 1219, File 726-6-15, Part 1.

102 Letter from W.H. Tidmarsh to W.A. Found, 9 May 1921, NAC, RG 23, Vol. 1219, File 726-6-15, Part 1.

103 Letter from C.J. Tidmarsh to A.P. Knight, 14 April 1922, NAC, RG 23, Vol. 1219, File 726-6-16, Part 1.

104 Letter from A.P. Knight to W.A. Found, 24 April 1922, NAC, Rg 23, Vol. 1219, File 726-6-16, Part 1.

105 Letter from A.P. Knight to W.A. Found, 21 January 1924. NAC, RG 23, Vol. 1489, File 769-4-4, Part 1.

106 J.J. Cowie, 'Observations on Doctor Knight's Report on the Sanitation and Grading of Lobster Canneries for 1922,' 30 December 1922, NAC, RG 23, Vol. 1219, File 726-6-16, Part 1. Some of this opposition may have stemmed from envy of those with scientific qualifications; he told Found: 'With all due deference to what scientists lay down as desirable or perhaps necessary, in the methods of canning lobsters, such methods must be governed largely by what is practicable, consistent with due cleanliness.'

107 'Graded Products Must Be Offered: Dairymen Told by Minister of Agriculture the Essentials of Success,' *Toronto Globe*, 4 January 1923. Article preserved in Found's Departmental files, NAC, RG 23, Vol. 1219, File 726-6-16, Part 1.

108 Letter from A.P. Knight to W.A. Found, 30 January 1923, NAC, RG 23, Vol. 1219, File 726-6-16, Part 1.

109 Letter from A.P. Knight to W.A. Found, 17 June 1923, NAC, RG 23, Vol. 1219, File 726-6-16, Part 2.

110 Meggs, *Salmon*, 137–8.

111 'Inspecting the Fish Canneries in Maritimes,' *Moncton Daily Times*, 20 June 1925, NAC, RG 23, Vol. 1220, File 726-6-16, Part 5.

112 Letter from A.G. McLeod to Ward Fisher, 13 January 1928, Huntsman Collection, University of Toronto Archives, Accession No. B78-0010, Box 26.

113 Letter from R.G. McKay to G.F. Pearson, 10 February 1928, Huntsman Collection, University of Toronto Archives, Accession No. B78-0010, Box 26.

114 Ernest Hess, 'Report on Course for Lobster Cannery Foremen 1933,' NAC, RG 23, Vol. 1489, File 769-4-4, Part 6.

115 Letter from W.A. Found to A.P. Knight, 10 February 1926, NAC, RG 23, Vol. 1220, File 726-6-16, Part 5.

116 Letter from J.J. Cowie to G.B. Reed, 9 May 1930, NAC, RG 23, Vol. 1220, File 726-6-16, Part 6.

117 Ibid.

118 Ernest Hess, 'Report on the Grading of Lobster Canneries,' June–July 1930. NAC, RG 23, Vol. 1220, File 726-6-16, Part 6.

119 Form letter from the Supervisor of Fisheries to fisheries inspectors, 1932, NAC, RG 23, Vol. 1220, File 726-6-16, Part 8.

120 Letter from A.P. Knight to J.P. McMurrich, 5 December 1926, Huntsman Collection, University of Toronto Archives, Accession No. B78-0010, Box 13.

121 Letter from W.A. Found to J.P. McMurrich, 18 January 1929, Huntsman Collection, University of Toronto Archives, Accession No. B78-0010, Box 13.

5. Rescuing Canada's Sinking Atlantic Fishing Industry, 1924–39

1 'Sense or Science,' *Punch*, 17 October 1923.

2 Bates, *Report on the Canadian Atlantic Sea-Fishery*, 62.

3 Ibid.

4 Ibid., 11.

5 Grant, *The Canadian Atlantic Fishery* 17–23.

6 Cowie, 'The Non-Progression of the Atlantic Fisheries of Canada,' 167; Howard, 'The Selling End of the Fish Game,' 202.

7 Bates, *Report on the Canadian Atlantic Sea-Fishery*, 30.

8 Grant, *The Canadian Atlantic Fishery*, 31–2.

9 Bates, *Report on the Canadian Atlantic Sea-Fishery*, 13.

10 Ibid., 50, and see table 55.

11 Forbes, *Maritime Rights*, 54.

12 Bates, *Report on the Canadian Atlantic Sea-Fishery*, 11.

13 Ibid., 29–30, 64, 81–2.

14 McEvoy, *The Fisherman's Problem*; 140.
15 Bates, *Report on the Canadian Atlantic Sea-Fishery*, 31.
16 *Canadian Fisherman* 3 (March 1916): 89; Grant, *The Canadian Atlantic Fishery*, 89–90; and Hazen, 'Canada's Fisheries,' 329–30.
17 Donnell, 'The Canadian Fresh Sea Fish Trade,' 142–3.
18 Prince, 'The Fishing Industries of Canada,' 17.
19 'Annual Banquet,' 90. Maritimers should note that a similar service for the West Coast fisheries cost up to $10,000 more than this per year.
20 Donnell, 'The Canadian Fresh Sea Fish Trade,' 143.
21 'An Expert Talks on Pickled Fish,' 260. Incidentally, this practice was still inveighed against as late as 1996.
22 Grant, *The Canadian Atlantic Fishery*, 33; J.J. Cowie, 'Packing Canadian Herring,' 82; 'Our Opportunity in Herring Packing,' 388.
23 *Report of the Royal Commission Investigating the Fisheries of the Maritime Provinces and the Magdalen Islands* (Ottawa: King's Printer, 1928), 70. Hereafter referenced as Maclean Commission, *Report*.
24 Bates, *Report on the Canadian Atlantic Sea-Fishery*, 16, 108, 26.
25 Pross and McCorquodale, *Economic Resurgence and the Constitutional Agenda*, 49–50, 55–6.
26 Ommer, 'What's Wrong with Canadian Fish?' 25–6.
27 Felt, 'On the Backs of Fish,' 48–9.
28 Ommer, 'What's Wrong with Canadian Fish?' 35–6.
29 Ibid., 37–8.
30 Ibid., 38.
31 Felt, 'On the Backs of Fish,' 54; Ommer, 'What's Wrong with Canadian Fish?' 42.
32 Dewar, *Industry in Trouble*, 54–76.
33 John J. Cowie, 'The Non-Progression of the Atlantic Fisheries of Canada,' 169–70; Donnell, 'The Canadian Fresh Sea Fish Trade,' 143; Hazen, 'Canada's Fisheries,' 328; Hazen's remarks in 'Annual Banquet,' 91.
34 Howard, 'The Selling End of the Fish Game,' 202–3.
35 Paulus, 'The Reasons for the Small Consumption of Fish,' 337.
36 A.G. Huntsman, 'What Fish Will Canadians Buy? Are Canadians Discriminating Fish Eaters?' manuscript, c. 1945, Huntsman Collection, University of Toronto Archives, Accession No. B79-0048, Box 9.
37 'The Gospel of Clean Fish,' 44.
38 'Memorandum: A Plan for the Inspection and Grading of Sea Fish,' c. 1945, Huntsman Collection, University of Toronto Archives, Accession No. B78-0010, Box 84.
39 McKay, 'Technical Education for Fishermen,' 103.

40 'Education for Fishermen,' 130–1.
41 McKay, 'Technical Education for Fishermen,' 102.
42 Prince, 'What Kind of Education do Fishermen Need,' 25, 27; 'Annual Banquet,' 89; Maclean Commission, *Report*, 76.
43 Grant, *The Canadian Atlantic Fishery*, 137.
44 W.A. Found, 'Extract from Memorandum March 7th, 1918 ... For File 729-3-13,' NAC, RG 23, Vol. 1318, File 729-3-13, Part 1.
45 Čuješ, *Fishermen's Co-operatives in Nova Scotia*, 1972, 22.
46 Grant, *The Canadian Atlantic Fishery*, 33. The ailing postwar New England fishery also suffered from lack of compulsory inspection. In 1976, Congress had to 'introduce bills to increase inspection and grading of fish to protect consumers from food poisoning' (Dewar, *Industry in Trouble*, 61–2).
47 Maclean Commission, *Report*, 41; Grant, *The Canadian Atlantic Fishery*, 131; and 'Report of a Committee Appointed to Investigate the Possibility of Compulsory Inspection of Fish in the Maritime Provinces' (1944), Huntsman Collection, University of Toronto Archives, Accession No. B78-0010, Box 84.
48 Grant, *The Canadian Atlantic Fishery*, 63, 131.
49 Ibid., 109–11.
50 Ibid., 39.
51 Bates, *Report on the Canadian Atlantic Sea-Fishery*, 8.
52 Letter from A. Johnston to A.P. Knight, 26 November 1923, Huntsman Collection, University of Toronto Archives, Accession No. B78-1101, Box 15.
53 Maclean Commission, *Report*, 17, 81-2.
54 Čuješ, *Fishermen's Co-operatives in Nova Scotia*, 16.
55 Bates, *Report on the Canadian Atlantic Sea-Fishery* 107; Čuješ, *Fishermen's Co-operatives in Nova Scotia*, 20, 28.
56 'Royal Fisheries Commission,' 4 page summary of recommendations concerning the Biological Board, n.d., Huntsman Collection, University of Toronto Archives, Accession No. B78-0010, Box 26.
57 Grant, *The Canadian Atlantic Fishery*, 127; Maclean Commission, *Report*, 38.
58 Letter from J.G. Robichaud to J.J. Cowie, 24 September 1928; and letter from L.P. Gaudet to W.A. Found, 11 October 1930. NAC, RG 23, Vol. 1318, File 729-3-13, Parts 2 and 3.
59 'Improvements in the Methods of Curing Cod'; 'New Method in Codfish Preparation,; and letter from George R. Earl to Found, 15 January 1931, NAC, RG 23, Vol. 1320, File 729-3-13, Part 1.
60 Letter from R.A. Merchant to J.J. Cowie, 19 September 1930, NAC, RG 23, Vol. 1318, File 729-3-13, Part 3.
61 Letter from G.R. Earl to J.J. Cowie, 29 September 1930, NAC, RG 23, Vol. 1320, File 729-3-13, Part 1.

62 'Minutes of meeting of the Advisory Sub-Committee on Education held at the Fisheries Experimental Station, Halifax, N.S., on December 12th., 1927,' Huntsman Collection, University of Toronto Archives, Accession No. B78-0010, Box 14.
63 Maclean Commission, *Report*, 75.
64 'Minutes of meeting of the Advisory Sub-Committee on Education, December 12th, 1927,' Huntsman Collection, University of Toronto Archives, Accession No. B78-0010, Box 14.
65 'Minutes of meeting of the Advisory Sub-Committee on Education, held at the Fisheries Experimental Station, Halifax, on Friday, January 13, 1928,' Huntsman Collection, University of Toronto Archives, Accession No. B78-0010, Box 14.
66 H.R. Chipman, 'Report on Course for Fishermen, 1930,' NAC, RG 23, Vol. 1489, File 769-4-4, Part 3.
67 Letter from Frank Young to W.A. Found, 7 December 1928; and letter from G.A. Barker to W.A. Found, 24 December 1928, NAC, RG 23, Vol. 1489, File 769-4-4, Part 2.
68 Letters from A.G. Huntsman to W.A. Found, 8 November 1928 and from W.A. Found to A.G. Huntsman, 26 November 1928, NAC, RG 23, Vol. 1489, File 769-4-4, Part 2; and 'Minutes of a Meeting of the Advisory Sub-Committee on Education held at the Fisheries Experimental Station, Halifax, N.S., on Monday, December 16, 1929,' Huntsman Collection, University of Toronto Archives, Accession No. B78-0010, Box 14.
69 'Minutes of a meeting of the Advisory Sub-Committee on Education held at the Fisheries Experimental Station on Friday, December 7, 1928,' Huntsman Collection, University of Toronto Archives, Accession No. B78-0010, Box 14.
70 Letters from Reggie Cheverie to J.J. Cowie, 16 November 1930; and from J.F. Sutherland to A.H. Leim, 17 November 1930, NAC, RG 23, Vol. 1489, File 769-4-4, Part 4.
71 Letters to A.H. Leim from Ernest Arsenault, 19 November 1930; and from George A. LeClair, 17 November 1930, NAC, RG 23, Vol. 1489, File 769-4-4, Part 4.
72 Letters from G.R. Earl to J.J. Cowie, 15 March, 5 April, and 1 June 1932; and 30 August 1933, NAC, RG 23, Vol. 1320, File 729-3-13, Part 3.
73 Letter from S.A. Beatty to J.J. Cowie, 14 November 1933; 'Biological Board of Canada: Fisheries Experimental Station (Atlantic): Report on Course for Fishermen, 1934'; 'Biological Board of Canada: Atlantic Fisheries Experimental Station: Report on Course for Fishermen, 1935'; and Letter from J.J. Cowie to D.H. Sutherland, 30 November 1934, NAC, RG 23, Vol. 1489, File 769-4-4, Part 6.

74 'Co-operation Improves Product'; 25–7.

75 'Fishermen Facing Future with Increased Courage.'

76 'Better Sales Are Resulting for Fishermen.' Thanks to Earl's work, it was possible to blame bad fish rather than fishermen when bad hake started turning up in the 1938 pack. D.B. Finn accompanied Earl to a fishing village where this was a problem, secured some fresh-caught fish, put them on ice, and four hours later at the Experimental Station confirmed that they were teeming with bacteria (letter from G.R. Earl to J.J. Cowie, 18 August 1939, NAC, RG 23, Vol. 1320, File 729-3-13, Part 5).

77 Letter from A.G. Huntsman to A. Johnston, 4 April 1925, NAC, RG 23, Vol. 1489, File 769-4-4, Part 1.

78 J.J. Cowie, 'Memorandum for Deputy Minister Re Establishment of Fisheries Experimental Station or School on the Atlantic Coast,' 22 November 1923, Huntsman Collection, University of Toronto Archives, Accession No. B78-0010, Box 15.

79 'Report on the United States Fisheries Products Laboratory at Washington, D.C.,' 14 February 1924, Huntsman Collection, University of Toronto Archives, Accession No. B78-0010, Box 15.

80 Grant, *The Canadian Atlantic Fishery*, 98–9; and Bates, *Report on the Canadian Atlantic Sea-Fishery*, 13.

81 A.G. Huntsman, 'What Fish Will Canadians Buy?'; Bates, *Report on the Canadian Atlantic Sea-Fishery*, 77.

82 A.G. Huntsman, 'Jacketed Cold Storage,' manuscript, n.d., Huntsman Collection, University of Toronto Archives, Accession No. B79-0048, Box 14.

83 Ricker, *The Fisheries Research Board*, 16.

84 Letter from A.G. Huntsman to J.J. Cowie, 14 March 1927, NAC, RG 23, Vol. 1489, File 769-4-4, Part 2.

85 Huntsman, 'What Fish Will Canadians Buy?'

86 A.G. Huntsman, 'Rapid Freezing,' manuscript, ca. 1928, Huntsman Collection, University of Toronto Archives, Accession No. B78-0010, Box 26.

87 Huntsman, 'What Fish Will Canadians Buy?'

88 A.G. Huntsman, 'Investigations of Fish Handling,' n.d., Huntsman Collection, University of Toronto Archives, Accession No. B79-0048, Box 1.

89 Letters from G.R. Earl to J.J. Cowie, 9 April 1934 and 1 May 1934, NAC, RG 23, Vol. 4320, File 729-3-13, Part 4.

90 Huntsman, 'What Fish Will Canadians Buy?'; Farstad, *Fisheries Development in Newfoundland*, 18.

91 Maclean Commission, *Report*, 73–4.

92 Dewar, *Industry in Trouble*, 62–5.

93 Letter from D.B. Finn to A.G. Huntsman, 23 April 1928, Huntsman Collec-

tion, University of Toronto Archives, Accession No. B78-0010, Box 13; and letter from J.P. McMurrich to W.A. Found, 1 May 1931, NAC, RG 23, Vol. 1467, File 769-1-2, Part 7.

94 A.T. Cameron, 'Memorandum on Technical Education of Fishermen in the Maritime Provinces,' 25 October 1938, NAC, RG 23, Vol. 1467, File 769-1-2, Part 10.

95 Ibid.; Letter from W.A. Found to A.T. Cameron, 2 November 1938, NAC, RG 23, Vol. 1467, File 769-1-2, Part 10. For later programs for the technical training of fishermen, see Patton, *Industrial Development and the Atlantic Fishery*, 54–64.

96 Barrett, 'Capital and the State in Atlantic Canada,' 79.

97 Dewar, *Industry in Trouble*, 86.

98 Ibid., 98, 103–4.

99 Barrett, 'Capital and the State in Atlantic Canada,' 98n.12.

6. Huxley's Red Herring

1 See below, and Mitchell, *Thomas Henry Huxley*, 279.

2 M. Graham, 'Science and the British Fisheries,' 1.

3 Kingsland, *Modelling Nature*, 4–5.

4 Cushing, *The Provident Sea*, 300.

5 Jester, 'Fisheries and the State,' 133–4.

6 Wesley C. Williams, 'Huxley, Thomas Henry,' *Dictionary of Scientific Biography* (New York: Charles Scribner's Sons, 1981), 589.

7 G. Parsons, 'Property, Profit, Pollution Conflicts,'

8 Jester, 'Fisheries and the State,' 135.

9 Ibid., 136.

10 Cushing, *The Provident Sea*, 114–16; Jester, 'Fisheries and the State,' 135–8.

11 Huxley, *Life and Letters*, 289.

12 Desmond, *Huxley*, 514.

13 Jester, 'Fisheries and the State,' 31–8.

14 Ibid., 49.

15 Ibid., 80–94.

16 Ibid., 105.

17 Ibid. 111–12; see also 146–53.

18 L. Huxley, *Life and Letters*, 292.

19 Jester, 'Fisheries and the State,' 55–6.

20 Desmond, *Huxley*, 515, 523.

21 Letter to John Donnelly, 1883, Imperial College of Science and Technology, Huxley papers, LVII, fol. 75; quoted in Jester, 'Fisheries and the State,' 59.

22 Huxley, *Life and Letters*, 303–4.
23 Desmond, *Huxley*, 533–4; Jester, 'Fisheries and the State,' 56–7.
24 Quoted from an account by Walpole in Huxley, *Life and Letters*, 298. Walpole believed that Huxley suffered from depression at this time; Huxley confessed that he had never enjoyed the inspectorate after Walpole left it.
25 Desmond, *Huxley*, 540.
26 Jester, 'Fisheries and the State,' 186.
27 Huxley, 'Professor Huxley and the Proposed Fishery Board.'
28 Ibid.
29 Jester, 'Fisheries and the State,' 195n.
30 Ibid., 200–5, 219–20.
31 T.H. Huxley, 'Inaugural Address.' 88.
32 Cushing, *The Provident Sea*, 117.
33 Huxley, 'Inaugural Address,' 88.
34 Southward and Roberts, 'One Hundred Years of Marine Research,' 466.
35 Lankester, quoted in ibid. 466.
36 Desmond, *Huxley*, 533.
37 Southward and Roberts, 'One Hundred Years of Marine Research,' 474.
38 Ibid., 468.
39 Graham, 'Science and the British Fisheries,' 1–2.
40 Cushing, *The Provident Sea*, 192–3, 202.
41 Ibid., 193, 203–4.
42 Graham, 'Science and the British Fisheries,' 3.
43 Jenkins, *The Sea Fisheries*, 252–3.
44 Ricker, *The Fisheries Research Board*, 18.
45 Ibid.
46 Hayes, 'The Fisheries Research Board of Canada,' 36.
47 Ricker, *Fisheries Research Board*, 14.
48 Hardy, 'Explanation,' xxii.
49 Jenkins, *The Sea Fisheries*, 10.
50 Gunther, *William Carmichael M'Intosh*, 82.
51 Ibid., 111.
52 Walter Garstang, 'The Impoverishment of the Sea,' 8.
53 Commission of Conservation, *Report*, 101.
54 Gunther, *William Carmichael M'Intosh*, 113.
55 Huntsman, 'Fisheries Management and Research,' 45.
56 Huntsman, 'Fisheries Research in Canada,' 121–2.
57 Huntsman, 'Fisheries Management and Research,' a paper presented to the Research Council of Ontario Advisory Committee of Fish and Wildlife; Report No. 3-6-52, unpublished mimeograph (December 1952), 6.

58 McEvoy, *The Fisherman's Problem*, 188.

59 Ibid., 189–90.

60 Ibid., 223–4.

61 J.C. Stevenson, interview with D.B. Finn, 17 October 1972, at the FAO Headquarters in Rome. NAC, RG 23, vol.430, transcript, 40.

62 'Annual Banquet' 85.

63 Garstang, 'The Impoverishment of the Sea', 5; Huxley, 'Inaugural Address,' 90.

64 Michael Graham, 'The Theory of Fishing,' paper presented to the British Association Section D, at Brighton, 13 September 1948, Huntsman Collection, University of Toronto Archives, Accession No. B78-0010, Box 17.

65 Needler, 'The Seventy-Fifth Anniversary,' 219.

66 *Report of the Royal Commission Investigating the Fisheries of the Maritime Provinces and the Magdalen Islands*, 88.

67 Grant, *The Canadian Atlantic Fishery*, 93.

68 See chapter 7, and Johnstone, *The Aquatic Explorers*, 140–1.

69 North American Council on Fishery Investigations, *Proceedings 1921–1930*, 3–4.

70 'Report of the Proceedings of the Annual Convention of the Canadian Fisheries Association, Halifax, 6–8 August 1918.' ICES asked Canada to join it several times, but the Ministry of Marine and Fisheries declined. Prince, a protégé of M'Intosh, shared M'Intosh's dislike of ICES and perhaps counselled against joining. See NAC, RG 23, Vol. 1107, File 721-32-2.

71 Meeting of the International Committee on Marine Fishery Investigations, Montreal, 23 June 1921, Huntsman Collection, University of Toronto Archives, Accession No. B78-0010, Box 8.

72 Although the link is not made explicitly in the literature, the Annual Report of the Commissioner of Fisheries in 1924 pushed for the creation of a permanent director and a more important role for the Woods Hole biological laboratory in order that work be carried on in a more consistent fashion. In this and the next year, the important work cited consisted primarily of North American Council investigations. The correspondence between the scientists concerning North American Council work is entirely scientific in its content, with no commentary on international political disputes in relation to the fisheries. The organization helped forge strong ties between scientists such as Huntsman and Bigelow. Annual and other meetings served as mini-scientific conferences for the participants. See Rich, 'Progress in Biological Inquiries, July 1 to December 24, 1924,' 45; and 'Progress in Biological Inquiries, 1925,' 43.

73 Smith, *Scaling Fisheries*, 182–3.

74 Meeting of the International Committee on Marine Fishery Investigations, Montreal, 23 June, 1921, Huntsman Collection, University of Toronto Archives, Accession No. B78-0010, Box 8.
75 North American Council, *Proceedings 1934–1936*, 4.
76 'Henry Bryant Bigelow,' unsigned biography, Huntsman Collection, University of Toronto Archives, Accession No. B78-0010 Box 20; and see Burstyn, 'Reviving American Oceanography,' 57–66.
77 North American Council, *Proceedings 1931–1933*, 11.
78 Cushing, *Fisheries Biology*, 204.
79 Cushing, *The Provident Sea*, 196.
80 Ibid., 202.
81 Parrish, 'The Cod, Haddock and Hake,' 255.
82 Cushing, *The Provident Sea*, 202.
83 Howell, *Ocean Research and the Great Fisheries* 25.
84 North American Council, *Proceedings 1921–1930*, 19–20.
85 North American Committee on Fisheries Investigations: Minutes of the Tenth Meeting, ... Friday, 6 November 1925, Huntsman Collection, University of Toronto Archives, Accession No. B78-0010, Box 13.
86 North American Council, Proceedings 1931–1933, 6–7.
87 North American Council on Fisheries Investigations: Minutes of the Seventeenth Meeting, ... November 6th and 7th, 1930, Huntsman Collection, University of Toronto Archives, Accession No. B78-0010, Box 13.
88 North American Council, *Proceedings 1921–1930*, 34.
89 North American Council, *Proceedings 1931–1933*, 15, 23.
90 North American Council, *Proceedings 1921–1930*, 21.
91 Smith, *Scaling Fisheries*, 184–5.
92 Ibid. 186–7.
93 North American Council, *Proceedings 1921–1930*, 14.
94 North American Council, *Proceedings 1931–1933*, 33–4.
95 North American Council, *Proceedings 1934–1936*, 20.
96 Letter from E.D. Le Danois to the Honourable Ernest Lapointe, Ministre de la Marine et des Pêcheries, 14 June 1922, NAC, RG 23, Vol. 1107, File 726-32-2, Part 3.
97 North American Council, *Proceedings 1921–1930*, 27.
98 Ibid., 29–30.
99 Howell, *Ocean Research and the Great Fisheries*, 122. Le Danois's papers 'à l'océanographie et á la biologie marine [ont été] grâce à la clarté de son style et á l'esprit de synthèse qui l'a toujours animé' (R. Letaconnoux, 'Edouard Le Danois').
100 North American Council, *Proceedings 1934–1936*, 24–5, 35–54.

7. An Environmental Assessment

1 All correspondence and archival material in the following section is, unless otherwise indicated, from the Huntsman Collection, University of Toronto Archives, Accession No. B78-0010, Boxes 32 and 52.
2 *Boston Evening Transcript,* 1 May 1929.
3 'Act to incorporate the Canadian Dexter P. Cooper Company' (16–17 George V, c. 23, assented to 15 June 1926).
4 Letter from A.G. Huntsman to Captain Fred. Anderson, 23 February 1927; Johnstone, *The Aquatic Explorers,* 121; Letter from Henry B. Hachey to A.G. Huntsman, 31 March 1928.
5 International Passamaquoddy Fisheries Board, *Passamaquoddy Fisheries Investigations,* 8.
6 Huntsman, *The Passamaquoddy Bay Power Project,* 24.
7 Ibid., 37–43.
8 Letter from Lewis Radcliffe to A.G. Huntsman, 27 March 1928.
9 North American Council on Fishery Investigations, *Proceedings 1921–1930,* 31; North American Council, 'Minutes of the Fifteenth Meeting,' 1928; Letter from Bigelow to Huntsman, 29 October 1928.
10 North American Council, *Proceedings 1921–1930,* 32.
11 Letters from Huntsman to Bigelow, 14 December 1928 and 5 January 1929; from Bigelow to Huntsman, 17 and 19 December 1928.
12 In chapter 4 of *Science in Action,* Bruno Latour thoroughly discusses the importance of allying as many diverse interests as possible, outside of science as well as within science itself, to ensure the successful support of scientific projects.
13 Letter from Bigelow to Huntsman, 8 February 1929; *Boston Evening Transcript,* 1 May 1929.
14 Letter from A.G. Huntsman to H.B. Bigelow, 23 February 1929,
15 Letter from Huntsman to Bigelow, 8 July 1929; Huntsman, 'The Passamaquoddy Power Project,' 1; Huntsman, 'International Passamaquoddy Fishery Investigations,' 357.
16 Letter from H.B. Bigelow to A.G. Huntsman, 25 May 1931.
17 Golley, A History of the Ecosystem Concept in Ecology, 42, 47–8.
18 Letter from Huntsman to W.A. Found, 19 June 1931; Braarud, 'Haaken Hasberg Gran,' 122–3.
19 E.L. Mills, *Biological Oceanography,* 152–7.
20 Letter from W.A. Found to Huntsman, 3 July 1931.
21 Letter from A.E. Parr to O.E. Sette, 15 October 1930; Johnstone, *The Aquatic Explorers,* 144–5.

22 Letter from Bigelow to Huntsman, 8 June 1930.
23 Wimpenny, 'Michael Graham,' ix.
24 Letters from Michael Graham to Huntsman, 14 June 1931; from Huntsman to H.G. Maurice, 17 July 1931.
25 Letters from W.A. Found to C.J. Fish, 25 July 1932; from H.H. Gran to Huntsman, 29 February 1932; from Huntsman to Gran, 15 March 1932.
26 Letter from Charles Fish to A.G. Huntsman, 23 March 1932.
27 Letters from Huntsman to C.J. Fish, 15 March 1932; from R. McGonigle to Fish, 28 March 1932; from Michael Graham to Huntsman, n.d., and 3 May 1932; from H.H. Gran to W.A. Found, 10 May 1932.
28 International Passamaquoddy Fisheries Commission, 'Report on the Scientific Investigations,' 1933 (manuscript): 3, 14–16. Huntsman Collection, University of Toronto Archives, Accession No. B78-0010, Box 32.
29 Ibid., 19.
30 Ibid., 3–4.
31 Ibid., 24–5.
32 Ibid., 32–3.
33 Braarud, 'Haaken Hasberg Gran,' E.L. Mills, 'Saint Simon and the Oceanographers,' 12.
34 W.F. Thompson, quoted in Smith, *Scaling Fisheries*, 182–3, 205.
35 Cushing, *The Provident Sea*, 193.
36 Ibid., 208.
37 North American Council, Proceedings 1931–1933, 11.
38 Smith, *Scaling Fisheries*, 186–92.
39 North American Council, *Proceedings 1921–1930*, 20–1, 24–5.
40 North American Council: 'Minutes of the Twenty-first Meeting,' 1934. Huntsman Collection, University of Toronto Archives, Accession No. B78-0010 Box 8, File 4.
41 Ibid.
42 Smith, *Scaling Fisheries*, 205–8; Huntsman, 'The Canadian Plaice,' 32.
43 Smith, *Scaling Fisheries*, 206–7, 205.
44 Ibid., 210–12.
45 Ibid., 213–14; McEvoy, *The Fisherman's Problem*, 208–9.
46 Beverton and Holt, 'The Theory of Fishing,' 412–3.
47 McEvoy, *The Fisherman's Problem*, 158–9.
48 Cushing, *The Provident Sea*, 199.
49 Ibid., 198–9, 201.
50 Parrish, 'The Cod, Haddock and Hake,' 264.
51 Ibid., 200.
52 Wimpenny, 'Michael Graham,' ix-xii.

53 A.G. Huntsman, 'Factors in Fishing,' 4–5.
54 Howell, *Ocean Research and the Great Fisheries,* 56.
55 Huntsman, 'Factors in Fishing,' 2.
56 Cushing, *The Provident Sea,* 199.
57 Letter from A.G. Huntsman to Stewart Bates, 20 January 1948. Huntsman Collection, University of Toronto Archives, Accession No. B78-0048, Box 1.
58 North American Council, 'Minutes of the Twenty-fourth Meeting,' 1937. Huntsman Collection, University of Toronto Archives, Accession No. B78-0010, Box 8, File 4.
59 McEvoy, *The Fisherman's Problem,* 164–5.
60 North American Council on Fisheries Investigations: Minutes of the Twenty-fifth Meeting, ... October 6, 1938, Huntsman Collection, University of Toronto Archives, Accession No. B78-1101, Box 8.
61 Letter from A.G. Huntsman to H.F.S. Paisley, 17 September 1940, Huntsman Collection, University of Toronto Archives, Accession no. B78-0010, Box 24.
62 Letter from A.G. Huntsman to S. Bates, 20 January 1948, Huntsman Collection, University of Toronto Archives, Accession No. B78-0048, Box 1.
63 Needler, 'The Atlantic Biological Station, St. Andrews, New Brunswick,' mimeograph, ca. 1942, Huntsman Collection, University of Toronto Archives, Accession No. B78-0010, Box 17.
64 Cushing, *The Provident Sea,* 206–7.
65 Ibid., 212, 217–8.
66 Ibid., 292–3, 227.
67 Johnstone, *The Aquatic Explorers,* 192.
68 Ricker, *The Fisheries Research Board,* 10.
69 Regier and Bronson, 'New Perspectives on Sustainable Development,' 111–20.
70 Larkin, 'An Epitaph for the Concept of Maximum Sustained Yield,' 3.
71 Regier, 'Indicators of Ecosystem Integrity.'
72 Ecosystem Principles Advisory Panel, *Ecosystem-Based Fishery Management: A Report to Congress by the Ecosystem Principles Advisory Panel* (Washington: U.S. Department of Commerce, 1999), 9–10.

8. Ebb Tide at the Atlantic Biological Station

1 A.G. Huntsman, 'The Fisheries Research Board of Canada,' Huntsman Collection, University of Toronto Archives, Accession No. B79-0048, Box 6.
2 Hughes, 'Professions,' 4. See also Vollmer and Mills, 'Professionalization and Technological Change,' 21.
3 For a description of some of this equipment, see Mills, 'Problems of Deep-Sea Biology.

4 de Vecchi, 'Science and Government,' 405; see also de Vecchi, 'Science and Scientists in Government, 1878–1896 – Part I,' and 'Science and Scientists in Government, 1878–1896 – Part II.'
5 Vollmer and Mills, 'The Social Context of Professionalization,' 46.
6 Beer and Lewis, 'Aspects of the Professionalization of Science,' 110.
7 Mendelsohn, 'The Emergence of Science as a Profession 7.
8 Allen, *The Naturalist in Britain,* 83.
9 Ibid., 32.
10 Mendelsohn, 'The Emergence of Science as a Profession,' 17.
11 Hatch, 'Introduction,' 3.
12 Freidson, 'The Theory of Professions,' 24–5.
13 Beer and Lewis, 'Aspects of the Professionalization of Science,' 111, 112–13.
14 Morrell, 'Professionalisation,' 982.
15 Ibid., 981, 983.
16 A.M. Carr Saunders, '(General)'; and Vollmer and Mills, 'The Historical Development of Professional Associations,' 4–5, 153.
17 Veysey, 'Higher Education as a Profession,' 18–19.
18 *A History of European Thought,* 170, 202. The strong German tradition of technical schools and training in the applied sciences was a construct of the nineteenth century, in which 'danger lies in the direction of being contented with practical usefulness' (Ibid., 166n).
19 Hachey, 'History of the Fisheries Research Board of Canada,' 80.
20 Reingold, 'Definitions and Speculations,' 35.
21 Ricker, *The Fisheries Research Board,* 4.
22 J.L. Hart, 'Fisheries Research Board of Canada,' 1129–30.
23 Letters from E.E. Prince to A.G. Huntsman, 15 February 1916; and from A.G. Huntsman to E.E. Prince, 19 February 1916, Huntsman Collection, University of Toronto Archives, Accession No. B78-0010 Box 9.
24 'Biological Stations of Canada,' 25.
25 Ibid., 29–30. While 'vessels and equipment were poor ... and results often sketchy ... the general picture was revealed' (Needler, 'The Seventy-Fifth Anniversary of Two Canadian Biological Stations,' 219).
26 Edward E. Prince, 'Department of Marine and Fisheries: The Biological Board of Canada.' Address delivered at Research Conference Ottawa, 20-2 February 1923, Huntsman Collection, University of Toronto Archives, Accession No. B78-0010, Box 72.
27 Hart, 'Fisheries Research Board of Canada,' 1130–1.
28 Biological Board of Canada, *Annual Announcement of the Atlantic and Pacific Biological Stations for 1921.* Printed circular.

29 Biological Board of Canada, *1922 Annual Announcement of the Atlantic and Pacific Biological Stations.* Printed circular.
30 See Biological Board, *1921 Annual Announcement.* Board publications consisted of the *Contributions to Canadian Biology* (results of original investigations); the *Bulletins of the Biological Board of Canada* (popular presentations of scientific fisheries facts); the *Studies from the Biological Stations* (reprints of articles based on work done at the Stations, published in other scientific journals); and various leaflets and pamphlets.
31 A.G. Huntsman, 'Recent Work from the Atlantic Biological Station,' ca. 1921; and 'untitled,' 1927 manuscript, Huntsman Collection, University of Toronto, Accession No. B78-0010, Boxes 76 and 14.
32 Needler, 'The Seventy-Fifth Anniversary of Two Canadian Biological Stations,' 219.
33 Ricker, *The Fisheries Research Board,* 5–6.
34 A.G. Huntsman, 'Recent Work,' Huntsman Collection, University of Toronto Archives, Accession No. B78-0010, Box 76.
35 W.A. Found, 'Memorandum Re Bill to Amend the Biological Board Act,' 9 June 1923, NAC, RG 23, Vol. 1466, File 769-1-2, Part 2.
36 Stevenson. *The First Seventy-Five Years.*
37 Ibid.
38 He was awarded three LLDs: by the University of Michigan in 1912, by the University of Cincinnati in 1923, and by the University of Toronto in 1931.
39 Stevenson. *The First Seventy-Five Years.*
40 A.G. Huntsman, 'The Fisheries Research Board of Canada,' Huntsman Collection, University of Toronto Archives, Accession No. B79-0048, Box 6; and Hachey, 'History of the Fisheries Research Board of Canada,' 1965, 80.
41 Letter from A.G. Huntsman to W.A. Found, 13 December 1920, NAC, RG 23, Vol. 1277, File 769-2-2, Part 1.
42 Letter from A.G. Huntsman to J.J. Cowie, 28 December 1929; and A.G. Huntsman, 'Continuous Operation of Atlantic Biological Station, St. Andrews, N.B.,' 1929. NAC, RG 23, Vol. 1477, File 769-2-1, Part 1.
43 *Report of the Royal Commission,* 1928, 79.
44 R. M'Gonigle, 'Report on the Fire at the Atlantic Biological Station, March 9, 1932; Causes, Etc.,' NAC, RG 23, Vol. 1497, File 768-2-20, Part 1.
45 Letter from C.B. Wilson to A.H. Leim, 10 November 1923; and letter from A.G. Huntsman to A. Wetmore, 10 March 1939, Huntsman Collection, University of Toronto Archives, Accession No. B78-0010, Boxes 73, 54.
46 Berger, *Science, God, and Nature,* 20–4. Also see Waiser, *The Field Naturalist*; and Chartrand, Duchesne, and Gingras, *Histoire des sciences au Québec,* 188–92.

47 'Canadian Atlantic Fauna,' list of proposed contributors, Huntsman Collection, University of Toronto Archives, Accession No. B78-0010, Box 23.

48 Letters from A.R. Cooper to A.G. Huntsman, 18 August 1923; and from A.G. Huntsman to A.R. Cooper, 3 October 1923, Huntsman Collection, University of Toronto, Accession No. B78-0010, Box 23.

49 Letter from H.B. Bigelow to A.G. Huntsman, 29 June 1931, Huntsman Collection, University of Toronto Archives, Accession No. B78-0010, Box 52.

50 Maurice also served as President to the International Council for the Exploration of the Sea (ICES) from 1920 to his retirement in 1938. His diplomacy played an important role in revitalizing ICES following the First World War (see Dobson, 'Henry Gascoyne Maurice,' 3–6).

51 Letter from H.G. Maurice to A.G. Huntsman, 22 June 1931. Huntsman Collection, University of Toronto Archives, Accession No. B78-0010, Box 52.

52 Letters from A.G. Huntsman to Henry G. Maurice, 22 June 1931, 17 July 1931, and 23 June 1931, Huntsman Collection, University of Toronto Archives, Accession No. B78-0010, Box 52.

53 Letter from J.P. McMurrich to A. Johnston, 7 June 1927, Huntsman Collection, University of Toronto Archives, Accession No. B78-0010, Box 12.

54 Letter from A.H. Hutchinson to J. Dybhaven, 3 October 1929, and letter from D.B. Finn to J. Dybhaven, 18 October 1929, Huntsman Collection, University of Toronto Archives, Accession No. B78-0010, Box 13.

55 Finn was then 'loaned' to the Department of Fisheries to become head of the salt fish board, and went on to become the first scientist–Deputy Minister of Fisheries. In 1946 he left this position to become head of the Fisheries Section of UN's Food and Agriculture Organization and never returned to the board. See Johnstone, *The Aquatic Explorers*, 169.

56 Reingold, 'Definitions and Speculations,' 48. Pauly indicates that 'the desire of university administrators to build medical schools based on laboratory science was the major force behind the creation of graduate biology programs from 1870 to 1900' ('The Appearance of Academic Biology,' 373). This was also true at the University of Toronto.

57 Winsor notes the importance of job funding in professionalization in *Reading the Shape of Nature*, 196.

58 Deason, 'A Survey of Academic Qualifications,' 137–8.

59 A.G. Huntsman, untitled, 1933 manuscript, Huntsman Collection, University of Toronto Archives, Accession No. B78-0010, Box 25.

60 Letter from P.F. Elson to A.G. Huntsman, 25 February 1934; letter from A.W.H. Needler to A.G. Huntsman, 9 October 1937; and C.J. Kerswill, 'Progress Report for the Term 1938–39,' Huntsman Collection, University of Toronto Archives, Accession No. B78-0010, Boxes 57, 25, and 18.

61 Johnstone, *The Aquatic Explorers*, 193, 249.

62 Letter from A.G. Huntsman to Hilary B. Moore, 15 April 1941, Huntsman Collection, University of Toronto Archives, Accession No. B78-0010, Box 57.

63 Letter from Bill Hoar to A.G. Huntsman, 19 December 1945; and letters from R.V. Truitt to A.G. Huntsman, 13 October 1938 and 25 October 1938, Huntsman Collection, University of Toronto Archives, Accession No. B78-0010, Boxes 57 and 58.

64 Letter from D.B. Finn to J.P. McMurrich, 25 August 1929; and from J.P. McMurrich to D.B. Finn, 21 September 1929, Huntsman Collection, University of Toronto Archives, Accession No. B78-0010, Box 13.

65 E.E. Prince's 1914 Preface to 'The Fishes of the Georgian Bay,' quoted in Hachey, *History of the Fisheries Research Board of Canada*, 191.

66 G.B. Reed, quoted in Hachey, *History*, 434. Cameron's biographical information is from Hachey, 431–5.

67 Hachey, *History*, 82 (from Minutes of the Executive Meeting of January 1931).

68 Hachey, *History*, 84.

69 Ibid., 85–6.

70 Clemens, 'Education and Fish,' 39.

71 Hachey, *History*, 87.

72 Hart, *Fisheries Research Board of Canada*, 1130.

73 Hachey, *History*, 89, 149–50.

74 Ibid., 98, 96; and Needler, 'Two Canadian Biological Stations,' 219–20.

75 M'Gonigle, 'Report on the Fire,' and letters from A.G. Huntsman to A.H. Whitman, 15 March 1932, and from J.J. Cowie to A.G. Huntsman, 18 March 1932, NAC, RG 23, Vol. 1479, File 769-2-20, Part 1.

76 Letter from A.H. Whitman to A.G. Huntsman, 22 March 1932, NAC, RG 23, Vol. 1479, File 769-2-20, Part 1.

77 A.T. Cameron, 'Memorandum re ration of total expenditure to total salary expenditure of the personnel paid by the Biological Board,' n.d., NAC, RG 23, Vol. 1527, File 769-81-1, Part 2.

78 Stevenson, *The First Seventy-Five Years*. Stevenson notes that scientists trained by Huntsman who later figured large in the Fisheries Research Board included B.E. Bailey, S.A. Beatty, H.N. Brocklesby, N.M. Carter, W.A. Clemens, D.B. Finn, H.B. Hachey, J.L. Hart, E. Hess, R.A. McKenzie, A.W.H. Needler, A.L. Pritchard, W.E. Ricker, M.W. Smith, W. Templeman, A.L. Tester, J.P. Tully, H.C. White, and O.C. Young.

79 Reginald Hayes interview, quoted in Johnstone, *The Aquatic Explorers*. 141.

80 Letter from John Bonsall Porter to A.G. Huntsman, 7 March 1936. Huntsman Collection, University of Toronto Archives, Accession No. B78-0010, Box 79.

81 A.G. Huntsman, 'To the Chairman and members of the Fisheries Research Board of Canada,' memorandum, n.d., Huntsman Collection, University of Toronto Archives, Accession No. B79-0048, Box 14.
82 However, from the beginning, the fisheries experimental stations received equivalent funding, and both shared a high priority in the government scheme of things. For example, in 1926, both stations received $33,500.
83 Hachey, *History*, 172–3.
84 Ibid., 174–5.
85 Ibid., 175.
86 Ibid., 176.
87 Ibid., 179.
88 Needler, 'The Atlantic Biological Station, St. Andrews, New Brunswick,' mimeograph, circa 1942, Huntsman Collection, University of Toronto Archives, Accession No. B78-0010, Box 17.
89 Letter from J.P. McMurrich to A. Johnston, 8 February 1927, NAC, RG 23, Vol. 1467, File 769-1-2, Part 4.
90 Letter from Henry O'Malley to J.J. Cowie, 27 December 1930; and 'Department of Commerce: Bureau of Fisheries, Washington: Biological Positions in the Bureau of Fisheries,' information sheet sent to W.A. Found on 29 March 1930, NAC, RG 23, Vol. 1467, File 769-1-2, Part 6.
91 Letter from George Hogarth to J.J. Cowie, 21 January 1931, NAC, RG 23, Vol. 1467, File 769-1-2, Part 6.
92 Letter from Henry Maurice to J.J. Cowie, 9 January 1931; and Great Britain: Ministry of Agriculture and Fisheries, 'Memorandum re. Organisation of Scientific Marine Research,' NAC, RG 23, Vol. 1467, File 769-1-2, Part 6.
93 Letter from W.A. Found to W.S. Edwards, 7 February 1931, and J.J. Cowie, 'Memorandum re Proposed Changes in Constitution of the Biological Board,' 26 February 1931. NAC, RG 23, Vol. 1467, File 769-1-2, Part 6.
94 'Memorandum re Attached Draft Bill with Respect to the Biological Board of Canada,' March 1931; and 'Draft of a Bill to Repeal the Biological Board Act and Create the Fisheries Research Board of Canada,' NAC, RG 23, Vol. 1467, File 769-1-2, Part 6.
95 Indeed, this loss of power did not occur until 1972, when the board, shortly before its demise in 1973, was integrated into the Department of the Environment and became purely advisory. See Anderson, 'Policy Determination of Government Scientific Organizations,' 105–6; and Anderson 'The Demise of the Fisheries Research Board,' 151–6.
96 Hart, *Fisheries Research Board of Canada*, 1130–1.
97 Hachey, *History*, 164, 171–2.
98 Ibid., 332–3.

99 Letter from A.G. Huntsman to J.P. McMurrich, 8 June 1927, Huntsman Collection, University of Toronto Archives, Accession No. B78-0010, Box 70.

100 Ibid.

101 Letter from A.S. MacKenzie to A.G. Huntsman, 22 July 1927, Huntsman Collection, University of Toronto Archives, Accession No. B78-0010, Box 12.

102 Ibid., and James Nelson Gowanloch, 'Preliminary Summary of Requirements for Biological Station,' circa 1927, Huntsman Collection, University of Toronto Archives, Accession No. B78-0010, Box 70.

103 Letter from A. Johnston to G.F. Pearson, 1 September 1927; and 'P.C. 1705: Certified to be a True Copy of a Minute of a Meeting of the Committee of the Privy Council, approved by His Excellency the Governor General on the 31st August, 1927' (mimeograph copy), Huntsman Collection, University of Toronto Archives, Accession No. B78-0010, Box 70.

104 Letter from J.P. McMurrich to A.G. Huntsman, 5 July 1927, Huntsman Collection, University of Toronto Archives, Accession No. B78-0010 Box 70.

105 'Minutes of Executive Meeting of the Biological Board of Canada Held at Ottawa, Saturday November 26th, 1927,' Huntsman Collection, University of Toronto Archives, Accession No. B78-0010, Box 14.

106 Grant, *The Canadian Atlantic Fishery*, 128–9; and letter from A.H. Leim to W.L. Harper, 25 October 1932, NAC, RG 23, Vol. 1489, File 769-4-4, Part 5.

107 Henry Roper, 'Two Scandals in Academe,' 127–45.

108 'Icelanders to Receive Degrees.'

109 Ricker, *The Fisheries Research Board*, 20–1.

110 This ran from 1936 until 1966, when administration was transferred to the Halifax Laboratory, which operated it as a technological station until 1970. Ibid., 25.

111 Towards the end of its history, the Fisheries Research Board's Atlantic and Pacific Oceanographic Groups broke down, and personnel moved to new stations working in the same area: the FRB Marine Ecology Laboratory in Dartmouth, Nova Scotia, in 1965, and the Pacific Environment Institute in West Vancouver, British Columbia, in 1971.

112 Dr F. Ronald Hayes, interviewed by J.C. Stevenson in Halifax, July 1972, transcript, NAC, RG 23, Vol. 23, File 7, 1972, 38–40. Ricker noted that Kask 'gave away physical oceanography to the Department of Mines and Marine Resources.' (Notes based on a telephone interview of W.E. Ricker by J.C. Stevenson, 19 January 1973, transcript, NAC, RG 23, Vol. 430, File 14, 1) According to E.L. Mills, in a personal note, 'Kask was jockeying for position in a new political environment, one in which DMTS [Department of Mines and Technical Services] was becoming a player, and believed that [the]

FRB would be stronger by yielding physical oceanography and concentrating on its strengths. This was a mis-calculation – but only in hindsight.'

113 Dr F. Ronald Hayes, interviewed by J.C. Stevenson in Halifax, July 1972, transcript, NAC, RG 23, Vol. 23 File 7, 1972, 38–40.

114 Dr J.L. Hart, interviewed by J.C. Stevenson in St Andrews, N.B., July 1972, transcript, NAC, RG 23, Vol. 430, File 6, 1972, 26.

115 Notes based on a telephone interview of Ricker by J.C. Stevenson, 10 January 1973, transcript, NAC, RG 23, Vol. 430, File 14, 2.

116 Dr J.L. Hart, interviewed by J.C. Stevenson in St Andrews, N.B., July 1972, transcript, NAC, RG 23, Vol. 430, File 6, 1972, 27.

117 Although I grew up in New Brunswick, and nursed a strong interest in marine science from an early age, I had no idea there was a marine biological station in my province until I was in second-year university and signed up for a field course at the Huntsman Marine Laboratory. However, I was aware of the station in British Columbia.

118 Interview with Dr W.R. Martin, interviewed by J.C. Stevenson in Alta Vista 6 February 1973, transcript, NAC, RG 23, Vol. 430, File 12, 35.

119 Ibid.

120 Scientists here in the 1970s were working on trying the genetically select enhanced strains of salmon to enhance the declining Atlantic salmon stocks. See Johnstone, 295.

Epilogue

1 Healy, 'Comment.' This paper was written in response to a paper by Hutchings, Walters, and Haedrich, 'Is Scientific Inquiry Incompatible with Government Information Control?'

2 Former public works minister Alfonse Gagliano's testimony before the House of Commons Public Accounts Committee, 18 March 2004. Excerpted from 'Gagliano's Defense,' *National Post,* 20 March 2004, and from 'Alfonso Gagliano, the Hon. Spokesman.' *Globe and Mail,* 19 March 2004.

3 Ecosystem Principles Advisory Panel, *Ecosystem-Based Fishery Management: A Report to Congress by the Ecosystem Principles Advisory Panel* (Washington: U.S. Department of Commerce, National Marine Fisheries Service and National Oceanic and Atmospheric Administration, 1999), 19.

4 Larkin, 'An Epitaph.'

5 Hays, *Conservation and the Gospel of Efficiency.*

6 See, for example, the description of the origins of such conservation organizations in Foster, *Working for Wildlife,* 16–42.

7 This is an observation recognized by current scientists themselves. See Preikshot, 'Reinventing the formulation of policy,' 116.
8 'Memorandum Concerning Trawlers' (1927?), University of Toronto Archives, Accession No. B78-1101, Box 26, File 8.
9 Symes, 'Fishing in Troubled Waters,' 6.
10 Kurlansky, *Cod*, 167–9.
11 Buckworth, 'World Fisheries in Crisis? 10.
12 D.B. Finn, interview by J.C. Stevenson, 17 October 1972, NAC, RG 23, Vol. 430, 40.
13 Gilbert, *Fish for Tomorrow*, 7.
14 See U.S. National Research Council, *Upstream.*
15 Tim D. Smith, *Scaling Fisheries*, 1994, 212–13.
16 Ibid., 285, 310.
17 H.D. Smith, 'Introduction,' xi.
18 Berill, *The Plundered Seas*, 166–8.
19 Charles, 'Beyond the Status Quo,' 104; and Hutchings and Myers, What Can Be Learned from the Collapse of a Renewable Resource? 2126–7.
20 Percy, 'Managing Fundy's Fisheries.' <http://www.auracom.com/_bofep/Publications/Fundy%20issues/fisherie.htm>.
21 Symes, 'Fishing in Troubled Waters,' 6.
22 Percy. 'Managing Fundy's Fisheries.'
23 Symes, 'Fishing in Troubled Waters,' 7.
24 Neis, Ripley, and Hutchings, 'The "Nature" of Cod,' 168–9.
25 Gray, 'Fisheries Science and Fishers' Knowledge,' <www.efep.org/TSGENSUS.pdf>.
26 Neis, Ripley, and Hutchings, 'The "Nature" of Cod,' 170.
27 Harris, *Lament for an Ocean* 155–6; Percy, 'Managing Fundy's Fisheries.'
28 Frances Anderson, 'The Demise of the Fisheries Research Board of Canada': 152–3.
29 Ibid., 153.
30 Ibid, 152–5.
31 Larkin, 'From Rubber Boots to Recombinant DNA,' 484.
32 Transcript of interview with F.R. Hayes, by J.C. Stevenson, July 1972, NAC, RG 23, Vol 430, File 7, 52.
33 See Hutchings, Walters, and Haedrich, 'Is Scientific Inquiry Incompatible.'
34 Finlayson, *Fishing for Truth*, 93–5.
35 Ibid., 87–91.
36 Ransom Myer's term for the mentality prevailing in CAFSAC. Interview, Halifax, 27 June 2003.
37 Harris, *Lament for an Ocean*, 98.

38 Ibid., 99–100.
39 Finlayson, *Fishing for Truth*, 118, 52.
40 Harris, *Lament for an Ocean*, 67–8.
41 Ibid., 196; Hutchings and Myers, 'What Can Be Learned,' 2127, 2144.
42 Hutchings, 'Spatial and Temporal Variation in the Density of Northern Cod,' 1996, 960; Hutchings and Myers, 'What Can Be Learned,' 2127–8.
43 Harris, *Lament for an Ocean* 191, 303.
44 Hutchings, Walters, and Haedrich, 'Is Scientific Inquiry Incompatible,' 1198–9.
45 Harris, *Lament for an Ocean*, 289,
46 Interview with Dr. Ransom A. Myers, Halifax, 27 June 2003.
47 Doubleday, Atkinson, and Baird, 'Comment,' 1424.
48 Hutchings, Haedrich, and Walters, 'Reply,' 1430.
49 See Harris, *Lament for an Ocean*, for a description of federal officials' attempts to censor or repress opposing views, such as by giving scientists gag orders forbidding them to write about the controversy itself or to speak to the press. This culminated in a lawsuit against the *Ottawa Citizen* for publishing 'defamatory' articles about this censorship, and against Ransom Myers, presumably for giving the interviews. This galvanized Canadian scientific support for whistleblowers.
50 Gray, 'Fisheries Science and Fishers' Knowledge,' <www.efep.org/TSGENSUS>. The embedded quotations by Paul Corten are from 'The Widening Gap Between Fisheries Biology and Fisheries Management in the European Union,' *Fisheries Research* 27 (1996): 1–15.
51 Hutchings, Walters, and Haedrich, 'Is Scientific Inquiry Incompatible,' 1203–4, 1206, 1208.
52 Biological Board of Canada, 'Minutes of a Meeting of the Executive Committee held at Ottawa on November 3rd, 1934, at 9:30 a.m.' Pacific Biological Station Scientific Archives FP. Italics in quotation are mine.
53 W.E. Ricker, transcript of interview by J.C. Stevenson and K. Johnstone, 19 January 1973, NAC, RG 23, Vol. 430, File 14.
54 Kerr and Ryder, 'The Laurentian Great Lakes Experience.'
55 Harris, *Lament for an Ocean*, 109, 115–24.
56 Morgan, De Blois, and Rose, 'An Observation on the Reaction of Atlantic Cod.'
57 Interview with Ransom Myers, Halifax, 27 June 2003.
58 Douma, 'The Precautionary Principle.'
59 Ecosystem Principles Advisory Panel, *Ecosystem-Based Fishery Management*, 15.
60 FAO, 'Guidelines on the Precautionary Approach.'
61 Ecosystem Principles Advisory Panel, *Ecosystem-Based Fishery Management*, 1–2.

62 Ibid., 3–4.
63 Ibid., 18–19.
64 Ibid., 29.
65 Ibid., 19; Buckworth, 'World Fisheries in Crisis?' 12.
66 Walters, 'Designing Fisheries Management Systems,' 280–1.
67 Pomeroy and Berkes, 'Two to Tango,' 1997, 474.
68 Peacock and Hansen, 'Community Management in Groundfish.'
69 Ibid.
70 Parravano, 'Fisheries Co-Management.'
71 CCBRM website, 'What We Do.'
72 Björnsson, 'Can Fisheries Yield Be Enhanced,' 2091–2.
73 Pauly et al., 'Speaking for Themselves,' 412.
74 International Council for the Exploration of the Sea, press release 'ICES advises zero catches.'
75 'E.U. Agrees Cod Recovery Plan and 2004 Quotas.'
76 Pitcher and Pauly outline the possible future of our oceans – with, for example, jellyfish replacing the food fishes in many waters – should conservation and ecosystem-rebuilding goals not be embraced by fisheries managers, governments, and indeed by the general population. See 'Rebuilding Ecosystems, not Sustainability, as the Proper Goal of Fishery Management,' in Pitcher, et al., eds., 311–329; and McEvoy, *The Fisherman's Problem*, 232–3.
77 State of California Department of Fish & Game, press release, 'Pacific Sardine Resource Now Officially Recovered,' 15 January 1999, <www.dfg.ca.gov/news/news99/99001.html>.
78 Smith, *Scaling Fisheries*, 3–5, 332–4.
79 Hubbard, 'Missing Links? Elements of Ecology in Fisheries Biology,' unpublished paper given at the 2003 Canadian Society for the History and Philosophy of Science, Halifax. Pitcher and Pauly note that 'although the ecosystem concept was integral to the ideas of pioneers of fisheries science, it was forgotten early in the era of domination of single-species population dynamics'; see their 'Rebuilding Ecosystems.' 313 See also Gauldie, 'Comment,' 2059–61.
80 Myers accuses his target of implying that 'eggs are not in fact required to produce baby fish, and that there is absolutely no need to conserve spawners to ensure future recruits.' See Myers, 'Comment and Reanalysis,' 978.
81 On this point, the Huntsman Collection at the University of Toronto Archives has boxes and boxes of salmon scales and data from Huntsman's Atlantic salmon studies dating back as far as the 1930s.
82 Priekshot, 'Reinventing the Formulation of policy,' 115–16.
83 See, for example, Myers, 'Stock and Recruitment,' Myers et al., 'What Is the Carrying Capacity for Fish in the Ocean?'
84 Author's interview with W.E. Ricker, Pacific Biological Station, July 1993.

Bibliography

Archival Sources

National Archives of Canada, Ottawa
Record Group (RG) 23 – Department of Marine and Fisheries, Fisheries Branch, Biological Board.
Pacific Biological Station Archives, Nanaimo, British Columbia
University of Toronto Archives, University of Toronto, Toronto
William J.K. Harkness files
Huntsman Collection
A.B. Macallum files
J.P. McMurrich files
James Mavor collection

Unpublished Sources

Anderson, Frances. 'Policy Determination of Government Scientific Organizations: A Case Study of the Fisheries Research Board of Canada 1963–1973.' Unpublished PhD dissertation, Université de Montréal, 1988.

Creet, M. 'Watch Hollis Godfrey at the 1917 Conference of Canadian Universities.' Paper presented at the Seventh Kingston Conference, Ottawa, 19 October 1991.

de Vecchi, Vittorio. 'Science and Government in Nineteenth-Century Canada.' PhD dissertation, University of Toronto, 1978.

Hayes, F. Ronald. 'The Fisheries Research Board of Canada and Its Place in Canada's Scientific Development 1968–1978.' Manuscript outlining future development approved by the Fisheries Research Board at its annual meeting, January 1968. (Originally a restricted document.)

International Passamaquoddy Fisheries Commission. 'Report on the Scientific

Investigations' (1933) (manuscript). Huntsman Collection, University of Toronto Archives, Accession No. B78-0010, Box 32.

Jester, Roger E. 'Fisheries and the State: A Study of Some Aspects of Resource Conservation and Government Policy in England, 1860–1902.' Unpublished PhD thesis, New York University, 1971.

Kesteven, G.L. 'An Examination of Certain Aspects of the Methodology and Theory of Fisheries Biology.' Revised mimeograph edition, New Haven, Connecticut: CT: Bingham Oceanographic Laboratory, 1950. Huntsman Collection, University of Toronto Archives, Accession no. B78-0010, Box 118.

Lillie, Frank Rattray. 'James Playfair McMurrich, 1859–1939.' (1939). From McMurrich files, University of Toronto Archives, Accession No. B73-0013.

Parsons, Gill. 'Property, Profit, Pollution Conflicts in Estuarine Water Management 1800–1915.' Unpublished thesis, Lancaster University, 1993.

Regier, Henry A. 1990. 'Indicators of Ecosystem Integrity.' Paper presented at the International Symposium on Ecological Indicators, Fort Lauderdale, FL, 14–18 October. 1990.

Regier, Henry A., and Elizabeth A. Bronson. 'New Perspectives on Sustainable Development and Barriers to Relevant Information,' Paper from Workshop on Breaking Barriers to Environmental Information, Institute for Research on Environment and Economy, University of Ottawa, 5–6 March 1990.

Published Sources

Acheson, T.W. 'The National Policy and the Industrialization of the Maritimes, 1880–1940.' Pp. 176–201 in *Atlantic Canada after Confederation: The Acadiensis Reader.* Volume 2. Ed. P.A. Buckner and David Frank. Fredericton: Acadiensis Press, 1985.

'Afternoon Meeting.' *Daily Mail and Empire.* 20 August 1897.

Alexander, David. 'Economic Growth in the Atlantic Region, 1880–1940.' Pp. 146–76 in *Atlantic Canada after Confederation: The Acadiensis Reader.* Volume 2. Ed. P.A. Buckner and David Frank. Fredericton: Acadiensis Press, 1985.

– 'Newfoundland's Traditional Economy and Development to 1934.' Pp. 11–33 *Atlantic Canada after Confederation: The Acadiensis Reader.* Volume 2. Ed. P.A. Buckner and David Frank. Fredericton: Acadiensis Press, 1985.

'Alfonso Gagliano, the Hon. Spokesman.' *Globe and Mail.* 19 March 2004.

Allard, Dean C. 'The Fish Commission Laboratory and Its Influence on the Founding of the Marine Biological Laboratory.' *Journal of the History of Biology* 23 (1990): 251–70.

Allard, Dean Conrad, Jr. *Spencer Fullerton Baird and the U.S. Fish Commission: A Study in the History of American Science.* New York: Arno Press, 1978.

Allardyce, Gilbert. 'The Vexed Question of Sawdust River Pollution in Nineteenth-Century New Brunswick.' *Dalhousie Review* 52 (1972): 177–90.

Allen, David Elliston. *The Naturalist in Britain: A Social History.* Markham, ON: Penguin 1978.

Allen, E.J. 'First Report of the Council of the Marine Biological Association of the United Kingdom on Work Carried out in Connection with the International Fishery Investigations.' *Journal of the Marine Biological Association of the United Kingdom* 7 (1905): 383–90.

Allen, E.J. and H.W. Harvey. 'The Laboratory of the Marine Biological Association at Plymouth.' *Journal of The Marine Biological Association of the United Kingdom,* New Series 15 (1928): 735–47.

Alter, O. *The Reluctant Patron: Science and the State in Britain, 1850–1920.* New York: Berg, 1987.

Anderson, Frances. 'The Demise of the Fisheries Research Board of Canada: A Case Study of Canadian Research Policy,' *Scientia Canadensis* 8, 27 (1984): 151–6.

Anderson, K.A. 'Johan Hjort, 1869–1948.' *Journal du Conseil International pour l'Exploration de la Mer* 16 (1949): 4.

'Annual Banquet: Canadian Fisheries Association.' *Canadian Fisherman* 3 (March 1916): 84–101.

Anstey, T.H. *One Hundred Harvests: Research Branch, Agriculture Canada 1886–1936.* Ottawa: Agriculture Canada, 1986.

Arnon, I. *Organisation and Administration of Agricultural Research.* New York: Elsevier, 1968.

B., G.C. 'Lankester, Edwin Ray, 1847–1929.' *Journal of the Marine Biological Association of the United Kingdom* New Series 16 (1929–30): 365–71.

Bailey, J.W. *Loring Woart Bailey: The Story of a Man of Science.* Saint John, NB, 1925.

Barber, Bernard. 'Some Problems in the Sociology of the Professions.' Pp. 15–34 in *The Professions in America.* Ed. Kenneth S. Lynn and the Editors of *Dædalus.* Boston: Houghton Mifflin, 1965.

Barrett, Gene, and Anthony Davis. 'Floundering in Troubled Waters: The Political Economy of the Atlantic Fishery and the Task Force on Atlantic Fisheries.' *Journal of Canadian Studies* 19 (1984): 125–56.

Barrett, L. Gene. 'Capital and the State in Atlantic Canada: The Structural Context of Fishery Policy between 1939 and 1977.' In *Atlantic Fisheries and Coastal Communities: Fisheries Decision-Making Case Studies.* Ed. Cynthia Lamson and Arthur J. Hanson. Halifax: Dalhousie Ocean Studies Program, 1984.

– 'Underdevelopment and Social Movements in the Nova Scotia Fishing Industry to 1938.' Pp. 127–60 in *Underdevelopment and Social Movements in Atlantic*

Canada. Ed. Robert J. Brym and R. James Sacouman. Toronto: New Hogtown Press, 1979.

Bates, Stewart. *Report on the Canadian Atlantic Sea-Fishery: Royal Commission on Provincial Development and Rehabilitation, IX.* Halifax: King's Printer, 1944.

Beard, Harry R. 'University Training for the Fisheries.' *Canadian Fisheries Manual* (1930): 40–2.

Beer, John J., and W. David Lewis. 'Aspects of the Professionalization of Science.' Pp. 110–30 in *The Professions in America*. Ed. Kenneth S. Lynn and the Editors of *Dædalus*. Boston: Houghton Mifflin, (1965): 110–30.

Behrman, Daniel. *The New World of Oceans: Men and Oceanography*. Toronto: Little, Brown. 1969.

Ben-David, J. 'Scientific Growth: A Sociological View.' *Minerva* 2 (1964): 455–76.

Benson, Keith R. 'Laboratories on the New England Shore: The "Somewhat Different Direction" of American Marine Biology.' *New England Quarterly* 61 (1988): 55–78.

Benson, Keith R., and Philip F. Rehbock, eds. *Oceanographic History: The Pacific and Beyond*. Seattle: University of Washington Press, 2002.

Benson, Norman G. 'The American Fisheries Society, 1920–1970.' Pp. 13–24 in *A Century of Fisheries in North America*. Ed. Norman G. Benson. Washington, DC: American Fisheries Society, 1970.

– ed. *A Century of Fisheries In North America*. Washington, DC: American Fisheries Society, 1970.

Berger, Carl. *Science, God, and Nature in Victorian Canada*. Toronto: University of Toronto Press, 1983.

– *The Sense of Power: Studies in the Ideas of Canadian Imperialism, 1867–1914*. Toronto: University of Toronto Press, 1970.

Berill, Michael. *The Plundered Seas: Can the World's Fish Be Saved?* Toronto: Graystone, 1997.

'Better Sales Are Resulting for Fishermen.' *Halifax Herald*, 8 August 1933.

Beverton, R.J.H., and S.J. Holt. 'The Theory of Fishing.' Pp. 372–411 in *Sea Fisheries: Their Investigation in the United Kingdom*. Edited by Michael Graham. London: Edward Arnold, 1956.

Bigelow, Henry B. *Oceanography: Its Scope, Problems, and Economic Importance*. Boston: Houghton Mifflin, 1931.

'Biological Stations of Canada.' Sessional Paper No. 29, *Fifty-Fifth Annual Report of the Fisheries Branch, Department of the Marine and Fisheries, For the Year 1921–22*. Ottawa: King's Printer, 1922.

Bjerkan, Paul. 'Results of the Hydrographic Observations Made by Dr Johan

Hjort in the Canadian Atlantic Waters during the Year 1915.' Pp. 349–404 in Canada, Canadian Fisheries Expedition, 1914–15, *Investigations in the Gulf of St Lawrence and Atlantic Waters of Canada.* Ottawa: King's Printer, 1919.

Björnsson, Björn. 'Can Fisheries Yield Be Enhanced by Large-Scale Feeding of a Predatory Fish Stock? A Case Study of the Icelandic Cod Stock.' *Canadian Journal of Fisheries and Aquatic Sciences* 58 (2001): 2091–104.

Bledstein, Burton J. *The Culture of Professionalism: The Middle Class and the Development of Higher Education in America.* New York: W.W. Norton, 1976.

Blegvad, H. *Report of the Danish Biological Station to the Ministry of Agriculture and Fisheries* XLV. Copenhagen: C.A. Reitzel, 1944.

Bocking, Steven. 'Fisheries and Fundamental Science: Donald Rawson's Studies of Lake Productivity.' *Scientia Canadensis* 38/39 (1990): 38–50.

Bowler, Peter. *The Fontana History of the Environmental Sciences.* London: Fontana, 1992.

Braarud, Trygve. 'Haaken Hasberg Gran, 1970–1955.' *Journal du Conseil International pour l'Exploration de la Mer* 21 (1955): 121–4.

Brannigan, A. *The Social Basis of Scientific Discoveries.* Cambridge: Cambridge University Press, 1982.

Brattström, Hans. 'The Biological Stations of the Bergens Museum and the University of Bergen 1892–1967.' *Sarsia* 29 (1967): 7–80.

'The British Association for the Advancement of Science.' *Toronto Daily Mail and Empire,* 16 August 1897.

Brosco, Jeffrey P. 'Henry Bryant Bigelow, the U.S. Bureau of Fisheries, and Intensive Area Study.' *Social Studies of Science* 19 (1989): 239–64.

Buckworth, Rik. C. 'World Fisheries in Crisis? We Must Respond!' Pp. 3–18 in *Reinventing Fisheries Management.* Ed. Tony J. Pitcher, Paul J.B. Hart, and Daniel Pauly. Boston: Kluwer, 1998.

Bunge, Mario. 'Towards a Philosophy of Technology.' Pp. 28–47 in *Philosophical Problems of Science and Technology.* Ed. Alex C. Michalos. Boston: Allyn and Bacon, 1974.

Burstyn, Harold L. 'Reviving American Oceanography: Frank Lillie, Wickliffe Rose, and the Founding of the Woods Hole Oceanographic Institution.' Pp. 57–66 in *Oceanography: The Past.* Ed. Mary Sears and Daniel Merriman. New York: Springer-Verlag, 1980.

Calderwood, W.L. 'On Recent Investigations of the Marine Biological Association (Fishery and Physical).' *Report of the British Association for the Advancement of Science* (1891): 685.

Canada: Canadian Fisheries Expedition, 1914–15. *Investigations in the Gulf of St Lawrence and Atlantic Waters of Canada.* Ottawa: King's Printer, 1919.

Canada: Department of Marine and Fisheries. *Eat More Fish: Recipes and Hints for the Cook.* Publicity Division Bulletin No. 3. Ottawa: F.A. Acland, 1932.

Canada, Department of the Naval Service. *Instructions to Fishery Officers.* Ottawa: Government Printing Bureau, 1914.

'The Canadian Biological Station – First Report of the Committee consisting of Professor E.E. Prince, (Chairman), Dr T. Wesley Mills, Dr A.B. Macallum, Professor John Macoun, Professor E.W. McBride, Mr W.T. Thistleton-Dyer, and Professor D.P. Penhallow (Secretary), on the Establishment of a Biological Station in the Gulf of St. Lawrence.' *Report of the British Association for the Advancement of Science* (1898): 582–583.

Cannon, Susan Faye. *Science in Culture: The Early Victorian Period.* New York: Dawson and Science History Publications, 1978.

Cardwell, D.S.L. *The Organisation of Science in England.* London: Heinemann, 1972.

Carlander, Kenneth D. 'Fishery Education and Training.' Pp. 57–70 in *A Century of Fisheries in North America.* Ed. Norman G. Bensen. Washington, DC: American Fisheries Society, 1970.

Carr Saunders, A.M. '(General)'. Pp. 3–9 in *Professionalization.* Ed. Howard M. Vollmer and Donald L. Mills. Englewood Cliffs, NJ: Prentice-Hall, 1966.

– *Professions: Their Organization and Place in Society.* Oxford: Clarendon Press, 1928.

CCBRM. 'What We Do: Atlantic Canada' <www.stfx.ca/institutes/ccbm/text/marine_resources.html>.

Chambers, Captain Ernest J. ca. *The Canadian Marine: A History of the Department of Marine and Fisheries.* Toronto: Canadian Marine and Fisheries History, ca. 1906.

Charles, A.T. 'Beyond the Status Quo: Rethinking Fishery Management.' Pp. 101–12 in *Reinventing Fisheries Management.* Ed. Tony J. Pitcher, Paul J.B. Hart, and Daniel Pauley. Boston: Kluwer, 1998.

Chartrand, Luc, Raymond Duchesne, and Yves Gingras. *Histoire des sciences au Québec.* Montreal: Boréal, 1987.

Christensen, Jorgen Møller. 'A Century of Fisheries Research and Management.' *Dana* 8 (1989): 5–16.

Clark, A.M. 'William Leadbetter Calderwood, I.S.O.' *Royal Society of Edinburgh Yearbook for 1951* (1952): 78.

Clarke, Burton R. '(College Teaching).' Pp. 283–91 in *Professionalization.* Ed. Howard M. Vollmer and Donald L. Mills. Englewood Cliffs, NJ: Prentice-Hall, 1966.

Clemens, W.A. 'Education and Fish: An Autobiography.' Fisheries Research Board of Canada Manuscript Report Series No. 974. Ottawa: Fisheries Research Board of Canada, 1968.

– 'Reminiscences of a Director.' *Journal of the Fisheries Research Board of Canada* 15 (1958): 779–96.

Clement, Wallace. 'Canada's Coastal Fisheries: Formation of Unions, Co-operatives and Associations.' *Journal of Canadian Studies* 19 (Spring 1984): 5–33.

Clepper, Henry. *Origins of American Conservation.* New York: Ronald Press, 1966.

Commission of Conservation. *Report of the Fourth Annual Meeting, 1913.* Toronto, n.p., 1913.

Collins, James P., John Beatty, and Jane Maienschein. 'Introduction: Between Ecology and Evolutionary Biology.' *Journal of the History of Biology* 19 (1986): 169–80.

'Co-operation Improves Product: Use Greater Care in Fish Preparation.' *Halifax Herald,* 2 September 1932.

Cotgrove, Steven, and Steven Box. *Science, Industry and Society.* London: Allen and Unwin, 1970.

Coull, James R. *The Fisheries of Europe: An Economic Geography.* London: G. Bell and Sons, 1972.

Cowie, J.J. 'Drift Net Fishing for Herring.' *Canadian Fisherman* 3 (February 1916): 42–5.

– 'The Non-Progression of the Atlantic Fisheries of Canada.' *Canadian Fisherman* 1 (June 1914): 167–71.

– 'Packing Canadian Herring for Foreign Markets.' *Canadian Fisherman* 3 (March 1916): 80–2.

Craigie, E. Horne. *A History of the Department of Zoology at the University of Toronto up to 1962.* Toronto: University of Toronto Press, 1966.

Crowcroft, Peter. *Elton's Ecologists: A History of the Bureau of Animal Population.* Chicago: University of Chicago Press, 1991.

Čuješ, Rudolf. *Fishermen's Co-operatives in Nova Scotia.* Antigonish: St Francis Xavier University, 1972.

Cunningham, J.T. *The Natural History of the Marketable Marine Fishes of the British Islands.* London: Macmillan, 1896.

– 'On the Growth of Food-fishes and Their Distribution at Different Ages.' *Report of the British Association for the Advancement of Science* (1891): 685–6.

Currie, R.I. 'Marine Science: Two Hundred Years of the Biological Sciences in Scotland.' *Proceedings of the Royal Society of Edinburgh* (B) 84 (1983): 231–50.

Cushing, D.H. *Fisheries Biology: A Study in Population Dynamics.* Madison: University of Wisconsin Press, 1981.

– *The Provident Sea.* New York: Cambridge University Press, 1988.

Cutting, C.L. *Fish Saving: A History of Fish Processing from Ancient to Modern times.* New York: Philosophical Library, 1956.

Daniels, George R., ed. *Nineteenth-Century American Science: A Reappraisal.* Evanston: Northwestern University Press, 1972.

Dannevig, Alf. 'Canadian Fish Eggs and Larvae.' Canada, Canadian Fisheries Expedition, 1914–15. Pp. 1–74 in *Investigations in the Gulf of St. Lawrence and Atlantic Waters of Canada.* Ottawa: King's Printer, 1919.

Davis, Celia. 'Professionals in Bureaucracies: The Conflict Thesis Revisited.' Pp. 177–94 in *The Sociology of the Professions: Lawyers, Doctors and Others.* Ed. Robert Dingwall and Philip Lewis. London: Macmillan, 1983.

Deacon, Margaret. *Scientists and the Sea, 1650–1900.* Aldershot: Ashgate 1997.

Deason, H.J. 'A Survey of Academic Qualifications for Fishery Biologists and of Institutional Facilities for Training Fishery Biologists.' *Transactions of the American Fisheries Society* 70 (1941): 128–42.

Denton, E.J., and A.J. Southward. 'Frederick Stratten Russell, 1897–1938. Elected F.R.S. 1938.' *Biographical Memoirs of Fellows of the Royal Society* 32 (1939): 463–93.

Desmond, Adrian. *Huxley: From Devil's Disciple to Evolution's High Priest.* Toronto: Penguin, 1997.

de Vecchi, Vittorio. 'The Dawning of a National Scientific Community in Canada, 1878–1896.' *Scientia Canadensis* 8 (1984): 32–58.

– 'Science and Scientists in Government, 1878–1896–Part I.' *Scientia Canadensis* 8 (1984): 112–142.

– 'Science and Scientists in Government, 1878–1896 – Part II.' *Scientia Canadensis* 9 (1985): 97–113.

Dewar, Margaret E. *Industry in Trouble: The Federal Government and the New England Fisheries.* Philadelphia: Temple University Press, 1983.

Dobson, A.T.A. 'Henry Gascoyne Maurice, 1874–1950.' *Journal du Conseil International Pour l'Exploration de la Mer* 17 (1950): 3–6.

'Dr E.E. Prince Retires after 30 Years.' *Canadian Fisherman* 11 (October 1924): 286–7.

Doern, G.B. *Science and Politics in Canada.* Montreal: McGill-Queen's University Press, 1972.

Dohrn, Anton. 'Report of the Committee, consisting of Dr Anton Dohrn, Professor Rolleston, and Mr P.L. Sclater, Appointed for the Purpose of Promoting the Foundation of Zoological Stations in Different Parts of the World.' *Report of the British Association for the Advancement of Science* (1871): 192.

– 'Report of the Committee, consisting of Dr Rolleston, Dr Sclater, Dr Anton Dohrn, Professor Huxley, Professor Wyville Thomson, and E. Ray Lankester, for the Foundation of Zoological Stations in Different Parts of the Globe.' *British Association for the Advancement of Science, Report* (1873): 408–12.

Donnell, Allen. 'The Canadian Fresh Sea Fish Trade.' *Canadian Fisherman* 2 (May 1915): 142–3.

Doubleday, William G., D.B. Atkinson, D.B., J. Baird. 'Comment: Scientific Inquiry and Fish Stock Assessment in the Canadian Department of Fisheries and Oceans.' *Canadian Journal of Fisheries and Aquatic Sciences* 54 (1997): 1422–6.

Douma, Wybe Th. 'The Precautionary Principle,' *Úfljótur* 49 (1996): 417–30.

Dunlop, Jean. *The British Fisheries Society 1786–1893.* Edinburgh: John Donald, 1978.

Ebert, J.D. 'Carnegie Institution of Washington and Marine Biology: Naples, Woods Hole, and Tortugas.' *Biological Bulletin* 168 (Suppl.) (1985): 172–82.

Ebert, James D. 'Evolving Institutional Patterns for Excellence: A Brief Comparison of the Organization and Management of the Cold Spring Harbour Laboratory and the Marine Biological laboratory.' *Biological Bulletin* 168 (Suppl.) (1985): 80–7.

Ecosystem Principles Advisory Panel. *Ecosystem-Based Fishery Management: A Report to Congress by the Ecosystem Principles Advisory Panel.* Washington: U.S. Department of Commerce, National Marine Fisheries Service and National Oceanic and Atmospheric Administration, 1999.

'Education for Fishermen.' *Canadian Fisherman* 1 (May 1914): 130–1.

'Edward Ernest Prince.' *Proceedings of the Royal Society of Canada Series III* 31 (1937): XX–XXIII.

Egerton, Frank N. 'Essay Review: A Worldwide Inventory of the History of Ecology.' *Journal of the History of Biology* 16 (1983): 171–5.

– 'The History of Ecology: Achievements and Opportunities, Part One.' *Journal of the History of Biology* 16 (1983): 259–310.

– 'The History of Ecology: Achievements and Opportunities, Part Two.' *Journal of the History of Biology* 18 (1985): 103–43.

Eggleston, Wilfrid. *National Research in Canada: The NRC, 1916–66.* Toronto: Clarke Irwin, 1978.

Elliott, Philip. *The Sociology of the Professions.* New York: Macmillan, 1982.

'E.U. Agrees Cod Recovery Plan and 2004 Quotas.' *Bridges Weekly Trade Digest* (2004) <www.flmnh.ufl.edu/fish/InNews/cod2004.htm>.

Everhart, W. Harry, and William D. Youngs. *Principles of Fishery Science.* 2nd ed. Ithaca, NY: Cornell University Press, 1981.

'An Expert Talks on Pickled Fish'. *Canadian Fisherman* 3 (August 1916): 260.

FAO (Food and Agriculture Organization of the United Nations). Precautionary Approach to Fisheries. Part 1: Guidelines on the Precautionary Approach to Capture Fisheries and Species Introductions. Rome: FAO, 1995.

Farstad, Nelvin. *Fisheries Development in Newfoundland: Aspects of Development, Location and Infrastructure.* Bergen: Universitetsforlaget, 1972.

Felt, Lawrence F. 'On the Backs of Fish: Newfoundland and Iceland's Experiences with Fishery-Induced Capital Goods Production in the Twentieth Century.' Pp. 45–72 in *A Question of Survival: The Fisheries and Newfoundland Society.* Edited by Peter R. Sinclair. St John's, NF: Institute of Social and Economic Research, Memorial University of Newfoundland, 1988.

Finlayson, Alan Christopher. *Fishing for Truth: A Sociological Analysis of Northern Cod Stock Assessments, 1977–1990.* St John's: Institute of Social and Economic Research, Memoria University, 1994.

'Fishermen Facing Future with Increased Courage: Greater Cheer Found as Plentiful Supply Is Stocked in Larders.' *Halifax Herald*, 8 February 1936.

Fishery Board for Scotland. *Tenth Annual Report of the Fishery Board For Scotland, 1891.* Glasgow: James Hedderwick and Sons, 1892.

Forbes, Ernest R. *Maritime Rights: The Maritime Rights Movement, 1919–1927: A Study in Canadian Regionalism.* Montreal: McGill-Queen's University Press, 1979.

– ed. *Challenging the Regional Stereotype: Essays on the 20th Century Maritimes.* Fredericton: Acadiensis, 1989.

Foster, Janet. *Working for Wildlife: The Beginning of Preservation in Canada.* Toronto: University of Toronto Press, 1978.

'Freezing Fish in Brine as Soon as Caught.' *Canadian Fisherman* 2 (January 1915): 5.

Freidson, Eliot. 'The Theory of Professions: State of the Art.' Pp. 19–37 in *The Sociology of The Professions.* Ed. Robert Dingwall and Philip Lewis. London: Macmillan, 1983.

Friedman, Robert Marc. *Appropriating the Weather: Vilhelm Bjerknes and the Construction of a Modern Meteorology.* Ithaca: Cornell University Press, 1989.

Friedrich, Hermann. *Marine Biology: An Introduction to its Problems and Results.* Seattle: University of Washington Press, 1969.

Fulton, T.W. 'The Scientific Work of the Fishery Board for Scotland.' *Journal of the Marine Biological Association of the United Kingdom* 1 (1889): 75–91.

'Gagliano's Defence.' *National Post*, 20 March 2004.

Galland, J.A. *The Management of Marine Fishes.* Seattle: University of Washington Press, 1974.

Garstang, W. 'Sea-fishery Investigations and the Balance of Life.' *Nature* 104 (1919): 48–9.

Garstang, Walter. 'The Impoverishment of the Sea. A Critical Summary of the Experimental and Statistical Evidence Bearing upon the Alleged Depletion of the Fishing Grounds.' *Journal of the Marine Biological Association of the United Kingdom* 6, New Series (1900–3): 1–69.

Gauldie, R.W. 'Comment – Fisheries Management Science: A Plea for Conceptual Change.' *Canadian Journal of Fisheries and Aquatic Sciences* 52 (1995): 2059–61.

Gilbert, DeWitt. *Fish for Tomorrow.* Seattle: University of Washington School of Fisheries, 1988.

Gillis, R. Peter. 'Rivers of Sawdust: The Battle over Industrial Pollution in Canada, 1865–1903.' *Journal of Canadian Studies* 21 (Spring 1986): 84–103.

Gingras, Yves. *Physics and the Rise of Scientific Research in Canada.* Trans. Peter Keating. Montreal: McGill-Queen's University Press, 1991.

Golley, Frank Benjamin. *A History of the Ecosystem Concept in Ecology.* New Haven, CT: Yale University Press, 1993.

Goodrich, E.S. 'Edwin Ray Lankester, 1847–1929.' *Proceedings of the Royal Society* (1930): x–xv.

'The Gospel of Clean Fish.' *Canadian Fisherman* 1 (February 1914): 43–4.

Gough, Joseph. *Fisheries Management in Canada, 1880–1910.* Canadian Manuscript Report of Fisheries and Aquatic Sciences 2105. Ottawa: Supply and Services Canada, 1991.

'Government Daily Bait Reports of Great Value to Fishermen.' *Canadian Fisherman* 2 (February 1915): 61–2.

Graham, Herbert W. 'Management of the Groundfish Fisheries of the Northwest Atlantic.' Pp. 249–63 in *A Century of Fisheries In North America.* Ed. Norman G. Benson. Washington, DC: American Fisheries Society, 1970.

Graham, Michael. *The Fish Gate.* London: Faber and Faber, 1943.

– 'Science and the British Fisheries.' Pp. 1–9 in *Sea Fisheries: Their Investigation in the United Kingdom.* Ed. Michael Graham. London: Edward Arnold, 1956.

Gran, H.H. 'Quantitative Investigations as to Phytoplankton and Pelagic Protozoa in the Gulf of St. Lawrence and Outside the Same.' Pp. 486–501 in Canada, Canadian Fisheries Expedition, 1914–15, *Investigations in the Gulf of St. Lawrence and Atlantic Waters of Canada.* Ottawa: King's Printer, 1919.

Grant, Ruth Fulton. *The Canadian Atlantic Fishery.* Toronto: Ryerson Press, 1934.

Gray, Tim. 'Fisheries Science and Fishers' Knowledge.' *European Fisheries Ecosystem Plan* (2002) <www.efep.org/TSGENSUS.pdf>.

Groeben, Christiane. 'Anton Dohrn: The Statesman of Darwinism.' *Biological Bulletin* 168 (Suppl.) (1985): 4–25.

Gulland, J.A. *Fish Population Dynamics: The Implications for Management.* Toronto: John Wiley and Sons, 1988.

Gunther, A.E. *The Life of William Carmichael M'Intosh, M.D., F.R.S. of St Andrews (1838–1931): A Pioneer in Marine Biology.* Edinburgh: Scottish Academic Press, 1977.

Hachey, H.B. *History of the Fisheries Research Board of Canada.* Ottawa: Fisheries Research Board of Canada, 1965.

Hardy, A.C. 'Explanation.' *Hull Bulletin of Marine Ecology* 1 (1939–44): vii–xxx.

– 'Walter Garstang, 1868–1949.' *Journal du Conseil International Pour l'Exploration de la Mer* 17 (1950): 7–12.

– 'Walter Garstang, 1868–1949.' *Journal of the Marine Biological Association of the United Kingdom,* New Series 29 (1950–1): 561–6.

Harkness, W.J.K., Justin W. Leonard, and Paul R. Needham. 'Fishery Research at Mid-Century.' *Transactions of the American Fisheries Society* 83 (1954): 212–6.

Harris, Michael. *Lament for an Ocean: The Collapse of the Atlantic Cod Fishery. A True Crime Story.* Toronto: McClelland and Stewart, 1998.

Hart, J.L. 'A.G. Huntsman: An Appreciation.' *Journal of the Fisheries Research Board of Canada* 22 (1965): 255–8.

– 'Fisheries Research Board of Canada: Biological Station, St Andrews N.B., 1908–1958: Fifty Years of Research in Aquatic Biology.' *Journal of the Fisheries Research Board of Canada* 15 (1958): 1127–61.

Hatch, Nathan O. 'Introduction.' In *The Professions in American History.* Ed. Nathan O. Hatch. Notre Dame, IN: University of Notre Dame Press, 1988.

Hayward, Brian. 'The Co-op Strategy.' *Journal of Canadian Studies* 19 (Spring 1984): 48–64.

Hayes, F. Ronald. *The Chaining of Prometheus.* Toronto: University of Toronto Press, 1971.

– *The Fisheries Research Board of Canada and Its Place in Canada's Scientific Development, 1968–1978.* Ottawa: Fisheries Research Board of Canada, 1968.

Hays, Samuel P. *Conservation and the Gospel of Efficiency: The Progressive Conservation Movement 1890–1920.* Pittsburgh: University of Pittsburgh Press, 1999.

Hazen, J.D. 'Canada's Fisheries – Past, Present, and Future.' *Canadian Fisherman* 2 (October 1915): 328–32.

Healy, M.C. 'Comment: The Interplay of Politics, Policy, and Science.' *Canadian Journal of Fisheries and Aquatic Science* 54 (1999): 1428.

Herdman, W.A. *Founders of Oceanography and Their Work.* London: Edward Arnold, 1923.

Herwig, W., O. Pettersson and P.P.C. Hoek. 'The Aims of the International Council for the Exploration of the Sea' (excerpted from the 'General Report on the work of the Period July 1902–July 1904'). *Rapports et Procès Verbaux des Réunions* (1905). Reprinted in *Study of the Sea: The Development of Marine Research under the Auspices of the International Council for the Exploration of the Sea.* Ed. E.M. Thomasson, Farnham, U.K.: Fishing News Books, 1981.

Hjort, Johan. 'Fluctuations in the Great Fisheries of Northern Europe (1914).' Reprinted in *Study of the Sea: The Development of Marine Research under the Auspices of the International Council for the Exploration of the Sea.* Ed. E.M. Thomasson Farnham, U.K. Fishing News Books, 1981.

– 'The Formative Years of European Marine Science.' (n.d.). Reprinted in *Study of the Sea: The Development of Marine Research Under the Auspices of the International Council for the Exploration of the Sea.* Ed. E.M. Thomasson. Farnham, U.K.: Fishing News Books, 1981.

– 'Introduction to the Canadian Fisheries Expedition, 1914–15.' Pp. i–xv in Canada, Canadian Fisheries Expedition, 1914–15, *Investigations in the Gulf of St Lawrence and Atlantic Waters of Canada.* Ottawa: King's Printer, 1919.

– 'International Exploration of the Sea.' *Rapports et Procès-Verbaux des Réunions, Conseil International pour l'Exploration de la Mer* 105 (1945): 3–19.

Hoek, P.P.C. 'Migration and Overfishing' (n.d.). Reprinted in *Study of the Sea: The Development of Marine Research Under the Auspices of the International Council for the Exploration of the Sea.* Ed. E.M. Thomasson. Farnham, U.K.: Fishing News Books, 1981.

Holt, E.W.L. 'An Examination of the Present State of the Grimsby Trawl Fishery, with Special Reference to the Destruction of Immature Fish.' *Journal of the Marine Biological Association of the United Kingdom* 3 (1895): 339–448.

Howard, A.E. 'The Selling End of the Fish Game.' *Canadian Fisherman* 1 (July 1914): 202–3.

Howell, G.C.L. *Ocean Research and the Great Fisheries.* Oxford: Clarendon Press, 1921.

Hubbard, Jennifer. 'The Commission of Conservation and the Canadian Atantic Fisheries.' *Scientia Canadensis* 34 (1988): 22–52.

– 'Home Sweet Home? A.G. Huntsman and the Homing Behaviour of Canadian Atlantic Salmon.' *Acadiensis* 19 (1990): 40–71.

Hughes, Everett C. *Men and Their Work.* Glencoe, IL: The Free Press, 1958.

– 'Professions.' Pp. 1–14 in *The Professions in America.* Ed. Kenneth S. Lynn and the Editors of *Daedalus.* Boston: Houghton Mifflin, 1965. 1–14.

Huntsman, A.G. 'Arthur Willey: 1867–1942.' *Proceedings of the Royal Society of Canada* (Series 3), Appendix B (1948): 95–8.

– 'The Canadian Plaice.' Biological Board of Canada, Bulletin No. 1 (1918).

– 'Edward Ernest Prince 1858–1936.' *Canadian Field Naturalist* 59 (1945): 1–2.

– 'Factors in Fishing.' Fisheries Research Board, 1951. (mimeograph).

– 'Fisheries Research in Canada.' *Science* 98 (6 August 1943): 117–22.

– 'Fisheries Management and Research.' *Journal du Conseil International pour l'Exploration de la Mer* 19 (1953): 44–55.

– 'Fishing and Assessing Populations.' Yale University, Peabody Museum of Natural History, Bingham Oceanographic Collection, Bulletin No. 11 (May) (1948): 16–9.

– 'Growth of Young Herring (So-Called Sardines) of the Bay of Fundy.' Pp. 165–72 in Canada, Canadian Fisheries Expedition, 1914–15, *Investigations in the Gulf of St Lawrence and Atlantic Waters of Canada.* Ottawa: King's Printer, 1919.

– 'International Passamaquoddy Fishery Investigations.' *Extrait du Journal du Conseil International Pour L'Exploration de la Mer* 13 (1938): 357–69.

– 'James Playfair McMurrich: 1859–1939.' *Proceedings of the Royal Society of Canada* 33 (1939): 143–6.

– 'Migration and Conservation of Atlantic Salmon for Canada's Maritime Provinces.' *American Association for the Advancement of Science,* Publication No. 8 (1939): 32–44.

– *The Passamaquoddy Bay Power Project and Its Effect on the Fisheries.* Saint John: Telegraph Journal, 1928 (pamphlet).

– 'Research on the Use and Increase of Fish Stocks.' Proceedings of the United Nations Scientific Conference on the Conservation and Utilization of Resources, 17 August–6 September, 1949, Lake Success, New York. Vol. 7. *Wildlife and Fish Resources.* New York: United Nations, Department of Economic Affairs, 1949.

– 'The Scale Method of Calculating the Rate of Growth in Fishes.' *Transactions of the Royal Society of Canada Series 3,* Vol. 12, Section 4 (1918): 47–52.

– 'Some Quantitative and Qualitative Plankton Studies of the Eastern Canadian Plankton.' Pp. 405–85 in Canada, Canadian Fisheries Expedition, 1914–15, *Investigations in the Gulf of St Lawrence and Atlantic Waters of Canada.* Ottawa: King's Printer, 1919.

Huntsman A.G., and C.M. Fraser. 'List of Publications Based on Results Obtained at the Biological Stations of Canada, 1901–1921.' *Contributions to Canadian Biology* 12 (1921): 167–83.

Hutchings, Jeffrey. 'Spatial and Temporal variation in the Density of northern Cod and a Review of Hypotheses for the Stock's Collapse.' *Canadian Journal of Fisheries and Aquatic Science* 15 (1996): 943–62.

Hutchings, Jeffrey A., Richard L. Haedrich, and Carl Walters. 'Reply: Scientific Inquiry and Fish Stock Assessment in the Canadian Department of Fisheries and Oceans; and Reply: The Interplay of Policy, Politics and Science.' *Canadian Journal of Fisheries and Aquatic Sciences* 54 (1997): 1430–1.

Hutchings Jeffrey, and Ransom Myers. 'What Can Be Learned from the Collapse of a Renewable Resource? Atlantic Cod, *Gadua Morua,* of Newfoundland and Labrador.' *Canadian Journal of Fisheries and Aquatic Sciences* 51 (1994): 2126–46.

Hutchings, Jeffrey A., Carl Walters, and Richard L. Haedrich. 'Is Scientific Inquiry Incompatible with Government Information Control?' *Canadian Journal of Fisheries and Aquatic Sciences* 54 (1997): 1198–210.

Huxley, Leonard. *Life and Letters of Thomas Henry Huxley.* vol. 2. London: Macmillan, 1913.

Huxley, T.H. 'Inaugural Address: Fisheries Exhibition, London; 1883.' Pp. 80–90 in *The Scientific Memoirs of Thomas Henry Huxley* (Supplementary Vol.) Ed. Sir Michael Foster and E. Ray Lankester. London: Macmillan, 1903.

– 'Professor Huxley and the Proposed Fishery Board.' *Times* (London), 30 March 1886.

'Icelanders to Receive Degrees.' *Charlottetown Patriot,* 15 May 1933.

'Improvements in the Methods of Curing Cod.' *Charlottetown Evening Patriot,* 9 September 1929.

'International Conference for the Exploration of the Sea, Stockholm, 1899.' *Journal of the Marine Biological Association of the United Kingdom* 6, New Series (1900–3): 101–14.

International Council for the Exploration of the Sea. Press release: 'ICES Advises Zero Catches of Cod and Other Fish Stocks' (2003). <www.ices.dk/aboutus/pressrelease/ACFMautumn2003.pdf>.

International Passamaquoddy Fisheries Board. *Passamaquoddy Fisheries Investigations: Report to the International Joint Commission.* Washington/ Ottawa: The Board, 1959.

Jackson, F. Slater. 'The Canadian Marine Biological Station.' *Canadian Record of Science* 8 (1901): 308–14.

Jarrell, R.A. and N.R. Ball. *Science, Technology and Canadian History.* Waterloo, ON: Wilfrid Laurier University Press, 1980.

Jenkins, J.T. *The Sea Fisheries.* London: Constable, 1920.

Johnson, Terence J. *Professions and Power.* London: Macmillan, 1972.

Johnstone, Kenneth. *The Aquatic Explorers: A History of the Fisheries Research Board of Canada.* Toronto: University of Toronto Press, 1977.

Kellog, Charles E., and David C. Knapp. *The College of Agriculture: Science in the Public Service.* Toronto: McGraw-Hill, 1966.

Kerr, J.G. 'The Scottish Marine Biological Association.' *Notes and Records of the Royal Society of London* 7 (1949): 81–96.

Kerr, S.R., and R.A. Ryder. 'The Laurentian Great Lakes Experience: A Prognosis for the Fisheries of Atlantic Canada.' *Canadian Journal of Fisheries and Aquatic Sciences* 54 (1997): 1190–7.

Kingsland, Sharon E. 'Mathematical Figments, Biological Facts: Population Ecology in the Thirties.' *Journal of the History of Biology* 19 (1986): 235–56.

– *Modeling Nature: Episodes in the History of Population Ecology.* Chicago: University of Chicago Press, 1985.

Klugh, A. Brooker. 'The Makers of Queen's: A.P. Knight, M.A. M.D., F.R.C.S.' *Queen's Review* 2 (1928): 97–9.

Knight, A.P. 'Histology of the Flexor Tendon in the Crushing Claw of the Lobster.' *Contributions to Canadian Biology* (1918–20): 185–8.

– *Official Report on Standardization in Lobster Canning.* Ottawa: King's Printer, 1921.

– *Report upon the Conditions in and around Twenty-Three Lobster Canning Factories in Prince Edward Island and New Brunswick during the Summer of 1920* Ottawa: King's Printer, 1921.

Knight, David. *The Age of Science: The Scientific World View in the Nineteenth Century.* Oxford: Basil Blackwell, 1986.

Kofoid, Charles. *The Biological Stations of Europe.* Washington, DC: U.S. Bureau of Education, 1910.

Kurlansky, Mark. *Cod: A Biography of the Fish That Changed the World* Toronto: Knopf, 1998.

L., J.B. 'Archibald Byron Macallum 1858–1934.' *Obituary Notices of the Royal Society* No. 3 (December 1934): 288.

– 'Archibald Byron Macallum 1858–1934.' *Proceedings of the Royal Society* (1934): 287–91.

Lackey, Robert T., and Larry A. Nielson. *Fisheries Management.* Toronto: John Wiley and Sons, 1980.

Lamson, Cynthia, and Arthur J., Hanson, eds. *Atlantic Fisheries and Coastal Communities: Fisheries Decision-Making Case Studies.* Halifax: Dalhousie Ocean Studies Program, 1984.

Lankester, E. Ray. 'Address to Section D. – Biology.' *Transactions of the British Association for the Advancement of Science* (1883): 512–28.

– 'The Marine Biological Association.' *Nature* 30 (1884): 123.

Larkin, P.A. 'An Epitaph for the Concept of Maximum Sustained Yield.' *Transactions of the American Fisheries Society* 106 (1977): 1–11.

– 'From Rubber Boots to Recombinant DNA: Sockeye Research in the Past 35 Years.' Pp. 482–7 in *Sockeye Salmon (Oncorynchus nerka): Population Biology and Future Management.* Ed. H.D. Smith, L. Margolis, and C.C. Wood. Ottawa: Department of Fisheries and Oceans, 1987.

Larson, Magali Sarfatti. *The Rise of Professionalism: A Sociological Analysis.* London: University of California Press, 1977.

Latour, Bruno. *Science in Action.* Milton Keynes, U.K.: Open University Press, 1987.

Layton, Edwin. 'Mirror Image Twins: The Communities of Science and Technology in 19th-Century America.' *Technology and Culture* 12 (1971): 562–80.

– 'Technology as Knowledge.' *Technology and Culture* 15 (1974): 31–41.

Lea, Einar. 'Report on "Age and Growth of the Herring in Canadian Waters."' Pp. 75–166 in Canada, Canadian Fisheries Expedition, 1914–15, *Investigations in the Gulf of St Lawrence and Atlantic Waters of Canada.* Ottawa: King's Printer, 1919.

Lee, A.J. *The Directorate of Fisheries Research: Its Origins and Development.* Lowestoft, U.K.: Ministry of Agriculture and Fisheries, 1992.

Letaconnoux, R. 'Edouard Le Danois (1997–1968).' *Journal du Conseil International pour l'Exploration de la Mer* 33 (1970): 122–3.

Levere, T.H., and R.A. Jarrell. *A Curious Field Book: Science and Society in Canadian History.* Toronto: Oxford University Press, 1974.

Lillie, Frank Rattray. *The Woods Hole Marine Biological Laboratory.* Chicago: University of Chicago Press, 1944.

Lucas, C.E. 'The Scottish Home Department's Marine Laboratory in Aberdeen.' *Proceedings of the Royal Society of Edinburgh* (B) 66 (1956): 222–34.

Lussenhop, John. 'Victor Hensen and the Development of Sampling Methods in Ecology.' *Journal of the History of Biology* 7 (1974): 319–37.

Macallum, A.B. 'President's Address.' *Proceedings of the Royal Society of Canada* 11 (1917): LXXI.

Macleod, Don. 'A Maturing of Purpose: Recent Publications in the History of Technology and the Physical Sciences of Canada.' *Acadiensis* 20 (1990): 225–48.

Macleod, R.M. 'Science and the Treasury: Principles, Personalities and Policies, 1870–1885.' In *The Patronage of Science in the Nineteenth Century.* Ed. G. L'E. Turner. Leyden: Noordhoft International, 1976.

– 'The Support of Victorian Science: The Endowment of Research Movement in Great Britain, 1868–1900.' *Minerva* 9 (1971): 197–230.

Maienschein, Jane. 'Agassiz, Hyatt, Whitman, and the Birth of the Marine Biological Laboratory.' *Biological Bulletin* 168 (Suppl.) (1985): 26–34.

– *One Hundred Years Exploring Life, 1888–1988: The Marine Biological laboratory at Woods Hole.* Boston, MA: Jones and Bartlett, 1988.

– ed. *Defining Biology: Lectures from the 1890s.* Cambridge: Harvard University Press, 1986.

Manchester, Lorne. *Canada's Fisheries.* Toronto: McGraw-Hill, 1970.

Masters, D.C. 'The Scottish Tradition in Higher Education.' In *The Scottish Tradition in Canada.* Ed. by W. Stanford Reid. Toronto: McClelland and Stewart, 1979.

McEvoy, Arthur F. *The Fisherman's Problem: Ecology and Law in the California Fisheries, 1850–1980.* New York: Cambridge University Press, 1986.

McHugh, J.L. 'Trends in Fishery Research.' Pp. 25–56 in *A Century of Fisheries in North America.* Ed. Norman G. Bensen. Washington, DC: American Fisheries Society, 1970.

M'Intosh, W.C. 'The Fisheries and the International Council.' *Nature* 103 (1919): 355–8, 376–8.

– 1897. *The Life Histories of the British Marine Food Fishes.* London: C.J. Clay and Sons.

– *The Resources of the Sea.* 2nd ed. Cambridge: Cambridge University Press, 1921.

McKay, Colin. 'Fisheries Administration in England.' *Canadian Fisherman* 4 (September 1917): 388.

– 'Technical Education for Fishermen.' *Canadian Fisherman* 1 (April 1914): 102–3.

McKillop, A.B. *A Disciplined Intelligence: Critical Inquiry and Canadian Thought in the Victorian Era.* Montreal: McGill-Queen's University Press, 1979.

– 'The Research Ideal and the University of Toronto.' In *Contours of Canadian Thought.* Ed. A.B. McKillop. Toronto: University of Toronto Press, 1987.

'Many Talks Were Given to Fishers.' *Halifax Herald.* 1 July 1937.

Meggs, G. *Salmon: The Decline of the British Columbia Fishery.* Toronto: Douglas and McIntyre, 1991.

Mendelsohn, Everett. 'The Emergence of Science as a Profession in Nineteenth-Century Europe.' Pp. 3–48 in *The Management of Scientists.* Ed. Karl Hill. Boston: Beacon, 1964.

Merton, Robert. 'Singletons and Multiples in Science.' Pp. 343–70 in *The Sociology of Science: Theoretical and Empirical Investigation.* Ed. Norman W. Storer. Chicago: University of Chicago Press, 1973.

Merz, John Theodore. *A History of European Thought in the Nineteenth Century.* Vol. 1. New York: Dover, 1965.

Metz, C.B. 'The Naples Zoological Station and the Marine Biological Laboratory: One Hundred Years of Biology.' *Biological Bulletin* 168 (Supplement): (1985) 1–207.

Mill, John Stuart. *August Comte and Positivism.* Ann Arbor: University of Michigan Press, 1965.

Mills, E.L. *Biological Oceanography: An Early History, 1870–1960.* Ithaca, NY: Cornell University Press, 1989.

– 'Canadian Fisheries Expedition.' Pp. 341–2 in *The Canadian Encyclopedia.* 2nd ed. Vol. 1. Edmonton: Hurtig, 1988.

– 'The Ocean Regarded as a Pasture: Kiel, Plymouth and the Explanation of the Marine Plankton Cycle.' *Deutsche Hydrographische Zeitschrift Ergänzungsheft. Reihe* B (22) (1990): 20–9.

– 'Problems of Deep-Sea Biology: An Historical Perspective.' Pp. 1–79 in *The Sea.* Vol. 8. *Deep-Sea Biology.* Ed. G.T. Rowe. New York: John Wiley and Sons, 1983.

– 'Saint Simon and the Oceanographers: Patterns of Change in the Study of Plankton Dynamics 1897–1946.' Pp. 3–18 in *Selected Works of Gordon A. Riley.* Ed. J.S. Wroblewski. Halifax: Dalhousie University Press, 1982.

– 'Useful in Many Capacities: An Early Career in American Physical Oceanography.' *Historical Studies of the Physical and Biological Sciences* 20, 2 (1990): 265–311.

Mitchell, P. Chalmers. *Thomas Henry Huxley: A Sketch of His Life and Work* London: Knickerbocker Press, 1900.

Monroy, Alberto, and Christiane Groeben. 'The "New" Embryology at the Zoological Station and at the Marine Biological Laboratory.' *Biological Bulletin* 168 (Suppl.) (1985): 35–43.

Moore, Wilbert E. *The Professions: Roles and Rules.* New York: Russell Sage Foundation, 1970.

Morgan, H.J., ed. 'Prince, Edward Ernest.' In *Canadian Men and Women of Science.* 2nd ed. Toronto: William Briggs, 1912.

Morgan, M.J., E.M. De Blois, and G.A. Rose. 'An Observation on the reaction of Atlantic Cod (*Gadus morhua*) in a Spawning Shoal to Bottom Trawling.' *Canadian Journal of Fisheries and Aquatic Sciences* 54 (Suppl. 1) (1997): 217–23.

Morrell, J.B. 'Professionalisation.' Pp. 980–9 in *Companion to the History of Modern Science.* Ed. R.C. Olby, G.N. Cantor, J.R.R. Christie, and M.J.S. Hodge. New York: Routledge, 1990.

Myers, Ransom A. 'Comment and Reanalysis: Paradigms for Recruitment Studies.' *Canadian Journal of Fisheries and Aquatic Sciences* 54 (1997): 978–81.

– 'Stock and Recruitment: Generalizations about Maximum Reproductive Rate, Density Dependence, and Variability Using Meta-analytic Approaches.' *ICES Journal of Marine Science* (2001): 937–51.

Myers, Ransom A., Brian R. MacKenzie, Kieth G. Bowen, and Nicholas J. Barrowman. 'What Is the Carrying Capacity for Fish in the Ocean? A Meta-analysis of Population Dynamics of North Atlantic cod.' *Canadian Journal of Fisheries and Aquatic Sciences* 58 (2001): 1464–76.

Needler, A.W.H. 'Archibald Gowanlock Huntsman 1883–1973.' *Proceedings of the Royal Society of Canada,* series 4, vol. 13 (1974).

– *Oyster Farming on the Atlantic Coast of Canada.* Ottawa: F.A. Acland, 1932.

– 'The Seventy-Fifth Anniversary of Two Canadian Biological Stations.' *Canadian Journal of Fisheries and Aquatic Sciences* 41 (1984): 216–24.

Neis, Barbara, Paul Ripley, and Jeffrey Hutchings. 'The "Nature" of Cod (*Gadus Morhua*): Perceptions of Stock Structure and Cod Behaviour by Fishermen, "Experts" and Scientists from the Nineteenth Century to the Present.' Pp. 125–85 in *Marine Resources and Human Societies in the North Atlantic Since 1500: Papers Presented at the Conference Entitled 'Marine Resources and Human Societies in the North Atlantic Since 1500.'* Ed. Daniel Vickers. St. John's: Institute of Social and Economic Research, 1997.

'The New Biological Station Built at Departure Bay.' *Nanaimo Free Press,* 10 June 1908.

'New Method in Codfish Preparation: Island Fishermen and Packers Greatly Interested in Suggested Improvements by Visiting Ottawa Official.' *Charlottetown Guardian,* 16 August 1929.

Nicholson, Heather Johnston. 'Autonomy and Accountability of Basic Research.' *Minerva* 15 (1977): 32–61.

North American Council on Fishery Investigations. *Proceedings.* Vol. 1. *1921–1930.* Ottawa: King's Printer, 1932.

– *Proceedings.* Vol. 2. *1931–1933.* Ottawa: King's Printer, 1935.

– *Proceedings.* Vol. 3. *1934–1936.* Ottawa: King's Printer, 1939.

'Old Sea Song.' *Canadian Fisherman* 3 (1916): 325.

Ommer, Rosemary E. 'What's Wrong with Canadian Fish?' Pp. 23–44 in *A Question of Survival: The Fisheries and Newfoundland Society.* Ed. Peter R. Sinclair. St. John's: Institute of Social and Economic Research, Memorial University of Newfoundland, 1988.

Oppenheimer, J.M. 'Some Historical Backgrounds for the Establishment of the Statione Zoologica at Naples.' Pp. 179–87 in *Oceanography: The Past.* Ed. M. Sears and D. Merriman. New York: Springer-Verlag, 1980.

Oreskes, N. 'Weighing the Earth from a Submarine: The Gravity Measuring Cruise of the U.S.S. S-21.' Pp. 53–68 in *The Earth, the Heavens, and the Carnegie Institution of Washington.* Ed. G. Good. *History of Geophysics* 5. Washington, DC: American Geophysical Union, 1994.

'Our Opportunity in Herring Packing.' *Canadian Fisherman* 2 (December 1915): 388–9.

Parravano, Pietro, 'Fisheries Co-Management: Empowerment and Responsibility for Fishermen.' *Fishermen's News* (April 2000). The Pacific Coast Federation of Fishermen's Associations <www.pcffa.org/fn-apr00.htm>.

Parrish, B.B. 'The Cod, Haddock and Hake.' Pp. 251–331 in *Sea Fisheries: Their Investigation in the United Kingdom.* Ed. Michael Graham. London: Edward Arnold, 1956.

Parsons, Talcott. 'Professions.' Pp. 536–47 in *International Encyclopedia of the Social Sciences.* Vol. 12. Ed. David L. Sills. New York: Macmillan and Free Press 1968.

Patton, Donald J. *Industrial Development and the Atlantic Fishery: Opportunities for Manufacturing and Skilled Workers in the 1980s.* Toronto: James Lorimer, 1981.

Paulus, J.A. 1915. 'The Reasons for the Small Consumption of Fish in Canada.' *Canadian Fisherman* 2 (October 1915): 337.

Pauly, Daniel, Hart, B.J. Paul, and Tony J. Pitcher. 'Speaking for Themselves: New Acts, New Actors and a New Deal in a Reinvented Fisheries Management.' Pp. 409–15 in *Reinventing Fisheries Management.* Ed. Tony J. Pitcher, Paul J.B. Hart, and Daniel Pauly. Boston: Kluwer, 1998.

Pauly, Philip. 'The Appearance of Academic Biology in Late Nineteenth-Century America.' *Journal for the History of Biology* 17 (Fall 1984): 369–97.

– 'Summer Resort and Scientific Discipline: Woods Hole and the Structure of American Biology, 1882–1925.' Pp. 121–50 in *The Development of American Biol-*

ogy. Ed. Ronald Rainger, Keith R. Benson, and Jane Maienschein. Philadelphia: University of Pennsylvania Press, 1988.

Peacock F.G., and J. Hansen. 'Community Management in Groundfish: A New Approach to Property Rights,' and 'Co-Management and Rights-Based Management.' (2003) <www.fao.org/ docrep/003/X8985E/x8985e06.htm>.

Percy, J.A. 'Managing Fundy's Fisheries: Who Should Write the Rules?' Bay of Fundy Ecosystem Partnership, *Fundy Issues* 20 (Spring 2001) <www.auracom.com/_bofep/ Publications/Fundy%20issues/fisherie.htm>.

Pitcher, Tony J., Paul J.B. Hart, and Daniel Pauly, eds. *Reinventing Fisheries Management*. Boston: Kluwer, 1998.

Pitcher, Tony J., and Daniel Pauly. 'Rebuilding Ecosystems, not Sustainability, as the Proper Goal of Fishery Management.' Pp. 311–29 in *Reinventing Fisheries Management*. Ed. Tony J. Pitcher, Paul J.B. Hart, and Daniel Pauly. Boston: Kluwer, 1988.

Pomeroy Robert A., and Fikret Berkes. 'Two to Tango: The Role of Government in Fisheries Co-management.' *Marine Policy* 21 (1997): 465–80.

Popper, Karl. 'Normal Science and Its Dangers.' Pp. 51–8 in *Criticism and the Growth of Knowledge*. Ed. Imre Lakatos and Alan Musgrave. Cambridge: Cambridge University Press, 1970.

Preikshot, David B. 'Reinventing the Formulation of policy in Future Fisheries.' Pp. 113–22 in *Reinventing Fisheries Management*. Ed. Tony J. Pitcher, Paul J.B. Hart, and Daniel Pauly. Boston: Kluwer, 1998.

Prince, E.E. 'The Fishing Industries of Canada.' *Canadian Fisherman* 1 (January 1914): 12–19.

Prince, E.E. 'What Kind of Education Do Fishermen Need.' *Transactions of the American Fisheries Society* 54 (1924): 25–8.

Prince, Edward E. 'St Andrews Marine Laboratory: An Hour in a Scientific Laboratory on the Scottish Coast.' *Macmillan's English Illustrated Magazine* (1889): 747–57.

'Prof. E.E. Prince Passes on after Brilliant Career.' *Ottawa Citizen*, 12 October 1936.

Pross, A. Paul, and Susan McCorquodale. *Economic Resurgence and the Constitutional Agenda: The Case of the East Coast Fisheries*. Kingston, ON: Institute of Intergovernmental Relations, Queen's University, 1987.

Regier, H.A., and W.L. Hartman. 'Lake Erie's Fish Community: 150 Years of Cultural Stresses.' *Science* 180 (22 June 1973): 1248–55.

Reingold, Nathan. 'Alexander Dallas Bache: Science and Technology in the American Idiom.' Pp. 110–26 in *Science American Style*. Ed. Nathan Reingold. New Brunswick, NJ: Rutgers University Press, 1991.

– American Indifference to Basic Research: A Reappraisal.' Pp. 38–62 in *Nine-*

teenth-Century American Science: A Reappraisal. Ed. George R. Daniels. Evanston, IL: Northwestern University Press. 1972.

– 'Cleveland Abbe at Pulkovo: Theory and Practice in the Nineteenth-Century Physical Sciences.' Pp. 96–109 In *Science American Style.* Ed. Nathan Reingold. New Brunswick, NJ: Rutgers University Press, 1991.

– 'Definitions and Speculations: The Professionalization of Science in America in the Nineteenth Century.' Pp. 33–69 in *The Pursuit of Knowledge in the Early American Republic: American Scientific and Learned Societies from Colonial Times to the Civil War.* Ed. Alexandra Oleson and Sanborn C. Brown. Baltimore: Johns Hopkins University Press, 1976.

'Report of the Committee, consisting of Professor Ray Lankester, Mr. P.L. Sclater, Professor M. Foster, Mr. A. Sedgwick, Professor A.M. Marshall, Professor A.C. Hadddon, Professor Moseley, and Mr. Percy Sladen (Secretary), appointed for the purpose of arranging for the occupaiton of a Table at the Zoological Station at Naples.' *Transactions of the British Association for the Advancement of Science* (1888).

'Report of the Proceedings of the Annual Convention of the Canadian Fisheries Association, Halifax, 6–8 August 1918,' *Canadian Fisherman* 4 (August 1918): 940.

Report of the Royal Commission Investigating the Fisheries of the Maritime Provinces and the Magdalen Islands. Ottawa: King's Printer, 1928.

'Report on Sample Shipment of Canadian Fish Forwarded to Great Britain by the Canadian Fisheries Association and the Department of Marine and Fisheries of Canada.' *Canadian Fisherman* 3 (March 1916): 118.

Rich, Willis. 'Progress in Biological Inquiries, July 1 to December 24, 1924.' In *Report of the United States Commissioner of Fisheries, 1924.* Washington: Government Printing Office, 1925.

– 'Progress in Biological Inquiries, 1925.' In *Report of the United States Commissioner of Fisheries, 1925.* Washington, DC: Government Printing Office, 1926.

Ricker, W.E. *The Fisheries Research Board: Seventy-Five Years of Achievements.* Ottawa: Environment Canada, 1975.

Rigby, Margaret S., and A.G. Huntsman. *Materials Relating to the History of the Fisheries Research Board of Canada (Formerly the Biological Board of Canada) for the Period 1898–1924.* Ottawa: Fisheries Research Board of Canada, 1958.

Roper, H. 'Two Scandals in Academe.' *Collection of the Royal Nova Scotia Historical Society* 43 (1991): 127–45.

Rosenberg, Charles E. *No Other Gods: On Science and American Social Thought.* Baltimore: Johns Hopkins University Press, 1976.

– 'Science, Technology and Economic Growth: The Case of the Agricultural Experiment Station Scientist, 1875–1914.' In *Nineteenth-Century American Sci-*

ence: A Reappraisal. Ed. George H. Daniels. Evanston, IL: Northwestern University Press, 1972.

Rossiter, M.W. *The Emergence of Agricultural Science: Justus Liebig and the Americans, 1840–1880.* New Haven, CT: Yale University Press, 1975.

Royce, W.F. 'A History of Marine Fishery Management.' *Reviews in Aquatic Sciences* 1 (1989): 27–44.

Russell, E.J. *A History of Agricultural Science in Great Britain, 1620–1954.* London: Allen and Unwin, 1966.

Russell, F.S. 'The Plymouth Laboratory of the Marine Biological Association of the United Kingdom.' *Journal of the Marine Biological Association of the United Kingdom* 27 (1948): 761–74.

Russell-Hunter, W.D. 'From Woods Hole to the World: The *Biological Bulletin.*' *Biological Bulletin* 168 (Suppl.) (1985): 200–2.

Ruud, Johan T. 'The Herring Scale Dispute.' Pp. 62–4 in *Study of the Sea: The Development of Marine Research under the Auspices of the International Council for the Exploration of the Sea.* Ed. E.M. Thomasson. Farnham, U.K.: Fishing News Books, 1981.

Sacouman, R. James. 'Underdevelopment and the Structural Origins of Antigonish Movement Co-operatives in Eastern Nova Scotia.' Pp. 231–50 in *Atlantic Canada after Confederation: The Acadiensis Reader.* Vol. 2. Ed. P.A. Buckner and David Frank. Fredericton: Acadiensis Press, 1985.

Sandström, W.J. 'The Hydrodynamics of Canadian Atlantic Waters.' In *Canada, Canadian Fisheries Expedition, 1914–45. Investigations in the Gulf of St Lawrence and Atlantic Waters of Canada.* Ottawa: King's Printer, 1919.

Schlee, Susan. *The Edge of an Unfamiliar World: A History of Oceanography.* New York: Dutton, 1973.

Schuter, B.J., and H.A. Regier. 'The Ecology of Fish and Populations: Dealing with Interactions Between Levels.' Pp. 33–49 in *Proceedings of the National Workshop on Effects of Habitat Alteration on Salmonid Stocks.* Ed. C.D. Levings, L.B. Holtby, and M.A. Henderson. Ottawa: Canadian Special Publication of Fisheries and Aquatic Sciences, 1989.

Seaborn, Edwin. *The March of Medicine in Western Ontario.* Toronto: Ryerson Press, 1944.

Sears, M., and D. Merriman, eds. *Oceanography: The Past.* New York: Springer-Verlag, 1980.

'The Second International Conference for the Exploration of the Sea, Christiania, 1901.' *Journal of the Marine Biological Association of the United Kingdom* 6, New Series (1900–3): 389–412.

'Sense and Science,' *Punch,* 17 October 1923.

Shortland, Michael, ed. *Science and Nature: Essays in the History of the Environmental Sciences.* Oxford: British Society for the History of Science, 1993.

Simpson, A.C. 'The Pelagic Phase.' Pp. 207–50 in *Sea Fisheries: Their Investigation in the United Kingdom.* Edited by Michael Graham. London: Edward Arnold, 1956.

Sidney, H. 'The Presidency: A Force, Fame, and Fishing.' *Time,* 11 March 1991, 57.

Sinclair, Peter R., ed. *A Question of Survival: The Fisheries and Newfoundland Society.* St. John's. Institute of Social and Economic Research, Memoral University of Newfoundland, 1988.

Skud, Bernard E. 'Management of North American Herring Stocks.' Pp. 195–208 in *A Century of Fisheries in North America.* Ed. Norman G. Bensen. Washington, DC: American Fisheries Society, 1970.

Smallman, B.N., H.M. Good, and A.S. West. *Queen's Biology: An Academic History of Innocence Lost and Fame Gained, 1858–1965.* Montreal and Kingston: Queen's University Press, 1991.

Smith, H.D. 'Introduction.' Pp. x–xii in *Sockeye Salmon (Onchorynchus nerka): Population Biology and Future Management.* Ed. H.D. Smith, L. Margolis, and C.C. Wood. Ottawa: Canadian Special Publications of Fisheries and Aquatic Sciences, 1987.

Smith, Tim D. *Scaling Fisheries: The Science of Measuring the Effects of Fishing, 1855–1955.* Cambridge: Cambridge University Press, 1994.

Southward, A.J., and E.K. Roberts. 'One Hundred Years of Marine Research at Plymouth.' *Journal of the Marine Biological Association of the United Kingdom* 67 (1987): 465–506.

State of California Department of Fish and Game. 'Pacific Sardine Resource Now Officially Recovered.' Press Release, 1999. <www.dfg.ca.gov/news/news99/99001.html>.

Staudenmaier, John M. *Technology's Storytellers: Reweaving the Human Fabric.* Cambridge, MA: The MIT Press, 1985.

Steedman, R.J., and H.A. Regier. 'Ecosystem Science for the Great Lakes: Perspectives on Degredation and Rehabilitative Transformations.' *Canadian Journal of Fisheries and Aquatic Sciences* 44 (1987): 95–103.

Stevenson, J.C. *The First Seventy-Five Years: A Résumé of the History of the Fisheries Research Board of Canada from 1898 to 1973.* Ottawa: Fisheries Research Board of Canada Miscellaneous Special Publication, 1973.

Sutton, John R. 'Organizational Autonomy and Professional Norms in Science: A Case Study of the Lawrence Livermore Laboratory.' *Social Studies of Science* 14 (1984): 197–224.

Symes, David. 'Fishing in Troubled Waters.' Pp. 3–18 in *Fisheries Management in Crisis.* Ed. Kevin Crean and David Symes. Mississauga, ON: Fishing New Books, 1996.

Thistle, M. *The Inner Ring.* Toronto: University of Toronto Press, 1965.

Thomasson, E.M. 'Introduction to Johan Hjort, "The Formative Years of European Marine Science."' In *Study of the Sea: The Development of Marine Research under the Auspices of the International Council for the Exploration of the Sea.* Farnham, U.K.: Fishing News Books, 1981.

– ed. *Study of the Sea: The Development of Marine Research under the Auspices of the International Council for the Exploration of the Sea.* Farnham, U.K.: Fishing News Books, 1981.

Thompson, Paul E. 'The First Fifty Years: The Exciting Ones.' Pp. 1–13 in *A Century of Fisheries in North America.* Ed. Norman G. Benson. Washington, DC: American Fisheries Society. 1970.

Tomas, Carlo R. 'Marine Botany and Ecology at Statione Zoologica.' *Biological Bulletin* 168 (Suppl.) (1985): 168–71.

Tory, H.M. *The History of Science in Canada.* Toronto: Ryerson Press, 1939.

Turner, Steven, Edward Kerwin, and David Woolwine. 'Careers and Creativity in Nineteenth-Century Physiology: Zloczower *Redux.*' *ISIS* 75 (1984): 523–9.

U.S. National Research Council. *Upstream: Salmon and Society in the Pacific Northwest.* Washington, DC: National Academy Press, 1966.

VanderZwaag, David L. *The Fish Feud: The U.S. and Canadian Boundary Dispute.* Toronto: D.C. Heath, 1983.

Veysey, Laurence. 'Higher Education as a Profession: Changes and Continuities.' Pp. 15–32 in *The Professions in American History.* Ed. Nathan O. Hatch. Notre Dame, IN: University of Notre Dame Press, 1988.

Vollmer, Howard M., and Donald L. Mills. 'Professionalization and Technological Change.' Pp. 21–2 in *Professionalization.* Ed. Howard M. Vollmer and Donald L. Mills. Englewood Cliffs, NJ: Prentice-Hall, 1966.

– 'The Social Context of Professionalization.' Pp. 46–8 in *Professionalization.* Ed. Howard M. Vollmer and Donald L. Mills. Englewood Cliffs, NJ: Prentice-Hall, 1966.

– 'The Historical Development of Professional Associations.' Pp. 153–4 in *Professionalization.* Ed. Howard M. Vollmer and Donald L. Mills. Englewood Cliffs, NJ: Prentice-Hall, 1961.

Waiser, W.A. *The Field Naturalist: John Macoun, the Geological Survey, and Natural Science.* Toronto: University of Toronto Press, 1989.

Walters, Carl. 'Designing Fisheries Management Systems That Do Not Depend upon Accurate Stock Assessment.' Pp. 279–88 in *Reinventing Fisheries Management.* Ed. Tony J. Pitcher, Paul J.B. Hart, and Daniel Pauly. Boston: Kluwer, 1998.

Weir, Gary E. *An Ocean in Common: American Naval Officers, Scientists, and the Ocean Environment.* College Station: Texas A&M University Press, 2001.

Went, A.E.J. 'Seventy Years Agrowing: A History of the International Council for the Exploration of the Sea 1902–1972.' *Rapports et procés-verbaux du Conseil International Pour l'Exploration de la Mer. Rapports et procès-verbaux des réunions* 165 (1972): 1–252.

'Who's Who in the Fishing World [W.A. Found].' *Canadian Fisherman* 1 (March 1914): 84.

Wimpenny, R.S. 'Michael Graham.' In *Sea Fisheries Research*. Ed. F.R. Harden Jones. London: Elek House, 1974.

Winsor, M.P. *Reading the Shape of Nature: Comparative Zoology at the Agassiz Museum*. Chicago: University of Chicago Press, 1991.

– *Starfish, Jellyfish and the Order of Life: Issues in Nineteenth-Century Science*. New Haven, CT: Yale University Press, 1976.

Worster, Donald, ed. *The Ends of the Earth: Perspectives on Modern Environmental History*. New York: Cambridge University Press, 1988.

Wright, R. Ramsay. 'Report on the Occupation of a Table at Naples: f. The Methods of Preservation of Specimens Used at the Zoological Station.' *Report of the British Association for the Advancement of Science (1900)*: 388.

Wynn, Graeme. 'The Maritimes: The Geography of Fragmentation and Underdevelopment.' Pp. 157–213 in *Heartland and Hinterland: A Geography of Canada*. Ed. L.D. McCann. Toronto: Prentice-Hall, 1981.

Wynne-Edwards, V.C. 'The Late Professor Arthur Willey: 1967–1942.' *McGill News* (Spring 1943): 11–3.

Young, E.G. *The Development of Biochemistry in Canada*. Toronto: University of Toronto Press, 1976.

Zaslow, Morris. *Reading the Rocks: The Story of the Geological Survey of Canada, 1842–1972*. Toronto: Macmillan, 1975.

Zeller, Suzanne. *Inventing Canada: Early Victorian Science and the Idea of a Transcontinental Nation*. Toronto: University of Toronto Press, 1987.

'Zoological Station.' *Toronto Daily Mail and Empire*, 17 August 1897.

Index

HANDBOOK OF INFRARED STANDARDS II

With Spectral Coverage of
1.4 μm–4 μm and 6.2 μm–7.7 μm

HANDBOOK OF INFRARED STANDARDS II

With Spectral Coverage of 1.4 μm–4 μm and 6.2 μm–7.7 μm

Guy Guelachvili

Laboratoire de Physique
Moléculaire et Applications
CNRS, Bât 350
Université Paris-Sud
Orsay Cedex, France

K. Narahari Rao

Department of Physics
The Ohio State University
Columbus, Ohio, U.S.A.

ACADEMIC PRESS, INC.

Harcourt Brace & Company, Publishers
Boston San Diego New York
London Sydney Tokyo Toronto

This book is printed on acid-free paper. ∞

ACADEMIC PRESS, INC.
1250 Sixth Avenue, San Diego, CA 92101

United Kingdom Edition published by
ACADEMIC PRESS INC. (LONDON) LTD.
24-28 Oval Road, London NW1 7DX

Library of Congress Cataloging-in-Publication Data

Guelachvili, Guy.
Handbook of infrared standards II : with spectral coverage of 1.4 μm- 4 μm and 6.2 μm-7.7 μm / Guy Guelachvili and K. Narahari Rao.
p. cm.
Includes bibliographical references.
ISBN 0-12-305362-5 (acid-free)
1. Infrared radiation—Standards—Handbooks, manuals, etc. I. Rao, K. Narahari. II. Title. III. Title: Handbook of infrared standards 2. IV. Title: Handbook of infrared standards two.
QC457.G828 1993
543′.08583—dc20 92-47468
CIP

PRINTED IN THE UNITED STATES OF AMERICA

93 94 95 96 MV 9 8 7 6 5 4 3 2 1

CONTENTS

INTRODUCTION

Numerous studies of molecular infrared spectra are being made routinely in several laboratories of the world making use of commercially developed high resolution spectrometers. In 1986 a volume of the *Handbook of Infrared Standards*[1] was published for use in wavenumber calibrations of the day-to-day experimental spectroscopic investigations. It provides maps of selected molecular spectra along with their measured wavenumbers in the region between 3 and 2600 μm. This wavelength coverage was mainly dictated by our desire to bring out a book of reasonable dimensions. The present effort, designated as *Handbook of Infrared Standards II*, gives the same type of information about closely spaced spectral lines occurring in the nearer infrared mainly between 1.4 and 4 μm. In this spectral region, in addition to the interest on high resolution studies of overtone and combination bands employing Fourier spectrometers, tunable sources of one type or another are either operating or being developed and it is believed that all researchers working in this area may find this handbook useful. Complete listings and specifications of the diode lasers operable in the region 1.4–4 μm can be obtained from industries dealing with electro-optic products[2]. The present undertaking combined with the books published previously[3–5]

[1]G. Guelachvili and K. Narahari Rao, *Handbook of Infrared Standards*, Academic Press, Inc., Orlando, Florida, 1986; the spectral coverage included in this book is between 3 and 2600 μm.

[2]See for instance, *1992 Laser Focus World Buyers Guide*, PennWell Publishing Company, 1421 South Sheridan, Tulsa, Oklahoma 74101 USA.

[3]A. G. Maki and J. S. Wells, *Wavenumber Calibration Tables From Heterodyne Frequency Measurements*, National Institute of Standards and Technology Publication 821, U.S. Department of Commerce, Washington, D.C. (1991); it gives computer-generated spectral maps and calculated values of the wavenumbers of the stronger transitions in about the same spectral region as in Ref. 1.

[4]A. R. H. Cole, *Tables of Wavenumbers for the Calibration of Infrared Spectrometers*, second Edition, prepared under the auspices of the Commission on Molecular Structure and Spectroscopy of IUPAC, Pergamon Press, N.Y. (1977).

[5]Earlier books: (a) *Tables of Wavenumbers for the Calibration of Infrared Spectrometers*, prepared under the auspices of the Commission on Molecular Structure and Spectroscopy of IUPAC, Butterworths, Washington (1961). (b) K. Narahari Rao, C. J. Humphreys and D. H. Rank, *Wavelength Standards in the Infrared*, Academic Press, N.Y. (1966).

may fulfill some of the needs of the researcher in infrared spectroscopy. Surely, there will be other attempts forthcoming as time goes by.

Molecular assignment information presented in this handbook has been taken from the references given as superscripts after the species indicated below:

C_2H_2[6,7]; OCS[8]; NH_3[9,10]; CH_3Cl[11,12]; CO[13,14,15]; N_2O[16,17]; H_2O[18,19]; HDO, D_2O[20].

Now, it may be recalled that at the meeting of the 17th Conférence Générale des Poids et Measures on October 20, 1983, in Paris, a new definition[21] of the meter was adopted "to open the way to the major improvements in the precision of measurements." The new definition of the meter does not depend on a particular radiation. It states that the "meter is the length of the path traveled by light in vacuum during a time interval of 1/(299 792 458) of a second." That is to say, this definition fixes the value of the speed of light at 299 792 458 m s^{-1} exactly. "The new definition has become practicable with the development of techniques for the measurement of frequencies. . . . in terms of the cesium standard that is used in the definition of the second.*" In order to enable investigators to pin down the absolute accuracy of their data, we have herein summarized frequency measurements that have been made since the publication of the previous handbook.[1]

The arrangement of material in this book is as follows. In Section I we present spectral maps and wavenumber data for some easily available molecular species recorded with a Fourier transform spectrometer (FTS). Section II summarizes several heterodyne frequency measurements while Section III gives an update of the precise frequencies for OsO_4 lines determined by using saturation absorption resonances of

[6]R. D'Cunha, Y. A. Sarma, G. Guelachvili, R. Farrenq, Q. Kou, V. Malathy Devi, C. Benner, and K. Narahari Rao, *J. Mol. Spectrosc.* 148, 213–225 (1991).

[7]W. J. Lafferty and A. S. Pine, *J. Mol. Spectrosc.* 141, 223–230 (1990).

[8]André Fayt and A. Belafhal (Private Communication, 1992).

[9]S. Urban, N. Tu, K. Narahari Rao, and G. Guelachvili, *J. Mol. Spectrosc.* 133, 312–330 (1989).

[10]G. Guelachvili, A. H. Abdullah, N. Tu, K. Narahari Rao, Š. Urban, and D. Papoušek, *J. Mol. Spectrosc.* 133, 345–364 (1989).

[11]M. Morillon-Chapey, G. Guelachvili, and P. Jensen, *Can. J. Phys.* 62, 247–253 (1984).

[12]N. Bensari-Zizi, C. Alamichel, and G. Guelachvili, *Can. J. Phys*, 59, 994–1004 (1981).

[13]G. Guelachvili, *J. Mol. Spectrosc.* 75, 251–269 (1979).

[14]G. Guelachvili, *Opt. Comm.* 8, 171–175 (1973).

[15]R. Farrenq, G. Guelachvili, A. J. Sauval, N. Grevesse, and C. B. Farmer, *J. Mol. Spectrosc.* 149, 375–390 (1991).

[16]C. Amiot and G. Guelachvili, *J. Mol. Spectrosc.* 51, 475–471 (1974).

[17]C. Amiot and G. Guelachvili, *J. Mol. Spectrosc.* 59, 171–190 (1976).

[18]J.-M. Flaud, C. Camy-Peyret, and R.A. Toth, *Water Vapor Line Parameters from Microwave to Medium Infrared*, Pergamon Press (1981).

[19]C. Camy-Peyret, J.-M. Flaud, and J.-P. Maillard, *J. Phys.-Letters* 41, 23–26 (1980).

[20]T. Oshima and H. Sasada, *J. Mol. Spectrosc.* 136, 250–263 (1989).

[21]See, for instance, *Metrologia*, 19, 163–177 (1984).

*One second is identified as the duration of 9 192 631 770 periods of the radiation corresponding to the transition between the two hyperfine levels of the ground state of the cesium − 133 atom; F=4, $m_F=0$ to F=3, $m_F=0$ transition of the fundamental state $^2S_{1/2}$ of ^{133}Cs.—*NBS Spec. Publ.* (U.S.) (1977).

OsO_4 and CO_2. To be sure, our interest in these handbooks has always been to address ourselves to the development and availability of sufficient numbers of standards in the entire infrared. To that extent, in Section IV of the present book we have included FTS generated spectral maps and wavenumber data for the CH_3Cl molecule in the region of 1309–1594 cm^{-1} for supplementing water vapor standards[1] that may not be suitable in certain specific experiments.

Finally, we have provided summary maps so one can, at a glance, find out what information has been made available in these handbooks from FTS techniques. The summary map for this handbook appears on its inside front cover page; the same type of map for the previous handbook appears on the inside of the back cover page for convenient cross-reference.

Before concluding this introduction we are indeed very gratified that numerous scientists around the world assisted us through discussions and other ways in bringing out this book. We are particularly grateful to Jean COLLET and Alain UBELMANN who are in charge of the high resolution interferometer at LPMA in Orsay, France. We express our sincere thanks to:

In Belgium: A. BELAFHAL and André FAYT
In Czechoslovakia: Stepan URBAN
In Finland: Rauno ANTTILA
In France: Claude ALAMICHEL, Claude AMIOT, Christian BORDÉ, Claude CAMY-PERET, Pierre CÉREZ, Christian CHARDONNET, Jean-Marie FLAUD, Nicole HUSSON, Qingli KOU, Jean-Yves MANDIN, and Mireille MORILLON-CHAPEY
In Germany: Manfred WINNEWISSER and Brenda WINNEWISSER
In India: Romola D'CUNHA
In Japan: Eizi HIROTA, Tetsuya OHSHIMA, Hiroyuki SASADA
In Morocco: Najate BEN SARI ZIZI
In USA: Arlan MANTZ and Paul ORMSBY

We also benefitted by meetings related to the subject "Unified wavenumber standards" of the IUPAC: 1992 members of which are:

M. BIRK, C. BORDÉ, J. BRAULT, L. R. BROWN, B. CARLI, A. R. H. COLE, K. M. EVENSON, A. FAYT, G. GUELACHVILI, D. HAUSAMANN, J. W. C. JOHNS, J. KAUPPINEN, Q. KOU, A. G. MAKI, K. N. RAO, W. URBAN, A. VALENTIN, J. VERGES, G. WAGNER, M. WINNEWISSER, and B. WINNEWISSER.

Finally, we appreciate the interest shown by Bureau National de Métrologie.

SECTION I

MAPS AND WAVENUMBER TABLES FOR THE NEW SPECTRAL COVERAGE: 1.4 μm–4 μm

The spectra for this handbook have all been recorded with the high resolution Fourier interferometer at the Laboratoire de Physique Moleculaire et Applications (LPMA) in Orsay[22], France. The apodized spectral resolution used in most cases was about 5.5×10^{-3} cm^{-1}. The spectra were calibrated with respect to the ν_3 band of $^{12}C^{16}O_2$ and the $1 - 0$ and $2 - 0$ bands of $^{12}C^{16}O$; the absolute accuracy of the spectral positions is believed to be better than 0.001 cm^{-1}. The pressure and path length information of the gases appears on the right hand top corner of each of the spectral maps.

The format of this handbook is essentially the same as in prior efforts of this type. That is, in pages 2–571, one sees a spectral map on one side and wavenumber data on the other. Most of the entries here are self-explanatory. Each of the spectral lines is identified either by a symbol or by a vertical bar, and a serial number has been given to it. The wavenumbers and transition assignments appearing on the page opposite each map are listed against serial numbers, which correspond exactly to the numbers on the map.

[22]G. Guelachvili, *Applied Opt.* 17, 1322–1326 (1978).

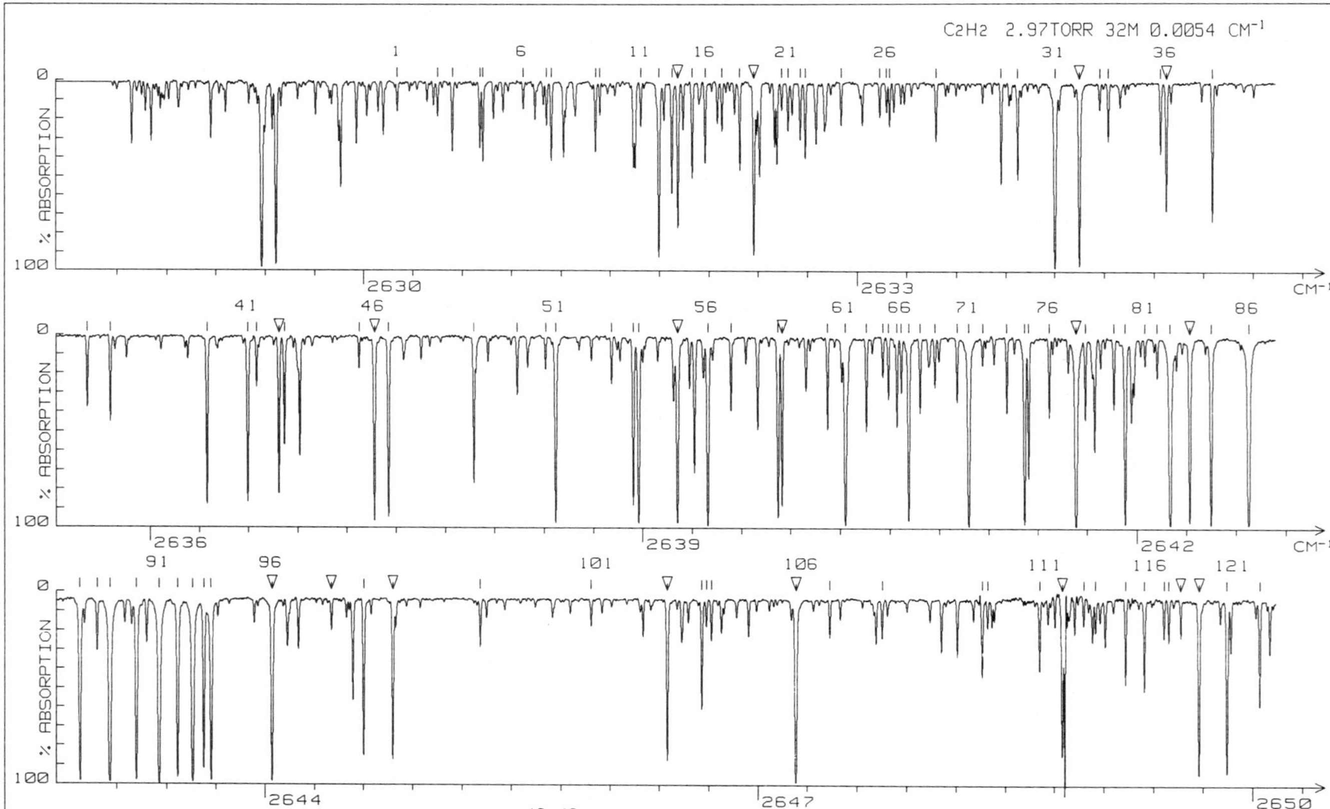

ISOTOPIC IDENTIFICATION – '∇' indicates $H^{12}C^{12}CH$; 'I': unidentified.

BAND 1= $(0101^{1}1^{-1})(\Sigma_u^-) \leftarrow (0001^{1}0)\ (\Pi_g)$; 2= $(0101^{-1}1^{1})(\Sigma_u^+) \leftarrow (0001^{1}0)\ (\Pi_g)$; 3= $(0010^{0}0^{0})(\Sigma_u^+) \leftarrow (0001^{1}0)(\Pi_g)$;
4= $(0101^{1}1^{1})(\Delta_u)e \leftarrow (0001^{1}0)e(\Pi_g)$; 5= $(0101^{1}1^{1})(\Delta_u)f \leftarrow (0001^{1}0)f(\Pi_g)$;

LINE NO.	WAVENUMBER CM^{-1}, VAC.	ROTATIONAL BAND ASSIGNMENTS			ISOTOPIC SPECIES
1	2630. 20538 (67)				
2	2630. 45051 (54)				
3	2630. 53888 (23)				
4	2630. 70931 (25)				
5	2630. 72865 (21)				
6	2630. 97181 (67)				
7	2631. 11137 (43)				
8	2631. 14020 (22)				
9	2631. 40895 (24)				
10	2631. 43734 (56)				
11	2631. 68702 (35)				
12	2631. 79748 (12)				
13	2631. 87567 (22)				
14	2631. 91184 (13)	P	(16)	2	$H^{12}C^{12}CH$
15	2631. 99783 (18)				
16	2632. 07638 (21)				
17	2632. 17770 (32)				
18	2632. 28741 (20)				
19	2632. 37459 (13)	P	(21)	3	$H^{12}C^{12}CH$
20	2632. 54450 (67)				
21	2632. 58302 (31)				
22	2632. 65618 (28)				
23	2632. 68663 (22)				
24	2632. 90315 (35)				
25	2633. 13785 (51)				
26	2633. 17862 (60)				
27	2633. 19697 (36)				
28	2633. 47713 (27)				
29	2633. 87185 (17)				
30	2633. 97391 (18)				
31	2634. 20308 (11)				
32	2634. 35139 (11)	P	(15)	2	$H^{12}C^{12}CH$
33	2634. 47289 (65)				
34	2634. 52422 (27)				
35	2634. 84123 (22)				
36	2634. 87608 (15)	P	(20)	3	$H^{12}C^{12}CH$
37	2635. 15426 (14)				
38	2635. 62325 (23)				
39	2635. 76217 (23)				
40	2636. 34792 (12)				
41	2636. 59551 (12)				
42	2636. 64885 (54)				
43	2636. 78423 (13)	P	(14)	2	$H^{12}C^{12}CH$
44	2636. 81752 (21)				
45	2637. 27367 (54)				
46	2637. 36711 (11)	P	(19)	3	$H^{12}C^{12}CH$
47	2637. 45258 (11)				
48	2637. 97148 (16)				
49	2638. 23618 (46)				
50	2638. 40709 (56)				
51	2638. 46795 (11)				
52	2638. 80861 (39)				
53	2638. 94207 (12)				
54	2638. 97486 (11)				
55	2639. 21033 (11)	P	(13)	2	$H^{12}C^{12}CH$
56	2639. 39381 (11)				
57	2639. 53408 (23)				
58	2639. 82312 (12)				
59	2639. 84772 (12)	P	(18)	3	$H^{12}C^{12}CH$
60	2640. 12288 (19)				

- "LINE NO.": serial number identifying the spectral line in the map on the opposite page.
- "WAVENUMBER": experimental value.
- Uncertainty in the last digit of the wavenumber is given in parentheses.
- "BAND" : identification on the opposite page.

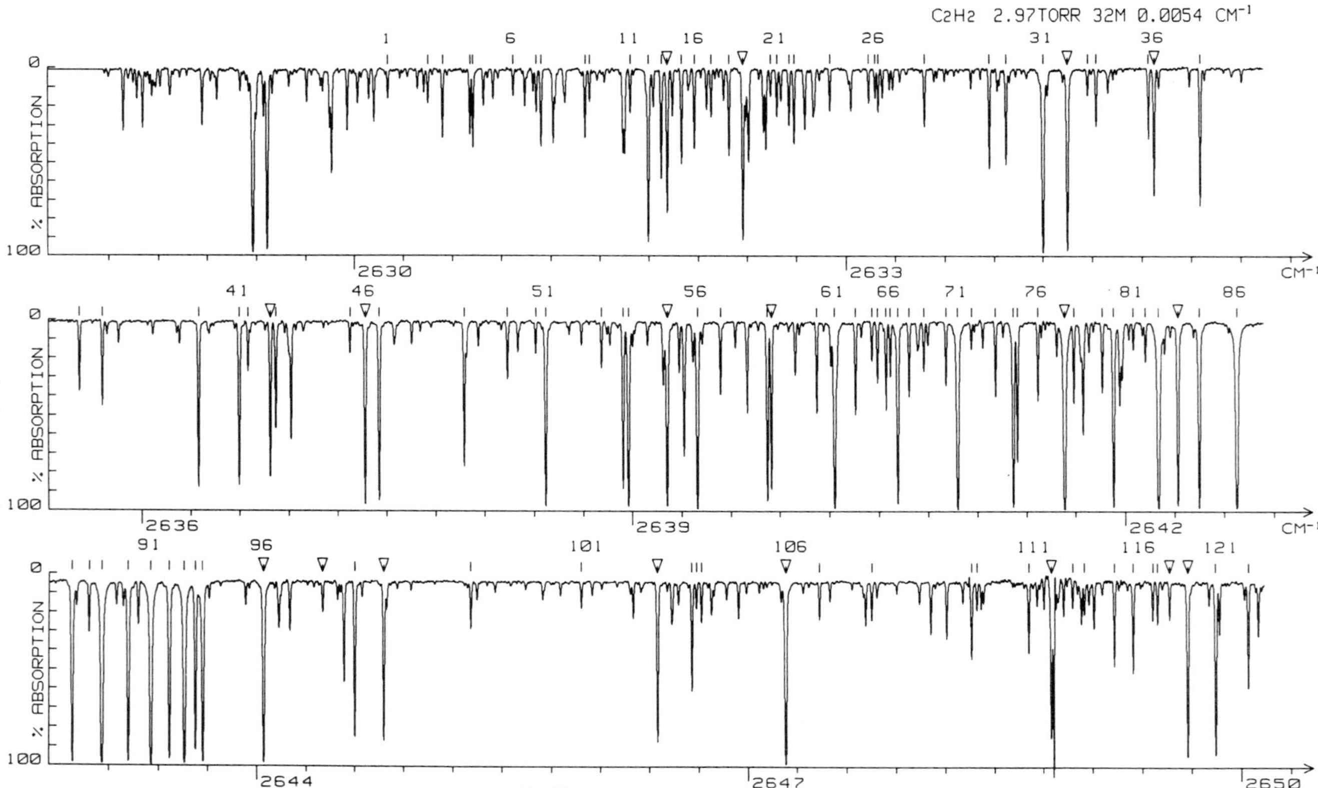

ISOTOPIC IDENTIFICATION – '∇' indicates $H^{12}C^{12}CH$; 'ı': unidentified.

BAND 1= $(0101^{1}1^{-1})(\Sigma_u^-) \leftarrow (0001^{1}0)$ (Π_g); 2= $(0101^{-1}1^{1})(\Sigma_u^+) \leftarrow (0001^{1}0)$ (Π_g); 3= $(0010^{0}0^{0})(\Sigma_u^+) \leftarrow (0001^{1}0)(\Pi_g)$;

4= $(0101^{1}1^{1})(\Delta u)e \leftarrow (0001^{1}0)e(\Pi_g)$; 5= $(0101^{1}1^{1})(\Delta u)f \leftarrow (0001^{1}0)f(\Pi_g)$;

LINE NO.	WAVENUMBER CM^{-1}, VAC.	ROTATIONAL ASSIGNMENTS	BAND	ISOTOPIC SPECIES
61	2640.22998(11)			
62	2640.35572(18)			
63	2640.45494(38)			
64	2640.48897(32)			
65	2640.54303(20)			
66	2640.56995(31)			
67	2640.61439(12)			
68	2640.68229(27)			
69	2640.77177(39)			
70	2640.90511(26)			
71	2640.97624(11)			
72	2641.06028(66)			
73	2641.20653(22)			
74	2641.31561(11)			
75	2641.34114(14)			
76	2641.46902(21)			
77	2641.63217(11)	P (12)	2	$H^{12}C^{12}CH$
78	2641.68775(25)			
79	2641.85903(23)			
80	2641.92667(11)			
81	2642.04678(66)			
82	2642.11992(48)			
83	2642.19842(11)			
84	2642.31852(11)	P (17)	3	$H^{12}C^{12}CH$
85	2642.44750(11)			
86	2642.67403(11)			
87	2642.87794(11)			
88	2642.98193(37)			
89	2643.05919(11)			
90	2643.21784(11)			
91	2643.35382(11)			
92	2643.46718(11)			
93	2643.55787(11)			
94	2643.62583(12)			
95	2643.67122(11)			
96	2644.04320(11)	P (11)	2	$H^{12}C^{12}CH$
97	2644.40388(61)	P (18)	1	$H^{12}C^{12}CH$
98	2644.60239(13)			
99	2644.77978(12)	P (16)	3	$H^{12}C^{12}CH$
100	2645.31105(32)			
101	2645.98454(63)			
102	2646.45030(12)	P (10)	2	$H^{12}C^{12}CH$
103	2646.65773(16)			
104	2646.68568(63)			
105	2646.71610(43)			
106	2647.23183(11)	P (15)	3	$H^{12}C^{12}CH$
107	2647.43692(46)			
108	2647.75192(39)			
109	2648.35972(21)			
110	2648.39258(63)			
111	2648.70854(27)			
112	2648.85144(11)	P (9)	2	$H^{12}C^{12}CH$
113	2648.97898(60)			
114	2649.05094(57)			
115	2649.23004(19)			
116	2649.34423(19)			
117	2649.46206(45)			
118	2649.49090(41)			
119	2649.56376(46)	P (16)	1	$H^{12}C^{12}CH$
120	2649.67549(12)	P (14)	3	$H^{12}C^{12}CH$
121	2649.84475(27)			
122	2650.04197(16)			

- "LINE NO.": serial number identifying the spectral line in the map on the opposite page.
- "WAVENUMBER": experimental value.
- Uncertainty in the last digit of the wavenumber is given in parentheses.
- "BAND" : identification on the opposite page.

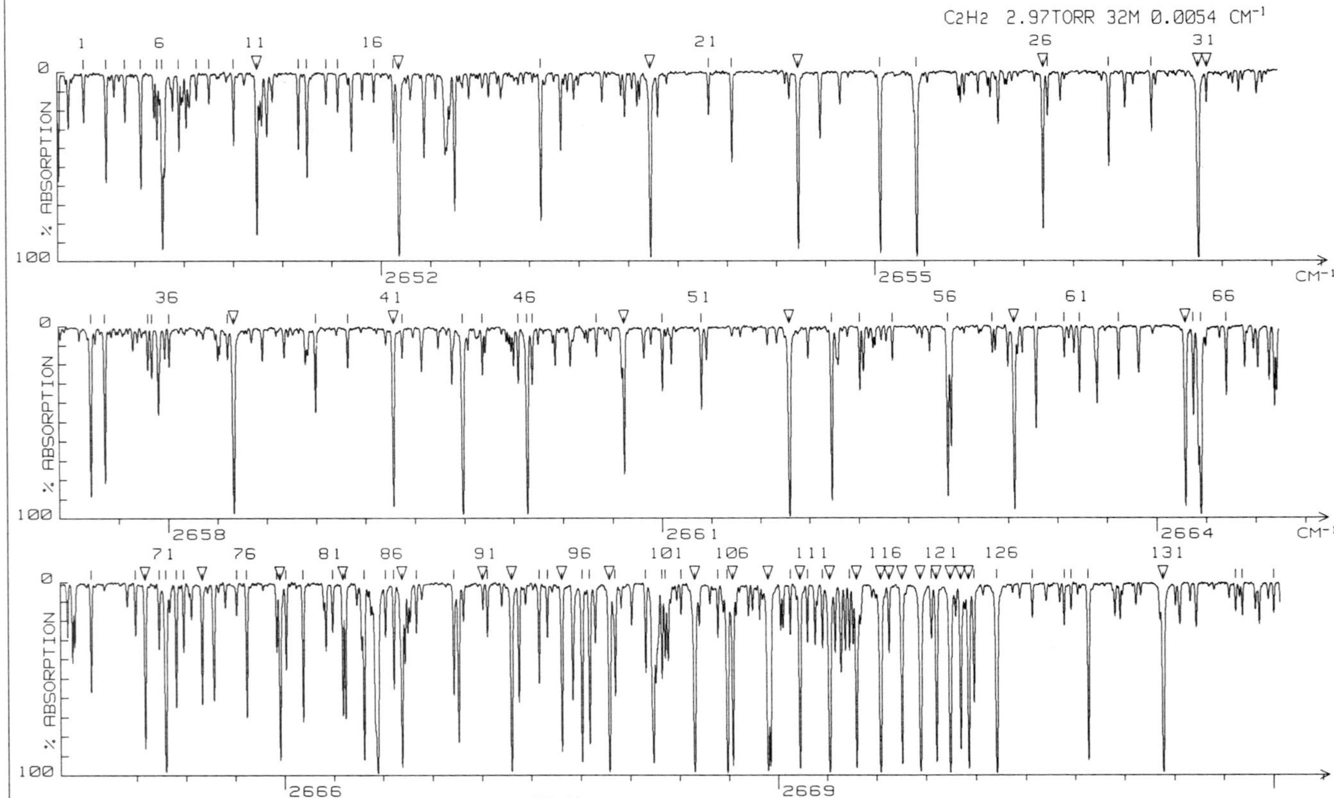

ISOTOPIC IDENTIFICATION – '∇' indicates $H^{12}C^{12}CH$; 'I': unidentified.

BAND 1= $(0101^1 1^{-1})(\Sigma_u^-) \leftarrow (0001^1 0)\ (\Pi_g)$; 2= $(0101^{-1} 1^1)(\Sigma_u^+) \leftarrow (0001^1 0)\ (\Pi_g)$; 3= $(0010^0 0^0)(\Sigma_u^+) \leftarrow (0001^1 0)(\Pi_g)$;
4= $(0101^1 1^1)(\Delta_u)e \leftarrow (0001^1 0)e(\Pi_g)$; 5= $(0101^1 1^1)(\Delta_u)f \leftarrow (0001^1 0)f(\Pi_g)$;

LINE NO.	WAVENUMBER CM^{-1}, VAC.	ROTATIONAL ASSIGNMENTS	BAND	ISOTOPIC SPECIES
1	2650.19600(38)			
2	2650.33070(16)			
3	2650.44604(33)			
4	2650.54221(24)			
5	2650.63988(42)			
6	2650.67266(25)			
7	2650.77246(34)			
8	2650.88133(68)			
9	2650.95755(60)			
10	2651.10526(23)			
11	2651.24662(14)	P (8)	2	$H^{12}C^{12}CH$
12	2651.49966(21)			
13	2651.55282(17)			
14	2651.67041(56)			
15	2651.74086(47)			
16	2651.95989(64)			
17	2652.07892(41)			
18	2652.11103(11)	P (13)	3	$H^{12}C^{12}CH$
19	2652.97211(13)			
20	2653.63605(11)	P (7)	2	$H^{12}C^{12}CH$
21	2653.99028(43)			
22	2654.13191(19)			
23	2654.53843(12)	P (12)	3	$H^{12}C^{12}CH$
24	2655.03846(11)			
25	2655.25833(11)			
26	2656.01975(13)	P (6)	2	$H^{12}C^{12}CH$
27	2656.04735(43)			
28	2656.41851(19)			
29	2656.67858(44)			
30	2656.95842(11)	P (11)	3	$H^{12}C^{12}CH$
31	2657.00985(58)	Pff (14)	5	$H^{12}C^{12}CH$
32	2657.53095(12)			
33	2657.61710(13)			
34	2657.87834(44)			
35	2657.90236(34)			
36	2658.00743(46)			
37	2658.36315(64)			
38	2658.39797(25)	P (5)	2	$H^{12}C^{12}CH$
39	2658.89763(20)			
40	2659.09497(45)			
41	2659.37104(12)	P (10)	3	$H^{12}C^{12}CH$
42	2659.42523(59)			
43	2659.79015(11)			
44	2659.91413(33)			
45	2660.12983(35)			
46	2660.18360(11)			
47	2660.21496(58)			
48	2660.60369(66)			
49	2660.77052(16)	P (4)	2	$H^{12}C^{12}CH$
50	2661.00343(31)			
51	2661.24033(21)			
52	2661.77678(11)	P (9)	3	$H^{12}C^{12}CH$
53	2662.03574(12)			
54	2662.20707(25)			
55	2662.40636(53)			
56	2662.73769(32)			
57	2663.00736(70)			
58	2663.13746(12)	P (3)	2	$H^{12}C^{12}CH$
59	2663.27154(21)			
60	2663.44369(60)			

- "LINE NO.": serial number identifying the spectral line in the map on the opposite page.
- "WAVENUMBER": experimental value.
- Uncertainty in the last digit of the wavenumber is given in parentheses.
- "BAND" : identification on the opposite page.

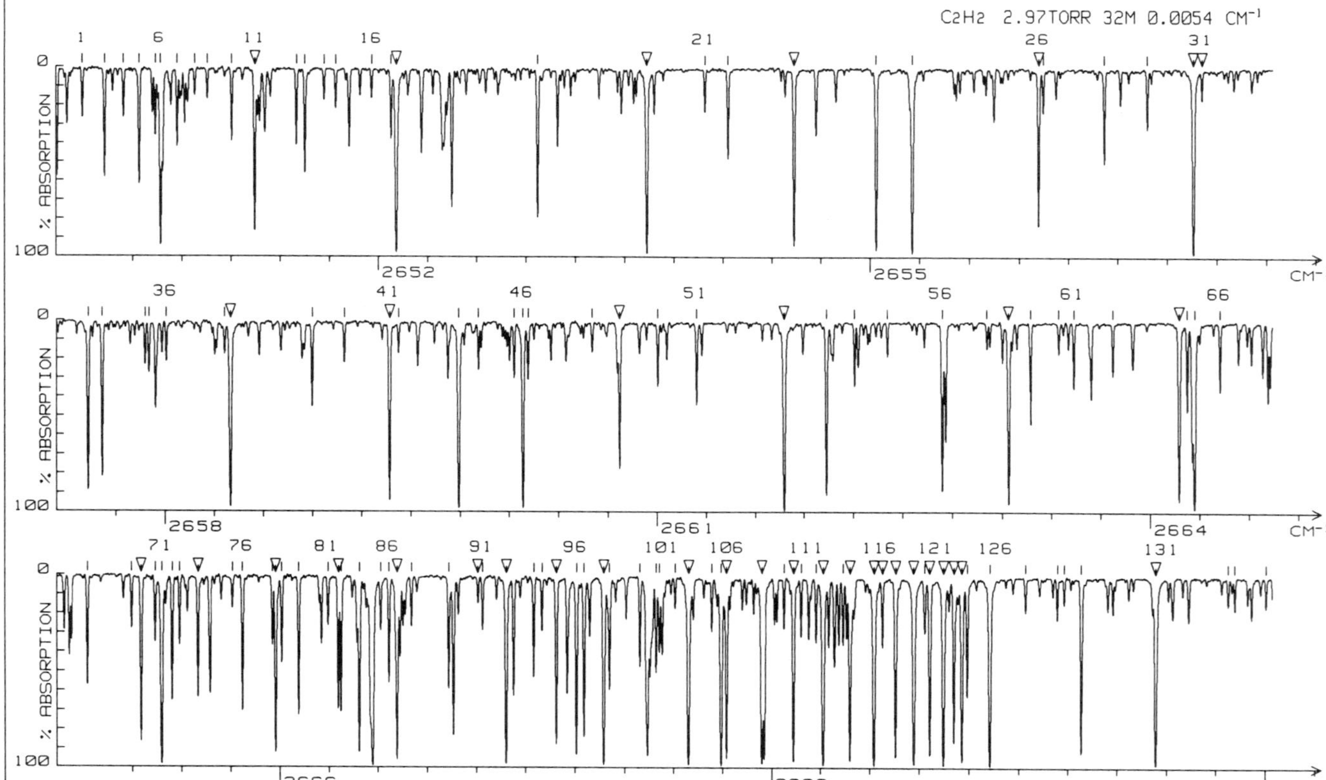

ISOTOPIC IDENTIFICATION – '∇' indicates $H^{12}C^{12}CH$; '|': unidentified.

BAND 1= $(0101^1 1^{-1})(\Sigma_u^-) \leftarrow (0001^1 0)\ (\Pi_g)$; 2= $(0101^{-1} 1^1)(\Sigma_u^+) \leftarrow (0001^1 0)\ (\Pi_g)$; 3= $(0010^0 0^0)(\Sigma_u^+) \leftarrow (0001^1 0)(\Pi_g)$;
4= $(0101^1 1^1)(\Delta u)e \leftarrow (0001^1 0)e(\Pi_g)$; 5= $(0101^1 1^1)(\Delta u)f \leftarrow (0001^1 0)f(\Pi_g)$;

LINE NO.	WAVENUMBER CM^{-1}, VAC.	ROTATIONAL BAND ASSIGNMENTS		ISOTOPIC SPECIES	LINE NO.	WAVENUMBER CM^{-1}, VAC.	ROTATIONAL BAND ASSIGNMENTS		ISOTOPIC SPECIES
61	2663.53525(25)				91	2667.23748(36)			
62	2663.77286(35)				92	2667.38306(11)	Q (18)	2	$H^{12}C^{12}CH$
63	2664.17578(25)	P (8)	3	$H^{12}C^{12}CH$	93	2667.54924(18)			
64	2664.22409(20)				94	2667.59894(30)			
65	2664.26778(14)				95	2667.68782(14)	Q (17)	2	$H^{12}C^{12}CH$
66	2664.42370(24)				96	2667.80934(12)			
67	2664.82633(17)				97	2667.85535(13)			
68	2665.09514(31)				98	2667.97415(11)	Q (16)	2	$H^{12}C^{12}CH$
69	2665.15440(14)	Q (24)	2	$H^{12}C^{12}CH$	99	2668.00862(24)			
70	2665.23929(43)				100	2668.19468(24)			
71	2665.27975(11)				101	2668.29415(29)			
72	2665.34250(15)				102	2668.31368(43)			
73	2665.38734(28)				103	2668.41041(57)			
74	2665.49918(16)	P (2)	2	$H^{12}C^{12}CH$	104	2668.49237(11)	Q (14)	2	$H^{12}C^{12}CH$
75	2665.70937(60)				105	2668.63380(35)			
76	2665.77216(15)				106	2668.69097(11)			
77	2665.95453(41)				107	2668.72465(12)	Q (13)	2	$H^{12}C^{12}CH$
78	2665.97476(12)	Q (22)	2	$H^{12}C^{12}CH$	108	2668.93929(13)	Q (12)	2	$H^{12}C^{12}CH$
79	2666.01177(24)				109	2669.07614(36)			
80	2666.11491(14)				110	2669.13633(33)	Q (11)	2	$H^{12}C^{12}CH$
81	2666.29353(43)				111	2669.18241(33)			
82	2666.35565(18)	Q (21)	2	$H^{12}C^{12}CH$	112	2669.27360(41)			
83	2666.37265(18)				113	2669.31660(11)	Q (10)	2	$H^{12}C^{12}CH$
84	2666.48619(13)				114	2669.43707(31)			
85	2666.61718(37)				115	2669.47964(12)	Q (9)	2	$H^{12}C^{12}CH$
86	2666.66789(17)				116	2669.62606(11)	Q (8)	2	$H^{12}C^{12}CH$
87	2666.71717(12)	Q (20)	2	$H^{12}C^{12}CH$	117	2669.67691(29)	P (8)	1	$H^{12}C^{12}CH$
88	2666.80324(39)				118	2669.75562(12)	Q (7)	2	$H^{12}C^{12}CH$
89	2667.02972(24)				119	2669.86871(11)	Q (6)	2	$H^{12}C^{12}CH$
90	2667.20876(69)	P (9)	1	$H^{12}C^{12}CH$	120	2669.93445(77)			

- "LINE NO.": serial number identifying the spectral line in the map on the opposite page.
- "WAVENUMBER": experimental value.
- Uncertainty in the last digit of the wavenumber is given in parentheses.
- "BAND" : identification on the opposite page.

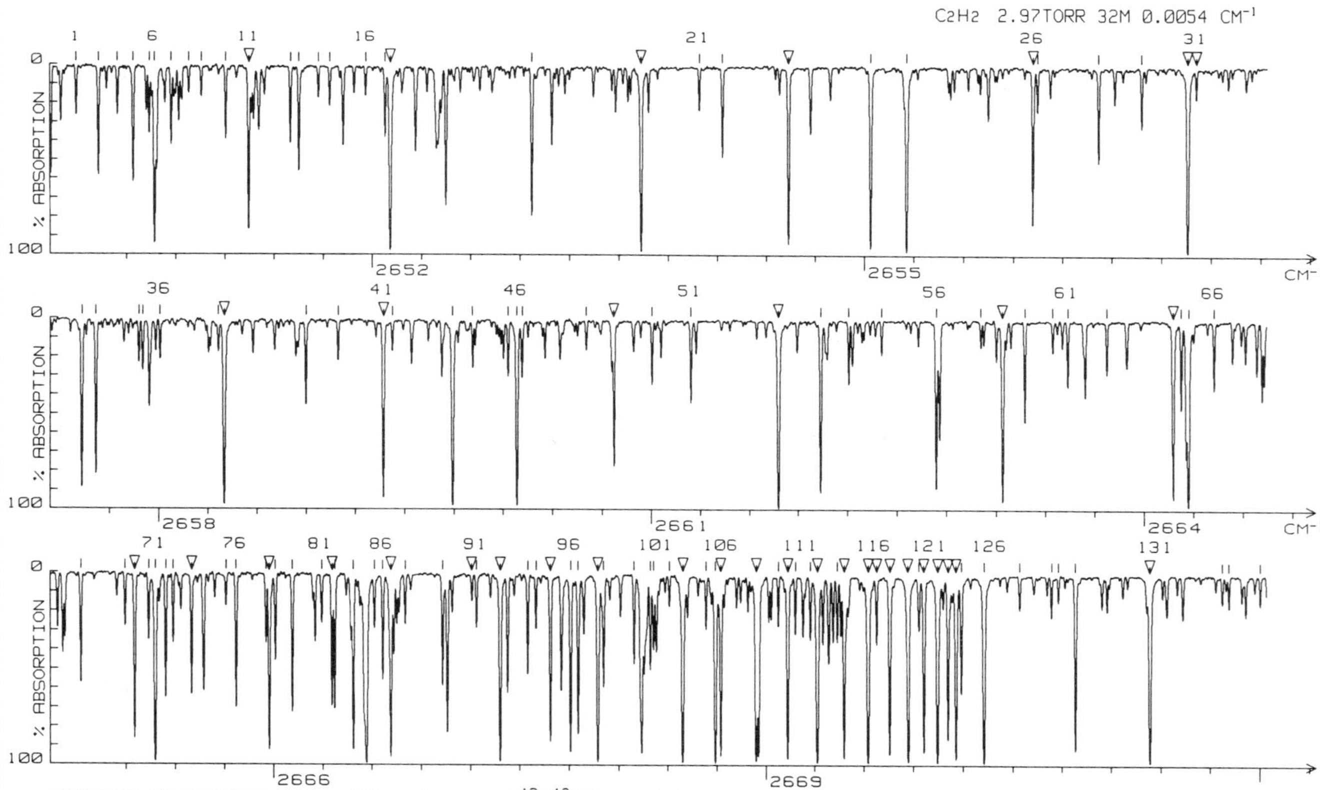

ISOTOPIC IDENTIFICATION – '∇' indicates $H^{12}C^{12}CH$; 'ı': unidentified.
BAND 1= $(0101^1 1^{-1})(\Sigma_u^-) \leftarrow (0001^1 0)$ (Π_g); 2= $(0101^{-1} 1^1)(\Sigma_u^+) \leftarrow (0001^1 0)$ (Π_g); 3= $(0010^0 0^0)(\Sigma_u^+) \leftarrow (0001^1 0)(\Pi_g)$;
4= $(0101^1 1^1)(\Delta_u)e \leftarrow (0001^1 0)e(\Pi_g)$; 5= $(0101^1 1^1)(\Delta_u)f \leftarrow (0001^1 0)f(\Pi_g)$;

LINE NO.	WAVENUMBER CM^{-1}, VAC.	ROTATIONAL BAND ASSIGNMENTS	ISOTOPIC SPECIES	LINE NO.	WAVENUMBER CM^{-1}, VAC.	ROTATIONAL BAND ASSIGNMENTS	ISOTOPIC SPECIES
121	2669.96533(12)	Q (5) 2	$H^{12}C^{12}CH$				
122	2670.04576(15)	Q (4) 2	$H^{12}C^{12}CH$				
123	2670.10994(14)	Q (3) 2	$H^{12}C^{12}CH$				
124	2670.15801(11)	Q (2) 2	$H^{12}C^{12}CH$				
125	2670.18997(16)						
126	2670.32663(11)						
127	2670.54068(57)						
128	2670.73579(44)						
129	2670.77741(66)						
130	2670.88205(12)						
131	2671.33476(11)	P (5) 3	$H^{12}C^{12}CH$				
132	2671.77452(72)						
133	2671.81531(57)						
134	2672.00426(61)						

- "LINE NO.": serial number identifying the spectral line in the map on the opposite page.
- "WAVENUMBER": experimental value.
- Uncertainty in the last digit of the wavenumber is given in parentheses.
- "BAND" : identification on the opposite page.

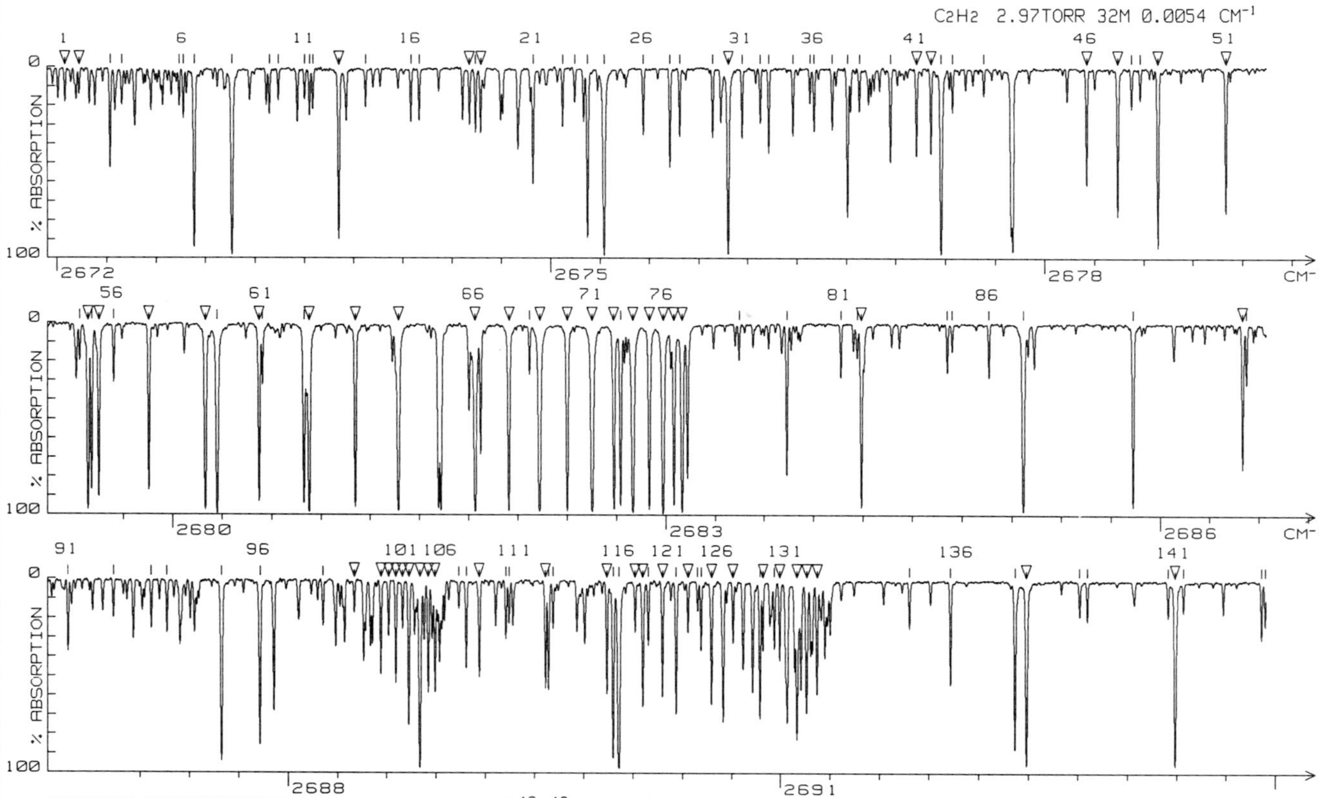

ISOTOPIC IDENTIFICATION - '∇' indicates $H^{12}C^{12}CH$; 'I': unidentified.

BAND 1= $(0101^{1}1^{-1})(\Sigma_u^-) \leftarrow (0001^{1}0)\ (\Pi_g)$; 2= $(0101^{-1}1^{1})(\Sigma_u^+) \leftarrow (0001^{1}0)\ (\Pi_g)$; 3= $(0010^{0}0^{0})(\Sigma_u^+) \leftarrow (0001^{1}0)(\Pi_g)$;

4= $(0101^{1}1^{1})(\Delta_u)e \leftarrow (0001^{1}0)e(\Pi_g)$; 5= $(0101^{1}1^{1})(\Delta_u)f \leftarrow (0001^{1}0)f(\Pi_g)$;

LINE NO.	WAVENUMBER CM^{-1}, VAC.	ROTATIONAL ASSIGNMENTS	BAND	ISOTOPIC SPECIES	LINE NO.	WAVENUMBER CM^{-1}, VAC.	ROTATIONAL ASSIGNMENTS	BAND	ISOTOPIC SPECIES
1	2672.04601 (55)	Pff (8)	5	$H^{12}C^{12}CH$	31	2676.16469 (23)			
2	2672.13171 (72)	P (7)	1	$H^{12}C^{12}CH$	32	2676.27620 (29)			
3	2672.31772 (18)				33	2676.32893 (20)			
4	2672.38853 (59)				34	2676.47513 (24)			
5	2672.73787 (50)				35	2676.57749 (52)			
6	2672.76524 (32)				36	2676.60310 (31)			
7	2672.83149 (12)				37	2676.71281 (26)			
8	2673.05944 (11)				38	2676.80535 (14)			
9	2673.28457 (42)				39	2676.87717 (45)			
10	2673.33839 (57)				40	2677.06800 (19)			
11	2673.49850 (61)				41	2677.22437 (20)	R (2)	2	$H^{12}C^{12}CH$
12	2673.52742 (42)				42	2677.31109 (20)	Q (25)	3	$H^{12}C^{12}CH$
13	2673.54782 (39)				43	2677.37290 (11)			
14	2673.70882 (12)	P (4)	3	$H^{12}C^{12}CH$	44	2677.44194 (43)			
15	2673.87130 (48)				45	2677.62895 (67)			
16	2674.14742 (30)				46	2678.25099 (16)	Q (23)	3	$H^{12}C^{12}CH$
17	2674.19775 (31)				47	2678.43935 (13)	P (2)	3	$H^{12}C^{12}CH$
18	2674.50666 (29)	Pff (7)	5	$H^{12}C^{12}CH$	48	2678.52234 (40)			
19	2674.54275 (26)				49	2678.57524 (57)			
20	2674.57361 (43)	P (6)	1	$H^{12}C^{12}CH$	50	2678.68596 (12)	Q (22)	3	$H^{12}C^{12}CH$
21	2674.89061 (16)				51	2679.09796 (14)	Q (21)	3	$H^{12}C^{12}CH$
22	2675.07082 (28)				52	2679.43739 (53)			
23	2675.14489 (55)				53	2679.48737 (12)	Q (20)	3	$H^{12}C^{12}CH$
24	2675.22303 (12)				54	2679.50883 (14)			
25	2675.32399 (11)				55	2679.55251 (12)	R (3)	2	$H^{12}C^{12}CH$
26	2675.56272 (24)				56	2679.64071 (28)			
27	2675.72262 (18)				57	2679.85461 (13)	Q (19)	3	$H^{12}C^{12}CH$
28	2675.78147 (24)				58	2680.20014 (11)	Q (18)	3	$H^{12}C^{12}CH$
29	2675.98214 (23)				59	2680.27147 (11)			
30	2676.07705 (11)	P (3)	3	$H^{12}C^{12}CH$	60	2680.52419 (12)	Q (17)	3	$H^{12}C^{12}CH$

- "LINE NO.": serial number identifying the spectral line in the map on the opposite page.
- "WAVENUMBER": experimental value.
- Uncertainty in the last digit of the wavenumber is given in parentheses.
- "BAND" : identification on the opposite page.

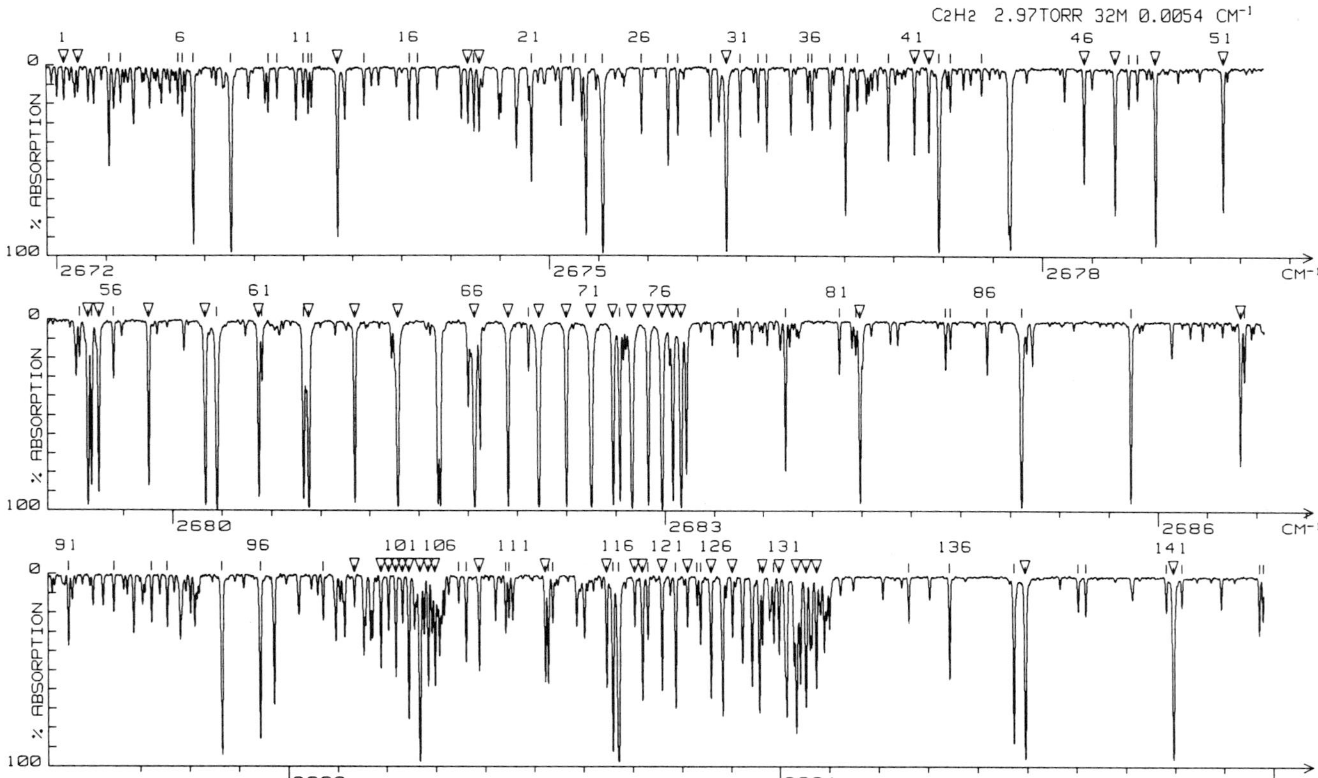

ISOTOPIC IDENTIFICATION – '∇' indicates $H^{12}C^{12}CH$; 'I': unidentified.

BAND 1= $(0101^{1}1^{-1})(\Sigma_u^-) \leftarrow (0001^{1}0)$ (Π_g); 2= $(0101^{-1}1^{1})(\Sigma_u^+) \leftarrow (0001^{1}0)$ (Π_g); 3= $(0010^{0}0^{0})(\Sigma_u^+) \leftarrow (0001^{1}0)(\Pi_g)$;
4= $(0101^{1}1^{1})(\Delta_u)e \leftarrow (0001^{1}0)e(\Pi_g)$; 5= $(0101^{1}1^{1})(\Delta_u)f \leftarrow (0001^{1}0)f(\Pi_g)$;

LINE NO.	WAVENUMBER CM^{-1}, VAC.	ROTATIONAL ASSIGNMENTS	BAND	ISOTOPIC SPECIES
61	2680. 54436 (44)			
62	2680. 79588 (12)			
63	2680. 82746 (11)	Q (16)	3	$H^{12}C^{12}CH$
64	2681. 11014 (12)	Q (15)	3	$H^{12}C^{12}CH$
65	2681. 37293 (11)	Q (14)	3	$H^{12}C^{12}CH$
66	2681. 83999 (11)	Q (12)	3	$H^{12}C^{12}CH$
67	2682. 04510 (11)	Q (11)	3	$H^{12}C^{12}CH$
68	2682. 17033 (33)			
69	2682. 23187 (11)	Q (10)	3	$H^{12}C^{12}CH$
70	2682. 40047 (11)	Q (9)	3	$H^{12}C^{12}CH$
71	2682. 55142 (11)	Q (8)	3	$H^{12}C^{12}CH$
72	2682. 68480 (11)	Q (7)	3	$H^{12}C^{12}CH$
73	2682. 72638 (12)			
74	2682. 80101 (11)	Q (6)	3	$H^{12}C^{12}CH$
75	2682. 90018 (11)	Q (5)	3	$H^{12}C^{12}CH$
76	2682. 98255 (11)	Q (4)	3	$H^{12}C^{12}CH$
77	2683. 04827 (12)	Q (3)	3	$H^{12}C^{12}CH$
78	2683. 09744 (11)	Q (2)	3	$H^{12}C^{12}CH$
79	2683. 44507 (51)			
80	2683. 73925 (14)			
81	2684. 06615 (36)			
82	2684. 16522 (68)			
83	2684. 19080 (12)	R (5)	2	$H^{12}C^{12}CH$
84	2684. 71160 (33)			
85	2684. 74141 (67)			
86	2684. 96131 (30)			
87	2685. 16876 (11)			
88	2685. 83366 (12)			
89	2686. 50076 (17)	R (6)	2	$H^{12}C^{12}CH$
90	2686. 52442 (34)			

LINE NO.	WAVENUMBER CM^{-1}, VAC.	ROTATIONAL ASSIGNMENTS	BAND	ISOTOPIC SPECIES
91	2686. 66488 (24)			
92	2686. 94025 (49)			
93	2687. 16947 (40)			
94	2687. 26593 (36)			
95	2687. 59850 (12)			
96	2687. 82982 (15)			
97	2688. 21198 (59)			
98	2688. 40442 (57)	Q (18)	1	$H^{12}C^{12}CH$
99	2688. 56699 (19)	Q (15)	1	$H^{12}C^{12}CH$
100	2688. 61407 (29)	Q (14)	1	$H^{12}C^{12}CH$
101	2688. 65832 (18)	Q (13)	1	$H^{12}C^{12}CH$
102	2688. 69920 (41)	Q (12)	1	$H^{12}C^{12}CH$
103	2688. 73696 (14)	Q (11)	1	$H^{12}C^{12}CH$
104	2688. 80407 (11)	Q (9)	1	$H^{12}C^{12}CH$
105	2688. 85517 (17)	Q (7)	1	$H^{12}C^{12}CH$
106	2688. 89611 (24)	Q (5)	1	$H^{12}C^{12}CH$
107	2689. 04156 (68)			
108	2689. 08703 (31)			
109	2689. 16529 (19)	Qef (18)	4	$H^{12}C^{12}CH$
110	2689. 32827 (29)			
111	2689. 34724 (45)			
112	2689. 56885 (25)	Qef (16)	4	$H^{12}C^{12}CH$
113	2689. 58740 (24)			
114	2689. 61515 (34)			
115	2689. 94221 (16)	Qef (14)	4	$H^{12}C^{12}CH$
116	2689. 98043 (12)			
117	2690. 01477 (23)			
118	2690. 11586 (32)	Qef (13)	4	$H^{12}C^{12}CH$
119	2690. 16280 (20)	R (2)	3	$H^{12}C^{12}CH$
120	2690. 19510 (26)			

- "LINE NO.": serial number identifying the spectral line in the map on the opposite page.
- "WAVENUMBER": experimental value.
- Uncertainty in the last digit of the wavenumber is given in parentheses.
- "BAND" : identification on the opposite page.

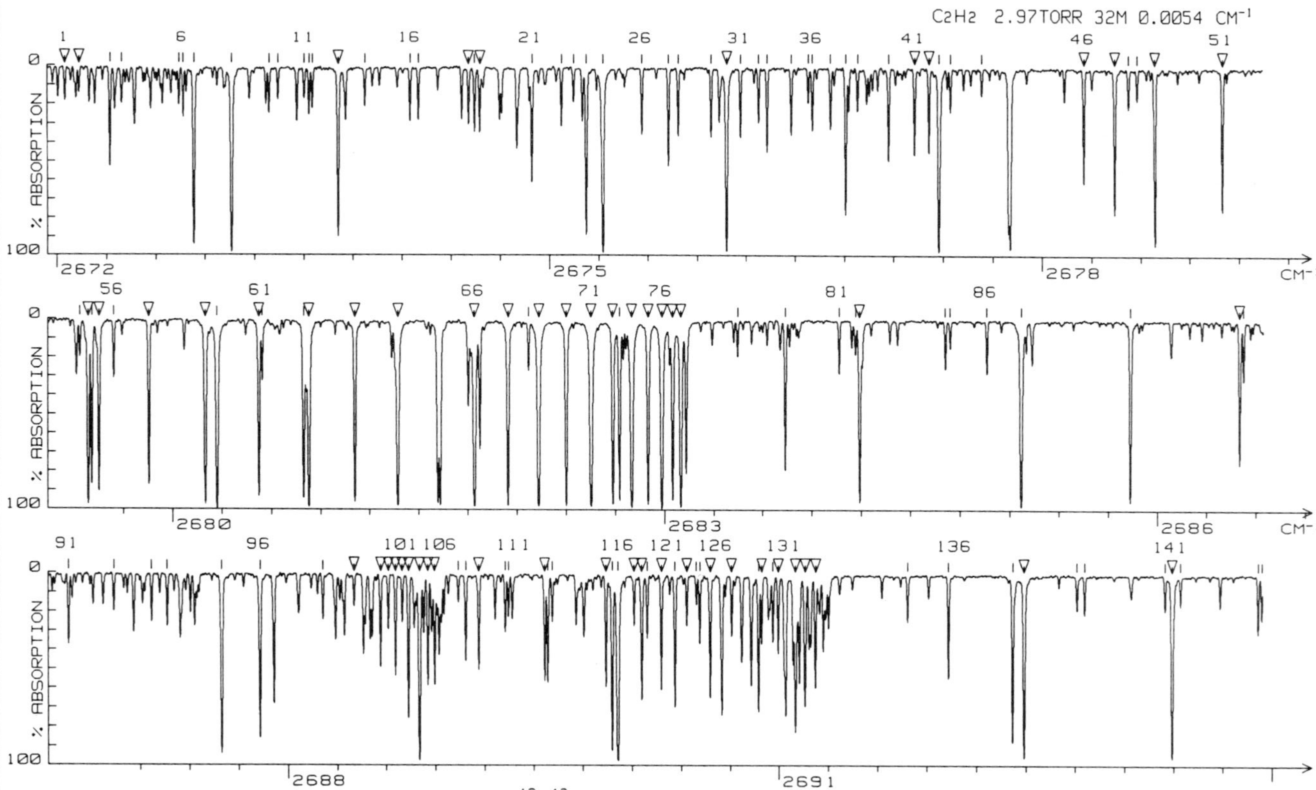

ISOTOPIC IDENTIFICATION – '∇' indicates $H^{12}C^{12}CH$; '|': unidentified.

BAND 1= $(0101^{1}1^{-1})(\Sigma_u^-) \leftarrow (0001^{1}0)\ (\Pi_g)$; 2= $(0101^{-1}1^{1})(\Sigma_u^+) \leftarrow (0001^{1}0)\ (\Pi_g)$; 3= $(0010^{0}0^{0})(\Sigma_u^+) \leftarrow (0001^{1}0)(\Pi_g)$;
4= $(0101^{1}1^{1})(\Delta_u)e \leftarrow (0001^{1}0)e(\Pi_g)$; 5= $(0101^{1}1^{1})(\Delta_u)f \leftarrow (0001^{1}0)f(\Pi_g)$;

LINE NO.	WAVENUMBER CM^{-1}, VAC.	ROTATIONAL ASSIGNMENTS	BAND	ISOTOPIC SPECIES
121	2690. 28008 (16)	Qef (12)	4	$H^{12}C^{12}CH$
122	2690. 36265 (15)			
123	2690. 43397 (37)	Qef (11)	4	$H^{12}C^{12}CH$
124	2690. 49287 (62)			
125	2690. 51373 (24)			
126	2690. 57741 (20)	Qef (10)	4	$H^{12}C^{12}CH$
127	2690. 70944 (27)	Qef (9)	4	$H^{12}C^{12}CH$
128	2690. 87441 (14)			
129	2690. 89328 (40)	Qfe (15)	5	$H^{12}C^{12}CH$
130	2690. 96556 (25)			
131	2690. 99760 (26)	Qfe (13)	5	$H^{12}C^{12}CH$
132	2691. 10170 (20)	R (8)	2	$H^{12}C^{12}CH$
133	2691. 15992 (26)	Qfe (9)	5	$H^{12}C^{12}CH$
134	2691. 22298 (18)	Qfe (7)	5	$H^{12}C^{12}CH$
135	2691. 78463 (57)			
136	2692. 03280 (17)			
137	2692. 42007 (13)			
138	2692. 48933 (11)	R (3)	3	$H^{12}C^{12}CH$
139	2692. 81294 (55)			
140	2692. 85808 (46)			
141	2693. 34831 (52)			
142	2693. 39151 (12)	R (9)	2	$H^{12}C^{12}CH$
143	2693. 44464 (61)			
144	2693. 91807 (33)			
145	2693. 94089 (43)			

LINE NO.	WAVENUMBER CM^{-1}, VAC.	ROTATIONAL ASSIGNMENTS	BAND	ISOTOPIC SPECIES

- "LINE NO.": serial number identifying the spectral line in the map on the opposite page.
- "WAVENUMBER": experimental value.
- Uncertainty in the last digit of the wavenumber is given in parentheses.
- "BAND" : identification on the opposite page.

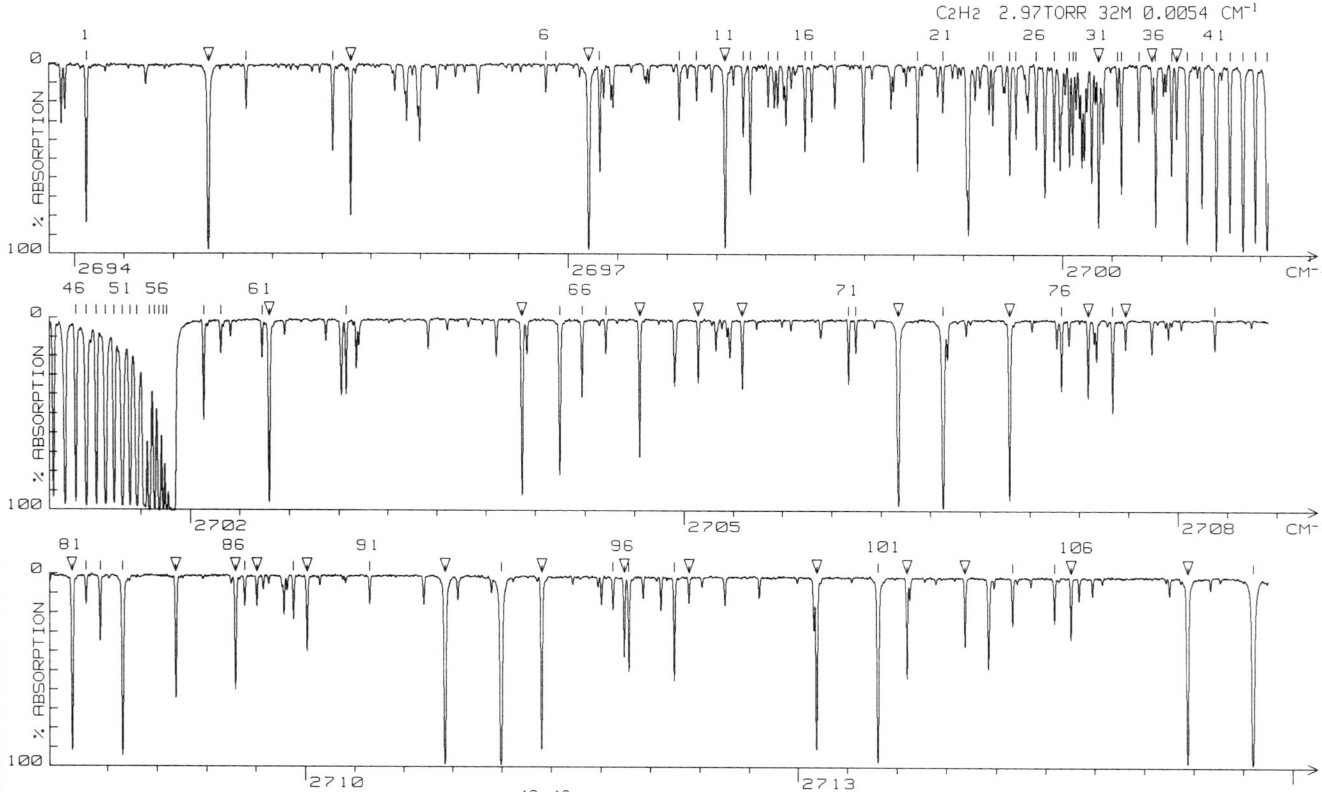

ISOTOPIC IDENTIFICATION – '∇' indicates $H^{12}C^{12}CH$; 'ı': unidentified.

BAND 1= $(0101^{1}1^{-1})(\Sigma_u^-) \leftarrow (0001^{1}0)\ (\Pi_g)$; 2= $(0101^{-1}1^{1})(\Sigma_u^+) \leftarrow (0001^{1}0)\ (\Pi_g)$; 3= $(0010^{0}0^{0})(\Sigma_u^+) \leftarrow (0001^{1}0)(\Pi_g)$;

4= $(0101^{1}1^{1})(\Delta_u)e \leftarrow (0001^{1}0)e(\Pi_g)$; 5= $(0101^{1}1^{1})(\Delta_u)f \leftarrow (0001^{1}0)f(\Pi_g)$;

LINE NO.	WAVENUMBER CM^{-1}, VAC.	ROTATIONAL BAND ASSIGNMENTS		ISOTOPIC SPECIES	LINE NO.	WAVENUMBER CM^{-1}, VAC.	ROTATIONAL BAND ASSIGNMENTS		ISOTOPIC SPECIES
1	2694.07123 (13)				31	2700.21943 (13)	R (12)	2	$H^{12}C^{12}CH$
2	2694.81115 (11)	R (4)	3	$H^{12}C^{12}CH$	32	2700.33191 (43)			
3	2695.04039 (42)				33	2700.35632 (15)			
4	2695.56526 (20)				34	2700.46204 (22)			
5	2695.67467 (14)	R (10)	2	$H^{12}C^{12}CH$	35	2700.54353 (58)	R (4)	1	$H^{12}C^{12}CH$
6	2696.86592 (65)				36	2700.56393 (13)			
7	2697.12311 (11)	R (5)	3	$H^{12}C^{12}CH$	37	2700.66181 (16)			
8	2697.19070 (17)				38	2700.69184 (23)	Ree (3)	4	$H^{12}C^{12}CH$
9	2697.67193 (29)				39	2700.75580 (12)			
10	2697.77746 (51)				40	2700.84600 (14)			
11	2697.95076 (12)	R (11)	2	$H^{12}C^{12}CH$	41	2700.93237 (11)			
12	2698.06269 (40)				42	2701.01487 (13)			
13	2698.10572 (15)				43	2701.09369 (11)			
14	2698.21457 (43)				44	2701.16873 (12)			
15	2698.27265 (44)				45	2701.24004 (11)			
16	2698.43858 (20)				46	2701.30765 (12)			
17	2698.47958 (34)				47	2701.37158 (11)			
18	2698.61935 (36)				48	2701.43185 (11)			
19	2698.79407 (22)				49	2701.48845 (11)			
20	2699.12292 (17)				50	2701.54145 (11)			
21	2699.27536 (33)				51	2701.59084 (11)			
22	2699.55314 (38)				52	2701.63666 (11)			
23	2699.57500 (26)				53	2701.67888 (11)			
24	2699.67723 (17)				54	2701.75263 (11)			
25	2699.71591 (23)				55	2701.78419 (11)			
26	2699.84035 (29)				56	2701.81218 (11)			
27	2699.94849 (27)				57	2701.83667 (11)			
28	2700.04032 (18)				58	2701.85772 (11)			
29	2700.06100 (20)				59	2702.08358 (21)			
30	2700.08036 (45)				60	2702.18673 (55)			

- "LINE NO.": serial number identifying the spectral line in the map on the opposite page.
- "WAVENUMBER": experimental value.
- Uncertainty in the last digit of the wavenumber is given in parentheses.
- "BAND" : identification on the opposite page.

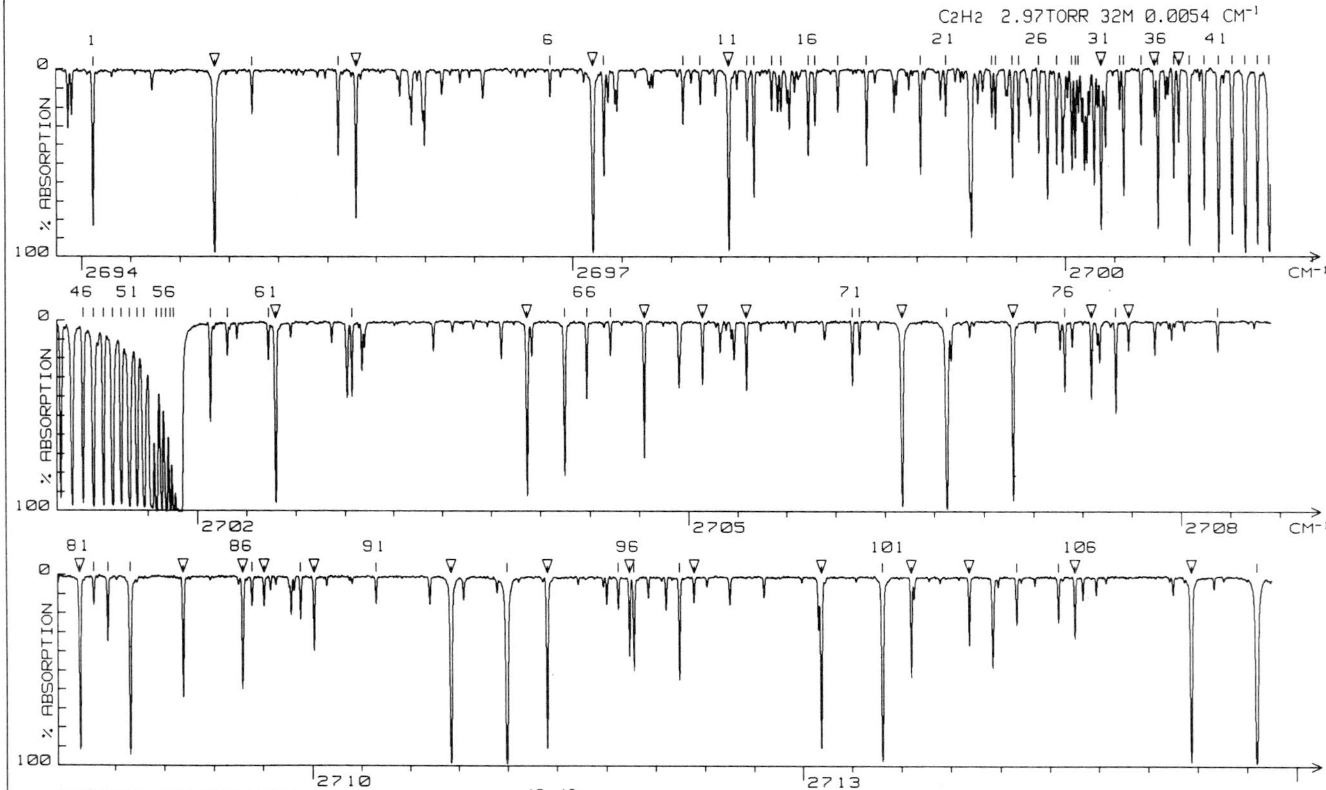

ISOTOPIC IDENTIFICATION – '▽' indicates $H^{12}C^{12}CH$; 'ı': unidentified.

BAND 1= $(0101^{1}1^{-1})(\Sigma_u^-) \leftarrow (0001^{1}0)\ (\Pi_g)$; 2= $(0101^{-1}1^{1})(\Sigma_u^+) \leftarrow (0001^{1}0)\ (\Pi_g)$; 3= $(0010^{0}0^{0})(\Sigma_u^+) \leftarrow (0001^{1}0)(\Pi_g)$;
4= $(0101^{1}1^{1})(\Delta_u)e \leftarrow (0001^{1}0)e(\Pi_g)$; 5= $(0101^{1}1^{1})(\Delta_u)f \leftarrow (0001^{1}0)f(\Pi_g)$;

LINE NO.	WAVENUMBER CM^{-1}, VAC.	ROTATIONAL BAND ASSIGNMENTS		ISOTOPIC SPECIES
61	2702.43613(48)			
62	2702.48011(12)	R (13)	2	$H^{12}C^{12}CH$
63	2702.94747(23)			
64	2704.02199(12)	R (8)	3	$H^{12}C^{12}CH$
65	2704.25051(13)			
66	2704.38303(22)			
67	2704.52712(52)			
68	2704.73283(15)	R (14)	2	$H^{12}C^{12}CH$
69	2705.08794(26)	R (6)	1	$H^{12}C^{12}CH$
70	2705.35544(24)	Ree (5)	4	$H^{12}C^{12}CH$
71	2706.00486(44)			
72	2706.04853(53)			
73	2706.30668(11)	R (9)	3	$H^{12}C^{12}CH$
74	2706.57796(11)			
75	2706.97750(12)	R (15)	2	$H^{12}C^{12}CH$
76	2707.29446(25)			
77	2707.45691(22)	Rff (6)	5	$H^{12}C^{12}CH$
78	2707.60540(19)			
79	2707.68285(65)	Ree (6)	4	$H^{12}C^{12}CH$
80	2708.22450(59)			
81	2708.58327(12)	R (10)	3	$H^{12}C^{12}CH$
82	2708.66934(63)			
83	2708.75452(26)			
84	2708.89253(12)			
85	2709.21348(16)	R (16)	2	$H^{12}C^{12}CH$
86	2709.57569(17)	R (8)	1	$H^{12}C^{12}CH$
87	2709.63181(64)			
88	2709.70451(63)	Rff (7)	5	$H^{12}C^{12}CH$
89	2709.92675(43)			
90	2710.00831(37)	Ree (7)	4	$H^{12}C^{12}CH$
91	2710.39181(65)			
92	2710.85145(11)	R (11)	3	$H^{12}C^{12}CH$
93	2711.19407(11)			
94	2711.44072(12)	R (17)	2	$H^{12}C^{12}CH$
95	2711.86980(59)			
96	2711.93775(22)	Rff (8)	5	$H^{12}C^{12}CH$
97	2711.96627(20)			
98	2712.24271(18)			
99	2712.33086(69)	Ree (8)	4	$H^{12}C^{12}CH$
100	2713.11073(12)	R (12)	3	$H^{12}C^{12}CH$
101	2713.48254(11)			
102	2713.65883(18)	R (18)	2	$H^{12}C^{12}CH$
103	2714.01029(24)	R (10)	1	$H^{12}C^{12}CH$
104	2714.29584(34)			
105	2714.55226(42)			
106	2714.65225(27)	Ree (9)	4	$H^{12}C^{12}CH$
107	2715.36071(11)	R (13)	3	$H^{12}C^{12}CH$
108	2715.75796(11)			

- "LINE NO.": serial number identifying the spectral line in the map on the opposite page.
- "WAVENUMBER": experimental value.
- Uncertainty in the last digit of the wavenumber is given in parentheses.
- "BAND" : identification on the opposite page.

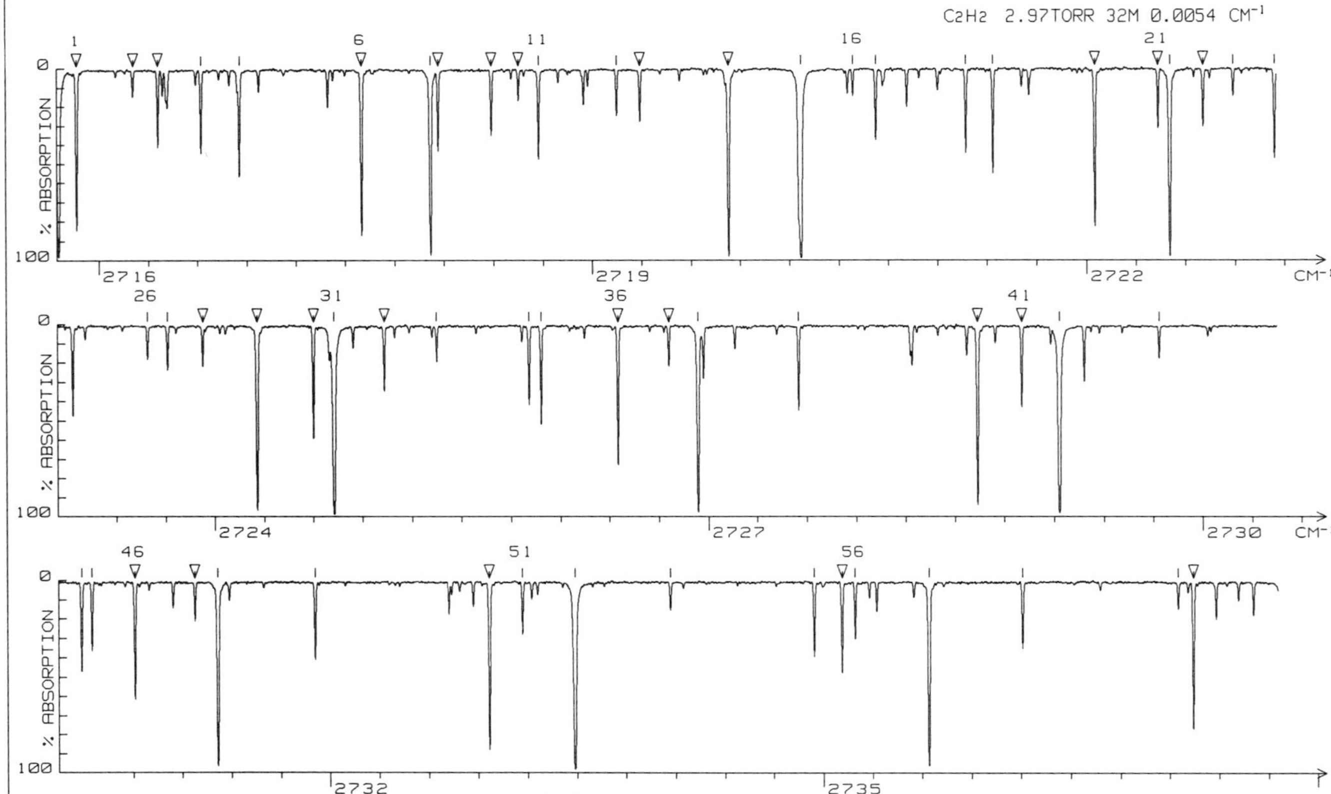

ISOTOPIC IDENTIFICATION – '∇' indicates $H^{12}C^{12}CH$; 'ı': unidentified.

BAND 1= $(0101^{1}1^{-1})(\Sigma_u^-) \leftarrow (0001^{1}0)$ (Π_g); 2= $(0101^{-1}1^{1})(\Sigma_u^+) \leftarrow (0001^{1}0)$ (Π_g); 3= $(0010^{0}0^{0})(\Sigma_u^+) \leftarrow (0001^{1}0)(\Pi_g)$;

4= $(0101^{1}1^{1})(\Delta_u)e \leftarrow (0001^{1}0)e(\Pi_g)$; 5= $(0101^{1}1^{1})(\Delta_u)f \leftarrow (0001^{1}0)f(\Pi_g)$;

LINE NO.	WAVENUMBER CM^{-1}, VAC.	ROTATIONAL BAND ASSIGNMENTS		ISOTOPIC SPECIES	LINE NO.	WAVENUMBER CM^{-1}, VAC.	ROTATIONAL BAND ASSIGNMENTS		ISOTOPIC SPECIES
1	2715.86758(13)	R (19)	2	$H^{12}C^{12}CH$	31	2724.72795(11)			
2	2716.20575(69)	R (11)	1	$H^{12}C^{12}CH$	32	2725.03408(30)	Rff (14)	5	$H^{12}C^{12}CH$
3	2716.36106(22)	Rff (10)	5	$H^{12}C^{12}CH$	33	2725.35303(53)			
4	2716.62348(25)				34	2725.91169(26)			
5	2716.85525(24)				35	2725.98399(18)			
6	2717.60082(13)	R (14)	3	$H^{12}C^{12}CH$	36	2726.45391(14)	R (18)	3	$H^{12}C^{12}CH$
7	2718.02025(11)				37	2726.76099(46)	R (24)	2	$H^{12}C^{12}CH$
8	2718.06662(25)	R (20)	2	$H^{12}C^{12}CH$	38	2726.93736(11)			
9	2718.38728(25)	R (12)	1	$H^{12}C^{12}CH$	39	2727.55222(21)			
10	2718.55114(57)	Rff (11)	5	$H^{12}C^{12}CH$	40	2728.63789(12)	R (19)	3	$H^{12}C^{12}CH$
11	2718.67294(20)				41	2728.90771(21)	R (25)	2	$H^{12}C^{12}CH$
12	2719.15047(34)				42	2729.13342(11)			
13	2719.29150(37)	Ree (11)	4	$H^{12}C^{12}CH$	43	2729.73754(57)			
14	2719.83079(11)	R (15)	3	$H^{12}C^{12}CH$	44	2730.48810(20)			
15	2720.26933(12)				45	2730.54988(29)			
16	2720.58716(70)				46	2730.80901(16)	R (20)	3	$H^{12}C^{12}CH$
17	2720.72660(24)				47	2731.17691(47)	R (18)	1	$H^{12}C^{12}CH$
18	2721.27109(21)				48	2731.31613(11)			
19	2721.43757(17)				49	2731.90893(22)			
20	2722.04989(13)	R (16)	3	$H^{12}C^{12}CH$	50	2732.96705(13)	R (21)	3	$H^{12}C^{12}CH$
21	2722.43472(32)	R (22)	2	$H^{12}C^{12}CH$	51	2733.16740(32)			
22	2722.50527(11)				52	2733.48543(12)			
23	2722.70759(34)	R (14)	1	$H^{12}C^{12}CH$	53	2734.06633(70)			
24	2722.88750(67)				54	2734.94326(23)			
25	2723.14014(20)				55	2735.11150(23)	R (22)	3	$H^{12}C^{12}CH$
26	2723.59197(57)				56	2735.19087(30)			
27	2723.71561(43)				57	2735.64129(11)			
28	2723.93004(46)	Ree (13)	4	$H^{12}C^{12}CH$	58	2736.20981(25)			
29	2724.25782(12)	R (17)	3	$H^{12}C^{12}CH$	59	2737.15038(68)			
30	2724.60324(16)	R (23)	2	$H^{12}C^{12}CH$	60	2737.24202(14)	R (23)	3	$H^{12}C^{12}CH$

- "LINE NO.": serial number identifying the spectral line in the map on the opposite page.
- "WAVENUMBER": experimental value.
- Uncertainty in the last digit of the wavenumber is given in parentheses.
- "BAND" : identification on the opposite page.

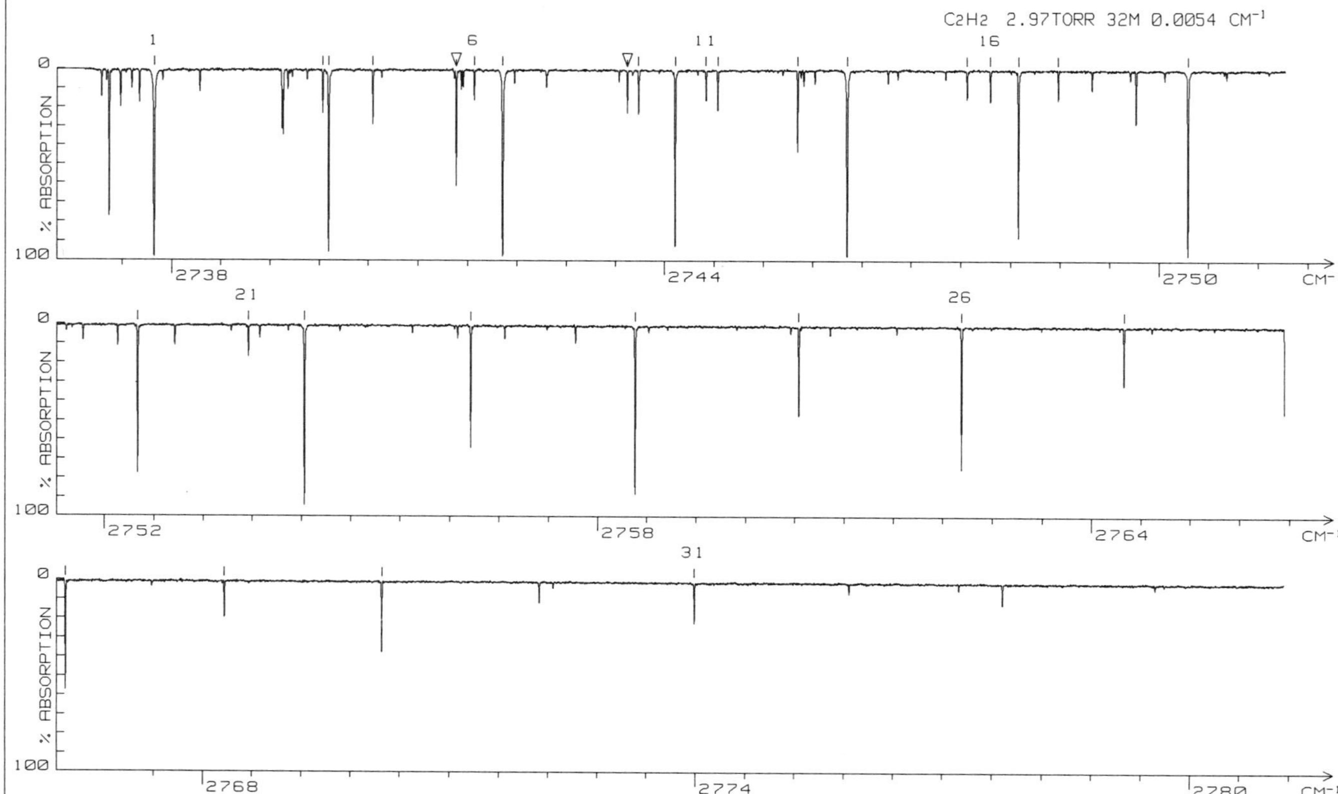

ISOTOPIC IDENTIFICATION – '∇' indicates $H^{12}C^{12}CH$; 'ı': unidentified.

BAND 1= $(0101^{1}1^{-1})(\Sigma_u^-) \leftarrow (0001^{1}0)\ (\Pi_g)$; 2= $(0101^{-1}1^{1})(\Sigma_u^+) \leftarrow (0001^{1}0)\ (\Pi_g)$; 3= $(0010^{0}0^{0})(\Sigma_u^+) \leftarrow (0001^{1}0)(\Pi_g)$;
4= $(0101^{1}1^{1})(\Delta_u)e \leftarrow (0001^{1}0)e(\Pi_g)$; 5= $(0101^{1}1^{1})(\Delta_u)f \leftarrow (0001^{1}0)f(\Pi_g)$;

LINE NO.	WAVENUMBER CM^{-1}, VAC.	ROTATIONAL BAND ASSIGNMENTS			ISOTOPIC SPECIES
1	2737.78368(11)				
2	2739.83966(36)				
3	2739.91247(12)				
4	2740.45464(35)				
5	2741.45988(16)	R	(25)	3	$H^{12}C^{12}CH$
6	2741.67875(61)				
7	2742.02778(11)				
8	2743.54662(43)	R	(26)	3	$H^{12}C^{12}CH$
9	2743.68235(66)				
10	2744.12942(12)				
11	2744.50256(59)				
12	2744.64318(47)				
13	2745.61839(21)				
14	2746.21741(11)				
15	2747.67509(64)				
16	2747.95557(58)				
17	2748.29167(13)				
18	2748.77509(59)				
19	2750.35224(12)				
20	2752.39898(14)				
21	2753.75187(57)				
22	2754.43193(12)				
23	2756.45098(16)				
24	2758.45618(13)				
25	2760.44729(20)				
26	2762.42450(14)				
27	2764.38756(33)				
28	2766.33668(17)				
29	2768.27150(52)				
30	2770.19233(24)				
31	2773.99105(46)				

- "LINE NO.": serial number identifying the spectral line in the map on the opposite page.
- "WAVENUMBER": experimental value.
- Uncertainty in the last digit of the wavenumber is given in parentheses.
- "BAND" : identification on the opposite page.

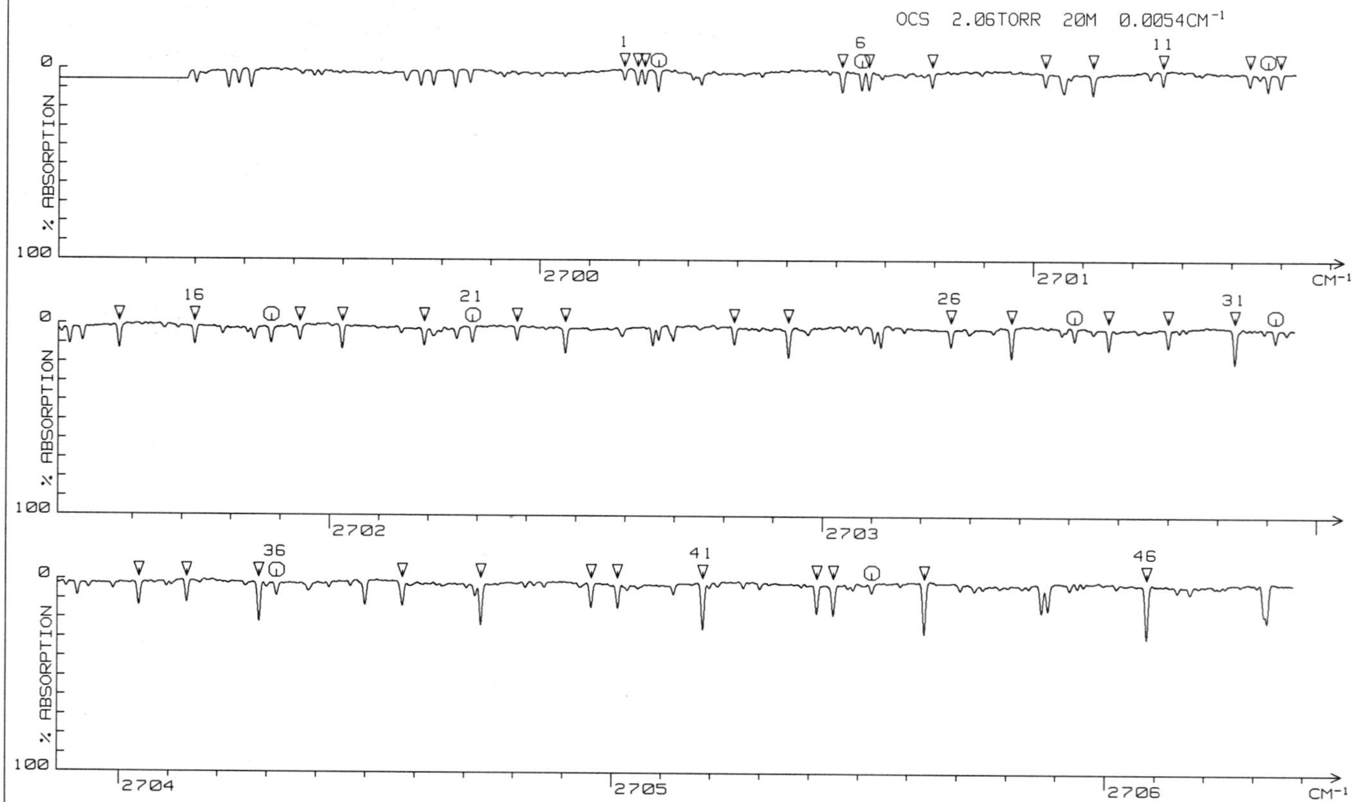

ISOTOPIC IDENTIFICATION – '▽' indicates $^{16}O^{12}C^{32}S$; '⊕': $^{16}O^{12}C^{34}S$; '◇': $^{16}O^{13}C^{32}S$; 'Y': $^{16}O^{12}C^{33}S$;

'✧' indicates $^{18}O^{12}C^{32}S$; '⍖': $^{16}O^{13}C^{34}S$; '⍏': Other OCS species; 'I': not OCS.

LINE NO.	WAVENUMBER UNC. CM^{-1}, VAC.	ROTATIONAL BAND ASSIGNMENTS	ISOTOPIC SPECIES
1	2700.16978(70)	P 58 2301F-0101F	16012C32S
2	2700.19623(52)	P 13 3200 -1000	16012C32S
3	2700.21090(54)	P 72 2200 -0000	16012C32S
4	2700.23818(39)	P 20 2200 -0000	16012C34S
4		P 35 2400 -0200	16012C32S
4		P 22 2501E-0301E	16012C32S
5	2700.61119(43)	P 12 3200 -1000	16012C32S
5		P 57 2301F-0101F	16012C32S
6	2700.65046(45)	P 19 2200 -0000	16012C34S
7	2700.66503(45)	P 71 2200 -0000	16012C32S
8	2700.79377(53)	P 55 2301E-0101E	16012C32S
9	2701.02285(59)	P 11 3200 -1000	16012C32S
10	2701.11915(41)	P 70 2200 -0000	16012C32S
10		P 41 2400 -0200	16012C32S
10		P 20 2501E-0301E	16012C32S
11	2701.26032(53)	P 54 2301E-0101E	16012C32S
12	2701.43483(66)	P 10 3200 -1000	16012C32S
13	2701.47153(46)	P 17 2200 -0000	16012C34S
14	2701.49759(56)	P 55 2301F-0101F	16012C32S
14		P 25 2503 -0303	16012C32S
15	2701.57224(34)	P 69 2200 -0000	16012C32S
15		P 43 2200 -0000	16012C33S
16	2701.72607(44)	P 53 2301E-0101E	16012C32S
16		P 32 2400 -0200	16012C32S
17	2701.88084(47)	P 16 2200 -0000	16012C34S
18	2701.93911(54)	P 54 2301F-0101F	16012C32S
19	2702.02520(32)	P 68 2200 -0000	16012C32S
20	2702.19108(45)	P 52 2301E-0101E	16012C32S
21	2702.28928(48)	P 15 2200 -0000	16012C34S
22	2702.38000(51)	P 53 2301F-0101F	16012C32S
23	2702.47792(31)	P 67 2200 -0000	16012C32S
24	2702.82033(43)	P 52 2301F-0101F	16012C32S
25	2702.93018(27)	P 66 2200 -0000	16012C32S

LINE NO.	WAVENUMBER UNC. CM^{-1}, VAC.	ROTATIONAL BAND ASSIGNMENTS	ISOTOPIC SPECIES
26	2703.25985(43)	P 51 2301F-0101F	16012C32S
27	2703.38199(26)	P 65 2200 -0000	16012C32S
28	2703.51000(56)	P 12 2200 -0000	16012C34S
29	2703.57931(37)	P 49 2301E-0101E	16012C32S
30	2703.69887(41)	P 50 2301F-0101F	16012C32S
31	2703.83332(23)	P 64 2200 -0000	16012C32S
31		R 8 3301E-1101E	16012C32S
32	2703.91511(56)	P 11 2200 -0000	16012C34S
33	2704.04015(35)	P 48 2301E-0101E	16012C32S
33		P 53 2202E-0000	16012C32S
34	2704.13717(35)	P 49 2301F-0101F	16012C32S
35	2704.28432(21)	P 63 2200 -0000	16012C32S
36	2704.31962(61)	P 10 2200 -0000	16012C34S
37	2704.57490(36)	P 48 2301F-0101F	16012C32S
37		P 26 2400 -0200	16012C32S
38	2704.73481(21)	P 62 2200 -0000	16012C32S
39	2704.95789(32)	P 46 2301E-0101E	16012C32S
40	2705.01209(33)	P 47 2301F-0101F	16012C32S
41	2705.18487(19)	P 61 2200 -0000	16012C32S
42	2705.41559(28)	P 45 2301E-0101E	16012C32S
42		R 8 2301E-0101E	16012C34S
43	2705.44897(28)	P 46 2301F-0101F	16012C32S
43		P 34 2200 -0000	16012C33S
43		R 12 3301F-1101F	16012C32S
44	2705.52743(84)	P 7 2200 -0000	16012C34S
45	2705.63449(18)	P 60 2200 -0000	16012C32S
46	2706.08363(17)	P 59 2200 -0000	16012C32S

- "LINE NO.": serial number identifying the spectral line in the map on the opposite page.
- "WAVENUMBER": experimental value.
- Uncertainty in the last digit of the wavenumber is given in parentheses.

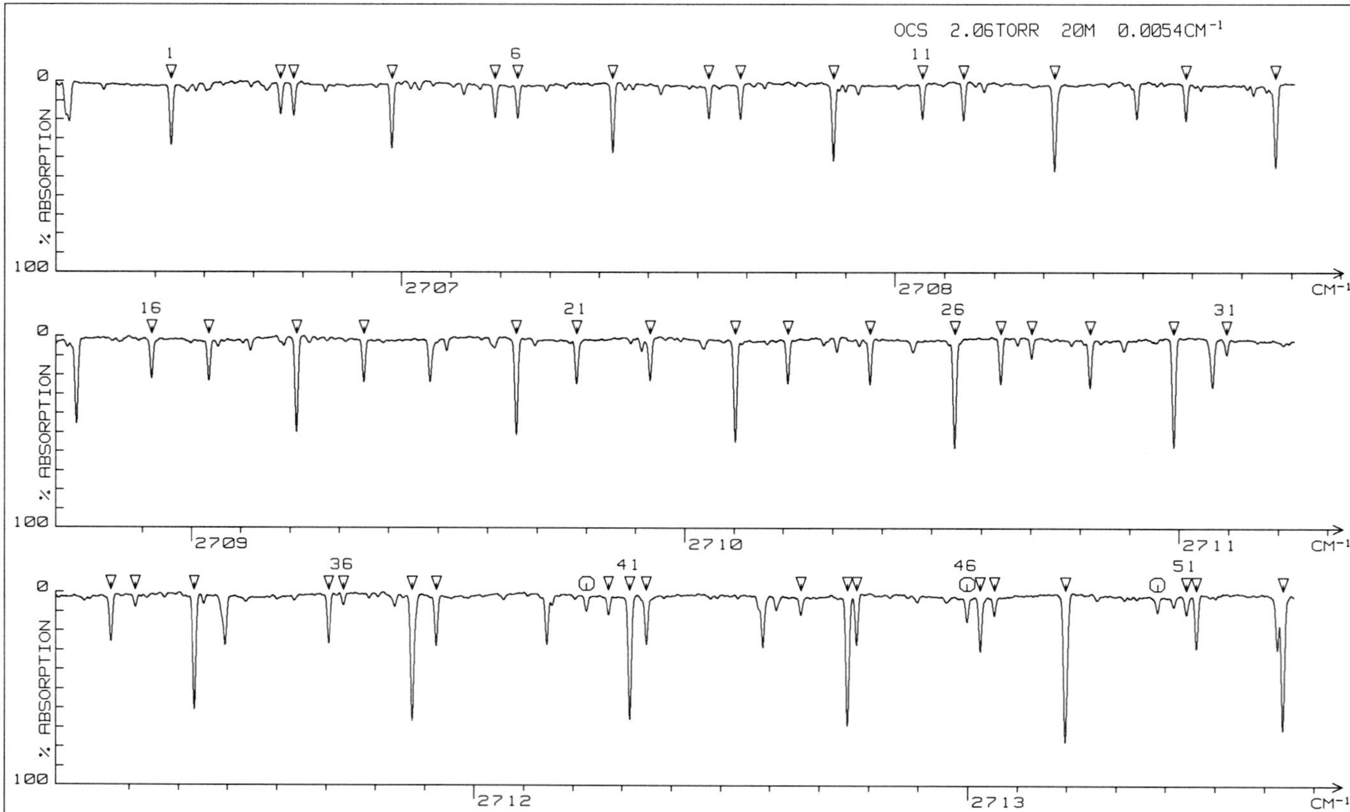

ISOTOPIC IDENTIFICATION – '▽' indicates $^{16}O^{12}C^{32}S$; 'Ⓘ': $^{16}O^{12}C^{34}S$; '◇': $^{16}O^{13}C^{32}S$; 'Y': $^{16}O^{12}C^{33}S$;

'✧' indicates $^{18}O^{12}C^{32}S$; '⍖': $^{16}O^{13}C^{34}S$; '⍏': Other OCS species; 'I': not OCS.

LINE NO.	WAVENUMBER CM^{-1}, VAC.	UNC.	ROTATIONAL ASSIGNMENTS	BAND ASSIGNMENTS	ISOTOPIC SPECIES
1	2706.53221	(16)	P 58	2200 -0000	16012C32S
2	2706.75472	(26)	P 43	2301F-0101F	16012C32S
3	2706.78128	(25)	P 42	2301E-0101E	16012C32S
4	2706.98041	(15)	P 57	2200 -0000	16012C32S
5	2707.18880	(25)	P 42	2301F-0101F	16012C32S
6	2707.23446	(25)	P 41	2301E-0101E	16012C32S
7	2707.42808	(14)	P 56	2200 -0000	16012C32S
8	2707.62229	(24)	P 41	2301F-0101F	16012C32S
9	2707.68644	(23)	P 40	2301E-0101E	16012C32S
10	2707.87514	(14)	P 55	2200 -0000	16012C32S
11	2708.05521	(23)	P 40	2301F-0101F	16012C32S
12	2708.13739	(22)	P 39	2301E-0101E	16012C32S
13	2708.32187	(13)	P 54	2200 -0000	16012C32S
13			R 6	3200 -1000	16012C32S
13			P 24	2402E-0202E	16012C32S
14	2708.58730	(23)	P 38	2301E-0101E	16012C32S
15	2708.76778	(13)	P 53	2200 -0000	16012C32S
16	2708.91887	(22)	P 38	2301F-0101F	16012C32S
16			P 42	2202E-0000	16012C32S
17	2709.03595	(21)	P 37	2301E-0101E	16012C32S
18	2709.21332	(13)	P 52	2200 -0000	16012C32S
18			R 18	2301E-0101E	16012C34S
19	2709.35006	(21)	P 37	2301F-0101F	16012C32S
20	2709.65824	(13)	P 51	2200 -0000	16012C32S
21	2709.78022	(20)	P 36	2301F-0101F	16012C32S
21			R 23	3301F-1101F	16012C32S
22	2709.93004	(21)	P 35	2301E-0101E	16012C32S
23	2710.10256	(12)	P 50	2200 -0000	16012C32S
24	2710.20991	(20)	P 35	2301F-0101F	16012C32S
25	2710.37551	(19)	P 34	2301E-0101E	16012C32S
26	2710.54642	(12)	P 49	2200 -0000	16012C32S
27	2710.63895	(19)	P 34	2301F-0101F	16012C32S

LINE NO.	WAVENUMBER CM^{-1}, VAC.	UNC.	ROTATIONAL ASSIGNMENTS	BAND ASSIGNMENTS	ISOTOPIC SPECIES
28	2710.70199	(39)	R 12	3200 -1000	16012C32S
28			R 22	2301E-0101E	16012C34S
28			R 49	2200 -0000	16013C32S
29	2710.81987	(19)	P 33	2301E-0101E	16012C32S
30	2710.98958	(12)	P 48	2200 -0000	16012C32S
31	2711.09586	(49)	R 13	3200 -1000	16012C32S
32	2711.26298	(19)	P 32	2301E-0101E	16012C32S
33	2711.31206	(67)	P 17	2402 -0202	16012C32S
34	2711.43219	(12)	P 47	2200 -0000	16012C32S
35	2711.70502	(18)	P 31	2301E-0101E	16012C32S
36	2711.73489	(69)	P 16	2402 -0202	16012C32S
37	2711.87420	(20)	P 46	2200 -0000	16012C32S
38	2711.92244	(18)	P 31	2301F-0101F	16012C32S
39	2712.22636	(51)	R 9	2200 -0000	16012C34S
39			R 30	3301E-1101E	16012C32S
40	2712.27215	(43)	R 16	3200 -1000	16012C32S
41	2712.31554	(12)	P 45	2200 -0000	16012C32S
42	2712.34892	(19)	P 30	2301F-0101F	16012C32S
43	2712.66258	(45)	R 17	3200 -1000	16012C32S
44	2712.75625	(11)	P 44	2200 -0000	16012C32S
45	2712.77480	(18)	P 29	2301F-0101F	16012C32S
46	2712.99765	(33)	R 11	2200 -0000	16012C34S
46			P 13	2402 -0202	16012C32S
47	2713.02439	(17)	P 28	2301E-0101E	16012C32S
48	2713.05245	(41)	R 18	3200 -1000	16012C32S
49	2713.19691	(12)	P 43	2200 -0000	16012C32S
49			P 28	2301F-0101F	16012C32S
50	2713.38266	(52)	R 12	2200 -0000	16012C34S
51	2713.44169	(46)	R 19	3200 -1000	16012C32S
51			P 15	2200 -0000	16012C33S
52	2713.46190	(17)	P 27	2301E-0101E	16012C32S
53	2713.63573	(17)	P 42	2200 -0000	16012C32S

- "LINE NO.": serial number identifying the spectral line in the map on the opposite page.
- "WAVENUMBER": experimental value.
- Uncertainty in the last digit of the wavenumber is given in parentheses.

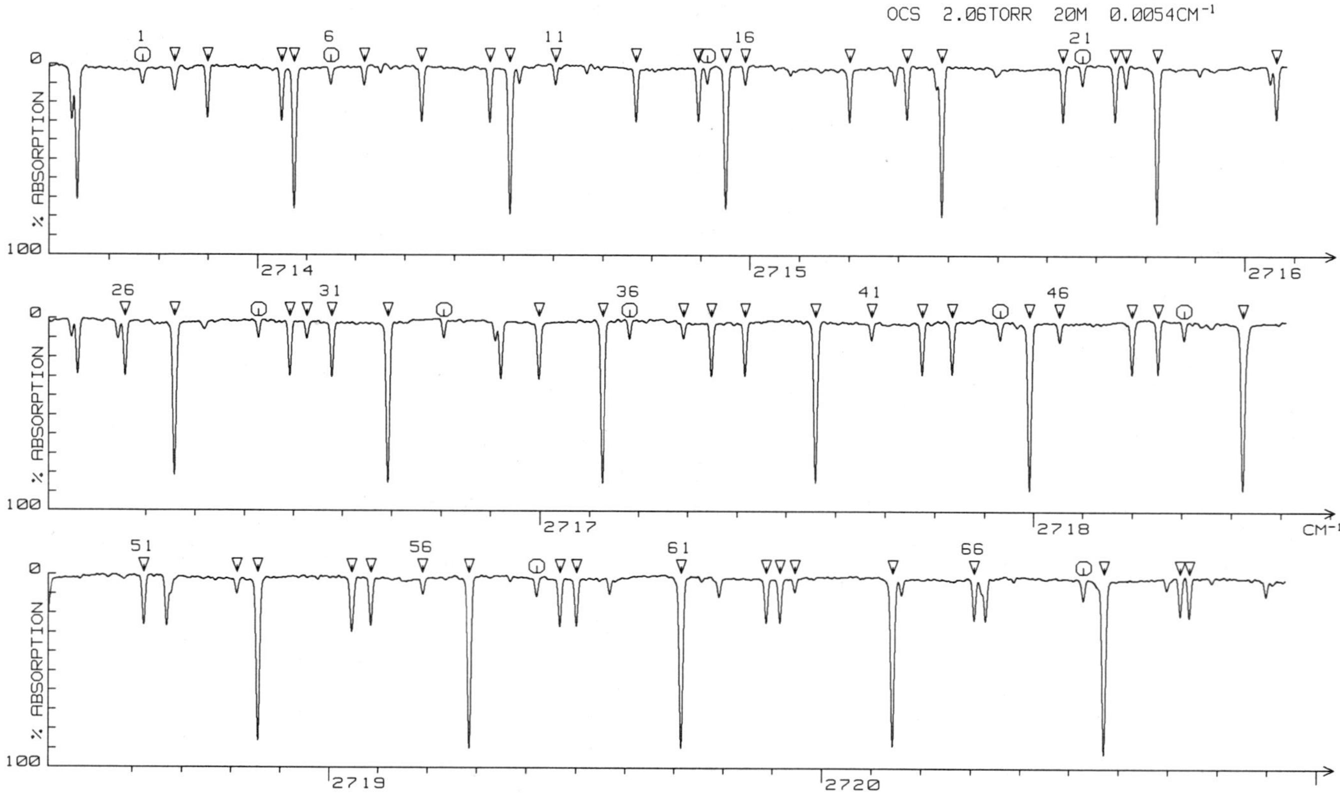

ISOTOPIC IDENTIFICATION – '▽' indicates $^{16}O^{12}C^{32}S$; 'Ⓘ': $^{16}O^{12}C^{34}S$; '◊': $^{16}O^{13}C^{32}S$; 'Y': $^{16}O^{12}C^{33}S$;
'✧' indicates $^{18}O^{12}C^{32}S$; '⫛': $^{16}O^{13}C^{34}S$; '✦': Other OCS species; 'I': not OCS.

LINE NO.	WAVENUMBER UNC. CM^{-1}, VAC.	ROTATIONAL BAND ASSIGNMENTS	ISOTOPIC SPECIES
1	2713.76638(51)	R 13 2200 -0000	16012C34S
2	2713.83109(40)	R 20 3200 -1000	16012C32S
2		P 17 2010 -0200	16012C32S
2		P 11 2402 -0202	16012C32S
3	2713.89828(17)	P 26 2301E-0101E	16012C32S
4	2714.04851(17)	P 26 2301F-0101F	16012C32S
5	2714.07451(11)	P 41 2200 -0000	16012C32S
6	2714.14903(47)	R 14 2200 -0000	16012C34S
6		P 4 2400 -0200	16012C32S
7	2714.21723(40)	R 21 3200 -1000	16012C32S
8	2714.33364(16)	P 25 2301E-0101E	16012C32S
8		R 32 2301E-0101E	16012C34S
9	2714.47180(17)	P 25 2301F-0101F	16012C32S
10	2714.51262(11)	P 40 2200 -0000	16012C32S
11	2714.60404(39)	R 22 3200 -1000	16012C32S
12	2714.76766(17)	P 24 2301E-0101E	16012C32S
13	2714.89451(17)	P 24 2301F-0101F	16012C32S
14	2714.91213(47)	R 16 2200 -0000	16012C34S
15	2714.95002(11)	P 39 2200 -0000	16012C32S
16	2714.98994(40)	R 23 3200 -1000	16012C32S
16		R 33 2301F-0101F	16012C34S
17	2715.20064(17)	P 23 2301E-0101E	16012C32S
18	2715.31647(17)	P 23 2301F-0101F	16012C32S
19	2715.38670(11)	P 38 2200 -0000	16012C32S
20	2715.63243(16)	P 22 2301E-0101E	16012C32S
21	2715.67208(40)	R 18 2200 -0000	16012C34S
22	2715.73772(17)	P 22 2301F-0101F	16012C32S
22		R 35 2301F-0101F	16012C34S
23	2715.75980(35)	R 25 3200 -1000	16012C32S
23		R 36 2301E-0101E	16012C34S
24	2715.82273(11)	P 37 2200 -0000	16012C32S
25	2716.06318(17)	P 21 2301E-0101E	16012C32S
26	2716.15846(17)	P 21 2301F-0101F	16012C32S
27	2716.25805(11)	P 36 2200 -0000	16012C32S
28	2716.42832(44)	R 20 2200 -0000	16012C34S
29	2716.49265(17)	P 20 2301E-0101E	16012C32S
30	2716.52702(45)	R 27 3200 -1000	16012C32S
31	2716.57845(17)	P 20 2301F-0101F	16012C32S
32	2716.69262(11)	P 35 2200 -0000	16012C32S
33	2716.80517(41)	R 21 2200 -0000	16012C34S
34	2716.99781(17)	P 19 2301F-0101F	16012C32S
35	2717.12652(11)	P 34 2200 -0000	16012C32S
36	2717.18127(41)	R 22 2200 -0000	16012C34S
37	2717.29135(46)	R 29 3200 -1000	16012C32S
38	2717.34823(17)	P 18 2301E-0101E	16012C32S
39	2717.41663(18)	P 18 2301F-0101F	16012C32S
40	2717.55945(11)	P 33 2200 -0000	16012C32S
40		R 23 2200 -0000	16012C34S
41	2717.67230(45)	R 30 3200 -1000	16012C32S
42	2717.77432(18)	P 17 2301E-0101E	16012C32S
43	2717.83457(17)	P 17 2301F-0101F	16012C32S
44	2717.93110(45)	R 24 2200 -0000	16012C34S
45	2717.99191(11)	P 32 2200 -0000	16012C32S
46	2718.05284(42)	R 31 3200 -1000	16012C32S
47	2718.19926(18)	P 16 2301E-0101E	16012C32S
48	2718.25202(17)	P 16 2301F-0101F	16012C32S
49	2718.30439(40)	R 25 2200 -0000	16012C34S
50	2718.42362(11)	P 31 2200 -0000	16012C32S
51	2718.62297(18)	P 15 2301E-0101E	16012C32S
52	2718.81136(51)	R 33 3200 -1000	16012C32S
53	2718.85450(11)	P 30 2200 -0000	16012C32S
54	2719.04626(19)	P 14 2301E-0101E	16012C32S
54		R 27 2200 -0000	16012C34S
55	2719.08484(19)	P 14 2301F-0101F	16012C32S
56	2719.19006(50)	R 34 3200 -1000	16012C32S

- "LINE NO.": serial number identifying the spectral line in the map on the opposite page.
- "WAVENUMBER": experimental value.
- Uncertainty in the last digit of the wavenumber is given in parentheses.

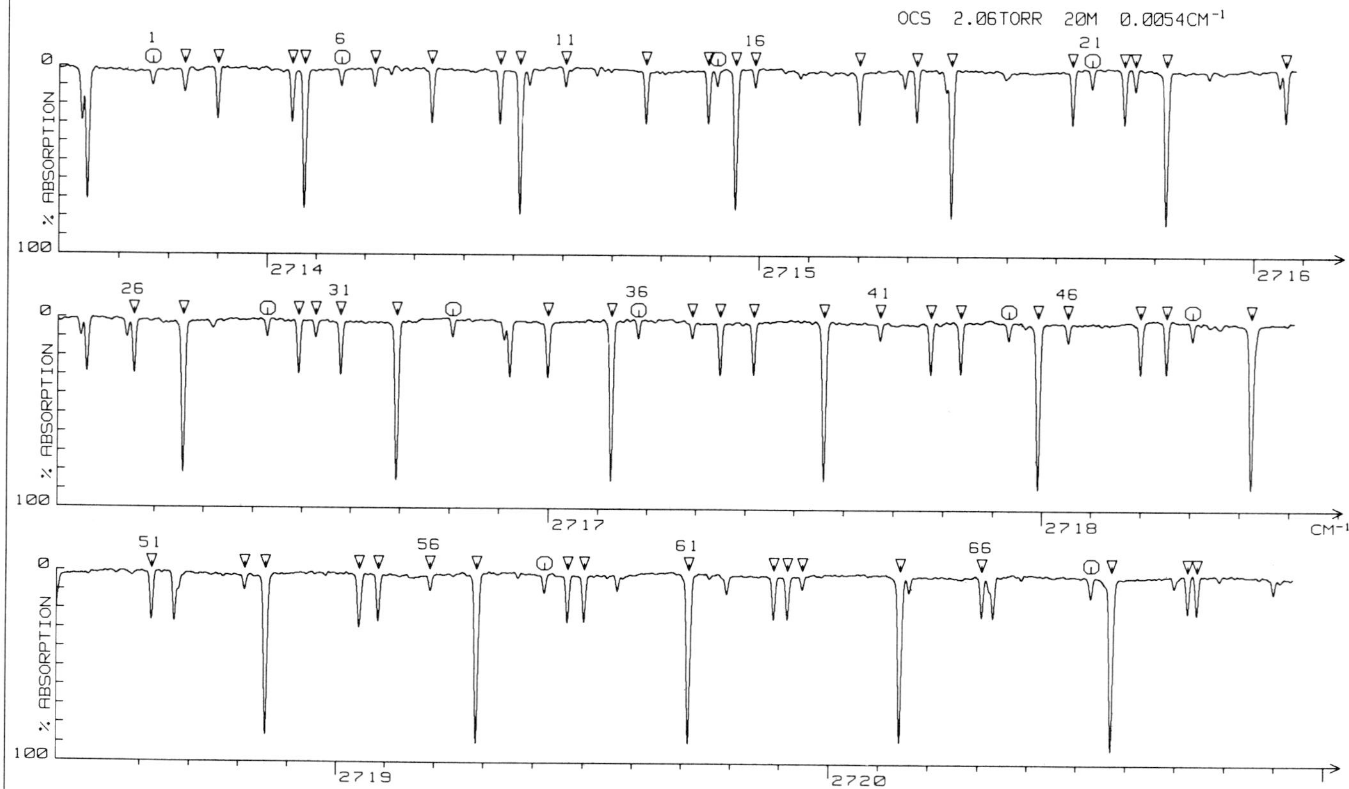

ISOTOPIC IDENTIFICATION – '▽' indicates $^{16}O^{12}C^{32}S$; 'Ⓘ': $^{16}O^{12}C^{34}S$; '◇': $^{16}O^{13}C^{32}S$; 'Y': $^{16}O^{12}C^{33}S$;
'✧' indicates $^{18}O^{12}C^{32}S$; '↓': $^{16}O^{13}C^{34}S$; '↑': Other OCS species; 'I': not OCS.

LINE NO.	WAVENUMBER UNC. CM^{-1}, VAC.	ROTATIONAL BAND ASSIGNMENTS	ISOTOPIC SPECIES
57	2719.28458(11)	P 29 2200 -0000	16O12C32S
58	2719.42031(41)	R 28 2200 -0000	16O12C34S
59	2719.46698(19)	P 13 2301E-0101E	16O12C32S
60	2719.50028(19)	P 13 2301F-0101F	16O12C32S
61	2719.71390(11)	P 28 2200 -0000	16O12C32S
62	2719.88731(20)	P 12 2301E-0101E	16O12C32S
63	2719.91502(21)	P 12 2301F-0101F	16O12C32S
63		R 48 2301E-0101E	16O12C34S
64	2719.94489(53)	R 36 3200 -1000	16O12C32S
65	2720.14247(11)	P 27 2200 -0000	16O12C32S
66	2720.30642(21)	P 11 2301E-0101E	16O12C32S
67	2720.52855(39)	R 31 2200 -0000	16O12C34S
67		R 11 2400 -0200	16O12C32S
68	2720.57022(11)	P 26 2200 -0000	16O12C32S
69	2720.72429(22)	P 10 2301E-0101E	16O12C32S
70	2720.74244(21)	P 10 2301F-0101F	16O12C32S

- "LINE NO.": serial number identifying the spectral line in the map on the opposite page.
- "WAVENUMBER": experimental value.
- Uncertainty in the last digit of the wavenumber is given in parentheses.

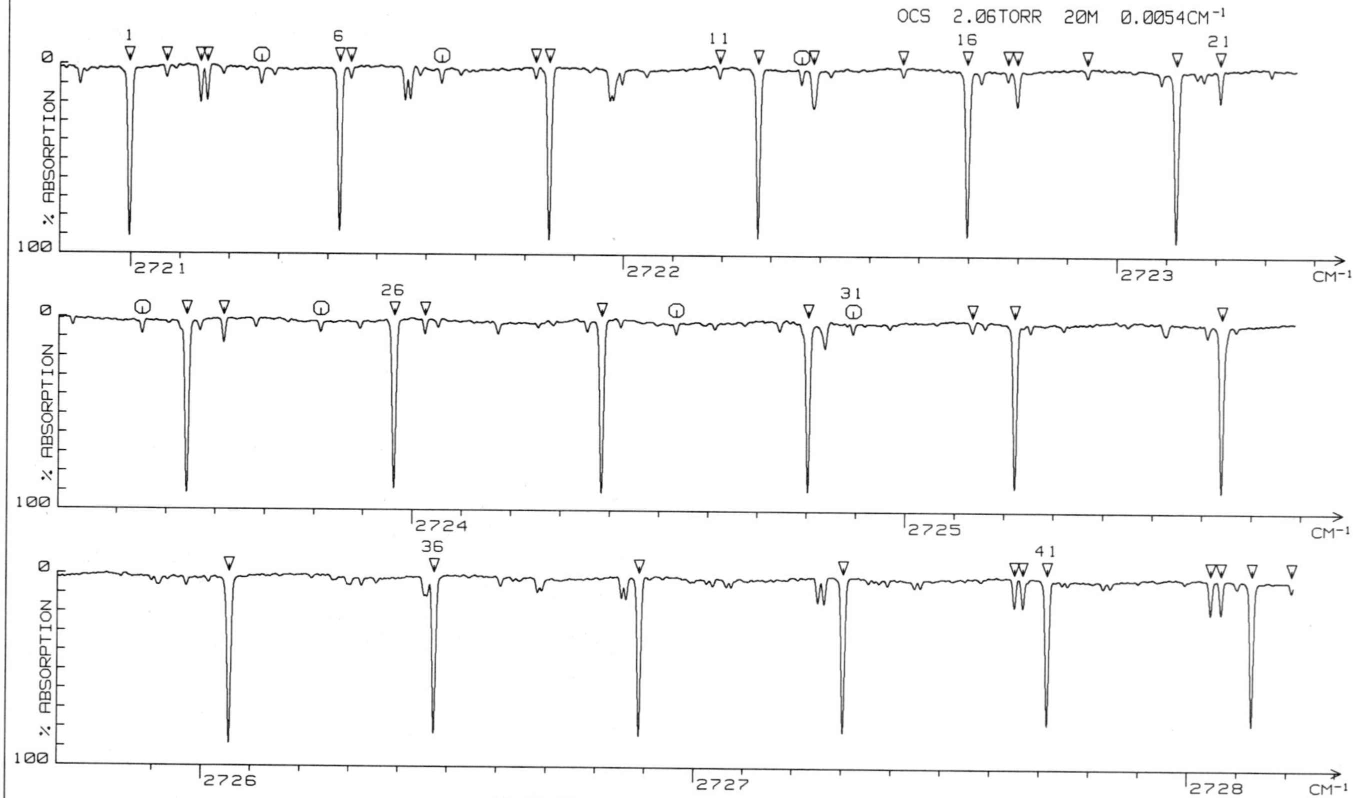

ISOTOPIC IDENTIFICATION – '▽' indicates $^{16}O^{12}C^{32}S$; 'Ⓘ': $^{16}O^{12}C^{34}S$; '◇': $^{16}O^{13}C^{32}S$; 'Y': $^{16}O^{12}C^{33}S$;
'✧' indicates $^{18}O^{12}C^{32}S$; '⍖': $^{16}O^{13}C^{34}S$; '⍏': Other OCS species; 'I': not OCS.

LINE NO.	WAVENUMBER CM^{-1}, VAC.	UNC.	ROTATIONAL BAND ASSIGNMENTS	ISOTOPIC SPECIES
1	2720.99723	(11)	P 25 2200 -0000	16012C32S
2	2721.07191	(67)	R 39 3200 -1000	16012C32S
2			R 29 2501E-0301E	16012C32S
3	2721.14112	(23)	P 9 2301E-0101E	16012C32S
4	2721.15511	(24)	P 9 2301F-0101F	16012C32S
5	2721.26428	(48)	R 33 2200 -0000	16012C34S
5			R 52 2301E-0101E	16012C34S
6	2721.42337	(11)	P 24 2200 -0000	16012C32S
7	2721.44668	(64)	R 40 3200 -1000	16012C32S
8	2721.63060	(50)	R 34 2200 -0000	16012C34S
9	2721.82083	(68)	R 41 3200 -1000	16012C32S
10	2721.84873	(11)	P 23 2200 -0000	16012C32S
11	2722.19395	(68)	R 42 3200 -1000	16012C32S
12	2722.27328	(11)	P 22 2200 -0000	16012C32S
13	2722.36097	(49)	R 36 2200 -0000	16012C34S
14	2722.38615	(33)	P 6 2301 -0101	16012C32S
14			R 9 2402 -0202	16012C32S
15	2722.56709	(82)	R 43 3200 -1000	16012C32S
16	2722.69704	(11)	P 21 2200 -0000	16012C32S
17	2722.77870	(72)	R 10 2402 -0202	16012C32S
18	2722.79763	(28)	P 5 2301 -0101	16012C32S
19	2722.93907	(84)	R 44 3200 -1000	16012C32S
20	2723.11996	(11)	P 20 2200 -0000	16012C32S
21	2723.20807	(25)	P 4 2301 -0101	16012C32S
22	2723.45089	(56)	R 39 2200 -0000	16012C34S
23	2723.54210	(11)	P 19 2200 -0000	16012C32S
24	2723.61711	(33)	P 3 2301 -0101	16012C32S
25	2723.81282	(72)	R 40 2200 -0000	16012C34S
26	2723.96329	(11)	P 18 2200 -0000	16012C32S
27	2724.02567	(51)	P 2 2301 -0101	16012C32S
28	2724.38373	(11)	P 17 2200 -0000	16012C32S
29	2724.53445	(71)	R 42 2200 -0000	16012C34S
30	2724.80335	(11)	P 16 2200 -0000	16012C32S
31	2724.89382	(68)	R 43 2200 -0000	16012C34S
32	2725.13583	(90)	R 16 2402 -0202	16012C32S
33	2725.22209	(11)	P 15 2200 -0000	16012C32S
34	2725.63998	(11)	P 14 2200 -0000	16012C32S
35	2726.05696	(11)	P 13 2200 -0000	16012C32S
36	2726.47322	(11)	P 12 2200 -0000	16012C32S
37	2726.88854	(11)	P 11 2200 -0000	16012C32S
38	2727.30301	(11)	P 10 2200 -0000	16012C32S
39	2727.64968	(27)	R 6 2301E-0101E	16012C32S
40	2727.66657	(27)	R 6 2301F-0101F	16012C32S
41	2727.71659	(11)	P 9 2200 -0000	16012C32S
42	2728.04631	(24)	R 7 2301E-0101E	16012C32S
43	2728.06752	(24)	R 7 2301F-0101F	16012C32S
44	2728.12935	(11)	P 8 2200 -0000	16012C32S
45	2728.20945	(71)	R 24 2402E-0202E	16012C32S
45			R 18 2010 -0200	16012C32S

- "LINE NO.": serial number identifying the spectral line in the map on the opposite page.
- "WAVENUMBER": experimental value.
- Uncertainty in the last digit of the wavenumber is given in parentheses.

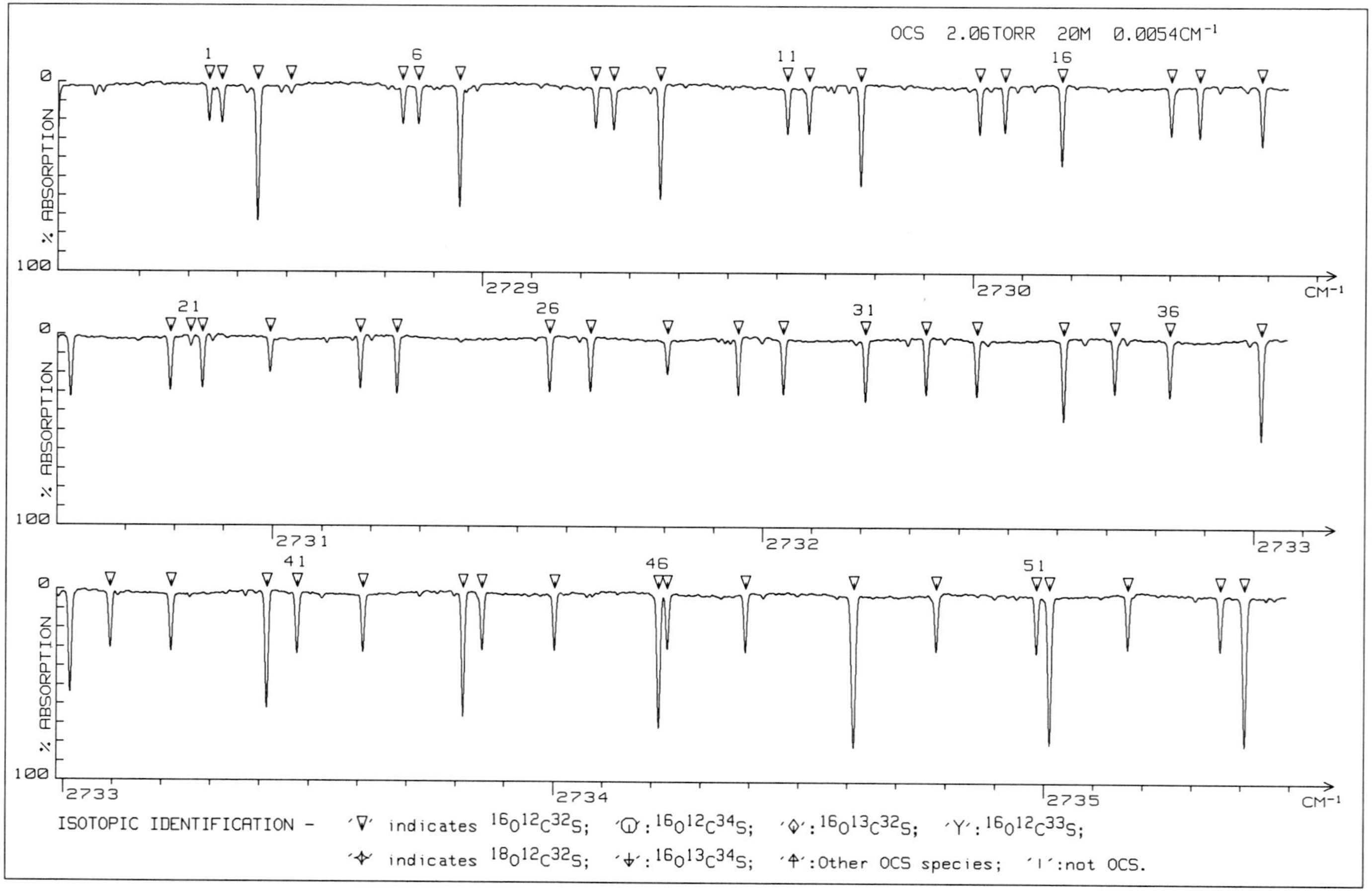
OCS 2.06TORR 20M 0.0054CM-1
1
6
11
16
21
26
31
36
41
46
51
0
% ABSORPTION
100
2729
2730
2731
2732
2733
2734
2735
CM-1
ISOTOPIC IDENTIFICATION - '∇' indicates 16O12C32S; 16O12C34S; 16O13C32S; 'Y': 16O12C33S;
indicates 18O12C32S; 16O13C34S; Other OCS species; 'I': not OCS.

LINE NO.	WAVENUMBER CM^{-1}, VAC.	UNC.	ROTATIONAL ASSIGNMENTS	BAND	ISOTOPIC SPECIES
1	2728.44198	(22)	R 8	2301E-0101E	16012C32S
2	2728.46808	(23)	R 8	2301F-0101F	16012C32S
3	2728.54126	(11)	P 7	2200 -0000	16012C32S
4	2728.60826	(105)	R 25	2402F-0202F	16012C32S
4			R 34	2400 -0200	16012C32S
5	2728.83625	(22)	R 9	2301E-0101E	16012C32S
6	2728.86792	(22)	R 9	2301F-0101F	16012C32S
7	2728.95225	(12)	P 6	2200 -0000	16012C32S
8	2729.22949	(21)	R 10	2301E-0101E	16012C32S
9	2729.26688	(21)	R 10	2301F-0101F	16012C32S
10	2729.36241	(12)	P 5	2200 -0000	16012C32S
11	2729.62146	(19)	R 11	2301E-0101E	16012C32S
12	2729.66532	(19)	R 11	2301F-0101F	16012C32S
13	2729.77155	(12)	P 4	2200 -0000	16012C32S
14	2730.01234	(19)	R 12	2301E-0101E	16012C32S
15	2730.06299	(19)	R 12	2301F-0101F	16012C32S
16	2730.17991	(14)	P 3	2200 -0000	16012C32S
17	2730.40191	(19)	R 13	2301E-0101E	16012C32S
18	2730.46013	(17)	R 13	2301F-0101F	16012C32S
18			R 30	2402E-0202E	16012C32S
19	2730.58737	(16)	P 2	2200 -0000	16012C32S
20	2730.79030	(18)	R 14	2301E-0101E	16012C32S
21	2730.83216	(75)	R 31	2402E-0202E	16012C32S
21			R 27	2010 -0200	16012C32S
22	2730.85630	(17)	R 14	2301F-0101F	16012C32S
23	2730.99399	(24)	P 1	2200 -0000	16012C32S
24	2731.17761	(17)	R 15	2301E-0101E	16012C32S
25	2731.25189	(16)	R 15	2301F-0101F	16012C32S
25			R 32	2402F-0202F	16012C32S
26	2731.56382	(17)	R 16	2301E-0101E	16012C32S
26			R 33	2402E-0202E	16012C32S
27	2731.64684	(17)	R 16	2301F-0101F	16012C32S
27			R 30	2010 -0200	16012C32S
28	2731.80454	(24)	R 0	2200 -0000	16012C32S
29	2731.94846	(16)	R 17	2301E-0101E	16012C32S
30	2732.04100	(17)	R 17	2301F-0101F	16012C32S
31	2732.20851	(16)	R 1	2200 -0000	16012C32S
32	2732.33214	(16)	R 18	2301E-0101E	16012C32S
33	2732.43454	(17)	R 18	2301F-0101F	16012C32S
34	2732.61157	(13)	R 2	2200 -0000	16012C32S
35	2732.71462	(17)	R 19	2301E-0101E	16012C32S
36	2732.82736	(16)	R 19	2301F-0101F	16012C32S
37	2733.01375	(12)	R 3	2200 -0000	16012C32S
38	2733.09588	(16)	R 20	2301E-0101E	16012C32S
39	2733.21950	(17)	R 20	2301F-0101F	16012C32S
40	2733.41498	(12)	R 4	2200 -0000	16012C32S
41	2733.47608	(16)	R 21	2301E-0101E	16012C32S
41			R 38	2402F-0202F	16012C32S
42	2733.61090	(16)	R 21	2301F-0101F	16012C32S
43	2733.81541	(12)	R 5	2200 -0000	16012C32S
44	2733.85483	(17)	R 22	2301E-0101E	16012C32S
45	2734.00158	(16)	R 22	2301F-0101F	16012C32S
46	2734.21486	(11)	R 6	2200 -0000	16012C32S
47	2734.23255	(17)	R 23	2301E-0101E	16012C32S
48	2734.39159	(16)	R 23	2301F-0101F	16012C32S
49	2734.61315	(17)	R 7	2200 -0000	16012C32S
49			R 24	2301E-0101E	16012C32S
50	2734.78093	(17)	R 24	2301F-0101F	16012C32S
51	2734.98438	(16)	R 25	2301E-0101E	16012C32S
52	2735.01129	(11)	R 8	2200 -0000	16012C32S
53	2735.16949	(17)	R 25	2301F-0101F	16012C32S
54	2735.35843	(17)	R 26	2301E-0101E	16012C32S
55	2735.40814	(11)	R 9	2200 -0000	16012C32S

- "LINE NO.": serial number identifying the spectral line in the map on the opposite page.
- "WAVENUMBER": experimental value.
- Uncertainty in the last digit of the wavenumber is given in parentheses.

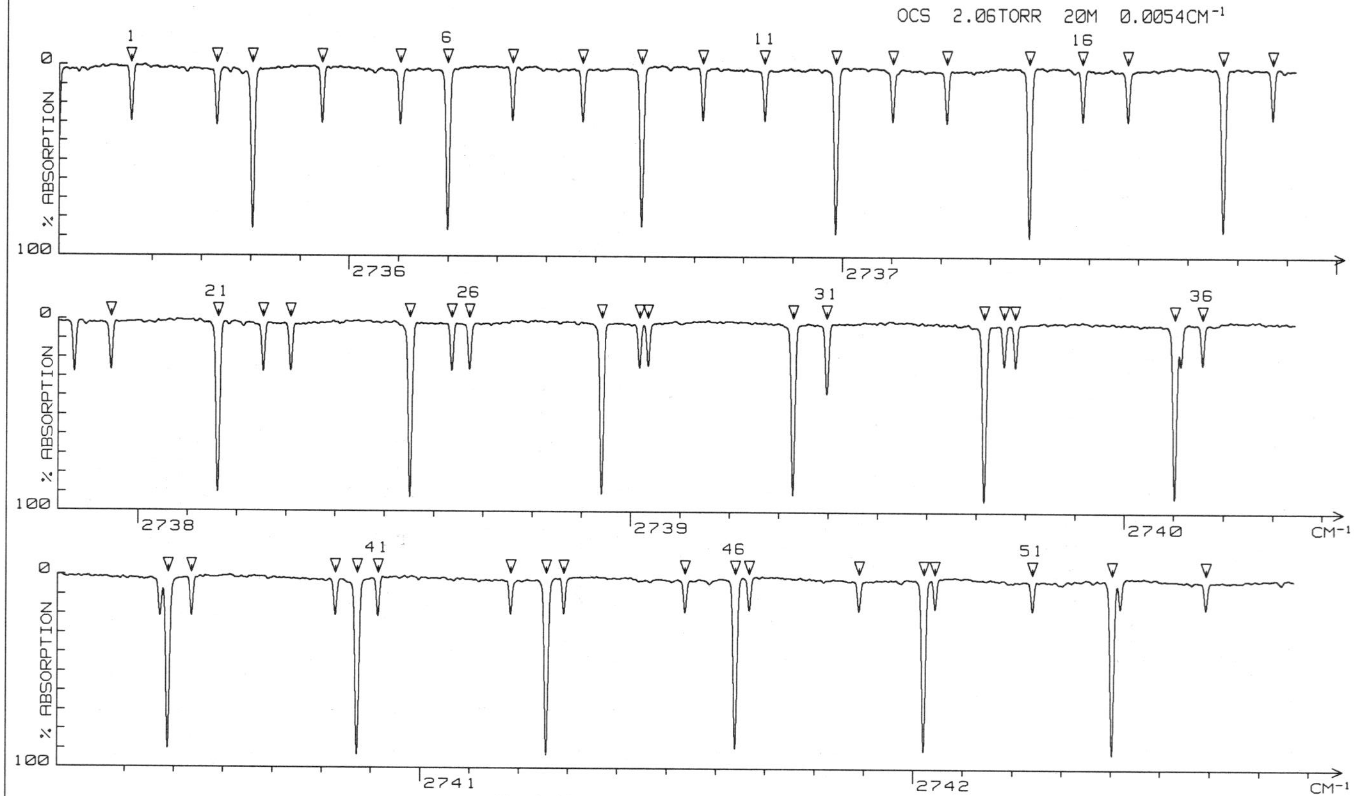

ISOTOPIC IDENTIFICATION – '∇' indicates $^{16}O^{12}C^{32}S$; '⦶': $^{16}O^{12}C^{34}S$; '◊': $^{16}O^{13}C^{32}S$; 'Y': $^{16}O^{12}C^{33}S$;
'✧' indicates $^{18}O^{12}C^{32}S$; '⫛': $^{16}O^{13}C^{34}S$; '⍏': Other OCS species; 'I': not OCS.

LINE NO.	WAVENUMBER CM^{-1}, VAC.	UNC.	ROTATIONAL BAND ASSIGNMENTS	ISOTOPIC SPECIES
1	2735.55742	(17)	R 26 2301F-0101F	16012C32S
2	2735.73146	(16)	R 27 2301E-0101E	16012C32S
3	2735.80408	(11)	R 10 2200 -0000	16012C32S
4	2735.94462	(16)	R 27 2301F-0101F	16012C32S
5	2736.10323	(17)	R 28 2301E-0101E	16012C32S
5			R 46 2402E-0202E	16012C32S
6	2736.19921	(11)	R 11 2200 -0000	16012C32S
7	2736.33105	(17)	R 28 2301F-0101F	16012C32S
8	2736.47372	(17)	R 29 2301E-0101E	16012C32S
9	2736.59343	(11)	R 12 2200 -0000	16012C32S
10	2736.71690	(17)	R 29 2301F-0101F	16012C32S
11	2736.84306	(18)	R 30 2301E-0101E	16012C32S
12	2736.98673	(11)	R 13 2200 -0000	16012C32S
13	2737.10202	(17)	R 30 2301F-0101F	16012C32S
14	2737.21119	(18)	R 31 2301E-0101E	16012C32S
15	2737.37919	(11)	R 14 2200 -0000	16012C32S
16	2737.48641	(17)	R 31 2301F-0101F	16012C32S
17	2737.57818	(18)	R 32 2301E-0101E	16012C32S
18	2737.77075	(11)	R 15 2200 -0000	16012C32S
19	2737.87010	(18)	R 32 2301F-0101F	16012C32S
20	2737.94396	(19)	R 33 2301E-0101E	16012C32S
21	2738.16140	(11)	R 16 2200 -0000	16012C32S
22	2738.25298	(19)	R 33 2301F-0101F	16012C32S
23	2738.30859	(19)	R 34 2301E-0101E	16012C32S
24	2738.55125	(11)	R 17 2200 -0000	16012C32S
25	2738.63530	(19)	R 34 2301F-0101F	16012C32S
26	2738.67198	(20)	R 35 2301E-0101E	16012C32S
27	2738.94023	(11)	R 18 2200 -0000	16012C32S
28	2739.01690	(20)	R 35 2301F-0101F	16012C32S
29	2739.03430	(20)	R 36 2301E-0101E	16012C32S
30	2739.32829	(11)	R 19 2200 -0000	16012C32S
31	2739.39655	(15)	R 36 2301F-0101F	16012C32S
31			R 37 2301E-0101E	16012C32S
32	2739.71553	(11)	R 20 2200 -0000	16012C32S
33	2739.75513	(21)	R 38 2301E-0101E	16012C32S
34	2739.77786	(21)	R 37 2301F-0101F	16012C32S
35	2740.10187	(13)	R 21 2200 -0000	16012C32S
36	2740.15730	(21)	R 38 2301F-0101F	16012C32S
37	2740.48743	(11)	R 22 2200 -0000	16012C32S
38	2740.53603	(22)	R 39 2301F-0101F	16012C32S
39	2740.82768	(24)	R 41 2301E-0101E	16012C32S
40	2740.87205	(11)	R 23 2200 -0000	16012C32S
41	2740.91398	(22)	R 40 2301F-0101F	16012C32S
42	2741.18296	(25)	R 42 2301E-0101E	16012C32S
43	2741.25588	(11)	R 24 2200 -0000	16012C32S
44	2741.29133	(23)	R 41 2301F-0101F	16012C32S
45	2741.53683	(26)	R 43 2301E-0101E	16012C32S
46	2741.63883	(11)	R 25 2200 -0000	16012C32S
47	2741.66792	(25)	R 42 2301F-0101F	16012C32S
48	2741.88965	(27)	R 44 2301E-0101E	16012C32S
49	2742.02091	(11)	R 26 2200 -0000	16012C32S
50	2742.04373	(26)	R 43 2301F-0101F	16012C32S
51	2742.24135	(27)	R 45 2301E-0101E	16012C32S
52	2742.40221	(11)	R 27 2200 -0000	16012C32S
53	2742.59192	(30)	R 46 2301E-0101E	16012C32S

- "LINE NO.": serial number identifying the spectral line in the map on the opposite page.
- "WAVENUMBER": experimental value.
- Uncertainty in the last digit of the wavenumber is given in parentheses.

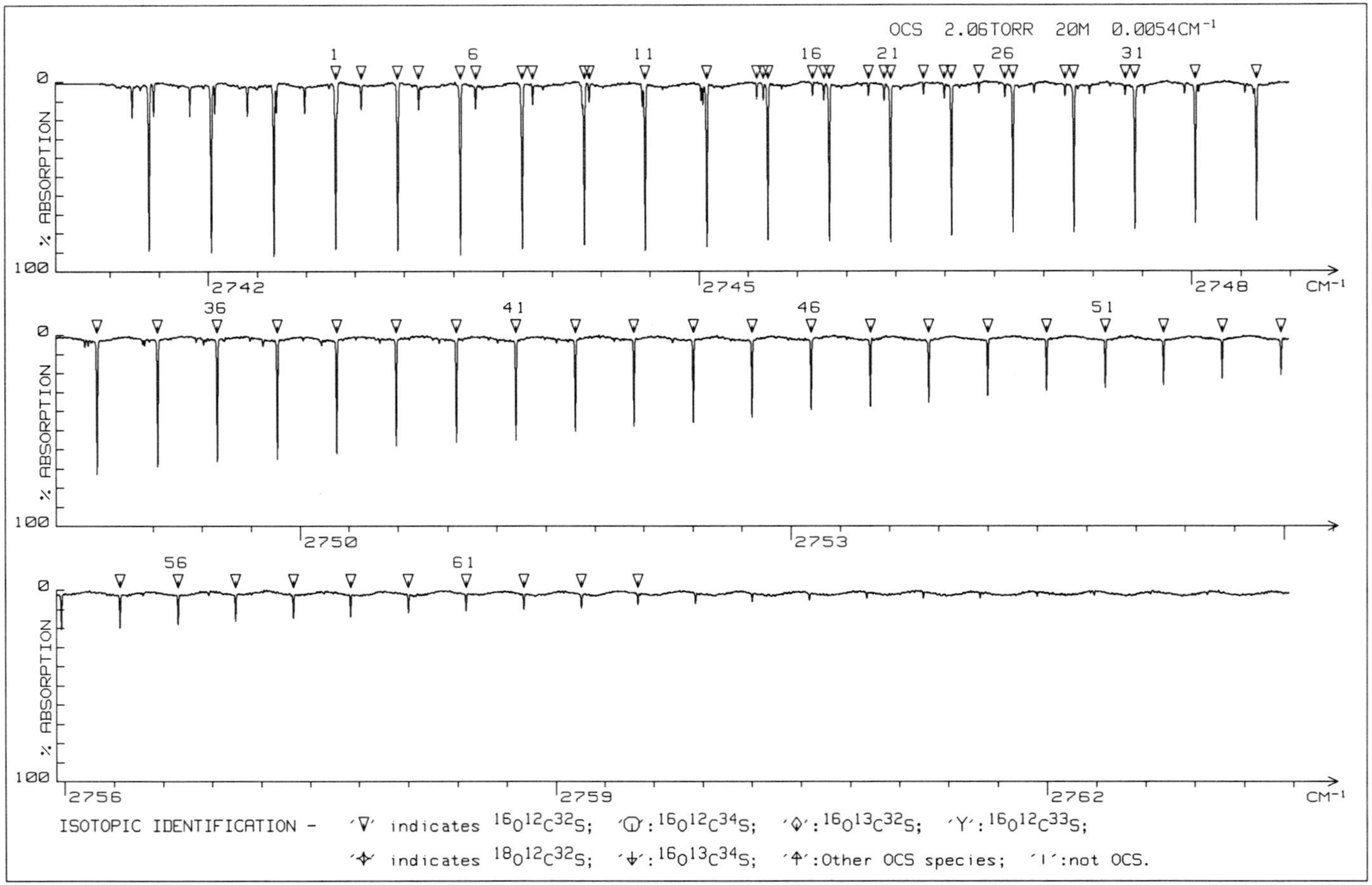
OCS 2.06TORR 20M 0.0054CM⁻¹
% ABSORPTION
0
100
2742
2745
2748
2750
2753
2756
2759
2762
CM⁻¹
1
6
11
16
21
26
31
36
41
46
51
56
61
ISOTOPIC IDENTIFICATION - '∇' indicates 16O12C32S; 16O12C34S; 16O13C32S; 'Y': 16O12C33S;
indicates 18O12C32S; 16O13C34S; Other OCS species; 'I': not OCS.

LINE NO.	WAVENUMBER CM^{-1}, VAC.	UNC.	ROTATIONAL BAND ASSIGNMENTS	ISOTOPIC SPECIES
1	2742.78264	(13)	R 28 2200 -0000	16012C32S
2	2742.94091	(34)	R 47 2301E-0101E	16012C32S
3	2743.16241	(11)	R 29 2200 -0000	16012C32S
3			R 46 2301F-0101F	16012C32S
4	2743.28925	(34)	R 48 2301E-0101E	16012C32S
5	2743.54100	(11)	R 30 2200 -0000	16012C32S
5			R 47 2301F-0101F	16012C32S
6	2743.63641	(34)	R 49 2301E-0101E	16012C32S
7	2743.91896	(13)	R 31 2200 -0000	16012C32S
8	2743.98242	(40)	R 50 2301E-0101E	16012C32S
9	2744.29618	(11)	R 32 2200 -0000	16012C32S
10	2744.32711	(44)	R 51 2301E-0101E	16012C32S
11	2744.67248	(11)	R 33 2200 -0000	16012C32S
12	2745.04814	(11)	R 34 2200 -0000	16012C32S
13	2745.35448	(48)	R 54 2301E-0101E	16012C32S
14	2745.39484	(46)	R 52 2301F-0101F	16012C32S
15	2745.42292	(11)	R 35 2200 -0000	16012C32S
16	2745.69476	(56)	R 55 2301E-0101E	16012C32S
17	2745.76328	(48)	R 53 2301F-0101F	16012C32S
18	2745.79690	(11)	R 36 2200 -0000	16012C32S
19	2746.03394	(61)	R 56 2301E-0101E	16012C32S
20	2746.13135	(49)	R 54 2301F-0101F	16012C32S
21	2746.17011	(11)	R 37 2200 -0000	16012C32S
22	2746.37208	(66)	R 57 2301E-0101E	16012C32S
23	2746.49876	(56)	R 55 2301F-0101F	16012C32S
24	2746.54256	(11)	R 38 2200 -0000	16012C32S
25	2746.70889	(69)	R 58 2301E-0101E	16012C32S
26	2746.86525	(55)	R 56 2301F-0101F	16012C32S
27	2746.91422	(11)	R 39 2200 -0000	16012C32S
28	2747.23107	(64)	R 57 2301F-0101F	16012C32S
29	2747.28514	(11)	R 40 2200 -0000	16012C32S
30	2747.59619	(73)	R 58 2301F-0101F	16012C32S

LINE NO.	WAVENUMBER CM^{-1}, VAC.	UNC.	ROTATIONAL BAND ASSIGNMENTS	ISOTOPIC SPECIES
31	2747.65532	(11)	R 41 2200 -0000	16012C32S
32	2748.02472	(11)	R 42 2200 -0000	16012C32S
33	2748.39341	(11)	R 43 2200 -0000	16012C32S
34	2748.76138	(12)	R 44 2200 -0000	16012C32S
35	2749.12863	(12)	R 45 2200 -0000	16012C32S
36	2749.49516	(12)	R 46 2200 -0000	16012C32S
37	2749.86098	(12)	R 47 2200 -0000	16012C32S
38	2750.22610	(12)	R 48 2200 -0000	16012C32S
39	2750.59055	(12)	R 49 2200 -0000	16012C32S
40	2750.95432	(12)	R 50 2200 -0000	16012C32S
41	2751.31745	(12)	R 51 2200 -0000	16012C32S
42	2751.67989	(13)	R 52 2200 -0000	16012C32S
43	2752.04165	(13)	R 53 2200 -0000	16012C32S
44	2752.40283	(13)	R 54 2200 -0000	16012C32S
45	2752.76338	(14)	R 55 2200 -0000	16012C32S
46	2753.12327	(14)	R 56 2200 -0000	16012C32S
47	2753.48247	(15)	R 57 2200 -0000	16012C32S
48	2753.84120	(15)	R 58 2200 -0000	16012C32S
49	2754.19930	(16)	R 59 2200 -0000	16012C32S
50	2754.55675	(17)	R 60 2200 -0000	16012C32S
51	2754.91368	(18)	R 61 2200 -0000	16012C32S
52	2755.26997	(19)	R 62 2200 -0000	16012C32S
53	2755.62578	(21)	R 63 2200 -0000	16012C32S
54	2755.98098	(23)	R 64 2200 -0000	16012C32S
55	2756.33566	(25)	R 65 2200 -0000	16012C32S
56	2756.68973	(27)	R 66 2200 -0000	16012C32S
57	2757.04327	(29)	R 67 2200 -0000	16012C32S
58	2757.39636	(33)	R 68 2200 -0000	16012C32S
59	2757.74883	(35)	R 69 2200 -0000	16012C32S
60	2758.10087	(42)	R 70 2200 -0000	16012C32S
61	2758.45228	(46)	R 71 2200 -0000	16012C32S
62	2758.80311	(52)	R 72 2200 -0000	26012C32S
63	2759.15378	(55)	R 73 2200 -0000	16012C32S
64	2759.50349	(63)	R 74 2200 -0000	16012C32S

- "LINE NO.": serial number identifying the spectral line in the map on the opposite page.
- "WAVENUMBER": experimental value.
- Uncertainty in the last digit of the wavenumber is given in parentheses.

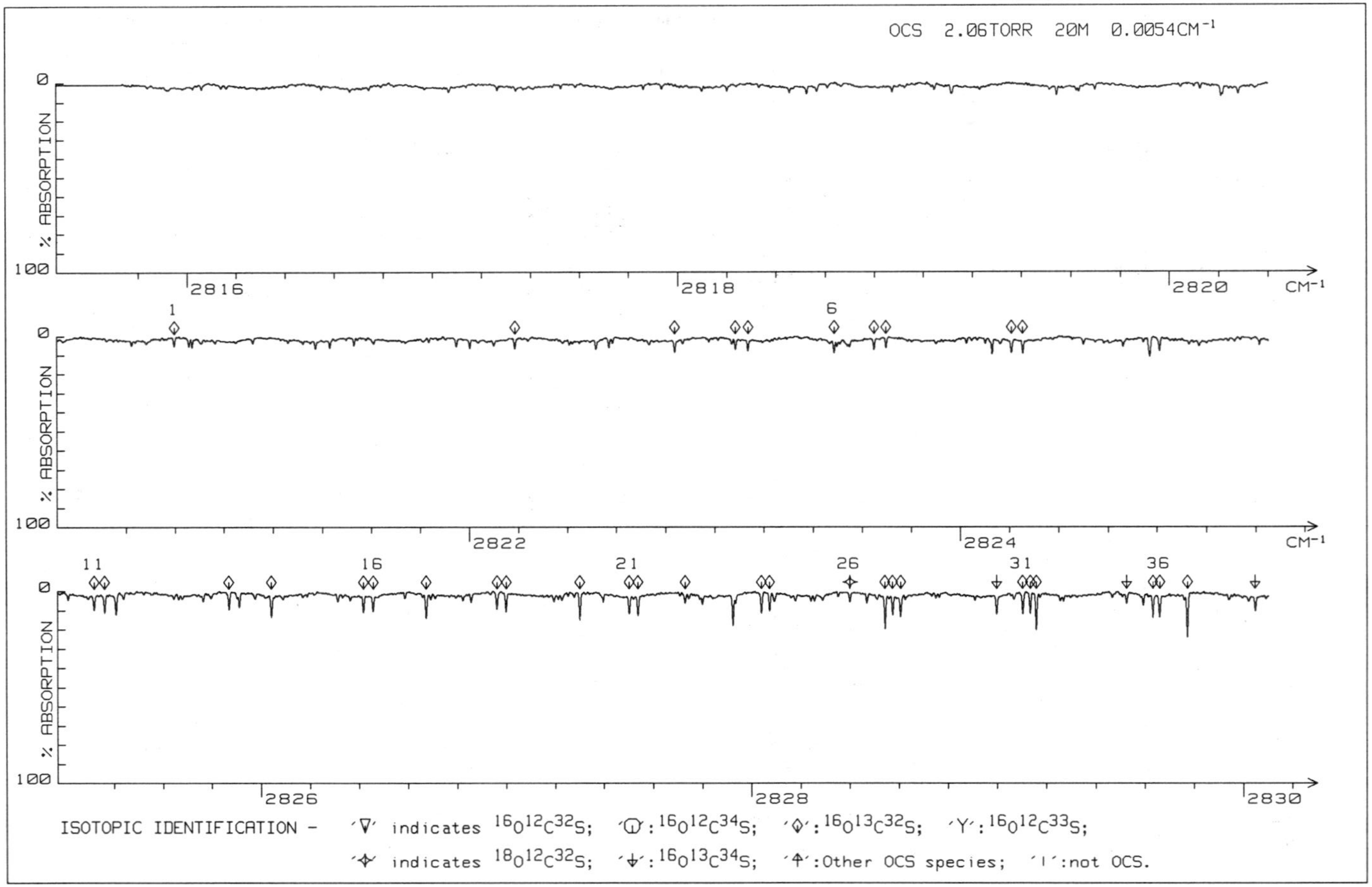

OCS 2.06TORR 20M 0.0054CM⁻¹
0
% ABSORPTION
100
2816
2818
2820
CM⁻¹
1
6
2822
2824
11
16
21
26
31
36
2826
2828
2830
ISOTOPIC IDENTIFICATION – ′▽′ indicates 16O12C32S; ′Ⓤ′: 16O12C34S; ′◇′: 16O13C32S; ′Y′: 16O12C33S;
′✧′ indicates 18O12C32S; ′↓′: 16O13C34S; ′↑′: Other OCS species; ′I′: not OCS.

LINE NO.	WAVENUMBER CM^{-1}, VAC.	UNC.	ROTATIONAL BAND ASSIGNMENTS	ISOTOPIC SPECIES
1	2820.79695	(86)	P 54 1111E-0101E	16O13C32S
2	2822.18575	(73)	P 74 1010 -0000	16O13C32S
3	2822.83780	(66)	P 73 1010 -0000	16O13C32S
4	2823.08459	(74)	P 50 1111E-0101E	16O13C32S
5	2823.13668	(70)	P 50 1111F-0101F	16O13C32S
6	2823.48605	(65)	P 72 1010 -0000	16O13C32S
7	2823.64846	(72)	P 49 1111E-0101E	16O13C32S
8	2823.69819	(70)	P 49 1111F-0101F	16O13C32S
9	2824.20899	(66)	P 48 1111E-0101E	16O13C32S
10	2824.25655	(61)	P 48 1111F-0101F	16O13C32S
11	2825.32034	(56)	P 46 1111E-0101E	16O13C32S
12	2825.36322	(50)	P 46 1111F-0101F	16O13C32S
12			P 71 1010 -0000	18O12C32S
13	2825.87068	(49)	P 45 1111E-0101E	16O13C32S
13			P 20 1210 -0200	16O13C32S
14	2826.04333	(39)	P 68 1010 -0000	16O13C32S
15	2826.41837	(48)	P 44 1111E-0101E	16O13C32S
16	2826.45750	(55)	P 44 1111F-0101F	16O13C32S
17	2826.67395	(35)	P 67 1010 -0000	16O13C32S
18	2826.96237	(49)	P 43 1111E-0101E	16O13C32S
19	2826.99964	(45)	P 43 1111F-0101F	16O13C32S
19			P 48 1010 -0000	18O12C34S
20	2827.30085	(30)	P 66 1010 -0000	16O13C32S
21	2827.50331	(48)	P 42 1111E-0101E	16O13C32S
22	2827.53882	(44)	P 42 1111F-0101F	16O13C32S
23	2827.72936	(87)	P 15 1212 -0202	16O13C32S
23			P 42 1111E-0101E	18O12C32S
24	2828.04091	(40)	P 41 1111E-0101E	16O13C32S
24			P 46 1010 -0000	18O12C34S
25	2828.07445	(44)	P 41 1111F-0101F	16O13C32S
26	2828.40080	(85)	P 66 1010 -0000	18O12C32S
27	2828.54404	(27)	P 64 1010 -0000	16O13C32S
28	2828.57514	(43)	P 40 1111E-0101E	16O13C32S
29	2828.60693	(43)	P 40 1111F-0101F	16O13C32S
29			P 14 1210 -0200	16O13C32S
30	2828.99872	(44)	P 43 1010 -0000	16O13C34S
30			P 65 1010 -0000	18O12C32S
31	2829.10613	(39)	P 39 1111E-0101E	16O13C32S
32	2829.13599	(38)	P 39 1111F-0101F	16O13C32S
33	2829.16044	(24)	P 63 1010 -0000	16O13C32S
34	2829.52519	(80)	P 42 1010 -0000	16O13C34S
35	2829.63389	(38)	P 38 1111E-0101E	16O13C32S
36	2829.66198	(41)	P 38 1111F-0101F	16O13C32S
37	2829.77319	(21)	P 62 1010 -0000	16O13C32S
37			P 38 1111F-0101F	18O12C32S
38	2830.04843	(58)	P 41 1010 -0000	16O13C34S
38			P 13 1111 -0101	16O13C34S

- "LINE NO.": serial number identifying the spectral line in the map on the opposite page.
- "WAVENUMBER": experimental value.
- Uncertainty in the last digit of the wavenumber is given in parentheses.

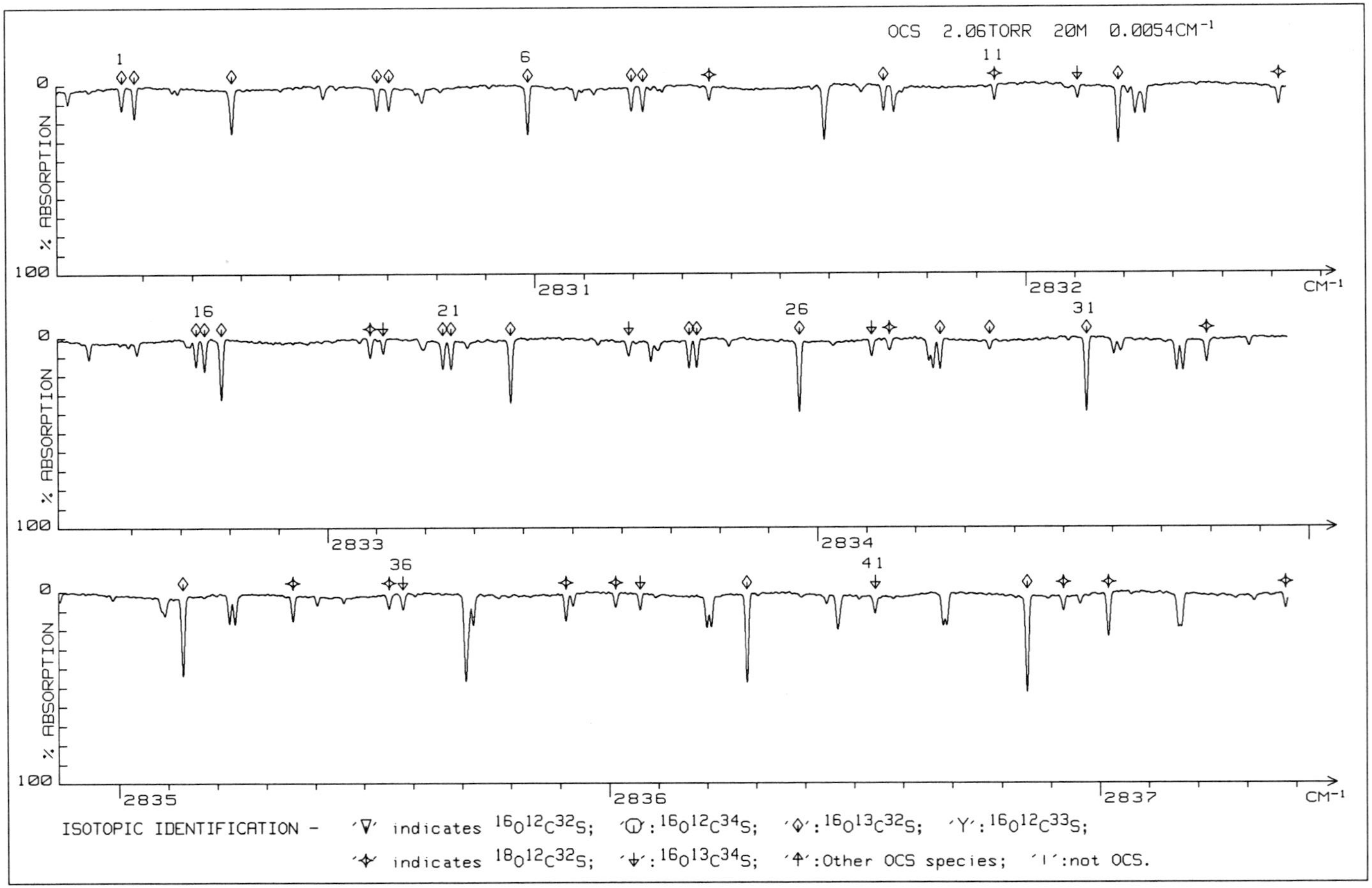

OCS 2.06TORR 20M 0.0054CM⁻¹
% ABSORPTION
0
100
1
6
11
16
21
26
31
36
41
2831
2832
2833
2834
2835
2836
2837
CM⁻¹
ISOTOPIC IDENTIFICATION – 'V' indicates $^{16}O^{12}C^{32}S$; 'Ѡ': $^{16}O^{12}C^{34}S$; '◊': $^{16}O^{13}C^{32}S$; 'Y': $^{16}O^{12}C^{33}S$;
'✧' indicates $^{18}O^{12}C^{32}S$; '↓': $^{16}O^{13}C^{34}S$; '↑': Other OCS species; 'I': not OCS.

LINE NO.	WAVENUMBER CM^{-1}, VAC.	UNC.	ROTATIONAL BAND ASSIGNMENTS	ISOTOPIC SPECIES
1	2830.15859	(36)	P 37 1111E-0101E	16013C32S
2	2830.18466	(26)	P 37 1111F-0101F	16013C32S
2			P 63 1010 -0000	18012C32S
3	2830.38221	(21)	P 61 1010 -0000	16013C32S
3			P 9 1212 -0202	16013C32S
4	2830.67971	(36)	P 36 1111E-0101E	16013C32S
5	2830.70428	(38)	P 36 1111F-0101F	16013C32S
6	2830.98745	(19)	P 60 1010 -0000	16013C32S
7	2831.19772	(36)	P 35 1111E-0101E	16013C32S
8	2831.22082	(34)	P 35 1111F-0101F	16013C32S
9	2831.35648	(58)	P 61 1010 -0000	18012C32S
10	2831.71240	(33)	P 34 1111E-0101E	16013C32S
11	2831.93774	(50)	P 60 1010 -0000	18012C32S
12	2832.10707	(65)	P 37 1010 -0000	16013C34S
13	2832.18964	(18)	P 58 1010 -0000	16013C32S
14	2832.51532	(46)	P 59 1010 -0000	18012C32S
14			P 5 1210 -0200	16013C32S
15	2832.73223	(31)	P 32 1111E-0101E	16013C32S
16	2832.75026	(27)	P 32 1111F-0101F	16013C32S
16			P 42 2010 -1000	16013C32S
17	2832.78460	(16)	P 57 1010 -0000	16013C32S
18	2833.08976	(44)	P 58 1010 -0000	18012C32S
19	2833.11618	(52)	P 35 1010 -0000	16013C34S
20	2833.23719	(31)	P 31 1111E-0101E	16013C32S
21	2833.25404	(30)	P 31 1111F-0101F	16013C32S
22	2833.37615	(16)	P 56 1010 -0000	16013C32S
23	2833.61547	(62)	P 34 1010 -0000	16013C34S
24	2833.73891	(30)	P 30 1111E-0101E	16013C32S
25	2833.75435	(31)	P 30 1111F-0101F	16013C32S
26	2833.96425	(15)	P 55 1010 -0000	16013C32S
27	2834.11189	(53)	P 33 1010 -0000	16013C34S
27			Q 4 1212 -0202	16013C32S
28	2834.14849	(92)	P 29 1111 -0101	18012C32S
29	2834.25134	(30)	P 29 1111F-0101F	16013C32S
30	2834.35122	(90)	P 39 2010 -1000	16013C32S
31	2834.54874	(15)	P 54 1010 -0000	16013C32S
32	2834.79305	(37)	P 55 1010 -0000	18012C32S
33	2835.12974	(14)	P 53 1010 -0000	16013C32S
34	2835.35400	(33)	P 54 1010 -0000	18012C32S
35	2835.55156	(63)	P 26 1111 -0101	18012C32S
36	2835.58004	(55)	P 30 1010 -0000	16013C34S
37	2835.91184	(32)	P 53 1010 -0000	18012C32S
38	2836.01279	(58)	P 25 1111 -0101	18012C32S
39	2836.06291	(48)	P 29 1010 -0000	16013C34S
40	2836.28120	(14)	P 51 1010 -0000	16013C32S
41	2836.54244	(49)	P 28 1010 -0000	16013C34S
42	2836.85176	(13)	P 50 1010 -0000	16013C32S
42			P 28 1010 -0000	18012C34S
43	2836.92560	(57)	P 23 1111 -0101	18012C32S
44	2837.01768	(22)	P 51 1010 -0000	18012C32S
44			P 27 1010 -0000	16013C34S
45	2837.37732	(58)	P 22 1111 -0101	18012C32S

- "LINE NO.": serial number identifying the spectral line in the map on the opposite page.
- "WAVENUMBER": experimental value.
- Uncertainty in the last digit of the wavenumber is given in parentheses.

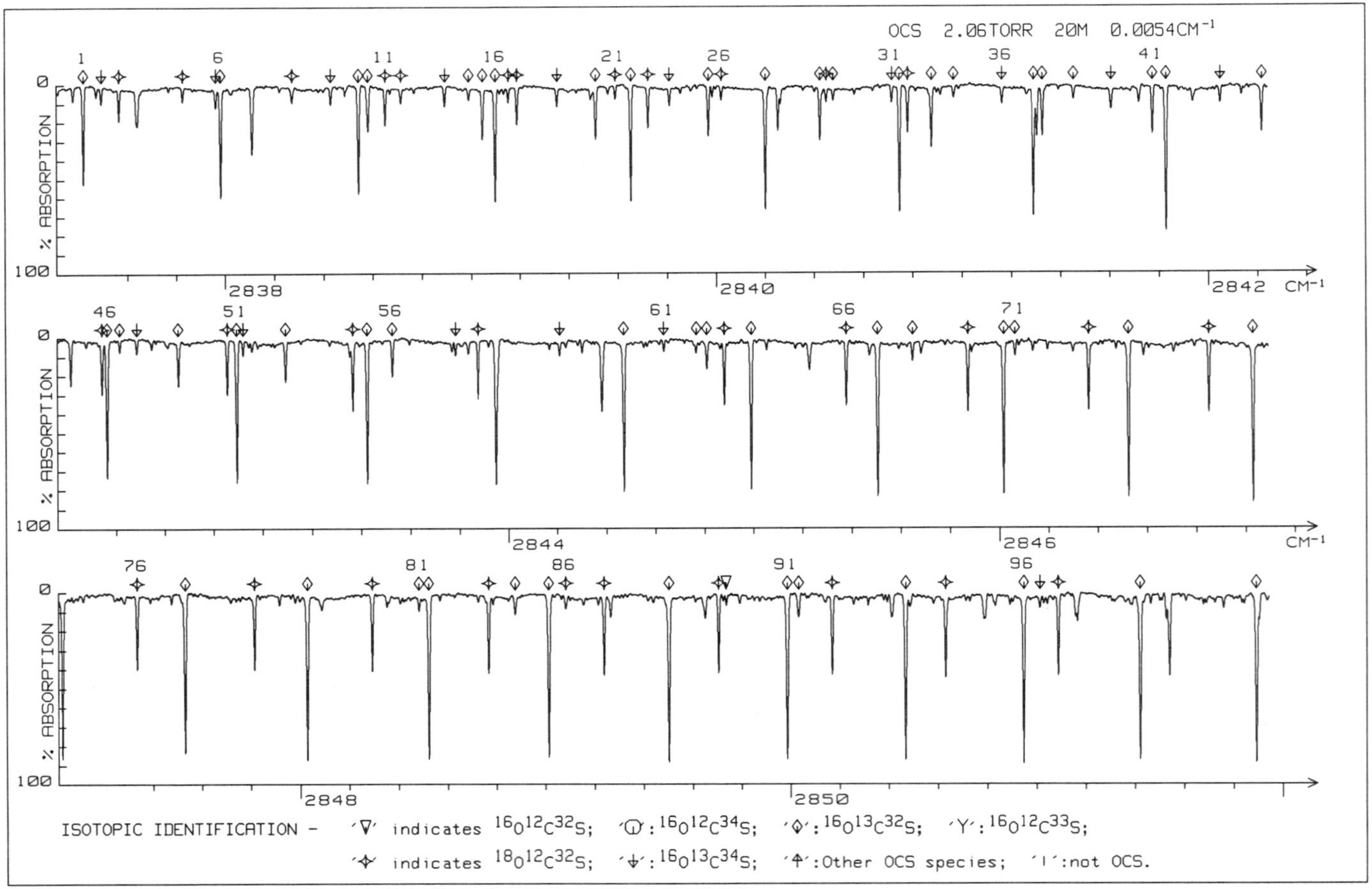
OCS 2.06TORR 20M 0.0054CM⁻¹
% ABSORPTION
0
100
2838
2840
2842
2844
2846
2848
2850
CM⁻¹
ISOTOPIC IDENTIFICATION – '▽' indicates 16O12C32S; 'Ⓘ': 16O12C34S; '◇': 16O13C32S; 'Y': 16O12C33S;
'✧' indicates 18O12C32S; '↓': 16O13C34S; '↑': Other OCS species; 'I': not OCS.

LINE NO.	WAVENUMBER CM^{-1}, VAC.	UNC.	ROTATIONAL BAND ASSIGNMENTS	ISOTOPIC SPECIES
1	2837.41867(	13)	P 49 1010 -0000	16013C32S
2	2837.49190(	48)	P 26 1010 -0000	16013C34S
3	2837.56481(	27)	P 50 1010 -0000	18012C32S
4	2837.82566(	55)	P 21 1111 -0101	18012C32S
5	2837.96168(	46)	P 25 1010 -0000	16013C34S
6	2837.98202(	13)	P 48 1010 -0000	16013C32S
6			P 32 2010 -1000	16013C32S
7	2838.27117(	60)	P 20 1111 -0101	18012C32S
8	2838.42846(	47)	P 24 1010 -0000	16013C34S
9	2838.54204(	13)	P 47 1010 -0000	16013C32S
10	2838.58005(	26)	P 20 1111 -0101	16013C32S
11	2838.64996(	24)	P 48 1010 -0000	18012C32S
12	2838.71320(	57)	P 19 1111 -0101	18012C32S
13	2838.89159(	42)	P 23 1010 -0000	16013C34S
14	2838.98875(	77)	P 30 2010 -1000	16013C32S
15	2839.04579(	22)	P 19 1111 -0101	16013C32S
16	2839.09851(	13)	P 46 1010 -0000	16013C32S
17	2839.15194(	56)	P 18 1111 -0101	18012C32S
18	2839.18771(	24)	P 47 1010 -0000	18012C32S
19	2839.35204(	48)	P 22 1010 -0000	16013C34S
20	2839.50829(	20)	P 18 1111 -0101	16013C32S
21	2839.58763(	60)	P 17 1111 -0101	18012C32S
22	2839.65144(	12)	P 45 1010 -0000	16013C32S
23	2839.72191(	23)	P 46 1010 -0000	18012C32S
24	2839.80910(	48)	P 21 1010 -0000	16013C34S
25	2839.96766(	20)	P 17 1111 -0101	16013C32S
26	2840.02050(	65)	P 16 1111 -0101	18012C32S
27	2840.20097(	12)	P 44 1010 -0000	16013C32S
28	2840.42373(	19)	P 16 1111 -0101	16013C32S
29	2840.44931(	67)	P 15 1111 -0101	18012C32S
30	2840.47732(	74)	P 27 2010 -1000	16013C32S
31	2840.71416(	50)	P 19 1010 -0000	16013C34S
32	2840.74691(	12)	P 43 1010 -0000	16013C32S
33	2840.78046(	21)	P 44 1010 -0000	18012C32S
34	2840.87657(	17)	P 15 1111F-0101F	16013C32S
34			P 15 1111E-0101E	16013C32S
34			P 14 1111E-0101E	18012C32S
34			P 14 1111F-0101F	18012C32S
35	2840.96658(	85)	P 26 2010 -1000	16013C32S
36	2841.16192(	54)	P 18 1010 -0000	16013C34S
37	2841.28944(	20)	P 42 1010 -0000	16013C32S
38	2841.32654(	20)	P 14 1111 -0101	16013C32S
39	2841.45301(	69)	P 25 2010 -1000	16013C32S
40	2841.60590(	48)	P 17 1010 -0000	16013C34S
40			R 19 1212 -0202	16013C32S
41	2841.77301(	20)	P 13 1111 -0101	16013C32S
42	2841.82795(	12)	P 41 1010 -0000	16013C32S
42			P 42 1010 -0000	18012C32S
43	2842.04805(	55)	P 16 1010 -0000	16013C34S
44	2842.21635(	21)	P 12 1111 -0101	16013C32S
45	2842.34317(	19)	P 41 1010 -0000	18012C32S
46	2842.36409(	12)	P 40 1010 -0000	16013C32S
47	2842.41628(	66)	P 23 2010 -1000	16013C32S
48	2842.48661(	57)	P 15 1010 -0000	16013C34S
49	2842.65659(	22)	P 11 1111 -0101	16013C32S
50	2842.85727(	18)	P 40 1010 -0000	18012C32S
51	2842.89603(	12)	P 39 1010 -0000	16013C32S
51			P 22 2010 -1000	16013C32S
52	2842.92230(	52)	P 14 1010 -0000	16013C34S
53	2843.09352(	24)	P 10 1111 -0101	16013C32S

- "LINE NO.": serial number identifying the spectral line in the map on the opposite page.
- "WAVENUMBER": experimental value.
- Uncertainty in the last digit of the wavenumber is given in parentheses.

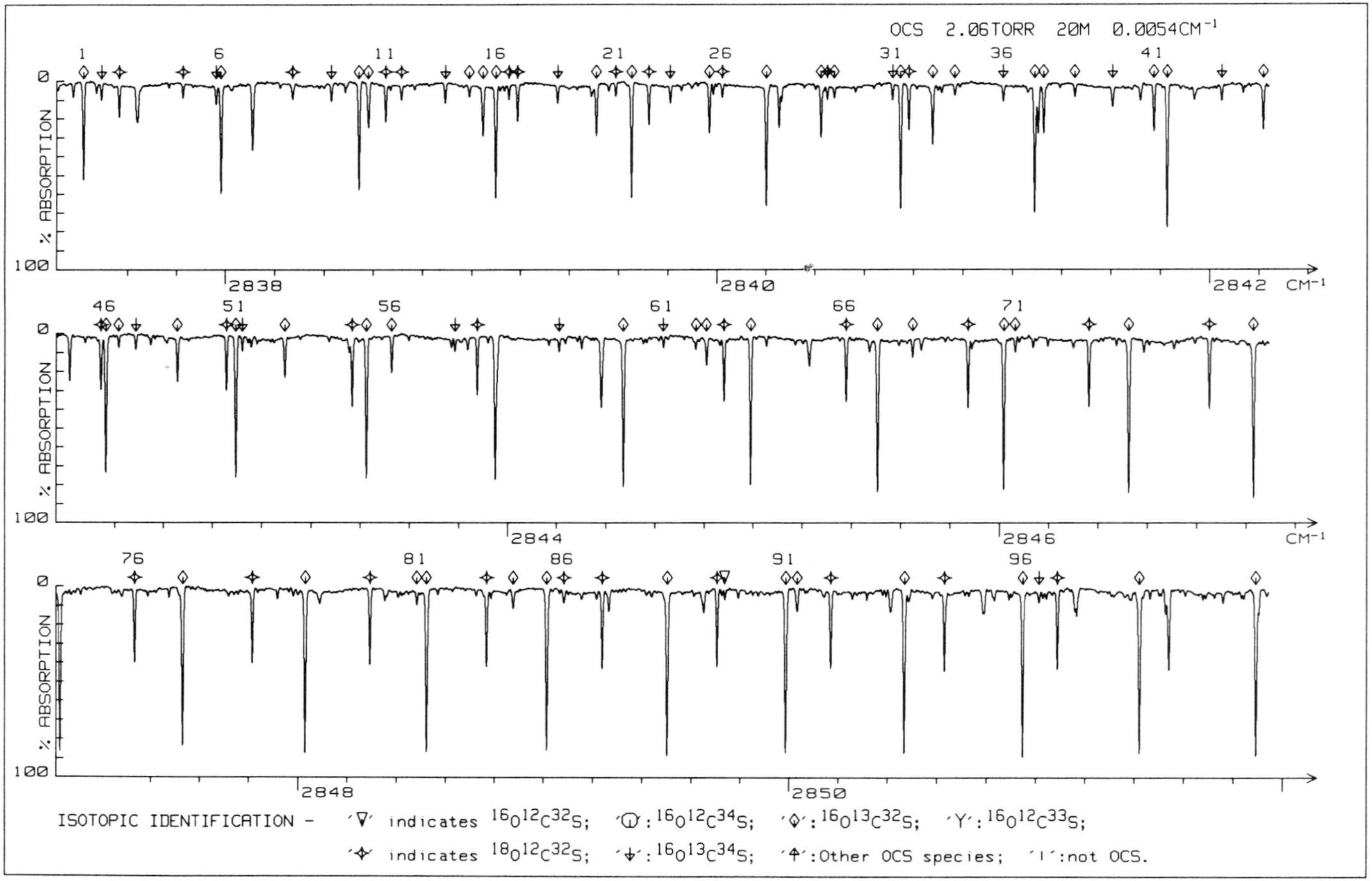
OCS 2.06TORR 20M 0.0054CM⁻¹
% ABSORPTION
0
100
2838
2840
2842
CM⁻¹
2844
2846
2848
2850
ISOTOPIC IDENTIFICATION – 'Ꮩ' indicates 16O12C32S; 'Ⓘ': 16O12C34S; '◊': 16O13C32S; 'Y': 16O12C33S;
'✧' indicates 18O12C32S; '↓': 16O13C34S; '↑': Other OCS species; 'I': not OCS.

LINE NO.	WAVENUMBER CM^{-1}, VAC.	UNC.	ROTATIONAL	BAND ASSIGNMENTS	ISOTOPIC SPECIES
54	2843.36769	(16)	P 39	1010 -0000	18012C32S
54			P 21	2010 -1000	16013C32S
54			R 23	1210 -0200	16013C32S
54			P 8	1111E-0101E	18012C32S
55	2843.42482	(12)	P 38	1010 -0000	16013C32S
56	2843.52733	(25)	P 9	1111 -0101	16013C32S
57	2843.78428	(60)	P 12	1010 -0000	16013C34S
58	2843.87589	(17)	P 38	1010 -0000	18012C32S
59	2844.21113	(63)	P 11	1010 -0000	16013C34S
59			P 11	1010 -0000	18012C34S
60	2844.47183	(12)	P 36	1010 -0000	16013C32S
61	2844.63435	(70)	P 10	1010 -0000	16013C34S
62	2844.76560	(78)	P 18	2010 -1000	16013C32S
63	2844.80961	(34)	P 6	1111 -0101	16013C32S
64	2844.88091	(17)	P 36	1010 -0000	18012C32S
65	2844.99018	(12)	P 35	1010 -0000	16013C32S
66	2845.37840	(16)	P 35	1010 -0000	18012C32S
67	2845.50511	(12)	P 34	1010 -0000	16013C32S
68	2845.64840	(47)	P 4	1111 -0101	16013C32S
69	2845.87259	(16)	P 34	1010 -0000	18012C32S
70	2846.01671	(12)	P 33	1010 -0000	16013C32S
71	2846.06264	(57)	P 3	1111 -0101	16013C32S
72	2846.36346	(16)	P 33	1010 -0000	18012C32S
73	2846.52477	(12)	P 32	1010 -0000	16013C32S
74	2846.85093	(15)	P 32	1010 -0000	18012C32S
75	2847.02963	(12)	P 31	1010 -0000	16013C32S
76	2847.33513	(15)	P 31	1010 -0000	18012C32S
77	2847.53091	(12)	P 30	1010 -0000	16013C32S
78	2847.81597	(16)	P 30	1010 -0000	18012C32S
79	2848.02888	(12)	P 29	1010 -0000	16013C32S
80	2848.29346	(15)	P 29	1010 -0000	18012C32S
81	2848.48399	(64)	R 2	1111 -0101	16013C32S
81			P 6	2010 -1000	18012C32S
82	2848.52352	(12)	P 28	1010 -0000	16013C32S
83	2848.76759	(15)	P 28	1010 -0000	18012C32S
84	2848.87637	(56)	R 3	1111 -0101	16013C32S
84			P 5	2010 -1000	18012C32S
85	2849.01479	(12)	P 27	1010 -0000	16013C32S
86	2849.08395	(82)	R 6	1111 -0101	18012C32S
87	2849.23845	(15)	P 27	1010 -0000	18012C32S
88	2849.50273	(12)	P 26	1010 -0000	16013C32S
89	2849.70591	(15)	P 26	1010 -0000	18012C32S
90	2849.73745	(81)	P 55	1313 -0303	16012C32S
91	2849.98738	(12)	P 25	1010 -0000	16013C32S
92	2850.03373	(56)	R 6	1111 -0101	16013C32S
93	2850.17007	(15)	P 25	1010 -0000	18012C32S
94	2850.46863	(12)	P 24	1010 -0000	16013C32S
95	2850.63086	(15)	P 24	1010 -0000	18012C32S
96	2850.94664	(12)	P 23	1010 -0000	16013C32S
97	2851.01376	(90)	R 5	1010 -0000	16013C34S
98	2851.08837	(15)	P 23	1010 -0000	18012C32S
99	2851.42135	(12)	P 22	1010 -0000	16013C32S
100	2851.89280	(12)	P 21	1010 -0000	16013C32S
100			R 11	1111E-0101E	16013C32S

- "LINE NO.": serial number identifying the spectral line in the map on the opposite page.
- "WAVENUMBER": experimental value.
- Uncertainty in the last digit of the wavenumber is given in parentheses.

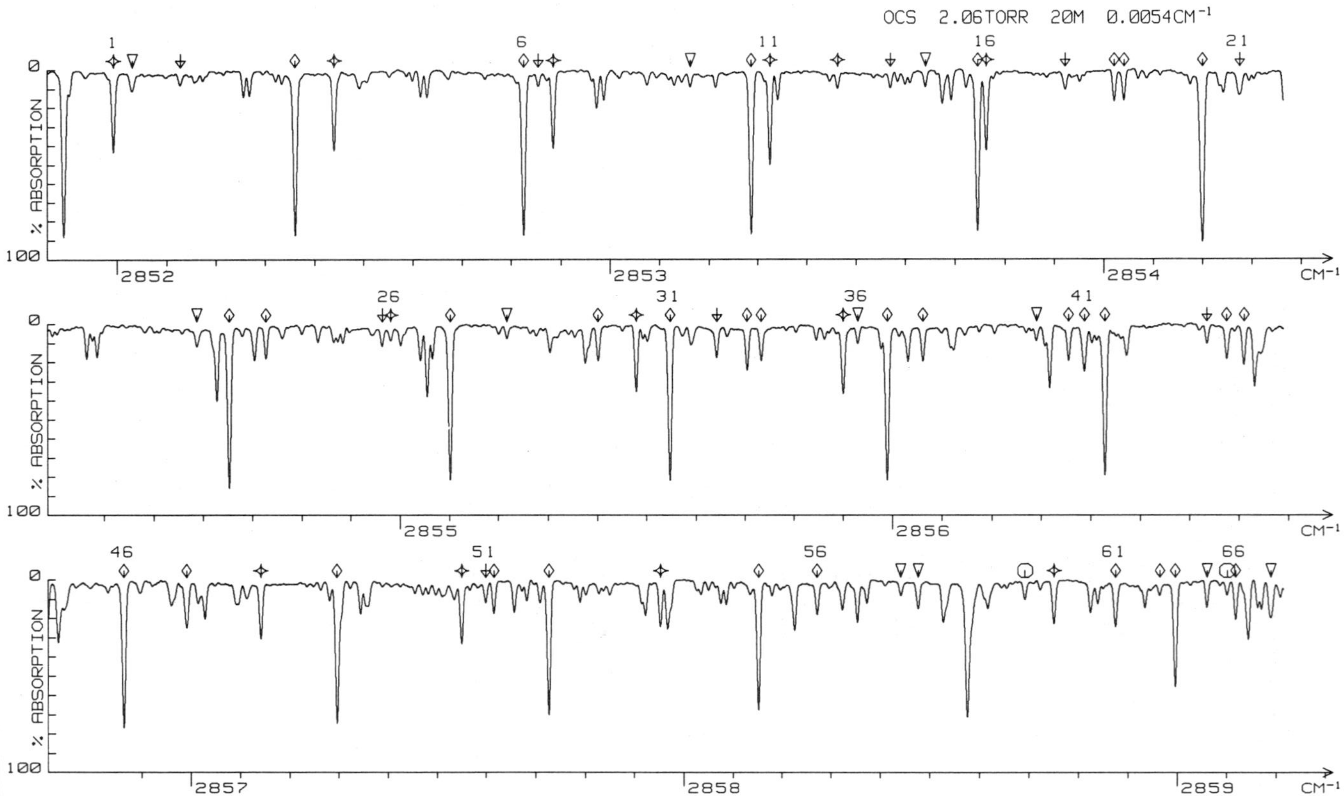

ISOTOPIC IDENTIFICATION – '∇' indicates $^{16}O^{12}C^{32}S$; '(I)': $^{16}O^{12}C^{34}S$; '◊': $^{16}O^{13}C^{32}S$; 'Y': $^{16}O^{12}C^{33}S$;
'✧' indicates $^{18}O^{12}C^{32}S$; '↓': $^{16}O^{13}C^{34}S$; '↑': Other OCS species; 'I': not OCS.

LINE NO.	WAVENUMBER CM^{-1}, VAC.	UNC.	ROTATIONAL BAND ASSIGNMENTS	ISOTOPIC SPECIES
1	2851.99322(	15)	P 21 1010 -0000	18012C32S
2	2852.03114(	67)	P 51 1313 -0303	16012C32S
2			P 73 1210 -0200	16012C32S
2			P 92 1111E-0101E	16012C32S
3	2852.12762(	82)	R 8 1010 -0000	16013C34S
3			P 1 2010 -1000	16013C32S
4	2852.36084(	12)	P 20 1010 -0000	16013C32S
5	2852.44065(	15)	P 20 1010 -0000	18012C32S
6	2852.82571(	12)	P 19 1010 -0000	16013C32S
7	2852.85500(	66)	R 10 1010 -0000	16013C34S
8	2852.88477(	15)	P 19 1010 -0000	18012C32S
9	2853.16253(	68)	P 49 1313 -0303	16012C32S
10	2853.28734(	12)	P 18 1010 -0000	16013C32S
11	2853.32549(	14)	P 18 1010 -0000	18012C32S
11			R 15 1111E-0101E	16013C32S
12	2853.46326(	56)	R 19 1111F-0101F	18012C32S
12			P 90 1111E-0101E	16012C32S
13	2853.57022(	58)	R 12 1010 -0000	16013C34S
14	2853.64121(	58)	P 49 1311E-0301E	16012C32S
14			R 14 1010 -0000	18012C34S
14			P 52 1212 -0202	16012C34S
15	2853.74570(	12)	P 17 1010 -0000	16013C32S
16	2853.76286(	16)	P 17 1010 -0000	18012C32S
16			R 20 1111E-0101E	18012C32S
17	2853.92282(	57)	R 13 1010 -0000	16013C34S
17			P 52 1210 -0200	16012C34S
18	2854.02174(	30)	R 17 1111E-0101E	16013C32S
19	2854.04102(	30)	R 17 1111F-0101F	16013C32S
20	2854.20025(	12)	P 16 1010 -0000	16013C32S
20			P 16 1010 -0000	18012C32S
21	2854.27571(	66)	R 14 1010 -0000	16013C34S
21			P 47 1313 -0303	16012C32S
21			R 16 1010 -0000	18012C34S

LINE NO.	WAVENUMBER CM^{-1}, VAC.	UNC.	ROTATIONAL BAND ASSIGNMENTS	ISOTOPIC SPECIES
22	2854.58765(	62)	P 69 1210 -0200	16012C32S
22			R 17 1010 -0000	18012C34S
22			P 44 1400 -0000	16013C32S
23	2854.65279(	12)	P 15 1010 -0000	16013C32S
24	2854.72759(	29)	R 19 1111F-0101F	16013C32S
25	2854.96357(	49)	R 16 1010 -0000	16013C34S
26	2854.98077(	78)	R 24 1111E-0101E	18012C32S
26			P 21 1414 -0404	16012C32S
27	2855.10149(	12)	P 14 1010 -0000	16013C32S
28	2855.21766(	72)	P 68 1210 -0200	16012C32S
29	2855.40126(	28)	R 21 1111F-0101F	16013C32S
30	2855.47924(	16)	P 13 1010 -0000	18012C32S
31	2855.54695(	12)	P 13 1010 -0000	16013C32S
32	2855.64141(	31)	R 18 1010 -0000	16013C34S
32			R 7 2010 -1000	16013C32S
33	2855.70454(	24)	R 22 1111E-0101E	16013C32S
33			P 87 1111F-0101F	16012C32S
34	2855.73334(	29)	R 22 1111F-0101F	16013C32S
35	2855.90010(	17)	P 12 1010 -0000	18012C32S
35			P 45 1311F-0301F	16012C32S
36	2855.92961(	52)	P 44 1313 -0303	16012C32S
37	2855.98929(	12)	P 12 1010 -0000	16013C32S
38	2856.06197(	29)	R 23 1111F-0101F	16013C32S
39	2856.29029(	62)	P 86 1111E-0101E	16012C32S
40	2856.35532(	29)	R 24 1111E-0101E	16013C32S
41	2856.38709(	27)	R 24 1111F-0101F	16013C32S
41			P 39 1400 -0000	16013C32S
41			R 23 1010 -0000	18012C34S
41			R 9 2010 -1000	16013C32S
42	2856.42844(	12)	P 11 1010 -0000	16013C32S
43	2856.63549(	47)	R 21 1010 -0000	16013C34S
44	2856.67564(	30)	R 25 1111E-0101E	16013C32S
44			R 24 1010 -0000	18012C34S

- "LINE NO.": serial number identifying the spectral line in the map on the opposite page.
- "WAVENUMBER": experimental value.
- Uncertainty in the last digit of the wavenumber is given in parentheses.

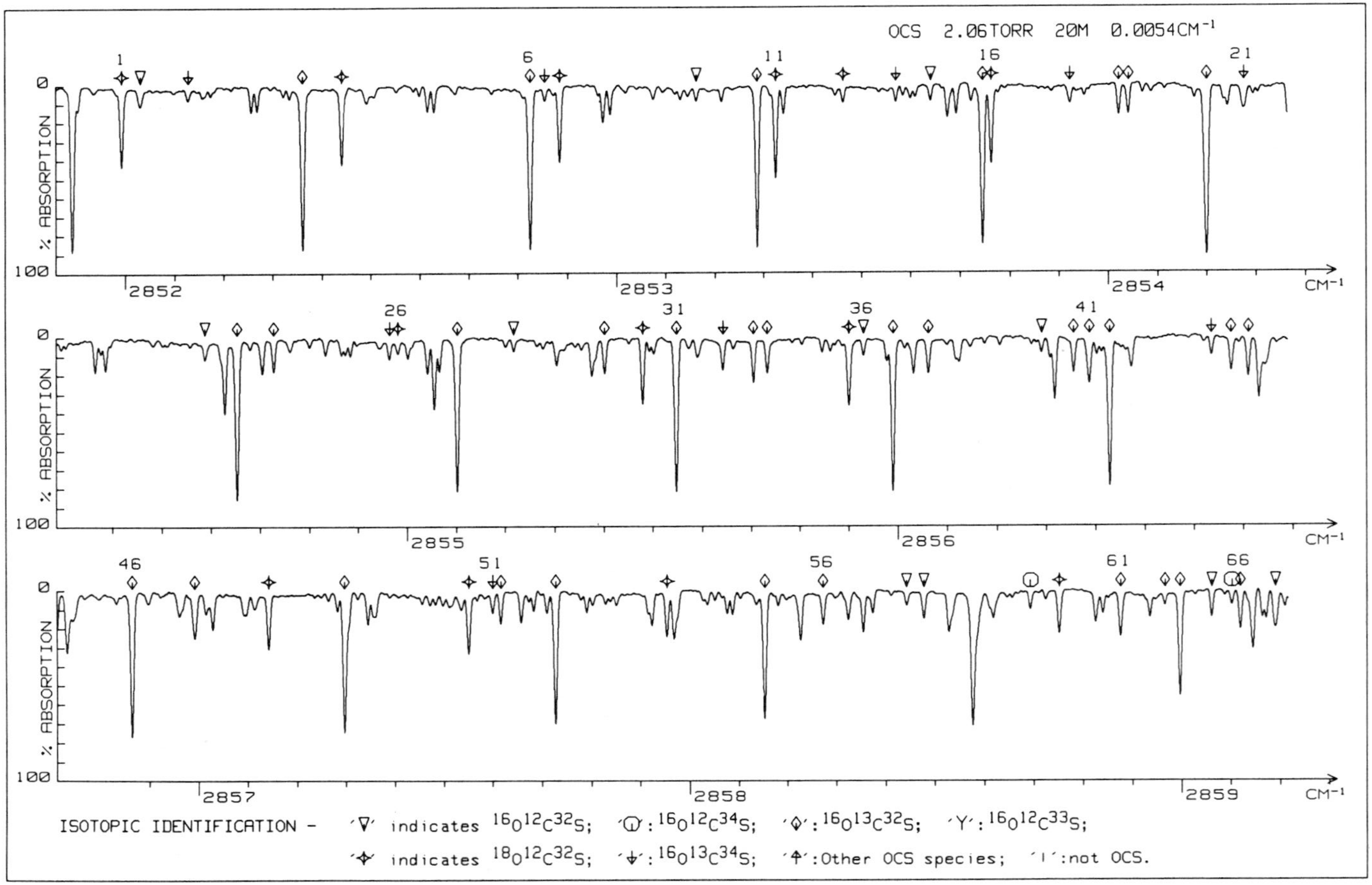
OCS 2.06TORR 20M 0.0054CM⁻¹
% ABSORPTION
0
100
2852
2853
2854
2855
2856
2857
2858
2859
CM⁻¹
1
6
11
16
21
26
31
36
41
46
51
56
61
66
ISOTOPIC IDENTIFICATION - '▽' indicates $^{16}O^{12}C^{32}S$; 'Ⓘ': $^{16}O^{12}C^{34}S$; '◊': $^{16}O^{13}C^{32}S$; 'Y': $^{16}O^{12}C^{33}S$;
'✧' indicates $^{18}O^{12}C^{32}S$; '↓': $^{16}O^{13}C^{34}S$; '↑': Other OCS species; 'I': not OCS.

LINE NO.	WAVENUMBER UNC. CM^{-1}, VAC.	ROTATIONAL BAND ASSIGNMENTS	ISOTOPIC SPECIES
45	2856.71019(26)	R 25 1111F-0101F	16013C32S
45		R 30 1111E-0101E	18012C32S
46	2856.86435(12)	P 10 1010 -0000	16013C32S
47	2856.99145(29)	R 26 1111E-0101E	16013C32S
47		P 85 1111E-0101E	16012C32S
47		R 31 1111E-0101E	18012C32S
47		P 43 1311F-0301F	16012C32S
48	2857.14211(19)	P 9 1010 -0000	18012C32S
49	2857.29709(19)	P 9 1010 -0000	16013C32S
49		R 32 1111F-0101F	18012C32S
50	2857.55011(19)	P 8 1010 -0000	18012C32S
50		P 41 1313 -0303	16012C32S
51	2857.59994(46)	R 24 1010 -0000	16013C34S
52	2857.61646(28)	R 28 1111E-0101E	16013C32S
53	2857.72661(12)	P 8 1010 -0000	16013C32S
54	2857.95349(23)	P 7 1010 -0000	18012C32S
55	2858.15313(12)	P 7 1010 -0000	16013C32S
56	2858.27276(31)	R 30 1111F-0101F	16013C32S
57	2858.44282(65)	P1 2 1010 -0000	16012C32S
58	2858.47795(36)	P 83 1111F-0101F	16012C32S
58		P 47 2210 -1200	16012C32S
59	2858.69253(51)	P 66 1111E-0101E	16012C34S
60	2858.75129(24)	P 5 1010 -0000	18012C32S
60		P 66 1111F-0101F	16012C34S
61	2858.87544(25)	R 32 1111F-0101F	16013C32S
61		R 16 2010 -1000	16013C32S
61		R 38 1111F-0101F	18012C32S
61		R 32 1010 -0000	18012C34S
62	2858.96510(81)	P 32 1400 -0000	16013C32S
63	2858.99658(13)	P 5 1010 -0000	16013C32S
64	2859.06097(34)	P 82 1111E-0101E	16012C32S
65	2859.10177(67)	P 42 1212 -0202	16012C34S
66	2859.11895(26)	R 33 1111E-0101E	16013C32S
66		P 39 1311E-0301E	16012C32S
66		P 88 2010 -1000	16012C32S
67	2859.18993(39)	P 61 1212E-0202E	16012C32S
67		P 61 1212F-0202F	16012C32S

- "LINE NO.": serial number identifying the spectral line in the map on the opposite page.
- "WAVENUMBER": experimental value.
- Uncertainty in the last digit of the wavenumber is given in parentheses.

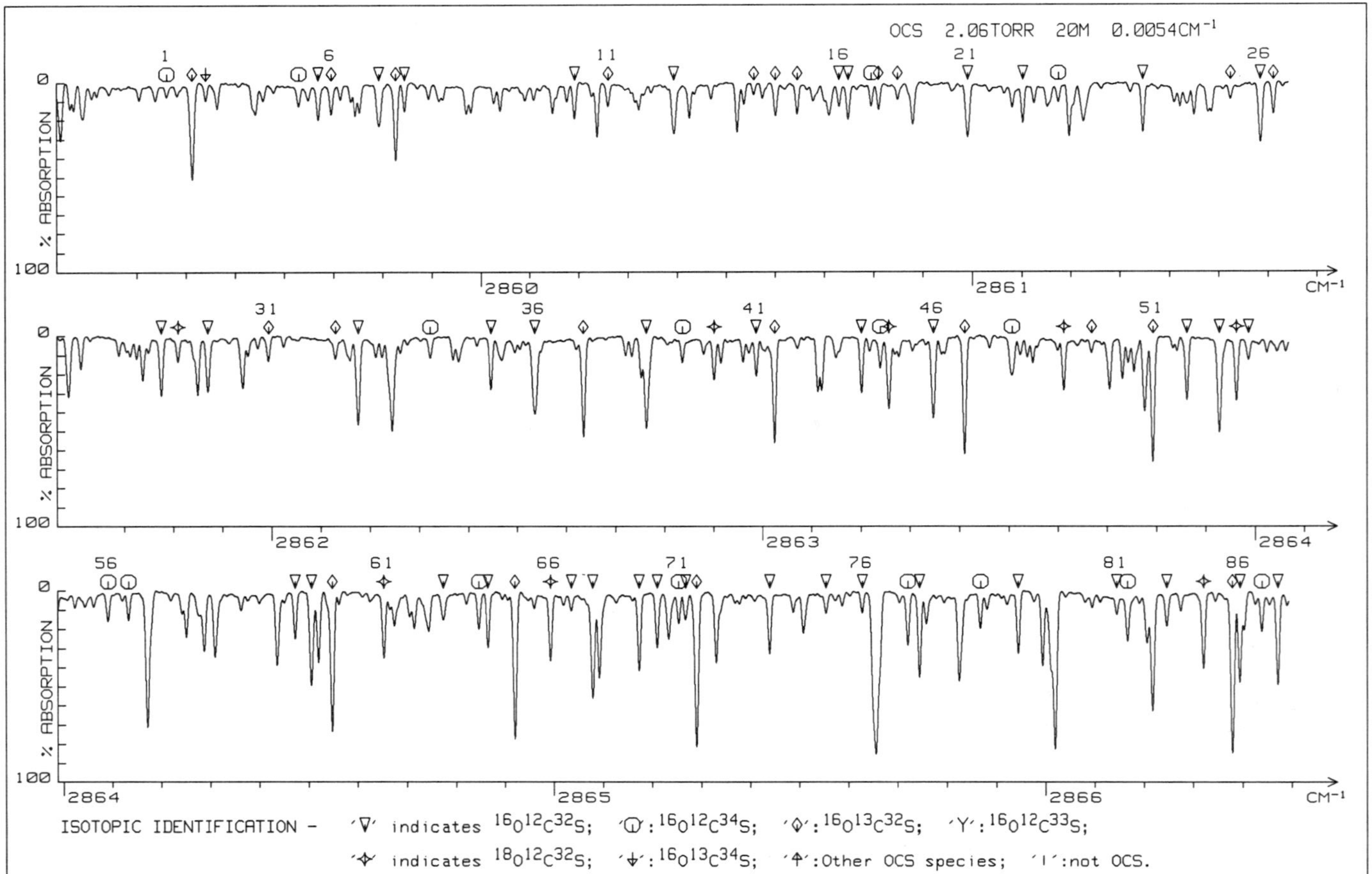
OCS 2.06TORR 20M 0.0054CM^{-1}
% ABSORPTION
0
100
2860
2861
2862
2863
2864
2865
2866
CM^{-1}
ISOTOPIC IDENTIFICATION - '∇' indicates $^{16}O^{12}C^{32}S$; 'Ⓘ': $^{16}O^{12}C^{34}S$; '◊': $^{16}O^{13}C^{32}S$; 'Y': $^{16}O^{12}C^{33}S$;
'✧' indicates $^{18}O^{12}C^{32}S$; '↓': $^{16}O^{13}C^{34}S$; '↑': Other OCS species; 'I': not OCS.

LINE NO.	WAVENUMBER CM^{-1}, VAC.	UNC.	ROTATIONAL ASSIGNMENTS	BAND ASSIGNMENTS	ISOTOPIC SPECIES
1	2859.36140	(81)	P 65	1111F-0101F	16012C34S
2	2859.41321	(15)	P 4	1010 -0000	16013C32S
2			R 34	1111E-0101E	16013C32S
2			P 86	1010 -0000	16012C34S
3	2859.44063	(49)	R 30	1010 -0000	16013C34S
4	2859.62992	(47)	P 41	1212 -0202	16012C34S
4			R 41	1111F-0101F	18012C32S
5	2859.66968	(29)	P 37	1313 -0303	16012C32S
5			P 38	1311F-0301F	16012C32S
6	2859.69599	(32)	R 35	1111E-0101E	16013C32S
7	2859.79322	(33)	P 60	1212 -0202	16012C32S
8	2859.82734	(15)	P 3	1010 -0000	16013C32S
8			P 87	2010 -1000	16012C32S
8			R 26	2010 -1000	18012C32S
9	2859.84509	(32)	P 81	1111F-0101F	16012C32S
10	2860.19232	(27)	P 36	1313 -0303	16012C32S
10			P 37	1311F-0301F	16012C32S
11	2860.26008	(39)	R 37	1111E-0101E	16013C32S
12	2860.39413	(26)	P 59	1212 -0202	16012C32S
13	2860.55677	(71)	R 21	2010 -1000	16013C32S
14	2860.60198	(31)	R 38	1111F-0101F	16013C32S
14			R 34	1010 -0000	16013C34S
15	2860.64568	(29)	P 1	1010 -0000	16013C32S
16	2860.73097	(42)	P 99	1010 -0000	16012C32S
17	2860.74918	(27)	P 59	1210 -0200	16012C32S
17			R 46	1111E-0101E	18012C32S
18	2860.79605	(40)	P 84	1010 -0000	16012C34S
18			P 65	2111F-1101F	16012C32S
19	2860.81093	(36)	R 39	1111E-0101E	16013C32S
20	2860.84958	(62)	P 27	1400 -0000	16013C32S
21	2860.99150	(24)	P 58	1212 -0202	16012C32S
22	2861.10255	(25)	P 79	1111E-0101E	16012C32S
23	2861.17564	(54)	P 62	1111F-0101F	16012C34S
24	2861.34735	(21)	P 58	1210 -0200	16012C32S
24			R 41	1111E-0101E	16013C32S
25	2861.52481	(64)	R 24	2010 -1000	16013C32S
26	2861.58544	(20)	P 57	1212 -0202	16012C32S
27	2861.61162	(29)	R 42	1111E-0101E	16013C32S
27			P 25	1400 -0000	16013C32S
28	2861.77557	(19)	P 78	1111E-0101E	16012C32S
28			P 61	1111F-0101F	16012C34S
29	2861.80949	(39)	R 2	1010 -0000	18012C32S
30	2861.86995	(21)	P 78	1111F-0101F	16012C32S
30			R 43	1111E-0101E	16013C32S
31	2861.99372	(38)	P 24	1400 -0000	16013C32S
31			P 37	1210 -0200	16012C34S
32	2862.12944	(50)	R 44	1111E-0101E	16013C32S
33	2862.17623	(16)	P 56	1212E-0202E	16012C32S
33			P 56	1212F-0202F	16012C32S
33			R 3	1010 -0000	18012C32S
34	2862.32226	(48)	P 60	1111E-0101E	16012C34S
35	2862.44631	(20)	P 77	1111E-0101E	16012C32S
36	2862.53661	(21)	P 77	1111F-0101F	16012C32S
36			P 56	1210 -0200	16012C32S
36			R 4	1010 -0000	18012C32S
37	2862.63582	(14)	R 3	1010 -0000	16013C32S
37			R 46	1111E-0101E	16013C32S
37			P 62	2111F-1101F	16012C32S
38	2862.76328	(42)	P 55	1212E-0202E	16012C32S
38			P 55	1212F-0202F	16012C32S
38			P 32	1311E-0301E	16012C32S
38			P 22	1400 -0000	16013C32S
39	2862.83648	(40)	P 81	1010 -0000	16012C34S
39			R 49	1010 -0000	18012C34S
40	2862.90107	(25)	R 5	1010 -0000	18012C32S

- "LINE NO.": serial number identifying the spectral line in the map on the opposite page.
- "WAVENUMBER": experimental value.
- Uncertainty in the last digit of the wavenumber is given in parentheses.

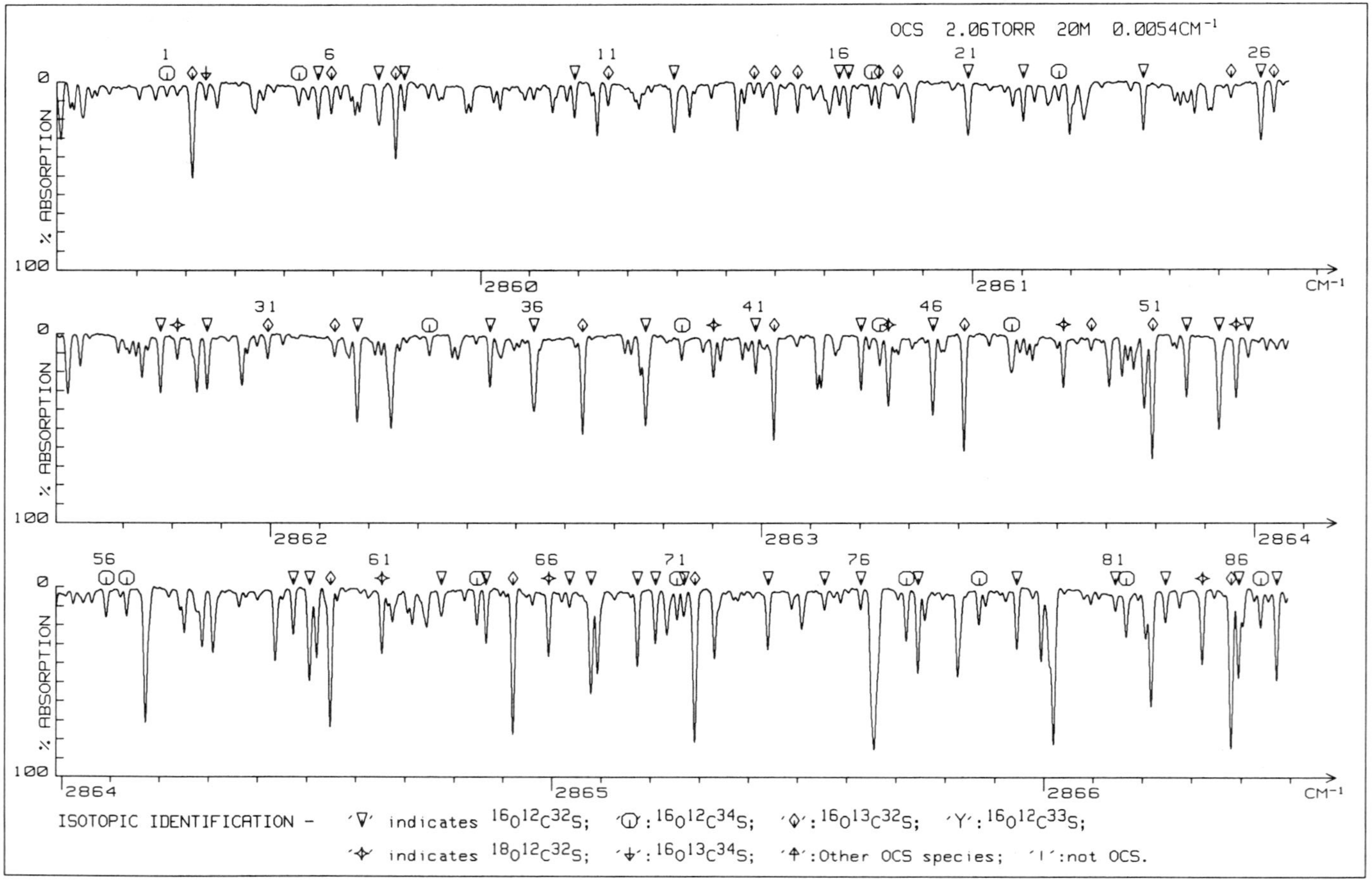
OCS 2.06TORR 20M 0.0054CM⁻¹
% ABSORPTION
0
100
2860
2861
2862
2863
2864
2865
2866
CM⁻¹
ISOTOPIC IDENTIFICATION – '▽' indicates 16O12C32S; 16O12C34S; 16O13C32S; 'Y': 16O12C33S;
indicates 18O12C32S; 16O13C34S; Other OCS species; 'I': not OCS.

LINE NO.	WAVENUMBER CM^{-1}, VAC.	UNC.	ROTATIONAL BAND ASSIGNMENTS	ISOTOPIC SPECIES
41	2862.98693(	26)	P 96 1010 -0000	16012C32S
41			P 52 1010 -0000	17012C32S
42	2863.02480(	13)	R 4 1010 -0000	16013C32S
43	2863.20129(	19)	P 76 1111F-0101F	16012C32S
44	2863.23941(	35)	P 34 1212 -0202	16012C34S
44			P 61 2111F-1101F	16012C32S
45	2863.25698(	17)	R 6 1010 -0000	18012C32S
45			P 30 1313 -0303	16012C32S
45			R 44 1010 -0000	16013C34S
46	2863.34733(	16)	P 54 1212 -0202	16012C32S
47	2863.41035(	13)	R 5 1010 -0000	16013C32S
48	2863.50613(	39)	P 58 1111E-0101E	16012C34S
48			P 80 1010 -0000	16012C34S
48			R 45 1010 -0000	16013C34S
49	2863.61185(	20)	R 7 1010 -0000	18012C32S
50	2863.66828(	70)	R 31 2010 -1000	16013C32S
51	2863.79267(	13)	R 6 1010 -0000	16013C32S
52	2863.86218(	17)	P 75 1111F-0101F	16012C32S
53	2863.92763(	15)	P 53 1212 -0202	16012C32S
53			P 19 1400 -0000	16013C32S
54	2863.96215(	18)	R 8 1010 -0000	18012C32S
54			R 32 2010 -1000	16013C32S
55	2863.98645(	52)	P 81 2010 -1000	16012C32S
55			R 47 1010 -0000	16013C34S
56	2864.09133(	33)	P 57 1111E-0101E	16012C34S
57	2864.13320(	33)	P 57 1111F-0101F	16012C34S
58	2864.47253(	21)	P 94 1010 -0000	16012C32S
59	2864.50537(	36)	P 52 1212 -0202	16012C32S
60	2864.54774(	12)	R 8 1010 -0000	16013C32S
61	2864.65293(	17)	R 10 1010 -0000	18012C32S
62	2864.77379(	47)	P 28 1311 -0301	16012C32S
63	2864.84595(	27)	P 78 1010 -0000	16012C34S
63			R 55 1111F-0101F	16013C32S
64	2864.86510(	19)	P 52 1210 -0200	16012C32S
65	2864.92056(	12)	R 9 1010 -0000	16013C32S
66	2864.99330(	17)	R 11 1010 -0000	18012C32S
67	2865.03558(	58)	P 58 2111F-1101F	16012C32S
68	2865.07922(	32)	P 51 1212 -0202	16012C32S
68			P 34 2212 -1202	16012C32S
69	2865.17371(	16)	P 73 1111F-0101F	16012C32S
70	2865.21002(	20)	P 93 1010 -0000	16012C32S
71	2865.25388(	31)	P 55 1111E-0101E	16012C34S
72	2865.26769(	40)	P 27 1311 -0301	16012C32S
73	2865.29029(	12)	R 10 1010 -0000	16013C32S
73			P 55 1111F-0101F	16012C34S
74	2865.43985(	18)	P 51 1210 -0200	16012C32S
75	2865.55491(	49)	P 57 2111E-1101E	16012C32S
75			R 54 1010 -0000	16013C34S
76	2865.62761(	41)	P 57 2111F-1101F	16012C32S
77	2865.72115(	20)	P 29 1212 -0202	16012C34S
77			P 25 1313 -0303	16012C32S
78	2865.74473(	15)	P 72 1111E-0101E	16012C32S
79	2865.86741(	27)	P 54 1111F-0101F	16012C34S
80	2865.94389(	17)	P 92 1010 -0000	16012C32S
81	2866.14505(	51)	P 56 2111E-1101E	16012C32S
82	2866.16707(	24)	P 76 1010 -0000	16012C34S
82			R 62 1111E-0101E	16013C32S
83	2866.24590(	31)	P 25 1311 -0301	16012C32S
84	2866.32094(	15)	R 15 1010 -0000	18012C32S
85	2866.37953(	12)	R 13 1010 -0000	16013C32S
86	2866.39408(	19)	P 71 1111E-0101E	16012C32S
87	2866.43873(	26)	P 53 1111F-0101F	16012C34S
88	2866.47141(	14)	P 71 1111F-0101F	16012C32S

- "LINE NO.": serial number identifying the spectral line in the map on the opposite page.
- "WAVENUMBER": experimental value.
- Uncertainty in the last digit of the wavenumber is given in parentheses.

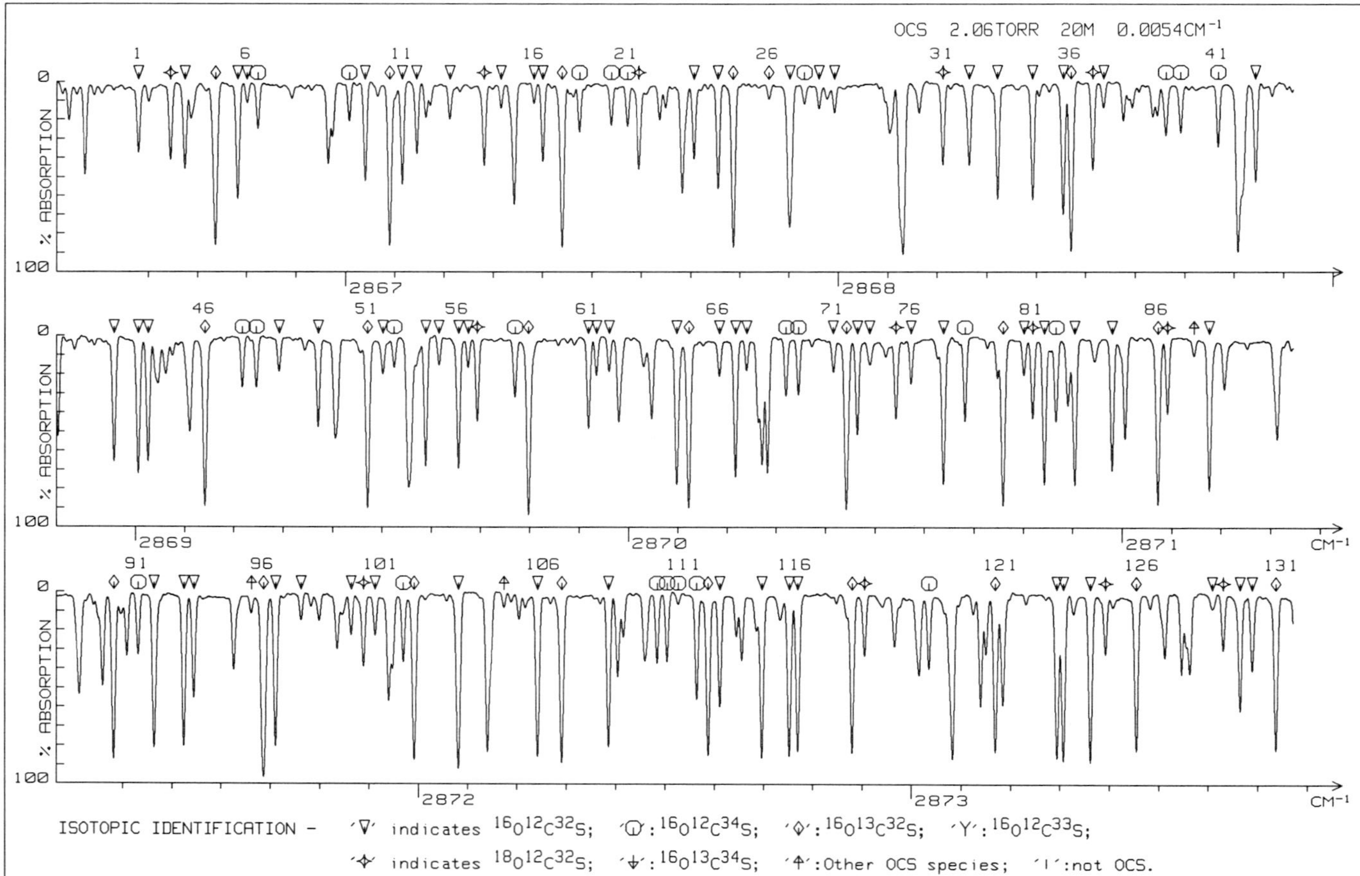
OCS 2.06TORR 20M 0.0054CM-1
% ABSORPTION
0
100
2867
2868
2869
2870
2871
2872
2873
CM-1
ISOTOPIC IDENTIFICATION - '▽' indicates 16O12C32S; 'Ⓘ': 16O12C34S; '◊': 16O13C32S; 'Y': 16O12C33S;
'✧' indicates 18O12C32S; '⏚': 16O13C34S; '↑': Other OCS species; 'I': not OCS.

LINE NO.	WAVENUMBER CM^{-1}, VAC.	UNC.	ROTATIONAL ASSIGNMENTS	BAND ASSIGNMENTS	ISOTOPIC SPECIES
1	2866.57952	(17)	P 49	1210 -0200	16012C32S
2	2866.64446	(15)	R 16	1010 -0000	18012C32S
3	2866.67374	(38)	P 91	1010 -0000	16012C32S
3			P 77	2010 -1000	16012C32S
3			P 12	1400 -0000	16013C32S
4	2866.73598	(16)	R 14	1010 -0000	16013C32S
4			P 55	2111E-1101E	16012C32S
5	2866.78179	(13)	P 48	1212 -0202	16012C32S
6	2866.80112	(44)	P 55	2111F-1101F	16012C32S
7	2866.82257	(22)	P 75	1010 -0000	16012C34S
8	2867.00675	(24)	P 52	1111F-0101F	16012C34S
9	2867.03990	(14)	P 70	1111E-0101E	16012C32S
10	2867.08957	(12)	R 15	1010 -0000	16013C32S
11	2867.11496	(13)	P 70	1111F-0101F	16012C32S
12	2867.14431	(16)	P 48	1210 -0200	16012C32S
13	2867.21130	(28)	P 23	1311 -0301	16012C32S
14	2867.28129	(15)	R 18	1010 -0000	18012C32S
15	2867.31626	(37)	P 54	2111E-1101E	16012C32S
15			R 63	1010 -0000	16013C34S
16	2867.38284	(43)	P 54	2111F-1101F	16012C32S
17	2867.40068	(15)	P 90	1010 -0000	16012C32S
18	2867.43969	(12)	R 16	1010 -0000	16013C32S
19	2867.47456	(20)	P 74	1010 -0000	16012C34S
20	2867.53902	(23)	P 51	1111E-0101E	16012C34S
21	2867.57160	(24)	P 51	1111F-0101F	16012C34S
22	2867.59490	(15)	R 19	1010 -0000	18012C32S
22			P 29	2212 -1202	16012C32S
23	2867.70595	(16)	P 47	1210 -0200	16012C32S
24	2867.75527	(13)	P 69	1111F-0101F	16012C32S
25	2867.78665	(12)	R 17	1010 -0000	16013C32S
26	2867.85884	(63)	P 9	1400 -0000	16013C32S
27	2867.90127	(14)	P 46	1212 -0202	16012C32S
27			R 20	1010 -0000	18012C32S
28	2867.93096	(48)	P 25	1210 -0200	16012C34S
28			P 29	2210 -1200	16012C32S
29	2867.96072	(39)	P 53	2111F-1101F	16012C32S
30	2867.99179	(34)	P 75	2010 -1000	16012C32S
31	2868.21124	(15)	R 21	1010 -0000	18012C32S
32	2868.26416	(39)	P 46	1210 -0200	16012C32S
33	2868.32115	(13)	P 68	1111E-0101E	16012C32S
34	2868.39207	(13)	P 68	1111F-0101F	16012C32S
35	2868.45458	(13)	P 45	1212 -0202	16012C32S
36	2868.47060	(12)	R 19	1010 -0000	16013C32S
36			P 52	2111E-1101E	16012C32S
37	2868.51438	(14)	R 22	1010 -0000	18012C32S
37			P 42	1010 -0000	17012C32S
38	2868.53595	(36)	P 52	2111F-1101F	16012C32S
39	2868.66125	(22)	P 49	1111E-0101E	16012C34S
40	2868.69104	(21)	P 49	1111F-0101F	16012C34S
41	2868.76810	(17)	P 72	1010 -0000	16012C34S
42	2868.84312	(14)	P 88	1010 -0000	16012C32S
43	2868.95666	(13)	P 67	1111E-0101E	16012C32S
44	2869.00557	(12)	P 44	1212 -0202	16012C32S
45	2869.02539	(13)	P 67	1111F-0101F	16012C32S
46	2869.14108	(12)	R 21	1010 -0000	16013C32S
47	2869.21769	(20)	P 48	1111E-0101E	16012C34S
48	2869.24581	(21)	P 48	1111F-0101F	16012C34S
49	2869.29180	(28)	P 73	2010 -1000	16012C32S
50	2869.37069	(14)	P 44	1210 -0200	16012C32S
51	2869.47132	(12)	R 22	1010 -0000	16013C32S
52	2869.50200	(30)	P 17	1313 -0303	16012C32S
53	2869.52487	(33)	P 21	1212 -0202	16012C34S
54	2869.58863	(12)	P 66	1111E-0101E	16012C32S
55	2869.61579	(31)	P 50	2111E-1101E	16012C32S
56	2869.65529	(12)	P 66	1111F-0101F	16012C32S

- "LINE NO.": serial number identifying the spectral line in the map on the opposite page.
- "WAVENUMBER": experimental value.
- Uncertainty in the last digit of the wavenumber is given in parentheses.

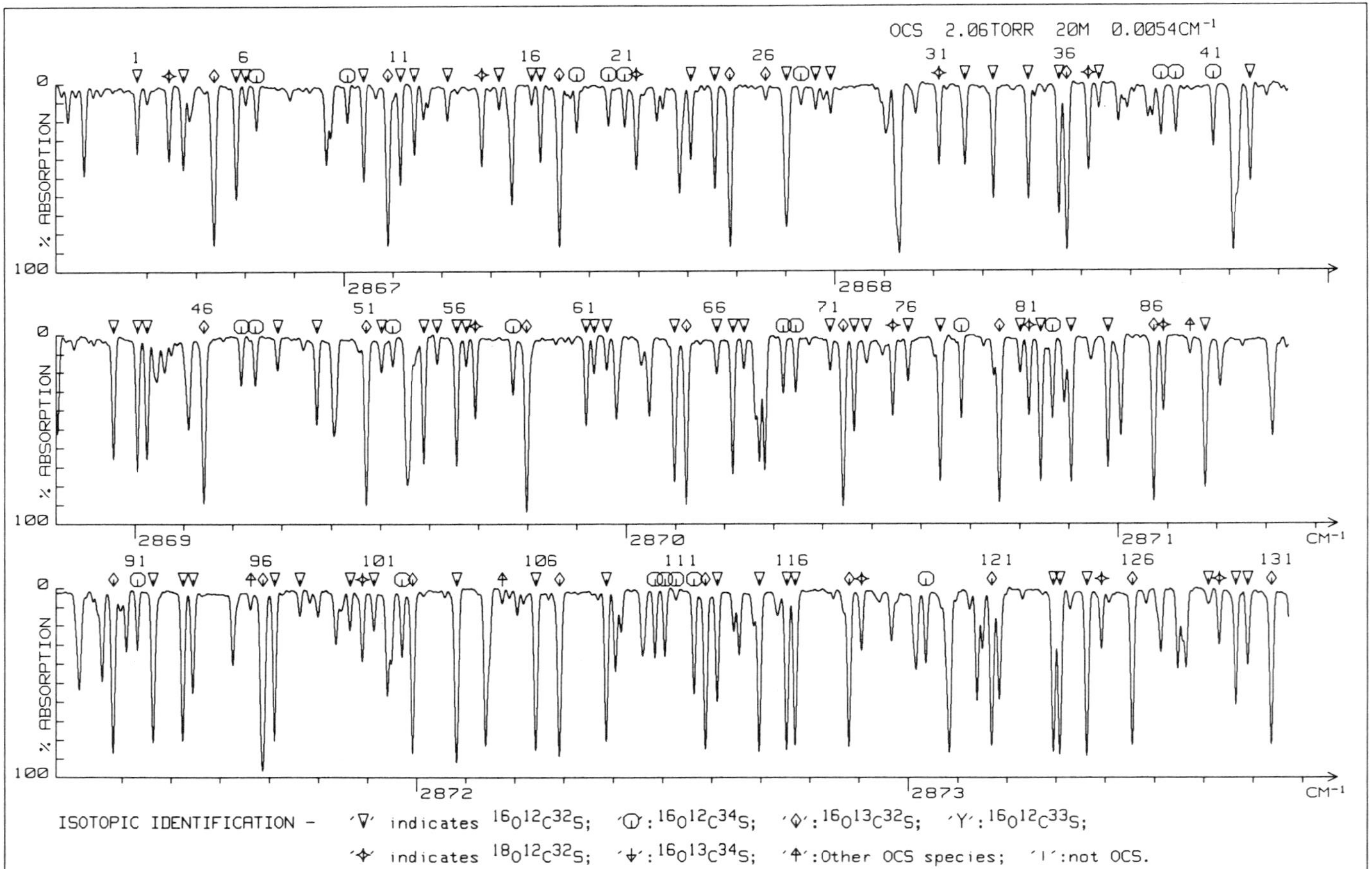
OCS 2.06TORR 20M 0.0054CM⁻¹
% ABSORPTION
2867
2868
2869
2870
2871
2872
2873
CM⁻¹
ISOTOPIC IDENTIFICATION - '▽' indicates $^{16}O^{12}C^{32}S$; 'Ⓘ': $^{16}O^{12}C^{34}S$; '◊': $^{16}O^{13}C^{32}S$; 'Y': $^{16}O^{12}C^{33}S$;
'✧' indicates $^{18}O^{12}C^{32}S$; '⏚': $^{16}O^{13}C^{34}S$; '↑': Other OCS species; 'I': not OCS.

LINE NO.	WAVENUMBER UNC. CM^{-1}, VAC.	ROTATIONAL BAND ASSIGNMENTS	ISOTOPIC SPECIES
57	2869.67499(31)	P 50 2111F-1101F	16012C32S
58	2869.69364(15)	R 26 1010 -0000	18012C32S
59	2869.77055(19)	P 47 1111E-0101E	16012C34S
60	2869.79814(12)	R 23 1010 -0000	16013C32S
60		P 47 1111F-0101F	16012C34S
61	2869.91910(14)	P 43 1210 -0200	16012C32S
62	2869.93508(28)	P 72 2010 -1000	16012C32S
63	2869.96070(28)	P 16 1313 -0303	16012C32S
64	2870.09781(12)	P 42 1212 -0202	16012C32S
64		P 39 1010 -0000	17012C32S
65	2870.12184(12)	R 24 1010 -0000	16013C32S
66	2870.18411(28)	P 49 2111E-1101E	16012C32S
66		P 78 1010 -0000	16012C33S
67	2870.21722(12)	P 65 1111E-0101E	16012C32S
68	2870.23957(30)	P 49 2111F-1101F	16012C32S
69	2870.32009(18)	P 46 1111E-0101E	16012C34S
70	2870.34534(19)	P 46 1111F-0101F	16012C34S
71	2870.41589(29)	P 15 1313 -0303	16012C32S
72	2870.44202(12)	R 25 1010 -0000	16013C32S
73	2870.46413(14)	P 42 1210 -0200	16012C32S
74	2870.48935(54)	P 16 1311 -0301	16012C32S
75	2870.54237(15)	R 29 1010 -0000	18012C32S
76	2870.57333(23)	P 71 2010 -1000	16012C32S
77	2870.63880(12)	P 41 1212 -0202	16012C32S
78	2870.68188(15)	P 69 1010 -0000	16012C34S
79	2870.75868(15)	R 26 1010 -0000	16013C32S
80	2870.80078(27)	P 48 2111F-1101F	16012C32S
81	2870.81856(15)	R 30 1010 -0000	18012C32S
82	2870.84227(12)	P 64 1111E-0101E	16012C32S
83	2870.86674(15)	P 45 1111E-0101E	16012C34S
83		P 14 1313 -0303	16012C32S
84	2870.90480(18)	P 64 1111F-0101F	16012C32S
85	2870.97972(12)	P 85 1010 -0000	16012C32S
86	2871.07209(12)	R 27 1010 -0000	16013C32S
87	2871.09120(15)	R 31 1010 -0000	18012C32S
88	2871.14539(53)	P 37 1010 -0000	17012C32S
88		P 29 1212 -0202	16012C33S
89	2871.17672(12)	P 40 1212 -0202	16012C32S
89		P 18 1210 -0200	16012C34S
90	2871.38208(12)	R 28 1010 -0000	16013C32S
91	2871.43154(18)	P 44 1111F-0101F	16012C34S
92	2871.46403(12)	P 63 1111E-0101E	16012C32S
93	2871.52443(12)	P 63 1111F-0101F	16012C32S
94	2871.54435(14)	P 40 1210 -0200	16012C32S
95	2871.66208(52)	P 36 1010 -0000	17012C32S
96	2871.68710(12)	R 29 1010 -0000	16013C32S
96		P 84 1010 -0000	16012C32S
97	2871.71121(12)	P 39 1212 -0202	16012C32S
98	2871.76261(35)	P 12 1313 -0303	16012C32S
99	2871.86315(25)	P 46 2111E-1101E	16012C32S
100	2871.88901(16)	R 34 1010 -0000	18012C32S
101	2871.91275(25)	P 46 2111F-1101F	16012C32S
102	2871.96975(17)	P 43 1111F-0101F	16012C34S
103	2871.99168(12)	R 30 1010 -0000	16013C32S
104	2872.08137(12)	P 62 1111E-0101E	16012C32S
104		P 39 1210 -0200	16012C32S
105	2872.17475(59)	P 35 1010 -0000	17012C32S
106	2872.24258(12)	P 38 1212 -0202	16012C32S
107	2872.29131(12)	R 31 1010 -0000	16013C32S
107		P 12 1311E-0301E	16012C32S
107		P 12 1311F-0301F	16012C32S
108	2872.38606(12)	P 83 1010 -0000	16012C32S
109	2872.48447(16)	P 42 1111E-0101E	16012C34S
110	2872.50456(16)	P 42 1111F-0101F	16012C34S

- "LINE NO.": serial number identifying the spectral line in the map on the opposite page.
- "WAVENUMBER": experimental value.
- Uncertainty in the last digit of the wavenumber is given in parentheses.

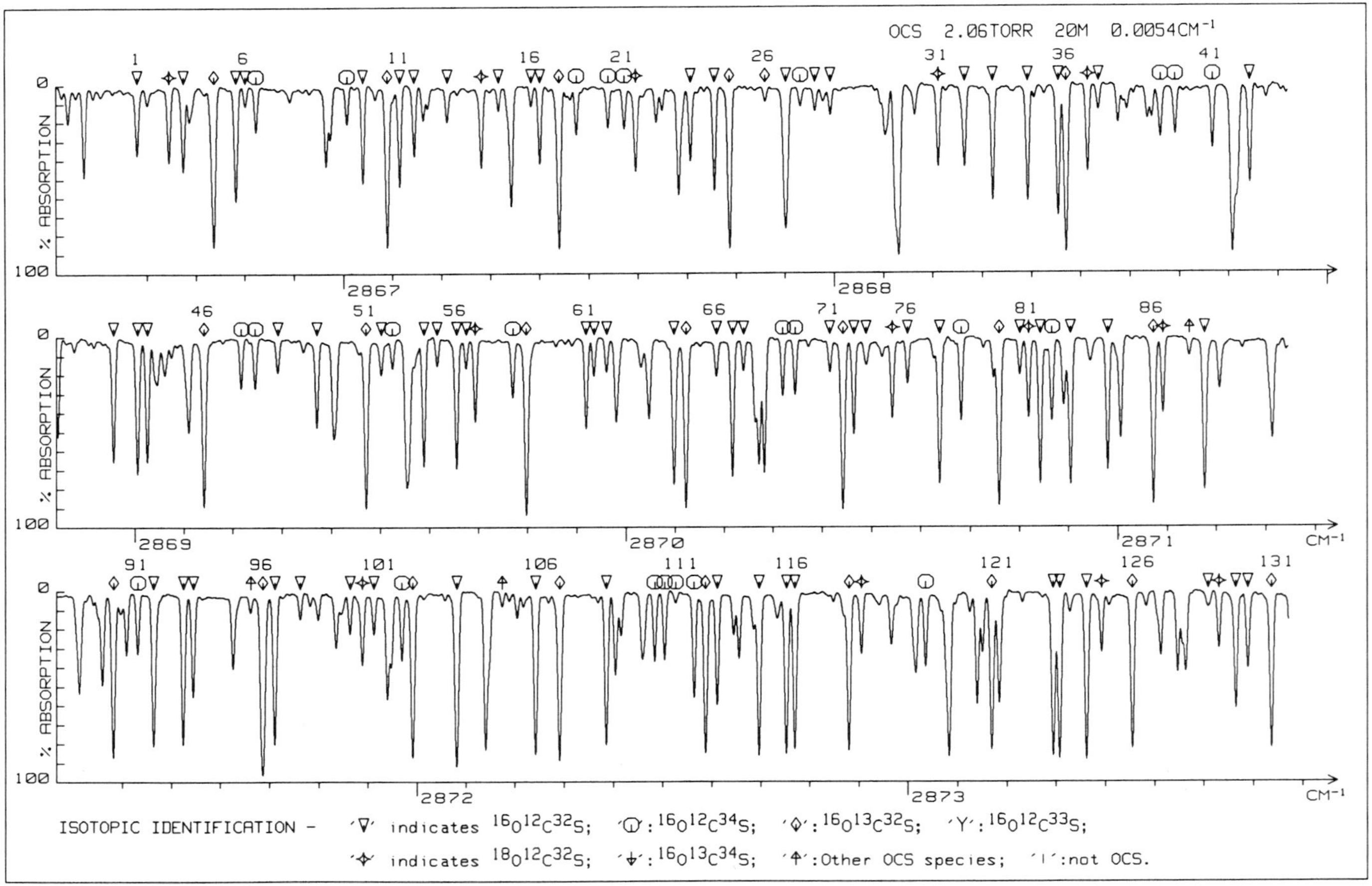
OCS 2.06TORR 20M 0.0054CM⁻¹
% ABSORPTION
2867
2868
2869
2870
2871
2872
2873
CM⁻¹
ISOTOPIC IDENTIFICATION – '▽' indicates $^{16}O^{12}C^{32}S$; '$^{16}O^{12}C^{34}S$; '◊': $^{16}O^{13}C^{32}S$; 'Y': $^{16}O^{12}C^{33}S$;
'✧' indicates $^{18}O^{12}C^{32}S$; '': $^{16}O^{13}C^{34}S$; '': Other OCS species; '|': not OCS.

LINE NO.	WAVENUMBER CM^{-1}, VAC.	UNC.	ROTATIONAL	BAND ASSIGNMENTS		ISOTOPIC SPECIES
111	2872.52696	(86)	P 15	1210	-0200	16012C34S
112	2872.56408	(13)	P 66	1010	-0000	16012C34S
113	2872.58743	(12)	R 32	1010	-0000	16013C32S
114	2872.61137	(13)	P 38	1210	-0200	16012C32S
115	2872.69680	(15)	P 61	1111E	-0101E	16012C32S
116	2872.75337	(12)	P 61	1111F	-0101F	16012C32S
117	2872.77055	(12)	P 37	1212	-0202	16012C32S
118	2872.88006	(12)	R 33	1010	-0000	16013C32S
119	2872.90511	(17)	R 38	1010	-0000	18012C32S
120	2873.03614	(16)	P 41	1111F	-0101F	16012C34S
121	2873.16941	(12)	R 34	1010	-0000	16013C32S
121			P 10	1311	-0301	16012C32S
122	2873.29521	(16)	P 36	1212	-0202	16012C32S
123	2873.30812	(15)	P 60	1111E	-0101E	16012C32S
124	2873.36267	(12)	P 60	1111F	-0101F	16012C32S
125	2873.39277	(19)	R 40	1010	-0000	18012C32S
126	2873.45500	(12)	R 35	1010	-0000	16013C32S
127	2873.60967	(89)	P 9	1311	-0301	16012C32S
128	2873.63136	(19)	R 41	1010	-0000	18012C32S
129	2873.66518	(13)	P 36	1210	-0200	16012C32S
130	2873.68946	(16)	P 66	2010	-1000	16012C32S
130			P 32	1010	-0000	17012C32S
131	2873.73712	(12)	R 36	1010	-0000	16013C32S

- "LINE NO.": serial number identifying the spectral line in the map on the opposite page.
- "WAVENUMBER": experimental value.
- Uncertainty in the last digit of the wavenumber is given in parentheses.

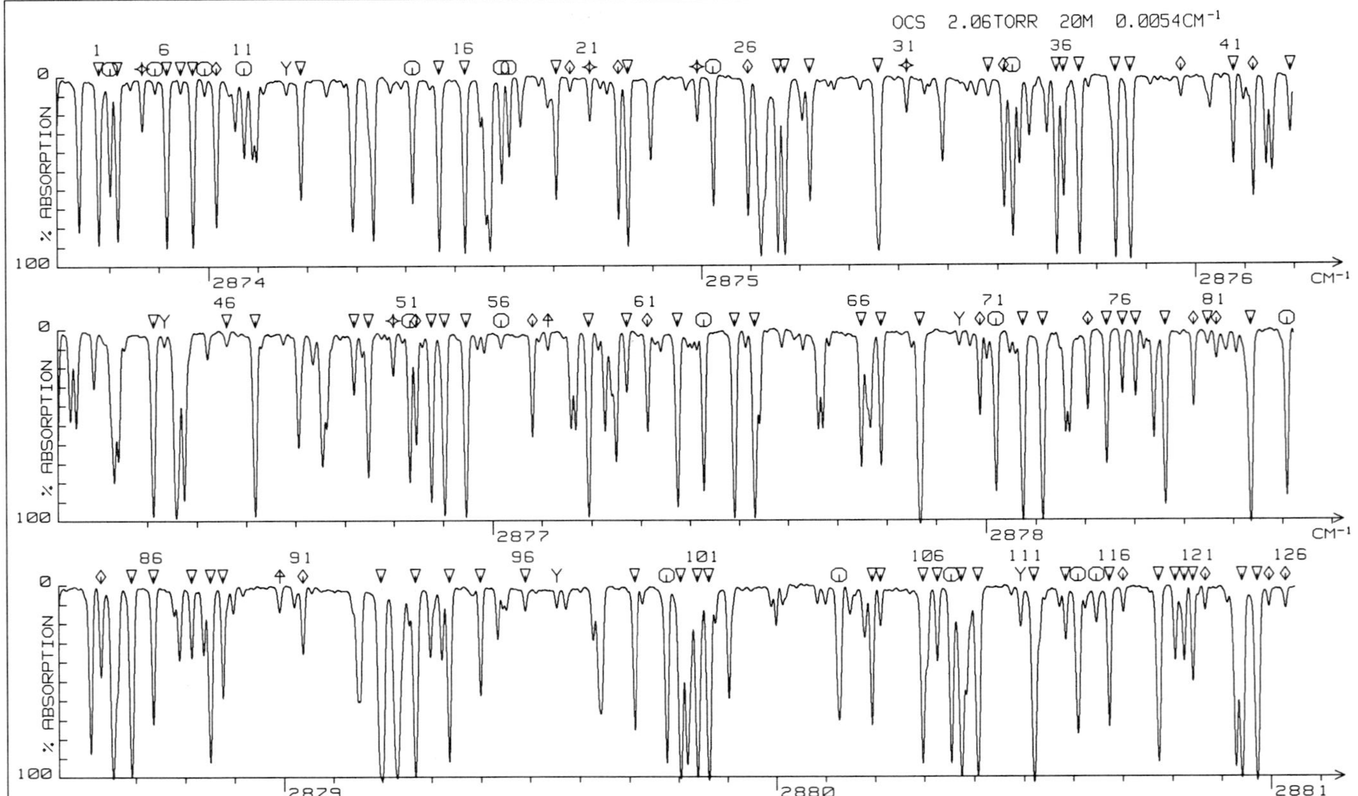

ISOTOPIC IDENTIFICATION – '▽' indicates $^{16}O^{12}C^{32}S$; '⏀': $^{16}O^{12}C^{34}S$; '◊': $^{16}O^{13}C^{32}S$; 'Y': $^{16}O^{12}C^{33}S$;
'✧' indicates $^{18}O^{12}C^{32}S$; '⏚': $^{16}O^{13}C^{34}S$; '↑': Other OCS species; 'I': not OCS.

LINE NO.	WAVENUMBER CM^{-1}, VAC.	UNC.	ROTATIONAL BAND ASSIGNMENTS	ISOTOPIC SPECIES
1	2873.77797	(12)	P 81 1010 -0000	16012C32S
2	2873.80134	(13)	P 64 1010 -0000	16012C34S
3	2873.81665	(12)	P 35 1212 -0202	16012C32S
4	2873.86664	(20)	R 42 1010 -0000	18012C32S
5	2873.89294	(60)	P 44 2400 -1000	16012C34S
6	2873.91595	(12)	P 59 1111E-0101E	16012C32S
7	2873.94427	(62)	P 7 1313 -0303	16012C32S
8	2873.96851	(12)	P 59 1111F-0101F	16012C32S
9	2873.99265	(52)	P 11 1212 -0202	16012C34S
10	2874.01577	(12)	R 37 1010 -0000	16013C32S
11	2874.07283	(16)	P 39 1111E-0101E	16012C34S
11			P 49 1111E-0101E	16012C33S
12	2874.15854	(62)	P 72 1010 -0000	16012C33S
13	2874.18738	(13)	P 35 1210 -0200	16012C32S
13			P 31 1010 -0000	17012C32S
14	2874.41467	(13)	P 63 1010 -0000	16012C34S
15	2874.46854	(12)	P 80 1010 -0000	16012C32S
16	2874.52039	(12)	P 58 1111E-0101E	16012C32S
17	2874.59508	(13)	P 38 1111E-0101E	16012C34S
17			P 41 2111E-1101E	16012C32S
18	2874.61072	(15)	P 38 1111F-0101F	16012C34S
19	2874.70578	(22)	P 34 1210 -0200	16012C32S
19			P 10 1210 -0200	16012C34S
20	2874.73406	(63)	R 7 1400 -0000	16013C32S
21	2874.77335	(23)	R 46 1010 -0000	18012C32S
22	2874.83053	(12)	R 40 1010 -0000	16013C32S
23	2874.84966	(12)	P 33 1212 -0202	16012C32S
24	2874.99181	(23)	R 47 1010 -0000	18012C32S
24			P 21 1212 -0202	16012C33S
25	2875.02438	(12)	P 62 1010 -0000	16012C34S
26	2875.09484	(12)	R 41 1010 -0000	16013C32S
27	2875.15539	(13)	P 79 1010 -0000	16012C32S
28	2875.16988	(13)	P 57 1111F-0101F	16012C32S
28			P 40 2111F-1101F	16012C32S
29	2875.22126	(13)	P 33 1210 -0200	16012C32S
30	2875.35939	(14)	P 32 1212 -0202	16012C32S
30			R 42 1010 -0000	16013C32S
31	2875.41739	(27)	R 49 1010 -0000	18012C32S
32	2875.58230	(63)	P 12 2212 -1202	16012C32S
33	2875.61323	(13)	R 43 1010 -0000	16013C32S
34	2875.63060	(12)	P 61 1010 -0000	16012C34S
34			P 36 1111E-0101E	16012C34S
35	2875.71883	(12)	P 56 1111E-0101E	16012C32S
36	2875.73338	(26)	P 32 1210 -0200	16012C32S
37	2875.76570	(12)	P 56 1111F-0101F	16012C32S
38	2875.83868	(13)	P 78 1010 -0000	16012C32S
39	2875.86887	(12)	P 31 1212 -0202	16012C32S
39			R 44 1010 -0000	16013C32S
40	2875.97103	(46)	R 10 1400 -0000	16013C32S
40			P 7 1210 -0200	16012C34S
41	2876.07710	(14)	P 62 2010 -1000	16012C32S
42	2876.11717	(13)	R 45 1010 -0000	16013C32S
43	2876.19247	(19)	P 38 2111E-1101E	16012C32S
44	2876.31285	(12)	P 55 1111E-0101E	16012C32S
45	2876.33534	(74)	P 45 1111F-0101F	16012C33S
46	2876.46162	(84)	P 10 2212 -1202	16012C32S
47	2876.51840	(11)	P 77 1010 -0000	16012C32S
48	2876.71870	(18)	P 37 2111E-1101E	16012C32S
49	2876.74842	(13)	P 30 1210 -0200	16012C32S
49			P 37 2111F-1101F	16012C32S
50	2876.79961	(24)	R 56 1010 -0000	18012C32S
50			R 12 1400 -0000	16013C32S
50			P 5 1210 -0200	16012C34S
50			Q 6 1313 -0303	16012C32S

- "LINE NO.": serial number identifying the spectral line in the map on the opposite page.
- "WAVENUMBER": experimental value.
- Uncertainty in the last digit of the wavenumber is given in parentheses.

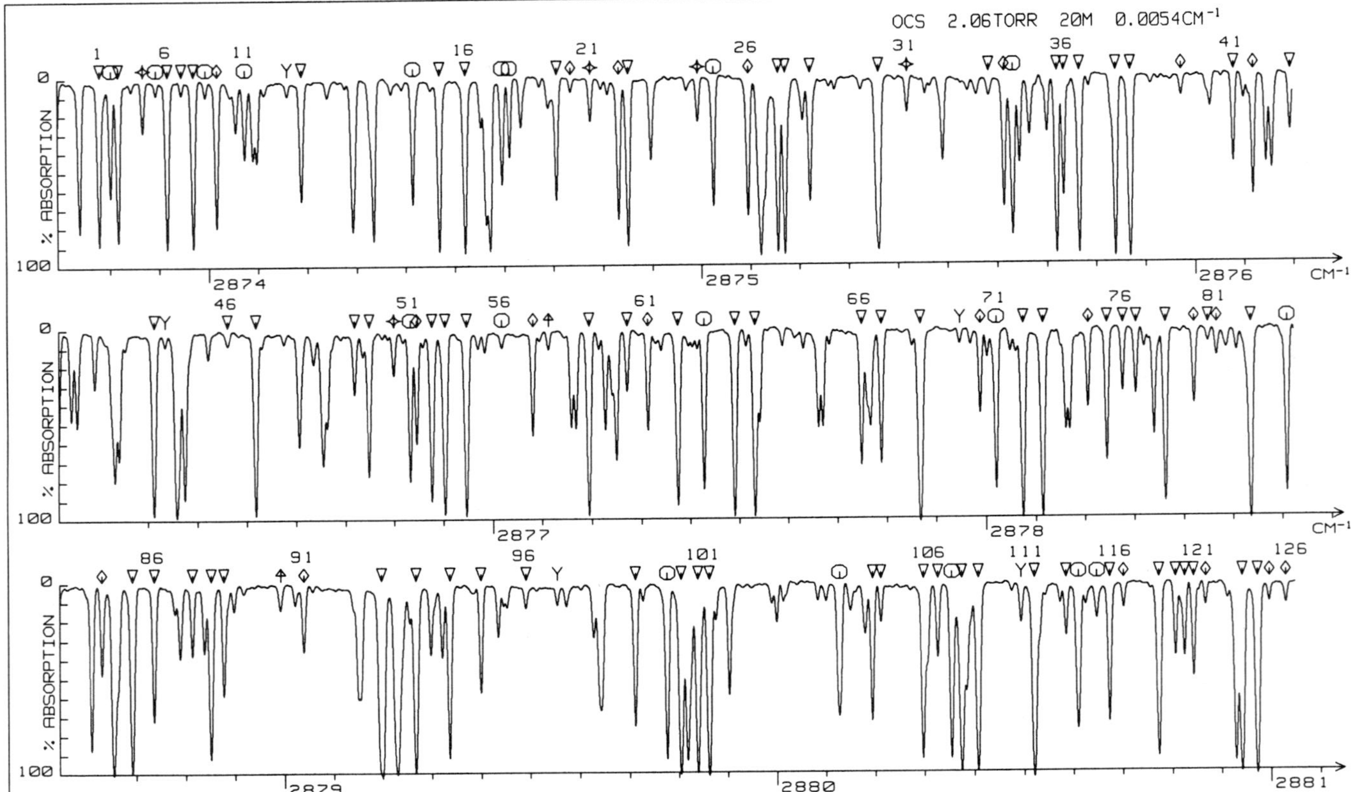

ISOTOPIC IDENTIFICATION – '▽' indicates $^{16}O^{12}C^{32}S$; 'Ⓘ': $^{16}O^{12}C^{34}S$; '◊': $^{16}O^{13}C^{32}S$; 'Y': $^{16}O^{12}C^{33}S$;
'✧' indicates $^{18}O^{12}C^{32}S$; '⍖': $^{16}O^{13}C^{34}S$; '↑': Other OCS species; 'I': not OCS.

LINE NO.	WAVENUMBER CM^{-1}, VAC.	UNC.	ROTATIONAL BAND ASSIGNMENTS	ISOTOPIC SPECIES
51	2876.83284(	17)	P 59 1010 -0000	16012C34S
52	2876.84605(	26)	R 48 1010 -0000	16013C32S
53	2876.87664(	12)	P 29 1212 -0202	16012C32S
54	2876.90345(	12)	P 54 1111E-0101E	16012C32S
55	2876.94673(	11)	P 54 1111F-0101F	16012C32S
56	2877.01812(	65)	P 36 2400 -1000	16012C34S
57	2877.08187(	13)	R 49 1010 -0000	16013C32S
58	2877.11324(	49)	P 25 1010 -0000	17012C32S
59	2877.19449(	11)	P 76 1010 -0000	16012C32S
60	2877.27132(	18)	P 36 2111F-1101F	16012C32S
61	2877.31392(	14)	R 50 1010 -0000	16013C32S
62	2877.37501(	12)	P 28 1212 -0202	16012C32S
62			P 67 1010 -0000	16012C33S
63	2877.42861(	12)	P 58 1010 -0000	16012C34S
64	2877.49073(	11)	P 53 1111E-0101E	16012C32S
65	2877.53213(	12)	P 53 1111F-0101F	16012C32S
66	2877.74886(	20)	P 28 1210 -0200	16012C32S
67	2877.78907(	13)	P 59 2010 -1000	16012C32S
67			P 35 2111F-1101F	16012C32S
67			P 44 3010 -2000	16012C32S
68	2877.86812(	12)	P 75 1010 -0000	16012C32S
68			P 27 1212 -0202	16012C32S
69	2877.94736(	62)	P 42 1111E-0101E	16012C33S
70	2877.98858(	15)	R 53 1010 -0000	16013C32S
71	2878.02078(	12)	P 57 1010 -0000	16012C34S
72	2878.07442(	11)	P 52 1111E-0101E	16012C32S
73	2878.11411(	11)	P 52 1111F-0101F	16012C32S
74	2878.20616(	16)	R 54 1010 -0000	16013C32S
75	2878.24455(	12)	P 27 1210 -0200	16012C32S
76	2878.27649(	17)	P 34 2111E-1101E	16012C32S
77	2878.30302(	17)	P 34 2111F-1101F	16012C32S
78	2878.36238(	12)	P 26 1212 -0202	16012C32S
79	2878.42026(	16)	R 55 1010 -0000	16013C32S
80	2878.44923(	73)	P 62 2402E-1000	16012C32S
80			Q 7 1212 -0202	16012C34S
81	2878.46670(	38)	R 16 1400 -0000	16013C32S
81			R 3 1313 -0303	16012C32S
82	2878.53588(	12)	P 74 1010 -0000	16012C32S
83	2878.60956(	12)	P 56 1010 -0000	16012C34S
84	2878.63069(	14)	R 56 1010 -0000	16013C32S
84			P 65 1010 -0000	16012C33S
84			R 67 1010 -0000	18012C32S
85	2878.69267(	11)	P 51 1111F-0101F	16012C32S
86	2878.73693(	12)	P 26 1210 -0200	16012C32S
87	2878.81392(	16)	P 33 2111F-1101F	16012C32S
88	2878.85118(	14)	P 25 1212 -0202	16012C32S
89	2878.87735(	27)	P 57 2010 -1000	16012C32S
90	2878.99216(	40)	P 21 1010 -0000	17012C32S
90			R 3 1311F-0301F	16012C32S
90			R 3 1311E-0301E	16012C32S
91	2879.03946(	18)	R 58 1010 -0000	16013C32S
91			P 40 1111F-0101F	16012C33S
92	2879.19871(	15)	P 73 1010 -0000	16012C32S
92			P 55 1010 -0000	16012C34S
93	2879.26791(	11)	P 50 1111F-0101F	16012C32S
94	2879.33659(	12)	P 24 1212 -0202	16012C32S
95	2879.39967(	13)	P 56 2010 -1000	16012C32S
96	2879.49136(	43)	P 59 2402E-1000	16012C32S
97	2879.55484(	51)	P 39 1111E-0101E	16012C33S
98	2879.71158(	12)	P 24 1210 -0200	16012C32S
98			P 40 2010 -1000	16012C34S
99	2879.77650(	12)	P 54 1010 -0000	16012C34S
100	2879.80503(	12)	P 49 1111E-0101E	16012C32S
100			P 31 2111E-1101E	16012C32S
101	2879.83961(	11)	P 49 1111F-0101F	16012C32S

- "LINE NO.": serial number identifying the spectral line in the map on the opposite page.
- "WAVENUMBER": experimental value.
- Uncertainty in the last digit of the wavenumber is given in parentheses.

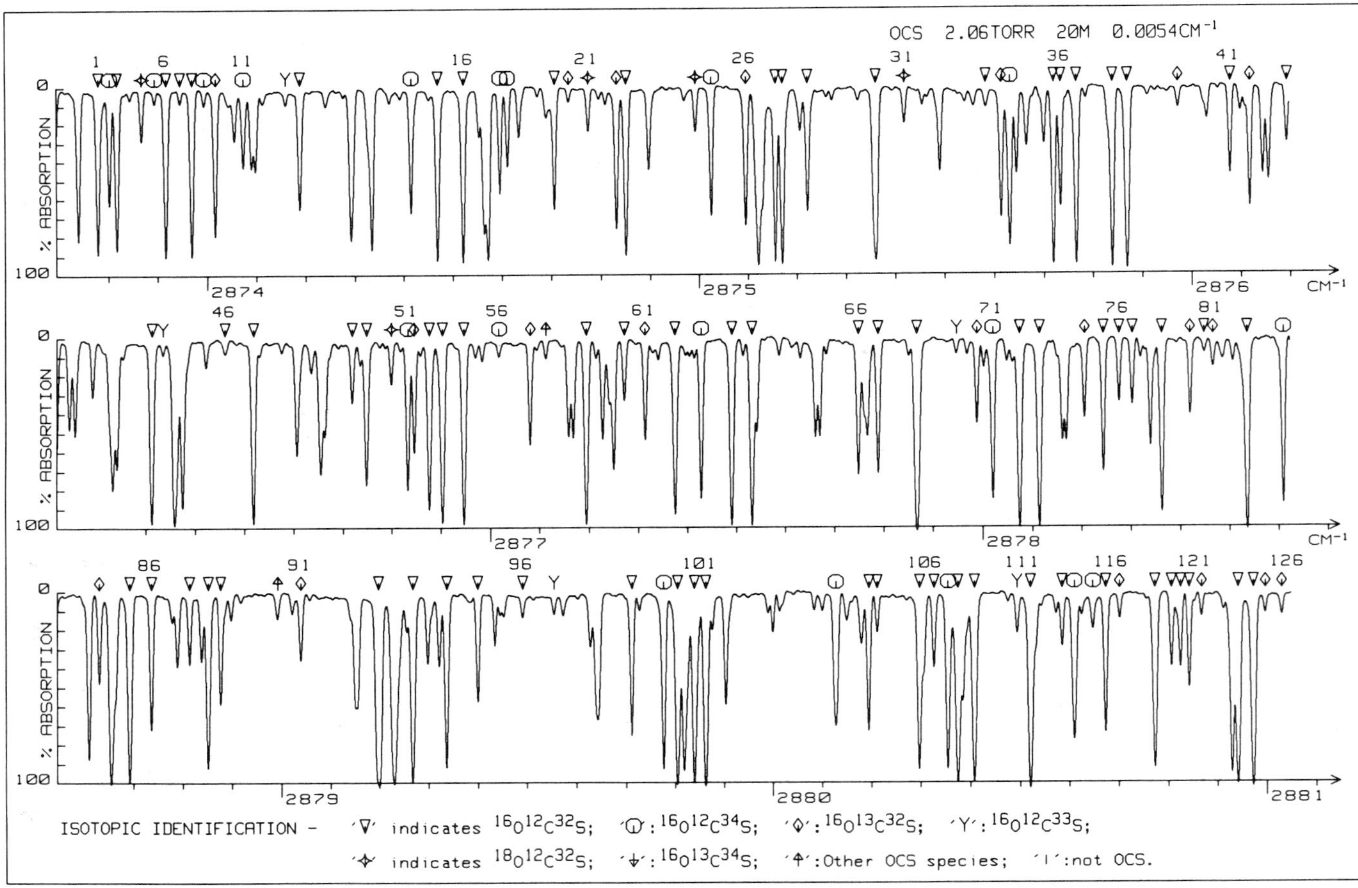
OCS 2.06TORR 20M 0.0054CM⁻¹
% ABSORPTION
0
100
2874
2875
2876
2877
2878
2879
2880
2881
CM⁻¹
ISOTOPIC IDENTIFICATION - '▽' indicates 16O12C32S; 'Ⓘ': 16O12C34S; '◊': 16O13C32S; 'Y': 16O12C33S;
'✧' indicates 18O12C32S; '⍖': 16O13C34S; '↑': Other OCS species; 'I': not OCS.

LINE NO.	WAVENUMBER CM^{-1}, VAC.	UNC.	ROTATIONAL	BAND ASSIGNMENTS	ISOTOPIC SPECIES
102	2879.86276	(11)	P 72	1010 -0000	16012C32S
103	2880.12784	(14)	P 27	1111 -0101	16012C34S
104	2880.19437	(12)	P 23	1210 -0200	16012C32S
105	2880.21185	(27)	P 57	2402E-1000	16012C32S
106	2880.29779	(69)	P 22	1212 -0202	16012C32S
107	2880.32698	(16)	P 30	2111F-1101F	16012C32S
108	2880.35493	(14)	P 53	1010 -0000	16012C34S
108			R 65	1010 -0000	16013C32S
109	2880.37508	(12)	P 48	1111E-0101E	16012C32S
110	2880.40799	(11)	P 48	1111F-0101F	16012C32S
111	2880.49395	(25)	P 62	1010 -0000	16012C33S
112	2880.52082	(11)	P 71	1010 -0000	16012C32S
113	2880.58522	(21)	P 56	2402E-1000	16012C32S
114	2880.61010	(13)	P 26	1111E-0101E	16012C34S
114			P 26	1111F-0101F	16012C34S
114			P 37	1111E-0101E	16012C33S
115	2880.64754	(33)	P 38	2010 -1000	16012C34S
115			Q 3	2212 -1202	16012C32S
116	2880.67363	(12)	P 22	1210 -0200	16012C32S
117	2880.70198	(37)	R 67	1010 -0000	16013C32S
118	2880.77322	(12)	P 21	1212 -0202	16012C32S
119	2880.80632	(40)	P 29	2111E-1101E	16012C32S
120	2880.82469	(43)	P 29	2111F-1101F	16012C32S
121	2880.84291	(14)	P 53	2010 -1000	16012C32S
122	2880.86798	(39)	R 68	1010 -0000	16013C32S
123	2880.94163	(12)	P 47	1111E-0101E	16012C32S
124	2880.97282	(11)	P 47	1111F-0101F	16012C32S
124			P 55	2402E-1000	16012C32S
125	2880.99691	(50)	R 22	1400 -0000	16013C32S
126	2881.03060	(46)	R 69	1010 -0000	16013C32S

- "LINE NO.": serial number identifying the spectral line in the map on the opposite page.
- "WAVENUMBER": experimental value.
- Uncertainty in the last digit of the wavenumber is given in parentheses.

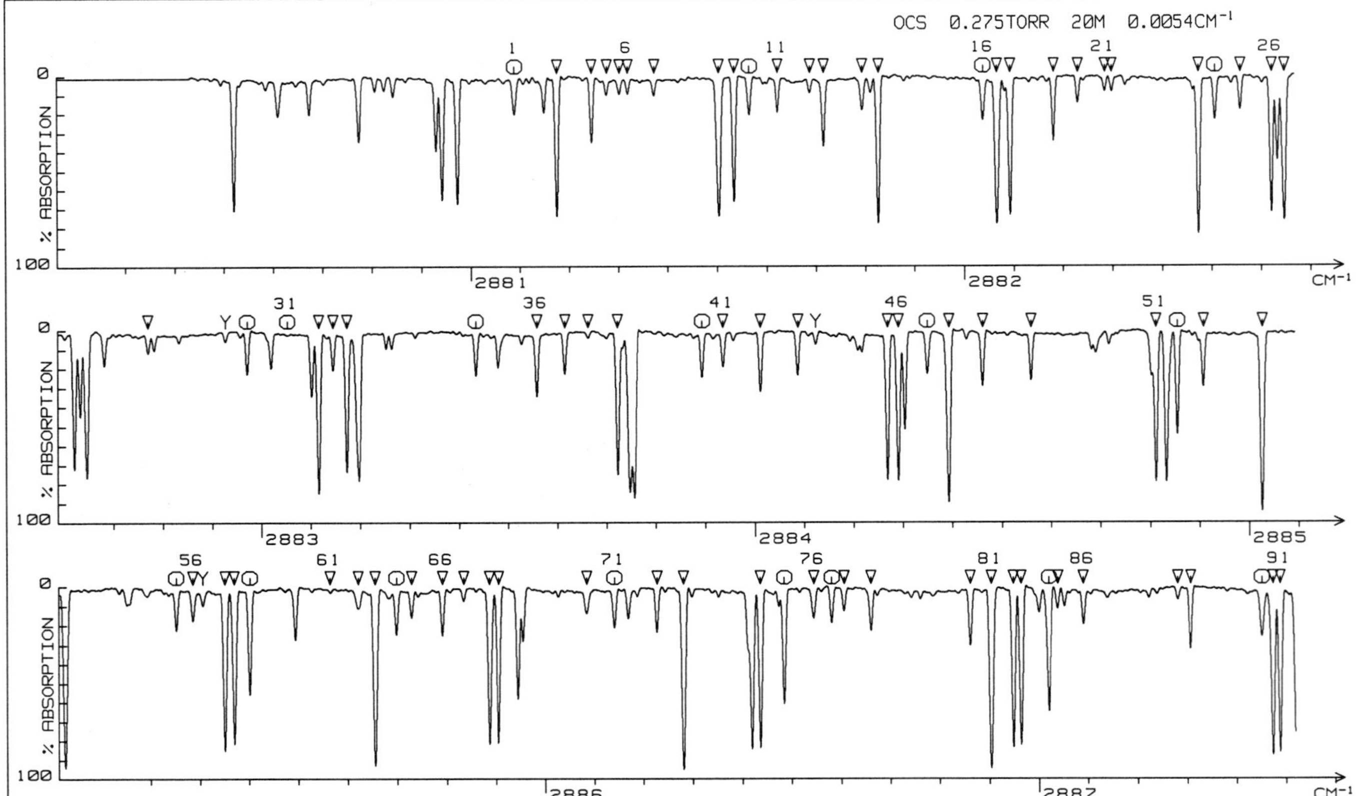

ISOTOPIC IDENTIFICATION – '▽' indicates $^{16}O^{12}C^{32}S$; '⦶': $^{16}O^{12}C^{34}S$; '◇': $^{16}O^{13}C^{32}S$; 'Y': $^{16}O^{12}C^{33}S$;
'✧' indicates $^{18}O^{12}C^{32}S$; '⊕': $^{16}O^{13}C^{34}S$; '✦': Other OCS species; 'I': not OCS.

LINE NO.	WAVENUMBER UNC. CM^{-1}, VAC.	ROTATIONAL BAND ASSIGNMENTS	ISOTOPIC SPECIES
1	2881.08922(13)	P 25 1111 -0101	16012C34S
2	2881.17529(11)	P 70 1010 -0000	16012C32S
3	2881.24581(12)	P 20 1212 -0202	16012C32S
4	2881.27624(15)	P 52 2402E-1000	16012C32S
5	2881.30201(16)	P 28 2111E-1101E	16012C32S
5		R 9 1311F-0301F	16012C32S
6	2881.31916(15)	P 28 2111F-1101F	16012C32S
7	2881.37301(15)	P 54 2402E-1000	16012C32S
8	2881.50374(12)	P 46 1111E-0101E	16012C32S
8		P 51 1010 -0000	16012C34S
9	2881.53447(11)	P 46 1111F-0101F	16012C32S
10	2881.56492(13)	P 24 1111 -0101	16012C34S
11	2881.62227(12)	P 20 1210 -0200	16012C32S
12	2881.68769(16)	P 51 2402E-1000	16012C32S
13	2881.71519(12)	P 19 1212 -0202	16012C32S
13		P 60 1010 -0000	16012C33S
14	2881.79342(13)	P 53 2402E-1000	16012C32S
14		P 27 2111E-1101E	16012C32S
15	2881.82617(11)	P 69 1010 -0000	16012C32S
16	2882.03751(13)	P 23 1111 -0101	16012C34S
16		R 11 1311F-0301F	16012C32S
17	2882.06559(12)	P 45 1111E-0101E	16012C32S
17		P 50 1010 -0000	16012C34S
18	2882.09248(11)	P 45 1111F-0101F	16012C32S
18		P 19 1210 -0200	16012C32S
19	2882.18079(12)	P 18 1212 -0202	16012C32S
19		P 34 1111F-0101F	16012C33S
20	2882.23002(13)	P 52 2010 -1000	16012C32S
21	2882.28423(43)	P 26 2111E-1101E	16012C32S
22	2882.29833(16)	P 26 2111F-1101F	16012C32S
23	2882.47338(11)	P 68 1010 -0000	16012C32S
24	2882.50654(12)	P 22 1111 -0101	16012C34S
25	2882.55773(12)	P 18 1210 -0200	16012C32S
26	2882.62081(24)	P 44 1111E-0101E	16012C32S
27	2882.64690(25)	P 44 1111F-0101F	16012C32S
27		P 17 1212 -0202	16012C32S
28	2882.77070(34)	P 25 2111E-1101E	16012C32S
28		R 13 1311F-0301F	16012C32S
29	2882.92828(20)	P 58 1010 -0000	16012C33S
30	2882.97236(12)	P 21 1111 -0101	16012C34S
31	2883.05348(54)	R 11 1212 -0202	16012C34S
32	2883.11700(11)	P 67 1010 -0000	16012C32S
33	2883.14608(12)	P 50 2010 -1000	16012C32S
34	2883.17368(11)	P 43 1111E-0101E	16012C32S
35	2883.43486(12)	P 20 1111 -0101	16012C34S
36	2883.55897(12)	P 15 1212 -0202	16012C32S
36		P 46 2402E-1000	16012C32S
37	2883.61573(12)	P 49 2010 -1000	16012C32S
38	2883.66245(27)	R 17 1313 -0303	16012C32S
39	2883.72309(11)	P 42 1111E-0101E	16012C32S
40	2883.89410(12)	P 19 1111 -0101	16012C34S
41	2883.93632(13)	P 15 1210 -0200	16012C32S
42	2884.01167(12)	P 14 1212 -0202	16012C32S
43	2884.08770(12)	P 48 2010 -1000	16012C32S
44	2884.12413(17)	P 56 1010 -0000	16012C33S
44		R 7 2210 -1200	16012C32S
45	2884.26915(11)	P 41 1111E-0101E	16012C32S
46	2884.29135(11)	P 41 1111F-0101F	16012C32S
47	2884.35040(12)	P 18 1111 -0101	16012C34S
48	2884.39321(11)	P 65 1010 -0000	16012C32S
48		P 14 1210 -0200	16012C32S
49	2884.46122(15)	P 13 1212 -0202	16012C32S
50	2884.55932(12)	P 47 2010 -1000	16012C32S
51	2884.81177(21)	P 40 1111E-0101E	16012C32S
52	2884.85462(12)	P 45 1010 -0000	16012C34S

- "LINE NO.": serial number identifying the spectral line in the map on the opposite page.
- "WAVENUMBER": experimental value.
- Uncertainty in the last digit of the wavenumber is given in parentheses.

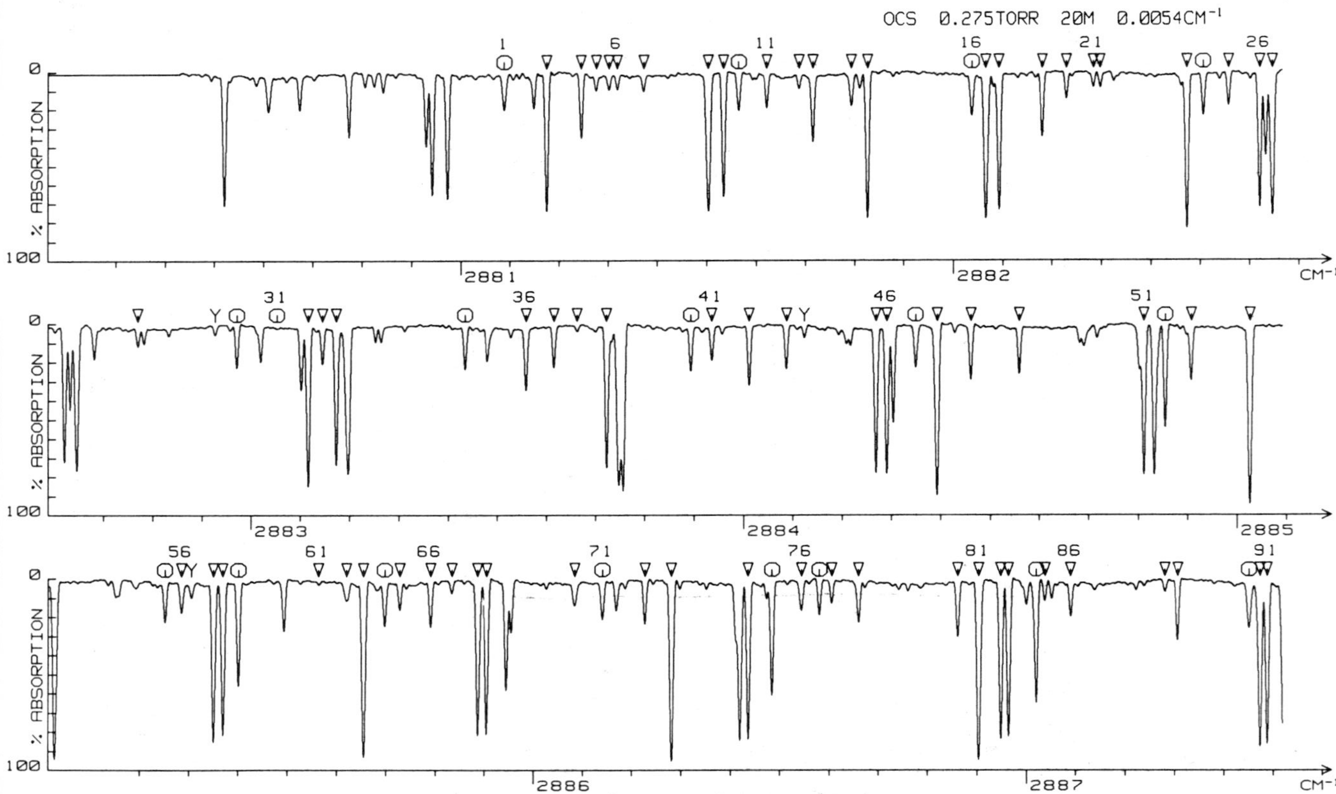

ISOTOPIC IDENTIFICATION – '▽' indicates $^{16}O^{12}C^{32}S$; 'Ⓘ': $^{16}O^{12}C^{34}S$; '◇': $^{16}O^{13}C^{32}S$; 'Y': $^{16}O^{12}C^{33}S$;
'✧' indicates $^{18}O^{12}C^{32}S$; '⏚': $^{16}O^{13}C^{34}S$; '✝': Other OCS species; 'I': not OCS.

LINE NO.	WAVENUMBER CM^{-1}, VAC.	UNC.	ROTATIONAL BAND ASSIGNMENTS	ISOTOPIC SPECIES
53	2884.90764	(15)	P 12 1212 -0202	16012C32S
54	2885.02658	(11)	P 64 1010 -0000	16012C32S
54			P 46 2010 -1000	16012C32S
55	2885.25194	(12)	P 16 1111 -0101	16012C34S
56	2885.28554	(13)	P 12 1210 -0200	16012C32S
57	2885.30571	(16)	P 54 1010 -0000	16012C33S
57			R 32 1400 -0000	16013C32S
58	2885.35087	(11)	P 39 1111E-0101E	16012C32S
58			P 11 1212 -0202	16012C32S
59	2885.37024	(11)	P 39 1111F-0101F	16012C32S
60	2885.40146	(11)	P 44 1010 -0000	16012C34S
61	2885.56435	(50)	R 21 1311F-0301F	16012C32S
62	2885.62111	(54)	P 19 2111E-1101E	16012C32S
62			P 19 2111F-1101F	16012C32S
63	2885.65551	(11)	P 63 1010 -0000	16012C32S
64	2885.69806	(17)	P 15 1111F-0101F	16012C34S
64			P 15 1111E-0101E	16012C34S
64			R 23 1313 -0303	16012C32S
65	2885.72847	(13)	P 11 1210 -0200	16012C32S
66	2885.79050	(12)	P 10 1212 -0202	16012C32S
67	2885.83369	(17)	P 55 2400 -1000	16012C32S
67			P 27 2010 -1000	16012C34S
68	2885.88672	(11)	P 38 1111E-0101E	16012C32S
69	2885.90474	(11)	P 38 1111F-0101F	16012C32S
70	2886.08384	(15)	P 18 2111E-1101E	16012C32S
70			P 18 2111F-1101F	16012C32S
70			R 19 1210 -0200	16012C34S
71	2886.14053	(13)	P 14 1111 -0101	16012C34S
72	2886.22715	(12)	P 9 1212 -0202	16012C32S
72			R 23 1311F-0301F	16012C32S
73	2886.28113	(11)	P 62 1010 -0000	16012C32S
74	2886.43582	(11)	P 37 1111F-0101F	16012C32S
75	2886.48450	(12)	P 42 1010 -0000	16012C34S

LINE NO.	WAVENUMBER CM^{-1}, VAC.	UNC.	ROTATIONAL BAND ASSIGNMENTS	ISOTOPIC SPECIES
76	2886.54383	(14)	P 17 2111E-1101E	16012C32S
76			P 17 2111F-1101F	16012C32S
76			P 53 2400 -1000	16012C32S
77	2886.58014	(13)	P 13 1111 -0101	16012C34S
78	2886.60487	(14)	P 9 1210 -0200	16012C32S
79	2886.66002	(18)	P 8 1212 -0202	16012C32S
79			P 25 1111 -0101	16012C33S
80	2886.86093	(12)	P 42 2010 -1000	16012C32S
81	2886.90326	(11)	P 61 1010 -0000	16012C32S
82	2886.94791	(11)	P 36 1111E-0101E	16012C32S
83	2886.96347	(11)	P 36 1111F-0101F	16012C32S
84	2887.01973	(14)	P 41 1010 -0000	16012C34S
84			P 12 1111 -0101	16012C34S
85	2887.03806	(14)	P 8 1210 -0200	16012C32S
86	2887.09025	(13)	P 7 1212 -0202	16012C32S
87	2887.27996	(17)	P 51 2400 -1000	16012C32S
88	2887.30594	(12)	P 41 2010 -1000	16012C32S
89	2887.45065	(20)	P 11 1111 -0101	16012C34S
89			P 15 2111 -1101	16012C32S
90	2887.47326	(11)	P 35 1111E-0101E	16012C32S
91	2887.48776	(11)	P 35 1111F-0101F	16012C32S

- "LINE NO.": serial number identifying the spectral line in the map on the opposite page.
- "WAVENUMBER": experimental value.
- Uncertainty in the last digit of the wavenumber is given in parentheses.

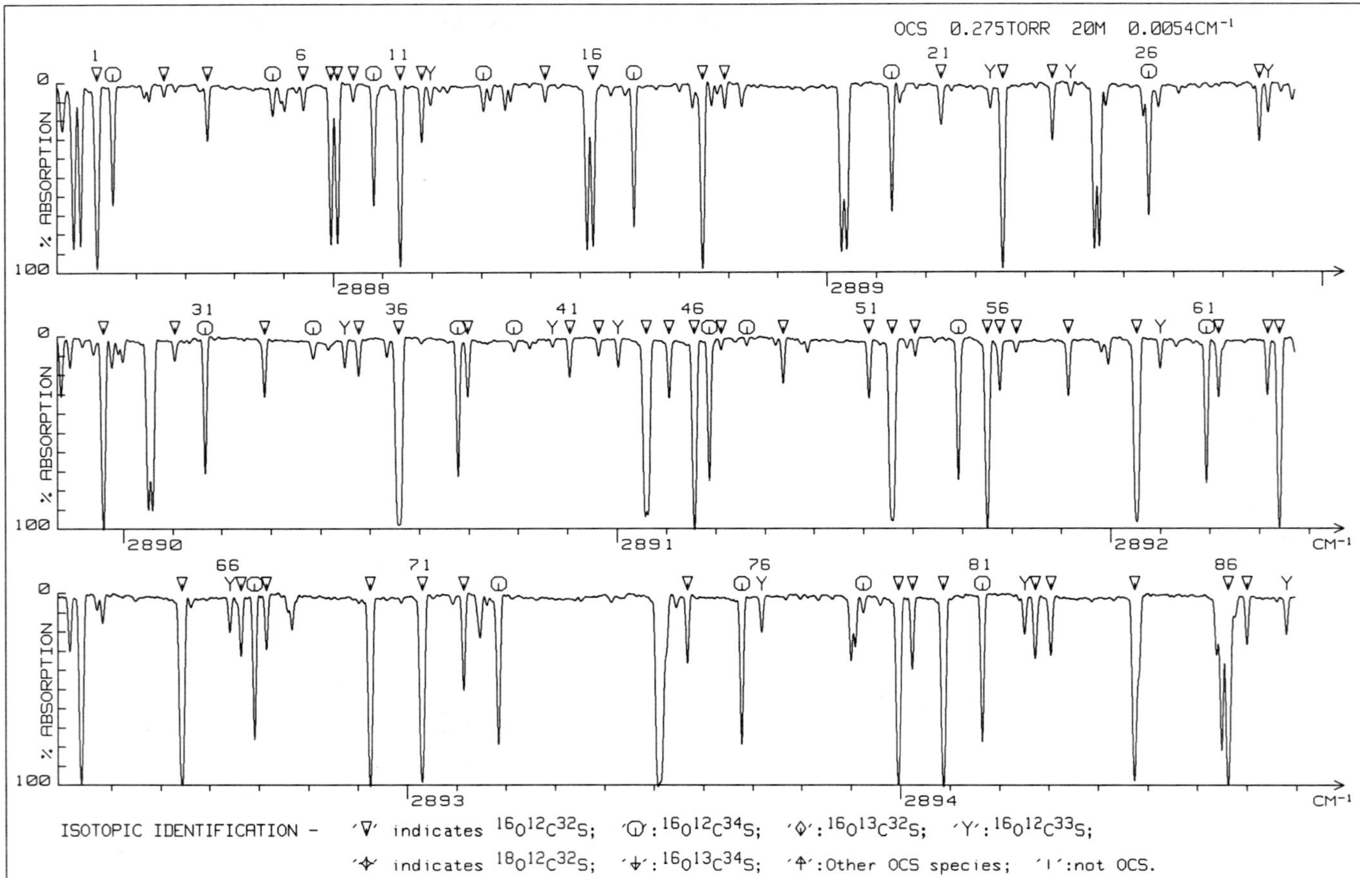

OCS 0.275TORR 20M 0.0054CM-1
% ABSORPTION
0
100
2888
2889
2890
2891
2892
2893
2894
CM-1
ISOTOPIC IDENTIFICATION – '▽' indicates $^{16}O^{12}C^{32}S$; 'Ⓘ': $^{16}O^{12}C^{34}S$; '◊': $^{16}O^{13}C^{32}S$; 'Y': $^{16}O^{12}C^{33}S$;
'✧' indicates $^{18}O^{12}C^{32}S$; '⫛': $^{16}O^{13}C^{34}S$; '✝': Other OCS species; 'I': not OCS.

LINE NO.	WAVENUMBER CM^{-1}, VAC.	UNC.	ROTATIONAL ASSIGNMENTS	BAND	ISOTOPIC SPECIES
1	2887.52142	(11)	P 60	1010 -0000	16012C32S
2	2887.55365	(11)	P 40	1010 -0000	16012C34S
3	2887.65691	(16)	P 50	2400 -1000	16012C32S
4	2887.74561	(12)	P 40	2010 -1000	16012C32S
5	2887.87862	(13)	P 10	1111 -0101	16012C34S
6	2887.94046	(13)	P 5	1212 -0202	16012C32S
7	2887.99557	(18)	P 34	1111E-0101E	16012C32S
8	2888.00882	(18)	P 34	1111F-0101F	16012C32S
9	2888.04083	(16)	P 49	2400 -1000	16012C32S
9			R 26	1212 -0202	16012C34S
10	2888.08299	(12)	P 39	1010 -0000	16012C34S
11	2888.13639	(11)	P 59	1010 -0000	16012C32S
12	2888.18015	(12)	P 39	2400 -1000	16012C32S
13	2888.19821	(14)	P 49	1010 -0000	16012C33S
14	2888.30478	(26)	P 9	1111 -0101	16012C34S
15	2888.43038	(14)	P 48	2400 -1000	16012C32S
16	2888.52631	(18)	P 33	1111F-0101F	16012C32S
17	2888.60915	(11)	P 38	1010 -0000	16012C34S
17			P 38	2400 -1000	16012C32S
18	2888.74767	(11)	P 58	1010 -0000	16012C32S
19	2888.79268	(25)	P 12	2111 -1101	16012C32S
19			R 21	2212 -1202	16012C32S
20	2889.13124	(11)	P 37	1010 -0000	16012C34S
21	2889.23140	(13)	P 46	2400 -1000	16012C32S
21			P 11	2111 -1101	16012C32S
21			R 29	1210 -0200	16012C34S
22	2889.33036	(29)	P 47	1010 -0000	16012C33S
23	2889.35530	(11)	P 57	1010 -0000	16012C32S
24	2889.45595	(15)	P 36	2400 -1000	16012C32S
25	2889.49362	(18)	P 19	1111 -0101	16012C33S
25			P 19	2010 -1000	16012C34S
25			R 3	1010 -0000	17012C32S
26	2889.65026	(12)	P 36	1010 -0000	16012C34S

LINE NO.	WAVENUMBER CM^{-1}, VAC.	UNC.	ROTATIONAL ASSIGNMENTS	BAND	ISOTOPIC SPECIES
27	2889.87318	(15)	P 35	2400 -1000	16012C32S
28	2889.89122	(22)	P 46	1010 -0000	16012C33S
28			Q 8	1212 -0202	16012C32S
29	2889.95929	(11)	P 56	1010 -0000	16012C32S
30	2890.10391	(14)	P 9	2111 -1101	16012C32S
30			R 24	2210 -1200	16012C32S
30			R 25	2212F-1202F	16012C32S
30			R 25	2212E-1202E	16012C32S
31	2890.16575	(11)	P 35	1010 -0000	16012C34S
32	2890.28690	(15)	P 34	2400 -1000	16012C32S
33	2890.38498	(17)	P 4	1111 -0101	16012C34S
33			P 17	2010 -1000	16012C34S
34	2890.44851	(13)	P 45	1010 -0000	16012C33S
34			R 37	1311E-0301E	16012C32S
35	2890.47703	(12)	P 43	2400 -1000	16012C32S
36	2890.55866	(13)	P 55	1010 -0000	16012C32S
36			P 29	1111E-0101E	16012C32S
36			P 29	1111F-0101F	16012C32S
37	2890.67776	(11)	P 34	1010 -0000	16012C34S
38	2890.69745	(15)	P 33	2400 -1000	16012C32S
39	2890.79222	(20)	P 3	1111 -0101	16012C34S
39			R 0	1210 -0200	16012C32S
39			R 16	1210 -0200	16012C33S
39			R 17	1212 -0202	16012C33S
40	2890.86988	(23)	P 16	1111 -0101	16012C33S
41	2890.90446	(12)	P 42	2400 -1000	16012C32S
42	2890.96264	(15)	P 7	2111E-1101E	16012C32S
42			P 7	2111F-1101F	16012C32S
42			R 35	1210 -0200	16012C34S
42			R 36	1212E-0202E	16012C34S
43	2891.00221	(13)	P 44	1010 -0000	16012C33S
43			R 39	1311E-0301E	16012C32S

- "LINE NO.": serial number identifying the spectral line in the map on the opposite page.
- "WAVENUMBER": experimental value.
- Uncertainty in the last digit of the wavenumber is given in parentheses.

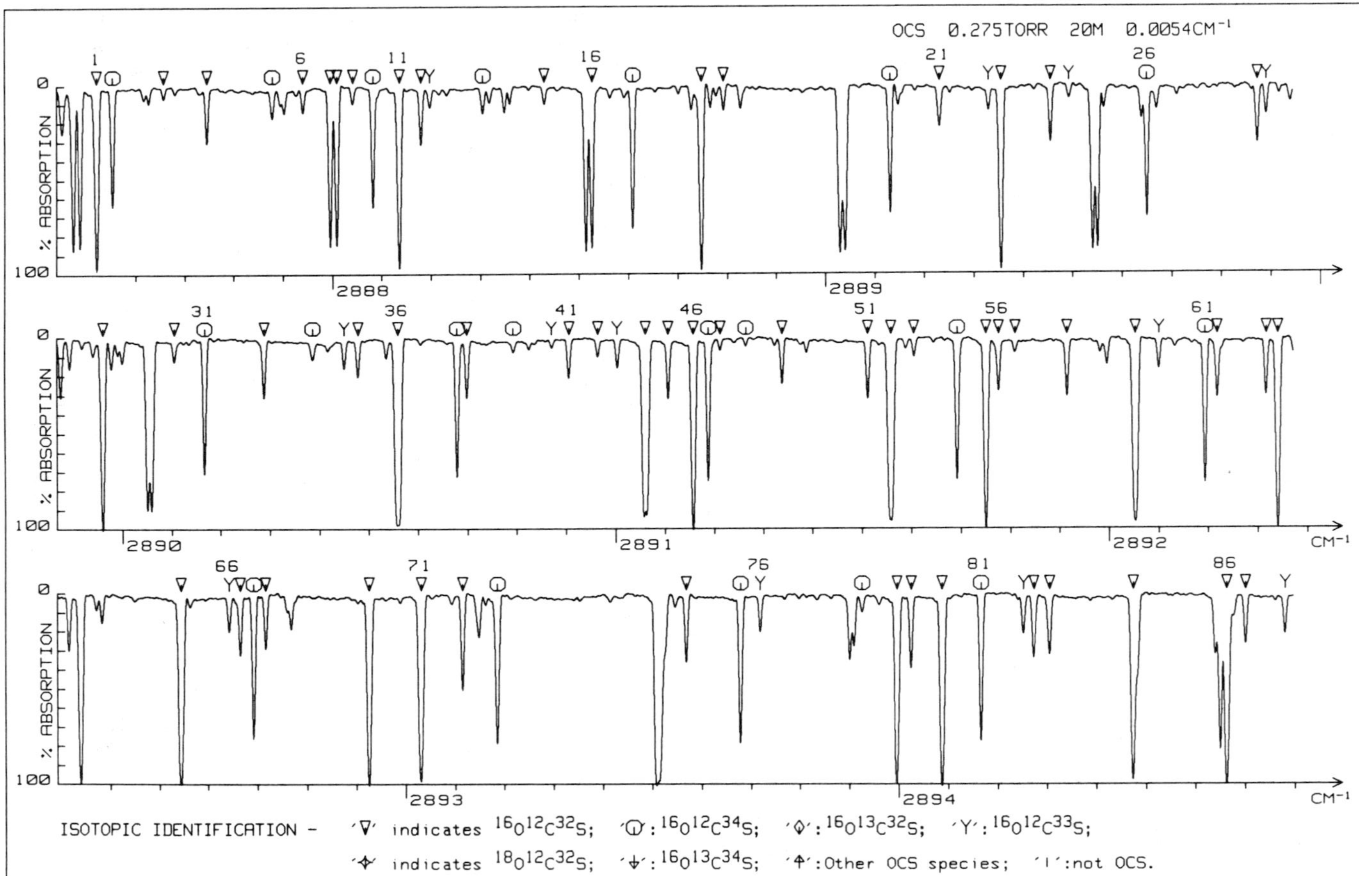
OCS 0.275TORR 20M 0.0054CM⁻¹
% ABSORPTION
0
100
2888
2889
2890
2891
2892
2893
2894
CM⁻¹
1
6
11
16
21
26
31
36
41
46
51
56
61
66
71
76
81
86
ISOTOPIC IDENTIFICATION – '▽' indicates 16O12C32S; 'Ⓘ': 16O12C34S; 'Ⓘ': 16O13C32S; 'Y': 16O12C33S;
'✧' indicates 18O12C32S; '✧': 16O13C34S; '✧': Other OCS species; 'I': not OCS.

LINE NO.	WAVENUMBER CM^{-1}, VAC.	UNC.	ROTATIONAL BAND ASSIGNMENTS	ISOTOPIC SPECIES
44	2891.05967	(13)	P 28 1111E-0101E	16012C32S
44			P 28 1111F-0101F	16012C32S
45	2891.10549	(12)	P 32 2400 -1000	16012C32S
46	2891.15662	(11)	P 54 1010 -0000	16012C32S
47	2891.18640	(11)	P 33 1010 -0000	16012C34S
48	2891.21042	(50)	R 2 1212 -0202	16012C32S
49	2891.26234	(22)	P 15 2010 -1000	16012C34S
50	2891.33723	(12)	P 41 2400 -1000	16012C32S
51	2891.51080	(12)	P 31 2400 -1000	16012C32S
52	2891.55738	(12)	P 27 1111E-0101E	16012C32S
52			P 27 1111F-0101F	16012C32S
53	2891.60442	(15)	R 3 1212 -0202	16012C32S
54	2891.69146	(11)	P 32 1010 -0000	16012C34S
55	2891.74984	(11)	P 53 1010 -0000	16012C32S
56	2891.77508	(16)	P 40 2400 -1000	16012C32S
57	2891.80804	(19)	P 5 2111E-1101E	16012C32S
57			P 5 2111F-1101F	16012C32S
57			R 42 1311E-0301E	16012C32S
57			R 19 1210 -0200	16012C33S
58	2891.91410	(15)	P 30 2400 -1000	16012C32S
58			R 42 1311F-0301F	16012C32S
59	2892.05180	(12)	P 26 1111E-0101E	16012C32S
59			P 26 1111F-0101F	16012C32S
60	2892.09892	(13)	P 42 1010 -0000	16012C33S
61	2892.19303	(11)	P 31 1010 -0000	16012C34S
62	2892.21735	(15)	P 39 2010 -1000	16012C32S
62			P 13 1111 -0101	16012C33S
63	2892.31560	(15)	P 29 2400 -1000	16012C32S
64	2892.33954	(11)	P 52 1010 -0000	16012C32S
65	2892.54285	(12)	P 25 1111 -0101	16012C32S
66	2892.64193	(13)	P 41 1010 -0000	16012C33S
66			P 3 2111 -1101	16012C32S
67	2892.66413	(15)	P 38 2010 -1000	16012C32S
68	2892.69127	(11)	P 30 1010 -0000	16012C34S
69	2892.71557	(12)	P 28 2400 -1000	16012C32S
70	2892.92563	(11)	P 51 1010 -0000	16012C32S
71	2893.03046	(11)	P 24 1111 -0101	16012C32S
72	2893.11422	(12)	P 27 2400 -1000	16012C32S
72			P 37 2010 -1000	16012C32S
73	2893.18558	(11)	P 29 1010 -0000	16012C34S
73			P 40 1010 -0000	16012C33S
74	2893.56768	(12)	P 36 2010 -1000	16012C32S
75	2893.67729	(11)	P 28 1010 -0000	16012C34S
76	2893.71778	(13)	P 39 1010 -0000	16012C33S
77	2893.92470	(45)	R 4 1111 -0101	16012C34S
77			R 7 2400 -1000	16012C34S
78	2893.99548	(11)	P 22 1111 -0101	16012C32S
79	2894.02381	(12)	P 35 2010 -1000	16012C32S
80	2894.08708	(11)	P 49 1010 -0000	16012C32S
81	2894.16517	(11)	P 27 1010 -0000	16012C34S
82	2894.25055	(12)	P 38 1010 -0000	16012C33S
82			P 8 2010 -1000	16012C34S
83	2894.27183	(12)	R 10 1212 -0202	16012C32S
83			R 9 1210 -0200	16012C32S
84	2894.30387	(12)	P 24 2400 -1000	16012C32S
84			R 5 1111 -0101	16012C34S
85	2894.47303	(14)	P 21 1111 -0101	16012C32S
86	2894.66241	(12)	P 48 1010 -0000	16012C32S
87	2894.69995	(12)	P 23 2400 -1000	16012C32S
88	2894.77938	(12)	P 37 1010 -0000	16012C33S

- "LINE NO.": serial number identifying the spectral line in the map on the opposite page.
- "WAVENUMBER": experimental value.
- Uncertainty in the last digit of the wavenumber is given in parentheses.

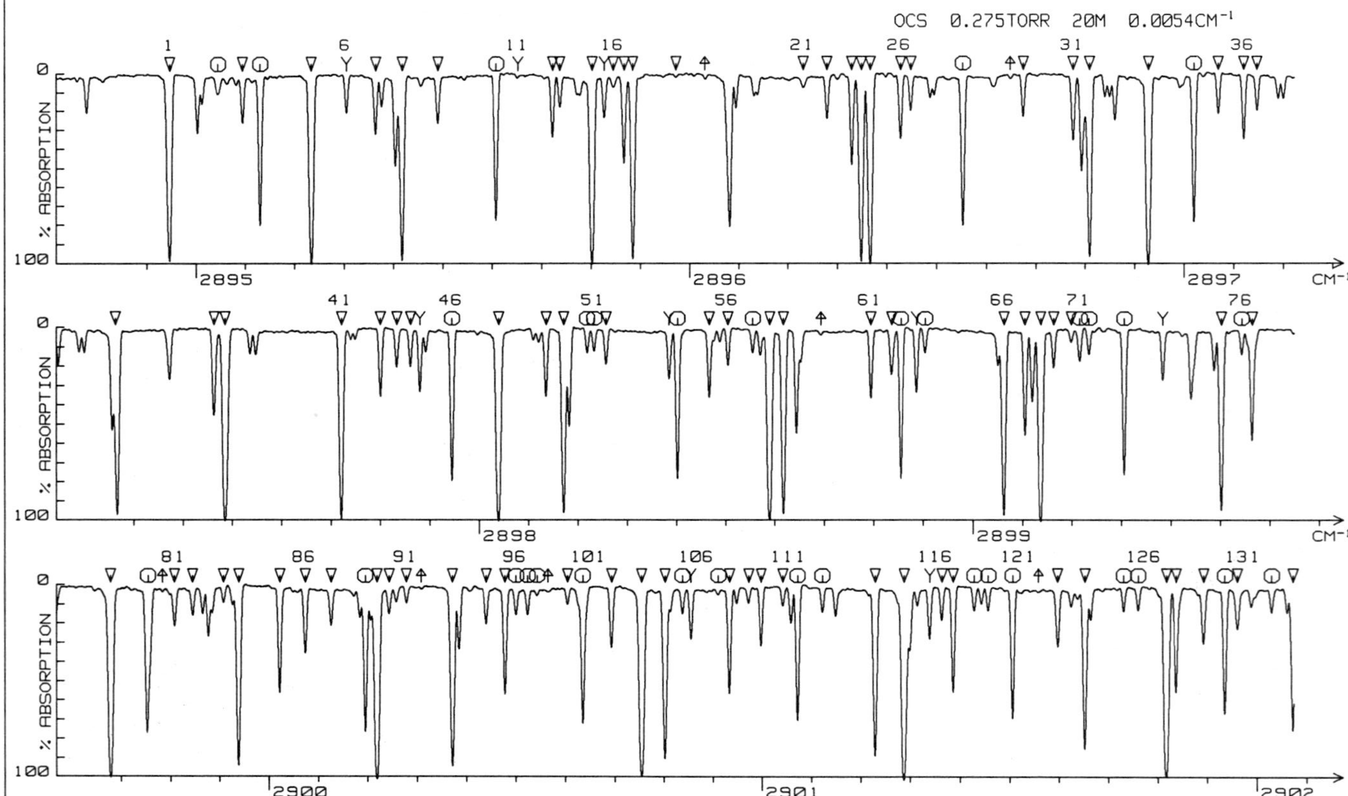

ISOTOPIC IDENTIFICATION – '▽' indicates $^{16}O^{12}C^{32}S$; '①': $^{16}O^{12}C^{34}S$; '◇': $^{16}O^{13}C^{32}S$; 'Y': $^{16}O^{12}C^{33}S$;

'✧' indicates $^{18}O^{12}C^{32}S$; '⫛': $^{16}O^{13}C^{34}S$; '⍏': Other OCS species; 'I': not OCS.

LINE NO.	WAVENUMBER UNC. CM^{-1}, VAC.	ROTATIONAL BAND ASSIGNMENTS	ISOTOPIC SPECIES
1	2894.94658(13)	P 20 1111 -0101	16012C32S
2	2895.04488(18)	R 7 1111 -0101	16012C34S
3	2895.09508(12)	P 22 2400 -1000	16012C32S
4	2895.13059(11)	P 25 1010 -0000	16012C34S
5	2895.23412(11)	P 47 1010 -0000	16012C32S
6	2895.30490(12)	P 36 1010 -0000	16012C33S
7	2895.36418(15)	R 13 1212 -0202	16012C32S
8	2895.41771(13)	P 19 1111 -0101	16012C32S
9	2895.49004(12)	P 21 2400 -1000	16012C32S
9		P 5 2010 -1000	16012C34S
10	2895.60815(11)	P 24 1010 -0000	16012C34S
11	2895.65198(52)	P 5 1111 -0101	16012C33S
12	2895.72200(12)	R 14 1212 -0202	16012C32S
12		R 21 1010 -0000	17012C32S
13	2895.73729(23)	R 13 1210 -0200	16012C32S
14	2895.80227(11)	P 46 1010 -0000	16012C32S
15	2895.82692(18)	P 35 1010 -0000	16012C33S
16	2895.84506(26)	R 4 2111F-1101F	16012C32S
16		R 4 2111E-1101E	16012C32S
16		R 33 1212E-0202E	16012C33S
16		R 33 1212F-0202F	16012C33S
17	2895.86646(13)	P 31 2010 -1000	16012C32S
18	2895.88497(11)	P 18 1111 -0101	16012C32S
18		P 20 2400 -1000	16012C32S
19	2895.97280(61)	P 7 3010 -2000	16012C32S
20	2896.03193(44)	R 22 1010 -0000	17012C32S
21	2896.23157(24)	R 5 2111 -1101	16012C32S
22	2896.27953(12)	P 19 2400 -1000	16012C32S
23	2896.32961(13)	P 30 2010 -1000	16012C32S
24	2896.34874(11)	P 17 1111 -0101	16012C32S
24		P 34 1010 -0000	16012C33S
25	2896.36687(11)	P 45 1010 -0000	16012C32S
26	2896.42794(12)	R 16 1212 -0202	16012C32S

LINE NO.	WAVENUMBER UNC. CM^{-1}, VAC.	ROTATIONAL BAND ASSIGNMENTS	ISOTOPIC SPECIES
27	2896.44893(13)	R 15 1210 -0200	16012C32S
28	2896.55296(11)	P 22 1010 -0000	16012C34S
29	2896.64785(47)	R 24 1010 -0000	17012C32S
30	2896.67440(12)	P 18 2400 -1000	16012C32S
31	2896.77597(14)	R 17 1212 -0202	16012C32S
32	2896.80945(13)	P 16 1111 -0101	16012C32S
33	2896.92790(11)	P 44 1010 -0000	16012C32S
34	2897.02023(11)	P 21 1010 -0000	16012C34S
35	2897.06943(12)	P 17 2400 -1000	16012C32S
36	2897.12057(12)	R 18 1212 -0202	16012C32S
37	2897.14727(13)	R 17 1210 -0200	16012C32S
38	2897.26276(20)	P 15 1111 -0101	16012C32S
39	2897.46288(13)	R 19 1212 -0202	16012C32S
39		P 16 2400 -1000	16012C32S
40	2897.48529(11)	P 43 1010 -0000	16012C32S
40		P 20 1010 -0000	16012C34S
41	2897.72025(11)	P 14 1111 -0101	16012C32S
41		P 27 2010 -1000	16012C32S
42	2897.80001(12)	R 20 1212 -0202	16012C32S
43	2897.83259(13)	R 19 1210 -0200	16012C32S
43		R 28 1010 -0000	17012C32S
44	2897.86032(13)	P 15 2400 -1000	16012C32S
45	2897.87963(14)	P 31 1010 -0000	16012C33S
45		R 15 1111E-0101E	16012C34S
46	2897.94452(11)	P 19 1010 -0000	16012C34S
47	2898.03925(11)	P 42 1010 -0000	16012C32S
48	2898.13476(14)	R 21 1212 -0202	16012C32S
49	2898.17070(14)	P 13 1111E-0101E	16012C32S
49		P 13 1111F-0101F	16012C32S
49		R 20 1210 -0200	16012C32S
50	2898.21823(27)	R 16 1111E-0101E	16012C34S
51	2898.23215(30)	R 16 1111F-0101F	16012C34S

- "LINE NO.": serial number identifying the spectral line in the map on the opposite page.
- "WAVENUMBER": experimental value.
- Uncertainty in the last digit of the wavenumber is given in parentheses.

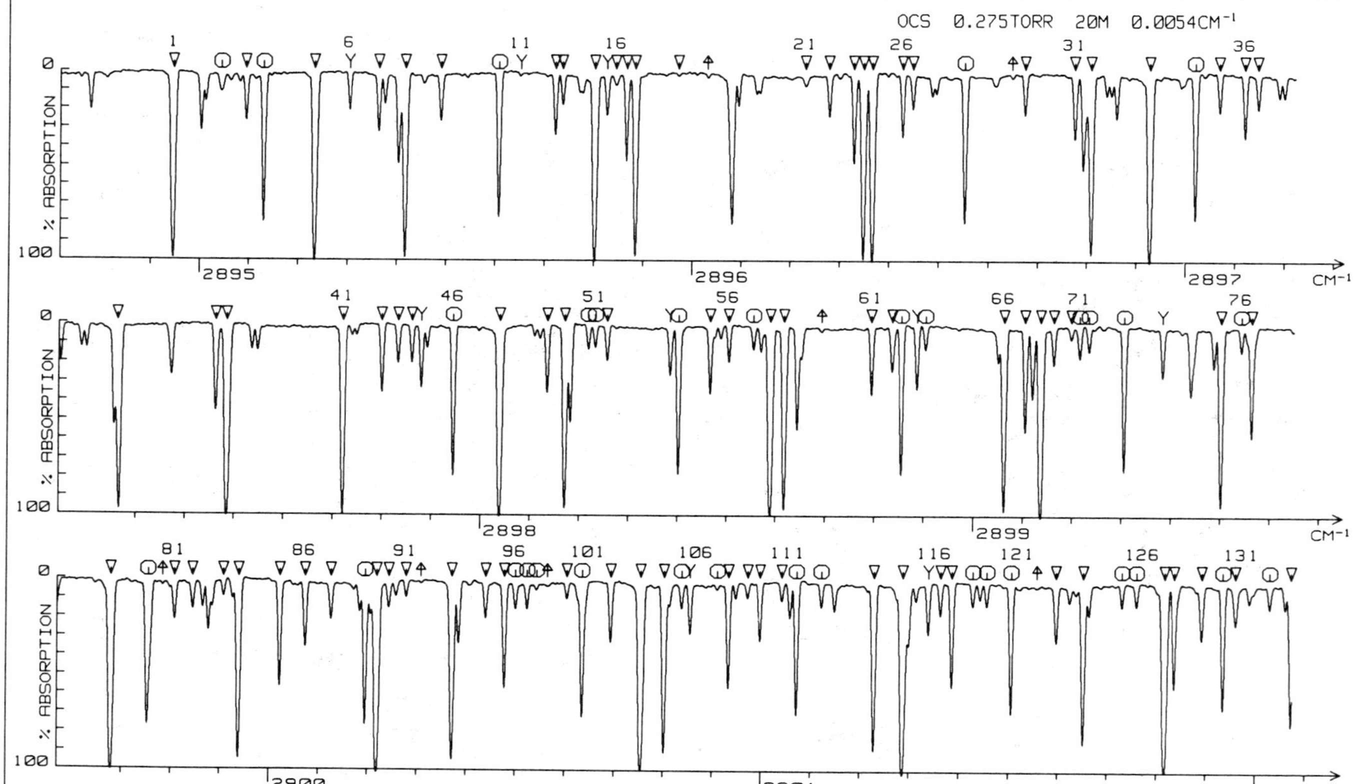

ISOTOPIC IDENTIFICATION – '▽' indicates $^{16}O^{12}C^{32}S$; 'ⓘ': $^{16}O^{12}C^{34}S$; '◊': $^{16}O^{13}C^{32}S$; 'Y': $^{16}O^{12}C^{33}S$;
'✧' indicates $^{18}O^{12}C^{32}S$; '⏚': $^{16}O^{13}C^{34}S$; '↑': Other OCS species; 'I': not OCS.

LINE NO.	WAVENUMBER CM^{-1}, VAC.	UNC.	ROTATIONAL BAND ASSIGNMENTS	ISOTOPIC SPECIES
52	2898.25627	(13)	P 14 2400 -1000	16012C32S
53	2898.38457	(18)	P 30 1010 -0000	16012C33S
54	2898.40153	(11)	P 18 1010 -0000	16012C34S
55	2898.46617	(14)	R 22 1212 -0202	16012C32S
56	2898.50429	(13)	R 21 1210 -0200	16012C32S
57	2898.55421	(28)	R 17 1111E-0101E	16012C34S
58	2898.58947	(11)	P 41 1010 -0000	16012C32S
59	2898.61767	(11)	P 12 1111 -0101	16012C32S
60	2898.69241	(50)	R 31 1010 -0000	17012C32S
61	2898.79428	(12)	R 23 1212 -0202	16012C32S
62	2898.83591	(19)	R 22 1210 -0200	16012C32S
62			R 12 2111E-1101E	16012C32S
63	2898.85505	(11)	P 17 1010 -0000	16012C34S
64	2898.88603	(12)	P 29 1010 -0000	16012C33S
64			R 18 1111E-0101E	16012C34S
65	2898.90348	(14)	R 18 1111F-0101F	16012C34S
66	2899.06137	(11)	P 11 1111 -0101	16012C32S
67	2899.10445	(12)	P 24 2010 -1000	16012C32S
68	2899.13628	(11)	P 40 1010 -0000	16012C32S
69	2899.16273	(19)	R 23 1210 -0200	16012C32S
70	2899.19811	(19)	R 13 2111E-1101E	16012C32S
71	2899.21546	(14)	R 19 1111E-0101E	16012C34S
71			R 13 2111F-1101F	16012C32S
72	2899.23407	(14)	R 19 1111F-0101F	16012C34S
72			R 20 2400 -1000	16012C34S
73	2899.30528	(11)	P 16 1010 -0000	16012C34S
74	2899.38313	(12)	P 28 1010 -0000	16012C33S
75	2899.50162	(11)	P 10 1111 -0101	16012C32S
76	2899.54254	(27)	R 20 1111E-0101E	16012C34S
77	2899.56355	(12)	P 23 2010 -1000	16012C32S
77			R 20 1111F-0101F	16012C34S
78	2899.67937	(11)	P 39 1010 -0000	16012C32S
79	2899.75441	(16)	P 15 1010 -0000	16012C34S
79			R 26 1212 -0202	16012C32S
80	2899.78379	(76)	R 35 1010 -0000	17012C32S
81	2899.80806	(20)	R 25 1210 -0200	16012C32S
82	2899.84484	(13)	P 10 2400 -1000	16012C32S
83	2899.90788	(16)	R 15 2111E-1101E	16012C32S
84	2899.93846	(11)	P 9 1111 -0101	16012C32S
85	2900.02181	(11)	P 22 2010 -1000	16012C32S
86	2900.07337	(12)	R 27 1212 -0202	16012C32S
87	2900.12562	(13)	R 26 1210 -0200	16012C32S
88	2900.19550	(11)	P 14 1010 -0000	16012C34S
89	2900.21905	(11)	P 38 1010 -0000	16012C32S
90	2900.24319	(27)	P 9 2400 -1000	16012C32S
91	2900.27815	(16)	R 16 2111F-1101F	16012C32S
92	2900.30823	(67)	R 37 1010 -0000	17012C32S
93	2900.37159	(11)	P 8 1111 -0101	16012C32S
93			P 26 1010 -0000	16012C33S
94	2900.43996	(13)	R 27 1210 -0200	16012C32S
95	2900.47814	(12)	P 21 2010 -1000	16012C32S
96	2900.50034	(14)	R 23 1111E-0101E	16012C34S
97	2900.52365	(14)	R 23 1111F-0101F	16012C34S
98	2900.54289	(33)	R 7 2010 -1000	16012C34S
99	2900.56445	(50)	R 38 1010 -0000	17012C32S
99			Q 40 1111E-0101F	16012C32S
100	2900.60436	(16)	R 17 2111E-1101E	16012C32S
101	2900.63550	(12)	P 13 1010 -0000	16012C34S
102	2900.69291	(12)	R 29 1212 -0202	16012C32S
103	2900.75495	(11)	P 37 1010 -0000	16012C32S
104	2900.80205	(11)	P 7 1111 -0101	16012C32S
105	2900.83751	(14)	R 24 1111F-0101F	16012C34S
106	2900.85479	(12)	P 25 1010 -0000	16012C33S

- "LINE NO.": serial number identifying the spectral line in the map on the opposite page.
- "WAVENUMBER": experimental value.
- Uncertainty in the last digit of the wavenumber is given in parentheses.

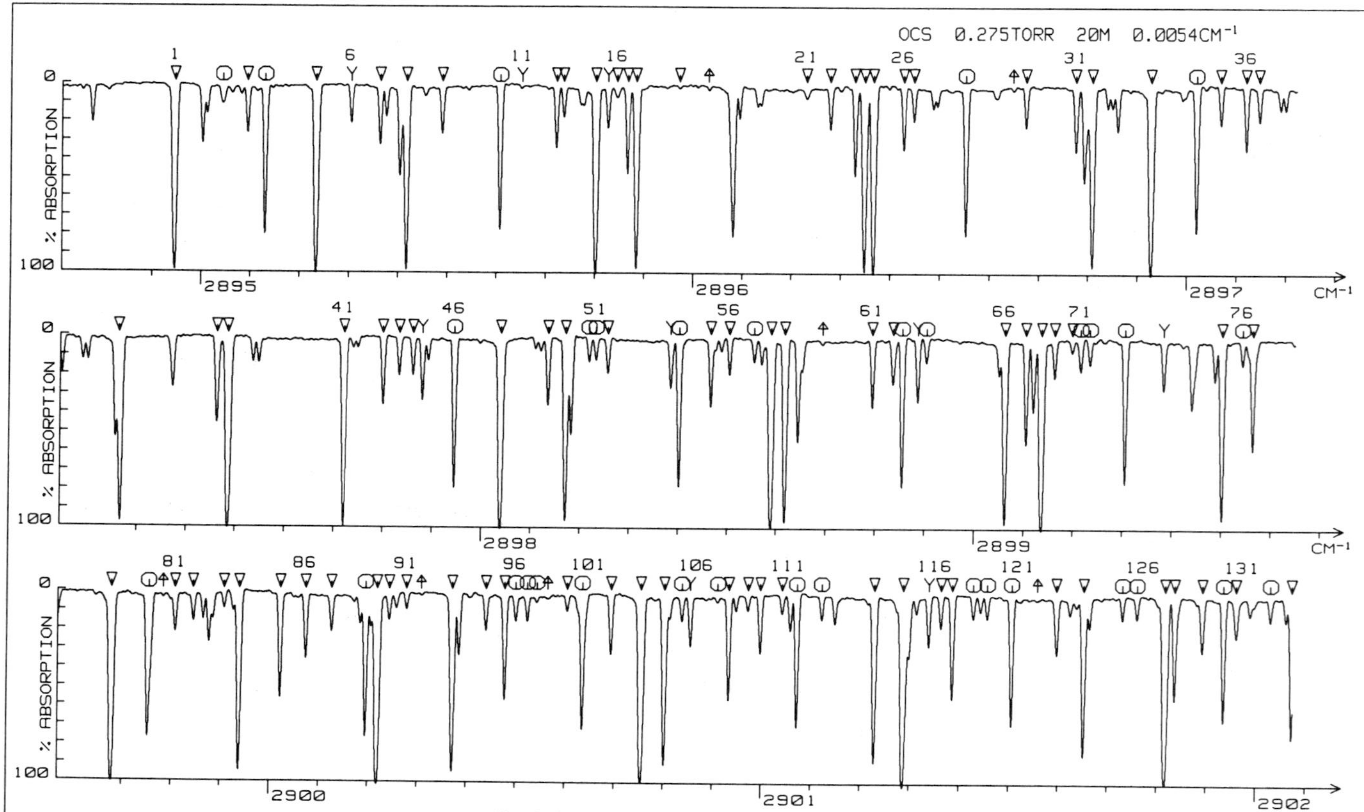

ISOTOPIC IDENTIFICATION – '▽' indicates $^{16}O^{12}C^{32}S$; '⦶': $^{16}O^{12}C^{34}S$; '◇': $^{16}O^{13}C^{32}S$; 'Y': $^{16}O^{12}C^{33}S$;
'✧' indicates $^{18}O^{12}C^{32}S$; '⍖': $^{16}O^{13}C^{34}S$; '✢': Other OCS species; 'ı': not OCS.

LINE NO.	WAVENUMBER UNC. CM^{-1}, VAC.	ROTATIONAL BAND ASSIGNMENTS	ISOTOPIC SPECIES
107	2900.91045(31)	R 8 2010 -1000	16012C34S
108	2900.93287(12)	P 20 2010 -1000	16012C32S
109	2900.97191(16)	R 18 2111F-1101F	16012C32S
110	2900.99774(12)	R 30 1212 -0202	16012C32S
111	2901.04223(15)	P 7 2400 -1000	16012C32S
112	2901.07199(12)	P 12 1010 -0000	16012C34S
113	2901.12198(14)	R 25 1111E-0101E	16012C34S
114	2901.22876(11)	P 6 1111 -0101	16012C32S
115	2901.28762(12)	P 36 1010 -0000	16012C32S
116	2901.33843(12)	P 24 1010 -0000	16012C33S
117	2901.36281(13)	R 30 1210 -0200	16012C32S
118	2901.38564(12)	P 19 2010 -1000	16012C32S
119	2901.42786(14)	R 26 1111E-0101E	16012C34S
120	2901.45571(27)	R 26 1111F-0101F	16012C34S
121	2901.50516(11)	P 11 1010 -0000	16012C34S
122	2901.55804(73)	R 42 1010 -0000	17012C32S
122		Q 33 1111E-0101F	16012C32S
123	2901.59711(12)	R 32 1212 -0202	16012C32S
124	2901.65215(11)	P 5 1111E-0101E	16012C32S
124		P 5 1111F-0101F	16012C32S
124		R 20 2111F-1101F	16012C32S
125	2901.73025(14)	R 27 1111E-0101E	16012C34S
126	2901.75976(14)	R 27 1111F-0101F	16012C34S
127	2901.81650(11)	P 35 1010 -0000	16012C32S
128	2901.83645(12)	P 18 2010 -1000	16012C32S
129	2901.89179(16)	R 33 1212 -0202	16012C32S
129		R 10 1111F-0101F	16012C33S
130	2901.93492(11)	P 10 1010 -0000	16012C34S
131	2901.96028(13)	R 32 1210 -0200	16012C32S
131		R 21 2111E-1101E	16012C32S
132	2902.02938(14)	R 28 1111E-0101E	16012C34S
132		R 44 1010 -0000	17012C32S
133	2902.07208(12)	P 4 1111 -0101	16012C32S

- "LINE NO.": serial number identifying the spectral line in the map on the opposite page.
- "WAVENUMBER": experimental value.
- Uncertainty in the last digit of the wavenumber is given in parentheses.

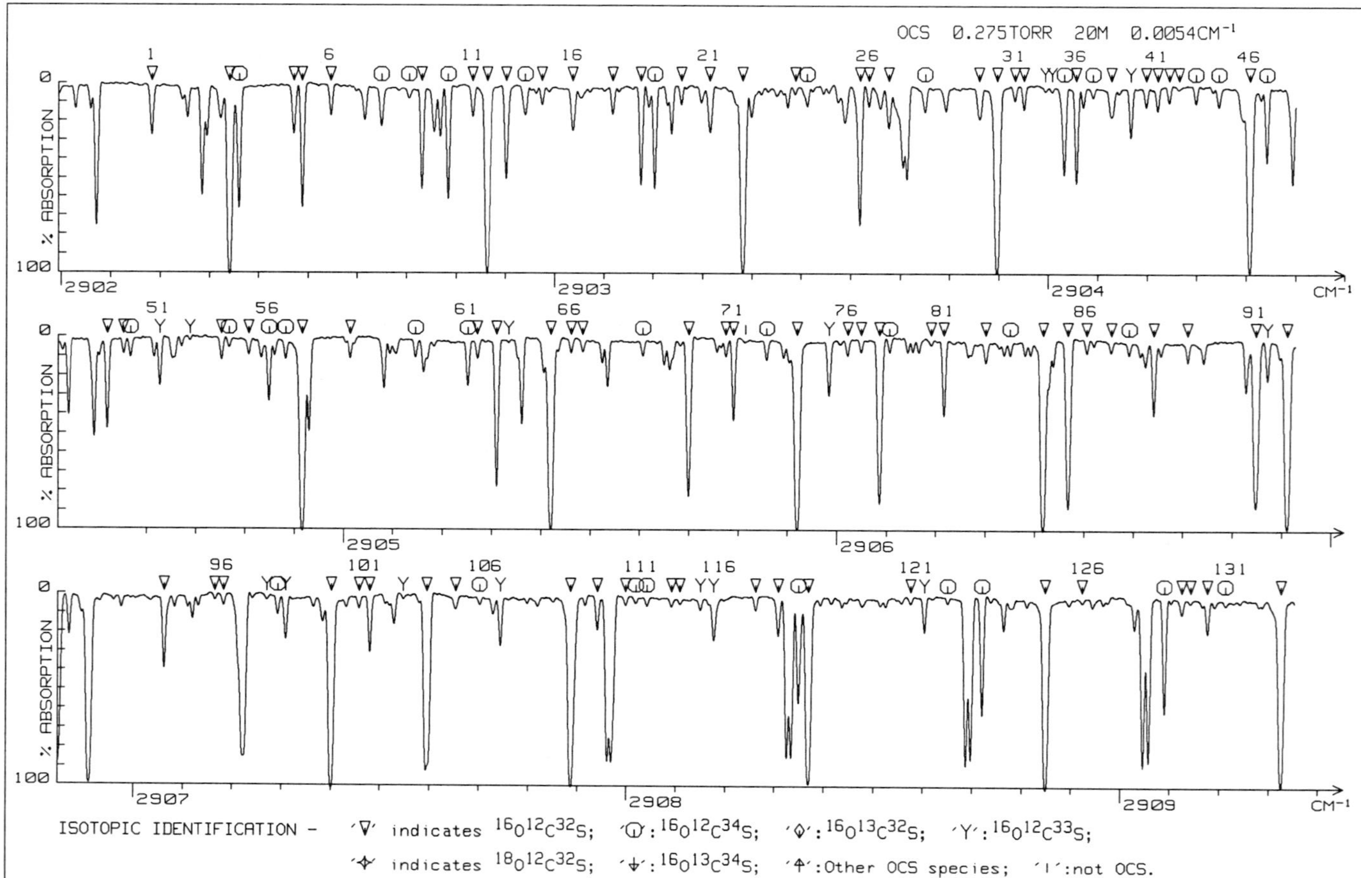
OCS 0.275TORR 20M 0.0054CM⁻¹
% ABSORPTION
0
100
2902
2903
2904
2905
2906
2907
2908
2909
CM⁻¹
ISOTOPIC IDENTIFICATION - '▽' indicates 16O12C32S; 'Ⓘ': 16O12C34S; '◊': 16O13C32S; 'Y': 16O12C33S;
'✧' indicates 18O12C32S; '⇣': 16O13C34S; '⇡': Other OCS species; 'I': not OCS.

LINE NO.	WAVENUMBER UNC. CM^{-1}, VAC.	ROTATIONAL BAND ASSIGNMENTS	ISOTOPIC SPECIES
1	2902.18326(12)	R 34 1212 -0202	16012C32S
2	2902.34187(11)	P 34 1010 -0000	16012C32S
3	2902.36117(33)	P 9 1010 -0000	16012C34S
3		R 29 1111F-0101F	16012C34S
4	2902.47130(18)	R 35 1212 -0202	16012C32S
5	2902.48865(12)	P 3 1111 -0101	16012C32S
6	2902.54633(13)	R 34 1210 -0200	16012C32S
6		R 28 2400 -1000	16012C34S
7	2902.65020(13)	R 30 1111F-0101F	16012C34S
7		R 23 2111F-1101F	16012C32S
7		P 3 2400 -1000	16012C32S
8	2902.70588(21)	R 13 2010 -1000	16012C34S
8		R 9 3010 -2000	16012C32S
9	2902.73171(12)	P 16 2010 -1000	16012C32S
10	2902.78428(12)	P 8 1010 -0000	16012C34S
11	2902.83384(22)	R 35 1210 -0200	16012C32S
11		Q 21 1111E-0101F	16012C32S
11		Q 24 1111F-0101E	16012C32S
12	2902.86367(11)	P 33 1010 -0000	16012C32S
13	2902.90222(13)	P 2 1111 -0101	16012C32S
13		R 31 1111E-0101E	16012C34S
14	2902.94042(15)	R 31 1111F-0101F	16012C34S
14		R 24 2111E-1101E	16012C32S
14		R 48 1010 -0000	17012C32S
15	2902.97491(34)	R 24 2111F-1101F	16012C32S
15		Q 22 1111F-0101E	16012C32S
16	2903.03727(13)	R 37 1212E-0202E	16012C32S
16		R 37 1212F-0202F	16012C32S
16		Q 21 1111F-0101E	16012C32S
17	2903.11811(13)	R 36 1210 -0200	16012C32S
18	2903.17596(12)	P 15 2010 -1000	16012C32S
19	2903.20385(12)	P 7 1010 -0000	16012C34S
20	2903.25773(32)	R 25 2111E-1101E	16012C32S
20		Q 15 1111E-0101F	16012C32S
20		P 54 1400 -0000	16012C34S
21	2903.31529(20)	R 38 1212E-0202E	16012C32S
21		R 38 1212F-0202F	16012C32S
21		R 14 1111F-0101F	16012C33S
21		Q 14 1111E-0101F	16012C32S
22	2903.38195(11)	P 32 1010 -0000	16012C32S
23	2903.48883(31)	Q 12 1111F-0101E	16012C32S
24	2903.51271(18)	R 33 1111F-0101F	16012C34S
25	2903.61886(26)	P 14 2010 -1000	16012C32S
25		P 6 1010 -0000	16012C34S
25		R 26 2111F-1101F	16012C32S
26	2903.63684(34)	Q 6 1111E-0101F	16012C32S
26		Q 7 1111F-0101E	16012C32S
27	2903.67667(22)	R 38 1210 -0200	16012C32S
27		Q 4 1111E-0101F	16012C32S
27		Q 5 1111F-0101E	16012C32S
28	2903.75100(15)	R 34 1111E-0101E	16012C34S
28		R 16 2010 -1000	16012C34S
29	2903.86091(30)	R 40 1212 -0202	16012C32S
30	2903.89670(11)	P 31 1010 -0000	16012C32S
31	2903.93213(43)	R 27 2111F-1101F	16012C32S
32	2903.95028(14)	R 39 1210 -0200	16012C32S
33	2903.99393(36)	R 16 1111E-0101E	16012C33S
33		R 53 1010 -0000	17012C32S
34	2904.00786(38)	R 16 1111F-0101F	16012C33S
35	2904.03283(51)	P 5 1010 -0000	16012C34S
36	2904.05733(12)	P 13 2010 -1000	16012C32S
37	2904.09054(22)	R 17 2010 -1000	16012C34S
38	2904.12828(36)	R 41 1212E-0202E	16012C32S
38		R 41 1212F-0202F	16012C32S

- "LINE NO.": serial number identifying the spectral line in the map on the opposite page.
- "WAVENUMBER": experimental value.
- Uncertainty in the last digit of the wavenumber is given in parentheses.

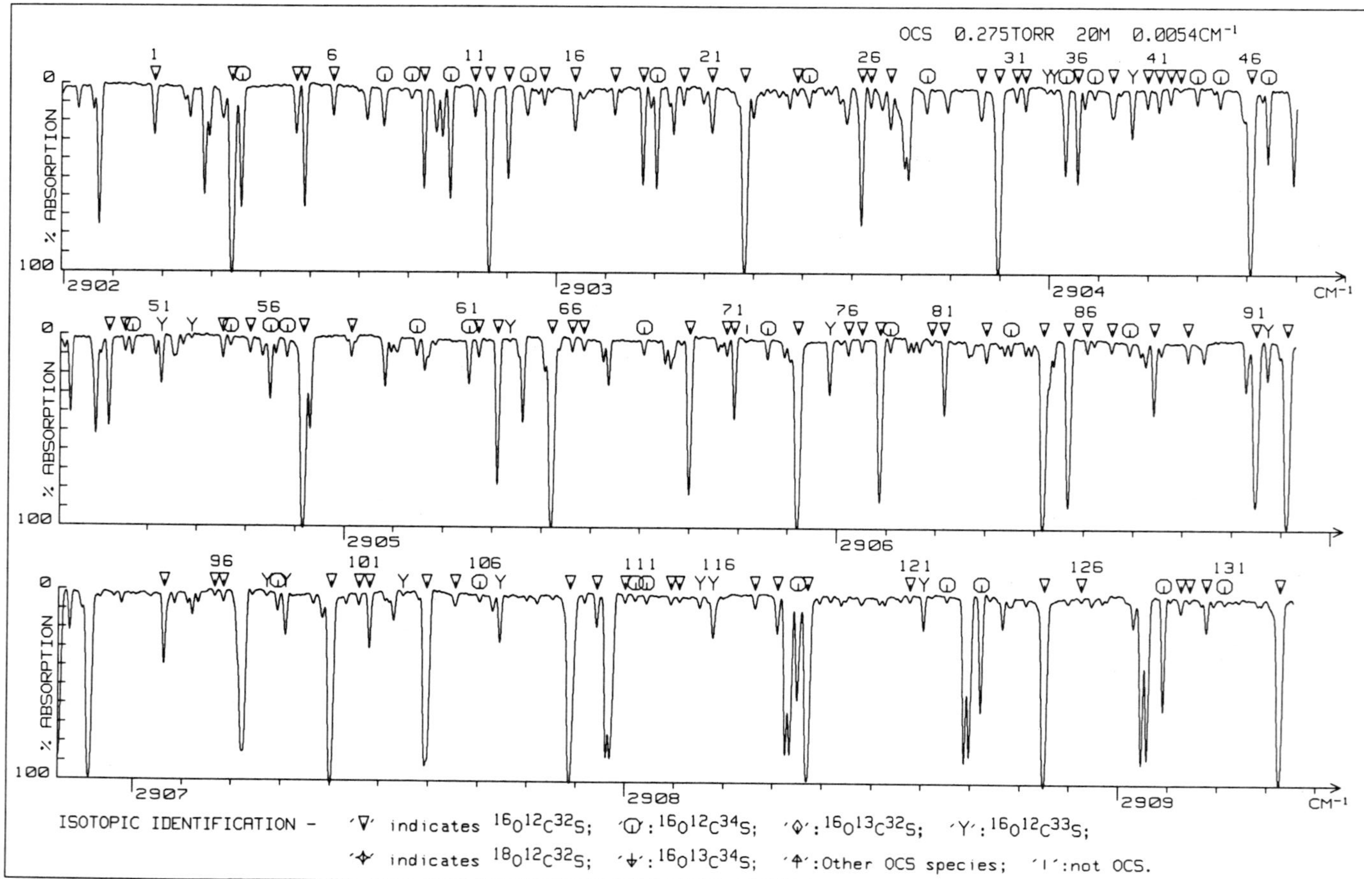
OCS 0.275TORR 20M 0.0054CM⁻¹
% ABSORPTION
0
100
2902
2903
2904
2905
2906
2907
2908
2909
CM⁻¹
ISOTOPIC IDENTIFICATION - '▽' indicates $^{16}O^{12}C^{32}S$; 'Ⓘ': $^{16}O^{12}C^{34}S$; '◇': $^{16}O^{13}C^{32}S$; 'Y': $^{16}O^{12}C^{33}S$;
'✧' indicates $^{18}O^{12}C^{32}S$; '⍖': $^{16}O^{13}C^{34}S$; '⍏': Other OCS species; 'I': not OCS.

LINE NO.	WAVENUMBER UNC. CM^{-1}, VAC.	ROTATIONAL BAND ASSIGNMENTS	ISOTOPIC SPECIES
39	2904.16710(12)	P 18 1010 -0000	16012C33S
40	2904.19792(15)	R 28 2111E-1101E	16012C32S
40		R 32 2400 -1000	16012C34S
40		R 54 1010 -0000	17012C32S
41	2904.22105(14)	R 40 1210 -0200	16012C32S
42	2904.24459(16)	R 28 2111F-1101F	16012C32S
43	2904.26448(44)	R 0 2400 -1000	16012C32S
44	2904.29892(15)	R 36 1111E-0101E	16012C34S
45	2904.34517(17)	R 36 1111F-0101F	16012C34S
45		R 17 1111F-0101F	16012C33S
46	2904.40785(11)	P 30 1010 -0000	16012C32S
47	2904.44217(13)	P 4 1010 -0000	16012C34S
48	2904.52075(12)	R 1 1111 -0101	16012C32S
49	2904.55355(44)	R 29 2111F-1101F	16012C32S
50	2904.56736(35)	R 37 1111E-0101E	16012C34S
51	2904.62637(16)	P 17 1010 -0000	16012C33S
52	2904.68750(37)	R 18 1111F-0101F	16012C33S
53	2904.75233(14)	R 42 1210 -0200	16012C32S
54	2904.76744(25)	R 19 2010 -1000	16012C34S
55	2904.80747(16)	R 30 2111E-1101E	16012C32S
56	2904.84814(15)	P 3 1010 -0000	16012C34S
57	2904.88187(35)	R 38 1111F-0101F	16012C34S
58	2904.91566(12)	P 29 1010 -0000	16012C32S
59	2905.01299(33)	R 43 1210 -0200	16012C32S
60	2905.14537(15)	R 39 1111F-0101F	16012C34S
61	2905.25078(12)	P 2 1010 -0000	16012C34S
62	2905.27018(15)	R 44 1210 -0200	16012C32S
63	2905.30989(11)	R 3 1111 -0101	16012C32S
64	2905.33404(43)	R 20 1111E-0101E	16012C33S
65	2905.41958(11)	P 28 1010 -0000	16012C32S
66	2905.46072(45)	R 32 2111F-1101F	16012C32S
67	2905.48411(19)	R 3 2400 -1000	16012C32S
68	2905.60657(16)	R 41 1111E-0101E	16012C34S

LINE NO.	WAVENUMBER UNC. CM^{-1}, VAC.	ROTATIONAL BAND ASSIGNMENTS	ISOTOPIC SPECIES
69	2905.69924(11)	R 4 1111 -0101	16012C32S
69		R 33 2111E-1101E	16012C32S
70	2905.77445(19)	R 46 1210 -0200	16012C32S
71	2905.79005(12)	P 9 2010 -1000	16012C32S
72	2905.81422(74)		CH4
73	2905.85794(15)	R 42 1111E-0101E	16012C34S
73		R 18 3010 -2000	16012C32S
74	2905.92013(11)	P 27 1010 -0000	16012C32S
75	2905.98399(12)	P 14 1010 -0000	16012C33S
75		R 34 2111E-1101E	16012C32S
75		R 22 1111E-0101E	16012C33S
76	2906.02146(16)	R 47 1210 -0200	16012C32S
77	2906.04851(17)	R 34 2111F-1101F	16012C32S
77		R 6 2010 -1000	16012C33S
78	2906.08547(12)	R 5 1111 -0101	16012C32S
79	2906.10571(51)	R 43 1111E-0101E	16012C34S
80	2906.19044(34)	R 19 3010 -2000	16012C32S
81	2906.21665(12)	P 8 2010 -1000	16012C32S
82	2906.30112(15)	R 5 2400 -1000	16012C32S
82		R 23 1111E-0101E	16012C33S
83	2906.35012(19)	R 44 1111E-0101E	16012C34S
84	2906.41732(12)	P 26 1010 -0000	16012C32S
85	2906.46805(12)	R 6 1111 -0101	16012C32S
86	2906.50543(16)	R 49 1210 -0200	16012C32S
86		P 44 1400 -0000	16012C34S
87	2906.55481(18)	R 36 2111E-1101E	16012C32S
88	2906.59090(17)	R 45 1111E-0101E	16012C34S
89	2906.64062(14)	P 7 2010 -1000	16012C32S
89		R 24 1111F-0101F	16012C33S
90	2906.70993(15)	R 6 2400 -1000	16012C32S
91	2906.84726(27)	R 7 1111F-0101F	16012C32S
91		R 7 1111E-0101E	16012C32S

- "LINE NO.": serial number identifying the spectral line in the map on the opposite page.
- "WAVENUMBER": experimental value.
- Uncertainty in the last digit of the wavenumber is given in parentheses.

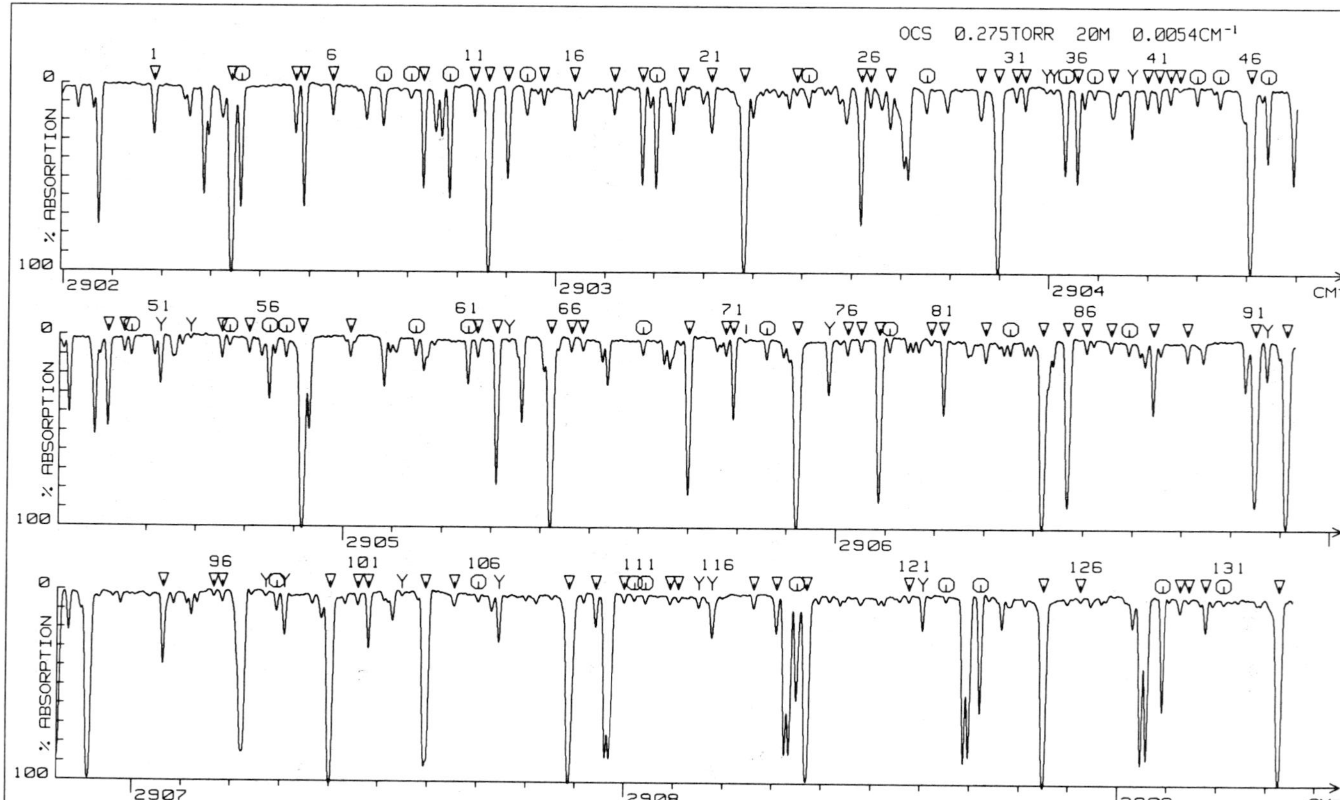

ISOTOPIC IDENTIFICATION – '▽' indicates $^{16}O^{12}C^{32}S$; '⏀': $^{16}O^{12}C^{34}S$; '◇': $^{16}O^{13}C^{32}S$; 'Y': $^{16}O^{12}C^{33}S$;
'✧' indicates $^{18}O^{12}C^{32}S$; '⍖': $^{16}O^{13}C^{34}S$; '⍏': Other OCS species; 'I': not OCS.

LINE NO.	WAVENUMBER UNC. CM^{-1}, VAC.	ROTATIONAL BAND ASSIGNMENTS	ISOTOPIC SPECIES
92	2906.87101(13)	P 12 1010 -0000	16012C33S
93	2906.91066(11)	P 25 1010 -0000	16012C32S
94	2907.06205(14)	P 6 2010 -1000	16012C32S
94		R 47 1111E-0101E	16012C34S
94		R 26 2010 -1000	16012C34S
95	2907.16582(30)	R 22 3010 -2000	16012C32S
95		R 9 2010 -1000	16012C33S
96	2907.18325(19)	R 38 2111F-1101F	16012C32S
96		P 42 1400 -0000	16012C34S
97	2907.27107(36)	R 26 1111F-0101F	16012C33S
98	2907.29294(37)	R 48 1111E-0101E	16012C34S
98		R 54 1212E-0202E	16012C32S
99	2907.30911(13)	P 11 1010 -0000	16012C33S
99		R 54 1212F-0202F	16012C32S
100	2907.40070(11)	P 24 1010 -0000	16012C32S
101	2907.45860(21)	R 39 2111F-1101F	16012C32S
102	2907.48046(12)	P 5 2010 -1000	16012C32S
102		R 23 3010 -2000	16012C32S
103	2907.54811(41)	R 27 1111E-0101E	16012C33S
104	2907.59554(13)	R 9 1111F-0101F	16012C32S
104		R 9 1111E-0101E	16012C32S
104		R 3 1010 -0000	16012C34S
105	2907.65326(21)	R 40 2111E-1101E	16012C32S
105		R 54 1210 -0200	16012C32S
106	2907.70234(23)	R 28 2010 -1000	16012C34S
106		P 67 1402E-0000	16012C32S
107	2907.74450(12)	P 10 1010 -0000	16012C33S
107		R 50 1111E-0101E	16012C34S
107		R 56 1212F-0202F	16012C32S
108	2907.88724(12)	P 23 1010 -0000	16012C32S
109	2907.94106(14)	R 9 2400 -1000	16012C32S
109		R 57 1212E-0202E	16012C32S
110	2907.99863(22)	R 41 2111F-1101F	16012C32S

LINE NO.	WAVENUMBER UNC. CM^{-1}, VAC.	ROTATIONAL BAND ASSIGNMENTS	ISOTOPIC SPECIES
111	2908.01959(24)	R 29 2010 -1000	16012C34S
112	2908.04217(23)	R 51 1111F-0101F	16012C34S
112		P 66 1402E-0000	16012C32S
113	2908.09207(24)	R 56 1210 -0200	16012C32S
114	2908.10916(33)	R 25 3010 -2000	16012C32S
115	2908.14978(20)	R 29 1111E-0101E	16012C33S
115		R 58 1212E-0202E	16012C32S
116	2908.17725(14)	P 9 1010 -0000	16012C33S
116		R 42 2111E-1101E	16012C32S
116		R 52 1111E-0101E	16012C34S
117	2908.26253(17)	R 42 2111F-1101F	16012C32S
117		R 52 1111F-0101F	16012C34S
117		R 12 2010 -1000	16012C33S
118	2908.30861(22)	P 3 2010 -1000	16012C32S
118		R 57 1210 -0200	16012C32S
119	2908.34968(13)	R 5 1010 -0000	16012C34S
119		R 10 2400 -1000	16012C32S
120	2908.37005(11)	P 22 1010 -0000	16012C32S
121	2908.57690(29)	R 60 1212F-0202F	16012C32S
121		R 43 2400 -1000	16012C34S
122	2908.60448(13)	P 8 1010 -0000	16012C33S
122		R 54 1111E-0101E	16012C34S
123	2908.65138(26)	R 31 2010 -1000	16012C34S
124	2908.72077(12)	R 6 1010 -0000	16012C34S
124		P 2 2010 -1000	16012C32S
125	2908.84947(11)	P 21 1010 -0000	16012C32S
126	2908.92343(33)	R 60 1210 -0200	16012C32S
127	2909.08962(11)	R 7 1010 -0000	16012C34S
128	2909.12560(17)	R 61 1210 -0200	16012C32S
129	2909.14296(47)	R 63 1212E-0202E	16012C32S
130	2909.17699(21)	R 12 2400 -1000	16012C32S
131	2909.21293(36)	R 57 1111E-0101E	16012C34S
132	2909.32538(11)	P 20 1010 -0000	16012C32S

- "LINE NO.": serial number identifying the spectral line in the map on the opposite page.
- "WAVENUMBER": experimental value.
- Uncertainty in the last digit of the wavenumber is given in parentheses.

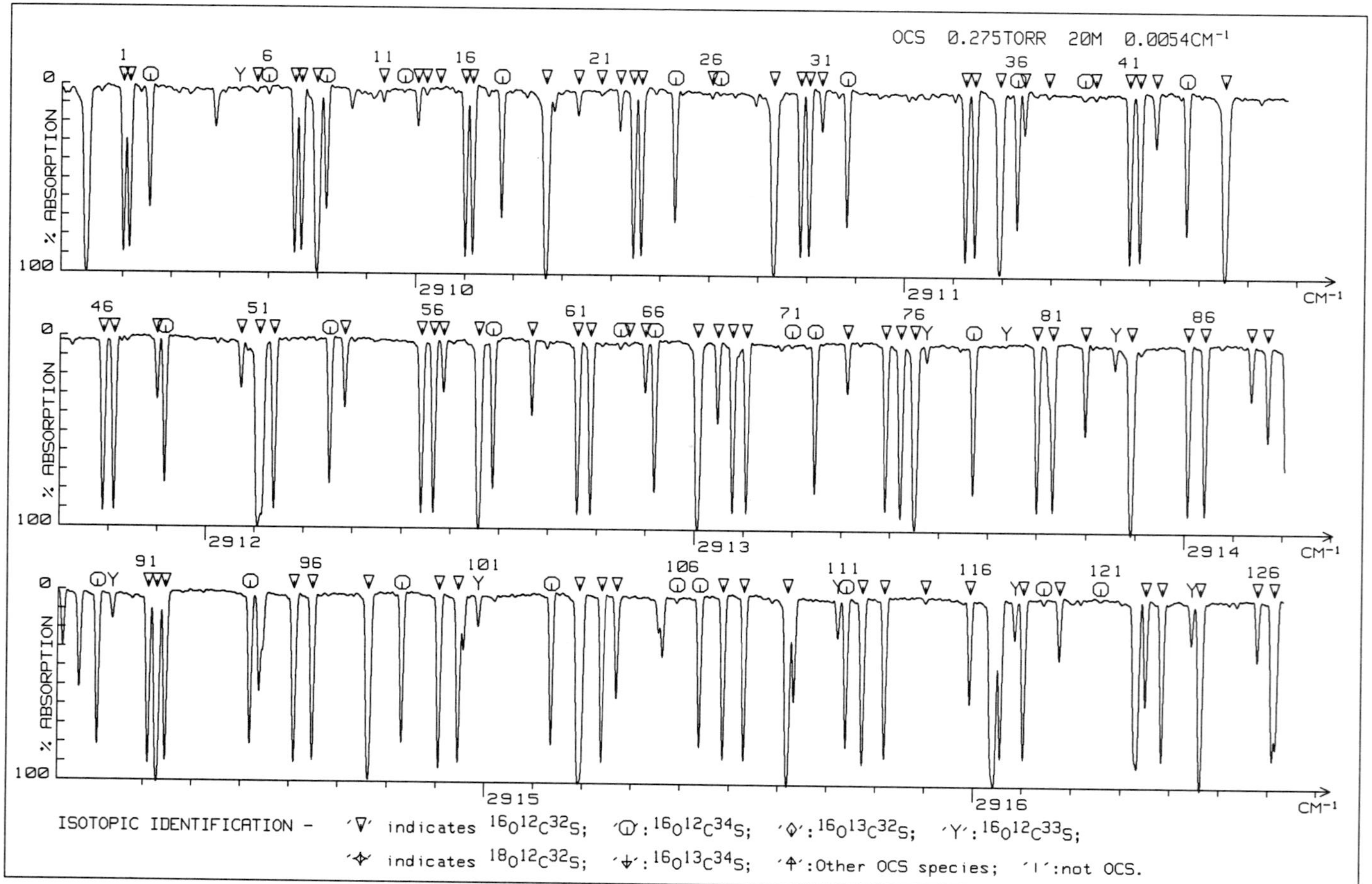
OCS 0.275TORR 20M 0.0054CM⁻¹
% ABSORPTION
0
100
2910
2911
2912
2913
2914
2915
2916
CM⁻¹
ISOTOPIC IDENTIFICATION - 'V' indicates 16O12C32S; 'Ⓘ': 16O12C34S; '◇': 16O13C32S; 'Y': 16O12C33S;
'✧' indicates 18O12C32S; '⏚': 16O13C34S; '⯊': Other OCS species; 'I': not OCS.

LINE NO.	WAVENUMBER CM^{-1}, VAC.	UNC.	ROTATIONAL	BAND ASSIGNMENTS	ISOTOPIC SPECIES
1	2909.40092	(17)	R 14	1111E-0101E	16012C32S
2	2909.41337	(18)	R 14	1111F-0101F	16012C32S
3	2909.45433	(12)	R 8	1010 -0000	16012C34S
3			P 6	1010 -0000	16012C33S
4	2909.63787	(44)	R 34	1111F-0101F	16012C33S
5	2909.67465	(34)	R 48	2111E-1101E	16012C32S
6	2909.69876	(24)	R 59	1111F-0101F	16012C34S
6			R 64	1210 -0200	16012C32S
6			R 66	1212E-0202E	16012C32S
6			P 55	1400 -0000	16012C33S
7	2909.75235	(17)	R 15	1111E-0101E	16012C32S
8	2909.76607	(18)	R 15	1111F-0101F	16012C32S
9	2909.79769	(11)	P 19	1010 -0000	16012C32S
10	2909.81647	(12)	R 9	1010 -0000	16012C34S
11	2909.93266	(47)	R 0	2010 -1000	16012C32S
12	2909.97572	(47)	R 61	1111E-0101E	16012C34S
13	2910.00307	(13)	R 14	2400 -1000	16012C32S
14	2910.02055	(28)	R 49	2111F-1101F	16012C32S
14			R 47	2010 -1000	16012C34S
15	2910.04819	(71)	R 68	1212E-0202E	16012C32S
16	2910.10025	(11)	R 16	1111E-0101E	16012C32S
17	2910.11525	(11)	R 16	1111F-0101F	16012C32S
18	2910.17478	(12)	R 10	1010 -0000	16012C34S
19	2910.26651	(11)	P 18	1010 -0000	16012C32S
20	2910.33161	(14)	R 1	2010 -1000	16012C32S
20			R 63	1111E-0101E	16012C34S
21	2910.37853	(32)	R 51	2111E-1101E	16012C32S
21			R 18	2010 -1000	16012C33S
22	2910.41677	(18)	R 15	2400 -1000	16012C32S
22			R 37	1111E-0101E	16012C33S
23	2910.44481	(11)	R 17	1111E-0101E	16012C32S
24	2910.46113	(11)	R 17	1111F-0101F	16012C32S
25	2910.52967	(11)	R 11	1010 -0000	16012C34S

LINE NO.	WAVENUMBER CM^{-1}, VAC.	UNC.	ROTATIONAL	BAND ASSIGNMENTS	ISOTOPIC SPECIES
26	2910.60486	(39)	R 52	2111E-1101E	16012C32S
27	2910.62135	(68)	R 64	1111F-0101F	16012C34S
28	2910.73183	(11)	P 17	1010 -0000	16012C32S
29	2910.78592	(11)	R 18	1111E-0101E	16012C32S
30	2910.80358	(11)	R 18	1111F-0101F	16012C32S
31	2910.82995	(13)	R 16	2400 -1000	16012C32S
31			R 53	2111E-1101E	16012C32S
32	2910.88112	(11)	R 12	1010 -0000	16012C34S
33	2911.12334	(11)	R 19	1111E-0101E	16012C32S
33			R 3	2010 -1000	16012C32S
34	2911.14269	(11)	R 19	1111F-0101F	16012C32S
35	2911.19365	(11)	P 16	1010 -0000	16012C32S
36	2911.22913	(12)	R 13	1010 -0000	16012C34S
37	2911.24397	(20)	R 17	2400 -1000	16012C32S
38	2911.29367	(42)	R 36	3010 -2000	16012C32S
38			R 68	1111F-0101F	16012C34S
39	2911.36586	(53)	P 30	1400 -0000	16012C34S
40	2911.38802	(46)	R 55	2111F-1101F	16012C32S
41	2911.45776	(11)	R 20	1111E-0101E	16012C32S
42	2911.47833	(11)	R 20	1111F-0101F	16012C32S
43	2911.51113	(16)	R 4	2010 -1000	16012C32S
43			R 40	2010 -1000	16012C34S
44	2911.57375	(12)	R 14	1010 -0000	16012C34S
45	2911.65208	(11)	P 15	1010 -0000	16012C32S
46	2911.78850	(11)	R 21	1111E-0101E	16012C32S
47	2911.81064	(11)	R 21	1111F-0101F	16012C32S
48	2911.89911	(15)	R 5	2010 -1000	16012C32S
49	2911.91491	(12)	R 15	1010 -0000	16012C34S
50	2912.07157	(17)	R 19	2400 -1000	16012C32S
51	2912.10996	(14)	P 14	1010 -0000	16012C32S
52	2912.13945	(11)	R 22	1111F-0101F	16012C32S
53	2912.25260	(11)	R 16	1010 -0000	16012C34S

- "LINE NO.": serial number identifying the spectral line in the map on the opposite page.
- "WAVENUMBER": experimental value.
- Uncertainty in the last digit of the wavenumber is given in parentheses.

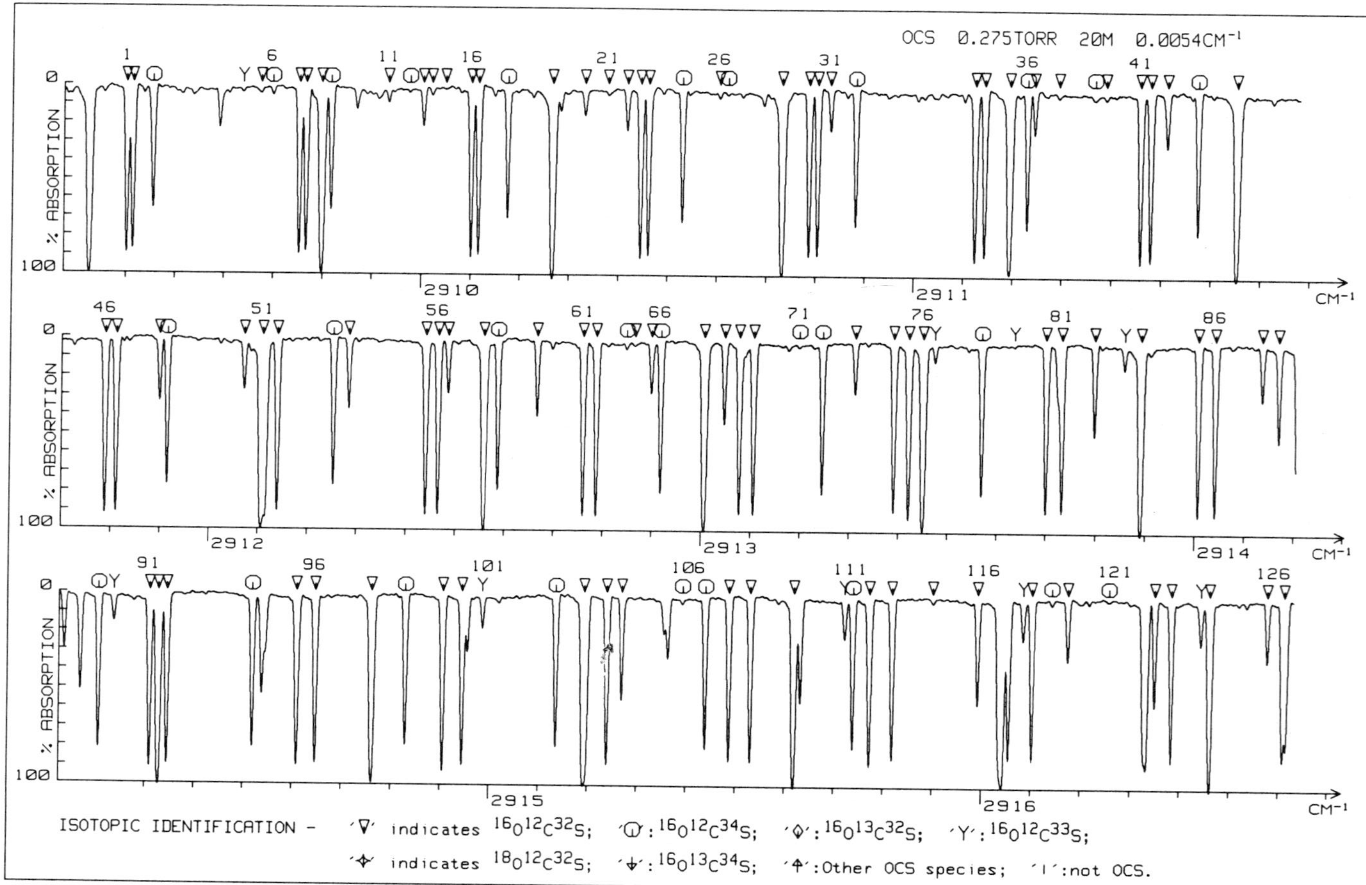
OCS 0.275TORR 20M 0.0054CM⁻¹
% ABSORPTION
0
100
2910
2911
2912
2913
2914
2915
2916
CM⁻¹
ISOTOPIC IDENTIFICATION - '∇' indicates $^{16}O^{12}C^{32}S$; 'Ⓘ': $^{16}O^{12}C^{34}S$; '◊': $^{16}O^{13}C^{32}S$; 'Y': $^{16}O^{12}C^{33}S$;
'✧' indicates $^{18}O^{12}C^{32}S$; '⍖': $^{16}O^{13}C^{34}S$; '↑': Other OCS species; 'I': not OCS.

LINE NO.	WAVENUMBER CM^{-1}, VAC.	UNC.	ROTATIONAL ASSIGNMENTS	BAND ASSIGNMENTS	ISOTOPIC SPECIES
54	2912.28422	(12)	R 6	2010 -1000	16012C32S
55	2912.43961	(11)	R 23	1111E-0101E	16012C32S
56	2912.46487	(11)	R 23	1111F-0101F	16012C32S
57	2912.48516	(16)	R 20	2400 -1000	16012C32S
58	2912.55808	(11)	P 13	1010 -0000	16012C32S
59	2912.58686	(11)	R 17	1010 -0000	16012C34S
60	2912.66635	(12)	R 7	2010 -1000	16012C32S
61	2912.75999	(11)	R 24	1111E-0101E	16012C32S
62	2912.78690	(11)	R 24	1111F-0101F	16012C32S
63	2912.84772	(34)	R 44	2010 -1000	16012C34S
63			R 42	3010 -2000	16012C32S
64	2912.86495	(84)	R 63	2111E-1101E	16012C32S
64			P 68	1400 -0000	16012C32S
65	2912.89862	(16)	R 21	2400 -1000	16012C32S
66	2912.91768	(11)	R 18	1010 -0000	16012C34S
67	2913.00583	(11)	P 12	1010 -0000	16012C32S
68	2913.04601	(13)	R 8	2010 -1000	16012C32S
69	2913.07688	(11)	R 25	1111E-0101E	16012C32S
70	2913.10549	(11)	R 25	1111F-0101F	16012C32S
71	2913.19787	(56)	R 45	2010 -1000	16012C34S
71			R 64	2111F-1101F	16012C32S
72	2913.24503	(11)	R 19	1010 -0000	16012C34S
73	2913.31153	(12)	R 22	2400 -1000	16012C32S
74	2913.39033	(11)	R 26	1111E-0101E	16012C32S
75	2913.42109	(11)	R 26	1111F-0101F	16012C32S
75			R 9	2010 -1000	16012C32S
76	2913.45005	(11)	P 11	1010 -0000	16012C32S
77	2913.47370	(36)	R 3	1010 -0000	16012C33S
78	2913.56888	(11)	R 20	1010 -0000	16012C34S
79	2913.63508	(77)	R 50	1111E-0101E	16012C33S
80	2913.70034	(11)	R 27	1111E-0101E	16012C32S

LINE NO.	WAVENUMBER CM^{-1}, VAC.	UNC.	ROTATIONAL ASSIGNMENTS	BAND ASSIGNMENTS	ISOTOPIC SPECIES
81	2913.73251	(17)	R 27	1111F-0101F	16012C32S
82	2913.79739	(12)	R 10	2010 -1000	16012C32S
83	2913.85720	(29)	R 4	1010 -0000	16012C33S
83			R 51	1111E-0101E	16012C33S
84	2913.89073	(11)	P 10	1010 -0000	16012C32S
84			R 21	1010 -0000	16012C34S
85	2914.00683	(11)	R 28	1111E-0101E	16012C32S
86	2914.04079	(11)	R 28	1111F-0101F	16012C32S
87	2914.13552	(12)	R 24	2400 -1000	16012C32S
88	2914.16924	(12)	R 11	2010 -1000	16012C32S
89	2914.20623	(11)	R 22	1010 -0000	16012C34S
90	2914.23703	(14)	R 5	1010 -0000	16012C33S
91	2914.30992	(11)	R 29	1111E-0101E	16012C32S
92	2914.32814	(11)	P 9	1010 -0000	16012C32S
93	2914.34571	(11)	R 29	1111F-0101F	16012C32S
94	2914.51969	(12)	R 23	1010 -0000	16012C34S
95	2914.60966	(11)	R 30	1111E-0101E	16012C32S
95			R 6	1010 -0000	16012C33S
96	2914.64714	(11)	R 30	1111F-0101F	16012C32S
97	2914.76185	(11)	P 8	1010 -0000	16012C32S
98	2914.82967	(11)	R 24	1010 -0000	16012C34S
99	2914.90548	(11)	R 31	1111E-0101E	16012C32S
99			R 13	2010 -1000	16012C32S
100	2914.94520	(18)	R 31	1111F-0101F	16012C32S
101	2914.98638	(13)	R 7	1010 -0000	16012C33S
102	2915.13617	(11)	R 25	1010 -0000	16012C34S
103	2915.19385	(13)	P 7	1010 -0000	16012C32S
103			R 32	1111E-0101E	16012C32S
104	2915.23967	(11)	R 32	1111F-0101F	16012C32S
105	2915.27002	(12)	R 14	2010 -1000	16012C32S
106	2915.39324	(32)	P 19	1400 -0000	16012C34S
107	2915.43923	(11)	R 26	1010 -0000	16012C34S

- "LINE NO.": serial number identifying the spectral line in the map on the opposite page.
- "WAVENUMBER": experimental value.
- Uncertainty in the last digit of the wavenumber is given in parentheses.

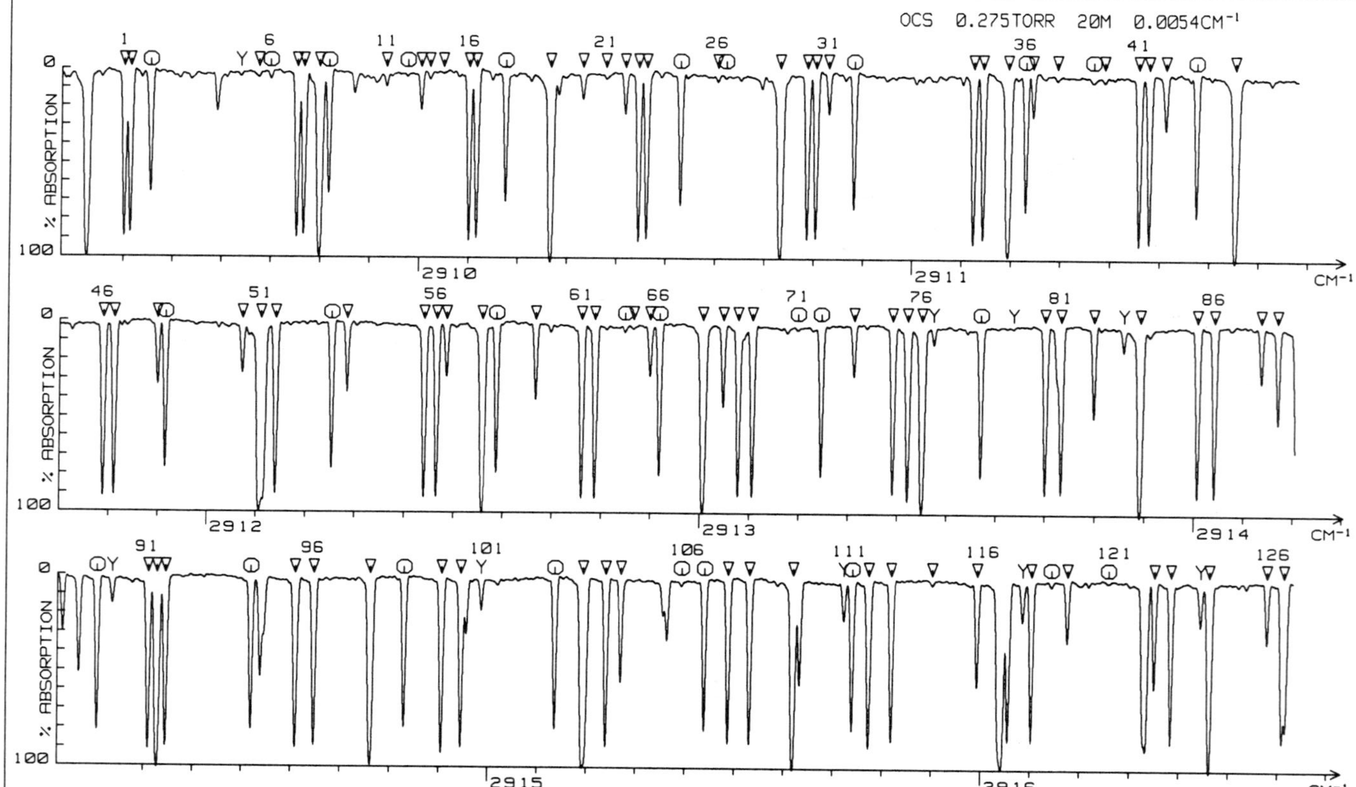

ISOTOPIC IDENTIFICATION – '▽' indicates $^{16}O^{12}C^{32}S$; 'Ⓘ': $^{16}O^{12}C^{34}S$; '◇': $^{16}O^{13}C^{32}S$; 'Y': $^{16}O^{12}C^{33}S$;
'✧' indicates $^{18}O^{12}C^{32}S$; '⏚': $^{16}O^{13}C^{34}S$; '✦': Other OCS species; 'I': not OCS.

LINE NO.	WAVENUMBER UNC. CM^{-1}, VAC.	ROTATIONAL BAND ASSIGNMENTS	ISOTOPIC SPECIES
108	2915.48724(11)	R 33 1111E-0101E	16O12C32S
109	2915.53082(11)	R 33 1111F-0101F	16O12C32S
110	2915.61887(12)	P 6 1010 -0000	16O12C32S
111	2915.72188(22)	R 9 1010 -0000	16O12C33S
112	2915.73873(12)	R 27 1010 -0000	16O12C34S
113	2915.77255(11)	R 34 1111E-0101E	16O12C32S
113		R 28 2400 -1000	16O12C32S
114	2915.81856(11)	R 34 1111F-0101F	16O12C32S
115	2915.90114(39)	P 58 1400 -0000	16O12C32S
116	2915.99213(12)	R 16 2010 -1000	16O12C32S
117	2916.08443(13)	R 10 1010 -0000	16O12C33S
118	2916.10275(11)	R 35 1111F-0101F	16O12C32S
119	2916.14348(38)	P 17 1400 -0000	16O12C34S
120	2916.17539(12)	R 29 2400 -1000	16O12C32S
121	2916.25788(77)	R 53 2400 -1000	16O12C34S
121		R 63 1111E-0101E	16O12C33S
122	2916.34997(12)	R 17 2010 -1000	16O12C32S
123	2916.38350(11)	R 36 1111F-0101F	16O12C32S
124	2916.44356(21)	R 11 1010 -0000	16O12C33S
125	2916.46182(11)	P 4 1010 -0000	16O12C32S
126	2916.57803(12)	R 30 2400 -1000	16O12C32S
127	2916.61222(19)	R 37 1111E-0101E	16O12C32S
127		R 30 1010 -0000	16O12C34S

- "LINE NO.": serial number identifying the spectral line in the map on the opposite page.
- "WAVENUMBER": experimental value.
- Uncertainty in the last digit of the wavenumber is given in parentheses.

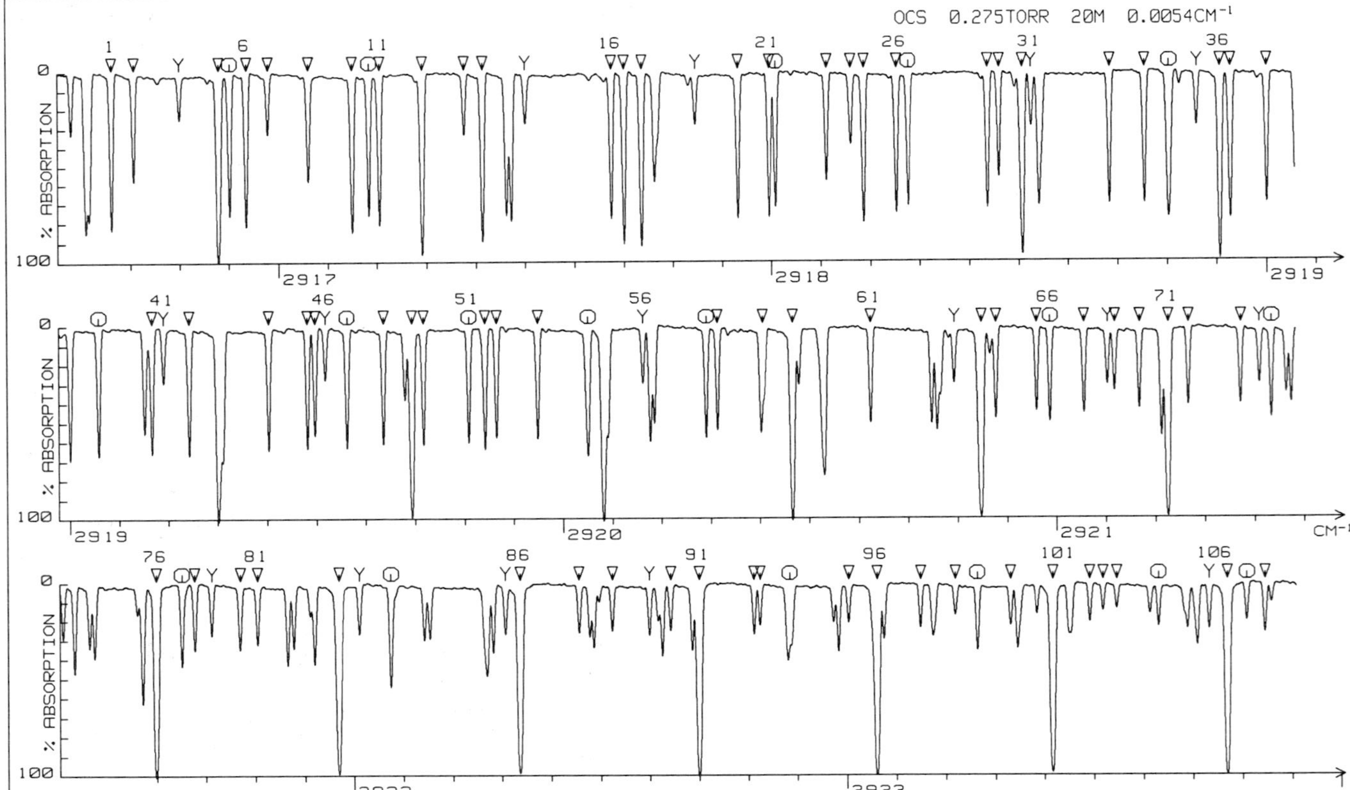

ISOTOPIC IDENTIFICATION – '▽' indicates $^{16}O^{12}C^{32}S$; 'Ⓘ': $^{16}O^{12}C^{34}S$; '◊': $^{16}O^{13}C^{32}S$; 'Y': $^{16}O^{12}C^{33}S$;
'✧' indicates $^{18}O^{12}C^{32}S$; '⇟': $^{16}O^{13}C^{34}S$; '⇞': Other OCS species; 'I': not OCS.

LINE NO.	WAVENUMBER CM^{-1}, VAC.	UNC.	ROTATIONAL	BAND ASSIGNMENTS	ISOTOPIC SPECIES
1	2916.66078	(11)	R 37	1111F-0101F	16012C32S
2	2916.70582	(12)	R 18	2010 -1000	16012C32S
3	2916.79897	(13)	R 12	1010 -0000	16012C33S
4	2916.87855	(11)	P 3	1010 -0000	16012C32S
4			R 38	1111E-0101E	16012C32S
5	2916.90164	(12)	R 31	1010 -0000	16012C34S
6	2916.93466	(11)	R 38	1111F-0101F	16012C32S
7	2916.97803	(15)	R 31	2400 -1000	16012C32S
8	2917.05977	(12)	R 19	2010 -1000	16012C32S
9	2917.14883	(11)	R 39	1111E-0101E	16012C32S
9			R 13	1010 -0000	16012C33S
10	2917.18356	(11)	R 32	1010 -0000	16012C34S
11	2917.20503	(12)	R 39	1111F-0101F	16012C32S
12	2917.29076	(11)	P 2	1010 -0000	16012C32S
13	2917.37529	(12)	R 32	2400 -1000	16012C32S
14	2917.41272	(11)	R 40	1111E-0101E	16012C32S
14			R 20	2010 -1000	16012C32S
15	2917.50008	(13)	R 14	1010 -0000	16012C33S
15			P 53	1400 -0000	16012C32S
16	2917.67421	(12)	R 41	1111E-0101E	16012C32S
17	2917.69990	(11)	P 1	1010 -0000	16012C32S
18	2917.73604	(11)	R 41	1111F-0101F	16012C32S
18			R 34	1010 -0000	16012C34S
19	2917.84482	(12)	R 15	1010 -0000	16012C33S
20	2917.93188	(11)	R 42	1111E-0101E	16012C32S
21	2917.99537	(24)	R 42	1111F-0101F	16012C32S
22	2918.00827	(26)	R 35	1010 -0000	16012C34S
23	2918.11151	(12)	R 22	2010 -1000	16012C32S
24	2918.15995	(12)	R 34	2400 -1000	16012C32S
24			P 51	1400 -0000	16012C32S
25	2918.18598	(11)	R 43	1111E-0101E	16012C32S
25			R 16	1010 -0000	16012C33S
26	2918.25172	(11)	R 43	1111F-0101F	16012C32S

LINE NO.	WAVENUMBER CM^{-1}, VAC.	UNC.	ROTATIONAL	BAND ASSIGNMENTS	ISOTOPIC SPECIES
27	2918.27613	(11)	R 36	1010 -0000	16012C34S
28	2918.43647	(11)	R 44	1111E-0101E	16012C32S
29	2918.45920	(12)	R 23	2010 -1000	16012C32S
30	2918.50680	(12)	R 0	1010 -0000	16012C32S
30			R 44	1111F-0101F	16012C32S
31	2918.52458	(13)	R 17	1010 -0000	16012C33S
32	2918.68351	(11)	R 45	1111E-0101E	16012C32S
33	2918.75413	(11)	R 45	1111F-0101F	16012C32S
34	2918.80296	(12)	R 38	1010 -0000	16012C34S
34			R 24	2010 -1000	16012C32S
35	2918.85910	(12)	R 18	1010 -0000	16012C33S
36	2918.90655	(11)	R 1	1010 -0000	16012C32S
37	2918.92749	(12)	R 46	1111E-0101E	16012C32S
37			R 36	2400 -1000	16012C32S
38	2919.00011	(11)	R 46	1111F-0101F	16012C32S
39	2919.05831	(11)	R 39	1010 -0000	16012C34S
40	2919.16706	(31)	R 47	1111E-0101E	16012C32S
41	2919.19037	(12)	R 19	1010 -0000	16012C33S
41			P 9	1400 -0000	16012C34S
42	2919.24254	(12)	R 47	1111F-0101F	16012C32S
43	2919.40346	(12)	R 48	1111E-0101E	16012C32S
44	2919.48157	(12)	R 48	1111F-0101F	16012C32S
45	2919.49677	(12)	R 26	2010 -1000	16012C32S
45			P 47	1400 -0000	16012C32S
46	2919.51782	(12)	R 20	1010 -0000	16012C33S
47	2919.56188	(12)	R 41	1010 -0000	16012C34S
48	2919.63631	(12)	R 49	1111E-0101E	16012C32S
49	2919.69343	(11)	R 3	1010 -0000	16012C32S
50	2919.71693	(12)	R 49	1111F-0101F	16012C32S
51	2919.80838	(12)	R 42	1010 -0000	16012C34S
52	2919.84148	(12)	R 27	2010 -1000	16012C32S
52			R 21	1010 -0000	16012C33S

- "LINE NO.": serial number identifying the spectral line in the map on the opposite page.
- "WAVENUMBER": experimental value.
- Uncertainty in the last digit of the wavenumber is given in parentheses.

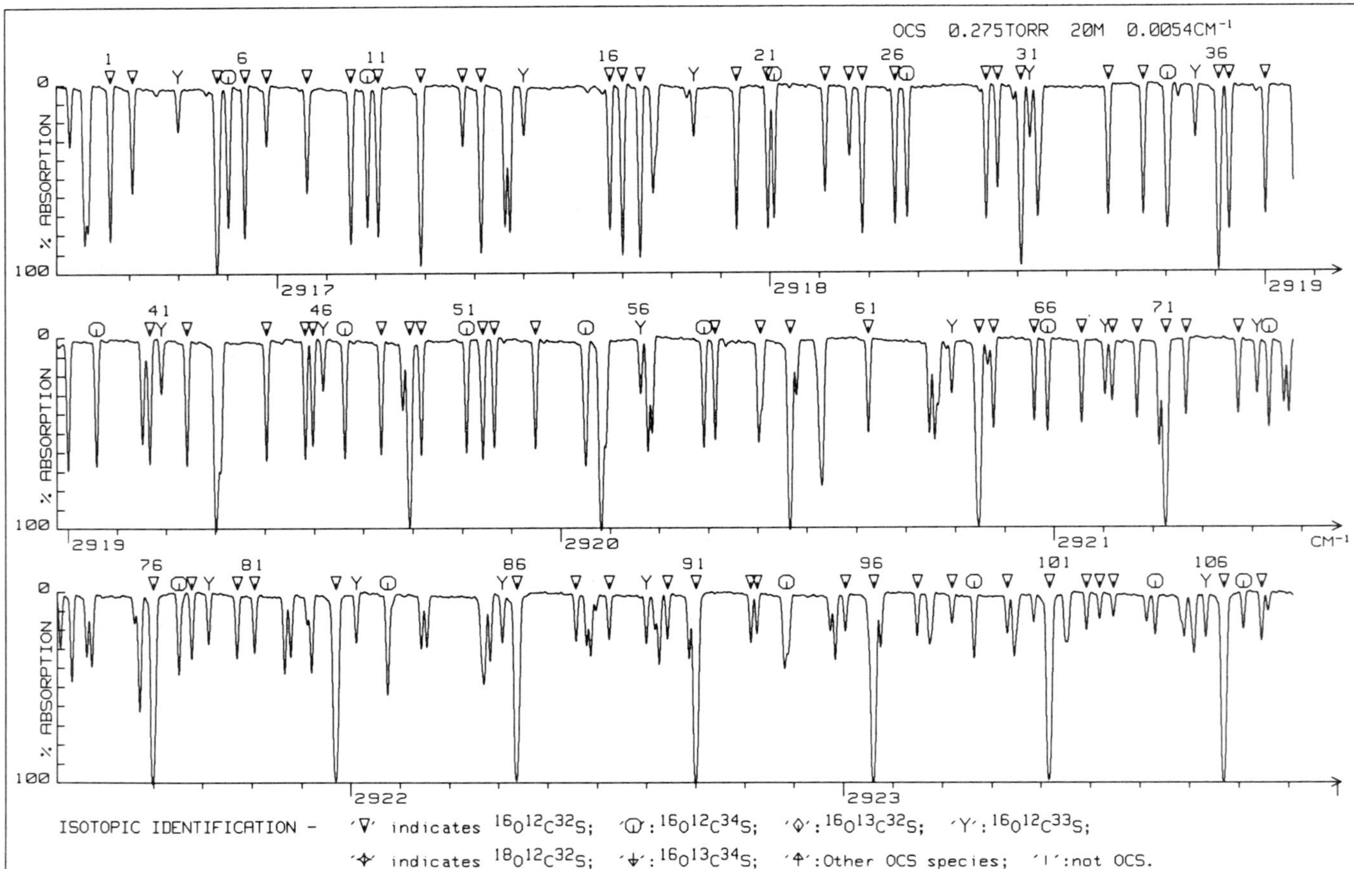
OCS 0.275TORR 20M 0.0054CM⁻¹
% ABSORPTION
0
100
2917
2918
2919
2920
2921
CM⁻¹
2922
2923
ISOTOPIC IDENTIFICATION - '∇' indicates 16O12C32S; 'ⓘ': 16O12C34S; '◊': 16O13C32S; 'Y': 16O12C33S;
'✧' indicates 18O12C32S; '⍖': 16O13C34S; '⍏': Other OCS species; 'I': not OCS.

LINE NO.	WAVENUMBER CM^{-1}, VAC.	UNC.	ROTATIONAL BAND ASSIGNMENTS	ISOTOPIC SPECIES
53	2919.86565	12	R 50 1111E-0101E	16012C32S
54	2919.94887	12	R 50 1111F-0101F	16012C32S
55	2920.05053	36	R 43 1010 -0000	16012C34S
55			R 39 2010 -1000	16012C32S
56	2920.16244	17	R 22 1010 -0000	16012C33S
57	2920.29057	12	R 44 1010 -0000	16012C34S
58	2920.31372	12	R 52 1111E-0101E	16012C32S
59	2920.40525	16	R 52 1111F-0101F	16012C32S
59			R 40 2010 -1000	16012C32S
60	2920.46636	12	R 5 1010 -0000	16012C32S
61	2920.62355	12	R 53 1111F-0101F	16012C32S
62	2920.79297	13	R 24 1010 -0000	16012C33S
63	2920.84740	12	R 6 1010 -0000	16012C32S
64	2920.87650	13	R 30 2010 -1000	16012C32S
65	2920.95920	12	R 55 1111E-0101E	16012C32S
66	2920.98704	12	R 47 1010 -0000	16012C34S
67	2921.05571	12	R 55 1111F-0101F	16012C32S
68	2921.10290	16	R 25 1010 -0000	16012C33S
69	2921.11736	15	R 42 2010 -1000	16012C32S
70	2921.16726	12	R 56 1111E-0101E	16012C32S
71	2921.22501	12	R 7 1010 -0000	16012C32S
72	2921.26648	13	R 56 1111F-0101F	16012C32S
73	2921.37168	12	R 57 1111E-0101E	16012C32S
74	2921.40930	12	R 26 1010 -0000	16012C33S
75	2921.43341	12	R 49 1010 -0000	16012C34S
76	2921.59912	11	R 8 1010 -0000	16012C32S
77	2921.65128	13	R 50 1010 -0000	16012C34S
78	2921.67737	12	R 58 1111F-0101F	16012C32S
79	2921.71219	12	R 27 1010 -0000	16012C33S
80	2921.76988	12	R 59 1111E-0101E	16012C32S
81	2921.80493	12	R 44 2010 -1000	16012C32S
82	2921.96973	12	R 9 1010 -0000	16012C32S

LINE NO.	WAVENUMBER CM^{-1}, VAC.	UNC.	ROTATIONAL BAND ASSIGNMENTS	ISOTOPIC SPECIES
83	2922.01153	17	R 28 1010 -0000	16012C33S
84	2922.07517	13	R 52 1010 -0000	16012C34S
84			R 60 1111F-0101F	16012C32S
85	2922.30739	17	R 29 1010 -0000	16012C33S
86	2922.33681	11	R 10 1010 -0000	16012C32S
87	2922.45651	12	R 62 1111F-0101F	16012C32S
88	2922.52335	12	R 63 1111E-0101E	16012C32S
89	2922.59952	12	R 30 1010 -0000	16012C33S
90	2922.64244	13	R 63 1111F-0101F	16012C32S
91	2922.70030	12	R 11 1010 -0000	16012C32S
92	2922.81207	17	R 47 2010 -1000	16012C32S
93	2922.82473	19	R 64 1111F-0101F	16012C32S
94	2922.88435	18	R 56 1010 -0000	16012C34S
94			R 31 1010 -0000	16012C33S
95	2923.00349	13	R 65 1111F-0101F	16012C32S
96	2923.06028	12	R 12 1010 -0000	16012C32S
97	2923.14888	13	R 48 2010 -1000	16012C32S
98	2923.21939	13	R 67 1111E-0101E	16012C32S
99	2923.26375	12	R 58 1010 -0000	16012C34S
99			R 51 2010 -1000	16012C32S
100	2923.33044	22	P 36 1400 -0000	16012C32S
101	2923.41664	12	R 13 1010 -0000	16012C32S
102	2923.49176	13	R 49 2010 -1000	16012C32S
103	2923.51817	14	R 68 1111F-0101F	16012C32S
104	2923.54580	14	R 69 1111E-0101E	16012C32S
105	2923.63025	13	R 60 1010 -0000	16012C34S
106	2923.73270	18	R 34 1010 -0000	16012C33S
107	2923.76952	12	R 14 1010 -0000	16012C32S
108	2923.80820	20	R 61 1010 -0000	16012C34S
109	2923.84465	18	R 50 2010 -1000	16012C32S
109			R 70 1111F-0101F	16012C32S

- "LINE NO.": serial number identifying the spectral line in the map on the opposite page.
- "WAVENUMBER": experimental value.
- Uncertainty in the last digit of the wavenumber is given in parentheses.

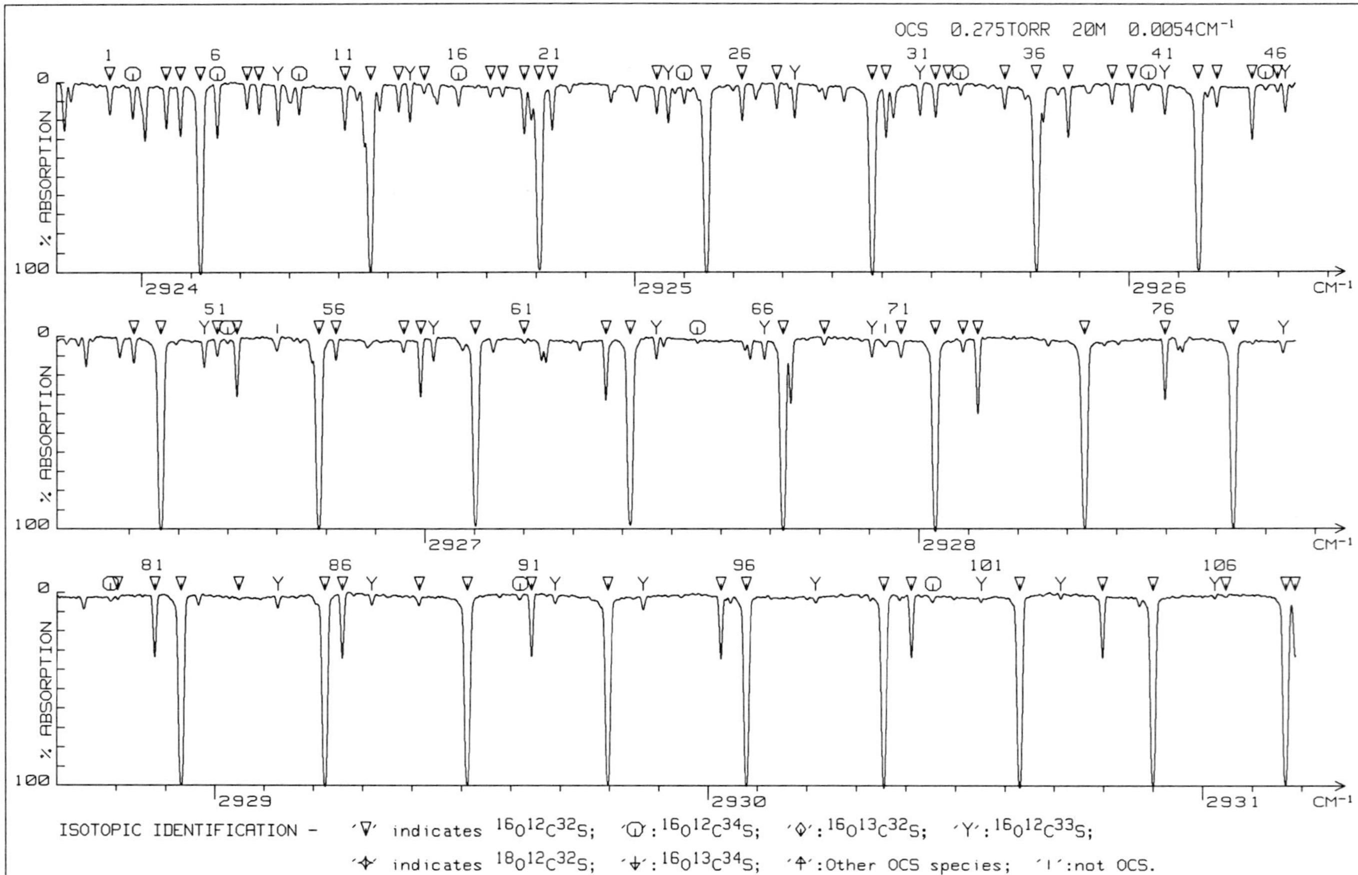
OCS 0.275TORR 20M 0.0054CM⁻¹
% ABSORPTION
CM⁻¹
2924
2925
2926
2927
2928
2929
2930
2931
ISOTOPIC IDENTIFICATION – '∇' indicates 16O12C32S; 16O12C34S; 16O13C32S; 'Y': 16O12C33S;
indicates 18O12C32S; 16O13C34S; Other OCS species; 'I': not OCS.

LINE NO.	WAVENUMBER CM^{-1}, VAC.	UNC.	ROTATIONAL BAND ASSIGNMENTS	ISOTOPIC SPECIES
1	2923.93685	(13)	R 53 2010 -1000	16012C32S
2	2923.98254	(13)	R 62 1010 -0000	16012C34S
3	2924.05017	(13)	P 34 1400 -0000	16012C32S
4	2924.07859	(17)	R 39 2400 -1000	16012C32S
5	2924.11886	(11)	R 15 1010 -0000	16012C32S
6	2924.15390	(16)	R 63 1010 -0000	16012C34S
6			R 72 1111F-0101F	16012C32S
6			R 73 1111E-0101E	16012C32S
7	2924.21413	(14)	R 51 2402E-1000	16012C32S
8	2924.23889	(13)	R 54 2010 -1000	16012C32S
9	2924.27780	(13)	R 36 1010 -0000	16012C33S
10	2924.32024	(13)	R 64 1010 -0000	16012C34S
11	2924.41254	(13)	P 33 1400 -0000	16012C32S
12	2924.46477	(12)	R 16 1010 -0000	16012C32S
13	2924.52225	(14)	R 55 2010 -1000	16012C32S
14	2924.54498	(13)	R 37 1010 -0000	16012C33S
15	2924.57420	(21)	R 76 1111E-0101E	16012C32S
16	2924.64332	(14)	R 66 1010 -0000	16012C34S
17	2924.70709	(22)	R 77 1111E-0101E	16012C32S
17			R 4 1400 -0000	16012C34S
18	2924.73256	(21)	R 76 1111F-0101F	16012C32S
19	2924.77634	(17)	P 32 1400 -0000	16012C32S
20	2924.80688	(12)	R 17 1010 -0000	16012C32S
21	2924.83284	(18)	R 41 2400 -1000	16012C32S
21			R 78 1111E-0101E	16012C32S
22	2925.04483	(13)	R 57 2010 -1000	16012C32S
23	2925.06853	(13)	R 39 1010 -0000	16012C33S
24	2925.10068	(40)	R 69 1010 -0000	16012C34S
25	2925.14554	(12)	R 18 1010 -0000	16012C32S
26	2925.21789	(13)	R 42 2400 -1000	16012C32S
27	2925.28862	(14)	R 58 2010 -1000	16012C32S
28	2925.32487	(21)	R 40 1010 -0000	16012C33S
29	2925.48073	(12)	R 19 1010 -0000	16012C32S
30	2925.50890	(17)	P 30 1400 -0000	16012C32S

LINE NO.	WAVENUMBER CM^{-1}, VAC.	UNC.	ROTATIONAL BAND ASSIGNMENTS	ISOTOPIC SPECIES
31	2925.57771	(13)	R 41 1010 -0000	16012C33S
32	2925.60885	(13)	R 43 2400 -1000	16012C32S
32			R 83 1111F-0101F	16012C32S
33	2925.63456	(51)	R 85 1111E-0101E	16012C32S
34	2925.65912	(21)	R 73 1010 -0000	16012C34S
35	2925.74852	(15)	R 60 2010 -1000	16012C32S
36	2925.81229	(12)	R 20 1010 -0000	16012C32S
37	2925.87747	(13)	P 29 1400 -0000	16012C32S
38	2925.96642	(15)	R 61 2010 -1000	16012C32S
39	2926.00658	(14)	R 44 2400 -1000	16012C32S
39			R 89 1111E-0101E	16012C32S
40	2926.03948	(27)	R 76 1010 -0000	16012C34S
41	2926.07233	(14)	R 43 1010 -0000	16012C33S
42	2926.14027	(12)	R 21 1010 -0000	16012C32S
43	2926.17702	(36)	R 62 2010 -1000	16012C32S
43			R 91 1111E-0101E	16012C32S
44	2926.24754	(12)	P 28 1400 -0000	16012C32S
45	2926.27451	(32)	R 78 1010 -0000	16012C34S
46	2926.29843	(31)	R 56 2402E-1000	16012C32S
46			R 90 1111F-0101F	16012C32S
47	2926.31429	(14)	R 44 1010 -0000	16012C33S
48	2926.41016	(14)	R 45 2400 -1000	16012C32S
49	2926.46465	(12)	R 22 1010 -0000	16012C32S
50	2926.55252	(14)	R 45 1010 -0000	16012C33S
51	2926.57936	(16)	R 64 2010 -1000	16012C32S
52	2926.59997	(45)	R 81 1010 -0000	16012C34S
53	2926.61900	(12)	P 27 1400 -0000	16012C32S
54	2926.70085	(22)		CH4
55	2926.78554	(12)	R 23 1010 -0000	16012C32S
56	2926.82021	(32)	R 46 2400 -1000	16012C32S
57	2926.95749	(18)	R 66 2010 -1000	16012C32S
58	2926.99203	(12)	P 26 1400 -0000	16012C32S
59	2927.01824	(14)	R 47 1010 -0000	16012C33S

- "LINE NO.": serial number identifying the spectral line in the map on the opposite page.
- "WAVENUMBER": experimental value.
- Uncertainty in the last digit of the wavenumber is given in parentheses.

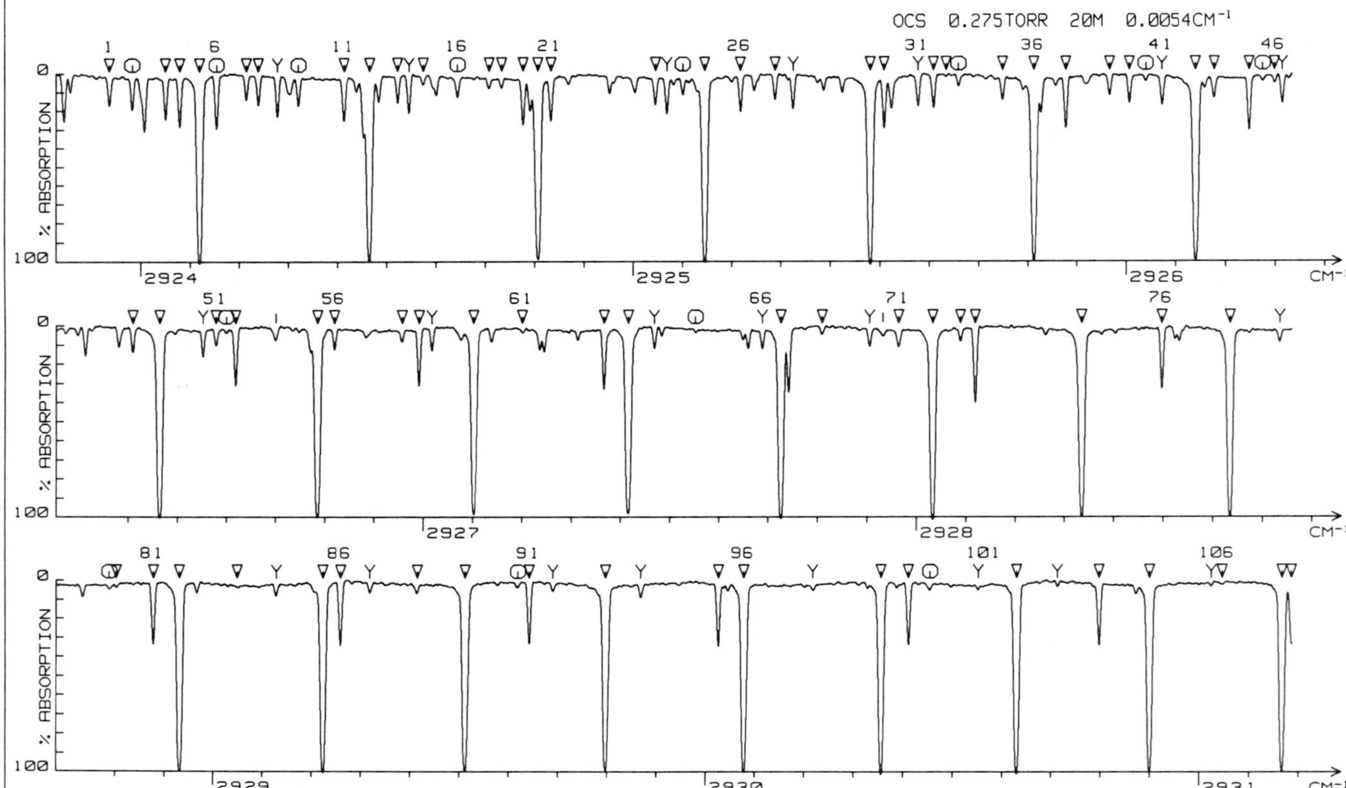

ISOTOPIC IDENTIFICATION – '∇' indicates $^{16}O^{12}C^{32}S$; 'Ⓘ': $^{16}O^{12}C^{34}S$; '◊': $^{16}O^{13}C^{32}S$; 'Y': $^{16}O^{12}C^{33}S$;

'✧' indicates $^{18}O^{12}C^{32}S$; '⍖': $^{16}O^{13}C^{34}S$; '⍏': Other OCS species; 'I': not OCS.

LINE NO.	WAVENUMBER UNC. CM^{-1}, VAC.	ROTATIONAL BAND ASSIGNMENTS	ISOTOPIC SPECIES
60	2927.10287(12)	R 24 1010 -0000	16012C32S
61	2927.20129(57)	R 58 2402E-1000	16012C32S
62	2927.36653(15)	P 25 1400 -0000	16012C32S
63	2927.41656(12)	R 25 1010 -0000	16012C32S
64	2927.46952(32)	R 49 1010 -0000	16012C33S
65	2927.55264(41)	R 11 1400 -0000	16012C34S
66	2927.68926(35)	R 50 1010 -0000	16012C33S
67	2927.72666(12)	R 26 1010 -0000	16012C32S
68	2927.80964(25)	R 71 2010 -1000	16012C32S
69	2927.90588(15)	R 51 1010 -0000	16012C33S
70	2927.93276(41)		CH4
71	2927.96456(18)	R 72 2010 -1000	16012C32S
71		R 12 1400 -0000	16012C34S
72	2928.03319(12)	R 27 1010 -0000	16012C32S
73	2928.08906(18)	R 49 2400 -1000	16012C32S
74	2928.11918(12)	P 23 1400 -0000	16012C32S
74		R 52 1010 -0000	16012C33S
74		R 73 2010 -1000	16012C32S
74		R 60 2402E-1000	16012C32S
75	2928.33601(12)	R 28 1010 -0000	16012C32S
76	2928.49803(12)	P 22 1400 -0000	16012C32S
77	2928.63533(12)	R 29 1010 -0000	16012C32S
78	2928.73518(18)	R 55 1010 -0000	16012C33S
79	2928.78874(36)	R 14 1400 -0000	16012C34S
80	2928.80290(58)	R 78 2010 -1000	16012C32S
81	2928.87802(12)	P 21 1400 -0000	16012C32S
82	2928.93100(12)	R 30 1010 -0000	16012C32S
83	2929.04834(52)	R 62 2402E-1000	16012C32S
83		R 80 2010 -1000	16012C32S
84	2929.12789(21)	R 57 1010 -0000	16012C33S
85	2929.22314(12)	R 31 1010 -0000	16012C32S
86	2929.25924(15)	P 20 1400 -0000	16012C32S

LINE NO.	WAVENUMBER UNC. CM^{-1}, VAC.	ROTATIONAL BAND ASSIGNMENTS	ISOTOPIC SPECIES
87	2929.31865(22)	R 58 1010 -0000	16012C33S
88	2929.41447(25)	R 52 2400 -1000	16012C32S
89	2929.51160(12)	R 32 1010 -0000	16012C32S
90	2929.61836(37)	R 16 1400 -0000	16012C34S
90		P 29 1501E-0101E	16012C32S
91	2929.64191(12)	P 19 1400 -0000	16012C32S
92	2929.68927(25)	R 60 1010 -0000	16012C33S
93	2929.79644(12)	R 33 1010 -0000	16012C32S
94	2929.86861(18)	R 61 1010 -0000	16012C33S
94		R 53 2400 -1000	16012C32S
95	2930.02584(15)	P 18 1400 -0000	16012C32S
96	2930.07761(12)	R 34 1010 -0000	16012C32S
97	2930.21772(31)	R 63 1010 -0000	16012C33S
98	2930.35521(11)	R 35 1010 -0000	16012C32S
99	2930.41104(12)	P 17 1400 -0000	16012C32S
100	2930.45409(34)	R 18 1400 -0000	16012C34S
101	2930.55143(37)	R 65 1010 -0000	16012C33S
102	2930.62923(11)	R 36 1010 -0000	16012C32S
103	2930.71279(41)	R 66 1010 -0000	16012C33S
104	2930.79750(15)	P 16 1400 -0000	16012C32S
105	2930.89952(12)	R 37 1010 -0000	16012C32S
106	2931.02424(48)	R 68 1010 -0000	16012C33S
107	2931.04584(80)	P 26 1501F-0101F	16012C32S
108	2931.16616(12)	R 38 1010 -0000	16012C32S
109	2931.18528(16)	P 15 1400 -0000	16012C32S

- "LINE NO.": serial number identifying the spectral line in the map on the opposite page.
- "WAVENUMBER": experimental value.
- Uncertainty in the last digit of the wavenumber is given in parentheses.

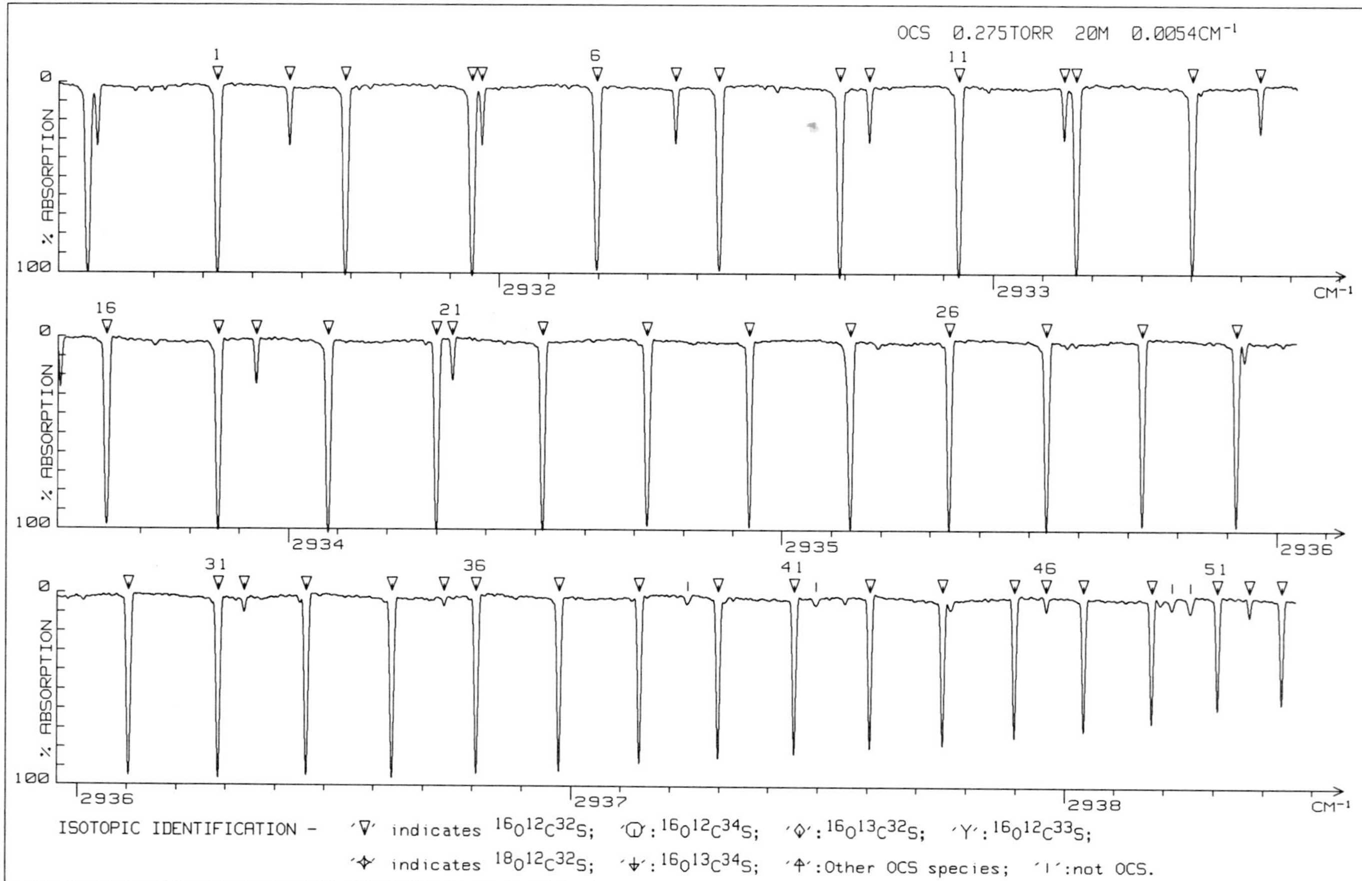

OCS 0.275TORR 20M 0.0054CM⁻¹
% ABSORPTION
0
100
2932
2933
2934
2935
2936
2937
2938
CM⁻¹
ISOTOPIC IDENTIFICATION – '∇' indicates 16O12C32S; 16O12C34S; 16O13C32S; 'Y': 16O12C33S;
indicates 18O12C32S; 16O13C34S; Other OCS species; 'I': not OCS.

LINE NO.	WAVENUMBER UNC. CM^{-1}, VAC.	ROTATIONAL BAND ASSIGNMENTS	ISOTOPIC SPECIES
1	2931.42918(11)	R 39 1010 -0000	16012C32S
2	2931.57431(12)	P 14 1400 -0000	16012C32S
3	2931.68860(11)	R 40 1010 -0000	16012C32S
4	2931.94430(11)	R 41 1010 -0000	16012C32S
5	2931.96456(16)	P 13 1400 -0000	16012C32S
6	2932.19633(11)	R 42 1010 -0000	16012C32S
7	2932.35616(12)	P 12 1400 -0000	16012C32S
8	2932.44476(11)	R 43 1010 -0000	16012C32S
9	2932.68944(11)	R 44 1010 -0000	16012C32S
10	2932.74890(12)	P 11 1400 -0000	16012C32S
11	2932.93049(11)	R 45 1010 -0000	16012C32S
12	2933.14284(16)	P 10 1400 -0000	16012C32S
13	2933.16795(11)	R 46 1010 -0000	16012C32S
14	2933.40160(11)	R 47 1010 -0000	16012C32S
15	2933.53803(13)	P 9 1400 -0000	16012C32S
16	2933.63158(11)	R 48 1010 -0000	16012C32S
17	2933.85793(11)	R 49 1010 -0000	16012C32S
18	2933.93430(13)	P 8 1400 -0000	16012C32S
19	2934.08052(11)	R 50 1010 -0000	16012C32S
20	2934.29951(11)	R 51 1010 -0000	16012C32S
21	2934.33178(13)	P 7 1400 -0000	16012C32S
22	2934.51475(11)	R 52 1010 -0000	16012C32S
23	2934.72636(12)	R 53 1010 -0000	16012C32S
24	2934.93412(11)	R 54 1010 -0000	16012C32S
25	2935.13799(13)	R 55 1010 -0000	16012C32S
26	2935.33864(11)	R 56 1010 -0000	16012C32S
27	2935.53525(11)	R 57 1010 -0000	16012C32S
28	2935.72831(11)	R 58 1010 -0000	16012C32S
29	2935.91757(11)	R 59 1010 -0000	16012C32S
30	2936.10308(11)	R 60 1010 -0000	16012C32S
31	2936.28486(11)	R 61 1010 -0000	16012C32S
32	2936.33715(18)	P 2 1400 -0000	16012C32S
33	2936.46292(11)	R 62 1010 -0000	16012C32S
34	2936.63725(11)	R 63 1010 -0000	16012C32S
35	2936.74156(26)	P 1 1400 -0000	16012C32S
36	2936.80788(11)	R 64 1010 -0000	16012C32S
37	2936.97471(11)	R 65 1010 -0000	16012C32S
38	2937.13785(11)	R 66 1010 -0000	16012C32S
39	2937.23523(30)		CH4
40	2937.29718(12)	R 67 1010 -0000	16012C32S
41	2937.45284(11)	R 68 1010 -0000	16012C32S
42	2937.49571(29)		CH4
43	2937.60465(11)	R 69 1010 -0000	16012C32S
44	2937.75277(12)	R 70 1010 -0000	16012C32S
45	2937.89715(11)	R 71 1010 -0000	16012C32S
46	2937.96232(17)	R 1 1400 -0000	16012C32S
47	2938.03773(12)	R 72 1010 -0000	16012C32S
48	2938.17462(12)	R 73 1010 -0000	16012C32S
49	2938.21570(23)		CH4
50	2938.25243(21)		CH4
51	2938.30769(12)	R 74 1010 -0000	16012C32S
52	2938.37152(15)	R 2 1400 -0000	16012C32S
53	2938.43697(12)	R 75 1010 -0000	16012C32S

- "LINE NO.": serial number identifying the spectral line in the map on the opposite page.
- "WAVENUMBER": experimental value.
- Uncertainty in the last digit of the wavenumber is given in parentheses.

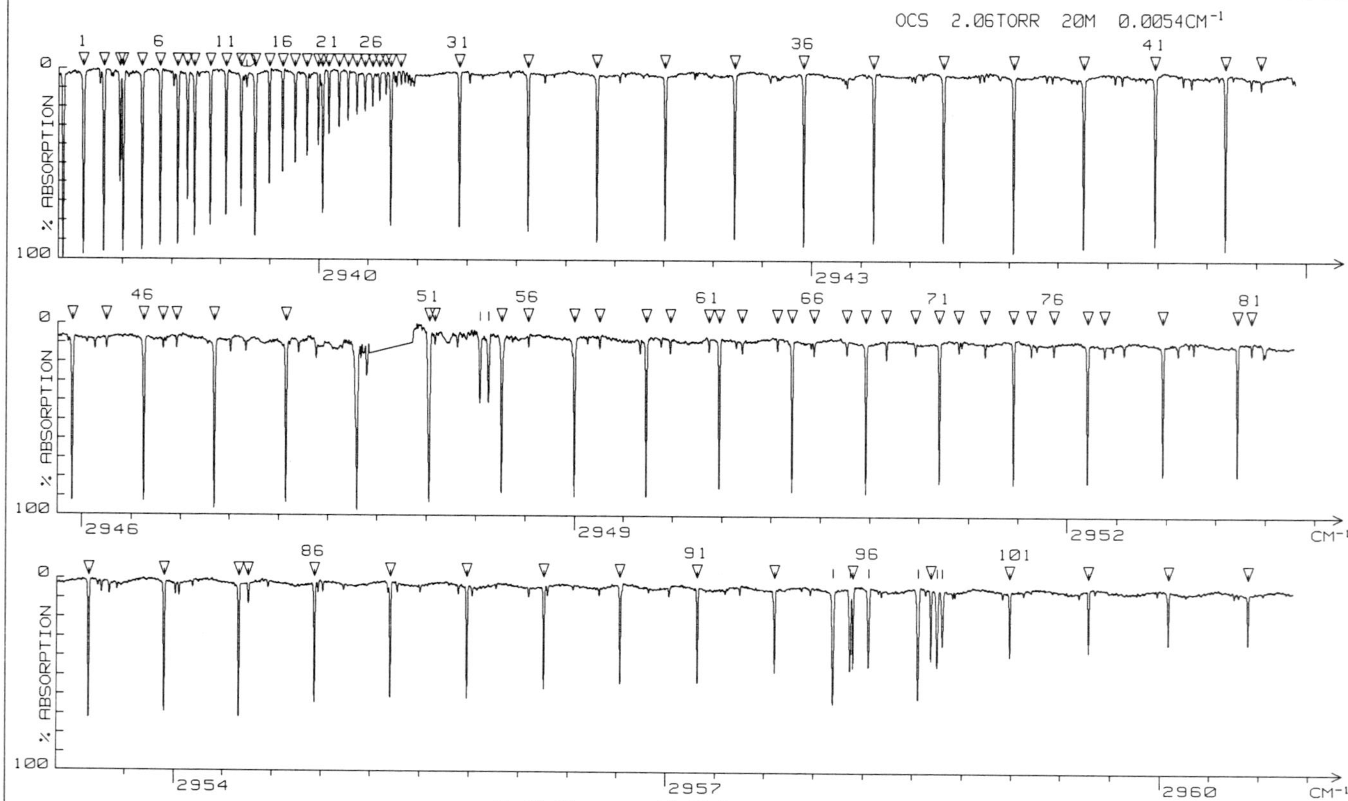

ISOTOPIC IDENTIFICATION – '∇' indicates $^{16}O^{12}C^{32}S$; 'Ⓘ': $^{16}O^{12}C^{34}S$; '◇': $^{16}O^{13}C^{32}S$; 'Y': $^{16}O^{12}C^{33}S$;
'✧' indicates $^{18}O^{12}C^{32}S$; '⫛': $^{16}O^{13}C^{34}S$; '✦': Other OCS species; 'I': not OCS.

LINE NO.	WAVENUMBER CM^{-1}, VAC.	UNC.	ROTATIONAL BAND ASSIGNMENTS	ISOTOPIC SPECIES
1	2938.56253	(12)	R 76 1010 -0000	16012C32S
2	2938.68425	(12)	R 77 1010 -0000	16012C32S
3	2938.78190	(14)	R 3 1400 -0000	16012C32S
4	2938.80228	(12)	R 78 1010 -0000	16012C32S
5	2938.91652	(12)	R 79 1010 -0000	16012C32S
6	2939.02690	(12)	R 80 1010 -0000	16012C32S
7	2939.13359	(12)	R 81 1010 -0000	16012C32S
8	2939.19337	(13)	R 4 1400 -0000	16012C32S
9	2939.23651	(12)	R 82 1010 -0000	16012C32S
10	2939.33561	(12)	R 83 1010 -0000	16012C32S
11	2939.43089	(13)	R 84 1010 -0000	16012C32S
12	2939.52244	(13)	R 85 1010 -0000	16012C32S
13	2939.55659	(70)	R 39 1400 -0000	16012C34S
14	2939.60797	(13)	R 5 1400 -0000	16012C32S
14			R 86 1010 -0000	16012C32S
15	2939.69404	(14)	R 87 1010 -0000	16012C32S
16	2939.77421	(14)	R 88 1010 -0000	16012C32S
17	2939.85057	(15)	R 89 1010 -0000	16012C32S
18	2939.92311	(16)	R 90 1010 -0000	16012C32S
19	2939.99187	(17)	R 91 1010 -0000	16012C32S
20	2940.01991	(13)	R 6 1400 -0000	16012C32S
21	2940.05681	(18)	R 92 1010 -0000	16012C32S
22	2940.11797	(21)	R 93 1010 -0000	16012C32S
22			R 24 1400 -0000	16012C33S
23	2940.17520	(22)	R 94 1010 -0000	16012C32S
24	2940.22888	(26)	R 95 1010 -0000	16012C32S
25	2940.27854	(28)	R 96 1010 -0000	16012C32S
26	2940.32431	(32)	R 97 1010 -0000	16012C32S
27	2940.36656	(36)	R 98 1010 -0000	16012C32S
28	2940.40488	(45)	R 99 1010 -0000	16012C32S
29	2940.43508	(13)	R 7 1400 -0000	16012C32S
29			R1 0 1010 -0000	16012C32S
30	2940.49656	(73)	R1 2 1010 -0000	16012C32S

LINE NO.	WAVENUMBER CM^{-1}, VAC.	UNC.	ROTATIONAL BAND ASSIGNMENTS	ISOTOPIC SPECIES
31	2940.85110	(13)	R 8 1400 -0000	16012C32S
32	2941.26848	(12)	R 9 1400 -0000	16012C32S
33	2941.68704	(12)	R 10 1400 -0000	16012C32S
34	2942.10685	(12)	R 11 1400 -0000	16012C32S
35	2942.52784	(12)	R 12 1400 -0000	16012C32S
36	2942.95002	(12)	R 13 1400 -0000	16012C32S
37	2943.37341	(12)	R 14 1400 -0000	16012C32S
38	2943.79802	(12)	R 15 1400 -0000	16012C32S
39	2944.22396	(12)	R 16 1400 -0000	16012C32S
40	2944.65114	(12)	R 17 1400 -0000	16012C32S
41	2945.07957	(12)	R 18 1400 -0000	16012C32S
42	2945.50925	(12)	R 19 1400 -0000	16012C32S
43	2945.72309	(78)	R 10 1501F-0101F	16012C32S
44	2945.94028	(12)	R 20 1400 -0000	16012C32S
45	2946.14598	(82)	R 11 1501F-0101F	16012C32S
46	2946.37255	(12)	R 21 1400 -0000	16012C32S
47	2946.48917	(88)	R 12 1501E-0101E	16012C32S
48	2946.57100	(78)	R 12 1501F-0101F	16012C32S
49	2946.80627	(12)	R 22 1400 -0000	16012C32S
50	2947.24129	(12)	R 23 1400 -0000	16012C32S
51	2948.11528	(88)	R 25 1400 -0000	16012C32S
52	2948.14827	(78)	R 16 1501E-0101E	16012C32S
53	2948.42208	(23)		CH4
54	2948.47465	(23)		CH4
55	2948.55453	(12)	R 26 1400 -0000	16012C32S
56	2948.71599	(73)	R 17 1501F-0101F	16012C32S
57	2948.99508	(12)	R 27 1400 -0000	16012C32S
58	2949.14899	(76)	R 18 1501F-0101F	16012C32S
59	2949.43711	(12)	R 28 1400 -0000	16012C32S
60	2949.58379	(70)	R 19 1501F-0101F	16012C32S

- "LINE NO.": serial number identifying the spectral line in the map on the opposite page.
- "WAVENUMBER": experimental value.
- Uncertainty in the last digit of the wavenumber is given in parentheses.

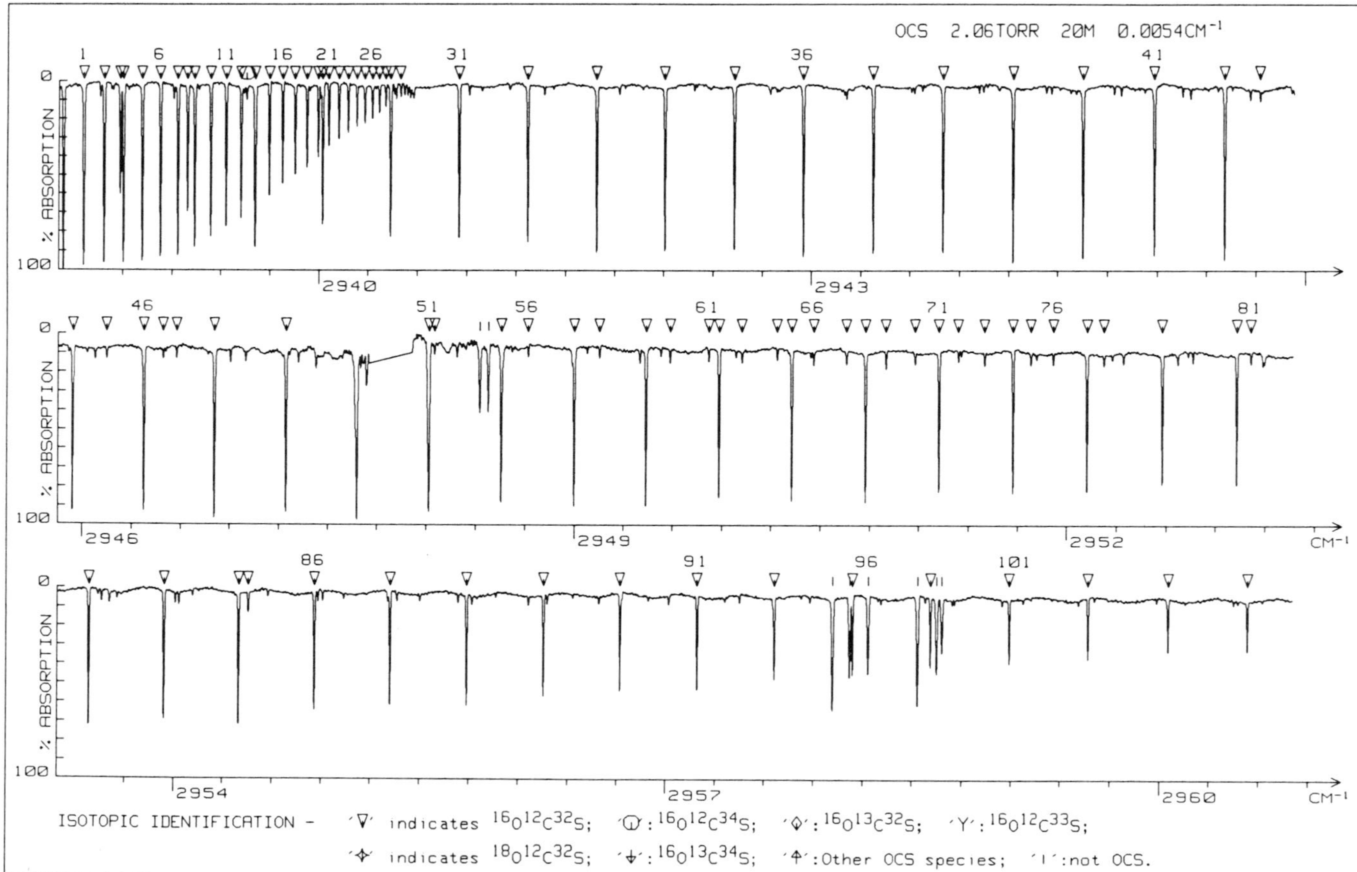
OCS 2.06TORR 20M 0.0054CM-1
% ABSORPTION
0
100
2940
2943
2946
2949
2952
2954
2957
2960
CM-1
ISOTOPIC IDENTIFICATION - 'V' indicates 16O12C32S; ':16O12C34S; ':16O13C32S; 'Y':16O12C33S;
' indicates 18O12C32S; ':16O13C34S; ':Other OCS species; 'I':not OCS.

LINE NO.	WAVENUMBER UNC. CM^{-1}, VAC.	ROTATIONAL BAND ASSIGNMENTS	ISOTOPIC SPECIES
61	2949.81773(69)	R 20 1501E-0101E	16012C32S
62	2949.88051(13)	R 29 1400 -0000	16012C32S
63	2950.01980(77)	R 20 1501F-0101F	16012C32S
64	2950.23638(78)	R 21 1501E-0101E	16012C32S
65	2950.32542(13)	R 30 1400 -0000	16012C32S
66	2950.45759(71)	R 21 1501F-0101F	16012C32S
66		R 47 1400 -0000	16012C33S
67	2950.65569(74)	R 22 1501E-0101E	16012C32S
68	2950.77192(13)	R 31 1400 -0000	16012C32S
69	2950.89625(52)	R 22 1501F-0101F	16012C32S
69		R 42 1402E-0000	16012C32S
69		R 63 1400 -0000	16012C34S
70	2951.07577(76)	R 23 1501E-0101E	16012C32S
71	2951.21990(13)	R 32 1400 -0000	16012C32S
72	2951.33655(78)	R 23 1501F-0101F	16012C32S
73	2951.49593(71)	R 24 1501E-0101E	16012C32S
74	2951.66935(13)	R 33 1400 -0000	16012C32S
75	2951.77851(75)	R 24 1501F-0101F	16012C32S
76	2951.91734(80)	R 25 1501E-0101E	16012C32S
77	2952.12047(13)	R 34 1400 -0000	16012C32S
78	2952.22189(83)	R 25 1501F-0101F	16012C32S
79	2952.57316(13)	R 35 1400 -0000	16012C32S
80	2953.02745(13)	R 36 1400 -0000	16012C32S
81	2953.11263(76)	R 27 1501F-0101F	16012C32S
82	2953.48341(13)	R 37 1400 -0000	16012C32S
83	2953.94092(14)	R 38 1400 -0000	16012C32S
84	2954.40029(13)	R 39 1400 -0000	16012C32S
85	2954.45839(60)	R 30 1501F-0101F	16012C32S
85		R 31 1501E-0101E	16012C32S
86	2954.86131(14)	R 40 1400 -0000	16012C32S
87	2955.32407(14)	R 41 1400 -0000	16012C32S
88	2955.78860(14)	R 42 1400 -0000	16012C32S
89	2956.25498(15)	R 43 1400 -0000	16012C32S
90	2956.72308(15)	R 44 1400 -0000	16012C32S

LINE NO.	WAVENUMBER UNC. CM^{-1}, VAC.	ROTATIONAL BAND ASSIGNMENTS	ISOTOPIC SPECIES
91	2957.19318(15)	R 45 1400 -0000	16012C32S
91		R 36 1501F-0101F	16012C32S
92	2957.66492(17)	R 46 1400 -0000	16012C32S
93	2958.01777(18)		CH4
94	2958.12036(22)		CH4
95	2958.13880(17)	R 47 1400 -0000	16012C32S
96	2958.23338(23)		CH4
97	2958.53689(18)		CH4
98	2958.61443(18)	R 48 1400 -0000	16012C32S
99	2958.65141(23)		CH4
100	2958.68301(31)		CH4
101	2959.09212(19)	R 49 1400 -0000	16012C32S
102	2959.57174(20)	R 50 1400 -0000	16012C32S
103	2960.05332(23)	R 51 1400 -0000	16012C32S
104	2960.53690(23)	R 52 1400 -0000	16012C32S

- "LINE NO.": serial number identifying the spectral line in the map on the opposite page.
- "WAVENUMBER": experimental value.
- Uncertainty in the last digit of the wavenumber is given in parentheses.

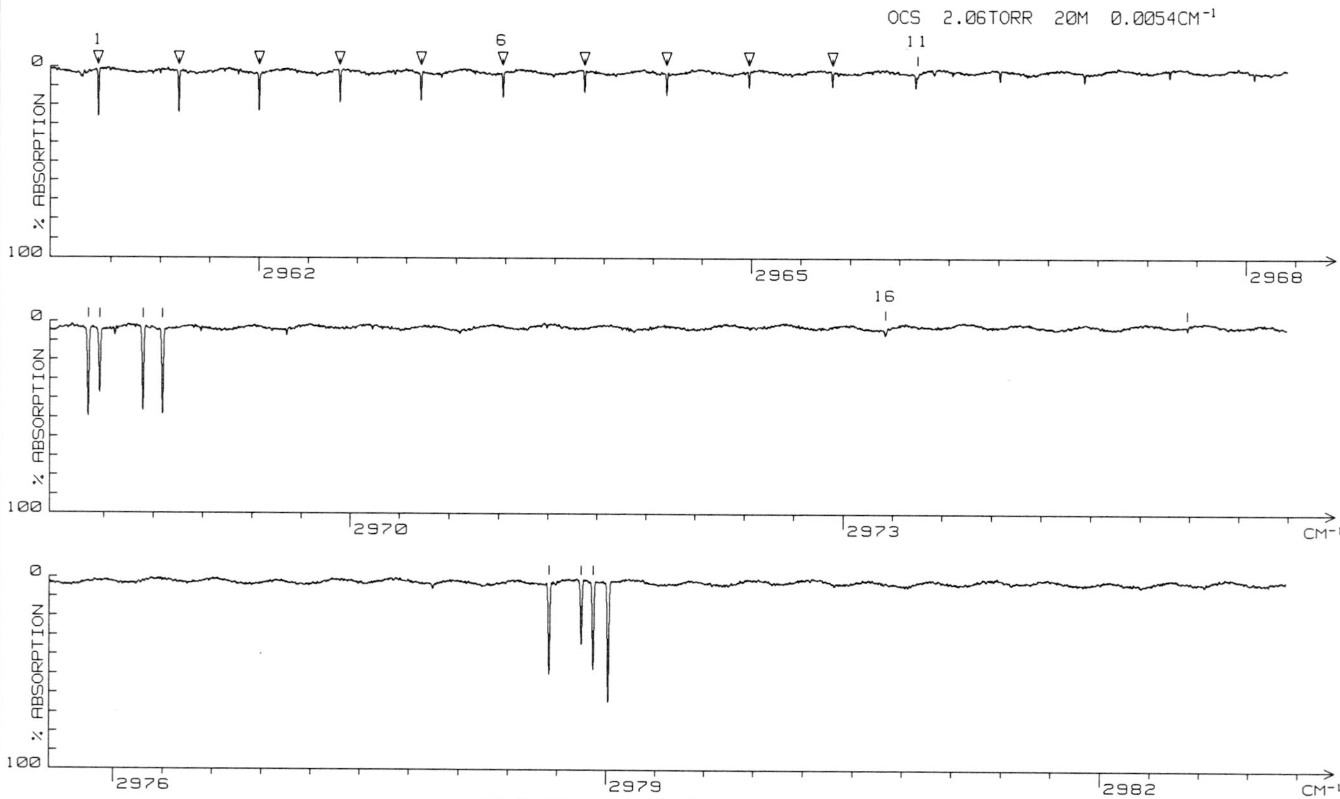

ISOTOPIC IDENTIFICATION – '▽' indicates $^{16}O^{12}C^{32}S$; 'Ⓘ': $^{16}O^{12}C^{34}S$; '◊': $^{16}O^{13}C^{32}S$; 'Y': $^{16}O^{12}C^{33}S$;
'✧' indicates $^{18}O^{12}C^{32}S$; '⫰': $^{16}O^{13}C^{34}S$; '↑': Other OCS species; 'I': not OCS.

LINE NO.	WAVENUMBER UNC. CM^{-1}, VAC.	ROTATIONAL BAND ASSIGNMENTS	ISOTOPIC SPECIES
1	2961.02274(26)	R 53 1400 -0000	16012C32S
2	2961.51034(29)	R 54 1400 -0000	16012C32S
3	2962.00021(30)	R 55 1400 -0000	16012C32S
4	2962.49231(34)	R 56 1400 -0000	16012C32S
5	2962.98621(40)	R 57 1400 -0000	16012C32S
6	2963.48249(45)	R 58 1400 -0000	16012C32S
7	2963.98097(49)	R 59 1400 -0000	16012C32S
8	2964.48143(48)	R 60 1400 -0000	16012C32S
9	2964.98427(63)	R 61 1400 -0000	16012C32S
10	2965.48962(70)	R 62 1400 -0000	16012C32S
11	2966.00692(93)		HOH
12	2968.40374(21)		CH4
13	2968.47412(28)		CH4
14	2968.73682(23)		CH4
15	2968.85574(22)		CH4
16	2973.25297(39)		HOH
17	2975.08567(83)		HOH
18	2978.65105(21)		CH4
19	2978.84860(27)		CH4
20	2978.92044(21)		CH4

- "LINE NO.": serial number identifying the spectral line in the map on the opposite page.
- "WAVENUMBER": experimental value.
- Uncertainty in the last digit of the wavenumber is given in parentheses.

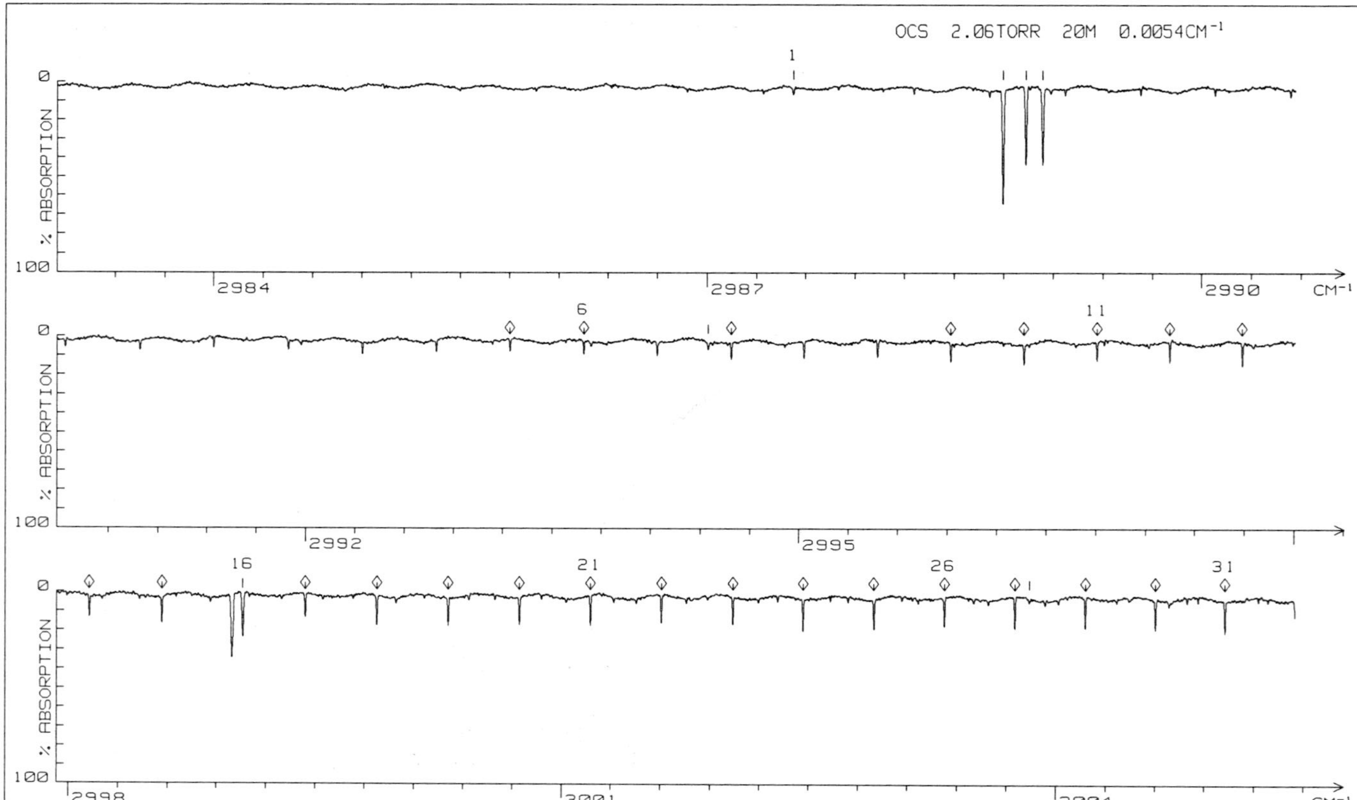

ISOTOPIC IDENTIFICATION – '▽' indicates $^{16}O^{12}C^{32}S$; '◎': $^{16}O^{12}C^{34}S$; '◇': $^{16}O^{13}C^{32}S$; 'Y': $^{16}O^{12}C^{33}S$;
'✧' indicates $^{18}O^{12}C^{32}S$; '⏚': $^{16}O^{13}C^{34}S$; '↑': Other OCS species; 'I': not OCS.

LINE NO.	WAVENUMBER CM^{-1}, VAC.	UNC.	ROTATIONAL BAND ASSIGNMENTS	ISOTOPIC SPECIES
1	2987.52547	(36)		HOH
2	2988.79589	(18)		CH4
3	2988.93311	(23)		CH4
4	2989.03414	(23)		CH4
5	2993.24430	(86)	P 46 0210 -0000	16013C32S
6	2993.69260	(73)	P 45 0210 -0000	16013C32S
7	2994.44700	(37)		HOH
8	2994.58686	(65)	P 43 0210 -0000	16013C32S
9	2995.92102	(55)	P 40 0210 -0000	16013C32S
10	2996.36408	(52)	P 39 0210 -0000	16013C32S
11	2996.80632	(54)	P 38 0210 -0000	16013C32S
12	2997.24730	(48)	P 37 0210 -0000	16013C32S
13	2997.68783	(47)	P 36 0210 -0000	16013C32S
14	2998.12719	(51)	P 35 0210 -0000	16013C32S
15	2998.56549	(41)	P 34 0210 -0000	16013C32S
16	2999.06093	(40)		CH4
17	2999.43932	(49)	P 32 0210 -0000	16013C32S
18	2999.87528	(38)	P 31 0210 -0000	16013C32S
19	3000.30977	(41)	P 30 0210 -0000	16013C32S
20	3000.74320	(42)	P 29 0210 -0000	16013C32S
21	3001.17616	(38)	P 28 0210 -0000	16013C32S
22	3001.60791	(41)	P 27 0210 -0000	16013C32S
23	3002.03879	(40)	P 26 0210 -0000	16013C32S
24	3002.46860	(35)	P 25 0210 -0000	16013C32S
25	3002.89756	(38)	P 24 0210 -0000	16013C32S
26	3003.32527	(39)	P 23 0210 -0000	16013C32S
27	3003.75218	(36)	P 22 0210 -0000	16013C32S
28	3003.83868	(51)		HOH
29	3004.17802	(36)	P 21 0210 -0000	16013C32S
30	3004.60280	(38)	P 20 0210 -0000	16013C32S
31	3005.02672	(35)	P 19 0210 -0000	16013C32S

- "LINE NO.": serial number identifying the spectral line in the map on the opposite page.
- "WAVENUMBER": experimental value.
- Uncertainty in the last digit of the wavenumber is given in parentheses.

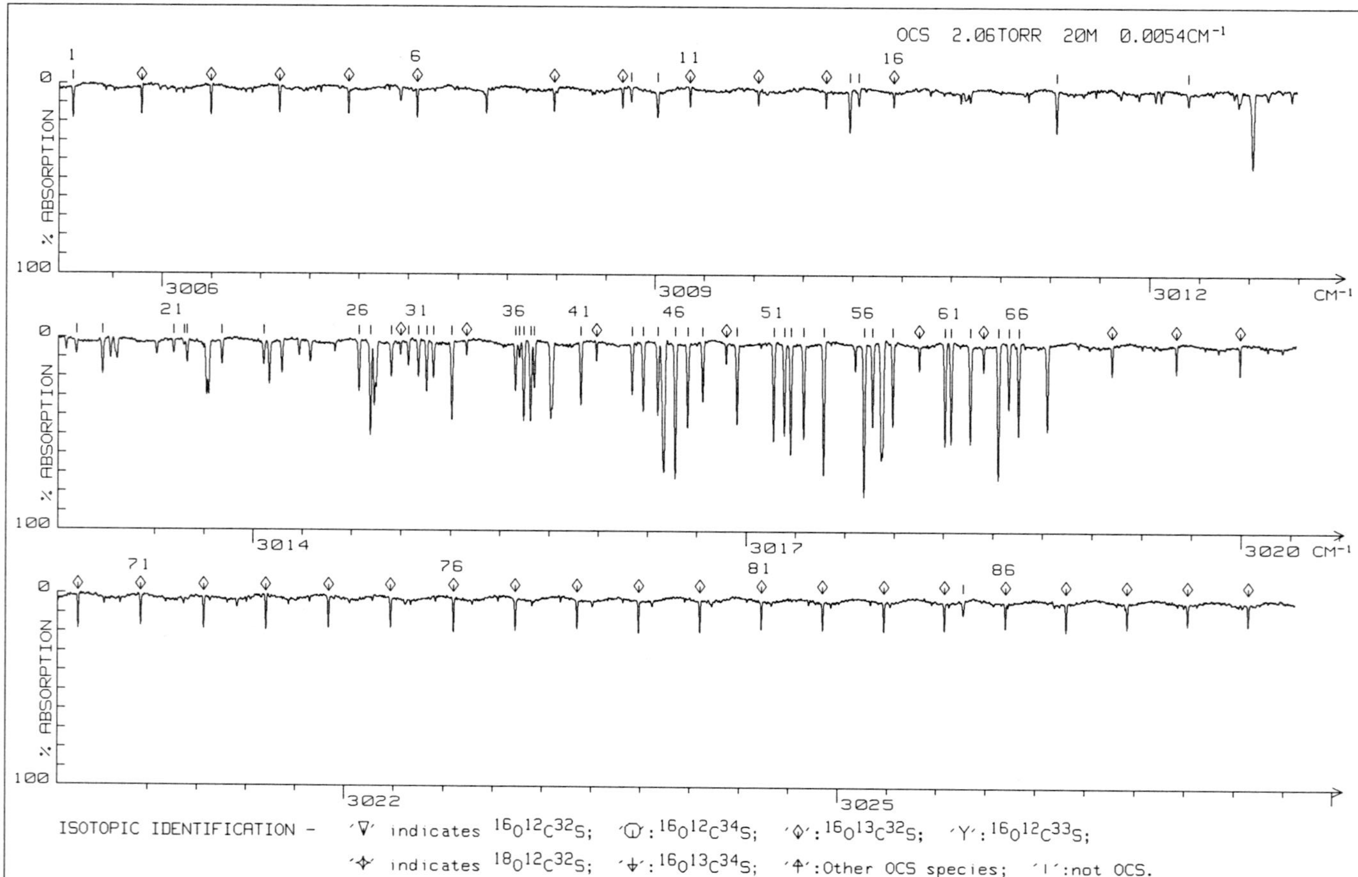
OCS 2.06TORR 20M 0.0054CM-1
% ABSORPTION
0
100
3006
3009
3012
CM-1
3014
3017
3020 CM-1
3022
3025
ISOTOPIC IDENTIFICATION -
'∇' indicates 16O12C32S; 'Φ': 16O12C34S; '◊': 16O13C32S; 'Y': 16O12C33S;
'◊' indicates 18O12C32S; '↓': 16O13C34S; '↑': Other OCS species; 'I': not OCS.

LINE NO.	WAVENUMBER UNC. CM^{-1}, VAC.	ROTATIONAL BAND ASSIGNMENTS	ISOTOPIC SPECIES	LINE NO.	WAVENUMBER UNC. CM^{-1}, VAC.	ROTATIONAL BAND ASSIGNMENTS	ISOTOPIC SPECIES
1	3005.44968(38)		HOH	31	3015.00197(46)		CH4
2	3005.87158(39)	P 17 0210 -0000	16013C32S	32	3015.05092(35)		CH4
3	3006.29238(39)	P 16 0210 -0000	16013C32S	33	3015.09207(48)		CH4
4	3006.71238(41)	P 15 0210 -0000	16013C32S	34	3015.20369(25)		CH4
5	3007.13110(45)	P 14 0210 -0000	16013C32S	35	3015.29173(75)	R 5 0210 -0000	16013C32S
6	3007.54898(40)	P 13 0210 -0000	16013C32S	36	3015.59152(38)		CH4
7	3008.38144(52)	P 11 0210 -0000	16013C32S	37	3015.61596(23)		HOH
8	3008.79633(52)	P 10 0210 -0000	16013C32S	38	3015.64227(25)		CH4
9	3008.84989(103)		CH4	39	3015.68413(87)		CH4
10	3009.01200(62)		CH4	40	3015.70684(38)		CH4
11	3009.21009(55)	P 9 0210 -0000	16013C32S	41	3015.98968(29)		CH4
12	3009.62267(71)	P 8 0210 -0000	16013C32S	42	3016.08465(56)	R 7 0210 -0000	16013C32S
13	3010.03449(59)	P 7 0210 -0000	16013C32S	43	3016.30231(35)		CH4
14	3010.17863(40)		CH4	44	3016.36958(27)		CH4
15	3010.23313(21)		HOH	45	3016.45870(27)		CH4
16	3010.44519(72)	P 6 0210 -0000	16013C32S	46	3016.56589(41)		CH4
17	3011.43521(40)		CH4	47	3016.64110(23)		CH4
18	3012.22946(112)		CH4	48	3016.73362(32)		CH4
19	3012.91730(136)		CH4	49	3016.87386(55)	R 9 0210 -0000	16013C32S
20	3013.07895(51)		CH4	50	3016.94025(24)		CH4
21	3013.51193(124)		CH4	51	3017.16377(21)		CH4
22	3013.57378(76)		HOH	52	3017.22847(22)		CH4
23	3013.59282(74)		CH4	53	3017.26711(19)		CH4
24	3013.80681(70)		CH4	54	3017.34633(21)		CH4
25	3014.06097(77)		CH4	55	3017.46748(42)		CH4
26	3014.64224(35)		CH4	56	3017.71223(34)		CH4
27	3014.71220(22)		CH4	57	3017.76362(24)		CH4
28	3014.83861(51)		CH4	58	3017.88616(23)		CH4
29	3014.89383(88)	R 4 0210 -0000	16013C32S	59	3018.04955(46)	R 12 0210 -0000	16013C32S
30	3014.94168(62)		CH4	60	3018.20599(20)		CH4

- "LINE NO.": serial number identifying the spectral line in the map on the opposite page.
- "WAVENUMBER": experimental value.
- Uncertainty in the last digit of the wavenumber is given in parentheses.

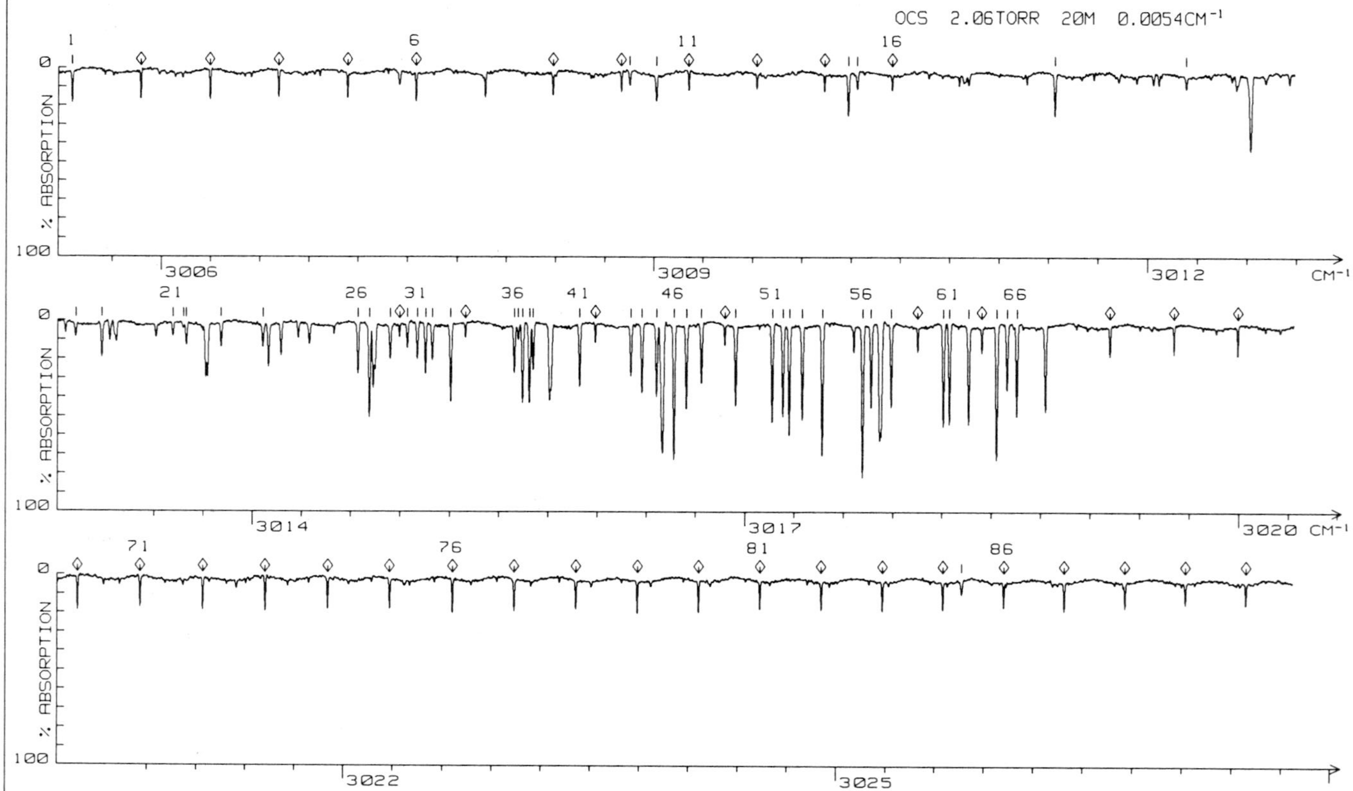

ISOTOPIC IDENTIFICATION – '▽' indicates $^{16}O^{12}C^{32}S$; '◯': $^{16}O^{12}C^{34}S$; '◊': $^{16}O^{13}C^{32}S$; 'Y': $^{16}O^{12}C^{33}S$;
'✧' indicates $^{18}O^{12}C^{32}S$; '↓': $^{16}O^{13}C^{34}S$; '↑': Other OCS species; 'I': not OCS.

LINE NO.	WAVENUMBER CM^{-1}, VAC.	UNC.	ROTATIONAL BAND ASSIGNMENTS	ISOTOPIC SPECIES
61	3018.24287	(20)		CH4
62	3018.35920	(21)		CH4
63	3018.43901	(46)	R 13 0210 -0000	16013C32S
64	3018.52918	(36)		CH4
65	3018.59180	(28)		CH4
66	3018.65066	(21)		CH4
67	3019.21550	(41)	R 15 0210 -0000	16013C32S
68	3019.60211	(39)	R 16 0210 -0000	16013C32S
69	3019.98771	(38)	R 17 0210 -0000	16013C32S
70	3020.37225	(35)	R 18 0210 -0000	16013C32S
71	3020.75575	(38)	R 19 0210 -0000	16013C32S
72	3021.13827	(38)	R 20 0210 -0000	16013C32S
73	3021.51967	(36)	R 21 0210 -0000	16013C32S
74	3021.90025	(38)	R 22 0210 -0000	16013C32S
75	3022.27947	(40)	R 23 0210 -0000	16013C32S
76	3022.65767	(35)	R 24 0210 -0000	16013C32S
77	3023.03520	(39)	R 25 0210 -0000	16013C32S
78	3023.41132	(42)	R 26 0210 -0000	16013C32S
79	3023.78678	(36)	R 27 0210 -0000	16013C32S
80	3024.16089	(38)	R 28 0210 -0000	16013C32S
81	3024.53410	(40)	R 29 0210 -0000	16013C32S
82	3024.90621	(42)	R 30 0210 -0000	16013C32S
83	3025.27757	(41)	R 31 0210 -0000	16013C32S
84	3025.64782	(43)	R 32 0210 -0000	16013C32S
85	3025.76165	(25)		HOH
86	3026.01697	(46)	R 33 0210 -0000	16013C32S
87	3026.38527	(41)	R 34 0210 -0000	16013C32S
87			R 35 0311E-0101E	16013C32S
88	3026.75234	(48)	R 35 0210 -0000	16013C32S
89	3027.11838	(52)	R 36 0210 -0000	16013C32S
90	3027.48377	(52)	R 37 0210 -0000	16013C32S

- "LINE NO.": serial number identifying the spectral line in the map on the opposite page.
- "WAVENUMBER": experimental value.
- Uncertainty in the last digit of the wavenumber is given in parentheses.

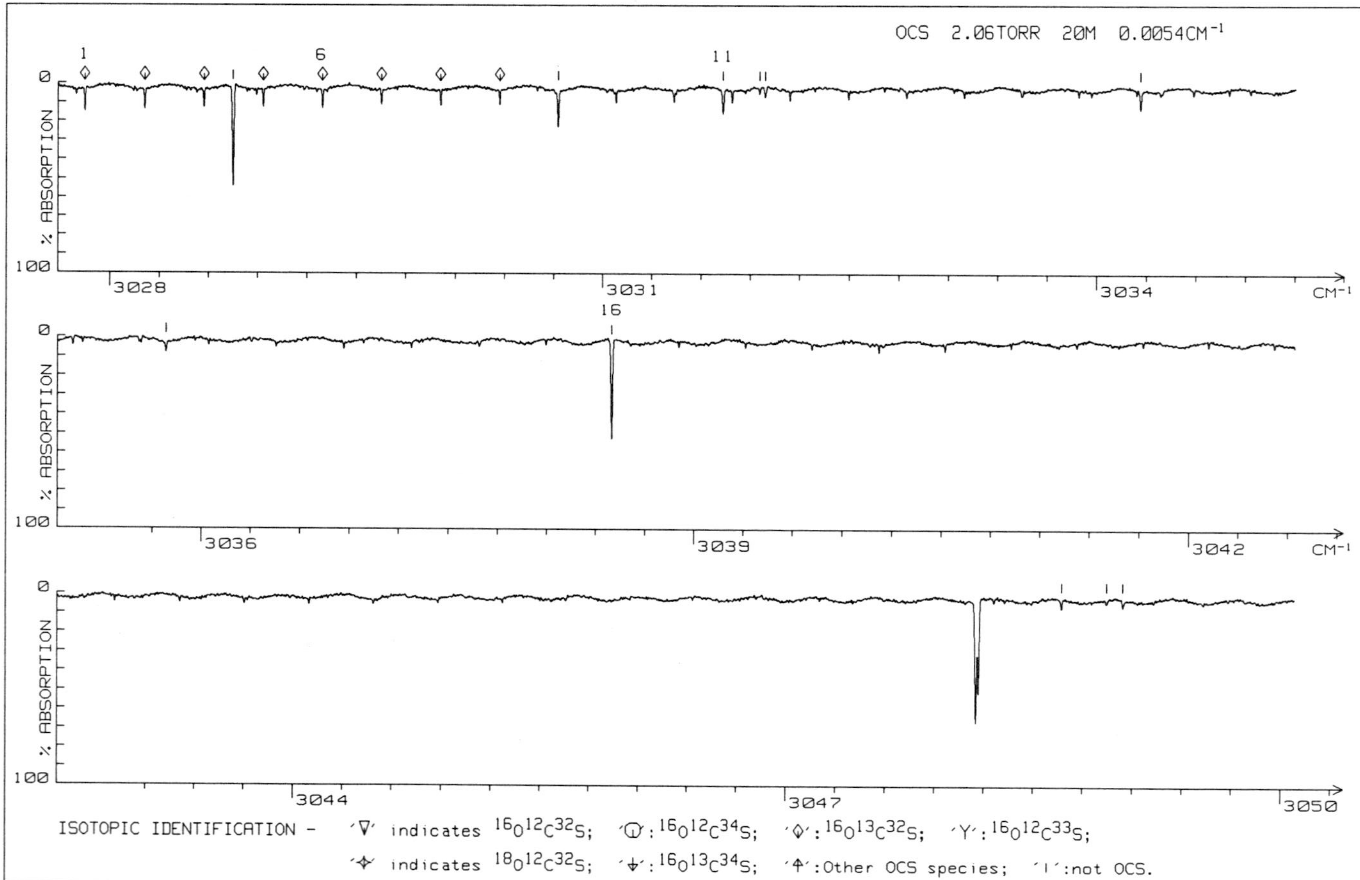
OCS 2.06TORR 20M 0.0054CM⁻¹
% ABSORPTION
CM⁻¹
3028
3031
3034
3036
3039
3042
3044
3047
3050
ISOTOPIC IDENTIFICATION - '∇' indicates ¹⁶O¹²C³²S; '⦶':¹⁶O¹²C³⁴S; '◊':¹⁶O¹³C³²S; 'Y':¹⁶O¹²C³³S;
'✧' indicates ¹⁸O¹²C³²S; '⊥':¹⁶O¹³C³⁴S; '↑':Other OCS species; 'I':not OCS.

LINE NO.	WAVENUMBER CM^{-1}, VAC.	UNC.	ROTATIONAL BAND ASSIGNMENTS	ISOTOPIC SPECIES
1	3089.77633	(14)	P 14 0210 -0000	16012C32S
2	3089.82900	(41)	R 28 1210 -1000	16012C32S
2			P 5 0311 -0101	16012C34S
3	3089.90132	(40)	P 9 0412 -0202	16012C32S
4	3089.95776	(16)	P 12 0311E-0101E	16012C32S
5	3089.98007	(16)	P 12 0311F-0101F	16012C32S
6	3090.00606	(32)	P 7 0210 -0000	16012C34S
7	3090.04233	(44)	P 10 0410 -0200	16012C32S
8	3090.19607	(13)	P 13 0210 -0000	16012C32S
9	3090.31870	(38)	P 8 0412 -0202	16012C32S
10	3090.37944	(16)	P 11 0311E-0101E	16012C32S
11	3090.39752	(28)	P 11 0311F-0101F	16012C32S
12	3090.46126	(54)	P 9 0410 -0200	16012C32S
13	3090.61495	(14)	P 12 0210 -0000	16012C32S
14	3090.73488	(52)	P 7 0412 -0202	16012C32S
15	3091.03234	(14)	P 11 0210 -0000	16012C32S
16	3091.14922	(62)	P 6 0412 -0202	16012C32S
17	3091.29491	(45)	P 7 0410 -0200	16012C32S
17			P 7 0210 -0000	16012C33S
18	3091.34266	(48)	R 32 1210 -1000	16012C32S
18			R 33 1311E-1101E	16012C32S
19	3091.44883	(13)	P 10 0210 -0000	16012C32S
20	3091.86455	(14)	P 9 0210 -0000	16012C32S
21	3092.05434	(63)	P 7 0311E-0101E	16012C32S
21			P 7 0311F-0101F	16012C32S
22	3092.09464	(75)	R 34 1210 -1000	16012C32S
23	3092.27876	(15)	P 8 0210 -0000	16012C32S
24	3092.46766	(19)	P 6 0311E-0101E	16012C32S
24			P 6 0311F-0101F	16012C32S
24			R 35 1210 -1000	16012C32S
25	3092.53499	(87)	P 4 0410 -0200	16012C32S
26	3092.69206	(14)	P 7 0210 -0000	16012C32S
27	3092.84364	(59)	R 36 1210 -1000	16012C32S
28	3092.88004	(17)	P 5 0311 -0101	16012C32S
29	3093.10456	(14)	P 6 0210 -0000	16012C32S
30	3093.21698	(77)	R 37 1210 -1000	16012C32S
31	3093.29128	(20)	P 4 0311 -0101	16012C32S
32	3093.51551	(16)	P 5 0210 -0000	16012C32S
33	3093.70097	(21)	P 3 0311 -0101	16012C32S
34	3093.92537	(14)	P 4 0210 -0000	16012C32S
35	3094.11000	(26)	P 2 0311 -0101	16012C32S
36	3094.33462	(15)	P 3 0210 -0000	16012C32S
37	3094.37164	(60)	R 3 0210 -0000	16012C34S
38	3094.70246	(86)	R 41 1210 -1000	16012C32S
38			Q 15 0311E-0101F	16012C32S
39	3094.74238	(18)	P 2 0210 -0000	16012C32S
40	3094.76190	(54)	R 4 0210 -0000	16012C34S
40			Q 13 0311E-0101F	16012C32S
41	3095.10447	(42)		CH4
42	3095.14952	(18)	P 1 0210 -0000	16012C32S
42			R 5 0210 -0000	16012C34S
43	3095.17973	(72)		CH4
44	3095.54021	(40)	R 6 0210 -0000	16012C34S
45	3095.62748	(81)	R 4 0412 -0202	16012C32S
46	3095.92759	(40)	R 7 0210 -0000	16012C34S
46			R 7 0511E-0301E	16012C32S
47	3095.94577	(36)		HOH
48	3096.31363	(34)	R 8 0210 -0000	16012C34S
49	3096.36301	(16)	R 1 0210 -0000	16012C32S
50	3096.42558	(50)	R 6 0412 -0202	16012C32S
51	3096.58481	(66)	R 5 0410 -0200	16012C32S
52	3096.69932	(30)	R 9 0210 -0000	16012C34S
53	3096.76563	(15)	R 2 0210 -0000	16012C32S
54	3096.82243	(52)	R 7 0412 -0202	16012C32S

- "LINE NO.": serial number identifying the spectral line in the map on the opposite page.
- "WAVENUMBER": experimental value.
- Uncertainty in the last digit of the wavenumber is given in parentheses.

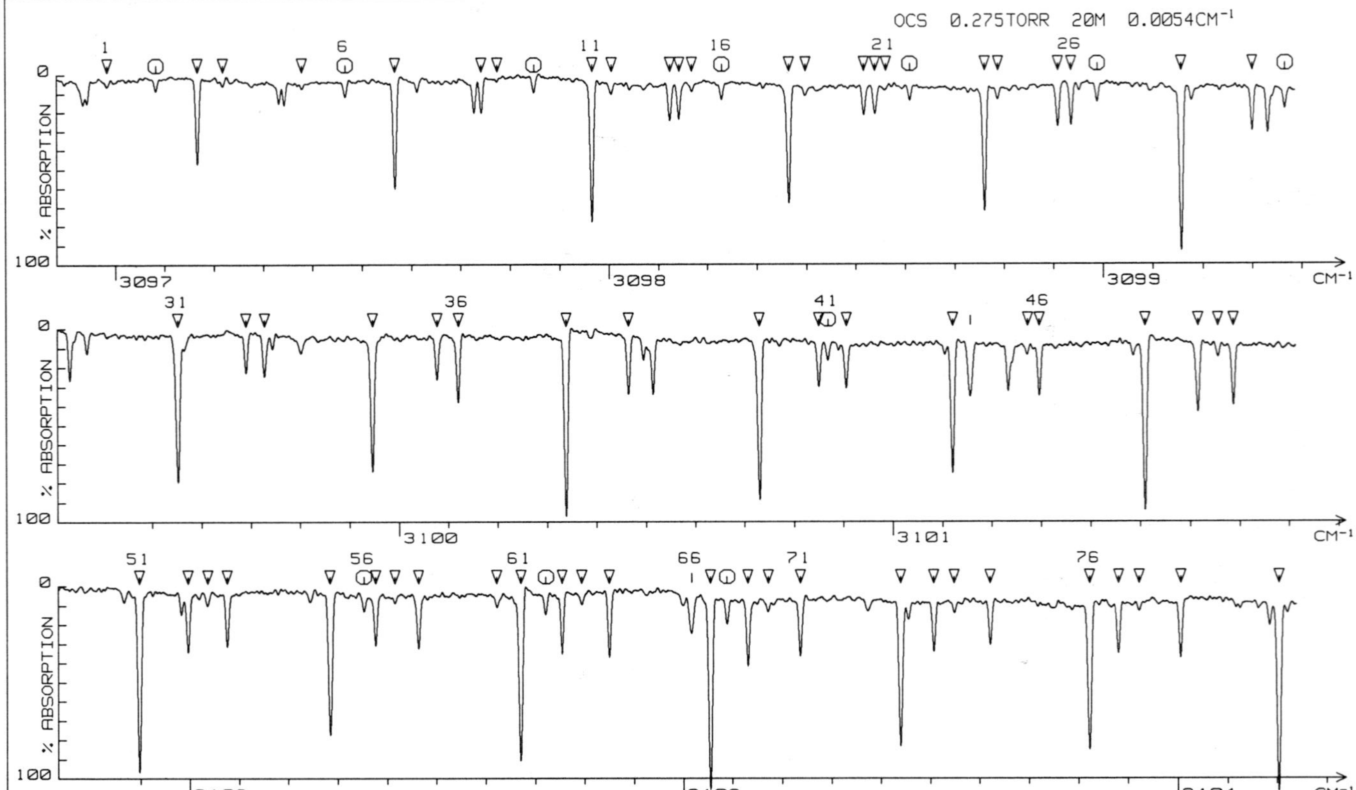

ISOTOPIC IDENTIFICATION – '▽' indicates $^{16}O^{12}C^{32}S$; 'Ⓘ': $^{16}O^{12}C^{34}S$; '◇': $^{16}O^{13}C^{32}S$; 'Y': $^{16}O^{12}C^{33}S$;
'✧' indicates $^{18}O^{12}C^{32}S$; '⫛': $^{16}O^{13}C^{34}S$; '⍏': Other OCS species; 'I': not OCS.

LINE NO.	WAVENUMBER UNC. CM^{-1}, VAC.	ROTATIONAL BAND ASSIGNMENTS	ISOTOPIC SPECIES
1	3096.98307(74)	R 6 0410 -0200	16012C32S
2	3097.08347(36)	R 10 0210 -0000	16012C34S
3	3097.16696(17)	R 3 0210 -0000	16012C32S
4	3097.21784(58)	R 8 0412 -0202	16012C32S
5	3097.37912(69)	R 7 0410 -0200	16012C32S
6	3097.46641(31)	R 11 0210 -0000	16012C34S
7	3097.56697(15)	R 4 0210 -0000	16012C32S
8	3097.74278(41)	R 6 0311F-0101F	16012C32S
9	3097.77417(54)	R 8 0410 -0200	16012C32S
10	3097.84868(27)	R 12 0210 -0000	16012C34S
11	3097.96631(14)	R 5 0210 -0000	16012C32S
12	3098.00525(40)	R 10 0412 -0202	16012C32S
12		R 50 1210 -1000	16012C32S
13	3098.12356(20)	R 7 0311E-0101E	16012C32S
14	3098.14193(21)	R 7 0311F-0101F	16012C32S
15	3098.16838(63)	R 9 0410 -0200	16012C32S
16	3098.22989(33)	R 13 0210 -0000	16012C34S
17	3098.36447(16)	R 6 0210 -0000	16012C32S
18	3098.39766(49)	R 11 0412 -0202	16012C32S
19	3098.51708(23)	R 8 0311E-0101E	16012C32S
20	3098.53949(22)	R 8 0311F-0101F	16012C32S
21	3098.56085(67)	R 10 0410 -0200	16012C32S
22	3098.60973(33)	R 14 0210 -0000	16012C34S
23	3098.76119(15)	R 7 0210 -0000	16012C32S
24	3098.78771(40)	R 12 0412 -0202	16012C32S
25	3098.90915(18)	R 9 0311E-0101E	16012C32S
26	3098.93627(18)	R 9 0311F-0101F	16012C32S
27	3098.98880(26)	R 15 0210 -0000	16012C34S
28	3099.15719(14)	R 8 0210 -0000	16012C32S
29	3099.30021(19)	R 10 0311E-0101E	16012C32S
30	3099.36686(28)	R 16 0210 -0000	16012C34S
31	3099.55197(17)	R 9 0210 -0000	16012C32S
32	3099.68970(21)	R 11 0311E-0101E	16012C32S
33	3099.72761(25)	R 11 0311F-0101F	16012C32S
33		R 13 0410 -0200	16012C32S
34	3099.94574(16)	R 10 0210 -0000	16012C32S
35	3100.07733(19)	R 12 0311E-0101E	16012C32S
36	3100.12018(18)	R 12 0311F-0101F	16012C32S
36		R 18 0210 -0000	16012C34S
36		R 14 0410 -0200	16012C32S
37	3100.33818(14)	R 11 0210 -0000	16012C32S
38	3100.46422(17)	R 13 0311E-0101E	16012C32S
39	3100.72991(16)	R 12 0210 -0000	16012C32S
40	3100.84961(21)	R 14 0311E-0101E	16012C32S
41	3100.86817(34)	R 20 0210 -0000	16012C34S
42	3100.90584(22)	R 14 0311F-0101F	16012C32S
43	3101.12018(17)	R 13 0210 -0000	16012C32S
44	3101.15609(32)		CH4
45	3101.27250(50)	R 17 0410 -0200	16012C32S
46	3101.29643(19)	R 15 0311F-0101F	16012C32S
46		R 24 0311E-0101E	16012C34S
47	3101.50929(15)	R 14 0210 -0000	16012C32S
48	3101.61487(18)	R 16 0311E-0101E	16012C32S
48		R 22 0210 -0000	16012C34S
49	3101.65554(38)	R 18 0410 -0200	16012C32S
49		R 25 0311E-0101E	16012C34S
49		R 22 0511E-0301E	16012C32S
50	3101.68659(17)	R 16 0311F-0101F	16012C32S
51	3101.89777(15)	R 15 0210 -0000	16012C32S
52	3101.99682(31)	R 17 0311E-0101E	16012C32S
53	3102.03615(55)	R 19 0410 -0200	16012C32S
54	3102.07566(21)	R 17 0311F-0101F	16012C32S
55	3102.28491(17)	R 16 0210 -0000	16012C32S

- "LINE NO.": serial number identifying the spectral line in the map on the opposite page.
- "WAVENUMBER": experimental value.
- Uncertainty in the last digit of the wavenumber is given in parentheses.

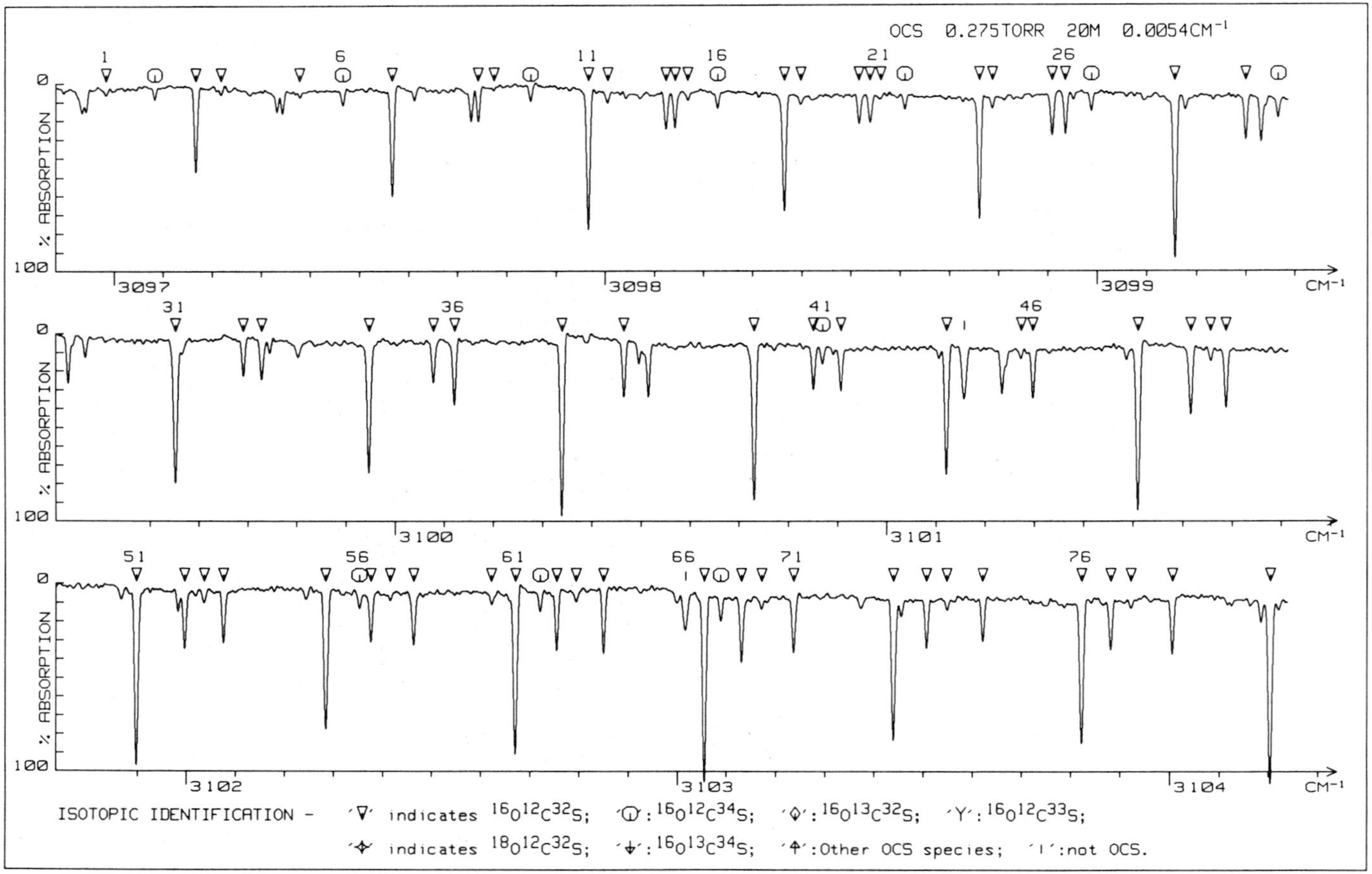
OCS 0.275TORR 20M 0.0054CM⁻¹
% ABSORPTION
0
100
3097
3098
3099
3100
3101
3102
3103
3104
CM⁻¹
ISOTOPIC IDENTIFICATION – '▽' indicates $^{16}O^{12}C^{32}S$; 'Ⓘ': $^{16}O^{12}C^{34}S$; '◇': $^{16}O^{13}C^{32}S$; 'Y': $^{16}O^{12}C^{33}S$;
'✧' indicates $^{18}O^{12}C^{32}S$; '⏚': $^{16}O^{13}C^{34}S$; '⍏': Other OCS species; 'I': not OCS.

LINE NO.	WAVENUMBER CM^{-1}, VAC.	UNC.	ROTATIONAL BAND ASSIGNMENTS	ISOTOPIC SPECIES
56	3102.35277	(35)	R 24 0210 -0000	16012C34S
57	3102.37634	(21)	R 18 0311E-0101E	16012C32S
57			R 27 0311E-0101E	16012C34S
58	3102.41582	(63)	R 20 0410 -0200	16012C32S
59	3102.46328	(21)	R 18 0311F-0101F	16012C32S
60	3102.62260	(58)	R 22 0412 -0202	16012C32S
61	3102.67065	(16)	R 17 0210 -0000	16012C32S
62	3102.72137	(28)	R 25 0210 -0000	16012C34S
63	3102.75441	(18)	R 19 0311E-0101E	16012C32S
64	3102.79438	(43)	R 21 0410 -0200	16012C32S
65	3102.85029	(18)	R 19 0311F-0101F	16012C32S
66	3103.01622	(44)		HOH
67	3103.05576	(15)	R 18 0210 -0000	16012C32S
68	3103.08922	(28)	R 26 0210 -0000	16012C34S
68			R 29 0311E-0101E	16012C34S
69	3103.13146	(19)	R 20 0311E-0101E	16012C32S
70	3103.17196	(51)	R 22 0410 -0200	16012C32S
71	3103.23652	(20)	R 20 0311F-0101F	16012C32S
72	3103.43986	(17)	R 19 0210 -0000	16012C32S
73	3103.50683	(22)	R 21 0311E-0101E	16012C32S
74	3103.54792	(59)	R 23 0410 -0200	16012C32S
75	3103.62120	(23)	R 21 0311F-0101F	16012C32S
76	3103.82222	(16)	R 20 0210 -0000	16012C32S
77	3103.88063	(20)	R 22 0311E-0101E	16012C32S
78	3103.92241	(49)	R 24 0410 -0200	16012C32S
79	3104.00498	(19)	R 22 0311F-0101F	16012C32S
80	3104.20410	(15)	R 21 0210 -0000	16012C32S

- "LINE NO.": serial number identifying the spectral line in the map on the opposite page.
- "WAVENUMBER": experimental value.
- Uncertainty in the last digit of the wavenumber is given in parentheses.

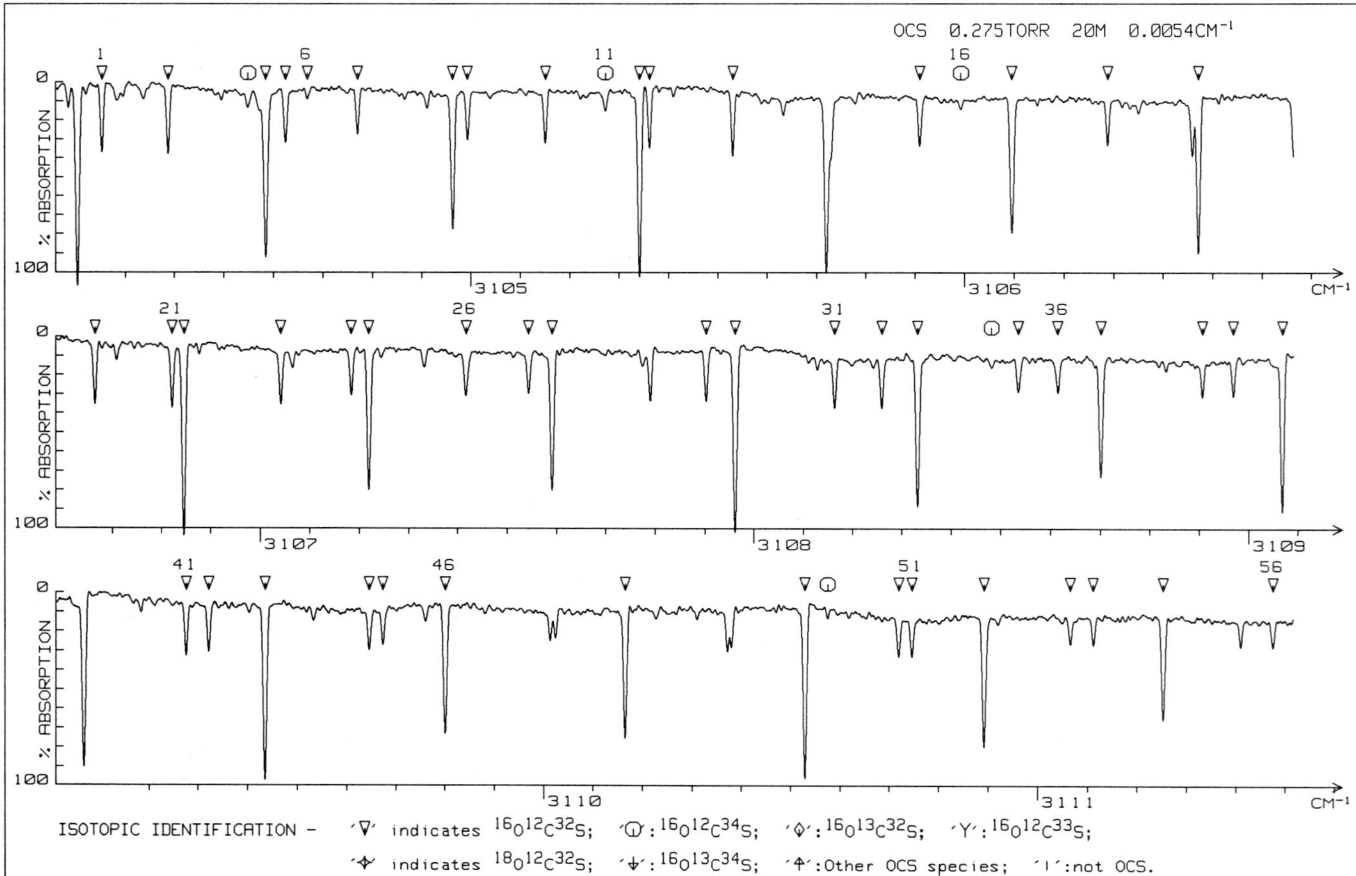
OCS 0.275TORR 20M 0.0054CM⁻¹
% ABSORPTION
0
100
3105
3106
3107
3108
3109
3110
3111
CM⁻¹
1
6
11
16
21
26
31
36
41
46
51
56
ISOTOPIC IDENTIFICATION - '∇' indicates 16O12C32S; '⊕': 16O12C34S; '◊': 16O13C32S; 'Y': 16O12C33S;
'✧' indicates 18O12C32S; '↓': 16O13C34S; '↑': Other OCS species; 'I': not OCS.

LINE NO.	WAVENUMBER CM^{-1}, VAC.	UNC.	ROTATIONAL ASSIGNMENTS	BAND	ISOTOPIC SPECIES
1	3104.25339	(18)	R 23	0311E-0101E	16012C32S
2	3104.38815	(19)	R 23	0311F-0101F	16012C32S
3	3104.54934	(39)	R 30	0210 -0000	16012C34S
4	3104.58503	(17)	R 22	0210 -0000	16012C32S
5	3104.62471	(22)	R 24	0311E-0101E	16012C32S
6	3104.66891	(67)	R 26	0410 -0200	16012C32S
7	3104.77012	(23)	R 24	0311F-0101F	16012C32S
8	3104.96419	(17)	R 23	0210 -0000	16012C32S
9	3104.99416	(22)	R 25	0311E-0101E	16012C32S
10	3105.15090	(20)	R 25	0311F-0101F	16012C32S
11	3105.27322	(31)	R 32	0210 -0000	16012C34S
11			R 28	0210 -0000	16012C33S
12	3105.34261	(16)	R 24	0210 -0000	16012C32S
13	3105.36250	(31)	R 26	0311E-0101E	16012C32S
14	3105.53113	(19)	R 26	0311F-0101F	16012C32S
15	3105.91013	(24)	R 27	0311F-0101F	16012C32S
16	3105.99292	(47)	R 34	0210 -0000	16012C34S
16			R 34	0511E-0301E	16012C32S
17	3106.09635	(19)	R 26	0210 -0000	16012C32S
17			R 28	0311E-0101E	16012C32S
18	3106.28785	(23)	R 28	0311F-0101F	16012C32S
19	3106.47163	(42)	R 27	0210 -0000	16012C32S
20	3106.66495	(19)	R 29	0311F-0101F	16012C32S
21	3106.82194	(33)	R 30	0311E-0101E	16012C32S
22	3106.84582	(15)	R 28	0210 -0000	16012C32S
23	3107.04113	(22)	R 30	0311F-0101F	16012C32S
23			R 34	0412E-0202E	16012C32S
24	3107.18351	(25)	R 31	0311E-0101E	16012C32S
25	3107.21901	(19)	R 29	0210 -0000	16012C32S
26	3107.41700	(34)	R 31	0311F-0101F	16012C32S
26			R 38	0210 -0000	16012C34S
26			R 35	0412F-0202F	16012C32S
27	3107.54332	(25)	R 32	0311E-0101E	16012C32S
28	3107.59085	(19)	R 30	0210 -0000	16012C32S
29	3107.90181	(21)	R 33	0311E-0101E	16012C32S
30	3107.96189	(16)	R 31	0210 -0000	16012C32S
31	3108.16321	(22)	R 33	0311F-0101F	16012C32S
32	3108.25920	(23)	R 34	0311E-0101E	16012C32S
33	3108.33201	(17)	R 32	0210 -0000	16012C32S
34	3108.48101	(63)	R 41	0210 -0000	16012C34S
35	3108.53523	(32)	R 34	0311F-0101F	16012C32S
36	3108.61462	(33)	R 35	0311E-0101E	16012C32S
37	3108.70067	(21)	R 33	0210 -0000	16012C32S
38	3108.90605	(25)	R 35	0311F-0101F	16012C32S
39	3108.96880	(24)	R 36	0311E-0101E	16012C32S
40	3109.06831	(18)	R 34	0210 -0000	16012C32S
41	3109.27619	(22)	R 36	0311F-0101F	16012C32S
41			R 39	0210 -0000	16012C33S
42	3109.32204	(23)	R 37	0311E-0101E	16012C32S
43	3109.43534	(17)	R 35	0210 -0000	16012C32S
44	3109.64561	(56)	R 37	0311F-0101F	16012C32S
45	3109.67360	(31)	R 38	0311E-0101E	16012C32S
46	3109.80111	(21)	R 36	0210 -0000	16012C32S
47	3110.16544	(19)	R 37	0210 -0000	16012C32S
48	3110.52919	(16)	R 38	0210 -0000	16012C32S
49	3110.57530	(66)	R 47	0210 -0000	16012C34S
50	3110.71973	(27)	R 41	0311E-0101E	16012C32S
51	3110.74659	(27)	R 40	0311F-0101F	16012C32S
52	3110.89206	(21)	R 39	0210 -0000	16012C32S
53	3111.06581	(39)	R 42	0311E-0101E	16012C32S
54	3111.11157	(39)	R 41	0311F-0101F	16012C32S
55	3111.25354	(25)	R 40	0210 -0000	16012C32S
56	3111.47545	(31)	R 42	0311F-0101F	16012C32S

- "LINE NO.": serial number identifying the spectral line in the map on the opposite page.
- "WAVENUMBER": experimental value.
- Uncertainty in the last digit of the wavenumber is given in parentheses.

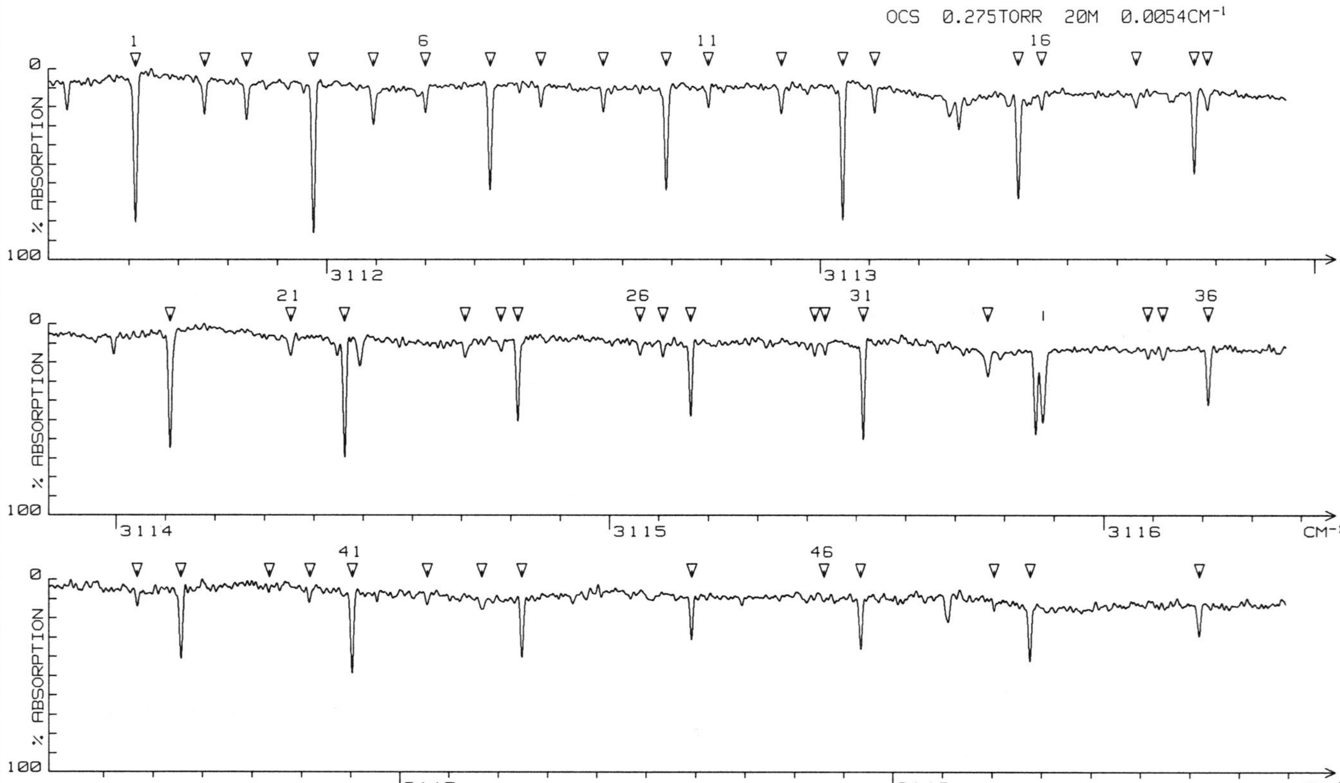

ISOTOPIC IDENTIFICATION – '▽' indicates $^{16}O^{12}C^{32}S$; ':' $^{16}O^{12}C^{34}S$; ':' $^{16}O^{13}C^{32}S$; 'Y': $^{16}O^{12}C^{33}S$;
'✧' indicates $^{18}O^{12}C^{32}S$; ':' $^{16}O^{13}C^{34}S$; ':' Other OCS species; 'I': not OCS.

LINE NO.	WAVENUMBER CM^{-1}, VAC.	UNC.	ROTATIONAL BAND ASSIGNMENTS	ISOTOPIC SPECIES
1	3111.61379(	20)	R 41 0210 -0000	16012C32S
2	3111.75327(	28)	R 44 0311E-0101E	16012C32S
2			R 46 0210 -0000	16012C33S
3	3111.83873(	28)	R 43 0311F-0101F	16012C32S
4	3111.97371(	18)	R 42 0210 -0000	16012C32S
5	3112.09517(	37)	R 45 0311E-0101E	16012C32S
6	3112.20101(	38)	R 44 0311F-0101F	16012C32S
7	3112.33234(	26)	R 43 0210 -0000	16012C32S
8	3112.43528(	42)	R 46 0311E-0101E	16012C32S
9	3112.56197(	42)	R 45 0311F-0101F	16012C32S
10	3112.68944(	22)	R 44 0210 -0000	16012C32S
11	3112.77417(	37)	R 47 0311E-0101E	16012C32S
12	3112.92204(	31)	R 46 0311F-0101F	16012C32S
12			R 49 0410 -0200	16012C32S
13	3113.04613(	18)	R 45 0210 -0000	16012C32S
14	3113.11203(	34)	R 48 0311E-0101E	16012C32S
15	3113.40202(	23)	R 46 0210 -0000	16012C32S
16	3113.44831(	43)	R 49 0311E-0101E	16012C32S
17	3113.63923(	48)	R 48 0311F-0101F	16012C32S
18	3113.75633(	26)	R 47 0210 -0000	16012C32S
19	3113.78296(	55)	R 50 0311E-0101E	16012C32S
20	3114.10977(	26)	R 48 0210 -0000	16012C32S
21	3114.35269(	39)	R 50 0311F-0101F	16012C32S
22	3114.46244(	24)	R 49 0210 -0000	16012C32S
23	3114.70794(	53)	R 51 0311F-0101F	16012C32S
24	3114.77945(	64)	R 53 0311E-0101E	16012C32S
25	3114.81410(	26)	R 50 0210 -0000	16012C32S
26	3115.06163(	60)	R 52 0311F-0101F	16012C32S
27	3115.10876(	70)	R 54 0311E-0101E	16012C32S
28	3115.16455(	26)	R 51 0210 -0000	16012C32S
29	3115.41494(	45)	R 53 0311F-0101F	16012C32S
30	3115.43653(	54)	R 55 0311E-0101E	16012C32S
31	3115.51412(	21)	R 52 0210 -0000	16012C32S
32	3115.76560(	50)	R 54 0311F-0101F	16012C32S
32			R 56 0311E-0101E	16012C32S
33	3115.87743(	54)		HOH
34	3116.08879(	85)	R 57 0311E-0101E	16012C32S
35	3116.11872(	84)	R 55 0311F-0101F	16012C32S
36	3116.21079(	32)	R 54 0210 -0000	16012C32S
37	3116.46835(	70)	R 56 0311F-0101F	16012C32S
38	3116.55729(	27)	R 55 0210 -0000	16012C32S
39	3116.73533(	69)	R 59 0311E-0101E	16012C32S
40	3116.81771(	64)	R 57 0311F-0101F	16012C32S
41	3116.90335(	23)	R 56 0210 -0000	16012C32S
42	3117.05618(	88)	R 60 0311E-0101E	16012C32S
43	3117.16633(	86)	R 58 0311F-0101F	16012C32S
44	3117.24849(	30)	R 57 0210 -0000	16012C32S
45	3117.59195(	40)	R 58 0210 -0000	16012C32S
46	3117.85967(	87)	R 60 0311F-0101F	16012C32S
47	3117.93507(	27)	R 59 0210 -0000	16012C32S
48	3118.20479(	91)	R 61 0311F-0101F	16012C32S
49	3118.27760(	29)	R 60 0210 -0000	16012C32S
50	3118.61879(	45)	R 61 0210 -0000	16012C32S

- "LINE NO.": serial number identifying the spectral line in the map on the opposite page.
- "WAVENUMBER": experimental value.
- Uncertainty in the last digit of the wavenumber is given in parentheses.

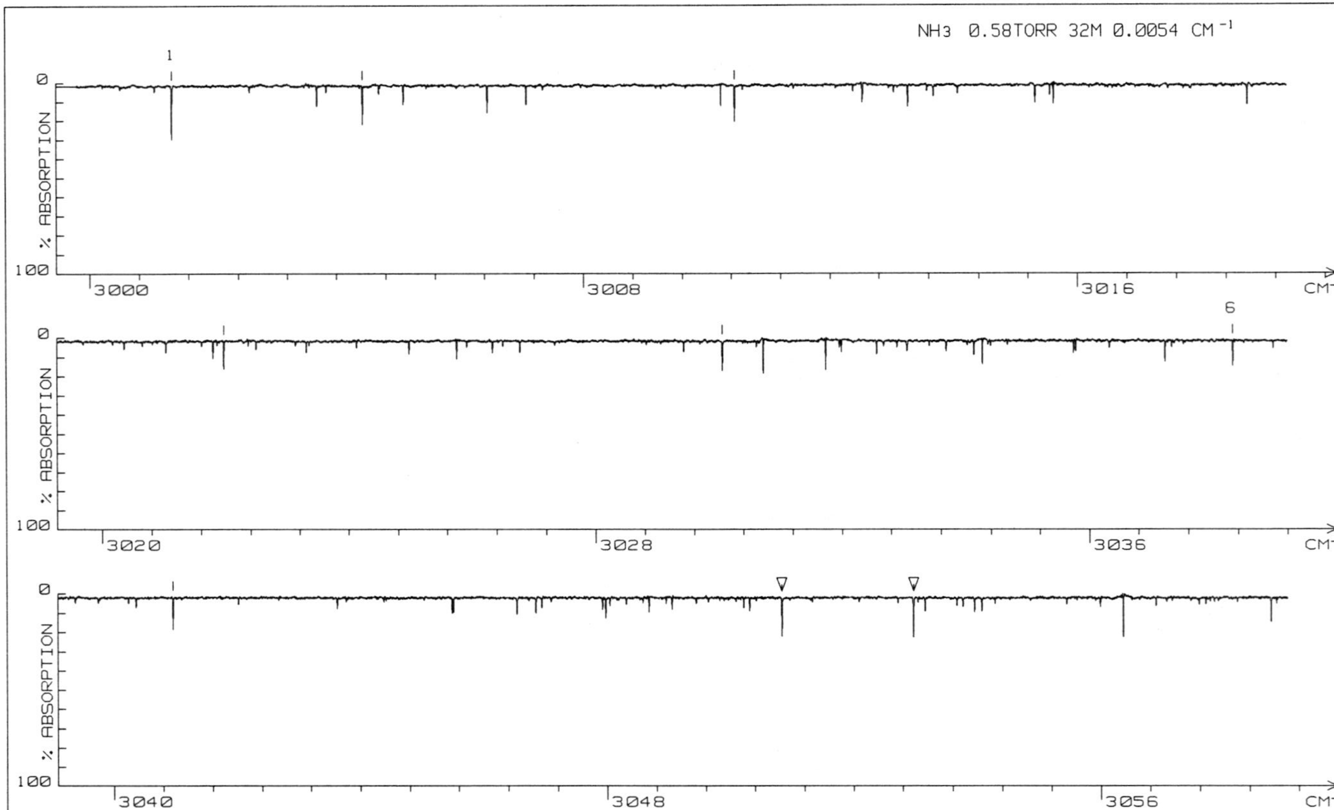

ISOTOPIC IDENTIFICATION – '∇' indicates $^{14}NH_3$; '|': unidentified.

LINE NO.	WAVENUMBER CM^{-1}, VAC.	ROTATIONAL ASSIGNMENTS	BAND	ISOTOPIC SPECIES
1	3001.32751(42)			
2	3004.42408(56)			
3	3010.45871(61)			
4	3021.97232(82)			
5	3030.06435(79)			
6	3038.31389(86)			
7	3040.95692(73)			
8	3050.83376(60)	s^rP(8, 3)	2NU4 pe	$^{14}NH_3$
9	3052.97083(59)	a^rP(8, 3)	2NU4 pe	$^{14}NH_3$

- "LINE NO.": serial number identifying the spectral line in the map on the opposite page.
- "WAVENUMBER": experimental value.
- Uncertainty in the last digit of the wavenumber is given in parentheses.
- "pa" : parallel ; "pe" : perpendicular.

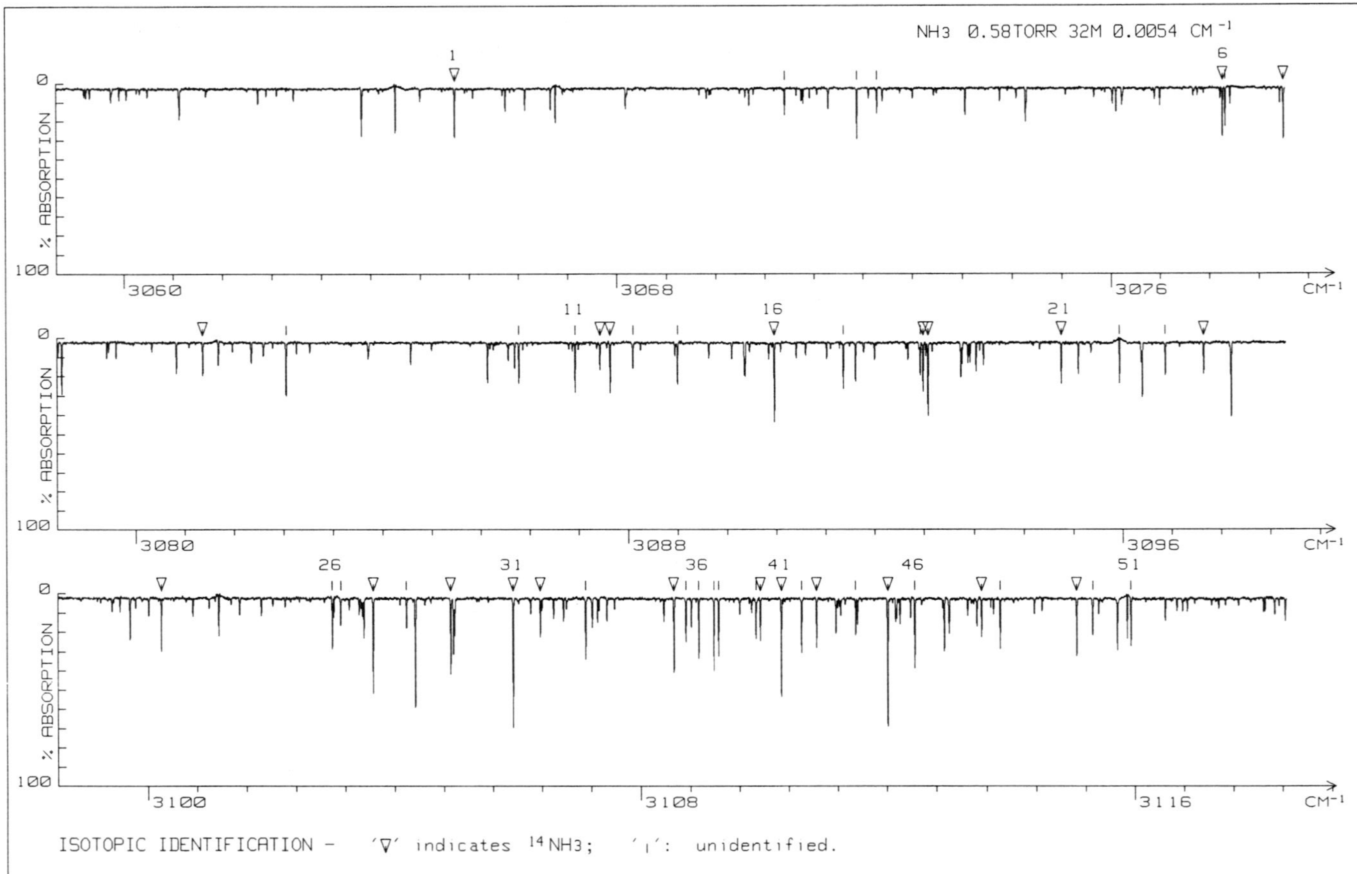
NH3 0.58TORR 32M 0.0054 CM⁻¹
% ABSORPTION
0
100
3060
3068
3076
3080
3088
3096
3100
3108
3116
CM⁻¹
1
6
11
16
21
26
31
36
41
46
51
ISOTOPIC IDENTIFICATION - '∇' indicates 14NH3; '|': unidentified.

LINE NO.	WAVENUMBER CM^{-1}, VAC.	ROTATIONAL ASSIGNMENTS	BAND	ISOTOPIC SPECIES
1	3065.36685(47)	$a^{r}P$(7, 3)	2NU4 pe	$^{14}NH_3$
2	3070.73027(83)			
3	3071.91204(45)			
4	3072.23345(87)			
5	3077.81146(48)	$s^{r}P$(6, 3)	2NU4 pe	$^{14}NH_3$
6	3077.85873(57)			
7	3078.80463(46)	$a^{r}P$(6, 3)	2NU4 pe	$^{14}NH_3$
8	3081.09361(67)	$s^{r}P$(7, 2)	2NU4 pe	$^{14}NH_3$
9	3082.43818(45)			
10	3086.22152(56)			
11	3087.13418(48)			
12	3087.54233(81)	$a^{r}P$(8, 1)	2NU4 pe	$^{14}NH_3$
13	3087.70511(48)	aP(7, 1)	2NU4 pa	$^{14}NH_3$
14	3088.07281(88)			
15	3088.79432(53)			
16	3090.36327(32)	aP(7, 3)	2NU4 pa	$^{14}NH_3$
17	3091.49006(54)			
18	3092.74116(71)			
19	3092.78742(52)	$s^{r}P$(5, 3)	2NU4 pe	$^{14}NH_3$
20	3092.86658(34)	sP(7, 0)	2NU4 pa	$^{14}NH_3$
21	3095.00124(59)	$s^{r}P$(6, 2)	2NU4 pe	$^{14}NH_3$
22	3095.94544(32)			
23	3096.69358(71)			
24	3097.31466(75)	$s^{r}P$(7, 1)	2NU4 pe	$^{14}NH_3$
25	3100.22118(47)	aP(6, 4)	2NU4 pa	$^{14}NH_3$
26	3102.99184(46)			
27	3103.12631(86)			
28	3103.65427(26)	$a^{r}P$(8, 0)	2NU4 pe	$^{14}NH_3$
29	3104.19223(76)			
30	3104.91024(34)	aP(6, 1)	2NU4 pa	$^{14}NH_3$
31	3105.93122(22)	aP(6, 2)	2NU4 pa	$^{14}NH_3$
31		aP(6, 0)	2NU4 pa	$^{14}NH_3$
32	3106.37752(62)	sP(6, 4)	2NU4 pa	$^{14}NH_3$
33	3107.11390(41)			
34	3108.53817(35)	sP(6, 1)	2NU4 pa	$^{14}NH_3$
35	3108.72986(61)			
36	3108.93757(41)			
37	3109.18302(34)			
38	3109.26335(43)			
39	3109.86627(60)			
40	3109.93940(56)	$s^{r}P$(5, 2)	2NU4 pe	$^{14}NH_3$
41	3110.28311(28)	sP(6, 3)	2NU4 pa	$^{14}NH_3$
42	3110.61333(45)			
43	3110.85635(50)	$a^{r}P$(5, 2)	2NU4 pe	$^{14}NH_3$
44	3111.49832(64)			
45	3112.01619(23)	$s^{r}P$(7, 0)	2NU4 pe	$^{14}NH_3$
46	3112.45626(37)			
47	3113.52313(61)	$s^{p}P$(8, 1)	2NU4 pe	$^{14}NH_3$
48	3113.82445(47)			
49	3115.05385(42)	sP(6, 2)	2NU4 pa	$^{14}NH_3$
50	3115.31333(66)			
51	3115.94176(49)			

- "LINE NO.": serial number identifying the spectral line in the map on the opposite page.
- "WAVENUMBER": experimental value.
- Uncertainty in the last digit of the wavenumber is given in parentheses.
- "pa" : parallel ; "pe" : perpendicular.

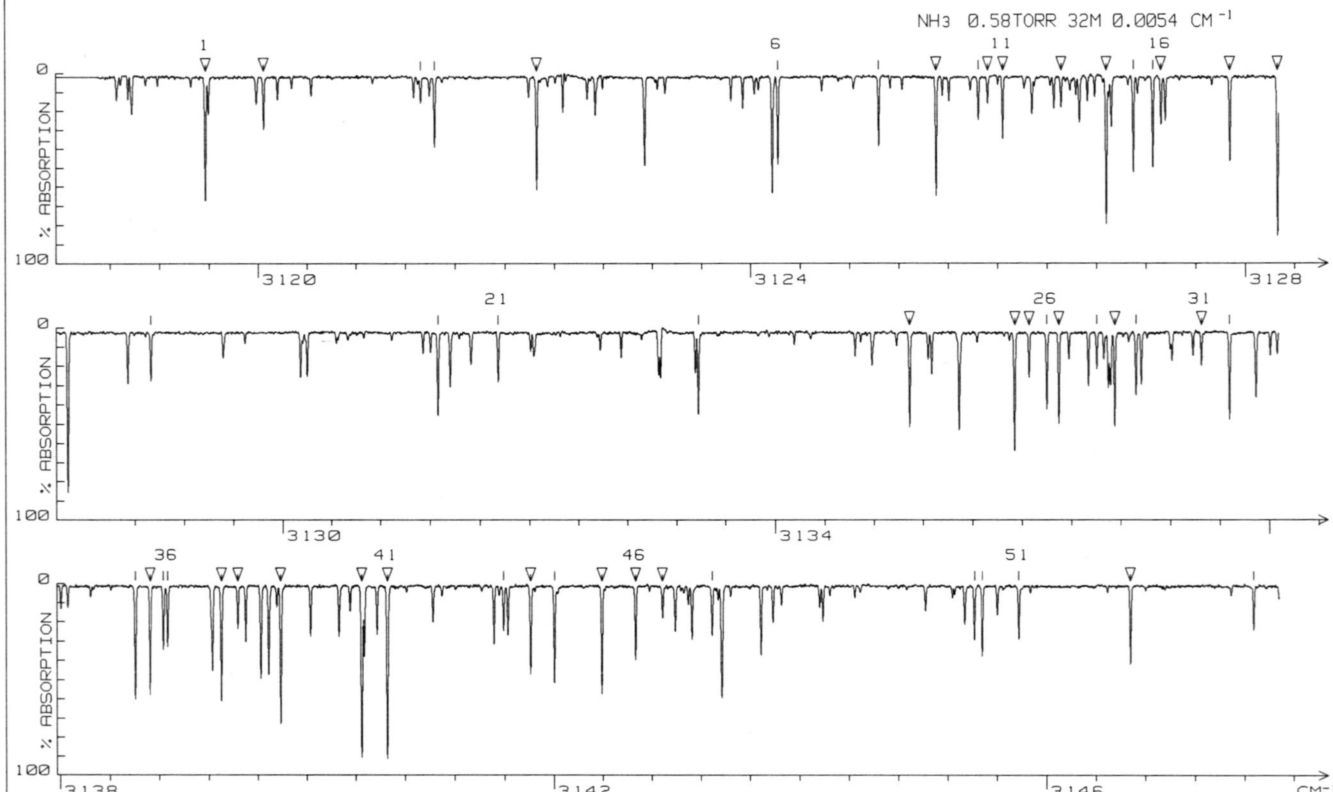

ISOTOPIC IDENTIFICATION - '∇' indicates $^{14}NH_3$; '|': unidentified.

LINE NO.	WAVENUMBER CM^{-1}, VAC.	ROTATIONAL ASSIGNMENTS	BAND	ISOTOPIC SPECIES
1	3119.57088(24)	aP(5, 3)	2NU4 pa	$^{14}NH_3$
2	3120.04507(47)	sP(5, 4)	2NU4 pa	$^{14}NH_3$
3	3121.32426(92)			
4	3121.43482(36)			
5	3122.25743(60)	aP(5, 1)	2NU4 pa	$^{14}NH_3$
6	3124.22193(30)			
7	3125.03675(36)			
8	3125.50231(23)	s^{r}P(5, 1)	2NU4 pe	$^{14}NH_3$
8		sP(5, 1)	2NU4 pa	$^{14}NH_3$
9	3125.84370(58)			
10	3125.91850(86)	s^{r}P(4, 2)	2NU4 pe	$^{14}NH_3$
11	3126.04237(39)	s^{P}P(7, 1)	2NU4 pe	$^{14}NH_3$
12	3126.51447(78)	a^{r}P(4, 2)	2NU4 pe	$^{14}NH_3$
13	3126.87992(20)	sP(5, 0)	2NU4 pa	$^{14}NH_3$
14	3127.10060(27)			
15	3127.25590(75)			
16	3127.32088(50)	a^{r}P(5, 1)	2NU4 pe	$^{14}NH_3$
17	3127.87689(31)	sP(5, 2)	2NU4 pa	$^{14}NH_3$
18	3128.26285(19)	a^{r}P(6, 0)	2NU4 pe	$^{14}NH_3$
19	3128.92950(52)			
20	3131.26531(32)			
21	3131.74775(50)			
22	3133.37577(33)			
23	3135.08988(29)	aP(10, 6)	NU1	$^{14}NH_3$
24	3135.93960(25)	aP(4, 3)	2NU4 pa	$^{14}NH_3$
25	3136.05765(54)	aP(10, 4)	NU1	$^{14}NH_3$
26	3136.20055(34)			
27	3136.29778(30)	aP(10, 3)	NU1	$^{14}NH_3$
28	3136.60474(64)			
29	3136.74991(29)	aP(10, 0)	NU1	$^{14}NH_3$
30	3136.92366(41)			
31	3137.44828(71)	sP(10, 8)	NU1	$^{14}NH_3$
32	3137.67581(31)			
33	3138.60422(26)			
34	3138.72428(25)	s^{P}P(7, 2)	2NU4 pe	$^{14}NH_3$
34		aP(4, 2)	2NU4 pa	$^{14}NH_3$
35	3138.83189(41)			
36	3138.86654(42)			
37	3139.30080(26)	sP(4, 3)	2NU4 pa	$^{14}NH_3$
38	3139.43581(58)	sP(10, 4)	NU1	$^{14}NH_3$
39	3139.78389(21)	aP(4, 1)	2NU4 pa	$^{14}NH_3$
40	3140.44921(30)	aP(4, 0)	2NU4 pa	$^{14}NH_3$
41	3140.65753(19)	s^{r}P(5, 0)	2NU4 pe	$^{14}NH_3$
42	3141.59105(56)			
43	3141.80911(81)	s^{r}P(4, 1)	2NU4 pe	$^{14}NH_3$
44	3142.00299(71)			
45	3142.39130(25)	sP(4, 2)	2NU4 pa	$^{14}NH_3$
46	3142.66665(36)	a^{P}P(6, 1)	2NU4 pe	$^{14}NH_3$
47	3142.88649(79)	a^{r}P(4, 1)	2NU4 pe	$^{14}NH_3$
48	3143.29265(54)			
49	3145.41566(48)			
50	3145.48042(38)			
51	3145.77477(47)			
52	3146.68180(34)	s^{P}P(7, 3)	2NU4 pe	$^{14}NH_3$
53	3147.67199(56)			

- "LINE NO.": serial number identifying the spectral line in the map on the opposite page.
- "WAVENUMBER": experimental value.
- Uncertainty in the last digit of the wavenumber is given in parentheses.
- "pa" : parallel ; "pe" : perpendicular.

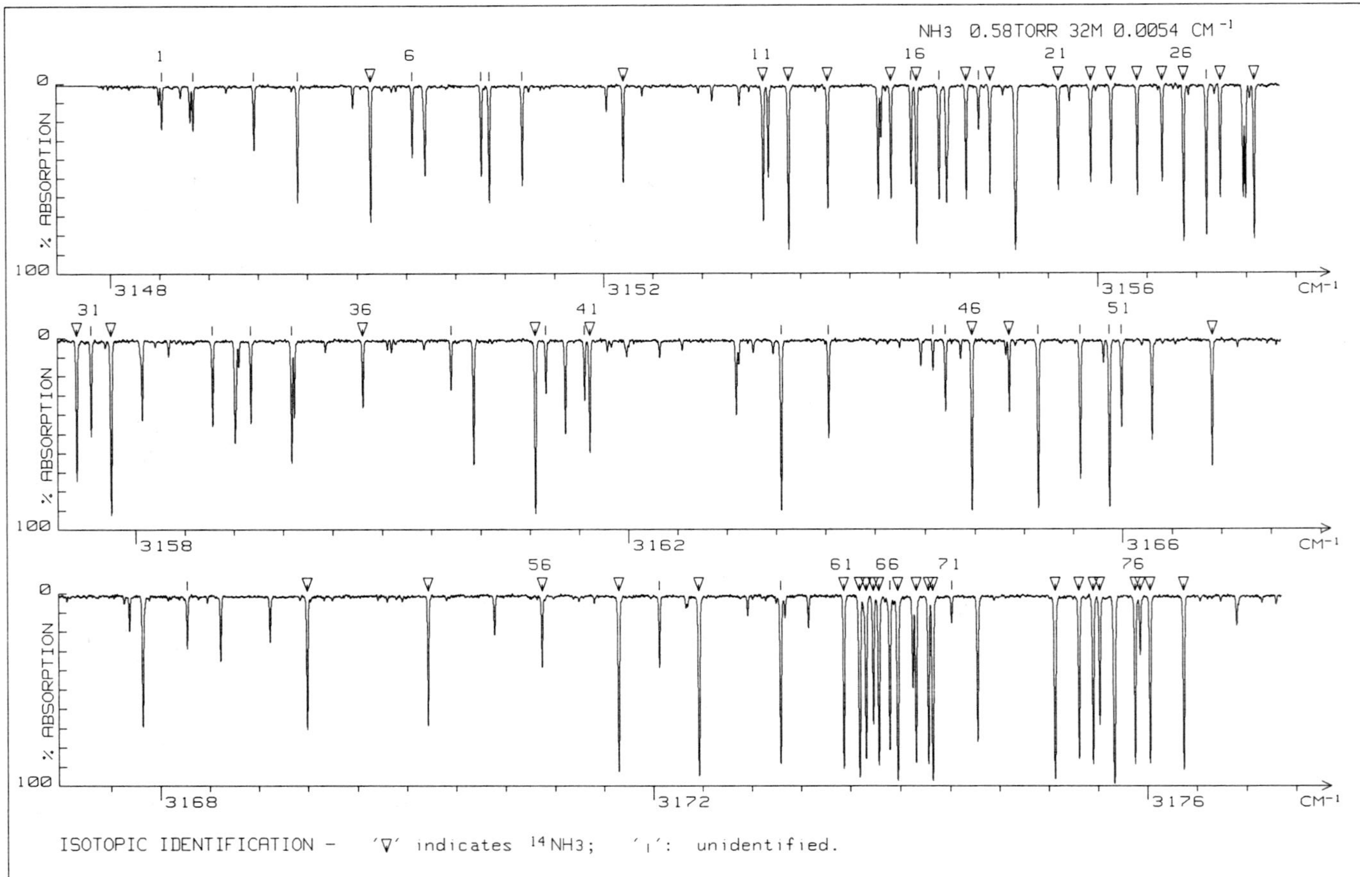
NH3 0.58TORR 32M 0.0054 CM⁻¹
% ABSORPTION
0
100
3148
3152
3156
3158
3162
3166
3168
3172
3176
CM⁻¹
ISOTOPIC IDENTIFICATION - '▽' indicates ¹⁴NH3; '|': unidentified.

LINE NO.	WAVENUMBER CM^{-1}, VAC.	ROTATIONAL ASSIGNMENTS	BAND	ISOTOPIC SPECIES
1	3148.42227(57)			
2	3148.67549(56)			
3	3149.17142(40)			
4	3149.52434(30)			
5	3150.11147(22)	aPP(7, 3)	2NU4 pe	$^{14}NH_3$
6	3150.45038(36)			
7	3151.01079(31)			
8	3151.07528(24)			
9	3151.34362(66)			
10	3152.16375(28)	sPP(6, 2)	2NU4 pe	$^{14}NH_3$
11	3153.29996(43)	aP(9, 7)	NU1	$^{14}NH_3$
12	3153.50723(22)	aP(9, 6)	NU1	$^{14}NH_3$
13	3153.82441(23)	aP(9, 5)	NU1	$^{14}NH_3$
14	3154.33554(24)	sPP(5, 1)	2NU4 pe	$^{14}NH_3$
15	3154.49564(29)			
16	3154.53782(22)	aP(9, 3)	NU1	$^{14}NH_3$
17	3154.71847(27)			
18	3154.93405(24)	aP(9, 1)	NU1	$^{14}NH_3$
19	3155.03468(56)			
20	3155.12828(27)	sP(9, 8)	NU1	$^{14}NH_3$
21	3155.68482(25)	aPP(6, 2)	2NU4 pe	$^{14}NH_3$
22	3155.94706(64)	aP(3, 2)	2NU4 pa	$^{14}NH_3$
23	3156.11283(29)	sP(9, 5)	NU1	$^{14}NH_3$
24	3156.32618(26)	sP(9, 4)	NU1	$^{14}NH_3$
25	3156.52568(28)	aPP(5, 1)	2NU4 pe	$^{14}NH_3$
26	3156.69941(20)	sP(9, 3)	NU1	$^{14}NH_3$
27	3156.88292(40)			
28	3156.99538(24)	sP(9, 2)	NU1	$^{14}NH_3$
29	3157.26674(20)	sP(9, 0)	NU1	$^{14}NH_3$
30	3157.52963(21)	aP(3, 1)	2NU4 pa	$^{14}NH_3$
31	3157.64537(28)			
32	3157.80653(19)	arP(4, 0)	2NU4 pe	$^{14}NH_3$
33	3158.62927(33)			
34	3158.94024(33)			
35	3159.27084(25)			
36	3159.84670(39)	aPP(7, 4)	2NU4 pe	$^{14}NH_3$
37	3160.56387(52)			
38	3161.24538(19)	sP(3, 0)	2NU4 pa	$^{14}NH_3$
39	3161.33141(48)			
40	3161.64684(43)			
41	3161.68853(26)	sPP(7, 4)	2NU4 pe	$^{14}NH_3$
42	3163.24325(22)			
43	3163.62822(29)			
44	3164.46784(76)			
45	3164.56666(38)			
46	3164.77748(22)	aPP(6, 3)	2NU4 pe	$^{14}NH_3$
47	3165.08270(38)	arQ(6, 5)	2NU4 pe	$^{14}NH_3$
48	3165.31722(19)			
49	3165.65925(24)			
50	3165.89584(20)			
51	3165.99365(30)			
52	3166.72728(25)	sPP(5, 2)	2NU4 pe	$^{14}NH_3$
53	3168.21848(48)			
54	3169.19596(24)	aPP(5, 2)	2NU4 pe	$^{14}NH_3$
55	3170.17323(24)	sPP(4, 1)	2NU4 pe	$^{14}NH_3$
56	3171.09649(38)	sPP(7, 5)	2NU4 pe	$^{14}NH_3$
57	3171.71900(19)	aPP(4, 1)	2NU4 pe	$^{14}NH_3$
58	3172.05077(37)			
59	3172.37140(19)	sPP(6, 4)	2NU4 pe	$^{14}NH_3$
59		sPP(7, 6)	2NU4 pe	$^{14}NH_3$
60	3173.03562(20)			

- "LINE NO.": serial number identifying the spectral line in the map on the opposite page.
- "WAVENUMBER": experimental value.
- Uncertainty in the last digit of the wavenumber is given in parentheses.
- "pa" : parallel ; "pe" : perpendicular.

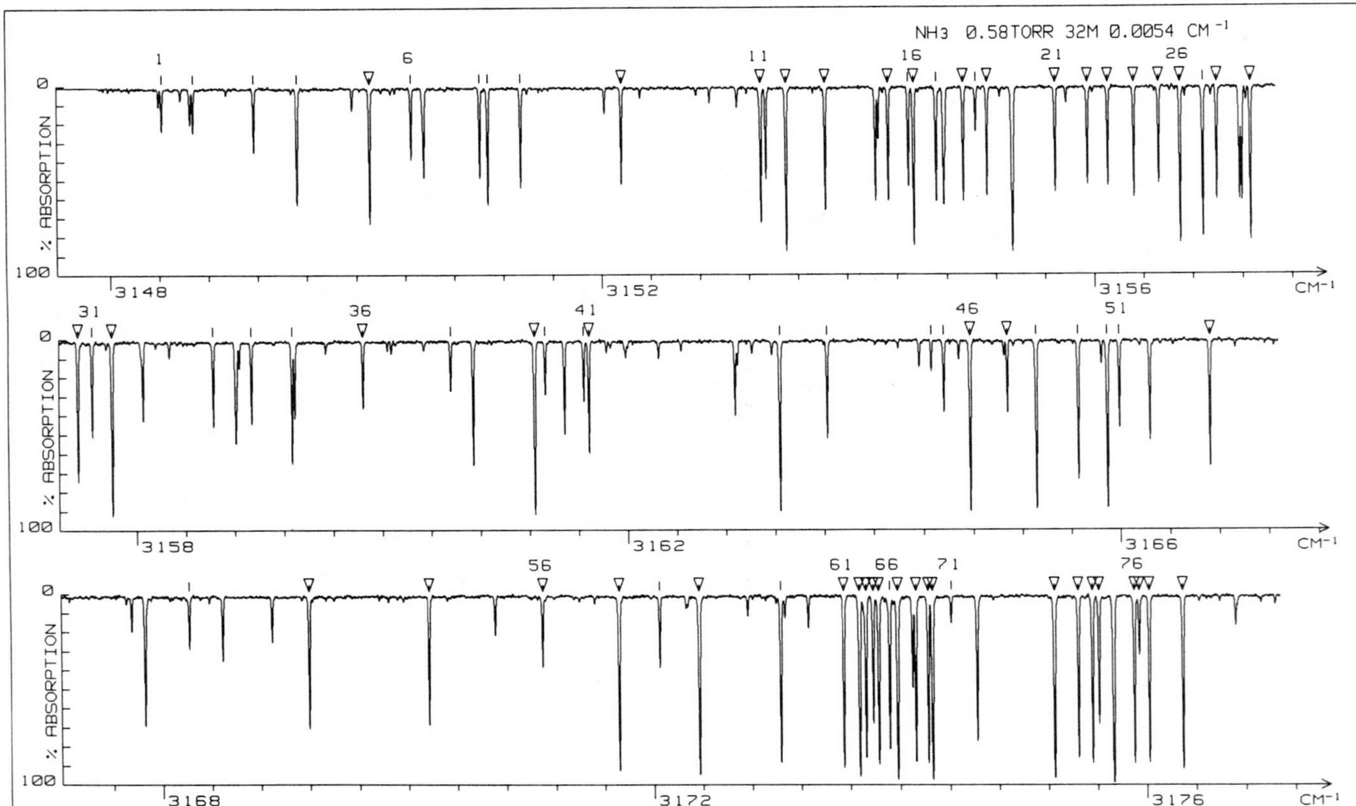

ISOTOPIC IDENTIFICATION – '∇' indicates $^{14}NH_3$; '|': unidentified.

LINE NO.	WAVENUMBER CM^{-1}, VAC.	ROTATIONAL ASSIGNMENTS	BAND	ISOTOPIC SPECIES
61	3173.54885(19)	$^{s}{}^{r}P$(3, 0)	2NU4 pe	$^{14}NH_3$
62	3173.67751(26)	aP(8, 6)	NU1	$^{14}NH_3$
63	3173.72837(20)	aP(8, 5)	NU1	$^{14}NH_3$
64	3173.78702(23)	aP(8, 7)	NU1	$^{14}NH_3$
65	3173.83160(33)	aP(8, 4)	NU1	$^{14}NH_3$
66	3173.91882(37)			
67	3173.98229(18)	aP(8, 3)	NU1	$^{14}NH_3$
68	3174.12774(20)	aP(8, 2)	NU1	$^{14}NH_3$
69	3174.22706(19)	$^{a}{}^{P}P$(7, 6)	2NU4 pe	$^{14}NH_3$
69		aP(8, 1)	NU1	$^{14}NH_3$
70	3174.26234(18)	aP(8, 0)	NU1	$^{14}NH_3$
71	3174.41201(95)			
72	3175.25213(19)	sP(8, 6)	NU1	$^{14}NH_3$
73	3175.44709(20)	sP(8, 5)	NU1	$^{14}NH_3$
74	3175.56038(20)	sP(8, 4)	NU1	$^{14}NH_3$
75	3175.61446(23)	sP(8, 7)	NU1	$^{14}NH_3$
76	3175.90311(19)	sP(8, 2)	NU1	$^{14}NH_3$
77	3175.94377(44)	$^{s}{}^{P}P$(5, 3)	2NU4 pe	$^{14}NH_3$
77		aP(2, 1)	2NU4 pa	$^{14}NH_3$
78	3176.02307(19)	sP(8, 1)	NU1	$^{14}NH_3$
79	3176.29550(19)	aP(2, 0)	2NU4 pa	$^{14}NH_3$

- "LINE NO.": serial number identifying the spectral line in the map on the opposite page.
- "WAVENUMBER": experimental value.
- Uncertainty in the last digit of the wavenumber is given in parentheses.
- "pa" : parallel ; "pe" : perpendicular.

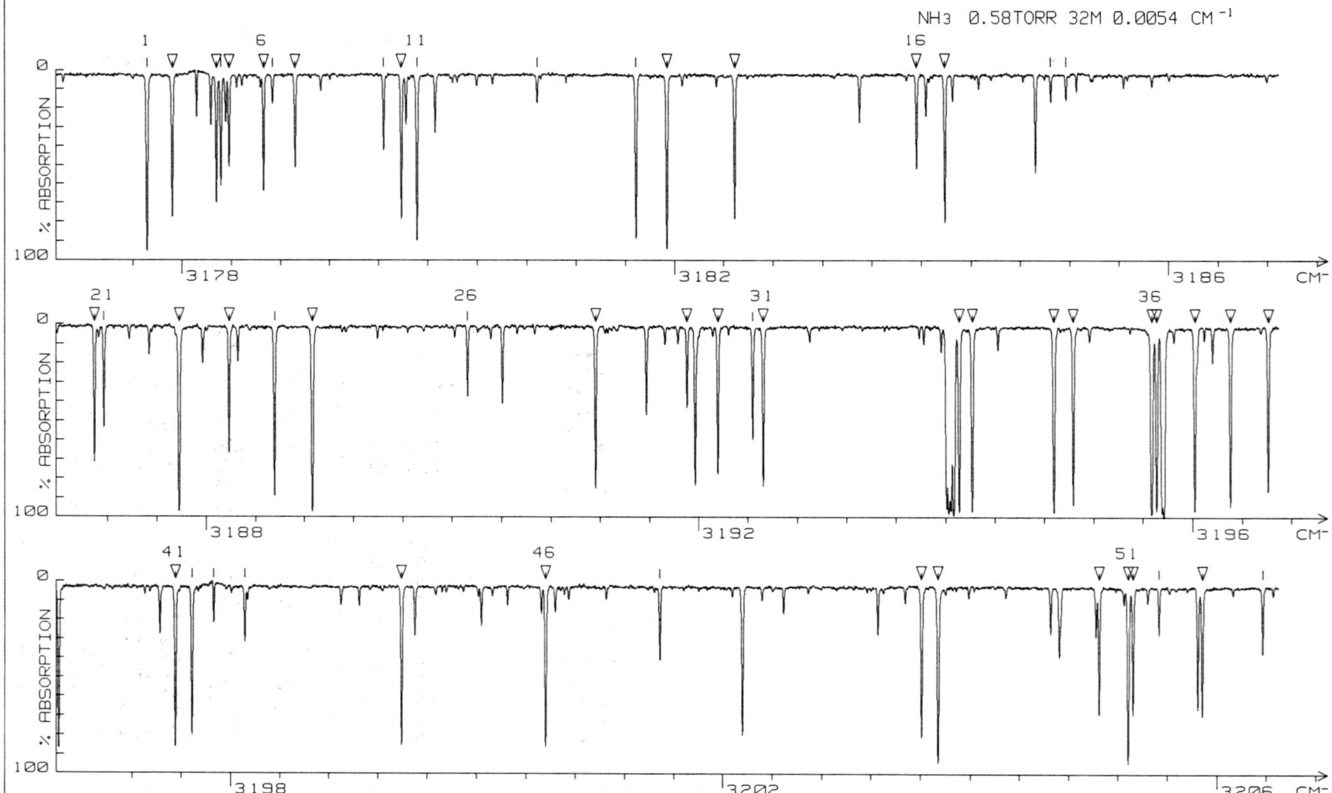

ISOTOPIC IDENTIFICATION – '▽' indicates $^{14}NH_3$; '|': unidentified.

LINE NO.	WAVENUMBER CM^{-1}, VAC.	ROTATIONAL ASSIGNMENTS	BAND	ISOTOPIC SPECIES
1	3177.71540(22)			
2	3177.92330(21)	s^{P}P(6, 5)	2NU4 pe	$^{14}NH_3$
3	3178.28014(25)	a^{P}P(6, 4)	2NU4 pe	$^{14}NH_3$
4	3178.31720(26)			
5	3178.38317(31)	s^{r}Q(5, 4)	2NU4 pe	$^{14}NH_3$
6	3178.66174(26)	sP(2, 1)	2NU4 pa	$^{14}NH_3$
6		s^{P}P(7, 7)	2NU4 pe	$^{14}NH_3$
7	3178.73459(88)			
8	3178.92306(29)	a^{r}Q(5, 4)	2NU4 pe	$^{14}NH_3$
9	3179.64269(36)			
10	3179.78499(21)	a^{P}P(6, 5)	2NU4 pe	$^{14}NH_3$
11	3179.91081(19)			
12	3180.88488(88)			
13	3181.68724(20)			
14	3181.93670(19)	a^{P}P(5, 3)	2NU4 pe	$^{14}NH_3$
15	3182.49285(23)	s^{P}P(4, 2)	2NU4 pe	$^{14}NH_3$
16	3183.96325(28)	s^{r}Q(6, 4)	2NU4 pe	$^{14}NH_3$
17	3184.18973(21)	a^{P}P(4, 2)	2NU4 pe	$^{14}NH_3$
18	3185.04770(86)			
19	3185.17047(92)			
20	3187.09475(24)	s^{P}P(3, 1)	2NU4 pe	$^{14}NH_3$
21	3187.17206(28)			
22	3187.78274(24)	s^{P}P(6, 6)	2NU4 pe	$^{14}NH_3$
23	3188.18628(23)	a^{P}P(3, 1)	2NU4 pe	$^{14}NH_3$
24	3188.55475(20)			
25	3188.86805(18)	a^{P}P(6, 6)	2NU4 pe	$^{14}NH_3$
26	3190.12786(38)			
27	3191.16491(20)	a^{P}P(5, 4)	2NU4 pe	$^{14}NH_3$
28	3191.90495(34)	a^{r}Q(7, 4)	2NU4 pe	$^{14}NH_3$
29	3192.16070(23)	a^{r}P(2, 0)	2NU4 pe	$^{14}NH_3$
30	3192.44315(27)			

LINE NO.	WAVENUMBER CM^{-1}, VAC.	ROTATIONAL ASSIGNMENTS	BAND	ISOTOPIC SPECIES
31	3192.52641(20)	a^{r}Q(4, 3)	2NU4 pe	$^{14}NH_3$
32	3194.10861(18)	aP(7, 1)	NU1	$^{14}NH_3$
33	3194.21302(19)	aP(7, 6)	NU1	$^{14}NH_3$
34	3194.87877(19)	a^{P}P(4, 3)	2NU4 pe	$^{14}NH_3$
35	3195.03449(19)	s^{P}P(5, 5)	2NU4 pe	$^{14}NH_3$
36	3195.67143(19)	sP(7, 4)	NU1	$^{14}NH_3$
36		sP(7, 3)	NU1	$^{14}NH_3$
37	3195.71249(19)	sP(7, 2)	NU1	$^{14}NH_3$
38	3196.01807(24)	sP(7, 6)	NU1	$^{14}NH_3$
39	3196.30443(19)	a^{P}P(5, 5)	2NU4 pe	$^{14}NH_3$
40	3196.61343(20)	s^{r}Q(5, 3)	2NU4 pe	$^{14}NH_3$
41	3197.55565(20)	a^{r}Q(5, 3)	2NU4 pe	$^{14}NH_3$
42	3197.68996(21)			
43	3197.86548(60)			
44	3198.11760(47)			
45	3199.39410(22)	s^{P}P(3, 2)	2NU4 pe	$^{14}NH_3$
46	3200.56157(20)	a^{P}P(3, 2)	2NU4 pe	$^{14}NH_3$
47	3201.48770(38)			
48	3203.61229(23)	a^{r}Q(6, 3)	2NU4 pe	$^{14}NH_3$
49	3203.74471(23)	s^{P}P(4, 4)	2NU4 pe	$^{14}NH_3$
50	3205.04933(23)	s^{P}P(2, 1)	2NU4 pe	$^{14}NH_3$
51	3205.28611(19)	a^{P}P(4, 4)	2NU4 pe	$^{14}NH_3$
52	3205.32581(25)	s^{r}Q(3, 2)	2NU4 pe	$^{14}NH_3$
53	3205.53622(55)			
54	3205.88379(25)	a^{r}Q(3, 2)	2NU4 pe	$^{14}NH_3$
55	3206.37343(39)			

- "LINE NO.": serial number identifying the spectral line in the map on the opposite page.
- "WAVENUMBER": experimental value.
- Uncertainty in the last digit of the wavenumber is given in parentheses.
- "pa" : parallel ; "pe" : perpendicular.

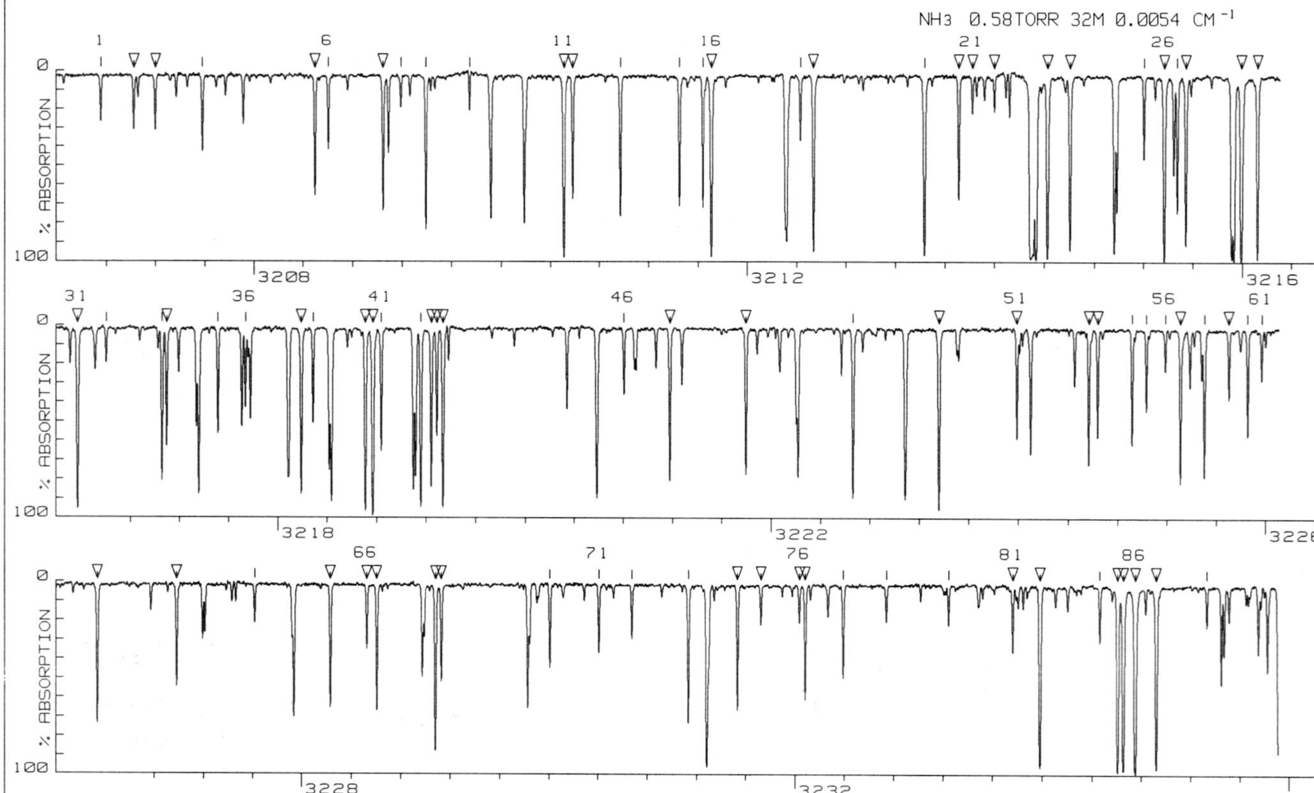

ISOTOPIC IDENTIFICATION – '▽' indicates $^{14}NH_3$; '|': unidentified.

LINE NO.	WAVENUMBER CM^{-1}, VAC.	ROTATIONAL ASSIGNMENTS	BAND	ISOTOPIC SPECIES
1	3206.75587(58)			
2	3207.02210(47)	a^rP(10, 3)	NU3	$^{14}NH_3$
3	3207.19990(49)	s^rP(10, 3)	NU3	$^{14}NH_3$
4	3207.58131(36)			
5	3208.49973(25)	s^rQ(7, 3)	2NU4 pe	$^{14}NH_3$
6	3208.60669(36)			
7	3209.04762(24)	s^rQ(4, 2)	2NU4 pe	$^{14}NH_3$
8	3209.19144(79)			
9	3209.39394(21)			
10	3209.74685(44)			
11	3210.51103(18)	s^pP(3, 3)	2NU4 pe	$^{14}NH_3$
12	3210.58486(23)	a^rQ(7, 3)	2NU4 pe	$^{14}NH_3$
13	3210.97426(24)			
14	3211.45124(44)			
15	3211.64196(23)			
16	3211.71126(19)	a^pP(3, 3)	2NU4 pe	$^{14}NH_3$
17	3212.43104(42)			
18	3212.53669(19)	aQ(5, 5)	2NU4 pa	$^{14}NH_3$
19	3213.43561(24)			
20	3213.71201(25)	s^rQ(5, 2)	2NU4 pe	$^{14}NH_3$
21	3213.82233(65)	a^rP(10, 2)	NU3	$^{14}NH_3$
22	3214.00228(65)	s^rP(10, 2)	NU3	$^{14}NH_3$
23	3214.42680(22)	aP(6, 4)	NU1	$^{14}NH_3$
24	3214.60530(19)	aP(6, 5)	NU1	$^{14}NH_3$
25	3215.20470(32)			
26	3215.36971(18)	aQ(3, 3)	2NU4 pa	$^{14}NH_3$
27	3215.47415(24)			
28	3215.54346(20)	aQ(2, 2)	2NU4 pa	$^{14}NH_3$
28		aQ(1, 1)	2NU4 pa	$^{14}NH_3$
29	3215.99126(39)	a^rP(9, 5)	NU3	$^{14}NH_3$
29		sP(6, 3)	NU1	$^{14}NH_3$
30	3216.12453(19)	sP(6, 4)	NU1	$^{14}NH_3$

LINE NO.	WAVENUMBER CM^{-1}, VAC.	ROTATIONAL ASSIGNMENTS	BAND	ISOTOPIC SPECIES
31	3216.37640(19)	sP(6, 5)	NU1	$^{14}NH_3$
32	3216.60522(75)			
33	3217.06044(21)			
34	3217.09879(25)	aQ(2, 1)	2NU4 pa	$^{14}NH_3$
35	3217.51445(26)			
36	3217.73803(37)			
37	3218.19438(23)	a^pP(2, 2)	2NU4 pe	$^{14}NH_3$
38	3218.28960(30)			
39	3218.70975(23)	aQ(4, 3)	2NU4 pa	$^{14}NH_3$
40	3218.77034(18)	sQ(3, 3)	2NU4 pa	$^{14}NH_3$
41	3218.83844(24)			
42	3219.15814(19)			
43	3219.24552(21)	aQ(6, 5)	2NU4 pa	$^{14}NH_3$
44	3219.29042(29)	s^rQ(6, 2)	2NU4 pe	$^{14}NH_3$
45	3219.33952(19)	sQ(4, 4)	2NU4 pa	$^{14}NH_3$
46	3220.80812(41)			
47	3221.17897(23)	s^rQ(3, 1)	2NU4 pe	$^{14}NH_3$
48	3221.79833(22)	sQ(3, 2)	2NU4 pa	$^{14}NH_3$
49	3222.66492(32)			
50	3223.36062(19)	sQ(4, 3)	2NU4 pa	$^{14}NH_3$
51	3223.99067(57)	s^pP(1, 1)	2NU4 pe	$^{14}NH_3$
52	3224.56428(24)	s^rQ(4, 1)	2NU4 pe	$^{14}NH_3$
52		sQ(4, 1)	2NU4 pa	$^{14}NH_3$
52		aQ(5, 2)	2NU4 pa	$^{14}NH_3$
53	3224.63545(28)	a^pP(1, 1)	2NU4 pe	$^{14}NH_3$
54	3224.91729(27)			
55	3225.03399(33)			
56	3225.18753(65)			
57	3225.30905(36)	sQ(5, 4)	2NU4 pa	$^{14}NH_3$
58	3225.50284(22)			
59	3225.70091(38)	s^rQ(7, 2)	2NU4 pe	$^{14}NH_3$
60	3225.85293(29)			

- "LINE NO.": serial number identifying the spectral line in the map on the opposite page.
- "WAVENUMBER": experimental value.
- Uncertainty in the last digit of the wavenumber is given in parentheses.
- "pa" : parallel ; "pe" : perpendicular.

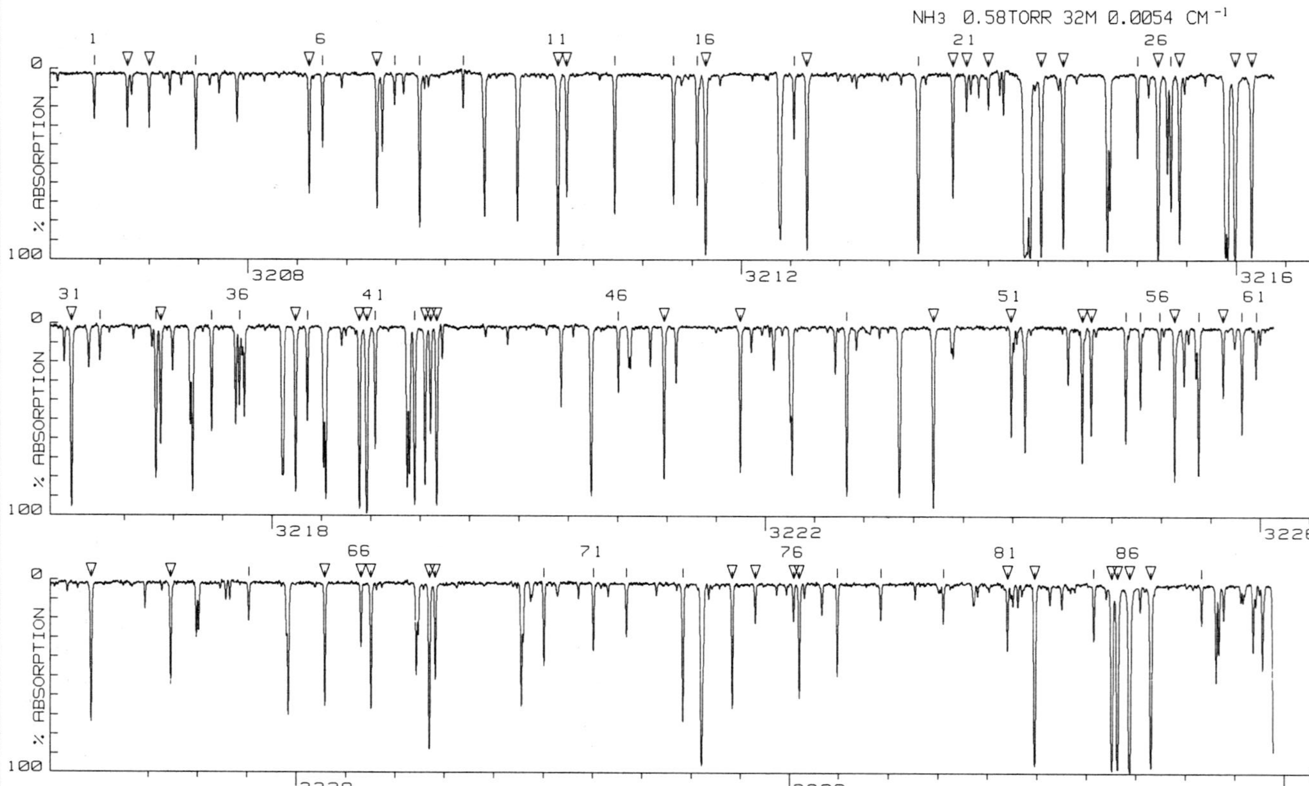

ISOTOPIC IDENTIFICATION - '∇' indicates $^{14}NH_3$; '|': unidentified.

LINE NO.	WAVENUMBER CM^{-1}, VAC.	ROTATIONAL BAND ASSIGNMENTS	ISOTOPIC SPECIES
61	3225.96867(52)		
62	3226.33959(22)	a^rQ(4, 1) 2NU4 pe	$^{14}NH_3$
63	3226.98504(27)	sQ(4, 2) 2NU4 pa	$^{14}NH_3$
64	3227.61857(67)		
65	3228.23447(25)	a^rP(10, 0) NU3	$^{14}NH_3$
66	3228.52716(43)	a^rQ(7, 2) 2NU4 pe	$^{14}NH_3$
67	3228.60875(25)	aQ(6, 3) 2NU4 pa	$^{14}NH_3$
68	3229.08495(23)	sQ(5, 3) 2NU4 pa	$^{14}NH_3$
69	3229.13257(28)	s^rP(9, 3) NU3	$^{14}NH_3$
70	3230.00746(34)		
71	3230.41028(38)		
72	3230.67779(48)		
73	3231.13454(22)		
74	3231.53667(25)	a^rQ(5, 1) 2NU4 pe	$^{14}NH_3$
75	3231.72561(65)	a^rP(8, 6) NU3	$^{14}NH_3$
76	3232.03566(73)	s^rP(8, 6) NU3	$^{14}NH_3$
77	3232.08272(27)	sQ(6, 4) 2NU4 pa	$^{14}NH_3$
78	3232.39327(31)		
79	3232.74264(74)		
80	3233.24504(63)		
81	3233.76468(39)	sQ(5, 2) 2NU4 pa	$^{14}NH_3$
82	3233.98416(19)	a^rQ(2, 0) 2NU4 pe	$^{14}NH_3$
83	3234.46144(49)		
84	3234.60807(28)	aP(5, 1) NU1	$^{14}NH_3$
85	3234.65524(21)	aP(5, 2) NU1	$^{14}NH_3$
86	3234.75232(20)	aP(5, 3) NU1	$^{14}NH_3$
87	3234.92369(19)	aP(5, 4) NU1	$^{14}NH_3$
88	3235.33208(63)		

- "LINE NO.": serial number identifying the spectral line in the map on the opposite page.
- "WAVENUMBER": experimental value.
- Uncertainty in the last digit of the wavenumber is given in parentheses.
- "pa" : parallel ; "pe" : perpendicular.

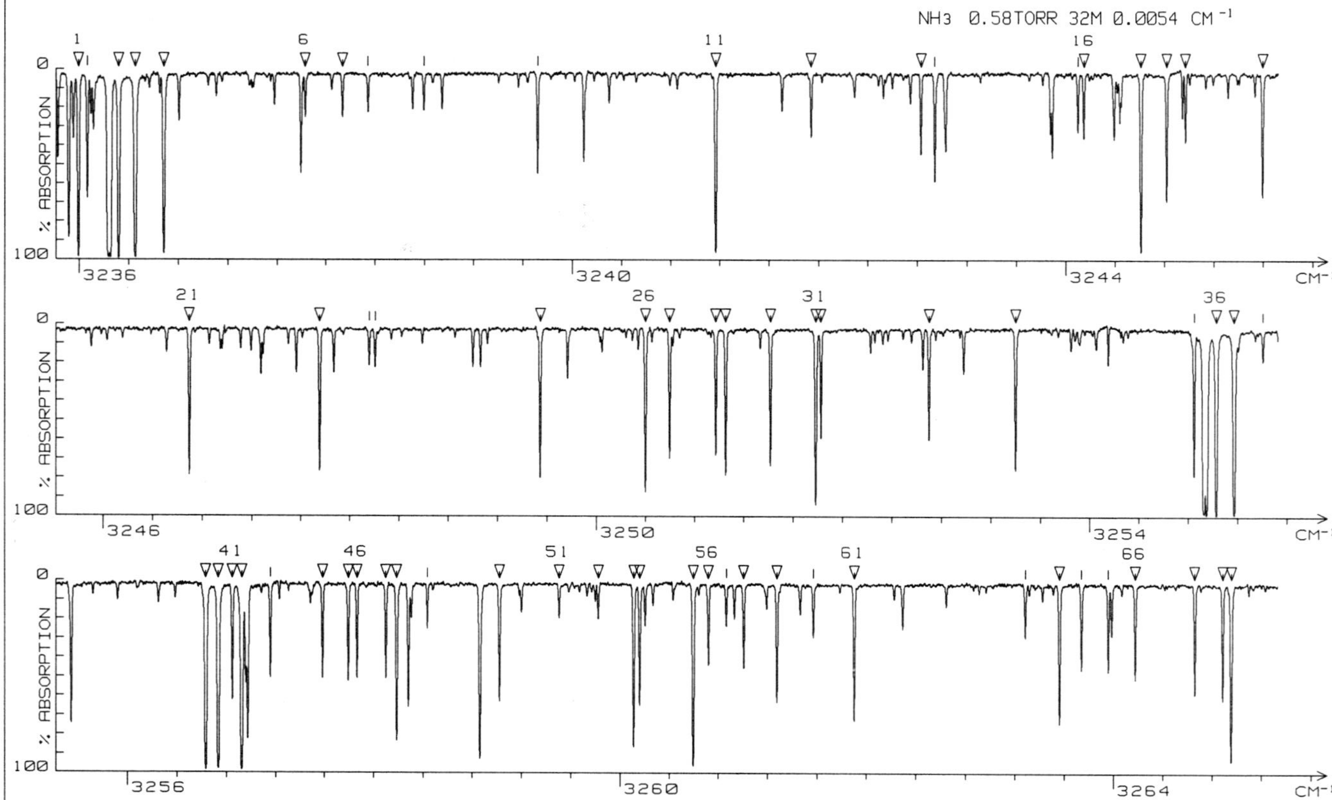

ISOTOPIC IDENTIFICATION – '∇' indicates $^{14}NH_3$; '|': unidentified.

LINE NO.	WAVENUMBER CM^{-1}, VAC.	ROTATIONAL ASSIGNMENTS	BAND		ISOTOPIC SPECIES
1	3235.99228(19)	s^{r}Q(3, 0)	2NU4	pe	$^{14}NH_3$
2	3236.06468(26)				
3	3236.31808(18)	sP(5, 2)	NU1		$^{14}NH_3$
4	3236.45080(20)	sP(5, 3)	NU1		$^{14}NH_3$
5	3236.68181(22)	sP(5, 4)	NU1		$^{14}NH_3$
6	3237.83023(60)	a^{r}P(8, 5)	NU3		$^{14}NH_3$
7	3238.13552(59)	s^{r}P(8, 5)	NU3		$^{14}NH_3$
8	3238.34229(68)				
9	3238.79268(69)				
10	3239.71657(54)				
11	3241.16642(19)	a^{r}Q(4, 0)	2NU4	pe	$^{14}NH_3$
12	3241.93710(40)	s^{r}Q(7, 1)	2NU4	pe	$^{14}NH_3$
13	3242.82695(33)	a^{r}P(9, 1)	NU3		$^{14}NH_3$
14	3242.93794(81)				
15	3244.09538(41)				
16	3244.14257(111)	a^{r}P(8, 4)	NU3		$^{14}NH_3$
17	3244.60814(19)	s^{r}Q(5, 0)	2NU4	pe	$^{14}NH_3$
18	3244.81511(23)	s^{P}Q(1, 1)	2NU4	pe	$^{14}NH_3$
19	3244.96849(38)	a^{r}Q(7, 1)	2NU4	pe	$^{14}NH_3$
20	3245.59292(26)	a^{P}Q(1, 1)	2NU4	pe	$^{14}NH_3$
21	3246.69267(22)	s^{P}Q(2, 1)	2NU4	pe	$^{14}NH_3$
22	3247.75544(22)	a^{P}Q(2, 1)	2NU4	pe	$^{14}NH_3$
23	3248.15690(70)				
24	3248.20556(72)				
25	3249.54289(23)	s^{P}Q(3, 1)	2NU4	pe	$^{14}NH_3$
26	3250.39829(23)	s^{r}P(9, 0)	NU3		$^{14}NH_3$
27	3250.59463(25)	a^{r}P(8, 3)	NU3		$^{14}NH_3$
28	3250.97324(23)	s^{r}P(8, 3)	NU3		$^{14}NH_3$
29	3251.05272(21)	a^{P}Q(3, 1)	2NU4	pe	$^{14}NH_3$
30	3251.41576(76)	a^{r}Q(6, 0)	2NU4	pe	$^{14}NH_3$
31	3251.78039(24)	a^{r}R(0, 0)	2NU4	pe	$^{14}NH_3$
32	3251.82266(28)	a^{P}P(10, 3)	NU3		$^{14}NH_3$
33	3252.69845(27)	s^{P}P(10, 3)	NU3		$^{14}NH_3$
34	3253.39760(23)	s^{P}Q(4, 1)	2NU4	pe	$^{14}NH_3$
35	3254.84572(24)				
36	3255.02404(20)	aP(4, 2)	NU1		$^{14}NH_3$
37	3255.16878(19)	aP(4, 3)	NU1		$^{14}NH_3$
38	3255.40359(85)				
39	3256.63156(24)	sP(4, 1)	NU1		$^{14}NH_3$
40	3256.73251(20)	sP(4, 2)	NU1		$^{14}NH_3$
41	3256.84459(25)	aR(1, 1)	2NU4	pa	$^{14}NH_3$
42	3256.92145(19)	s^{r}Q(7, 0)	2NU4	pe	$^{14}NH_3$
42		sP(4, 3)	NU1		$^{14}NH_3$
43	3257.15369(28)				
44	3257.58394(30)	a^{r}P(8, 2)	NU3		$^{14}NH_3$
45	3257.79304(29)	s^{r}P(8, 2)	NU3		$^{14}NH_3$
46	3257.86432(29)	a^{P}P(9, 1)	NU3		$^{14}NH_3$
47	3258.09834(29)	s^{P}P(9, 1)	NU3		$^{14}NH_3$
48	3258.18568(24)	s^{r}R(1, 1)	2NU4	pe	$^{14}NH_3$
49	3258.43389(58)				
50	3259.02026(25)	s^{P}Q(2, 2)	2NU4	pe	$^{14}NH_3$
51	3259.50062(77)	a^{r}P(7, 5)	NU3		$^{14}NH_3$
52	3259.81882(74)	s^{r}P(7, 5)	NU3		$^{14}NH_3$
53	3260.10878(21)	sR(1, 1)	2NU4	pa	$^{14}NH_3$
54	3260.15814(26)	a^{P}Q(2, 2)	2NU4	pe	$^{14}NH_3$
55	3260.59352(19)	sR(1, 0)	2NU4	pa	$^{14}NH_3$
56	3260.71995(79)	a^{P}P(10, 4)	NU3		$^{14}NH_3$
57	3260.86343(68)				
58	3261.00492(58)	s^{P}P(10, 4)	NU3		$^{14}NH_3$
59	3261.27403(69)	a^{P}Q(5, 1)	2NU4	pe	$^{14}NH_3$
60	3261.56741(47)				

- "LINE NO.": serial number identifying the spectral line in the map on the opposite page.
- "WAVENUMBER": experimental value.
- Uncertainty in the last digit of the wavenumber is given in parentheses.
- "pa" : parallel ; "pe" : perpendicular.

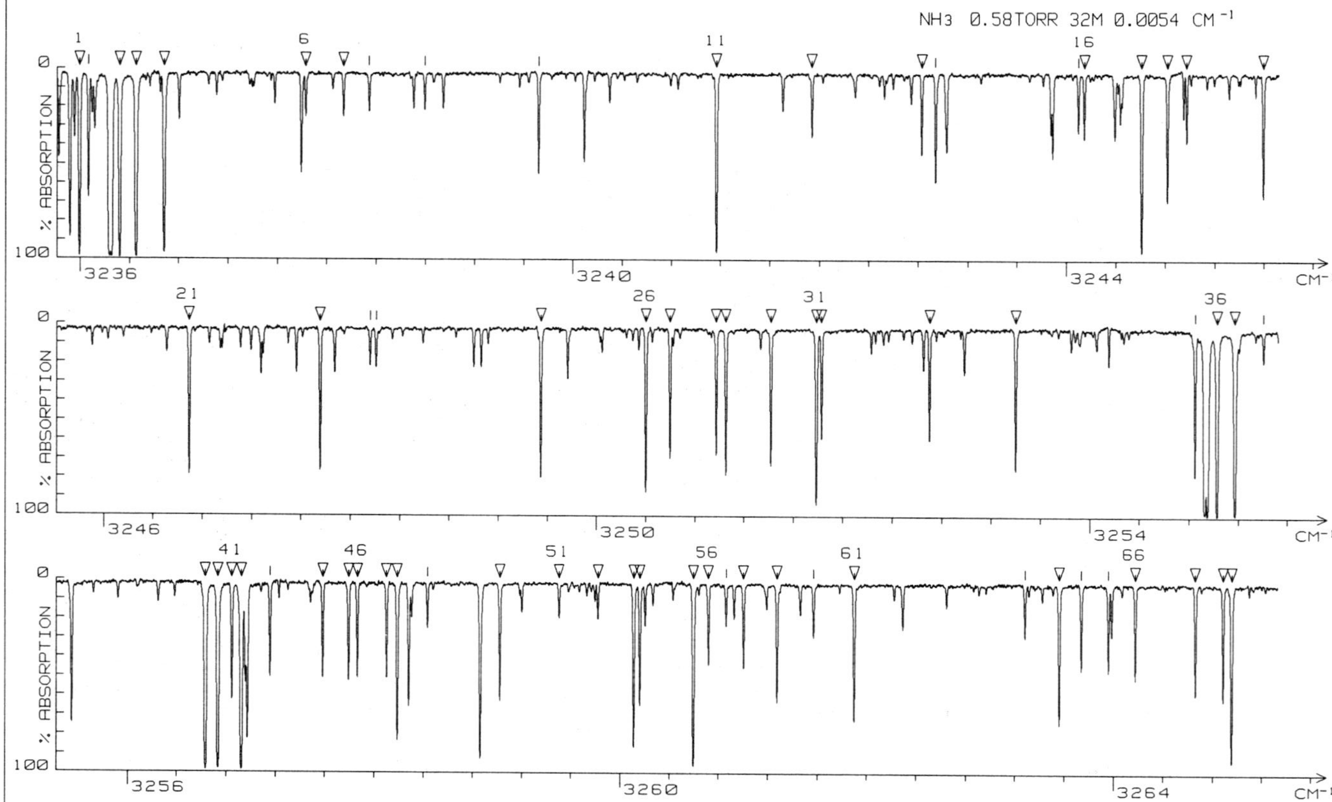

ISOTOPIC IDENTIFICATION – '∇' indicates $^{14}NH_3$; 'ı': unidentified.

LINE NO.	WAVENUMBER CM^{-1}, VAC.	ROTATIONAL BAND ASSIGNMENTS	ISOTOPIC SPECIES
61	3261.89989(22)	$s^{P}Q$(3, 2) 2NU4 pe	$^{14}NH_3$
62	3263.28199(50)		
63	3263.55896(25)	$a^{P}Q$(3, 2) 2NU4 pe	$^{14}NH_3$
64	3263.73750(32)		
65	3263.95519(33)		
66	3264.17660(31)	$s^{P}Q$(6, 1) 2NU4 pe	$^{14}NH_3$
67	3264.65846(27)	$a^{r}P$(8, 1) NU3	$^{14}NH_3$
68	3264.88259(26)	$s^{r}P$(8, 1) NU3	$^{14}NH_3$
69	3264.95181(24)	$s^{r}R$(2, 2) 2NU4 pe	$^{14}NH_3$

- "LINE NO.": serial number identifying the spectral line in the map on the opposite page.
- "WAVENUMBER": experimental value.
- Uncertainty in the last digit of the wavenumber is given in parentheses.
- "pa" : parallel ; "pe" : perpendicular.

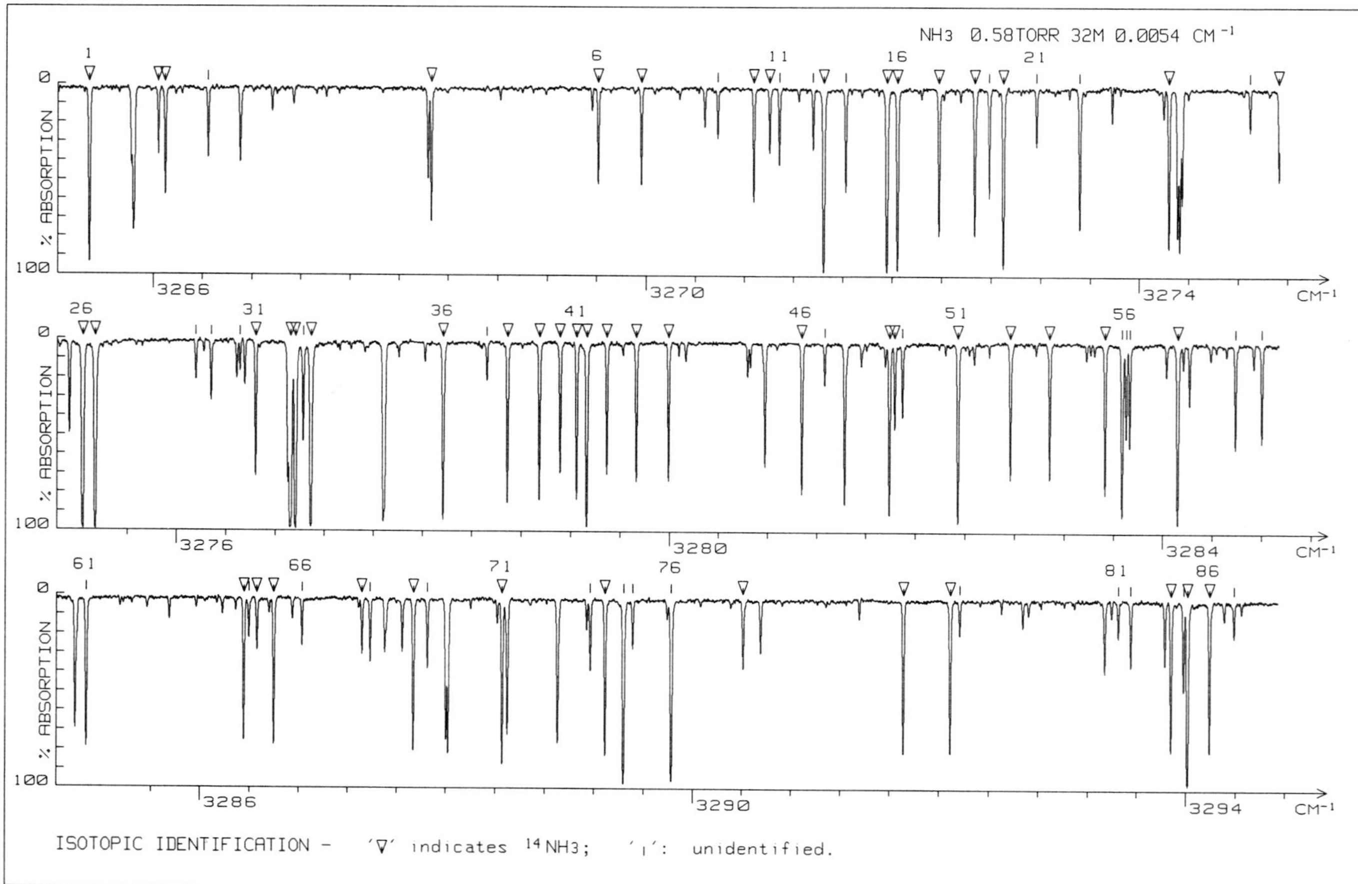
NH3 0.58TORR 32M 0.0054 CM⁻¹
% ABSORPTION
0
100
3266
3270
3274
3276
3280
3284
3286
3290
3294
CM⁻¹
ISOTOPIC IDENTIFICATION - '∇' indicates ¹⁴NH3; 'ı': unidentified.

LINE NO.	WAVENUMBER CM^{-1}, VAC.	ROTATIONAL ASSIGNMENTS	BAND	ISOTOPIC SPECIES
1	3265.48030(20)	a^rR(2, 2)	2NU4 pe	$^{14}NH_3$
2	3266.04057(39)	s^rP(7, 4)	NU3	$^{14}NH_3$
3	3266.09867(27)	s^PP(9, 2)	NU3	$^{14}NH_3$
4	3266.44535(39)			
5	3268.25954(23)	a^PQ(4, 2)	2NU4 pe	$^{14}NH_3$
6	3269.61146(67)	a^PP(10, 5)	NU3	$^{14}NH_3$
7	3269.95698(28)	s^PP(10, 5)	NU3	$^{14}NH_3$
8	3270.58134(47)			
9	3270.87441(24)	s^PQ(5, 2)	2NU4 pe	$^{14}NH_3$
10	3271.00250(72)	s^PQ(7, 1)	2NU4 pe	$^{14}NH_3$
11	3271.08149(32)			
12	3271.35606(39)			
13	3271.44122(19)	s^rR(3, 3)	2NU4 pe	$^{14}NH_3$
13		s^PQ(3, 3)	2NU4 pe	$^{14}NH_3$
14	3271.62149(30)			
15	3271.95658(26)	a^rR(3, 3)	2NU4 pe	$^{14}NH_3$
16	3272.04227(19)	a^rP(8, 0)	NU3	$^{14}NH_3$
17	3272.37757(22)	a^rP(7, 3)	NU3	$^{14}NH_3$
18	3272.66451(40)	s^rP(7, 3)	NU3	$^{14}NH_3$
19	3272.78154(27)			
20	3272.89676(19)	s^rR(1, 0)	2NU4 pe	$^{14}NH_3$
21	3273.16343(88)			
22	3273.51170(25)			
23	3274.23805(36)	s^PP(9, 3)	NU3	$^{14}NH_3$
24	3274.89200(113)			
25	3275.12912(69)	s^PQ(4, 3)	2NU4 pe	$^{14}NH_3$
26	3275.23698(20)	aP(3, 1)	NU1	$^{14}NH_3$
27	3275.33717(19)	aP(3, 2)	NU1	$^{14}NH_3$
28	3276.15572(94)			
29	3276.27888(39)			
30	3276.51136(57)			
31	3276.63808(22)	s^PQ(6, 2)	2NU4 pe	$^{14}NH_3$
32	3276.92238(25)	sP(3, 0)	NU1	$^{14}NH_3$
33	3276.96277(21)	sP(3, 1)	NU1	$^{14}NH_3$
34	3277.02833(79)			
35	3277.09053(20)	sP(3, 2)	NU1	$^{14}NH_3$
36	3278.16813(20)	a^rR(4, 4)	2NU4 pe	$^{14}NH_3$
37	3278.52005(98)			
38	3278.68599(21)	aR(2, 1)	2NU4 pa	$^{14}NH_3$
39	3278.94413(21)	a^PP(10, 6)	NU3	$^{14}NH_3$
40	3279.11173(23)	a^rP(7, 2)	NU3	$^{14}NH_3$
41	3279.24825(21)	s^PP(10, 6)	NU3	$^{14}NH_3$
42	3279.33039(19)	aR(2, 0)	2NU4 pa	$^{14}NH_3$
43	3279.49249(24)	s^rP(7, 2)	NU3	$^{14}NH_3$
44	3279.72871(22)	a^PP(8, 1)	NU3	$^{14}NH_3$
45	3279.98968(23)	s^PP(8, 1)	NU3	$^{14}NH_3$
46	3281.07548(24)	a^PQ(4, 3)	2NU4 pe	$^{14}NH_3$
46		s^rR(5, 5)	2NU4 pe	$^{14}NH_3$
47	3281.25952(52)			
48	3281.78866(23)	a^rR(2, 1)	2NU4 pe	$^{14}NH_3$
49	3281.83162(31)	a^PQ(6, 2)	2NU4 pe	$^{14}NH_3$
50	3281.89427(39)			
51	3282.34216(19)	sR(2, 1)	2NU4 pa	$^{14}NH_3$
52	3282.76444(23)	a^PP(9, 4)	NU3	$^{14}NH_3$
53	3283.08069(23)	s^PP(9, 4)	NU3	$^{14}NH_3$
54	3283.52815(21)	a^PQ(5, 3)	2NU4 pe	$^{14}NH_3$
55	3283.66667(20)			
56	3283.69931(33)			
57	3283.73111(30)			
58	3284.12161(25)	a^rR(5, 5)	2NU4 pe	$^{14}NH_3$
59	3284.58538(28)			
60	3284.79972(29)			

- "LINE NO.": serial number identifying the spectral line in the map on the opposite page.
- "WAVENUMBER": experimental value.
- Uncertainty in the last digit of the wavenumber is given in parentheses.
- "pa" : parallel ; "pe" : perpendicular.

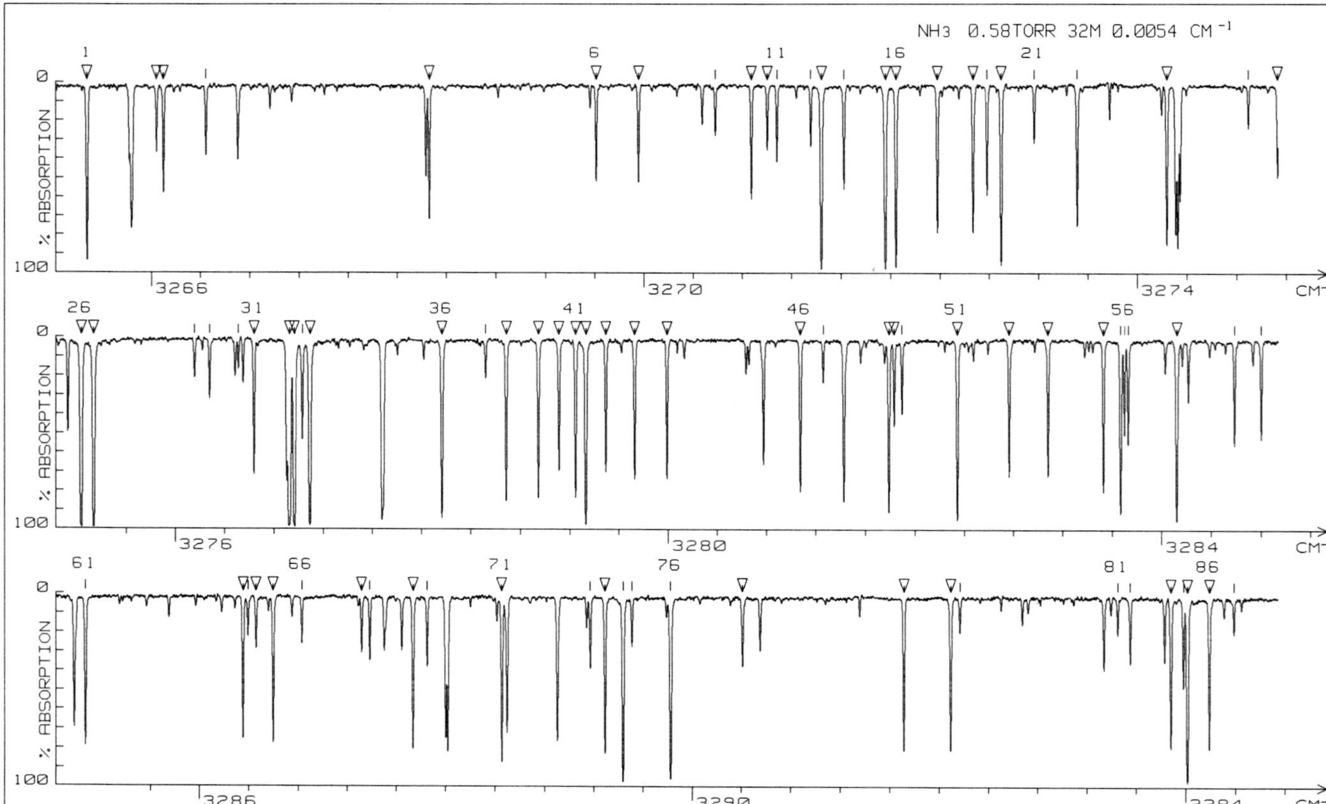

ISOTOPIC IDENTIFICATION – '∇' indicates $^{14}NH_3$; 'ı': unidentified.

LINE NO.	WAVENUMBER CM^{-1}, VAC.	ROTATIONAL ASSIGNMENTS	BAND	ISOTOPIC SPECIES
61	3285.07573(22)			
62	3286.35273(22)	a^rP(7, 1)	NU3	$^{14}NH_3$
63	3286.39298(110)			
64	3286.45849(47)	s^pR(1, 1)	2NU4 pe	$^{14}NH_3$
65	3286.59647(21)	s^pQ(4, 4)	2NU4 pe	$^{14}NH_3$
65		s^rP(7, 1)	NU3	$^{14}NH_3$
66	3286.82918(45)			
67	3287.31580(44)	a^rP(6, 4)	NU3	$^{14}NH_3$
68	3287.38331(44)			
69	3287.73560(22)	a^pP(8, 2)	NU3	$^{14}NH_3$
70	3287.84928(37)			
71	3288.45464(23)	s^rR(3, 2)	2NU4 pe	$^{14}NH_3$
72	3289.17024(35)			
73	3289.28947(24)	a^rR(3, 2)	2NU4 pe	$^{14}NH_3$
74	3289.43754(24)			
75	3289.51063(49)			
76	3289.82357(19)			
77	3290.40979(43)	a^pQ(4, 4)	2NU4 pe	$^{14}NH_3$
78	3291.71286(21)	a^pP(9, 5)	NU3	$^{14}NH_3$
79	3292.09119(21)	s^pP(9, 5)	NU3	$^{14}NH_3$
80	3292.16655(94)			
81	3293.44746(97)			
82	3293.55199(65)			
83	3293.87881(24)	a^rP(6, 3)	NU3	$^{14}NH_3$
84	3293.97932(75)			
85	3294.01212(19)	s^rP(7, 0)	NU3	$^{14}NH_3$
86	3294.18870(22)	s^rP(6, 3)	NU3	$^{14}NH_3$
87	3294.38527(98)			

- "LINE NO.": serial number identifying the spectral line in the map on the opposite page.
- "WAVENUMBER": experimental value.
- Uncertainty in the last digit of the wavenumber is given in parentheses.
- "pa" : parallel ; "pe" : perpendicular.

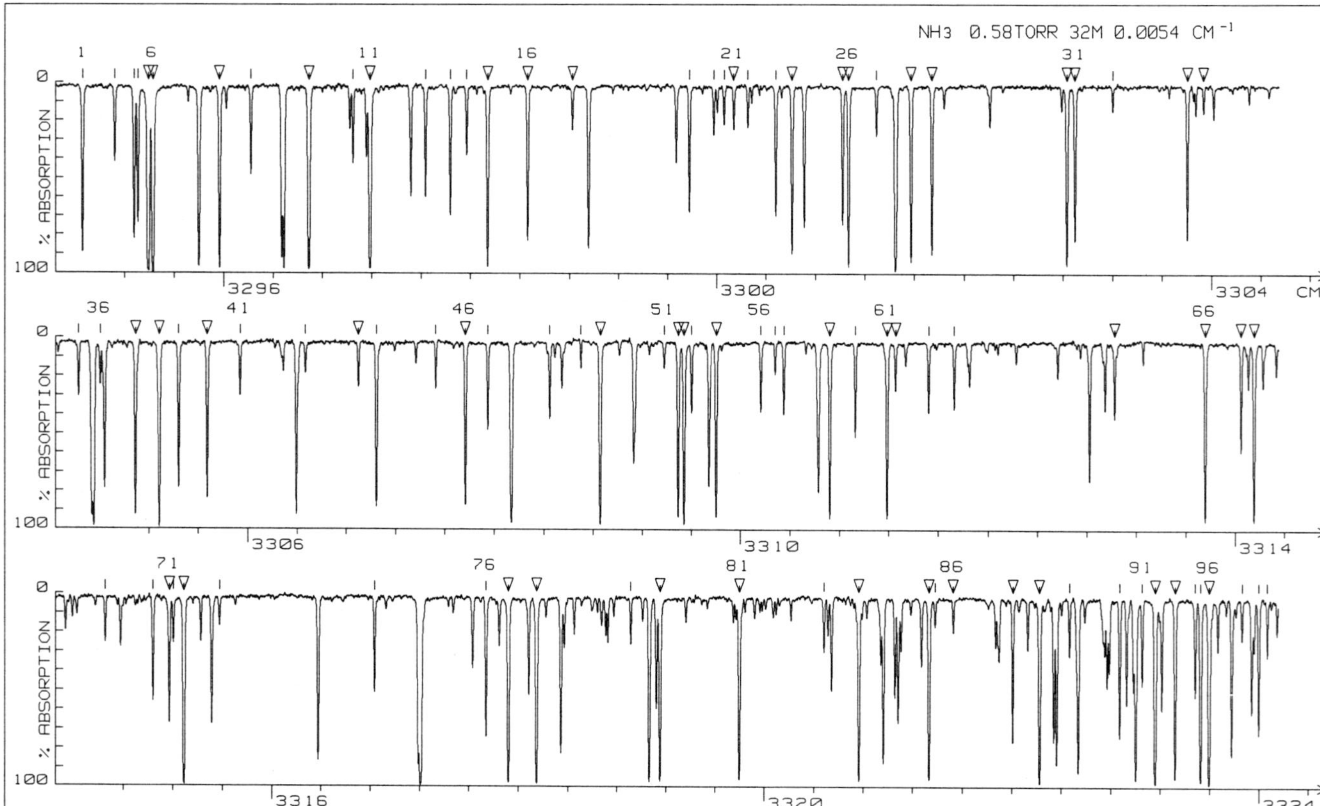

ISOTOPIC IDENTIFICATION – '∇' indicates $^{14}NH_3$; '|': unidentified.

LINE NO.	WAVENUMBER CM^{-1}, VAC.	ROTATIONAL ASSIGNMENTS	BAND	ISOTOPIC SPECIES
1	3294.85945(20)			
2	3295.11948(66)			
3	3295.27472(38)			
4	3295.30615(46)			
5	3295.38841(20)	aP(2, 0)	NU1	$^{14}NH_3$
6	3295.42604(22)	aP(2, 1)	NU1	$^{14}NH_3$
7	3295.96552(24)	aPP(8, 3)	NU3	$^{14}NH_3$
8	3296.21811(33)			
9	3296.69100(31)	a^{r}R(2, 0)	2NU4 pe	$^{14}NH_3$
9		a^{r}R(4, 3)	2NU4 pe	$^{14}NH_3$
10	3297.04921(32)			
11	3297.18647(19)	sP(2, 1)	NU1	$^{14}NH_3$
12	3297.63400(26)			
13	3297.83862(23)			
14	3297.96987(41)			
15	3298.13978(19)	aR(3, 3)	2NU4 pa	$^{14}NH_3$
15		aPQ(6, 4)	2NU4 pe	$^{14}NH_3$
16	3298.46158(23)	aPP(10, 8)	NU3	$^{14}NH_3$
17	3298.82457(64)	aPQ(5, 5)	2NU4 pe	$^{14}NH_3$
18	3299.77607(24)			
19	3299.97584(56)			
20	3300.05966(95)			
21	3300.13937(51)	sPQ(6, 4)	2NU4 pe	$^{14}NH_3$
22	3300.25364(49)			
23	3300.47939(24)			
24	3300.60922(19)	aR(3, 1)	2NU4 pa	$^{14}NH_3$
25	3301.01725(25)	s^{r}P(6, 2)	NU3	$^{14}NH_3$
26	3301.06620(24)	aPP(9, 6)	NU3	$^{14}NH_3$
27	3301.29320(79)			
28	3301.57306(20)	aR(3, 2)	2NU4 pa	$^{14}NH_3$
29	3301.74251(23)	sPP(7, 1)	NU3	$^{14}NH_3$
30	3302.83009(19)	sR(3, 3)	2NU4 pa	$^{14}NH_3$

LINE NO.	WAVENUMBER CM^{-1}, VAC.	ROTATIONAL ASSIGNMENTS	BAND	ISOTOPIC SPECIES
31	3302.89491(21)	s^{r}R(5, 4)	2NU4 pe	$^{14}NH_3$
32	3303.19825(104)			
33	3303.80213(24)	a^{r}R(5, 4)	2NU4 pe	$^{14}NH_3$
34	3303.93396(91)	s^{r}R(3, 1)	2NU4 pe	$^{14}NH_3$
34		sR(3, 1)	2NU4 pa	$^{14}NH_3$
35	3304.63217(42)			
36	3304.80708(93)			
37	3305.09083(29)	sPP(8, 4)	NU3	$^{14}NH_3$
38	3305.28420(63)	sR(3, 0)	2NU4 pa	$^{14}NH_3$
39	3305.44161(25)			
40	3305.67271(21)	a^{r}R(3, 1)	2NU4 pe	$^{14}NH_3$
41	3305.93582(43)			
42	3306.46777(85)			
43	3306.89853(63)	aPQ(6, 5)	2NU4 pe	$^{14}NH_3$
44	3307.04454(23)			
45	3307.52555(44)			
46	3307.76603(20)	a^{r}P(6, 1)	NU3	$^{14}NH_3$
47	3307.94720(48)			
48	3308.44549(31)			
49	3308.69914(53)			
50	3308.85606(25)	aPP(10, 9)	NU3	$^{14}NH_3$
51	3309.37703(114)			
52	3309.49150(23)	aPP(7, 2)	NU3	$^{14}NH_3$
53	3309.53937(19)	sPP(10, 9)	NU3	$^{14}NH_3$
54	3309.60066(36)			
55	3309.80227(19)	sPP(7, 2)	NU3	$^{14}NH_3$
56	3310.16498(58)			
57	3310.28004(79)			
58	3310.34931(57)			
59	3310.72031(19)	aPP(9, 7)	NU3	$^{14}NH_3$
60	3310.92900(31)			

- "LINE NO.": serial number identifying the spectral line in the map on the opposite page.
- "WAVENUMBER": experimental value.
- Uncertainty in the last digit of the wavenumber is given in parentheses.
- "pa" : parallel ; "pe" : perpendicular.

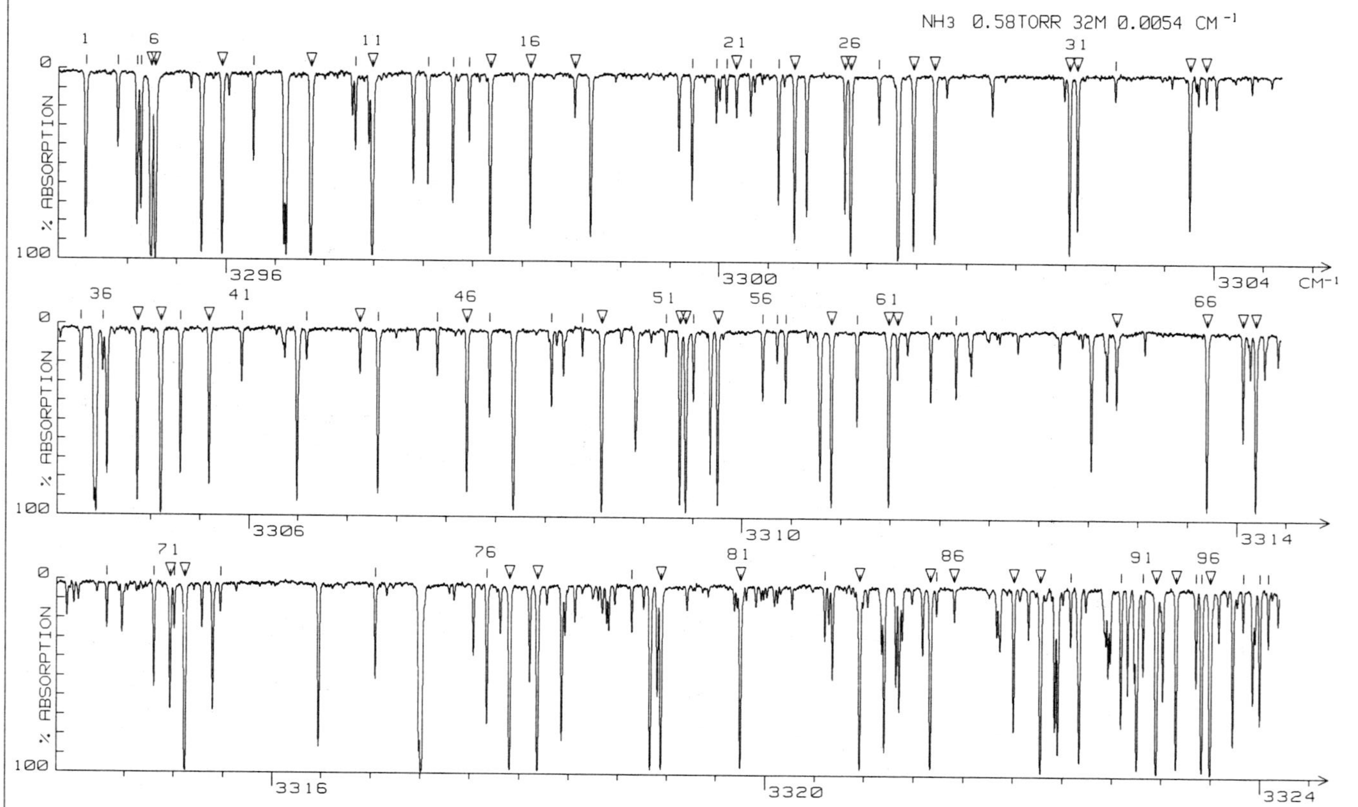

ISOTOPIC IDENTIFICATION – '∇' indicates $^{14}NH_3$; '|': unidentified.

LINE NO.	WAVENUMBER CM^{-1}, VAC.	ROTATIONAL ASSIGNMENTS	BAND	ISOTOPIC SPECIES
61	3311.18701(20)	s^{P}P(9, 7)	NU3	$^{14}NH_3$
62	3311.25523(78)	s^{P}Q(6, 6)	2NU4 pe	$^{14}NH_3$
63	3311.52233(34)			
64	3311.72550(61)			
65	3313.02191(37)	a^{P}Q(6, 6)	2NU4 pe	$^{14}NH_3$
66	3313.75535(19)	a^{P}P(8, 5)	NU3	$^{14}NH_3$
67	3314.04519(27)	a^{r}R(4, 2)	2NU4 pe	$^{14}NH_3$
68	3314.15066(24)	s^{P}P(8, 5)	NU3	$^{14}NH_3$
69	3314.65397(47)			
70	3315.03647(47)			
71	3315.17166(25)	a^{r}P(5, 3)	NU3	$^{14}NH_3$
72	3315.20472(62)			
73	3315.28944(19)	a^{r}P(6, 0)	NU3	$^{14}NH_3$
74	3315.57548(74)			
75	3316.83504(27)			
76	3317.73943(25)			
77	3317.92102(19)	a^{P}P(7, 3)	NU3	$^{14}NH_3$
78	3318.14639(25)	s^{P}P(7, 3)	NU3	$^{14}NH_3$
79	3318.91231(71)			
80	3319.15097(25)	a^{P}P(10, 10)	NU3	$^{14}NH_3$
81	3319.79905(19)	s^{P}P(10, 10)	NU3	$^{14}NH_3$
82	3320.48029(50)			
83	3320.76539(19)	a^{P}P(9, 8)	NU3	$^{14}NH_3$
84	3321.33397(19)	s^{P}P(9, 8)	NU3	$^{14}NH_3$
85	3321.38242(91)			
86	3321.52636(90)	s^{P}R(2, 2)	2NU4 pe	$^{14}NH_3$
87	3322.00668(22)	a^{r}P(5, 2)	NU3	$^{14}NH_3$
88	3322.22420(26)	aR(4, 3)	2NU4 pa	$^{14}NH_3$
89	3322.46580(37)			
90	3322.87059(23)			

LINE NO.	WAVENUMBER CM^{-1}, VAC.	ROTATIONAL ASSIGNMENTS	BAND	ISOTOPIC SPECIES
91	3323.05406(29)			
92	3323.15965(19)	a^{P}R(2, 2)	2NU4 pe	$^{14}NH_3$
92		a^{P}P(8, 6)	NU3	$^{14}NH_3$
93	3323.31871(19)	s^{P}P(6, 1)	NU3	$^{14}NH_3$
94	3323.48126(81)			
95	3323.52531(19)			
96	3323.59478(28)	s^{P}P(8, 6)	NU3	$^{14}NH_3$
97	3323.86059(77)			
98	3323.99615(26)			
99	3324.06735(36)			

- "LINE NO.": serial number identifying the spectral line in the map on the opposite page.
- "WAVENUMBER": experimental value.
- Uncertainty in the last digit of the wavenumber is given in parentheses.
- "pa" : parallel ; "pe" : perpendicular.

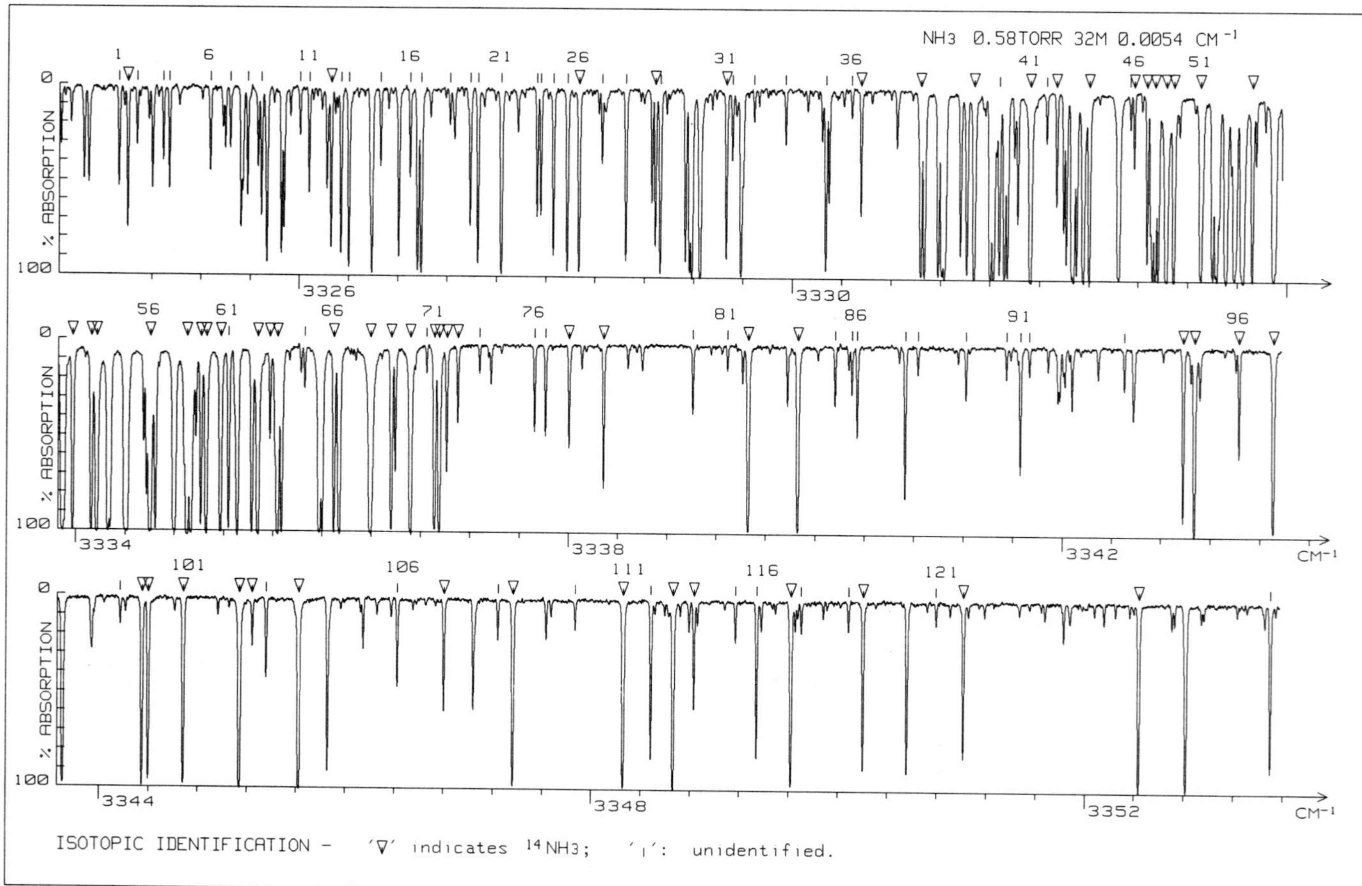

NH3 0.58TORR 32M 0.0054 CM⁻¹
% ABSORPTION
0
100
3326
3330
3334
3338
3342
3344
3348
3352
CM⁻¹
ISOTOPIC IDENTIFICATION - '∇' indicates ¹⁴NH3; '|': unidentified.

LINE NO.	WAVENUMBER CM^{-1}, VAC.	ROTATIONAL BAND ASSIGNMENTS	ISOTOPIC SPECIES
1	3324.53449(49)		
2	3324.60380(23)	sR(4, 4) 2NU4 pa	$^{14}NH_3$
3	3324.67855(58)		
4	3324.89449(31)		
5	3324.94423(26)		
6	3325.27349(56)		
7	3325.43505(51)		
8	3325.57678(28)		
9	3325.69067(25)		
10	3326.00496(59)		
11	3326.07806(30)		
12	3326.25628(21)	sR(4, 1) 2NU4 pa	$^{14}NH_3$
13	3326.33677(21)		
14	3326.40081(27)		
15	3326.65838(53)		
16	3326.89765(45)		
17	3326.98881(22)		
18	3327.22207(74)		
19	3327.38651(46)		
20	3327.44816(20)		
21	3327.63689(19)		
22	3327.92559(26)		
23	3327.95670(27)		
24	3328.05867(20)		
25	3328.17228(25)		
26	3328.27052(19)	sR(4, 3) 2NU4 pa	$^{14}NH_3$
27	3328.45789(40)		
28	3328.64947(20)		
29	3328.88833(21)	s^{r}R(4, 1) 2NU4 pe	$^{14}NH_3$
30	3328.92777(23)		

LINE NO.	WAVENUMBER CM^{-1}, VAC.	ROTATIONAL BAND ASSIGNMENTS	ISOTOPIC SPECIES
31	3329.46605(20)	s^{r}P(5, 1) NU3	$^{14}NH_3$
32	3329.51467(36)		
33	3329.69060(58)		
34	3329.94439(37)		
35	3330.26971(26)		
36	3330.48136(97)		
37	3330.55538(23)	a^{r}R(4, 1) 2NU4 pe	$^{14}NH_3$
38	3331.03887(22)	aQ(8, 6) NU1	$^{14}NH_3$
39	3331.47019(21)	aQ(8, 2) NU1	$^{14}NH_3$
40	3331.67428(26)		
41	3331.92461(21)	aQ(7, 6) NU1	$^{14}NH_3$
41		sPP(9, 9) NU3	$^{14}NH_3$
42	3332.05323(67)		
43	3332.13456(46)	aQ(9, 4) NU1	$^{14}NH_3$
44	3332.39912(22)	aQ(6, 4) NU1	$^{14}NH_3$
45	3332.73502(100)		
46	3332.76713(31)	sPR(3, 1) 2NU4 pe	$^{14}NH_3$
47	3332.87291(30)	sR(4, 2) 2NU4 pa	$^{14}NH_3$
48	3332.93884(23)	sQ(8, 6) NU1	$^{14}NH_3$
49	3333.02578(29)	aQ(6, 6) NU1	$^{14}NH_3$
50	3333.08702(20)	aQ(5, 3) NU1	$^{14}NH_3$
51	3333.30404(26)	aQ(5, 4) NU1	$^{14}NH_3$
51		sQ(8, 7) NU1	$^{14}NH_3$
52	3333.71894(19)	aQ(4, 2) NU1	$^{14}NH_3$
53	3333.97418(19)	sQ(6, 3) NU1	$^{14}NH_3$
54	3334.12409(22)	sQ(6, 4) NU1	$^{14}NH_3$
55	3334.16877(20)	aQ(4, 4) NU1	$^{14}NH_3$
56	3334.59895(21)	aQ(3, 3) NU1	$^{14}NH_3$

- "LINE NO.": serial number identifying the spectral line in the map on the opposite page.
- "WAVENUMBER": experimental value.
- Uncertainty in the last digit of the wavenumber is given in parentheses.
- "pa" : parallel ; "pe" : perpendicular.

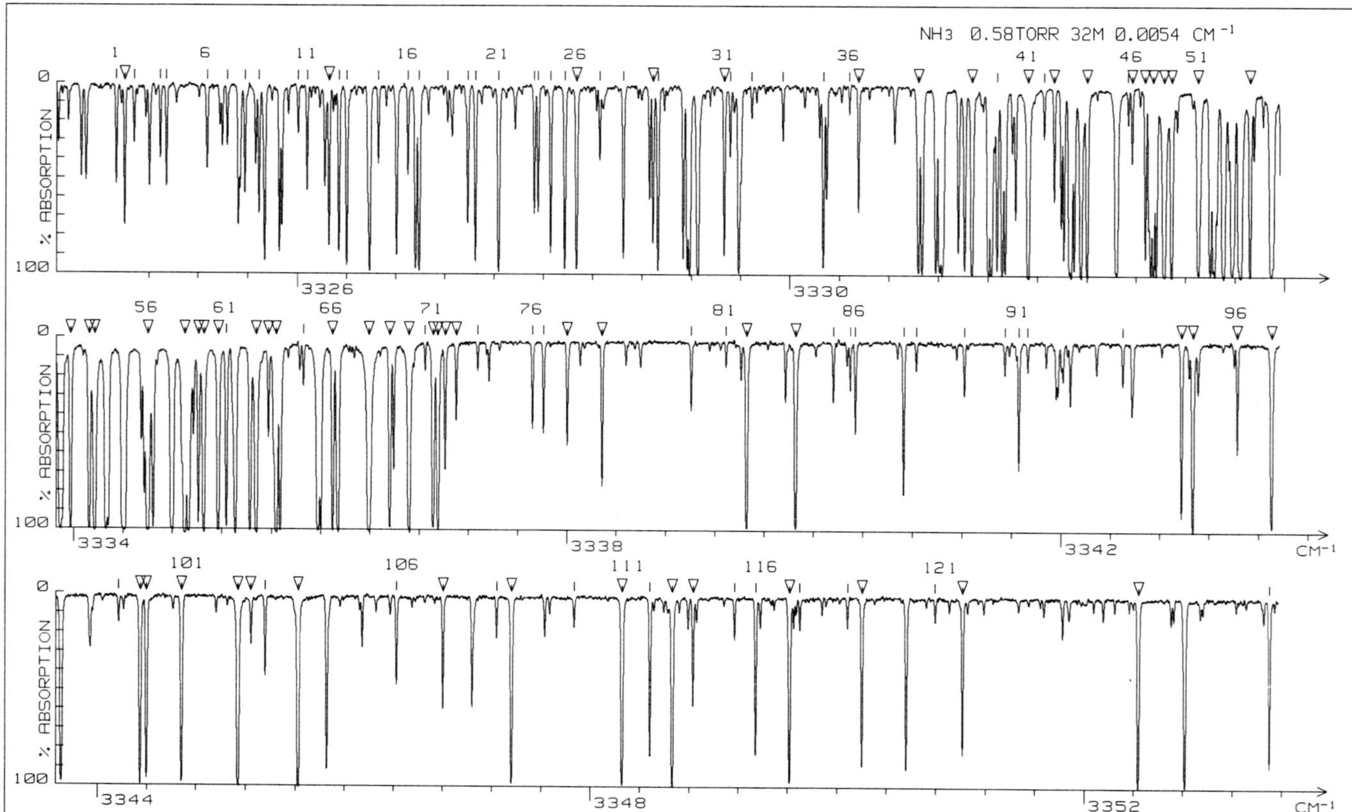

ISOTOPIC IDENTIFICATION – '∇' indicates $^{14}NH_3$; '|': unidentified.

LINE NO.	WAVENUMBER CM^{-1}, VAC.	ROTATIONAL BAND ASSIGNMENTS	ISOTOPIC SPECIES
57	3334.90191(21)	sQ(9, 6) NU1	$^{14}NH_3$
57		sQ(6, 6) NU1	$^{14}NH_3$
57		aPR(3, 1) 2NU4 pe	$^{14}NH_3$
57		aQ(10, 6) NU1	$^{14}NH_3$
58	3335.01342(28)	aQ(10, 8) NU1	$^{14}NH_3$
58		sQ(9, 8) NU1	$^{14}NH_3$
59	3335.05626(21)	sQ(5, 4) NU1	$^{14}NH_3$
60	3335.17187(21)	sQ(9, 7) NU1	$^{14}NH_3$
60		aQ(1, 1) NU1	$^{14}NH_3$
61	3335.23741(19)		
62	3335.47622(20)	aQ(10, 10) NU1	$^{14}NH_3$
62		sQ(5, 5) NU1	$^{14}NH_3$
63	3335.57439(34)	sQ(9, 4) NU1	$^{14}NH_3$
64	3335.63634(20)	sQ(4, 3) NU1	$^{14}NH_3$
65	3335.85710(79)		
66	3336.09512(19)	sPP(7, 5) NU3	$^{14}NH_3$
67	3336.39092(25)	sQ(3, 3) NU1	$^{14}NH_3$
68	3336.56083(19)	sQ(2, 1) NU1	$^{14}NH_3$
69	3336.71669(27)	sQ(2, 2) NU1	$^{14}NH_3$
70	3336.84657(82)		
71	3336.91122(23)	s^{r}P(5, 0) NU3	$^{14}NH_3$
72	3336.95229(32)	sQ(1, 1) NU1	$^{14}NH_3$
73	3337.01178(25)	sQ(10, 3) NU1	$^{14}NH_3$
74	3337.10124(47)	aQ(10, 3) NU1	$^{14}NH_3$
75	3337.27576(88)		
76	3337.72056(28)		
77	3337.80931(45)		
78	3338.00155(76)	s^{r}R(5, 2) 2NU4 pe	$^{14}NH_3$
79	3338.28526(21)	aR(5, 5) 2NU4 pa	$^{14}NH_3$
80	3339.00513(48)		

LINE NO.	WAVENUMBER CM^{-1}, VAC.	ROTATIONAL BAND ASSIGNMENTS	ISOTOPIC SPECIES
81	3339.28917(109)		
82	3339.45797(19)	aPP(6, 3) NU3	$^{14}NH_3$
83	3339.85630(21)	sPP(6, 3) NU3	$^{14}NH_3$
84	3340.15940(34)		
85	3340.29627(42)		
86	3340.33971(44)		
87	3340.72789(22)		
88	3340.82965(95)		
89	3341.21774(55)		
90	3341.54538(80)		
91	3341.65848(24)		
92	3341.73066(59)		
93	3342.49831(44)		
94	3342.97846(23)	aR(5, 4) 2NU4 pa	$^{14}NH_3$
95	3343.06961(19)	aPP(8, 8) NU3	$^{14}NH_3$
95		a^{r}P(4, 2) NU3	$^{14}NH_3$
96	3343.42824(29)	s^{r}P(4, 2) NU3	$^{14}NH_3$
97	3343.70587(19)	sPP(8, 8) NU3	$^{14}NH_3$
98	3344.17059(95)		
99	3344.34528(24)	aPP(5, 1) NU3	$^{14}NH_3$
100	3344.39631(25)	aR(5, 1) 2NU4 pa	$^{14}NH_3$
101	3344.68438(19)	sPP(5, 1) NU3	$^{14}NH_3$
102	3345.13846(21)	aPP(7, 6) NU3	$^{14}NH_3$
103	3345.24235(43)	sPR(3, 2) 2NU4 pe	$^{14}NH_3$
104	3345.35621(41)		
105	3345.62161(20)	sPP(7, 6) NU3	$^{14}NH_3$
106	3346.42378(80)		
107	3346.80304(34)	s^{r}R(6, 3) 2NU4 pe	$^{14}NH_3$
108	3347.23865(55)		
109	3347.35977(19)	aR(5, 3) 2NU4 pa	$^{14}NH_3$
110	3347.86594(75)		

- "LINE NO.": serial number identifying the spectral line in the map on the opposite page.
- "WAVENUMBER": experimental value.
- Uncertainty in the last digit of the wavenumber is given in parentheses.
- "pa" : parallel ; "pe" : perpendicular.

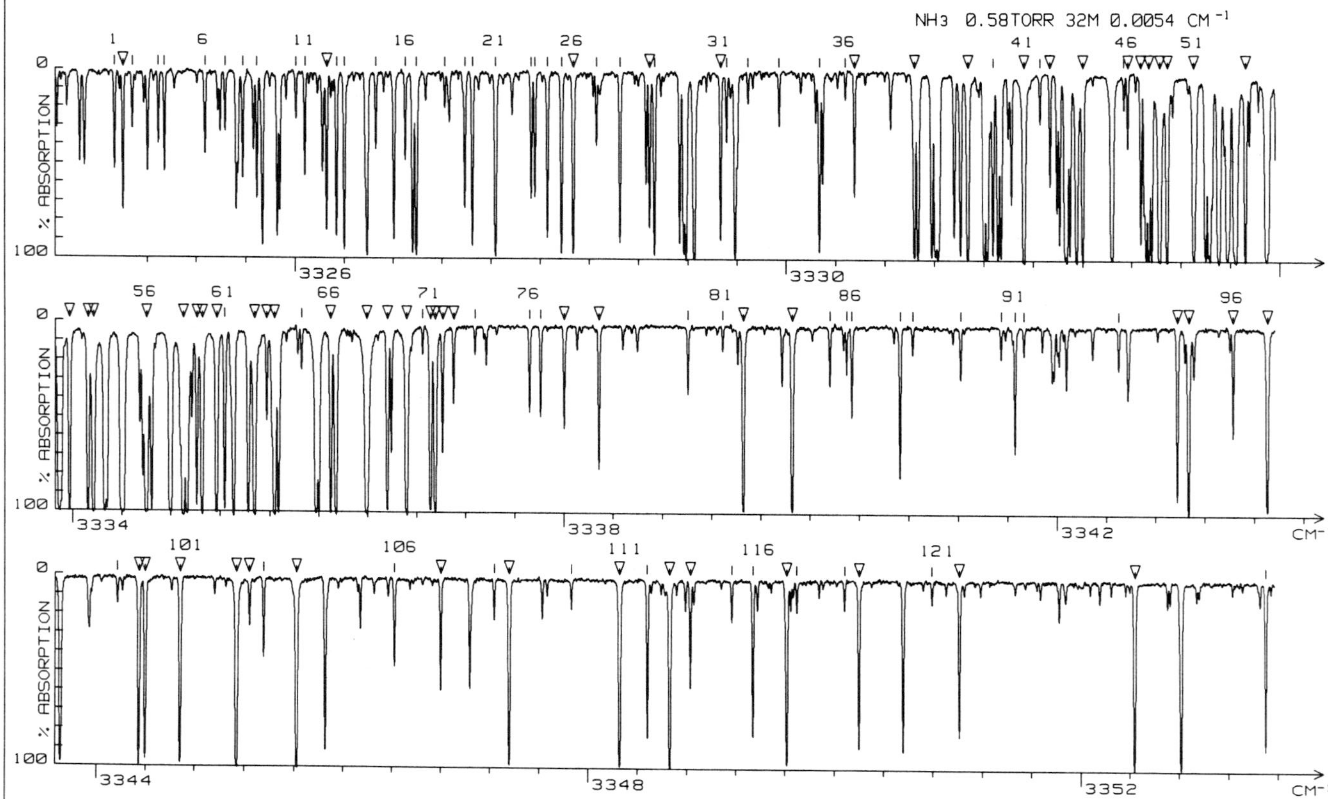

ISOTOPIC IDENTIFICATION – '∇' indicates $^{14}NH_3$; 'ı': unidentified.

LINE NO.	WAVENUMBER CM^{-1}, VAC.	ROTATIONAL BAND ASSIGNMENTS	ISOTOPIC SPECIES
111	3348.25817 (19)	$^aP^P$(6, 4) NU3	$^{14}NH_3$
112	3348.48271 (22)		
113	3348.66317 (19)	$^sP^P$(6, 4) NU3	$^{14}NH_3$
114	3348.83019 (24)	$^{a^r}R$(6, 3) 2NU4 pe	$^{14}NH_3$
115	3349.16917 (64)		
116	3349.34170 (20)		
117	3349.61790 (24)	^sR(5, 0) 2NU4 pa	$^{14}NH_3$
118	3349.70039 (77)		
119	3350.08289 (54)		
120	3350.20311 (21)	$^{a^r}P$(4, 1) NU3	$^{14}NH_3$
121	3350.79363 (95)		
122	3351.01445 (20)	^sR(5, 4) 2NU4 pa	$^{14}NH_3$
123	3352.43855 (18)	$^aP^P$(5, 2) NU3	$^{14}NH_3$
124	3353.49854 (26)		

- "LINE NO.": serial number identifying the spectral line in the map on the opposite page.
- "WAVENUMBER": experimental value.
- Uncertainty in the last digit of the wavenumber is given in parentheses.
- "pa" : parallel ; "pe" : perpendicular.

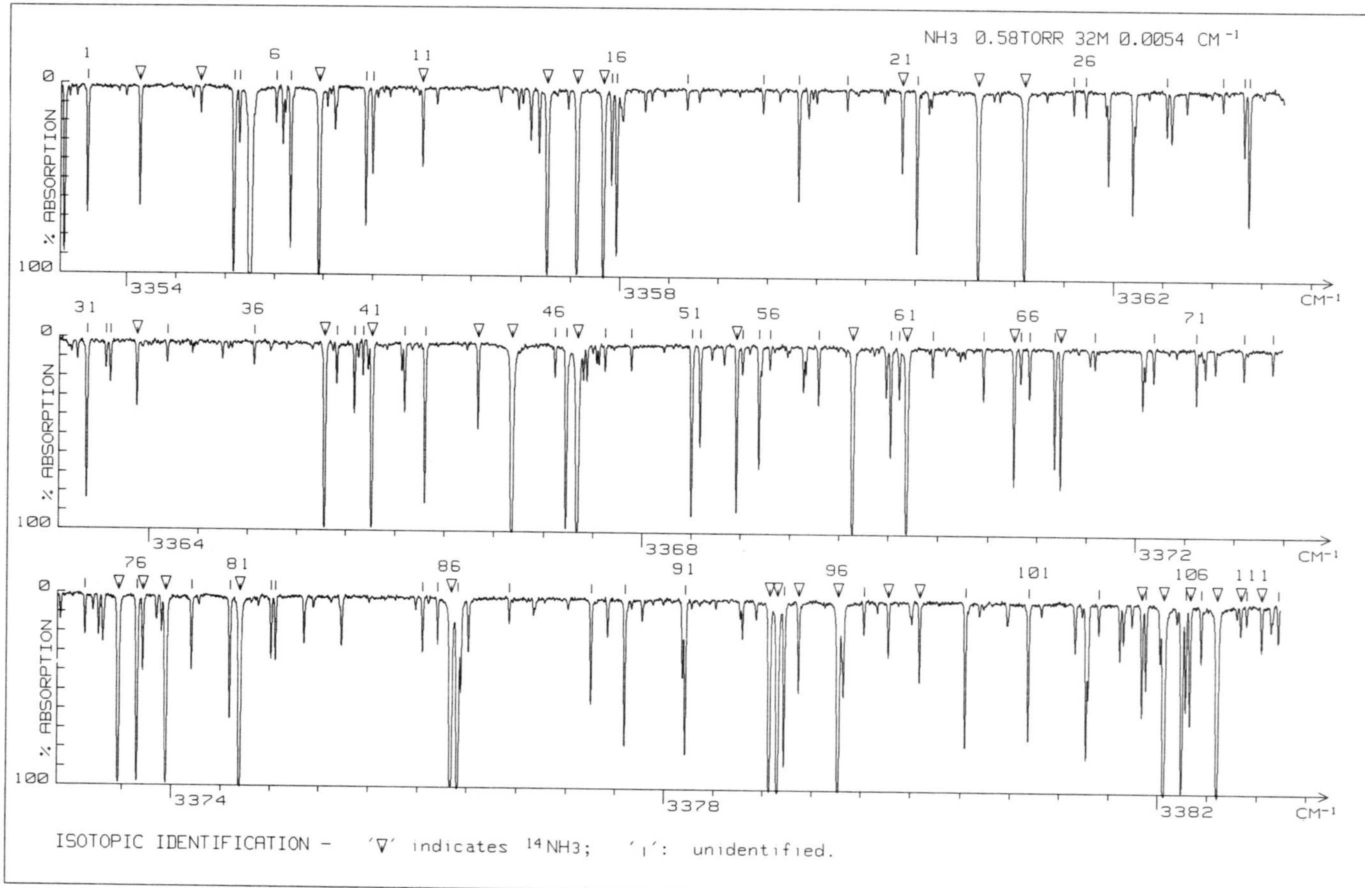
NH3 0.58TORR 32M 0.0054 CM -1
% ABSORPTION
0
100
3354
3358
3362
3364
3368
3372
3374
3378
3382
CM-1
ISOTOPIC IDENTIFICATION - '∇' indicates 14NH3; 'ı': unidentified.

LINE NO.	WAVENUMBER CM^{-1}, VAC.	ROTATIONAL BAND ASSIGNMENTS	ISOTOPIC SPECIES
1	3353.68533 (23)		
2	3354.10517 (31)	$s^{r}R$(5, 1) 2NU4 pe	$^{14}NH_3$
3	3354.59849 (86)	$s^{P}R$(3, 3) 2NU4 pe	$^{14}NH_3$
4	3354.86660 (24)		
5	3354.91050 (58)		
6	3355.20853 (64)		
7	3355.33052 (23)		
8	3355.56616 (20)	$s^{P}P$(7, 7) NU3	$^{14}NH_3$
9	3355.94421 (22)		
10	3355.99859 (40)		
11	3356.40185 (26)	$a^{r}R$(5, 1) 2NU4 pe	$^{14}NH_3$
12	3357.40657 (19)	$a^{P}P$(6, 5) NU3	$^{14}NH_3$
13	3357.64938 (20)	$a^{r}P$(4, 0) NU3	$^{14}NH_3$
14	3357.86309 (21)	$s^{P}P$(6, 5) NU3	$^{14}NH_3$
15	3357.92942 (41)		
16	3357.97083 (20)		
17	3358.54307 (95)		
18	3359.16103 (89)		
19	3359.45189 (36)		
20	3359.84606 (83)		
21	3360.29246 (38)	$a^{P}R$(4, 1) 2NU4 pe	$^{14}NH_3$
22	3360.41268 (21)		
23	3360.90180 (20)	$a^{P}P$(5, 3) NU3	$^{14}NH_3$
24	3361.27574 (21)	$s^{P}P$(5, 3) NU3	$^{14}NH_3$
25	3361.67343 (52)		
26	3361.77079 (83)		
27	3362.42846 (37)		
28	3362.88126 (108)		
29	3363.05602 (40)		
30	3363.09350 (22)		
31	3363.49709 (20)		
32	3363.64901 (69)		
33	3363.68413 (49)		
34	3363.89783 (31)	$s^{r}R$(6, 2) 2NU4 pe	$^{14}NH_3$
35	3364.14589 (91)		
36	3364.84831 (76)		
37	3365.42890 (19)	$a^{P}P$(4, 1) NU3	$^{14}NH_3$
38	3365.52376 (46)		
39	3365.66758 (43)		
40	3365.73747 (38)		
41	3365.80929 (19)	$s^{P}P$(4, 1) NU3	$^{14}NH_3$
42	3366.07361 (41)		
43	3366.23913 (19)		
44	3366.67091 (34)	$a^{r}R$(6, 2) 2NU4 pe	$^{14}NH_3$
45	3366.94368 (20)	$a^{P}P$(6, 6) NU3	$^{14}NH_3$
46	3367.28623 (63)		
47	3367.37771 (19)		
48	3367.47069 (20)	$s^{P}P$(6, 6) NU3	$^{14}NH_3$
49	3367.69405 (62)		
50	3367.90860 (89)		
51	3368.39925 (21)		
52	3368.46741 (24)		
53	3368.76750 (21)	$s^{r}R$(5, 0) 2NU4 pe	$^{14}NH_3$
54	3368.81397 (84)		
55	3368.94862 (57)		
56	3369.03587 (82)		
57	3369.43123 (40)		
58	3369.71223 (19)	$a^{P}P$(5, 4) NU3	$^{14}NH_3$
59	3370.01828 (37)		
60	3370.08753 (57)		

- "LINE NO.": serial number identifying the spectral line in the map on the opposite page.
- "WAVENUMBER": experimental value.
- Uncertainty in the last digit of the wavenumber is given in parentheses.
- "pa" : parallel ; "pe" : perpendicular.

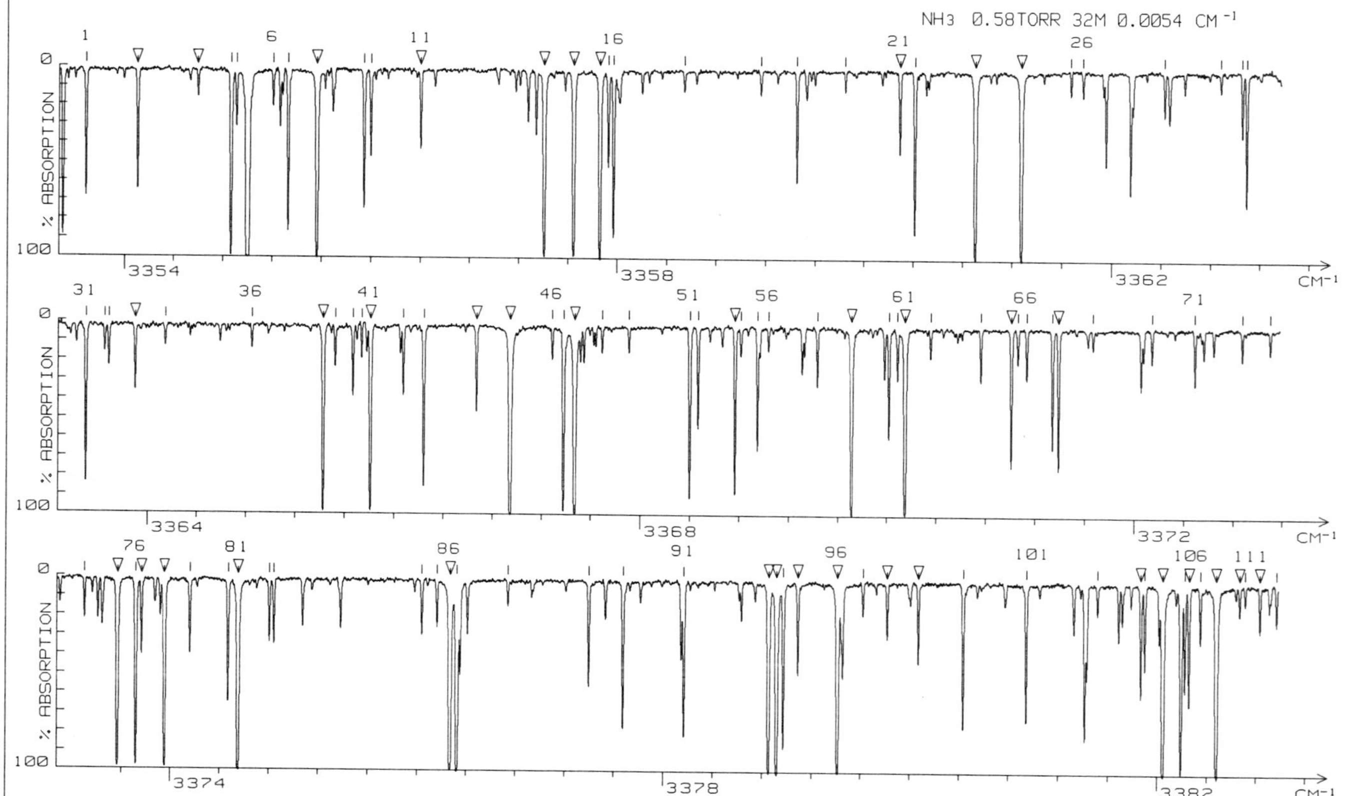

ISOTOPIC IDENTIFICATION – '∇' indicates $^{14}NH_3$; 'I': unidentified.

LINE NO.	WAVENUMBER CM^{-1}, VAC.	ROTATIONAL ASSIGNMENTS	BAND	ISOTOPIC SPECIES
61	3370.14972(20)	$s^{P}P$(5, 4)	NU3	$^{14}NH_3$
62	3370.35509(73)			
63	3370.76070(35)			
64	3371.01031(22)	$a^{r}P$(3, 1)	NU3	$^{14}NH_3$
65	3371.06297(76)			
66	3371.13405(37)			
67	3371.34096(29)			
68	3371.38888(21)	$s^{r}P$(3, 1)	NU3	$^{14}NH_3$
69	3371.66461(99)			
70	3372.14189(52)			
71	3372.48766(33)			
72	3372.87046(65)			
73	3373.10261(82)			
74	3373.30016(49)			
75	3373.56814(25)	$a^{P}P$(4, 2)	NU3	$^{14}NH_3$
76	3373.71825(24)			
77	3373.76521(39)	$a^{r}Q$(9, 8)	NU3	$^{14}NH_3$
78	3373.95422(21)	$s^{P}P$(4, 2)	NU3	$^{14}NH_3$
79	3374.15979(30)			
80	3374.47052(27)			
81	3374.55193(20)	aR(1, 1)	NU1	$^{14}NH_3$
82	3374.80582(37)			
83	3374.84263(29)			
84	3376.03821(47)			
85	3376.16149(55)			
86	3376.27037(20)	sR(1, 0)	NU1	$^{14}NH_3$
87	3376.32653(20)			
88	3376.73941(68)			
89	3377.40061(24)			
90	3377.67590(24)			
91	3378.16765(40)			
92	3378.85286(18)	$s^{r}P$(3, 0)	NU3	$^{14}NH_3$
93	3378.91826(23)	$a^{P}P$(5, 5)	NU3	$^{14}NH_3$
94	3378.97174(23)			
95	3379.08930(37)	$a^{r}Q$(9, 7)	NU3	$^{14}NH_3$
96	3379.41170(21)	$s^{P}P$(5, 5)	NU3	$^{14}NH_3$
97	3379.62213(51)			
98	3379.81949(34)	$s^{P}R$(4, 3)	2NU4 pe	$^{14}NH_3$
99	3380.07130(26)	$s^{r}R$(6, 1)	2NU4 pe	$^{14}NH_3$
100	3380.43484(21)			
101	3380.94579(21)			
102	3381.51684(65)			
103	3381.86904(26)	$a^{r}Q$(8, 7)	NU3	$^{14}NH_3$
104	3381.90110(35)			
105	3382.04673(20)	$a^{P}P$(4, 3)	NU3	$^{14}NH_3$
106	3382.22325(28)			
107	3382.25686(28)	$s^{r}Q$(8, 7)	NU3	$^{14}NH_3$
108	3382.34764(32)			
109	3382.47850(20)	$s^{P}P$(4, 3)	NU3	$^{14}NH_3$
110	3382.66679(71)	$a^{P}R$(4, 3)	2NU4 pe	$^{14}NH_3$
111	3382.71211(83)			
112	3382.83252(36)	$s^{P}R$(5, 1)	2NU4 pe	$^{14}NH_3$
113	3382.97043(66)			

- "LINE NO.": serial number identifying the spectral line in the map on the opposite page.
- "WAVENUMBER": experimental value.
- Uncertainty in the last digit of the wavenumber is given in parentheses.
- "pa" : parallel ; "pe" : perpendicular.

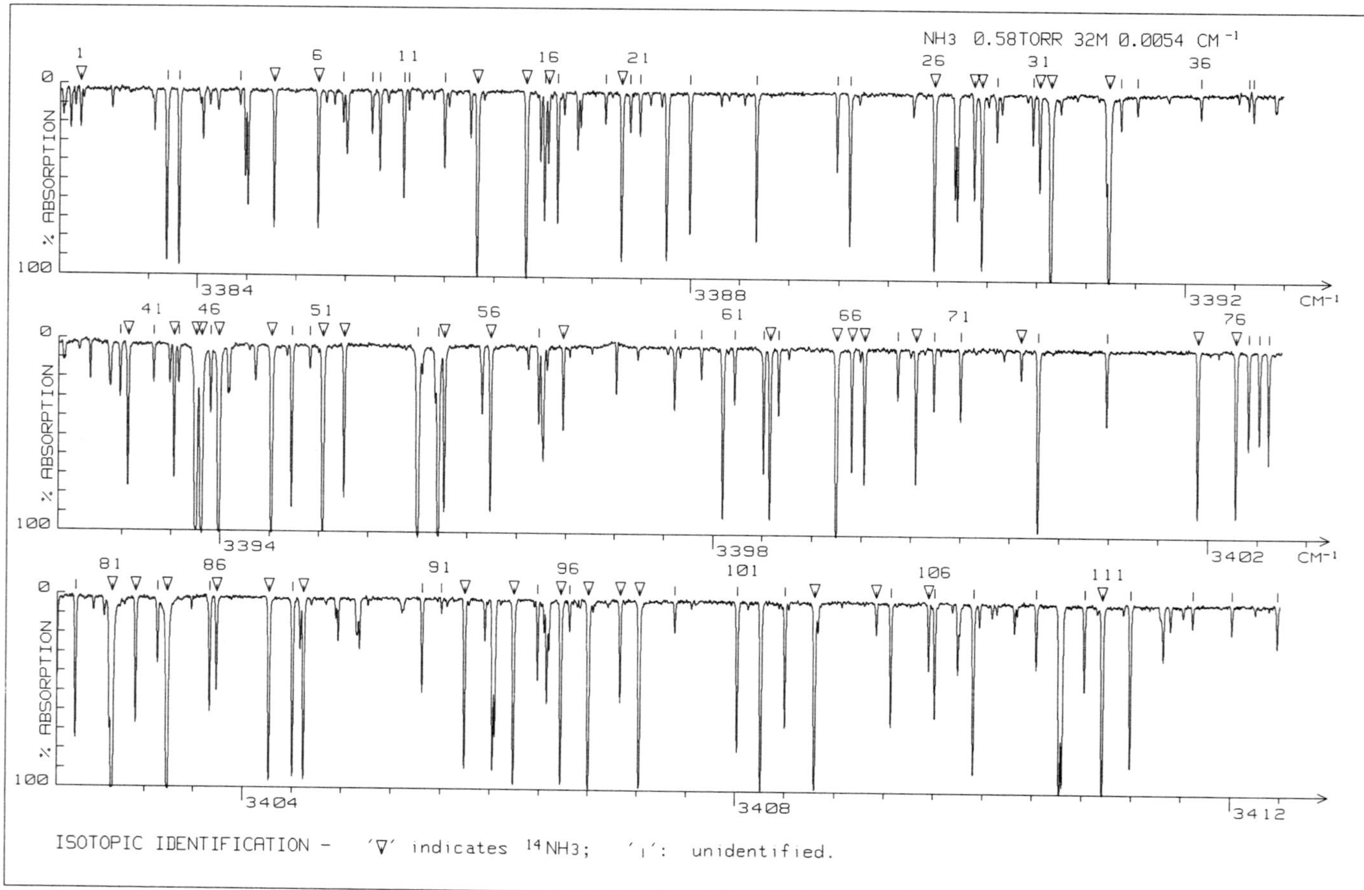
NH3 0.58TORR 32M 0.0054 CM⁻¹
% ABSORPTION
0
100
3384
3388
3392
CM⁻¹
3394
3398
3402
3404
3408
3412
ISOTOPIC IDENTIFICATION - '∇' indicates ¹⁴NH3; '|': unidentified.

LINE NO.	WAVENUMBER CM^{-1}, VAC.	ROTATIONAL BAND ASSIGNMENTS	ISOTOPIC SPECIES
1	3383.05078(51)	a^rR(6, 1) 2NU4 pe	$^{14}NH_3$
2	3383.75045(25)		
3	3383.85205(19)		
4	3384.34402(103)		
5	3384.61992(21)	a^rQ(9, 6) NU3	$^{14}NH_3$
6	3384.97901(20)	s^rQ(9, 6) NU3	$^{14}NH_3$
7	3385.18095(76)		
8	3385.41476(36)		
9	3385.47848(25)		
10	3385.67088(32)		
11	3385.70975(48)		
12	3386.00035(32)		
13	3386.26509(19)	a^pP(3, 1) NU3	$^{14}NH_3$
14	3386.66061(19)	s^pP(3, 1) NU3	$^{14}NH_3$
15	3386.80863(36)		
16	3386.83975(39)	a^pR(5, 1) 2NU4 pe	$^{14}NH_3$
17	3386.91444(21)		
18	3387.30734(44)		
19	3387.44135(19)	a^rQ(8, 6) NU3	$^{14}NH_3$
20	3387.51002(41)		
21	3387.58809(39)		
22	3387.99466(21)		
23	3388.53685(21)		
24	3389.18206(26)		
25	3389.28880(26)		
26	3389.97365(19)	a^rQ(7, 6) NU3	$^{14}NH_3$
27	3390.29599(25)	a^rQ(9, 5) NU3	$^{14}NH_3$
28	3390.35709(29)	s^rQ(7, 6) NU3	$^{14}NH_3$
29	3390.47984(45)		
30	3390.76854(33)		
31	3390.82440(24)	s^rQ(9, 5) NU3	$^{14}NH_3$
32	3390.91605(23)	a^pP(4, 4) NU3	$^{14}NH_3$
33	3391.38457(21)	s^pP(4, 4) NU3	$^{14}NH_3$
34	3391.47847(45)		
35	3391.61180(86)		
36	3392.12172(70)		
37	3392.50738(59)		
38	3392.54569(68)		
39	3393.18394(41)		
40	3393.24874(47)	a^rQ(8, 5) NU3	$^{14}NH_3$
41	3393.45527(48)		
42	3393.62560(22)	s^rQ(8, 5) NU3	$^{14}NH_3$
43	3393.66333(52)		
44	3393.80423(20)	aR(2, 0) NU1	$^{14}NH_3$
45	3393.84921(20)	aR(2, 1) NU1	$^{14}NH_3$
46	3393.92256(34)		
47	3393.98981(21)	aR(2, 2) NU1	$^{14}NH_3$
48	3394.41537(19)	a^pP(3, 2) NU3	$^{14}NH_3$
49	3394.57807(21)		
50	3394.72705(89)		
51	3394.83324(19)	s^pP(3, 2) NU3	$^{14}NH_3$
52	3395.00632(19)	s^pR(5, 2) 2NU4 pe	$^{14}NH_3$
53	3395.59922(20)		
54	3395.76558(21)		
55	3395.81604(21)	a^rQ(7, 5) NU3	$^{14}NH_3$
56	3396.18769(21)	s^rQ(7, 5) NU3	$^{14}NH_3$
57	3396.57662(37)		
58	3396.77660(25)	s^rQ(9, 4) NU3	$^{14}NH_3$
59	3397.68514(34)		
60	3397.90103(43)		

- "LINE NO.": serial number identifying the spectral line in the map on the opposite page.
- "WAVENUMBER": experimental value.
- Uncertainty in the last digit of the wavenumber is given in parentheses.
- "pa" : parallel ; "pe" : perpendicular.

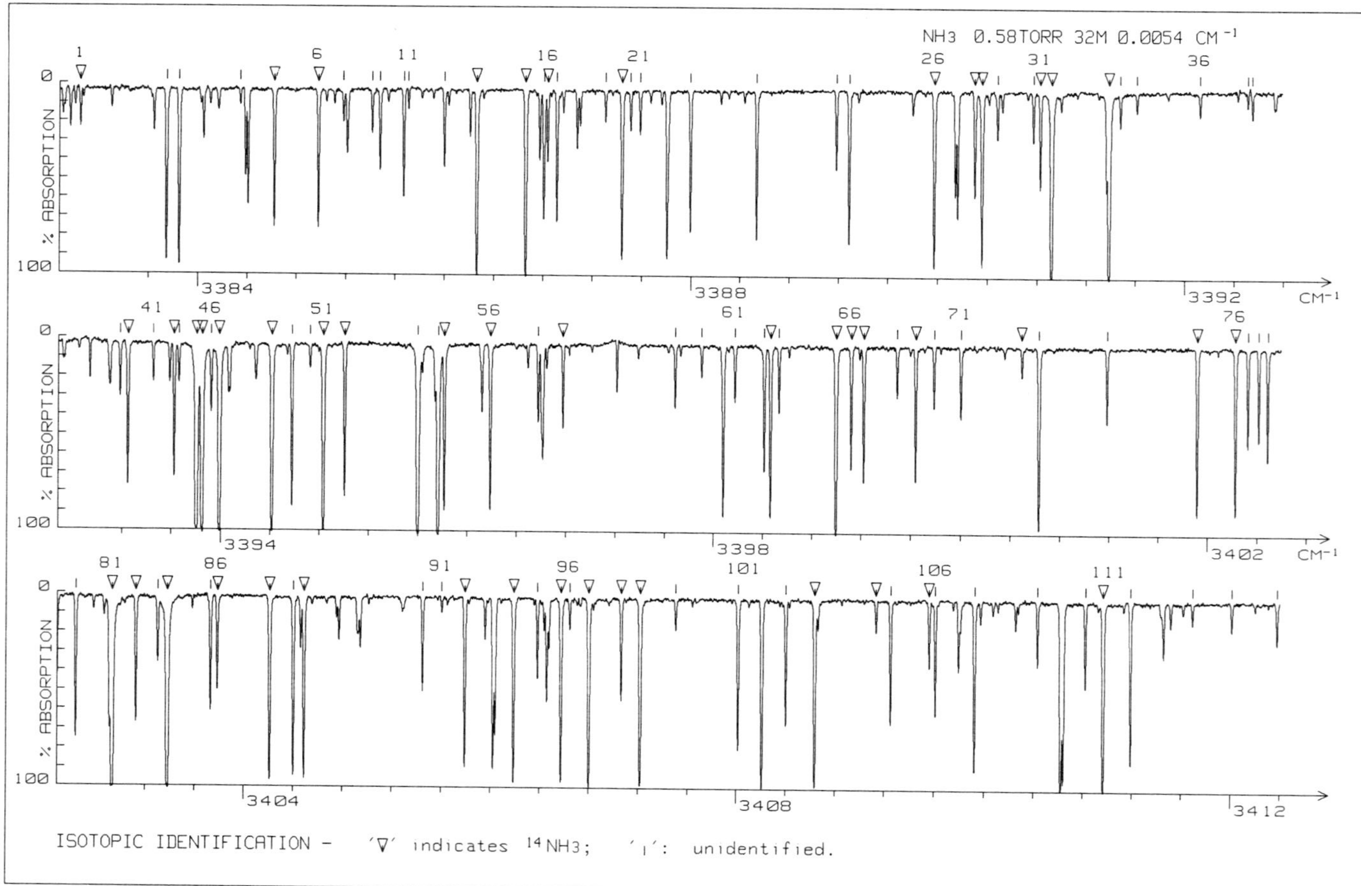
NH3 0.58TORR 32M 0.0054 CM⁻¹
% ABSORPTION
0
100
CM⁻¹
1
6
11
16
21
26
31
36
41
46
51
56
61
66
71
76
81
86
91
96
101
106
111
3384
3388
3392
3394
3398
3402
3404
3408
3412
ISOTOPIC IDENTIFICATION - '∇' indicates ¹⁴NH3; '|': unidentified.

LINE NO.	WAVENUMBER CM^{-1}, VAC.	ROTATIONAL BAND ASSIGNMENTS	ISOTOPIC SPECIES
61	3398.17493(36)		
62	3398.41257(21)		
63	3398.46295(19)	$s^{r}Q$(6, 5) NU3	$^{14}NH_3$
64	3398.53247(35)		
65	3398.99547(19)	$a^{r}P$(2, 0) NU3	$^{14}NH_3$
66	3399.11968(20)	$a^{r}R$(6, 0) 2NU4 pe	$^{14}NH_3$
67	3399.22284(21)	$a^{r}Q$(8, 4) NU3	$^{14}NH_3$
68	3399.49157(41)		
69	3399.64016(21)	$s^{r}Q$(8, 4) NU3	$^{14}NH_3$
70	3399.78645(29)		
71	3400.00100(27)		
72	3400.49225(58)	$a^{P}R$(5, 2) 2NU4 pe	$^{14}NH_3$
73	3400.63027(18)		
74	3401.18241(27)		
75	3401.91689(19)	$a^{r}Q$(7, 4) NU3	$^{14}NH_3$
76	3402.22540(19)	$s^{r}Q$(7, 4) NU3	$^{14}NH_3$
77	3402.32551(22)		
78	3402.41229(32)		
79	3402.48866(22)		
80	3402.64139(20)		
81	3402.93561(26)	$a^{P}P$(3, 3) NU3	$^{14}NH_3$
82	3403.12729(21)	$s^{r}Q$(9, 3) NU3	$^{14}NH_3$
83	3403.30485(47)		
84	3403.38490(26)	$s^{P}P$(3, 3) NU3	$^{14}NH_3$
85	3403.73183(28)		
86	3403.78687(32)	$s^{P}R$(5, 3) 2NU4 pe	$^{14}NH_3$
87	3404.21189(21)	$a^{r}Q$(6, 4) NU3	$^{14}NH_3$
88	3404.40303(21)		
89	3404.49189(23)	$s^{r}Q$(6, 4) NU3	$^{14}NH_3$
90	3405.45109(23)		
91	3405.60846(105)		
92	3405.79722(19)	$a^{r}Q$(8, 3) NU3	$^{14}NH_3$
93	3406.19215(19)	$a^{r}Q$(5, 4) NU3	$^{14}NH_3$
94	3406.38616(24)		
95	3406.57626(20)	$s^{r}Q$(5, 4) NU3	$^{14}NH_3$
96	3406.64752(55)		
97	3406.80360(19)	$a^{P}P$(2, 1) NU3	$^{14}NH_3$
98	3407.06410(22)	$a^{P}R$(5, 3) 2NU4 pe	$^{14}NH_3$
99	3407.21901(19)	$s^{P}P$(2, 1) NU3	$^{14}NH_3$
100	3407.50564(49)		
101	3408.01594(21)		
102	3408.40330(20)		
103	3408.63908(24)	$s^{r}Q$(7, 3) NU3	$^{14}NH_3$
104	3409.13687(57)	$s^{P}R$(6, 1) 2NU4 pe	$^{14}NH_3$
105	3409.25483(21)		
106	3409.56349(33)	$a^{r}Q$(9, 2) NU3	$^{14}NH_3$
107	3409.61289(21)		
108	3409.92490(19)		
109	3410.43348(33)		
110	3410.82200(24)		
111	3410.96790(19)	$s^{r}Q$(6, 3) NU3	$^{14}NH_3$
112	3411.19206(20)		
113	3411.69348(80)		
114	3412.01017(70)		
115	3412.37908(48)		

- "LINE NO.": serial number identifying the spectral line in the map on the opposite page.
- "WAVENUMBER": experimental value.
- Uncertainty in the last digit of the wavenumber is given in parentheses.
- "pa" : parallel ; "pe" : perpendicular.

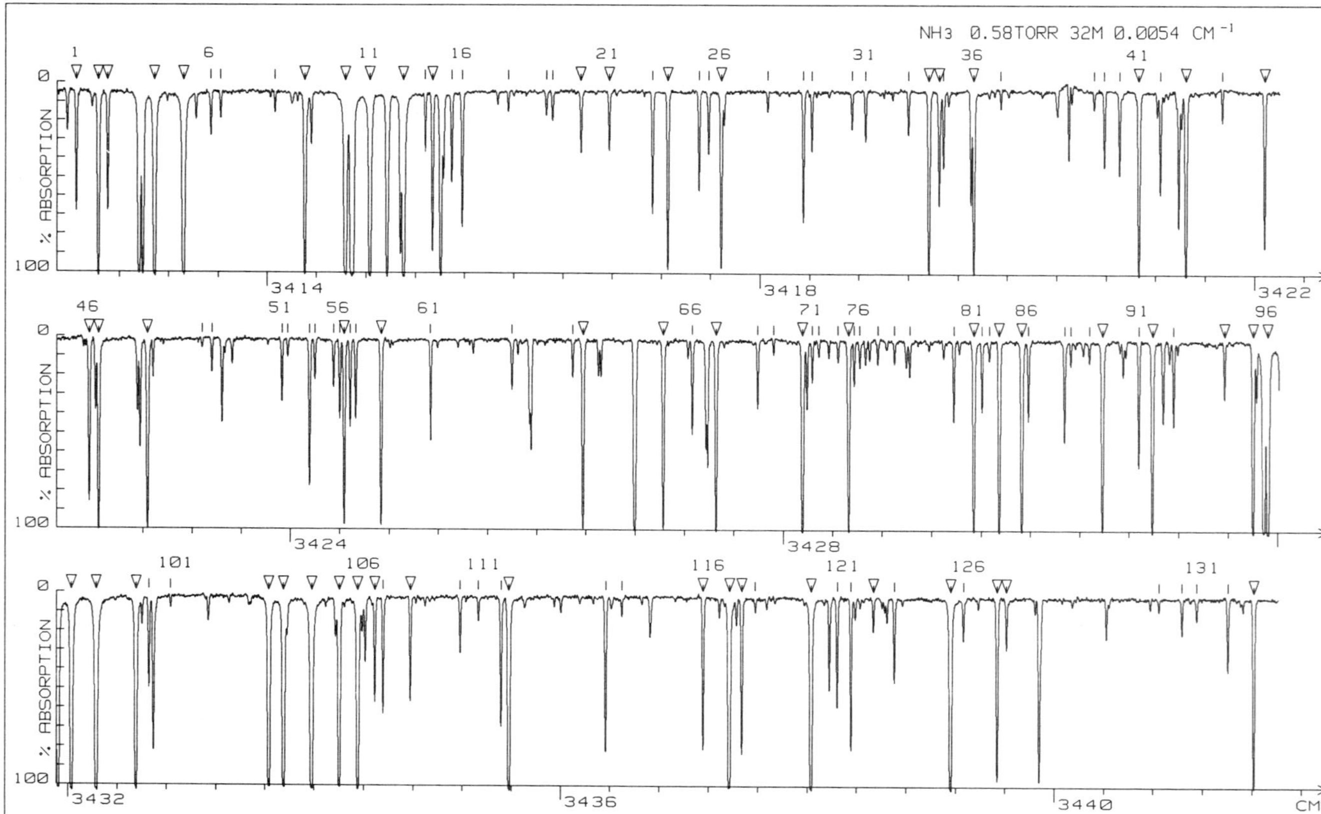

ISOTOPIC IDENTIFICATION – '∇' indicates $^{14}NH_3$; '|': unidentified.

LINE NO.	WAVENUMBER CM^{-1}, VAC.	ROTATIONAL BAND ASSIGNMENTS	ISOTOPIC SPECIES	LINE NO.	WAVENUMBER CM^{-1}, VAC.	ROTATIONAL BAND ASSIGNMENTS	ISOTOPIC SPECIES
1	3412.45118(20)	a^rQ(8, 2) NU3	$^{14}NH_3$	31	3418.85585(38)		
2	3412.62970(22)	a^rQ(5, 3) NU3	$^{14}NH_3$	32	3419.20122(33)		
3	3412.70606(21)	s^rQ(8, 2) NU3	$^{14}NH_3$	33	3419.36765(19)	a^rQ(5, 2) NU3	$^{14}NH_3$
4	3413.08818(26)	aR(3, 2) NU1	$^{14}NH_3$	34	3419.45003(28)	a^rQ(8, 1) NU3	$^{14}NH_3$
5	3413.32123(20)	aR(3, 3) NU1	$^{14}NH_3$	35	3419.48382(36)		
6	3413.54145(39)			36	3419.72798(22)	s^rQ(5, 2) NU3	$^{14}NH_3$
7	3413.62107(58)			37	3419.95062(56)		
8	3414.06547(74)			38	3420.70431(81)		
9	3414.31063(22)	a^rQ(4, 3) NU3	$^{14}NH_3$	39	3420.78636(24)		
10	3414.63697(20)	sR(3, 0) NU1	$^{14}NH_3$	40	3420.91204(23)		
11	3414.83330(20)	sR(3, 2) NU1	$^{14}NH_3$	41	3421.07039(19)	a^rQ(4, 2) NU3	$^{14}NH_3$
12	3415.10583(20)	sR(3, 3) NU1	$^{14}NH_3$	42	3421.24172(22)		
13	3415.28282(35)			43	3421.44688(19)	s^rQ(4, 2) NU3	$^{14}NH_3$
14	3415.34341(23)	s^rQ(7, 2) NU3	$^{14}NH_3$	44	3421.73943(34)		
15	3415.49909(25)			45	3422.08461(19)	a^rQ(7, 1) NU3	$^{14}NH_3$
16	3415.58410(20)			46	3422.36247(32)	s^rQ(7, 1) NU3	$^{14}NH_3$
17	3415.95685(84)			47	3422.43888(19)	a^rQ(3, 2) NU3	$^{14}NH_3$
18	3416.26678(67)			48	3422.83529(19)	s^rQ(3, 2) NU3	$^{14}NH_3$
19	3416.31729(46)			49	3423.27817(85)		
20	3416.55069(33)	a^pR(5, 4) 2NU4 pe	$^{14}NH_3$	50	3423.35654(42)		
21	3416.77917(27)	s^rQ(9, 1) NU3	$^{14}NH_3$	51	3423.92991(28)		
22	3417.13129(20)			52	3423.97586(90)		
23	3417.25562(18)	a^rQ(6, 2) NU3	$^{14}NH_3$	53	3424.15485(24)		
24	3417.50776(27)			54	3424.19730(36)		
25	3417.58750(28)			55	3424.34856(39)		
26	3417.68941(19)	s^rQ(6, 2) NU3	$^{14}NH_3$	56	3424.39828(36)		
27	3418.06300(68)			57	3424.43532(18)	a^rQ(6, 1) NU3	$^{14}NH_3$
28	3418.34945(20)			58	3424.48257(24)		
29	3418.41880(33)			59	3424.52587(30)		
30	3418.74742(45)			60	3424.73062(18)	s^rQ(6, 1) NU3	$^{14}NH_3$

- "LINE NO.": serial number identifying the spectral line in the map on the opposite page.
- "WAVENUMBER": experimental value.
- Uncertainty in the last digit of the wavenumber is given in parentheses.
- "pa" : parallel ; "pe" : perpendicular.

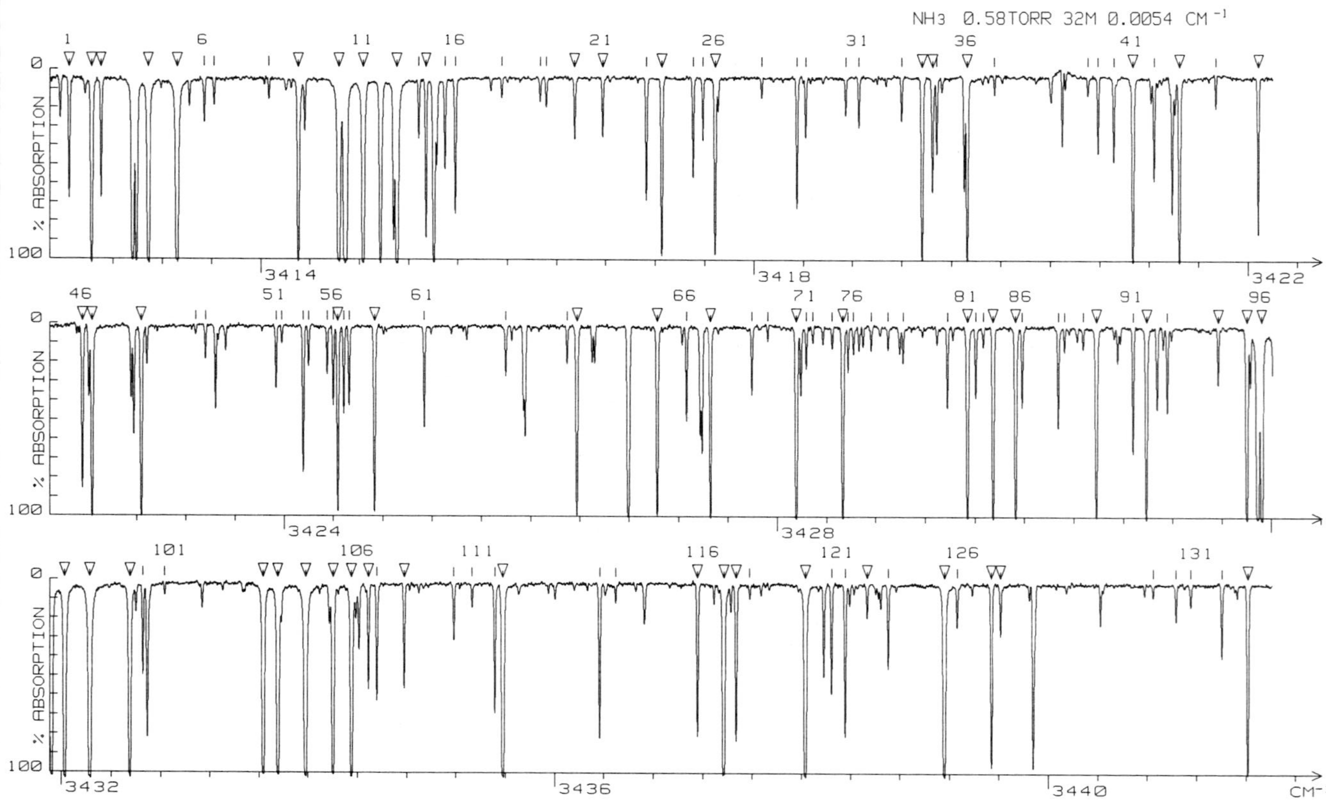

ISOTOPIC IDENTIFICATION - '∇' indicates $^{14}NH_3$; 'ı': unidentified.

LINE NO.	WAVENUMBER CM^{-1}, VAC.	ROTATIONAL BAND ASSIGNMENTS	ISOTOPIC SPECIES	LINE NO.	WAVENUMBER CM^{-1}, VAC.	ROTATIONAL BAND ASSIGNMENTS	ISOTOPIC SPECIES
61	3425.13355(26)			91	3430.87737(20)		
62	3425.79370(40)			92	3430.98693(19)	$s^{r}Q$(2, 1) NU3	$^{14}NH_3$
63	3426.29076(37)			93	3431.15535(24)		
64	3426.37383(19)	$a^{r}Q$(5, 1) NU3	$^{14}NH_3$	94	3431.56587(29)	$a^{P}Q$(9, 1) NU3	$^{14}NH_3$
65	3427.02506(19)	$a^{P}P$(1, 1) NU3	$^{14}NH_3$	95	3431.79736(22)	$a^{r}Q$(6, 0) NU3	$^{14}NH_3$
66	3427.26056(22)			96	3431.91919(25)	aR(4, 1) NU1	$^{14}NH_3$
67	3427.45574(19)	$s^{P}P$(1, 1) NU3	$^{14}NH_3$	97	3432.02626(25)	aR(4, 2) NU1	$^{14}NH_3$
68	3427.78957(26)			98	3432.22575(27)	aR(4, 3) NU1	$^{14}NH_3$
69	3427.91730(56)			99	3432.54815(20)	aR(4, 4) NU1	$^{14}NH_3$
70	3428.15111(23)	$a^{r}Q$(4, 1) NU3	$^{14}NH_3$	100	3432.65240(23)		
71	3428.22963(37)			101	3432.83043(80)		
72	3428.28339(92)			102	3433.63142(20)	sR(4, 1) NU1	$^{14}NH_3$
73	3428.43974(62)			103	3433.75155(25)	sR(4, 2) NU1	$^{14}NH_3$
74	3428.52812(19)	$s^{r}Q$(4, 1) NU3	$^{14}NH_3$	104	3433.97869(20)	sR(4, 3) NU1	$^{14}NH_3$
75	3428.57234(60)			105	3434.20228(20)	$s^{r}Q$(5, 0) NU3	$^{14}NH_3$
76	3428.61778(97)			106	3434.35091(20)	sR(4, 4) NU1	$^{14}NH_3$
77	3428.76110(75)			107	3434.48720(21)	$a^{P}Q$(8, 1) NU3	$^{14}NH_3$
78	3428.89617(71)			108	3434.55436(28)		
79	3429.01910(73)			109	3434.77529(21)	$s^{P}Q$(8, 1) NU3	$^{14}NH_3$
80	3429.37452(25)			110	3435.18252(34)		
81	3429.53616(19)	$a^{r}Q$(3, 1) NU3	$^{14}NH_3$	111	3435.32905(67)		
82	3429.60235(27)			112	3435.51345(24)		
83	3429.66342(74)			113	3435.57800(20)	$a^{r}Q$(4, 0) NU3	$^{14}NH_3$
84	3429.74599(19)	$s^{r}Q$(7, 0) NU3	$^{14}NH_3$	114	3436.36519(19)		
85	3429.92858(19)	$s^{r}Q$(3, 1) NU3	$^{14}NH_3$	115	3436.49524(68)		
86	3429.98079(30)			116	3437.15473(27)	$a^{P}Q$(7, 1) NU3	$^{14}NH_3$
87	3430.27233(23)			117	3437.36956(20)	$s^{r}Q$(3, 0) NU3	$^{14}NH_3$
88	3430.32173(64)			118	3437.46953(20)	$s^{P}Q$(7, 1) NU3	$^{14}NH_3$
89	3430.47248(77)			119	3437.57812(97)		
90	3430.57955(19)	$a^{r}Q$(2, 1) NU3	$^{14}NH_3$	120	3438.02877(20)	$a^{r}Q$(2, 0) NU3	$^{14}NH_3$

- "LINE NO.": serial number identifying the spectral line in the map on the opposite page.
- "WAVENUMBER": experimental value.
- Uncertainty in the last digit of the wavenumber is given in parentheses.
- "pa" : parallel ; "pe" : perpendicular.

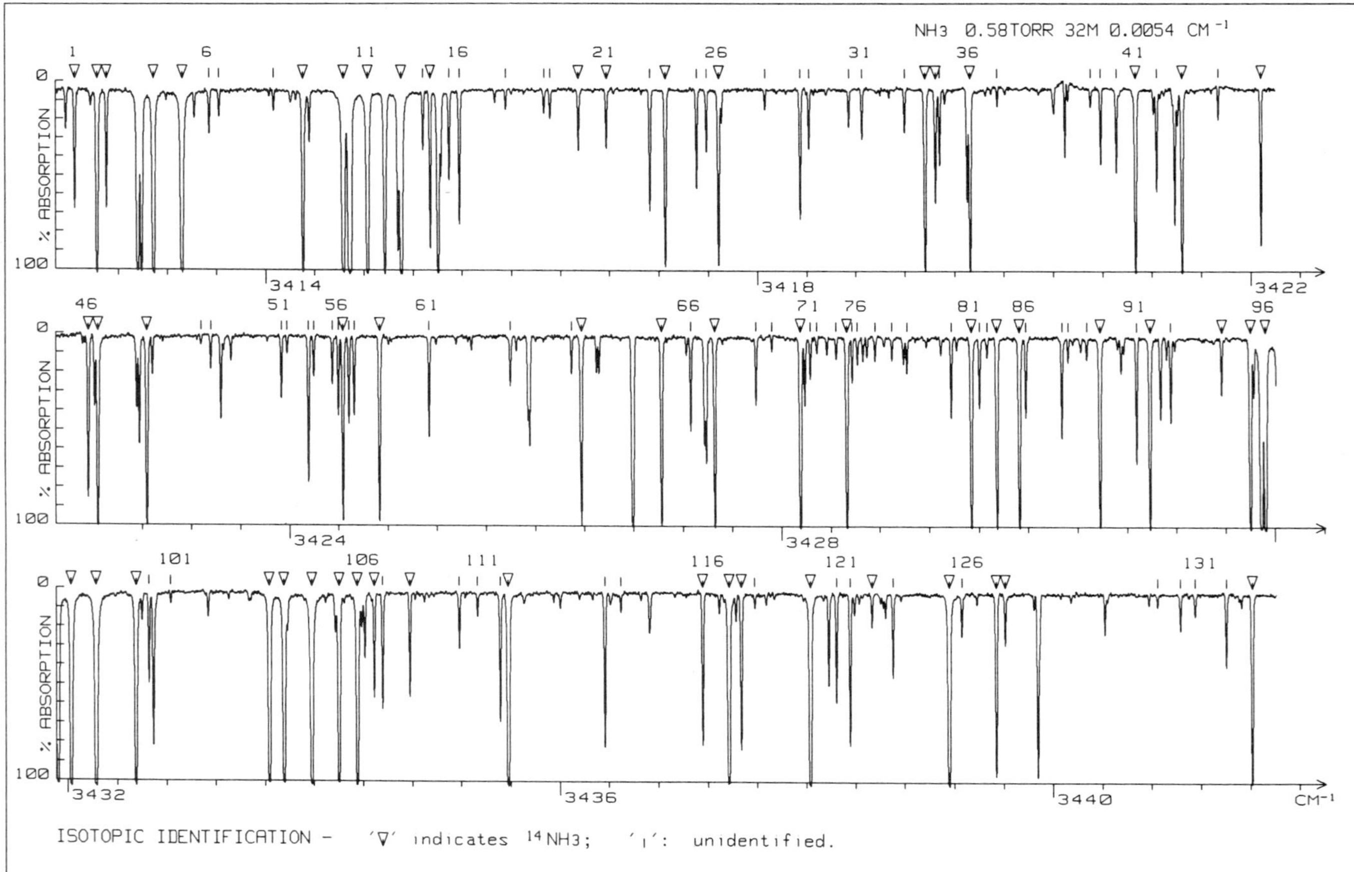
NH3 0.58TORR 32M 0.0054 CM⁻¹
% ABSORPTION
0
100
3414
3418
3422
3424
3428
3432
3436
3440
CM⁻¹
ISOTOPIC IDENTIFICATION - '∇' indicates ¹⁴NH3; 'ı': unidentified.

LINE NO.	WAVENUMBER CM^{-1}, VAC.	ROTATIONAL ASSIGNMENTS	BAND	ISOTOPIC SPECIES
121	3438.24141(20)			
122	3438.35325(20)			
123	3438.53536(49)	^sPQ(9, 2)	NU3	$^{14}NH_3$
124	3438.70444(29)			
125	3439.15681(24)	s^rQ(1, 0)	NU3	$^{14}NH_3$
126	3439.25928(35)			
127	3439.53678(20)	^aPQ(6, 1)	NU3	$^{14}NH_3$
128	3439.61213(34)	^aPQ(9, 2)	NU3	$^{14}NH_3$
129	3440.84625(100)			
130	3441.03049(49)			
131	3441.14943(60)			
132	3441.40197(26)			
133	3441.61117(18)	^aPQ(5, 1)	NU3	$^{14}NH_3$

- "LINE NO.": serial number identifying the spectral line in the map on the opposite page.
- "WAVENUMBER": experimental value.
- Uncertainty in the last digit of the wavenumber is given in parentheses.
- "pa" : parallel ; "pe" : perpendicular.

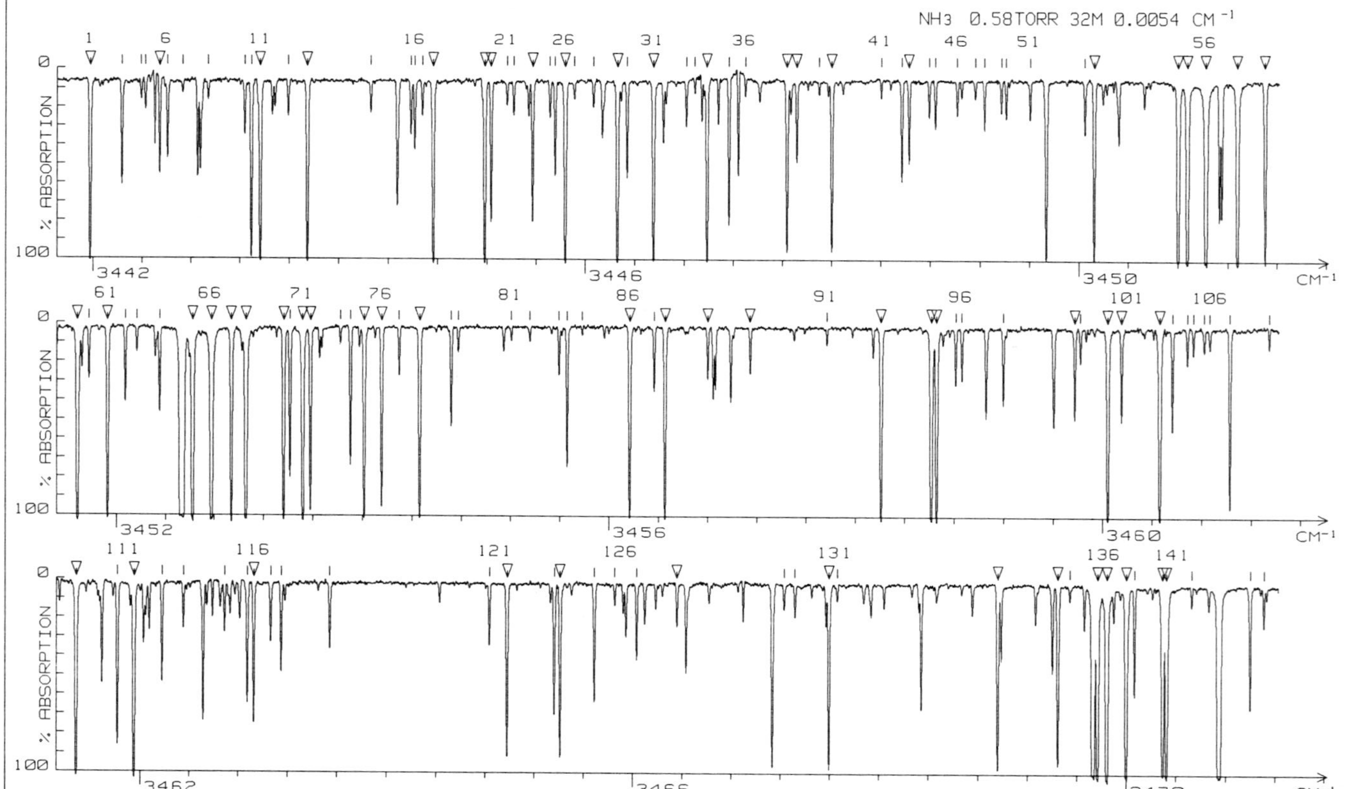

ISOTOPIC IDENTIFICATION - '∇' indicates $^{14}NH_3$; 'ı': unidentified.

LINE NO.	WAVENUMBER CM^{-1}, VAC.	ROTATIONAL BAND ASSIGNMENTS	ISOTOPIC SPECIES
1	3441.97473(18)	sPQ(5, 1) NU3	$^{14}NH_3$
2	3442.23416(21)		
3	3442.39078(81)		
4	3442.42549(46)		
5	3442.54395(25)	aPQ(8, 2) NU3	$^{14}NH_3$
6	3442.60847(25)		
7	3442.73038(94)		
8	3442.94047(71)		
9	3443.23796(47)		
10	3443.29070(18)		
11	3443.36397(19)	aPQ(4, 1) NU3	$^{14}NH_3$
12	3443.58955(36)		
13	3443.74645(18)	sPQ(4, 1) NU3	$^{14}NH_3$
14	3444.25692(41)		
15	3444.58011(47)		
16	3444.61090(29)		
17	3444.67304(37)		
18	3444.76192(18)	aPQ(3, 1) NU3	$^{14}NH_3$
19	3445.17909(22)	sPQ(3, 1) NU3	$^{14}NH_3$
20	3445.23105(23)	aPQ(7, 2) NU3	$^{14}NH_3$
21	3445.36231(53)		
22	3445.41629(37)		
23	3445.56959(20)	sPQ(7, 2) NU3	$^{14}NH_3$
24	3445.71044(50)		
25	3445.75115(26)		
26	3445.83406(19)	aPQ(2, 1) NU3	$^{14}NH_3$
27	3445.91041(70)		
28	3446.06480(58)		
29	3446.25848(19)	sPQ(2, 1) NU3	$^{14}NH_3$
30	3446.33678(22)		
31	3446.54946(19)	aPQ(1, 1) NU3	$^{14}NH_3$
32	3446.81499(32)		
33	3446.88471(55)		
34	3446.98489(19)	sPQ(1, 1) NU3	$^{14}NH_3$
35	3447.16342(19)		
36	3447.29532(72)		
37	3447.63504(29)	aPQ(6, 2) NU3	$^{14}NH_3$
38	3447.71407(22)	aPQ(9, 3) NU3	$^{14}NH_3$
39	3447.89715(95)		
40	3447.99925(18)	sPQ(6, 2) NU3	$^{14}NH_3$
41	3448.40076(51)		
42	3448.56734(21)		
43	3448.62616(23)	sPQ(9, 3) NU3	$^{14}NH_3$
44	3448.78868(36)		
45	3448.83908(32)		
46	3449.01383(38)		
47	3449.15806(83)		
48	3449.23299(41)		
49	3449.36977(54)		
50	3449.40838(34)		
51	3449.60171(34)		
52	3450.04467(36)		
53	3450.12144(18)	sPQ(5, 2) NU3	$^{14}NH_3$
54	3450.79911(20)	aR(5, 1) NU1	$^{14}NH_3$
55	3450.87371(20)	aR(5, 2) NU1	$^{14}NH_3$
56	3451.02163(20)	aR(5, 3) NU1	$^{14}NH_3$
57	3451.27549(20)	aR(5, 4) NU1	$^{14}NH_3$
58	3451.50231(19)	aPQ(4, 2) NU3	$^{14}NH_3$
59	3451.68068(20)	aR(5, 5) NU1	$^{14}NH_3$
60	3451.77219(35)		

- "LINE NO.": serial number identifying the spectral line in the map on the opposite page.
- "WAVENUMBER": experimental value.
- Uncertainty in the last digit of the wavenumber is given in parentheses.
- "pa" : parallel ; "pe" : perpendicular.

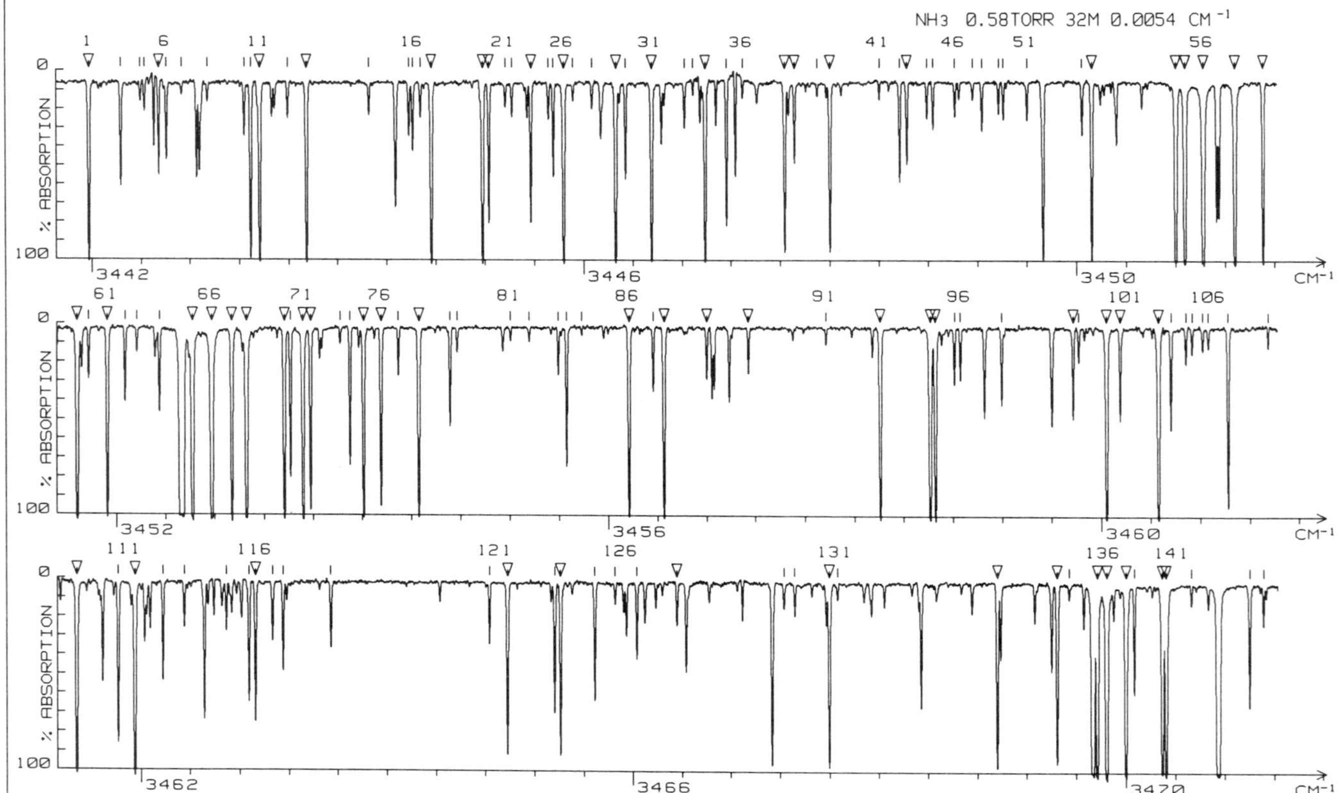

ISOTOPIC IDENTIFICATION – '∇' indicates $^{14}NH_3$; 'ı': unidentified.

LINE NO.	WAVENUMBER CM^{-1}, VAC.	ROTATIONAL ASSIGNMENTS	BAND	ISOTOPIC SPECIES	LINE NO.	WAVENUMBER CM^{-1}, VAC.	ROTATIONAL ASSIGNMENTS	BAND	ISOTOPIC SPECIES
61	3451.92510(19)	sPQ(4, 2)	NU3	$^{14}NH_3$	91	3457.76741(80)			
62	3452.06894(23)				92	3458.20897(18)	aPQ(5, 3)	NU3	$^{14}NH_3$
63	3452.16303(53)				93	3458.61507(22)	arR(0, 0)	NU3	$^{14}NH_3$
64	3452.34839(24)				94	3458.65821(19)	sPQ(5, 3)	NU3	$^{14}NH_3$
65	3452.62036(20)	sR(5, 2)	NU1	$^{14}NH_3$	95	3458.81001(30)			
66	3452.77605(20)	sR(5, 3)	NU1	$^{14}NH_3$	96	3458.85858(32)			
67	3452.93757(18)	aPQ(3, 2)	NU3	$^{14}NH_3$	97	3459.19471(36)			
68	3453.05570(20)	sR(5, 4)	NU1	$^{14}NH_3$	98	3459.77179(28)	aPQ(8, 4)	NU3	$^{14}NH_3$
69	3453.36130(24)	sPQ(3, 2)	NU3	$^{14}NH_3$	99	3459.81744(51)			
70	3453.41189(20)				100	3460.04048(19)	aPQ(4, 3)	NU3	$^{14}NH_3$
71	3453.51498(29)	sR(5, 5)	NU1	$^{14}NH_3$	101	3460.15004(28)	sPQ(8, 4)	NU3	$^{14}NH_3$
72	3453.57915(20)	aPQ(7, 3)	NU3	$^{14}NH_3$	102	3460.46129(21)	sPQ(4, 3)	NU3	$^{14}NH_3$
73	3453.82035(92)				103	3460.56111(27)			
74	3453.90129(20)				104	3460.68186(38)			
75	3454.01186(19)	aPQ(2, 2)	NU3	$^{14}NH_3$	105	3460.73054(64)			
76	3454.15306(18)	sPQ(7, 3)	NU3	$^{14}NH_3$	106	3460.81691(65)			
77	3454.29390(30)				107	3460.86181(65)			
78	3454.45937(19)	sPQ(2, 2)	NU3	$^{14}NH_3$	108	3461.02243(18)			
79	3454.71498(26)				109	3461.34441(41)			
80	3454.77211(60)				110	3461.47685(19)	aPQ(3, 3)	NU3	$^{14}NH_3$
81	3455.20421(105)				111	3461.81292(20)			
82	3455.35409(97)				112	3461.94787(21)	sPQ(3, 3)	NU3	$^{14}NH_3$
83	3455.59089(34)				113	3462.17581(21)			
84	3455.65769(19)				114	3462.35238(31)			
85	3455.78153(84)				115	3462.69062(30)			
86	3456.16649(24)	aPQ(6, 3)	NU3	$^{14}NH_3$	116	3462.87456(25)			
87	3456.36266(26)				117	3462.92633(20)	sPQ(7, 4)	NU3	$^{14}NH_3$
88	3456.44995(19)	sPQ(6, 3)	NU3	$^{14}NH_3$	118	3463.06365(31)			
89	3456.79768(30)	aPQ(9, 4)	NU3	$^{14}NH_3$	119	3463.14859(23)			
90	3457.14315(37)	sPQ(9, 4)	NU3	$^{14}NH_3$	120	3463.54326(26)			

- "LINE NO.": serial number identifying the spectral line in the map on the opposite page.
- "WAVENUMBER": experimental value.
- Uncertainty in the last digit of the wavenumber is given in parentheses.
- "pa" : parallel ; "pe" : perpendicular.

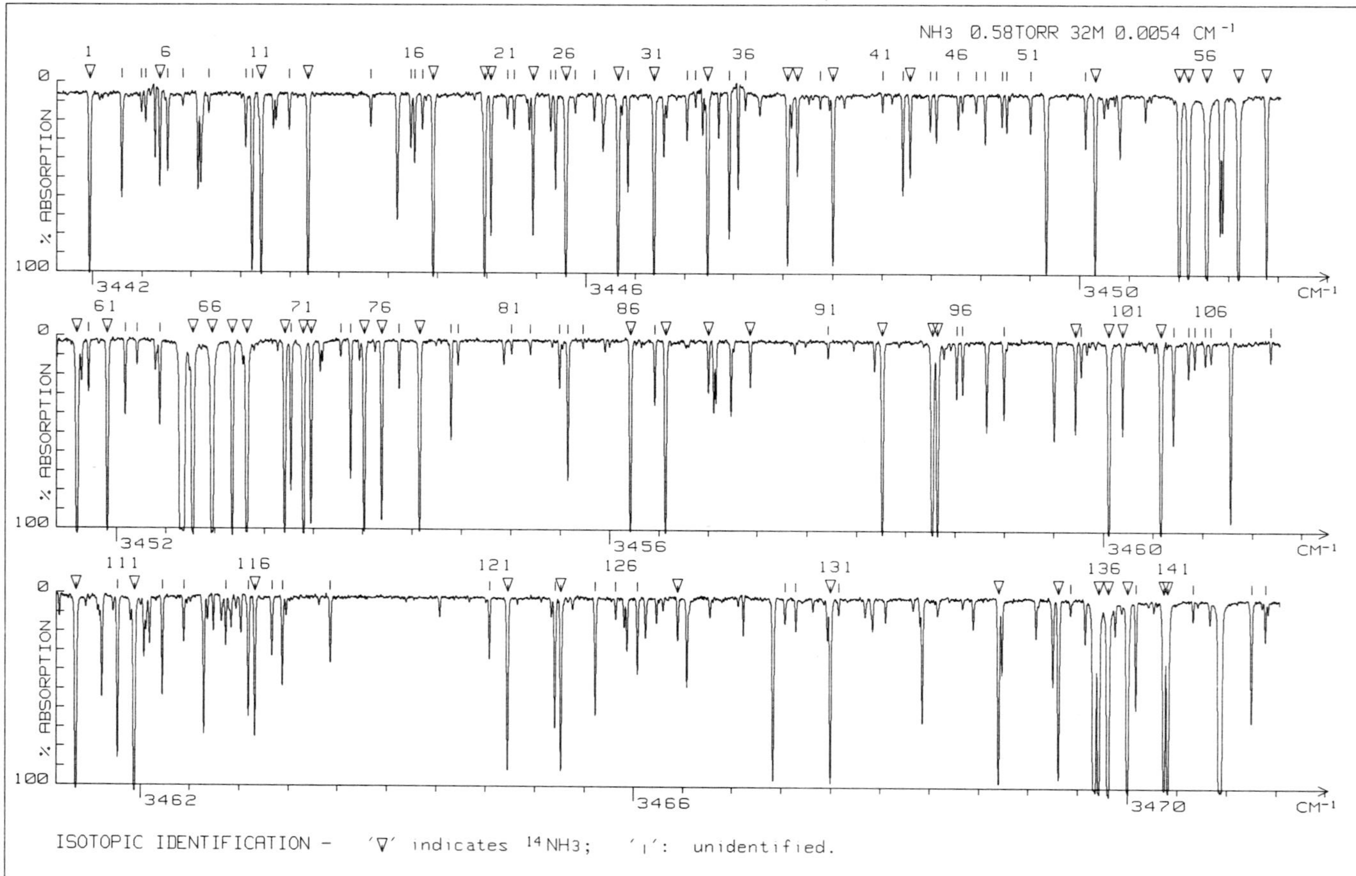
NH3 0.58TORR 32M 0.0054 CM⁻¹
% ABSORPTION
0
100
3442
3446
3450
3452
3456
3460
3462
3466
3470
CM⁻¹
ISOTOPIC IDENTIFICATION - '∇' indicates 14NH3; '|': unidentified.

LINE NO.	WAVENUMBER CM^{-1}, VAC.	ROTATIONAL ASSIGNMENTS	BAND	ISOTOPIC SPECIES
121	3464.83281(26)			
122	3464.97921(24)	a^PQ(6, 4)	NU3	$^{14}NH_3$
123	3465.36133(28)			
124	3465.40947(18)	s^PQ(6, 4)	NU3	$^{14}NH_3$
125	3465.68931(20)			
126	3465.85439(69)			
127	3466.03057(23)			
128	3466.35519(37)	s^PQ(9, 5)	NU3	$^{14}NH_3$
129	3467.22836(52)			
130	3467.31668(44)			
131	3467.59493(20)	s^PQ(5, 4)	NU3	$^{14}NH_3$
132	3467.66010(80)			
133	3468.95686(20)	a^PQ(4, 4)	NU3	$^{14}NH_3$
134	3469.44427(33)	s^PQ(4, 4)	NU3	$^{14}NH_3$
135	3469.54085(106)			
136	3469.76704(22)	aR(6, 2)	NU1	$^{14}NH_3$
137	3469.84177(20)	aR(6, 3)	NU1	$^{14}NH_3$
138	3469.99732(22)	aR(6, 4)	NU1	$^{14}NH_3$
139	3470.06493(20)			
140	3470.29439(22)	aR(6, 5)	NU1	$^{14}NH_3$
141	3470.32533(22)	a^rR(1, 1)	NU3	$^{14}NH_3$
142	3470.52691(59)			
143	3470.99774(57)			
144	3471.10839(63)			

- "LINE NO.": serial number identifying the spectral line in the map on the opposite page.
- "WAVENUMBER": experimental value.
- Uncertainty in the last digit of the wavenumber is given in parentheses.
- "pa" : parallel ; "pe" : perpendicular.

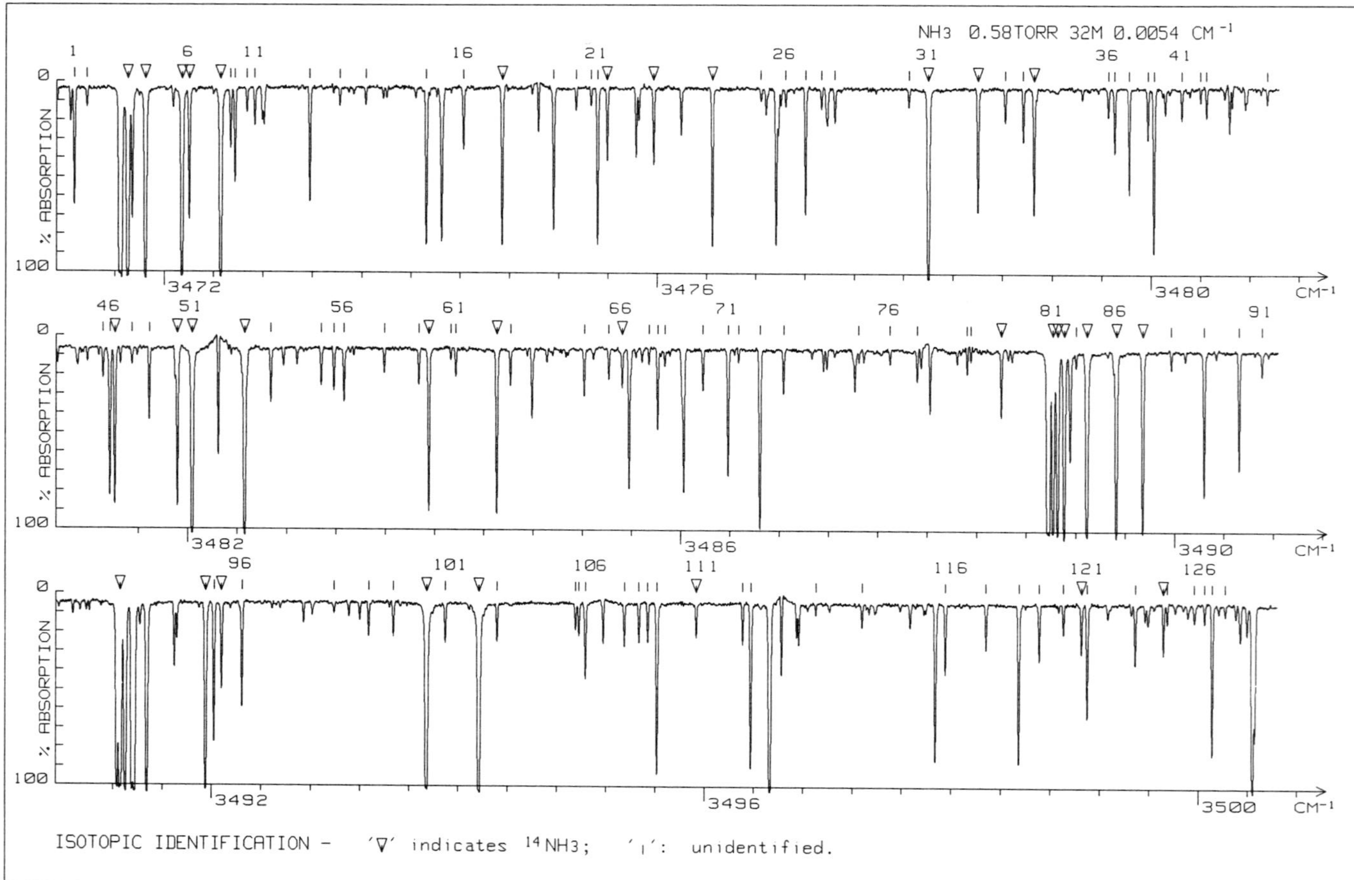
NH3 0.58TORR 32M 0.0054 CM⁻¹
% ABSORPTION
0
100
3472
3476
3480
3482
3486
3490
3492
3496
3500
CM⁻¹
ISOTOPIC IDENTIFICATION - '∇' indicates ¹⁴NH3; '|': unidentified.

LINE NO.	WAVENUMBER CM^{-1}, VAC.	ROTATIONAL BAND ASSIGNMENTS	ISOTOPIC SPECIES
1	3471.26948 (26)		
2	3471.37329 (89)		
3	3471.70404 (25)	sR(6, 3) NU1	$^{14}NH_3$
4	3471.84689 (20)	sR(6, 4) NU1	$^{14}NH_3$
5	3472.14377 (19)	sR(6, 5) NU1	$^{14}NH_3$
6	3472.20336 (50)	s^{P}Q(7, 5) NU3	$^{14}NH_3$
7	3472.45754 (21)	sR(6, 6) NU1	$^{14}NH_3$
8	3472.53750 (28)		
9	3472.57309 (21)		
10	3472.67109 (52)		
11	3472.73395 (35)		
12	3473.17872 (20)		
13	3473.42067 (72)		
14	3473.62842 (71)		
15	3474.12116 (19)		
16	3474.42596 (29)		
17	3474.73925 (20)	s^{P}Q(6, 5) NU3	$^{14}NH_3$
18	3475.15645 (20)		
19	3475.34029 (55)		
20	3475.46041 (73)		
21	3475.51476 (19)		
22	3475.59214 (29)	a^{P}Q(9, 6) NU3	$^{14}NH_3$
23	3475.96912 (24)	s^{P}Q(9, 6) NU3	$^{14}NH_3$
24	3476.44617 (20)	a^{P}Q(5, 5) NU3	$^{14}NH_3$
25	3476.84105 (86)		
26	3477.04035 (76)		
27	3477.20281 (19)		
28	3477.33220 (65)		
29	3477.44024 (40)		
30	3478.04201 (61)		
31	3478.20086 (19)	s^{r}R(1, 0) NU3	$^{14}NH_3$
32	3478.59760 (20)	a^{P}Q(8, 6) NU3	$^{14}NH_3$
33	3478.82108 (40)		
34	3478.96573 (31)		
35	3479.05140 (49)	s^{P}Q(8, 6) NU3	$^{14}NH_3$
36	3479.65047 (48)		
37	3479.70251 (28)		
38	3479.82110 (20)		
39	3479.97093 (28)		
40	3480.02001 (19)		
41	3480.24367 (45)		
42	3480.39572 (53)		
43	3480.44441 (45)		
44	3480.94020 (96)		
45	3481.31202 (81)		
46	3481.36744 (19)		
47	3481.40765 (19)	a^{P}Q(7, 6) NU3	$^{14}NH_3$
48	3481.54650 (73)		
49	3481.68649 (23)		
50	3481.91636 (36)	s^{P}Q(7, 6) NU3	$^{14}NH_3$
51	3482.03529 (22)	a^{r}R(2, 2) NU3	$^{14}NH_3$
52	3482.46143 (21)	s^{r}R(2, 2) NU3	$^{14}NH_3$
53	3482.67108 (27)		
54	3483.07949 (37)		
55	3483.18185 (33)		
56	3483.26261 (32)		
57	3483.58832 (50)		
58	3483.87046 (40)		
59	3483.95100 (20)	a^{P}Q(6, 6) NU3	$^{14}NH_3$
60	3484.13199 (76)		

- "LINE NO.": serial number identifying the spectral line in the map on the opposite page.
- "WAVENUMBER": experimental value.
- Uncertainty in the last digit of the wavenumber is given in parentheses.
- "pa" : parallel ; "pe" : perpendicular.

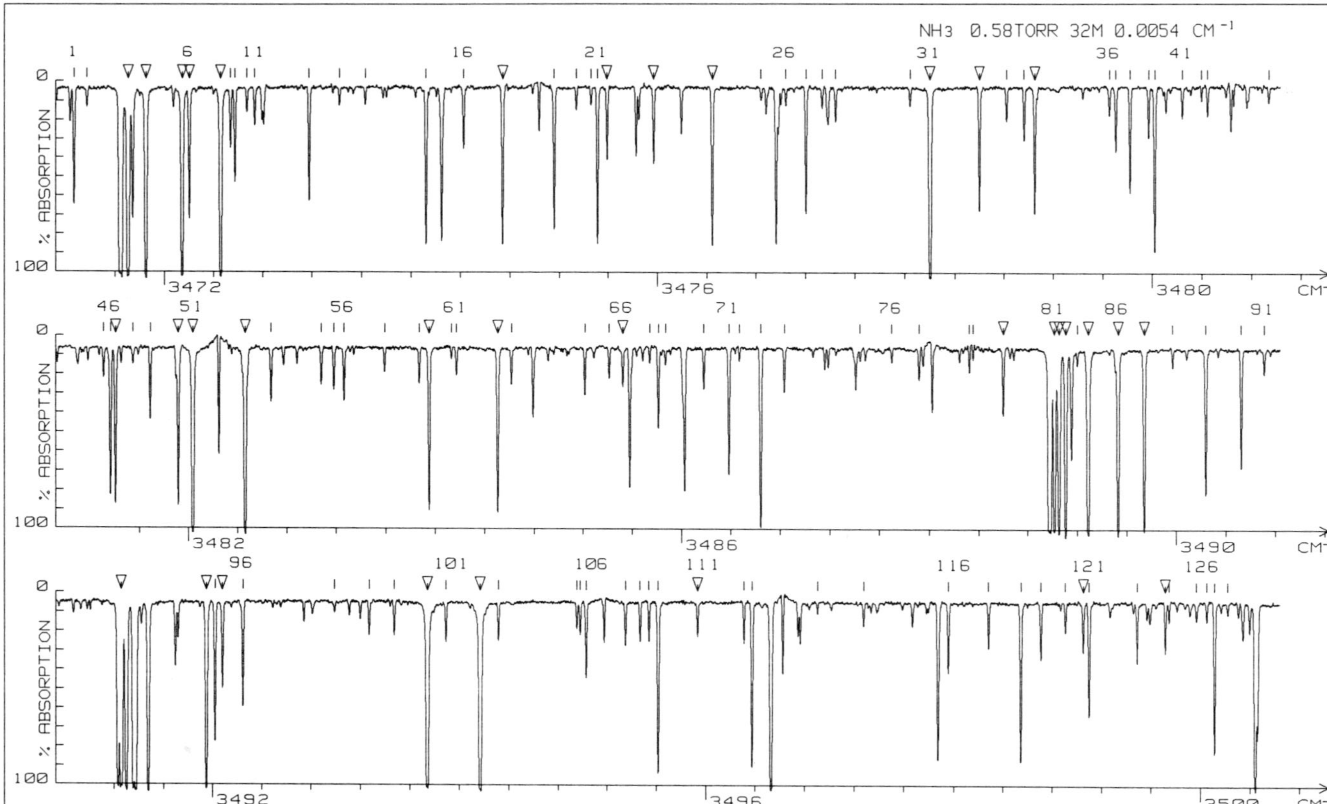

ISOTOPIC IDENTIFICATION – '▽' indicates $^{14}NH_3$; '|': unidentified.

LINE NO.	WAVENUMBER CM^{-1}, VAC.	ROTATIONAL BAND ASSIGNMENTS	ISOTOPIC SPECIES
61	3484.17382 (49)		
62	3484.50539 (20)	$s^{P}Q$(6, 6) NU3	$^{14}NH_3$
63	3484.61501 (37)		
64	3485.21512 (34)		
65	3485.41419 (44)		
66	3485.52386 (36)	$a^{P}Q$(9, 7) NU3	$^{14}NH_3$
67	3485.74091 (46)		
68	3485.81183 (23)		
69	3485.86955 (68)		
70	3486.17857 (33)		
71	3486.38480 (25)		
72	3486.47027 (60)		
73	3486.64303 (18)		
74	3486.83277 (36)		
75	3487.44573 (100)		
76	3487.69906 (83)		
77	3487.91912 (46)		
78	3488.32078 (36)		
79	3488.35058 (57)		
80	3488.59899 (30)	$a^{P}Q$(8, 7) NU3	$^{14}NH_3$
81	3489.01453 (32)	aR(7, 4) NU1	$^{14}NH_3$
82	3489.05225 (49)	aR(7, 5) NU1	$^{14}NH_3$
83	3489.10524 (18)	$a^{r}R$(2, 1) NU3	$^{14}NH_3$
84	3489.19845 (94)		
85	3489.28686 (22)	aR(7, 6) NU1	$^{14}NH_3$
86	3489.52652 (19)	$s^{r}R$(2, 1) NU3	$^{14}NH_3$
87	3489.74123 (19)	aR(7, 7) NU1	$^{14}NH_3$
88	3489.97000 (79)		
89	3490.23631 (20)		
90	3490.52095 (20)		

LINE NO.	WAVENUMBER CM^{-1}, VAC.	ROTATIONAL BAND ASSIGNMENTS	ISOTOPIC SPECIES
91	3490.70552 (48)		
92	3491.25818 (28)	sR(7, 6) NU1	$^{14}NH_3$
92		sR(7, 3) NU1	$^{14}NH_3$
93	3491.95105 (19)	sR(7, 7) NU1	$^{14}NH_3$
94	3492.01966 (20)		
95	3492.08123 (23)	$s^{P}Q$(7, 7) NU3	$^{14}NH_3$
96	3492.24637 (20)		
97	3492.98994 (94)		
98	3493.26873 (46)		
99	3493.47077 (38)		
100	3493.74071 (20)	$a^{r}R$(3, 3) NU3	$^{14}NH_3$
101	3493.89259 (34)		
102	3494.16900 (26)	$s^{r}R$(3, 3) NU3	$^{14}NH_3$
103	3494.31498 (35)		
104	3494.94998 (53)		
105	3494.97815 (38)		
106	3495.03028 (27)		
107	3495.34919 (31)		
108	3495.46768 (37)		
109	3495.53954 (37)		
110	3495.61485 (18)		
111	3495.93243 (45)	$a^{P}Q$(9, 8) NU3	$^{14}NH_3$
112	3496.31098 (31)		
113	3496.37656 (18)		
114	3496.90445 (96)		
115	3497.27762 (57)		
116	3497.95529 (27)		
117	3498.28171 (33)		
118	3498.54879 (19)		
119	3498.71179 (30)		
120	3498.90806 (46)		

- "LINE NO.": serial number identifying the spectral line in the map on the opposite page.
- "WAVENUMBER": experimental value.
- Uncertainty in the last digit of the wavenumber is given in parentheses.
- "pa" : parallel ; "pe" : perpendicular.

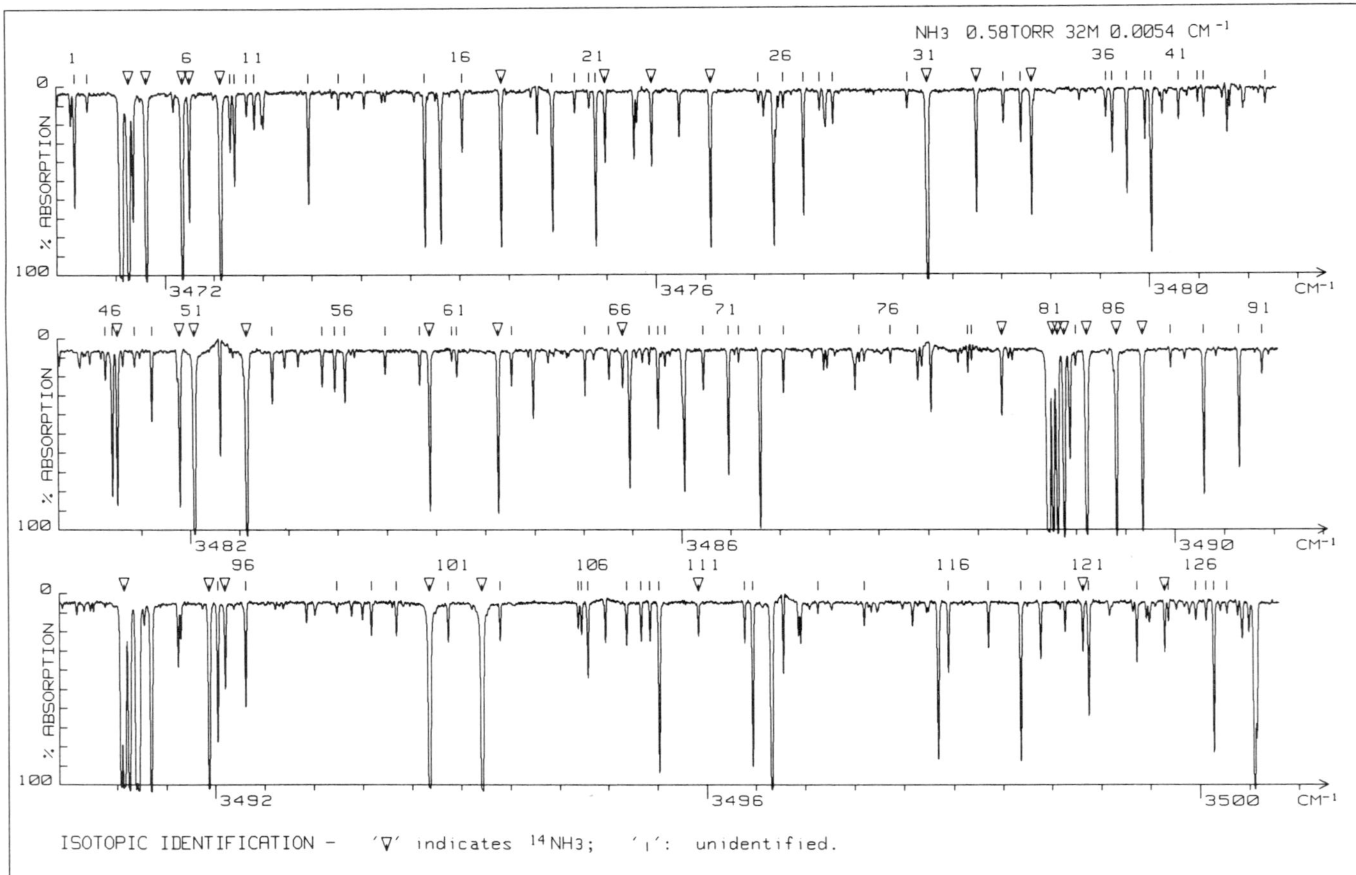
NH3 0.58TORR 32M 0.0054 CM⁻¹
% ABSORPTION
0
100
1
6
11
16
21
26
31
36
41
3472
3476
3480
CM⁻¹
46
51
56
61
66
71
76
81
86
91
3482
3486
3490
96
101
106
111
116
121
126
3492
3496
3500
ISOTOPIC IDENTIFICATION - '∇' indicates ¹⁴NH3; '|': unidentified.

LINE NO.	WAVENUMBER CM^{-1}, VAC.	ROTATIONAL BAND ASSIGNMENTS	ISOTOPIC SPECIES
121	3499.05203(32)	$a^{P}Q$(8, 8) NU3	$^{14}NH_3$
122	3499.09919(20)		
123	3499.48708(26)		
124	3499.71585(36)	$s^{P}Q$(8, 8) NU3	$^{14}NH_3$
125	3499.74699(38)		
126	3499.96602(96)		
127	3500.04947(63)		
128	3500.11230(19)		
129	3500.21779(113)		

- "LINE NO.": serial number identifying the spectral line in the map on the opposite page.
- "WAVENUMBER": experimental value.
- Uncertainty in the last digit of the wavenumber is given in parentheses.
- "pa" : parallel ; "pe" : perpendicular.

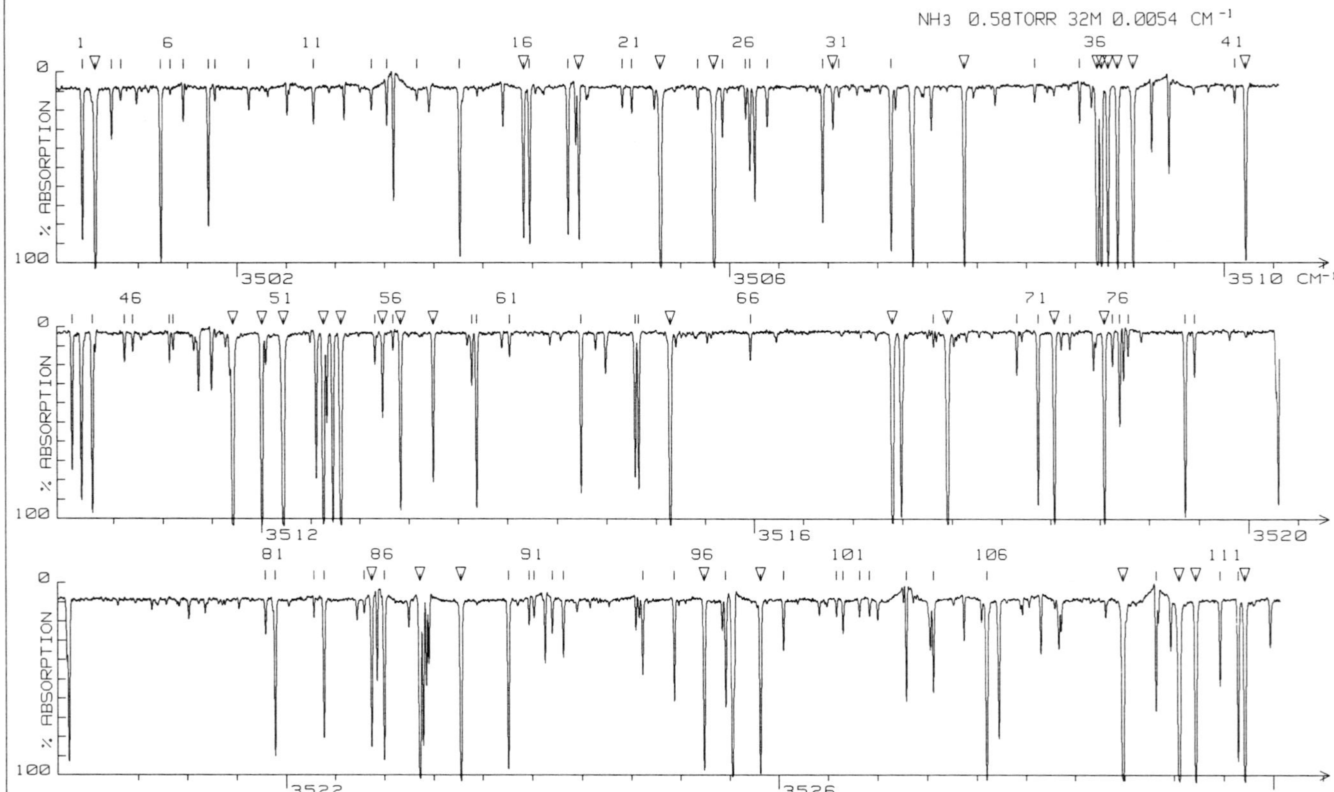

ISOTOPIC IDENTIFICATION – '∇' indicates $^{14}NH_3$; '|': unidentified.

LINE NO.	WAVENUMBER CM^{-1}, VAC.	ROTATIONAL BAND ASSIGNMENTS	ISOTOPIC SPECIES
1	3500.74954(20)		
2	3500.85385(19)	s^{r}R(3, 2) NU3	$^{14}NH_3$
3	3500.98637(27)		
4	3501.06038(82)		
5	3501.38255(18)		
6	3501.46285(89)		
7	3501.56829(24)		
8	3501.77167(19)		
9	3501.82598(49)		
10	3502.09844(58)		
11	3502.62610(39)		
12	3503.09633(61)		
13	3503.22085(32)		
14	3503.46571(89)		
15	3503.81534(20)		
16	3504.33106(23)	a^{P}R(2, 1) NU3	$^{14}NH_3$
17	3504.37922(18)		
18	3504.68754(19)		
19	3504.77690(22)	s^{P}R(2, 1) NU3	$^{14}NH_3$
20	3505.12882(65)		
21	3505.20512(42)		
22	3505.43721(20)	a^{r}R(4, 4) NU3	$^{14}NH_3$
23	3505.74258(54)		
24	3505.87081(19)	s^{r}R(4, 4) NU3	$^{14}NH_3$
25	3505.94470(33)		
26	3506.13456(43)		
27	3506.16732(21)		
28	3506.30907(65)		
29	3506.75787(19)		
30	3506.84071(33)	a^{P}Q(9, 9) NU3	$^{14}NH_3$

LINE NO.	WAVENUMBER CM^{-1}, VAC.	ROTATIONAL BAND ASSIGNMENTS	ISOTOPIC SPECIES
31	3506.89010(109)		
32	3507.31008(18)		
33	3507.89780(19)	s^{r}R(3, 1) NU3	$^{14}NH_3$
34	3508.46922(63)		
35	3508.83648(25)		
36	3508.97842(23)	aR(8, 0) NU1	$^{14}NH_3$
36		aR(8, 2) NU1	$^{14}NH_3$
36		aR(8, 1) NU1	$^{14}NH_3$
37	3509.01102(22)	aR(8, 3) NU1	$^{14}NH_3$
38	3509.06479(19)	aR(8, 5) NU1	$^{14}NH_3$
39	3509.14302(23)	aR(8, 4) NU1	$^{14}NH_3$
40	3509.27001(19)	aR(8, 6) NU1	$^{14}NH_3$
41	3510.08692(75)		
42	3510.17540(26)	aR(8, 7) NU1	$^{14}NH_3$
43	3510.46533(19)		
44	3510.63136(20)		
45	3510.88956(41)		
46	3510.95755(60)		
47	3511.25379(42)		
48	3511.28540(67)		
49	3511.76854(23)	a^{r}R(4, 3) NU3	$^{14}NH_3$
50	3512.00269(19)	sR(8, 6) NU1	$^{14}NH_3$
51	3512.17615(23)	s^{r}R(4, 3) NU3	$^{14}NH_3$
52	3512.50544(18)	sR(8, 3) NU1	$^{14}NH_3$
53	3512.64740(82)	sR(8, 1) NU1	$^{14}NH_3$
53		sR(8, 4) NU1	$^{14}NH_3$
54	3512.92486(77)		
55	3512.98706(23)	s^{P}R(2, 2) NU3	$^{14}NH_3$
56	3513.07150(46)		
57	3513.13319(18)	sR(8, 7) NU1	$^{14}NH_3$
58	3513.39560(19)	sR(8, 8) NU1	$^{14}NH_3$

- "LINE NO.": serial number identifying the spectral line in the map on the opposite page.
- "WAVENUMBER": experimental value.
- Uncertainty in the last digit of the wavenumber is given in parentheses.
- "pa" : parallel ; "pe" : perpendicular.

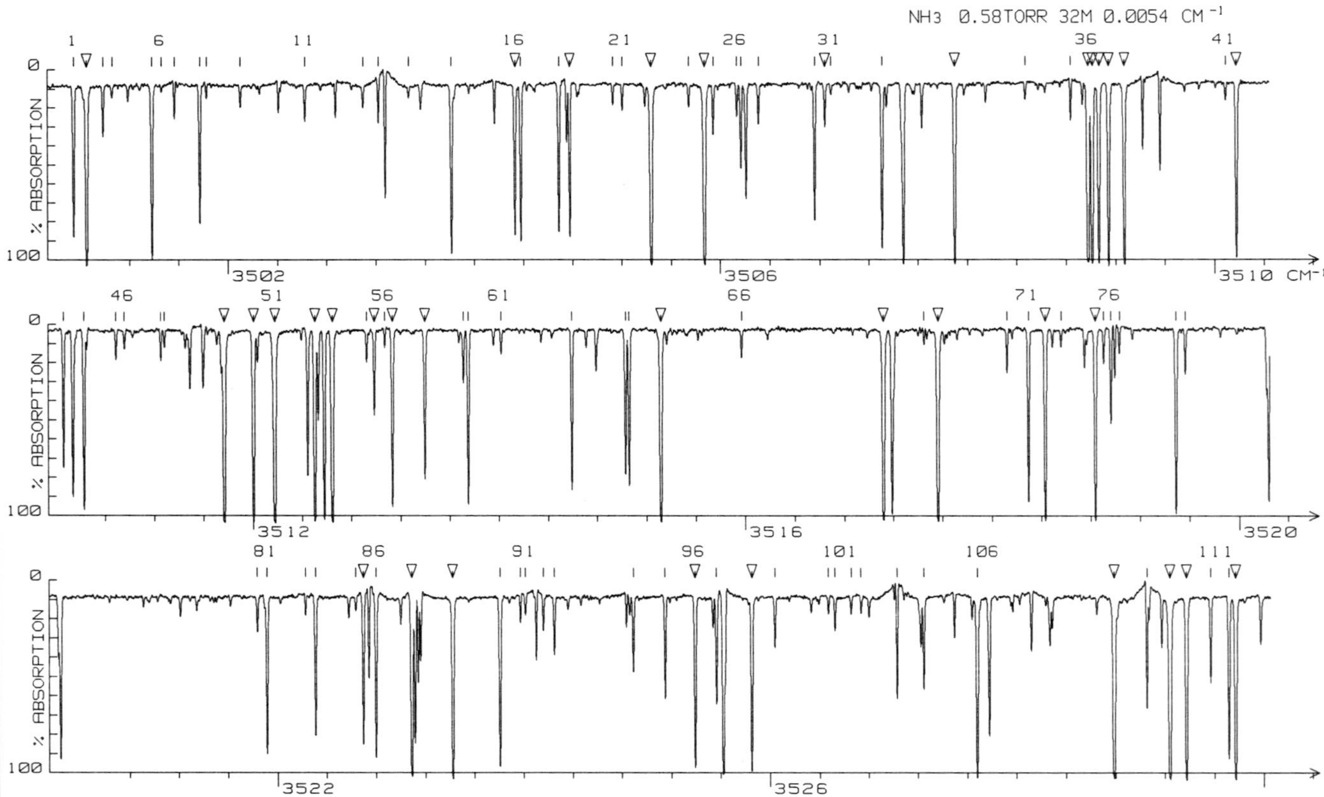

ISOTOPIC IDENTIFICATION - '∇' indicates $^{14}NH_3$; '|': unidentified.

LINE NO.	WAVENUMBER CM^{-1}, VAC.	ROTATIONAL BAND ASSIGNMENTS	ISOTOPIC SPECIES
59	3513.71083(31)		
60	3513.75037(18)		
61	3514.01803(50)		
62	3514.59169(18)		
63	3515.02995(24)		
64	3515.05962(21)		
65	3515.31529(18)	$s^{r}R(3, 0)$ NU3	$^{14}NH_3$
66	3515.97176(46)		
67	3517.12051(19)	$a^{r}R(5, 5)$ NU3	$^{14}NH_3$
68	3517.45032(49)		
69	3517.56301(19)	$s^{r}R(5, 5)$ NU3	$^{14}NH_3$
70	3518.12419(35)		
71	3518.29625(18)		
72	3518.43135(19)	$a^{r}R(4, 2)$ NU3	$^{14}NH_3$
73	3518.55815(73)		
74	3518.83619(19)	$s^{r}R(4, 2)$ NU3	$^{14}NH_3$
75	3518.90275(39)		
76	3518.96103(24)		
77	3519.02934(91)		
78	3519.48947(18)		
79	3519.56441(33)		
80	3521.83331(40)		
81	3521.91159(20)		
82	3522.22618(39)		
83	3522.31000(19)		
84	3522.63657(84)		
85	3522.69679(19)	$a^{p}R(3, 1)$ NU3	$^{14}NH_3$
86	3522.79912(22)		
87	3523.08859(21)	$a^{r}R(5, 4)$ NU3	$^{14}NH_3$
88	3523.42350(19)	$s^{r}R(5, 4)$ NU3	$^{14}NH_3$
89	3523.80801(18)		
90	3523.97309(27)		
91	3524.01391(59)		
92	3524.15957(38)		
93	3524.24926(26)		
94	3524.89165(27)		
95	3525.14859(20)		
96	3525.39255(18)	$a^{r}R(4, 1)$ NU3	$^{14}NH_3$
97	3525.56636(21)		
98	3525.85199(18)	$s^{r}R(4, 1)$ NU3	$^{14}NH_3$
99	3526.03974(26)		
100	3526.46867(59)		
101	3526.52088(37)		
102	3526.65198(57)		
103	3526.73473(87)		
104	3527.03107(17)		
105	3527.24866(30)		
106	3527.67854(19)		
107	3528.78647(21)	$a^{r}R(6, 6)$ NU3	$^{14}NH_3$
108	3529.05644(20)		
109	3529.24102(21)	$s^{r}R(6, 6)$ NU3	$^{14}NH_3$
110	3529.37392(18)	$a^{r}R(5, 3)$ NU3	$^{14}NH_3$
111	3529.56934(21)		
112	3529.71605(20)		
113	3529.76975(19)	$s^{r}R(5, 3)$ NU3	$^{14}NH_3$

- "LINE NO.": serial number identifying the spectral line in the map on the opposite page.
- "WAVENUMBER": experimental value.
- Uncertainty in the last digit of the wavenumber is given in parentheses.
- "pa" : parallel ; "pe" : perpendicular.

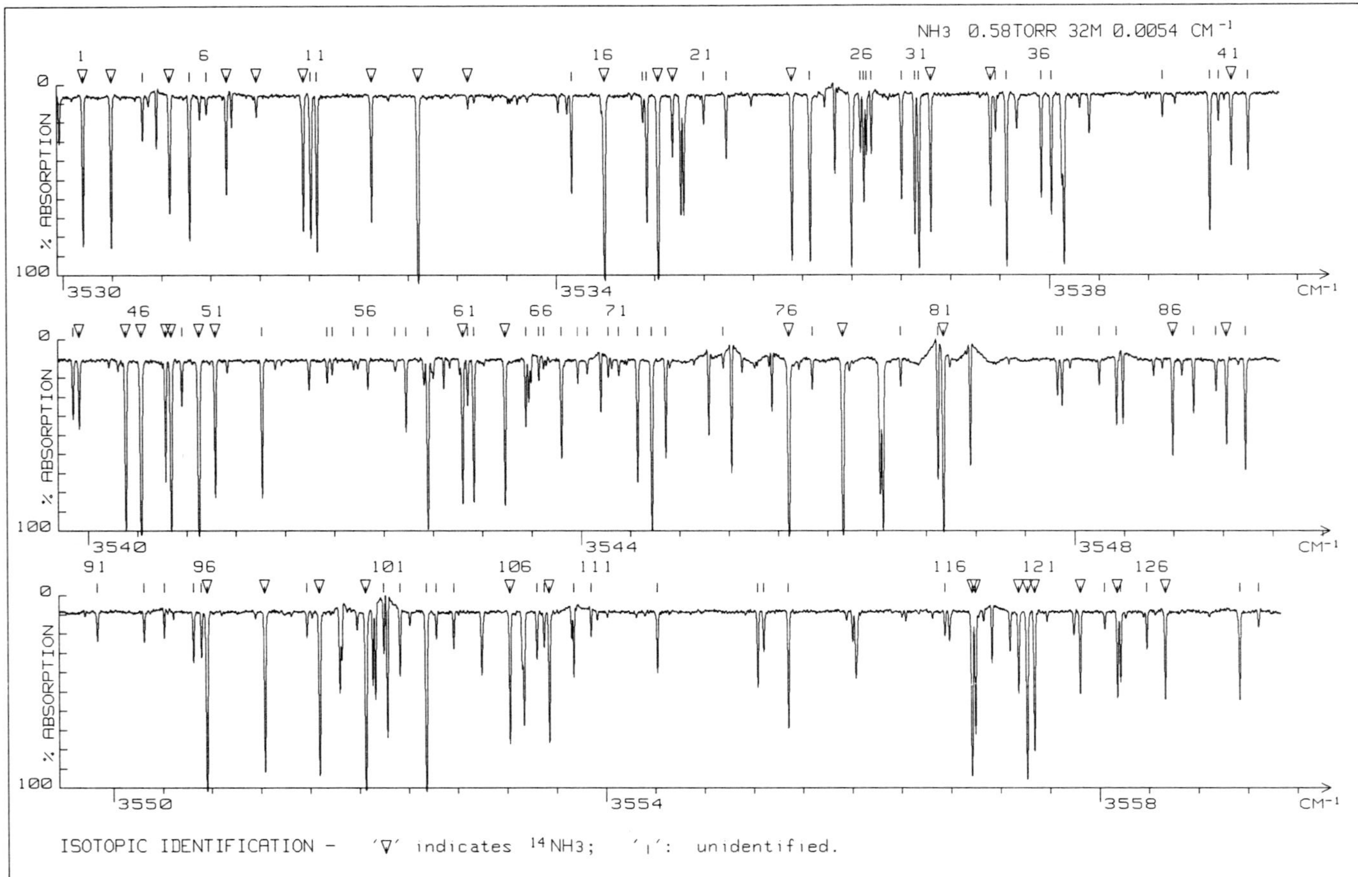
NH3 0.58TORR 32M 0.0054 CM⁻¹
0
% ABSORPTION
100
3530
3534
3538
CM⁻¹
3540
3544
3548
3550
3554
3558
ISOTOPIC IDENTIFICATION - '▽' indicates ¹⁴NH3; '|': unidentified.

LINE NO.	WAVENUMBER CM^{-1}, VAC.	ROTATIONAL BAND ASSIGNMENTS	ISOTOPIC SPECIES
1	3530.16618 (19)	aR (9, 1) NU1	$^{14}NH_3$
2	3530.39447 (19)	aR (9, 2) NU1	$^{14}NH_3$
3	3530.64841 (29)		
4	3530.86910 (27)	$a^{P}R$ (3, 2) NU3	$^{14}NH_3$
5	3531.03073 (19)		
6	3531.16975 (60)		
7	3531.33198 (20)	$s^{P}R$ (3, 2) NU3	$^{14}NH_3$
8	3531.57512 (53)	aR (9, 6) NU1	$^{14}NH_3$
9	3531.95481 (22)	aR (9, 7) NU1	$^{14}NH_3$
10	3532.01524 (28)		
11	3532.06495 (19)		
12	3532.50579 (19)	sR (9, 3) NU1	$^{14}NH_3$
12		aR (9, 8) NU1	$^{14}NH_3$
13	3532.88335 (19)	$a^{r}R$ (4, 0) NU3	$^{14}NH_3$
14	3533.29108 (72)	aR (9, 9) NU1	$^{14}NH_3$
15	3534.12831 (21)		
16	3534.39596 (19)	$a^{r}R$ (6, 5) NU3	$^{14}NH_3$
16		sR (9, 0) NU1	$^{14}NH_3$
16		sR (9, 4) NU1	$^{14}NH_3$
17	3534.70877 (94)		
18	3534.74053 (19)		
19	3534.83199 (19)	sR (9, 1) NU1	$^{14}NH_3$
19		$s^{r}R$ (6, 5) NU3	$^{14}NH_3$
20	3534.94980 (26)	sR (9, 5) NU1	$^{14}NH_3$
21	3535.20159 (52)		
22	3535.38534 (23)		
23	3535.91669 (18)	$a^{r}R$ (5, 2) NU3	$^{14}NH_3$
24	3536.06457 (22)		
25	3536.47693 (44)		
26	3536.50487 (26)		
27	3536.52624 (43)		
28	3536.56566 (30)		
29	3536.80773 (20)		
30	3536.91458 (22)		
31	3536.94915 (20)		
32	3537.04356 (19)	sR (9, 8) NU1	$^{14}NH_3$
33	3537.52865 (20)	sR (9, 9) NU1	$^{14}NH_3$
34	3537.56963 (37)		
35	3537.65855 (18)		
36	3537.93842 (21)		
37	3538.01801 (19)		
38	3538.91716 (46)		
39	3539.29719 (26)		
40	3539.36832 (88)		
41	3539.47022 (24)	$a^{P}R$ (3, 3) NU3	$^{14}NH_3$
42	3539.60681 (28)		
43	3539.88076 (24)		
44	3539.93052 (24)	$s^{P}R$ (3, 3) NU3	$^{14}NH_3$
45	3540.30853 (20)	$a^{r}R$ (6, 4) NU3	$^{14}NH_3$
46	3540.43143 (25)	$a^{r}R$ (7, 7) NU3	$^{14}NH_3$
47	3540.63001 (24)	$a^{P}R$ (4, 1) NU3	$^{14}NH_3$
48	3540.67644 (19)	$s^{r}R$ (6, 4) NU3	$^{14}NH_3$
49	3540.76351 (28)		
50	3540.90151 (19)	$s^{r}R$ (7, 7) NU3	$^{14}NH_3$
51	3541.03588 (19)	$s^{P}R$ (4, 1) NU3	$^{14}NH_3$
52	3541.41488 (19)		
53	3541.94465 (44)		
54	3541.98552 (66)		
55	3542.15770 (98)		
56	3542.27239 (39)		
57	3542.49543 (110)		
58	3542.58310 (28)		
59	3542.75875 (20)		

- "LINE NO.": serial number identifying the spectral line in the map on the opposite page.
- "WAVENUMBER": experimental value.
- Uncertainty in the last digit of the wavenumber is given in parentheses.
- "pa" : parallel ; "pe" : perpendicular.

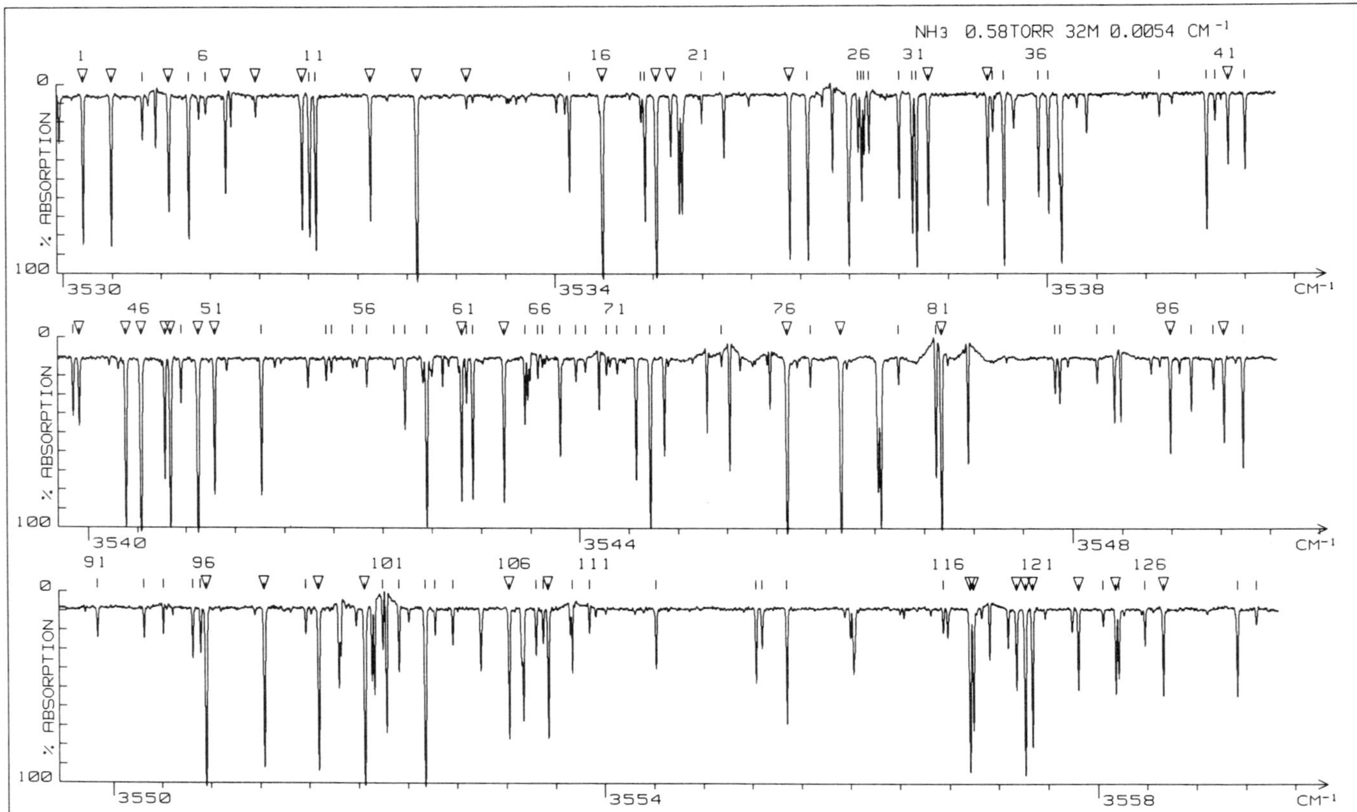

ISOTOPIC IDENTIFICATION – '∇' indicates $^{14}NH_3$; 'ı': unidentified.

LINE NO.	WAVENUMBER CM^{-1}, VAC.	ROTATIONAL BAND ASSIGNMENTS	ISOTOPIC SPECIES	LINE NO.	WAVENUMBER CM^{-1}, VAC.	ROTATIONAL BAND ASSIGNMENTS	ISOTOPIC SPECIES
60	3543.04324(21)	$a^{r}R(5, 1)$ NU3	$^{14}NH_3$	90	3549.38135(19)		
61	3543.08366(30)			91	3549.87319(37)		
62	3543.13317(19)			92	3550.25211(34)		
63	3543.38673(18)	$s^{r}R(5, 1)$ NU3	$^{14}NH_3$	93	3550.41239(31)		
64	3543.55453(37)			94	3550.65418(49)		
65	3543.65987(32)			95	3550.71849(52)		
66	3543.69784(75)			96	3550.76335(18)	$s^{r}R(5, 0)$ NU3	$^{14}NH_3$
67	3543.84206(25)			97	3551.23424(18)	$a^{r}R(7, 5)$ NU3	$^{14}NH_3$
68	3543.97556(44)			98	3551.57505(51)		
69	3544.05301(105)			99	3551.67820(18)	$s^{r}R(7, 5)$ NU3	$^{14}NH_3$
70	3544.22325(51)			100	3552.05202(23)	$a^{r}R(8, 8)$ NU3	$^{14}NH_3$
71	3544.30916(82)			101	3552.19766(44)		
72	3544.46129(20)			102	3552.32940(25)		
73	3544.57618(18)			103	3552.54248(19)		
74	3544.69048(22)			104	3552.62376(44)		
75	3545.15809(77)			105	3552.76778(33)		
76	3545.68929(24)	$a^{r}R(7, 6)$ NU3	$^{14}NH_3$	106	3553.22311(19)	$a^{r}R(6, 2)$ NU3	$^{14}NH_3$
77	3545.87951(35)			107	3553.43983(29)		
78	3546.12727(26)	$s^{r}R(7, 6)$ NU3	$^{14}NH_3$	108	3553.49741(31)		
79	3546.59774(41)			109	3553.54024(23)	$s^{r}R(6, 2)$ NU3	$^{14}NH_3$
80	3546.89800(25)			110	3553.73836(39)		
81	3546.94271(19)	$s^{r}R(6, 3)$ NU3	$^{14}NH_3$	111	3553.87993(31)		
82	3547.86417(31)			112	3554.41865(24)		
83	3547.90129(27)			113	3555.23277(22)		
84	3548.20119(44)			114	3555.28122(31)		
85	3548.34023(35)			115	3555.48288(19)		
86	3548.79500(20)	$a^{P}R(4, 2)$ NU3	$^{14}NH_3$	116	3556.75122(39)		
87	3548.96463(24)			117	3556.96864(22)	$a^{r}R(8, 7)$ NU3	$^{14}NH_3$
88	3549.14491(72)			118	3556.99686(52)	$a^{r}R(7, 4)$ NU3	$^{14}NH_3$
89	3549.22936(21)	$s^{P}R(4, 2)$ NU3	$^{14}NH_3$	119	3557.34746(23)	$a^{P}R(4, 3)$ NU3	$^{14}NH_3$

- "LINE NO.": serial number identifying the spectral line in the map on the opposite page.
- "WAVENUMBER": experimental value.
- Uncertainty in the last digit of the wavenumber is given in parentheses.
- "pa" : parallel ; "pe" : perpendicular.

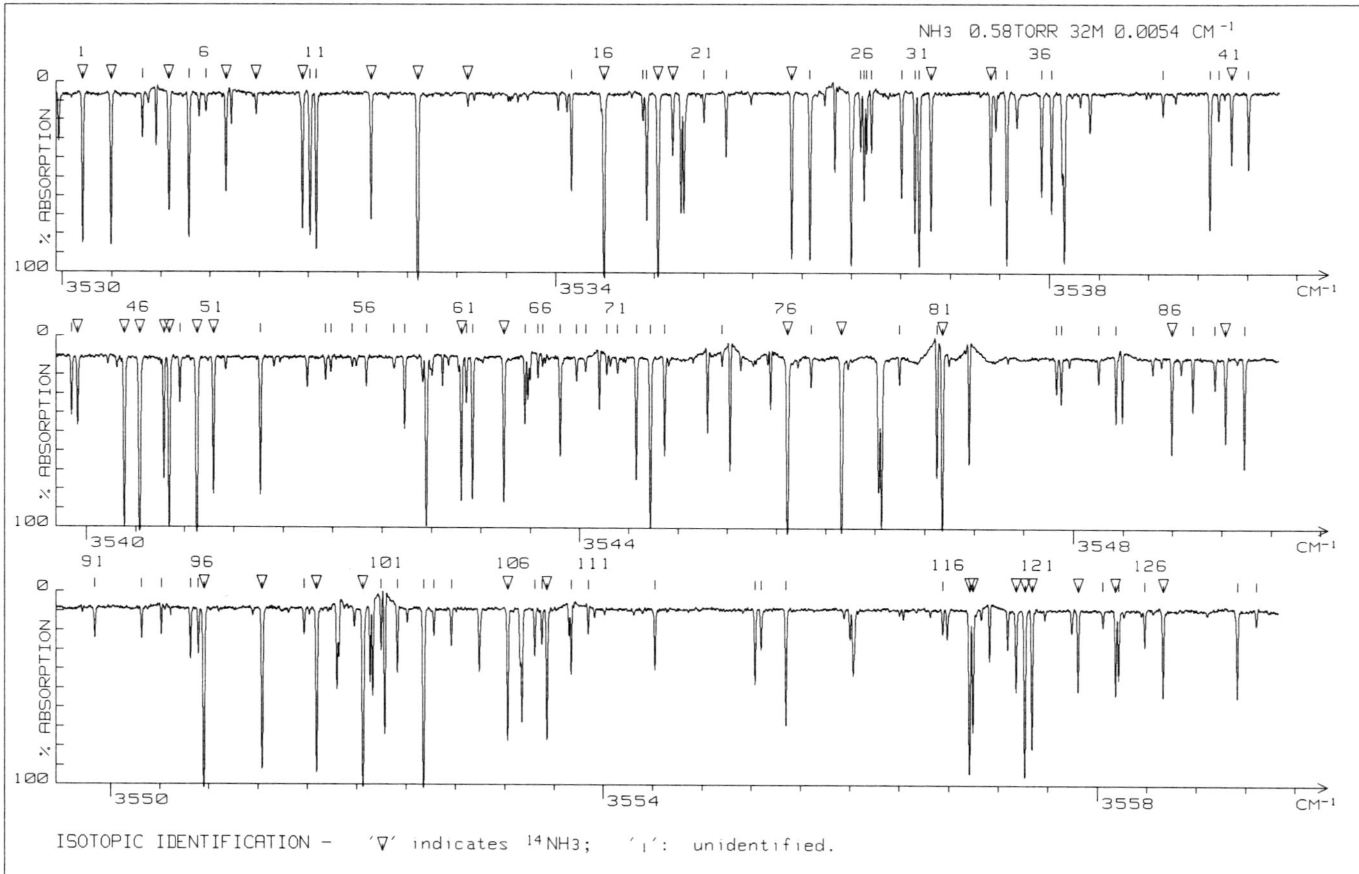
NH3 0.58TORR 32M 0.0054 CM⁻¹
% ABSORPTION
0
100
3530
3534
3538
3540
3544
3548
3550
3554
3558
CM⁻¹
1 6 11 16 21 26 31 36 41
46 51 56 61 66 71 76 81 86
91 96 101 106 111 116 121 126
ISOTOPIC IDENTIFICATION - '∇' indicates ¹⁴NH3; '|': unidentified.

LINE NO.	WAVENUMBER CM^{-1}, VAC.	ROTATIONAL BAND ASSIGNMENTS	ISOTOPIC SPECIES
120	3557.41670(18)	s^rR(8, 7) NU3	$^{14}NH_3$
121	3557.47518(19)	s^rR(7, 4) NU3	$^{14}NH_3$
122	3557.84335(25)	s^pR(4, 3) NU3	$^{14}NH_3$
123	3558.04302(72)		
124	3558.14482(32)	a^pR(5, 1) NU3	$^{14}NH_3$
125	3558.17050(37)		
126	3558.38380(36)		
127	3558.53266(21)	s^pR(5, 1) NU3	$^{14}NH_3$
128	3559.13299(31)		
129	3559.28557(60)		

- "LINE NO.": serial number identifying the spectral line in the map on the opposite page.
- "WAVENUMBER": experimental value.
- Uncertainty in the last digit of the wavenumber is given in parentheses.
- "pa" : parallel ; "pe" : perpendicular.

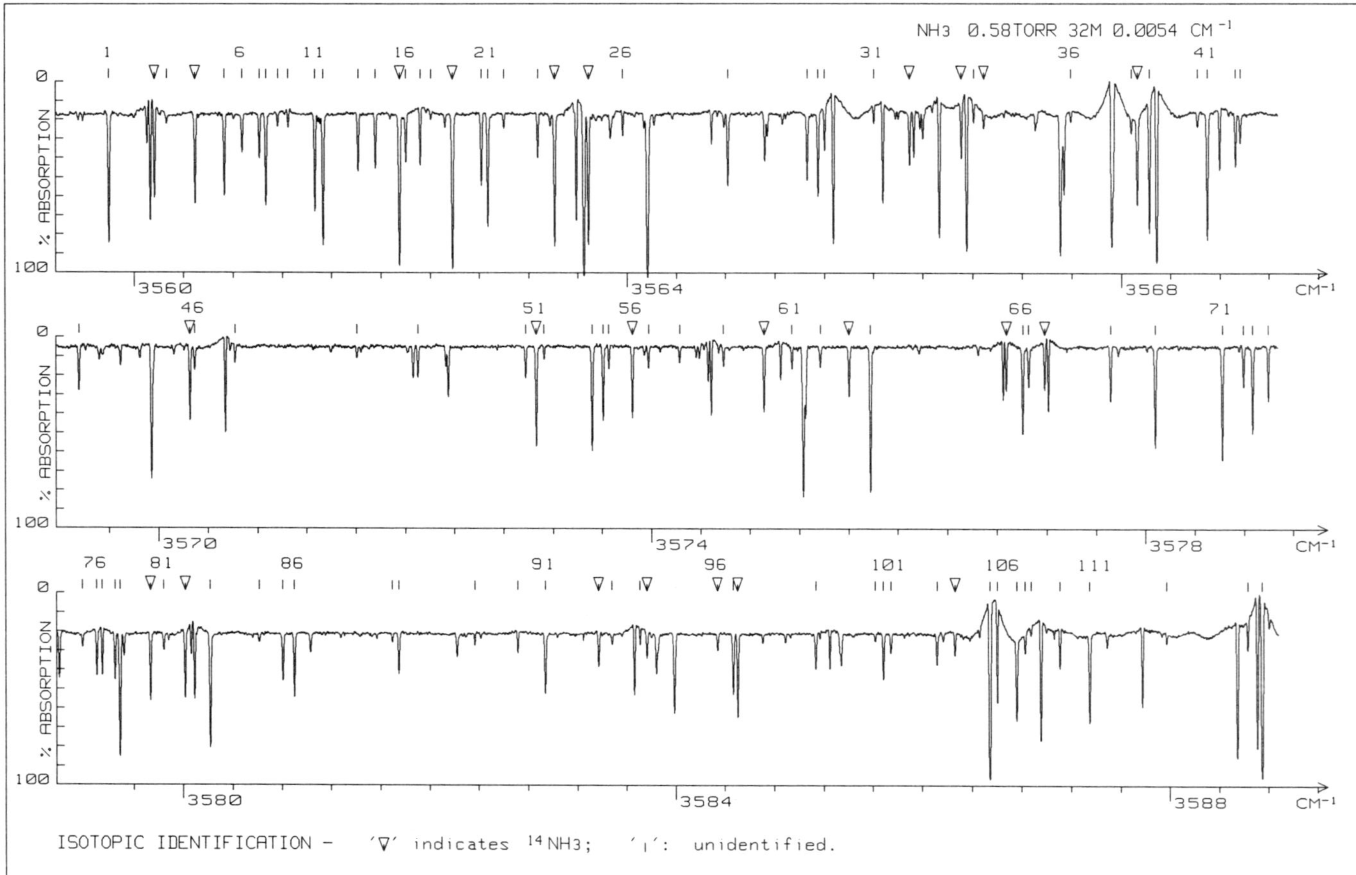
NH3 0.58TORR 32M 0.0054 CM⁻¹
% ABSORPTION
CM⁻¹
3560
3564
3568
3570
3574
3578
3580
3584
3588
ISOTOPIC IDENTIFICATION – '∇' indicates ¹⁴NH3; '|': unidentified.

LINE NO.	WAVENUMBER CM^{-1}, VAC.	ROTATIONAL BAND ASSIGNMENTS	ISOTOPIC SPECIES	LINE NO.	WAVENUMBER CM^{-1}, VAC.	ROTATIONAL BAND ASSIGNMENTS	ISOTOPIC SPECIES
1	3559.79870(19)			31	3566.00520(39)		
2	3560.16692(24)	a^rR(6, 1) NU3	$^{14}NH_3$	32	3566.29616(25)	a^pR(5, 2) NU3	$^{14}NH_3$
3	3560.26675(85)			33	3566.70963(29)	s^pR(5, 2) NU3	$^{14}NH_3$
4	3560.49677(20)	s^rR(6, 1) NU3	$^{14}NH_3$	34	3566.80931(65)		
5	3560.73149(21)			35	3566.88897(81)	s^pR(4, 4) NU3	$^{14}NH_3$
6	3560.87316(35)			36	3567.59276(80)		
7	3561.01411(28)			37	3568.08400(45)		
8	3561.06752(25)			38	3568.13337(21)	s^rR(8, 5) NU3	$^{14}NH_3$
9	3561.16377(46)			39	3568.23196(25)		
10	3561.24782(40)			40	3568.61841(70)		
11	3561.46912(19)			41	3568.69897(19)		
12	3561.53542(21)			42	3568.92672(27)		
13	3561.81849(23)			43	3568.96464(38)		
14	3561.95748(25)			44	3569.35591(29)		
15	3562.15147(18)	a^rR(8, 6) NU3	$^{14}NH_3$	45	3570.25614(26)	s^rR(7, 2) NU3	$^{14}NH_3$
16	3562.20320(47)			46	3570.29401(58)		
17	3562.32033(22)			47	3570.61867(67)		
18	3562.40525(81)			48	3571.61075(91)		
19	3562.58467(20)	s^rR(8, 6) NU3	$^{14}NH_3$	49	3572.10359(40)		
20	3562.81912(23)			50	3572.97899(38)		
21	3562.87191(20)			51	3573.06546(21)	a^rR(8, 4) NU3	$^{14}NH_3$
22	3563.00148(62)			52	3573.12776(53)		
23	3563.27537(26)			53	3573.52157(21)		
24	3563.41092(18)	a^rR(7, 3) NU3	$^{14}NH_3$	54	3573.61107(23)		
25	3563.68963(21)	s^rR(7, 3) NU3	$^{14}NH_3$	55	3573.65669(41)		
26	3563.96707(33)			56	3573.84609(23)	s^rR(8, 4) NU3	$^{14}NH_3$
27	3564.82257(27)			57	3573.97754(55)		
28	3565.46151(24)			58	3574.23080(72)		
29	3565.54881(21)			59	3574.58974(53)		
30	3565.60300(30)			60	3574.91688(29)	a^pR(5, 3) NU3	$^{14}NH_3$

- "LINE NO.": serial number identifying the spectral line in the map on the opposite page.
- "WAVENUMBER": experimental value.
- Uncertainty in the last digit of the wavenumber is given in parentheses.
- "pa" : parallel ; "pe" : perpendicular.

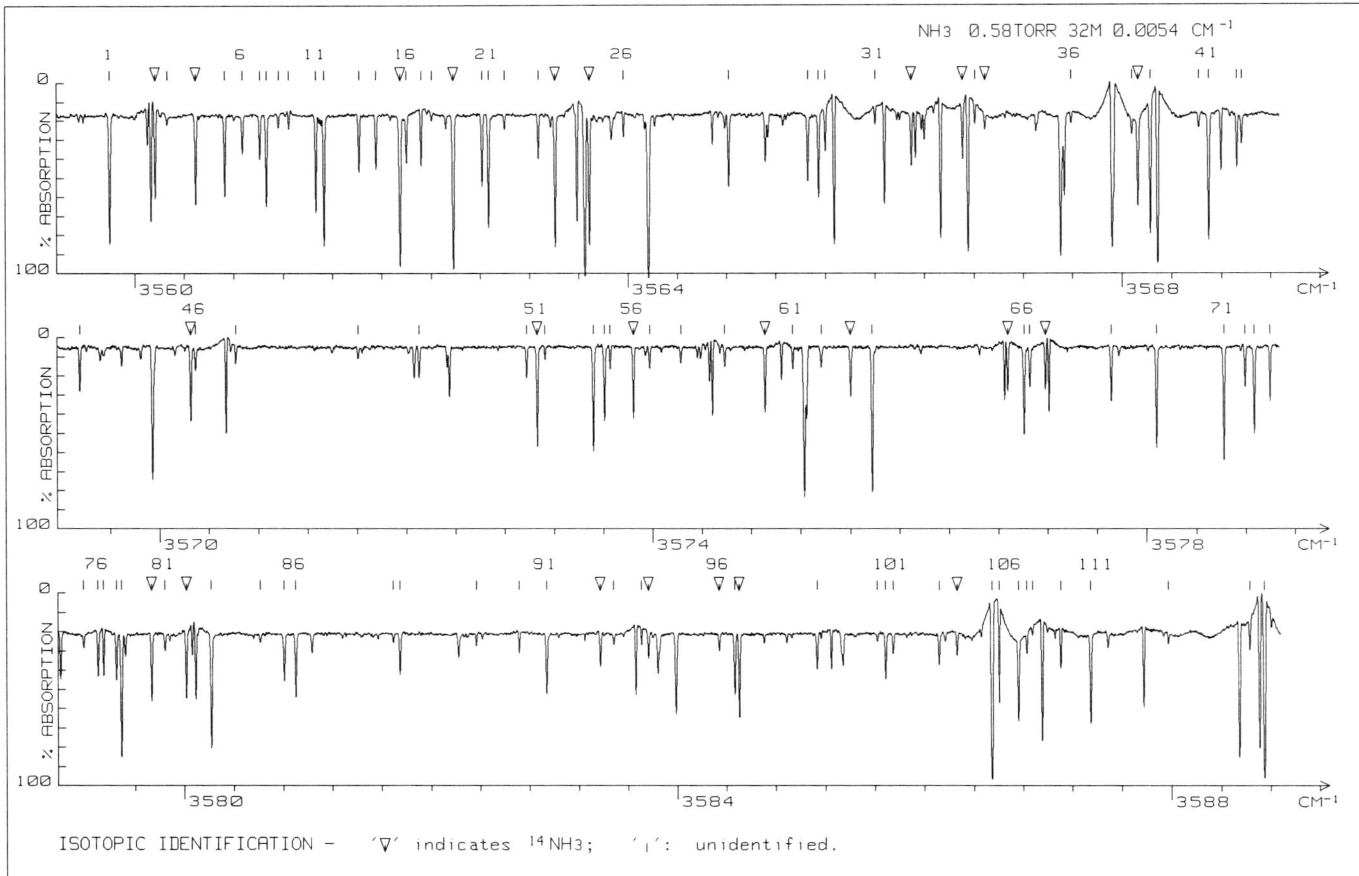
NH3 0.58TORR 32M 0.0054 CM⁻¹
% ABSORPTION
0
100
3560
3564
3568
3570
3574
3578
3580
3584
3588
CM⁻¹
1 6 11 16 21 26 31 36 41
46 51 56 61 66 71
76 81 86 91 96 101 106 111
ISOTOPIC IDENTIFICATION – '∇' indicates ¹⁴NH3; '|': unidentified.

LINE NO.	WAVENUMBER CM^{-1}, VAC.	ROTATIONAL BAND ASSIGNMENTS	ISOTOPIC SPECIES
61	3575.14128(51)		
62	3575.36853(53)		
63	3575.60357(27)	$s^{P}R$(6, 1) NU3	$^{14}NH_3$
64	3575.78075(20)		
65	3576.87675(51)	$a^{r}R$(7, 1) NU3	$^{14}NH_3$
66	3577.00876(22)		
67	3577.05672(35)		
68	3577.18359(27)	$s^{r}R$(7, 1) NU3	$^{14}NH_3$
69	3577.71617(28)		
70	3578.08090(21)		
71	3578.62201(20)		
72	3578.79466(33)		
73	3578.86863(26)		
74	3578.99725(33)		
75	3579.18528(71)		
76	3579.30167(33)		
77	3579.34571(21)		
78	3579.44949(46)		
79	3579.48972(19)		
80	3579.73509(25)	$a^{r}R$(8, 3) NU3	$^{14}NH_3$
81	3579.84330(65)		
82	3580.01799(25)	$s^{r}R$(8, 3) NU3	$^{14}NH_3$
83	3580.22068(20)		
84	3580.61463(102)		
85	3580.80776(32)		
86	3580.90121(28)		
87	3581.69529(94)		
88	3581.74828(27)		
89	3582.36927(51)		
90	3582.71709(34)		

LINE NO.	WAVENUMBER CM^{-1}, VAC.	ROTATIONAL BAND ASSIGNMENTS	ISOTOPIC SPECIES
91	3582.93863(28)		
92	3583.37638(34)	$a^{P}R$(6, 2) NU3	$^{14}NH_3$
93	3583.48396(88)		
94	3583.70993(42)		
95	3583.76628(42)	$s^{P}R$(6, 2) NU3	$^{14}NH_3$
96	3584.34089(66)	$s^{P}R$(5, 4) NU3	$^{14}NH_3$
97	3584.46880(28)		
98	3584.50564(20)	$s^{r}R$(7, 0) NU3	$^{14}NH_3$
99	3585.13323(32)		
100	3585.61917(103)		
101	3585.68728(28)		
102	3585.74777(49)		
103	3586.11957(36)		
104	3586.26289(40)	$a^{r}R$(8, 2) NU3	$^{14}NH_3$
105	3586.54380(22)		
106	3586.60401(20)		
107	3586.76149(22)		
108	3586.83018(49)		
109	3586.87655(94)		
110	3587.10837(23)		
111	3587.34968(21)		
112	3587.97572(58)		
113	3588.63247(43)		
114	3588.75038(22)		

- "LINE NO.": serial number identifying the spectral line in the map on the opposite page.
- "WAVENUMBER": experimental value.
- Uncertainty in the last digit of the wavenumber is given in parentheses.
- "pa" : parallel ; "pe" : perpendicular.

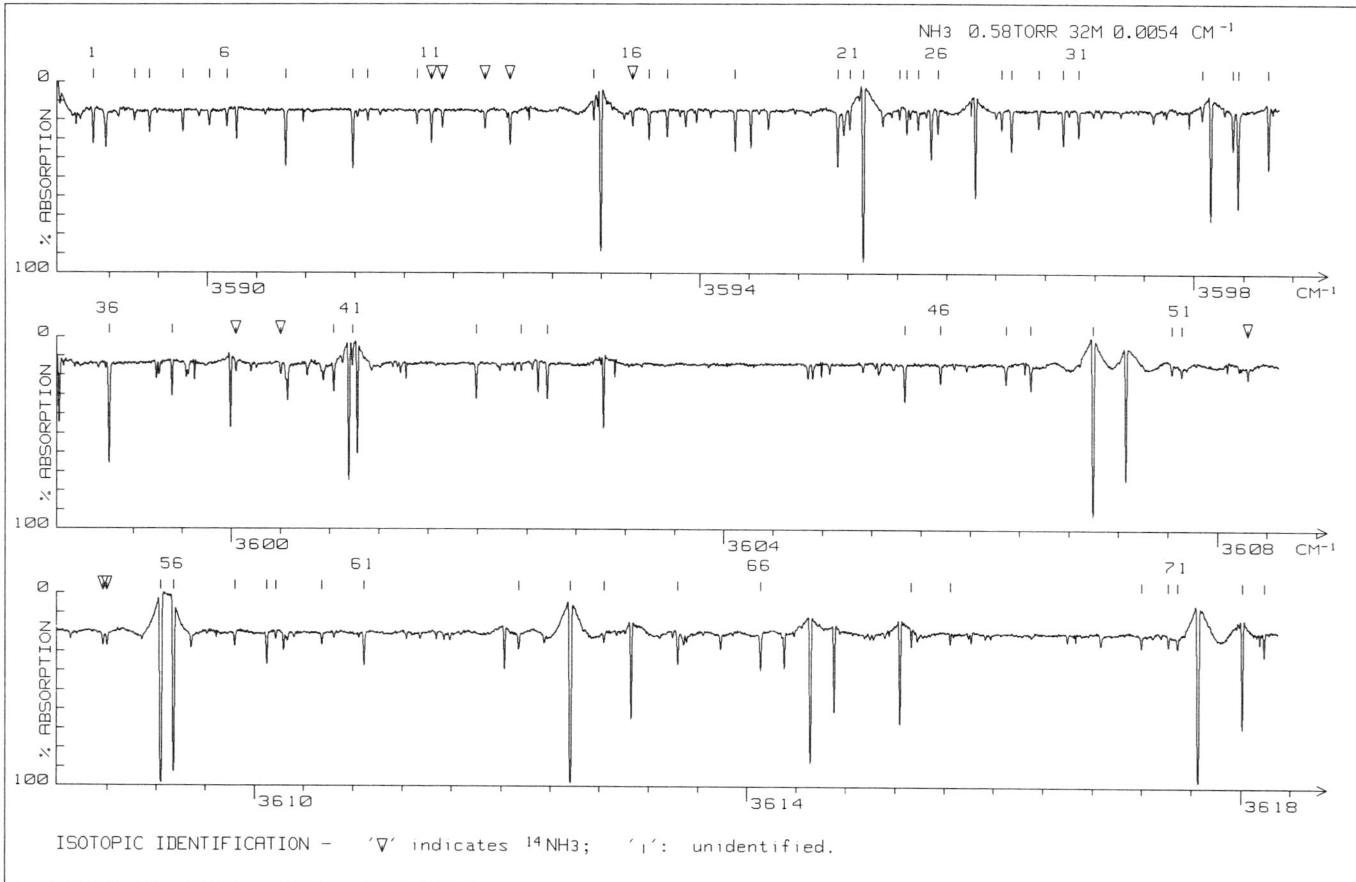
NH3 0.58TORR 32M 0.0054 CM⁻¹
% ABSORPTION
0
100
1
6
11
16
21
26
31
36
41
46
51
56
61
66
71
3590
3594
3598
3600
3604
3608
3610
3614
3618
CM⁻¹
ISOTOPIC IDENTIFICATION - '∇' indicates ¹⁴NH3; '|': unidentified.

LINE NO.	WAVENUMBER CM^{-1}, VAC.	ROTATIONAL BAND ASSIGNMENTS	ISOTOPIC SPECIES	LINE NO.	WAVENUMBER CM^{-1}, VAC.	ROTATIONAL BAND ASSIGNMENTS	ISOTOPIC SPECIES
1	3589.07248(32)			31	3597.06802(46)		
2	3589.40468(104)			32	3598.06687(82)		
3	3589.53267(49)			33	3598.31414(39)		
4	3589.80151(50)			34	3598.35532(21)		
5	3590.01938(77)			35	3598.60384(44)		
6	3590.15978(57)			36	3599.00580(21)		
7	3590.64126(31)			37	3599.52050(25)		
8	3591.18479(30)			38	3600.03912(98)	aPR(7, 2) NU3	14NH3
9	3591.30641(89)			39	3600.40372(90)	sPR(7, 2) NU3	14NH3
10	3591.70709(80)			40	3600.83528(44)		
11	3591.82436(39)	aPR(6, 3) NU3	14NH3	41	3600.98684(100)		
12	3591.91306(60)	aPR(7, 1) NU3	14NH3	42	3601.99182(44)		
13	3592.25486(61)	sPR(7, 1) NU3	14NH3	43	3602.35394(72)		
14	3592.45608(39)	sPR(6, 3) NU3	14NH3	44	3602.56508(41)		
15	3593.14131(54)			45	3605.46731(33)		
16	3593.45615(66)	srR(8, 1) NU3	14NH3	46	3605.75784(71)		
17	3593.58796(38)			47	3606.29334(66)		
18	3593.73331(42)			48	3606.49269(47)		
19	3594.28841(35)			49	3606.99415(16)		
20	3595.11991(26)			50	3607.63205(96)		
21	3595.21758(57)			51	3607.71201(96)		
22	3595.32668(16)			52	3608.24351(87)	sPR(8, 1) NU3	14NH3
23	3595.61814(91)			53	3608.76215(86)	aPR(7, 3) NU3	14NH3
24	3595.67792(45)			54	3608.79425(86)	sPR(7, 3) NU3	14NH3
25	3595.76910(57)			55	3609.23450(18)		
26	3595.92895(40)			56	3609.33952(20)		
27	3596.45056(63)			57	3609.83585(85)		
28	3596.52953(32)			58	3610.09664(42)		
29	3596.74699(54)			59	3610.16946(66)		
30	3596.94585(35)			60	3610.54552(85)		

- "LINE NO.": serial number identifying the spectral line in the map on the opposite page.
- "WAVENUMBER": experimental value.
- Uncertainty in the last digit of the wavenumber is given in parentheses.
- "pa" : parallel ; "pe" : perpendicular.

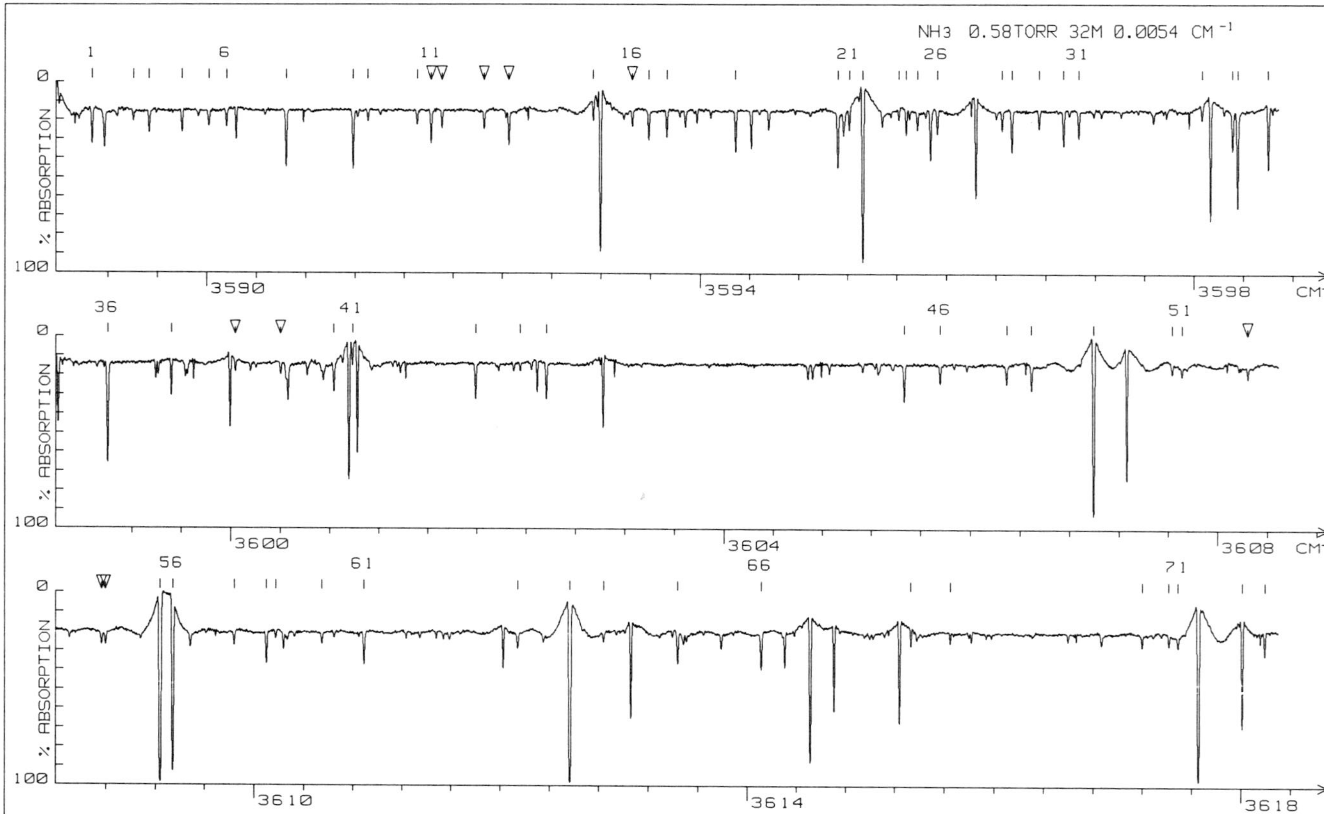

ISOTOPIC IDENTIFICATION – '∇' indicates $^{14}NH_3$; '|': unidentified.

LINE NO.	WAVENUMBER CM^{-1}, VAC.	ROTATIONAL BAND ASSIGNMENTS	ISOTOPIC SPECIES
61	3610.88784(36)		
62	3612.13994(76)		
63	3612.56325(18)		
64	3612.83639(95)		
65	3613.43410(37)		
66	3614.10866(40)		
67	3615.32886(42)		
68	3615.64528(62)		
69	3617.19594(85)		
70	3617.41307(81)		
71	3617.48767(111)		
72	3618.00677(21)		
73	3618.18718(33)		

- "LINE NO.": serial number identifying the spectral line in the map on the opposite page.
- "WAVENUMBER": experimental value.
- Uncertainty in the last digit of the wavenumber is given in parentheses.
- "pa" : parallel ; "pe" : perpendicular.

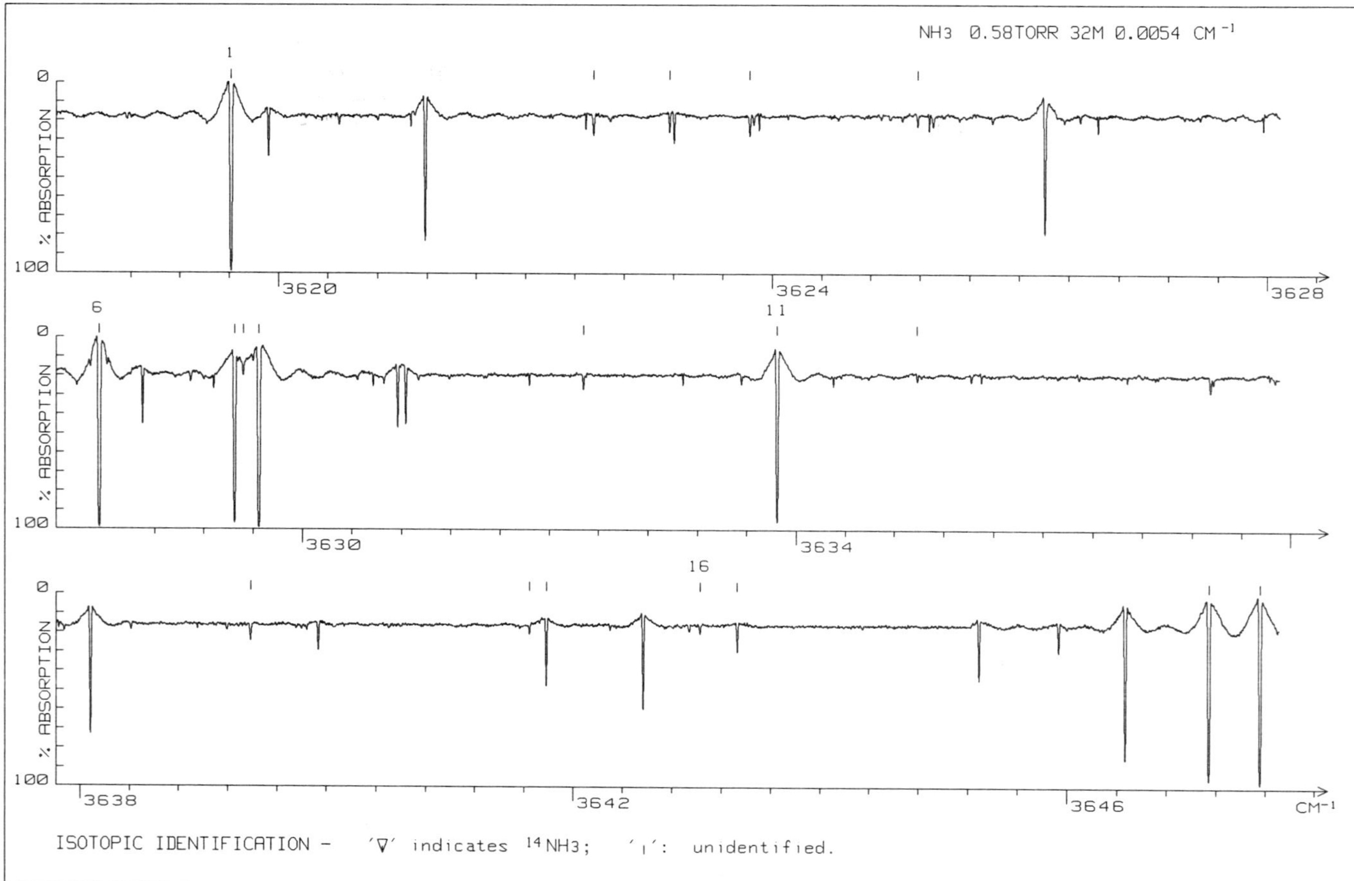
NH3 0.58TORR 32M 0.0054 CM⁻¹
% ABSORPTION
0
100
1
6
11
16
3620
3624
3628
3630
3634
3638
3642
3646
CM⁻¹
ISOTOPIC IDENTIFICATION - '∇' indicates ¹⁴NH3; '|': unidentified.

LINE NO.	WAVENUMBER CM^{-1}, VAC.	ROTATIONAL BAND ASSIGNMENTS	ISOTOPIC SPECIES
1	3619.61274(18)		
2	3622.55324(71)		
3	3623.16579(39)		
4	3623.81702(66)		
5	3625.17927(61)		
6	3628.34730(18)		
7	3629.44739(18)		
8	3629.51643(79)		
9	3629.64430(19)		
10	3632.27664(49)		
11	3633.84429(18)		
12	3634.98257(87)		
13	3639.38343(90)		
14	3641.64298(77)		
15	3641.77886(21)		
16	3643.02532(84)		
17	3643.32969(31)		
18	3647.13913(18)		
19	3647.55374(18)		

- "LINE NO.": serial number identifying the spectral line in the map on the opposite page.
- "WAVENUMBER": experimental value.
- Uncertainty in the last digit of the wavenumber is given in parentheses.
- "pa" : parallel ; "pe" : perpendicular.

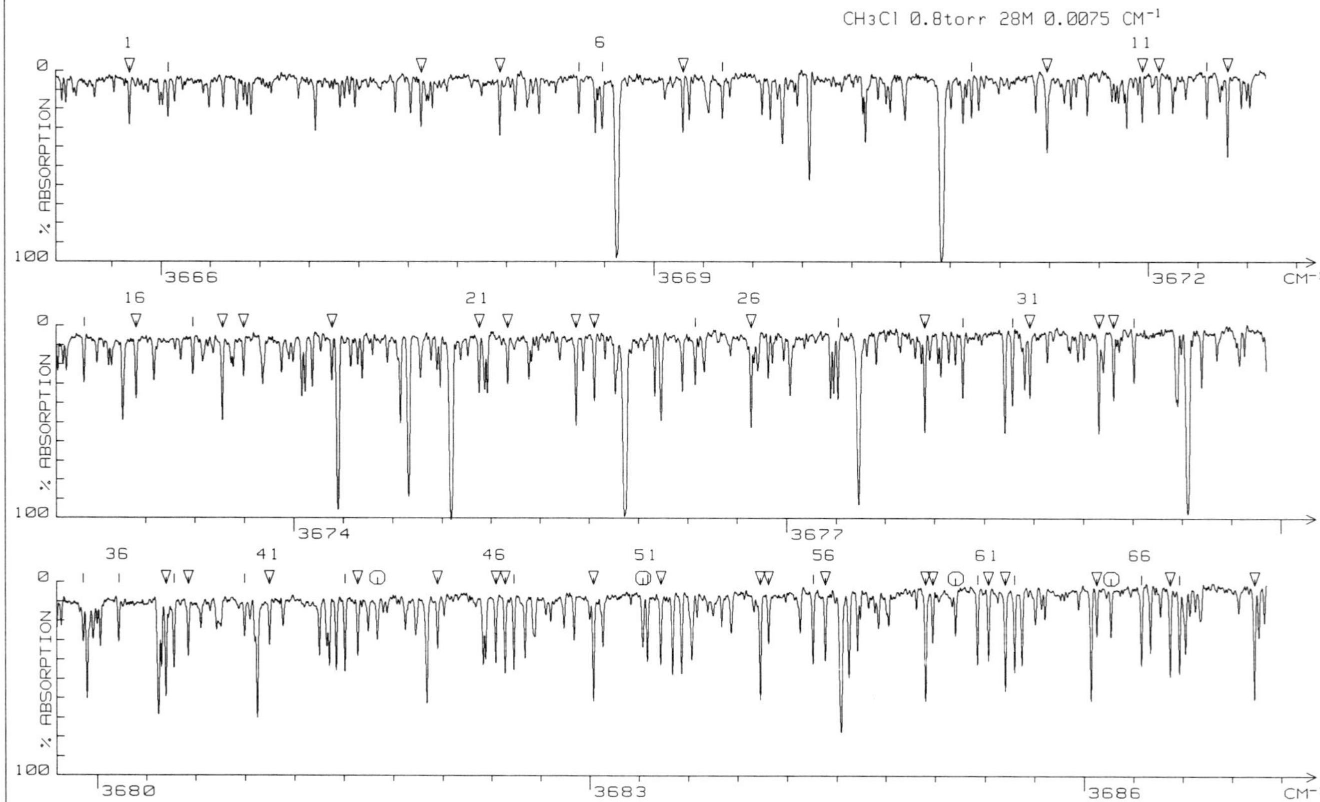

ISOTOPIC IDENTIFICATION – '▽' indicates $CH_3{}^{35}Cl$; '⓵': $CH_3{}^{37}Cl$; 'ı': unidentified.

LINE NO.	WAVENUMBER CM^{-1}, VAC.	UNC.	ROTATIONAL ASSIGNMENTS	BAND	ISOTOPIC SPECIES
1	3665.80923	67	${}^{Q}P_{3}$ (34)	NU1+NU3	$CH_3{}^{35}Cl$
2	3666.04558	81			
3	3667.57801	67	${}^{Q}P_{6}$ (31)	NU1+NU3	$CH_3{}^{35}Cl$
4	3668.06184	56	${}^{Q}P_{3}$ (32)	NU1+NU3	$CH_3{}^{35}Cl$
5	3668.54607	85			
6	3668.68768	81			
7	3669.17773	59	${}^{Q}P_{3}$ (31)	NU1+NU3	$CH_3{}^{35}Cl$
8	3669.41948	82			
9	3670.93013	70			
10	3671.38717	52	${}^{Q}P_{3}$ (29)	NU1+NU3	$CH_3{}^{35}Cl$
11	3671.96513	70	${}^{Q}P_{6}$ (27)	NU1+NU3	$CH_3{}^{35}Cl$
12	3672.06517	79	${}^{Q}P_{4}$ (28)	NU1+NU3	$CH_3{}^{35}Cl$
13	3672.35759	74			
14	3672.48183	43	${}^{Q}P_{3}$ (28)	NU1+NU3	$CH_3{}^{35}Cl$
15	3672.72519	71			
16	3673.04415	57	${}^{Q}P_{6}$ (26)	NU1+NU3	$CH_3{}^{35}Cl$
17	3673.39046	86			
18	3673.56874	41	${}^{Q}P_{3}$ (27)	NU1+NU3	$CH_3{}^{35}Cl$
19	3673.69703	80	${}^{Q}P_{5}$ (26)	NU1+NU3	$CH_3{}^{35}Cl$
20	3674.23198	75	${}^{Q}P_{4}$ (26)	NU1+NU3	$CH_3{}^{35}Cl$
21	3675.12915	59	${}^{Q}P_{1}$ (26)	NU1+NU3	$CH_3{}^{35}Cl$
22	3675.30385	71	${}^{Q}P_{4}$ (25)	NU1+NU3	$CH_3{}^{35}Cl$
23	3675.72059	41	${}^{Q}P_{3}$ (25)	NU1+NU3	$CH_3{}^{35}Cl$
24	3675.83437	54	${}^{Q}P_{5}$ (24)	NU1+NU3	$CH_3{}^{35}Cl$
			${}^{Q}P_{6}$ (18)	NU1+NU3	$CH_3{}^{37}Cl$
25	3676.44634	72			
26	3676.78605	40	${}^{Q}P_{3}$ (24)	NU1+NU3	$CH_3{}^{35}Cl$
27	3677.31165	48			
28	3677.84390	36	${}^{Q}P_{3}$ (23)	NU1+NU3	$CH_3{}^{35}Cl$
29	3678.07280	54			
30	3678.37248	47			
31	3678.47787	53	${}^{Q}P_{4}$ (22)	NU1+NU3	$CH_3{}^{35}Cl$
32	3678.89477	36	${}^{Q}P_{3}$ (22)	NU1+NU3	$CH_3{}^{35}Cl$
33	3678.98467	47	${}^{Q}P_{5}$ (21)	NU1+NU3	$CH_3{}^{35}Cl$
34	3679.11050	59			
35	3679.91397	86			
36	3680.13201	70			
37	3680.41888	35	${}^{Q}P_{1}$ (21)	NU1+NU3	$CH_3{}^{35}Cl$
38	3680.46971	47			
39	3680.55602	56	${}^{Q}P_{4}$ (20)	NU1+NU3	$CH_3{}^{35}Cl$
40	3680.89704	81			
41	3681.04770	67	${}^{Q}P_{5}$ (19)	NU1+NU3	$CH_3{}^{35}Cl$
42	3681.50703	42			
43	3681.58461	56	${}^{Q}P_{4}$ (19)	NU1+NU3	$CH_3{}^{35}Cl$
44	3681.70276	89	${}^{Q}P_{6}$ (12)	NU1+NU3	$CH_3{}^{37}Cl$
45	3682.06835	59	${}^{Q}P_{5}$ (18)	NU1+NU3	$CH_3{}^{35}Cl$
46	3682.42583	46	${}^{Q}P_{6}$ (17)	NU1+NU3	$CH_3{}^{35}Cl$
47	3682.48289	41	${}^{Q}P_{1}$ (19)	NU1+NU3	$CH_3{}^{35}Cl$
48	3682.53680	42			
49	3683.02418	33	${}^{Q}P_{3}$ (18)	NU1+NU3	$CH_3{}^{35}Cl$
50	3683.32366	71	${}^{Q}P_{5}$ (11)	NU1+NU3	$CH_3{}^{37}Cl$
51	3683.35187	47			
52	3683.43216	45	${}^{Q}P_{6}$ (16)	NU1+NU3	$CH_3{}^{35}Cl$
53	3684.03789	33	${}^{Q}P_{3}$ (17)	NU1+NU3	$CH_3{}^{35}Cl$
54	3684.08844	63	${}^{Q}P_{5}$ (16)	NU1+NU3	$CH_3{}^{35}Cl$
55	3684.35554	42			
56	3684.43036	48	${}^{Q}P_{6}$ (15)	NU1+NU3	$CH_3{}^{35}Cl$
57	3685.04451	31	${}^{Q}P_{3}$ (16)	NU1+NU3	$CH_3{}^{35}Cl$
58	3685.08707	57	${}^{Q}P_{5}$ (15)	NU1+NU3	$CH_3{}^{35}Cl$
59	3685.22495	66	${}^{Q}P_{3}$ (10)	NU1+NU3	$CH_3{}^{37}Cl$
60	3685.35564	41			

- "LINE NO.": serial number identifying the spectral line in the map on the opposite page.
- "WAVENUMBER": experimental value.
- Uncertainty in the last digit of the wavenumber is given in parentheses.

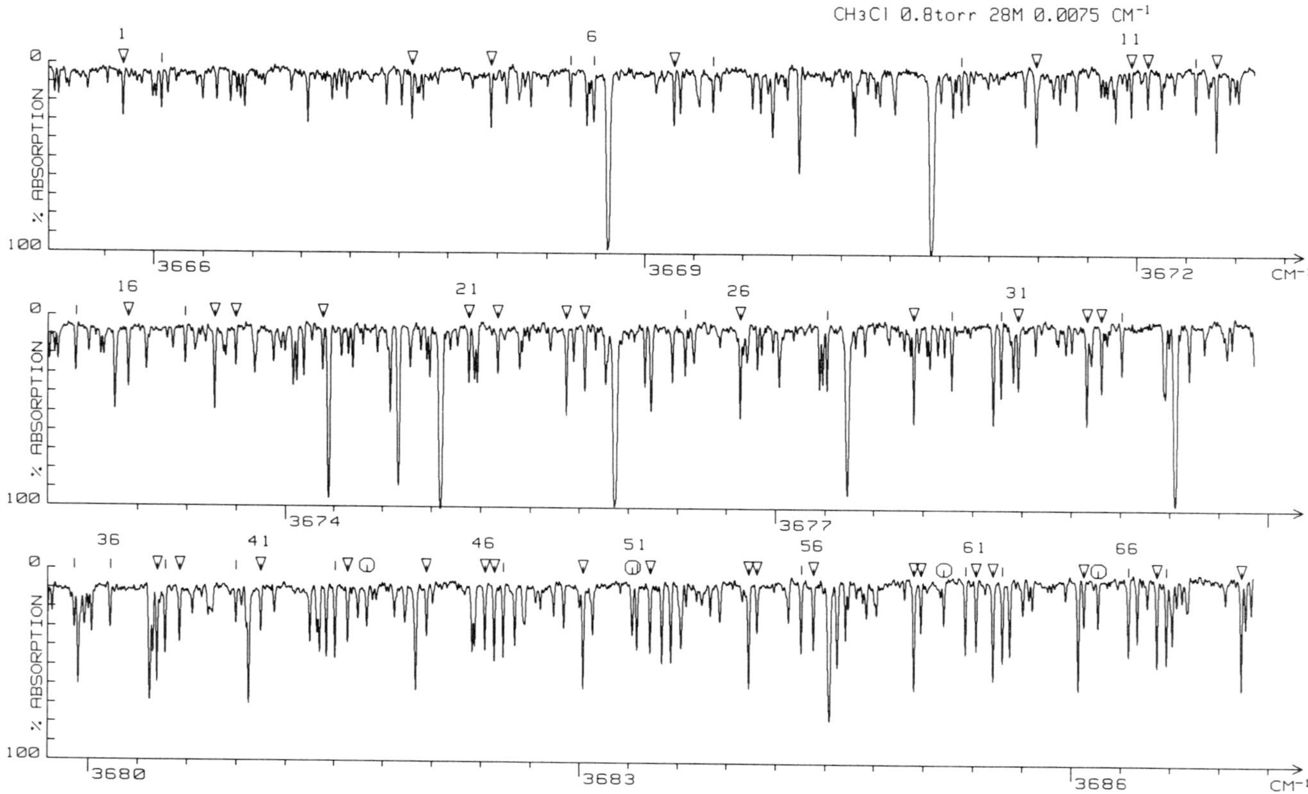

ISOTOPIC IDENTIFICATION – '∇' indicates $CH_3{}^{35}Cl$; 'ⓘ': $CH_3{}^{37}Cl$; 'ı': unidentified.

LINE NO.	WAVENUMBER CM^{-1}, VAC.	UNC.	ROTATIONAL ASSIGNMENTS	BAND	ISOTOPIC SPECIES
61	3685.42151	(43)	$^{Q}P_{6}$ (14)	NU1+NU3	$CH_3{}^{35}Cl$
62	3685.52489	(34)	$^{Q}P_{1}$ (16)	NU1+NU3	$CH_3{}^{35}Cl$
63	3685.58127	(37)			
64	3686.07827	(59)	$^{Q}P_{5}$ (14)	NU1+NU3	$CH_3{}^{35}Cl$
65	3686.16387	(63)	$^{Q}P_{3}$ (9)	NU1+NU3	$CH_3{}^{37}Cl$
66	3686.35084	(41)			
67	3686.52427	(37)	$^{Q}P_{1}$ (15)	NU1+NU3	$CH_3{}^{35}Cl$
68	3686.58078	(38)			
69	3687.03595	(31)	$^{Q}P_{3}$ (14)	NU1+NU3	$CH_3{}^{35}Cl$

- "LINE NO.": serial number identifying the spectral line in the map on the opposite page.
- "WAVENUMBER": experimental value.
- Uncertainty in the last digit of the wavenumber is given in parentheses.

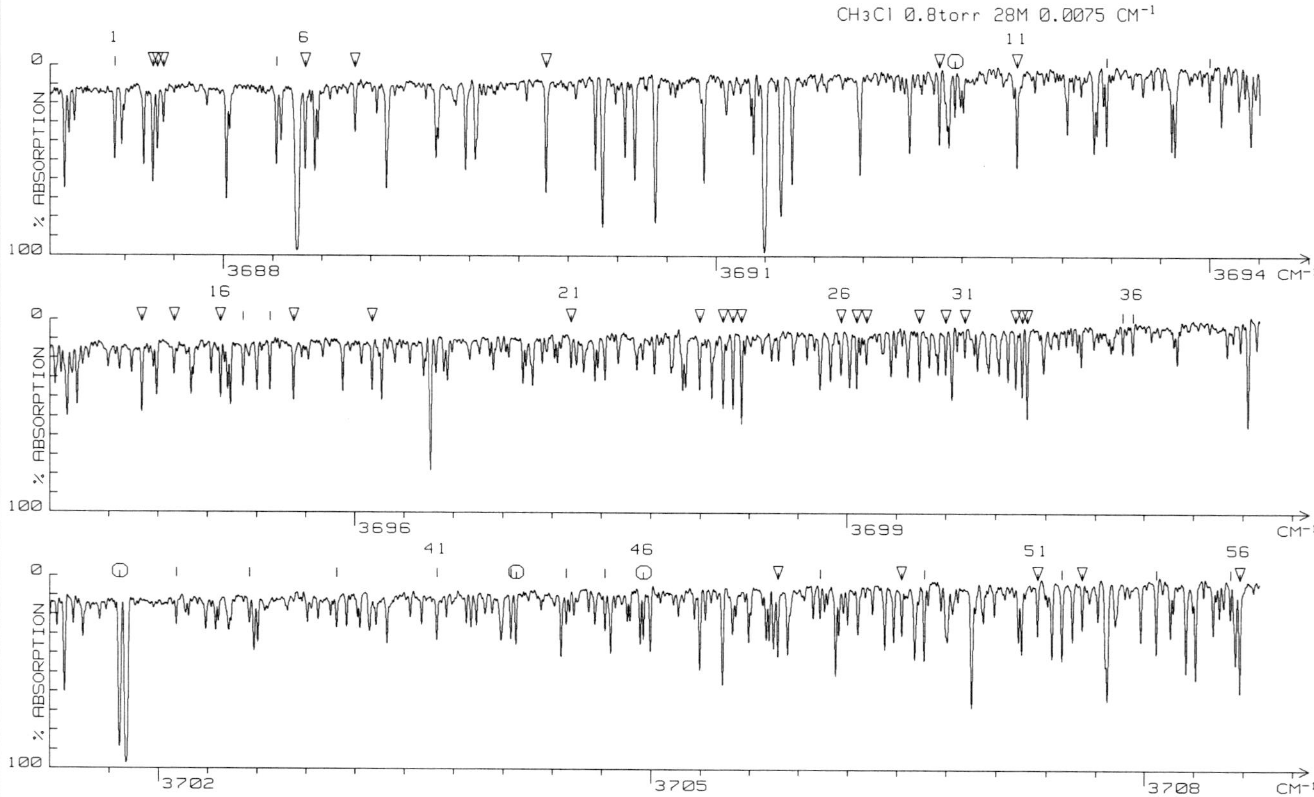

ISOTOPIC IDENTIFICATION – '▽' indicates $CH_3{}^{35}Cl$; 'ⓘ': $CH_3{}^{37}Cl$; 'ı': unidentified.

LINE NO.	WAVENUMBER CM^{-1}, VAC.	UNC.	ROTATIONAL ASSIGNMENTS	BAND	ISOTOPIC SPECIES
1	3687.34046	(41)			
2	3687.57379	(33)	$^{Q}P_{7}$ (11)	NU1+NU3	$CH_3{}^{35}Cl$
3	3687.60081	(44)	$^{Q}P_{4}$ (13)	NU1+NU3	$CH_3{}^{35}Cl$
4	3687.63701	(80)	$^{Q}P_{8}$ (10)	NU1+NU3	$CH_3{}^{35}Cl$
5	3688.32292	(37)			
6	3688.50015	(35)	$^{Q}P_{1}$ (13)	NU1+NU3	$CH_3{}^{35}Cl$
7	3688.80529	(60)	$^{Q}Q_{7}$ (48)	NU1+NU3	$CH_3{}^{35}Cl$
8	3689.96758	(30)	$^{Q}P_{3}$ (11)	NU1+NU3	$CH_3{}^{35}Cl$
9	3692.36396	(36)	$^{Q}P_{1}$ (9)	NU1+NU3	$CH_3{}^{35}Cl$
10	3692.45726	(55)	$^{Q}Q_{4}$ (19)	NU1+NU3	$CH_3{}^{37}Cl$
11	3692.83063	(32)	$^{Q}P_{3}$ (8)	NU1+NU3	$CH_3{}^{35}Cl$
12	3693.37000	(39)			
13	3693.99984	(78)			
14	3694.70213	(37)	$^{Q}P_{3}$ (6)	NU1+NU3	$CH_3{}^{35}Cl$
15	3694.89988	(78)	$^{Q}Q_{9}$ (15)	NU1+NU3	$CH_3{}^{35}Cl$
16	3695.18244	(43)	$^{Q}P_{1}$ (6)	NU1+NU3	$CH_3{}^{35}Cl$
17	3695.31853	(53)			
18	3695.48097	(44)			
19	3695.62705	(44)	$^{Q}P_{3}$ (5)	NU1+NU3	$CH_3{}^{35}Cl$
20	3696.10729	(48)	$^{Q}P_{1}$ (5)	NU1+NU3	$CH_3{}^{35}Cl$
21	3697.31813	(69)	$^{Q}Q_{7}$ (10)	NU1+NU3	$CH_3{}^{35}Cl$
22	3698.09693	(45)	$^{Q}Q_{6}$ (10)	NU1+NU3	$CH_3{}^{35}Cl$
23	3698.24225	(34)	$^{Q}Q_{6}$ (8)	NU1+NU3	$CH_3{}^{35}Cl$
24	3698.30227	(32)	$^{Q}Q_{6}$ (7)	NU1+NU3	$CH_3{}^{35}Cl$
25	3698.35548	(28)	$^{Q}Q_{6}$ (6)	NU1+NU3	$CH_3{}^{35}Cl$
26	3698.96273	(64)	$^{Q}Q_{5}$ (7)	NU1+NU3	$CH_3{}^{35}Cl$
27	3699.06084	(41)	$^{Q}Q_{5}$ (5)	NU1+NU3	$CH_3{}^{35}Cl$
28	3699.11939	(77)	$^{Q}Q_{4}$ (12)	NU1+NU3	$CH_3{}^{35}Cl$
29	3699.44180	(47)	$^{Q}Q_{3}$ (13)	NU1+NU3	$CH_3{}^{35}Cl$
			$^{Q}Q_{4}$ (8)	NU1+NU3	$CH_3{}^{35}Cl$
30	3699.60131	(51)	$^{Q}Q_{4}$ (5)	NU1+NU3	$CH_3{}^{35}Cl$
31	3699.71735	(71)	$^{Q}Q_{3}$ (10)	NU1+NU3	$CH_3{}^{35}Cl$
32	3700.02181	(39)	$^{Q}Q_{3}$ (5)	NU1+NU3	$CH_3{}^{35}Cl$
33	3700.05988	(35)	$^{Q}Q_{3}$ (4)	NU1+NU3	$CH_3{}^{35}Cl$
34	3700.09051	(29)	$^{Q}Q_{3}$ (3)	NU1+NU3	$CH_3{}^{35}Cl$
35	3700.67149	(77)			
36	3700.73271	(78)			
37	3701.76547	(26)	$^{Q}R_{6}$ (10)	NU1+NU3	$CH_3{}^{37}Cl$
38	3702.11098	(70)			
39	3702.55432	(86)			
40	3703.08795	(67)			
41	3703.70074	(51)			
42	3704.14803	(47)			
43	3704.17980	(43)	$^{Q}R_{3}$ (11)	NU1+NU3	$CH_3{}^{37}Cl$
44	3704.48579	(66)			
45	3704.72262	(53)			
46	3704.95610	(43)	$^{Q}R_{3}$ (12)	NU1+NU3	$CH_3{}^{37}Cl$
47	3705.77738	(33)	$^{Q}R_{1}$ (5)	NU1+NU3	$CH_3{}^{35}Cl$
48	3706.03415	(61)			
49	3706.53290	(44)	$^{Q}R_{4}$ (7)	NU1+NU3	$CH_3{}^{35}Cl$
50	3706.67028	(30)			
51	3707.35068	(38)	$^{Q}R_{4}$ (8)	NU1+NU3	$CH_3{}^{35}Cl$
52	3707.49566	(29)			
53	3707.61890	(44)	$^{Q}R_{5}$ (9)	NU1+NU3	$CH_3{}^{35}Cl$
54	3708.07370	(30)			
55	3708.52394	(52)			
56	3708.58121	(24)	$^{Q}R_{3}$ (9)	NU1+NU3	$CH_3{}^{35}Cl$

- "LINE NO.": serial number identifying the spectral line in the map on the opposite page.
- "WAVENUMBER": experimental value.
- Uncertainty in the last digit of the wavenumber is given in parentheses.

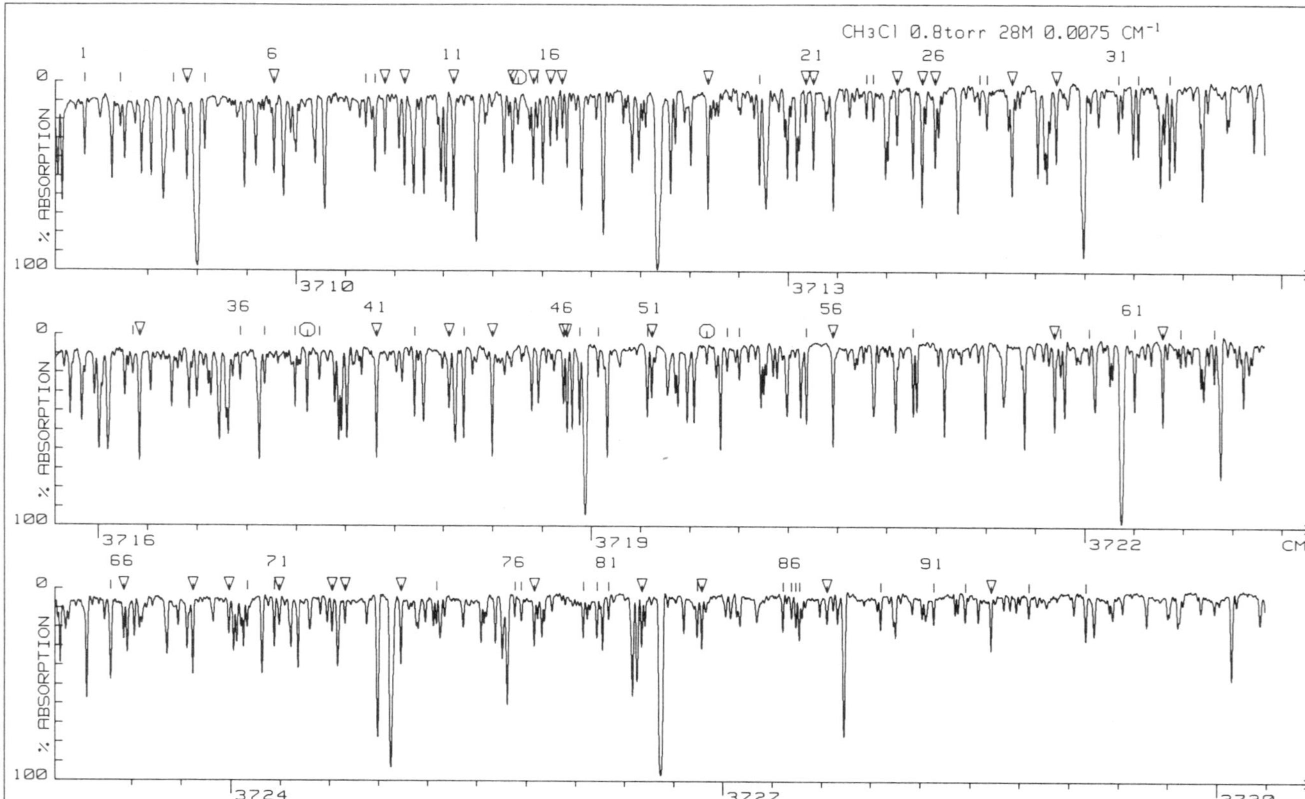

ISOTOPIC IDENTIFICATION – '∇' indicates $CH_3{}^{35}Cl$; 'Ⓘ': $CH_3{}^{37}Cl$; 'ı': unidentified.

LINE NO.	WAVENUMBER CM^{-1}, VAC.	UNC.	ROTATIONAL ASSIGNMENTS	BAND	ISOTOPIC SPECIES
1	3708.71876	(35)			
2	3708.93594	(70)			
3	3709.25974	(37)			
4	3709.34000	(29)	$^{Q}R_{6}$ (12)	NU1+NU3	$CH_3{}^{35}Cl$
5	3709.44845	(38)			
6	3709.86604	(27)	$^{Q}R_{1}$ (10)	NU1+NU3	$CH_3{}^{35}Cl$
7	3710.42660	(56)			
8	3710.48241	(27)			
9	3710.54290	(33)	$^{Q}R_{4}$ (12)	NU1+NU3	$CH_3{}^{35}Cl$
10	3710.66112	(25)	$^{Q}R_{1}$ (11)	NU1+NU3	$CH_3{}^{35}Cl$
11	3710.96503	(22)	$^{Q}R_{3}$ (12)	NU1+NU3	$CH_3{}^{35}Cl$
12	3711.32212	(31)	$^{Q}R_{4}$ (13)	NU1+NU3	$CH_3{}^{35}Cl$
13	3711.35503	(71)	$^{Q}R_{6}$ (23)	NU1+NU3	$CH_3{}^{37}Cl$
14	3711.44741	(27)	$^{Q}R_{1}$ (12)	NU1+NU3	$CH_3{}^{35}Cl$
15	3711.47273	(62)			
16	3711.55086	(37)	$^{Q}R_{5}$ (14)	NU1+NU3	$CH_3{}^{35}Cl$
17	3711.62330	(45)	$^{Q}R_{7}$ (16)	NU1+NU3	$CH_3{}^{35}Cl$
18	3712.51402	(21)	$^{Q}R_{3}$ (14)	NU1+NU3	$CH_3{}^{35}Cl$
19	3712.82779	(26)			
20	3713.10796	(59)	$^{Q}R_{7}$ (18)	NU1+NU3	$CH_3{}^{35}Cl$
21	3713.15549	(27)	$^{Q}R_{6}$ (17)	NU1+NU3	$CH_3{}^{35}Cl$
22	3713.48074	(61)			
23	3713.52277	(58)			
24	3713.66476	(37)	$^{Q}R_{8}$ (20)	NU1+NU3	$CH_3{}^{35}Cl$
25	3713.81675	(21)	$^{Q}R_{5}$ (17)	NU1+NU3	$CH_3{}^{35}Cl$
26	3713.89506	(28)	$^{Q}R_{6}$ (18)	NU1+NU3	$CH_3{}^{35}Cl$
27	3714.16116	(65)			
28	3714.20677	(50)			
29	3714.36047	(25)	$^{Q}R_{4}$ (17)	NU1+NU3	$CH_3{}^{35}Cl$
30	3714.62695	(28)	$^{Q}R_{6}$ (19)	NU1+NU3	$CH_3{}^{35}Cl$

LINE NO.	WAVENUMBER CM^{-1}, VAC.	UNC.	ROTATIONAL ASSIGNMENTS	BAND	ISOTOPIC SPECIES
31	3715.00576	(45)			
32	3715.12820	(29)			
33	3715.31809	(24)			
34	3716.20954	(82)			
35	3716.25050	(22)	$^{Q}R_{3}$ (19)	NU1+NU3	$CH_3{}^{35}Cl$
36	3716.85810	(63)			
37	3717.00693	(56)			
38	3717.19276	(31)			
39	3717.26782	(29)	$^{Q}R_{5}$ (31)	NU1+NU3	$CH_3{}^{37}Cl$
40	3717.34240	(62)			
41	3717.68966	(22)	$^{Q}R_{3}$ (21)	NU1+NU3	$CH_3{}^{35}Cl$
42	3717.92105	(28)			
43	3718.13558	(38)	$^{Q}R_{5}$ (23)	NU1+NU3	$CH_3{}^{35}Cl$
44	3718.22429	(24)			
45	3718.39743	(21)	$^{Q}R_{3}$ (22)	NU1+NU3	$CH_3{}^{35}Cl$
46	3718.82731	(33)	$^{Q}R_{5}$ (24)	NU1+NU3	$CH_3{}^{35}Cl$
			$^{Q}R_{3}$ (32)	NU1+NU3	$CH_3{}^{37}Cl$
47	3718.85013	(25)	$^{Q}R_{6}$ (25)	NU1+NU3	$CH_3{}^{35}Cl$
48	3718.92855	(26)			
49	3719.04115	(55)			
50	3719.34090	(29)			
51	3719.36782	(35)	$^{Q}R_{4}$ (24)	NU1+NU3	$CH_3{}^{35}Cl$
52	3719.70250	(85)	$^{Q}R_{5}$ (35)	NU1+NU3	$CH_3{}^{37}Cl$
53	3719.82855	(73)			
54	3719.90271	(50)			
55	3720.31010	(25)			
56	3720.47315	(22)	$^{Q}R_{3}$ (25)	NU1+NU3	$CH_3{}^{35}Cl$
57	3720.95994	(27)			
58	3721.81681	(23)	$^{Q}R_{3}$ (27)	NU1+NU3	$CH_3{}^{35}Cl$
59	3721.85321	(61)			
60	3722.02614	(88)			

- "LINE NO.": serial number identifying the spectral line in the map on the opposite page.
- "WAVENUMBER": experimental value.
- Uncertainty in the last digit of the wavenumber is given in parentheses.

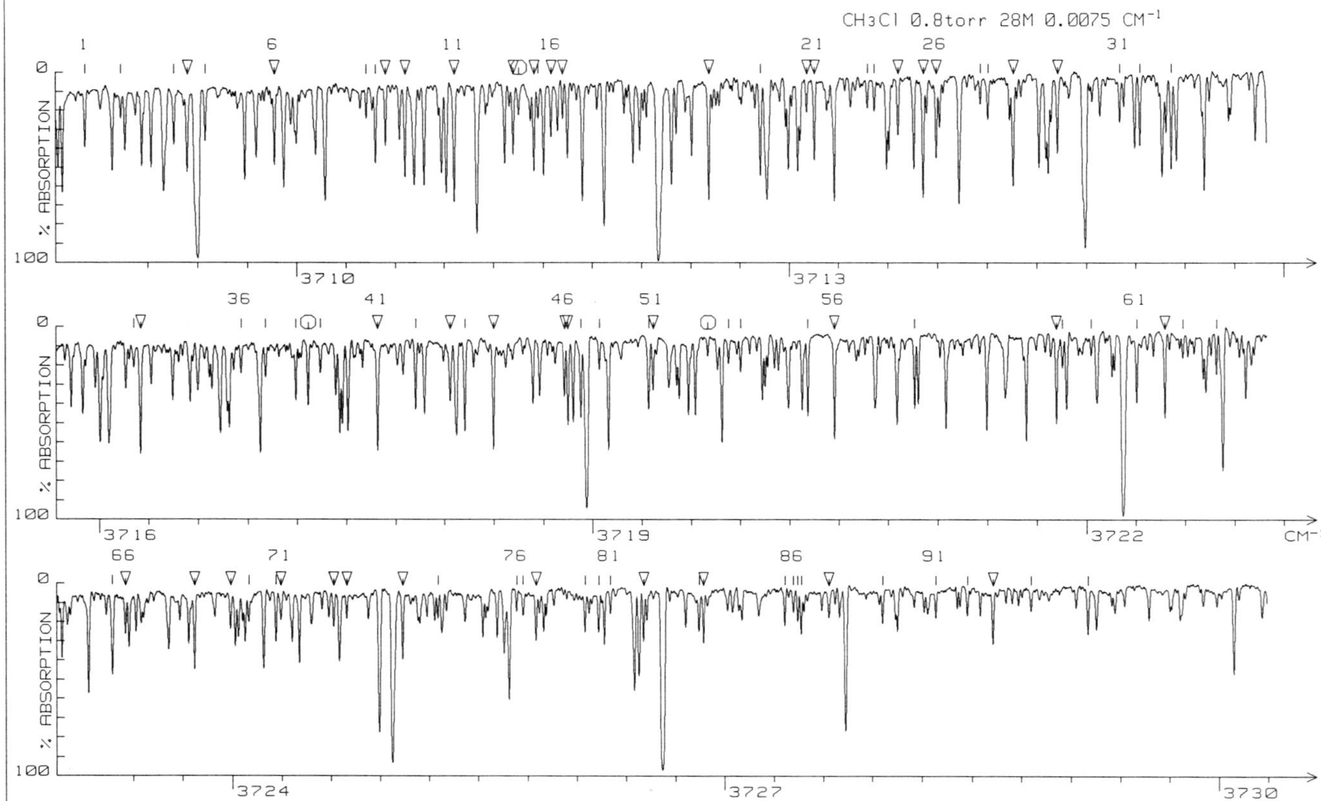

ISOTOPIC IDENTIFICATION - '∇' indicates $CH_3{}^{35}Cl$; 'ⓘ': $CH_3{}^{37}Cl$; 'ı': unidentified.

LINE NO.	WAVENUMBER CM^{-1}, VAC.	UNC.	ROTATIONAL ASSIGNMENTS	BAND	ISOTOPIC SPECIES
61	3722.30485	(29)			
62	3722.47644	(24)	$^{Q}R_{3}$ (28)	NU1+NU3	$CH_3{}^{35}Cl$
63	3722.58532	(71)			
64	3722.78821	(40)			
65	3723.27339	(28)			
66	3723.35177	(44)	$^{Q}R_{4}$ (30)	NU1+NU3	$CH_3{}^{35}Cl$
67	3723.77207	(25)	$^{Q}R_{3}$ (30)	NU1+NU3	$CH_3{}^{35}Cl$
68	3723.98758	(45)	$^{Q}R_{4}$ (31)	NU1+NU3	$CH_3{}^{35}Cl$
69	3724.09820	(66)			
70	3724.26256	(33)			
71	3724.29293	(69)	$^{Q}R_{9}$ (38)	NU1+NU3	$CH_3{}^{35}Cl$
72	3724.61580	(48)	$^{Q}R_{4}$ (32)	NU1+NU3	$CH_3{}^{35}Cl$
73	3724.69496	(64)	$^{Q}R_{5}$ (33)	NU1+NU3	$CH_3{}^{35}Cl$
74	3725.03504	(27)	$^{Q}R_{3}$ (32)	NU1+NU3	$CH_3{}^{35}Cl$
75	3725.25417	(61)			
76	3725.73094	(60)			
77	3725.77052	(72)			
78	3725.84819	(36)	$^{Q}R_{4}$ (34)	NU1+NU3	$CH_3{}^{35}Cl$
79	3726.14792	(42)			
80	3726.23091	(43)			
81	3726.30296	(64)			
82	3726.50419	(35)	$^{Q}R_{5}$ (36)	NU1+NU3	$CH_3{}^{35}Cl$
83	3726.84085	(45)			
84	3726.86933	(34)	$^{Q}R_{3}$ (35)	NU1+NU3	$CH_3{}^{35}Cl$
85	3727.36443	(42)			
86	3727.41703	(71)			
87	3727.44359	(50)			
88	3727.46427	(39)			
89	3727.63247	(66)	$^{Q}R_{4}$ (37)	NU1+NU3	$CH_3{}^{35}Cl$
90	3727.96175	(48)			
91	3728.28232	(59)			
92	3728.47123	(59)			
93	3728.62803	(34)	$^{Q}R_{8}$ (44)	NU1+NU3	$CH_3{}^{35}Cl$
94	3728.85894	(76)			
95	3729.20208	(41)			

- "LINE NO.": serial number identifying the spectral line in the map on the opposite page.
- "WAVENUMBER": experimental value.
- Uncertainty in the last digit of the wavenumber is given in parentheses.

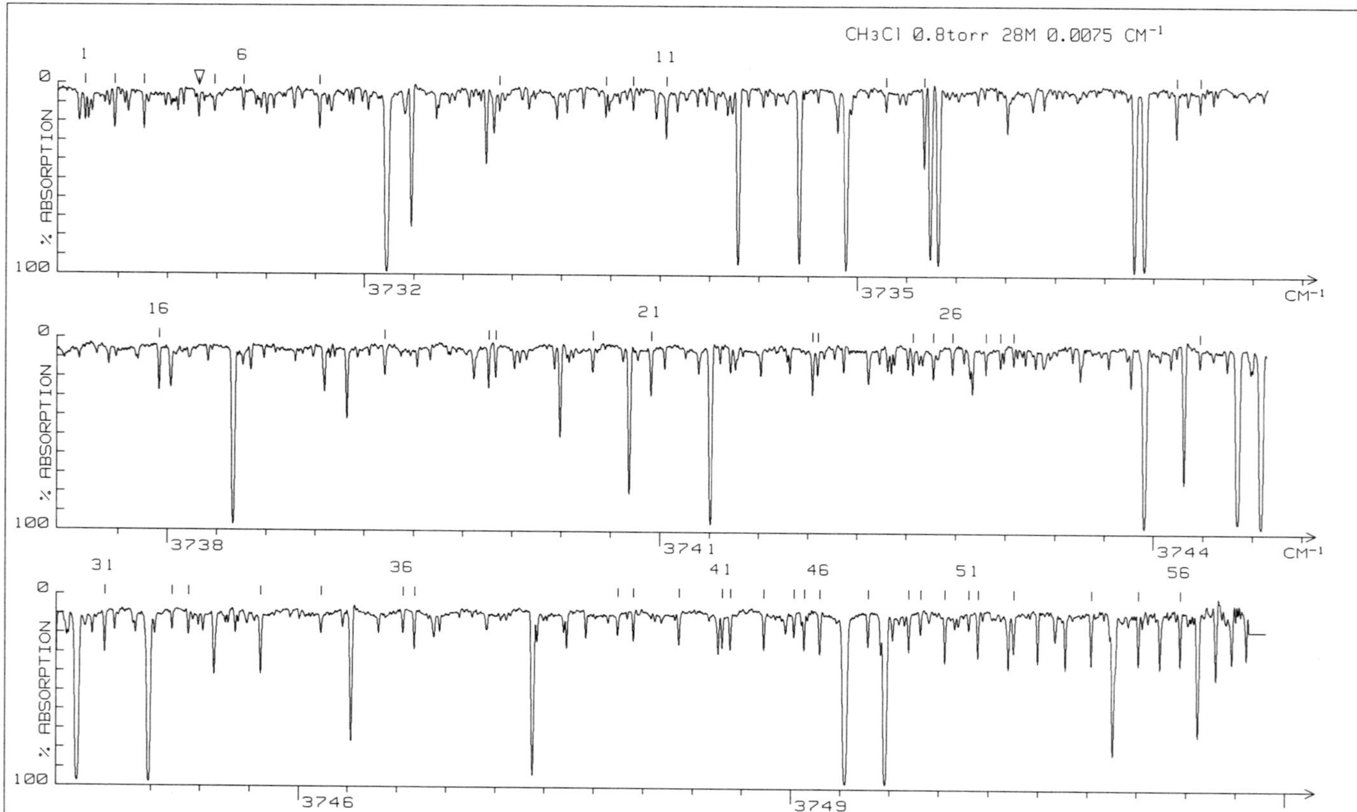

ISOTOPIC IDENTIFICATION – '∇' indicates $CH_3{}^{35}Cl$; 'Ⓘ': $CH_3{}^{37}Cl$; 'ı': unidentified.

LINE NO.	WAVENUMBER UNC. CM^{-1}, VAC.	ROTATIONAL ASSIGNMENTS	BAND	ISOTOPIC SPECIES	LINE NO.	WAVENUMBER UNC. CM^{-1}, VAC.	ROTATIONAL ASSIGNMENTS	BAND	ISOTOPIC SPECIES
1	3730.29687 (55)				31	3744.81985 (37)			
2	3730.47688 (49)				32	3745.23082 (71)			
3	3730.65428 (44)				33	3745.33006 (64)			
4	3730.98859 (61)	$^{Q}R_{4}$ (43)	NU1+NU3	$CH_3{}^{35}Cl$	34	3745.76792 (29)			
5	3731.08440 (106)				35	3746.13661 (87)			
6	3731.25872 (73)				36	3746.63686 (67)			
7	3731.72389 (44)				37	3746.70526 (39)			
8	3732.82045 (81)				38	3747.94545 (74)			
9	3733.46784 (57)				39	3748.04083 (47)			
10	3733.63382 (73)				40	3748.31617 (48)			
11	3733.83467 (37)				41	3748.58074 (45)			
12	3735.17324 (76)				42	3748.62917 (45)			
13	3735.40719 (27)				43	3748.83483 (39)			
14	3736.93480 (37)				44	3749.01729 (59)			
15	3737.07783 (67)				45	3749.07883 (40)			
16	3737.95007 (37)				46	3749.17426 (42)			
17	3739.32475 (62)				47	3749.46910 (44)			
18	3739.95576 (37)				48	3749.71834 (40)			
19	3739.99905 (52)				49	3749.78925 (66)			
20	3740.59223 (75)				50	3749.93813 (32)			
21	3740.94552 (37)				51	3750.08125 (79)			
22	3741.92730 (37)				52	3750.13574 (32)			
23	3741.95822 (65)				53	3750.35310 (43)			
24	3742.54075 (58)				54	3750.82297 (30)			
25	3742.66461 (57)				55	3751.11147 (30)			
26	3742.77980 (58)				56	3751.36414 (28)			
27	3742.98272 (66)								
28	3743.07112 (73)								
29	3743.15081 (76)								
30	3744.28038 (79)								

- "LINE NO.": serial number identifying the spectral line in the map on the opposite page.
- "WAVENUMBER": experimental value.
- Uncertainty in the last digit of the wavenumber is given in parentheses.

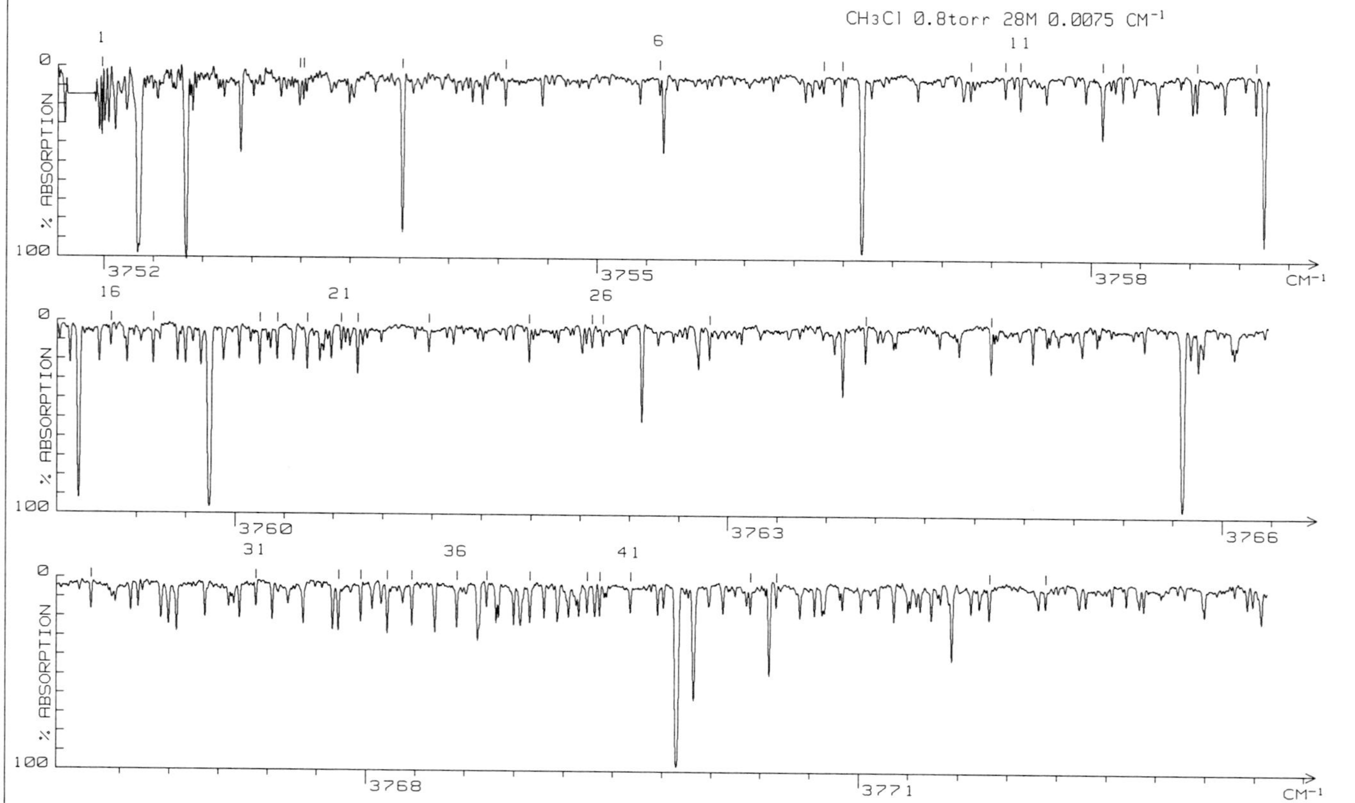

ISOTOPIC IDENTIFICATION – '∇' indicates $CH_3{}^{35}Cl$; 'Ⓘ': $CH_3{}^{37}Cl$; 'ı': unidentified.

LINE NO.	WAVENUMBER CM^{-1}, VAC.	UNC.	ROTATIONAL BAND ASSIGNMENTS	ISOTOPIC SPECIES
1	3751.98829	(20)		
2	3753.19262	(57)		
3	3753.21622	(60)		
4	3753.81882	(20)		
5	3754.44352	(53)		
6	3755.38057	(79)		
7	3756.37927	(87)		
8	3756.49283	(50)		
9	3757.26944	(69)		
10	3757.48174	(67)		
11	3757.57209	(45)		
12	3758.06805	(33)		
13	3758.19144	(55)		
14	3758.64097	(45)		
15	3758.99704	(40)		
16	3759.24306	(68)		
17	3759.50089	(44)		
18	3760.14990	(42)		
19	3760.25640	(47)		
20	3760.44045	(41)		
21	3760.64902	(61)		
22	3760.74757	(37)		
23	3761.18078	(69)		
24	3761.78860	(45)		
25	3762.17206	(86)		
26	3762.23736	(89)		
27	3762.88611	(48)		
28	3763.83954	(42)		
29	3764.59945	(39)		
30	3766.32484	(53)		
31	3767.32885	(58)		
32	3767.83312	(40)		
33	3767.96816	(40)		
34	3768.12919	(35)		
35	3768.27968	(39)		
36	3768.55430	(39)		
37	3768.73395	(60)		
38	3768.99481	(48)		
39	3769.34372	(59)		
40	3769.42071	(48)		
41	3769.60693	(58)		
42	3770.34109	(59)		
43	3770.49664	(57)		
44	3771.78982	(55)		
45	3772.13045	(70)		

- "LINE NO.": serial number identifying the spectral line in the map on the opposite page.
- "WAVENUMBER": experimental value.
- Uncertainty in the last digit of the wavenumber is given in parentheses.

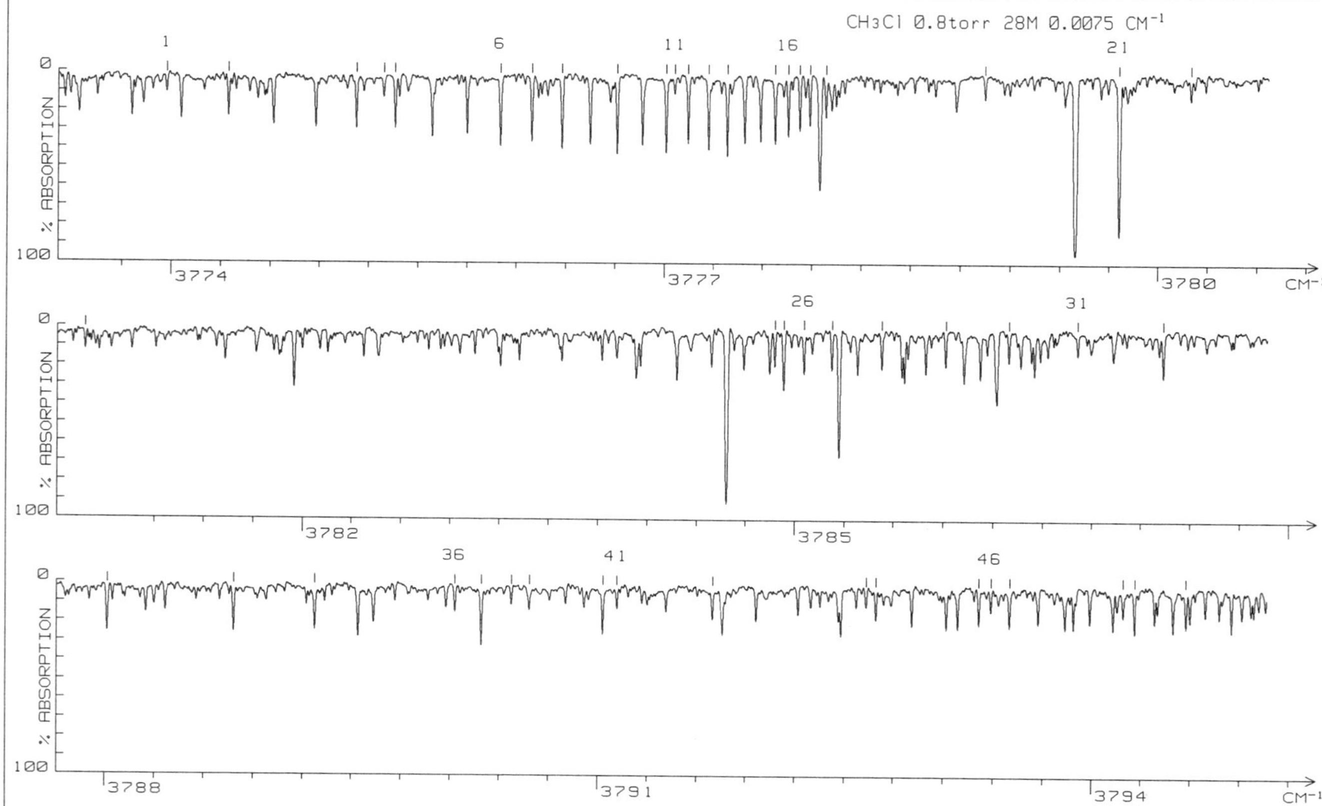

ISOTOPIC IDENTIFICATION – '∇' indicates $CH_3{}^{35}Cl$; 'Ⓘ': $CH_3{}^{37}Cl$; 'ı': unidentified.

LINE NO.	WAVENUMBER UNC. CM^{-1}, VAC.	ROTATIONAL BAND ASSIGNMENTS	ISOTOPIC SPECIES	LINE NO.	WAVENUMBER UNC. CM^{-1}, VAC.	ROTATIONAL BAND ASSIGNMENTS	ISOTOPIC SPECIES
1	3773.97133(85)			31	3786.71302(67)		
2	3774.34544(38)			32	3787.23620(37)		
3	3775.12431(31)			33	3788.01516(37)		
4	3775.29073(78)			34	3788.78296(37)		
5	3775.36000(31)			35	3789.27704(41)		
6	3776.00207(26)			36	3790.13075(55)		
7	3776.19390(26)			37	3790.29238(30)		
8	3776.37597(26)			38	3790.47590(76)		
9	3776.71229(24)			39	3790.58464(78)		
10	3777.00875(25)			40	3791.03138(35)		
11	3777.06150(88)			41	3791.11753(67)		
12	3777.14336(25)			42	3791.69794(55)		
13	3777.26833(26)			43	3792.63565(83)		
14	3777.38407(24)			44	3792.69599(49)		
15	3777.67630(26)			45	3793.31688(41)		
16	3777.75582(29)			46	3793.38817(64)		
17	3777.82622(29)			47	3793.50114(43)		
18	3777.88788(31)			48	3794.19364(63)		
19	3777.98438(39)			49	3794.26466(37)		
20	3778.94893(67)			50	3794.57743(43)		
21	3779.76240(20)						
22	3780.20075(64)						
23	3780.67962(71)						
24	3784.87795(42)						
25	3784.93203(32)						
26	3785.05634(40)						
27	3785.22387(38)						
28	3785.52944(40)						
29	3785.91433(42)						
30	3786.29522(48)						

- "LINE NO.": serial number identifying the spectral line in the map on the opposite page.
- "WAVENUMBER": experimental value.
- Uncertainty in the last digit of the wavenumber is given in parentheses.

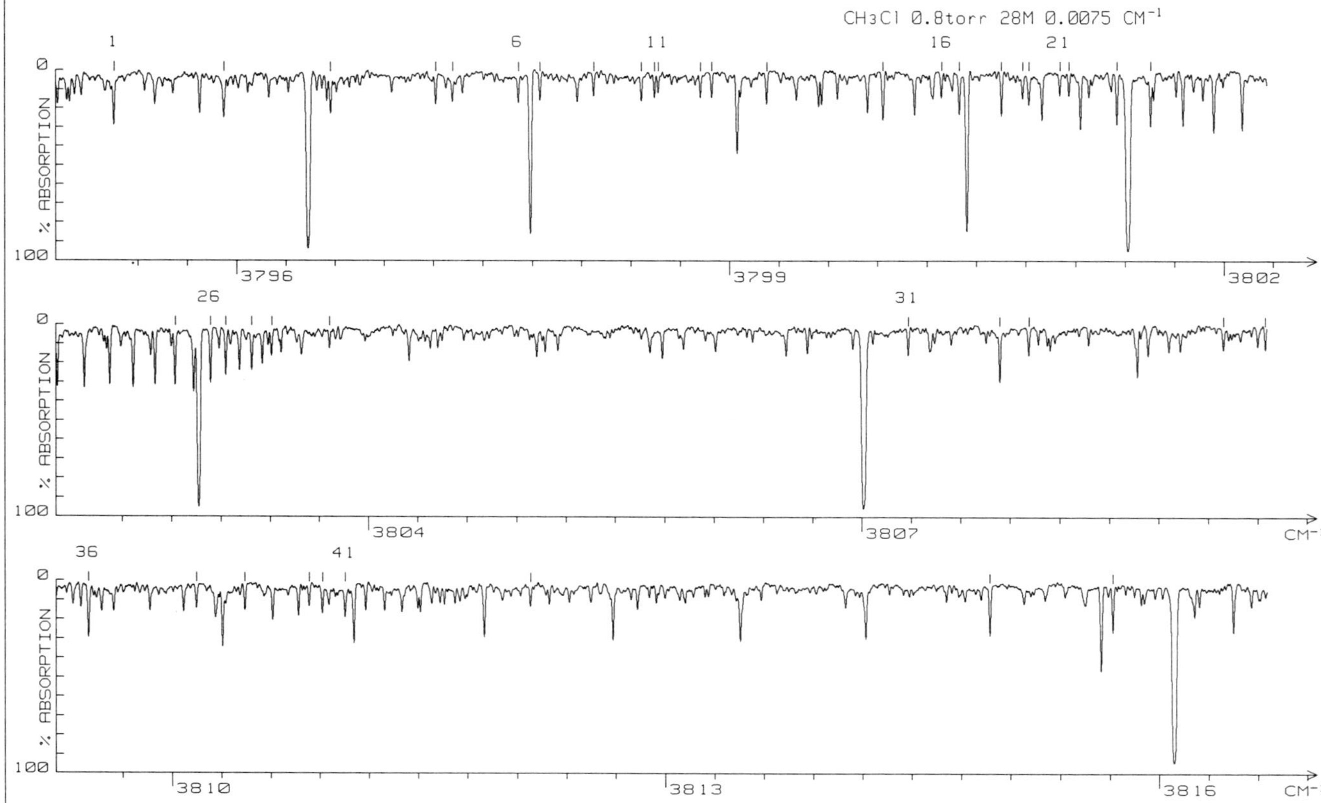

ISOTOPIC IDENTIFICATION – '∇' indicates $CH_3{}^{35}Cl$; 'Ⓘ': $CH_3{}^{37}Cl$; 'ı': unidentified.

LINE NO.	WAVENUMBER UNC. CM^{-1}, VAC.	ROTATIONAL BAND ASSIGNMENTS	ISOTOPIC SPECIES
1	3795.25488(34)		
2	3795.92205(46)		
3	3796.57441(44)		
4	3797.21671(47)		
5	3797.31830(61)		
6	3797.71545(54)		
7	3797.84717(50)		
8	3798.17533(60)		
9	3798.46381(55)		
10	3798.54302(57)		
11	3798.56615(72)		
12	3798.82301(59)		
13	3798.89328(54)		
14	3799.22491(45)		
15	3799.93399(38)		
16	3800.28817(57)		
17	3800.39697(38)		
18	3800.65451(35)		
19	3800.78226(63)		
20	3800.81984(57)		
21	3801.00474(63)		
22	3801.05939(62)		
23	3801.34936(32)		
24	3801.55600(32)		
25	3802.82240(29)		
26	3803.03760(30)		
27	3803.13067(33)		
28	3803.28823(40)		
29	3803.40835(54)		
30	3803.76393(70)		
31	3807.28282(53)		
32	3807.84103(31)		
33	3808.01576(54)		
34	3809.19304(71)		
35	3809.44790(60)		
36	3809.49458(32)		
37	3810.14718(63)		
38	3810.44184(55)		
39	3810.83632(68)		
40	3810.91626(57)		
41	3811.05376(58)		
42	3812.17943(71)		
43	3814.97536(31)		
44	3815.71818(33)		

- "LINE NO.": serial number identifying the spectral line in the map on the opposite page.
- "WAVENUMBER": experimental value.
- Uncertainty in the last digit of the wavenumber is given in parentheses.

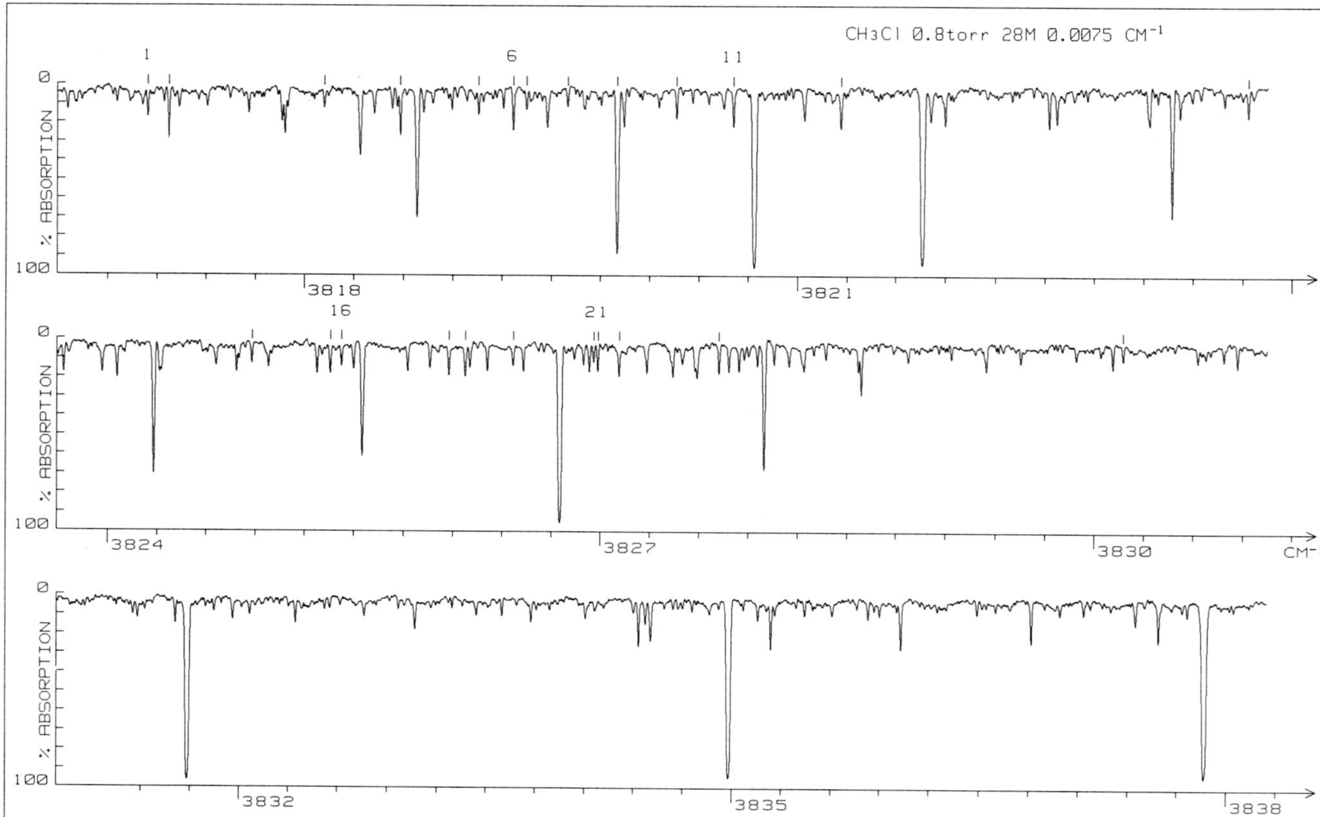

ISOTOPIC IDENTIFICATION - '∇' indicates $CH_3{}^{35}Cl$; 'Ⓘ': $CH_3{}^{37}Cl$; 'ı': unidentified.

LINE NO.	WAVENUMBER CM^{-1}, VAC.	UNC.	ROTATIONAL BAND ASSIGNMENTS	ISOTOPIC SPECIES
1	3817.04461	(51)		
2	3817.17267	(30)		
3	3818.12427	(78)		
4	3818.58220	(32)		
5	3819.05831	(59)		
6	3819.27059	(35)		
7	3819.35075	(75)		
8	3819.60000	(74)		
9	3819.90441	(20)		
10	3820.26479	(46)		
11	3820.61118	(39)		
12	3821.26308	(44)		
13	3823.73374	(53)		
14	3824.87999	(65)		
15	3825.36024	(52)		
16	3825.42834	(65)		
17	3826.07767	(52)		
18	3826.17751	(56)		
19	3826.46950	(75)		
20	3826.96058	(79)		
21	3826.98630	(48)		
22	3827.11657	(53)		
23	3827.72279	(45)		
24	3830.17578	(83)		

- "LINE NO.": serial number identifying the spectral line in the map on the opposite page.
- "WAVENUMBER": experimental value.
- Uncertainty in the last digit of the wavenumber is given in parentheses.

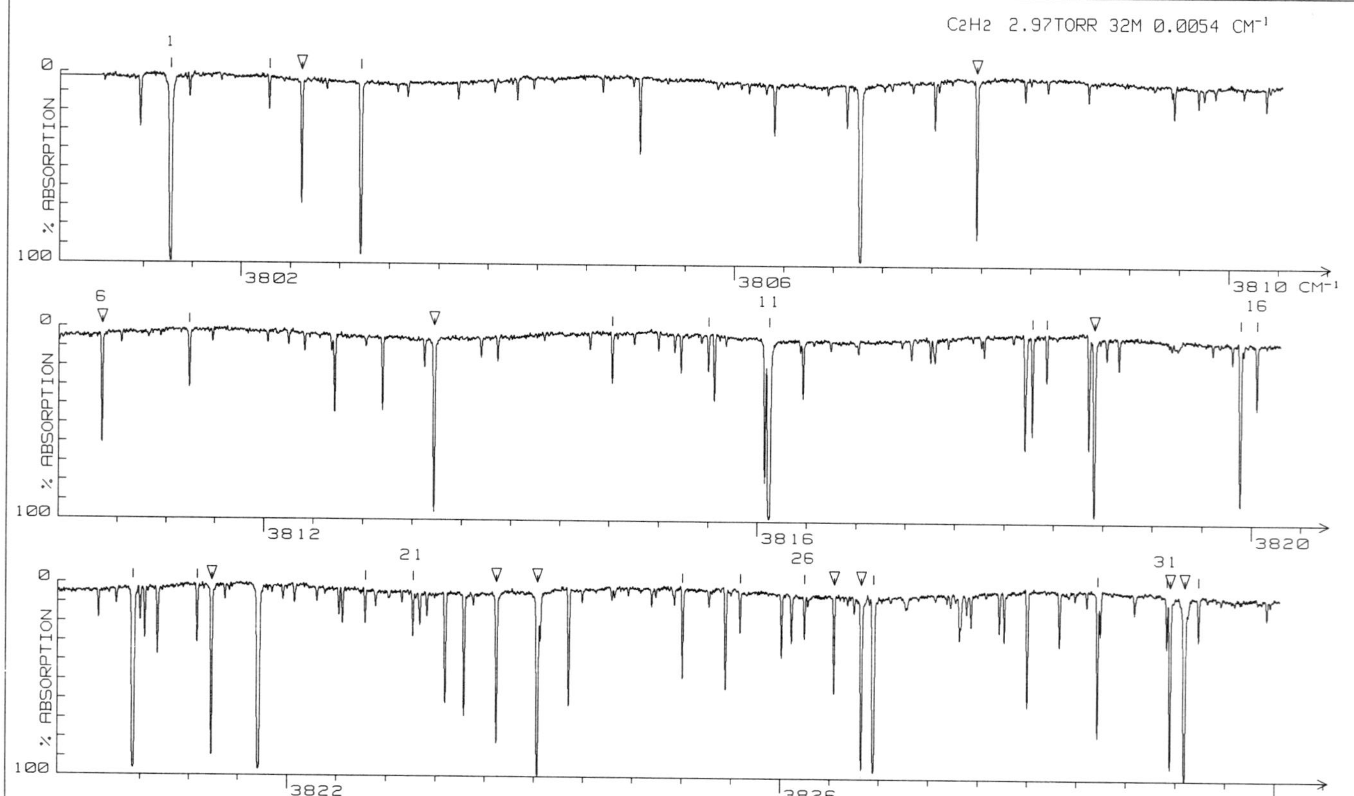

ISOTOPIC IDENTIFICATION – '∇' indicates $H^{12}C^{12}CH$; 'ı': unidentified.
BAND 1= $(0011^{1}0^{0})(\Pi_u) \leftarrow (0000^{0}0^{0})(\Sigma_g^{+})$; 2= $(0102^{2}1^{-1})(\Pi_u) \leftarrow (0000^{0}0^{0})(\Sigma_g^{+})$;

LINE NO.	WAVENUMBER CM^{-1}, VAC.	ROTATIONAL BAND ASSIGNMENTS		ISOTOPIC SPECIES
1	3801.41938(25)			
2	3802.22357(67)			
3	3802.48799(23)	P (31)	2	$H^{12}C^{12}CH$
4	3802.96632(17)			
5	3807.96002(22)	P (29)	2	$H^{12}C^{12}CH$
6	3810.67381(26)	P (28)	2	$H^{12}C^{12}CH$
7	3811.38760(43)			
8	3813.37290(20)	P (27)	2	$H^{12}C^{12}CH$
9	3814.82213(47)			
10	3815.59870(59)			
11	3816.09254(25)			
12	3818.22572(27)			
13	3818.34190(81)			
14	3818.72793(21)	P (25)	2	$H^{12}C^{12}CH$
15	3819.90492(22)			
16	3820.03910(36)			
17	3820.73918(27)			
18	3821.26546(43)			
19	3821.38409(21)	P (24)	2	$H^{12}C^{12}CH$
20	3822.62880(65)			
21	3823.01591(51)			
22	3823.69486(85)	P (29)	1	$H^{12}C^{12}CH$
23	3824.02623(22)	P (23)	2	$H^{12}C^{12}CH$
24	3825.20617(29)			
25	3825.67326(53)			
26	3826.19506(54)			
27	3826.43327(27)	P (28)	1	$H^{12}C^{12}CH$
28	3826.65458(37)	P (22)	2	$H^{12}C^{12}CH$
29	3826.75469(19)			
30	3828.56835(22)			
31	3829.12834(45)			
32	3829.15504(21)	P (27)	1	$H^{12}C^{12}CH$
33	3829.26904(19)	P (21)	2	$H^{12}C^{12}CH$
34	3829.38264(55)			

- "LINE NO.": serial number identifying the spectral line in the map on the opposite page.
- "WAVENUMBER": experimental value.
- Uncertainty in the last digit of the wavenumber is given in parentheses.
- "BAND" : identification on the opposite page.

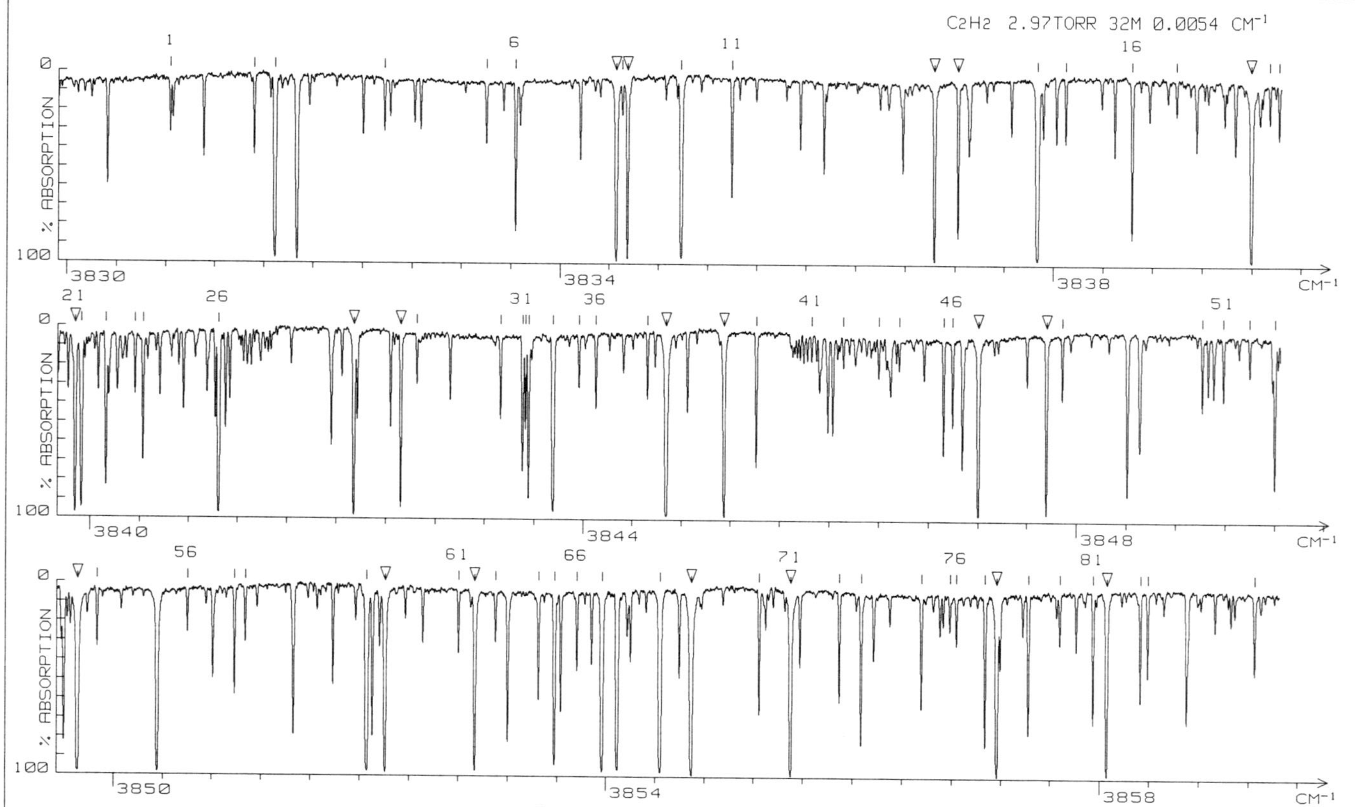

ISOTOPIC IDENTIFICATION – '∇' indicates $H^{12}C^{12}CH$; 'I': unidentified.

BAND $1=(0011^{1}0^{0})(\Pi_u)\leftarrow(0000^{0}0^{0})(\Sigma_g^{+})$; $2=(0102^{2}1^{-1})(\Pi_u)\leftarrow(0000^{0}0^{0})(\Sigma_g^{+})$;

LINE NO.	WAVENUMBER CM^{-1}, VAC.	ROTATIONAL ASSIGNMENTS	BAND	ISOTOPIC SPECIES
1	3830. 83826 (44)			
2	3831. 52008 (33)			
3	3831. 68698 (36)			
4	3832. 57764 (43)			
5	3833. 40236 (41)			
6	3833. 63643 (52)			
7	3834. 45767 (19)	P (19)	2	$H^{12}C^{12}CH$
8	3834. 50812 (78)			
9	3834. 54936 (20)	P (25)	1	$H^{12}C^{12}CH$
10	3834. 98383 (24)			
11	3835. 39824 (24)			
12	3837. 03196 (20)	P (18)	2	$H^{12}C^{12}CH$
13	3837. 22226 (22)	P (24)	1	$H^{12}C^{12}CH$
14	3837. 87003 (26)			
15	3838. 09757 (38)			
16	3838. 63145 (22)			
17	3838. 99119 (65)			
18	3839. 59309 (18)	P (17)	2	$H^{12}C^{12}CH$
19	3839. 74327 (56)			
20	3839. 82203 (42)			
21	3839. 87936 (19)	P (23)	1	$H^{12}C^{12}CH$
22	3839. 93009 (25)			
23	3840. 12660 (21)			
24	3840. 36059 (40)			
25	3840. 42546 (24)			
26	3841. 04561 (19)			
27	3842. 14136 (19)	P (16)	2	$H^{12}C^{12}CH$
28	3842. 52080 (45)	P (22)	1	$H^{12}C^{12}CH$
29	3842. 65013 (45)			
30	3843. 32865 (32)			

LINE NO.	WAVENUMBER CM^{-1}, VAC.	ROTATIONAL ASSIGNMENTS	BAND	ISOTOPIC SPECIES
31	3843. 50547 (28)			
32	3843. 52915 (29)			
33	3843. 55401 (23)			
34	3843. 75156 (22)			
35	3843. 96343 (46)			
36	3844. 10392 (34)			
37	3844. 52434 (38)			
38	3844. 67664 (19)	P (15)	2	$H^{12}C^{12}CH$
39	3845. 14679 (21)	P (21)	1	$H^{12}C^{12}CH$
40	3845. 40864 (24)			
41	3845. 85737 (80)			
42	3846. 11321 (72)			
43	3846. 39967 (83)			
44	3846. 56376 (70)			
45	3846. 91878 (25)			
46	3846. 99266 (30)			
47	3847. 19942 (20)	P (14)	2	$H^{12}C^{12}CH$
48	3847. 75785 (21)	P (20)	1	$H^{12}C^{12}CH$
49	3847. 88388 (37)			
50	3849. 01357 (33)			
51	3849. 18391 (39)			
52	3849. 39301 (60)			
53	3849. 59942 (98)			
54	3849. 70946 (18)	P (13)	2	$H^{12}C^{12}CH$
55	3849. 86751 (35)			
56	3850. 60275 (54)			
57	3850. 98782 (26)			
58	3851. 07380 (45)			
59	3852. 05836 (31)			
60	3852. 20711 (21)	P (12)	2	$H^{12}C^{12}CH$

- "LINE NO.": serial number identifying the spectral line in the map on the opposite page.
- "WAVENUMBER": experimental value.
- Uncertainty in the last digit of the wavenumber is given in parentheses.
- "BAND" : identification on the opposite page.

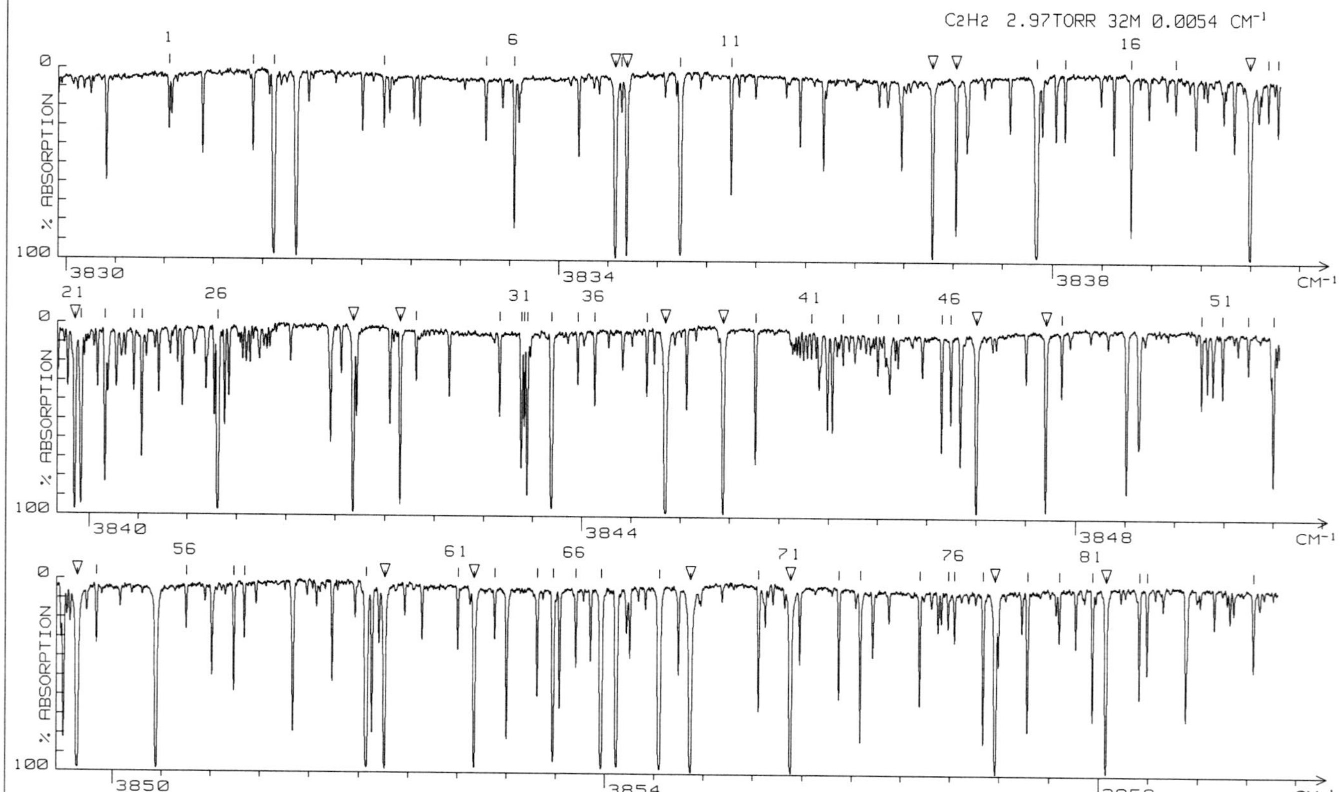

ISOTOPIC IDENTIFICATION – '∇' indicates $H^{12}C^{12}CH$; 'I': unidentified.

BAND $1 = (0011^1 0^0)(\Pi_u) \leftarrow (0000^0 0^0)(\Sigma_g^+)$; $2 = (0102^2 1^{-1})(\Pi_u) \leftarrow (0000^0 0^0)(\Sigma_g^+)$;

LINE NO.	WAVENUMBER CM^{-1}, VAC.	ROTATIONAL ASSIGNMENTS	BAND	ISOTOPIC SPECIES
61	3852.80270(39)			
62	3852.93550(24)	P (18)	1	$H^{12}C^{12}CH$
63	3853.10169(48)			
64	3853.44762(27)			
65	3853.57653(24)			
66	3853.76110(34)			
67	3853.96695(25)			
68	3854.43887(31)			
69	3854.69231(19)	P (11)	2	$H^{12}C^{12}CH$
70	3855.24609(25)			
71	3855.50277(19)	P (17)	1	$H^{12}C^{12}CH$
72	3855.89409(26)			
73	3856.06980(46)			
74	3856.55450(26)			
75	3856.78482(72)			
76	3856.83694(46)			
77	3857.06876(55)			
78	3857.16509(18)	P (10)	2	$H^{12}C^{12}CH$
79	3857.42563(19)			
80	3857.67904(47)			
81	3857.94717(24)			
82	3858.05602(20)	P (16)	1	$H^{12}C^{12}CH$
83	3858.32722(26)			
84	3858.38999(31)			
85	3859.24920(38)			

- "LINE NO.": serial number identifying the spectral line in the map on the opposite page.
- "WAVENUMBER": experimental value.
- Uncertainty in the last digit of the wavenumber is given in parentheses.
- "BAND" : identification on the opposite page.

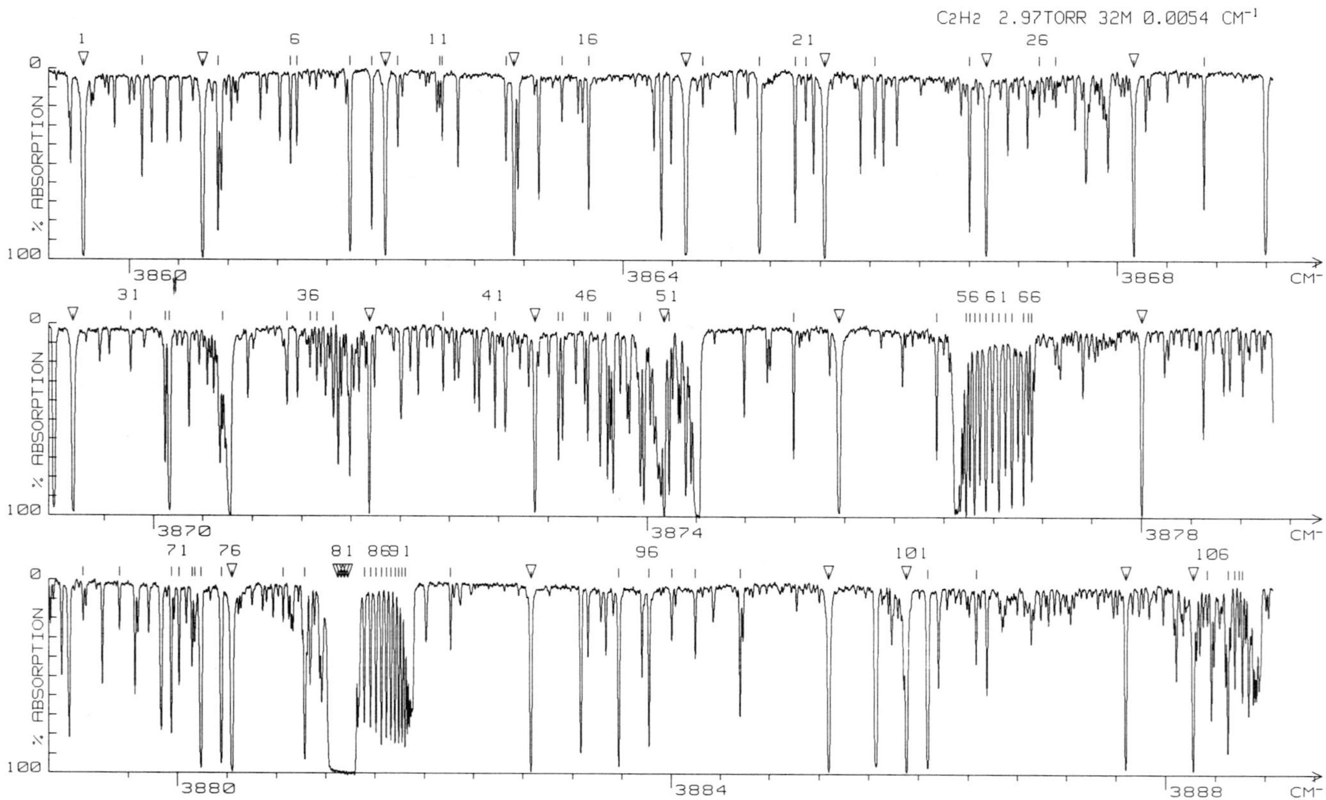

ISOTOPIC IDENTIFICATION - '∇' indicates $H^{12}C^{12}CH$; 'I': unidentified.

BAND 1= $(0011^{1}0^{0})(\Pi_u) \leftarrow (0000^{0}0^{0})(\Sigma_g^{+})$; 2= $(0102^{2}1^{-1})(\Pi_u) \leftarrow (0000^{0}0^{0})(\Sigma_g^{+})$;

LINE NO.	WAVENUMBER CM^{-1}, VAC.	ROTATIONAL BAND ASSIGNMENTS		ISOTOPIC SPECIES	LINE NO.	WAVENUMBER CM^{-1}, VAC.	ROTATIONAL BAND ASSIGNMENTS		ISOTOPIC SPECIES
1	3859.62615(18)	P (9)	2	$H^{12}C^{12}CH$	31	3869.81510(54)			
2	3860.10202(28)				32	3870.09333(22)			
3	3860.59550(19)	P (15)	1	$H^{12}C^{12}CH$	33	3870.13028(25)			
4	3860.71949(71)				34	3870.55919(48)			
5	3861.30490(30)				35	3871.08237(35)			
6	3861.35598(35)				36	3871.27033(64)			
7	3861.78854(20)				37	3871.32565(46)			
8	3861.96480(23)				38	3871.45398(45)			
9	3862.07488(23)	P (8)	2	$H^{12}C^{12}CH$	39	3871.75000(20)	P (4)	2	$H^{12}C^{12}CH$
10	3862.17584(35)				40	3872.34826(38)			
11	3862.51449(72)				41	3872.77218(29)			
12	3862.53846(38)				42	3873.09373(21)	P (10)	1	$H^{12}C^{12}CH$
13	3863.05633(32)				43	3873.28281(24)			
14	3863.12130(23)	P (14)	1	$H^{12}C^{12}CH$	44	3873.31573(27)			
15	3863.50631(50)				45	3873.49526(36)			
16	3863.72404(66)				46	3873.52093(27)			
17	3864.51156(19)	P (7)	2	$H^{12}C^{12}CH$	47	3873.68037(57)			
18	3864.64995(70)				48	3873.70240(28)			
19	3865.11235(29)				49	3873.94507(28)			
20	3865.39855(23)				50	3874.13892(22)	P (3)	2	$H^{12}C^{12}CH$
21	3865.48378(49)				51	3874.17972(63)			
22	3865.63383(18)	P (13)	1	$H^{12}C^{12}CH$	52	3875.19236(25)			
23	3866.04035(32)				53	3875.55525(18)	P (9)	1	$H^{12}C^{12}CH$
24	3866.80182(56)				54	3876.34344(25)			
25	3866.93628(21)	P (6)	2	$H^{12}C^{12}CH$	55	3876.57885(37)			
26	3867.37066(58)				56	3876.61020(24)			
27	3867.50473(79)				57	3876.64771(21)			
28	3868.13333(24)	P (12)	1	$H^{12}C^{12}CH$	58	3876.69057(24)			
29	3868.69706(24)				59	3876.73889(33)			
30	3869.34910(19)	P (5)	2	$H^{12}C^{12}CH$	60	3876.79035(22)			

- "LINE NO.": serial number identifying the spectral line in the map on the opposite page.
- "WAVENUMBER": experimental value.
- Uncertainty in the last digit of the wavenumber is given in parentheses.
- "BAND" : identification on the opposite page.

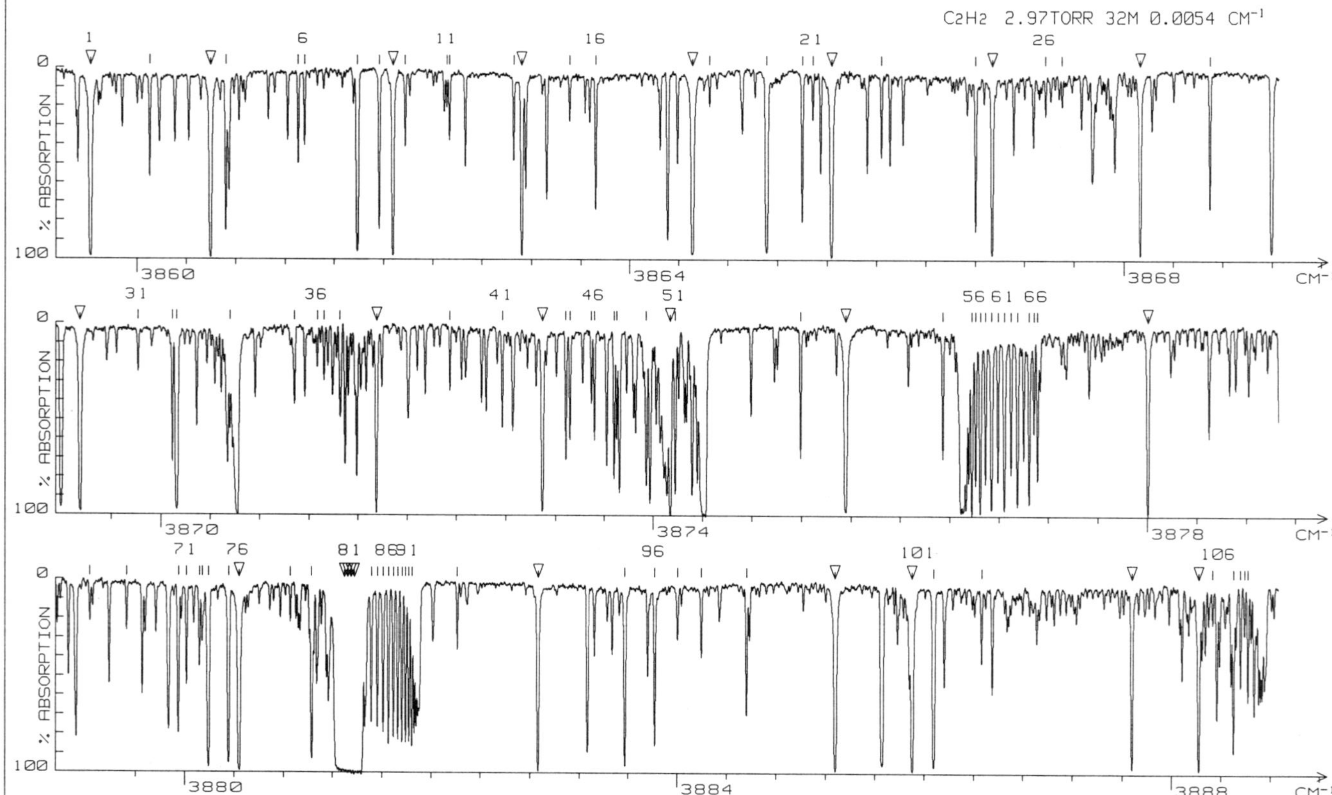

ISOTOPIC IDENTIFICATION - '∇' indicates $H^{12}C^{12}CH$; 'ı': unidentified.

BAND 1= $(0011^10^0)(\Pi_u) \leftarrow (0000^00^0)(\Sigma_g^+)$; 2= $(0102^21^{-1})(\Pi_u) \leftarrow (0000^00^0)(\Sigma_g^+)$;

LINE NO.	WAVENUMBER CM^{-1}, VAC.	ROTATIONAL BAND ASSIGNMENTS		ISOTOPIC SPECIES
61	3876.84478 (27)			
62	3876.89896 (24)			
63	3876.95218 (30)			
64	3877.04643 (23)			
65	3877.08416 (29)			
66	3877.11387 (48)			
67	3878.00433 (20)	P (8)	1	$H^{12}C^{12}CH$
68	3879.23394 (64)			
69	3879.52969 (51)			
70	3879.95018 (22)			
71	3880.01555 (28)			
72	3880.11949 (31)			
73	3880.14133 (51)			
74	3880.19213 (20)			
75	3880.35541 (18)			
76	3880.44114 (18)	P (7)	1	$H^{12}C^{12}CH$
77	3880.85505 (58)			
78	3881.02933 (25)			
79	3881.30064 (43)	Q (30)	2	$H^{12}C^{12}CH$
79		Q (9)	2	$H^{12}C^{12}CH$
80	3881.32687 (29)	Q (11)	2	$H^{12}C^{12}CH$
81	3881.35378 (27)	Q (13)	2	$H^{12}C^{12}CH$
82	3881.38011 (33)	Q (15)	2	$H^{12}C^{12}CH$
83	3881.51473 (25)			
84	3881.56415 (24)			
85	3881.61035 (23)			
86	3881.65316 (22)			
87	3881.69327 (23)			
88	3881.72978 (24)			
89	3881.76318 (55)			
90	3881.79286 (23)			
91	3881.81956 (24)			
92	3881.84364 (75)			
93	3882.21605 (38)			
94	3882.86590 (20)	P (6)	1	$H^{12}C^{12}CH$
95	3883.57591 (22)			
96	3883.82068 (74)			
97	3884.00527 (53)			
98	3884.19721 (37)			
99	3884.56482 (77)			
100	3885.27853 (19)	P (5)	1	$H^{12}C^{12}CH$
101	3885.90508 (19)	R (1)	2	$H^{12}C^{12}CH$
102	3886.07773 (21)			
103	3886.46547 (35)			
104	3887.67938 (23)	P (4)	1	$H^{12}C^{12}CH$
105	3888.22244 (23)	R (2)	2	$H^{12}C^{12}CH$
106	3888.33544 (69)			
107	3888.50195 (39)			
108	3888.55851 (28)			
109	3888.59095 (50)			
110	3888.61958 (27)			

- "LINE NO.": serial number identifying the spectral line in the map on the opposite page.
- "WAVENUMBER": experimental value.
- Uncertainty in the last digit of the wavenumber is given in parentheses.
- "BAND" : identification on the opposite page.

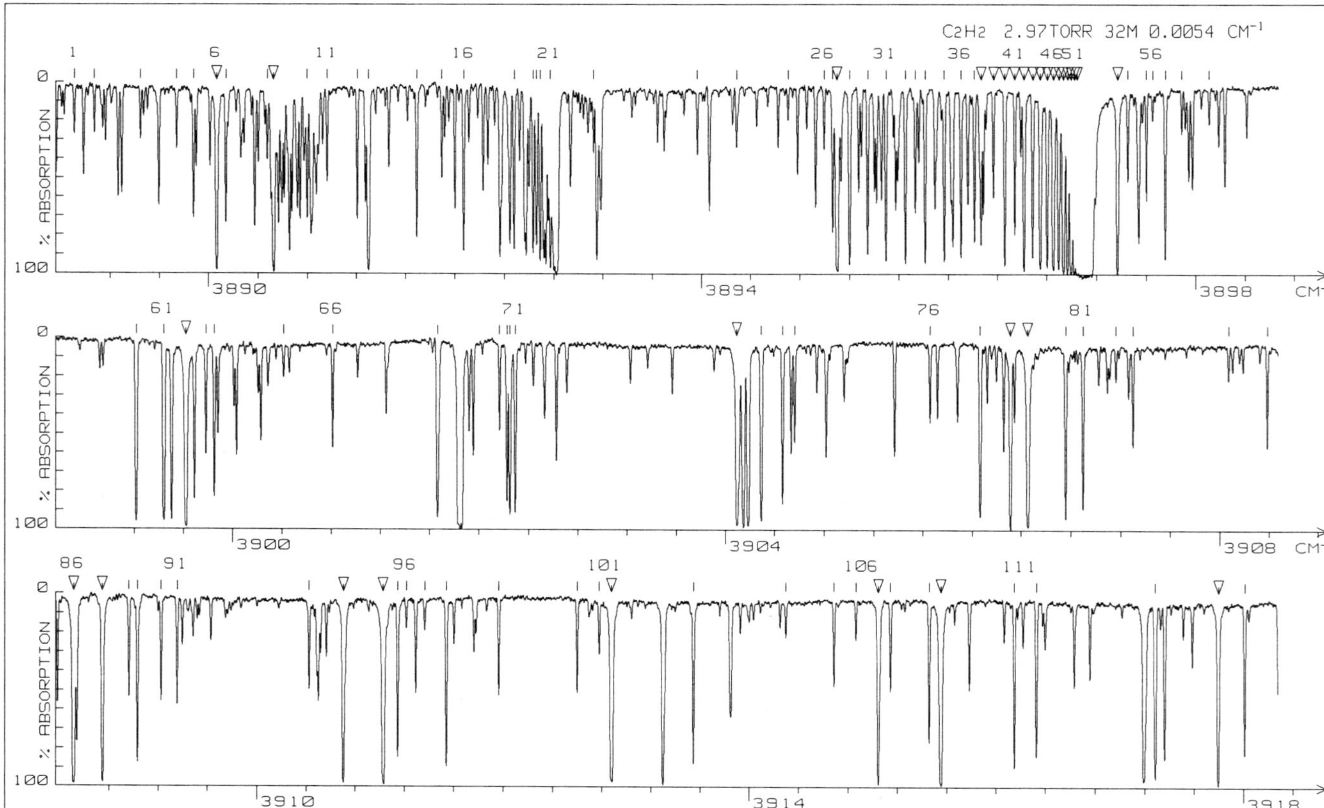

ISOTOPIC IDENTIFICATION – '∇' indicates $H^{12}C^{12}CH$; 'I': unidentified.

BAND $1=(0011^{1}0^{0})(\Pi_u)\leftarrow(0000^{0}0^{0})(\Sigma_g^+)$; $2=(0102^{2}1^{-1})(\Pi_u)\leftarrow(0000^{0}0^{0})(\Sigma_g^+)$;

LINE NO.	WAVENUMBER CM^{-1}, VAC.	ROTATIONAL BAND ASSIGNMENTS		ISOTOPIC SPECIES
1	3888.91127 (50)			
2	3889.07439 (49)			
3	3889.44673 (44)			
4	3889.74049 (41)			
5	3889.88068 (24)			
6	3890.06825 (21)	P (3)	1	$H^{12}C^{12}CH$
7	3890.14210 (72)			
8	3890.47550 (40)			
9	3890.52746 (19)	R (3)	2	$H^{12}C^{12}CH$
10	3890.79866 (26)			
11	3890.96207 (29)			
12	3891.20979 (68)			
13	3891.30040 (30)			
14	3891.69379 (23)			
15	3891.89662 (30)			
16	3892.07534 (22)			
17	3892.47922 (22)			
18	3892.63248 (62)			
19	3892.66016 (84)			
20	3892.68948 (23)			
21	3892.77197 (45)			
22	3893.12506 (46)			
23	3893.96754 (38)			
24	3894.28807 (52)			
25	3894.70649 (68)			
26	3894.99562 (40)			
27	3895.06436 (86)			
28	3895.10072 (18)	R (5)	2	$H^{12}C^{12}CH$
29	3895.20034 (24)			
30	3895.34591 (23)			
31	3895.49676 (23)			
32	3895.65449 (22)			
33	3895.73454 (26)			
34	3895.81399 (22)			
35	3895.96709 (21)			
36	3896.10324 (55)			
37	3896.21223 (23)			
38	3896.26623 (63)	Q (29)	1	$H^{12}C^{12}CH$
39	3896.36788 (27)	Q (28)	1	$H^{12}C^{12}CH$
40	3896.46021 (22)	Q (27)	1	$H^{12}C^{12}CH$
41	3896.54422 (23)	Q (26)	1	$H^{12}C^{12}CH$
42	3896.62010 (30)	Q (25)	1	$H^{12}C^{12}CH$
43	3896.68844 (22)	Q (24)	1	$H^{12}C^{12}CH$
44	3896.74990 (19)	Q (23)	1	$H^{12}C^{12}CH$
45	3896.80485 (22)	Q (22)	1	$H^{12}C^{12}CH$
46	3896.85362 (19)	Q (21)	1	$H^{12}C^{12}CH$
47	3896.89711 (21)	Q (20)	1	$H^{12}C^{12}CH$
48	3896.93537 (20)	Q (19)	1	$H^{12}C^{12}CH$
49	3896.96920 (25)	Q (18)	1	$H^{12}C^{12}CH$
50	3896.99888 (19)	Q (17)	1	$H^{12}C^{12}CH$
51	3897.02488 (54)	Q (16)	1	$H^{12}C^{12}CH$
52	3897.04741 (36)	Q (15)	1	$H^{12}C^{12}CH$
53	3897.36877 (21)	R (6)	2	$H^{12}C^{12}CH$
54	3897.45150 (31)			
55	3897.60179 (26)			
56	3897.65317 (83)			
57	3897.75520 (21)			
58	3897.88641 (52)			
59	3898.10923 (64)			
60	3899.21749 (24)			

- "LINE NO.": serial number identifying the spectral line in the map on the opposite page.
- "WAVENUMBER": experimental value.
- Uncertainty in the last digit of the wavenumber is given in parentheses.
- "BAND" : identification on the opposite page.

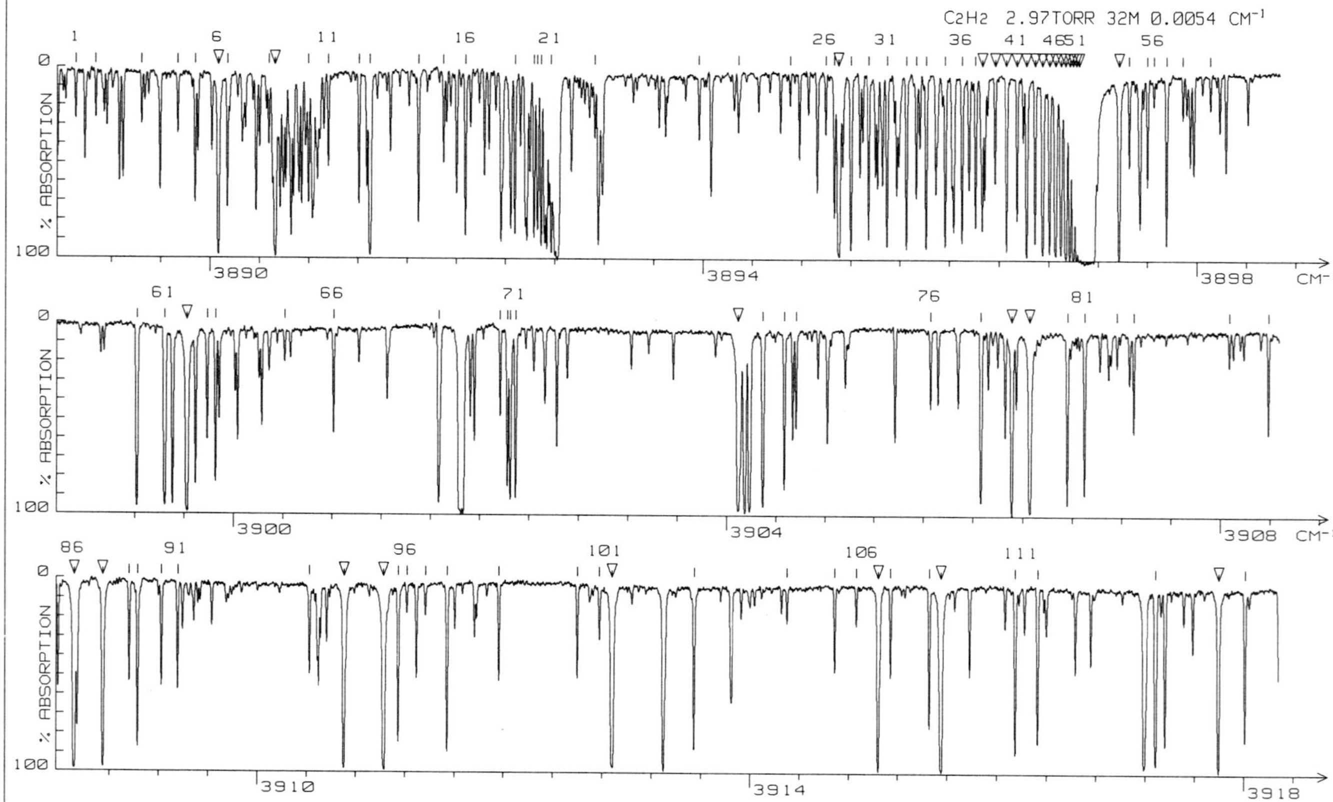

ISOTOPIC IDENTIFICATION - '∇' indicates $H^{12}C^{12}CH$; 'I': unidentified.
BAND 1= $(0011^{1}0^{0})(\Pi_u) \leftarrow (0000^{0}0^{0})(\Sigma_g^+)$; 2= $(0102^{2}1^{-1})(\Pi_u) \leftarrow (0000^{0}0^{0})(\Sigma_g^+)$;

LINE NO.	WAVENUMBER CM^{-1}, VAC.	ROTATIONAL ASSIGNMENTS	BAND	ISOTOPIC SPECIES
61	3899.44205 (22)			
62	3899.62406 (19)	R (7)	2	$H^{12}C^{12}CH$
63	3899.78690 (27)			
64	3899.85488 (60)			
65	3900.41383 (79)			
66	3900.81333 (27)			
67	3901.66727 (26)			
68	3902.16760 (29)			
69	3902.22721 (51)			
70	3902.25068 (20)			
71	3902.29411 (22)			
72	3904.09638 (19)	R (9)	2	$H^{12}C^{12}CH$
73	3904.29555 (44)			
74	3904.46838 (23)			
75	3904.56525 (29)			
76	3905.65932 (34)			
77	3906.06569 (21)			
78	3906.31293 (19)	R (10)	2	$H^{12}C^{12}CH$
79	3906.45694 (19)	R (3)	1	$H^{12}C^{12}CH$
80	3906.75934 (23)			
81	3906.89823 (22)			
82	3907.16206 (66)			
83	3907.29865 (28)			
84	3908.07116 (65)			
85	3908.38634 (27)			
86	3908.51628 (19)	R (11)	2	$H^{12}C^{12}CH$
87	3908.74992 (24)	R (4)	1	$H^{12}C^{12}CH$
88	3908.96364 (29)			
89	3909.03260 (56)			
90	3909.22488 (29)			
91	3909.35508 (27)			
92	3910.42513 (32)			
93	3910.70613 (22)	R (12)	2	$H^{12}C^{12}CH$
94	3911.03033 (19)	R (5)	1	$H^{12}C^{12}CH$
95	3911.14843 (22)			
96	3911.21815 (88)			
97	3911.36952 (82)			
98	3911.54347 (21)			
99	3911.96675 (29)			
100	3912.60107 (29)			
101	3912.78123 (50)			
102	3912.88264 (34)	R (13)	2	$H^{12}C^{12}CH$
103	3913.54659 (22)			
104	3914.30012 (63)			
105	3914.68724 (31)			
106	3914.86348 (56)			
107	3915.04502 (25)	R (14)	2	$H^{12}C^{12}CH$
108	3915.14535 (30)			
109	3915.45912 (24)			
110	3915.55316 (18)	R (7)	1	$H^{12}C^{12}CH$
111	3916.15004 (23)			
112	3916.32949 (24)			
113	3917.28654 (19)			
114	3917.79520 (20)	R (8)	1	$H^{12}C^{12}CH$
115	3918.00904 (23)			

- "LINE NO.": serial number identifying the spectral line in the map on the opposite page.
- "WAVENUMBER": experimental value.
- Uncertainty in the last digit of the wavenumber is given in parentheses.
- "BAND" : identification on the opposite page.

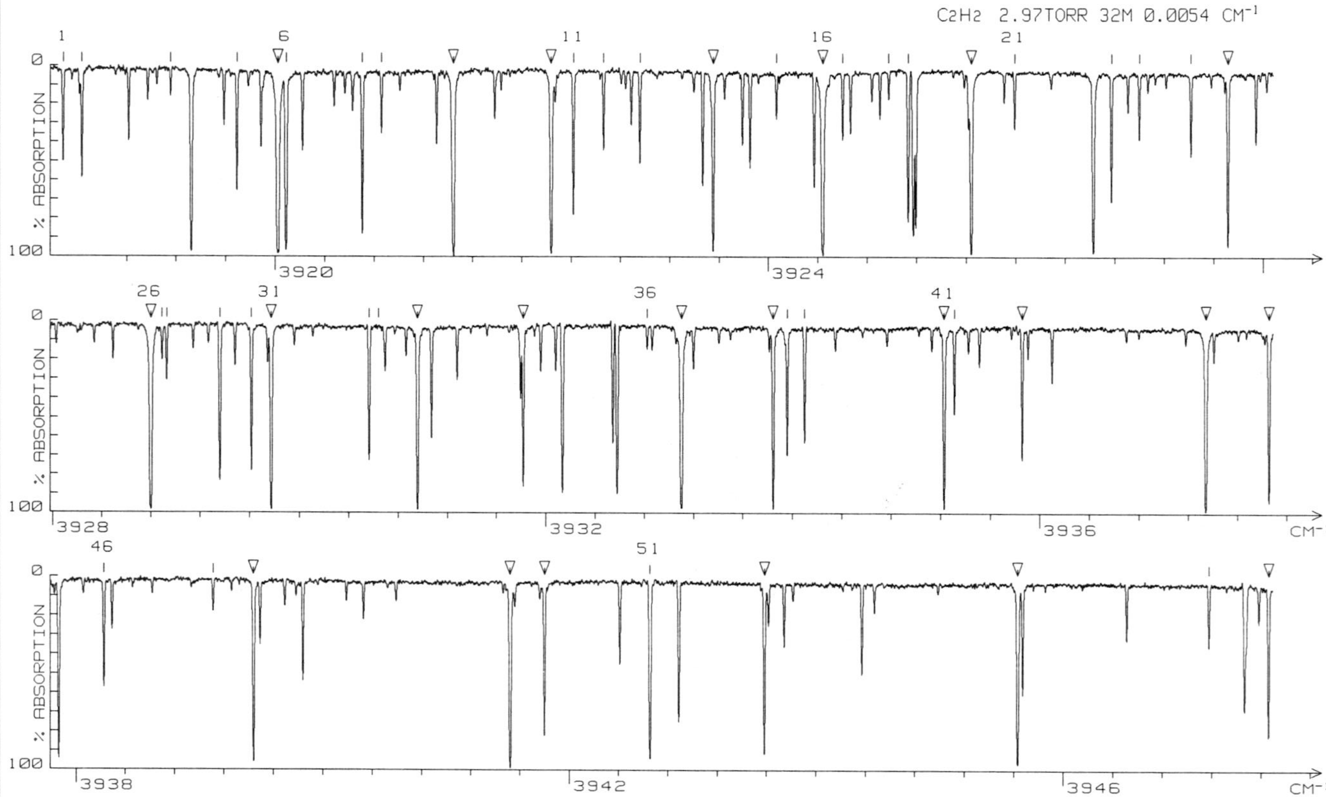

ISOTOPIC IDENTIFICATION - '∇' indicates $H^{12}C^{12}CH$; 'I': unidentified.
BAND 1= $(0011^{1}0^{0})(\Pi_u) \leftarrow (0000^{0}0^{0})(\Sigma_g^+)$; 2= $(0102^{2}1^{-1})(\Pi_u) \leftarrow (0000^{0}0^{0})(\Sigma_g^+)$;

LINE NO.	WAVENUMBER CM^{-1}, VAC.	ROTATIONAL BAND ASSIGNMENTS		BAND	ISOTOPIC SPECIES
1	3918.27977(30)				
2	3918.43471(27)				
3	3919.15751(76)				
4	3919.69412(26)				
5	3920.02390(18)	R	(9)	1	$H^{12}C^{12}CH$
6	3920.08948(18)				
7	3920.70536(22)				
8	3920.86030(40)				
9	3921.44681(44)	R	(17)	2	$H^{12}C^{12}CH$
10	3922.23919(20)	R	(10)	1	$H^{12}C^{12}CH$
11	3922.41845(24)				
12	3922.66583(33)				
13	3922.95968(30)				
14	3923.55177(38)	R	(18)	2	$H^{12}C^{12}CH$
15	3924.06885(50)				
16	3924.44065(19)	R	(11)	1	$H^{12}C^{12}CH$
17	3924.60291(38)				
18	3924.97663(74)				
19	3925.13499(20)				
20	3925.64183(22)	R	(19)	2	$H^{12}C^{12}CH$
21	3925.99244(43)				
22	3926.77344(24)				
23	3927.00025(37)				
24	3927.41673(31)				
25	3927.71656(22)	R	(20)	2	$H^{12}C^{12}CH$
26	3928.80101(19)	R	(13)	1	$H^{12}C^{12}CH$
27	3928.89249(69)				
28	3928.92978(46)				
29	3929.36170(25)				
30	3929.61754(24)				
31	3929.77601(23)	R	(21)	2	$H^{12}C^{12}CH$
32	3930.56683(19)				
33	3930.64305(59)				
34	3930.95951(22)	R	(14)	1	$H^{12}C^{12}CH$
35	3931.81989(47)	R	(22)	2	$H^{12}C^{12}CH$
36	3932.82415(98)				
37	3933.10292(20)	R	(15)	1	$H^{12}C^{12}CH$
38	3933.84776(21)	R	(23)	2	$H^{12}C^{12}CH$
39	3933.96066(25)				
40	3934.10128(21)				
41	3935.23086(21)	R	(16)	1	$H^{12}C^{12}CH$
42	3935.31201(32)				
43	3935.85954(24)	R	(24)	2	$H^{12}C^{12}CH$
44	3937.34330(22)	R	(17)	1	$H^{12}C^{12}CH$
45	3937.85520(21)	R	(25)	2	$H^{12}C^{12}CH$
46	3938.22475(27)				
47	3939.11234(83)				
48	3939.43964(22)	R	(18)	1	$H^{12}C^{12}CH$
49	3941.51969(23)	R	(19)	1	$H^{12}C^{12}CH$
50	3941.79647(23)	R	(27)	2	$H^{12}C^{12}CH$
51	3942.65321(21)				
52	3943.58288(22)	R	(20)	1	$H^{12}C^{12}CH$
53	3945.62918(27)	R	(21)	1	$H^{12}C^{12}CH$
54	3947.17488(37)				
55	3947.65795(83)	R	(22)	1	$H^{12}C^{12}CH$

- "LINE NO.": serial number identifying the spectral line in the map on the opposite page.
- "WAVENUMBER": experimental value.
- Uncertainty in the last digit of the wavenumber is given in parentheses.
- "BAND" : identification on the opposite page.

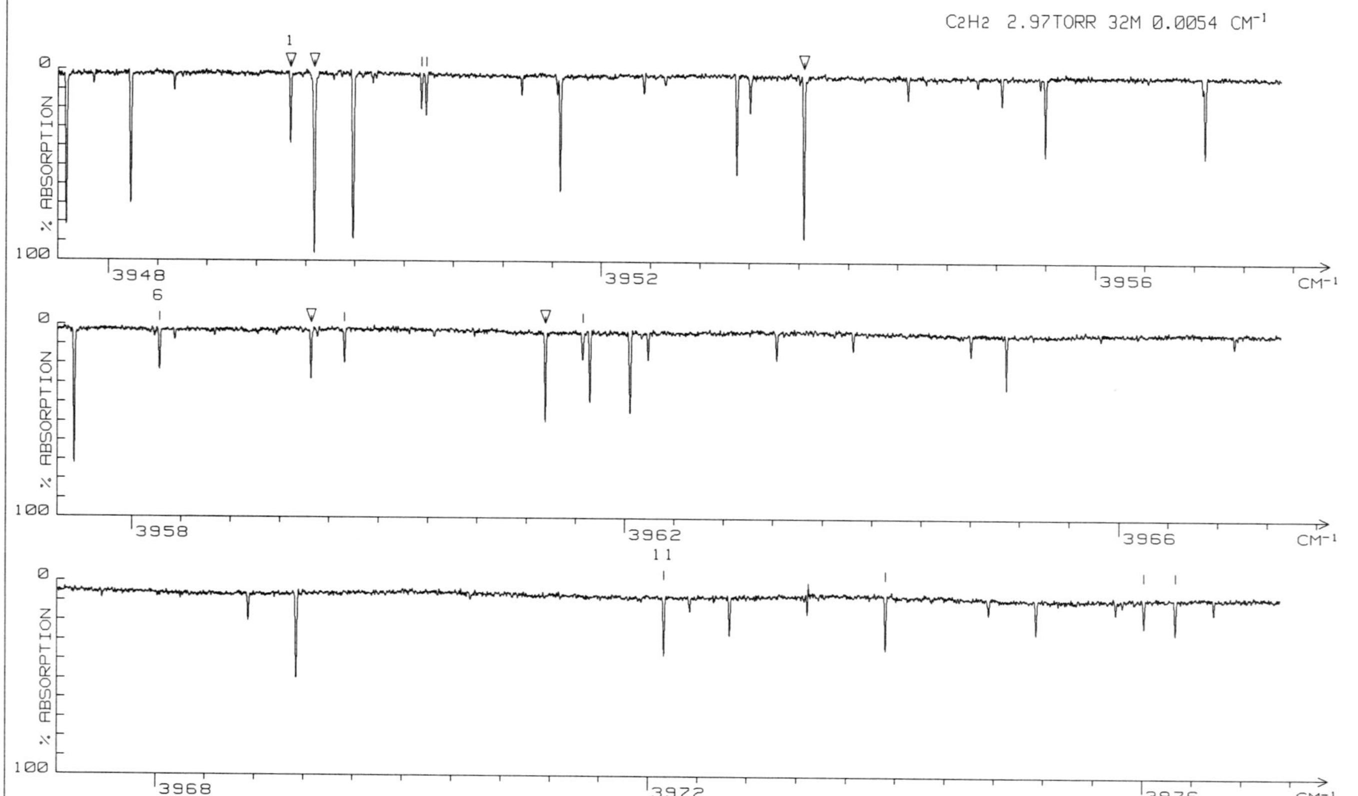

ISOTOPIC IDENTIFICATION – '∇' indicates $H^{12}C^{12}CH$; 'I': unidentified.

BAND 1= $(0011^10^0)(\Pi_u) \leftarrow (0000^00^0)(\Sigma_g^+)$; 2= $(0102^21^{-1})(\Pi_u) \leftarrow (0000^00^0)(\Sigma_g^+)$;

LINE NO.	WAVENUMBER CM^{-1}, VAC.	ROTATIONAL ASSIGNMENTS		BAND	ISOTOPIC SPECIES
1	3949.47346(35)	R	(31)	2	$H^{12}C^{12}CH$
2	3949.66917(21)	R	(23)	1	$H^{12}C^{12}CH$
3	3950.53530(61)				
4	3950.57407(56)				
5	3953.63737(22)	R	(25)	1	$H^{12}C^{12}CH$
6	3958.22063(45)				
7	3959.44963(49)	R	(28)	1	$H^{12}C^{12}CH$
8	3959.72476(75)				
9	3961.34900(31)	R	(29)	1	$H^{12}C^{12}CH$
10	3961.65432(74)				
11	3972.12436(34)				
12	3973.91899(37)				
13	3976.00923(79)				
14	3976.26439(60)				

- "LINE NO.": serial number identifying the spectral line in the map on the opposite page.
- "WAVENUMBER": experimental value.
- Uncertainty in the last digit of the wavenumber is given in parentheses.
- "BAND" : identification on the opposite page.

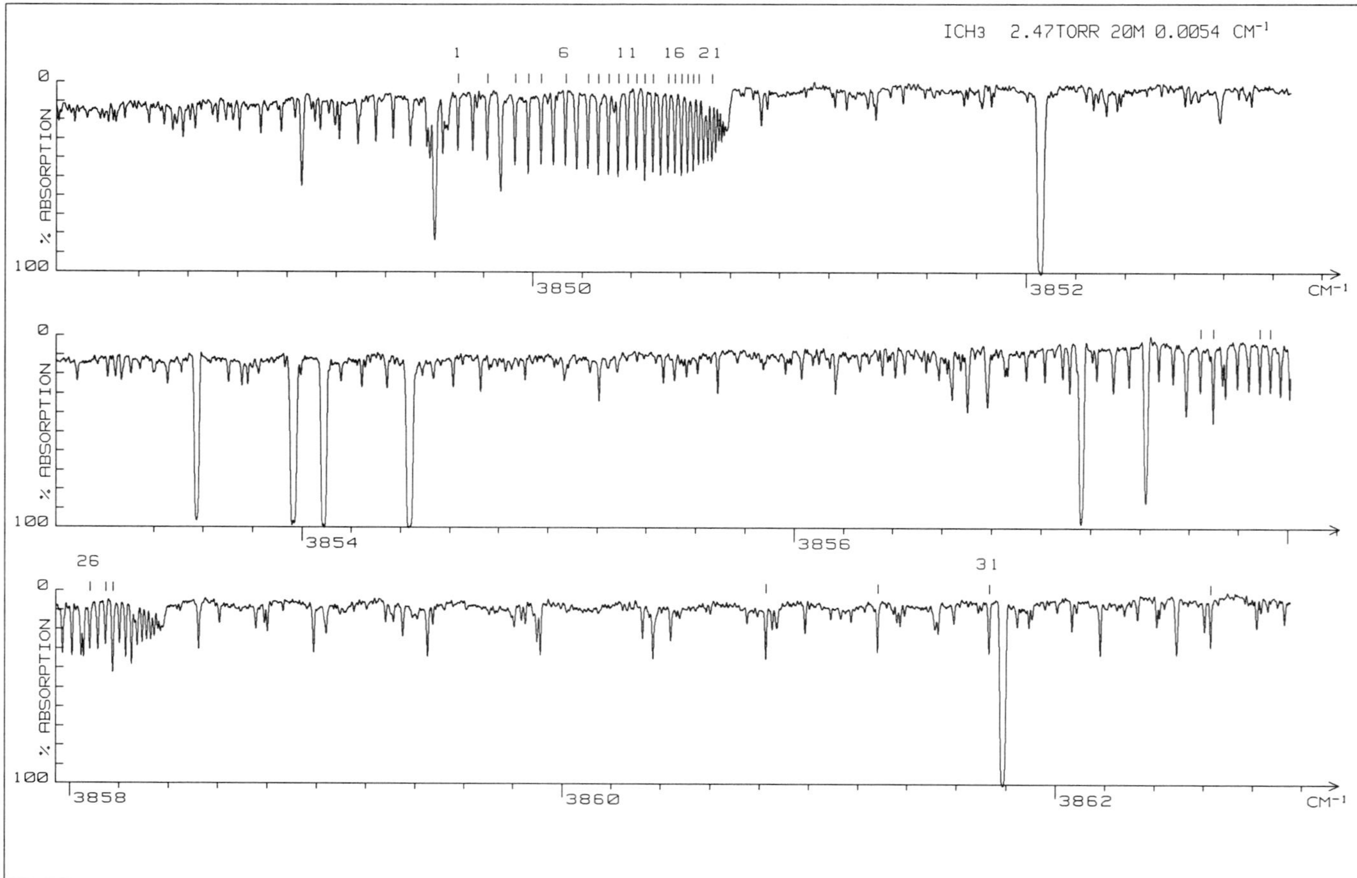

ICH3 2.47TORR 20M 0.0054 CM-1
1 6 11 16 21
0
% ABSORPTION
100
3850
3852
CM-1
0
% ABSORPTION
100
3854
3856
26
31
0
% ABSORPTION
100
3858
3860
3862
CM-1

LINE NO	WAVENUMBER CM^{-1}, VAC
1	3849.69419(51)
2	3849.81454(47)
3	3849.92745(45)
4	3849.98115(41)
5	3850.03304(46)
6	3850.13149(42)
7	3850.22285(44)
8	3850.26595(41)
9	3850.30741(41)
10	3850.34732(42)
11	3850.38499(42)
12	3850.42118(41)
13	3850.45548(38)
14	3850.48861(41)
15	3850.54933(41)
16	3850.57718(41)
17	3850.60317(40)
18	3850.62805(42)
19	3850.65093(44)
20	3850.67213(47)
21	3850.72528(60)
22	3857.64666(64)
23	3857.69885(39)
24	3857.88798(59)
25	3857.93044(62)
26	3858.08207(61)
27	3858.14614(63)
28	3858.17539(44)
29	3860.82684(54)
30	3861.28089(55)
31	3861.73276(54)
32	3862.62945(57)

- "LINE NO.": serial number identifying the spectral line in the map on the opposite page.
- "WAVENUMBER": experimental value.
- Uncertainty in the last digit of the wavenumber is given in parentheses.

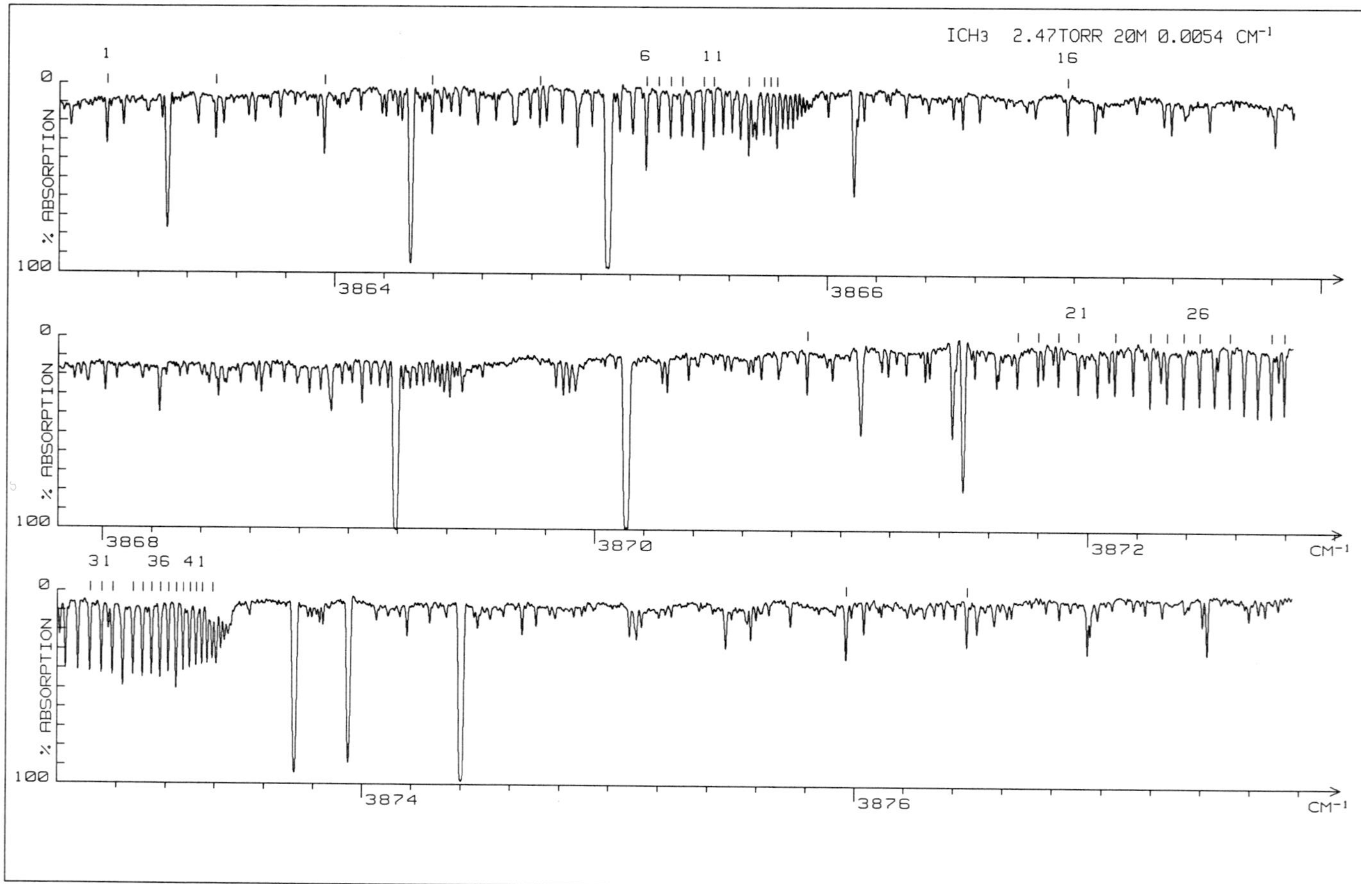
ICH3 2.47TORR 20M 0.0054 CM⁻¹
1
6
11
16
21
26
31
36
41
0
100
% ABSORPTION
3864
3866
3868
3870
3872
3874
3876
CM⁻¹

LINE NO	WAVENUMBER CM^{-1}, VAC	LINE NO	WAVENUMBER CM^{-1}, VAC
1	3863.07426(56)	31	3872.93912(35)
2	3863.51701(56)	32	3872.98339(36)
3	3863.95785(45)	33	3873.06672(36)
4	3864.39532(54)	34	3873.10548(35)
5	3864.83144(55)	35	3873.14224(36)
6	3865.26518(35)	36	3873.17710(38)
7	3865.31475(57)	37	3873.21040(36)
8	3865.36276(56)	38	3873.24232(30)
9	3865.40906(55)	39	3873.27113(37)
10	3865.49556(42)	40	3873.29849(39)
11	3865.53657(52)	41	3873.32421(43)
12	3865.67971(42)	42	3873.34802(41)
13	3865.74064(59)	43	3873.39008(55)
14	3865.76790(55)	44	3875.96467(52)
15	3865.79412(48)	45	3876.45460(54)
16	3866.97439(60)		
17	3870.86170(55)		
18	3871.71705(64)		
19	3871.79996(58)		
20	3871.88118(56)		
21	3871.96045(53)		
22	3872.10823(49)		
23	3872.25239(42)		
24	3872.32117(47)		
25	3872.38756(43)		
26	3872.45160(43)		
27	3872.57370(38)		
28	3872.74195(37)		
29	3872.79419(38)		
30	3872.89268(35)		

- "LINE NO.": serial number identifying the spectral line in the map on the opposite page.
- "WAVENUMBER": experimental value.
- Uncertainty in the last digit of the wavenumber is given in parentheses.

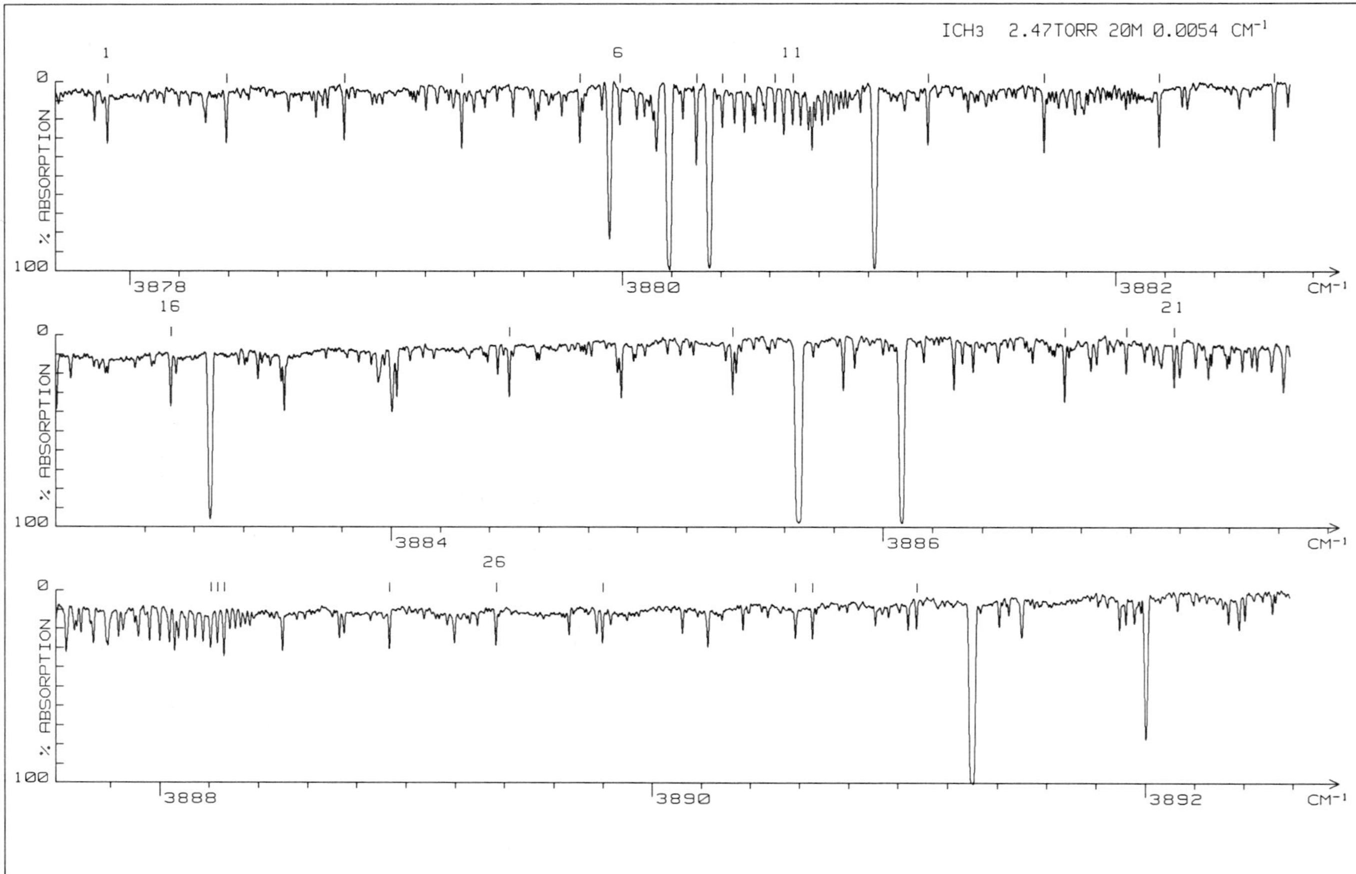
ICH3 2.47TORR 20M 0.0054 CM-1
1
6
11
16
21
26
% ABSORPTION
0
100
3878
3880
3882
3884
3886
3888
3890
3892
CM-1

LINE NO	WAVENUMBER CM^{-1}, VAC
1	3877.91124(45)
2	3878.39308(43)
3	3878.87300(41)
4	3879.35086(40)
5	3879.82724(40)
6	3879.99195(57)
7	3880.30221(32)
8	3880.40750(60)
9	3880.49868(52)
10	3880.62346(61)
11	3880.69535(60)
12	3881.24442(40)
13	3881.71220(36)
14	3882.17905(40)
15	3882.64354(39)
16	3883.10615(38)
17	3884.48177(38)
18	3885.38892(41)
19	3886.73449(36)
20	3886.98350(61)
21	3887.17799(43)
22	3888.20692(60)
23	3888.23449(56)
24	3888.26088(42)
25	3888.93334(49)
26	3889.36689(51)
27	3889.79813(54)
28	3890.58451(64)
29	3890.65338(55)
30	3891.07600(57)

- "LINE NO.": serial number identifying the spectral line in the map on the opposite page.
- "WAVENUMBER": experimental value.
- Uncertainty in the last digit of the wavenumber is given in parentheses.

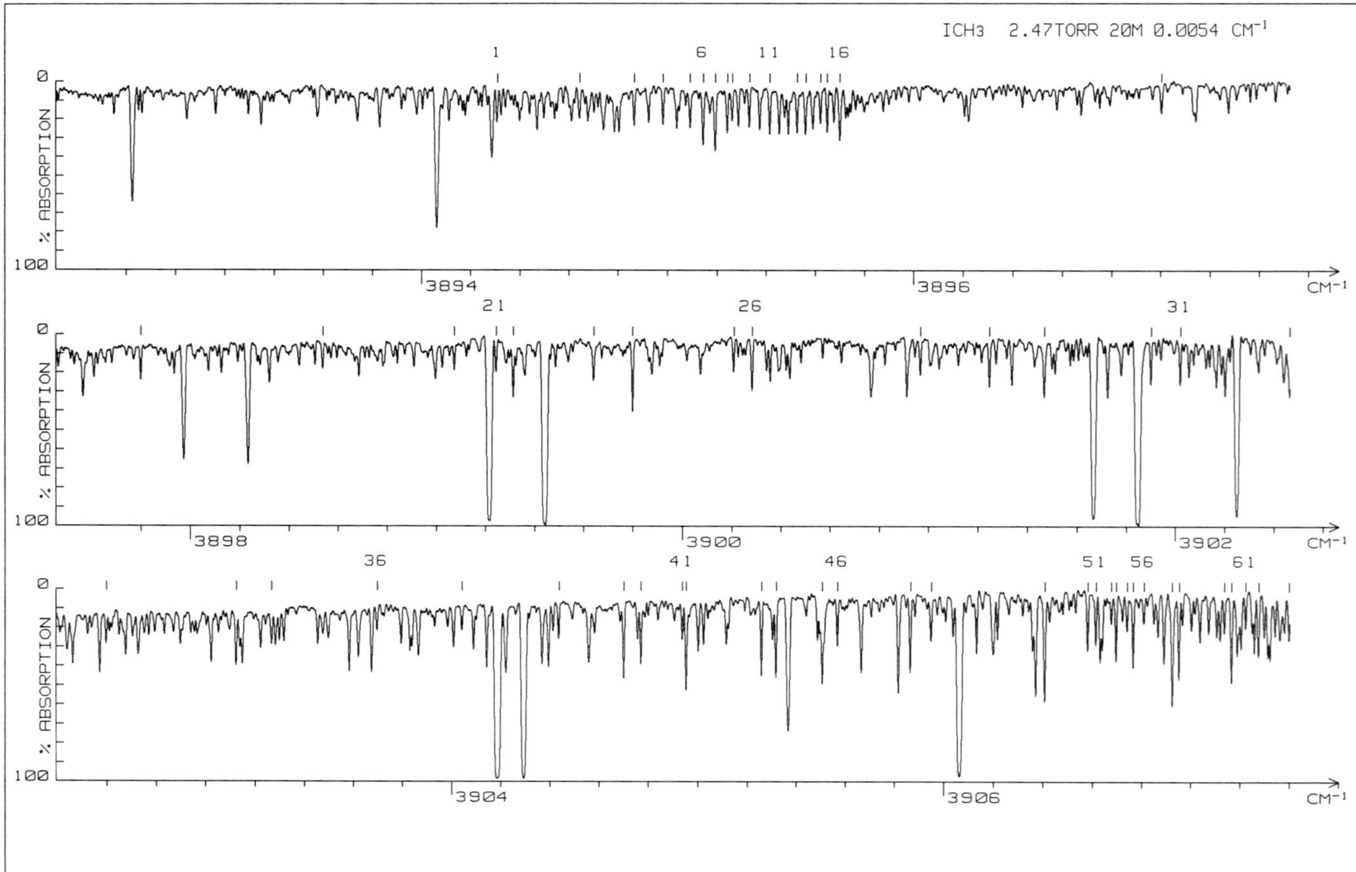
ICH3 2.47TORR 20M 0.0054 CM⁻¹
% ABSORPTION
0
100
1
6
11
16
3894
3896
21
26
31
3898
3900
3902
36
41
46
51
56
61
3904
3906
CM⁻¹

LINE NO	WAVENUMBER CM^{-1}, VAC	LINE NO	WAVENUMBER CM^{-1}, VAC	LINE NO	WAVENUMBER CM^{-1}, VAC
1	3894.30823(54)	31	3902.02106(35)	61	3907.22613(41)
2	3894.64456(68)	32	3902.46361(40)	62	3907.27813(30)
3	3894.86730(48)	33	3902.59815(48)	63	3907.40160(37)
4	3894.98343(46)	34	3903.12473(31)		
5	3895.09250(49)	35	3903.26851(54)		
6	3895.14462(38)	36	3903.69828(62)		
7	3895.19351(35)	37	3904.04362(63)		
8	3895.24199(43)	38	3904.43778(39)		
9	3895.26198(56)	39	3904.70380(52)		
10	3895.33266(44)	40	3904.77304(25)		
11	3895.41625(41)	41	3904.93933(45)		
12	3895.52724(45)	42	3904.95510(22)		
13	3895.56151(45)	43	3905.26146(22)		
14	3895.62262(51)	44	3905.32012(45)		
15	3895.65024(41)	45	3905.50841(48)		
16	3895.70130(42)	46	3905.57036(32)		
17	3897.00502(72)	47	3905.86683(22)		
18	3897.80007(47)	48	3905.95111(40)		
19	3898.53684(64)	49	3906.40960(20)		
20	3899.07348(57)	50	3906.58303(28)		
21	3899.24381(51)	51	3906.61570(30)		
22	3899.31353(34)	52	3906.67895(54)		
23	3899.64110(48)	53	3906.69879(26)		
24	3899.80036(25)	54	3906.74314(44)		
25	3900.20742(44)	55	3906.76828(23)		
26	3900.28356(35)	56	3906.81176(56)		
27	3900.96830(46)	57	3906.92865(21)		
28	3901.24677(36)	58	3906.95566(21)		
29	3901.46802(37)	59	3907.13958(41)		
30	3901.90089(36)	60	3907.16797(22)		

- "LINE NO.": serial number identifying the spectral line in the map on the opposite page.
- "WAVENUMBER": experimental value.
- Uncertainty in the last digit of the wavenumber is given in parentheses.

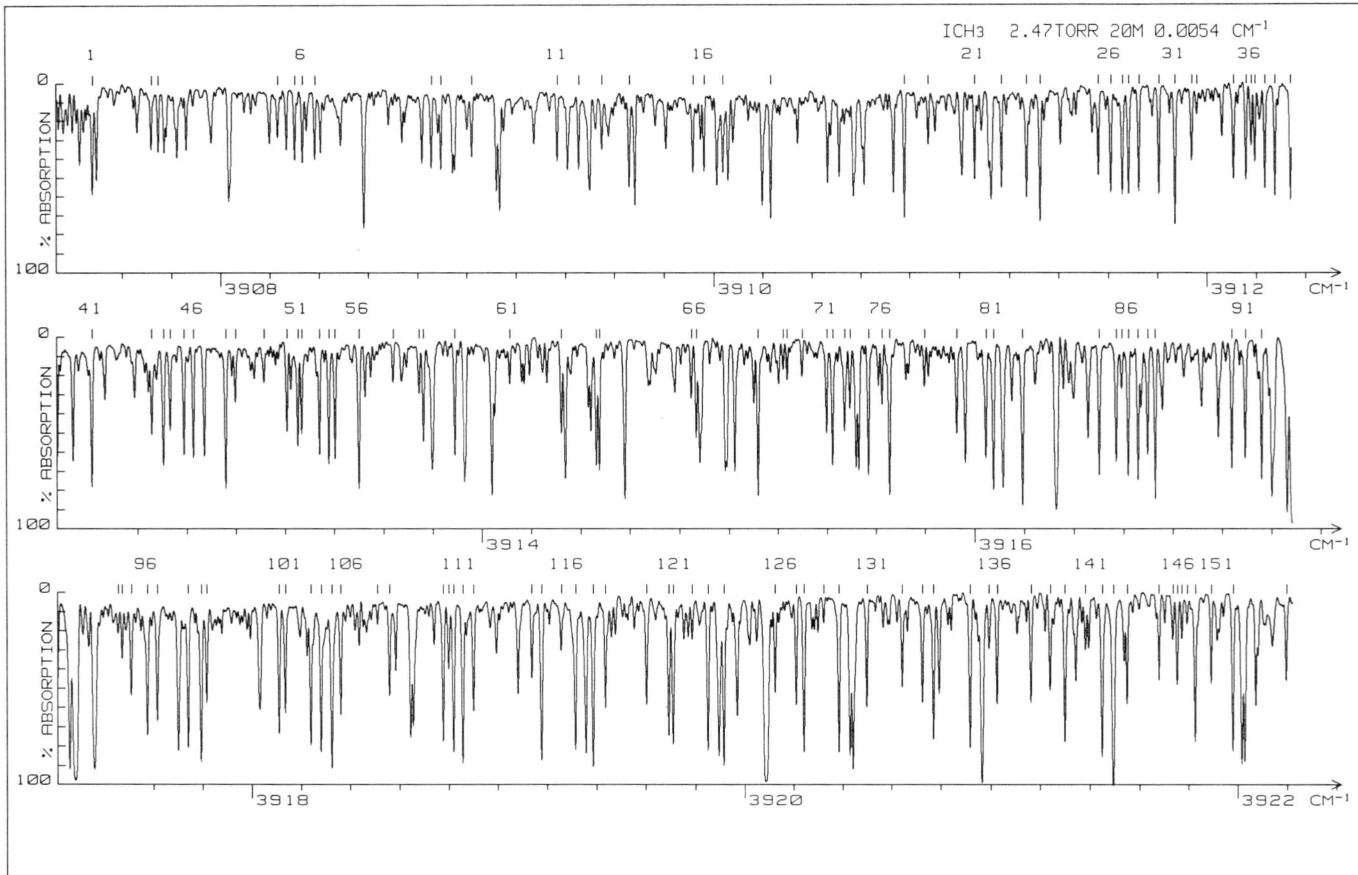
ICH3 2.47TORR 20M 0.0054 CM⁻¹
% ABSORPTION
CM⁻¹
3908
3910
3912
3914
3916
3918
3920
3922

LINE NO	WAVENUMBER CM^{-1}, VAC	LINE NO	WAVENUMBER CM^{-1}, VAC	LINE NO	WAVENUMBER CM^{-1}, VAC	LINE NO	WAVENUMBER CM^{-1}, VAC
1	3907.48154(36)	31	3911.87186(17)	61	3914.11290(38)	91	3917.09370(18)
2	3907.71922(28)	32	3911.93927(24)	62	3914.32134(33)	92	3917.16030(22)
3	3907.74760(27)	33	3911.96043(66)	63	3914.46385(25)	93	3917.45764(57)
4	3908.23322(39)	34	3912.10914(22)	64	3914.47625(24)	94	3917.47437(28)
5	3908.30251(51)	35	3912.15921(20)	65	3914.84725(28)	95	3917.51083(20)
6	3908.33386(24)	36	3912.17978(28)	66	3914.86736(33)	96	3917.57622(22)
7	3908.38382(25)	37	3912.19530(22)	67	3915.11951(16)	97	3917.61611(25)
8	3908.85689(23)	38	3912.23601(19)	68	3915.22164(53)	98	3917.74146(17)
9	3908.89550(47)	39	3912.27637(18)	69	3915.23797(36)	99	3917.79531(16)
10	3909.02266(28)	40	3912.34001(18)	70	3915.29851(41)	100	3917.81819(19)
11	3909.36855(25)	41	3912.41605(17)	71	3915.39811(20)	101	3918.11204(17)
12	3909.45528(23)	42	3912.65785(23)	72	3915.42260(18)	102	3918.13757(18)
13	3909.55116(33)	43	3912.70510(20)	73	3915.47180(25)	103	3918.24046(20)
14	3909.66125(20)	44	3912.73203(21)	74	3915.49382(29)	104	3918.28177(19)
15	3909.91706(23)	45	3912.78839(18)	75	3915.56929(17)	105	3918.32736(18)
16	3909.96173(22)	46	3912.82787(18)	76	3915.62621(26)	106	3918.36357(18)
17	3910.03831(28)	47	3912.95884(17)	77	3915.65692(19)	107	3918.51242(58)
18	3910.23355(18)	48	3912.99834(30)	78	3915.79858(33)	108	3918.56046(20)
19	3910.78058(18)	49	3913.11446(44)	79	3915.92807(22)	109	3918.77870(20)
20	3910.87666(32)	50	3913.20825(22)	80	3916.04504(25)	110	3918.80033(25)
21	3911.06265(21)	51	3913.25147(19)	81	3916.07446(17)	111	3918.82067(19)
22	3911.16955(20)	52	3913.26730(21)	82	3916.19327(18)	112	3918.85802(16)
23	3911.27181(30)	53	3913.33923(29)	83	3916.50183(17)	113	3918.90305(19)
24	3911.32672(17)	54	3913.37864(18)	84	3916.56992(18)	114	3919.13929(49)
25	3911.56067(41)	55	3913.40418(18)	85	3916.59204(39)	115	3919.17859(19)
26	3911.61110(35)	56	3913.50076(17)	86	3916.61824(18)	116	3919.25739(33)
27	3911.65726(20)	57	3913.63857(37)	87	3916.65767(21)	117	3919.31562(19)
28	3911.68245(19)	58	3913.74413(36)	88	3916.69668(19)	118	3919.38751(17)
29	3911.72367(19)	59	3913.76186(20)	89	3916.72845(16)	119	3919.43830(27)
30	3911.80663(19)	60	3913.88907(18)	90	3917.03968(18)	120	3919.60502(19)

- "LINE NO.": serial number identifying the spectral line in the map on the opposite page.
- "WAVENUMBER": experimental value.
- Uncertainty in the last digit of the wavenumber is given in parentheses.

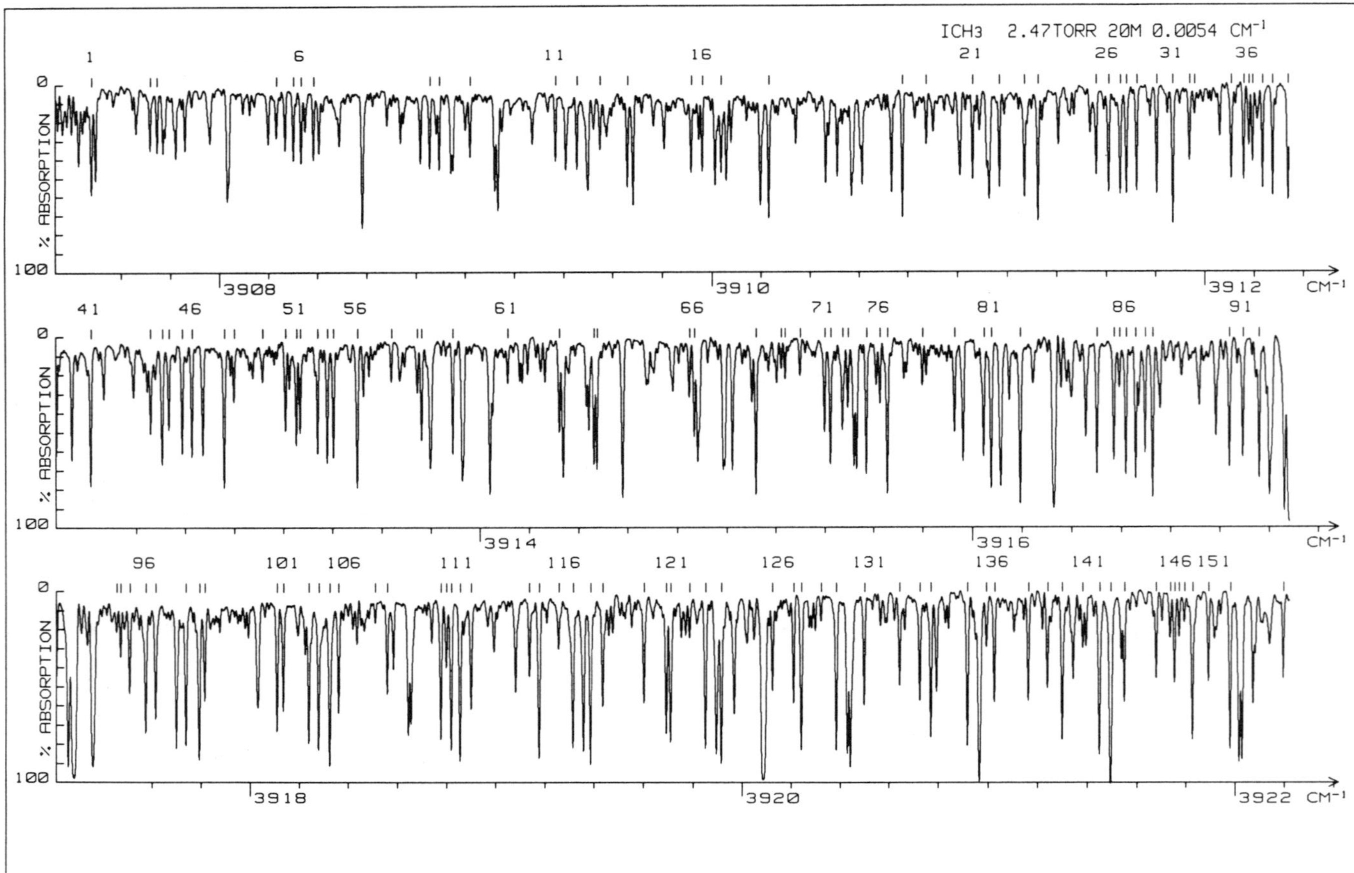
ICH3 2.47TORR 20M 0.0054 CM-1
% ABSORPTION
0
100
3908
3910
3912
3914
3916
3918
3920
3922
CM-1

LINE NO	WAVENUMBER CM^{-1}, VAC	LINE NO	WAVENUMBER CM^{-1}, VAC
121	3919.69430(17)	151	3921.89358(19)
122	3919.71169(17)	152	3921.98196(16)
123	3919.78769(39)	153	3922.19731(20)
124	3919.85158(16)		
125	3919.91605(17)		
126	3920.12584(22)		
127	3920.21184(18)		
128	3920.24320(16)		
129	3920.32308(51)		
130	3920.38589(16)		
131	3920.49950(18)		
132	3920.64570(20)		
133	3920.72761(19)		
134	3920.77253(17)		
135	3920.91915(16)		
136	3920.99489(32)		
137	3921.02629(18)		
138	3921.16430(18)		
139	3921.24213(33)		
140	3921.30116(20)		
141	3921.38386(33)		
142	3921.45102(16)		
143	3921.49590(19)		
144	3921.55109(26)		
145	3921.68132(34)		
146	3921.73744(36)		
147	3921.75555(20)		
148	3921.77474(34)		
149	3921.79547(52)		
150	3921.82865(20)		

- "LINE NO.": serial number identifying the spectral line in the map on the opposite page.
- "WAVENUMBER": experimental value.
- Uncertainty in the last digit of the wavenumber is given in parentheses.

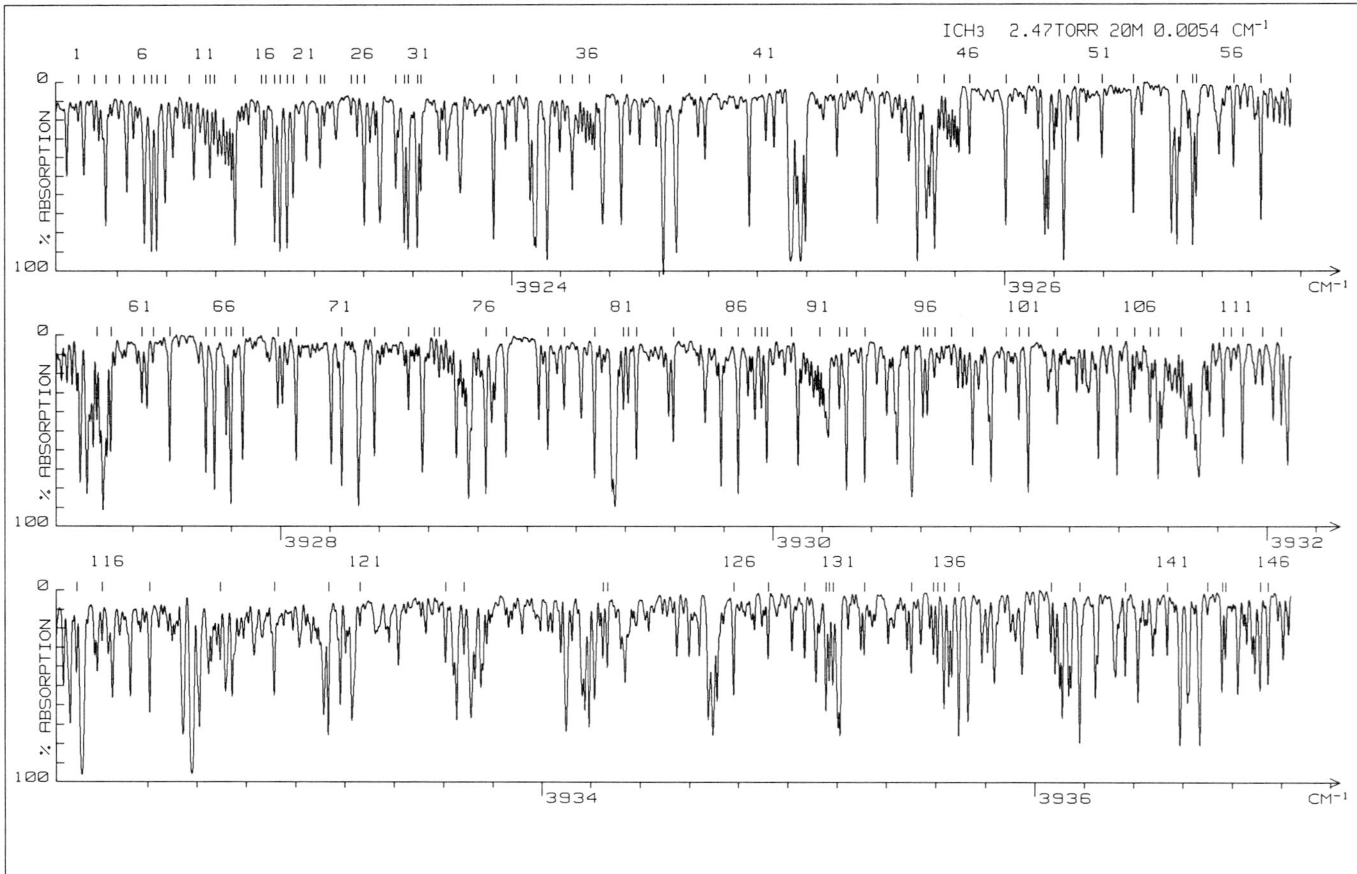
ICH3 2.47TORR 20M 0.0054 CM-1
1 6 11 16 21 26 31 36 41 46 51 56
61 66 71 76 81 86 91 96 101 106 111
116 121 126 131 136 141 146
0
% ABSORPTION
100
3924
3926
3928
3930
3932
3934
3936
CM-1

LINE NO	WAVENUMBER CM^{-1}, VAC	LINE NO	WAVENUMBER CM^{-1}, VAC	LINE NO	WAVENUMBER CM^{-1}, VAC	LINE NO	WAVENUMBER CM^{-1}, VAC
1	3922.24376(58)	31	3923.63301(28)	61	3927.43973(48)	91	3930.19331(28)
2	3922.30845(37)	32	3923.92721(16)	62	3927.48475(58)	92	3930.27185(25)
3	3922.35497(20)	33	3924.01964(30)	63	3927.55221(17)	93	3930.30095(19)
4	3922.40769(75)	34	3924.19870(29)	64	3927.69974(21)	94	3930.37423(19)
5	3922.46728(37)	35	3924.24901(20)	65	3927.73487(19)	95	3930.61147(46)
6	3922.51158(16)	36	3924.31730(34)	66	3927.78163(28)	96	3930.62924(22)
7	3922.54029(17)	37	3924.44825(17)	67	3927.80097(17)	97	3930.65793(52)
8	3922.56093(18)	38	3924.61813(15)	68	3927.84894(17)	98	3930.72665(59)
9	3922.59516(19)	39	3924.78713(22)	69	3927.99177(21)	99	3930.81081(17)
10	3922.69379(46)	40	3924.96864(16)	70	3928.06532(17)	100	3930.94531(30)
11	3922.75950(38)	41	3925.03603(28)	71	3928.25101(17)	101	3930.99787(21)
12	3922.77788(24)	42	3925.32417(22)	72	3928.38312(17)	102	3931.03597(19)
13	3922.79424(33)	43	3925.48778(16)	73	3928.52174(21)	103	3931.15320(23)
14	3922.87964(18)	44	3925.65009(16)	74	3928.62664(47)	104	3931.31894(18)
15	3922.98728(19)	45	3925.75703(33)	75	3928.64593(36)	105	3931.39489(22)
16	3923.00503(37)	46	3925.86001(41)	76	3928.83401(18)	106	3931.46594(34)
17	3923.04056(16)	47	3926.00568(20)	77	3928.91608(22)	107	3931.52941(36)
18	3923.06172(16)	48	3926.13730(30)	78	3929.08747(17)	108	3931.56263(21)
19	3923.09042(16)	49	3926.24113(17)	79	3929.15402(22)	109	3931.65378(28)
20	3923.11485(18)	50	3926.29893(25)	80	3929.27856(16)	110	3931.82572(19)
21	3923.16844(23)	51	3926.39477(21)	81	3929.39542(24)	111	3931.85711(55)
22	3923.22560(43)	52	3926.52240(16)	82	3929.41468(24)	112	3931.90292(17)
23	3923.24269(53)	53	3926.69941(18)	83	3929.44781(17)	113	3931.98360(35)
24	3923.35069(61)	54	3926.76273(17)	84	3929.59671(28)	114	3932.05997(21)
25	3923.37547(31)	55	3926.77695(24)	85	3929.79052(16)	115	3932.11289(24)
26	3923.40405(19)	56	3926.92784(20)	86	3929.86206(16)	116	3932.21652(34)
27	3923.53192(31)	57	3927.03819(16)	87	3929.93044(25)	117	3932.40996(25)
28	3923.56620(18)	58	3927.15702(30)	88	3929.95688(22)	118	3932.69585(47)
29	3923.58192(17)	59	3927.25676(36)	89	3929.97830(23)	119	3932.91525(21)
30	3923.61873(17)	60	3927.31347(25)	90	3930.07594(58)	120	3933.13501(19)

- "LINE NO.": serial number identifying the spectral line in the map on the opposite page.
- "WAVENUMBER": experimental value.
- Uncertainty in the last digit of the wavenumber is given in parentheses.

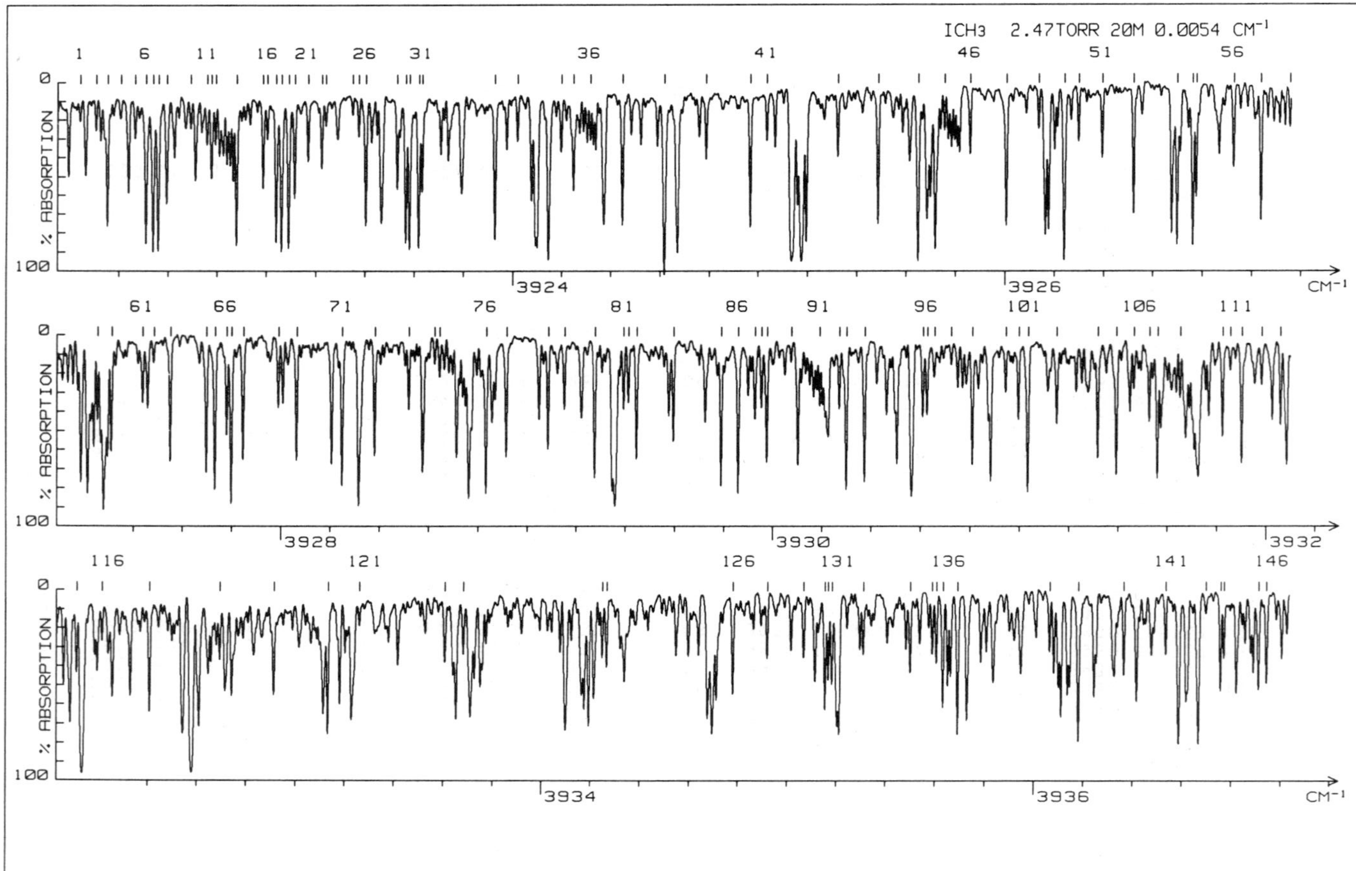
ICH3 2.47TORR 20M 0.0054 CM⁻¹
% ABSORPTION
0
100
3924
3926
3928
3930
3932
3934
3936
CM⁻¹

LINE NO	WAVENUMBER CM^{-1}, VAC
121	3933.26333(50)
122	3933.60941(26)
123	3933.68524(27)
124	3934.25074(23)
125	3934.26881(23)
126	3934.78263(18)
127	3934.92191(37)
128	3935.06992(26)
129	3935.15662(24)
130	3935.16979(32)
131	3935.18424(20)
132	3935.31129(24)
133	3935.50220(38)
134	3935.59245(45)
135	3935.60899(21)
136	3935.63534(17)
137	3935.69463(19)
138	3936.06598(42)
139	3936.18296(17)
140	3936.36761(20)
141	3936.53818(22)
142	3936.70056(70)
143	3936.75939(27)
144	3936.77315(40)
145	3936.91325(34)
146	3936.94577(19)

- "LINE NO.": serial number identifying the spectral line in the map on the opposite page.
- "WAVENUMBER": experimental value.
- Uncertainty in the last digit of the wavenumber is given in parentheses.

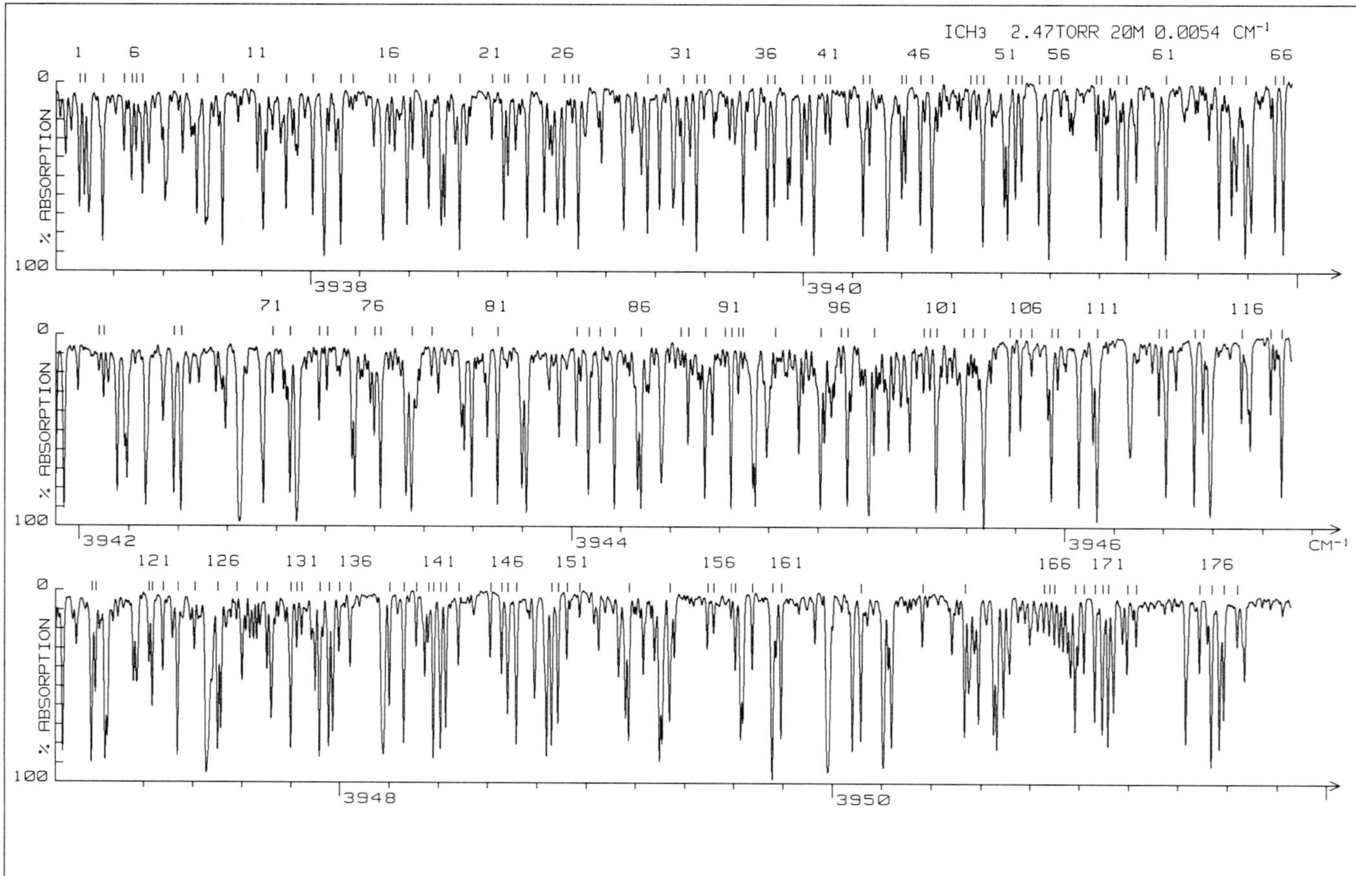
ICH3 2.47TORR 20M 0.0054 CM-1
% ABSORPTION
0
100
3938
3940
3942
3944
3946
3948
3950
CM-1
1 6 11 16 21 26 31 36 41 46 51 56 61 66
71 76 81 86 91 96 101 106 111 116
121 126 131 136 141 146 151 156 161 166 171 176

LINE NO	WAVENUMBER CM^{-1}, VAC	LINE NO	WAVENUMBER CM^{-1}, VAC	LINE NO	WAVENUMBER CM^{-1}, VAC	LINE NO	WAVENUMBER CM^{-1}, VAC
1	3937.06318(17)	31	3939.51228(19)	61	3941.46992(16)	91	3944.64889(15)
2	3937.08283(17)	32	3939.56639(15)	62	3941.68511(17)	92	3944.67824(26)
3	3937.15653(18)	33	3939.59636(37)	63	3941.73554(22)	93	3944.69685(56)
4	3937.24158(22)	34	3939.70101(29)	64	3941.78992(17)	94	3944.82824(42)
5	3937.27199(18)	35	3939.75823(16)	65	3941.90848(16)	95	3945.01156(17)
6	3937.29029(23)	36	3939.85567(16)	66	3941.94234(16)	96	3945.09463(35)
7	3937.31573(17)	37	3939.88314(21)	67	3942.08177(53)	97	3945.12056(16)
8	3937.47867(24)	38	3939.99431(16)	68	3942.10059(23)	98	3945.22969(17)
9	3937.53651(21)	39	3940.04428(16)	69	3942.38331(16)	99	3945.42913(24)
10	3937.64110(16)	40	3940.09035(27)	70	3942.41342(16)	100	3945.45385(29)
11	3937.78316(32)	41	3940.10936(45)	71	3942.78523(28)	101	3945.48017(16)
12	3937.89969(17)	42	3940.24250(16)	72	3942.85715(18)	102	3945.59065(16)
13	3938.00874(16)	43	3940.26870(20)	73	3942.97670(20)	103	3945.62556(39)
14	3938.12451(17)	44	3940.40049(23)	74	3943.00856(24)	104	3945.67116(15)
15	3938.17216(64)	45	3940.41625(25)	75	3943.12282(17)	105	3945.77657(17)
16	3938.32044(24)	46	3940.47492(16)	76	3943.20044(19)	106	3945.82042(28)
17	3938.34133(49)	47	3940.52092(16)	77	3943.22582(16)	107	3945.86600(36)
18	3938.41389(21)	48	3940.67416(30)	78	3943.35139(17)	108	3945.94723(17)
19	3938.47929(17)	49	3940.69944(39)	79	3943.43152(42)	109	3945.97236(29)
20	3938.60645(16)	50	3940.72544(18)	80	3943.59656(15)	110	3946.05926(15)
21	3938.73615(26)	51	3940.82556(17)	81	3943.70156(15)	111	3946.13250(15)
22	3938.78511(16)	52	3940.85915(17)	82	3944.02043(17)	112	3946.38352(19)
23	3938.80212(20)	53	3940.88371(26)	83	3944.06904(15)	113	3946.41292(17)
24	3938.88026(16)	54	3940.95404(18)	84	3944.11694(26)	114	3946.52645(15)
25	3938.94865(17)	55	3940.99594(16)	85	3944.17590(15)	115	3946.56207(25)
26	3939.02928(16)	56	3941.04384(45)	86	3944.28382(16)	116	3946.71595(18)
27	3939.06221(44)	57	3941.18671(20)	87	3944.44259(67)	117	3946.83255(18)
28	3939.08711(17)	58	3941.20628(16)	88	3944.47336(17)	118	3946.87720(15)
29	3939.36879(16)	59	3941.27490(21)	89	3944.54212(16)	119	3946.99239(16)
30	3939.41667(17)	60	3941.30920(17)	90	3944.62385(53)	120	3947.00834(26)

- "LINE NO.": serial number identifying the spectral line in the map on the opposite page.
- "WAVENUMBER": experimental value.
- Uncertainty in the last digit of the wavenumber is given in parentheses.

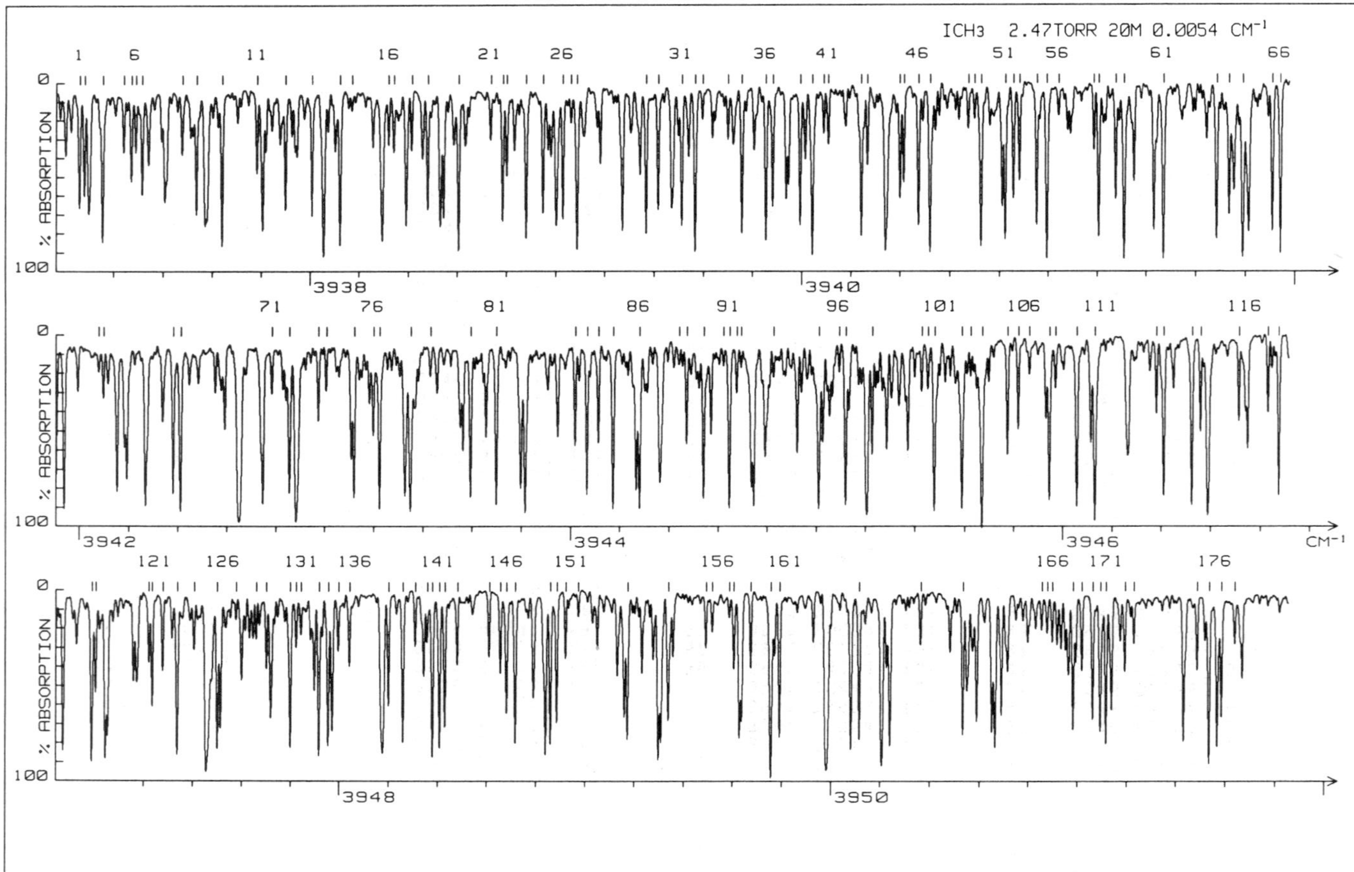
ICH3 2.47TORR 20M 0.0054 CM⁻¹
% ABSORPTION
0
100
3938
3940
3942
3944
3946
3948
3950
CM⁻¹

LINE NO	WAVENUMBER CM^{-1}, VAC	LINE NO	WAVENUMBER CM^{-1}, VAC
121	3947.22318(34)	151	3948.92187(20)
122	3947.23658(22)	152	3948.97427(40)
123	3947.27948(18)	153	3949.17514(18)
124	3947.33956(15)	154	3949.33996(21)
125	3947.40775(24)	155	3949.49252(22)
126	3947.50286(17)	156	3949.51847(33)
127	3947.57939(43)	157	3949.58910(49)
128	3947.66194(26)	158	3949.60764(19)
129	3947.70285(33)	159	3949.67614(18)
130	3947.80093(17)	160	3949.75786(16)
131	3947.82517(26)	161	3949.79288(16)
132	3947.84494(32)	162	3950.11634(16)
133	3947.91950(15)	163	3950.36558(22)
134	3947.95703(17)	164	3950.53494(16)
135	3947.99897(24)	165	3950.85673(33)
136	3948.04380(19)	166	3950.87907(31)
137	3948.20205(17)	167	3950.90003(26)
138	3948.26054(15)	168	3950.98497(19)
139	3948.31092(21)	169	3951.02077(18)
140	3948.35886(27)	170	3951.06429(16)
141	3948.38088(15)	171	3951.09497(17)
142	3948.40969(17)	172	3951.11824(16)
143	3948.43123(16)	173	3951.19457(31)
144	3948.48341(37)	174	3951.23163(26)
145	3948.61223(19)	175	3951.48900(19)
146	3948.65821(18)	176	3951.53712(16)
147	3948.68291(16)	177	3951.58726(21)
148	3948.71850(16)	178	3951.64023(27)
149	3948.86095(16)		
150	3948.88653(16)		

- "LINE NO.": serial number identifying the spectral line in the map on the opposite page.
- "WAVENUMBER": experimental value.
- Uncertainty in the last digit of the wavenumber is given in parentheses.

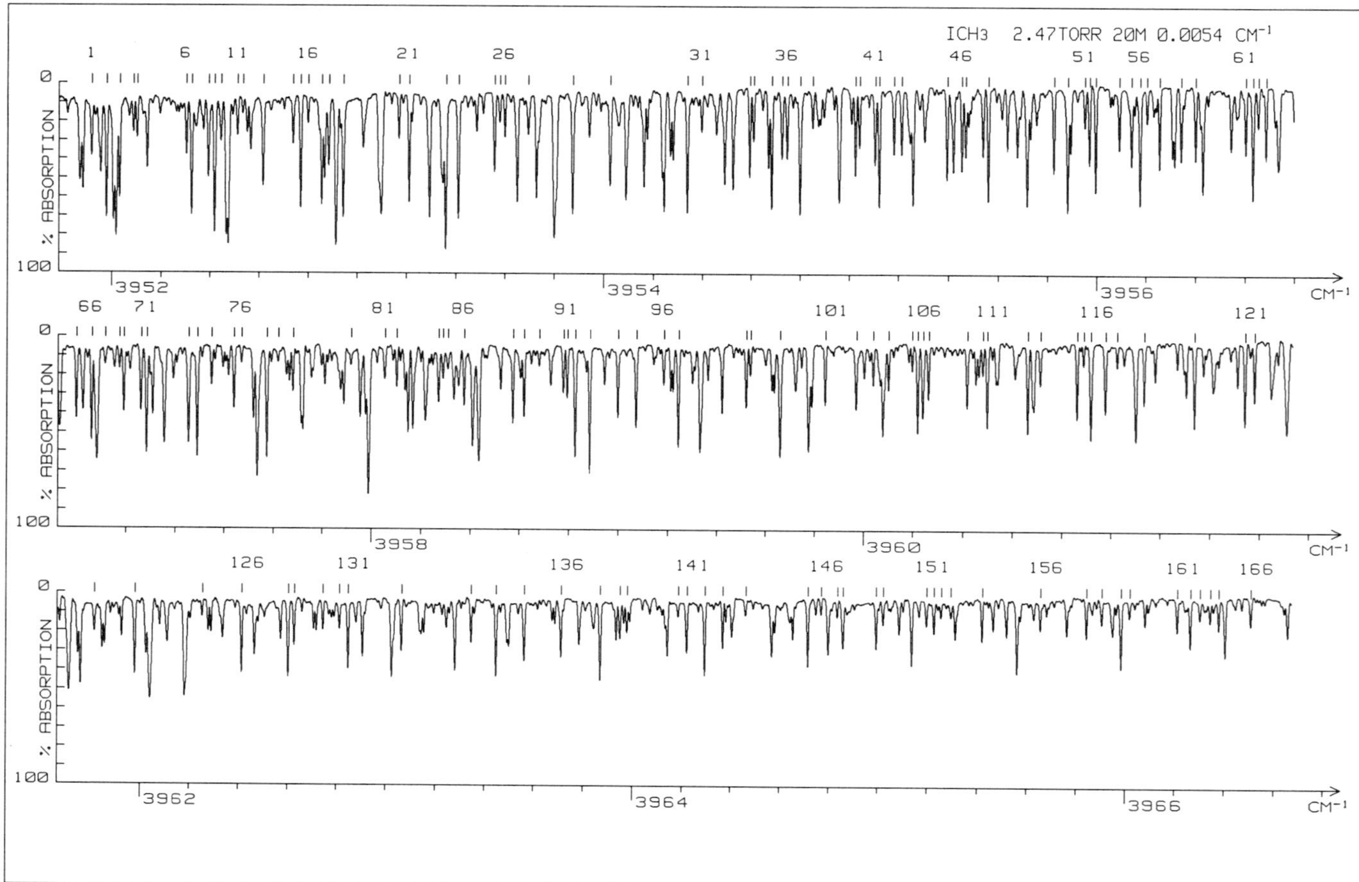
ICH3 2.47TORR 20M 0.0054 CM-1
% ABSORPTION
0
100
3952
3954
3956
3958
3960
3962
3964
3966
CM-1

LINE NO	WAVENUMBER CM^{-1}, VAC	LINE NO	WAVENUMBER CM^{-1}, VAC	LINE NO	WAVENUMBER CM^{-1}, VAC	LINE NO	WAVENUMBER CM^{-1}, VAC
1	3951.91880(40)	31	3954.39630(27)	61	3956.60181(19)	91	3958.79541(21)
2	3951.97896(16)	32	3954.59373(17)	62	3956.63052(16)	92	3958.82729(16)
3	3952.03223(22)	33	3954.60973(22)	63	3956.65194(28)	93	3958.88633(16)
4	3952.08795(29)	34	3954.68352(20)	64	3956.68292(18)	94	3959.00004(18)
5	3952.10312(25)	35	3954.72541(19)	65	3956.80082(18)	95	3959.07303(17)
6	3952.30425(21)	36	3954.74693(19)	66	3956.86166(17)	96	3959.18447(27)
7	3952.32543(16)	37	3954.79906(16)	67	3956.91633(65)	97	3959.24396(16)
8	3952.39402(18)	38	3954.85073(26)	68	3956.97437(44)	98	3959.52252(19)
9	3952.41945(16)	39	3955.02422(17)	69	3956.99245(19)	99	3959.53994(33)
10	3952.44524(44)	40	3955.04071(20)	70	3957.06170(19)	100	3959.66155(18)
11	3952.51258(28)	41	3955.10508(18)	71	3957.08465(16)	101	3959.84441(19)
12	3952.53401(54)	42	3955.12230(16)	72	3957.25648(18)	102	3959.97065(21)
13	3952.61637(17)	43	3955.18192(19)	73	3957.29194(16)	103	3960.04106(26)
14	3952.73766(23)	44	3955.21284(19)	74	3957.34993(25)	104	3960.10322(24)
15	3952.76885(16)	45	3955.39390(17)	75	3957.44147(20)	105	3960.19868(32)
16	3952.79889(33)	46	3955.45327(17)	76	3957.47173(54)	106	3960.22113(17)
17	3952.85646(22)	47	3955.46820(34)	77	3957.57460(16)	107	3960.24159(20)
18	3952.88471(31)	48	3955.55959(16)	78	3957.61975(72)	108	3960.26391(20)
19	3952.94338(19)	49	3955.82531(17)	79	3957.68017(23)	109	3960.41755(20)
20	3953.16941(22)	50	3955.88108(20)	80	3957.91760(61)	110	3960.48029(31)
21	3953.21068(21)	51	3955.95060(27)	81	3958.05518(35)	111	3960.49897(17)
22	3953.35808(17)	52	3955.96936(18)	82	3958.10361(24)	112	3960.66298(17)
23	3953.40955(16)	53	3955.99511(16)	83	3958.27182(22)	113	3960.71351(27)
24	3953.55652(18)	54	3956.09083(37)	84	3958.28997(37)	114	3960.86263(18)
25	3953.57858(35)	55	3956.14011(19)	85	3958.30926(60)	115	3960.88885(36)
26	3953.59984(24)	56	3956.17481(20)	86	3958.37619(21)	116	3960.91898(18)
27	3953.69517(32)	57	3956.20462(29)	87	3958.57554(33)	117	3960.97781(20)
28	3953.87409(16)	58	3956.25502(17)	88	3958.62131(19)	118	3961.02560(39)
29	3954.02552(19)	59	3956.34194(29)	89	3958.68218(53)	119	3961.13557(20)
30	3954.33725(16)	60	3956.39961(18)	90	3958.77947(22)	120	3961.33898(17)

- "LINE NO.": serial number identifying the spectral line in the map on the opposite page.
- "WAVENUMBER": experimental value.
- Uncertainty in the last digit of the wavenumber is given in parentheses.

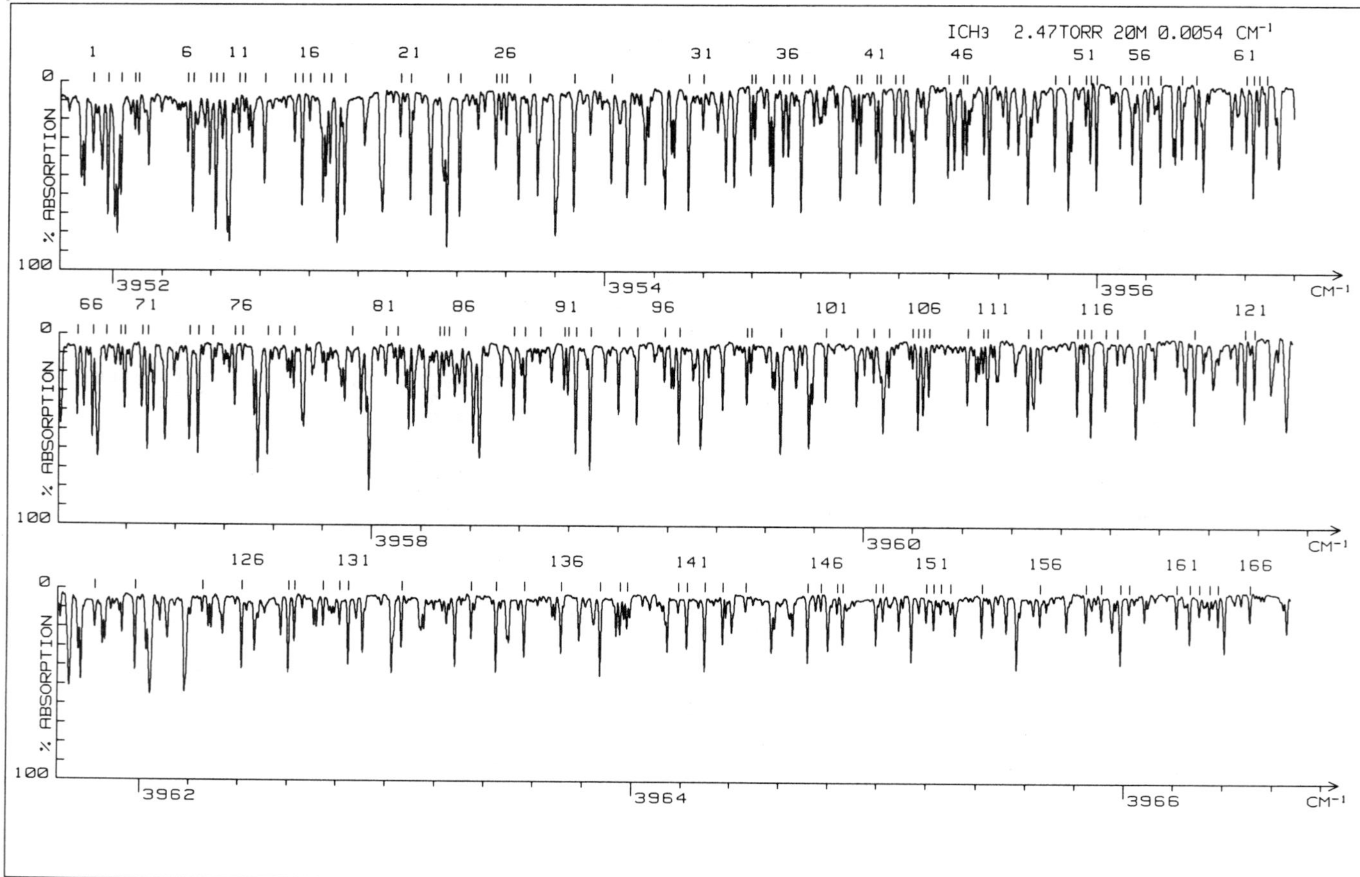
ICH3 2.47TORR 20M 0.0054 CM⁻¹
1 6 11 16 21 26 31 36 41 46 51 56 61
0
% ABSORPTION
100
3952 3954 3956 CM⁻¹
66 71 76 81 86 91 96 101 106 111 116 121
3958 3960 CM⁻¹
126 131 136 141 146 151 156 161 166
3962 3964 3966 CM⁻¹

LINE NO	WAVENUMBER CM^{-1}, VAC	LINE NO	WAVENUMBER CM^{-1}, VAC
121	3961.54232(17)	151	3965.19952(37)
122	3961.58072(20)	152	3965.22908(29)
123	3961.81566(36)	153	3965.25658(53)
124	3961.97939(18)	154	3965.29617(72)
125	3962.25423(48)	155	3965.42095(22)
126	3962.41518(18)	156	3965.65701(31)
127	3962.60362(18)	157	3965.84397(24)
128	3962.62807(22)	158	3965.90656(37)
129	3962.74730(29)	159	3965.98453(19)
130	3962.81405(36)	160	3966.01870(51)
131	3962.84933(18)	161	3966.21302(29)
132	3963.06531(20)	162	3966.26539(21)
133	3963.34715(49)	163	3966.30415(48)
134	3963.44886(18)	164	3966.34517(41)
135	3963.56410(21)	165	3966.38033(30)
136	3963.71292(20)	166	3966.50866(33)
137	3963.87153(18)		
138	3963.95104(28)		
139	3963.97908(31)		
140	3964.18557(60)		
141	3964.22102(22)		
142	3964.29480(18)		
143	3964.36868(21)		
144	3964.46320(59)		
145	3964.71766(18)		
146	3964.77085(47)		
147	3964.83674(45)		
148	3964.85983(26)		
149	3964.99622(22)		
150	3965.02426(37)		

- "LINE NO.": serial number identifying the spectral line in the map on the opposite page.
- "WAVENUMBER": experimental value.
- Uncertainty in the last digit of the wavenumber is given in parentheses.

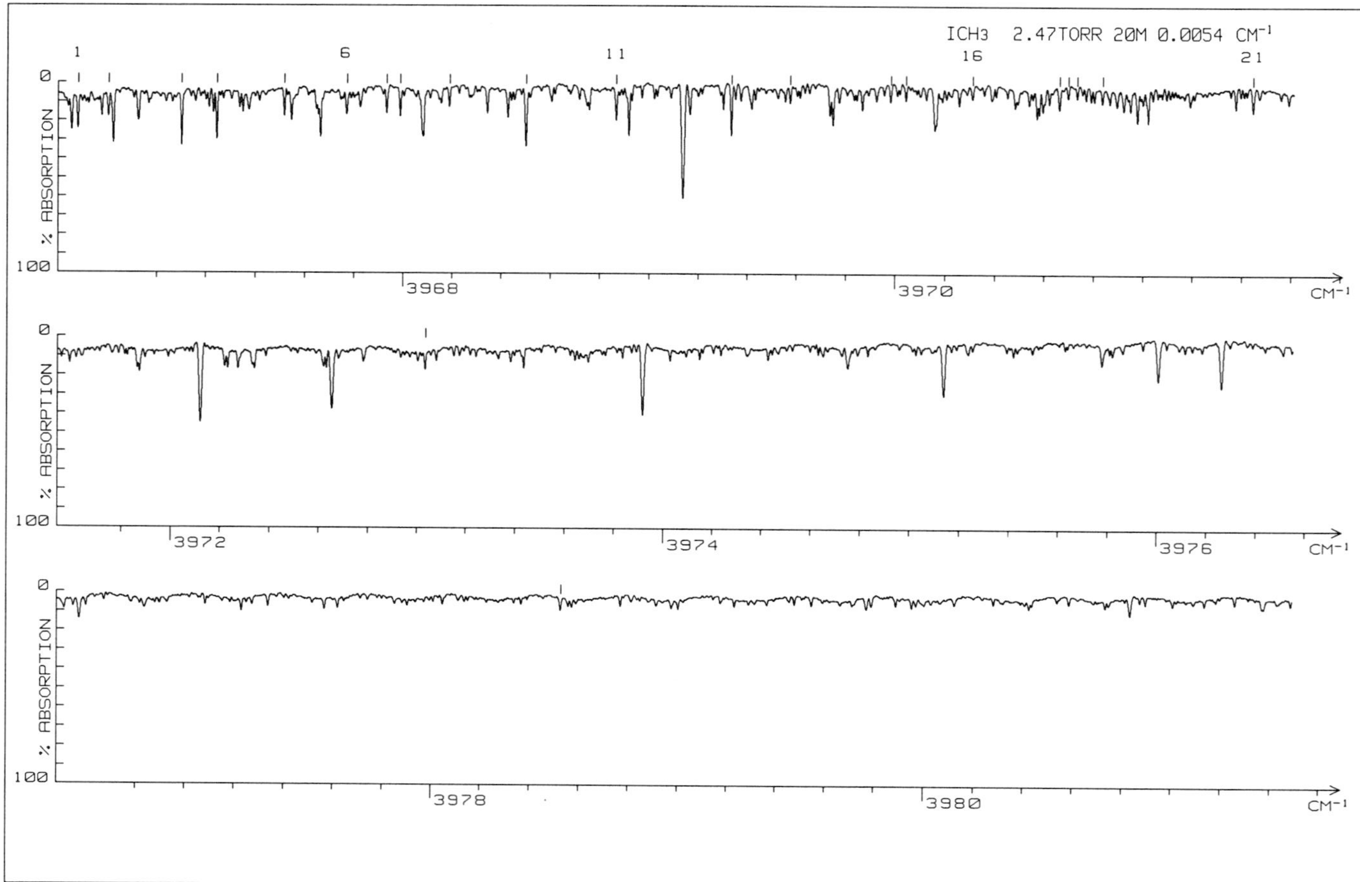
ICH3 2.47TORR 20M 0.0054 CM-1
1
6
11
16
21
0
% ABSORPTION
100
3968
3970
CM-1
3972
3974
3976
3978
3980

LINE NO	WAVENUMBER CM^{-1}, VAC
1	3966.68500(26)
2	3966.80738(34)
3	3967.10423(20)
4	3967.24771(20)
5	3967.51966(30)
6	3967.77332(42)
7	3967.93516(34)
8	3967.99055(31)
9	3968.19118(39)
10	3968.50332(21)
11	3968.86863(28)
12	3969.33641(20)
13	3969.57760(47)
14	3969.98485(48)
15	3970.04606(50)
16	3970.31205(70)
17	3970.66551(42)
18	3970.70087(63)
19	3970.73706(60)
20	3970.83769(56)
21	3971.44828(40)
22	3973.03499(44)
23	3978.52732(63)

- "LINE NO.": serial number identifying the spectral line in the map on the opposite page.
- "WAVENUMBER": experimental value.
- Uncertainty in the last digit of the wavenumber is given in parentheses.

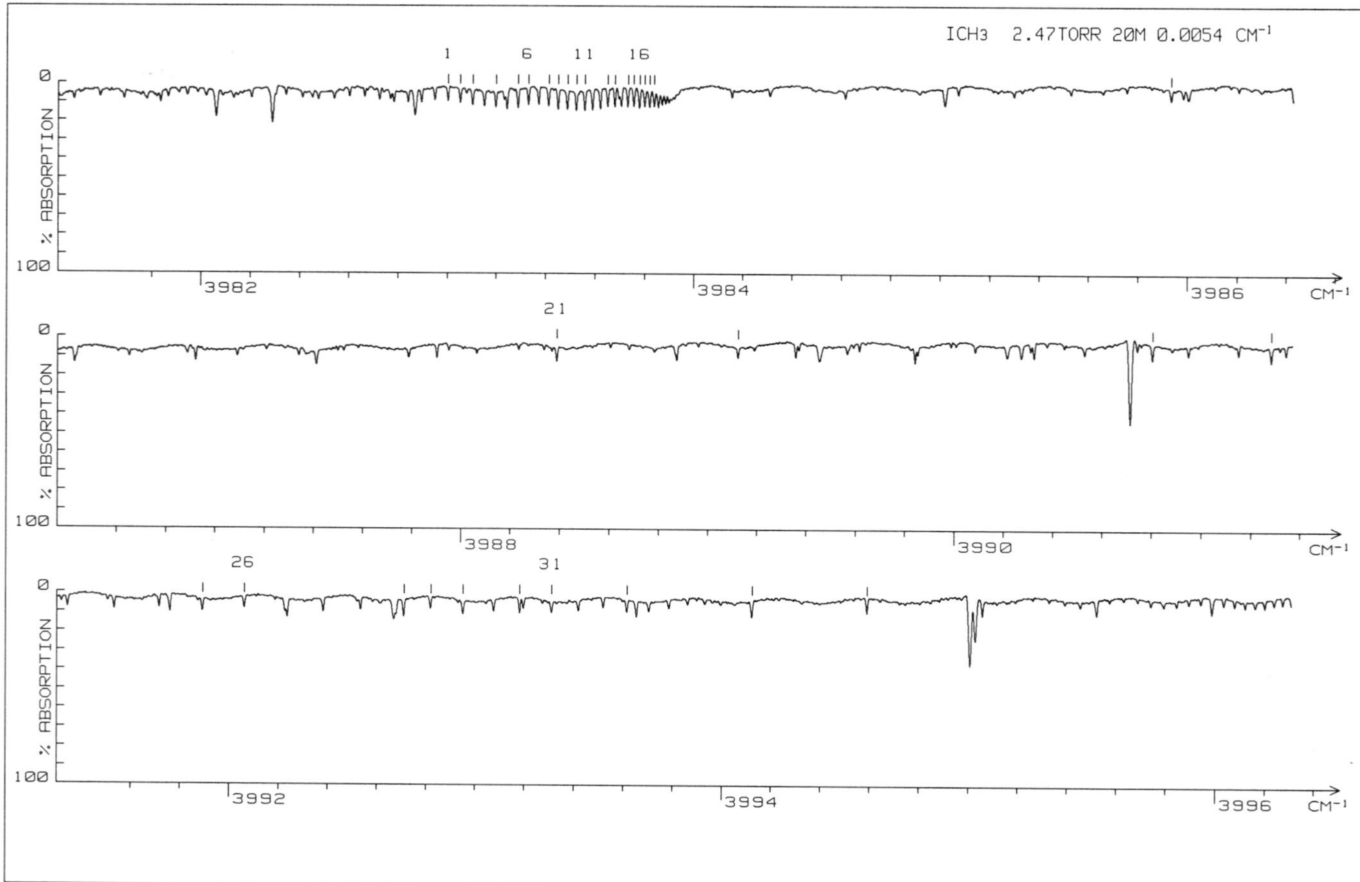
ICH3 2.47TORR 20M 0.0054 CM⁻¹
1
6
11
16
21
26
31
0
% ABSORPTION
100
3982
3984
3986
3988
3990
3992
3994
3996
CM⁻¹

LINE NO	WAVENUMBER CM^{-1}, VAC	LINE NO	WAVENUMBER CM^{-1}, VAC
1	3983.00374(51)	31	3993.31065(68)
2	3983.05501(50)	32	3993.61438(59)
3	3983.10498(47)	33	3994.12264(46)
4	3983.20014(45)	34	3994.59042(45)
5	3983.28990(40)		
6	3983.33243(43)		
7	3983.41345(42)		
8	3983.45190(38)		
9	3983.48889(39)		
10	3983.52451(38)		
11	3983.55854(38)		
12	3983.65207(41)		
13	3983.68064(38)		
14	3983.73291(39)		
15	3983.75727(41)		
16	3983.77992(39)		
17	3983.80138(46)		
18	3983.82102(45)		
19	3983.83943(48)		
20	3985.93710(73)		
21	3988.38907(52)		
22	3989.12422(63)		
23	3990.80417(47)		
24	3991.28262(49)		
25	3991.89150(60)		
26	3992.06140(65)		
27	3992.70934(44)		
28	3992.81794(60)		
29	3992.95075(58)		
30	3993.18196(48)		

- "LINE NO.": serial number identifying the spectral line in the map on the opposite page.
- "WAVENUMBER": experimental value.
- Uncertainty in the last digit of the wavenumber is given in parentheses.

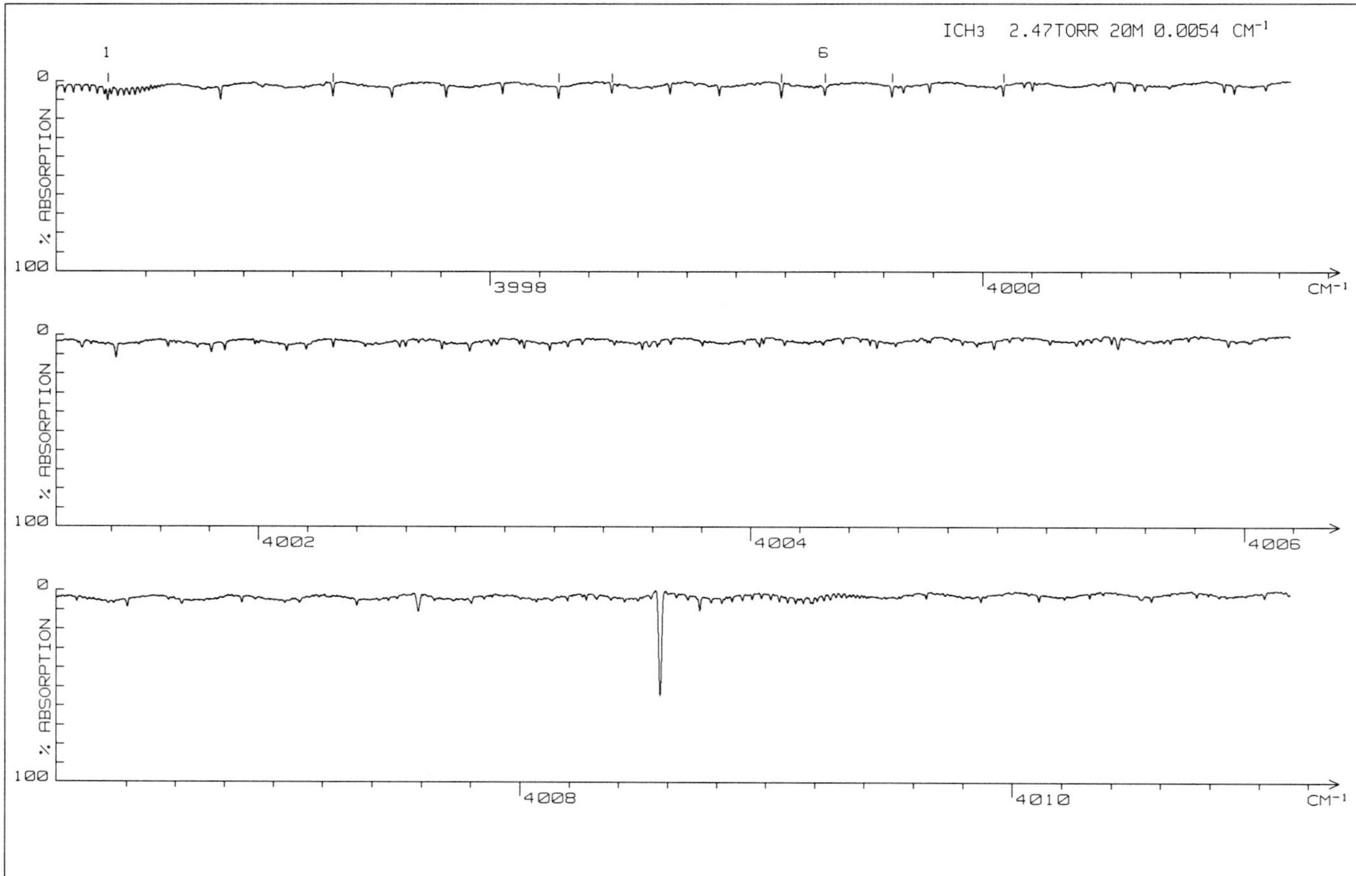
ICH3 2.47TORR 20M 0.0054 CM⁻¹
1
6
0
% ABSORPTION
100
3998
4000
CM⁻¹
0
% ABSORPTION
100
4002
4004
4006
0
% ABSORPTION
100
4008
4010
CM⁻¹

LINE NO	WAVENUMBER CM^{-1}, VAC
1	3996.44686(51)
2	3997.36533(53)
3	3998.27816(56)
4	3998.49557(64)
5	3999.18550(56)
6	3999.36225(69)
7	3999.63523(61)
8	4000.08414(57)

- "LINE NO.": serial number identifying the spectral line in the map on the opposite page.
- "WAVENUMBER": experimental value.
- Uncertainty in the last digit of the wavenumber is given in parentheses.

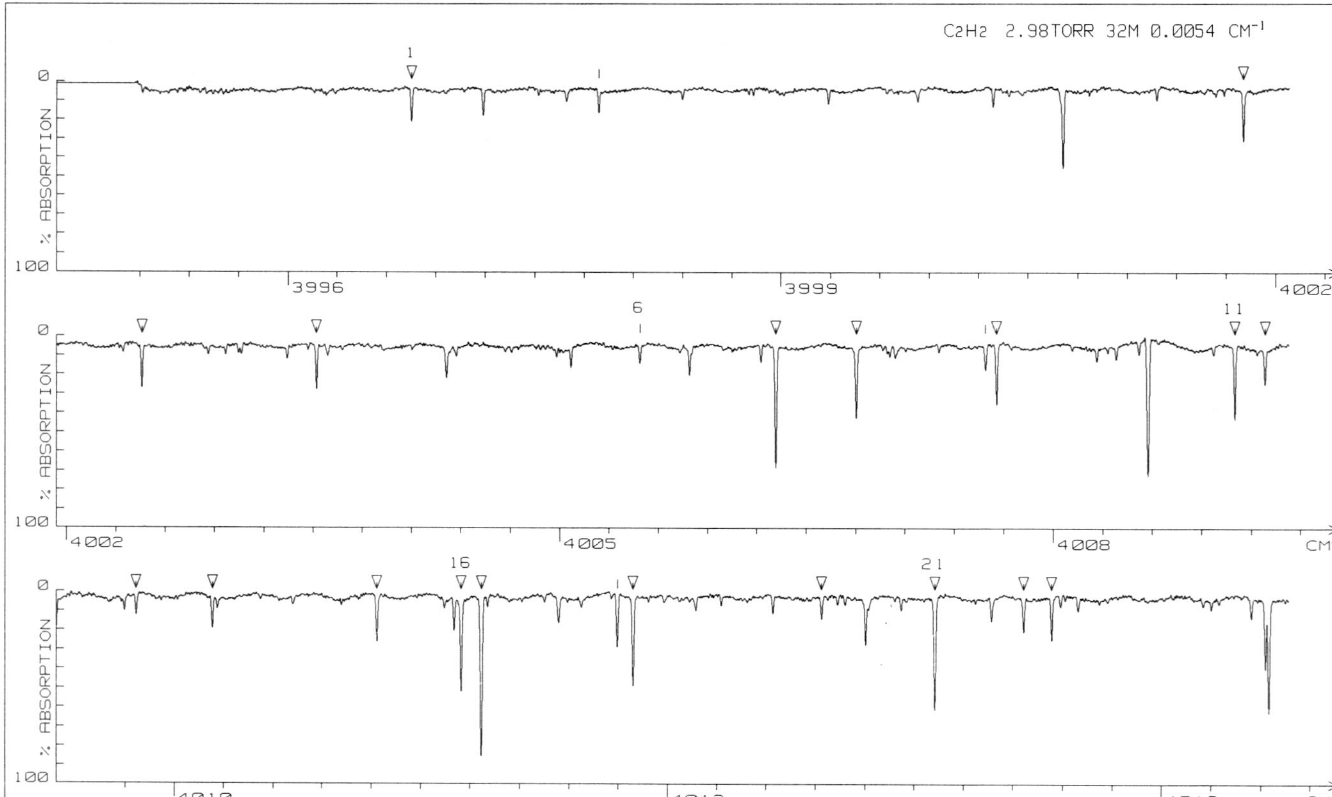

ISOTOPIC IDENTIFICATION – '▽' indicates $H^{12}C^{12}CH$; '◯': $H^{12}C^{13}CH$; 'I': unidentified.

BAND 1=$(1001^11^1)^{0+}(\Sigma_u^+) \leftarrow (0001^10^0)(\Pi_g)$; 2=$(1001^11^1)^{0-}(\Sigma_u^-) \leftarrow (0001^10^0)(\Pi_g)$; 3=$(1001^11^1)^2(\Delta_u) \leftarrow (0001^10^0)(\Pi_g)$; 4=$(1000^02^0)(\Sigma_g^+) \leftarrow (0000^01^1)(\Pi_u)$
5=$(1000^02^2)(\Delta_g) \leftarrow (0000^01^1)(\Pi_u)$; 6=$(1000^01^1)(\Pi_u) \leftarrow (0000^00^0)(\Sigma_g^+)$; 7=$(0100^03^1)(\Pi_u) \leftarrow (0000^00^0)(\Sigma_g^+)$; 8=$(1000^01^1)(\Pi) \leftarrow (0000^00^0)(\Sigma^+)$

LINE NO.	WAVENUMBER CM^{-1}, VAC.	ROTATIONAL ASSIGNMENTS		BAND	ISOTOPIC SPECIES
1	3996.75418(54)	P	(27)	1	$H^{12}C^{12}CH$
2	3997.89118(76)				
3	4001.80501(38)	P	(25)	1	$H^{12}C^{12}CH$
4	4002.45692(41)	P	(28)	4	$H^{12}C^{12}CH$
5	4003.51988(41)	P	(34)	6	$H^{12}C^{12}CH$
6	4005.48730(94)				
7	4006.31537(21)	P	(33)	6	$H^{12}C^{12}CH$
8	4006.80063(31)	P	(23)	1	$H^{12}C^{12}CH$
9	4007.58909(82)				
10	4007.65639(32)	P	(26)	4	$H^{12}C^{12}CH$
11	4009.09813(28)	P	(32)	6	$H^{12}C^{12}CH$
12	4009.28025(56)	P	(22)	1	$H^{12}C^{12}CH$
13	4009.76742(90)	Pff	(26)	3	$H^{12}C^{12}CH$
14	4010.23145(60)	P	(25)	4	$H^{12}C^{12}CH$
15	4011.23339(42)	P	(26)	2	$H^{12}C^{12}CH$
16	4011.74691(25)	P	(21)	1	$H^{12}C^{12}CH$
17	4011.86820(19)	P	(31)	6	$H^{12}C^{12}CH$
18	4012.69462(42)				
19	4012.79021(26)	P	(24)	4	$H^{12}C^{12}CH$
20	4013.93728(82)	P	(25)	2	$H^{12}C^{12}CH$
21	4014.62545(23)	P	(30)	6	$H^{12}C^{12}CH$
22	4015.16650(56)	Pff	(24)	3	$H^{12}C^{12}CH$
23	4015.33354(44)	P	(23)	4	$H^{12}C^{12}CH$

- "LINE NO.": serial number identifying the spectral line in the map on the opposite page.
- "WAVENUMBER": experimental value.
- Uncertainty in the last digit of the wavenumber is given in parentheses.
- "BAND" : identification on the opposite page.

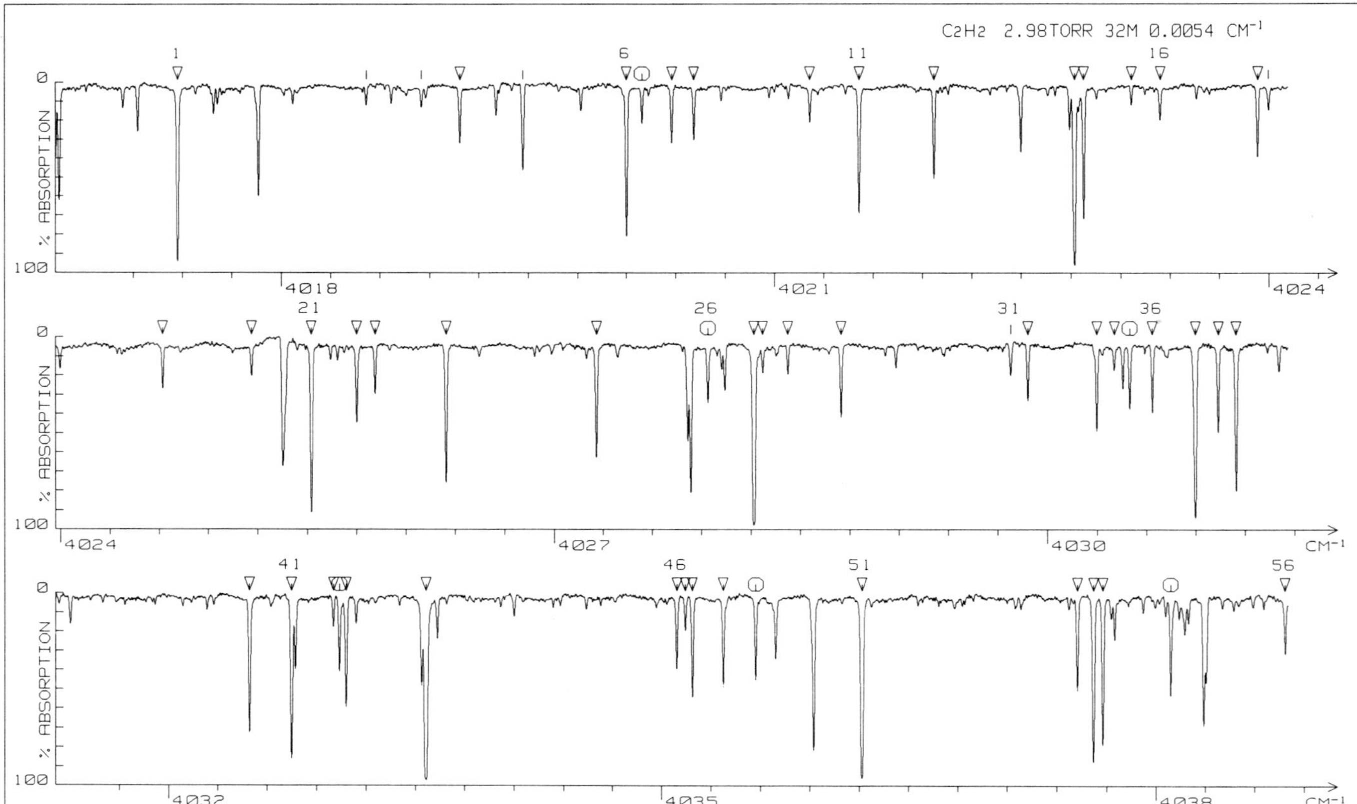

ISOTOPIC IDENTIFICATION – '∇' indicates $H^{12}C^{12}CH$; 'Ⓘ': $H^{12}C^{13}CH$; 'ı': unidentified.

BAND $1=(1001^11^1)^{0+}(\Sigma_u^+) \leftarrow (0001^10^0)(\Pi_g)$; $2=(1001^11^1)^{0-}(\Sigma_u^-) \leftarrow (0001^10^0)(\Pi_g)$; $3=(1001^11^1)^2(\Delta_u) \leftarrow (0001^10^0)(\Pi_g)$; $4=(1000^02^0)(\Sigma_g^+) \leftarrow (0000^01^1)(\Pi_u)$

$5=(1000^02^2)\ (\Delta_g) \leftarrow (0000^01^1)(\Pi_u)$; $6=(1000^01^1)\ (\Pi_u) \leftarrow (0000^00^0)(\Sigma_g^+)$; $7=(0100^03^1)(\Pi_u) \leftarrow (0000^00^0)(\Sigma_g^+)$; $8=(1000^01^1)\ (\Pi\ ) \leftarrow (0000^00^0)(\Sigma^+)$

LINE NO.	WAVENUMBER CM^{-1}, VAC.	ROTATIONAL ASSIGNMENTS	BAND	ISOTOPIC SPECIES
1	4017. 36976 (21)	P (29)	6	$H^{12}C^{12}CH$
2	4018. 51489 (96)			
3	4018. 84948 (94)			
4	4019. 08532 (37)	P (18)	1	$H^{12}C^{12}CH$
5	4019. 46667 (27)			
6	4020. 10112 (21)	P (28)	6	$H^{12}C^{12}CH$
7	4020. 19574 (52)	P (24)	8	$H^{12}C^{13}CH$
8	4020. 37503 (36)	P (21)	4	$H^{12}C^{12}CH$
9	4020. 51068 (39)	Pff (22)	3	$H^{12}C^{12}CH$
10	4021. 21445 (54)	Pee (23)	3	$H^{12}C^{12}CH$
11	4021. 51246 (22)	P (17)	1	$H^{12}C^{12}CH$
12	4021. 96990 (25)	P (22)	2	$H^{12}C^{12}CH$
13	4022. 81967 (18)	P (27)	6	$H^{12}C^{12}CH$
14	4022. 87466 (20)	P (20)	4	$H^{12}C^{12}CH$
15	4023. 16229 (90)	Pff (21)	3	$H^{12}C^{12}CH$
16	4023. 33844 (61)	Pee (24)	5	$H^{12}C^{12}CH$
17	4023. 93154 (31)	P (16)	1	$H^{12}C^{12}CH$
18	4023. 99738 (81)			
19	4024. 62126 (43)	P (21)	2	$H^{12}C^{12}CH$
20	4025. 16398 (64)	Pff (22)	5	$H^{12}C^{12}CH$
21	4025. 52529 (19)	P (26)	6	$H^{12}C^{12}CH$
22	4025. 80121 (30)	Pff (20)	3	$H^{12}C^{12}CH$
23	4025. 91336 (41)	Pee (21)	3	$H^{12}C^{12}CH$
24	4026. 34280 (21)	P (15)	1	$H^{12}C^{12}CH$
25	4027. 25919 (23)	P (20)	2	$H^{12}C^{12}CH$
26	4027. 94017 (37)	P (21)	8	$H^{12}C^{13}CH$
27	4028. 21760 (19)	P (25)	6	$H^{12}C^{12}CH$
28	4028. 27177 (75)	Pee (20)	3	$H^{12}C^{12}CH$
29	4028. 42468 (59)	Pff (19)	3	$H^{12}C^{12}CH$
30	4028. 74713 (28)	P (14)	1	$H^{12}C^{12}CH$
31	4029. 77959 (64)			
32	4029. 88352 (37)	P (19)	2	$H^{12}C^{12}CH$
33	4030. 29695 (27)	P (17)	4	$H^{12}C^{12}CH$
34	4030. 40284 (78)	Pee (21)	5	$H^{12}C^{12}CH$
35	4030. 49641 (33)	P (20)	8	$H^{12}C^{13}CH$
36	4030. 63505 (31)	Pee (19)	3	$H^{12}C^{12}CH$
37	4030. 89707 (23)	P (24)	6	$H^{12}C^{12}CH$
38	4031. 03566 (27)	Pff (18)	3	$H^{12}C^{12}CH$
39	4031. 14524 (20)	P (13)	1	$H^{12}C^{12}CH$
40	4032. 49511 (20)	P (18)	2	$H^{12}C^{12}CH$
41	4032. 74791 (29)	P (16)	4	$H^{12}C^{12}CH$
42	4033. 00284 (61)	Pee (18)	3	$H^{12}C^{12}CH$
43	4033. 04021 (29)	P (19)	8	$H^{12}C^{13}CH$
44	4033. 08070 (24)	Pff (19)	5	$H^{12}C^{12}CH$
45	4033. 56337 (22)	P (23)	6	$H^{12}C^{12}CH$
46	4035. 09297 (30)	P (17)	2	$H^{12}C^{12}CH$
47	4035. 14474 (53)	Pee (19)	5	$H^{12}C^{12}CH$
48	4035. 18855 (25)	P (15)	4	$H^{12}C^{12}CH$
49	4035. 37447 (26)	Pee (17)	3	$H^{12}C^{12}CH$
50	4035. 57128 (27)	P (18)	8	$H^{12}C^{13}CH$
51	4036. 21634 (19)	Pff (16)	3	$H^{12}C^{12}CH$
51		P (22)	6	$H^{12}C^{12}CH$
52	4037. 52387 (25)	Pee (18)	5	$H^{12}C^{12}CH$
53	4037. 61995 (21)	P (14)	4	$H^{12}C^{12}CH$
54	4037. 67715 (21)	P (16)	2	$H^{12}C^{12}CH$
55	4038. 08973 (64)	P (17)	8	$H^{12}C^{13}CH$
56	4038. 78585 (39)	Pff (15)	3	$H^{12}C^{12}CH$

- "LINE NO.": serial number identifying the spectral line in the map on the opposite page.
- "WAVENUMBER": experimental value.
- Uncertainty in the last digit of the wavenumber is given in parentheses.
- "BAND" : identification on the opposite page.

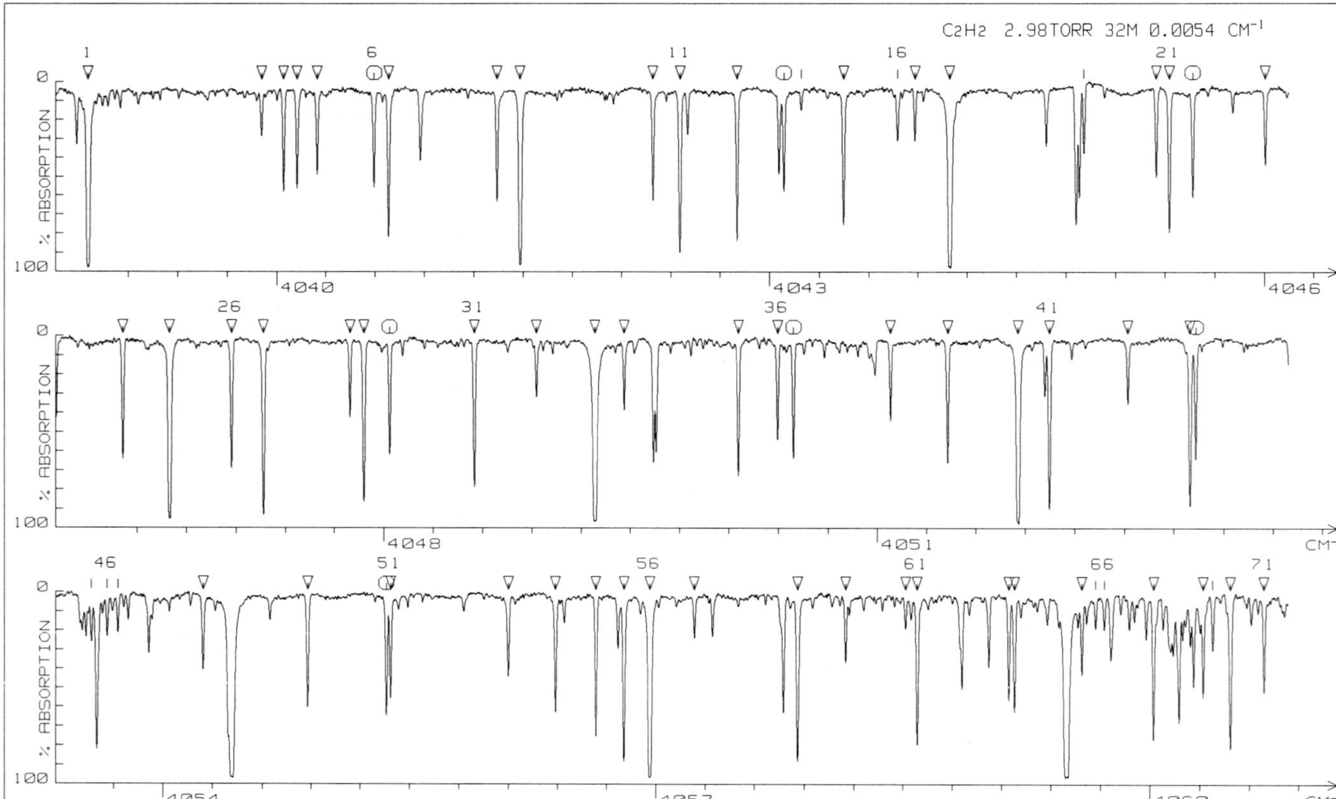

ISOTOPIC IDENTIFICATION - '▽' indicates $H^{12}C^{12}CH$; 'Ⓘ': $H^{12}C^{13}CH$; 'I': unidentified.

BAND 1=$(1001^11^1)^{0+}(\Sigma_u^+) \leftarrow (0001^10^0)(\Pi_g)$; 2=$(1001^11^1)^{0-}(\Sigma_u^-) \leftarrow (0001^10^0)(\Pi_g)$; 3=$(1001^11^1)^2(\Delta_u) \leftarrow (0001^10^0)(\Pi_g)$; 4=$(1000^02^0)(\Sigma_g^+) \leftarrow (0000^01^1)(\Pi_u)$
5=$(1000^02^2)(\Delta_g) \leftarrow (0000^01^1)(\Pi_u)$; 6=$(1000^01^1)(\Pi_u) \leftarrow (0000^00^0)(\Sigma_g^+)$; 7=$(0100^03^1)(\Pi_u) \leftarrow (0000^00^0)(\Sigma_g^+)$; 8=$(1000^01^1)(\Pi) \leftarrow (0000^00^0)(\Sigma^+)$

LINE NO.	WAVENUMBER CM^{-1}, VAC.	ROTATIONAL BAND ASSIGNMENTS		ISOTOPIC SPECIES
1	4038.85636 (23)	P (21)	6	$H^{12}C^{12}CH$
2	4039.90677 (45)	Pee (17)	5	$H^{12}C^{12}CH$
3	4040.04241 (23)	P (13)	4	$H^{12}C^{12}CH$
4	4040.12511 (67)	Pee (15)	3	$H^{12}C^{12}CH$
5	4040.24810 (26)	P (15)	2	$H^{12}C^{12}CH$
6	4040.59496 (24)	P (16)	8	$H^{12}C^{13}CH$
7	4040.68398 (21)	P (9)	1	$H^{12}C^{12}CH$
8	4041.34140 (23)	Pff (14)	3	$H^{12}C^{12}CH$
9	4041.48314 (19)	P (20)	6	$H^{12}C^{12}CH$
10	4042.29268 (23)	Pee (16)	5	$H^{12}C^{12}CH$
11	4042.45663 (21)	P (12)	4	$H^{12}C^{12}CH$
12	4042.80525 (19)	P (14)	2	$H^{12}C^{12}CH$
13	4043.08783 (23)	P (15)	8	$H^{12}C^{13}CH$
14	4043.19138 (91)			
15	4043.44748 (20)	Pff (15)	5	$H^{12}C^{12}CH$
16	4043.77774 (44)			
17	4043.88346 (38)	Pff (13)	3	$H^{12}C^{12}CH$
18	4044.09642 (25)	P (19)	6	$H^{12}C^{12}CH$
19	4044.90982 (32)			
20	4045.34865 (25)	P (13)	2	$H^{12}C^{12}CH$
21	4045.42794 (20)	P (7)	1	$H^{12}C^{12}CH$
22	4045.56746 (24)	P (14)	8	$H^{12}C^{13}CH$
23	4046.00544 (30)	Pff (14)	5	$H^{12}C^{12}CH$
24	4046.41199 (23)	Pff (12)	3	$H^{12}C^{12}CH$
25	4046.69657 (19)	P (18)	6	$H^{12}C^{12}CH$
26	4047.07154 (21)	Pee (14)	5	$H^{12}C^{12}CH$
27	4047.26314 (19)	Pee (12)	3	$H^{12}C^{12}CH$
27		P (10)	4	$H^{12}C^{12}CH$
28	4047.79465 (29)	P (6)	1	$H^{12}C^{12}CH$
29	4047.87837 (21)	P (12)	2	$H^{12}C^{12}CH$
30	4048.03484 (23)	P (13)	8	$H^{12}C^{13}CH$

LINE NO.	WAVENUMBER CM^{-1}, VAC.	ROTATIONAL BAND ASSIGNMENTS		ISOTOPIC SPECIES
31	4048.54896 (21)	Pff (13)	5	$H^{12}C^{12}CH$
32	4048.92653 (37)	Pff (11)	3	$H^{12}C^{12}CH$
33	4049.28301 (25)	P (17)	6	$H^{12}C^{12}CH$
34	4049.46273 (31)	Pee (13)	5	$H^{12}C^{12}CH$
35	4050.15866 (21)	P (5)	1	$H^{12}C^{12}CH$
36	4050.39437 (25)	P (11)	2	$H^{12}C^{12}CH$
37	4050.48921 (22)	P (12)	8	$H^{12}C^{13}CH$
38	4051.07919 (27)	Pff (12)	5	$H^{12}C^{12}CH$
39	4051.42719 (22)	Pff (10)	3	$H^{12}C^{12}CH$
40	4051.85621 (23)	Pee (12)	5	$H^{12}C^{12}CH$
40		P (16)	6	$H^{12}C^{12}CH$
41	4052.04516 (19)	P (8)	4	$H^{12}C^{12}CH$
42	4052.51937 (34)	P (4)	1	$H^{12}C^{12}CH$
43	4052.89648 (19)	P (10)	2	$H^{12}C^{12}CH$
44	4052.93052 (22)	P (11)	8	$H^{12}C^{13}CH$
45	4053.56379 (53)			
46	4053.66036 (53)			
47	4053.72566 (51)			
48	4054.24488 (30)	Pee (11)	5	$H^{12}C^{12}CH$
49	4054.87799 (23)	P (3)	1	$H^{12}C^{12}CH$
50	4055.35891 (21)	P (10)	8	$H^{12}C^{13}CH$
51	4055.38463 (25)	P (9)	2	$H^{12}C^{12}CH$
52	4056.09825 (30)	Pff (10)	5	$H^{12}C^{12}CH$
53	4056.38719 (23)	Pff (8)	3	$H^{12}C^{12}CH$
54	4056.63434 (20)	Pee (10)	5	$H^{12}C^{12}CH$
55	4056.80570 (19)	P (6)	4	$H^{12}C^{12}CH$
56	4056.96272 (23)	P (14)	6	$H^{12}C^{12}CH$
57	4057.23367 (48)	P (2)	1	$H^{12}C^{12}CH$
58	4057.85907 (21)	P (8)	2	$H^{12}C^{12}CH$
59	4058.14925 (32)	Q (22)	1	$H^{12}C^{12}CH$
60	4058.51530 (64)	Q (21)	1	$H^{12}C^{12}CH$

- "LINE NO.": serial number identifying the spectral line in the map on the opposite page.
- "WAVENUMBER": experimental value.
- Uncertainty in the last digit of the wavenumber is given in parentheses.
- "BAND" : identification on the opposite page.

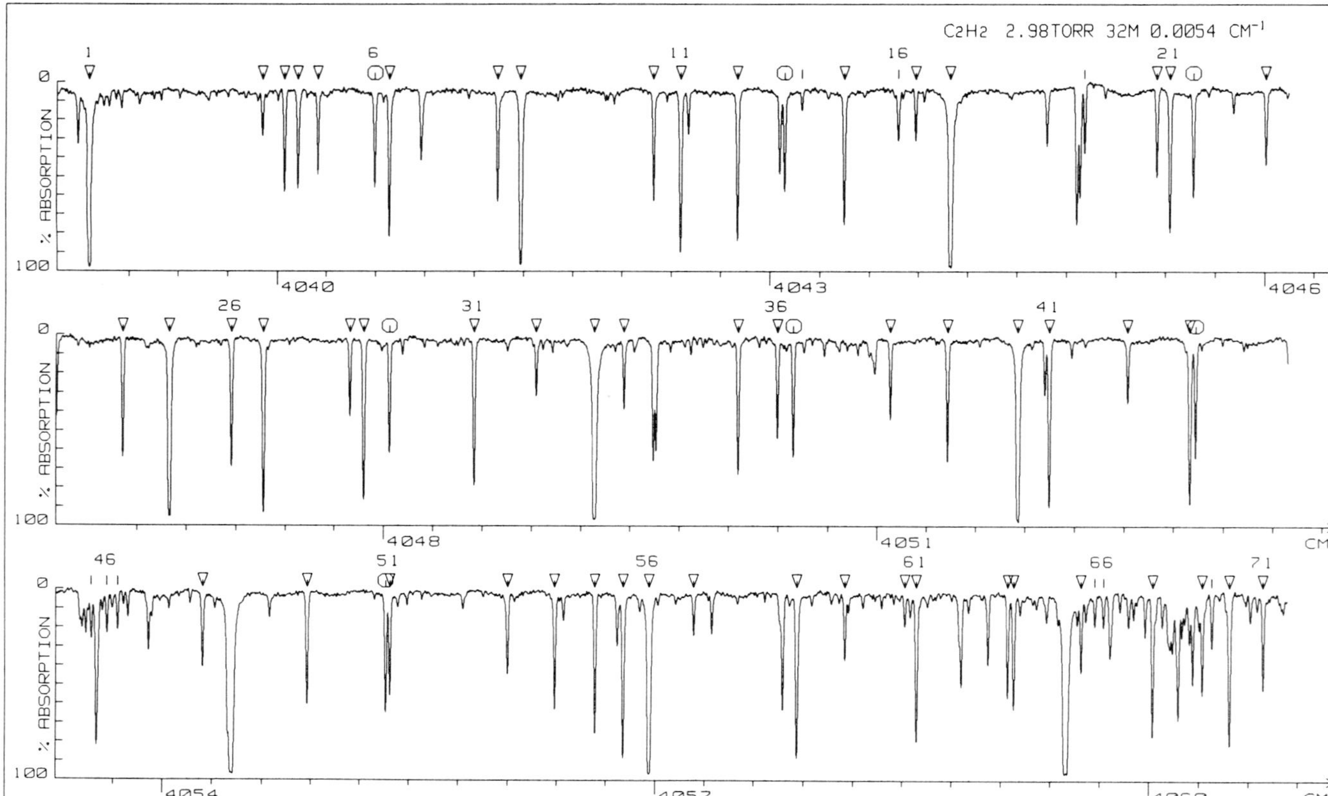

ISOTOPIC IDENTIFICATION - '∇' indicates $H^{12}C^{12}CH$; 'Ⓘ': $H^{12}C^{13}CH$; 'ı': unidentified.

BAND $1=(1001^11^1)^{0+}(\Sigma_u^+)\leftarrow(0001^10^0)(\Pi_g)$; $2=(1001^11^1)^{0-}(\Sigma_u^-)\leftarrow(0001^10^0)(\Pi_g)$; $3=(1001^11^1)^2(\Delta_u)\leftarrow(0001^10^0)(\Pi_g)$; $4=(1000^02^0)(\Sigma_g^+)\leftarrow(0000^01^1)(\Pi_u)$

$5=(1000^02^2)\ (\Delta_g)\leftarrow(0000^01^1)(\Pi_u)$; $6=(1000^01^1)\ (\Pi_u)\leftarrow(0000^00^0)(\Sigma_g^+)$; $7=(0100^03^1)(\Pi_u)\leftarrow(0000^00^0)(\Sigma_g^+)$; $8=(1000^01^1)\ (\Pi\ )\leftarrow(0000^00^0)(\Sigma^+)$

LINE NO.	WAVENUMBER CM^{-1}, VAC.	ROTATIONAL BAND ASSIGNMENTS		ISOTOPIC SPECIES	LINE NO.	WAVENUMBER CM^{-1}, VAC.	ROTATIONAL BAND ASSIGNMENTS	ISOTOPIC SPECIES
61	4058.58711 (21)	Pff (9)	5	$H^{12}C^{12}CH$				
62	4059.14249 (24)	Pee (7)	3	$H^{12}C^{12}CH$				
63	4059.17931 (24)	Q (19)	1	$H^{12}C^{12}CH$				
63		P (5)	4	$H^{12}C^{12}CH$				
64	4059.58640 (31)	P (1)	1	$H^{12}C^{12}CH$				
65	4059.66986 (62)							
66	4059.72215 (61)							
67	4060.02180 (41)	Q (16)	1	$H^{12}C^{12}CH$				
68	4060.31909 (63)	P (7)	2	$H^{12}C^{12}CH$				
69	4060.37763 (41)							
70	4060.48366 (33)	Q (14)	1	$H^{12}C^{12}CH$				
71	4060.68771 (26)	Q (13)	1	$H^{12}C^{12}CH$				

- "LINE NO.": serial number identifying the spectral line in the map on the opposite page.
- "WAVENUMBER": experimental value.
- Uncertainty in the last digit of the wavenumber is given in parentheses.
- "BAND" : identification on the opposite page.

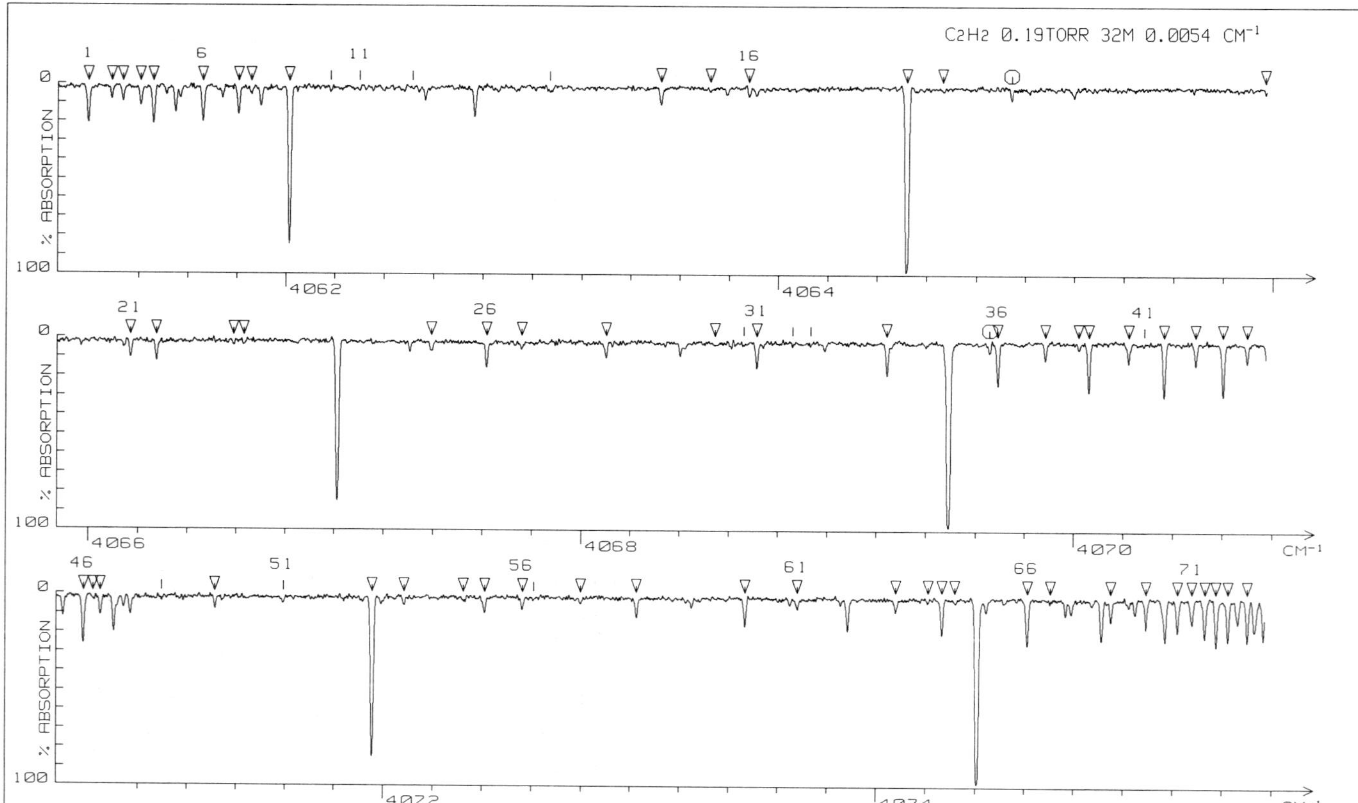

ISOTOPIC IDENTIFICATION - '∇' indicates $H^{12}C^{12}CH$; '⦶': $H^{12}C^{13}CH$; 'ı': unidentified.

BAND 1=$(1001^11^1)^{0+}(\Sigma_u^+) \leftarrow (0001^10^0)(\Pi_g)$; 2=$(1001^11^1)^{0-}(\Sigma_u^-) \leftarrow (0001^10^0)(\Pi_g)$; 3=$(1001^11^1)^2(\Delta_u) \leftarrow (0001^10^0)(\Pi_g)$; 4=$(1000^02^0)(\Sigma_g^+) \leftarrow (0000^01^1)(\Pi_u)$

5=$(1000^02^2)(\Delta_g) \leftarrow (0000^01^1)(\Pi_u)$; 6=$(1000^01^1)(\Pi_u) \leftarrow (0000^00^0)(\Sigma_g^+)$; 7=$(0100^03^1)(\Pi_u) \leftarrow (0000^00^0)(\Sigma_g^+)$; 8=$(1000^01^1)(\Pi) \leftarrow (0000^00^0)(\Sigma^+)$

LINE NO.	WAVENUMBER CM^{-1}, VAC.	ROTATIONAL BAND ASSIGNMENTS		ISOTOPIC SPECIES
1	4061.19769 (21)	Q (10)	1	$H^{12}C^{12}CH$
2	4061.29184 (25)	Pff (6)	3	$H^{12}C^{12}CH$
3	4061.33575 (23)	Q (9)	1	$H^{12}C^{12}CH$
4	4061.40754 (22)	Pee (8)	5	$H^{12}C^{12}CH$
5	4061.45872 (19)	Q (8)	1	$H^{12}C^{12}CH$
6	4061.66052 (19)	Q (6)	1	$H^{12}C^{12}CH$
7	4061.80658 (19)	Q (4)	1	$H^{12}C^{12}CH$
8	4061.85904 (30)	Q (3)	1	$H^{12}C^{12}CH$
9	4062.01466 (24)	P (12)	6	$H^{12}C^{12}CH$
10	4062.18317 (45)			
11	4062.30077 (42)			
12	4062.51392 (96)			
13	4063.07082 (76)			
14	4063.52285 (22)	Pff (7)	5	$H^{12}C^{12}CH$
15	4063.72280 (62)	Pff (5)	3	$H^{12}C^{12}CH$
16	4063.88084 (28)	Pee (5)	3	$H^{12}C^{12}CH$
17	4064.52026 (27)	P (11)	6	$H^{12}C^{12}CH$
18	4064.66455 (88)	Q (27)	4	$H^{12}C^{12}CH$
19	4064.94220 (57)	P (6)	8	$H^{12}C^{13}CH$
20	4065.96995 (36)	Pff (6)	5	$H^{12}C^{12}CH$
21	4066.16897 (22)	Pee (6)	5	$H^{12}C^{12}CH$
22	4066.27271 (22)	P (2)	4	$H^{12}C^{12}CH$
23	4066.58761 (33)	Q (23)	4	$H^{12}C^{12}CH$
24	4066.63096 (42)	R (1)	1	$H^{12}C^{12}CH$
25	4067.39329 (24)	Q (21)	4	$H^{12}C^{12}CH$
26	4067.61578 (35)	P (4)	2	$H^{12}C^{12}CH$
27	4067.75878 (37)	Q (20)	4	$H^{12}C^{12}CH$
28	4068.10030 (21)	Q (19)	4	$H^{12}C^{12}CH$
29	4068.54437 (39)	Pff (3)	3	$H^{12}C^{12}CH$
29		Pee (5)	5	$H^{12}C^{12}CH$
30	4068.66226 (66)			

LINE NO.	WAVENUMBER CM^{-1}, VAC.	ROTATIONAL BAND ASSIGNMENTS		ISOTOPIC SPECIES
31	4068.71462 (35)	Q (17)	4	$H^{12}C^{12}CH$
32	4068.85960 (64)			
33	4068.93167 (60)			
34	4069.24207 (29)	Q (15)	4	$H^{12}C^{12}CH$
35	4069.65492 (31)	P (4)	8	$H^{12}C^{13}CH$
36	4069.68844 (19)	Q (13)	4	$H^{12}C^{12}CH$
37	4069.88340 (21)	Q (12)	4	$H^{12}C^{12}CH$
38	4070.01978 (34)	P (3)	2	$H^{12}C^{12}CH$
39	4070.06038 (21)	Q (11)	4	$H^{12}C^{12}CH$
40	4070.22014 (22)	Q (10)	4	$H^{12}C^{12}CH$
41	4070.28377 (61)			
42	4070.36322 (21)	Q (9)	4	$H^{12}C^{12}CH$
43	4070.49029 (21)	Q (8)	4	$H^{12}C^{12}CH$
44	4070.60181 (22)	Q (7)	4	$H^{12}C^{12}CH$
45	4070.69843 (22)	Q (6)	4	$H^{12}C^{12}CH$
46	4070.78051 (19)	Q (5)	4	$H^{12}C^{12}CH$
47	4070.81982 (73)	Pff (4)	5	$H^{12}C^{12}CH$
48	4070.84814 (22)	Q (4)	4	$H^{12}C^{12}CH$
49	4071.09389 (53)			
50	4071.31330 (27)	R (3)	1	$H^{12}C^{12}CH$
51	4071.59189 (35)			
52	4071.95522 (23)	P (8)	6	$H^{12}C^{12}CH$
53	4072.08300 (28)	Qef (24)	3	$H^{12}C^{12}CH$
54	4072.32463 (48)	Qef (23)	3	$H^{12}C^{12}CH$
55	4072.40939 (23)	P (2)	2	$H^{12}C^{12}CH$
56	4072.56272 (24)	Qef (22)	3	$H^{12}C^{12}CH$
57	4072.60776 (69)			
58	4072.79691 (38)	Qef (21)	3	$H^{12}C^{12}CH$
59	4073.02629 (21)	Qef (20)	3	$H^{12}C^{12}CH$
60	4073.46820 (20)	Qef (18)	3	$H^{12}C^{12}CH$

- "LINE NO.": serial number identifying the spectral line in the map on the opposite page.
- "WAVENUMBER": experimental value.
- Uncertainty in the last digit of the wavenumber is given in parentheses.
- "BAND" : identification on the opposite page.

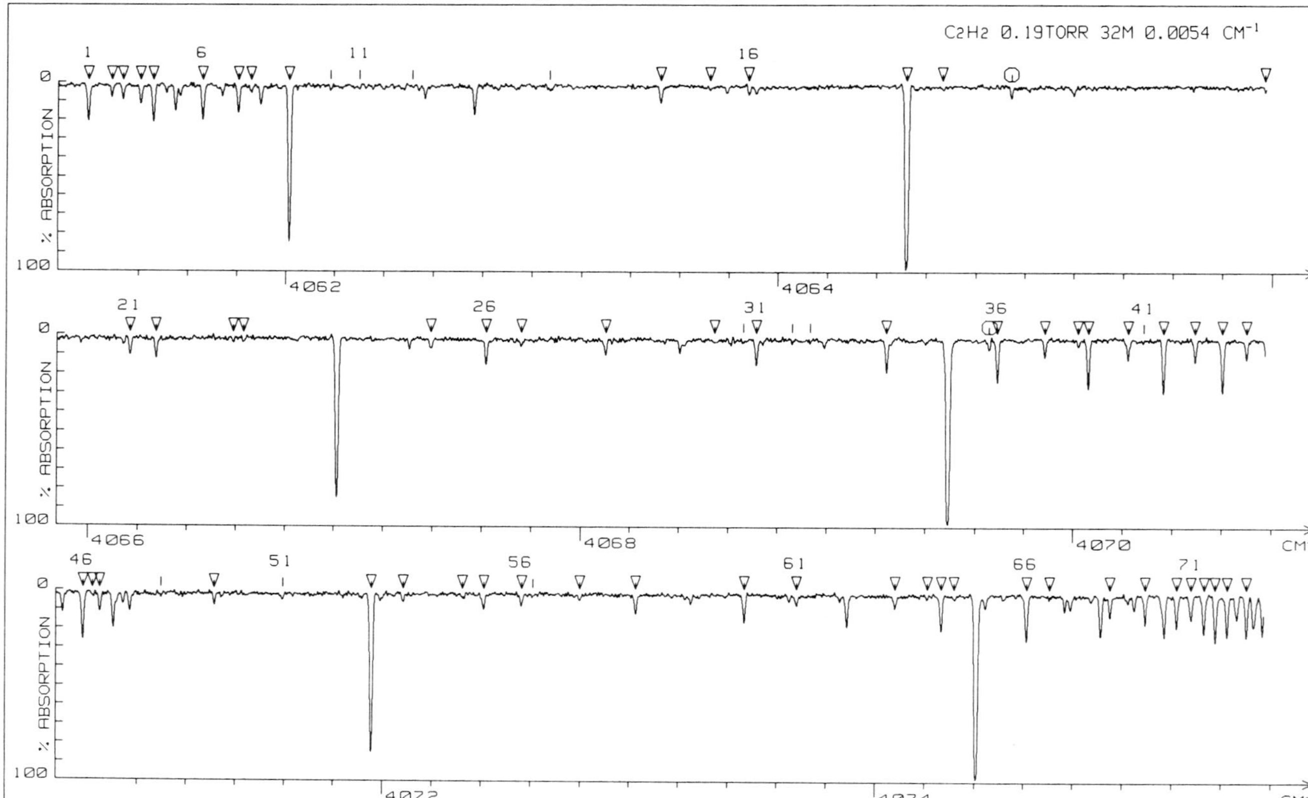

ISOTOPIC IDENTIFICATION - '∇' indicates $H^{12}C^{12}CH$; '⦶': $H^{12}C^{13}CH$; 'ı': unidentified.

BAND 1=$(1001^11^1)^{0+}(\Sigma_u^+) \leftarrow (0001^10^0)(\Pi_g)$; 2=$(1001^11^1)^{0-}(\Sigma_u^-) \leftarrow (0001^10^0)(\Pi_g)$; 3=$(1001^11^1)^2(\Delta_u) \leftarrow (0001^10^0)(\Pi_g)$; 4=$(1000^02^0)(\Sigma_g^+) \leftarrow (0000^01^1)(\Pi_u)$
5=$(1000^02^2)(\Delta_g) \leftarrow (0000^01^1)(\Pi_u)$; 6=$(1000^01^1)(\Pi_u) \leftarrow (0000^00^0)(\Sigma_g^+)$; 7=$(0100^03^1)(\Pi_u) \leftarrow (0000^00^0)(\Sigma_g^+)$; 8=$(1000^01^1)(\Pi) \leftarrow (0000^00^0)(\Sigma^+)$

LINE NO.	WAVENUMBER CM^{-1}, VAC.	ROTATIONAL ASSIGNMENTS	BAND	ISOTOPIC SPECIES
61	4073.67956 (26)	Qef (17)	3	$H^{12}C^{12}CH$
62	4074.07903 (24)	Qef (15)	3	$H^{12}C^{12}CH$
63	4074.20964 (41)	Qfe (27)	3	$H^{12}C^{12}CH$
64	4074.26593 (19)	Qef (14)	3	$H^{12}C^{12}CH$
65	4074.31675 (45)	Qfe (26)	3	$H^{12}C^{12}CH$
65		P (2)	8	$H^{12}C^{13}CH$
66	4074.61152 (22)	Qef (12)	3	$H^{12}C^{12}CH$
66		Qfe (23)	3	$H^{12}C^{12}CH$
67	4074.70609 (47)	Qfe (22)	3	$H^{12}C^{12}CH$
68	4074.95078 (22)	Qfe (19)	3	$H^{12}C^{12}CH$
69	4075.09315 (22)	Qfe (17)	3	$H^{12}C^{12}CH$
70	4075.21938 (19)	Qfe (15)	3	$H^{12}C^{12}CH$
71	4075.27752 (20)	Qef (7)	3	$H^{12}C^{12}CH$
71		Qfe (14)	3	$H^{12}C^{12}CH$
72	4075.32934 (20)	Qfe (13)	3	$H^{12}C^{12}CH$
73	4075.37588 (19)	Qef (6)	3	$H^{12}C^{12}CH$
73		Qfe (12)	3	$H^{12}C^{12}CH$
74	4075.42336 (22)	Qfe (11)	3	$H^{12}C^{12}CH$
75	4075.50207 (26)	Qfe (9)	3	$H^{12}C^{12}CH$

- "LINE NO.": serial number identifying the spectral line in the map on the opposite page.
- "WAVENUMBER": experimental value.
- Uncertainty in the last digit of the wavenumber is given in parentheses.
- "BAND" : identification on the opposite page.

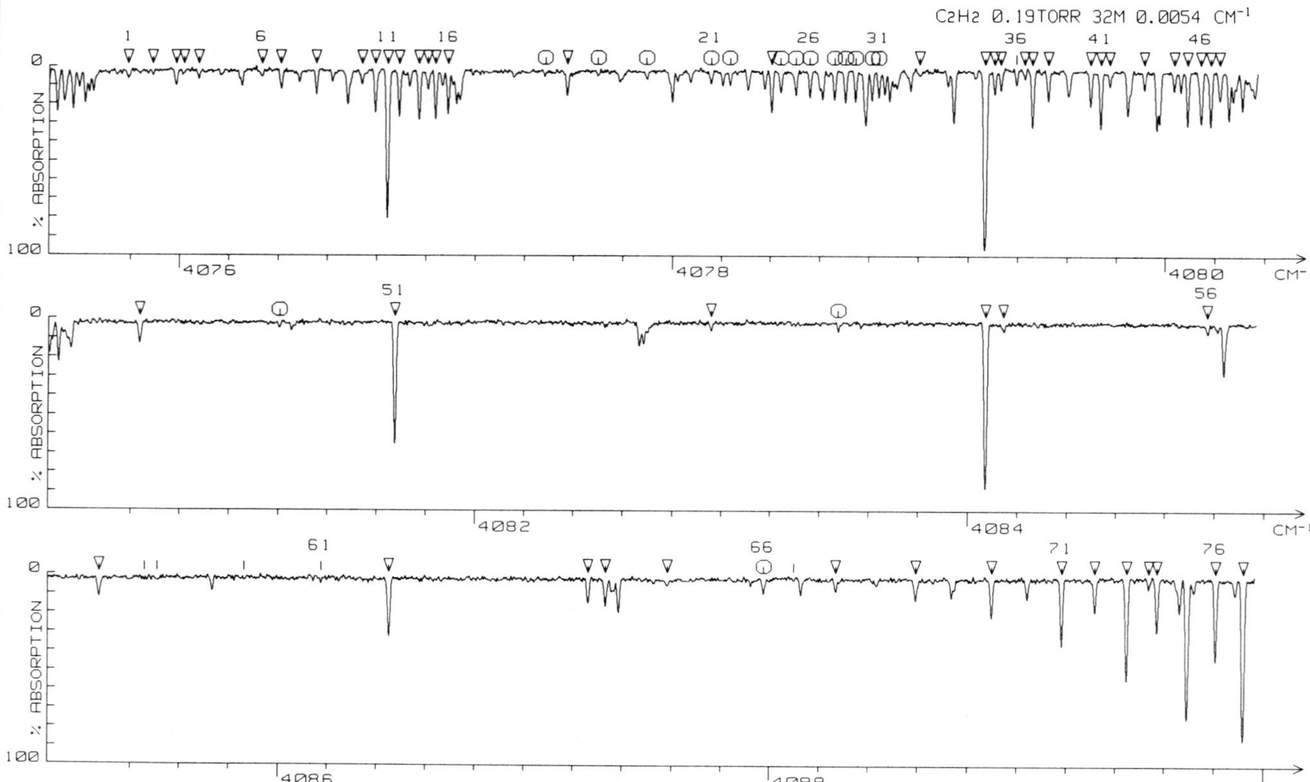

ISOTOPIC IDENTIFICATION - '∇' indicates $H^{12}C^{12}CH$; 'Ⓘ': $H^{12}C^{13}CH$; 'ı': unidentified.
BAND $1=(1001^{1}1^{1})^{0+}(\Sigma_u^+) \leftarrow (0001^{1}0^{0})(\Pi_g)$; $2=(1001^{1}1^{1})^{0-}(\Sigma_u^-) \leftarrow (0001^{1}0^{0})(\Pi_g)$; $3=(1001^{1}1^{1})^{2}(\Delta_u) \leftarrow (0001^{1}0^{0})(\Pi_g)$; $4=(1000^{0}2^{0})(\Sigma_g^+) \leftarrow (0000^{0}1^{1})(\Pi_u)$
$5=(1000^{0}2^{2})(\Delta_g) \leftarrow (0000^{0}1^{1})(\Pi_u)$; $6=(1000^{0}1^{1})(\Pi_u) \leftarrow (0000^{0}0^{0})(\Sigma_g^+)$; $7=(0100^{0}3^{1})(\Pi_u) \leftarrow (0000^{0}0^{0})(\Sigma_g^+)$; $8=(1000^{0}1^{1})(\Pi) \leftarrow (0000^{0}0^{0})(\Sigma^+)$

LINE NO.	WAVENUMBER CM^{-1}, VAC.	ROTATIONAL ASSIGNMENTS	BAND	ISOTOPIC SPECIES
1	4075.78916 (35)	Q (26)	2	$H^{12}C^{12}CH$
1		Qef (29)	5	$H^{12}C^{12}CH$
2	4075.88784 (38)	Q (25)	2	$H^{12}C^{12}CH$
3	4075.98295 (21)	R (5)	1	$H^{12}C^{12}CH$
3		Q (24)	2	$H^{12}C^{12}CH$
4	4076.01655 (78)	Qef (28)	5	$H^{12}C^{12}CH$
5	4076.07703 (27)	Q (23)	2	$H^{12}C^{12}CH$
6	4076.33235 (36)	Q (20)	2	$H^{12}C^{12}CH$
7	4076.40949 (20)	Q (19)	2	$H^{12}C^{12}CH$
8	4076.55279 (20)	Q (17)	2	$H^{12}C^{12}CH$
9	4076.73867 (23)	Q (14)	2	$H^{12}C^{12}CH$
10	4076.79302 (21)	Q (13)	2	$H^{12}C^{12}CH$
11	4076.84294 (23)	Q (12)	2	$H^{12}C^{12}CH$
11		P (6)	6	$H^{12}C^{12}CH$
12	4076.88989 (19)	Q (11)	2	$H^{12}C^{12}CH$
13	4076.97138 (19)	Q (9)	2	$H^{12}C^{12}CH$
14	4077.00599 (22)	Q (8)	2	$H^{12}C^{12}CH$
15	4077.03727 (19)	Q (7)	2	$H^{12}C^{12}CH$
16	4077.08766 (26)	Q (5)	2	$H^{12}C^{12}CH$
17	4077.48312 (60)	Q (26)	8	$H^{12}C^{13}CH$
18	4077.57217 (33)	Qef (21)	5	$H^{12}C^{12}CH$
19	4077.69766 (38)	Q (24)	8	$H^{12}C^{13}CH$
20	4077.89367 (30)	Q (22)	8	$H^{12}C^{13}CH$
·21	4078.15412 (23)	Q (19)	8	$H^{12}C^{13}CH$
22	4078.23209 (23)	Q (18)	8	$H^{12}C^{13}CH$
23	4078.40341 (19)	Qef (17)	5	$H^{12}C^{12}CH$
24	4078.44045 (21)	Q (15)	8	$H^{12}C^{13}CH$
25	4078.50123 (21)	Q (14)	8	$H^{12}C^{13}CH$
25		Qfe (27)	5	$H^{12}C^{12}CH$
26	4078.55804 (21)	Q (13)	8	$H^{12}C^{13}CH$
27	4078.65893 (21)	Q (11)	8	$H^{12}C^{13}CH$
28	4078.70314 (19)	Q (10)	8	$H^{12}C^{13}CH$
29	4078.74340 (31)	Q (9)	8	$H^{12}C^{13}CH$
30	4078.81107 (31)	Q (7)	8	$H^{12}C^{13}CH$

LINE NO.	WAVENUMBER CM^{-1}, VAC.	ROTATIONAL ASSIGNMENTS	BAND	ISOTOPIC SPECIES
31	4078.83905 (20)	Q (6)	8	$H^{12}C^{13}CH$
32	4079.00351 (49)	Qfe (23)	5	$H^{12}C^{12}CH$
33	4079.26601 (26)	P (5)	6	$H^{12}C^{12}CH$
34	4079.30349 (38)	Qef (12)	5	$H^{12}C^{12}CH$
35	4079.32901 (36)	Qfe (20)	5	$H^{12}C^{12}CH$
36	4079.39363 (37)			
37	4079.42727 (29)	Qfe (19)	5	$H^{12}C^{12}CH$
38	4079.45801 (22)	Qef (11)	5	$H^{12}C^{12}CH$
39	4079.52087 (21)	Qfe (18)	5	$H^{12}C^{12}CH$
40	4079.69246 (28)	Qfe (16)	5	$H^{12}C^{12}CH$
41	4079.73423 (19)	Qef (9)	5	$H^{12}C^{12}CH$
42	4079.77063 (23)	Qfe (15)	5	$H^{12}C^{12}CH$
43	4079.91254 (21)	Qfe (13)	5	$H^{12}C^{12}CH$
44	4080.03467 (38)	Qfe (11)	5	$H^{12}C^{12}CH$
45	4080.08820 (19)	Qfe (10)	5	$H^{12}C^{12}CH$
46	4080.14197 (26)	Qef (5)	5	$H^{12}C^{12}CH$
46		Qfe (9)	5	$H^{12}C^{12}CH$
47	4080.18073 (19)	Qfe (8)	5	$H^{12}C^{12}CH$
48	4080.21738 (24)	Qef (4)	5	$H^{12}C^{12}CH$
48		Qfe (7)	5	$H^{12}C^{12}CH$
49	4080.63707 (21)	R (7)	1	$H^{12}C^{12}CH$
50	4081.20621 (37)	R (0)	8	$H^{12}C^{13}CH$
51	4081.67531 (19)	P (4)	6	$H^{12}C^{12}CH$
52	4082.95724 (31)	R (8)	1	$H^{12}C^{12}CH$
53	4083.47619 (31)	R (1)	8	$H^{12}C^{13}CH$
54	4084.07079 (24)	P (3)	6	$H^{12}C^{12}CH$
55	4084.14285 (27)	R (2)	2	$H^{12}C^{12}CH$
56	4084.97141 (30)	Rff (3)	3	$H^{12}C^{12}CH$
56		P (22)	7	$H^{12}C^{12}CH$
57	4085.27246 (20)	R (9)	1	$H^{12}C^{12}CH$
58	4085.45849 (71)			
59	4085.50878 (87)			
60	4085.86179 (62)			

- "LINE NO.": serial number identifying the spectral line in the map on the opposite page.
- "WAVENUMBER": experimental value.
- Uncertainty in the last digit of the wavenumber is given in parentheses.
- "BAND" : identification on the opposite page.

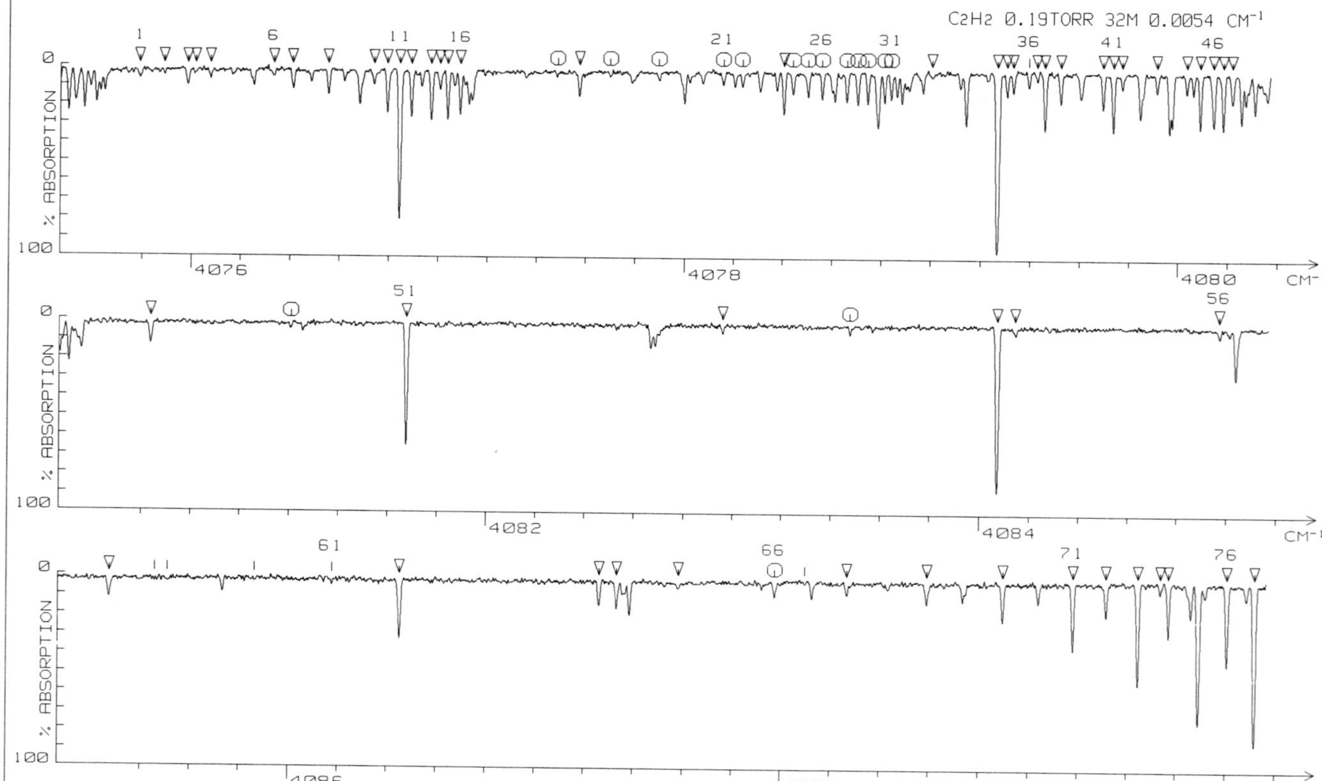

ISOTOPIC IDENTIFICATION - '▽' indicates $H^{12}C^{12}CH$; 'Ⓘ': $H^{12}C^{13}CH$; 'I': unidentified.

BAND 1=$(1001^11^1)^{0+}(\Sigma_u^+) \leftarrow (0001^10^0)(\Pi_g)$; 2=$(1001^11^1)^{0-}(\Sigma_u^-) \leftarrow (0001^10^0)(\Pi_g)$; 3=$(1001^11^1)^2(\Delta_u) \leftarrow (0001^10^0)(\Pi_g)$; 4=$(1000^02^0)(\Sigma_g^+) \leftarrow (0000^01^1)(\Pi_u)$

5=(1000^02^2) $(\Delta_g) \leftarrow (0000^01^1)(\Pi_u)$; 6=$(1000^01^1)$ $(\Pi_u) \leftarrow (0000^00^0)(\Sigma_g^+)$; 7=$(0100^03^1)(\Pi_u) \leftarrow (0000^00^0)(\Sigma_g^+)$; 8=$(1000^01^1)$ $(\Pi) \leftarrow (0000^00^0)(\Sigma^+)$

LINE NO.	WAVENUMBER CM^{-1}, VAC.	ROTATIONAL BAND ASSIGNMENTS			ISOTOPIC SPECIES
61	4086.17587 (101)				
62	4086.45192 (23)	R	(3)	2	$H^{12}C^{12}CH$
62		P	(2)	6	$H^{12}C^{12}CH$
63	4087.26157 (21)	Rff	(4)	3	$H^{12}C^{12}CH$
64	4087.33284 (31)	R	(6)	4	$H^{12}C^{12}CH$
65	4087.58199 (33)	R	(10)	1	$H^{12}C^{12}CH$
66	4087.97626 (24)	R	(3)	8	$H^{12}C^{13}CH$
67	4088.09784 (66)				
68	4088.26797 (25)	Q	(35)	6	$H^{12}C^{12}CH$
69	4088.59226 (38)	Q	(33)	6	$H^{12}C^{12}CH$
70	4088.89705 (19)	Q	(31)	6	$H^{12}C^{12}CH$
71	4089.18140 (19)	Q	(29)	6	$H^{12}C^{12}CH$
72	4089.31616 (19)	Q	(28)	6	$H^{12}C^{12}CH$
73	4089.44601 (19)	Q	(27)	6	$H^{12}C^{12}CH$
74	4089.53728 (28)	Rff	(5)	3	$H^{12}C^{12}CH$
75	4089.57102 (23)	Q	(26)	6	$H^{12}C^{12}CH$
76	4089.80641 (19)	Q	(24)	6	$H^{12}C^{12}CH$
77	4089.91680 (24)	Q	(23)	6	$H^{12}C^{12}CH$

- "LINE NO.": serial number identifying the spectral line in the map on the opposite page.
- "WAVENUMBER": experimental value.
- Uncertainty in the last digit of the wavenumber is given in parentheses.
- "BAND" : identification on the opposite page.

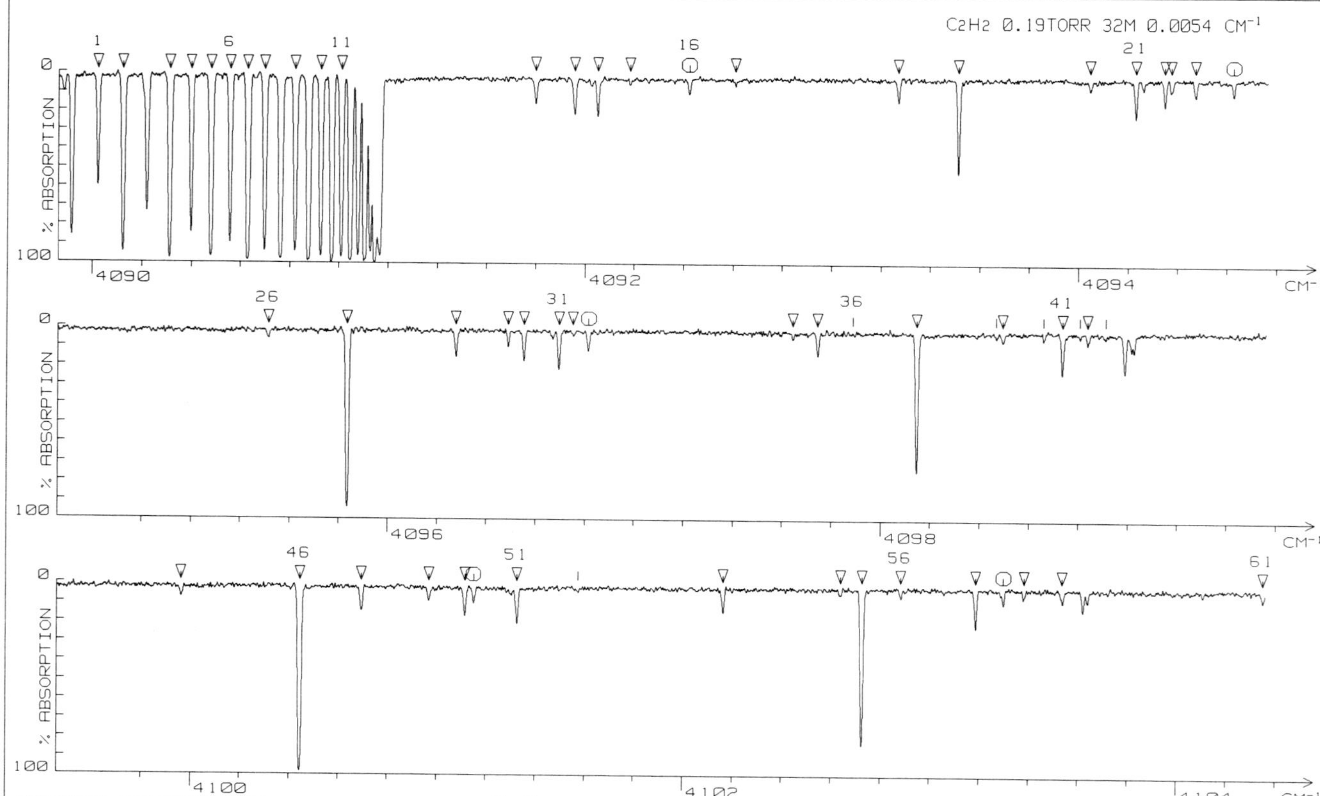

ISOTOPIC IDENTIFICATION - '∇' indicates $H^{12}C^{12}CH$; 'Ⓘ': $H^{12}C^{13}CH$; 'I': unidentified.

BAND 1=$(1001^11^1)^{0+}(\Sigma_u^+)\leftarrow(0001^10^0)(\Pi_g)$; 2=$(1001^11^1)^{0-}(\Sigma_u^-)\leftarrow(0001^10^0)(\Pi_g)$; 3=$(1001^11^1)^2(\Delta_u)\leftarrow(0001^10^0)(\Pi_g)$; 4=$(1000^02^0)(\Sigma_g^+)\leftarrow(0000^01^1)(\Pi_u)$
5=$(1000^02^2)(\Delta_g)\leftarrow(0000^01^1)(\Pi_u)$; 6=$(1000^01^1)(\Pi_u)\leftarrow(0000^00^0)(\Sigma_g^+)$; 7=$(0100^03^1)(\Pi_u)\leftarrow(0000^00^0)(\Sigma_g^+)$; 8=$(1000^01^1)(\Pi)\leftarrow(0000^00^0)(\Sigma^+)$

LINE NO.	WAVENUMBER CM^{-1}, VAC.	ROTATIONAL ASSIGNMENTS	BAND	ISOTOPIC SPECIES
1	4090.02271 (19)	Q (22)	6	$H^{12}C^{12}CH$
2	4090.12354 (25)	Q (21)	6	$H^{12}C^{12}CH$
3	4090.31119 (26)	Q (19)	6	$H^{12}C^{12}CH$
4	4090.39801 (23)	Q (18)	6	$H^{12}C^{12}CH$
5	4090.47997 (28)	Q (17)	6	$H^{12}C^{12}CH$
6	4090.55728 (24)	Q (16)	6	$H^{12}C^{12}CH$
7	4090.63012 (29)	Q (15)	6	$H^{12}C^{12}CH$
8	4090.69828 (25)	Q (14)	6	$H^{12}C^{12}CH$
9	4090.82061 (27)	Q (12)	6	$H^{12}C^{12}CH$
10	4090.92463 (27)	Q (10)	6	$H^{12}C^{12}CH$
11	4091.01057 (29)	R (5)	2	$H^{12}C^{12}CH$
11		Q (8)	6	$H^{12}C^{12}CH$
12	4091.79816 (21)	Rff (6)	3	$H^{12}C^{12}CH$
13	4091.95568 (20)	R (8)	4	$H^{12}C^{12}CH$
13		Rff (4)	5	$H^{12}C^{12}CH$
14	4092.04852 (27)	Ree (4)	5	$H^{12}C^{12}CH$
15	4092.18230 (35)	R (12)	1	$H^{12}C^{12}CH$
16	4092.42247 (23)	R (5)	8	$H^{12}C^{13}CH$
17	4092.60743 (41)	P (19)	7	$H^{12}C^{12}CH$
18	4093.26998 (22)	R (6)	2	$H^{12}C^{12}CH$
19	4093.51195 (19)	R (0)	6	$H^{12}C^{12}CH$
20	4094.04475 (28)	Rff (7)	3	$H^{12}C^{12}CH$
21	4094.22870 (19)	Rff (5)	5	$H^{12}C^{12}CH$
22	4094.34727 (21)	Ree (7)	3	$H^{12}C^{12}CH$
23	4094.37362 (56)	Ree (5)	5	$H^{12}C^{12}CH$
24	4094.47145 (21)	R (13)	1	$H^{12}C^{12}CH$
25	4094.62509 (23)	R (6)	8	$H^{12}C^{13}CH$
26	4095.51523 (30)	R (7)	2	$H^{12}C^{12}CH$
27	4095.83684 (25)	R (1)	6	$H^{12}C^{12}CH$
28	4096.27669 (20)	Rff (8)	3	$H^{12}C^{12}CH$
29	4096.49036 (22)	Rff (6)	5	$H^{12}C^{12}CH$
30	4096.55301 (20)	R (10)	4	$H^{12}C^{12}CH$

LINE NO.	WAVENUMBER CM^{-1}, VAC.	ROTATIONAL ASSIGNMENTS	BAND	ISOTOPIC SPECIES
31	4096.69516 (21)	Ree (6)	5	$H^{12}C^{12}CH$
32	4096.75259 (39)	R (14)	1	$H^{12}C^{12}CH$
33	4096.81398 (23)	R (7)	8	$H^{12}C^{13}CH$
34	4097.64525 (36)	P (17)	7	$H^{12}C^{12}CH$
35	4097.74594 (20)	R (8)	2	$H^{12}C^{12}CH$
36	4097.88616 (90)			
37	4098.14792 (22)	R (2)	6	$H^{12}C^{12}CH$
38	4098.46715 (47)			
39	4098.49476 (27)	Rff (9)	3	$H^{12}C^{12}CH$
40	4098.66159 (47)			
41	4098.73760 (21)	Rff (7)	5	$H^{12}C^{12}CH$
42	4098.80683 (47)			
43	4098.83876 (26)	R (11)	4	$H^{12}C^{12}CH$
44	4098.91085 (47)			
45	4099.96186 (29)	R (9)	2	$H^{12}C^{12}CH$
46	4100.44443 (26)	R (3)	6	$H^{12}C^{12}CH$
47	4100.69673 (21)	Rff (10)	3	$H^{12}C^{12}CH$
48	4100.96998 (23)	Rff (8)	5	$H^{12}C^{12}CH$
49	4101.11546 (20)	R (12)	4	$H^{12}C^{12}CH$
50	4101.15134 (21)	R (9)	8	$H^{12}C^{13}CH$
51	4101.32803 (27)	Ree (8)	5	$H^{12}C^{12}CH$
52	4101.57336 (87)			
53	4102.16340 (20)	R (10)	2	$H^{12}C^{12}CH$
54	4102.64027 (31)	P (15)	7	$H^{12}C^{12}CH$
55	4102.72707 (23)	R (4)	6	$H^{12}C^{12}CH$
56	4102.88477 (30)	Rff (11)	3	$H^{12}C^{12}CH$
57	4103.18786 (19)	Rff (9)	5	$H^{12}C^{12}CH$
58	4103.29926 (22)	R (10)	8	$H^{12}C^{13}CH$
59	4103.38246 (27)	R (13)	4	$H^{12}C^{12}CH$
60	4103.53957 (25)	R (17)	1	$H^{12}C^{12}CH$
61	4104.35013 (30)	R (11)	2	$H^{12}C^{12}CH$

- "LINE NO.": serial number identifying the spectral line in the map on the opposite page.
- "WAVENUMBER": experimental value.
- Uncertainty in the last digit of the wavenumber is given in parentheses.
- "BAND" : identification on the opposite page.

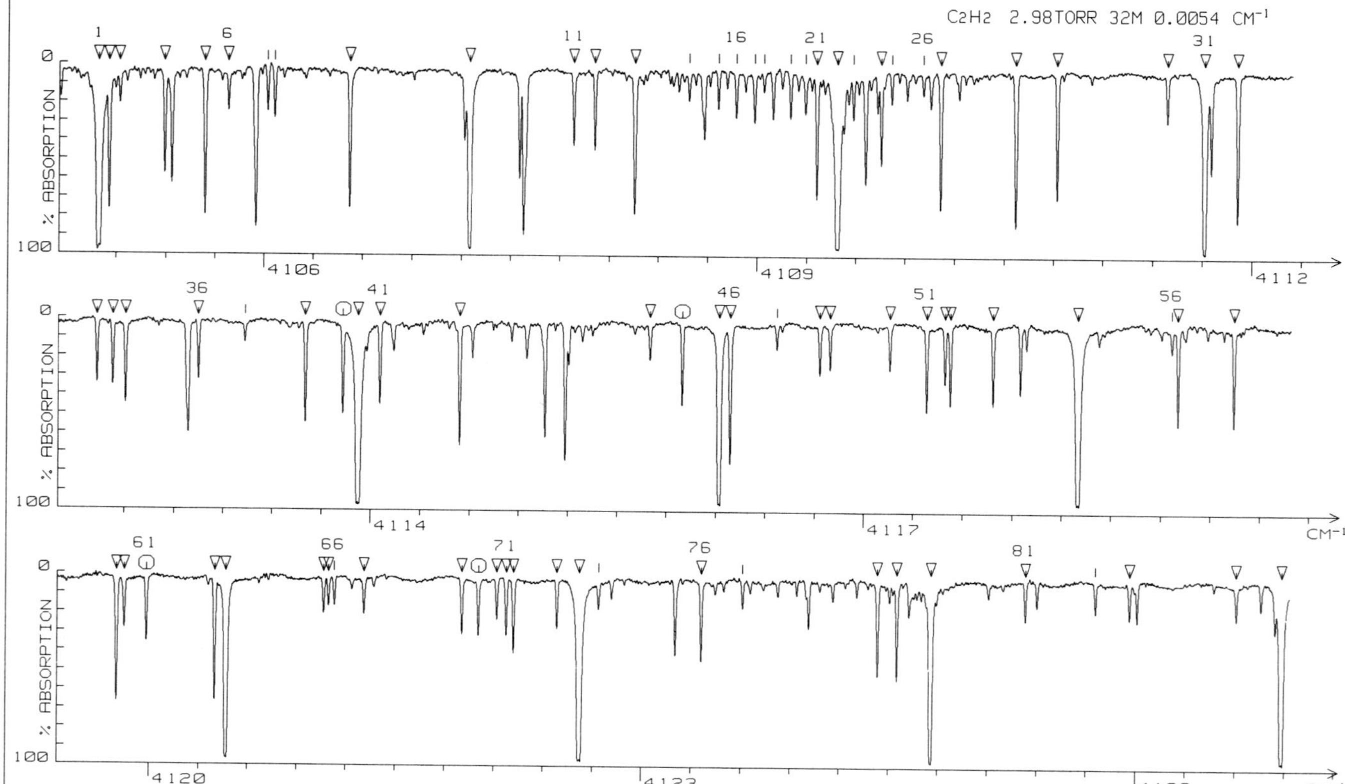

ISOTOPIC IDENTIFICATION - '∇' indicates $H^{12}C^{12}CH$; '⊙': $H^{12}C^{13}CH$; 'ı': unidentified.

BAND 1=$(1001^11^1)^{0+}(\Sigma_u^+) \leftarrow (0001^10^0)(\Pi_g)$; 2=$(1001^11^1)^{0-}(\Sigma_u^-) \leftarrow (0001^10^0)(\Pi_g)$; 3=$(1001^11^1)^2(\Delta_u) \leftarrow (0001^10^0)(\Pi_g)$; 4=$(1000^02^0)(\Sigma_g^+) \leftarrow (0000^01^1)(\Pi_u)$

5=$(1000^02^2)(\Delta_g) \leftarrow (0000^01^1)(\Pi_u)$; 6=$(1000^01^1)(\Pi_u) \leftarrow (0000^00^0)(\Sigma_g^+)$; 7=$(0100^03^1)(\Pi_u) \leftarrow (0000^00^0)(\Sigma_g^+)$; 8=$(1000^01^1)(\Pi) \leftarrow (0000^00^0)(\Sigma^+)$

LINE NO.	WAVENUMBER CM^{-1}, VAC.	ROTATIONAL ASSIGNMENTS	BAND	ISOTOPIC SPECIES
1	4104. 99502 (27)	R (5)	6	$H^{12}C^{12}CH$
2	4105. 05811 (21)	Rff (12)	3	$H^{12}C^{12}CH$
3	4105. 12163 (65)	P (14)	7	$H^{12}C^{12}CH$
4	4105. 39115 (25)	Rff (10)	5	$H^{12}C^{12}CH$
5	4105. 63842 (21)	R (14)	4	$H^{12}C^{12}CH$
6	4105. 78109 (52)	R (18)	1	$H^{12}C^{12}CH$
7	4106. 02146 (46)			
8	4106. 06382 (42)			
9	4106. 52204 (20)	R (12)	2	$H^{12}C^{12}CH$
10	4107. 24908 (23)	R (6)	6	$H^{12}C^{12}CH$
11	4107. 88333 (30)	R (15)	4	$H^{12}C^{12}CH$
12	4108. 01045 (28)	R (19)	1	$H^{12}C^{12}CH$
13	4108. 25574 (23)	Ree (13)	3	$H^{12}C^{12}CH$
13		Ree (11)	5	$H^{12}C^{12}CH$
14	4108. 58621 (71)			
15	4108. 76194 (49)			
16	4108. 86962 (44)			
17	4108. 98305 (43)			
18	4109. 04024 (91)			
19	4109. 20021 (42)			
20	4109. 29040 (53)			
21	4109. 36031 (23)	Rff (14)	3	$H^{12}C^{12}CH$
22	4109. 48848 (28)	R (7)	6	$H^{12}C^{12}CH$
23	4109. 58395 (48)			
24	4109. 75376 (63)	Rff (12)	5	$H^{12}C^{12}CH$
25	4109. 81883 (58)			
26	4110. 01053 (84)			
27	4110. 11554 (20)	R (16)	4	$H^{12}C^{12}CH$
28	4110. 56749 (20)	Ree (14)	3	$H^{12}C^{12}CH$
28		Ree (12)	5	$H^{12}C^{12}CH$
29	4110. 82151 (21)	R (14)	2	$H^{12}C^{12}CH$
30	4111. 48859 (41)	Rff (15)	3	$H^{12}C^{12}CH$
31	4111. 71363 (24)	R (8)	6	$H^{12}C^{12}CH$
32	4111. 91275 (21)	Rff (13)	5	$H^{12}C^{12}CH$
33	4112. 33481 (33)	R (17)	4	$H^{12}C^{12}CH$
34	4112. 43029 (34)	R (21)	1	$H^{12}C^{12}CH$
35	4112. 50720 (29)	P (11)	7	$H^{12}C^{12}CH$
36	4112. 94892 (36)	R (15)	2	$H^{12}C^{12}CH$
37	4113. 23279 (94)			
38	4113. 60218 (24)	Rff (16)	3	$H^{12}C^{12}CH$
39	4113. 83192 (26)	R (15)	8	$H^{12}C^{13}CH$
40	4113. 92440 (27)	R (9)	6	$H^{12}C^{12}CH$
41	4114. 05700 (29)	Rff (14)	5	$H^{12}C^{12}CH$
42	4114. 54026 (23)	R (18)	4	$H^{12}C^{12}CH$
43	4115. 70052 (58)	Rff (17)	3	$H^{12}C^{12}CH$
44	4115. 89641 (27)	R (16)	8	$H^{12}C^{13}CH$
45	4116. 12066 (23)	R (10)	6	$H^{12}C^{12}CH$
46	4116. 18644 (21)	Rff (15)	5	$H^{12}C^{12}CH$
47	4116. 47008 (71)			
48	4116. 73072 (41)	R (19)	4	$H^{12}C^{12}CH$
49	4116. 79363 (43)	R (23)	1	$H^{12}C^{12}CH$
50	4117. 15974 (43)	R (17)	2	$H^{12}C^{12}CH$
51	4117. 38285 (27)	P (9)	7	$H^{12}C^{12}CH$
52	4117. 49302 (36)	Ree (15)	5	$H^{12}C^{12}CH$
53	4117. 52600 (29)	Ree (17)	3	$H^{12}C^{12}CH$
54	4117. 78392 (29)	Rff (18)	3	$H^{12}C^{12}CH$
55	4118. 30227 (27)	Rff (16)	5	$H^{12}C^{12}CH$
55		R (11)	6	$H^{12}C^{12}CH$
56	4118. 86842 (84)			
57	4118. 90573 (24)	R (20)	4	$H^{12}C^{12}CH$
58	4119. 24268 (25)	R (18)	2	$H^{12}C^{12}CH$
59	4119. 80497 (23)	Ree (16)	5	$H^{12}C^{12}CH$
59		P (8)	7	$H^{12}C^{12}CH$
60	4119. 85082 (44)	Ree (18)	3	$H^{12}C^{12}CH$
60		Rff (19)	3	$H^{12}C^{12}CH$

- "LINE NO.": serial number identifying the spectral line in the map on the opposite page.
- "WAVENUMBER": experimental value.
- Uncertainty in the last digit of the wavenumber is given in parentheses.
- "BAND" : identification on the opposite page.

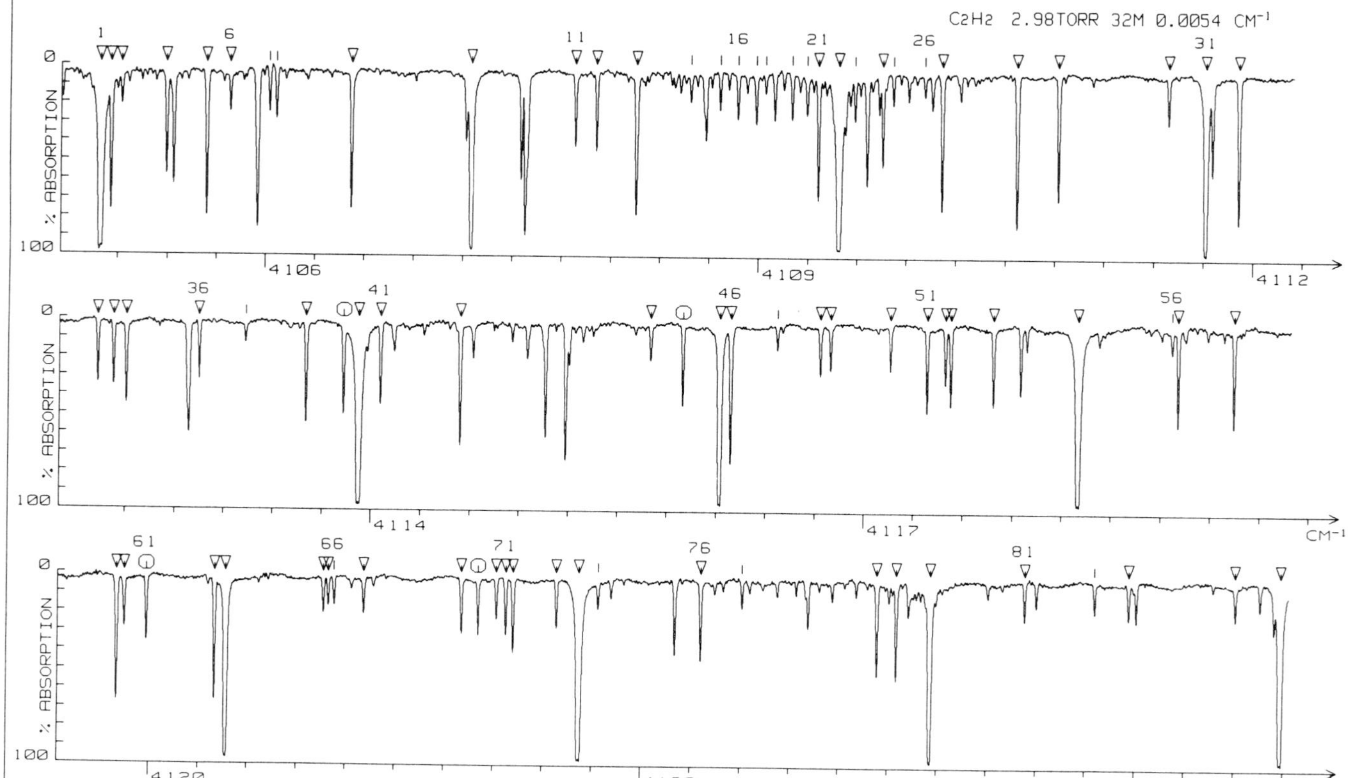

ISOTOPIC IDENTIFICATION – '∇' indicates $H^{12}C^{12}CH$; '⊕': $H^{12}C^{13}CH$; 'I': unidentified.

BAND 1=$(1001^11^1)^{0+}(\Sigma_u^+) \leftarrow (0001^10^0)(\Pi_g)$; 2=$(1001^11^1)^{0-}(\Sigma_u^-) \leftarrow (0001^10^0)(\Pi_g)$; 3=$(1001^11^1)^2(\Delta_u) \leftarrow (0001^10^0)(\Pi_g)$; 4=$(1000^02^0)(\Sigma_g^+) \leftarrow (0000^01^1)(\Pi_u)$

5=$(1000^02^2)(\Delta_g) \leftarrow (0000^01^1)(\Pi_u)$; 6=$(1000^01^1)(\Pi_u) \leftarrow (0000^00^0)(\Sigma_g^+)$; 7=$(0100^03^1)(\Pi_u) \leftarrow (0000^00^0)(\Sigma_g^+)$; 8=$(1000^01^1)(\Pi) \leftarrow (0000^00^0)(\Sigma^+)$

LINE NO.	WAVENUMBER CM^{-1}, VAC.	ROTATIONAL ASSIGNMENTS	BAND	ISOTOPIC SPECIES
61	4119.98411 (32)	R (18)	8	$H^{12}C^{13}CH$
62	4120.40038 (22)	Rff (17)	5	$H^{12}C^{12}CH$
63	4120.46941 (23)	R (12)	6	$H^{12}C^{12}CH$
64	4121.06435 (56)	R (21)	4	$H^{12}C^{12}CH$
65	4121.09448 (65)	R (25)	1	$H^{12}C^{12}CH$
66	4121.12956 (69)			
67	4121.31096 (56)	R (19)	2	$H^{12}C^{12}CH$
68	4121.90483 (37)	Rff (20)	3	$H^{12}C^{12}CH$
69	4122.00659 (37)	R (19)	8	$H^{12}C^{13}CH$
70	4122.11878 (49)	Ree (17)	5	$H^{12}C^{12}CH$
71	4122.17651 (37)	Ree (19)	3	$H^{12}C^{12}CH$
72	4122.22129 (29)	P (7)	7	$H^{12}C^{12}CH$
73	4122.48501 (42)	Rff (18)	5	$H^{12}C^{12}CH$
74	4122.62203 (27)	R (13)	6	$H^{12}C^{12}CH$
75	4122.73664 (77)			
76	4123.36359 (29)	R (20)	2	$H^{12}C^{12}CH$
77	4123.61263 (66)			
78	4124.43637 (26)	Ree (18)	5	$H^{12}C^{12}CH$
79	4124.55478 (25)	Rff (19)	5	$H^{12}C^{12}CH$
80	4124.75985 (23)	R (14)	6	$H^{12}C^{12}CH$
81	4125.32979 (48)	R (27)	1	$H^{12}C^{12}CH$
81		R (23)	4	$H^{12}C^{12}CH$
82	4125.75939 (58)			
83	4125.96399 (51)	Rff (22)	3	$H^{12}C^{12}CH$
84	4126.60945 (53)	Rff (20)	5	$H^{12}C^{12}CH$
85	4126.88312 (25)	R (15)	6	$H^{12}C^{12}CH$

- "LINE NO.": serial number identifying the spectral line in the map on the opposite page.
- "WAVENUMBER": experimental value.
- Uncertainty in the last digit of the wavenumber is given in parentheses.
- "BAND" : identification on the opposite page.

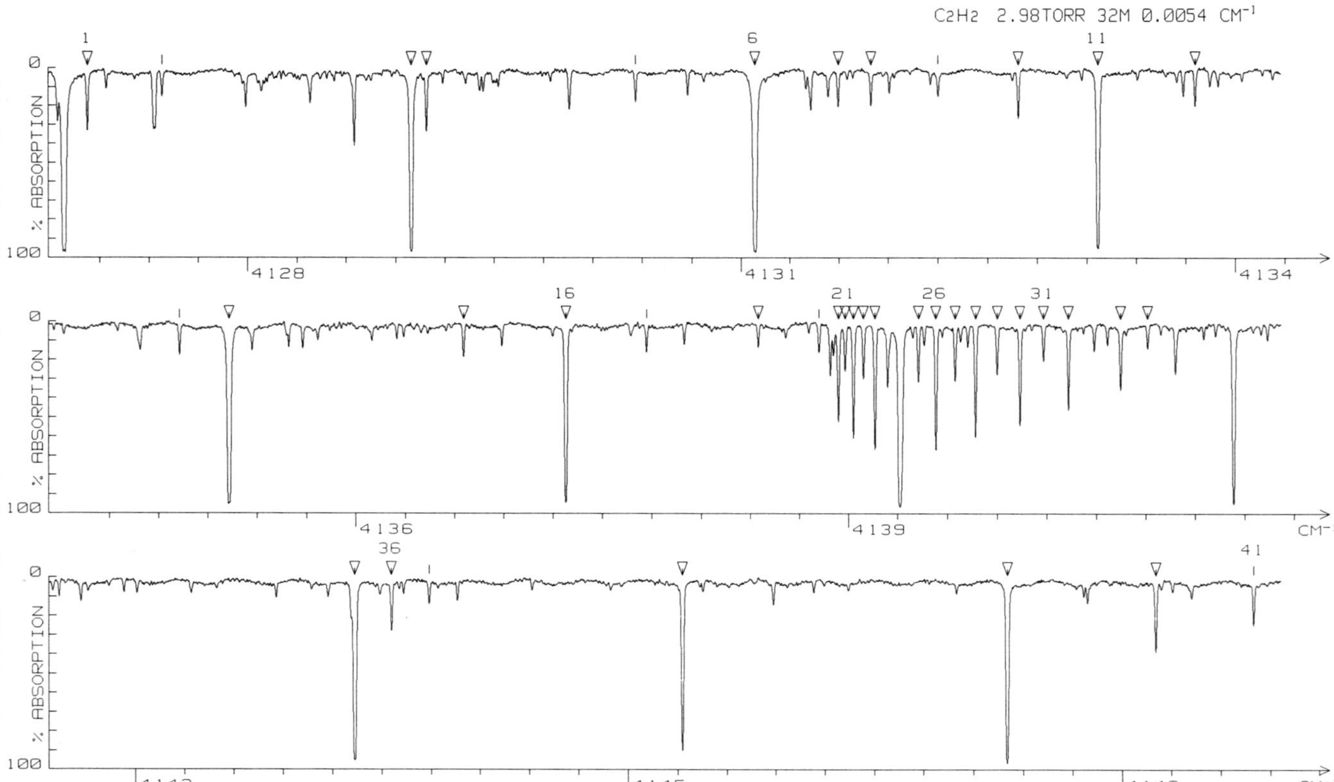

ISOTOPIC IDENTIFICATION - '▽' indicates $H^{12}C^{12}CH$; '⓵': $H^{12}C^{13}CH$; 'ı': unidentified.

BAND 1=$(1001^11^1)^{0+}(\Sigma_u^+)\leftarrow(0001^10^0)(\Pi_g)$; 2=$(1001^11^1)^{0-}(\Sigma_u^-)\leftarrow(0001^10^0)(\Pi_g)$; 3=$(1001^11^1)^2(\Delta_u)\leftarrow(0001^10^0)(\Pi_g)$; 4=$(1000^02^0)(\Sigma_g^+)\leftarrow(0000^01^1)(\Pi_u)$
5=$(1000^02^2)(\Delta_g)\leftarrow(0000^01^1)(\Pi_u)$; 6=$(1000^01^1)(\Pi_u)\leftarrow(0000^00^0)(\Sigma_g^+)$; 7=$(0100^03^1)(\Pi_u)\leftarrow(0000^00^0)(\Sigma_g^+)$; 8=$(1000^01^1)(\Pi)\leftarrow(0000^00^0)(\Sigma^+)$

LINE NO.	WAVENUMBER CM^{-1}, VAC.	ROTATIONAL ASSIGNMENTS	BAND	ISOTOPIC SPECIES
1	4127.02415(36)	P (5)	7	$H^{12}C^{12}CH$
2	4127.47863(70)			
3	4128.99182(19)	R (16)	6	$H^{12}C^{12}CH$
4	4129.08463(36)	Ree (20)	5	$H^{12}C^{12}CH$
5	4130.35908(60)			
6	4131.08574(25)	R (17)	6	$H^{12}C^{12}CH$
7	4131.59402(56)	R (26)	4	$H^{12}C^{12}CH$
8	4131.79174(55)	P (3)	7	$H^{12}C^{12}CH$
9	4132.19957(81)			
10	4132.68202(42)	Rff (23)	5	$H^{12}C^{12}CH$
11	4133.16491(19)	R (18)	6	$H^{12}C^{12}CH$
12	4133.75438(50)	Ree (22)	5	$H^{12}C^{12}CH$
13	4134.92632(63)			
14	4135.22921(24)	R (19)	6	$H^{12}C^{12}CH$
15	4136.65471(60)	Rff (25)	5	$H^{12}C^{12}CH$
16	4137.27864(23)	R (20)	6	$H^{12}C^{12}CH$
17	4137.76768(69)			
18	4138.44892(79)	Ree (24)	5	$H^{12}C^{12}CH$
19	4138.81924(73)			
20	4138.93862(27)	Q (3)	7	$H^{12}C^{12}CH$
21	4138.97927(45)	Q (4)	7	$H^{12}C^{12}CH$
22	4139.02959(24)	Q (5)	7	$H^{12}C^{12}CH$
23	4139.08969(38)	Q (6)	7	$H^{12}C^{12}CH$
24	4139.15953(23)	Q (7)	7	$H^{12}C^{12}CH$
25	4139.42455(38)	Q (10)	7	$H^{12}C^{12}CH$
26	4139.53082(22)	Q (11)	7	$H^{12}C^{12}CH$
27	4139.64538(41)	Q (12)	7	$H^{12}C^{12}CH$
28	4139.76803(24)	Q (13)	7	$H^{12}C^{12}CH$
29	4139.89822(43)	Q (14)	7	$H^{12}C^{12}CH$
30	4140.03561(24)	Q (15)	7	$H^{12}C^{12}CH$
31	4140.17980(52)	Q (16)	7	$H^{12}C^{12}CH$
32	4140.32986(28)	Q (17)	7	$H^{12}C^{12}CH$
33	4140.64643(34)	Q (19)	7	$H^{12}C^{12}CH$
34	4140.81010(81)	Ree (25)	5	$H^{12}C^{12}CH$
34		Q (20)	7	$H^{12}C^{12}CH$
35	4143.33760(22)	R (23)	6	$H^{12}C^{12}CH$
36	4143.55820(42)	R (1)	7	$H^{12}C^{12}CH$
37	4143.78636(82)			
38	4145.32742(19)	R (24)	6	$H^{12}C^{12}CH$
39	4147.30187(19)	R (25)	6	$H^{12}C^{12}CH$
40	4148.20288(32)	R (3)	7	$H^{12}C^{12}CH$
41	4148.79663(48)			

- "LINE NO.": serial number identifying the spectral line in the map on the opposite page.
- "WAVENUMBER": experimental value.
- Uncertainty in the last digit of the wavenumber is given in parentheses.
- "BAND" : identification on the opposite page.

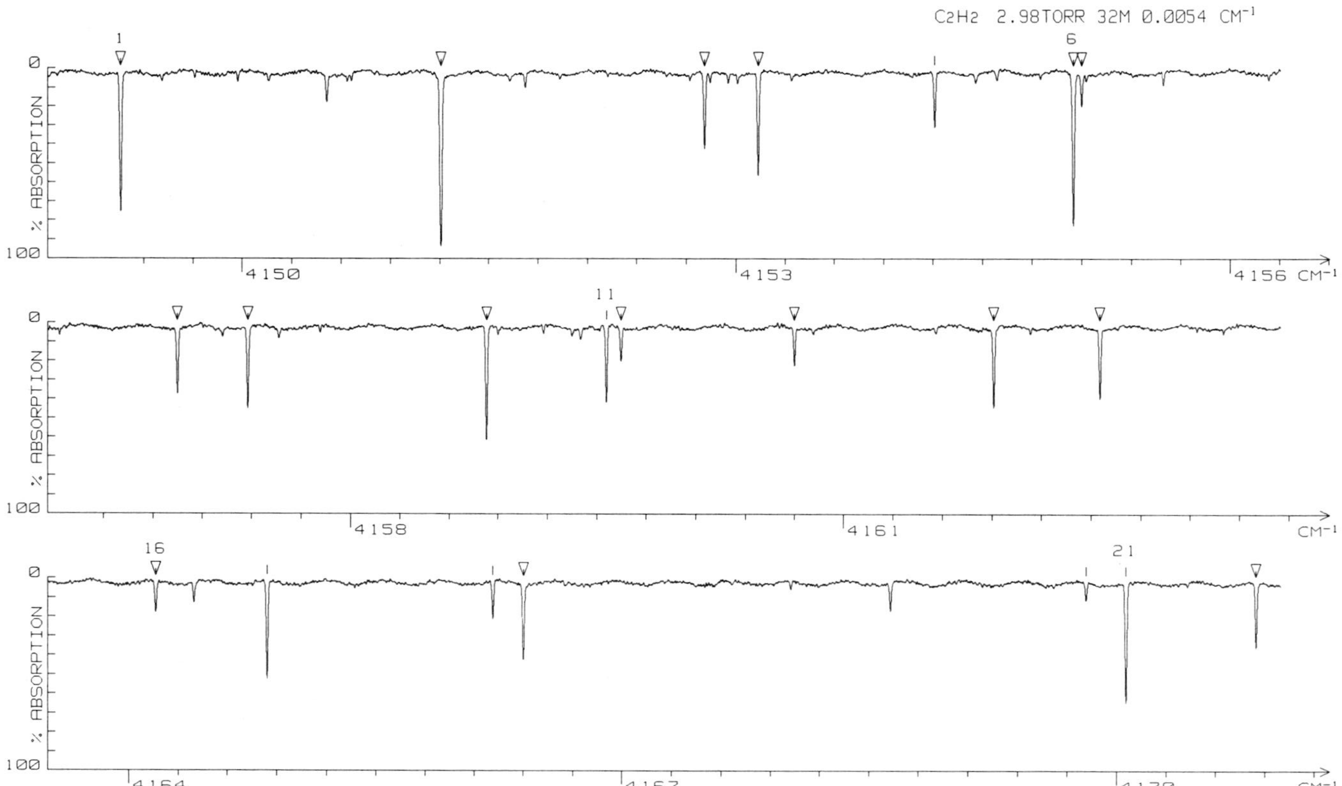

ISOTOPIC IDENTIFICATION – '∇' indicates $H^{12}C^{12}CH$; 'Ⓘ': $H^{12}C^{13}CH$; 'I': unidentified.
BAND 1=$(1001^11^1)^{0+}(\Sigma_u^+)\leftarrow(0001^10^0)(\Pi_g)$; 2=$(1001^11^1)^{0-}(\Sigma_u^-)\leftarrow(0001^10^0)(\Pi_g)$; 3=$(1001^11^1)^2(\Delta_u)\leftarrow(0001^10^0)(\Pi_g)$; 4=$(1000^02^0)(\Sigma_g^+)\leftarrow(0000^01^1)(\Pi_u)$
5=$(1000^02^2)(\Delta_g)\leftarrow(0000^01^1)(\Pi_u)$; 6=$(1000^01^1)(\Pi_u)\leftarrow(0000^00^0)(\Sigma_g^+)$; 7=$(0100^03^1)(\Pi_u)\leftarrow(0000^00^0)(\Sigma_g^+)$; 8=$(1000^01^1)(\Pi)\leftarrow(0000^00^0)(\Sigma^+)$

LINE NO.	WAVENUMBER CM^{-1}, VAC.	ROTATIONAL ASSIGNMENTS		BAND	ISOTOPIC SPECIES
1	4149.26171(20)	R	(26)	6	$H^{12}C^{12}CH$
2	4151.20634(21)	R	(27)	6	$H^{12}C^{12}CH$
3	4152.81068(28)	R	(5)	7	$H^{12}C^{12}CH$
4	4153.13578(24)	R	(28)	6	$H^{12}C^{12}CH$
5	4154.21162(29)				
6	4155.05000(21)	R	(29)	6	$H^{12}C^{12}CH$
7	4155.10034(59)	R	(6)	7	$H^{12}C^{12}CH$
8	4156.94912(33)	R	(30)	6	$H^{12}C^{12}CH$
9	4157.38064(28)	R	(7)	7	$H^{12}C^{12}CH$
10	4158.83302(23)	R	(31)	6	$H^{12}C^{12}CH$
11	4159.55968(24)				
12	4159.65076(59)	R	(8)	7	$H^{12}C^{12}CH$
13	4160.70180(51)	R	(32)	6	$H^{12}C^{12}CH$
14	4161.91124(29)	R	(9)	7	$H^{12}C^{12}CH$
15	4162.55506(31)	R	(33)	6	$H^{12}C^{12}CH$
16	4164.16092(66)	R	(10)	7	$H^{12}C^{12}CH$
17	4164.84106(25)				
18	4166.21559(50)				
19	4166.40023(31)	R	(11)	7	$H^{12}C^{12}CH$
20	4169.81458(100)				
21	4170.05537(22)				
22	4170.84578(33)	R	(13)	7	$H^{12}C^{12}CH$

- "LINE NO.": serial number identifying the spectral line in the map on the opposite page.
- "WAVENUMBER": experimental value.
- Uncertainty in the last digit of the wavenumber is given in parentheses.
- "BAND" : identification on the opposite page.

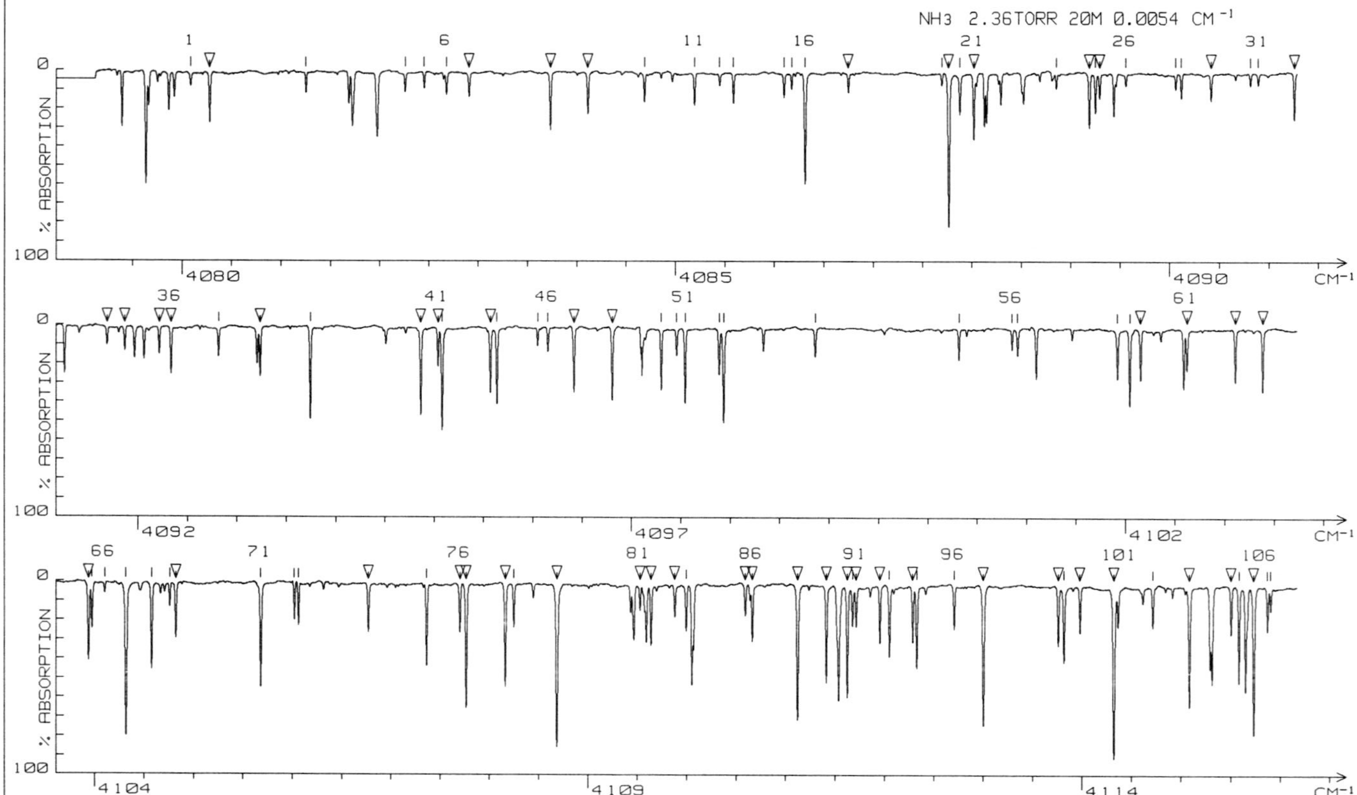

ISOTOPIC IDENTIFICATION – '∇' indicates $^{14}NH_3$; '|': unidentified.

LINE NO.	WAVENUMBER CM^{-1}, VAC.	ROTATIONAL ASSIGNMENTS	BAND		ISOTOPIC SPECIES
1	4080.08356 (99)				
2	4080.28034 (28)	sP (5, 2)	NU2+2NU4	pa	$^{14}NH_3$
3	4081.25310 (54)				
4	4082.25913 (63)				
5	4082.45305 (70)				
6	4082.67789 (58)				
7	4082.90680 (55)	$s^{r}Q$ (6, 4)	NU2+2NU4	pe	$^{14}NH_3$
8	4083.73308 (26)	$s^{P}P$ (3, 1)	NU2+2NU4	pe	$^{14}NH_3$
9	4084.11275 (34)	aP (10, 8)	NU1+NU2		$^{14}NH_3$
10	4084.68740 (43)				
11	4085.19424 (42)				
12	4085.44743 (102)				
13	4085.58640 (43)				
14	4086.10024 (51)				
15	4086.17652 (66)				
16	4086.31213 (19)				
17	4086.74751 (64)	aP (10, 7)	NU1+NU2		$^{14}NH_3$
18	4087.69359 (94)				
19	4087.76446 (17)	$s^{P}P$ (4, 3)	NU2+2NU4	pe	$^{14}NH_3$
20	4087.87823 (36)				
21	4088.02074 (25)	aP (10, 6)	NU1+NU2		$^{14}NH_3$
22	4088.85510 (78)				
23	4089.18899 (28)	$s^{r}Q$ (4, 3)	NU2+2NU4	pe	$^{14}NH_3$
24	4089.25084 (34)				
25	4089.29393 (51)	aP (10, 5)	NU1+NU2		$^{14}NH_3$
26	4089.55699 (105)				
27	4090.06210 (74)				
28	4090.11946 (48)				
29	4090.41598 (47)	aP (10, 4)	NU1+NU2		$^{14}NH_3$
30	4090.81171 (91)				
31	4090.89055 (105)				
32	4091.25440 (30)	aP (10, 3)	NU1+NU2		$^{14}NH_3$
33	4091.68606 (69)	$s^{r}Q$ (7, 4)	NU2+2NU4	pe	$^{14}NH_3$
34	4091.86565 (51)	aP (10, 2)	NU1+NU2		$^{14}NH_3$
35	4092.21337 (47)	aP (10, 1)	NU1+NU2		$^{14}NH_3$
36	4092.33108 (30)	aP (10, 0)	NU1+NU2		$^{14}NH_3$
37	4092.81270 (47)				
38	4093.23257 (30)	$a^{r}P$ (4, 1)	NU2+2NU4	pe	$^{14}NH_3$
39	4093.74094 (20)				
40	4094.86074 (21)	$s^{P}P$ (3, 2)	NU2+2NU4	pe	$^{14}NH_3$
41	4095.03511 (34)	$a^{P}P$ (8, 4)	NU2+2NU4	pe	$^{14}NH_3$
42	4095.07699 (19)				
43	4095.56810 (24)	$s^{r}Q$ (5, 3)	NU2+2NU4	pe	$^{14}NH_3$
44	4095.63382 (22)				
45	4096.04611 (66)				
46	4096.15051 (53)				
47	4096.41420 (25)	sP (4, 2)	NU2+2NU4	pa	$^{14}NH_3$
48	4096.80174 (23)	sP (4, 1)	NU2+2NU4	pa	$^{14}NH_3$
49	4097.29953 (25)				
50	4097.45556 (49)				
51	4097.54062 (22)				
52	4097.88480 (32)				
53	4097.93054 (21)				
54	4098.85797 (49)				
55	4100.31260 (42)				
56	4100.85029 (59)				
57	4100.90675 (50)				
58	4101.91726 (32)				
59	4102.04045 (24)				
60	4102.15047 (29)	$a^{P}P$ (6, 2)	NU2+2NU4	pe	$^{14}NH_3$

- "LINE NO.": serial number identifying the spectral line in the map on the opposite page.
- "WAVENUMBER": experimental value.
- Uncertainty in the last digit of the wavenumber is given in parentheses.
- "pa" : parallel ; "pe" : perpendicular.

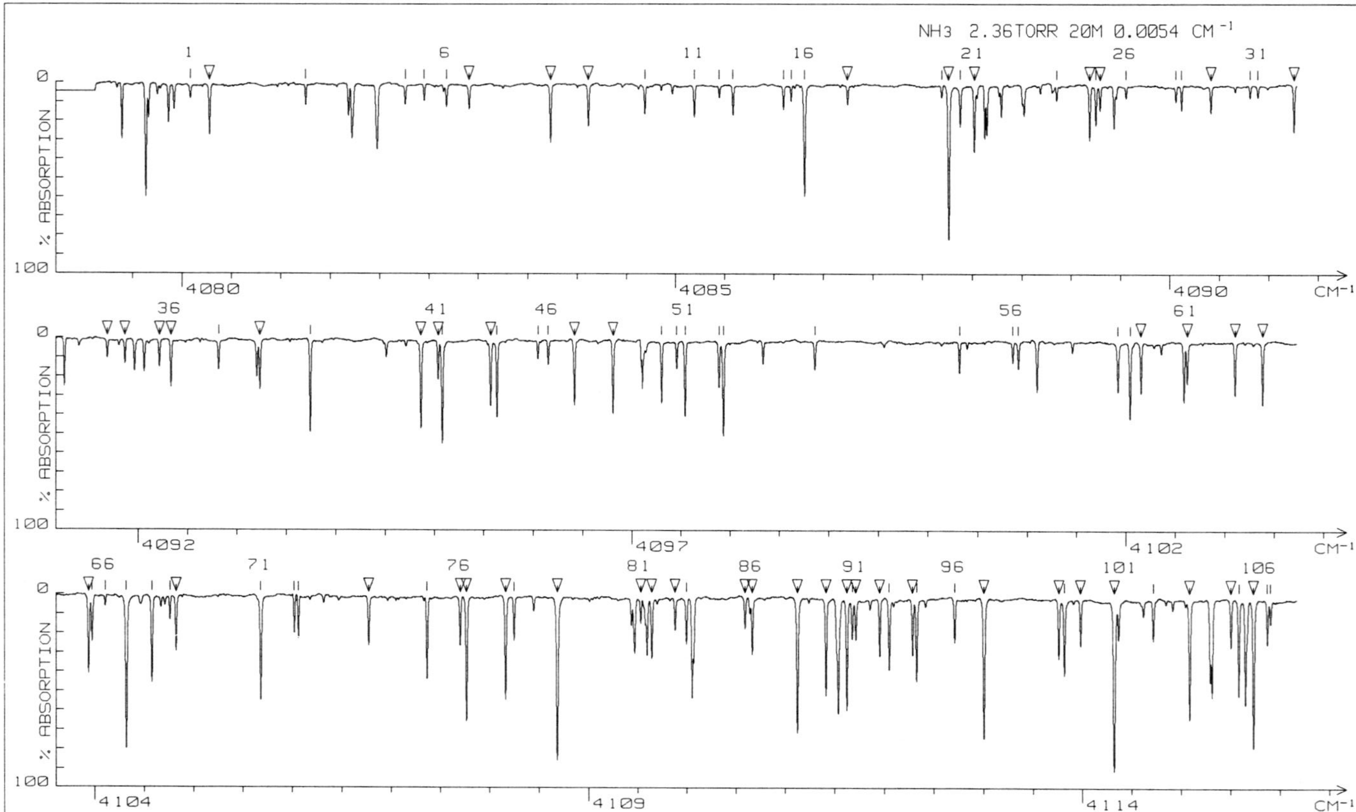

ISOTOPIC IDENTIFICATION – '∇' indicates $^{14}NH_3$; '|': unidentified.

LINE NO.	WAVENUMBER CM^{-1}, VAC.	ROTATIONAL ASSIGNMENTS	BAND ASSIGNMENTS		ISOTOPIC SPECIES
61	4102.61934(36)	s^{r}Q(3, 2)	NU2+2NU4	pe	$^{14}NH_3$
62	4103.10168(28)	s^{r}Q(6, 3)	NU2+2NU4	pe	$^{14}NH_3$
63	4103.37954(26)	a^{p}P(7, 3)	NU2+2NU4	pe	$^{14}NH_3$
64	4103.93990(23)	aP(9, 8)	NU1+NU2		$^{14}NH_3$
65	4103.97504(32)				
66	4104.10534(103)				
67	4104.32118(18)				
68	4104.57872(22)				
69	4104.76148(48)				
70	4104.82308(28)	a^{p}P(5, 1)	NU2+2NU4	pe	$^{14}NH_3$
71	4105.68231(20)				
72	4106.02147(34)				
73	4106.06373(30)				
74	4106.77408(31)	aP(9, 7)	NU1+NU2		$^{14}NH_3$
75	4107.36478(21)				
76	4107.69981(30)	s^{r}Q(4, 2)	NU2+2NU4	pe	$^{14}NH_3$
77	4107.76434(18)	a^{r}P(4, 0)	NU2+2NU4	pe	$^{14}NH_3$
78	4108.16213(20)	aP(9, 6)	NU1+NU2		$^{14}NH_3$
79	4108.24678(32)				
80	4108.68495(17)	aQ(6, 6)	NU2+2NU4	pa	$^{14}NH_3$
81	4109.52869(51)	sP(10, 2)	NU1+NU2		$^{14}NH_3$
82	4109.64136(28)	aP(9, 5)	NU1+NU2		$^{14}NH_3$
83	4109.88088(44)	sP(10, 7)	NU1+NU2		$^{14}NH_3$
84	4109.99574(33)				
85	4110.59122(42)	sP(10, 8)	NU1+NU2		$^{14}NH_3$
86	4110.66325(28)	aP(9, 4)	NU1+NU2		$^{14}NH_3$
87	4111.12450(18)	aQ(5, 5)	NU2+2NU4	pa	$^{14}NH_3$
88	4111.41291(21)	aP(9, 3)	NU1+NU2		$^{14}NH_3$
89	4111.62463(19)	a^{p}P(6, 3)	NU2+2NU4	pe	$^{14}NH_3$
90	4111.67771(37)				
91	4111.71451(37)	s^{r}Q(7, 3)	NU2+2NU4	pe	$^{14}NH_3$
92	4111.95332(27)	a^{p}P(7, 4)	NU2+2NU4	pe	$^{14}NH_3$
92		aP(9, 2)	NU1+NU2		$^{14}NH_3$
93	4112.04930(25)				
94	4112.28708(27)	aP(9, 1)	NU1+NU2		$^{14}NH_3$
95	4112.32828(23)				
96	4112.70723(32)				
97	4113.00405(25)	aQ(4, 4)	NU2+2NU4	pa	$^{14}NH_3$
98	4113.76242(27)	sP(3, 2)	NU2+2NU4	pa	$^{14}NH_3$
99	4113.81905(23)				
100	4113.98230(32)	s^{r}Q(5, 2)	NU2+2NU4	pe	$^{14}NH_3$
101	4114.32387(17)	aQ(3, 3)	NU2+2NU4	pa	$^{14}NH_3$
102	4114.71764(36)				
103	4115.08661(19)	aQ(2, 2)	NU2+2NU4	pa	$^{14}NH_3$
104	4115.50355(31)	s^{r}Q(2, 1)	NU2+2NU4	pe	$^{14}NH_3$
105	4115.58275(20)				
106	4115.73248(18)	sP(3, 0)	NU2+2NU4	pa	$^{14}NH_3$
107	4115.87133(32)				
108	4115.90410(58)				

- "LINE NO.": serial number identifying the spectral line in the map on the opposite page.
- "WAVENUMBER": experimental value.
- Uncertainty in the last digit of the wavenumber is given in parentheses.
- "pa" : parallel ; "pe" : perpendicular.

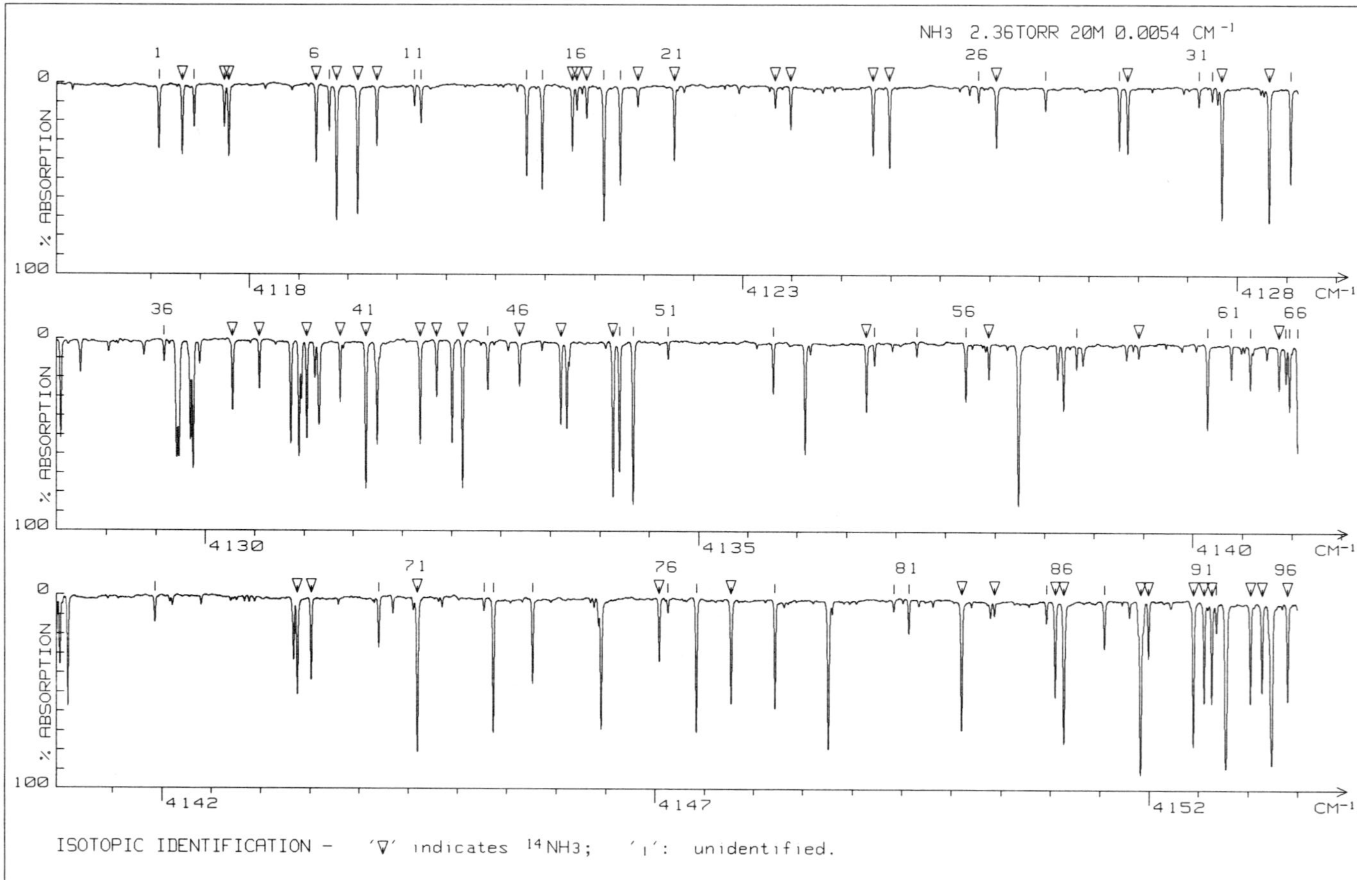
NH3 2.36TORR 20M 0.0054 CM⁻¹
% ABSORPTION
0
100
4118
4123
4128
4130
4135
4140
4142
4147
4152
CM⁻¹
1
6
11
16
21
26
31
36
41
46
51
56
61
66
71
76
81
86
91
96
ISOTOPIC IDENTIFICATION – '∇' indicates ¹⁴NH3; 'ı': unidentified.

LINE NO.	WAVENUMBER CM^{-1}, VAC.	ROTATIONAL ASSIGNMENTS	BAND		ISOTOPIC SPECIES
1	4117.08057 (26)				
2	4117.31542 (25)	$a^{P}P$(5, 2)	NU2+2NU4	pe	$^{14}NH_3$
3	4117.43552 (34)				
4	4117.74421 (34)	aQ(2, 1)	NU2+2NU4	pa	$^{14}NH_3$
5	4117.78966 (23)	aQ(5, 4)	NU2+2NU4	pa	$^{14}NH_3$
6	4118.67320 (23)	aQ(3, 2)	NU2+2NU4	pa	$^{14}NH_3$
7	4118.80518 (31)				
8	4118.88175 (18)	aQ(4, 3)	NU2+2NU4	pa	$^{14}NH_3$
9	4119.09803 (19)	aQ(7, 6)	NU2+2NU4	pa	$^{14}NH_3$
10	4119.29316 (26)	$s^{r}Q$(3, 1)	NU2+2NU4	pe	$^{14}NH_3$
11	4119.67283 (59)				
12	4119.74000 (40)				
13	4120.81321 (22)				
14	4120.96995 (20)				
15	4121.27158 (26)	$a^{P}P$(4, 1)	NU2+2NU4	pe	$^{14}NH_3$
16	4121.31796 (56)	$s^{r}Q$(8, 3)	NU2+2NU4	pe	$^{14}NH_3$
17	4121.41721 (41)	$s^{r}Q$(6, 2)	NU2+2NU4	pe	$^{14}NH_3$
17		aQ(6, 5)	NU2+2NU4	pa	$^{14}NH_3$
18	4121.59113 (19)				
19	4121.75978 (21)				
20	4121.93770 (66)	aQ(3, 1)	NU2+2NU4	pa	$^{14}NH_3$
21	4122.30377 (23)	$a^{P}P$(7, 5)	NU2+2NU4	pe	$^{14}NH_3$
22	4123.33384 (60)	aQ(6, 4)	NU2+2NU4	pa	$^{14}NH_3$
23	4123.48863 (33)	aQ(4, 2)	NU2+2NU4	pa	$^{14}NH_3$
24	4124.31744 (26)	$s^{r}Q$(4, 1)	NU2+2NU4	pe	$^{14}NH_3$
25	4124.48479 (22)	aQ(5, 3)	NU2+2NU4	pa	$^{14}NH_3$
26	4125.38374 (78)				
27	4125.56228 (27)	$a^{P}P$(6, 4)	NU2+2NU4	pe	$^{14}NH_3$
28	4126.06074 (59)				
29	4126.80583 (26)				
30	4126.88976 (25)	aP(8, 7)	NU1+NU2		$^{14}NH_3$
31	4127.61442 (69)				
32	4127.74847 (90)				
33	4127.84514 (19)	$a^{P}P$(7, 6)	NU2+2NU4	pe	$^{14}NH_3$
34	4128.31959 (19)	aP(8, 6)	NU1+NU2		$^{14}NH_3$
35	4128.53653 (20)				
36	4129.57793 (63)				
37	4130.27278 (25)	$a^{r}Q$(5, 4)	NU2+2NU4	pe	$^{14}NH_3$
38	4130.54417 (31)	$s^{r}Q$(5, 1)	NU2+2NU4	pe	$^{14}NH_3$
39	4131.02190 (21)	sP(9, 3)	NU1+NU2		$^{14}NH_3$
40	4131.36386 (27)	sP(9, 5)	NU1+NU2		$^{14}NH_3$
41	4131.62539 (19)	aP(8, 3)	NU1+NU2		$^{14}NH_3$
42	4132.17545 (20)	aP(8, 2)	NU1+NU2		$^{14}NH_3$
43	4132.34261 (29)	sP(9, 7)	NU1+NU2		$^{14}NH_3$
44	4132.60826 (18)	aP(8, 0)	NU1+NU2		$^{14}NH_3$
45	4132.86583 (32)				
46	4133.18602 (35)	sP(9, 8)	NU1+NU2		$^{14}NH_3$
47	4133.60093 (23)	sP(2, 1)	NU2+2NU4	pa	$^{14}NH_3$
48	4134.12773 (18)	$a^{P}P$(6, 5)	NU2+2NU4	pe	$^{14}NH_3$
49	4134.19537 (18)				
50	4134.33531 (18)				
51	4134.68778 (66)				
52	4135.75783 (29)				
53	4136.69471 (26)	$a^{r}Q$(6, 4)	NU2+2NU4	pe	$^{14}NH_3$
54	4136.77684 (59)				
55	4137.20794 (85)				
56	4137.70242 (27)				
57	4137.93381 (39)	$s^{r}Q$(6, 1)	NU2+2NU4	pe	$^{14}NH_3$
58	4138.81912 (50)				
59	4139.44598 (98)	$s^{r}Q$(8, 2)	NU2+2NU4	pe	$^{14}NH_3$
60	4140.14660 (23)				

- "LINE NO.": serial number identifying the spectral line in the map on the opposite page.
- "WAVENUMBER": experimental value.
- Uncertainty in the last digit of the wavenumber is given in parentheses.
- "pa" : parallel ; "pe" : perpendicular.

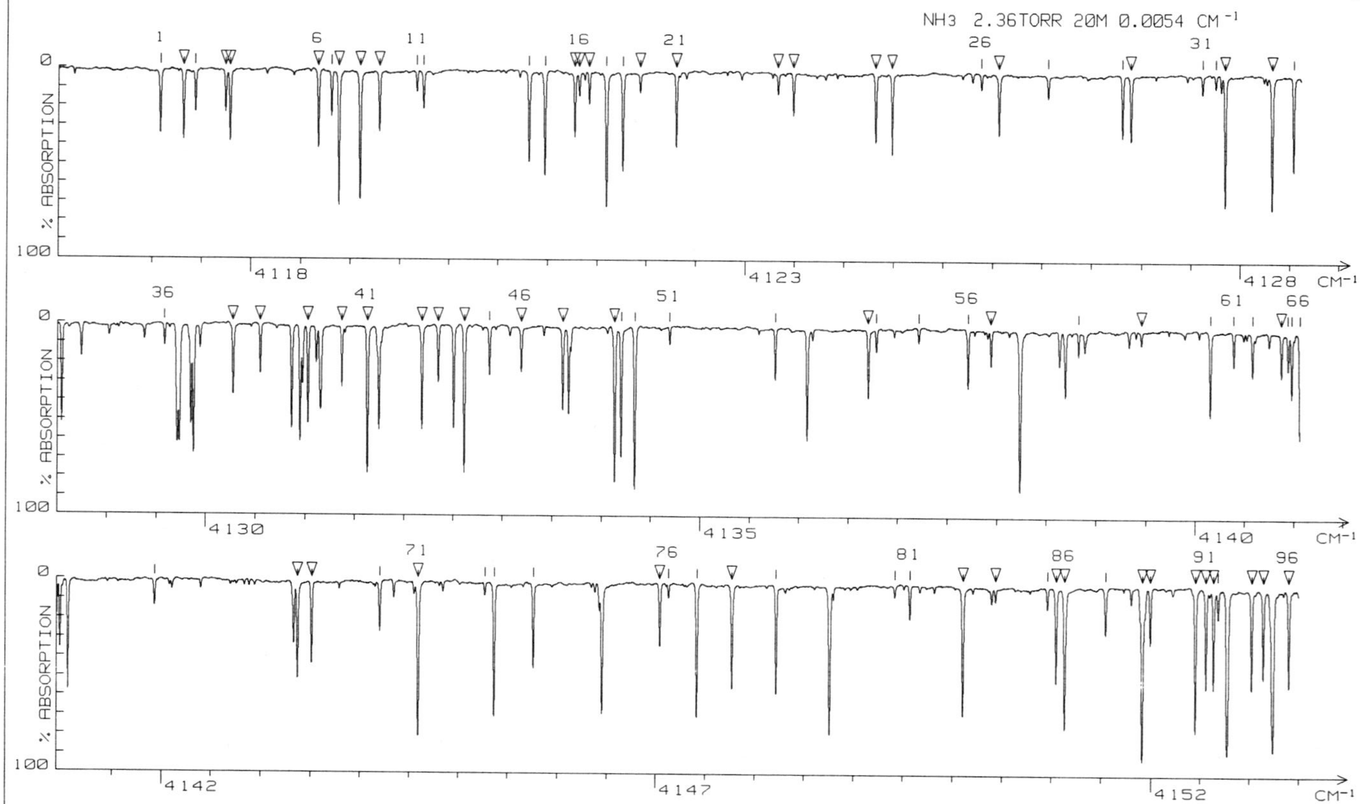

ISOTOPIC IDENTIFICATION – '∇' indicates $^{14}NH_3$; '|': unidentified.

LINE NO.	WAVENUMBER CM^{-1}, VAC.	ROTATIONAL ASSIGNMENTS	BAND ASSIGNMENTS		ISOTOPIC SPECIES
61	4140.38356(39)				
62	4140.57418(35)				
63	4140.86467(35)	$s^{P}Q$(1, 1)	NU2+2NU4	pe	$^{14}NH_3$
64	4140.93372(40)				
65	4140.96975(26)				
66	4141.05250(20)				
67	4141.92744(58)				
68	4143.36956(22)	$a^{r}Q$(4, 3)	NU2+2NU4	pe	$^{14}NH_3$
69	4143.50980(23)	$a^{r}P$(2, 0)	NU2+2NU4	pe	$^{14}NH_3$
70	4144.20066(32)				
71	4144.59158(19)	$a^{P}P$(4, 3)	NU2+2NU4	pe	$^{14}NH_3$
72	4145.27095(112)				
73	4145.36611(18)				
74	4145.76107(22)				
75	4147.04241(27)	$s^{P}Q$(3, 1)	NU2+2NU4	pe	$^{14}NH_3$
76	4147.12919(90)				
77	4147.41551(19)				
78	4147.76946(21)	$a^{r}Q$(5, 3)	NU2+2NU4	pe	$^{14}NH_3$
79	4148.22034(19)				
80	4149.41595(115)				
81	4149.57183(42)				
82	4150.10429(19)	aP(7, 5)	NU1+NU2		$^{14}NH_3$
83	4150.43311(97)	$s^{r}Q$(8, 1)	NU2+2NU4	pe	$^{14}NH_3$
84	4150.96037(64)				
85	4151.04889(22)	$a^{P}P$(3, 2)	NU2+2NU4	pe	$^{14}NH_3$
86	4151.13507(25)	aP(7, 4)	NU1+NU2		$^{14}NH_3$
87	4151.54464(34)				
88	4151.91122(19)	aP(7, 3)	NU1+NU2		$^{14}NH_3$
89	4151.99033(28)	$s^{P}Q$(4, 1)	NU2+2NU4	pe	$^{14}NH_3$
90	4152.44982(19)	aP(7, 2)	NU1+NU2		$^{14}NH_3$
91	4152.55795(20)	sP(8, 1)	NU1+NU2		$^{14}NH_3$
92	4152.63410(21)	sP(8, 2)	NU1+NU2		$^{14}NH_3$
93	4152.68106(49)				
94	4153.02129(21)	sP(8, 4)	NU1+NU2		$^{14}NH_3$
95	4153.13827(22)	$a^{r}Q$(6, 3)	NU2+2NU4	pe	$^{14}NH_3$
96	4153.39709(21)	sP(8, 5)	NU1+NU2		$^{14}NH_3$

- "LINE NO.": serial number identifying the spectral line in the map on the opposite page.
- "WAVENUMBER": experimental value.
- Uncertainty in the last digit of the wavenumber is given in parentheses.
- "pa" : parallel ; "pe" : perpendicular.

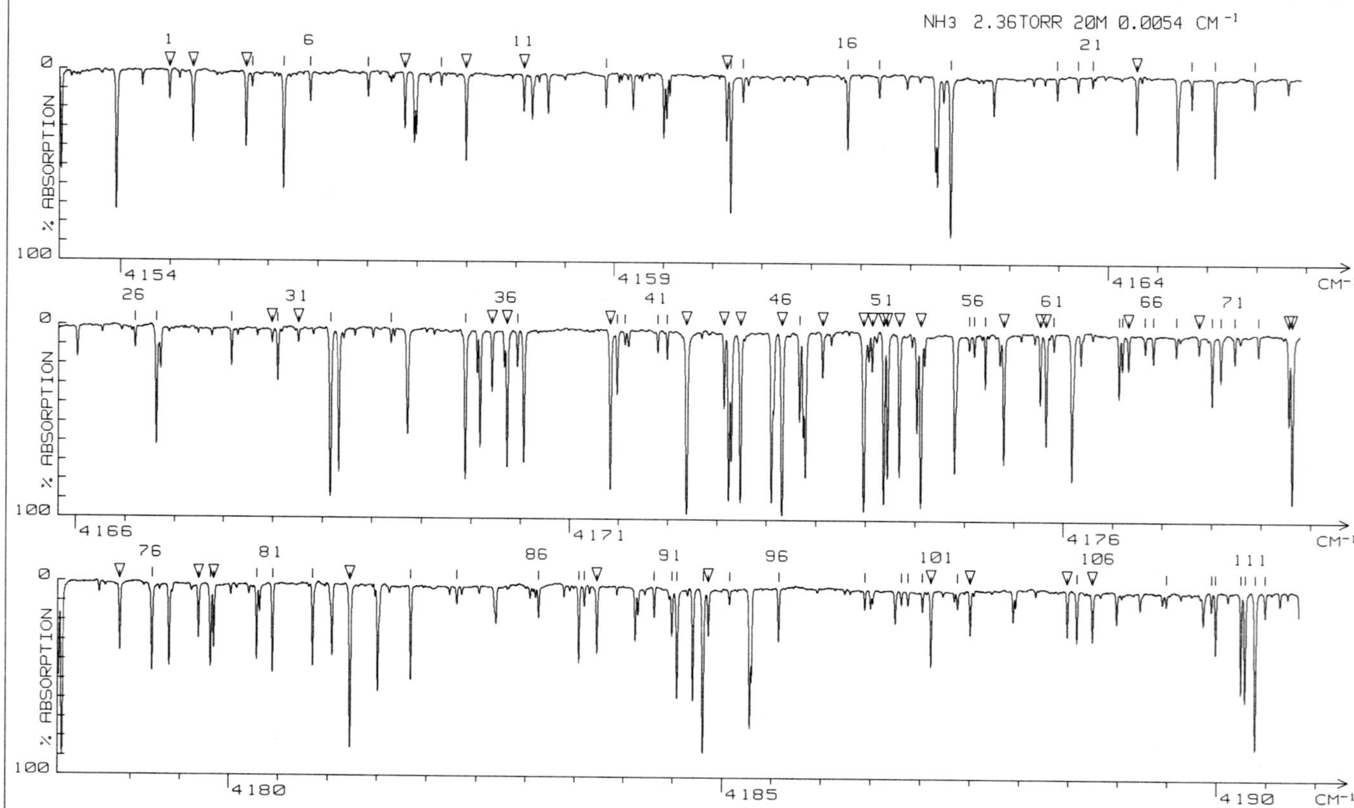

ISOTOPIC IDENTIFICATION – '∇' indicates $^{14}NH_3$; 'I': unidentified.

LINE NO.	WAVENUMBER CM^{-1}, VAC.	ROTATIONAL ASSIGNMENTS	BAND ASSIGNMENTS		ISOTOPIC SPECIES
1	4154.48683(49)	s^{P}Q(2, 2)	NU2+2NU4	pe	$^{14}NH_3$
2	4154.72590(25)	sP(8, 7)	NU1+NU2		$^{14}NH_3$
3	4155.26950(25)	s^{r}R(1, 1)	NU2+2NU4	pe	$^{14}NH_3$
4	4155.33005(96)				
5	4155.64994(21)				
6	4155.91761(48)				
7	4156.50111(59)				
8	4156.87160(30)	a^{r}Q(3, 2)	NU2+2NU4	pe	$^{14}NH_3$
9	4157.23978(101)				
10	4157.49018(23)	aR(1, 1)	NU2+2NU4	pa	$^{14}NH_3$
11	4158.07452(40)	s^{P}Q(3, 2)	NU2+2NU4	pe	$^{14}NH_3$
12	4158.90956(43)				
13	4160.13577(26)	a^{r}Q(4, 2)	NU2+2NU4	pe	$^{14}NH_3$
14	4160.17941(19)				
15	4160.30023(52)				
16	4161.36118(24)				
17	4161.67858(60)				
18	4162.39931(19)				
19	4163.47923(59)				
20	4163.68895(85)				
21	4163.83548(114)				
22	4164.28140(28)	a^{r}Q(5, 2)	NU2+2NU4	pe	$^{14}NH_3$
23	4164.84080(26)				
24	4165.07313(21)				
25	4165.47032(47)				
26	4166.59193(64)				
27	4166.80974(36)				
28	4167.57507(38)				
29	4167.98719(103)	a^{r}P(10, 7)	NU2+NU3		$^{14}NH_3$
30	4168.04365(33)				
31	4168.25104(110)	s^{P}Q(5, 2)	NU2+2NU4	pe	$^{14}NH_3$
32	4168.57435(19)				
33	4169.18586(93)				
34	4169.93577(20)				
35	4170.20802(29)	a^{r}Q(2, 1)	NU2+2NU4	pe	$^{14}NH_3$
36	4170.36037(19)	aP(6, 5)	NU1+NU2		$^{14}NH_3$
37	4170.46234(40)				
38	4171.41023(18)	aP(6, 4)	NU1+NU2		$^{14}NH_3$
39	4171.48112(28)				
40	4171.55894(99)				
41	4171.89135(66)				
42	4171.98736(49)				
43	4172.18876(20)	aP(6, 3)	NU1+NU2		$^{14}NH_3$
44	4172.56416(25)	a^{r}Q(3, 1)	NU2+2NU4	pe	$^{14}NH_3$
45	4172.72951(19)	aP(6, 2)	NU1+NU2		$^{14}NH_3$
46	4173.15336(20)	sQ(4, 4)	NU2+2NU4	pa	$^{14}NH_3$
46		aP(6, 0)	NU1+NU2		$^{14}NH_3$
47	4173.32909(24)				
48	4173.56204(36)	s^{r}P(10, 4)	NU2+NU3		$^{14}NH_3$
49	4173.97747(20)	sQ(3, 3)	NU2+2NU4	pa	$^{14}NH_3$
50	4174.06135(42)	a^{r}P(10, 6)	NU2+NU3		$^{14}NH_3$
51	4174.17628(18)	sP(7, 0)	NU1+NU2		$^{14}NH_3$
52	4174.21431(19)	sP(7, 1)	NU1+NU2		$^{14}NH_3$
53	4174.33484(19)	sP(7, 2)	NU1+NU2		$^{14}NH_3$
54	4174.55223(19)	sP(7, 3)	NU1+NU2		$^{14}NH_3$
55	4175.04035(100)				
56	4175.09209(68)				
57	4175.20289(19)				
58	4175.39051(20)	sP(7, 5)	NU1+NU2		$^{14}NH_3$
59	4175.76285(26)	a^{r}Q(4, 1)	NU2+2NU4	pe	$^{14}NH_3$
60	4175.82126(21)	sQ(3, 2)	NU2+2NU4	pa	$^{14}NH_3$

- "LINE NO.": serial number identifying the spectral line in the map on the opposite page.
- "WAVENUMBER": experimental value.
- Uncertainty in the last digit of the wavenumber is given in parentheses.
- "pa" : parallel ; "pe" : perpendicular.

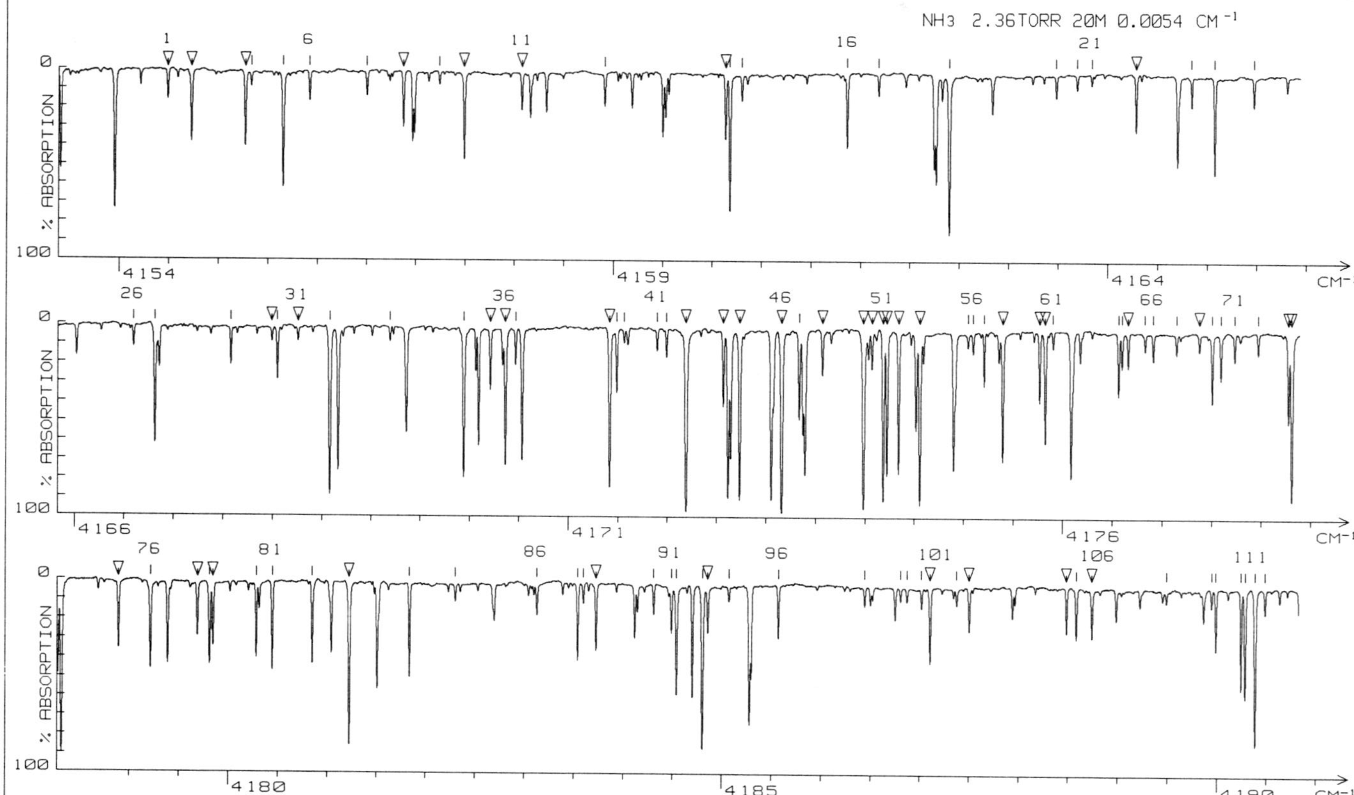

ISOTOPIC IDENTIFICATION – '∇' indicates $^{14}NH_3$; '|': unidentified.

LINE NO.	WAVENUMBER CM^{-1}, VAC.	ROTATIONAL ASSIGNMENTS	BAND ASSIGNMENTS		ISOTOPIC SPECIES
61	4175.89964 (82)				
62	4176.55743 (27)				
63	4176.59109 (43)				
64	4176.65260 (43)	$s^{P}Q$(5, 3)	NU2+2NU4	pe	$^{14}NH_3$
65	4176.82211 (75)				
66	4176.90390 (53)				
67	4177.14031 (66)				
68	4177.36808 (76)	sQ(4, 1)	NU2+2NU4	pa	$^{14}NH_3$
69	4177.49827 (27)				
70	4177.58554 (36)				
71	4177.72548 (52)				
72	4177.96357 (64)				
73	4178.27037 (34)	aR(2, 2)	NU2+2NU4	pa	$^{14}NH_3$
73		sQ(5, 5)	NU2+2NU4	pa	$^{14}NH_3$
74	4178.30529 (19)	sQ(4, 3)	NU2+2NU4	pa	$^{14}NH_3$
75	4178.89109 (27)	$s^{r}R$(2, 1)	NU2+2NU4	pe	$^{14}NH_3$
76	4179.21475 (24)				
77	4179.68692 (32)	$s^{r}P$(9, 6)	NU2+NU3		$^{14}NH_3$
78	4179.81184 (25)				
79	4179.84673 (28)	$a^{r}Q$(5, 1)	NU2+2NU4	pe	$^{14}NH_3$
80	4180.28258 (22)				
81	4180.44264 (23)				
82	4180.84696 (23)				
83	4181.22551 (19)	aR(2, 0)	NU2+2NU4	pa	$^{14}NH_3$
84	4181.83653 (22)				
85	4182.30084 (81)				
86	4183.12960 (51)				
87	4183.53969 (24)				
88	4183.59615 (71)				
89	4183.72342 (28)	$s^{P}Q$(5, 4)	NU2+2NU4	pe	$^{14}NH_3$
89		sQ(5, 4)	NU2+2NU4	pa	$^{14}NH_3$
90	4184.30783 (47)				
91	4184.48893 (38)				
92	4184.54221 (22)				
93	4184.80414 (18)				
94	4184.85815 (36)	$a^{r}Q$(6, 1)	NU2+2NU4	pe	$^{14}NH_3$
95	4185.07126 (100)				
96	4185.57052 (30)				
97	4186.43903 (69)				
98	4186.80598 (111)				
99	4186.87213 (109)				
100	4187.02000 (72)				
101	4187.10692 (26)	$s^{r}R$(3, 2)	NU2+2NU4	pe	$^{14}NH_3$
102	4187.37473 (88)				
103	4187.50336 (38)	$s^{r}P$(9, 5)	NU2+NU3		$^{14}NH_3$
104	4188.48612 (34)	$s^{r}P$(10, 2)	NU2+NU3		$^{14}NH_3$
105	4188.58338 (33)				
106	4188.74131 (36)	sQ(7, 7)	NU2+2NU4	pa	$^{14}NH_3$
107	4189.48744 (113)				
108	4189.94270 (72)				
109	4189.98665 (28)				
110	4190.24075 (19)				
111	4190.28250 (21)				
112	4190.38502 (19)				
113	4190.48519 (57)				

- "LINE NO.": serial number identifying the spectral line in the map on the opposite page.
- "WAVENUMBER": experimental value.
- Uncertainty in the last digit of the wavenumber is given in parentheses.
- "pa" : parallel ; "pe" : perpendicular.

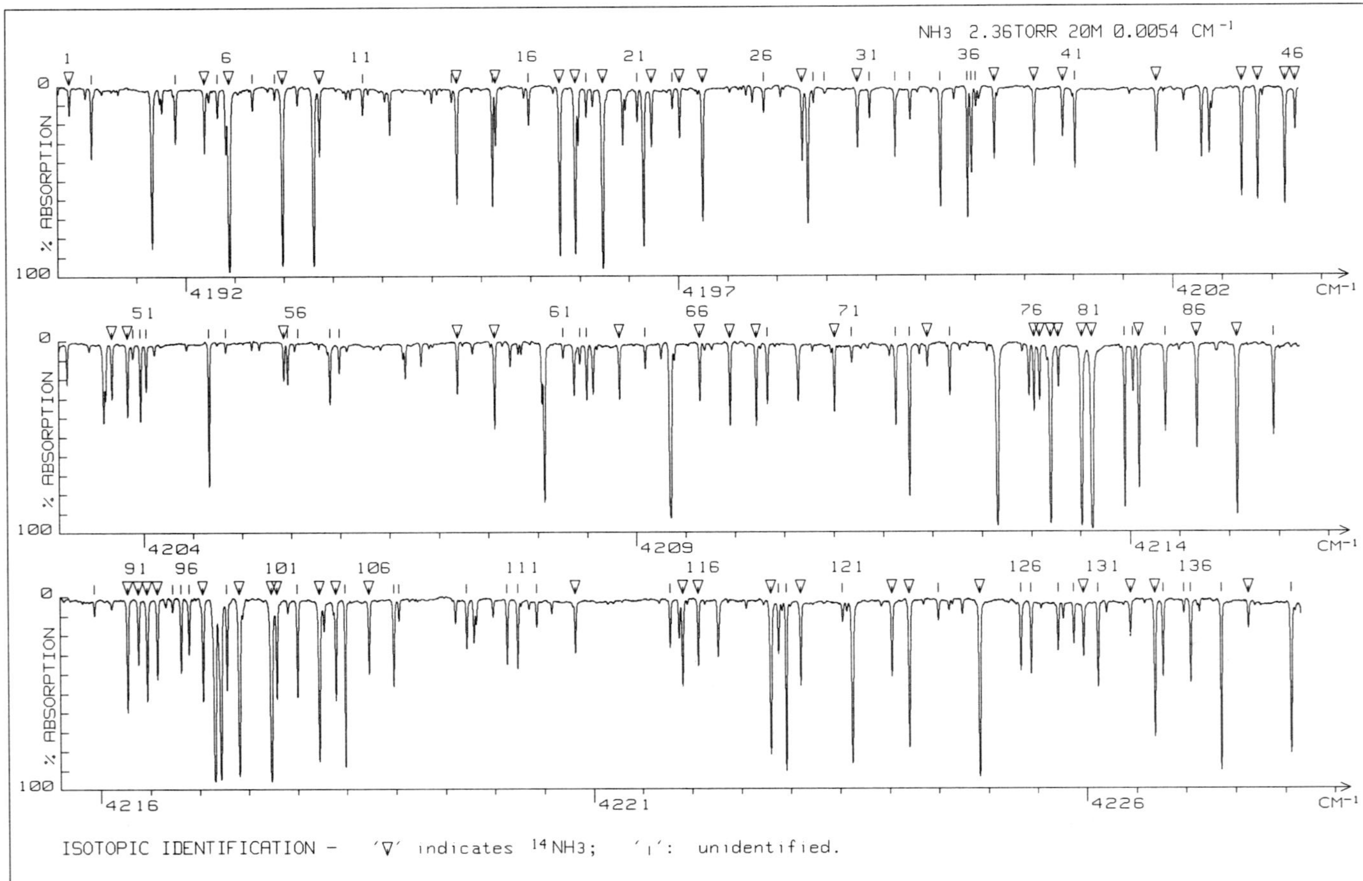
NH3 2.36TORR 20M 0.0054 CM⁻¹
% ABSORPTION
0
100
4192
4197
4202
4204
4209
4214
4216
4221
4226
CM⁻¹
ISOTOPIC IDENTIFICATION - '∇' indicates ¹⁴NH3; '|': unidentified.

LINE NO.	WAVENUMBER CM^{-1}, VAC.	ROTATIONAL ASSIGNMENTS	BAND		ISOTOPIC SPECIES
1	4190.82841(54)	$a^{r}Q$(7, 1)	NU2+2NU4	pe	$^{14}NH_3$
2	4191.05181(26)				
3	4191.89744(33)				
4	4192.19506(30)	$s^{r}R$(5, 5)	NU2+2NU4	pe	$^{14}NH_3$
5	4192.32720(55)				
6	4192.44489(18)	aP(5, 3)	NU1+NU2		$^{14}NH_3$
7	4192.68245(64)				
8	4192.90764(106)				
9	4192.98601(18)	aP(5, 2)	NU1+NU2		$^{14}NH_3$
10	4193.36436(26)	$a^{r}P$(10, 3)	NU2+NU3		$^{14}NH_3$
11	4193.80716(56)				
12	4194.70168(105)				
13	4194.75385(20)	$s^{r}R$(4, 3)	NU2+2NU4	pe	$^{14}NH_3$
14	4195.11819(18)				
15	4195.15013(31)	$s^{r}P$(9, 4)	NU2+NU3		$^{14}NH_3$
16	4195.48394(44)				
17	4195.79536(18)	sP(6, 1)	NU1+NU2		$^{14}NH_3$
18	4195.95198(19)	sP(6, 2)	NU1+NU2		$^{14}NH_3$
19	4196.06406(50)				
20	4196.22939(17)	sP(6, 3)	NU1+NU2		$^{14}NH_3$
21	4196.58777(47)				
22	4196.73242(32)	$a^{P}Q$(1, 1)	NU2+2NU4	pe	$^{14}NH_3$
23	4196.94239(39)				
24	4197.01583(34)	$a^{r}P$(9, 6)	NU2+NU3		$^{14}NH_3$
25	4197.24727(20)	sP(6, 5)	NU1+NU2		$^{14}NH_3$
26	4197.87045(59)				
27	4198.25681(26)	$a^{P}Q$(2, 1)	NU2+2NU4	pe	$^{14}NH_3$
28	4198.37093(56)				
29	4198.48347(59)				
30	4198.81701(31)	$s^{P}Q$(6, 5)	NU2+2NU4	pe	$^{14}NH_3$
30		sQ(6, 5)	NU2+2NU4	pa	$^{14}NH_3$

LINE NO.	WAVENUMBER CM^{-1}, VAC.	ROTATIONAL ASSIGNMENTS	BAND		ISOTOPIC SPECIES
31	4198.93798(49)				
32	4199.19667(27)				
33	4199.35045(53)				
34	4199.65543(21)				
35	4199.92814(18)				
36	4199.96701(24)				
37	4200.01231(82)				
38	4200.19800(26)	$a^{r}P$(10, 2)	NU2+NU3		$^{14}NH_3$
39	4200.60463(25)	$a^{P}Q$(3, 1)	NU2+2NU4	pe	$^{14}NH_3$
40	4200.89638(36)	$s^{r}P$(8, 6)	NU2+NU3		$^{14}NH_3$
41	4201.02055(24)				
42	4201.83853(29)	$s^{r}R$(5, 4)	NU2+2NU4	pe	$^{14}NH_3$
43	4202.69774(21)	$s^{r}P$(9, 3)	NU2+NU3		$^{14}NH_3$
44	4202.85792(21)	aR(3, 2)	NU2+2NU4	pa	$^{14}NH_3$
45	4203.12967(21)	$a^{r}R$(0, 0)	NU2+2NU4	pe	$^{14}NH_3$
46	4203.23348(39)	$a^{r}P$(9, 5)	NU2+NU3		$^{14}NH_3$
47	4203.68609(31)	$s^{r}R$(3, 1)	NU2+2NU4	pe	$^{14}NH_3$
48	4203.84182(27)	$a^{P}Q$(4, 1)	NU2+2NU4	pe	$^{14}NH_3$
49	4203.89515(118)				
50	4203.97130(25)				
51	4204.02944(35)				
52	4204.66996(17)				
53	4204.83888(56)				
54	4205.42995(43)	$s^{P}Q$(7, 5)	NU2+2NU4	pe	$^{14}NH_3$
55	4205.47051(39)				
56	4205.57502(73)				
57	4205.90081(31)				
58	4205.99513(52)				
59	4207.18913(32)	$a^{r}P$(10, 1)	NU2+NU3		$^{14}NH_3$
60	4207.56848(24)	aR(3, 1)	NU2+2NU4	pa	$^{14}NH_3$

- "LINE NO.": serial number identifying the spectral line in the map on the opposite page.
- "WAVENUMBER": experimental value.
- Uncertainty in the last digit of the wavenumber is given in parentheses.
- "pa" : parallel ; "pe" : perpendicular.

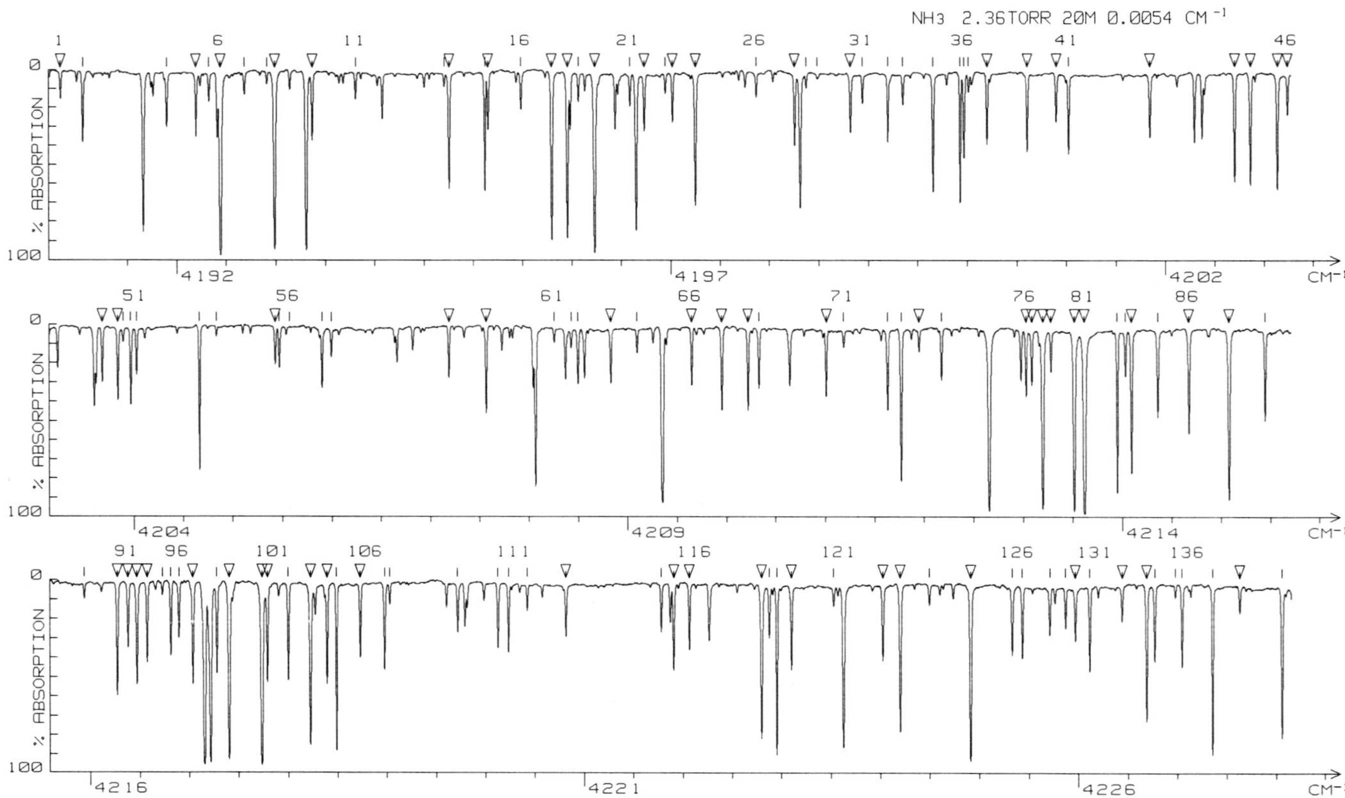

ISOTOPIC IDENTIFICATION – '∇' indicates $^{14}NH_3$; 'ı': unidentified.

LINE NO.	WAVENUMBER CM^{-1}, VAC.	ROTATIONAL ASSIGNMENTS	BAND		ISOTOPIC SPECIES
61	4208.25770(96)				
62	4208.42854(83)				
63	4208.49772(33)				
64	4208.83090(33)	s^{r}P(8, 5)	NU2+NU3		$^{14}NH_3$
65	4209.09819(66)				
66	4209.64817(31)	a^{r}P(9, 4)	NU2+NU3		$^{14}NH_3$
67	4209.95388(25)	a^{r}R(1, 1)	NU2+2NU4	pe	$^{14}NH_3$
68	4210.21871(24)	s^{r}P(9, 2)	NU2+NU3		$^{14}NH_3$
69	4210.33506(31)				
70	4211.01043(28)	s^{P}P(10, 1)	NU2+NU3		$^{14}NH_3$
71	4211.18784(94)				
72	4211.63144(27)				
73	4211.77167(19)				
74	4211.95334(69)	s^{P}Q(8, 5)	NU2+2NU4	pe	$^{14}NH_3$
75	4212.17930(33)				
76	4213.03049(29)	a^{P}Q(3, 2)	NU2+2NU4	pe	$^{14}NH_3$
77	4213.09051(33)	s^{r}R(4, 2)	NU2+2NU4	pe	$^{14}NH_3$
78	4213.19975(18)	aP(4, 2)	NU1+NU2		$^{14}NH_3$
79	4213.28186(40)	a^{P}Q(6, 1)	NU2+2NU4	pe	$^{14}NH_3$
80	4213.51698(21)	aP(4, 1)	NU1+NU2		$^{14}NH_3$
81	4213.62130(18)	aP(4, 0)	NU1+NU2		$^{14}NH_3$
82	4213.94887(16)				
83	4214.03163(37)				
84	4214.09228(20)	a^{r}P(10, 0)	NU2+NU3		$^{14}NH_3$
85	4214.35712(25)				
86	4214.67690(23)	sR(1, 1)	NU2+2NU4	pa	$^{14}NH_3$
87	4215.08078(20)	sR(1, 0)	NU2+2NU4	pa	$^{14}NH_3$
88	4215.44627(23)				
89	4215.93914(95)				
90	4216.27093(22)	a^{r}P(9, 3)	NU2+NU3		$^{14}NH_3$
91	4216.37911(30)	a^{P}Q(4, 2)	NU2+2NU4	pe	$^{14}NH_3$
92	4216.46790(22)	a^{r}R(2, 2)	NU2+2NU4	pe	$^{14}NH_3$
93	4216.57149(26)	s^{r}P(8, 4)	NU2+NU3		$^{14}NH_3$
94	4216.72558(114)				
95	4216.81185(24)				
96	4216.89228(32)				
97	4217.03468(23)	aR(4, 4)	NU2+2NU4	pa	$^{14}NH_3$
98	4217.28012(24)				
99	4217.40409(20)	sP(5, 2)	NU1+NU2		$^{14}NH_3$
100	4217.73087(19)	s^{P}R(2, 2)	NU2+2NU4	pe	$^{14}NH_3$
100		sP(5, 3)	NU1+NU2		$^{14}NH_3$
101	4217.78706(23)	s^{r}P(9, 1)	NU2+NU3		$^{14}NH_3$
102	4217.99675(23)				
103	4218.22042(19)	sP(5, 4)	NU1+NU2		$^{14}NH_3$
104	4218.38848(23)	s^{r}R(3, 0)	NU2+2NU4	pe	$^{14}NH_3$
105	4218.48601(16)				
106	4218.72861(26)	s^{P}P(10, 2)	NU2+NU3		$^{14}NH_3$
107	4218.97542(24)				
108	4219.02802(71)				
109	4219.71382(36)				
110	4220.12472(29)				
111	4220.23110(27)				
112	4220.42172(57)				
113	4220.81172(34)	a^{P}Q(5, 2)	NU2+2NU4	pe	$^{14}NH_3$
114	4221.77744(35)				
115	4221.90354(25)	s^{r}R(5, 3)	NU2+2NU4	pe	$^{14}NH_3$
116	4222.06189(28)	a^{P}P(10, 1)	NU2+NU3		$^{14}NH_3$
117	4222.79943(20)	a^{r}R(3, 3)	NU2+2NU4	pe	$^{14}NH_3$
118	4222.87712(36)				
119	4222.95451(18)				
120	4223.10002(25)	a^{r}P(9, 2)	NU2+NU3		$^{14}NH_3$

- "LINE NO.": serial number identifying the spectral line in the map on the opposite page.
- "WAVENUMBER": experimental value.
- Uncertainty in the last digit of the wavenumber is given in parentheses.
- "pa" : parallel ; "pe" : perpendicular.

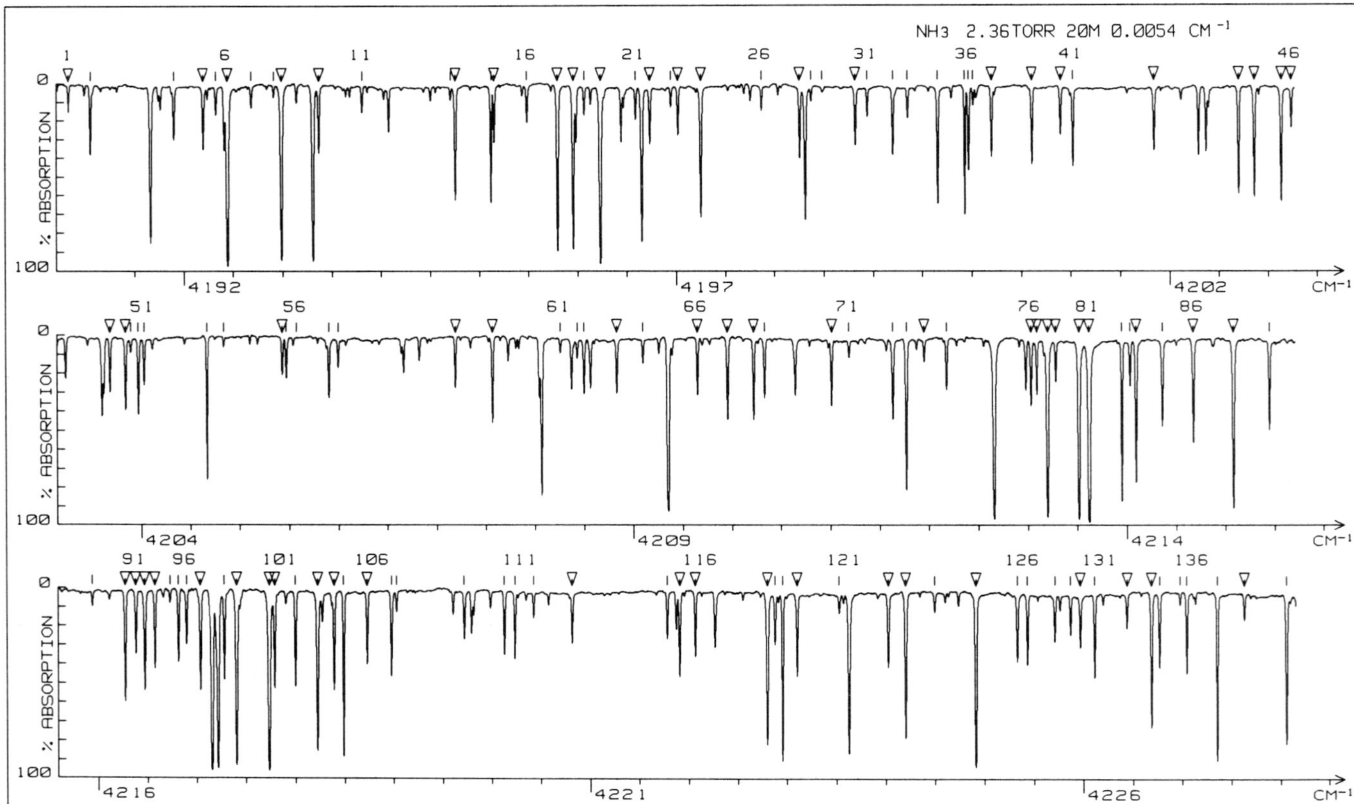

ISOTOPIC IDENTIFICATION – '∇' indicates $^{14}NH_3$; 'ı': unidentified.

LINE NO.	WAVENUMBER CM^{-1}, VAC.	ROTATIONAL ASSIGNMENTS	BAND		ISOTOPIC SPECIES
121	4223.52416(90)				
122	4224.02164(28)	a^pQ(3, 3)	NU2+2NU4	pe	$^{14}NH_3$
123	4224.20140(20)	s^rP(8, 3)	NU2+NU3		$^{14}NH_3$
124	4224.49421(80)				
125	4224.91001(18)	s^rP(9, 0)	NU2+NU3		$^{14}NH_3$
126	4225.32729(29)				
127	4225.43052(27)				
128	4225.70843(38)				
129	4225.86830(42)				
130	4225.96755(35)	a^rP(8, 5)	NU2+NU3		$^{14}NH_3$
131	4226.11229(24)				
132	4226.43675(47)	a^pQ(6, 2)	NU2+2NU4	pe	$^{14}NH_3$
133	4226.68369(20)	s^pP(10, 3)	NU2+NU3		$^{14}NH_3$
134	4226.76473(26)				
135	4226.97719(112)				
136	4227.04598(25)				
137	4227.35439(18)				
138	4227.62756(59)	s^pQ(9, 6)	NU2+2NU4	pe	$^{14}NH_3$
139	4228.05575(20)				

- "LINE NO.": serial number identifying the spectral line in the map on the opposite page.
- "WAVENUMBER": experimental value.
- Uncertainty in the last digit of the wavenumber is given in parentheses.
- "pa" : parallel ; "pe" : perpendicular.

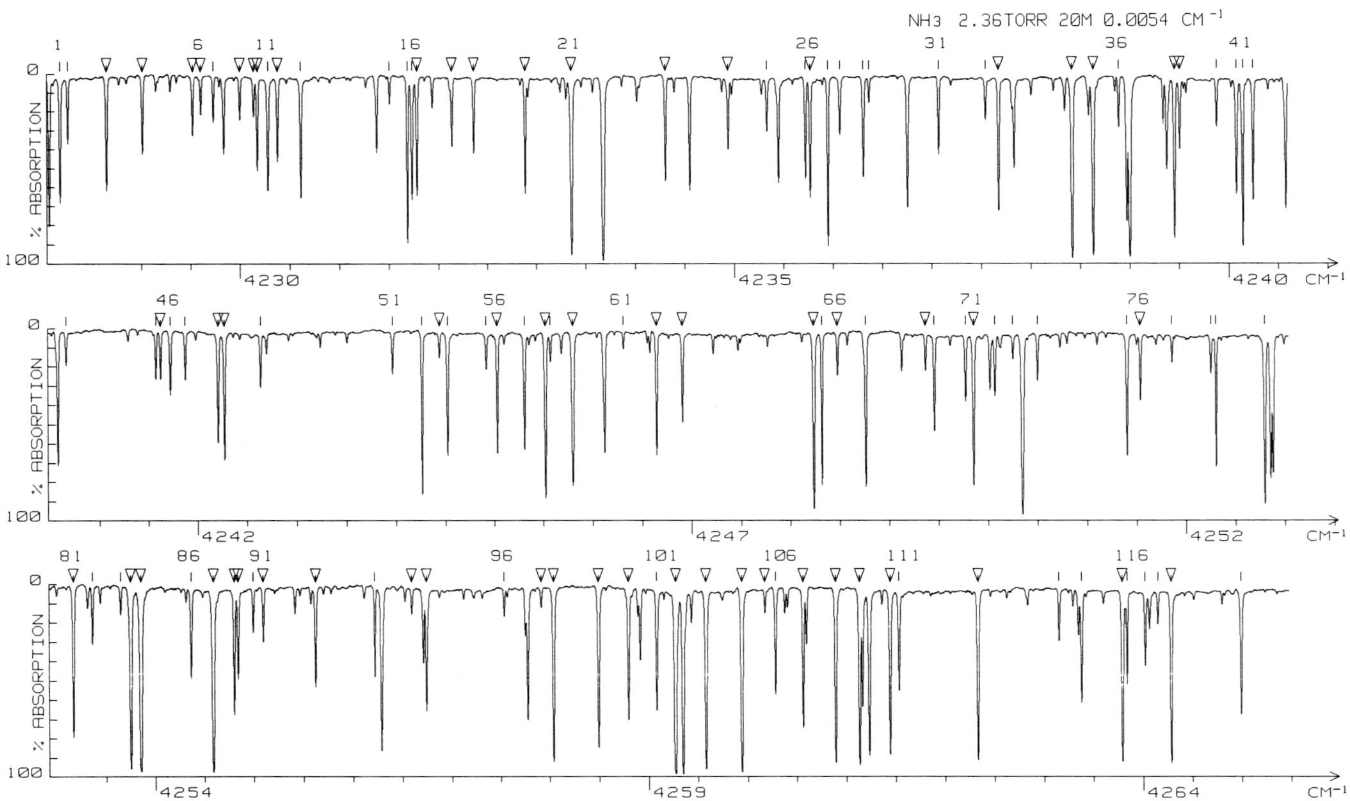

ISOTOPIC IDENTIFICATION – '∇' indicates $^{14}NH_3$; '|': unidentified.

LINE NO.	WAVENUMBER CM^{-1}, VAC.	ROTATIONAL ASSIGNMENTS	BAND		ISOTOPIC SPECIES	LINE NO.	WAVENUMBER CM^{-1}, VAC.	ROTATIONAL ASSIGNMENTS	BAND		ISOTOPIC SPECIES
1	4228.16961(21)					31	4237.06919(26)				
2	4228.24741(28)					32	4237.54117(40)				
3	4228.64197(22)	aR(4, 2)	NU2+2NU4	pa	$^{14}NH_3$	33	4237.67205(21)	a^{P}P(10, 3)	NU2+NU3		$^{14}NH_3$
4	4229.00952(26)	a^{P}Q(4, 3)	NU2+2NU4	pe	$^{14}NH_3$	34	4238.41703(18)	sP(4, 1)	NU1+NU2		$^{14}NH_3$
5	4229.51778(32)	a^{r}R(4, 4)	NU2+2NU4	pe	$^{14}NH_3$	35	4238.62952(18)	sP(4, 2)	NU1+NU2		$^{14}NH_3$
6	4229.60612(42)	s^{r}R(4, 1)	NU2+2NU4	pe	$^{14}NH_3$	36	4238.88752(34)				
7	4229.73117(36)					37	4239.45197(20)	s^{r}P(8, 1)	NU2+NU3		$^{14}NH_3$
8	4229.99921(40)	s^{r}P(7, 5)	NU2+NU3		$^{14}NH_3$	38	4239.50500(29)	a^{r}R(3, 2)	NU2+2NU4	pe	$^{14}NH_3$
9	4230.13705(45)	s^{r}R(6, 4)	NU2+2NU4	pe	$^{14}NH_3$	39	4239.87463(37)				
10	4230.17440(23)	a^{r}P(9, 1)	NU2+NU3		$^{14}NH_3$	40	4240.07600(23)				
11	4230.27888(23)					41	4240.14022(17)				
12	4230.37561(26)	a^{P}Q(5, 3)	NU2+2NU4	pe	$^{14}NH_3$	42	4240.24353(21)				
13	4230.61085(22)					43	4240.66193(50)				
14	4231.50598(54)					44	4241.57672(36)				
15	4231.68535(17)					45	4241.62495(37)	a^{P}Q(6, 3)	NU2+2NU4	pe	$^{14}NH_3$
16	4231.73035(22)					46	4241.72290(32)				
17	4231.78082(21)	s^{r}P(8, 2)	NU2+NU3		$^{14}NH_3$	47	4241.87543(36)				
18	4232.13334(87)	a^{r}R(2, 1)	NU2+2NU4	pe	$^{14}NH_3$	48	4242.21004(23)	aR(5, 4)	NU2+2NU4	pa	$^{14}NH_3$
19	4232.35962(27)	a^{r}P(8, 4)	NU2+NU3		$^{14}NH_3$	49	4242.27411(21)	s^{P}P(10, 5)	NU2+NU3		$^{14}NH_3$
20	4232.88621(21)	s^{P}P(9, 1)	NU2+NU3		$^{14}NH_3$	50	4242.63849(33)				
21	4233.35600(18)	aP(3, 2)	NU1+NU2		$^{14}NH_3$	51	4243.96644(33)				
22	4234.30432(22)	s^{P}P(10, 4)	NU2+NU3		$^{14}NH_3$	52	4244.26382(17)				
23	4234.93897(28)	aR(4, 1)	NU2+2NU4	pa	$^{14}NH_3$	53	4244.43845(66)	a^{P}Q(5, 4)	NU2+2NU4	pe	$^{14}NH_3$
24	4235.33188(34)					54	4244.52179(22)				
25	4235.72105(22)					55	4244.91978(49)				
26	4235.76938(22)	sR(2, 1)	NU2+2NU4	pa	$^{14}NH_3$	56	4245.03140(21)	a^{P}P(9, 1)	NU2+NU3		$^{14}NH_3$
27	4235.94736(17)					57	4245.30978(21)				
28	4236.06907(34)					58	4245.52395(20)	s^{r}P(7, 3)	NU2+NU3		$^{14}NH_3$
29	4236.30623(24)					59	4245.57569(59)				
30	4236.36504(60)					60	4245.80052(23)	a^{r}P(8, 2)	NU2+NU3		$^{14}NH_3$

- "LINE NO.": serial number identifying the spectral line in the map on the opposite page.
- "WAVENUMBER": experimental value.
- Uncertainty in the last digit of the wavenumber is given in parentheses.
- "pa" : parallel ; "pe" : perpendicular.

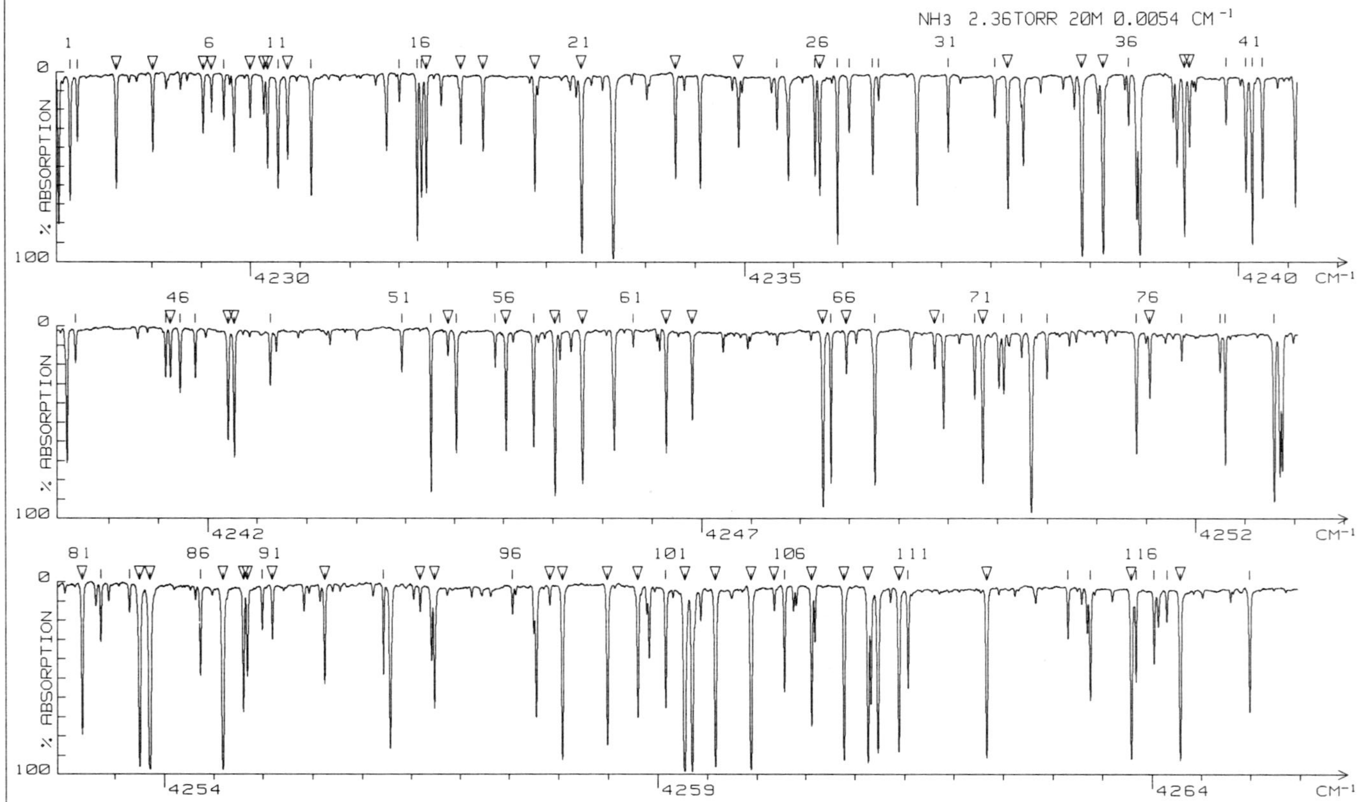

ISOTOPIC IDENTIFICATION – '∇' indicates $^{14}NH_3$; '|': unidentified.

LINE NO.	WAVENUMBER CM^{-1}, VAC.	ROTATIONAL ASSIGNMENTS	BAND ASSIGNMENTS		ISOTOPIC SPECIES
61	4246.31328(86)				
62	4246.64541(21)	a^{r}R(2, 0)	NU2+2NU4	pe	$^{14}NH_3$
63	4246.90827(24)	a^{r}R(4, 3)	NU2+2NU4	pe	$^{14}NH_3$
64	4248.23702(18)	s^{P}P(9, 3)	NU2+NU3		$^{14}NH_3$
65	4248.31804(17)				
66	4248.47634(43)	a^{r}P(7, 5)	NU2+NU3		$^{14}NH_3$
67	4248.76282(20)				
68	4249.36939(48)	s^{P}R(3, 3)	NU2+2NU4	pe	$^{14}NH_3$
69	4249.45869(24)				
70	4249.77277(31)				
71	4249.85424(21)	aR(5, 3)	NU2+2NU4	pa	$^{14}NH_3$
72	4250.06770(33)				
73	4250.24728(67)				
74	4250.49946(38)				
75	4251.40312(22)				
76	4251.53714(30)	aR(5, 1)	NU2+2NU4	pa	$^{14}NH_3$
77	4251.85959(57)				
78	4252.25019(46)				
79	4252.30260(17)				
80	4252.79660(19)				
81	4253.16529(19)	s^{r}P(7, 2)	NU2+NU3		$^{14}NH_3$
82	4253.35881(32)				
83	4253.64791(60)				
84	4253.75092(18)	aP(2, 1)	NU1+NU2		$^{14}NH_3$
85	4253.85322(18)	aP(2, 0)	NU1+NU2		$^{14}NH_3$
86	4254.36442(24)				
87	4254.59101(19)	s^{P}P(8, 1)	NU2+NU3		$^{14}NH_3$
88	4254.80333(22)	a^{P}P(10, 5)	NU2+NU3		$^{14}NH_3$
89	4254.84315(26)	a^{r}P(7, 4)	NU2+NU3		$^{14}NH_3$
90	4254.99425(40)				
91	4255.09611(32)	a^{r}R(3, 1)	NU2+2NU4	pe	$^{14}NH_3$
92	4255.62038(24)	aR(5, 2)	NU2+2NU4	pa	$^{14}NH_3$
93	4256.21743(18)				
94	4256.58997(68)	s^{r}R(5, 1)	NU2+2NU4	pe	$^{14}NH_3$
95	4256.73801(37)	sR(3, 1)	NU2+2NU4	pa	$^{14}NH_3$
96	4257.53474(60)				
97	4257.91061(88)	aR(6, 6)	NU2+2NU4	pa	$^{14}NH_3$
98	4258.03790(21)	sR(3, 0)	NU2+2NU4	pa	$^{14}NH_3$
99	4258.49056(21)	s^{P}P(10, 7)	NU2+NU3		$^{14}NH_3$
100	4258.79522(21)	sR(3, 2)	NU2+2NU4	pa	$^{14}NH_3$
101	4259.07851(23)				
102	4259.26897(18)	sP(3, 0)	NU1+NU2		$^{14}NH_3$
103	4259.57850(21)	sP(3, 2)	NU1+NU2		$^{14}NH_3$
104	4259.93804(19)	a^{r}P(8, 0)	NU2+NU3		$^{14}NH_3$
105	4260.17384(71)	a^{P}R(2, 1)	NU2+2NU4	pe	$^{14}NH_3$
106	4260.28126(23)				
107	4260.56096(21)	s^{r}P(7, 1)	NU2+NU3		$^{14}NH_3$
108	4260.88944(18)	a^{P}Q(6, 5)	NU2+2NU4	pe	$^{14}NH_3$
109	4261.13259(20)	a^{P}P(9, 3)	NU2+NU3		$^{14}NH_3$
110	4261.44134(19)	a^{r}P(7, 3)	NU2+NU3		$^{14}NH_3$
111	4261.53430(25)				
112	4262.33127(18)	s^{P}P(8, 2)	NU2+NU3		$^{14}NH_3$
113	4263.15070(36)				
114	4263.37739(24)				
115	4263.79016(21)	a^{P}P(10, 6)	NU2+NU3		$^{14}NH_3$
116	4263.83761(19)				
117	4264.01828(26)				
118	4264.14683(51)				
119	4264.28265(19)	s^{P}P(9, 5)	NU2+NU3		$^{14}NH_3$
120	4264.99042(23)				

- "LINE NO.": serial number identifying the spectral line in the map on the opposite page.
- "WAVENUMBER": experimental value.
- Uncertainty in the last digit of the wavenumber is given in parentheses.
- "pa" : parallel ; "pe" : perpendicular.

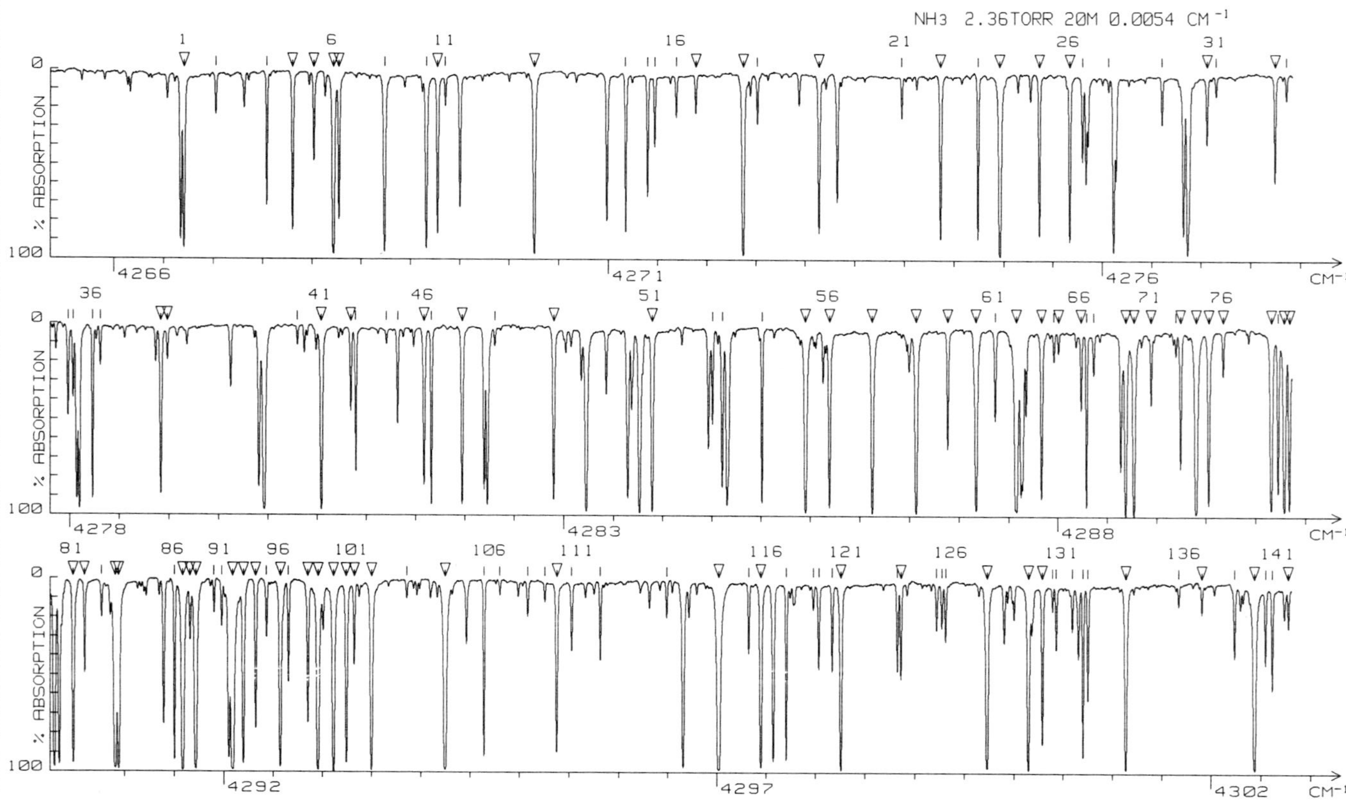

ISOTOPIC IDENTIFICATION – '▽' indicates $^{14}NH_3$; '|': unidentified.

LINE NO.	WAVENUMBER CM^{-1}, VAC.	ROTATIONAL ASSIGNMENTS	BAND		ISOTOPIC SPECIES
1	4266.70689(21)	$s^{P}P$(10, 8)	NU2+NU3		$^{14}NH_3$
2	4267.02980(45)				
3	4267.54237(17)				
4	4267.80347(20)	$a^{P}P$(8, 1)	NU2+NU3		$^{14}NH_3$
5	4268.02019(26)	$s^{r}R$(5, 0)	NU2+2NU4	pe	$^{14}NH_3$
6	4268.21823(18)	$s^{r}P$(7, 0)	NU2+NU3		$^{14}NH_3$
7	4268.27572(19)	$a^{r}P$(7, 2)	NU2+NU3		$^{14}NH_3$
8	4268.73893(22)				
9	4269.16179(22)				
10	4269.27635(21)	$a^{P}P$(9, 4)	NU2+NU3		$^{14}NH_3$
11	4269.35512(58)				
12	4270.25708(19)	$s^{P}P$(8, 3)	NU2+NU3		$^{14}NH_3$
13	4271.17717(17)				
14	4271.40292(21)				
15	4271.47387(28)				
16	4271.69056(42)				
17	4271.88915(47)	$a^{r}R$(5, 3)	NU2+2NU4	pe	$^{14}NH_3$
18	4272.37272(19)	$s^{P}P$(9, 6)	NU2+NU3		$^{14}NH_3$
19	4272.51646(39)				
20	4273.14096(19)	$a^{P}P$(10, 7)	NU2+NU3		$^{14}NH_3$
21	4273.97681(42)				
22	4274.36557(19)	$s^{r}P$(6, 2)	NU2+NU3		$^{14}NH_3$
23	4274.74115(17)				
24	4274.96087(18)	$s^{P}P$(10, 9)	NU2+NU3		$^{14}NH_3$
25	4275.36116(19)	$a^{r}P$(7, 1)	NU2+NU3		$^{14}NH_3$
26	4275.67468(19)	$a^{P}P$(8, 2)	NU2+NU3		$^{14}NH_3$
27	4275.80224(44)				
28	4276.06477(117)				
29	4276.60479(39)				
30	4277.06196(31)	$a^{r}P$(6, 4)	NU2+NU3		$^{14}NH_3$

LINE NO.	WAVENUMBER CM^{-1}, VAC.	ROTATIONAL ASSIGNMENTS	BAND		ISOTOPIC SPECIES
31	4277.15385(86)				
32	4277.74560(22)	sR(4, 1)	NU2+2NU4	pa	$^{14}NH_3$
33	4277.85939(67)				
34	4277.98175(26)				
35	4278.03243(81)				
36	4278.23460(17)				
37	4278.31071(43)				
38	4278.91875(19)	aQ(10, 9)	NU1+NU2		$^{14}NH_3$
39	4278.98391(58)	$s^{r}R$(7, 3)	NU2+2NU4	pe	$^{14}NH_3$
40	4280.29770(87)				
41	4280.54221(19)	$s^{P}P$(9, 7)	NU2+NU3		$^{14}NH_3$
42	4280.83833(28)	aQ(10, 8)	NU1+NU2		$^{14}NH_3$
43	4280.89225(21)				
44	4281.20208(118)				
45	4281.31498(24)				
46	4281.58414(22)	aQ(9, 8)	NU1+NU2		$^{14}NH_3$
47	4281.65754(17)				
48	4281.97194(19)	$s^{r}P$(6, 1)	NU2+NU3		$^{14}NH_3$
49	4282.30487(93)				
50	4282.89792(19)	$a^{P}P$(10, 8)	NU2+NU3		$^{14}NH_3$
51	4283.89425(18)	$s^{P}P$(7, 2)	NU2+NU3		$^{14}NH_3$
52	4284.50454(48)				
53	4284.60350(22)				
54	4285.00929(17)				
55	4285.45217(18)	aQ(7, 7)	NU1+NU2		$^{14}NH_3$
56	4285.69375(19)	aQ(8, 6)	NU1+NU2		$^{14}NH_3$
57	4286.12495(19)	$s^{P}P$(8, 5)	NU2+NU3		$^{14}NH_3$
58	4286.56767(19)	aQ(7, 6)	NU1+NU2		$^{14}NH_3$
59	4286.88385(23)	aQ(8, 5)	NU1+NU2		$^{14}NH_3$
60	4287.16994(19)	$a^{P}P$(9, 6)	NU2+NU3		$^{14}NH_3$

- "LINE NO.": serial number identifying the spectral line in the map on the opposite page.
- "WAVENUMBER": experimental value.
- Uncertainty in the last digit of the wavenumber is given in parentheses.
- "pa" : parallel ; "pe" : perpendicular.

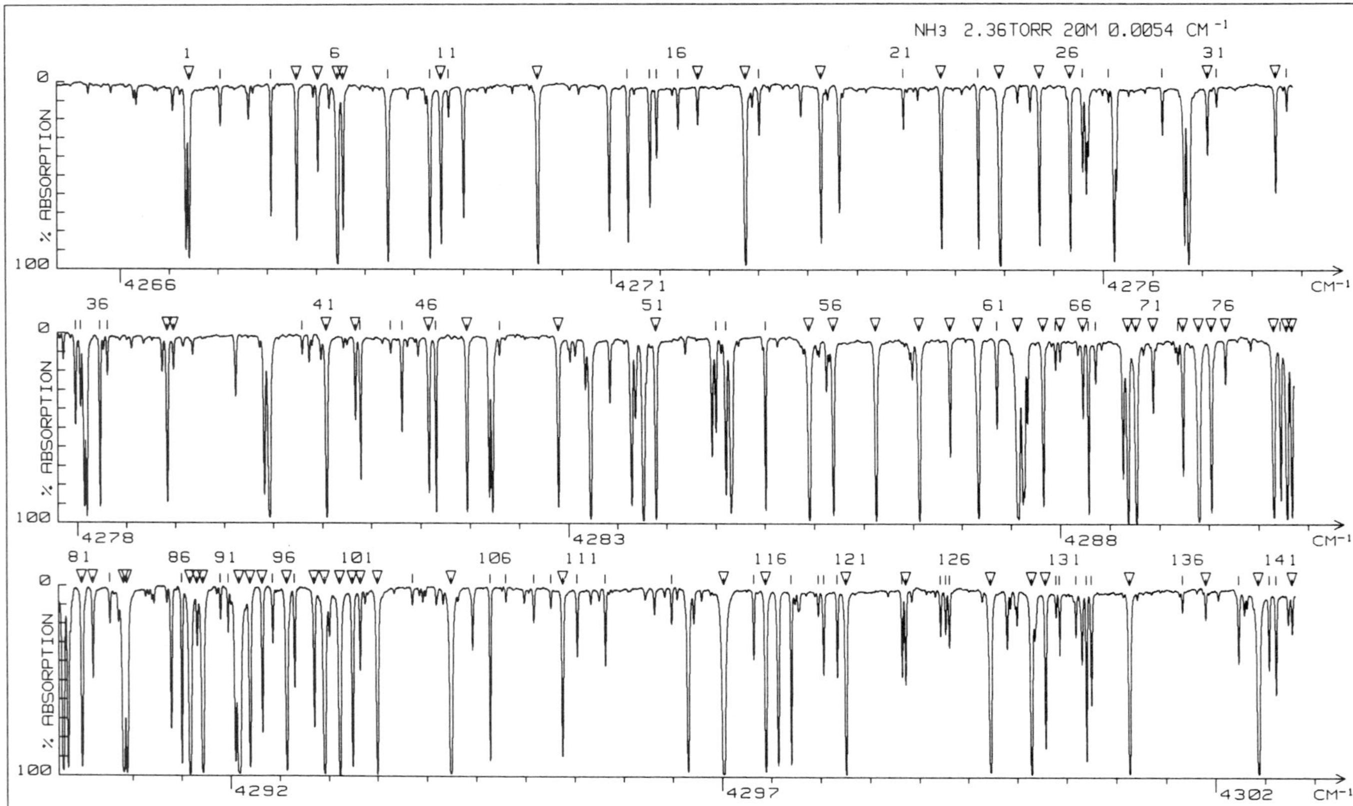

ISOTOPIC IDENTIFICATION – '∇' indicates $^{14}NH_3$; 'ı': unidentified.

LINE NO.	WAVENUMBER CM^{-1}, VAC.	ROTATIONAL ASSIGNMENTS	BAND	ISOTOPIC SPECIES
61	4287.36078(25)			
62	4287.57212(19)	aQ(9, 1)	NU1+NU2	$^{14}NH_3$
62		aQ(6, 6)	NU1+NU2	$^{14}NH_3$
63	4287.82931(19)	aQ(7, 5)	NU1+NU2	$^{14}NH_3$
64	4287.95801(64)			
65	4288.00510(82)	a^rR(5, 2)	NU2+2NU4 pe	$^{14}NH_3$
66	4288.23440(28)	aQ(8, 3)	NU1+NU2	$^{14}NH_3$
67	4288.29014(17)			
68	4288.36173(47)			
69	4288.68417(19)	aQ(6, 5)	NU1+NU2	$^{14}NH_3$
70	4288.76796(18)	s^pP(9, 8)	NU2+NU3	$^{14}NH_3$
71	4288.94046(30)	aQ(8, 1)	NU1+NU2	$^{14}NH_3$
72	4289.19022(83)			
73	4289.23929(22)	aQ(7, 3)	NU1+NU2	$^{14}NH_3$
74	4289.40009(18)	aQ(5, 5)	NU1+NU2	$^{14}NH_3$
75	4289.52738(19)	aQ(6, 4)	NU1+NU2	$^{14}NH_3$
76	4289.67077(42)	aQ(7, 2)	NU1+NU2	$^{14}NH_3$
77	4290.15459(19)	aQ(6, 3)	NU1+NU2	$^{14}NH_3$
78	4290.22365(19)			
79	4290.28666(19)	aQ(5, 4)	NU1+NU2	$^{14}NH_3$
80	4290.33741(19)	a^pP(7, 1)	NU2+NU3	$^{14}NH_3$
81	4290.47867(19)	a^rP(6, 2)	NU2+NU3	$^{14}NH_3$
82	4290.59378(26)	aQ(6, 2)	NU1+NU2	$^{14}NH_3$
83	4290.76610(54)			
84	4290.90438(19)	aQ(4, 4)	NU1+NU2	$^{14}NH_3$
85	4290.93957(19)	aQ(5, 3)	NU1+NU2	$^{14}NH_3$
86	4291.49977(17)			
87	4291.58369(18)	aQ(4, 3)	NU1+NU2	$^{14}NH_3$
88	4291.65581(40)	aQ(5, 1)	NU1+NU2	$^{14}NH_3$
89	4291.71353(18)	s^pP(7, 3)	NU2+NU3	$^{14}NH_3$
90	4291.89560(55)			
91	4291.97369(42)			
92	4292.08837(19)	aQ(3, 3)	NU1+NU2	$^{14}NH_3$
93	4292.19688(18)	a^pP(8, 4)	NU2+NU3	$^{14}NH_3$
94	4292.32235(20)	aQ(4, 1)	NU1+NU2	$^{14}NH_3$
95	4292.42883(38)			
96	4292.56914(19)	aQ(3, 2)	NU1+NU2	$^{14}NH_3$
97	4292.65103(25)			
98	4292.84990(20)	aQ(3, 1)	NU1+NU2	$^{14}NH_3$
99	4292.95252(18)	aQ(2, 2)	NU1+NU2	$^{14}NH_3$
100	4293.10956(18)	a^pP(10, 9)	NU2+NU3	$^{14}NH_3$
101	4293.24025(19)	aQ(2, 1)	NU1+NU2	$^{14}NH_3$
102	4293.31855(28)	a^rR(4, 0)	NU2+2NU4 pe	$^{14}NH_3$
103	4293.49670(19)	aQ(1, 1)	NU1+NU2	$^{14}NH_3$
104	4293.85272(101)			
105	4294.24177(19)	s^pP(8, 6)	NU2+NU3	$^{14}NH_3$
106	4294.63821(17)			
107	4294.79926(121)			
108	4295.08152(50)			
109	4295.25347(88)			
110	4295.37378(19)	s^rP(5, 2)	NU2+NU3	$^{14}NH_3$
111	4295.52243(33)			
112	4295.81163(28)			
113	4296.48619(50)			
114	4297.01462(19)	s^pP(9, 9)	NU2+NU3	$^{14}NH_3$
115	4297.32016(30)			
116	4297.44446(19)	s^pP(6, 1)	NU2+NU3	$^{14}NH_3$
117	4297.70507(17)			
118	4297.97547(76)			
119	4298.03001(28)			
120	4298.16654(27)			

- "LINE NO.": serial number identifying the spectral line in the map on the opposite page.
- "WAVENUMBER": experimental value.
- Uncertainty in the last digit of the wavenumber is given in parentheses.
- "pa" : parallel ; "pe" : perpendicular.

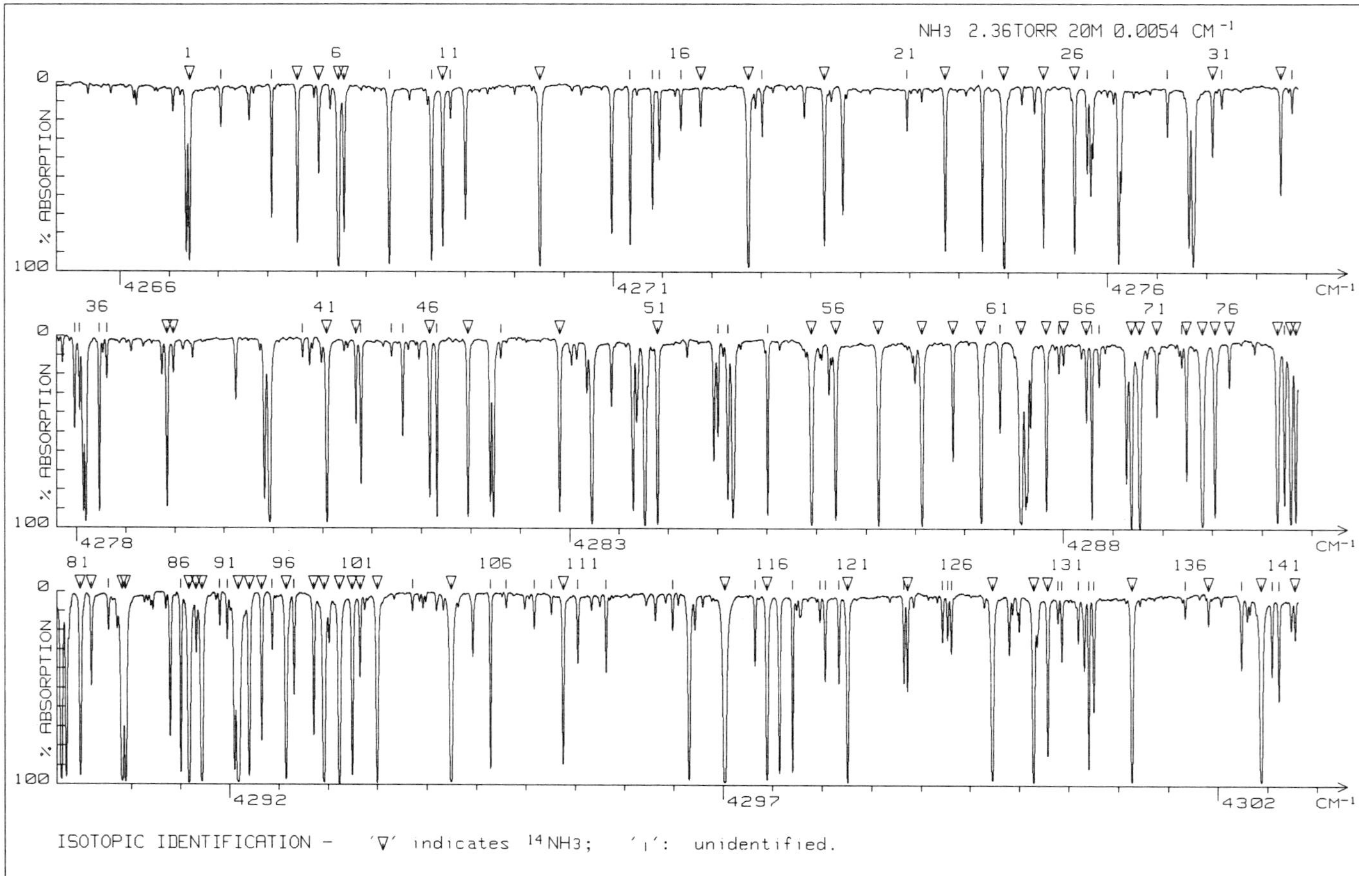
NH3 2.36TORR 20M 0.0054 CM⁻¹
% ABSORPTION
0
100
4266
4271
4276
4278
4283
4288
4292
4297
4302
CM⁻¹
ISOTOPIC IDENTIFICATION – '∇' indicates ¹⁴NH3; '|': unidentified.

LINE NO.	WAVENUMBER CM^{-1}, VAC.	ROTATIONAL ASSIGNMENTS	BAND		ISOTOPIC SPECIES
121	4298.25456(18)	^aPP(7, 2)	NU2+NU3		$^{14}NH_3$
122	4298.82246(27)				
123	4298.85848(26)	sR(5, 1)	NU2+2NU4	pa	$^{14}NH_3$
124	4299.21761(45)				
125	4299.26982(43)				
126	4299.30815(37)				
127	4299.72541(18)	^sPP(7, 4)	NU2+NU3		$^{14}NH_3$
128	4300.14092(19)	sP(1, 0)	NU1+NU2		$^{14}NH_3$
129	4300.28384(20)	sR(5, 0)	NU2+2NU4	pa	$^{14}NH_3$
130	4300.38818(74)				
131	4300.42861(30)				
132	4300.59254(43)				
133	4300.70033(17)				
134	4300.75086(24)				
135	4301.13137(19)	^aPP(8, 5)	NU2+NU3		$^{14}NH_3$
136	4301.66546(82)				
137	4301.90344(59)	aR(7, 1)	NU2+2NU4	pa	$^{14}NH_3$
138	4302.23219(32)				
139	4302.43325(18)	^sPP(8, 7)	NU2+NU3		$^{14}NH_3$
140	4302.54170(29)				
141	4302.61079(25)				
142	4302.77345(51)	sQ(10, 8)	NU1+NU2		$^{14}NH_3$

- "LINE NO.": serial number identifying the spectral line in the map on the opposite page.
- "WAVENUMBER": experimental value.
- Uncertainty in the last digit of the wavenumber is given in parentheses.
- "pa" : parallel ; "pe" : perpendicular.

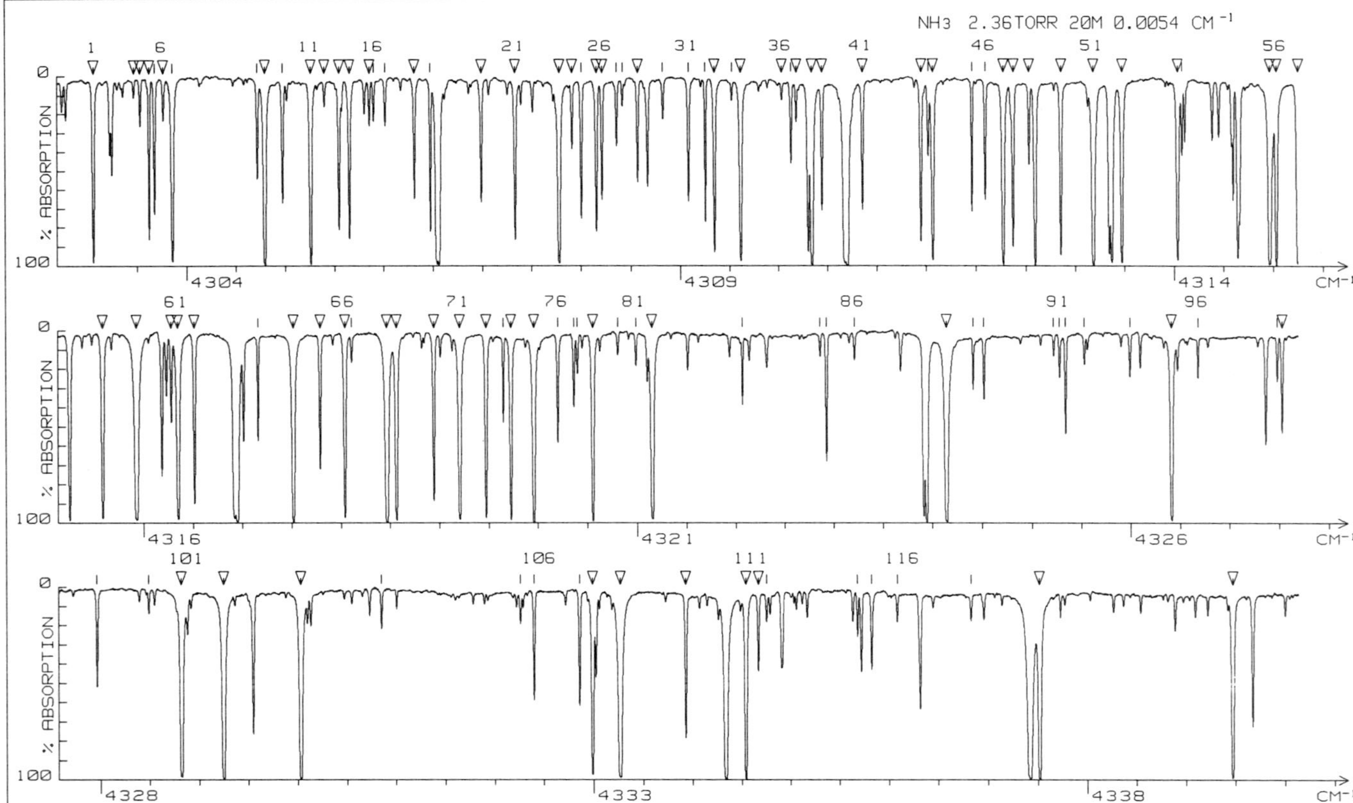

ISOTOPIC IDENTIFICATION – '∇' indicates $^{14}NH_3$; '|': unidentified.

LINE NO.	WAVENUMBER CM^{-1}, VAC.	ROTATIONAL ASSIGNMENTS	BAND ASSIGNMENTS		ISOTOPIC SPECIES
1	4303.05543 (18)	$s^{r}P$(5, 1)	NU2+NU3		$^{14}NH_3$
2	4303.46631 (98)	$a^{r}R$(5, 1)	NU2+2NU4	pe	$^{14}NH_3$
3	4303.53141 (42)	sQ(10, 6)	NU1+NU2		$^{14}NH_3$
4	4303.62373 (17)	$a^{P}P$(10, 10)	NU2+NU3		$^{14}NH_3$
5	4303.67796 (20)				
6	4303.76484 (49)	sQ(10, 7)	NU1+NU2		$^{14}NH_3$
7	4303.85742 (19)				
8	4304.71974 (26)				
9	4304.79450 (19)	$a^{r}P$(6, 0)	NU2+NU3		$^{14}NH_3$
10	4304.97416 (24)				
11	4305.25726 (18)	sQ(9, 1)	NU1+NU2		$^{14}NH_3$
11		$s^{P}P$(6, 2)	NU2+NU3		$^{14}NH_3$
12	4305.39696 (69)	sQ(9, 3)	NU1+NU2		$^{14}NH_3$
13	4305.54487 (20)	$a^{r}P$(5, 3)	NU2+NU3		$^{14}NH_3$
14	4305.64976 (22)	sQ(10, 9)	NU1+NU2		$^{14}NH_3$
15	4305.85621 (44)	sQ(9, 5)	NU1+NU2		$^{14}NH_3$
16	4305.90004 (45)				
17	4306.01820 (44)				
18	4306.31436 (24)	sQ(9, 6)	NU1+NU2		$^{14}NH_3$
19	4306.47521 (17)				
20	4306.98962 (22)	sQ(9, 7)	NU1+NU2		$^{14}NH_3$
21	4307.33620 (20)	sQ(10, 10)	NU1+NU2		$^{14}NH_3$
22	4307.78267 (18)	$s^{P}P$(7, 5)	NU2+NU3		$^{14}NH_3$
23	4307.91293 (32)	sQ(8, 3)	NU1+NU2		$^{14}NH_3$
23		$s^{r}R$(7, 1)	NU2+2NU4	pe	$^{14}NH_3$
24	4308.00562 (22)				
25	4308.15614 (22)	sQ(9, 8)	NU1+NU2		$^{14}NH_3$
26	4308.21217 (22)	sQ(8, 4)	NU1+NU2		$^{14}NH_3$
27	4308.35776 (31)				
28	4308.41655 (68)				
29	4308.57275 (24)	sR(5, 2)	NU2+2NU4	pa	$^{14}NH_3$
30	4308.82703 (50)				
31	4309.08437 (22)				
32	4309.25472 (17)				
33	4309.34677 (19)	sQ(8, 6)	NU1+NU2		$^{14}NH_3$
34	4309.52180 (87)				
35	4309.61400 (19)	sQ(9, 9)	NU1+NU2		$^{14}NH_3$
36	4310.03636 (101)	sQ(7, 1)	NU1+NU2		$^{14}NH_3$
37	4310.13016 (28)				
38	4310.18456 (47)	sQ(7, 2)	NU1+NU2		$^{14}NH_3$
39	4310.34325 (19)	$a^{P}P$(8, 6)	NU2+NU3		$^{14}NH_3$
40	4310.44595 (23)	sQ(7, 3)	NU1+NU2		$^{14}NH_3$
41	4310.85682 (23)	sQ(7, 4)	NU1+NU2		$^{14}NH_3$
42	4311.44960 (20)	sQ(7, 5)	NU1+NU2		$^{14}NH_3$
43	4311.52226 (46)				
44	4311.56830 (19)	sQ(8, 8)	NU1+NU2		$^{14}NH_3$
45	4311.96211 (17)				
46	4312.09519 (22)				
47	4312.27383 (18)	sQ(7, 6)	NU1+NU2		$^{14}NH_3$
48	4312.37154 (24)	sQ(6, 1)	NU1+NU2		$^{14}NH_3$
48		$a^{r}P$(5, 2)	NU2+NU3		$^{14}NH_3$
49	4312.53181 (29)	sQ(6, 2)	NU1+NU2		$^{14}NH_3$
50	4312.85580 (19)	sQ(6, 3)	NU1+NU2		$^{14}NH_3$
51	4313.18241 (19)	$s^{P}P$(6, 3)	NU2+NU3		$^{14}NH_3$
52	4313.47278 (18)	$a^{r}R$(6, 2)	NU2+2NU4	pe	$^{14}NH_3$
52		aR(0, 0)	NU1+NU2		$^{14}NH_3$
53	4314.03474 (19)	sQ(6, 5)	NU1+NU2		$^{14}NH_3$
54	4314.07873 (33)				
55	4314.96510 (19)	sQ(6, 6)	NU1+NU2		$^{14}NH_3$
56	4315.03121 (19)	sQ(5, 3)	NU1+NU2		$^{14}NH_3$
57	4315.24503 (19)	$a^{P}P$(7, 4)	NU2+NU3		$^{14}NH_3$
58	4315.58251 (19)	sQ(5, 4)	NU1+NU2		$^{14}NH_3$
59	4315.92718 (20)	$s^{P}P$(7, 6)	NU2+NU3		$^{14}NH_3$
60	4316.27904 (29)	sQ(4, 1)	NU1+NU2		$^{14}NH_3$

- "LINE NO.": serial number identifying the spectral line in the map on the opposite page.
- "WAVENUMBER": experimental value.
- Uncertainty in the last digit of the wavenumber is given in parentheses.
- "pa" : parallel ; "pe" : perpendicular.

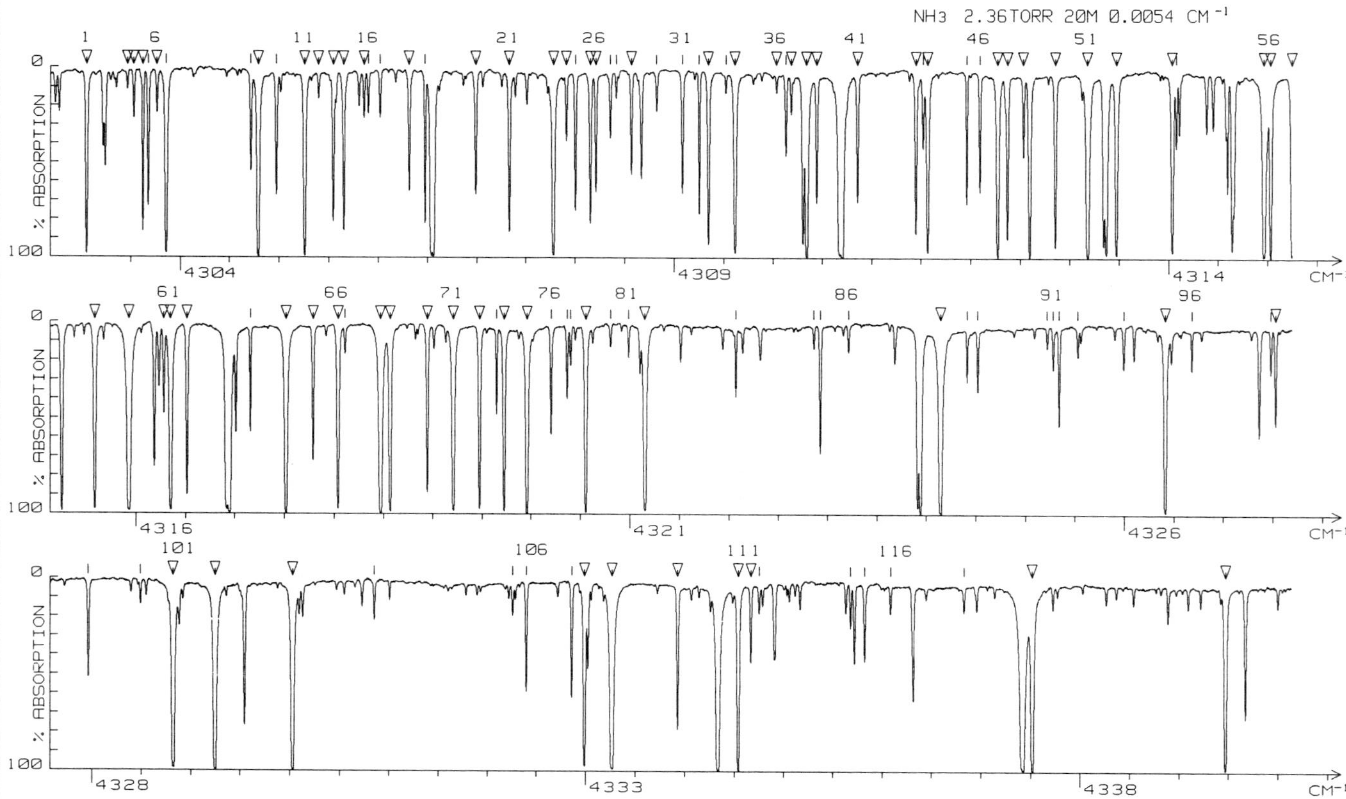

ISOTOPIC IDENTIFICATION – '∇' indicates $^{14}NH_3$; '|': unidentified.

LINE NO.	WAVENUMBER CM^{-1}, VAC.	ROTATIONAL ASSIGNMENTS	BAND ASSIGNMENTS	ISOTOPIC SPECIES
61	4316.34715(18)	sQ(5, 5)	NU1+NU2	$^{14}NH_3$
62	4316.51220(22)	sQ(4, 2)	NU1+NU2	$^{14}NH_3$
63	4317.15931(22)			
64	4317.51486(19)	sQ(4, 4)	NU1+NU2	$^{14}NH_3$
65	4317.78684(23)	sQ(3, 1)	NU1+NU2	$^{14}NH_3$
66	4318.03632(19)	sQ(3, 2)	NU1+NU2	$^{14}NH_3$
67	4318.10572(76)			
68	4318.46768(19)	sQ(3, 3)	NU1+NU2	$^{14}NH_3$
69	4318.56522(18)	sPP(5, 1)	NU2+NU3	$^{14}NH_3$
70	4318.94308(19)	sQ(2, 1)	NU1+NU2	$^{14}NH_3$
71	4319.20467(19)	sQ(2, 2)	NU1+NU2	$^{14}NH_3$
72	4319.47041(18)	arP(5, 1)	NU2+NU3	$^{14}NH_3$
73	4319.64907(24)			
74	4319.72564(19)	sQ(1, 1)	NU1+NU2	$^{14}NH_3$
75	4319.95873(18)	aPP(8, 7)	NU2+NU3	$^{14}NH_3$
76	4320.20313(25)			
77	4320.36464(30)			
78	4320.40120(51)			
79	4320.55381(18)	aPP(6, 2)	NU2+NU3	$^{14}NH_3$
80	4320.80401(88)			
81	4320.98717(60)			
82	4321.15250(19)	sPP(6, 4)	NU2+NU3	$^{14}NH_3$
83	4322.06606(20)			
84	4322.86255(82)			
85	4322.92694(21)			
86	4323.21011(72)			
87	4324.14300(19)	srR(7, 0)	NU2+2NU4 pe	$^{14}NH_3$
87		sPP(7, 7)	NU2+NU3	$^{14}NH_3$
88	4324.41015(32)			
89	4324.51862(35)			
90	4325.21930(86)			
91	4325.27952(46)			
92	4325.34014(24)			
93	4325.52899(67)			
94	4325.99225(44)			
95	4326.41018(18)	sPP(5, 2)	NU2+NU3	$^{14}NH_3$
96	4326.68106(39)			
97	4327.47776(42)			
98	4327.52629(27)	sPR(6, 3)	NU2+2NU4 pe	$^{14}NH_3$
99	4327.96073(26)			
100	4328.48723(77)			
101	4328.81650(19)	aPP(6, 3)	NU2+NU3	$^{14}NH_3$
102	4329.24037(19)	sPP(6, 5)	NU2+NU3	$^{14}NH_3$
103	4330.02535(19)	aPP(8, 8)	NU2+NU3	$^{14}NH_3$
104	4330.84830(49)			
105	4332.25877(65)			
106	4332.39690(25)			
107	4332.85700(22)			
108	4332.98576(19)	aR(1, 1)	NU1+NU2	$^{14}NH_3$
109	4333.26227(20)	aPP(7, 6)	NU2+NU3	$^{14}NH_3$
110	4333.92587(28)	arP(4, 2)	NU2+NU3	$^{14}NH_3$
111	4334.53569(18)	aPP(5, 1)	NU2+NU3	$^{14}NH_3$
112	4334.66288(31)	sR(6, 2)	NU2+2NU4 pa	$^{14}NH_3$
113	4334.74618(71)			
114	4335.67573(46)			
115	4335.81999(30)			
116	4336.07870(69)			
117	4336.82482(77)			
118	4337.51123(19)	aPP(6, 4)	NU2+NU3	$^{14}NH_3$
118		srQ(10, 8)	NU2+NU3	$^{14}NH_3$
119	4339.46864(18)	sPP(4, 1)	NU2+NU3	$^{14}NH_3$

- "LINE NO.": serial number identifying the spectral line in the map on the opposite page.
- "WAVENUMBER": experimental value.
- Uncertainty in the last digit of the wavenumber is given in parentheses.
- "pa" : parallel ; "pe" : perpendicular.

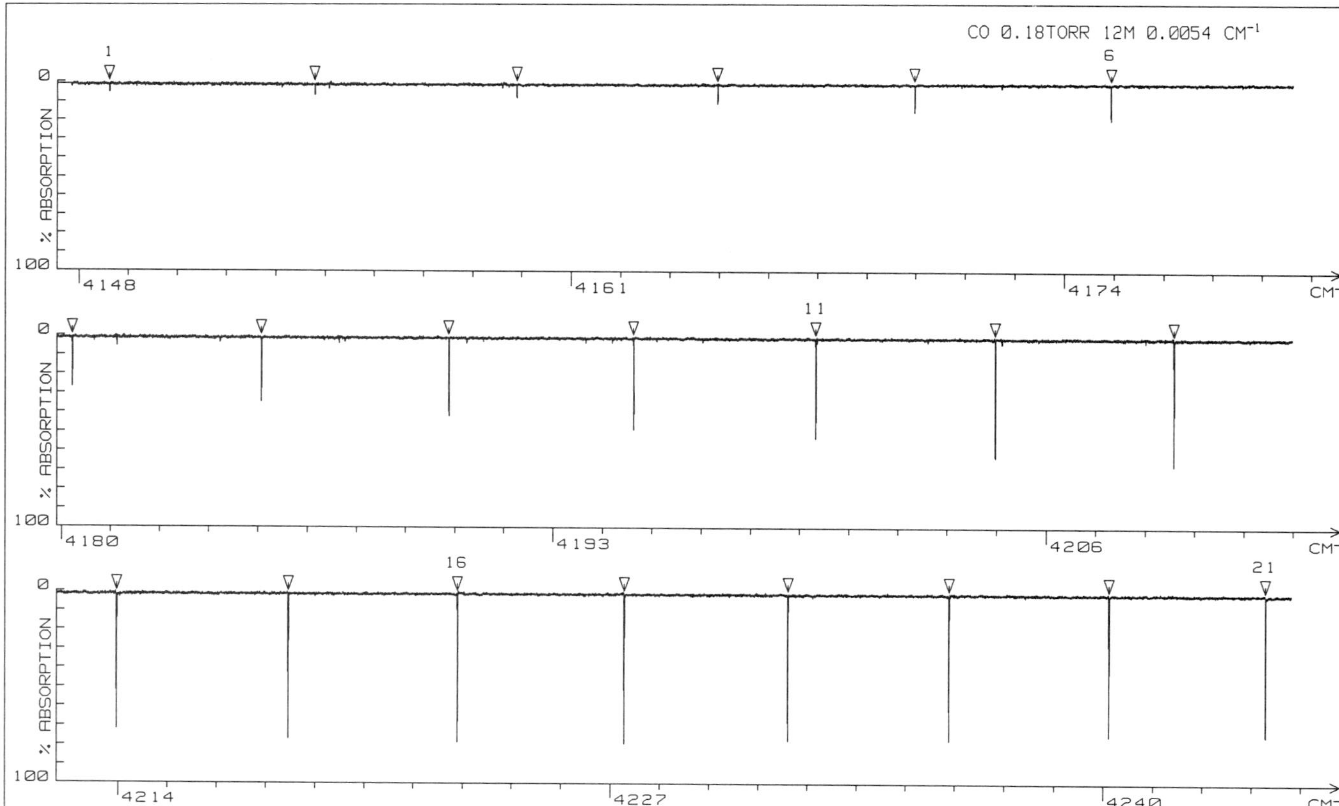

ISOTOPIC IDENTIFICATION – '∇' indicates $^{12}C^{16}O$

LINE NO.	WAVENUMBER CM^{-1}, VAC.	ROTATIONAL ASSIGNMENT	BAND	ISOTOPIC SPECIES
1	4148.7970	P(24)	2-0	$^{12}C^{16}O$
2	4154.2117	P(23)	2-0	$^{12}C^{16}O$
3	4159.5597	P(22)	2-0	$^{12}C^{16}O$
4	4164.8410	P(21)	2-0	$^{12}C^{16}O$
5	4170.0554	P(20)	2-0	$^{12}C^{16}O$
6	4175.2027	P(19)	2-0	$^{12}C^{16}O$
7	4180.2828	P(18)	2-0	$^{12}C^{16}O$
8	4185.2955	P(17)	2-0	$^{12}C^{16}O$
9	4190.2407	P(16)	2-0	$^{12}C^{16}O$
10	4195.1182	P(15)	2-0	$^{12}C^{16}O$
11	4199.9280	P(14)	2-0	$^{12}C^{16}O$
12	4204.6698	P(13)	2-0	$^{12}C^{16}O$
13	4209.3435	P(12)	2-0	$^{12}C^{16}O$
14	4213.9489	P(11)	2-0	$^{12}C^{16}O$
15	4218.4860	P(10)	2-0	$^{12}C^{16}O$
16	4222.9545	P(9)	2-0	$^{12}C^{16}O$
17	4227.3544	P(8)	2-0	$^{12}C^{16}O$
18	4231.6853	P(7)	2-0	$^{12}C^{16}O$
19	4235.9474	P(6)	2-0	$^{12}C^{16}O$
20	4240.1402	P(5)	2-0	$^{12}C^{16}O$
21	4244.2639	P(4)	2-0	$^{12}C^{16}O$

- "LINE NO.": serial number identifying the spectral line in the map on the opposite page.
- "WAVENUMBER": calculated value. Uncertainty: 0.001 CM-1

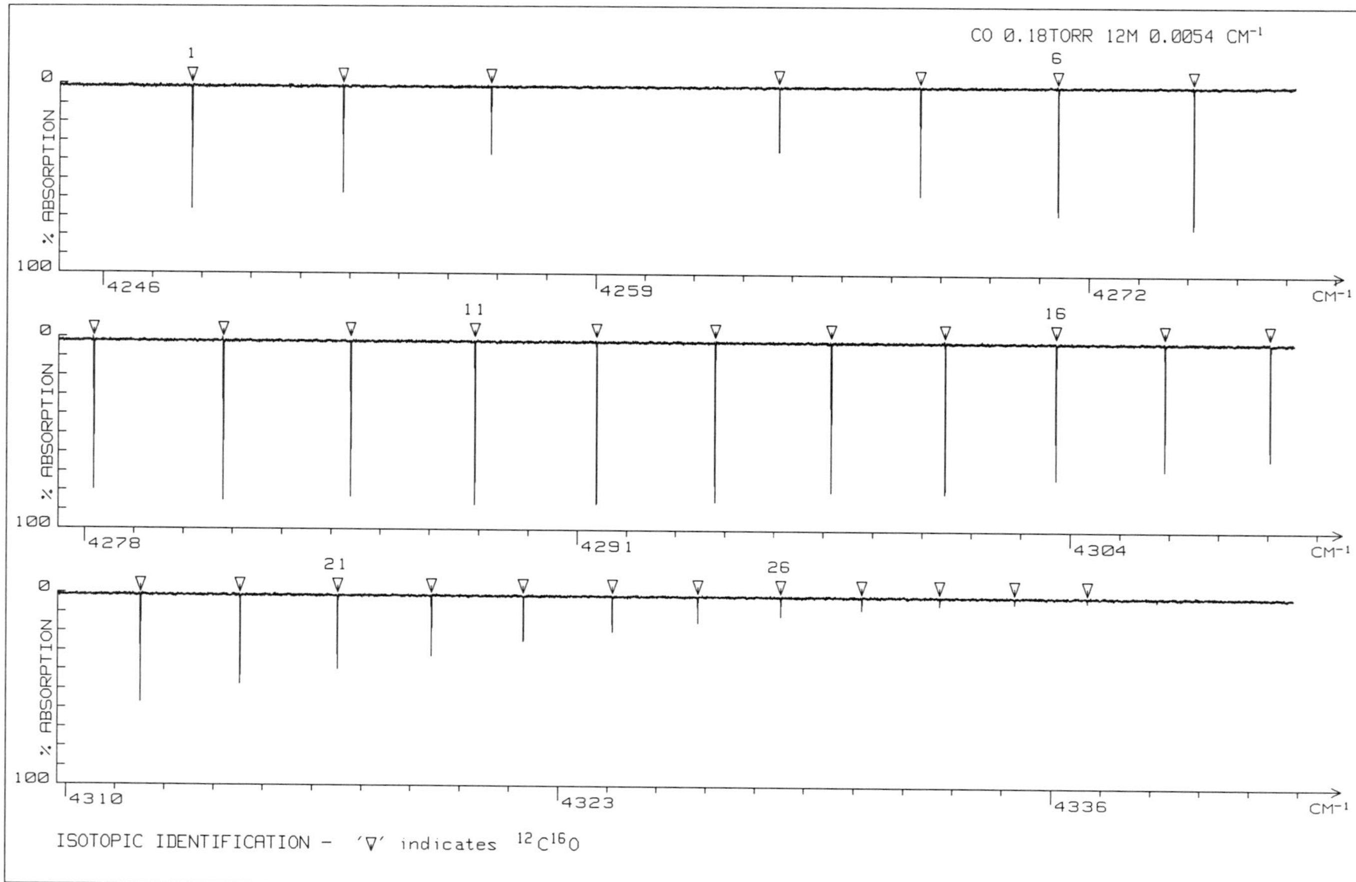
CO 0.18TORR 12M 0.0054 CM⁻¹
1
6
11
16
21
26
0
% ABSORPTION
100
4246
4259
4272
4278
4291
4304
4310
4323
4336
CM⁻¹
ISOTOPIC IDENTIFICATION - '∇' indicates 12C16O

LINE NO.	WAVENUMBER CM^{-1}, VAC.	ROTATIONAL ASSIGNMENT	BAND	ISOTOPIC SPECIES
1	4248.3180	P(3)	2-0	$^{12}C^{16}O$
2	4252.3027	P(2)	2-0	$^{12}C^{16}O$
3	4256.2176	P(1)	2-0	$^{12}C^{16}O$
4	4263.8376	R(0)	2-0	$^{12}C^{16}O$
5	4267.5425	R(1)	2-0	$^{12}C^{16}O$
6	4271.1770	R(2)	2-0	$^{12}C^{16}O$
7	4274.7411	R(3)	2-0	$^{12}C^{16}O$
8	4278.2347	R(4)	2-0	$^{12}C^{16}O$
9	4281.6574	R(5)	2-0	$^{12}C^{16}O$
10	4285.0093	R(6)	2-0	$^{12}C^{16}O$
11	4288.2901	R(7)	2-0	$^{12}C^{16}O$
12	4291.4998	R(8)	2-0	$^{12}C^{16}O$
13	4294.6381	R(9)	2-0	$^{12}C^{16}O$
14	4297.7050	R(10)	2-0	$^{12}C^{16}O$
15	4300.7002	R(11)	2-0	$^{12}C^{16}O$
16	4303.6236	R(12)	2-0	$^{12}C^{16}O$
17	4306.4752	R(13)	2-0	$^{12}C^{16}O$
18	4309.2547	R(14)	2-0	$^{12}C^{16}O$
19	4311.9619	R(15)	2-0	$^{12}C^{16}O$
20	4314.5968	R(16)	2-0	$^{12}C^{16}O$
21	4317.1593	R(17)	2-0	$^{12}C^{16}O$
22	4319.6490	R(18)	2-0	$^{12}C^{16}O$
23	4322.0660	R(19)	2-0	$^{12}C^{16}O$
24	4324.4100	R(20)	2-0	$^{12}C^{16}O$
25	4326.6810	R(21)	2-0	$^{12}C^{16}O$
26	4328.8787	R(22)	2-0	$^{12}C^{16}O$
27	4331.0030	R(23)	2-0	$^{12}C^{16}O$
28	4333.0538	R(24)	2-0	$^{12}C^{16}O$
29	4335.0310	R(25)	2-0	$^{12}C^{16}O$
30	4336.9343	R(26)	2-0	$^{12}C^{16}O$

- "LINE NO.": serial number identifying the spectral line in the map on the opposite page.
- "WAVENUMBER": calculated value. Uncertainty: 0.001 CM-1

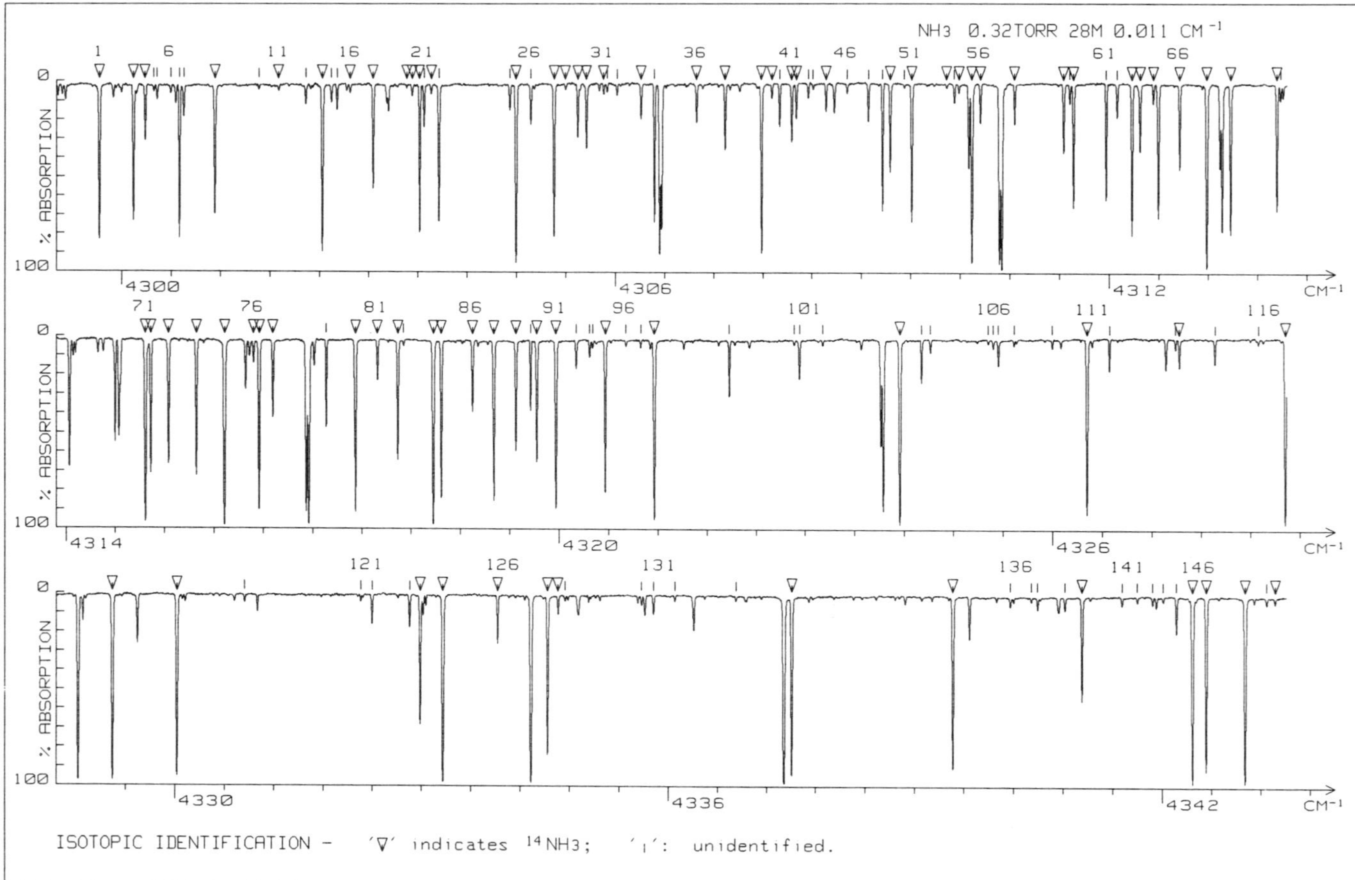
NH3 0.32TORR 28M 0.011 CM⁻¹
% ABSORPTION
0
100
4300
4306
4312
4314
4320
4326
4330
4336
4342
CM⁻¹
ISOTOPIC IDENTIFICATION - '∇' indicates ¹⁴NH3; '|': unidentified.

LINE NO.	WAVENUMBER CM^{-1}, VAC.	ROTATIONAL ASSIGNMENTS	BAND ASSIGNMENTS		ISOTOPIC SPECIES
1	4299.72541(18)	sPP(7, 4)	NU2+NU3		$^{14}NH_3$
2	4300.14092(19)	sP(1, 0)	NU1+NU2		$^{14}NH_3$
3	4300.28384(20)	sR(5, 0)	NU2+2NU4	pa	$^{14}NH_3$
4	4300.38818(74)				
5	4300.42861(30)				
6	4300.59254(43)				
7	4300.70033(17)				
8	4300.75086(24)				
9	4301.13137(19)	aPP(8, 5)	NU2+NU3		$^{14}NH_3$
10	4301.66546(82)				
11	4301.90344(59)	aR(7, 1)	NU2+2NU4	pa	$^{14}NH_3$
12	4302.23219(32)				
13	4302.43325(18)	sPP(8, 7)	NU2+NU3		$^{14}NH_3$
14	4302.54170(29)				
15	4302.61079(25)				
16	4302.77345(51)	sQ(10, 8)	NU1+NU2		$^{14}NH_3$
17	4303.05543(18)	srP(5, 1)	NU2+NU3		$^{14}NH_3$
18	4303.46631(98)	arR(5, 1)	NU2+2NU4	pe	$^{14}NH_3$
19	4303.53141(42)	sQ(10, 6)	NU1+NU2		$^{14}NH_3$
20	4303.62373(17)	aPP(10, 10)	NU2+NU3		$^{14}NH_3$
21	4303.67796(20)				
22	4303.76484(49)	sQ(10, 7)	NU1+NU2		$^{14}NH_3$
23	4303.85742(19)				
24	4304.71974(26)				
25	4304.79450(19)	arP(6, 0)	NU2+NU3		$^{14}NH_3$
26	4304.97416(24)				
27	4305.25726(18)	sQ(9, 1)	NU1+NU2		$^{14}NH_3$
27		sPP(6, 2)	NU2+NU3		$^{14}NH_3$
28	4305.39696(69)	sQ(9, 3)	NU1+NU2		$^{14}NH_3$
29	4305.54487(20)	arP(5, 3)	NU2+NU3		$^{14}NH_3$
30	4305.64976(22)	sQ(10, 9)	NU1+NU2		$^{14}NH_3$
31	4305.85621(44)	sQ(9, 5)	NU1+NU2		$^{14}NH_3$
32	4305.90004(45)				
33	4306.01820(44)				
34	4306.31436(24)	sQ(9, 6)	NU1+NU2		$^{14}NH_3$
35	4306.47521(17)				
36	4306.98962(22)	sQ(9, 7)	NU1+NU2		$^{14}NH_3$
37	4307.33620(20)	sQ(10, 10)	NU1+NU2		$^{14}NH_3$
38	4307.78267(18)	sPP(7, 5)	NU2+NU3		$^{14}NH_3$
39	4307.91293(32)	sQ(8, 3)	NU1+NU2		$^{14}NH_3$
39		srR(7, 1)	NU2+2NU4	pe	$^{14}NH_3$
40	4308.00562(22)				
41	4308.15614(22)	sQ(9, 8)	NU1+NU2		$^{14}NH_3$
42	4308.21217(22)	sQ(8, 4)	NU1+NU2		$^{14}NH_3$
43	4308.35776(31)				
44	4308.41655(68)				
45	4308.57275(24)	sR(5, 2)	NU2+2NU4	pa	$^{14}NH_3$
46	4308.82703(50)				
47	4309.08437(22)				
48	4309.25472(17)				
49	4309.34677(19)	sQ(8, 6)	NU1+NU2		$^{14}NH_3$
50	4309.52180(87)				
51	4309.61400(19)	sQ(9, 9)	NU1+NU2		$^{14}NH_3$
52	4310.03636(101)	sQ(7, 1)	NU1+NU2		$^{14}NH_3$
53	4310.13016(28)				
54	4310.18456(47)	sQ(7, 2)	NU1+NU2		$^{14}NH_3$
55	4310.34325(19)	aPP(8, 6)	NU2+NU3		$^{14}NH_3$
56	4310.44595(23)	sQ(7, 3)	NU1+NU2		$^{14}NH_3$
57	4310.85682(23)	sQ(7, 4)	NU1+NU2		$^{14}NH_3$
58	4311.44960(20)	sQ(7, 5)	NU1+NU2		$^{14}NH_3$
59	4311.52226(46)				
60	4311.56830(19)	sQ(8, 8)	NU1+NU2		$^{14}NH_3$

- "LINE NO.": serial number identifying the spectral line in the map on the opposite page.
- "WAVENUMBER": experimental value.
- Uncertainty in the last digit of the wavenumber is given in parentheses.
- "pa" : parallel ; "pe" : perpendicular.

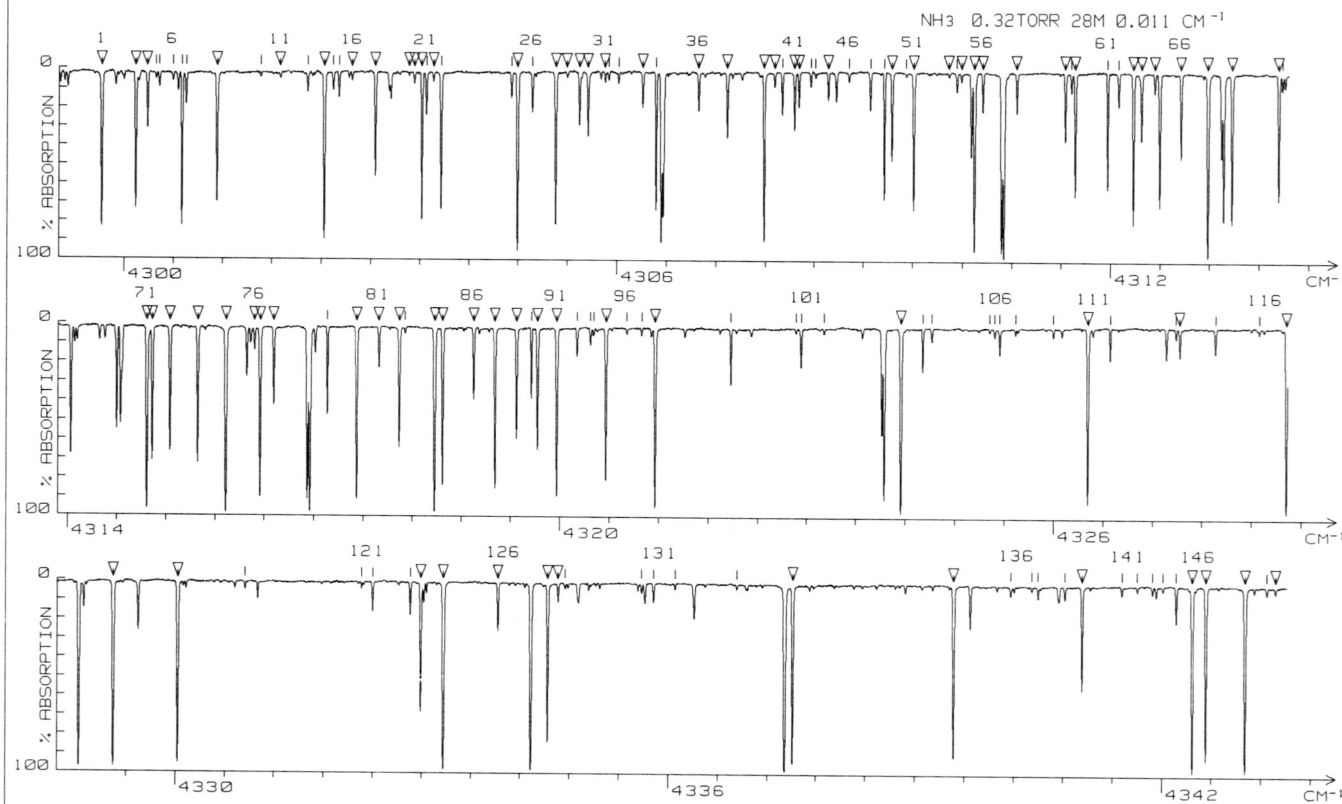

ISOTOPIC IDENTIFICATION – '▽' indicates $^{14}NH_3$; '|': unidentified.

LINE NO.	WAVENUMBER CM^{-1}, VAC.	ROTATIONAL ASSIGNMENTS	BAND		ISOTOPIC SPECIES
61	4311.96211(17)				
62	4312.09519(22)				
63	4312.27383(18)	sQ(7, 6)	NU1+NU2		$^{14}NH_3$
64	4312.37154(24)	sQ(6, 1)	NU1+NU2		$^{14}NH_3$
64		a^{r}P(5, 2)	NU2+NU3		$^{14}NH_3$
65	4312.53181(29)	sQ(6, 2)	NU1+NU2		$^{14}NH_3$
66	4312.85580(19)	sQ(6, 3)	NU1+NU2		$^{14}NH_3$
67	4313.18241(19)	s^{P}P(6, 3)	NU2+NU3		$^{14}NH_3$
68	4313.47278(18)	a^{r}R(6, 2)	NU2+2NU4	pe	$^{14}NH_3$
68		aR(0, 0)	NU1+NU2		$^{14}NH_3$
69	4314.03474(19)	sQ(6, 5)	NU1+NU2		$^{14}NH_3$
70	4314.07873(33)				
71	4314.96510(19)	sQ(6, 6)	NU1+NU2		$^{14}NH_3$
72	4315.03121(19)	sQ(5, 3)	NU1+NU2		$^{14}NH_3$
73	4315.24503(19)	a^{P}P(7, 4)	NU2+NU3		$^{14}NH_3$
74	4315.58251(19)	sQ(5, 4)	NU1+NU2		$^{14}NH_3$
75	4315.92718(20)	s^{P}P(7, 6)	NU2+NU3		$^{14}NH_3$
76	4316.27904(29)	sQ(4, 1)	NU1+NU2		$^{14}NH_3$
77	4316.34715(18)	sQ(5, 5)	NU1+NU2		$^{14}NH_3$
78	4316.51220(22)	sQ(4, 2)	NU1+NU2		$^{14}NH_3$
79	4317.15931(22)				
80	4317.51486(19)	sQ(4, 4)	NU1+NU2		$^{14}NH_3$
81	4317.78684(23)	sQ(3, 1)	NU1+NU2		$^{14}NH_3$
82	4318.03632(19)	sQ(3, 2)	NU1+NU2		$^{14}NH_3$
83	4318.10572(76)				
84	4318.46768(19)	sQ(3, 3)	NU1+NU2		$^{14}NH_3$
85	4318.56522(18)	s^{P}P(5, 1)	NU2+NU3		$^{14}NH_3$
86	4318.94308(19)	sQ(2, 1)	NU1+NU2		$^{14}NH_3$
87	4319.20467(19)	sQ(2, 2)	NU1+NU2		$^{14}NH_3$
88	4319.47041(18)	a^{r}P(5, 1)	NU2+NU3		$^{14}NH_3$
89	4319.64907(24)				
90	4319.72564(19)	sQ(1, 1)	NU1+NU2		$^{14}NH_3$

LINE NO.	WAVENUMBER CM^{-1}, VAC.	ROTATIONAL ASSIGNMENTS	BAND		ISOTOPIC SPECIES
91	4319.95873(18)	a^{P}P(8, 7)	NU2+NU3		$^{14}NH_3$
92	4320.20313(25)				
93	4320.36464(30)				
94	4320.40120(51)				
95	4320.55381(18)	a^{P}P(6, 2)	NU2+NU3		$^{14}NH_3$
96	4320.80401(88)				
97	4320.98717(60)				
98	4321.15250(19)	s^{P}P(6, 4)	NU2+NU3		$^{14}NH_3$
99	4322.06606(20)				
100	4322.86255(82)				
101	4322.92694(21)				
102	4323.21011(72)				
103	4324.14300(19)	s^{r}R(7, 0)	NU2+2NU4	pe	$^{14}NH_3$
103		s^{P}P(7, 7)	NU2+NU3		$^{14}NH_3$
104	4324.41015(32)				
105	4324.51862(35)				
106	4325.21930(86)				
107	4325.27952(46)				
108	4325.34014(24)				
109	4325.52899(67)				
110	4325.99225(44)				
111	4326.41018(18)	s^{P}P(5, 2)	NU2+NU3		$^{14}NH_3$
112	4326.68106(39)				
113	4327.47776(42)				
114	4327.52629(27)	s^{P}R(6, 3)	NU2+2NU4	pe	$^{14}NH_3$
115	4327.96073(26)				
116	4328.48723(77)				
117	4328.81650(19)	a^{P}P(6, 3)	NU2+NU3		$^{14}NH_3$
118	4329.24037(19)	s^{P}P(6, 5)	NU2+NU3		$^{14}NH_3$
119	4330.02535(19)	a^{P}P(8, 8)	NU2+NU3		$^{14}NH_3$
120	4330.84830(49)				

- "LINE NO.": serial number identifying the spectral line in the map on the opposite page.
- "WAVENUMBER": experimental value.
- Uncertainty in the last digit of the wavenumber is given in parentheses.
- "pa" : parallel ; "pe" : perpendicular.

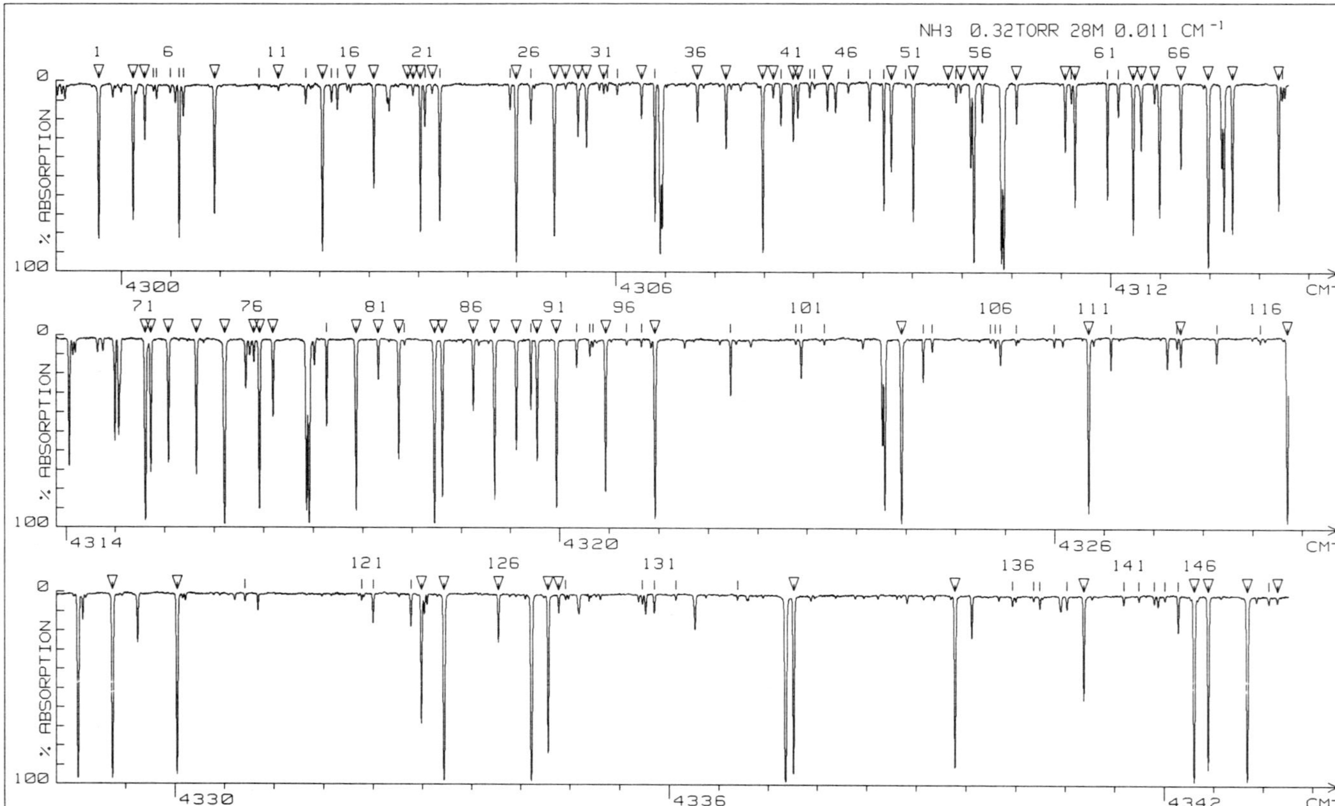

ISOTOPIC IDENTIFICATION – '∇' indicates $^{14}NH_3$; '|': unidentified.

LINE NO.	WAVENUMBER CM^{-1}, VAC.	ROTATIONAL ASSIGNMENTS	BAND		ISOTOPIC SPECIES
121	4332.25877(65)				
122	4332.39690(25)				
123	4332.85700(22)				
124	4332.98576(19)	aR(1, 1)	NU1+NU2		$^{14}NH_3$
125	4333.26227(20)	a^pP(7, 6)	NU2+NU3		$^{14}NH_3$
126	4333.92587(28)	a^rP(4, 2)	NU2+NU3		$^{14}NH_3$
127	4334.53569(18)	a^pP(5, 1)	NU2+NU3		$^{14}NH_3$
128	4334.66288(31)	sR(6, 2)	NU2+2NU4	pa	$^{14}NH_3$
129	4334.74618(71)				
130	4335.67573(46)				
131	4335.81999(30)				
132	4336.07870(69)				
133	4336.82482(77)				
134	4337.51123(19)	a^pP(6, 4)	NU2+NU3		$^{14}NH_3$
134		s^rQ(10, 8)	NU2+NU3		$^{14}NH_3$
135	4339.46864(18)	s^pP(4, 1)	NU2+NU3		$^{14}NH_3$
136	4340.16041(53)				
137	4340.41408(76)				
138	4340.48882(45)				
139	4340.82164(37)				
140	4341.02682(19)	a^rP(4, 1)	NU2+NU3		$^{14}NH_3$
141	4341.51088(56)				
142	4341.69611(65)				
143	4341.87972(58)				
144	4342.00801(71)				
145	4342.16885(24)				
146	4342.36679(19)	s^pP(5, 4)	NU2+NU3		$^{14}NH_3$
147	4342.53381(18)	a^pP(5, 2)	NU2+NU3		$^{14}NH_3$
148	4343.00708(19)	a^pP(7, 7)	NU2+NU3		$^{14}NH_3$
149	4343.26722(59)				
150	4343.37093(72)	a^rR(6, 0)	NU2+2NU4	pe	$^{14}NH_3$

- "LINE NO.": serial number identifying the spectral line in the map on the opposite page.
- "WAVENUMBER": experimental value.
- Uncertainty in the last digit of the wavenumber is given in parentheses.
- "pa" : parallel ; "pe" : perpendicular.

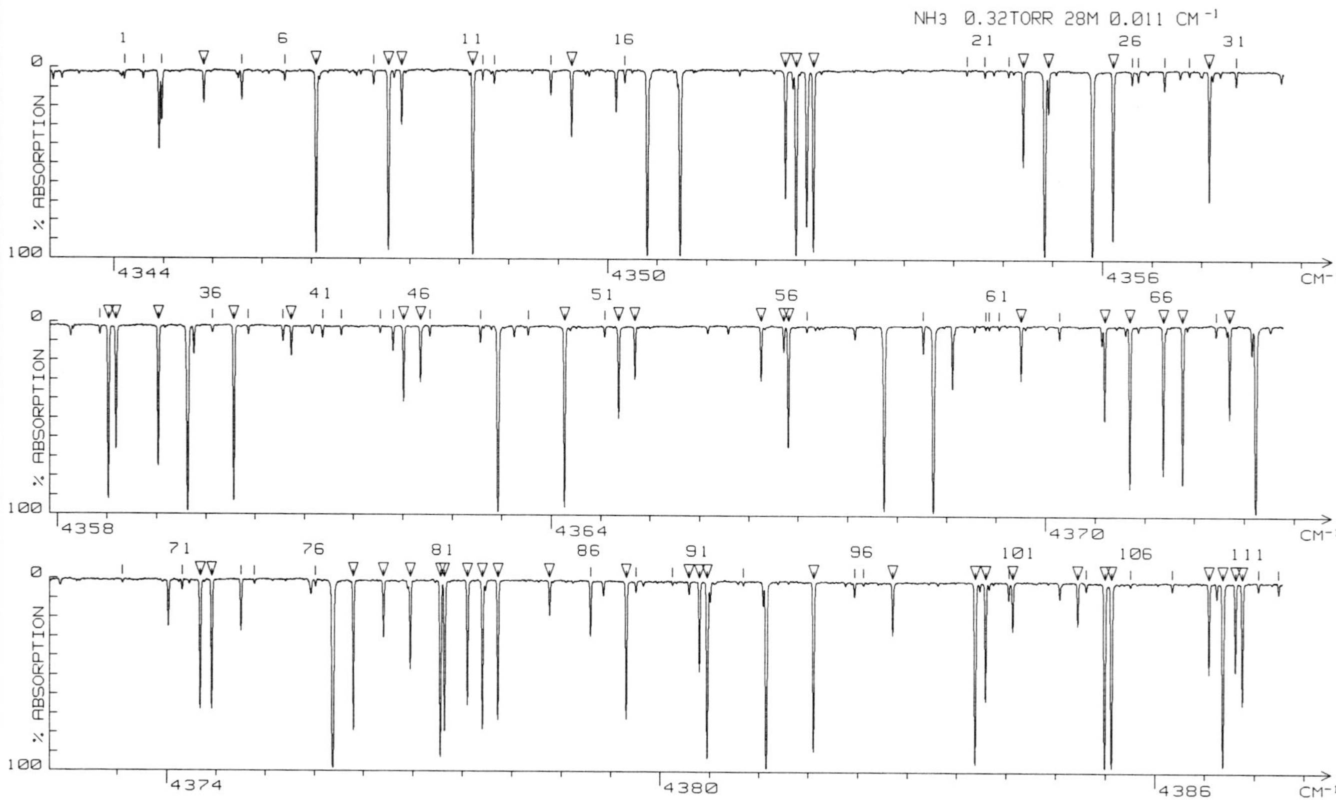

ISOTOPIC IDENTIFICATION – '▽' indicates $^{14}NH_3$; '|': unidentified.

LINE NO.	WAVENUMBER CM^{-1}, VAC.	ROTATIONAL ASSIGNMENTS	BAND		ISOTOPIC SPECIES
1	4344.12699(60)				
2	4344.35320(66)				
3	4344.57843(28)				
4	4345.09054(24)	s^rQ(10, 7)	NU2+NU3		$^{14}NH_3$
4		a^PR(5, 2)	NU2+2NU4	pe	$^{14}NH_3$
5	4345.55681(25)				
6	4346.07552(47)				
7	4346.45834(19)	a^PP(6, 5)	NU2+NU3		$^{14}NH_3$
8	4347.16218(37)				
9	4347.34152(19)	s^PP(4, 2)	NU2+NU3		$^{14}NH_3$
10	4347.50260(24)	s^rQ(9, 7)	NU2+NU3		$^{14}NH_3$
11	4348.36523(19)	a^rP(4, 0)	NU2+NU3		$^{14}NH_3$
12	4348.48794(45)				
13	4348.62833(38)				
14	4349.30896(26)				
15	4349.56106(20)	s^rQ(8, 7)	NU2+NU3		$^{14}NH_3$
16	4350.21137(41)				
17	4352.16560(19)	aR(2, 2)	NU1+NU2		$^{14}NH_3$
18	4352.29292(19)	s^rP(3, 0)	NU2+NU3		$^{14}NH_3$
19	4352.50231(19)	aR(2, 0)	NU1+NU2		$^{14}NH_3$
20	4354.36019(85)				
21	4354.57912(60)				
22	4354.86653(77)				
23	4355.04291(19)	s^rQ(9, 6)	NU2+NU3		$^{14}NH_3$
24	4355.35092(31)	a^rQ(10, 8)	NU2+NU3		$^{14}NH_3$
25	4356.12682(18)	a^PP(4, 1)	NU2+NU3		$^{14}NH_3$
26	4356.35530(39)				
27	4356.42969(44)				
28	4356.74798(28)				
29	4357.05233(76)				
30	4357.29265(19)	s^rQ(8, 6)	NU2+NU3		$^{14}NH_3$
31	4357.61544(34)				
32	4358.51502(59)				
33	4358.61695(18)	sR(1, 0)	NU1+NU2		$^{14}NH_3$
34	4358.70880(19)	sR(1, 1)	NU1+NU2		$^{14}NH_3$
35	4359.21772(19)	s^rQ(7, 6)	NU2+NU3		$^{14}NH_3$
36	4359.87893(72)				
37	4360.14199(19)	s^PP(3, 1)	NU2+NU3		$^{14}NH_3$
38	4360.31548(49)				
39	4360.73467(36)				
40	4360.83466(24)	a^rQ(10, 7)	NU2+NU3		$^{14}NH_3$
41	4361.21712(98)				
42	4361.44507(45)				
43	4361.92399(64)				
44	4362.07700(28)				
45	4362.20735(19)	a^rP(3, 1)	NU2+NU3		$^{14}NH_3$
46	4362.41179(23)	s^rQ(9, 5)	NU2+NU3		$^{14}NH_3$
47	4362.52396(93)				
48	4363.13732(34)				
49	4363.71638(49)				
50	4364.15820(19)	a^PP(4, 2)	NU2+NU3		$^{14}NH_3$
51	4364.64376(45)				
52	4364.81297(19)	s^rQ(8, 5)	NU2+NU3		$^{14}NH_3$
53	4365.01445(21)	a^rQ(9, 7)	NU2+NU3		$^{14}NH_3$
54	4366.54500(21)	a^rQ(10, 6)	NU2+NU3		$^{14}NH_3$
55	4366.82632(26)	s^rQ(10, 4)	NU2+NU3		$^{14}NH_3$
56	4366.88349(19)	s^rQ(7, 5)	NU2+NU3		$^{14}NH_3$
57	4367.10494(82)				
58	4368.51521(27)				
59	4369.27172(112)				
60	4369.31456(108)				

- "LINE NO.": serial number identifying the spectral line in the map on the opposite page.
- "WAVENUMBER": experimental value.
- Uncertainty in the last digit of the wavenumber is given in parentheses.
- "pa" : parallel ; "pe" : perpendicular.

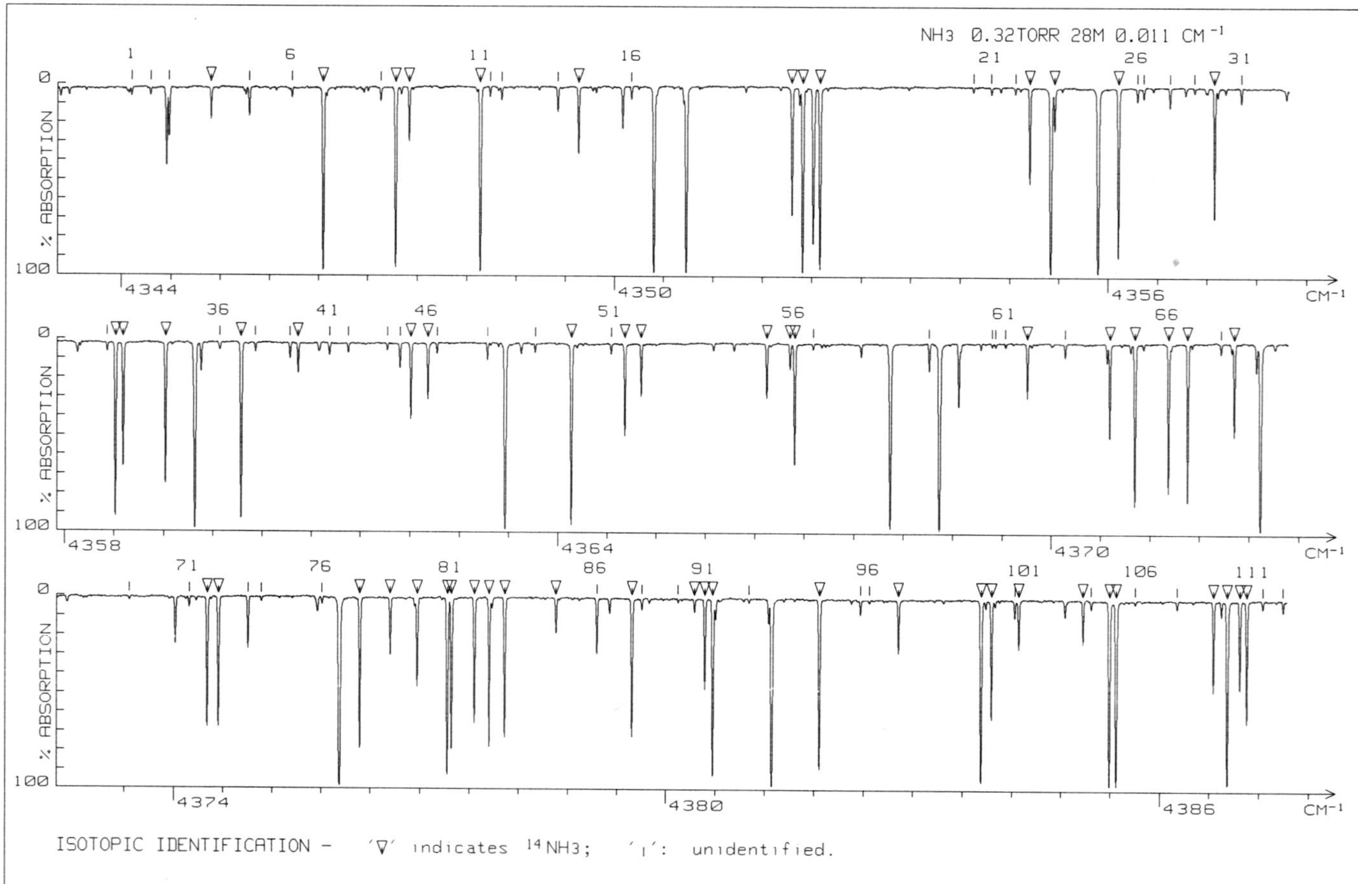
NH3 0.32TORR 28M 0.011 CM-1
% ABSORPTION
0
100
4344
4350
4356
4358
4364
4370
4374
4380
4386
CM-1
ISOTOPIC IDENTIFICATION - '∇' indicates 14NH3; '|': unidentified.

LINE NO.	WAVENUMBER CM^{-1}, VAC.	ROTATIONAL ASSIGNMENTS	BAND ASSIGNMENTS	ISOTOPIC SPECIES
61	4369.43526 (111)			
62	4369.69925 (23)	$s^{r}Q$(9, 4)	NU2+NU3	$^{14}NH_3$
63	4370.16648 (37)			
64	4370.70986 (19)	$a^{r}Q$(9, 6)	NU2+NU3	$^{14}NH_3$
65	4371.01372 (18)	aR(3, 3)	NU1+NU2	$^{14}NH_3$
66	4371.41906 (18)	aR(3, 2)	NU1+NU2	$^{14}NH_3$
67	4371.65564 (18)	aR(3, 1)	NU1+NU2	$^{14}NH_3$
68	4372.05814 (45)			
69	4372.21966 (19)	$s^{r}Q$(8, 4)	NU2+NU3	$^{14}NH_3$
70	4373.46439 (118)			
71	4374.18729 (43)			
72	4374.40725 (19)	$s^{r}Q$(7, 4)	NU2+NU3	$^{14}NH_3$
73	4374.54754 (19)	$a^{r}Q$(8, 6)	NU2+NU3	$^{14}NH_3$
74	4374.90882 (21)			
75	4375.07184 (114)			
76	4375.81092 (66)			
77	4376.27703 (18)	$s^{r}Q$(6, 4)	NU2+NU3	$^{14}NH_3$
78	4376.64491 (21)	$a^{r}Q$(9, 5)	NU2+NU3	$^{14}NH_3$
79	4376.96814 (21)	$s^{r}Q$(9, 3)	NU2+NU3	$^{14}NH_3$
80	4377.33362 (19)	$a^{p}P$(3, 1)	NU2+NU3	$^{14}NH_3$
81	4377.38478 (18)	sR(2, 1)	NU1+NU2	$^{14}NH_3$
82	4377.66247 (19)	sR(2, 2)	NU1+NU2	$^{14}NH_3$
83	4377.84400 (19)	$s^{r}Q$(5, 4)	NU2+NU3	$^{14}NH_3$
84	4378.03477 (19)	$a^{r}Q$(7, 6)	NU2+NU3	$^{14}NH_3$
85	4378.65813 (24)	$a^{r}Q$(10, 4)	NU2+NU3	$^{14}NH_3$
86	4379.15474 (21)			
87	4379.58873 (19)	$s^{r}Q$(8, 3)	NU2+NU3	$^{14}NH_3$
88	4379.70917 (52)			
89	4380.15120 (113)			
90	4380.35050 (41)	$s^{p}R$(8, 3)	NU2+2NU4 pe	$^{14}NH_3$
91	4380.47555 (19)	$a^{r}Q$(8, 5)	NU2+NU3	$^{14}NH_3$
92	4380.57158 (19)	$s^{p}P$(2, 1)	NU2+NU3	$^{14}NH_3$
93	4381.01073 (116)			
94	4381.86757 (18)	$s^{r}Q$(7, 3)	NU2+NU3	$^{14}NH_3$
95	4382.36140 (37)			
96	4382.46914 (136)			
97	4382.82319 (21)	$a^{r}Q$(9, 4)	NU2+NU3	$^{14}NH_3$
98	4383.82752 (19)	$s^{r}Q$(6, 3)	NU2+NU3	$^{14}NH_3$
99	4383.95359 (19)	$a^{r}Q$(7, 5)	NU2+NU3	$^{14}NH_3$
100	4384.23295 (30)			
101	4384.28218 (21)	$s^{r}Q$(9, 2)	NU2+NU3	$^{14}NH_3$
102	4385.06949 (22)	$a^{r}Q$(10, 3)	NU2+NU3	$^{14}NH_3$
103	4385.16894 (56)			
104	4385.39275 (19)	$a^{p}P$(3, 2)	NU2+NU3	$^{14}NH_3$
105	4385.47504 (19)	$s^{r}Q$(5, 3)	NU2+NU3	$^{14}NH_3$
106	4385.70243 (112)			
107	4386.20884 (62)			
108	4386.65595 (19)	$a^{r}Q$(8, 4)	NU2+NU3	$^{14}NH_3$
109	4386.82452 (19)	$s^{r}Q$(4, 3)	NU2+NU3	$^{14}NH_3$
110	4386.97372 (19)	$s^{r}Q$(8, 2)	NU2+NU3	$^{14}NH_3$
111	4387.05639 (19)	$a^{r}Q$(6, 5)	NU2+NU3	$^{14}NH_3$
112	4387.24735 (51)			
113	4387.49159 (43)			

- "LINE NO.": serial number identifying the spectral line in the map on the opposite page.
- "WAVENUMBER": experimental value.
- Uncertainty in the last digit of the wavenumber is given in parentheses.
- "pa" : parallel ; "pe" : perpendicular.

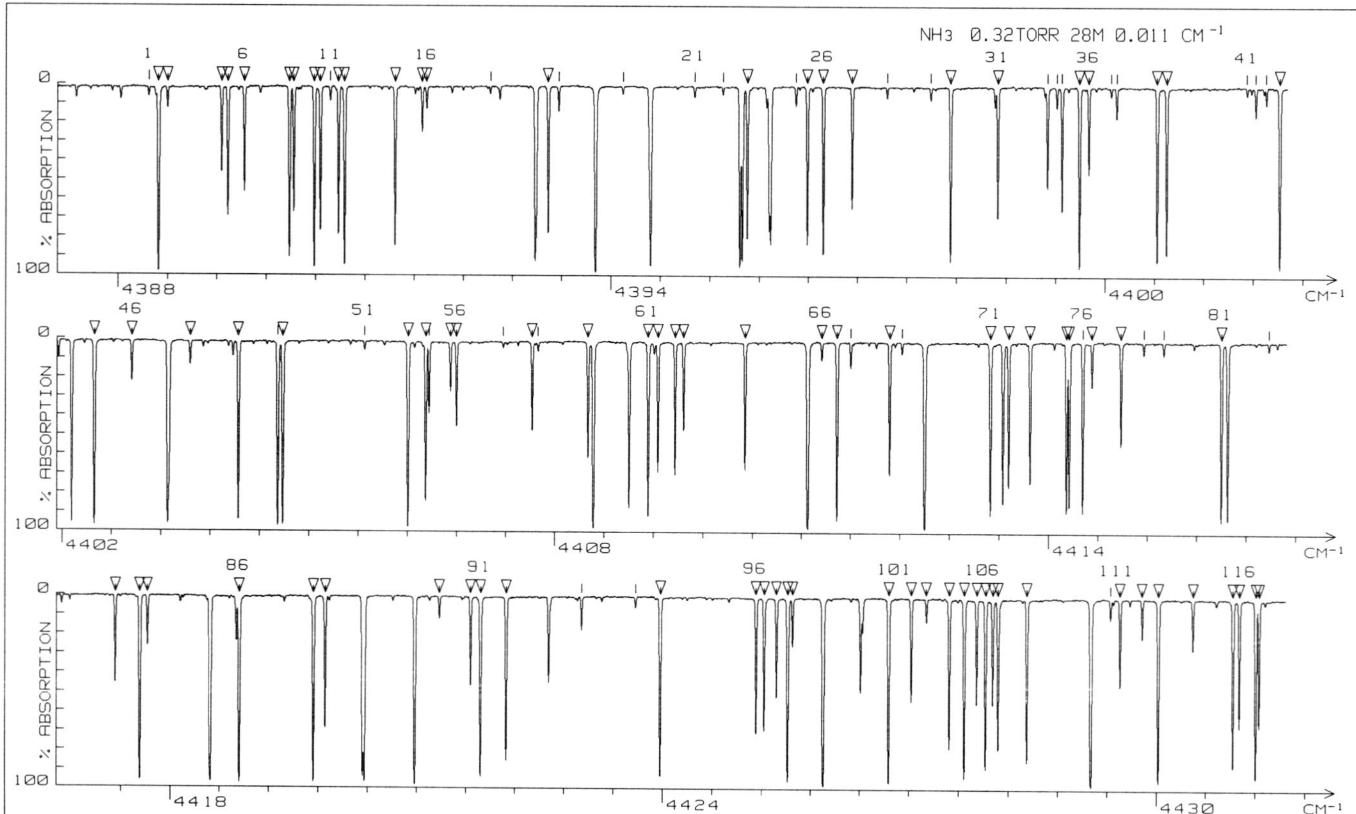

ISOTOPIC IDENTIFICATION – '∇' indicates $^{14}NH_3$; 'ı': unidentified.

LINE NO.	WAVENUMBER CM^{-1}, VAC.	ROTATIONAL BAND ASSIGNMENTS	ISOTOPIC SPECIES	LINE NO.	WAVENUMBER CM^{-1}, VAC.	ROTATIONAL BAND ASSIGNMENTS	ISOTOPIC SPECIES
1	4388.36992(51)			31	4398.69530(19)	s^{r}Q(6, 1) NU2+NU3	$^{14}NH_3$
2	4388.48745(19)	s^{P}P(2, 2) NU2+NU3	$^{14}NH_3$	32	4399.29684(19)		
3	4388.59556(28)	s^{r}Q(10, 1) NU2+NU3	$^{14}NH_3$	33	4399.41059(28)		
4	4389.25602(19)	a^{r}Q(9, 3) NU2+NU3	$^{14}NH_3$	34	4399.47429(18)		
5	4389.33138(19)	s^{r}Q(7, 2) NU2+NU3	$^{14}NH_3$	35	4399.68679(19)	a^{r}Q(6, 3) NU2+NU3	$^{14}NH_3$
6	4389.53167(20)	aR(4, 4) NU1+NU2	$^{14}NH_3$	36	4399.79943(19)	a^{r}Q(8, 2) NU2+NU3	$^{14}NH_3$
7	4390.07849(19)	aR(4, 3) NU1+NU2	$^{14}NH_3$	37	4400.07211(48)		
8	4390.13386(19)	a^{r}Q(7, 4) NU2+NU3	$^{14}NH_3$	38	4400.13605(24)		
9	4390.38043(19)	a^{r}P(2, 0) NU2+NU3	$^{14}NH_3$	39	4400.62794(19)	s^{r}Q(5, 1) NU2+NU3	$^{14}NH_3$
10	4390.45438(19)	aR(4, 2) NU1+NU2	$^{14}NH_3$	40	4400.74208(19)	s^{P}P(1, 1) NU2+NU3	$^{14}NH_3$
11	4390.57436(36)			41	4401.72151(56)		
12	4390.67465(19)	aR(4, 1) NU1+NU2	$^{14}NH_3$	42	4401.82735(28)		
13	4390.74736(19)	aR(4, 0) NU1+NU2	$^{14}NH_3$	43	4401.95512(32)		
14	4391.36247(18)	s^{r}Q(6, 2) NU2+NU3	$^{14}NH_3$	44	4402.11753(19)	s^{r}Q(4, 1) NU2+NU3	$^{14}NH_3$
15	4391.69312(21)	s^{r}Q(9, 1) NU2+NU3	$^{14}NH_3$	45	4402.39480(19)	a^{r}Q(5, 3) NU2+NU3	$^{14}NH_3$
16	4391.74767(28)	a^{r}Q(10, 2) NU2+NU3	$^{14}NH_3$	46	4402.84695(23)	a^{r}Q(9, 1) NU2+NU3	$^{14}NH_3$
17	4392.52862(60)			47	4403.55352(28)	s^{P}Q(10, 1) NU2+NU3	$^{14}NH_3$
18	4393.23488(19)	a^{r}Q(6, 4) NU2+NU3	$^{14}NH_3$	48	4404.14477(19)	s^{r}Q(2, 1) NU2+NU3	$^{14}NH_3$
19	4393.36293(28)			49	4404.62284(19)		
20	4394.14266(57)			50	4404.68361(19)	a^{r}Q(4, 3) NU2+NU3	$^{14}NH_3$
21	4395.01145(47)			51	4405.67675(57)		
22	4395.35544(55)			52	4406.20644(19)	a^{P}P(2, 2) NU2+NU3	$^{14}NH_3$
23	4395.64872(19)	sR(3, 1) NU1+NU2	$^{14}NH_3$	53	4406.41986(19)	a^{r}Q(6, 2) NU2+NU3	$^{14}NH_3$
24	4396.24592(30)			54	4406.46135(22)		
25	4396.38601(18)	sR(3, 3) NU1+NU2	$^{14}NH_3$	55	4406.72153(28)	s^{P}Q(9, 1) NU2+NU3	$^{14}NH_3$
26	4396.58118(18)	a^{r}Q(7, 3) NU2+NU3	$^{14}NH_3$	56	4406.79749(19)	a^{r}Q(8, 1) NU2+NU3	$^{14}NH_3$
27	4396.93232(19)	s^{r}Q(7, 1) NU2+NU3	$^{14}NH_3$	57	4407.36998(79)		
28	4397.35580(39)			58	4407.72394(20)	aR(5, 5) NU1+NU2	$^{14}NH_3$
29	4397.88591(37)			59	4407.79661(55)		
30	4398.12552(19)	a^{P}P(2, 1) NU2+NU3	$^{14}NH_3$	60	4408.40380(19)	aR(5, 4) NU1+NU2	$^{14}NH_3$

- "LINE NO.": serial number identifying the spectral line in the map on the opposite page.
- "WAVENUMBER": experimental value.
- Uncertainty in the last digit of the wavenumber is given in parentheses.
- "pa" : parallel ; "pe" : perpendicular.

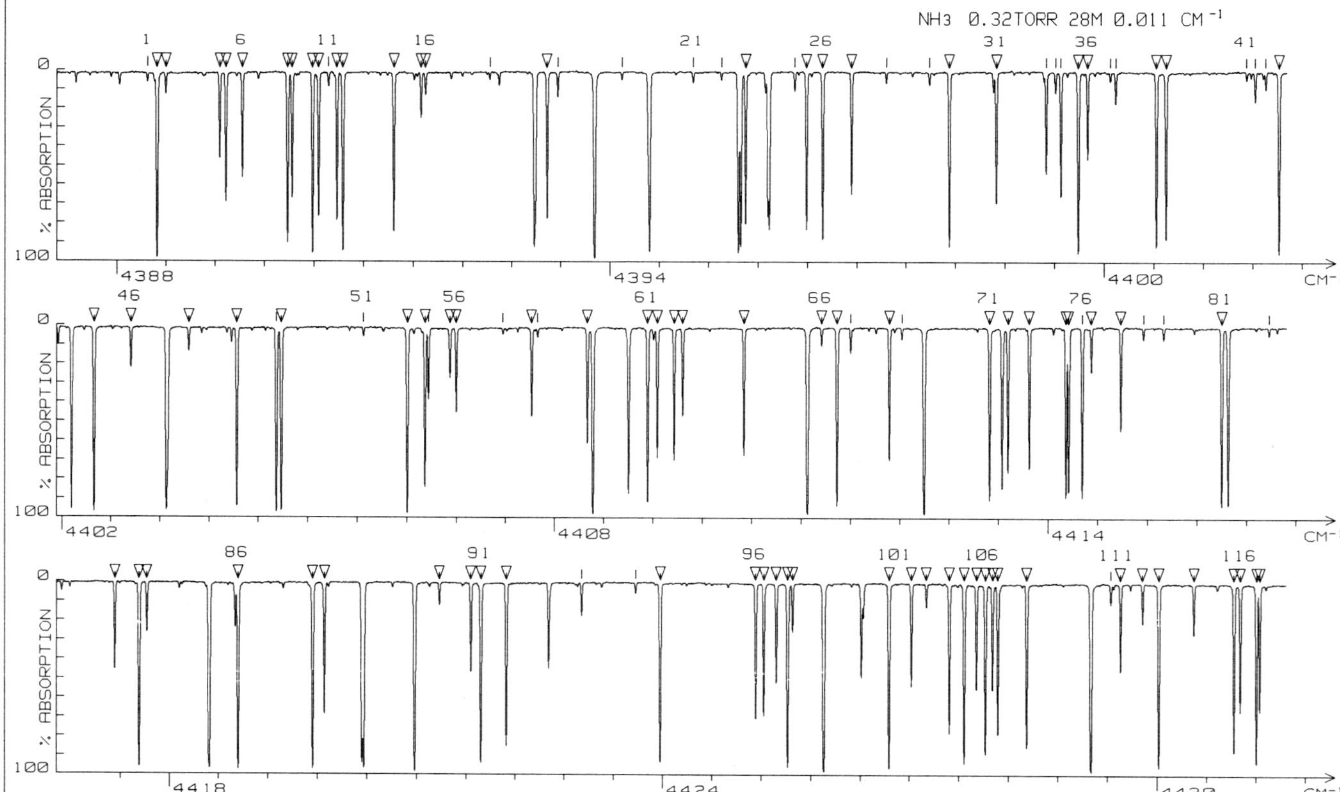

ISOTOPIC IDENTIFICATION – '∇' indicates $^{14}NH_3$; '|': unidentified.

LINE NO.	WAVENUMBER CM^{-1}, VAC.	ROTATIONAL ASSIGNMENTS	BAND	ISOTOPIC SPECIES
61	4409.13786(19)	a^{r}Q(5, 2)	NU2+NU3	$^{14}NH_3$
62	4409.25493(19)	aR(5, 2)	NU1+NU2	$^{14}NH_3$
63	4409.46045(19)	aR(5, 1)	NU1+NU2	$^{14}NH_3$
64	4409.56334(19)	s^{P}Q(8, 1)	NU2+NU3	$^{14}NH_3$
65	4410.30735(19)	a^{r}Q(7, 1)	NU2+NU3	$^{14}NH_3$
66	4411.24777(32)	s^{P}Q(10, 2)	NU2+NU3	$^{14}NH_3$
67	4411.43536(19)	a^{r}Q(4, 2)	NU2+NU3	$^{14}NH_3$
68	4411.60182(32)			
69	4412.07324(19)	s^{P}Q(7, 1)	NU2+NU3	$^{14}NH_3$
70	4412.22696(44)			
71	4413.29528(19)	a^{r}Q(3, 2)	NU2+NU3	$^{14}NH_3$
72	4413.51322(19)	sR(4, 1)	NU1+NU2	$^{14}NH_3$
73	4413.77071(19)	sR(4, 2)	NU1+NU2	$^{14}NH_3$
74	4414.21681(19)	sR(4, 3)	NU1+NU2	$^{14}NH_3$
75	4414.25090(19)	s^{P}Q(6, 1)	NU2+NU3	$^{14}NH_3$
76	4414.41410(18)			
77	4414.52456(22)	s^{P}Q(9, 2)	NU2+NU3	$^{14}NH_3$
78	4414.87710(20)	sR(4, 4)	NU1+NU2	$^{14}NH_3$
79	4415.15234(43)			
80	4415.39429(44)			
81	4416.10042(19)	s^{P}Q(5, 1)	NU2+NU3	$^{14}NH_3$
82	4416.67429(51)			
83	4417.32907(19)	s^{P}Q(8, 2)	NU2+NU3	$^{14}NH_3$
84	4417.62735(19)	s^{P}Q(4, 1)	NU2+NU3	$^{14}NH_3$
85	4417.72003(25)	a^{P}Q(9, 1)	NU2+NU3	$^{14}NH_3$
86	4418.83831(19)	s^{P}Q(3, 1)	NU2+NU3	$^{14}NH_3$
87	4419.73982(19)	s^{P}Q(2, 1)	NU2+NU3	$^{14}NH_3$
88	4419.88179(19)	s^{P}Q(7, 2)	NU2+NU3	$^{14}NH_3$
89	4421.27183(32)	a^{P}Q(10, 2)	NU2+NU3	$^{14}NH_3$
90	4421.65442(19)	a^{P}Q(8, 1)	NU2+NU3	$^{14}NH_3$
91	4421.77636(19)	a^{r}Q(2, 1)	NU2+NU3	$^{14}NH_3$
92	4422.09125(18)	s^{P}Q(6, 2)	NU2+NU3	$^{14}NH_3$
93	4423.01102(24)			
94	4423.66943(52)			
95	4423.96797(19)	s^{P}Q(5, 2)	NU2+NU3	$^{14}NH_3$
96	4425.12788(19)	s^{P}Q(8, 3)	NU2+NU3	$^{14}NH_3$
97	4425.22974(19)	a^{P}Q(7, 1)	NU2+NU3	$^{14}NH_3$
98	4425.38037(19)	aR(6, 6)	NU1+NU2	$^{14}NH_3$
99	4425.51837(19)	s^{P}Q(4, 2)	NU2+NU3	$^{14}NH_3$
100	4425.57762(22)	a^{P}Q(9, 2)	NU2+NU3	$^{14}NH_3$
101	4426.74847(19)	s^{P}Q(3, 2)	NU2+NU3	$^{14}NH_3$
102	4427.02548(19)	aR(6, 4)	NU1+NU2	$^{14}NH_3$
103	4427.20741(29)	s^{P}Q(10, 4)	NU2+NU3	$^{14}NH_3$
104	4427.48482(19)	aR(6, 3)	NU1+NU2	$^{14}NH_3$
105	4427.66469(19)	s^{P}Q(2, 2)	NU2+NU3	$^{14}NH_3$
106	4427.81482(19)	aR(6, 2)	NU1+NU2	$^{14}NH_3$
107	4427.92318(19)	s^{P}Q(7, 3)	NU2+NU3	$^{14}NH_3$
108	4428.00894(19)	aR(6, 1)	NU1+NU2	$^{14}NH_3$
109	4428.07366(19)	aR(6, 0)	NU1+NU2	$^{14}NH_3$
110	4428.42001(19)	a^{P}Q(6, 1)	NU2+NU3	$^{14}NH_3$
111	4429.43888(34)			
112	4429.55375(20)	a^{P}Q(8, 2)	NU2+NU3	$^{14}NH_3$
113	4429.82106(23)	a^{P}Q(10, 3)	NU2+NU3	$^{14}NH_3$
114	4430.01707(19)	s^{P}Q(6, 3)	NU2+NU3	$^{14}NH_3$
115	4430.44274(22)	s^{P}Q(9, 4)	NU2+NU3	$^{14}NH_3$
116	4430.92756(19)	sR(5, 0)	NU1+NU2	$^{14}NH_3$
117	4431.00487(19)	sR(5, 1)	NU1+NU2	$^{14}NH_3$
118	4431.20194(19)	a^{P}Q(5, 1)	NU2+NU3	$^{14}NH_3$
119	4431.24253(19)	sR(5, 2)	NU1+NU2	$^{14}NH_3$

- "LINE NO.": serial number identifying the spectral line in the map on the opposite page.
- "WAVENUMBER": experimental value.
- Uncertainty in the last digit of the wavenumber is given in parentheses.
- "pa" : parallel ; "pe" : perpendicular.

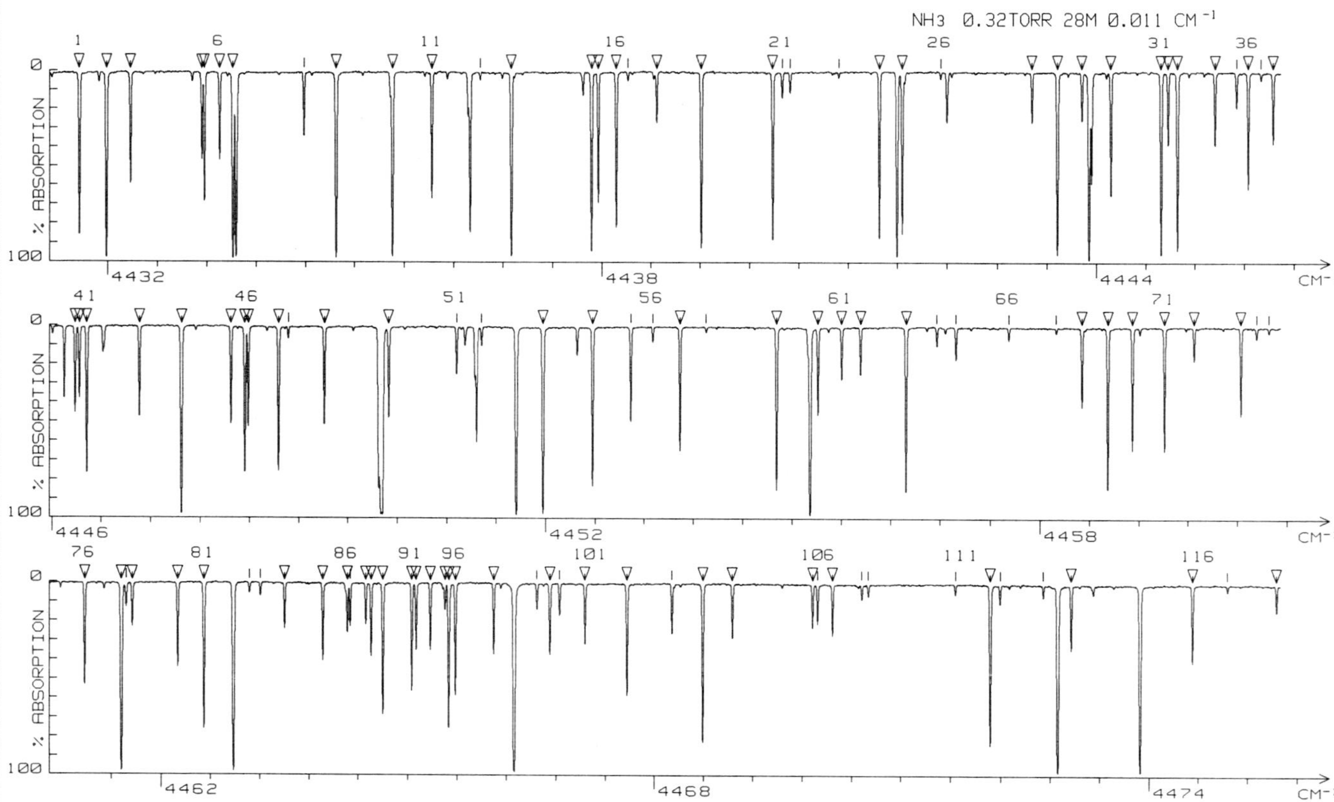

ISOTOPIC IDENTIFICATION – '∇' indicates $^{14}NH_3$; '|': unidentified.

LINE NO.	WAVENUMBER CM^{-1}, VAC.	ROTATIONAL ASSIGNMENTS	BAND ASSIGNMENTS	ISOTOPIC SPECIES
1	4431.65754(19)	sR(5, 3)	NU1+NU2	$^{14}NH_3$
2	4431.98433(19)	sPQ(5, 3)	NU2+NU3	$^{14}NH_3$
3	4432.27566(20)	sR(5, 4)	NU1+NU2	$^{14}NH_3$
4	4433.13448(21)	sR(5, 5)	NU1+NU2	$^{14}NH_3$
5	4433.17002(19)	aPQ(7, 2)	NU2+NU3	$^{14}NH_3$
6	4433.35466(20)	sPQ(8, 4)	NU2+NU3	$^{14}NH_3$
7	4433.51706(20)	sPQ(4, 3)	NU2+NU3	$^{14}NH_3$
8	4434.37828(21)			
9	4434.77436(19)	sPQ(3, 3)	NU2+NU3	$^{14}NH_3$
10	4435.45983(19)	aPQ(3, 1)	NU2+NU3	$^{14}NH_3$
11	4435.93354(19)	sPQ(7, 4)	NU2+NU3	$^{14}NH_3$
12	4436.52211(72)			
13	4436.90275(19)	aPQ(2, 1)	NU2+NU3	$^{14}NH_3$
14	4437.87145(19)	aPQ(1, 1)	NU2+NU3	$^{14}NH_3$
15	4437.95416(19)	aPQ(8, 3)	NU2+NU3	$^{14}NH_3$
16	4438.17620(19)	sPQ(6, 4)	NU2+NU3	$^{14}NH_3$
17	4438.32154(53)			
18	4438.67267(26)	sPQ(9, 5)	NU2+NU3	$^{14}NH_3$
19	4439.21493(19)	aPQ(5, 2)	NU2+NU3	$^{14}NH_3$
20	4440.08409(19)	sPQ(5, 4)	NU2+NU3	$^{14}NH_3$
21	4440.19910(28)			
22	4440.29379(36)			
23	4440.88527(105)			
24	4441.37815(19)	aPQ(7, 3)	NU2+NU3	$^{14}NH_3$
25	4441.66141(19)	sPQ(4, 4)	NU2+NU3	$^{14}NH_3$
26	4442.12606(82)			
27	4443.21546(27)	aR(7, 7)	NU1+NU2	$^{14}NH_3$
28	4443.52748(19)	aPQ(3, 2)	NU2+NU3	$^{14}NH_3$
29	4443.82384(28)	sPQ(10, 6)	NU2+NU3	$^{14}NH_3$
30	4444.17760(19)	sPQ(7, 5)	NU2+NU3	$^{14}NH_3$

LINE NO.	WAVENUMBER CM^{-1}, VAC.	ROTATIONAL ASSIGNMENTS	BAND ASSIGNMENTS	ISOTOPIC SPECIES
31	4444.78367(19)	aPQ(6, 3)	NU2+NU3	$^{14}NH_3$
32	4444.86974(21)	aR(7, 5)	NU1+NU2	$^{14}NH_3$
33	4444.98928(19)	aPQ(2, 2)	NU2+NU3	$^{14}NH_3$
34	4445.44534(20)	aR(7, 4)	NU1+NU2	$^{14}NH_3$
35	4445.70616(24)			
36	4445.84844(19)	aR(7, 3)	NU1+NU2	$^{14}NH_3$
37	4446.00249(70)			
38	4446.14814(20)	aR(7, 2)	NU1+NU2	$^{14}NH_3$
39	4446.28402(20)	aPQ(8, 4)	NU2+NU3	$^{14}NH_3$
40	4446.33633(20)	aR(7, 1)	NU1+NU2	$^{14}NH_3$
41	4446.42679(19)	sPQ(6, 5)	NU2+NU3	$^{14}NH_3$
42	4447.06098(20)	sPQ(9, 6)	NU2+NU3	$^{14}NH_3$
43	4447.56747(19)	aPQ(5, 3)	NU2+NU3	$^{14}NH_3$
44	4448.17222(19)	sR(6, 1)	NU1+NU2	$^{14}NH_3$
45	4448.34024(19)	sPQ(5, 5)	NU2+NU3	$^{14}NH_3$
46	4448.38160(20)	sR(6, 2)	NU1+NU2	$^{14}NH_3$
47	4448.74955(19)	sR(6, 3)	NU1+NU2	$^{14}NH_3$
48	4448.86738(55)			
49	4449.30767(19)	sR(6, 4)	NU1+NU2	$^{14}NH_3$
50	4450.09379(21)	sR(6, 5)	NU1+NU2	$^{14}NH_3$
51	4450.91494(28)			
52	4451.21946(34)			
53	4451.96449(19)	aPQ(3, 3)	NU2+NU3	$^{14}NH_3$
54	4452.56346(19)	sPQ(7, 6)	NU2+NU3	$^{14}NH_3$
55	4453.03503(20)			
56	4453.30352(40)			
57	4453.63668(19)	aPQ(6, 4)	NU2+NU3	$^{14}NH_3$
58	4453.95312(107)			
59	4454.81093(19)	sPQ(6, 6)	NU2+NU3	$^{14}NH_3$
60	4455.31177(20)	aPQ(8, 5)	NU2+NU3	$^{14}NH_3$

- "LINE NO.": serial number identifying the spectral line in the map on the opposite page.
- "WAVENUMBER": experimental value.
- Uncertainty in the last digit of the wavenumber is given in parentheses.
- "pa" : parallel ; "pe" : perpendicular.

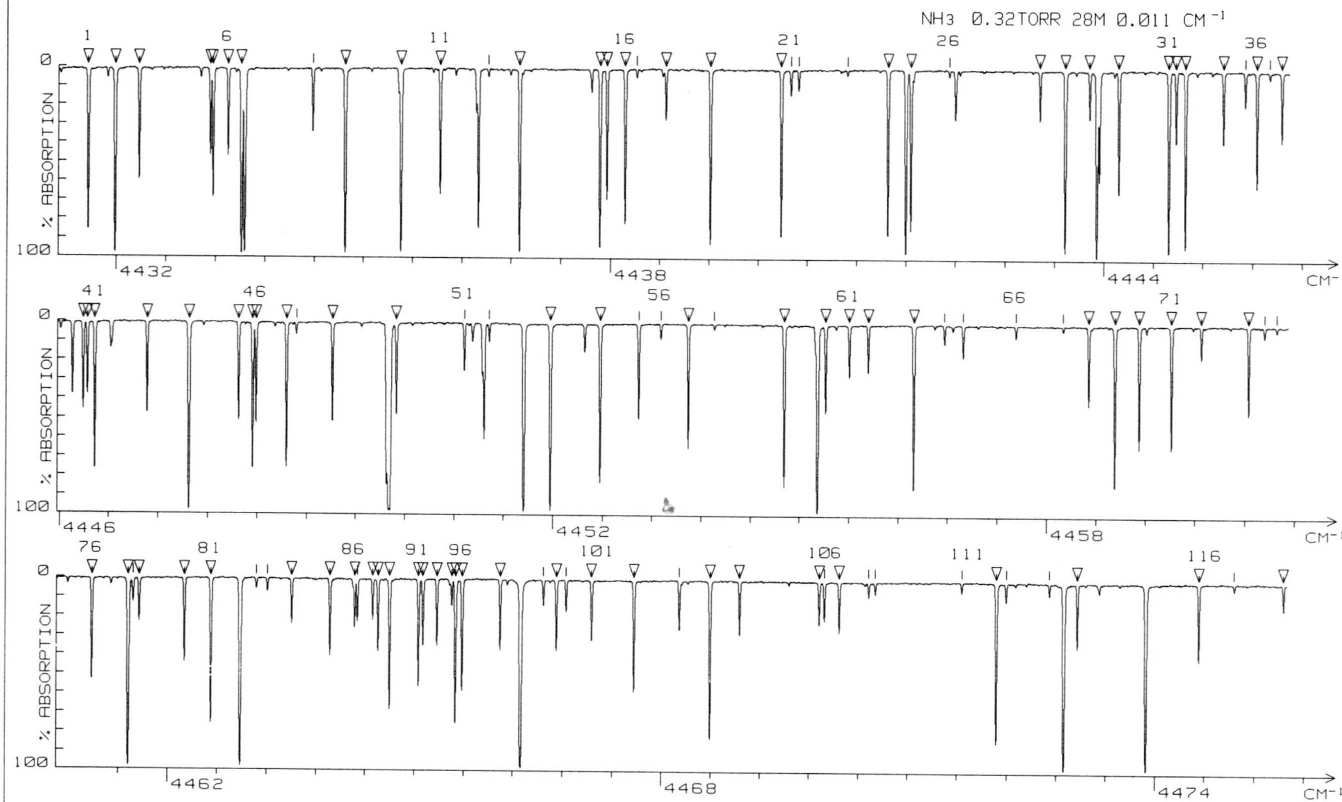

ISOTOPIC IDENTIFICATION – '∇' indicates $^{14}NH_3$; '|': unidentified.

LINE NO.	WAVENUMBER CM^{-1}, VAC.	ROTATIONAL ASSIGNMENTS	BAND	ISOTOPIC SPECIES
61	4455.59916 (27)	sPQ(9, 7)	NU2+NU3	$^{14}NH_3$
62	4455.82812 (29)	aPQ(10, 6)	NU2+NU3	$^{14}NH_3$
63	4456.38753 (19)	aPQ(5, 4)	NU2+NU3	$^{14}NH_3$
64	4456.75962 (33)			
65	4456.98528 (25)			
66	4457.61735 (52)			
67	4458.19327 (92)			
68	4458.50457 (20)	sPQ(8, 7)	NU2+NU3	$^{14}NH_3$
69	4458.82132 (19)	aPQ(4, 4)	NU2+NU3	$^{14}NH_3$
70	4459.11725 (19)	aPQ(7, 5)	NU2+NU3	$^{14}NH_3$
71	4459.50554 (19)	sPR(1, 1)	NU2+NU3	$^{14}NH_3$
72	4459.87110 (36)	aR(8, 8)	NU1+NU2	$^{14}NH_3$
73	4460.43856 (20)	aPQ(9, 6)	NU2+NU3	$^{14}NH_3$
74	4460.63002 (56)			
75	4460.77938 (103)			
76	4461.07881 (20)	sPQ(7, 7)	NU2+NU3	$^{14}NH_3$
76		sPQ(10, 8)	NU2+NU3	$^{14}NH_3$
77	4461.52218 (20)	arR(1, 1)	NU2+NU3	$^{14}NH_3$
78	4461.58026 (49)			
79	4461.65351 (31)	aR(8, 7)	NU1+NU2	$^{14}NH_3$
80	4462.20067 (20)	aR(8, 6)	NU1+NU2	$^{14}NH_3$
81	4462.52201 (19)	aPQ(6, 5)	NU2+NU3	$^{14}NH_3$
82	4463.07363 (64)			
83	4463.20715 (47)			
84	4463.50105 (23)	aR(8, 4)	NU1+NU2	$^{14}NH_3$
85	4463.96779 (20)	aR(8, 3)	NU1+NU2	$^{14}NH_3$
86	4464.27142 (28)	sPQ(9, 8)	NU2+NU3	$^{14}NH_3$
87	4464.30638 (30)			
88	4464.49437 (23)	aR(8, 1)	NU1+NU2	$^{14}NH_3$
88		aR(8, 2)	NU1+NU2	$^{14}NH_3$
89	4464.55913 (20)	aR(8, 0)	NU1+NU2	$^{14}NH_3$
90	4464.70152 (19)	aPQ(8, 6)	NU2+NU3	$^{14}NH_3$
91	4465.04878 (19)	sR(7, 0)	NU1+NU2	$^{14}NH_3$
92	4465.10313 (21)	sR(7, 1)	NU1+NU2	$^{14}NH_3$
93	4465.27281 (21)	sR(7, 2)	NU1+NU2	$^{14}NH_3$
94	4465.45105 (45)	aPQ(10, 7)	NU2+NU3	$^{14}NH_3$
95	4465.49802 (19)	aPQ(5, 5)	NU2+NU3	$^{14}NH_3$
96	4465.57887 (19)	sR(7, 3)	NU1+NU2	$^{14}NH_3$
97	4466.04664 (21)	sR(7, 4)	NU1+NU2	$^{14}NH_3$
98	4466.56975 (36)			
99	4466.72528 (21)	sR(7, 5)	NU1+NU2	$^{14}NH_3$
100	4466.84166 (26)			
101	4467.15017 (22)	sPQ(8, 8)	NU2+NU3	$^{14}NH_3$
102	4467.66830 (20)	sR(7, 6)	NU1+NU2	$^{14}NH_3$
103	4468.21636 (22)			
104	4468.59132 (19)	aPQ(7, 6)	NU2+NU3	$^{14}NH_3$
105	4468.95002 (22)	sR(7, 7)	NU1+NU2	$^{14}NH_3$
106	4469.92451 (30)	sPQ(10, 9)	NU2+NU3	$^{14}NH_3$
107	4469.98673 (24)			
108	4470.16794 (27)	aPQ(9, 7)	NU2+NU3	$^{14}NH_3$
109	4470.52323 (43)			
110	4470.60342 (50)			
111	4471.65799 (61)			
112	4472.07504 (19)	aPQ(6, 6)	NU2+NU3	$^{14}NH_3$
113	4472.19033 (34)			
114	4472.71562 (44)			
115	4473.05480 (22)	sPQ(9, 9)	NU2+NU3	$^{14}NH_3$
116	4474.53193 (21)	aPQ(8, 7)	NU2+NU3	$^{14}NH_3$
117	4474.95676 (92)			
118	4475.54892 (29)	aPQ(10, 8)	NU2+NU3	$^{14}NH_3$

- "LINE NO.": serial number identifying the spectral line in the map on the opposite page.
- "WAVENUMBER": experimental value.
- Uncertainty in the last digit of the wavenumber is given in parentheses.
- "pa" : parallel ; "pe" : perpendicular.

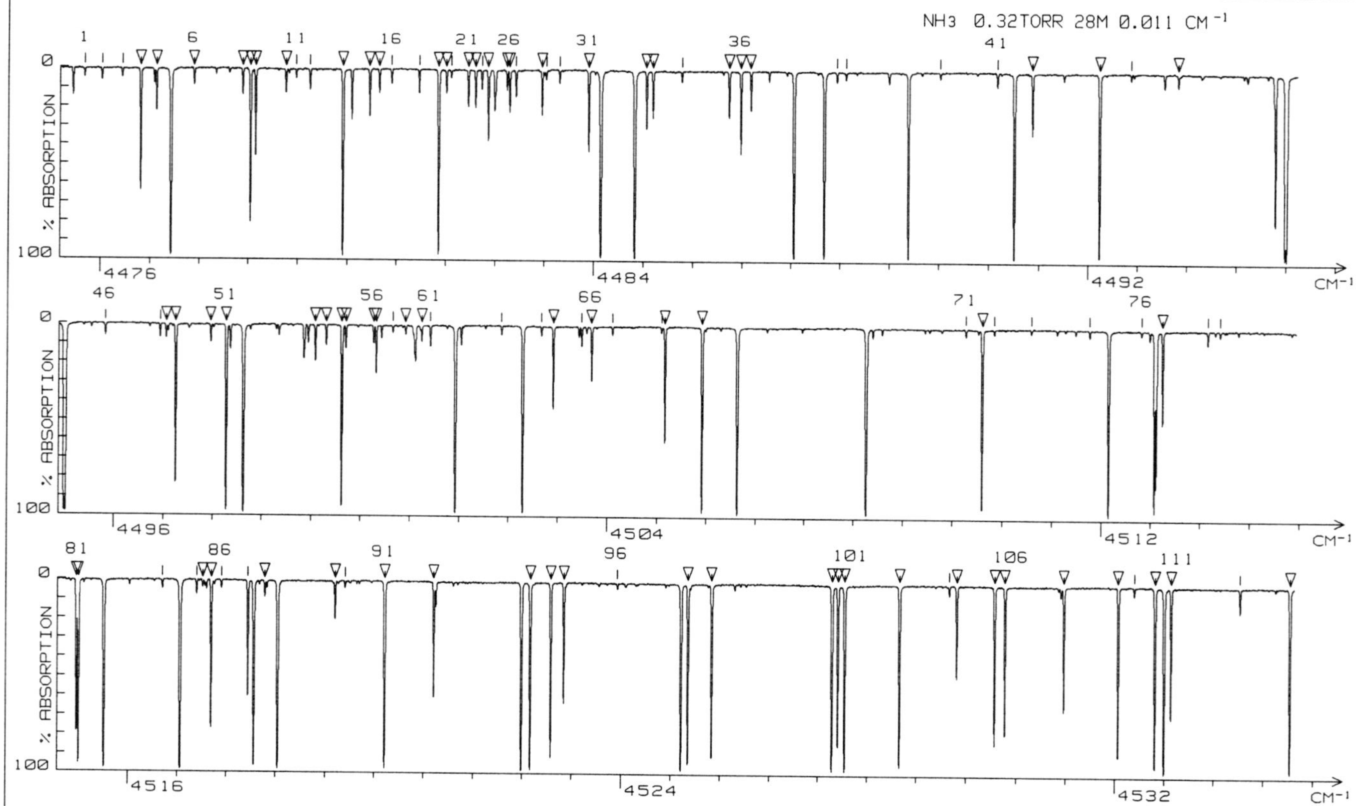

ISOTOPIC IDENTIFICATION – '∇' indicates $^{14}NH_3$; 'ı': unidentified.

LINE NO.	WAVENUMBER CM^{-1}, VAC.	ROTATIONAL ASSIGNMENTS	BAND ASSIGNMENTS	ISOTOPIC SPECIES
1	4475.73962(66)			
2	4476.02426(62)			
3	4476.35231(77)			
4	4476.64862(19)	aPR(1, 1)	NU2+NU3	$^{14}NH_3$
5	4476.90358(31)	aR(9, 9)	NU1+NU2	$^{14}NH_3$
6	4477.51416(41)	aR(9, 7)	NU1+NU2	$^{14}NH_3$
7	4478.30965(31)	aR(9, 8)	NU1+NU2	$^{14}NH_3$
8	4478.43618(19)	sPR(2, 1)	NU2+NU3	$^{14}NH_3$
9	4478.52098(20)	aPQ(7, 7)	NU2+NU3	$^{14}NH_3$
10	4479.01020(52)	sPQ(10, 10)	NU2+NU3	$^{14}NH_3$
11	4479.17615(64)			
12	4479.39866(37)			
13	4479.92898(20)	arR(2, 1)	NU2+NU3	$^{14}NH_3$
14	4480.36912(29)	aPQ(9, 8)	NU2+NU3	$^{14}NH_3$
15	4480.52634(31)	aR(9, 5)	NU1+NU2	$^{14}NH_3$
16	4480.72320(52)			
17	4481.17041(32)			
18	4481.48729(19)	srR(3, 1)	NU2+NU3	$^{14}NH_3$
19	4481.61198(32)	aR(9, 4)	NU1+NU2	$^{14}NH_3$
20	4481.68661(94)			
21	4481.96370(25)	sR(8, 1)	NU1+NU2	$^{14}NH_3$
22	4482.07973(25)	sR(8, 2)	NU1+NU2	$^{14}NH_3$
23	4482.18376(34)			
24	4482.28782(23)	sR(8, 3)	NU1+NU2	$^{14}NH_3$
25	4482.59586(33)	aR(9, 1)	NU1+NU2	$^{14}NH_3$
26	4482.64131(24)	sR(8, 4)	NU1+NU2	$^{14}NH_3$
27	4482.74130(28)			
28	4483.16507(30)	sR(8, 5)	NU1+NU2	$^{14}NH_3$
29	4483.23239(119)			
30	4483.44795(53)			
31	4483.91996(20)	sR(8, 6)	NU1+NU2	$^{14}NH_3$
32	4484.84902(22)	aPQ(8, 8)	NU2+NU3	$^{14}NH_3$
33	4484.95185(29)	sR(8, 7)	NU1+NU2	$^{14}NH_3$
34	4485.41879(56)			
35	4486.18520(29)	aPQ(10, 9)	NU2+NU3	$^{14}NH_3$
36	4486.37452(20)	sPR(2, 2)	NU2+NU3	$^{14}NH_3$
37	4486.53847(32)	sR(8, 8)	NU1+NU2	$^{14}NH_3$
38	4487.94790(58)			
39	4488.09599(79)			
40	4489.61966(97)			
41	4490.53256(42)			
42	4491.09443(22)	aPQ(9, 9)	NU2+NU3	$^{14}NH_3$
43	4492.18419(20)	srR(4, 2)	NU2+NU3	$^{14}NH_3$
44	4492.69266(94)			
45	4493.45676(61)	aR(10, 10)	NU1+NU2	$^{14}NH_3$
46	4495.85563(61)			
47	4496.74066(47)			
48	4496.83829(77)	aR(10, 8)	NU1+NU2	$^{14}NH_3$
49	4496.99697(19)	sPR(3, 1)	NU2+NU3	$^{14}NH_3$
50	4497.56474(40)	aR(10, 7)	NU1+NU2	$^{14}NH_3$
50		aPQ(10, 10)	NU2+NU3	$^{14}NH_3$
51	4497.82193(19)	arR(3, 1)	NU2+NU3	$^{14}NH_3$
52	4499.26303(26)	sR(9, 3)	NU1+NU2	$^{14}NH_3$
53	4499.43597(34)	sR(9, 4)	NU1+NU2	$^{14}NH_3$
54	4499.68999(19)	srR(4, 1)	NU2+NU3	$^{14}NH_3$
55	4499.75344(34)	sR(9, 5)	NU1+NU2	$^{14}NH_3$
56	4500.20737(43)	aR(10, 3)	NU1+NU2	$^{14}NH_3$
57	4500.25250(29)	sR(9, 6)	NU1+NU2	$^{14}NH_3$
58	4500.52269(84)			
59	4500.72945(70)	aR(10, 2)	NU1+NU2	$^{14}NH_3$
60	4500.99013(68)	aR(10, 0)	NU1+NU2	$^{14}NH_3$

- "LINE NO.": serial number identifying the spectral line in the map on the opposite page.
- "WAVENUMBER": experimental value.
- Uncertainty in the last digit of the wavenumber is given in parentheses.
- "pa" : parallel ; "pe" : perpendicular.

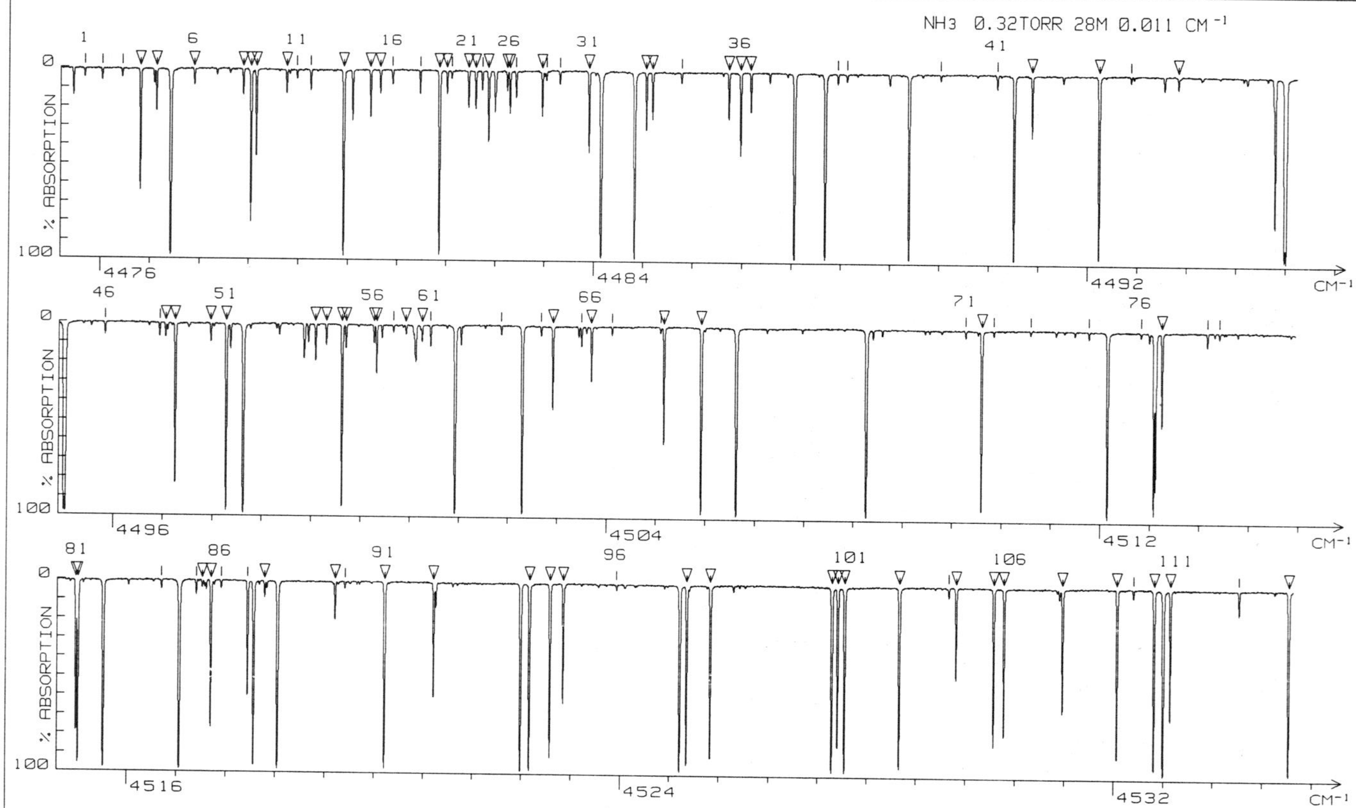

ISOTOPIC IDENTIFICATION – '∇' indicates $^{14}NH_3$; '|': unidentified.

LINE NO.	WAVENUMBER CM^{-1}, VAC.	ROTATIONAL ASSIGNMENTS	BAND	ISOTOPIC SPECIES
61	4501. 13026 (41)			
62	4502. 28590 (81)			
63	4502. 93395 (60)			
64	4503. 12390 (21)	a^{P}R(2, 2)	NU2+NU3	$^{14}NH_3$
65	4503. 58312 (47)			
66	4503. 74384 (27)	sR(9, 9)	NU1+NU2	$^{14}NH_3$
67	4504. 07756 (82)			
68	4504. 86596 (103)			
69	4504. 92532 (20)	s^{P}R(3, 2)	NU2+NU3	$^{14}NH_3$
70	4505. 52781 (20)	s^{r}R(6, 5)	NU2+NU3	$^{14}NH_3$
71	4509. 81212 (104)			
72	4510. 07324 (19)	s^{r}R(5, 2)	NU2+NU3	$^{14}NH_3$
73	4510. 26448 (109)			
74	4510. 85650 (132)			
75	4511. 80289 (91)			
76	4512. 64401 (102)			
77	4512. 98649 (20)	s^{P}R(3, 3)	NU2+NU3	$^{14}NH_3$
78	4513. 72007 (59)			
79	4513. 91719 (90)			
80	4515. 16238 (19)	s^{P}R(4, 1)	NU2+NU3	$^{14}NH_3$
81	4515. 19654 (19)	a^{r}R(4, 1)	NU2+NU3	$^{14}NH_3$
82	4516. 55596 (80)			
83	4517. 11247 (44)			
84	4517. 20916 (94)	sR(10, 1)	NU1+NU2	$^{14}NH_3$
85	4517. 35134 (19)	s^{r}R(5, 1)	NU2+NU3	$^{14}NH_3$
86	4517. 52194 (78)			
87	4517. 95294 (20)			
88	4518. 22611 (68)	sR(10, 8)	NU1+NU2	$^{14}NH_3$
89	4519. 36007 (34)	sR(10, 9)	NU1+NU2	$^{14}NH_3$
90	4519. 51917 (110)			

LINE NO.	WAVENUMBER CM^{-1}, VAC.	ROTATIONAL ASSIGNMENTS	BAND	ISOTOPIC SPECIES
91	4520. 17105 (20)	s^{r}R(6, 3)	NU2+NU3	$^{14}NH_3$
92	4520. 96680 (20)	a^{P}R(3, 2)	NU2+NU3	$^{14}NH_3$
93	4522. 53345 (20)	a^{r}R(6, 5)	NU2+NU3	$^{14}NH_3$
94	4522. 86559 (19)	s^{r}R(7, 5)	NU2+NU3	$^{14}NH_3$
95	4523. 07615 (20)	s^{P}R(4, 2)	NU2+NU3	$^{14}NH_3$
96	4523. 93477 (114)			
97	4525. 08099 (19)	a^{r}R(5, 2)	NU2+NU3	$^{14}NH_3$
98	4525. 46503 (19)	s^{r}R(8, 7)	NU2+NU3	$^{14}NH_3$
99	4527. 43234 (20)	a^{r}R(7, 7)	NU2+NU3	$^{14}NH_3$
100	4527. 52850 (19)	s^{r}R(6, 2)	NU2+NU3	$^{14}NH_3$
101	4527. 63959 (20)	s^{r}R(9, 9)	NU2+NU3	$^{14}NH_3$
102	4528. 52561 (19)	a^{r}R(6, 4)	NU2+NU3	$^{14}NH_3$
103	4529. 32868 (65)			
104	4529. 44894 (20)	a^{P}R(3, 3)	NU2+NU3	$^{14}NH_3$
105	4530. 05520 (19)	s^{r}R(7, 4)	NU2+NU3	$^{14}NH_3$
106	4530. 22059 (19)	a^{P}R(4, 1)	NU2+NU3	$^{14}NH_3$
107	4531. 16983 (20)	s^{P}R(4, 3)	NU2+NU3	$^{14}NH_3$
108	4532. 05169 (19)	a^{r}R(5, 1)	NU2+NU3	$^{14}NH_3$
109	4532. 31064 (102)			
110	4532. 64862 (19)	s^{r}R(8, 6)	NU2+NU3	$^{14}NH_3$
111	4532. 90696 (19)	s^{P}R(5, 1)	NU2+NU3	$^{14}NH_3$
112	4534. 02283 (40)			
113	4534. 82665 (20)	a^{r}R(6, 3)	NU2+NU3	$^{14}NH_3$

- "LINE NO.": serial number identifying the spectral line in the map on the opposite page.
- "WAVENUMBER": experimental value.
- Uncertainty in the last digit of the wavenumber is given in parentheses.
- "pa" : parallel ; "pe" : perpendicular.

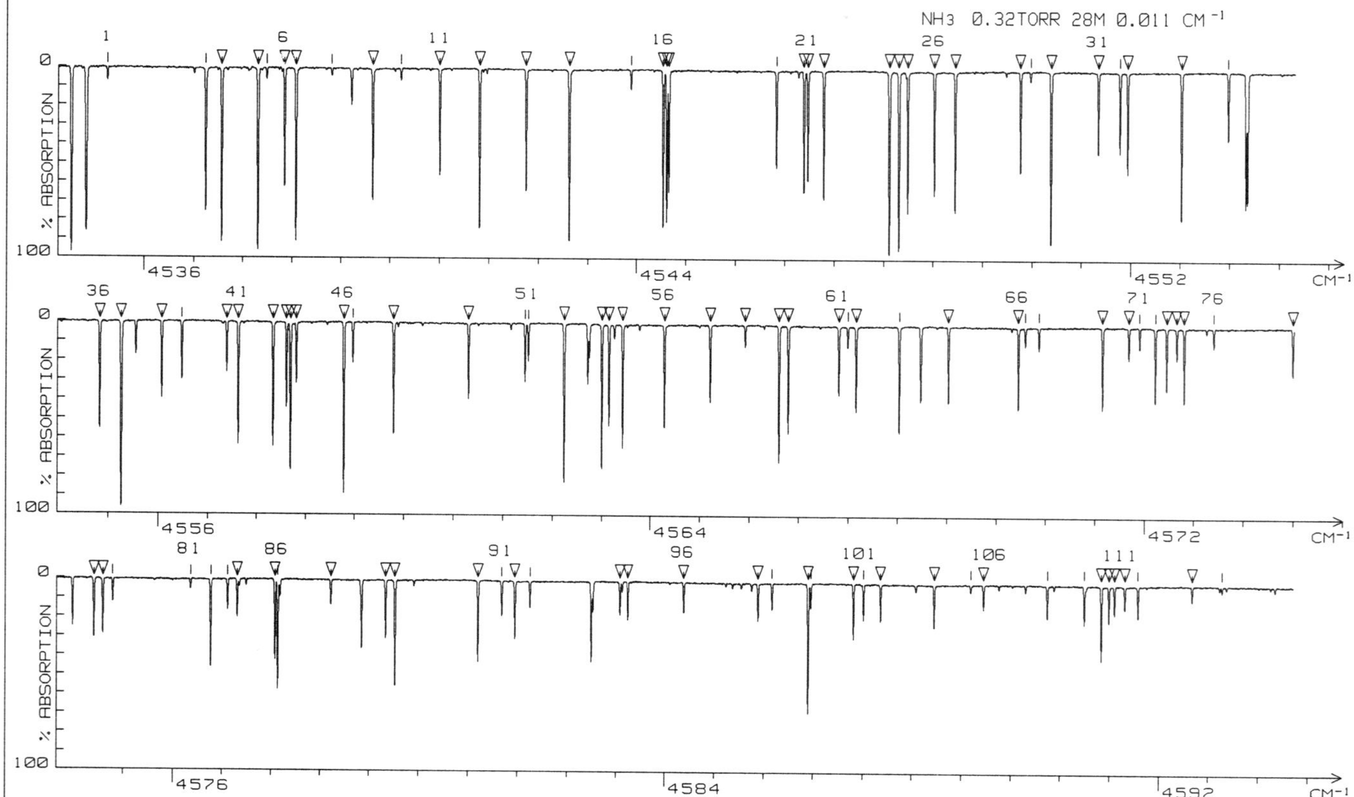

ISOTOPIC IDENTIFICATION – '∇' indicates $^{14}NH_3$; '|': unidentified.

LINE NO.	WAVENUMBER CM^{-1}, VAC.	ROTATIONAL ASSIGNMENTS	BAND	ISOTOPIC SPECIES
1	4535.41203(54)			
2	4536.98837(19)			
3	4537.25478(19)	s^rR(7, 3)	NU2+NU3	$^{14}NH_3$
3		s^rR(10, 10)	NU2+NU3	$^{14}NH_3$
4	4537.84634(20)	a^rR(8, 8)	NU2+NU3	$^{14}NH_3$
5	4537.98932(77)			
6	4538.27860(20)	a^pR(4, 2)	NU2+NU3	$^{14}NH_3$
7	4538.46163(19)	a^rR(7, 5)	NU2+NU3	$^{14}NH_3$
8	4539.05130(111)			
9	4539.72080(20)	s^rR(8, 5)	NU2+NU3	$^{14}NH_3$
10	4540.17890(58)			
11	4540.80198(20)	s^pR(5, 2)	NU2+NU3	$^{14}NH_3$
12	4541.44406(19)	a^rR(6, 2)	NU2+NU3	$^{14}NH_3$
13	4542.19910(20)	s^rR(9, 7)	NU2+NU3	$^{14}NH_3$
14	4542.89342(19)	a^rR(8, 7)	NU2+NU3	$^{14}NH_3$
15	4543.90918(40)			
16	4544.43018(19)	a^rR(7, 4)	NU2+NU3	$^{14}NH_3$
17	4544.48906(19)	s^rR(10, 9)	NU2+NU3	$^{14}NH_3$
18	4544.52411(20)	s^rR(7, 2)	NU2+NU3	$^{14}NH_3$
19	4546.26351(20)			
20	4546.70611(20)	a^pR(4, 3)	NU2+NU3	$^{14}NH_3$
21	4546.76889(20)	s^rR(8, 4)	NU2+NU3	$^{14}NH_3$
22	4547.02787(20)	a^pR(5, 1)	NU2+NU3	$^{14}NH_3$
23	4548.08504(20)	a^rR(9, 9)	NU2+NU3	$^{14}NH_3$
24	4548.24160(19)	a^rR(8, 6)	NU2+NU3	$^{14}NH_3$
25	4548.38989(19)	a^rR(6, 1)	NU2+NU3	$^{14}NH_3$
26	4548.81888(20)	s^pR(5, 3)	NU2+NU3	$^{14}NH_3$
27	4549.15257(19)	s^rR(9, 6)	NU2+NU3	$^{14}NH_3$
28	4550.20736(20)	s^pR(6, 1)	NU2+NU3	$^{14}NH_3$
29	4550.36859(91)			
30	4550.70652(19)	a^rR(7, 3)	NU2+NU3	$^{14}NH_3$
31	4551.47125(21)	s^rR(10, 8)	NU2+NU3	$^{14}NH_3$
32	4551.82110(21)			
33	4551.94421(20)	s^rR(7, 1)	NU2+NU3	$^{14}NH_3$
34	4552.82289(19)	a^rR(9, 8)	NU2+NU3	$^{14}NH_3$
35	4553.57620(22)			
36	4555.05965(20)	a^pR(5, 2)	NU2+NU3	$^{14}NH_3$
37	4555.40365(20)	a^rR(6, 0)	NU2+NU3	$^{14}NH_3$
38	4556.04948(22)	s^rR(9, 5)	NU2+NU3	$^{14}NH_3$
39	4556.37268(27)			
40	4557.10771(29)	s^pR(5, 4)	NU2+NU3	$^{14}NH_3$
41	4557.29487(20)	a^rR(7, 2)	NU2+NU3	$^{14}NH_3$
42	4557.86170(20)	a^rR(9, 7)	NU2+NU3	$^{14}NH_3$
43	4558.07885(21)	s^pR(6, 2)	NU2+NU3	$^{14}NH_3$
44	4558.14418(19)	a^rR(10, 10)	NU2+NU3	$^{14}NH_3$
45	4558.23840(24)	s^rR(10, 7)	NU2+NU3	$^{14}NH_3$
46	4559.01756(19)	s^rR(7, 0)	NU2+NU3	$^{14}NH_3$
47	4559.16873(34)			
48	4559.83101(20)	a^rR(8, 4)	NU2+NU3	$^{14}NH_3$
49	4561.03722(22)	s^rR(8, 2)	NU2+NU3	$^{14}NH_3$
50	4561.95155(24)			
51	4562.00470(36)			
52	4562.58018(19)	a^rR(10, 9)	NU2+NU3	$^{14}NH_3$
53	4563.19354(19)	a^rR(9, 6)	NU2+NU3	$^{14}NH_3$
54	4563.31238(20)	a^pR(6, 1)	NU2+NU3	$^{14}NH_3$
55	4563.53453(20)	a^pR(5, 3)	NU2+NU3	$^{14}NH_3$
56	4564.22376(20)	a^rR(7, 1)	NU2+NU3	$^{14}NH_3$
57	4564.96342(22)	s^rR(10, 6)	NU2+NU3	$^{14}NH_3$
58	4565.52652(40)	s^pR(5, 5)	NU2+NU3	$^{14}NH_3$
59	4566.07768(20)	a^rR(8, 3)	NU2+NU3	$^{14}NH_3$
60	4566.22672(20)	s^pR(6, 3)	NU2+NU3	$^{14}NH_3$

- "LINE NO.": serial number identifying the spectral line in the map on the opposite page.
- "WAVENUMBER": experimental value.
- Uncertainty in the last digit of the wavenumber is given in parentheses.
- "pa" : parallel ; "pe" : perpendicular.

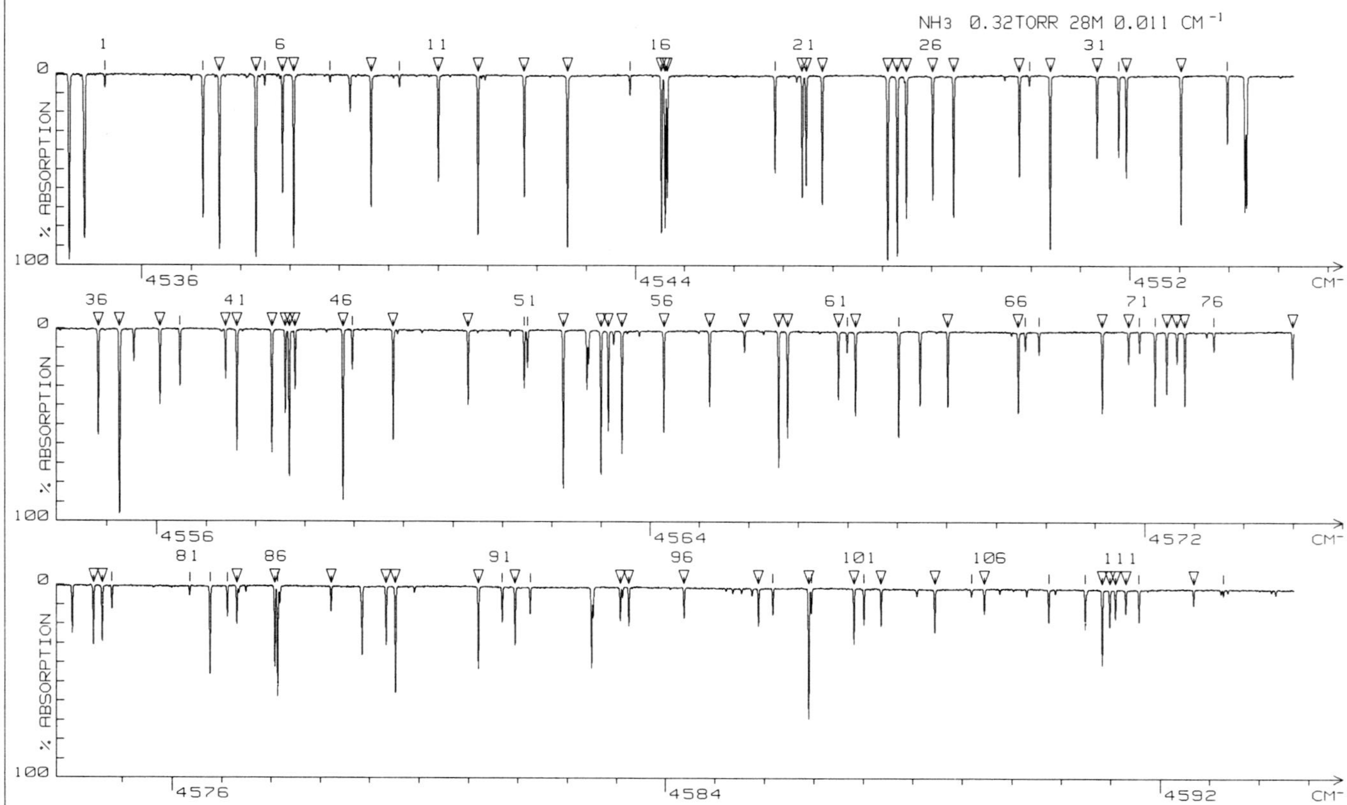

ISOTOPIC IDENTIFICATION - '∇' indicates $^{14}NH_3$; '|': unidentified.

LINE NO.	WAVENUMBER CM^{-1}, VAC.	ROTATIONAL ASSIGNMENTS	BAND	ISOTOPIC SPECIES
61	4567.04323(22)	$s^{P}R$(7, 1)	NU2+NU3	$^{14}NH_3$
62	4567.18406(42)			
63	4567.31902(21)	$a^{r}R$(10, 8)	NU2+NU3	$^{14}NH_3$
64	4568.02032(21)			
65	4568.81707(22)	$a^{r}R$(9, 5)	NU2+NU3	$^{14}NH_3$
66	4569.94663(21)	$s^{r}R$(9, 3)	NU2+NU3	$^{14}NH_3$
67	4570.06022(45)			
68	4570.27998(39)			
69	4571.31395(21)	$a^{P}R$(6, 2)	NU2+NU3	$^{14}NH_3$
70	4571.73751(31)	$s^{r}R$(10, 5)	NU2+NU3	$^{14}NH_3$
71	4571.91133(41)			
72	4572.16224(22)			
73	4572.34765(23)	$a^{r}R$(10, 7)	NU2+NU3	$^{14}NH_3$
74	4572.51292(32)	$a^{P}R$(5, 4)	NU2+NU3	$^{14}NH_3$
75	4572.64005(22)	$a^{r}R$(8, 2)	NU2+NU3	$^{14}NH_3$
76	4573.11702(43)			
77	4574.38445(30)	$s^{P}R$(6, 4)	NU2+NU3	$^{14}NH_3$
78	4574.73635(24)	$a^{r}R$(9, 4)	NU2+NU3	$^{14}NH_3$
79	4574.87937(24)	$s^{P}R$(7, 2)	NU2+NU3	$^{14}NH_3$
80	4575.03653(39)			
81	4576.28290(100)			
82	4576.61085(21)			
83	4576.89146(45)			
84	4577.04831(28)	$s^{r}R$(9, 2)	NU2+NU3	$^{14}NH_3$
85	4577.66170(23)	$a^{r}R$(10, 6)	NU2+NU3	$^{14}NH_3$
86	4577.70970(20)			
87	4578.57653(36)	$s^{r}R$(10, 4)	NU2+NU3	$^{14}NH_3$
88	4579.47022(24)	$a^{r}R$(8, 1)	NU2+NU3	$^{14}NH_3$
89	4579.62357(20)	$a^{P}R$(6, 3)	NU2+NU3	$^{14}NH_3$
90	4580.96118(22)	$a^{r}R$(9, 3)	NU2+NU3	$^{14}NH_3$

LINE NO.	WAVENUMBER CM^{-1}, VAC.	ROTATIONAL ASSIGNMENTS	BAND	ISOTOPIC SPECIES
91	4581.34623(37)			
92	4581.56018(25)	$a^{P}R$(5, 5)	NU2+NU3	$^{14}NH_3$
93	4581.80232(35)			
94	4583.26286(31)	$a^{r}R$(10, 5)	NU2+NU3	$^{14}NH_3$
95	4583.39876(29)	$s^{P}R$(8, 1)	NU2+NU3	$^{14}NH_3$
96	4584.30685(32)	$s^{r}R$(9, 1)	NU2+NU3	$^{14}NH_3$
97	4585.50567(29)	$s^{r}R$(10, 3)	NU2+NU3	$^{14}NH_3$
98	4585.73740(51)			
99	4586.32033(20)	$a^{r}R$(8, 0)	NU2+NU3	$^{14}NH_3$
100	4586.36381(51)			
101	4587.04916(37)	$a^{P}R$(7, 2)	NU2+NU3	$^{14}NH_3$
102	4587.21121(36)			
103	4587.48900(29)	$a^{r}R$(9, 2)	NU2+NU3	$^{14}NH_3$
104	4588.36265(32)	$a^{P}R$(6, 4)	NU2+NU3	$^{14}NH_3$
105	4588.95026(121)			
106	4589.15696(37)	$a^{r}R$(10, 4)	NU2+NU3	$^{14}NH_3$
107	4590.19347(39)			
108	4590.79357(35)			
109	4591.06880(22)	$s^{r}R$(9, 0)	NU2+NU3	$^{14}NH_3$
110	4591.19012(36)	$s^{P}R$(7, 4)	NU2+NU3	$^{14}NH_3$
111	4591.27955(34)	$s^{P}R$(8, 2)	NU2+NU3	$^{14}NH_3$
112	4591.44718(58)	$s^{P}R$(6, 6)	NU2+NU3	$^{14}NH_3$
113	4591.65874(32)			
114	4592.54058(67)	$s^{r}R$(10, 2)	NU2+NU3	$^{14}NH_3$
115	4593.02449(125)			

- "LINE NO.": serial number identifying the spectral line in the map on the opposite page.
- "WAVENUMBER": experimental value.
- Uncertainty in the last digit of the wavenumber is given in parentheses.
- "pa" : parallel ; "pe" : perpendicular.

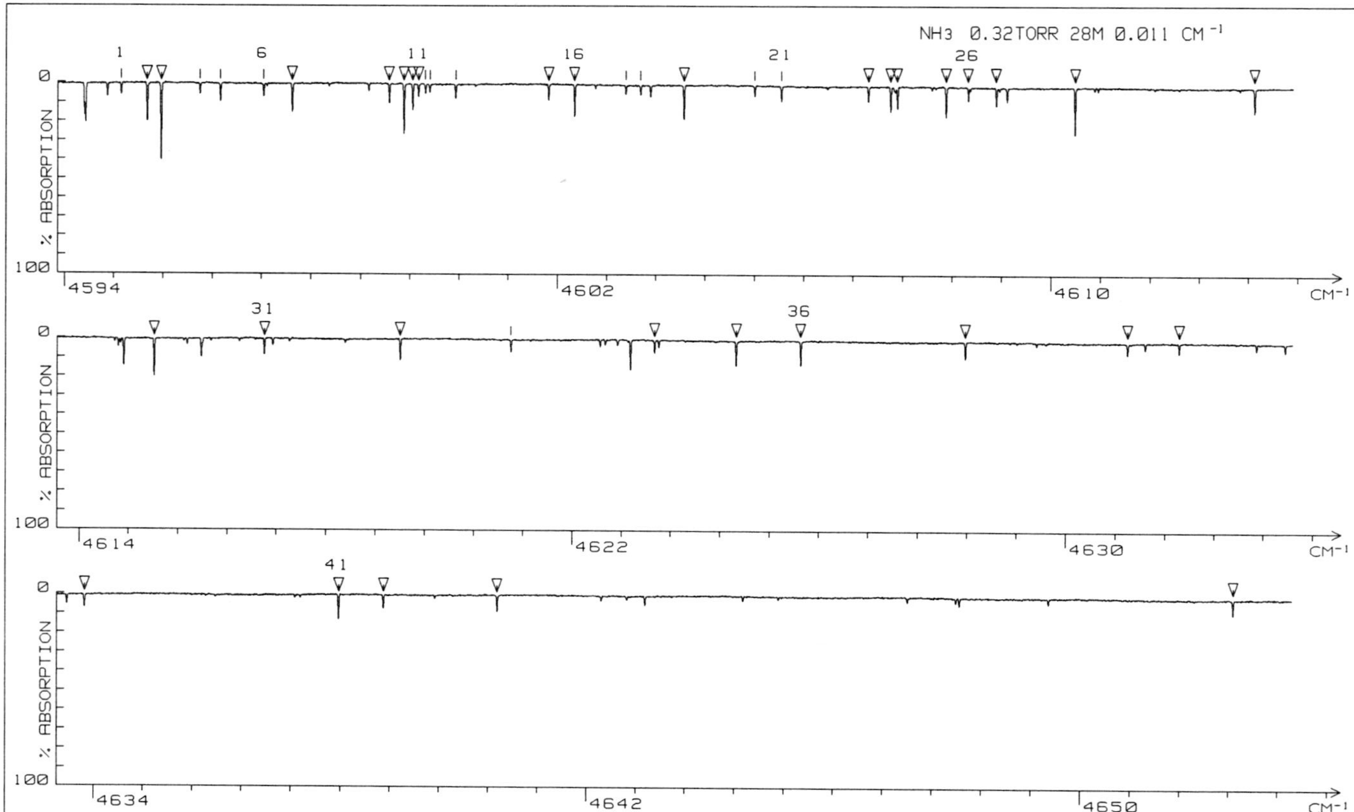

ISOTOPIC IDENTIFICATION – '∇' indicates $^{14}NH_3$; '|': unidentified.

LINE NO.	WAVENUMBER CM^{-1}, VAC.	ROTATIONAL ASSIGNMENTS	BAND	ISOTOPIC SPECIES
1	4594.92379(88)			
2	4595.34469(42)	$a^{r}R(10, 3)$	NU2+NU3	$^{14}NH_3$
3	4595.56803(22)	$a^{P}R(7, 3)$	NU2+NU3	$^{14}NH_3$
4	4596.19610(96)			
5	4596.52473(71)			
6	4597.23733(84)			
7	4597.69704(36)	$a^{P}R(6, 5)$	NU2+NU3	$^{14}NH_3$
8	4599.26484(45)	$s^{P}R(9, 1)$	NU2+NU3	$^{14}NH_3$
9	4599.50223(29)	$s^{P}R(8, 3)$	NU2+NU3	$^{14}NH_3$
10	4599.64413(39)	$s^{P}R(7, 5)$	NU2+NU3	$^{14}NH_3$
11	4599.73786(61)	$s^{r}R(10, 1)$	NU2+NU3	$^{14}NH_3$
12	4599.85130(114)			
13	4599.92416(131)			
14	4600.34149(72)			
15	4601.85475(50)	$a^{r}R(10, 2)$	NU2+NU3	$^{14}NH_3$
16	4602.27720(33)	$a^{P}R(8, 2)$	NU2+NU3	$^{14}NH_3$
17	4603.10662(109)			
18	4603.35190(85)			
19	4604.05819(37)	$a^{P}R(7, 4)$	NU2+NU3	$^{14}NH_3$
20	4605.20129(69)			
21	4605.63424(59)			
22	4607.04389(56)	$s^{P}R(9, 2)$	NU2+NU3	$^{14}NH_3$
23	4607.40409(41)	$a^{P}R(6, 6)$	NU2+NU3	$^{14}NH_3$
24	4607.51221(43)	$s^{P}R(8, 4)$	NU2+NU3	$^{14}NH_3$
25	4608.30004(41)	$s^{P}R(7, 6)$	NU2+NU3	$^{14}NH_3$
26	4608.66040(60)	$a^{r}R(10, 1)$	NU2+NU3	$^{14}NH_3$
27	4609.11369(62)	$a^{P}R(9, 1)$	NU2+NU3	$^{14}NH_3$
28	4610.38536(30)	$a^{P}R(8, 3)$	NU2+NU3	$^{14}NH_3$
29	4613.29767(40)	$a^{P}R(7, 5)$	NU2+NU3	$^{14}NH_3$
30	4615.22060(31)	$a^{r}R(10, 0)$	NU2+NU3	$^{14}NH_3$
31	4617.01311(54)	$a^{P}R(9, 2)$	NU2+NU3	$^{14}NH_3$
32	4619.20670(65)	$a^{P}R(8, 4)$	NU2+NU3	$^{14}NH_3$
33	4621.01143(87)			
34	4623.34521(100)	$s^{P}R(9, 4)$	NU2+NU3	$^{14}NH_3$
35	4624.66663(43)	$s^{P}R(8, 6)$	NU2+NU3	$^{14}NH_3$
36	4625.71286(59)	$a^{P}R(9, 3)$	NU2+NU3	$^{14}NH_3$
37	4628.37786(55)	$a^{P}R(8, 5)$	NU2+NU3	$^{14}NH_3$
38	4630.99736(74)	$s^{P}R(10, 3)$	NU2+NU3	$^{14}NH_3$
39	4631.83126(88)	$s^{P}R(9, 5)$	NU2+NU3	$^{14}NH_3$
40	4633.84667(104)	$a^{P}R(9, 4)$	NU2+NU3	$^{14}NH_3$
41	4637.97009(68)	$a^{P}R(8, 6)$	NU2+NU3	$^{14}NH_3$
42	4638.69979(68)	$a^{P}R(10, 3)$	NU2+NU3	$^{14}NH_3$
42		$s^{P}R(10, 4)$	NU2+NU3	$^{14}NH_3$
43	4640.54502(57)	$s^{P}R(9, 6)$	NU2+NU3	$^{14}NH_3$
44	4652.47683(66)	$a^{P}R(9, 6)$	NU2+NU3	$^{14}NH_3$

- "LINE NO.": serial number identifying the spectral line in the map on the opposite page.
- "WAVENUMBER": experimental value.
- Uncertainty in the last digit of the wavenumber is given in parentheses.
- "pa" : parallel ; "pe" : perpendicular.

N_2O 1.828TORR 40M 0.0054 CM-1

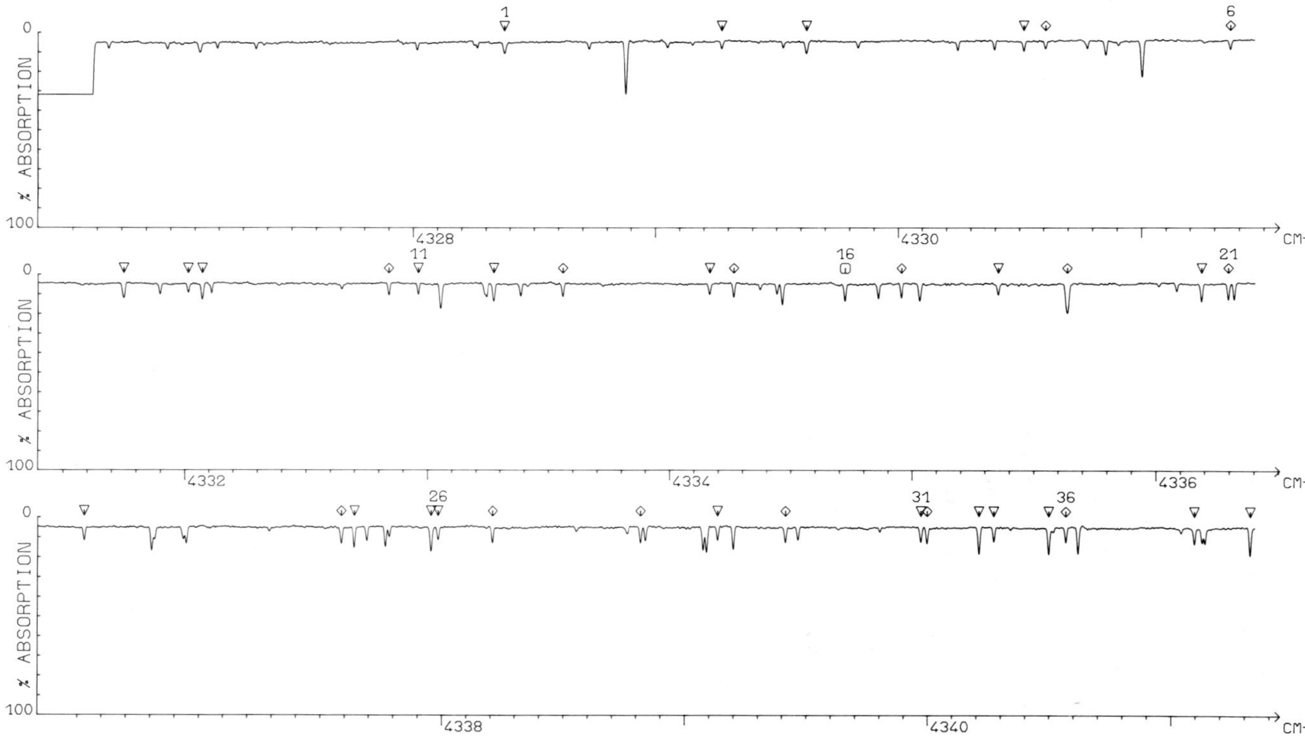

ISOTOPIC IDENTIFICATION – '▽' indicates $^{14}N^{14}N^{16}O$; '⊡': $^{15}N^{14}N^{16}O$; '◇': $^{14}N^{15}N^{16}O$; 'Y': $^{14}N^{14}N^{18}O$;
'✧' indicates $^{14}N^{14}N^{17}O$; '⊥': $^{15}N^{15}N^{16}O$; '△': $^{15}N^{14}N^{18}O$; '⊕': $^{14}N^{15}N^{18}O$.

LINE NO.	WAVENUMBER CM^{-1},VAC.	ROVIBRATIONAL ASSIGNMENT	ISOTOPIC SPECIES
1	4328.37963(73)	P31 0222E-0202E	14N14N16O
1		P31 0222F-0202F	14N14N16O
2	4329.27366(80)	P31 0220E-0200E	14N14N16O
3	4329.62308(64)	P30 0222E-0202E	14N14N16O
3		P30 0222F-0202F	14N14N16O
4	4330.51654(62)	P30 0220E-0200E	14N14N16O
5	4330.60690(82)	R 4 0020E-0000E	14N15N16O
6	4331.36523(67)	R 5 0020E-0000E	14N15N16O
7	4331.74916(48)	P29 0220E-0200E	14N14N16O
8	4332.01294(71)	P29 1020E-1000E	14N14N16O
9	4332.07042(39)	P28 0222E-0202E	14N14N16O
9		P28 0222F-0202F	14N14N16O
10	4332.84092(50)	R 7 0020E-0000E	14N15N16O
11	4332.96159(52)	P28 0220E-0200E	14N14N16O
12	4333.27437(36)	P27 0222E-0202E	14N14N16O
12		P27 0222F-0202F	14N14N16O
13	4333.55859(40)	R 8 0020E-0000E	14N15N16O
14	4334.16418(60)	P27 0220E-0200E	14N14N16O
15	4334.26341(37)	R 9 0020E-0000E	14N15N16O
16	4334.72206(34)	P37 0020E-0000E	15N14N16O
16		P47 0121E-0101E	14N14N16O
17	4334.95426(37)	R10 0020E-0000E	14N15N16O
18	4335.35330(57)	P26 0220E-0200E	14N14N16O
19	4335.63775(37)	R11 0020E-0000E	14N15N16O
19		R30 0220E-0200E	15N14N16O
20	4336.18755(34)	P46 0121E-0101E	14N14N16O
21	4336.29646(35)	R12 0020E-0000E	14N15N16O
22	4336.52899(41)	P25 0220E-0200E	14N14N16O
23	4337.58513(36)	R14 0020E-0000E	14N15N16O
24	4337.63923(30)	P45 0121E-0101E	14N14N16O
25	4337.95508(27)	P23 0222E-0202E	14N14N16O
25		P23 0222F-0202F	14N14N16O
26	4337.98407(50)	P24 1020E-1000E	14N14N16O
27	4338.20943(41)	R15 0020E-0000E	14N15N16O

LINE NO.	WAVENUMBER CM^{-1},VAC.	ROVIBRATIONAL ASSIGNMENT	ISOTOPIC SPECIES
28	4338.82007(34)	R16 0020E-0000E	14N15N16O
29	4339.13690(38)	P23 1020E-1000E	14N14N16O
30	4339.41732(36)	R17 0020E-0000E	14N15N16O
31	4339.97600(37)	P22 0220E-0200E	14N14N16O
32	4340.00150(35)	R18 0020E-0000E	14N15N16O
33	4340.21493(25)	P21 0222E-0202E	14N14N16O
33		P21 0222F-0202F	14N14N16O
34	4340.27596(44)	P22 1020E-1000E	14N14N16O
35	4340.50203(25)	P43 0121E-0101E	14N14N16O
36	4340.57214(35)	R19 0020E-0000E	14N15N16O
37	4341.09799(36)	P21 0220E-0200E	14N14N16O
38	4341.32473(24)	P20 0222E-0202E	14N14N16O

- "LINE NO.": serial number identifying the spectral line in the map on the opposite page.
- "WAVENUMBER": experimental value.
- Uncertainty in the last digit of the wavenumber is given in parentheses.

N_2O 1.828TORR 40M 0.0054 CM-1

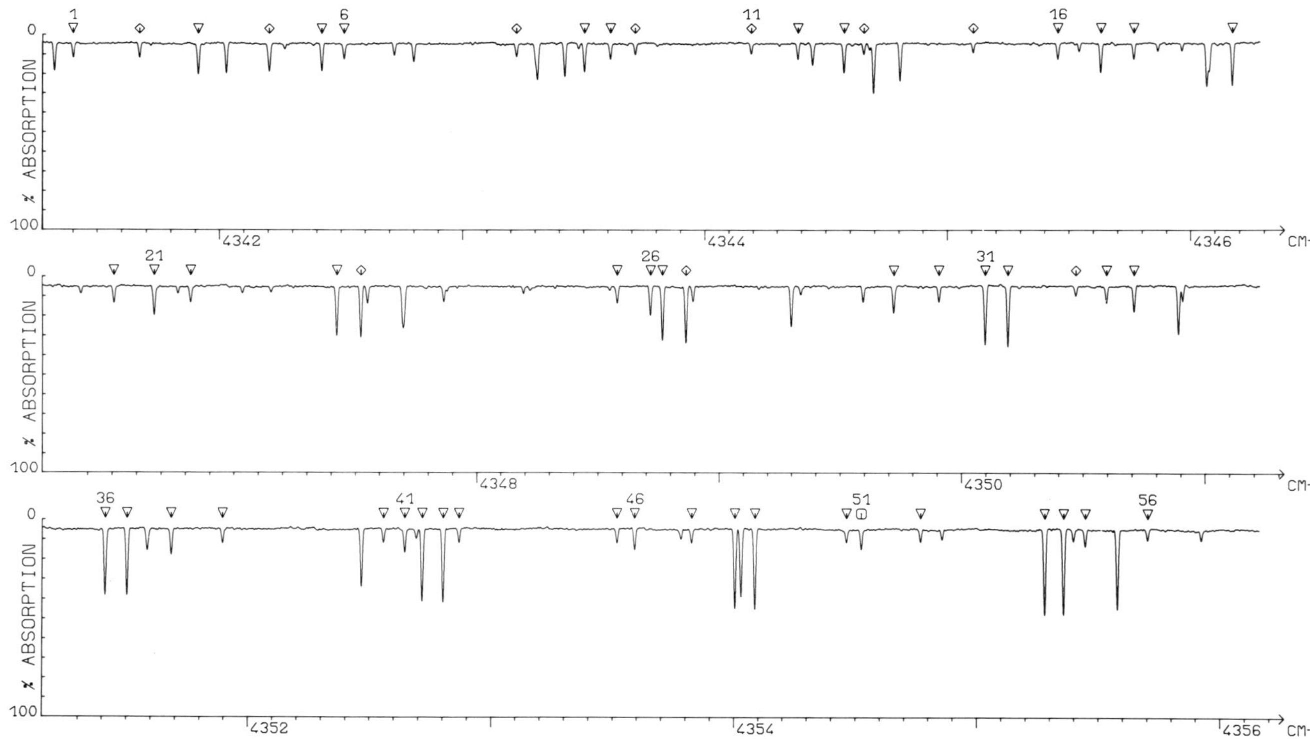

ISOTOPIC IDENTIFICATION – '▽' indicates $^{14}N^{14}N^{16}O$; '⊡': $^{15}N^{14}N^{16}O$; '◇': $^{14}N^{15}N^{16}O$; 'Y': $^{14}N^{14}N^{18}O$;
'✧' indicates $^{14}N^{14}N^{17}O$; '⊥': $^{15}N^{15}N^{16}O$; '△': $^{15}N^{14}N^{18}O$; '↑': $^{14}N^{15}N^{18}O$.

LINE NO.	WAVENUMBER CM^{-1}, VAC.	ROVIBRATIONAL ASSIGNMENT	ISOTOPIC SPECIES
1	4341.40109(40)	P21 1020E-1000E	14N14N160
2	4341.67313(38)	R21 0020E-0000E	14N15N160
3	4341.91304(23)	P42 0121E-0101E	14N14N160
4	4342.20499(40)	R22 0020E-0000E	14N15N160
5	4342.42089(25)	P19 0222E-0202E	14N14N160
5		P19 0222F-0202F	14N14N160
6	4342.51285(43)	P20 1020E-1000E	14N14N160
7	4343.22426(51)	R24 0020E-0000E	14N15N160
8	4343.50377(24)	P18 0222E-0202E	14N14N160
8		P18 0222F-0202F	14N14N160
9	4343.61047(35)	P19 1020E-1000E	14N14N160
10	4343.71431(55)	R25 0020E-0000E	14N15N160
11	4344.19098(58)	R26 0020E-0000E	14N15N160
12	4344.38442(33)	P18 0220E-0200E	14N14N160
13	4344.57334(24)	P17 0222E-0202E	14N14N160
13		P17 0222F-0202F	14N14N160
14	4344.65446(58)	R27 0020E-0000E	14N15N160
15	4345.10413(60)	R28 0020E-0000E	14N15N160
16	4345.45298(35)	P17 0220E-0200E	14N14N160
17	4345.62936(23)	P16 0222E-0202E	14N14N160
17		P16 0222F-0202F	14N14N160
18	4345.76435(41)	P17 1020E-1000E	14N14N160
19	4346.16984(20)	P39 0121E-0101E	14N14N160
20	4346.50836(39)	P16 0220E-0200E	14N14N160
21	4346.67183(24)	P15 0222E-0202E	14N14N160
21		P15 0222F-0202F	14N14N160
22	4346.82056(42)	P16 1020E-1000E	14N14N160
23	4347.42266(19)	P38 0121F-0101F	14N14N160
24	4347.52295(19)	R34 0020E-0000E	14N15N160
24		P38 0121E-0101E	14N14N160
25	4348.57830(34)	P14 0220E-0200E	14N14N160
26	4348.71657(23)	P13 0222E-0202E	14N14N160
26		P13 0222F-0202F	14N14N160
27	4348.76623(18)	P37 0121F-0101F	14N14N160

LINE NO.	WAVENUMBER CM^{-1}, VAC.	ROVIBRATIONAL ASSIGNMENT	ISOTOPIC SPECIES
28	4348.86296(18)	R38 0020E-0000E	14N15N160
28		P37 0121E-0101E	14N14N160
29	4349.71884(25)	P12 0222E-0202E	14N14N160
29		P12 0222F-0202F	14N14N160
30	4349.90615(41)	P13 1020E-1000E	14N14N160
31	4350.09635(18)	P36 0121F-0101F	14N14N160
32	4350.18942(18)	P36 0121E-0101E	14N14N160
33	4350.46883(65)	R44 0020E-0000E	14N15N160
34	4350.59470(38)	P12 0220E-0200E	14N14N160
35	4350.70738(25)	P11 0222E-0202E	14N14N160
35		P11 0222F-0202F	14N14N160
36	4351.41290(18)	P35 0121F-0101F	14N14N160
37	4351.50232(18)	P35 0121E-0101E	14N14N160
38	4351.68258(25)	P10 0222E-0202E	14N14N160
38		P10 0222F-0202F	14N14N160
39	4351.89444(44)	P11 1020E-1000E	14N14N160
39		R 8 0121E-0101E	15N14N160
40	4352.55726(44)	P10 0220E-0200E	14N14N160
41	4352.64431(27)	P 9 0222E-0202E	14N14N160
41		P 9 0222F-0202F	14N14N160
42	4352.71591(17)	P34 0121F-0101F	14N14N160
43	4352.80180(17)	P34 0121E-0101E	14N14N160
44	4352.86737(41)	P10 1020E-1000E	14N14N160
45	4353.51862(40)	P 9 0220E-0200E	14N14N160
46	4353.59255(29)	P 8 0222E-0202E	14N14N160
46		P 8 0222F-0202F	14N14N160
47	4353.82682(40)	P 9 1020E-1000E	14N14N160
48	4354.00539(17)	P33 0121F-0101F	14N14N160
49	4354.08782(17)	P33 0121E-0101E	14N14N160
50	4354.46636(51)	P 8 0220E-0200E	14N14N160
51	4354.52745(30)	R12 0121E-0101E	15N14N160
51		P 7 0222E-0202E	14N14N160
51		P 7 0222F-0202F	14N14N160

- "LINE NO.": serial number identifying the spectral line in the map on the opposite page.
- "WAVENUMBER": experimental value.
- Uncertainty in the last digit of the wavenumber is given in parentheses.

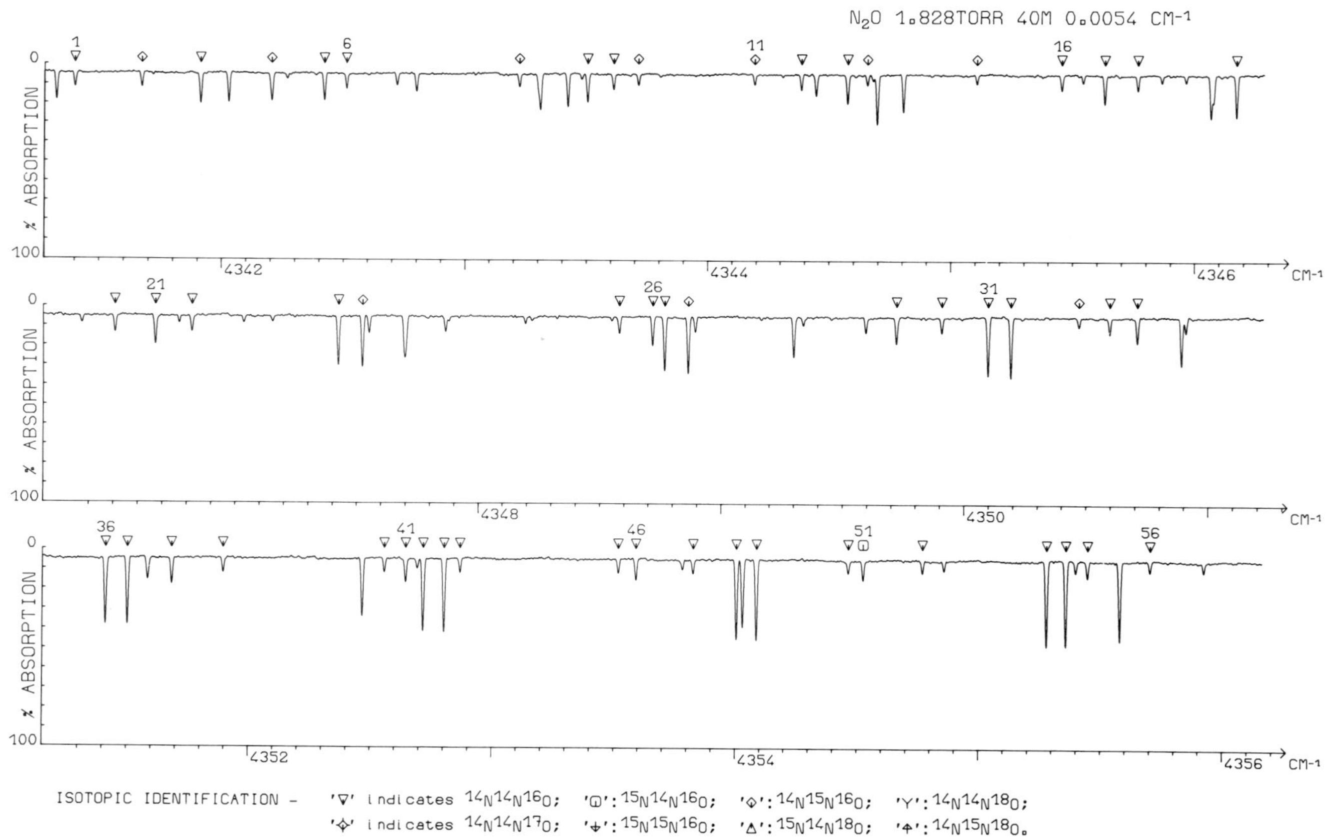
N2O 1.828TORR 40M 0.0054 CM-1
1
6
11
16
21
26
31
36
41
46
51
56
0
% ABSORPTION
100
4342
4344
4346
4348
4350
4352
4354
4356
CM-1
ISOTOPIC IDENTIFICATION – '▽' indicates 14N14N16O; '⊡': 15N14N16O; '◇': 14N15N16O; 'Y': 14N14N18O;
'✧' indicates 14N14N17O; '⏚': 15N15N16O; '△': 15N14N18O; '✢': 14N15N18O.

LINE NO.	WAVENUMBER CM^{-1}, VAC.	ROVIBRATIONAL ASSIGNMENT	ISOTOPIC SPECIES
52	4354.77244(47)	P 8 1020E-1000E	14N14N16O
53	4355.28131(17)	P32 0121F-0101F	14N14N16O
54	4355.36029(17)	P32 0121E-0101E	14N14N16O
55	4355.44875(33)	P 6 0222E-0202E	14N14N16O
55		P 6 0222F-0202F	14N14N16O
56	4355.70384(55)	P 7 1020E-1000E	14N14N16O

- "LINE NO.": serial number identifying the spectral line in the map on the opposite page.
- "WAVENUMBER": experimental value.
- Uncertainty in the last digit of the wavenumber is given in parentheses.

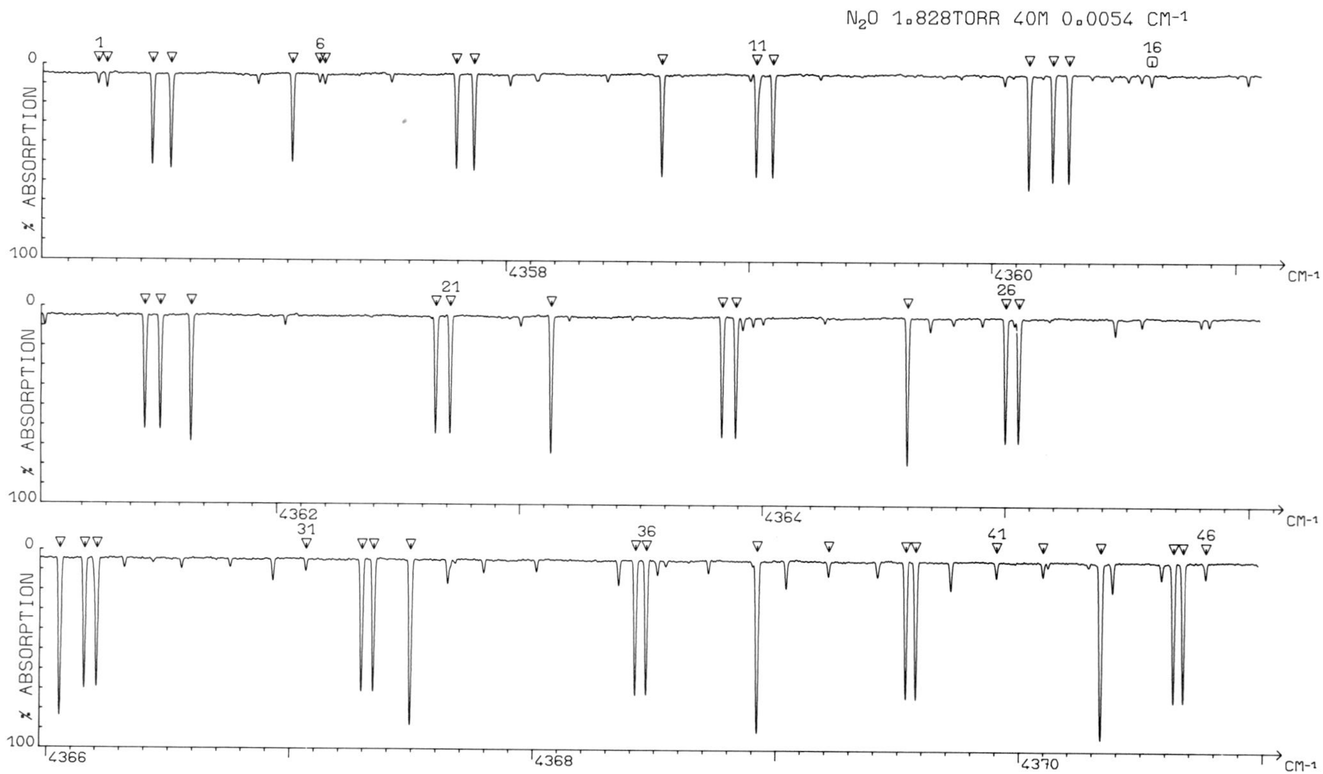

ISOTOPIC IDENTIFICATION – '▽' indicates $^{14}N^{14}N^{16}O$; 'ⓘ': $^{15}N^{14}N^{16}O$; '◇': $^{14}N^{15}N^{16}O$; 'Y': $^{14}N^{14}N^{18}O$;
'✧' indicates $^{14}N^{14}N^{17}O$; '⊥': $^{15}N^{15}N^{16}O$; '△': $^{15}N^{14}N^{18}O$; '✦': $^{14}N^{15}N^{18}O$.

LINE NO.	WAVENUMBER CM^{-1}, VAC.	ROVIBRATIONAL ASSIGNMENT	ISOTOPIC SPECIES
1	4356.32124(55)	P 6 0220E-0200E	14N14N160
2	4356.35651(37)	P 5 0222E-0202E	14N14N160
2		P 5 0222F-0202F	14N14N160
2		R15 0121E-0101E	15N14N160
3	4356.54374(17)	P31 0121F-0101F	14N14N160
4	4356.61953(17)	P31 0121E-0101E	14N14N160
5	4357.11748(17)	P51 0020E-0000E	14N14N160
6	4357.22863(68)	P 5 0220E-0200E	14N14N160
7	4357.25048(65)	P 4 0222E-0202E	14N14N160
7		P 4 0222F-0202F	14N14N160
8	4357.79262(17)	P30 0121F-0101F	14N14N160
9	4357.86489(17)	P30 0121E-0101E	14N14N160
10	4358.64060(17)	P50 0020E-0000E	14N14N160
11	4359.02782(18)	P29 0121F-0101F	14N14N160
12	4359.09689(17)	P29 0121E-0101E	14N14N160
13	4360.15025(16)	P49 0020E-0000E	14N14N160
14	4360.24960(17)	P28 0121F-0101F	14N14N160
15	4360.31540(17)	P28 0121E-0101E	14N14N160
16	4360.65209(54)	R23 0121E-0101E	15N14N160
17	4361.45773(16)	P27 0121F-0101F	14N14N160
18	4361.52042(16)	P27 0121E-0101E	14N14N160
19	4361.64615(16)	P48 0020E-0000E	14N14N160
20	4362.65229(16)	P26 0121F-0101F	14N14N160
21	4362.71192(16)	P26 0121E-0101E	14N14N160
22	4363.12860(16)	P47 0020E-0000E	14N14N160
23	4363.83328(16)	P25 0121F-0101F	14N14N160
24	4363.88988(16)	P25 0121E-0101E	14N14N160
25	4364.59728(16)	P46 0020E-0000E	14N14N160
26	4365.00067(16)	P24 0121F-0101F	14N14N160
27	4365.05431(16)	P24 0121E-0101E	14N14N160
28	4366.05245(16)	P45 0020E-0000E	14N14N160
29	4366.15445(16)	P23 0121F-0101F	14N14N160
30	4366.20511(17)	P23 0121E-0101E	14N14N160

LINE NO.	WAVENUMBER CM^{-1}, VAC.	ROVIBRATIONAL ASSIGNMENT	ISOTOPIC SPECIES
31	4367.06320(49)	R 6 0220E-0200E	14N14N160
32	4367.29454(16)	P22 0121F-0101F	14N14N160
33	4367.34260(16)	P22 0121E-0101E	14N14N160
34	4367.49404(16)	P44 0020E-0000E	14N14N160
35	4368.42126(16)	P21 0121F-0101F	14N14N160
36	4368.46641(16)	P21 0121E-0101E	14N14N160
37	4368.92202(16)	P43 0020E-0000E	14N14N160
38	4369.21719(36)	R 9 0220E-0200E	14N14N160
39	4369.53425(16)	P20 0121F-0101F	14N14N160
40	4369.57672(16)	P20 0121E-0101E	14N14N160
41	4369.90816(35)	R10 0220E-0200E	14N14N160
42	4370.09887(44)	R10 1020E-1000E	14N14N160
43	4370.33638(16)	P42 0020E-0000E	14N14N160
44	4370.63363(16)	P19 0121F-0101F	14N14N160
45	4370.67341(16)	P19 0121E-0101E	14N14N160
46	4370.76627(59)	R11 1020E-1000E	14N14N160

- "LINE NO.": serial number identifying the spectral line in the map on the opposite page.
- "WAVENUMBER": experimental value.
- Uncertainty in the last digit of the wavenumber is given in parentheses.

N_2O 1.828TORR 40M 0.0054 CM-1

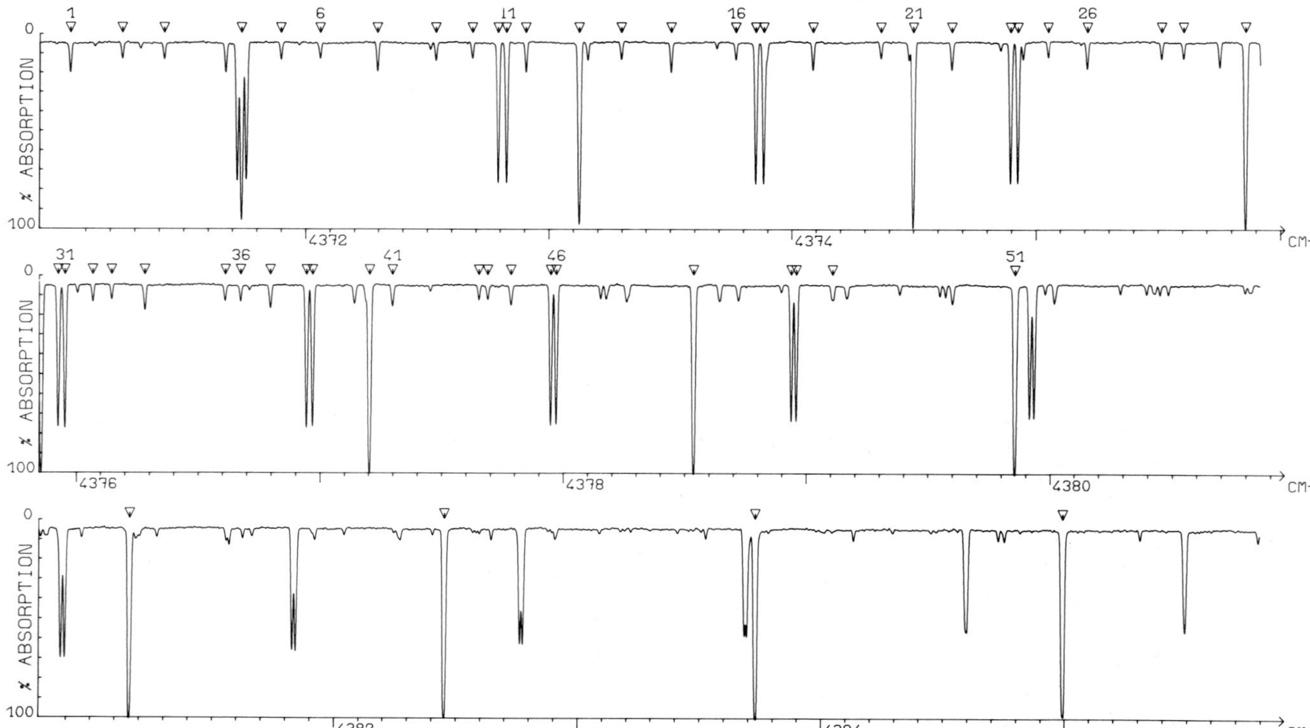

ISOTOPIC IDENTIFICATION – '▽' indicates $^{14}N^{14}N^{16}O$; '◫': $^{15}N^{14}N^{16}O$; '◇': $^{14}N^{15}N^{16}O$; 'Y': $^{14}N^{14}N^{18}O$;
'✧' indicates $^{14}N^{14}N^{17}O$; '⊥': $^{15}N^{15}N^{16}O$; '△': $^{15}N^{14}N^{18}O$; '✦': $^{14}N^{15}N^{18}O$.

LINE NO.	WAVENUMBER CM^{-1}, VAC.	ROVIBRATIONAL ASSIGNMENT	ISOTOPIC SPECIES
1	4371.03543(23)	R13 0222E-0202E	14N14N16O
1		R13 0222F-0202F	14N14N16O
2	4371.24902(34)	R12 0220E-0200E	14N14N16O
3	4371.41991(35)	R12 1020E-1000E	14N14N16O
4	4371.73711(17)	P41 0020E-0000E	14N14N16O
5	4371.89915(32)	R13 0220E-0200E	14N14N16O
6	4372.05948(35)	R13 1020E-1000E	14N14N16O
7	4372.29607(23)	R15 0222E-0202E	14N14N16O
7		R15 0222F-0202F	14N14N16O
8	4372.53568(32)	R14 0220E-0200E	14N14N16O
9	4372.68531(35)	R14 1020E-1000E	14N14N16O
10	4372.79162(16)	P17 0121F-0101F	14N14N16O
11	4372.82622(16)	P17 0121E-0101E	14N14N16O
12	4372.90616(23)	R16 0222E-0202E	14N14N16O
12		R16 0222F-0202F	14N14N16O
13	4373.12424(16)	P40 0020E-0000E	14N14N16O
14	4373.29677(37)	R15 1020E-1000E	14N14N16O
15	4373.50268(23)	R17 0222E-0202E	14N14N16O
15		R17 0222F-0202F	14N14N16O
16	4373.76809(32)	R16 0220E-0200E	14N14N16O
17	4373.85014(16)	P16 0121F-0101F	14N14N16O
18	4373.88228(16)	P16 0121E-0101E	14N14N16O
19	4374.08543(24)	R18 0222E-0202E	14N14N16O
19		R18 0222F-0202F	14N14N16O
20	4374.36402(33)	R17 0220E-0200E	14N14N16O
21	4374.49773(16)	P39 0020E-0000E	14N14N16O
22	4374.65475(25)	R19 0222E-0202E	14N14N16O
22		R19 0222F-0202F	14N14N16O
23	4374.89503(16)	P15 0121F-0101F	14N14N16O
24	4374.92477(16)	P15 0121E-0101E	14N14N16O
25	4375.04845(34)	R18 1020E-1000E	14N14N16O
26	4375.21038(25)	R20 0222E-0202E	14N14N16O
26		R20 0222F-0202F	14N14N16O
27	4375.51443(37)	R19 0220E-0200E	14N14N16O
28	4375.60432(35)	R19 1020E-1000E	14N14N16O
29	4375.85760(16)	P38 0020E-0000E	14N14N16O
30	4375.92633(16)	P14 0121F-0101F	14N14N16O
31	4375.95367(16)	P14 0121E-0101E	14N14N16O
31		R 2 0020E-0000E	15N14N16O
32	4376.06923(33)	R20 0220E-0200E	14N14N16O
33	4376.14602(37)	R20 1020E-1000E	14N14N16O
34	4376.28107(28)	R22 0222E-0202E	14N14N16O
34		R22 0222F-0202F	14N14N16O
35	4376.61045(36)	R21 0220E-0200E	14N14N16O
36	4376.67407(34)	R21 1020E-1000E	14N14N16O
37	4376.79591(29)	R23 0222E-0202E	14N14N16O
37		R23 0222F-0202F	14N14N16O
38	4376.94397(16)	P13 0121F-0101F	14N14N16O
39	4376.96897(16)	P13 0121E-0101E	14N14N16O
40	4377.20381(16)	P37 0020E-0000E	14N14N16O
41	4377.29747(37)	R24 0222E-0202E	14N14N16O
41		R24 0222F-0202F	14N14N16O
42	4377.65214(43)	R23 0220E-0200E	14N14N16O
43	4377.68821(41)	R23 1020E-1000E	14N14N16O
44	4377.78480(35)	R25 0222E-0202E	14N14N16O
44		R25 0222F-0202F	14N14N16O
45	4377.94797(16)	P12 0121F-0101F	14N14N16O
46	4377.97071(16)	P12 0121E-0101E	14N14N16O
47	4378.53639(16)	P36 0020E-0000E	14N14N16O
48	4378.93833(16)	P11 0121F-0101F	14N14N16O
49	4378.95889(16)	P11 0121E-0101E	14N14N16O
50	4379.10825(62)	R26 0220E-0200E	14N14N16O
51	4379.85531(16)	P35 0020E-0000E	14N14N16O
52	4381.16056(16)	P34 0020E-0000E	14N14N16O
53	4382.45217(16)	P33 0020E-0000E	14N14N16O
54	4383.73009(16)	P32 0020E-0000E	14N14N16O
55	4384.99435(16)	P31 0020E-0000E	14N14N16O

- "LINE NO.": serial number identifying the spectral line in the map on the opposite page.
- "WAVENUMBER": experimental value.
- Uncertainty in the last digit of the wavenumber is given in parentheses.

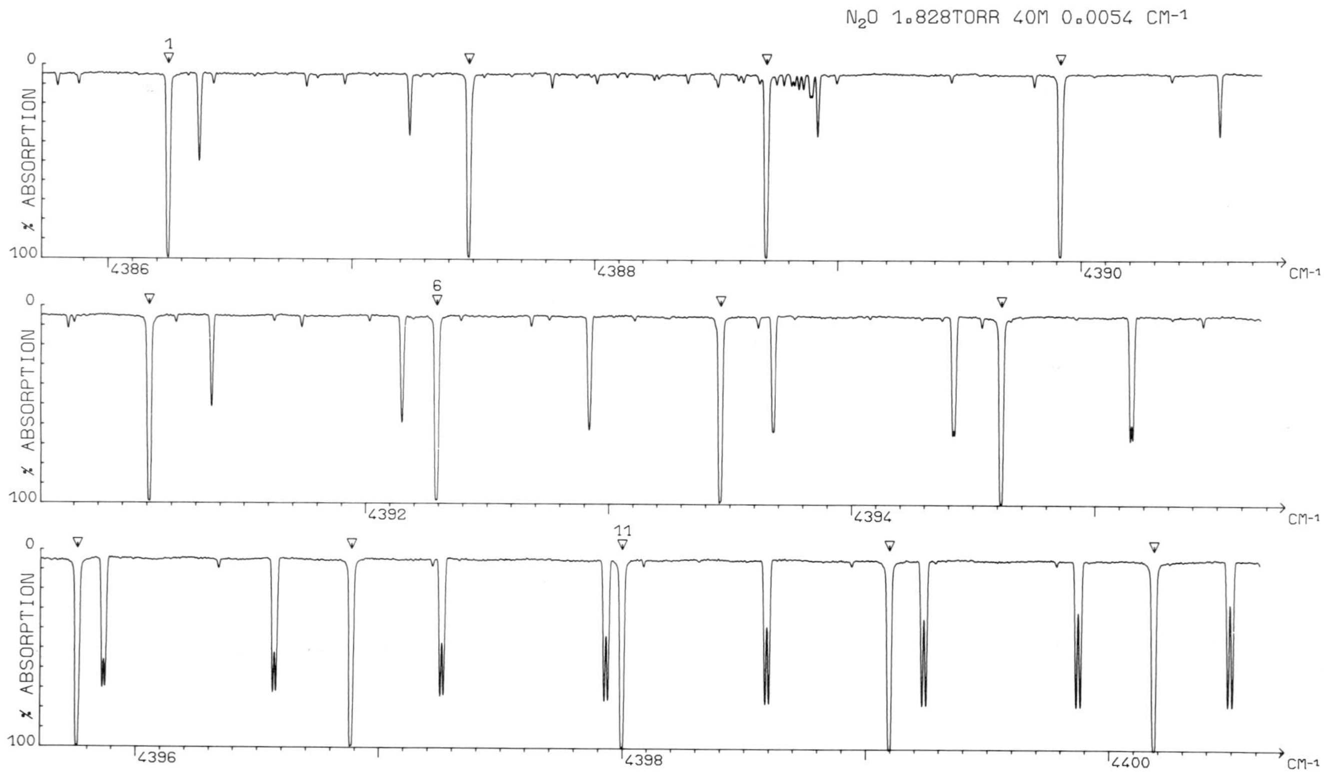

ISOTOPIC IDENTIFICATION – '▽' indicates $^{14}N^{14}N^{16}O$; '⌑': $^{15}N^{14}N^{16}O$; '◇': $^{14}N^{15}N^{16}O$; 'Y': $^{14}N^{14}N^{18}O$;
'✧' indicates $^{14}N^{14}N^{17}O$; '⏚': $^{15}N^{15}N^{16}O$; '△': $^{15}N^{14}N^{18}O$; '⚘': $^{14}N^{15}N^{18}O$.

LINE NO.	WAVENUMBER CM^{-1}, VAC.	ROVIBRATIONAL ASSIGNMENT	ISOTOPIC SPECIES
1	4386.24496(16)	P30 0020E-0000E	14N14N160
2	4387.48184(16)	P29 0020E-0000E	14N14N160
3	4388.70504(16)	P28 0020E-0000E	14N14N160
4	4389.91460(16)	P27 0020E-0000E	14N14N160
5	4391.11038(16)	P26 0020E-0000E	14N14N160
6	4392.29251(16)	P25 0020E-0000E	14N14N160
7	4393.46089(16)	P24 0020E-0000E	14N14N160
8	4394.61559(16)	P23 0020E-0000E	14N14N160
9	4395.75657(16)	P22 0020E-0000E	14N14N160
10	4396.88383(16)	P21 0020E-0000E	14N14N160
11	4397.99736(16)	P20 0020E-0000E	14N14N160
12	4399.09715(16)	P19 0020E-0000E	14N14N160
13	4400.18320(16)	P18 0020E-0000E	14N14N160

- "LINE NO.": serial number identifying the spectral line in the map on the opposite page.
- "WAVENUMBER": experimental value.
- Uncertainty in the last digit of the wavenumber is given in parentheses.

N_2O 1.828TORR 40M 0.0054 CM-1

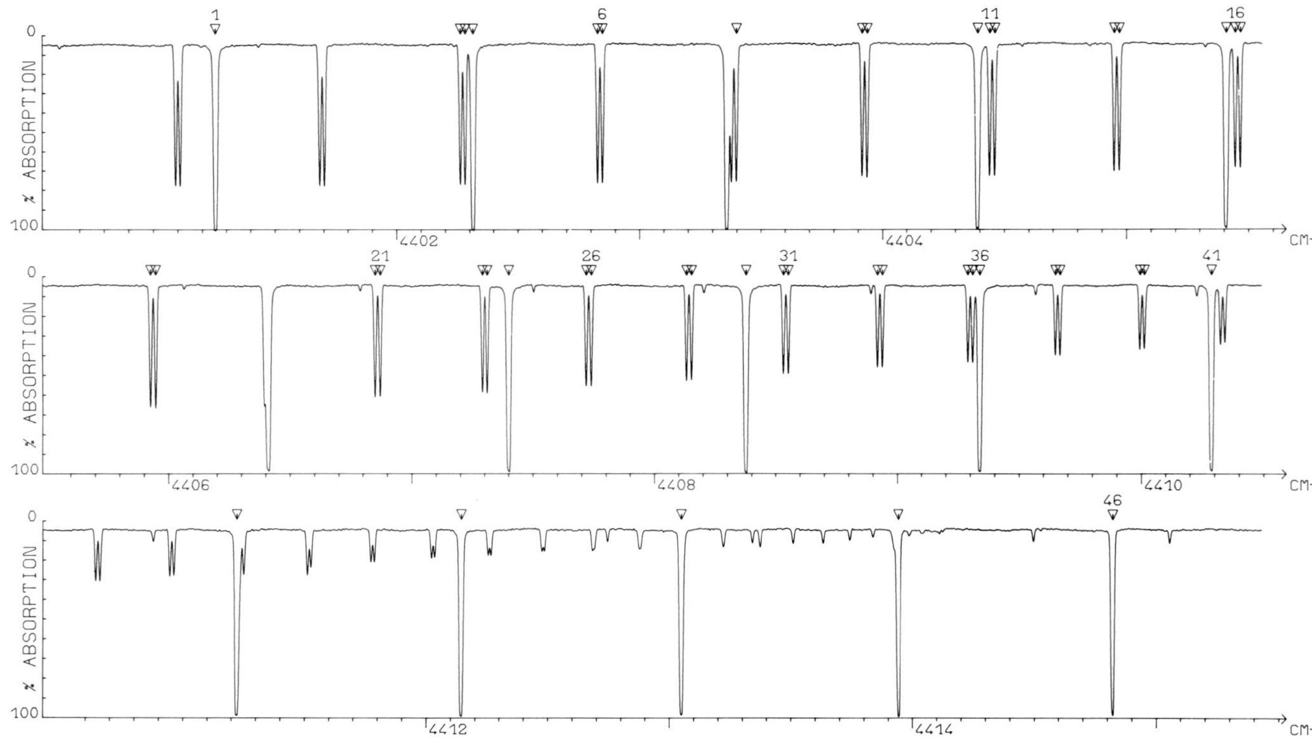

ISOTOPIC IDENTIFICATION – '▽' indicates $^{14}N^{14}N^{16}O$; '⊡': $^{15}N^{14}N^{16}O$; '◇': $^{14}N^{15}N^{16}O$; 'Y': $^{14}N^{14}N^{18}O$;
'◇' indicates $^{14}N^{14}N^{17}O$; '⏚': $^{15}N^{15}N^{16}O$; '△': $^{15}N^{14}N^{18}O$; '⏀': $^{14}N^{15}N^{18}O$.

LINE NO.	WAVENUMBER CM^{-1}, VAC.	ROVIBRATIONAL ASSIGNMENT	ISOTOPIC SPECIES
1	4401.25553(16)	P17 0020E-0000E	14N14N160
2	4402.26194(17)	R18 0121E-0101E	14N14N160
3	4402.28126(17)	R18 0121F-0101F	14N14N160
4	4402.31412(16)	P16 0020E-0000E	14N14N160
5	4402.82669(17)	R19 0121E-0101E	14N14N160
6	4402.84638(17)	R19 0121F-0101F	14N14N160
7	4403.39775(17)	R20 0121F-0101F	14N14N160
8	4403.91499(16)	R21 0121E-0101E	14N14N160
9	4403.93538(16)	R21 0121F-0101F	14N14N160
10	4404.38998(16)	P14 0020E-0000E	14N14N160
11	4404.43853(16)	R22 0121E-0101E	14N14N160
12	4404.45917(16)	R22 0121F-0101F	14N14N160
13	4404.94833(16)	R23 0121E-0101E	14N14N160
14	4404.96920(16)	R23 0121F-0101F	14N14N160
15	4405.40731(16)	P13 0020E-0000E	14N14N160
16	4405.44438(17)	R24 0121E-0101E	14N14N160
17	4405.46535(17)	R24 0121F-0101F	14N14N160
18	4405.92671(17)	R25 0121E-0101E	14N14N160
19	4405.94777(17)	R25 0121F-0101F	14N14N160
20	4406.85012(17)	R27 0121E-0101E	14N14N160
21	4406.87119(17)	R27 0121F-0101F	14N14N160
22	4407.29120(17)	R28 0121E-0101E	14N14N160
23	4407.31217(17)	R28 0121F-0101F	14N14N160
24	4407.40063(16)	P11 0020E-0000E	14N14N160
25	4407.71858(17)	R29 0121E-0101E	14N14N160
26	4407.73936(17)	R29 0121F-0101F	14N14N160
27	4408.13212(17)	R30 0121E-0101E	14N14N160
28	4408.15270(17)	R30 0121F-0101F	14N14N160
29	4408.37663(16)	P10 0020E-0000E	14N14N160
30	4408.53197(17)	R31 0121E-0101E	14N14N160
31	4408.55226(17)	R31 0121F-0101F	14N14N160
32	4408.91800(17)	R32 0121E-0101E	14N14N160
33	4408.93791(17)	R32 0121F-0101F	14N14N160
34	4409.29027(17)	R33 0121E-0101E	14N14N160
35	4409.30985(17)	R33 0121F-0101F	14N14N160
36	4409.33886(16)	P 9 0020E-0000E	14N14N160
37	4409.64882(18)	R34 0121E-0101E	14N14N160
38	4409.66784(18)	R34 0121F-0101F	14N14N160
39	4409.99348(21)	R35 0121E-0101E	14N14N160
40	4410.01202(19)	R35 0121F-0101F	14N14N160
41	4410.28731(16)	P 8 0020E-0000E	14N14N160
42	4411.22197(16)	P 7 0020E-0000E	14N14N160
43	4412.14278(16)	P 6 0020E-0000E	14N14N160
44	4413.04992(16)	P 5 0020E-0000E	14N14N160
45	4413.94313(16)	P 4 0020E-0000E	14N14N160
46	4414.82258(16)	P 3 0020E-0000E	14N14N160

- "LINE NO.": serial number identifying the spectral line in the map on the opposite page.
- "WAVENUMBER": experimental value.
- Uncertainty in the last digit of the wavenumber is given in parentheses.

N_2O 1.828TORR 40M 0.0054 CM-1

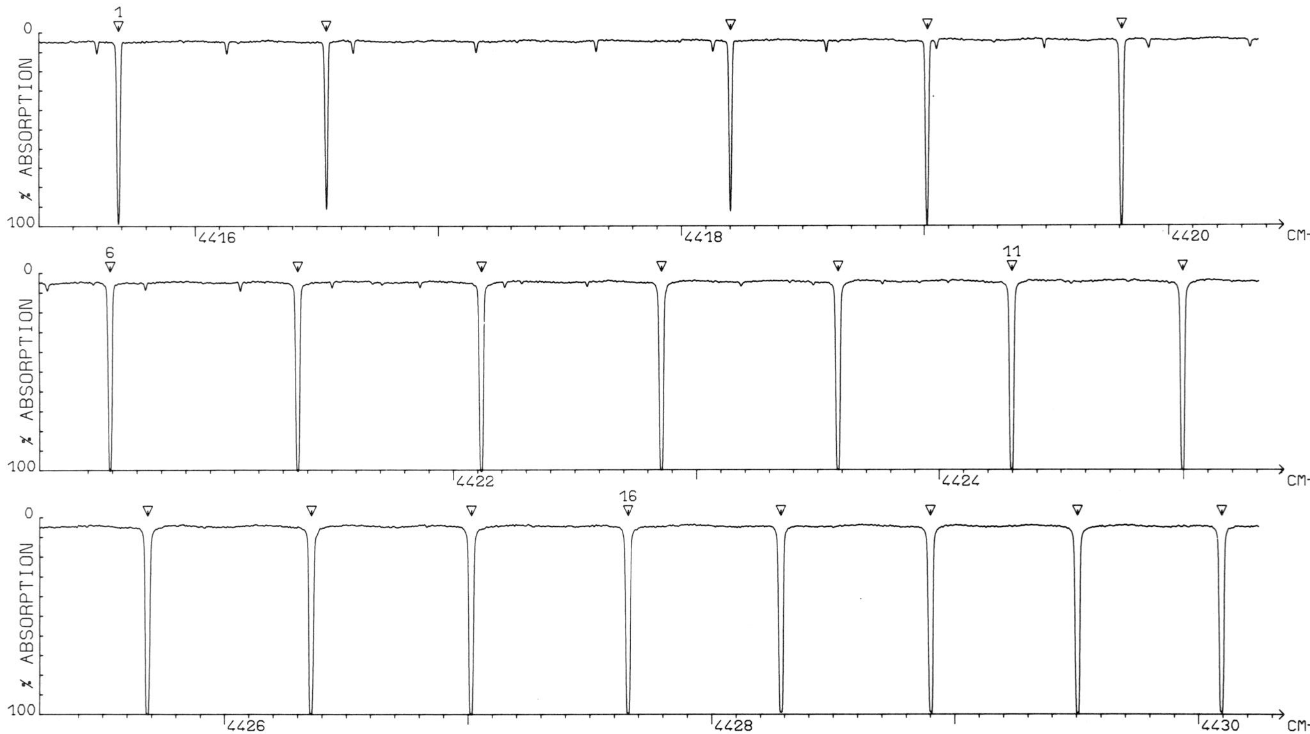

ISOTOPIC IDENTIFICATION – '▽' indicates $^{14}N^{14}N^{16}O$; '⊡': $^{15}N^{14}N^{16}O$; '◇': $^{14}N^{15}N^{16}O$; 'Y': $^{14}N^{14}N^{18}O$;
'✧' indicates $^{14}N^{14}N^{17}O$; '⊥': $^{15}N^{15}N^{16}O$; '△': $^{15}N^{14}N^{18}O$; '✦': $^{14}N^{15}N^{18}O$.

LINE NO.	WAVENUMBER CM^{-1}, VAC.	ROVIBRATIONAL ASSIGNMENT	ISOTOPIC SPECIES
1	4415.68819(16)	P 2 0020E-0000E	14N14N16O
2	4416.54005(16)	P 1 0020E-0000E	14N14N16O
3	4418.20224(16)	R 0 0020E-0000E	14N14N16O
4	4419.01264(16)	R 1 0020E-0000E	14N14N16O
5	4419.80919(16)	R 2 0020E-0000E	14N14N16O
6	4420.59191(16)	R 3 0020E-0000E	14N14N16O
7	4421.36077(16)	R 4 0020E-0000E	14N14N16O
8	4422.11583(16)	R 5 0020E-0000E	14N14N16O
9	4422.85702(16)	R 6 0020E-0000E	14N14N16O
10	4423.58435(16)	R 7 0020E-0000E	14N14N16O
11	4424.29784(16)	R 8 0020E-0000E	14N14N16O
12	4424.99745(16)	R 9 0020E-0000E	14N14N16O
13	4425.68319(16)	R10 0020E-0000E	14N14N16O
14	4426.35509(16)	R11 0020E-0000E	14N14N16O
15	4427.01311(16)	R12 0020E-0000E	14N14N16O
16	4427.65726(16)	R13 0020E-0000E	14N14N16O
17	4428.28754(16)	R14 0020E-0000E	14N14N16O
18	4428.90393(16)	R15 0020E-0000E	14N14N16O
19	4429.50642(16)	R16 0020E-0000E	14N14N16O
20	4430.09504(16)	R17 0020E-0000E	14N14N16O

- "LINE NO.": serial number identifying the spectral line in the map on the opposite page.
- "WAVENUMBER": experimental value.
- Uncertainty in the last digit of the wavenumber is given in parentheses.

N_2O 1.828TORR 40M 0.0054 CM-1

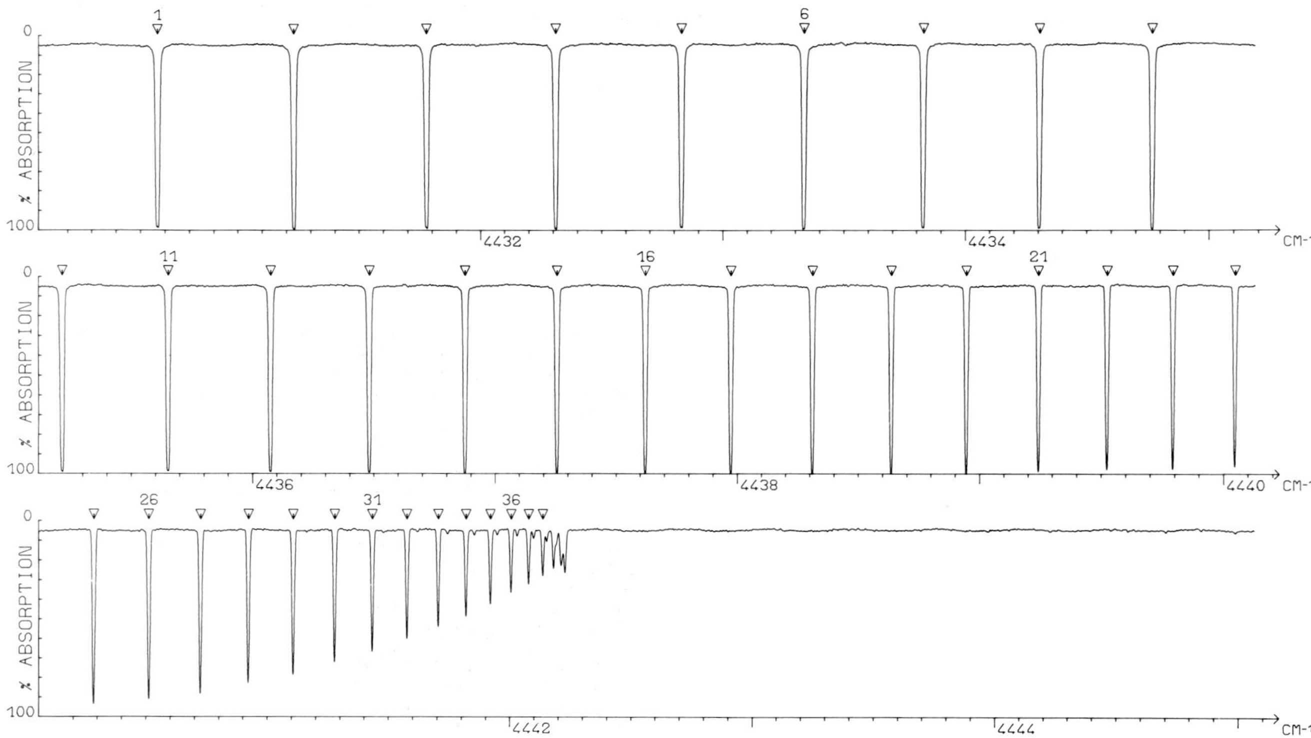

ISOTOPIC IDENTIFICATION – '▽' indicates $^{14}N^{14}N^{16}O$; '◫': $^{15}N^{14}N^{16}O$; '⟐': $^{14}N^{15}N^{16}O$; 'Y': $^{14}N^{14}N^{18}O$;
'⟡' indicates $^{14}N^{14}N^{17}O$; '⍖': $^{15}N^{15}N^{16}O$; '△': $^{15}N^{14}N^{18}O$; '⍏': $^{14}N^{15}N^{18}O$.

LINE NO.	WAVENUMBER CM^{-1}, VAC.	ROVIBRATIONAL ASSIGNMENT	ISOTOPIC SPECIES	LINE NO.	WAVENUMBER CM^{-1}, VAC.	ROVIBRATIONAL ASSIGNMENT	ISOTOPIC SPECIES
1	4430.66977(16)	R18 0020E-0000E	14N14N160	31	4441.43380(17)	R48 0020E-0000E	14N14N160
2	4431.23058(16)	R19 0020E-0000E	14N14N160	32	4441.57618(17)	R49 0020E-0000E	14N14N160
3	4431.77751(16)	R20 0020E-0000E	14N14N160	33	4441.70451(17)	R50 0020E-0000E	14N14N160
4	4432.31054(16)	R21 0020E-0000E	14N14N160	34	4441.81888(17)	R51 0020E-0000E	14N14N160
5	4432.82964(16)	R22 0020E-0000E	14N14N160	35	4441.91927(18)	R52 0020E-0000E	14N14N160
6	4433.33483(16)	R23 0020E-0000E	14N14N160	36	4442.00562(18)	R53 0020E-0000E	14N14N160
7	4433.82614(16)	R24 0020E-0000E	14N14N160	37	4442.07795(19)	R54 0020E-0000E	14N14N160
8	4434.30349(16)	R25 0020E-0000E	14N14N160	38	4442.13628(20)	R55 0020E-0000E	14N14N160
9	4434.76692(16)	R26 0020E-0000E	14N14N160				
10	4435.21647(16)	R27 0020E-0000E	14N14N160				
11	4435.65205(16)	R28 0020E-0000E	14N14N160				
12	4436.07371(16)	R29 0020E-0000E	14N14N160				
13	4436.48142(16)	R30 0020E-0000E	14N14N160				
14	4436.87520(16)	R31 0020E-0000E	14N14N160				
15	4437.25504(16)	R32 0020E-0000E	14N14N160				
16	4437.62094(16)	R33 0020E-0000E	14N14N160				
17	4437.97288(16)	R34 0020E-0000E	14N14N160				
18	4438.31088(16)	R35 0020E-0000E	14N14N160				
19	4438.63492(16)	R36 0020E-0000E	14N14N160				
20	4438.94500(16)	R37 0020E-0000E	14N14N160				
21	4439.24111(16)	R38 0020E-0000E	14N14N160				
22	4439.52328(16)	R39 0020E-0000E	14N14N160				
23	4439.79147(16)	R40 0020E-0000E	14N14N160				
24	4440.04567(16)	R41 0020E-0000E	14N14N160				
25	4440.28592(16)	R42 0020E-0000E	14N14N160				
26	4440.51218(16)	R43 0020E-0000E	14N14N160				
27	4440.72447(16)	R44 0020E-0000E	14N14N160				
28	4440.92280(16)	R45 0020E-0000E	14N14N160				
29	4441.10711(17)	R46 0020E-0000E	14N14N160				
30	4441.27745(17)	R47 0020E-0000E	14N14N160				

- "LINE NO.": serial number identifying the spectral line in the map on the opposite page.
- "WAVENUMBER": experimental value.
- Uncertainty in the last digit of the wavenumber is given in parentheses.

N_2O 1.152TORR 20M 0.0054 CM-1

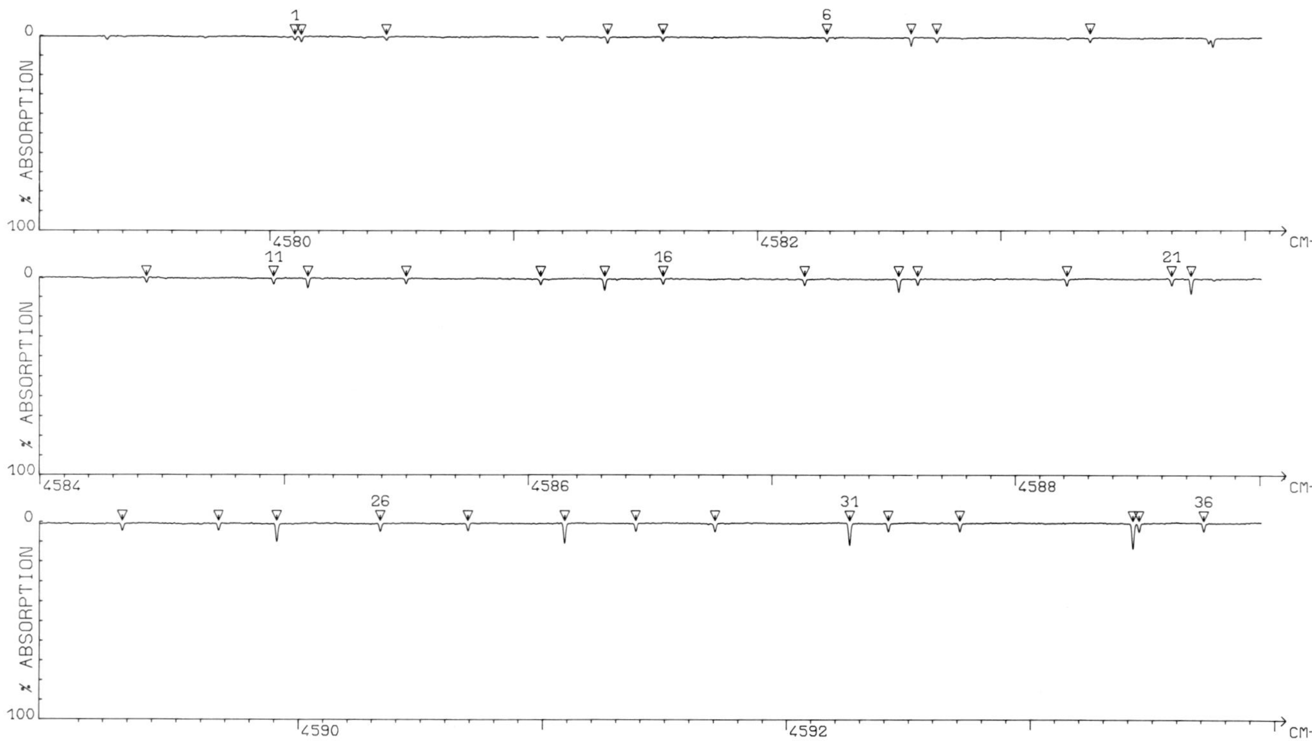

ISOTOPIC IDENTIFICATION – '▽' indicates $^{14}N^{14}N^{16}O$; '◫': $^{15}N^{14}N^{16}O$; '◇': $^{14}N^{15}N^{16}O$; 'Y': $^{14}N^{14}N^{18}O$;
'✧' indicates $^{14}N^{14}N^{17}O$; '⊥': $^{15}N^{15}N^{16}O$; '△': $^{15}N^{14}N^{18}O$; '⊕': $^{14}N^{15}N^{18}O$.

LINE NO.	WAVENUMBER CM^{-1}, VAC.	ROVIBRATIONAL ASSIGNMENT	ISOTOPIC SPECIES
1	4580.10193(60)	P33 1311F-0101F	14N14N160
2	4580.12753(36)	P48 1210E-0000E	14N14N160
3	4580.47807(51)	P32 1311E-0101E	14N14N160
4	4581.38341(32)	P47 1210E-0000E	14N14N160
5	4581.61127(42)	P31 1311E-0101E	14N14N160
6	4582.28457(41)	P31 1311F-0101F	14N14N160
7	4582.62959(27)	P46 1210E-0000E	14N14N160
8	4582.73511(39)	P30 1311E-0101E	14N14N160
9	4583.36381(40)	P30 1311F-0101F	14N14N160
10	4584.43511(60)	P29 1311F-0101F	14N14N160
11	4584.95424(34)	P28 1311E-0101E	14N14N160
12	4585.09375(24)	P44 1210E-0000E	14N14N160
13	4585.49783(35)	P28 1311F-0101F	14N14N160
14	4586.04919(33)	P27 1311E-0101E	14N14N160
15	4586.31162(23)	P43 1210E-0000E	14N14N160
16	4586.55257(34)	P27 1311F-0101F	14N14N160
17	4587.13470(32)	P26 1311E-0101E	14N14N160
18	4587.51996(22)	P42 1210E-0000E	14N14N160
19	4587.59928(32)	P26 1311F-0101F	14N14N160
20	4588.21051(30)	P25 1311E-0101E	14N14N160
20		P34 2111F-0101F	15N15N160
21	4588.63813(31)	P25 1311F-0101F	14N14N160
22	4588.71888(20)	P41 1210E-0000E	14N14N160
23	4589.27670(30)	P24 1311E-0101E	14N14N160
24	4589.66864(30)	P24 1311F-0101F	14N14N160
25	4589.90847(20)	P40 1210E-0000E	14N14N160
26	4590.33338(28)	P23 1311E-0101E	14N14N160
27	4590.69133(29)	P23 1311F-0101F	14N14N160
28	4591.08868(19)	P39 1210E-0000E	14N14N160
29	4591.38054(28)	P22 1311E-0101E	14N14N160
30	4591.70588(28)	P22 1311F-0101F	14N14N160
31	4592.25958(19)	P38 1210E-0000E	14N14N160
32	4592.41807(27)	P21 1311E-0101E	14N14N160
33	4592.71234(31)	P21 1311F-0101F	14N14N160
34	4593.42113(18)	P37 1210E-0000E	14N14N160
35	4593.44610(27)	P20 1311E-0101E	14N14N160
36	4593.71070(27)	P20 1311F-0101F	14N14N160

- "LINE NO.": serial number identifying the spectral line in the map on the opposite page.
- "WAVENUMBER": experimental value.
- Uncertainty in the last digit of the wavenumber is given in parentheses.

N_2O 1.152TORR 20M 0.0054 CM-1

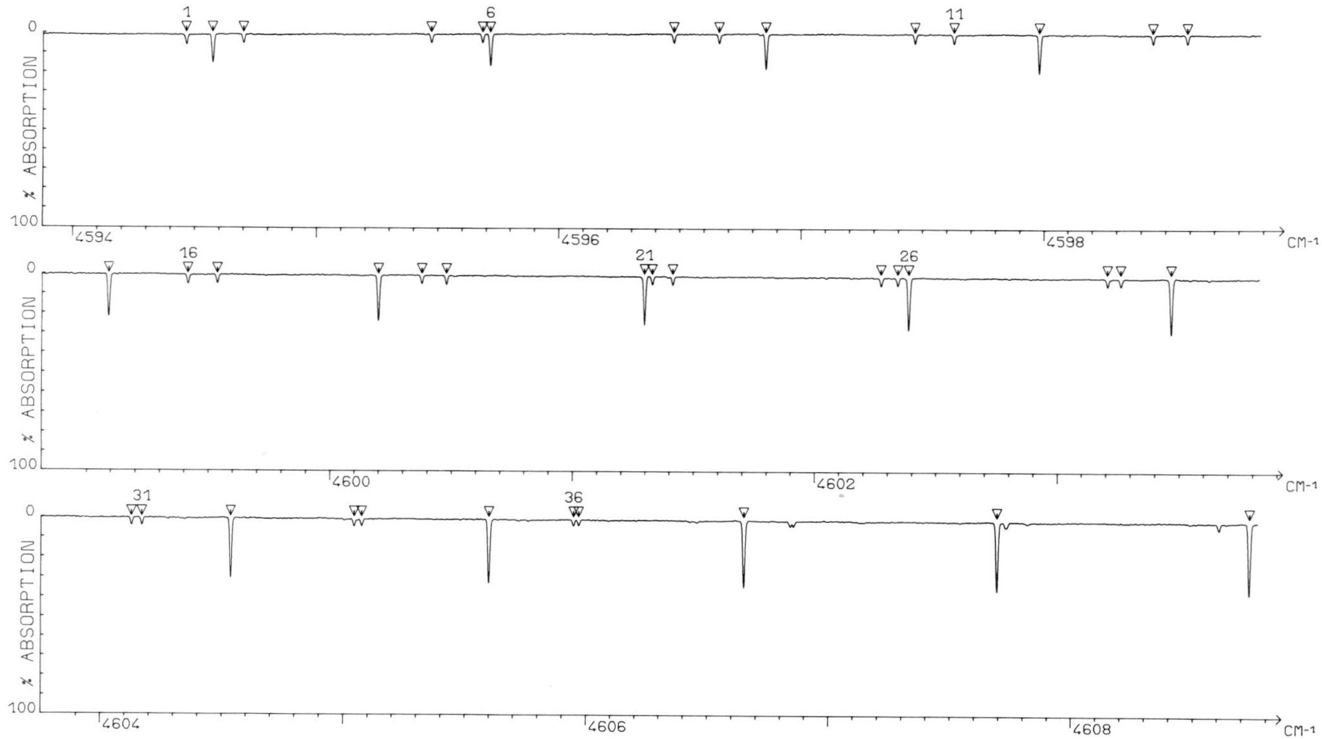

ISOTOPIC IDENTIFICATION – '▽' indicates $^{14}N^{14}N^{16}O$; '◫': $^{15}N^{14}N^{16}O$; '◇': $^{14}N^{15}N^{16}O$; 'Y': $^{14}N^{14}N^{18}O$;
'✧' indicates $^{14}N^{14}N^{17}O$; '⍖': $^{15}N^{15}N^{16}O$; '△': $^{15}N^{14}N^{18}O$; '⍏': $^{14}N^{15}N^{18}O$.

LINE NO.	WAVENUMBER CM^{-1}, VAC.	ROVIBRATIONAL ASSIGNMENT	ISOTOPIC SPECIES	LINE NO.	WAVENUMBER CM^{-1}, VAC.	ROVIBRATIONAL ASSIGNMENT	ISOTOPIC SPECIES
1	4594.46462(25)	P19 1311E-0101E	14N14N160	31	4604.16939(29)	P 9 1311F-0101F	14N14N160
2	4594.57355(18)	P36 1210E-0000E	14N14N160	32	4604.53469(17)	P27 1210E-0000E	14N14N160
3	4594.70121(27)	P19 1311F-0101F	14N14N160	33	4605.04129(35)	P 8 1311E-0101E	14N14N160
4	4595.47371(26)	P18 1311E-0101E	14N14N160	34	4605.07251(52)	P 8 1311F-0101F	14N14N160
5	4595.68392(27)	P18 1311F-0101F	14N14N160	35	4605.59658(17)	P26 1210E-0000E	14N14N160
6	4595.71659(18)	P35 1210E-0000E	14N14N160	36	4605.94612(34)	P 7 1311E-0101E	14N14N160
7	4596.47309(26)	P17 1311E-0101E	14N14N160	37	4605.96772(33)	P 7 1311F-0101F	14N14N160
8	4596.65854(25)	P17 1311F-0101F	14N14N160	38	4606.64960(17)	P25 1210E-0000E	14N14N160
9	4596.85062(18)	P34 1210E-0000E	14N14N160	39	4607.69380(17)	P24 1210E-0000E	14N14N160
10	4597.46302(26)	P16 1311E-0101E	14N14N160	40	4608.72919(17)	P23 1210E-0000E	14N14N160
11	4597.62504(26)	P16 1311F-0101F	14N14N160				
12	4597.97550(18)	P33 1210E-0000E	14N14N160				
13	4598.44355(26)	P15 1311E-0101E	14N14N160				
14	4598.58382(26)	P15 1311F-0101F	14N14N160				
15	4599.09128(18)	P32 1210E-0000E	14N14N160				
16	4599.41463(26)	P14 1311E-0101E	14N14N160				
17	4599.53447(26)	P14 1311F-0101F	14N14N160				
18	4600.19795(17)	P31 1210E-0000E	14N14N160				
19	4600.37610(26)	P13 1311E-0101E	14N14N160				
20	4600.47729(26)	P13 1311F-0101F	14N14N160				
21	4601.29558(17)	P30 1210E-0000E	14N14N160				
22	4601.32811(27)	P12 1311E-0101E	14N14N160				
23	4601.41216(26)	P12 1311F-0101F	14N14N160				
24	4602.27074(30)	P11 1311E-0101E	14N14N160				
25	4602.33917(28)	P11 1311F-0101F	14N14N160				
26	4602.38426(17)	P29 1210E-0000E	14N14N160				
27	4603.20370(27)	P10 1311E-0101E	14N14N160				
28	4603.25827(28)	P10 1311F-0101F	14N14N160				
29	4603.46392(17)	P28 1210E-0000E	14N14N160				
30	4604.12741(29)	P 9 1311E-0101E	14N14N160				

- "LINE NO.": serial number identifying the spectral line in the map on the opposite page.
- "WAVENUMBER": experimental value.
- Uncertainty in the last digit of the wavenumber is given in parentheses.

N_2O 1.152TORR 20M 0.0054 CM-1

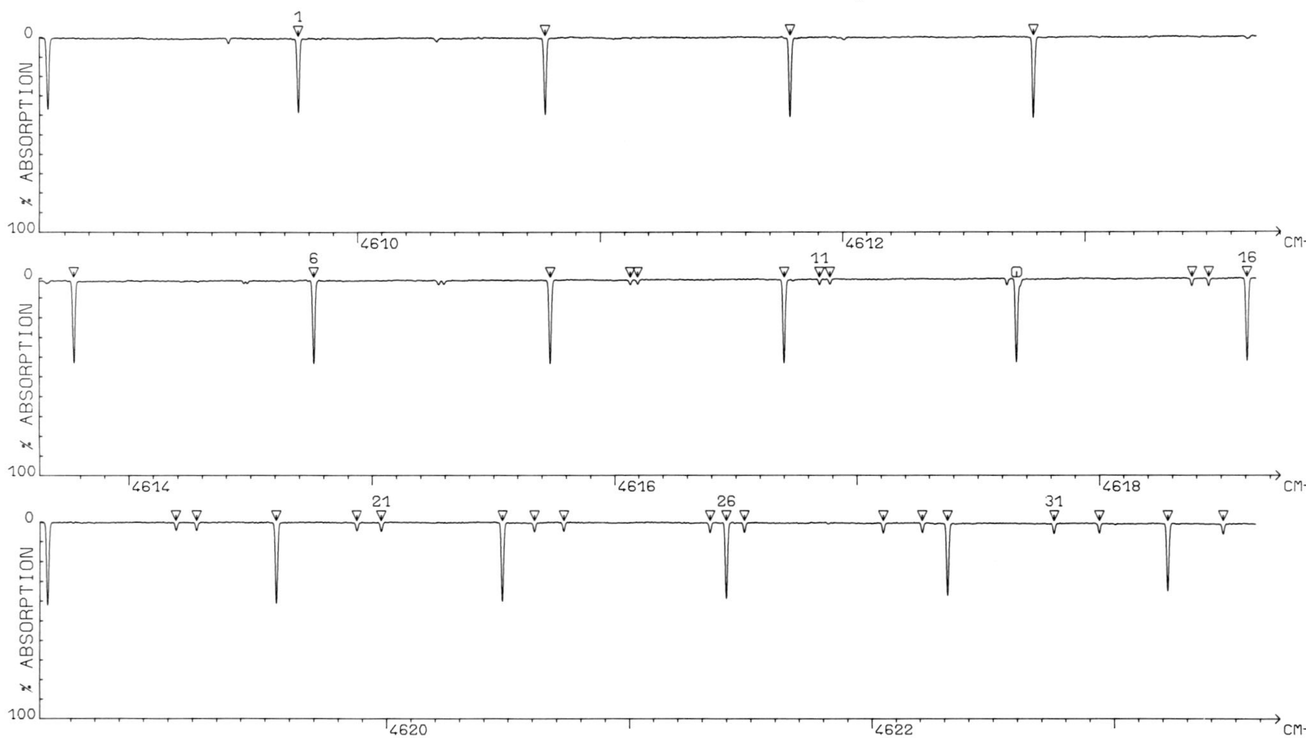

ISOTOPIC IDENTIFICATION – '▽' indicates $^{14}N^{14}N^{16}O$; '◫': $^{15}N^{14}N^{16}O$; '◇': $^{14}N^{15}N^{16}O$; 'Y': $^{14}N^{14}N^{18}O$;
'✧' indicates $^{14}N^{14}N^{17}O$; '⊥': $^{15}N^{15}N^{16}O$; '△': $^{15}N^{14}N^{18}O$; '♀': $^{14}N^{15}N^{18}O$.

LINE NO.	WAVENUMBER CM^{-1}, VAC.	ROVIBRATIONAL ASSIGNMENT	ISOTOPIC SPECIES
1	4609.75581(17)	P22 1210E-0000E	14N14N160
2	4610.77365(17)	P21 1210E-0000E	14N14N160
3	4611.78278(17)	P20 1210E-0000E	14N14N160
4	4612.78316(17)	P19 1210E-0000E	14N14N160
5	4613.77489(17)	P18 1210E-0000E	14N14N160
6	4614.75803(17)	P17 1210E-0000E	14N14N160
7	4615.73248(17)	P16 1210E-0000E	14N14N160
8	4616.06224(74)	R 4 1311E-0101E	14N14N160
9	4616.09431(49)	R 4 1311F-0101F	14N14N160
10	4616.69834(17)	P15 1210E-0000E	14N14N160
11	4616.84402(36)	R 5 1311E-0101E	14N14N160
12	4616.88653(36)	R 5 1311F-0101F	14N14N160
13	4617.65556(17)	R56 1210E-0000E	15N14N160
13		P14 1210E-0000E	14N14N160
14	4618.37842(30)	R 7 1311E-0101E	14N14N160
15	4618.44739(29)	R 7 1311F-0101F	14N14N160
16	4618.60422(17)	P13 1210E-0000E	14N14N160
17	4619.13118(32)	R 8 1311E-0101E	14N14N160
18	4619.21573(29)	R 8 1311F-0101F	14N14N160
19	4619.54437(17)	P12 1210E-0000E	14N14N160
20	4619.87432(28)	R 9 1311E-0101E	14N14N160
21	4619.97624(28)	R 9 1311F-0101F	14N14N160
22	4620.47592(17)	P11 1210E-0000E	14N14N160
23	4620.60789(26)	R10 1311E-0101E	14N14N160
24	4620.72853(26)	R10 1311F-0101F	14N14N160
25	4621.33204(25)	R11 1311E-0101E	14N14N160
26	4621.39896(17)	P10 1210E-0000E	14N14N160
27	4621.47298(26)	R11 1311F-0101F	14N14N160
28	4622.04663(25)	R12 1311E-0101E	14N14N160
29	4622.20931(26)	R12 1311F-0101F	14N14N160
30	4622.31344(17)	P 9 1210E-0000E	14N14N160
31	4622.75165(25)	R13 1311E-0101E	14N14N160
32	4622.93763(25)	R13 1311F-0101F	14N14N160
33	4623.21944(17)	P 8 1210E-0000E	14N14N160
34	4623.44684(25)	R14 1311E-0101E	14N14N160

- "LINE NO.": serial number identifying the spectral line in the map on the opposite page.
- "WAVENUMBER": experimental value.
- Uncertainty in the last digit of the wavenumber is given in parentheses.

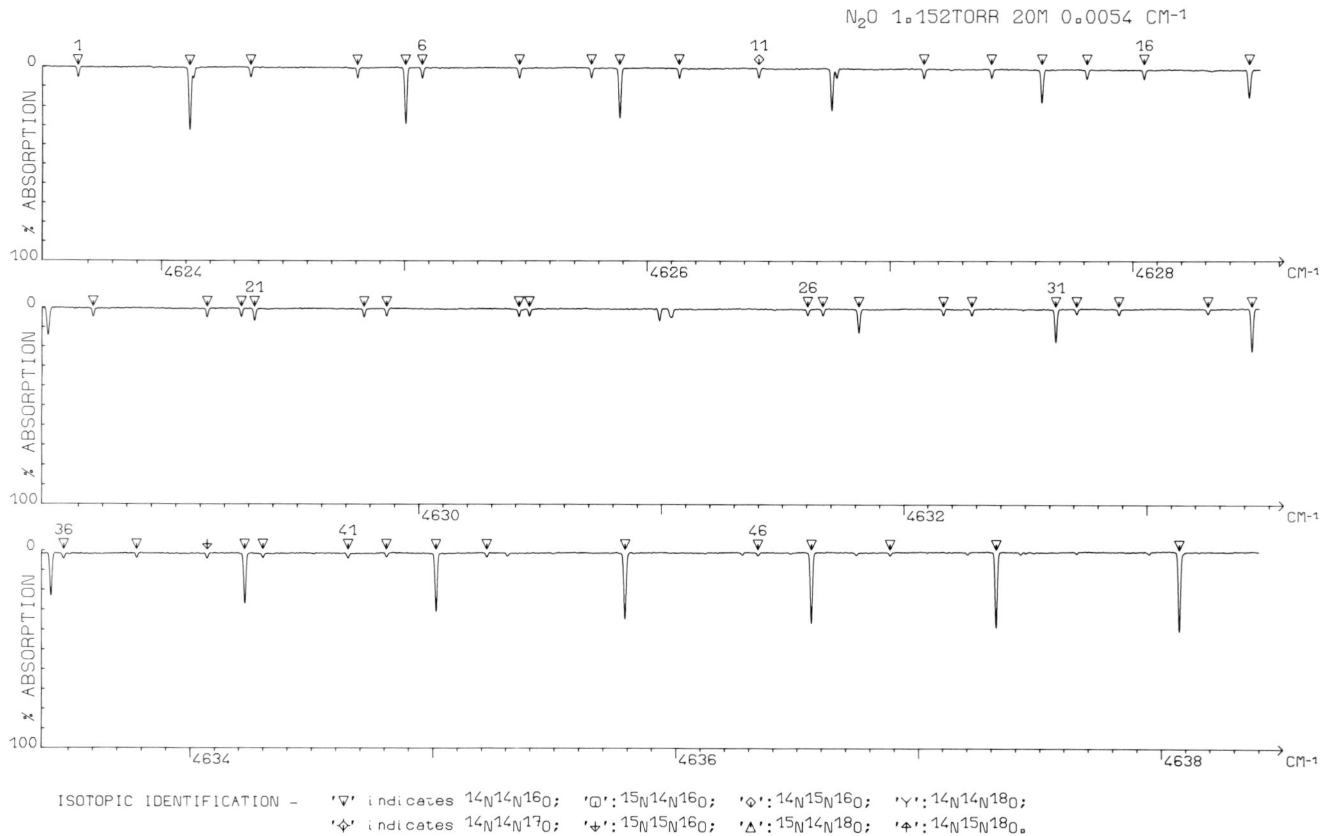
N_2O 1.152TORR 20M 0.0054 CM-1
1
6
11
16
21
26
31
36
41
46
0
% ABSORPTION
100
4624
4626
4628
4630
4632
4634
4636
4638
CM-1
ISOTOPIC IDENTIFICATION – '▽' indicates $^{14}N^{14}N^{16}O$; 'Ⓠ': $^{15}N^{14}N^{16}O$; '◇': $^{14}N^{15}N^{16}O$; 'Y': $^{14}N^{14}N^{18}O$;
'✧' indicates $^{14}N^{14}N^{17}O$; '⊥': $^{15}N^{15}N^{16}O$; '△': $^{15}N^{14}N^{18}O$; '↑': $^{14}N^{15}N^{18}O$.

LINE NO.	WAVENUMBER CM^{-1}, VAC.	ROVIBRATIONAL ASSIGNMENT	ISOTOPIC SPECIES	LINE NO.	WAVENUMBER CM^{-1}, VAC.	ROVIBRATIONAL ASSIGNMENT	ISOTOPIC SPECIES
1	4623.65794(25)	R14 1311F-0101F	14N14N160	31	4632.62453(18)	R 2 1210E-0000E	14N14N160
2	4624.11693(18)	P 7 1210E-0000E	14N14N160	32	4632.71055(37)	R29 1311E-0101E	14N14N160
3	4624.36989(25)	R15 1311F-0101F	14N14N160	33	4632.88332(35)	R28 1311F-0101F	14N14N160
4	4624.80844(25)	R16 1311E-0101E	14N14N160	34	4633.24984(41)	R30 1311E-0101E	14N14N160
5	4625.00590(17)	P 6 1210E-0000E	14N14N160	35	4633.42850(18)	R 3 1210E-0000E	14N14N160
6	4625.07390(24)	R16 1311F-0101F	14N14N160	36	4633.48016(36)	R29 1311F-0101F	14N14N160
7	4625.47479(25)	R17 1311E-0101E	14N14N160	37	4633.77893(40)	R31 1311E-0101E	14N14N160
8	4625.77008(27)	R17 1311F-0101F	14N14N160	38	4634.06834(39)	R15 2111E-0101E	15N15N160
9	4625.88643(17)	P 5 1210E-0000E	14N14N160	38		R30 1311F-0101F	14N14N160
10	4626.13148(25)	R18 1311E-0101E	14N14N160	39	4634.22391(18)	R 4 1210E-0000E	14N14N160
				40	4634.29797(45)	R32 1311E-0101E	14N14N160
11	4626.45775(25)	P42 2111E-0101E	14N15N160				
11		R18 1311F-0101F	14N14N160	41	4634.64826(40)	R31 1311F-0101F	14N14N160
12	4627.13747(25)	R19 1311F-0101F	14N14N160	42	4634.80686(60)	R33 1311E-0101E	14N14N160
13	4627.41559(26)	R20 1311E-0101E	14N14N160	42		P36 2111F-0101F	14N15N160
14	4627.62191(18)	P 3 1210E-0000E	14N14N160	43	4635.01088(17)	R 5 1210E-0000E	14N14N160
15	4627.80893(28)	R20 1311F-0101F	14N14N160	44	4635.21982(42)	R32 1311F-0101F	14N14N160
16	4628.04306(26)	R21 1311E-0101E	14N14N160	45	4635.78918(17)	R 6 1210E-0000E	14N14N160
17	4628.47587(19)	P 2 1210E-0000E	14N14N160	45		R18 2111E-0101E	15N15N160
18	4628.66089(27)	R22 1311E-0101E	14N14N160	46	4636.33709(64)	R34 1311F-0101F	14N14N160
19	4629.12718(27)	R22 1311F-0101F	14N14N160	46		R19 2111E-0101E	15N15N160
20	4629.26891(28)	R23 1311E-0101E	14N14N160	47	4636.55896(17)	R 7 1210E-0000E	14N14N160
				48	4636.88276(66)	R35 1311F-0101F	14N14N160
21	4629.32345(23)	P 1 1210E-0000E	14N14N160	49	4637.32016(17)	R 8 1210E-0000E	14N14N160
22	4629.77405(28)	R23 1311F-0101F	14N14N160	50	4638.07274(17)	R 9 1210E-0000E	14N14N160
23	4629.86721(29)	R24 1311E-0101E	14N14N160				
24	4630.41253(28)	R24 1311F-0101F	14N14N160				
25	4630.45555(30)	R25 1311E-0101E	14N14N160				
26	4631.60264(33)	R27 1311E-0101E	14N14N160				
27	4631.66466(31)	R26 1311F-0101F	14N14N160				
28	4631.81203(19)	R 1 1210E-0000E	14N14N160				
29	4632.16174(34)	R28 1311E-0101E	14N14N160				
30	4632.27814(33)	R27 1311F-0101F	14N14N160				

- "LINE NO.": serial number identifying the spectral line in the map on the opposite page.
- "WAVENUMBER": experimental value.
- Uncertainty in the last digit of the wavenumber is given in parentheses.

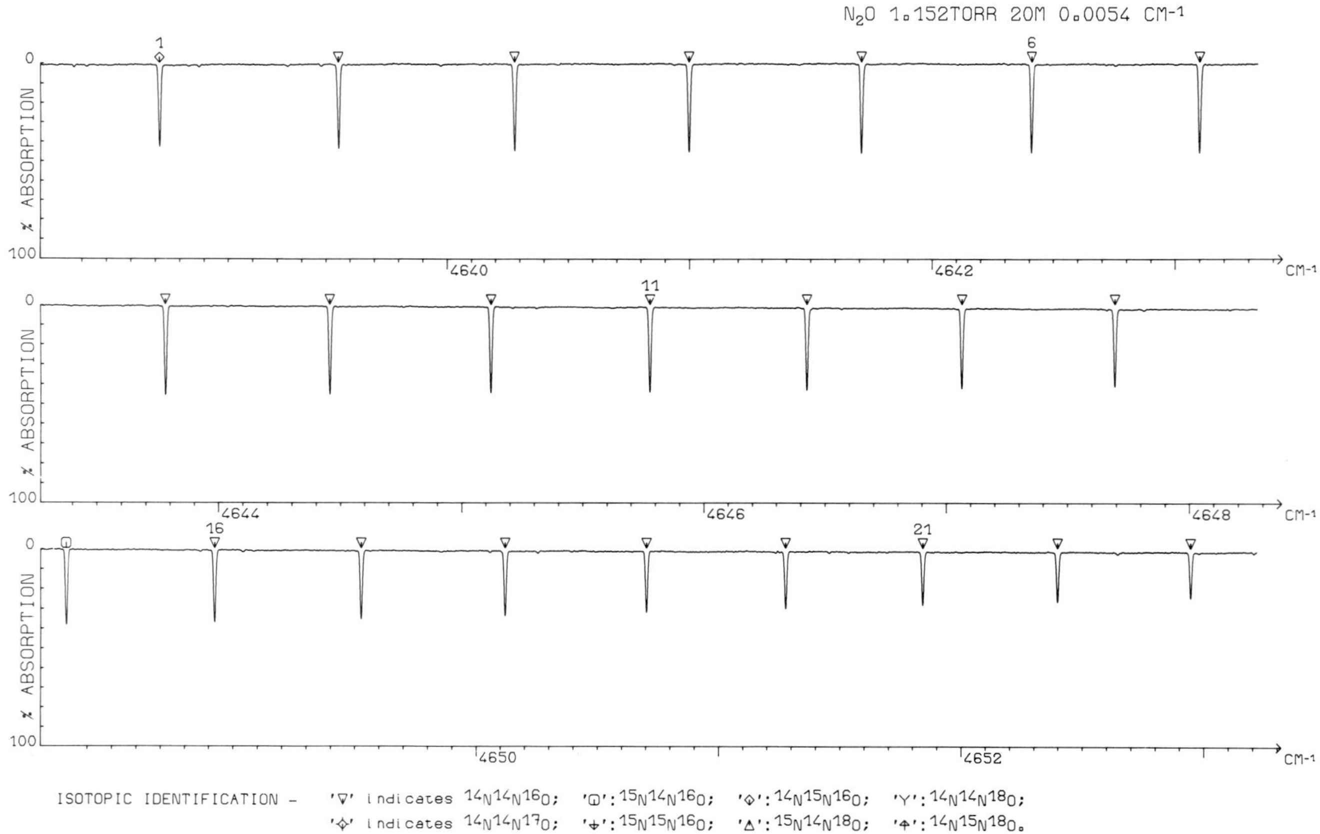
N_2O 1.152TORR 20M 0.0054 CM-1
0
% ABSORPTION
100
4640
4642
4644
4646
4648
4650
4652
CM-1
1
6
11
16
21
ISOTOPIC IDENTIFICATION - '▽' indicates $^{14}N^{14}N^{16}O$; '⊡': $^{15}N^{14}N^{16}O$; '◇': $^{14}N^{15}N^{16}O$; 'Y': $^{14}N^{14}N^{18}O$;
'◇' indicates $^{14}N^{14}N^{17}O$; '+': $^{15}N^{15}N^{16}O$; '△': $^{15}N^{14}N^{18}O$; '+': $^{14}N^{15}N^{18}O$.

LINE NO.	WAVENUMBER CM^{-1}, VAC.	ROVIBRATIONAL ASSIGNMENT	ISOTOPIC SPECIES
1	4638.81668(17)	P36 2010E-0000E	14N15N16O
1		R10 1210E-0000E	14N14N16O
2	4639.55206(17)	R11 1210E-0000E	14N14N16O
3	4640.27867(17)	R12 1210E-0000E	14N14N16O
4	4640.99667(17)	R13 1210E-0000E	14N14N16O
5	4641.70590(17)	R14 1210E-0000E	14N14N16O
6	4642.40637(17)	R15 1210E-0000E	14N14N16O
7	4643.09808(17)	R16 1210E-0000E	14N14N16O
7		P37 2111E-0101E	15N14N16O
8	4643.78103(17)	R17 1210E-0000E	14N14N16O
9	4644.45511(17)	R18 1210E-0000E	14N14N16O
10	4645.12033(17)	R19 1210E-0000E	14N14N16O
11	4645.77666(17)	R20 1210E-0000E	14N14N16O
12	4646.42410(17)	R21 1210E-0000E	14N14N16O
13	4647.06251(17)	R22 1210E-0000E	14N14N16O
14	4647.69192(17)	R23 1210E-0000E	14N14N16O
15	4648.31232(17)	P33 2111F-0101F	15N14N16O
15		R24 1210E-0000E	14N14N16O
16	4648.92364(17)	R25 1210E-0000E	14N14N16O
17	4649.52586(17)	R26 1210E-0000E	14N14N16O
18	4650.11886(17)	R27 1210E-0000E	14N14N16O
19	4650.70271(17)	R28 1210E-0000E	14N14N16O
20	4651.27733(17)	R29 1210E-0000E	14N14N16O
21	4651.84255(18)	R30 1210E-0000E	14N14N16O
22	4652.39870(18)	R31 1210E-0000E	14N14N16O
23	4652.94515(18)	R32 1210E-0000E	14N14N16O

- "LINE NO.": serial number identifying the spectral line in the map on the opposite page.
- "WAVENUMBER": experimental value.
- Uncertainty in the last digit of the wavenumber is given in parentheses.

N$_2$O 1.152TORR 20M 0.0054 CM-1

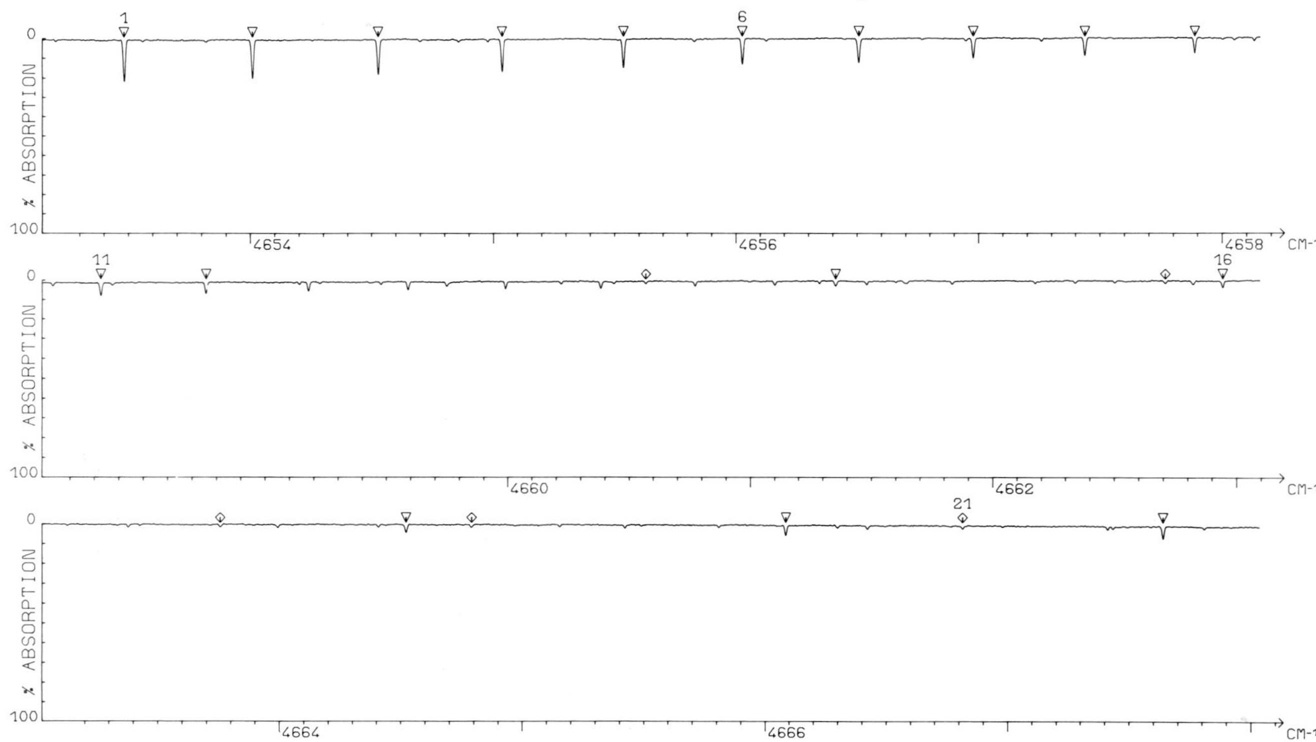

ISOTOPIC IDENTIFICATION – '▽' indicates $^{14}N^{14}N^{16}O$; '⊡': $^{15}N^{14}N^{16}O$; '◇': $^{14}N^{15}N^{16}O$; 'Y': $^{14}N^{14}N^{18}O$;
'✧' indicates $^{14}N^{14}N^{17}O$; '⏚': $^{15}N^{15}N^{16}O$; '△': $^{15}N^{14}N^{18}O$; '⏄': $^{14}N^{15}N^{18}O$.

LINE NO.	WAVENUMBER CM^{-1}, VAC.	ROVIBRATIONAL ASSIGNMENT	ISOTOPIC SPECIES
1	4653.48235(18)	R33 1210E-0000E	14N14N160
2	4654.01003(18)	R34 1210E-0000E	14N14N160
3	4654.52827(18)	R35 1210E-0000E	14N14N160
4	4655.03689(18)	R36 1210E-0000E	14N14N160
5	4655.53594(19)	R37 1210E-0000E	14N14N160
6	4656.02534(19)	R38 1210E-0000E	14N14N160
7	4656.50476(19)	R39 1210E-0000E	14N14N160
8	4656.97509(20)	R40 1210E-0000E	14N14N160
9	4657.43513(21)	R41 1210E-0000E	14N14N160
10	4657.88563(21)	R42 1210E-0000E	14N14N160
11	4658.32611(23)	R43 1210E-0000E	14N14N160
12	4658.75681(24)	R44 1210E-0000E	14N14N160
13	4660.56828(78)	P18 2010E-0000E	14N15N160
14	4661.35079(40)	P57 2010E-0000E	14N14N160
15	4662.70614(77)	P16 2010E-0000E	14N15N160
16	4662.94189(37)	P56 2010E-0000E	14N14N160
17	4663.75461(81)	P15 2010E-0000E	14N15N160
18	4664.52052(46)	P55 2010E-0000E	14N14N160
19	4664.78925(76)	P14 2010E-0000E	14N15N160
20	4666.08611(25)	P54 2010E-0000E	14N14N160
21	4666.81552(82)	P12 2010E-0000E	14N15N160
22	4667.63880(23)	P53 2010E-0000E	14N14N160

- "LINE NO.": serial number identifying the spectral line in the map on the opposite page.
- "WAVENUMBER": experimental value.
- Uncertainty in the last digit of the wavenumber is given in parentheses.

N_2O 1.152TORR 20M 0.0054 CM-1

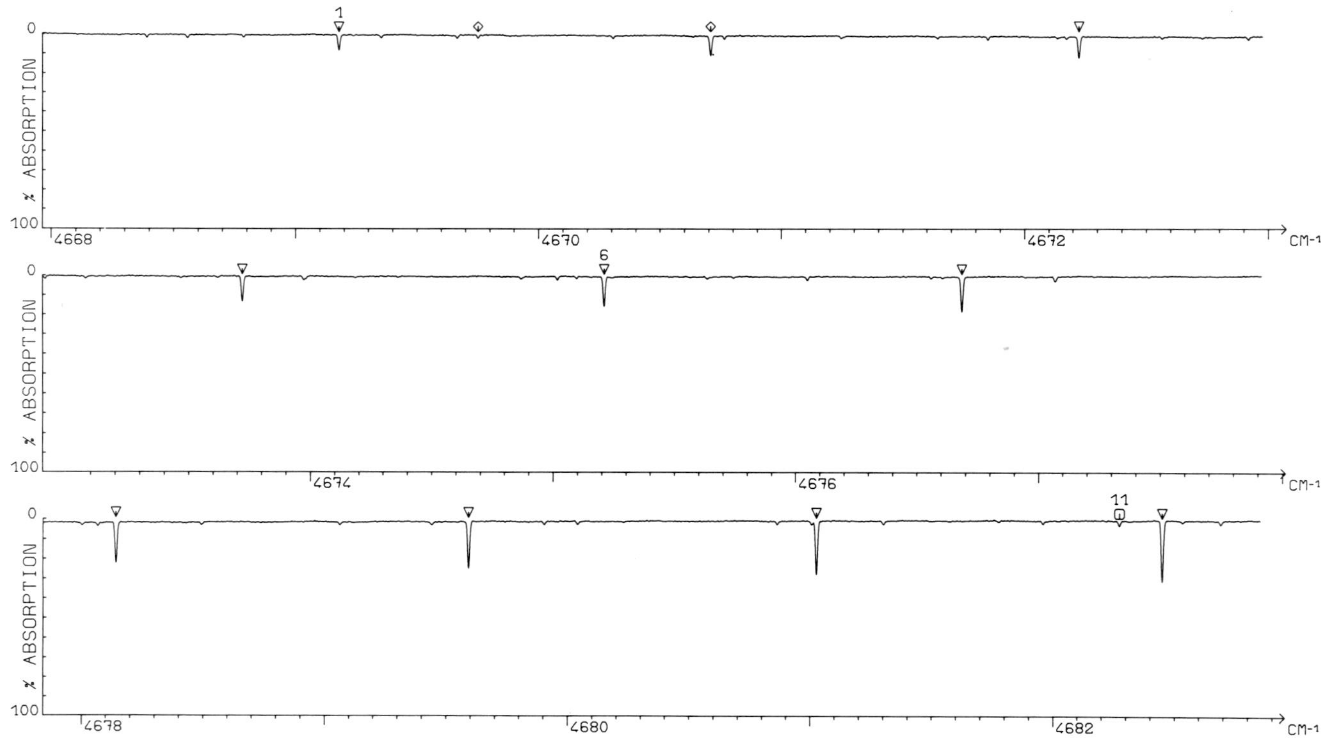

ISOTOPIC IDENTIFICATION – '▽' indicates $^{14}N^{14}N^{16}O$; '⊡': $^{15}N^{14}N^{16}O$; '◇': $^{14}N^{15}N^{16}O$; 'Y': $^{14}N^{14}N^{18}O$;
'✧' indicates $^{14}N^{14}N^{17}O$; '⊥': $^{15}N^{15}N^{16}O$; '△': $^{15}N^{14}N^{18}O$; '✦': $^{14}N^{15}N^{18}O$.

LINE NO.	WAVENUMBER CM^{-1} ,VAC.	ROVIBRATIONAL ASSIGNMENT	ISOTOPIC SPECIES
1	4669.17846(22)	P52 2010E-0000E	14N14N160
2	4669.75011(76)	P 9 2010E-0000E	14N15N160
3	4670.70495(20)	P 8 2010E-0000E	14N15N160
3		P51 2010E-0000E	14N14N160
4	4672.21897(20)	P50 2010E-0000E	14N14N160
5	4673.71960(19)	P49 2010E-0000E	14N14N160
6	4675.20725(19)	P48 2010E-0000E	14N14N160
7	4676.68171(18)	P47 2010E-0000E	14N14N160
8	4678.14305(18)	P46 2010E-0000E	14N14N160
9	4679.59122(18)	P45 2010E-0000E	14N14N160
10	4681.02619(18)	P44 2010E-0000E	14N14N160
11	4682.27253(65)	R 2 2010E-0000E	15N14N160
12	4682.44797(18)	P43 2010E-0000E	14N14N160

- "LINE NO.": serial number identifying the spectral line in the map on the opposite page.
- "WAVENUMBER": experimental value.
- Uncertainty in the last digit of the wavenumber is given in parentheses.

N_2O 1.152TORR 20M 0.0054 CM-1

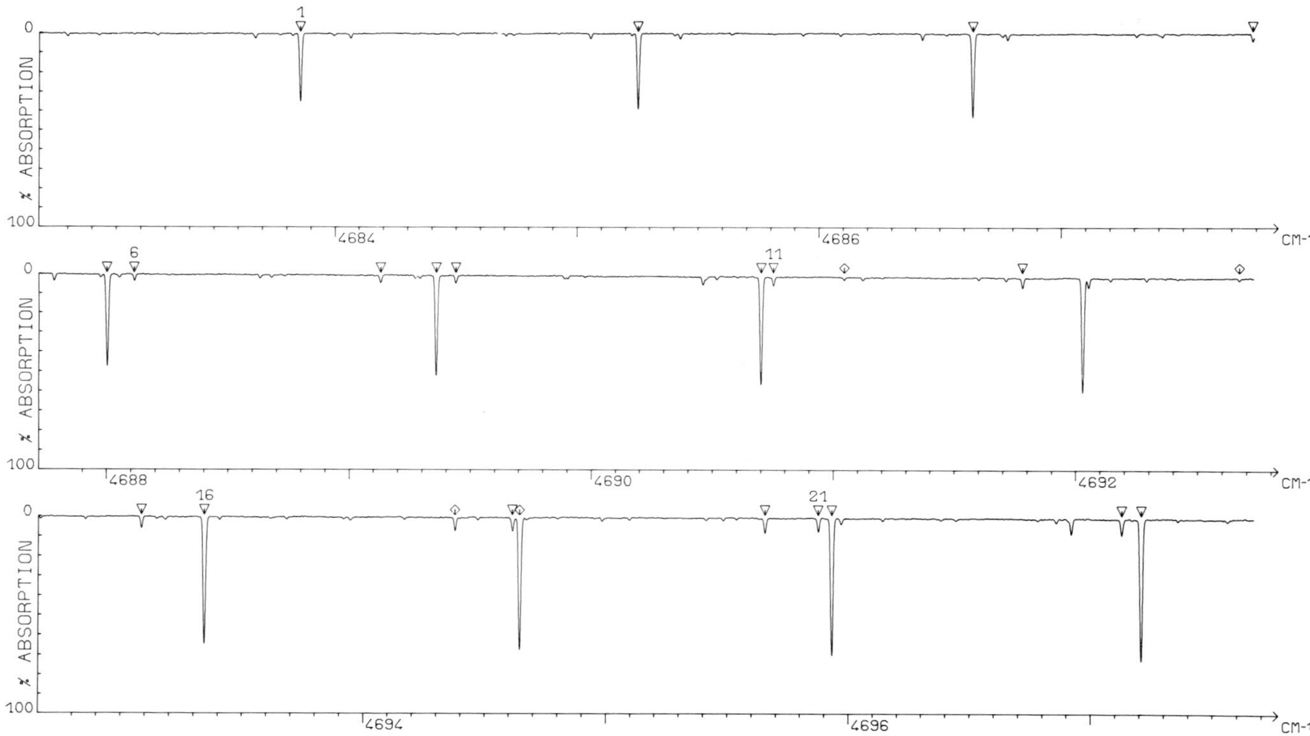

ISOTOPIC IDENTIFICATION – '▽' indicates $^{14}N^{14}N^{16}O$; '◫': $^{15}N^{14}N^{16}O$; '◇': $^{14}N^{15}N^{16}O$; 'Y': $^{14}N^{14}N^{18}O$;
'✧' indicates $^{14}N^{14}N^{17}O$; '⏚': $^{15}N^{15}N^{16}O$; '△': $^{15}N^{14}N^{18}O$; '✦': $^{14}N^{15}N^{18}O$.

LINE NO.	WAVENUMBER CM^{-1}, VAC.	ROVIBRATIONAL ASSIGNMENT	ISOTOPIC SPECIES
1	4683.85649(17)	P42 2010E-0000E	14N14N160
2	4685.25177(17)	P41 2010E-0000E	14N14N160
3	4686.63373(17)	P40 2010E-0000E	14N14N160
4	4687.78372(36)	P39 2111E-0101E	14N14N160
5	4688.00247(17)	P39 2010E-0000E	14N14N160
6	4688.11325(32)	P39 2111F-0101F	14N14N160
7	4689.12871(30)	P38 2111E-0101E	14N14N160
8	4689.35779(17)	P38 2010E-0000E	14N14N160
9	4689.43869(32)	P38 2111F-0101F	14N14N160
10	4690.69981(17)	P37 2010E-0000E	14N14N160
11	4690.75152(27)	P37 2111F-0101F	14N14N160
12	4691.04403(72)	R18 2010E-0000E	14N15N160
13	4691.77932(26)	P36 2111E-0101E	14N14N160
14	4692.66970(86)	R21 2010E-0000E	14N15N160
15	4693.08486(24)	P35 2111E-0101E	14N14N160
16	4693.34352(17)	P35 2010E-0000E	14N14N160
17	4694.37740(23)	R34 2111F-0101F	14N15N160
18	4694.61539(23)	P34 2111F-0101F	14N14N160
19	4694.64549(17)	R25 2010E-0000E	14N15N160
19		P34 2010E-0000E	14N14N160
20	4695.65688(22)	P33 2111E-0101E	14N14N160
20		R20 2111F-0101F	15N14N160
21	4695.87842(22)	P33 2111F-0101F	14N14N160
22	4695.93391(17)	P33 2010E-0000E	14N14N160
23	4697.12912(24)	P32 2111F-0101F	14N14N160
24	4697.20887(17)	P32 2010E-0000E	14N14N160

- "LINE NO.": serial number identifying the spectral line in the map on the opposite page.
- "WAVENUMBER": experimental value.
- Uncertainty in the last digit of the wavenumber is given in parentheses.

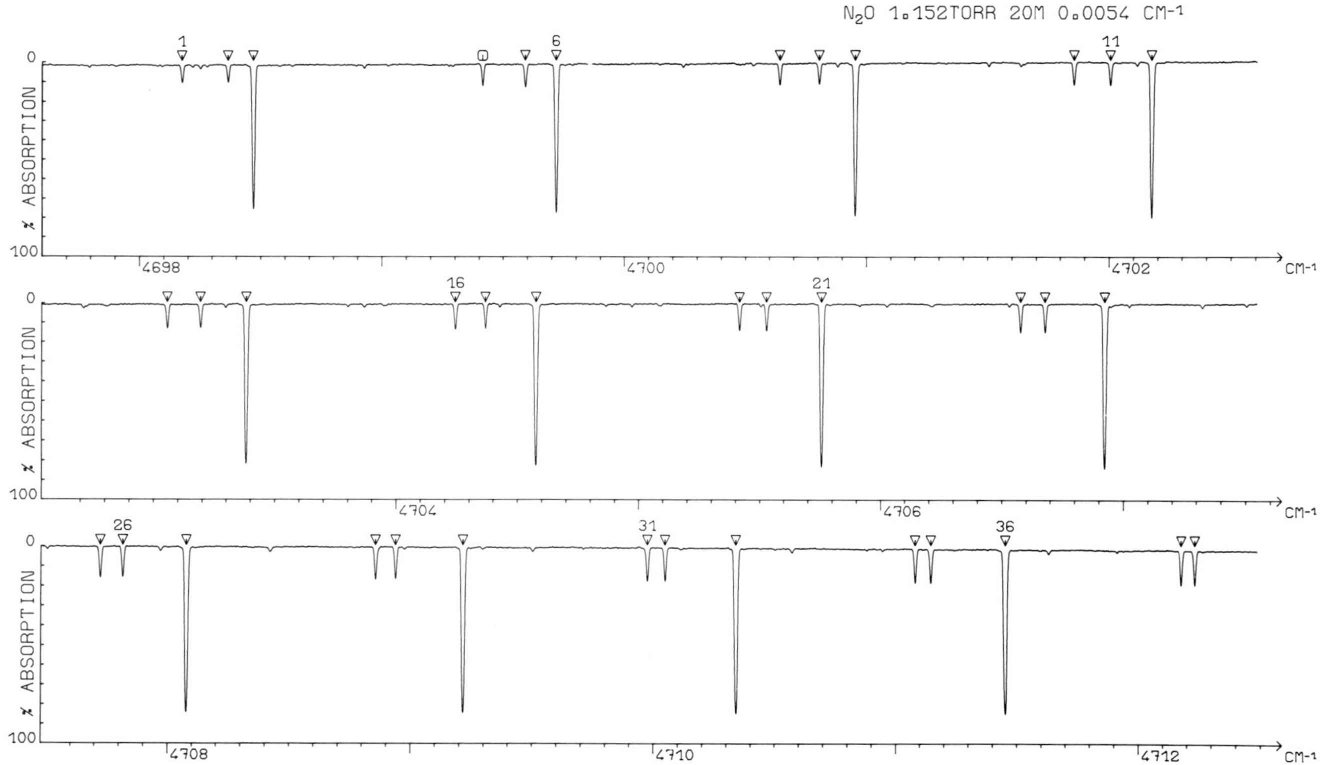

ISOTOPIC IDENTIFICATION – '▽' indicates $^{14}N^{14}N^{16}O$; '⊡': $^{15}N^{14}N^{16}O$; '◇': $^{14}N^{15}N^{16}O$; 'Y': $^{14}N^{14}N^{18}O$;
'✧' indicates $^{14}N^{14}N^{17}O$; '⊥': $^{15}N^{15}N^{16}O$; 'Δ': $^{15}N^{14}N^{18}O$; '♠': $^{14}N^{15}N^{18}O$.

LINE NO.	WAVENUMBER CM^{-1} ,VAC.	ROVIBRATIONAL ASSIGNMENT	ISOTOPIC SPECIES
1	4698.17573(21)	P31 2111E-0101E	14N14N160
2	4698.36669(21)	P31 2111F-0101F	14N14N160
3	4698.47035(17)	P31 2010E-0000E	14N14N160
4	4699.41513(20)	R32 2010E-0000E	15N14N160
4		P30 2111E-0101E	14N14N160
5	4699.59191(20)	P30 2111F-0101F	14N14N160
6	4699.71834(17)	P30 2010E-0000E	14N14N160
7	4700.64198(20)	P29 2111E-0101E	14N14N160
8	4700.80454(20)	P29 2111F-0101F	14N14N160
9	4700.95285(17)	P29 2010E-0000E	14N14N160
10	4701.85514(20)	P28 2111E-0101E	14N14N160
11	4702.00449(20)	P28 2111F-0101F	14N14N160
12	4702.17385(17)	P28 2010E-0000E	14N14N160
12		R35 2111E-0101E	15N14N160
13	4703.05512(19)	P27 2111E-0101E	14N14N160
14	4703.19189(19)	P27 2111F-0101F	14N14N160
15	4703.38128(17)	P27 2010E-0000E	14N14N160
16	4704.24192(19)	P26 2111E-0101E	14N14N160
17	4704.36647(19)	P26 2111F-0101F	14N14N160
18	4704.57520(17)	P26 2010E-0000E	14N14N160
19	4705.41535(19)	P25 2111E-0101E	14N14N160
20	4705.52850(19)	P25 2111F-0101F	14N14N160
21	4705.75556(17)	P25 2010E-0000E	14N14N160
22	4706.57550(19)	P24 2111E-0101E	14N14N160
23	4706.67777(19)	P24 2111F-0101F	14N14N160
24	4706.92235(17)	P24 2010E-0000E	14N14N160
25	4707.72238(19)	P23 2111E-0101E	14N14N160
26	4707.81428(19)	P23 2111F-0101F	14N14N160
27	4708.07554(17)	P23 2010E-0000E	14N14N160
28	4708.85594(19)	P22 2111E-0101E	14N14N160
29	4708.93802(19)	P22 2111F-0101F	14N14N160
30	4709.21516(17)	P22 2010E-0000E	14N14N160
31	4709.97602(18)	P21 2111E-0101E	14N14N160
32	4710.04906(18)	P21 2111F-0101F	14N14N160
33	4710.34113(17)	P21 2010E-0000E	14N14N160
34	4711.08307(18)	P20 2111E-0101E	14N14N160
35	4711.14725(18)	P20 2111F-0101F	14N14N160
36	4711.45348(17)	P20 2010E-0000E	14N14N160
37	4712.17664(18)	P19 2111E-0101E	14N14N160
38	4712.23267(18)	P19 2111F-0101F	14N14N160

- "LINE NO.": serial number identifying the spectral line in the map on the opposite page.
- "WAVENUMBER": experimental value.
- Uncertainty in the last digit of the wavenumber is given in parentheses.

*** ATTENTION *** CALCUL A VERIFIER

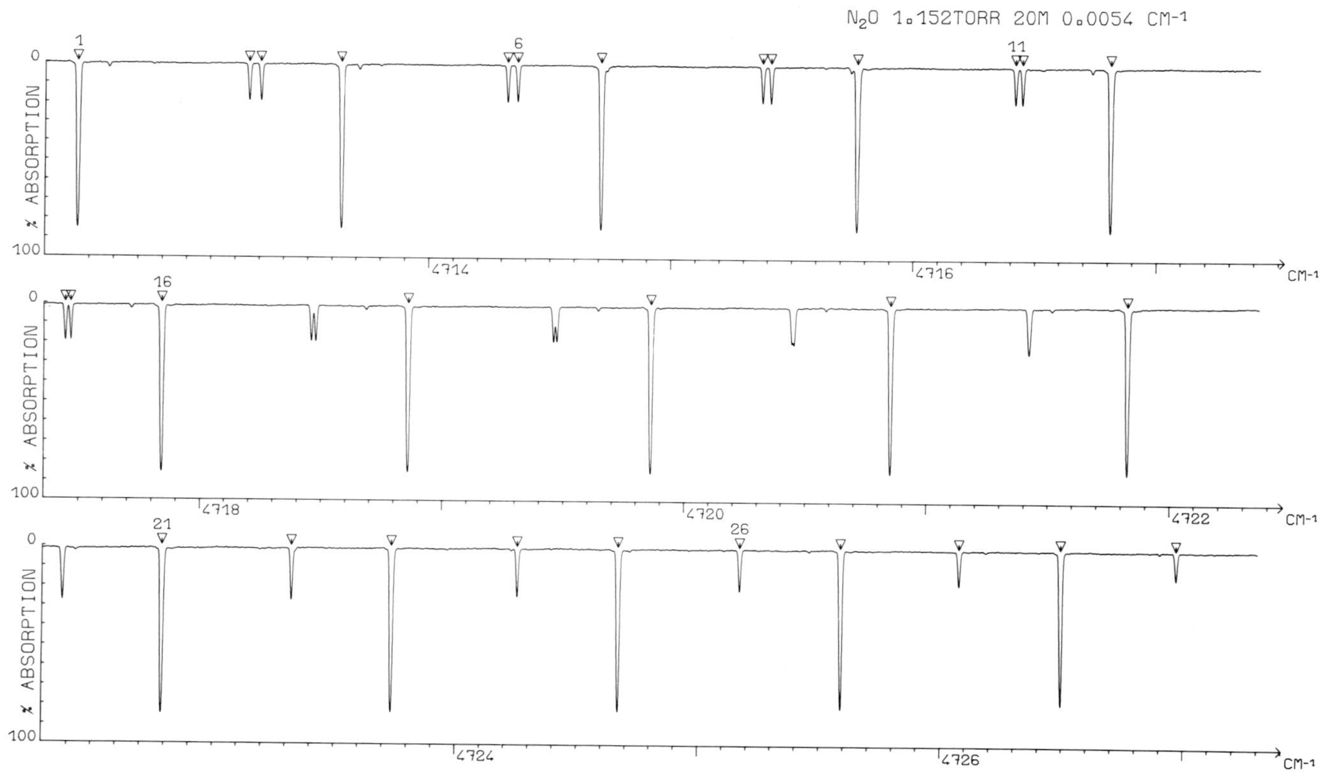

ISOTOPIC IDENTIFICATION – '▽' indicates $^{14}N^{14}N^{16}O$; '◫': $^{15}N^{14}N^{16}O$; '◇': $^{14}N^{15}N^{16}O$; 'Y': $^{14}N^{14}N^{18}O$;
'✧' indicates $^{14}N^{14}N^{17}O$; '⫫': $^{15}N^{15}N^{16}O$; '△': $^{15}N^{14}N^{18}O$; '⟠': $^{14}N^{15}N^{18}O$.

LINE NO.	WAVENUMBER CM^{-1}, VAC.	ROVIBRATIONAL ASSIGNMENT	ISOTOPIC SPECIES
1	4712.55223(17)	P19 2010E-0000E	14N14N160
2	4713.25681(18)	P18 2111E-0101E	14N14N160
3	4713.30527(18)	P18 2111F-0101F	14N14N160
4	4713.63731(17)	P18 2010E-0000E	14N14N160
5	4714.32364(18)	P17 2111E-0101E	14N14N160
6	4714.36502(18)	P17 2111F-0101F	14N14N160
7	4714.70877(17)	P17 2010E-0000E	14N14N160
8	4715.37709(18)	P16 2111E-0101E	14N14N160
9	4715.41198(18)	P16 2111F-0101F	14N14N160
10	4715.76654(17)	P16 2010E-0000E	14N14N160
11	4716.41717(18)	P15 2111E-0101E	14N14N160
12	4716.44601(18)	P15 2111F-0101F	14N14N160
13	4716.81063(17)	P15 2010E-0000E	14N14N160
14	4717.44379(18)	P14 2111E-0101E	14N14N160
15	4717.46715(18)	P14 2111F-0101F	14N14N160
16	4717.84104(17)	P14 2010E-0000E	14N14N160
17	4718.85774(17)	P13 2010E-0000E	14N14N160
18	4719.86079(17)	P12 2010E-0000E	14N14N160
19	4720.85011(17)	P11 2010E-0000E	14N14N160
20	4721.82568(17)	P10 2010E-0000E	14N14N160
21	4722.78751(17)	P 9 2010E-0000E	14N14N160
22	4723.32258(18)	P 8 2111F-0101F	14N14N160
22		P 8 2111E-0101E	14N14N160
23	4723.73564(17)	P 8 2010E-0000E	14N14N160
24	4724.25430(18)	P 7 2111F-0101F	14N14N160
24		P 7 2111E-0101E	14N14N160
25	4724.67005(17)	P 7 2010E-0000E	14N14N160
26	4725.17276(18)	P 6 2111E-0101E	14N14N160
26		P 6 2111F-0101F	14N14N160
27	4725.59062(17)	P 6 2010E-0000E	14N14N160
28	4726.07813(19)	P 5 2111E-0101E	14N14N160
28		P 5 2111F-0101F	14N14N160
29	4726.49752(17)	P 5 2010E-0000E	14N14N160
30	4726.97018(19)	P 4 2111E-0101E	14N14N160
30		P 4 2111F-0101F	14N14N160

- "LINE NO.": serial number identifying the spectral line in the map on the opposite page.
- "WAVENUMBER": experimental value.
- Uncertainty in the last digit of the wavenumber is given in parentheses.

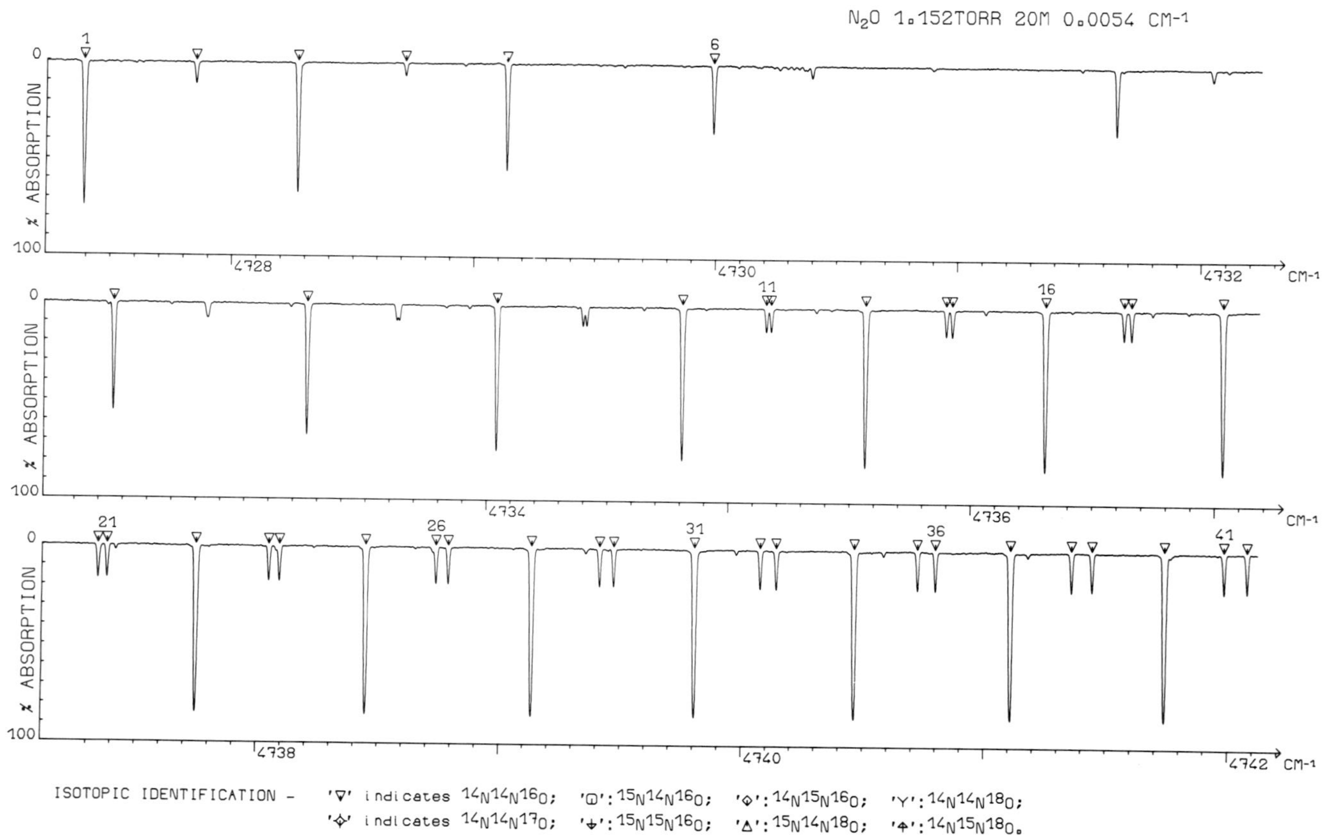
N2O 1.152TORR 20M 0.0054 CM-1
0
% ABSORPTION
100
1
6
11
16
21
26
31
36
41
4728
4730
4732
4734
4736
4738
4740
4742
CM-1
ISOTOPIC IDENTIFICATION - '▽' indicates 14N14N16O; 'Ⓞ': 15N14N16O; '◇': 14N15N16O; 'Y': 14N14N18O;
'◇' indicates 14N14N17O; '⊕': 15N15N16O; '△': 15N14N18O; '⊕': 14N15N18O.

LINE NO.	WAVENUMBER CM^{-1}, VAC.	ROVIBRATIONAL ASSIGNMENT	ISOTOPIC SPECIES
1	4727.39060(17)	P 4 2010E-0000E	14N14N16O
2	4727.84900(21)	P 3 2111E-0101E	14N14N16O
2		P 3 2111F-0101F	14N14N16O
3	4728.26994(17)	P 3 2010E-0000E	14N14N16O
4	4728.71472(24)	P 2 2111E-0101E	14N14N16O
4		P 2 2111F-0101F	14N14N16O
5	4729.13551(17)	P 2 2010E-0000E	14N14N16O
6	4729.98733(18)	P 1 2010E-0000E	14N14N16O
7	4732.46003(18)	R 1 2010E-0000E	14N14N16O
8	4733.25670(17)	R 2 2010E-0000E	14N14N16O
9	4734.03953(17)	R 3 2010E-0000E	14N14N16O
10	4734.80863(17)	R 4 2010E-0000E	14N14N16O
11	4735.15287(20)	R 5 2111E-0101E	14N14N16O
12	4735.17283(20)	R 5 2111F-0101F	14N14N16O
13	4735.56389(17)	R 5 2010E-0000E	14N14N16O
14	4735.89674(19)	R 6 2111E-0101E	14N14N16O
15	4735.92188(19)	R 6 2111F-0101F	14N14N16O
16	4736.30536(17)	R 6 2010E-0000E	14N14N16O
17	4736.62703(19)	R 7 2111E-0101E	14N14N16O
18	4736.65771(19)	R 7 2111F-0101F	14N14N16O
19	4737.03303(17)	R 7 2010E-0000E	14N14N16O
20	4737.34390(19)	R 8 2111E-0101E	14N14N16O
21	4737.38053(19)	R 8 2111F-0101F	14N14N16O
22	4737.74689(17)	R 8 2010E-0000E	14N14N16O
23	4738.04711(19)	R 9 2111E-0101E	14N14N16O
24	4738.09042(19)	R 9 2111F-0101F	14N14N16O
25	4738.44691(18)	R 9 2010E-0000E	14N14N16O
26	4738.73692(18)	R10 2111E-0101E	14N14N16O
27	4738.78728(18)	R10 2111F-0101F	14N14N16O
28	4739.13314(18)	R10 2010E-0000E	14N14N16O
29	4739.41315(18)	R11 2111E-0101E	14N14N16O
30	4739.47117(18)	R11 2111F-0101F	14N14N16O
31	4739.80555(18)	R11 2010E-0000E	14N14N16O
32	4740.07586(18)	R12 2111E-0101E	14N14N16O
33	4740.14199(18)	R12 2111F-0101F	14N14N16O
34	4740.46417(18)	R12 2010E-0000E	14N14N16O
35	4740.72506(18)	R13 2111E-0101E	14N14N16O
36	4740.79986(18)	R13 2111F-0101F	14N14N16O
37	4741.10898(18)	R13 2010E-0000E	14N14N16O
38	4741.36079(18)	R14 2111E-0101E	14N14N16O
39	4741.44472(18)	R14 2111F-0101F	14N14N16O
40	4741.73998(18)	R14 2010E-0000E	14N14N16O
41	4741.98299(18)	R15 2111E-0101E	14N14N16O
42	4742.07654(18)	R15 2111F-0101F	14N14N16O

- "LINE NO.": serial number identifying the spectral line in the map on the opposite page.
- "WAVENUMBER": experimental value.
- Uncertainty in the last digit of the wavenumber is given in parentheses.

N_2O 1.152TORR 20M 0.0054 CM-1

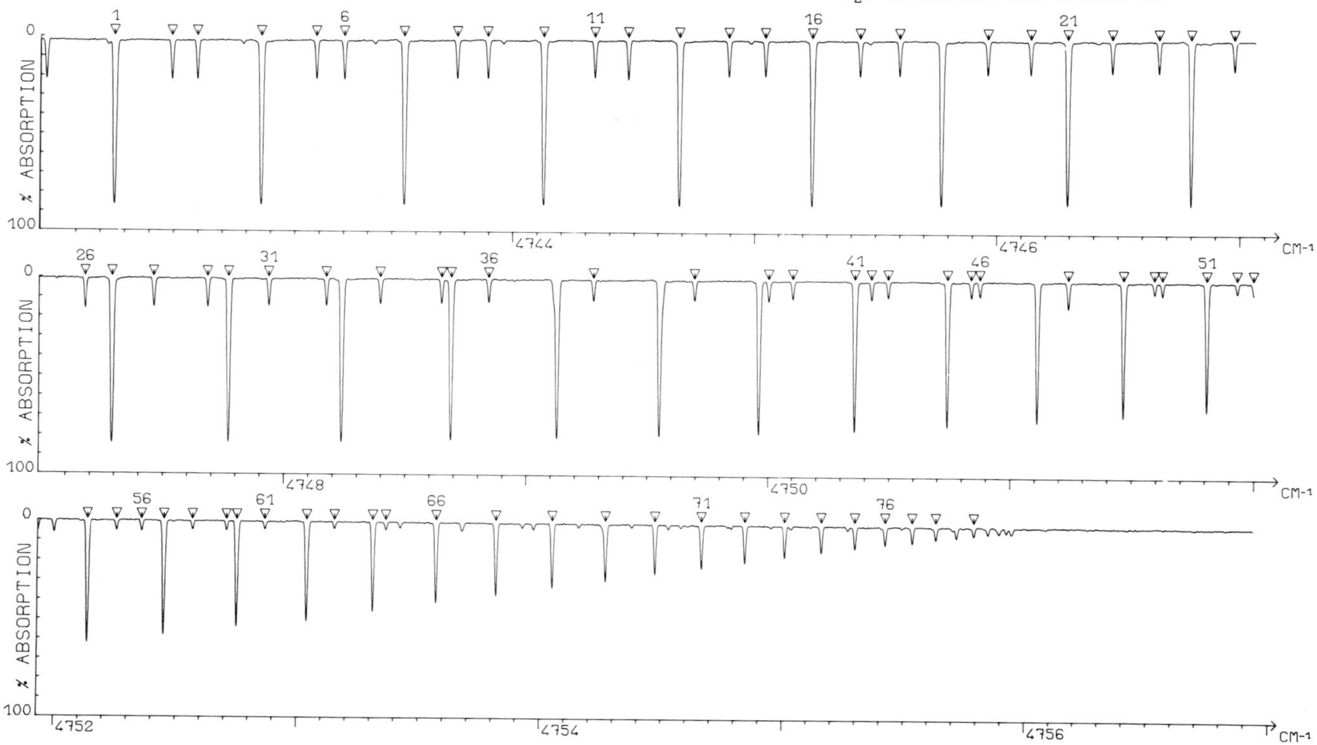

ISOTOPIC IDENTIFICATION – '▽' indicates $^{14}N^{14}N^{16}O$; '⊡': $^{15}N^{14}N^{16}O$; '◇': $^{14}N^{15}N^{16}O$; 'Y': $^{14}N^{14}N^{18}O$;
'✧' indicates $^{14}N^{14}N^{17}O$; '⊥': $^{15}N^{15}N^{16}O$; '△': $^{15}N^{14}N^{18}O$; '✦': $^{14}N^{15}N^{18}O$.

LINE NO.	WAVENUMBER CM^{-1}, VAC.	ROVIBRATIONAL ASSIGNMENT	ISOTOPIC SPECIES	LINE NO.	WAVENUMBER CM^{-1}, VAC.	ROVIBRATIONAL ASSIGNMENT	ISOTOPIC SPECIES
1	4742.35718(18)	R15 2010E-0000E	14N14N160	31	4747.93504(19)	R26 2111E-0101E	14N14N160
2	4742.59161(18)	R16 2111E-0101E	14N14N160	32	4748.17133(19)	R26 2111F-0101F	14N14N160
3	4742.69546(18)	R16 2111F-0101F	14N14N160	33	4748.39515(19)	R27 2111E-0101E	14N14N160
4	4742.96057(18)	R16 2010E-0000E	14N14N160	34	4748.64779(19)	R27 2111F-0101F	14N14N160
5	4743.18676(18)	R17 2111E-0101E	14N14N160	35	4748.68746(18)	R27 2010E-0000E	14N14N160
6	4743.30135(18)	R17 2111F-0101F	14N14N160	36	4748.84176(20)	R28 2111E-0101E	14N14N160
7	4743.55014(18)	R17 2010E-0000E	14N14N160	37	4749.27485(20)	R29 2111E-0101E	14N14N160
8	4743.76841(18)	R18 2111E-0101E	14N14N160	38	4749.69439(20)	R30 2111E-0101E	14N14N160
9	4743.89426(18)	R18 2111F-0101F	14N14N160	39	4750.00008(20)	R30 2111F-0101F	14N14N160
10	4744.12593(18)	R18 2010E-0000E	14N14N160	40	4750.10072(21)	R31 2111E-0101E	14N14N160
11	4744.33647(18)	R19 2111E-0101E	14N14N160	41	4750.35690(18)	R31 2010E-0000E	14N14N160
12	4744.47434(18)	R19 2111F-0101F	14N14N160	42	4750.42524(21)	R31 2111F-0101F	14N14N160
13	4744.68790(18)	R19 2010E-0000E	14N14N160	43	4750.49346(21)	R32 2111E-0101E	14N14N160
14	4744.89104(18)	R20 2111E-0101E	14N14N160	44	4750.73999(18)	R32 2010E-0000E	14N14N160
15	4745.04120(18)	R20 2111F-0101F	14N14N160	45	4750.83737(21)	R32 2111F-0101F	14N14N160
16	4745.23610(18)	R20 2010E-0000E	14N14N160	46	4750.87272(22)	R33 2111E-0101E	14N14N160
17	4745.43217(19)	R21 2111E-0101E	14N14N160	47	4751.23761(19)	R33 2111F-0101F	14N14N160
18	4745.59521(18)	R21 2111F-0101F	14N14N160	48	4751.46491(18)	R34 2010E-0000E	14N14N160
19	4745.95959(19)	R22 2111E-0101E	14N14N160	49	4751.59109(24)	R35 2111E-0101E	14N14N160
20	4746.13635(19)	R22 2111F-0101F	14N14N160	50	4751.62347(23)	R34 2111F-0101F	14N14N160
21	4746.29109(18)	R22 2010E-0000E	14N14N160	51	4751.80688(18)	R35 2010E-0000E	14N14N160
22	4746.47378(19)	R23 2111E-0101E	14N14N160	52	4751.92998(25)	R36 2111E-0101E	14N14N160
23	4746.66444(19)	R23 2111F-0101F	14N14N160	53	4751.99726(24)	R35 2111F-0101F	14N14N160
24	4746.79790(18)	R23 2010E-0000E	14N14N160	54	4752.13511(18)	R36 2010E-0000E	14N14N160
25	4746.97440(19)	R24 2111E-0101E	14N14N160	55	4752.25565(27)	R37 2111E-0101E	14N14N160
26	4747.17970(19)	R24 2111F-0101F	14N14N160	56	4752.35839(25)	R36 2111F-0101F	14N14N160
27	4747.29097(18)	R24 2010E-0000E	14N14N160	57	4752.44968(18)	R37 2010E-0000E	14N14N160
28	4747.46141(19)	R25 2111E-0101E	14N14N160	58	4752.56750(29)	R38 2111E-0101E	14N14N160
29	4747.68196(19)	R25 2111F-0101F	14N14N160	59	4752.70666(27)	R37 2111F-0101F	14N14N160
30	4747.77022(18)	R25 2010E-0000E	14N14N160	60	4752.75062(18)	R38 2010E-0000E	14N14N160

- "LINE NO.": serial number identifying the spectral line in the map on the opposite page.
- "WAVENUMBER": experimental value.
- Uncertainty in the last digit of the wavenumber is given in parentheses.

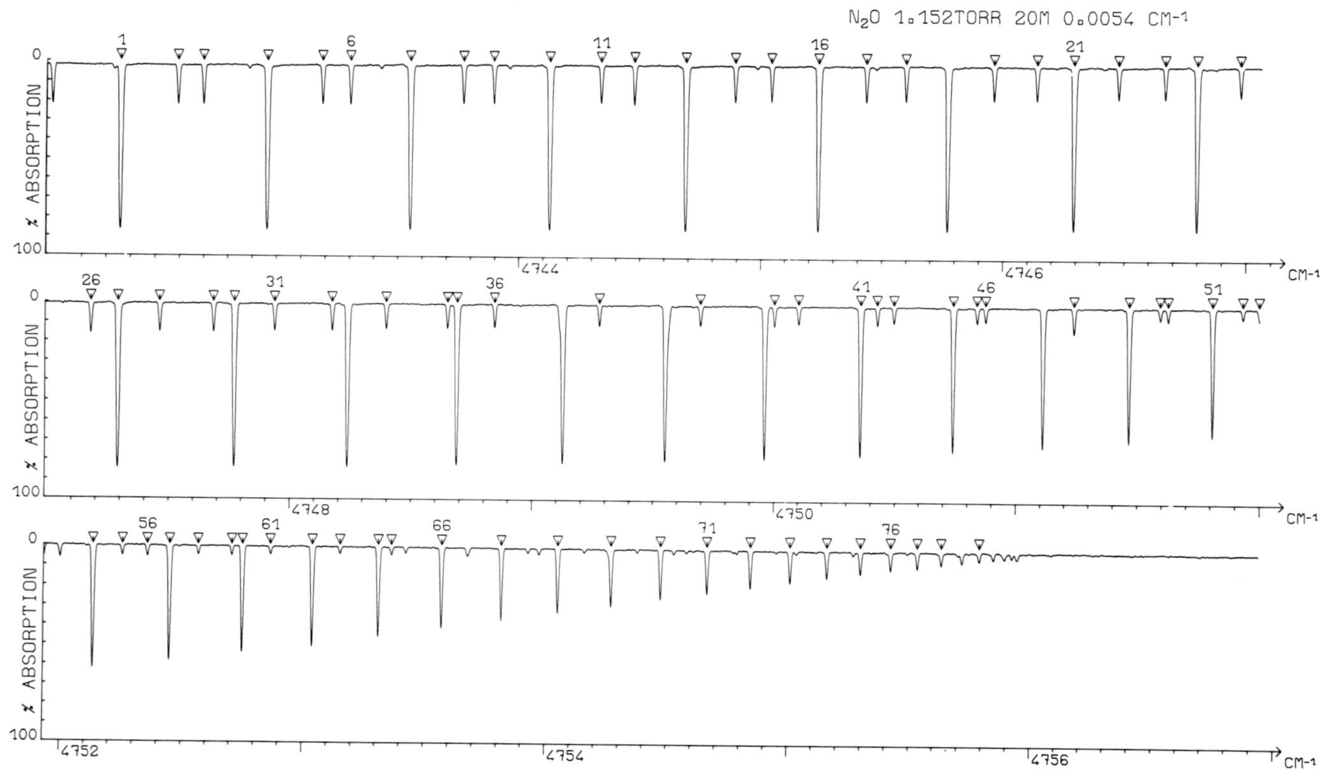

ISOTOPIC IDENTIFICATION – '▽' indicates $^{14}N^{14}N^{16}O$; '◻': $^{15}N^{14}N^{16}O$; '◇': $^{14}N^{15}N^{16}O$; 'Y': $^{14}N^{14}N^{18}O$;
'✧' indicates $^{14}N^{14}N^{17}O$; '⊥': $^{15}N^{15}N^{16}O$; '△': $^{15}N^{14}N^{18}O$; '✦': $^{14}N^{15}N^{18}O$.

LINE NO.	WAVENUMBER CM^{-1}, VAC.	ROVIBRATIONAL ASSIGNMENT	ISOTOPIC SPECIES
61	4752.86647(30)	R39 2111E-0101E	14N14N160
62	4753.03809(18)	R39 2010E-0000E	14N14N160
63	4753.15199(38)	R40 2111E-0101E	14N14N160
64	4753.31159(18)	R40 2010E-0000E	14N14N160
65	4753.36532(31)	R39 2111F-0101F	14N14N160
66	4753.57161(18)	R41 2010E-0000E	14N14N160
67	4753.81805(18)	R42 2010E-0000E	14N14N160
68	4754.05093(18)	R43 2010E-0000E	14N14N160
69	4754.27015(19)	R44 2010E-0000E	14N14N160
70	4754.47595(18)	R45 2010E-0000E	14N14N160
71	4754.66822(18)	R46 2010E-0000E	14N14N160
72	4754.84689(18)	R47 2010E-0000E	14N14N160
73	4755.01218(19)	R48 2010E-0000E	14N14N160
74	4755.16387(19)	R49 2010E-0000E	14N14N160
75	4755.30222(20)	R50 2010E-0000E	14N14N160
76	4755.42704(21)	R51 2010E-0000E	14N14N160
77	4755.53852(21)	R52 2010E-0000E	14N14N160
78	4755.63673(23)	R53 2010E-0000E	14N14N160
79	4755.79292(33)	R55 2010E-0000E	14N14N160

- "LINE NO.": serial number identifying the spectral line in the map on the opposite page.
- "WAVENUMBER": experimental value.
- Uncertainty in the last digit of the wavenumber is given in parentheses.

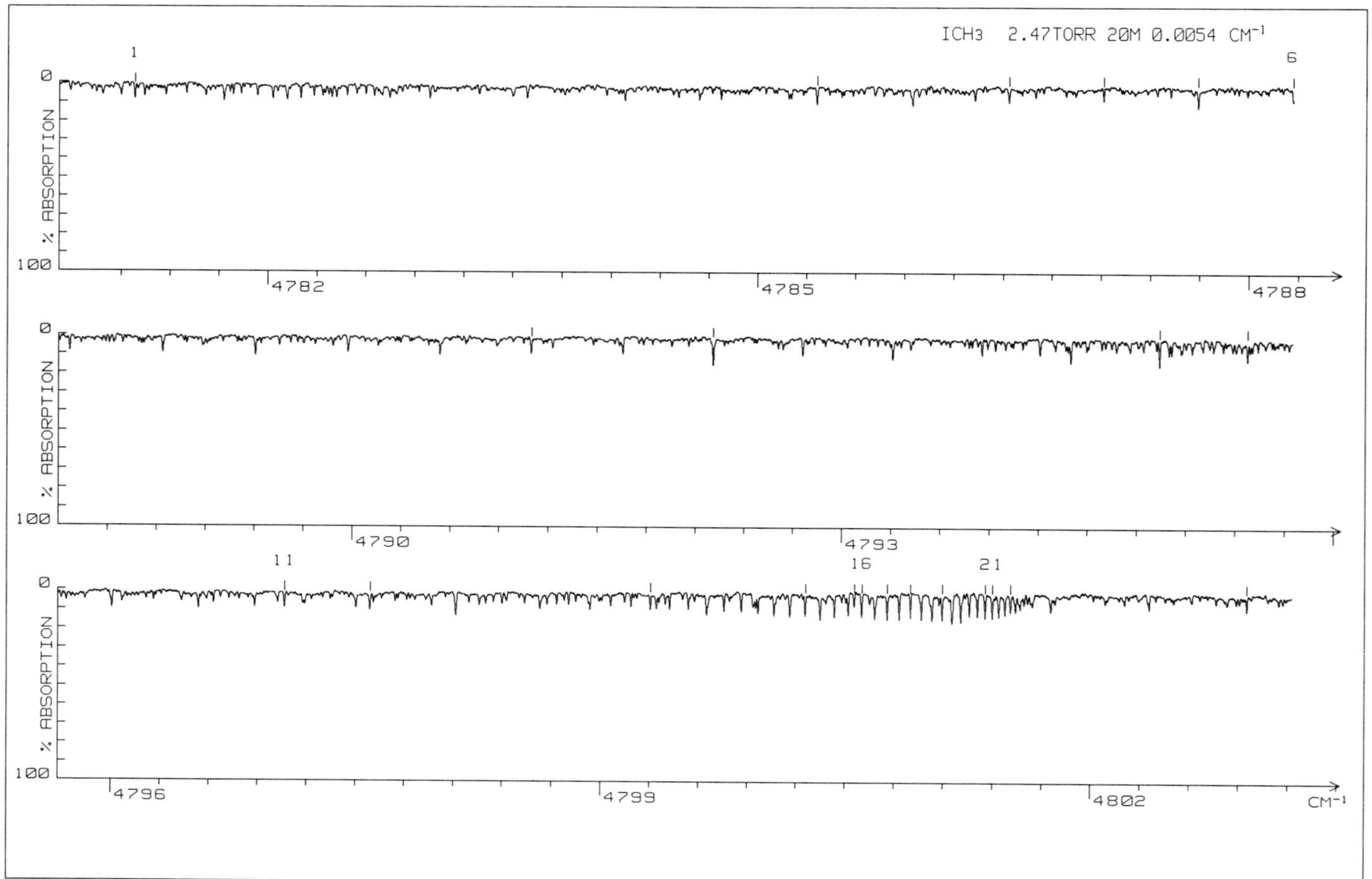
ICH3 2.47TORR 20M 0.0054 CM-1
1
6
0
% ABSORPTION
100
4782
4785
4788
4790
4793
11
16
21
4796
4799
4802
CM-1

LINE NO	WAVENUMBER CM^{-1}, VAC
1	4781.18204(60)
2	4785.36183(54)
3	4786.53233(67)
4	4787.11306(60)
5	4787.69105(47)
6	4788.26556(58)
7	4791.09719(55)
8	4792.20973(38)
9	4794.93901(37)
10	4795.47555(52)
11	4797.06865(60)
12	4797.59317(59)
13	4799.30889(58)
14	4800.26027(41)
15	4800.56052(58)
16	4800.60663(39)
17	4800.76172(40)
18	4800.90515(37)
19	4801.09768(37)
20	4801.35862(40)
21	4801.40171(42)
22	4801.51261(53)
23	4802.95583(53)

- "LINE NO.": serial number identifying the spectral line in the map on the opposite page.
- "WAVENUMBER": experimental value.
- Uncertainty in the last digit of the wavenumber is given in parentheses.

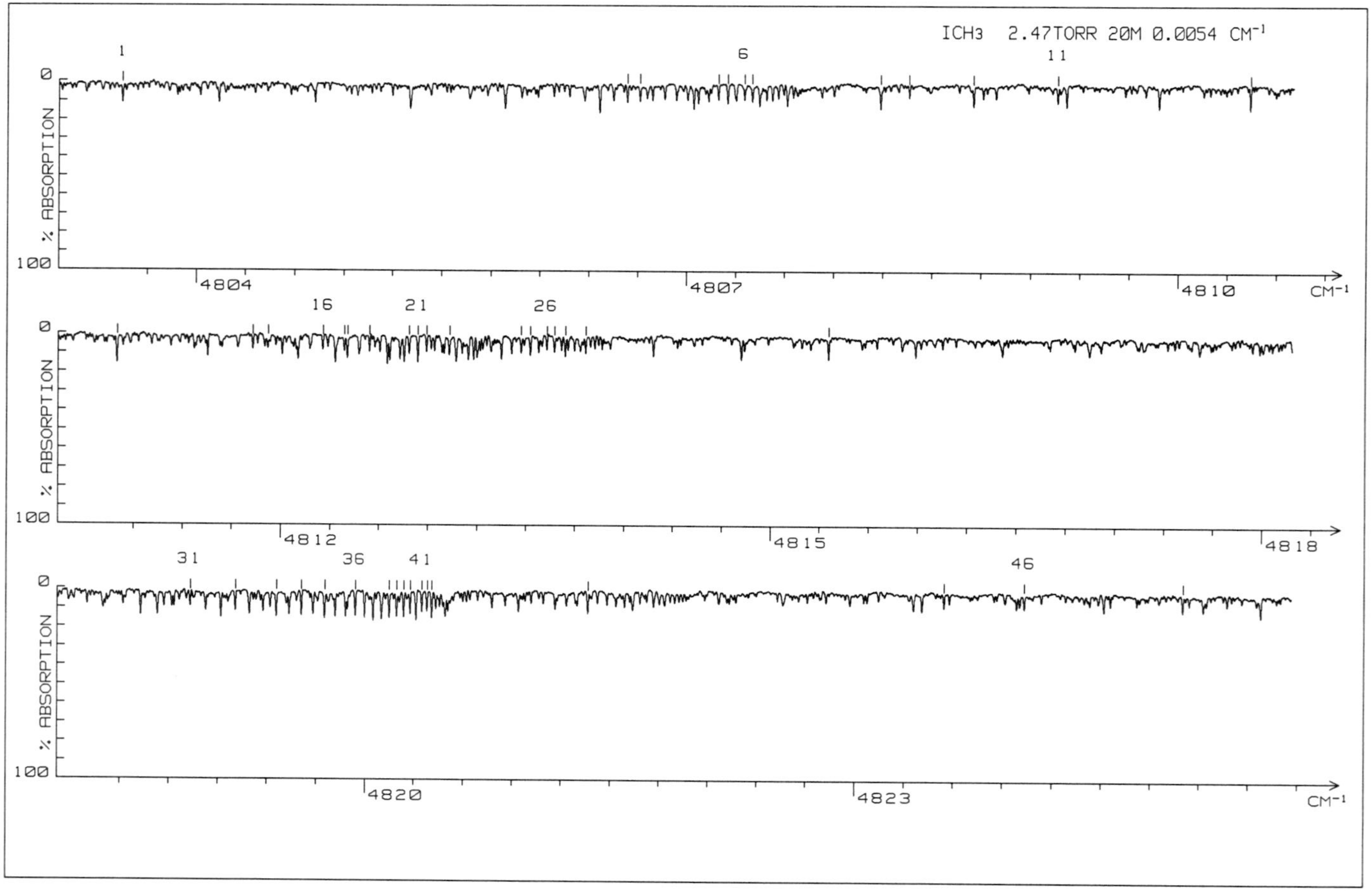

ICH3 2.47TORR 20M 0.0054 CM-1
1
6
11
0
% ABSORPTION
100
4804
4807
4810
CM-1
16
21
26
0
% ABSORPTION
100
4812
4815
4818
31
36
41
46
0
% ABSORPTION
100
4820
4823
CM-1

LINE NO	WAVENUMBER CM^{-1}, VAC	LINE NO	WAVENUMBER CM^{-1}, VAC
1	4803.54807(56)	31	4818.93228(58)
2	4806.63401(53)	32	4819.20875(47)
3	4806.71444(60)	33	4819.45860(40)
4	4807.19369(57)	34	4819.61065(37)
5	4807.25052(52)	35	4819.75118(40)
6	4807.35449(62)	36	4819.93955(39)
7	4807.40211(61)	37	4820.15082(39)
8	4808.18741(44)	38	4820.19632(41)
9	4808.36199(68)	39	4820.23919(38)
10	4808.75558(44)	40	4820.27890(41)
11	4809.26676(57)	41	4820.35002(43)
12	4810.44253(36)	42	4820.38112(45)
13	4810.99953(38)	43	4820.41025(38)
14	4811.83095(60)	44	4821.36503(43)
15	4811.92242(77)	45	4823.54653(59)
16	4812.25563(59)	46	4824.03423(62)
17	4812.38666(64)	47	4825.00092(52)
18	4812.40557(47)		
19	4812.54386(50)		
20	4812.78827(55)		
21	4812.84109(38)		
22	4812.89360(54)		
23	4813.03175(54)		
24	4813.47158(59)		
25	4813.52786(52)		
26	4813.62959(57)		
27	4813.67722(62)		
28	4813.74317(49)		
29	4813.86680(53)		
30	4815.35682(43)		

- "LINE NO.": serial number identifying the spectral line in the map on the opposite page.
- "WAVENUMBER": experimental value.
- Uncertainty in the last digit of the wavenumber is given in parentheses.

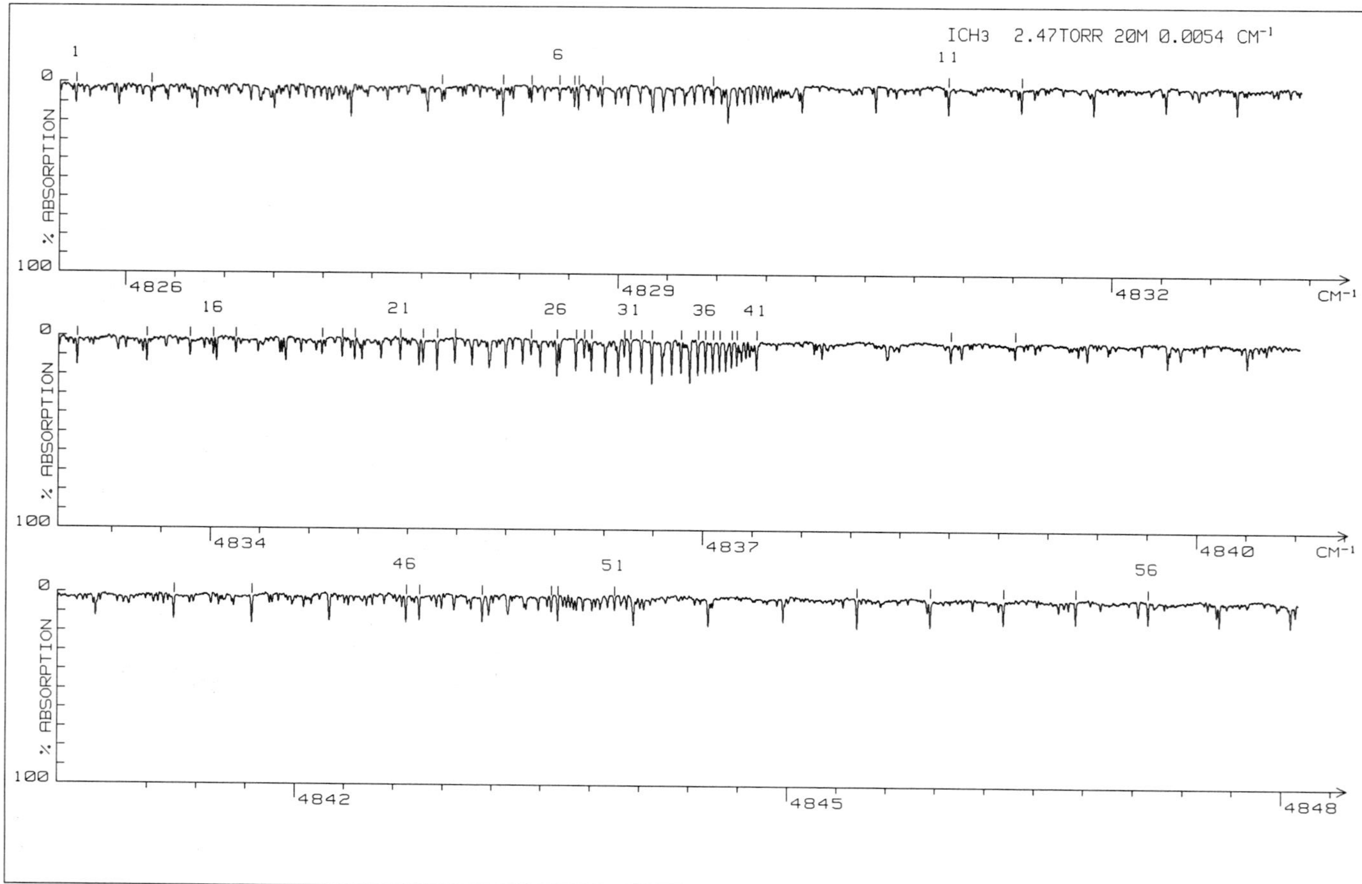

ICH3 2.47TORR 20M 0.0054 CM-1
1
6
11
16
21
26
31
36
41
46
51
56
0
% ABSORPTION
100
4826
4829
4832
4834
4837
4840
4842
4845
4848
CM-1

LINE NO	WAVENUMBER CM^{-1}, VAC	LINE NO	WAVENUMBER CM^{-1}, VAC
1	4825.69071(55)	31	4836.55696(30)
2	4826.15163(60)	32	4836.62587(31)
3	4827.91818(73)	33	4836.69095(27)
4	4828.29444(37)	34	4836.86754(33)
5	4828.46813(57)	35	4836.96963(31)
6	4828.63979(69)	36	4837.01606(33)
7	4828.72949(47)	37	4837.05975(35)
8	4828.75360(39)	38	4837.10030(35)
9	4828.89738(49)	39	4837.17285(39)
10	4829.57250(52)	40	4837.20483(41)
11	4831.00547(39)	41	4837.32406(40)
12	4831.44752(41)	42	4838.50765(58)
13	4833.18566(38)	43	4838.89698(59)
14	4833.61264(41)	44	4841.26229(44)
15	4833.87592(54)	45	4841.73715(41)
16	4834.01580(54)	46	4842.67574(38)
17	4834.15552(65)	47	4842.75650(41)
18	4834.67688(56)	48	4843.14135(40)
19	4834.80082(46)	49	4843.56429(61)
20	4834.87652(44)	50	4843.60364(39)
21	4835.15456(44)	51	4843.94871(63)
22	4835.29253(46)	52	4845.42181(34)
23	4835.37750(34)	53	4845.86963(38)
24	4835.48621(40)	54	4846.31307(41)
25	4835.95419(59)	55	4846.75380(40)
26	4836.11189(33)	56	4847.19135(41)
27	4836.22631(33)		
28	4836.27757(50)		
29	4836.32140(33)		
30	4836.52031(49)		

- "LINE NO.": serial number identifying the spectral line in the map on the opposite page.
- "WAVENUMBER": experimental value.
- Uncertainty in the last digit of the wavenumber is given in parentheses.

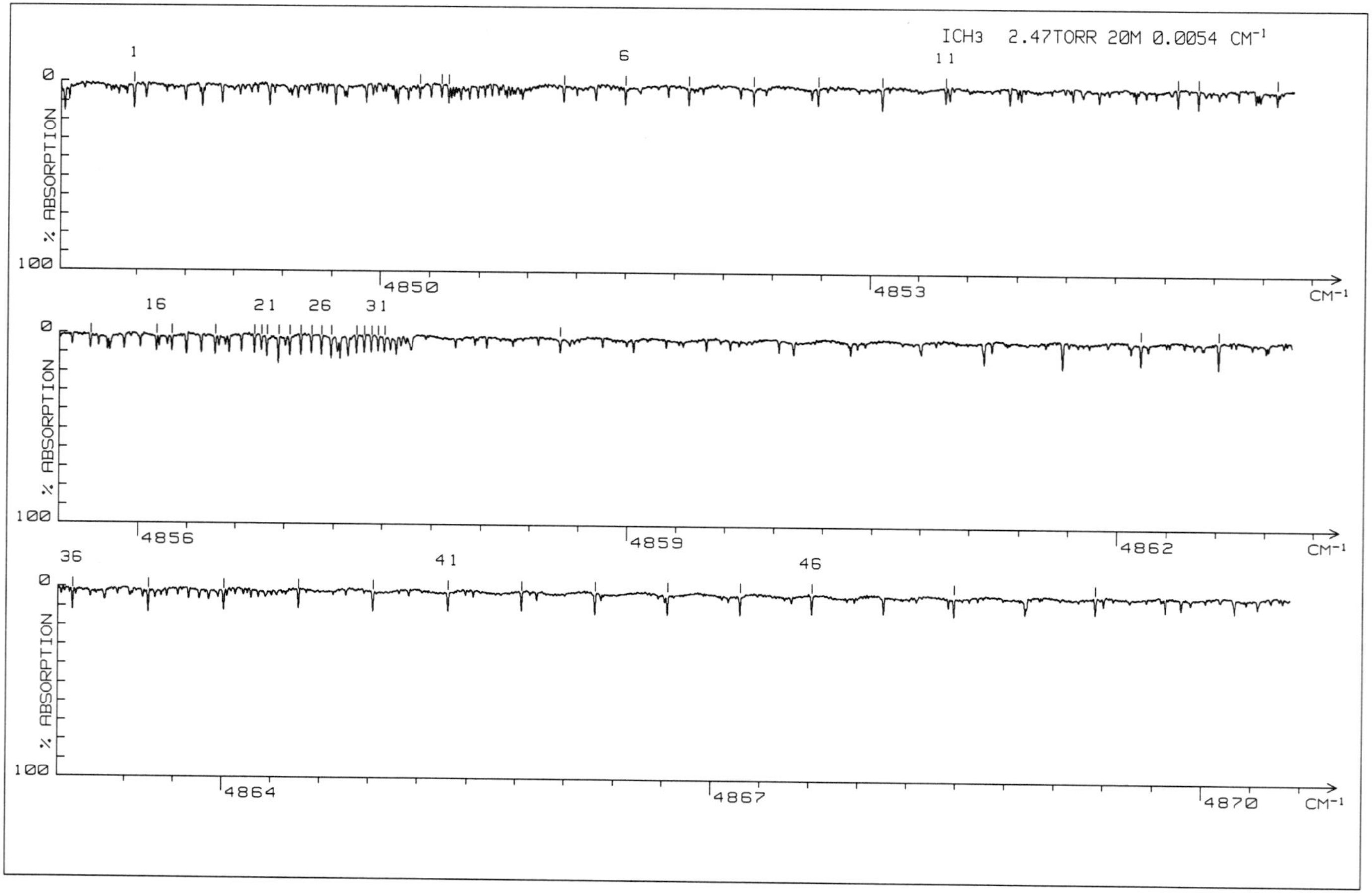
ICH3 2.47TORR 20M 0.0054 CM-1
1
6
11
0
% ABSORPTION
100
4850
4853
CM-1
16
21
26
31
0
% ABSORPTION
100
4856
4859
4862
CM-1
36
41
46
0
% ABSORPTION
100
4864
4867
4870
CM-1

LINE NO	WAVENUMBER CM^{-1}, VAC	LINE NO	WAVENUMBER CM^{-1}, VAC
1	4848.48122(43)	31	4857.47287(64)
2	4850.23741(69)	32	4857.51120(65)
3	4850.36773(68)	33	4858.58708(73)
4	4850.41040(55)	34	4862.14050(45)
5	4851.11826(61)	35	4862.61257(42)
6	4851.49660(55)	36	4863.08198(45)
7	4851.88692(54)	37	4863.54855(45)
8	4852.28129(50)	38	4864.01202(48)
9	4852.67557(51)	39	4864.47294(50)
10	4853.06750(47)	40	4864.93100(49)
11	4853.46005(57)	41	4865.38585(53)
12	4854.87453(54)	42	4865.83835(47)
13	4855.00040(49)	43	4866.28750(45)
14	4855.47970(72)	44	4866.73414(47)
15	4855.70112(68)	45	4867.17775(48)
16	4856.10869(59)	46	4867.61852(49)
17	4856.20360(58)	47	4868.49125(51)
18	4856.47151(52)	48	4869.35251(58)
19	4856.71283(48)		
20	4856.75511(63)		
21	4856.78824(48)		
22	4856.86077(38)		
23	4856.93035(49)		
24	4856.99720(47)		
25	4857.06105(48)		
26	4857.12239(50)		
27	4857.18111(46)		
28	4857.33902(53)		
29	4857.38622(58)		
30	4857.43103(53)		

- "LINE NO.": serial number identifying the spectral line in the map on the opposite page.
- "WAVENUMBER": experimental value.
- Uncertainty in the last digit of the wavenumber is given in parentheses.

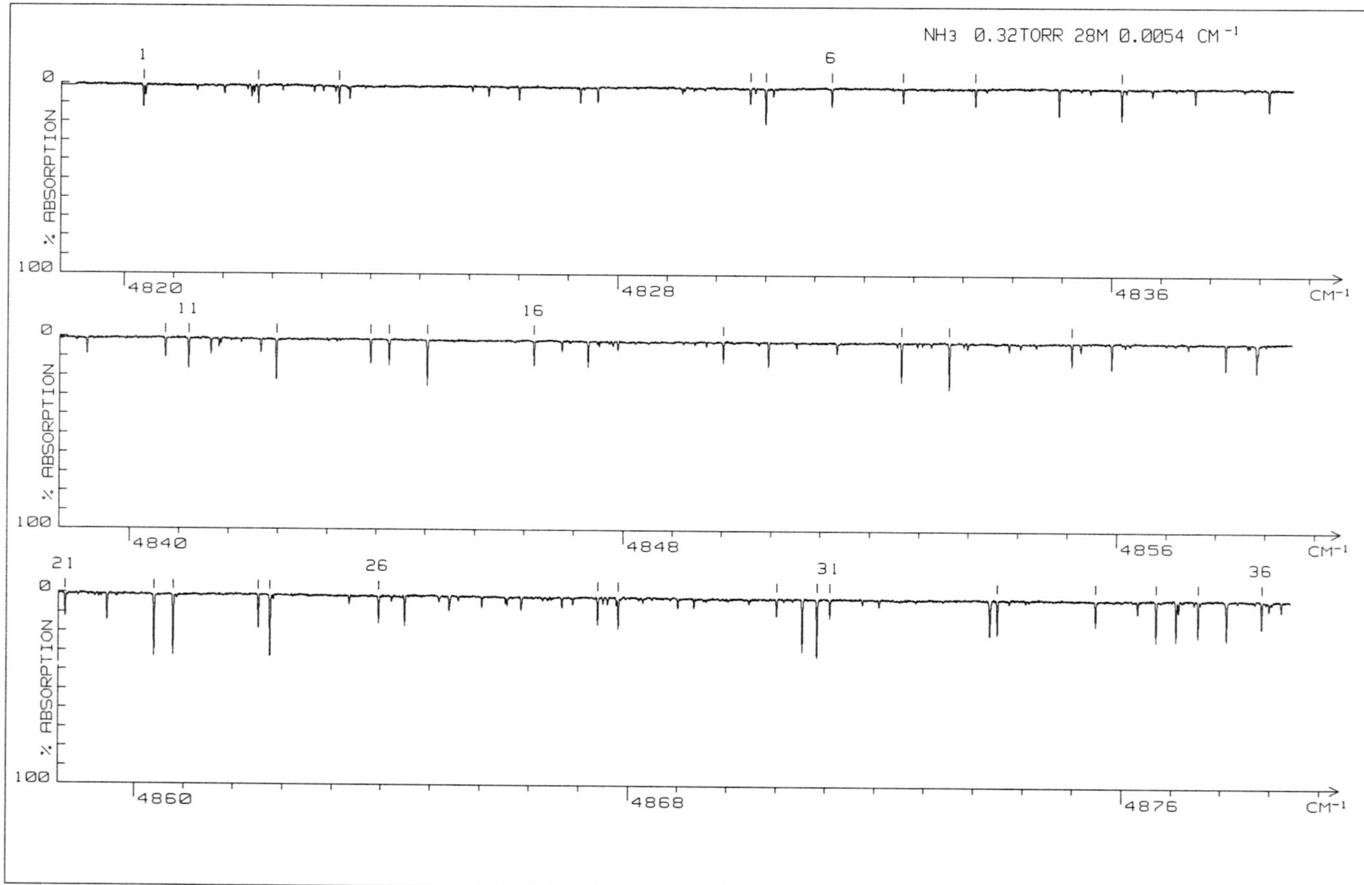
NH3 0.32TORR 28M 0.0054 CM⁻¹
1
6
11
16
21
26
31
36
0
% ABSORPTION
100
4820
4828
4836
4840
4848
4856
4860
4868
4876
CM⁻¹

LINE NO	WAVENUMBER CM^{-1}, VAC	LINE NO	WAVENUMBER CM^{-1}, VAC
1	4820.29455(69)	31	4871.26800(81)
2	4822.15960(82)	32	4873.98069(47)
3	4823.46536(79)	33	4875.56998(62)
4	4830.13491(96)	34	4876.54715(40)
5	4830.38577(44)	35	4877.23058(46)
6	4831.46201(80)	36	4878.25896(58)
7	4832.61360(101)		
8	4833.78707(81)		
9	4836.14719(51)		
10	4840.57295(84)		
11	4840.94955(54)		
12	4842.36980(43)		
13	4843.90444(66)		
14	4844.20563(62)		
15	4844.82788(39)		
16	4846.55465(64)		
17	4849.61877(71)		
18	4852.50749(43)		
19	4853.28283(37)		
20	4855.26430(65)		
21	4858.86085(69)		
22	4860.30846(32)		
23	4860.61807(32)		
24	4862.00029(49)		
25	4862.18657(31)		
26	4863.95266(58)		
27	4867.49825(58)		
28	4867.82896(53)		
29	4870.40200(90)		
30	4871.06039(33)		

- "LINE NO.": serial number identifying the spectral line in the map on the opposite page.
- "WAVENUMBER": experimental value.
- Uncertainty in the last digit of the wavenumber is given in parentheses.

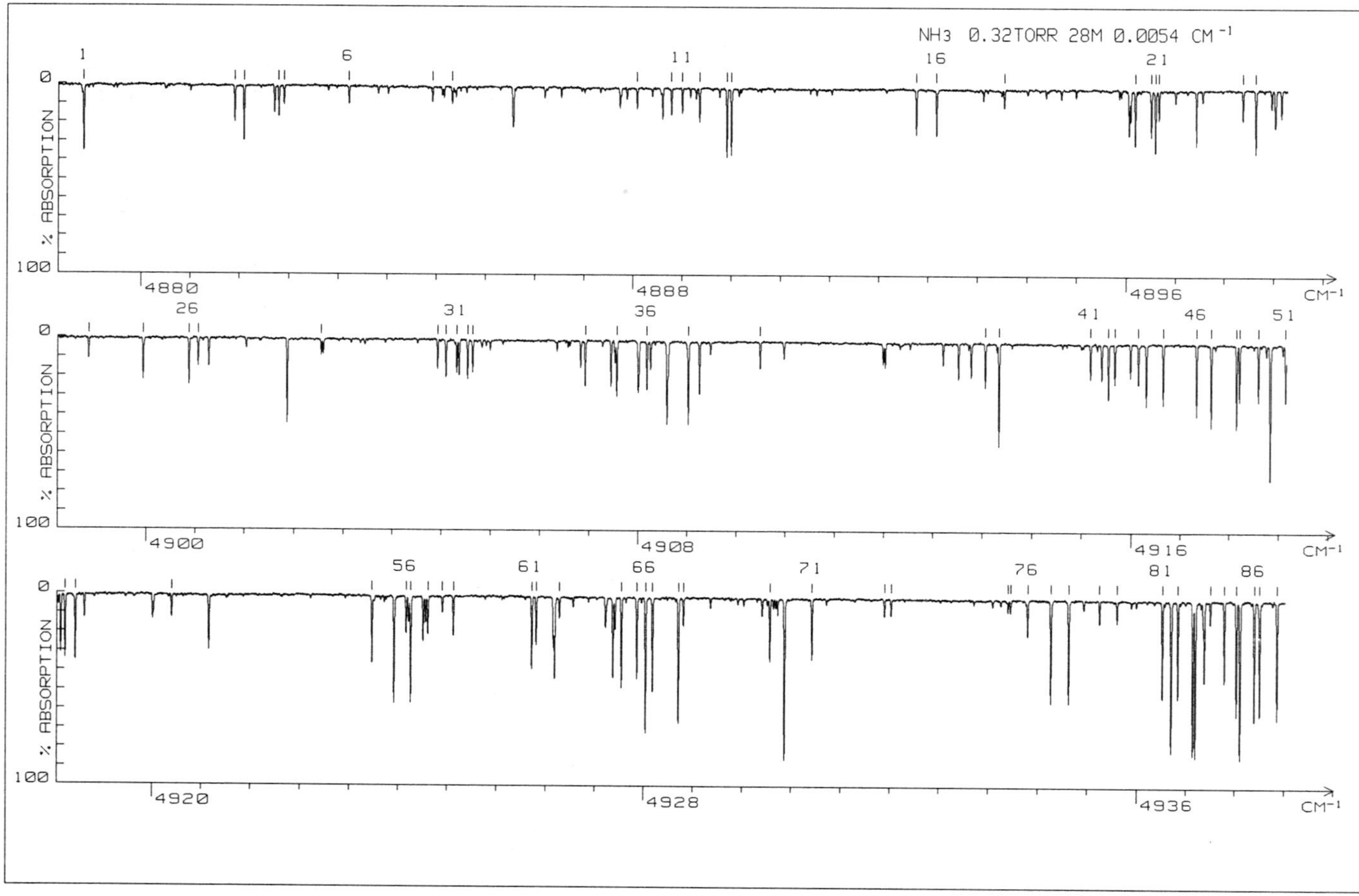
NH3 0.32TORR 28M 0.0054 CM⁻¹
% ABSORPTION
0
100
4880
4888
4896
4900
4908
4916
4920
4928
4936
CM⁻¹
1
6
11
16
21
26
31
36
41
46
51
56
61
66
71
76
81
86

LINE NO	WAVENUMBER CM^{-1}, VAC	LINE NO	WAVENUMBER CM^{-1}, VAC	LINE NO	WAVENUMBER CM^{-1}, VAC
1	4879.06139(30)	31	4905.04498(52)	61	4926.17645(29)
2	4881.52843(46)	32	4905.22154(46)	62	4926.24769(38)
3	4881.67640(34)	33	4905.30365(52)	63	4926.62421(76)
4	4882.23580(54)	34	4907.13012(39)	64	4927.63737(26)
5	4882.31949(82)	35	4907.64839(34)	65	4927.88659(28)
6	4883.37238(87)	36	4908.14193(38)	66	4928.03387(22)
7	4884.73554(107)	37	4908.81465(27)	67	4928.14653(26)
8	4885.05902(99)	38	4909.98195(61)	68	4928.56667(23)
9	4888.06437(77)	39	4913.64085(41)	69	4928.64742(61)
10	4888.62675(59)	40	4913.86304(24)	70	4930.05193(32)
11	4888.80212(61)	41	4915.33919(49)	71	4930.73988(32)
12	4889.08469(49)	42	4915.62643(35)	72	4931.91798(90)
13	4889.52845(29)	43	4915.72934(44)	73	4932.02477(92)
14	4889.60046(30)	44	4916.10724(45)	74	4933.91896(124)
15	4892.60479(39)	45	4916.51103(32)	75	4933.96230(124)
16	4892.93759(38)	46	4917.06312(29)	76	4934.23666(48)
17	4894.03288(90)	47	4917.30367(27)	77	4934.61271(25)
18	4896.13941(34)	48	4917.71400(27)	78	4934.90682(25)
19	4896.40021(38)	49	4917.76332(34)	79	4935.39992(66)
20	4896.46774(31)	50	4918.06663(34)	80	4935.68094(70)
21	4896.52662(55)	51	4918.50737(34)	81	4936.41488(25)
22	4897.90064(54)	52	4918.57968(32)	82	4936.66999(25)
23	4898.10848(31)	53	4918.74439(31)	83	4937.20688(69)
24	4899.05008(77)	54	4920.31803(72)	84	4937.43430(27)
25	4899.94331(44)	55	4923.57705(30)	85	4937.62776(24)
26	4900.69149(40)	56	4924.13160(47)	86	4937.91717(23)
27	4900.84002(60)	57	4924.20650(25)	87	4938.00407(24)
28	4902.83300(105)	58	4924.48650(47)	88	4938.28569(23)
29	4904.73239(57)	59	4924.72433(100)		
30	4904.86980(47)	60	4924.90355(44)		

- "LINE NO.": serial number identifying the spectral line in the map on the opposite page.
- "WAVENUMBER": experimental value.
- Uncertainty in the last digit of the wavenumber is given in parentheses.

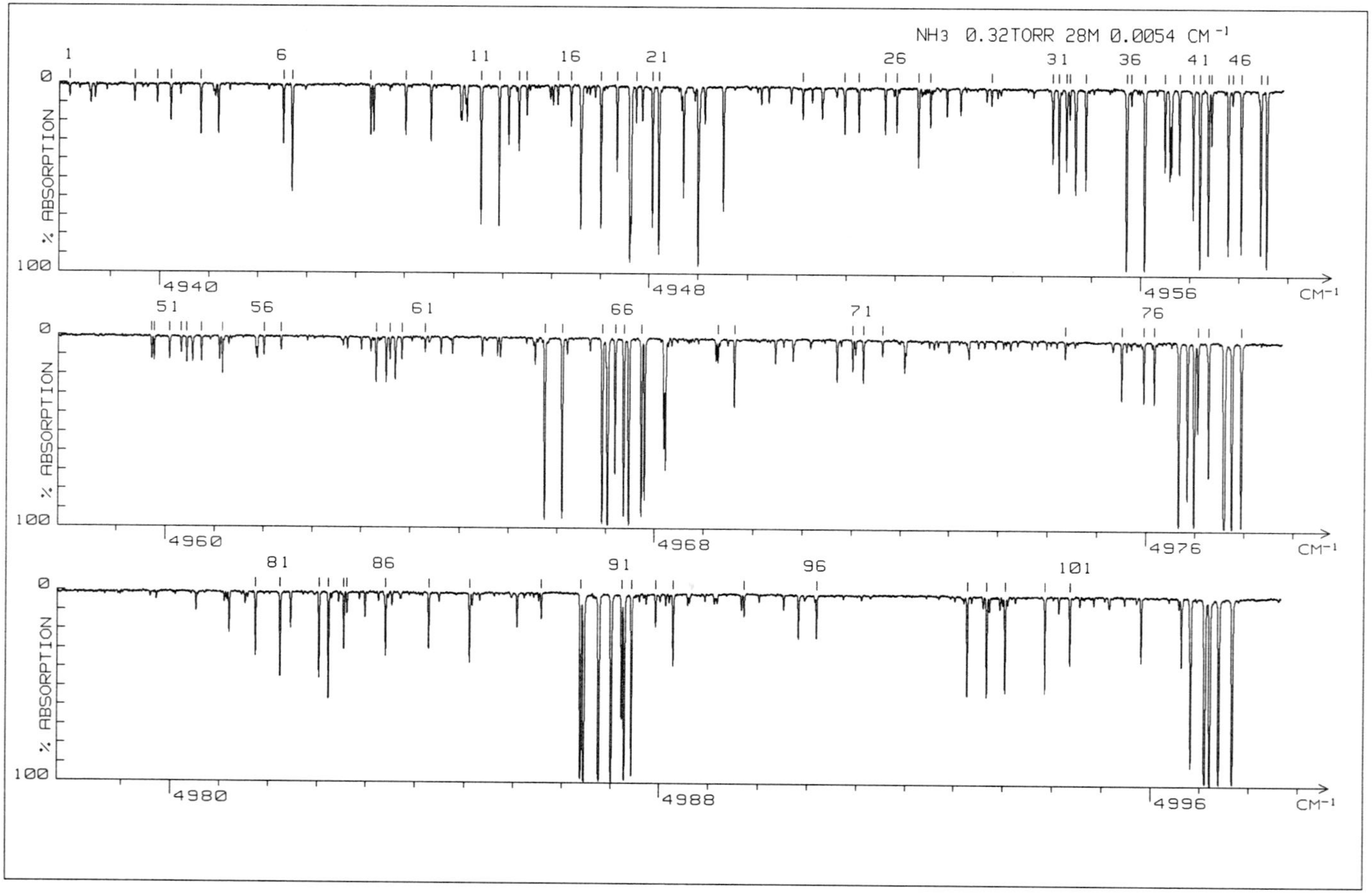
NH3 0.32TORR 28M 0.0054 CM⁻¹
1
6
11
16
21
26
31
36
41
46
51
56
61
66
71
76
81
86
91
96
101
0
% ABSORPTION
100
4940
4948
4956
4960
4968
4976
4980
4988
4996
CM⁻¹

LINE NO	WAVENUMBER CM^{-1}, VAC	LINE NO	WAVENUMBER CM^{-1}, VAC	LINE NO	WAVENUMBER CM^{-1}, VAC	LINE NO	WAVENUMBER CM^{-1}, VAC
1	4938.51244(128)	31	4954.66607(25)	61	4964.24043(104)	91	4987.37097(24)
2	4939.58319(95)	32	4954.78709(28)	62	4966.19406(22)	92	4987.53646(22)
3	4939.95008(87)	33	4954.84072(58)	63	4966.48050(22)	93	4987.93366(53)
4	4940.17068(47)	34	4955.09910(25)	64	4967.13632(23)	94	4988.21964(31)
5	4940.65755(37)	35	4955.76194(23)	65	4967.34832(23)	95	4989.36734(72)
6	4942.00993(33)	36	4955.83919(108)	66	4967.48981(22)	96	4990.54384(43)
7	4942.15417(25)	37	4956.05658(23)	67	4967.78576(22)	97	4993.01448(26)
8	4943.44040(37)	38	4956.38006(28)	68	4969.03039(78)	98	4993.33173(26)
9	4944.01288(36)	39	4956.61535(27)	69	4969.29538(31)	99	4993.62928(26)
10	4944.42024(34)	40	4956.84866(23)	70	4971.21181(55)	100	4994.27381(26)
11	4945.23169(23)	41	4956.95808(22)	71	4971.38570(43)	101	4994.67848(31)
12	4945.52962(23)	42	4957.09435(22)	72	4971.69722(111)		
13	4945.85358(33)	43	4957.14639(36)	73	4974.68259(94)		
14	4945.98244(56)	44	4957.41802(22)	74	4975.59773(35)		
15	4946.49060(84)	45	4957.48798(114)	75	4975.95865(34)		
16	4946.70808(43)	46	4957.63062(22)	76	4976.12761(33)		
17	4947.19847(22)	47	4957.94811(22)	77	4976.84294(27)		
18	4947.46266(27)	48	4958.04764(22)	78	4977.01765(23)		
19	4947.77980(47)	49	4959.77356(73)	79	4977.55161(23)		
20	4948.04990(23)	50	4959.81445(70)	80	4981.38455(32)		
21	4948.15583(22)	51	4960.06427(74)	81	4981.78722(28)		
22	4950.48574(54)	52	4960.25299(103)	82	4982.42169(27)		
23	4951.16449(40)	53	4960.34276(66)	83	4982.57728(25)		
24	4951.39489(41)	54	4960.58231(70)	84	4982.82688(35)		
25	4951.82608(39)	55	4960.92914(48)	85	4982.87977(81)		
26	4952.01388(41)	56	4961.60592(91)	86	4983.51693(33)		
27	4952.37837(30)	57	4961.88670(121)	87	4984.21533(36)		
28	4952.57592(45)	58	4963.44720(44)	88	4984.87529(31)		
29	4953.57459(92)	59	4963.67402(78)	89	4986.05216(64)		
30	4954.56657(29)	60	4963.86491(75)	90	4986.69659(23)		

- "LINE NO.": serial number identifying the spectral line in the map on the opposite page.
- "WAVENUMBER": experimental value.
- Uncertainty in the last digit of the wavenumber is given in parentheses.

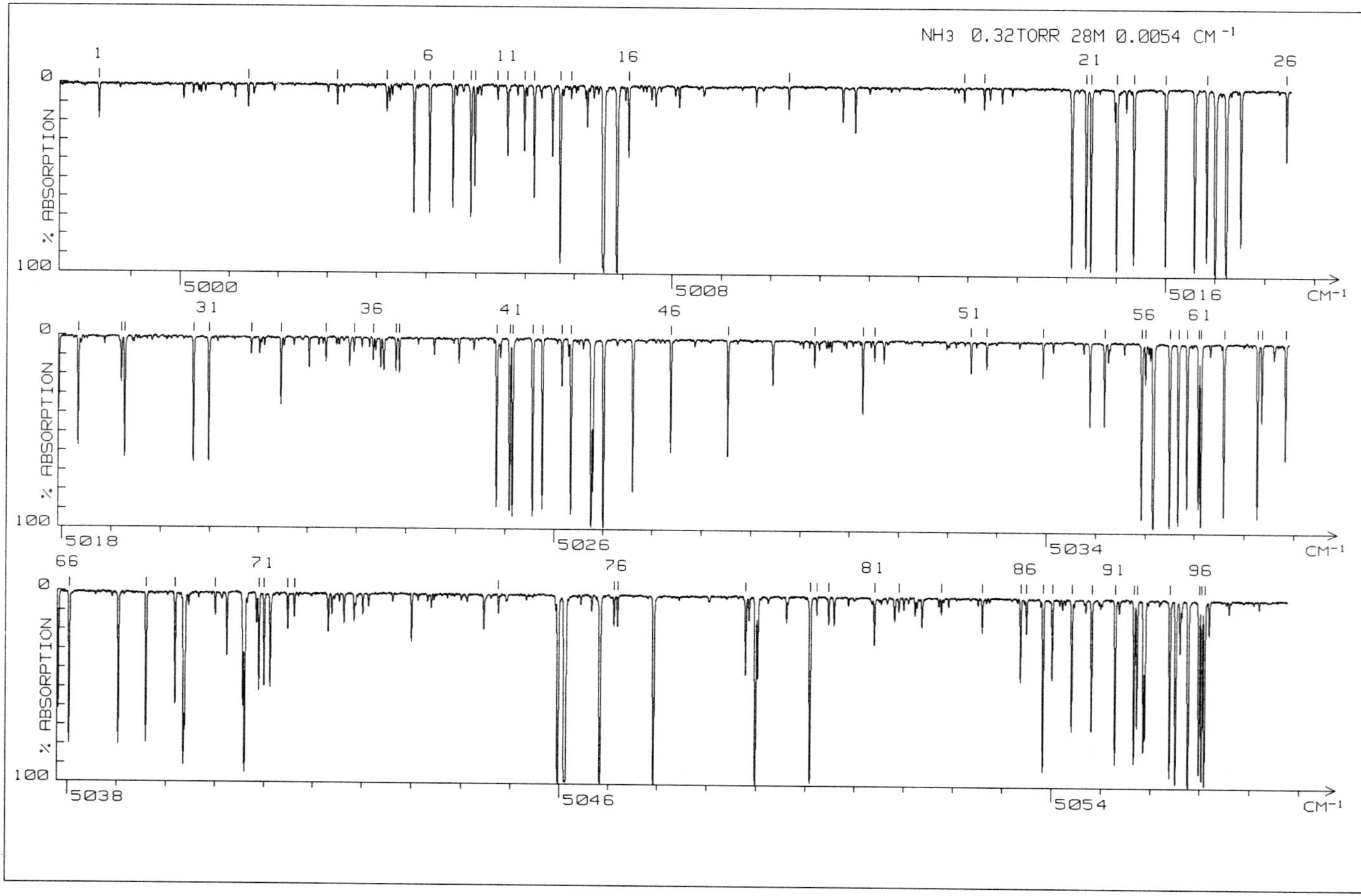

NH3 0.32TORR 28M 0.0054 CM⁻¹
% ABSORPTION
0
100
CM⁻¹
1
6
11
16
21
26
5000
5008
5016
31
36
41
46
51
56
61
5018
5026
5034
66
71
76
81
86
91
96
5038
5046
5054

LINE NO	WAVENUMBER CM^{-1}, VAC	LINE NO	WAVENUMBER CM^{-1}, VAC	LINE NO	WAVENUMBER CM^{-1}, VAC	LINE NO	WAVENUMBER CM^{-1}, VAC
1	4998.67623(51)	31	5020.37516(24)	61	5036.49625(23)	91	5055.02277(23)
2	5001.09922(74)	32	5021.06883(98)	62	5036.86303(22)	92	5055.32626(23)
3	5002.55597(86)	33	5021.55982(32)	63	5037.40050(22)	93	5055.37533(24)
4	5003.36077(74)	34	5022.28993(72)	64	5037.46908(29)	94	5055.91064(23)
5	5003.80144(24)	35	5022.74802(116)	65	5037.85058(25)	95	5056.38744(23)
6	5004.05332(24)	36	5023.05939(73)	66	5038.02598(23)	96	5056.43118(23)
7	5004.42547(24)	37	5023.42569(53)	67	5039.27148(23)	97	5056.47732(23)
8	5004.71387(24)	38	5023.48267(51)	68	5039.74592(25)		
9	5004.77877(26)	39	5025.05493(23)	69	5040.40178(73)		
10	5005.14944(125)	40	5025.26348(23)	70	5041.11195(27)		
11	5005.31072(31)	41	5025.31150(22)	71	5041.19564(27)		
12	5005.57881(32)	42	5025.63265(22)	72	5041.58703(51)		
13	5005.73715(25)	43	5025.79144(22)	73	5041.69399(75)		
14	5006.16858(23)	44	5026.11521(41)	74	5045.00024(98)		
15	5006.34491(129)	45	5026.26407(23)	75	5046.88764(59)		
16	5007.28830(31)	46	5027.89738(25)	76	5046.94717(57)		
17	5009.88245(77)	47	5028.81863(24)	77	5049.02629(30)		
18	5012.72985(112)	48	5030.21735(66)	78	5050.07064(23)		
19	5013.05603(78)	49	5031.00663(30)	79	5050.17961(94)		
20	5014.70470(23)	50	5031.19583(87)	80	5050.37513(64)		
21	5014.79084(23)	51	5032.76482(58)	81	5051.11674(46)		
22	5015.20229(23)	52	5033.01910(65)	82	5051.51056(127)		
23	5015.47624(22)	53	5033.93192(50)	83	5052.20450(111)		
24	5015.99571(22)	54	5034.92587(29)	84	5052.87170(52)		
25	5016.66542(22)	55	5035.52764(23)	85	5053.49166(28)		
26	5017.94248(30)	56	5035.58774(48)	86	5053.58676(53)		
27	5018.26055(25)	57	5035.98503(23)	87	5053.85945(23)		
28	5018.95448(42)	58	5036.12762(23)	88	5054.00670(29)		
29	5019.01154(24)	59	5036.27407(23)	89	5054.31865(24)		
30	5020.12703(24)	60	5036.45715(23)	90	5054.64534(24)		

- "LINE NO.": serial number identifying the spectral line in the map on the opposite page.
- "WAVENUMBER": experimental value.
- Uncertainty in the last digit of the wavenumber is given in parentheses.

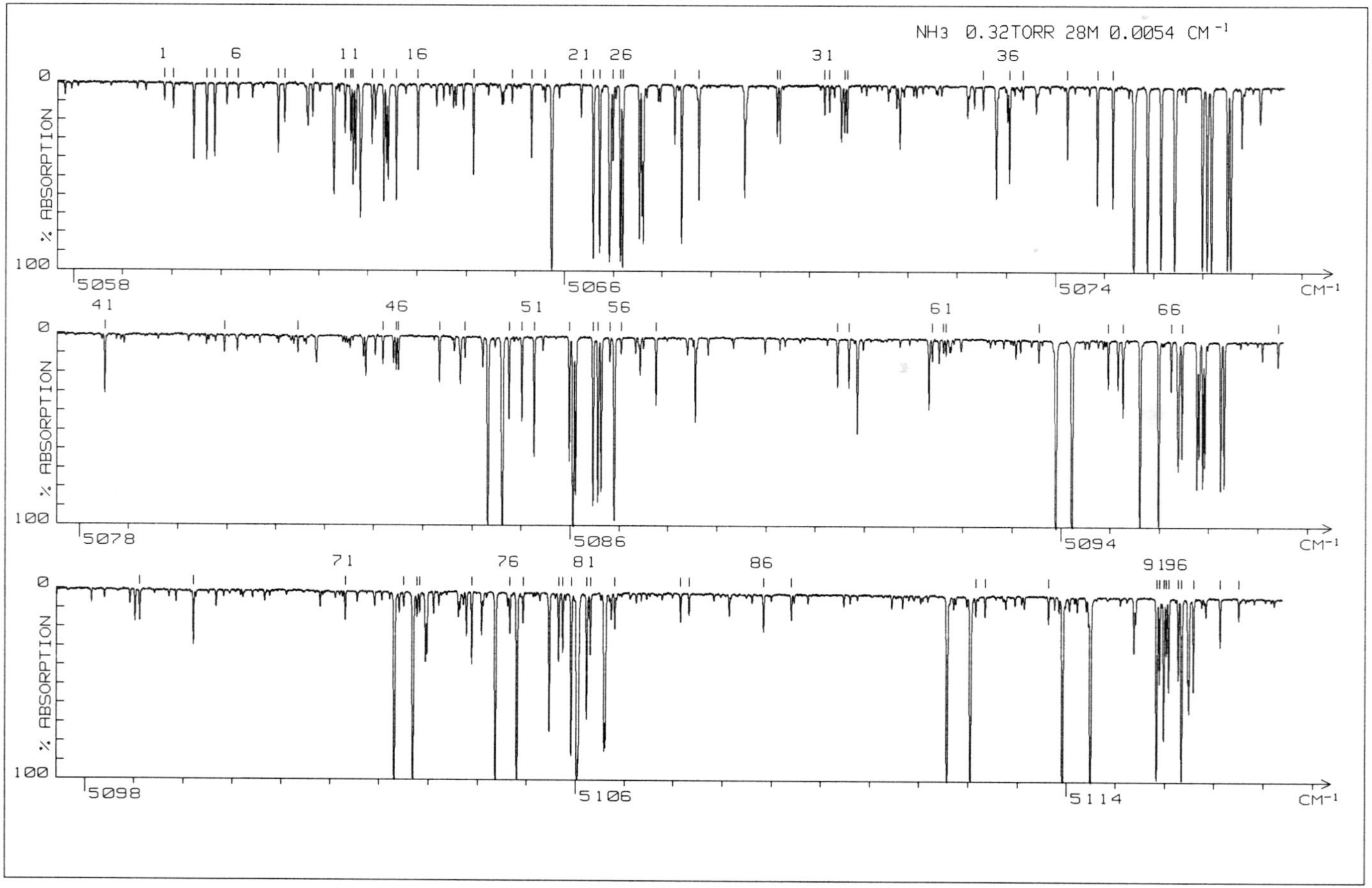
NH3 0.32TORR 28M 0.0054 CM⁻¹
% ABSORPTION
0
100
5058
5066
5074
5078
5086
5094
5098
5106
5114
CM⁻¹

LINE NO	WAVENUMBER CM^{-1}, VAC	LINE NO	WAVENUMBER CM^{-1}, VAC	LINE NO	WAVENUMBER CM^{-1}, VAC	LINE NO	WAVENUMBER CM^{-1}, VAC
1	5059.47177(94)	31	5070.23862(58)	61	5092.08355(129)	91	5115.46069(24)
2	5059.61463(68)	32	5070.31903(65)	62	5092.12596(107)	92	5115.50688(30)
3	5060.15662(30)	33	5070.56196(43)	63	5093.62293(83)	93	5115.57541(24)
4	5060.28757(31)	34	5070.60501(42)	64	5094.76580(42)	94	5115.61578(39)
5	5060.48342(81)	35	5072.81480(72)	65	5095.00535(31)	95	5115.65659(28)
6	5060.67444(117)	36	5073.24039(27)	66	5095.78396(41)	96	5115.81059(30)
7	5061.32982(32)	37	5073.45438(142)	67	5095.96128(25)	97	5115.86605(24)
8	5061.43492(48)	38	5074.18633(31)	68	5097.51694(73)	98	5116.06315(28)
9	5061.88972(53)	39	5074.67718(25)	69	5098.88880(64)	99	5116.49410(43)
10	5062.42170(40)	40	5074.93008(24)	70	5099.75701(40)	100	5116.79567(77)
11	5062.51192(36)	41	5078.41618(37)	71	5102.24110(69)		
12	5062.55142(27)	42	5080.35185(105)	72	5103.19766(121)		
13	5062.86604(34)	43	5081.55727(109)	73	5103.41078(80)		
14	5063.05746(25)	44	5082.95114(64)	74	5103.45437(93)		
15	5063.26137(25)	45	5083.16495(56)	75	5104.29699(32)		
16	5063.60920(28)	46	5083.20462(56)	76	5104.92344(51)		
17	5064.50742(27)	47	5083.86440(44)	77	5105.13935(61)		
18	5065.14161(87)	48	5084.27853(82)	78	5105.71780(33)		
19	5065.45840(31)	49	5085.00111(30)	79	5105.78404(37)		
20	5065.67774(100)	50	5085.21151(29)	80	5105.92218(23)		
21	5066.26410(56)	51	5085.41149(25)	81	5106.17241(25)		
22	5066.46140(23)	52	5085.98213(25)	82	5106.23209(36)		
23	5066.56558(23)	53	5086.36615(23)	83	5106.63061(56)		
24	5066.78175(31)	54	5086.44593(23)	84	5107.71396(66)		
25	5066.90340(23)	55	5086.63885(74)	85	5107.85316(87)		
26	5066.94350(23)	56	5086.82540(109)	86	5109.06198(57)		
27	5067.80036(35)	57	5087.40509(33)	87	5109.51894(76)		
28	5068.19584(25)	58	5090.35767(43)	88	5112.51871(95)		
29	5069.46100(39)	59	5090.54221(42)	89	5112.66847(87)		
30	5069.51002(36)	60	5091.90477(81)	90	5113.69304(67)		

- "LINE NO.": serial number identifying the spectral line in the map on the opposite page.
- "WAVENUMBER": experimental value.
- Uncertainty in the last digit of the wavenumber is given in parentheses.

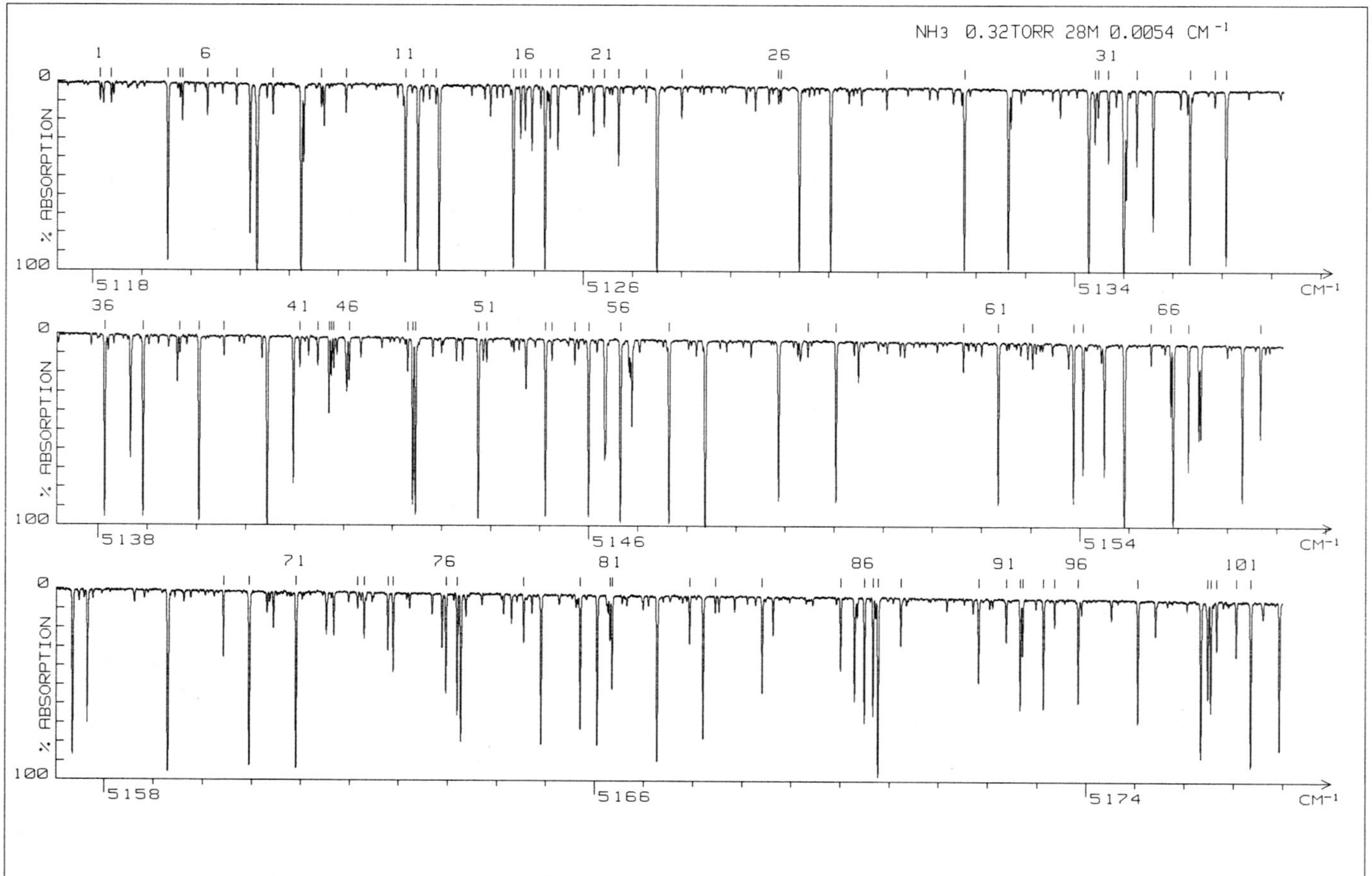
NH3 0.32TORR 28M 0.0054 CM⁻¹
% ABSORPTION
0
100
5118
5126
5134
5138
5146
5154
5158
5166
5174
CM⁻¹
1 6 11 16 21 26 31
36 41 46 51 56 61 66
71 76 81 86 91 96 101

LINE NO	WAVENUMBER CM^{-1}, VAC	LINE NO	WAVENUMBER CM^{-1}, VAC	LINE NO	WAVENUMBER CM^{-1}, VAC	LINE NO	WAVENUMBER CM^{-1}, VAC
1	5118.11695(108)	31	5134.54449(33)	61	5152.66457(24)	91	5172.69909(53)
2	5118.29578(95)	32	5135.01217(32)	62	5153.21663(81)	92	5172.92655(28)
3	5119.22056(24)	33	5135.88173(24)	63	5153.88771(25)	93	5172.96536(43)
4	5119.41688(119)	34	5136.28246(117)	64	5154.04151(26)	94	5173.30135(28)
5	5119.45569(49)	35	5136.46256(24)	65	5155.15506(90)	95	5173.48559(72)
6	5119.85673(60)	36	5138.10128(24)	66	5155.47160(33)	96	5173.87025(29)
7	5120.33518(91)	37	5138.72522(24)	67	5155.76393(26)	97	5174.84772(27)
8	5120.92785(63)	38	5139.32450(125)	68	5156.92700(29)	98	5175.98529(30)
9	5121.71581(83)	39	5139.63264(25)	69	5159.93460(37)	99	5176.03508(28)
10	5122.12112(69)	40	5140.03554(105)	70	5160.35868(25)	100	5176.12761(45)
11	5123.09921(24)	41	5141.27684(69)	71	5161.12335(25)	101	5176.44116(43)
12	5123.38513(109)	42	5141.56833(73)	72	5162.12532(141)	102	5176.67760(25)
13	5123.58868(106)	43	5141.75132(32)	73	5162.23948(50)		
14	5124.85508(24)	44	5141.78858(57)	74	5162.62806(42)		
15	5124.97030(40)	45	5141.83061(65)	75	5162.71363(34)		
16	5125.04665(49)	46	5142.08141(52)	76	5163.57112(30)		
17	5125.29804(87)	47	5143.04184(62)	77	5163.75327(27)		
18	5125.45146(41)	48	5143.12316(24)	78	5164.83844(49)		
19	5125.58076(36)	49	5143.17061(24)	79	5165.76395(26)		
20	5126.16211(43)	50	5144.18844(24)	80	5166.24593(55)		
21	5126.33663(50)	51	5144.32519(87)	81	5166.28109(31)		
22	5126.56938(31)	52	5145.28795(24)	82	5167.54420(48)		
23	5127.02063(116)	53	5145.39253(90)	83	5167.96049(118)		
24	5127.59897(62)	54	5145.76481(79)	84	5168.72413(30)		
25	5129.17740(135)	55	5145.99269(24)	85	5170.00658(36)		
26	5129.21678(123)	56	5146.50885(25)	86	5170.39219(27)		
27	5130.93612(90)	57	5147.30362(25)	87	5170.53277(27)		
28	5132.20922(24)	58	5149.56513(129)	88	5170.61449(25)		
29	5134.32707(40)	59	5150.02503(24)	89	5170.98360(48)		
30	5134.37982(72)	60	5152.09934(66)	90	5172.25486(33)		

- "LINE NO.": serial number identifying the spectral line in the map on the opposite page.
- "WAVENUMBER": experimental value.
- Uncertainty in the last digit of the wavenumber is given in parentheses.

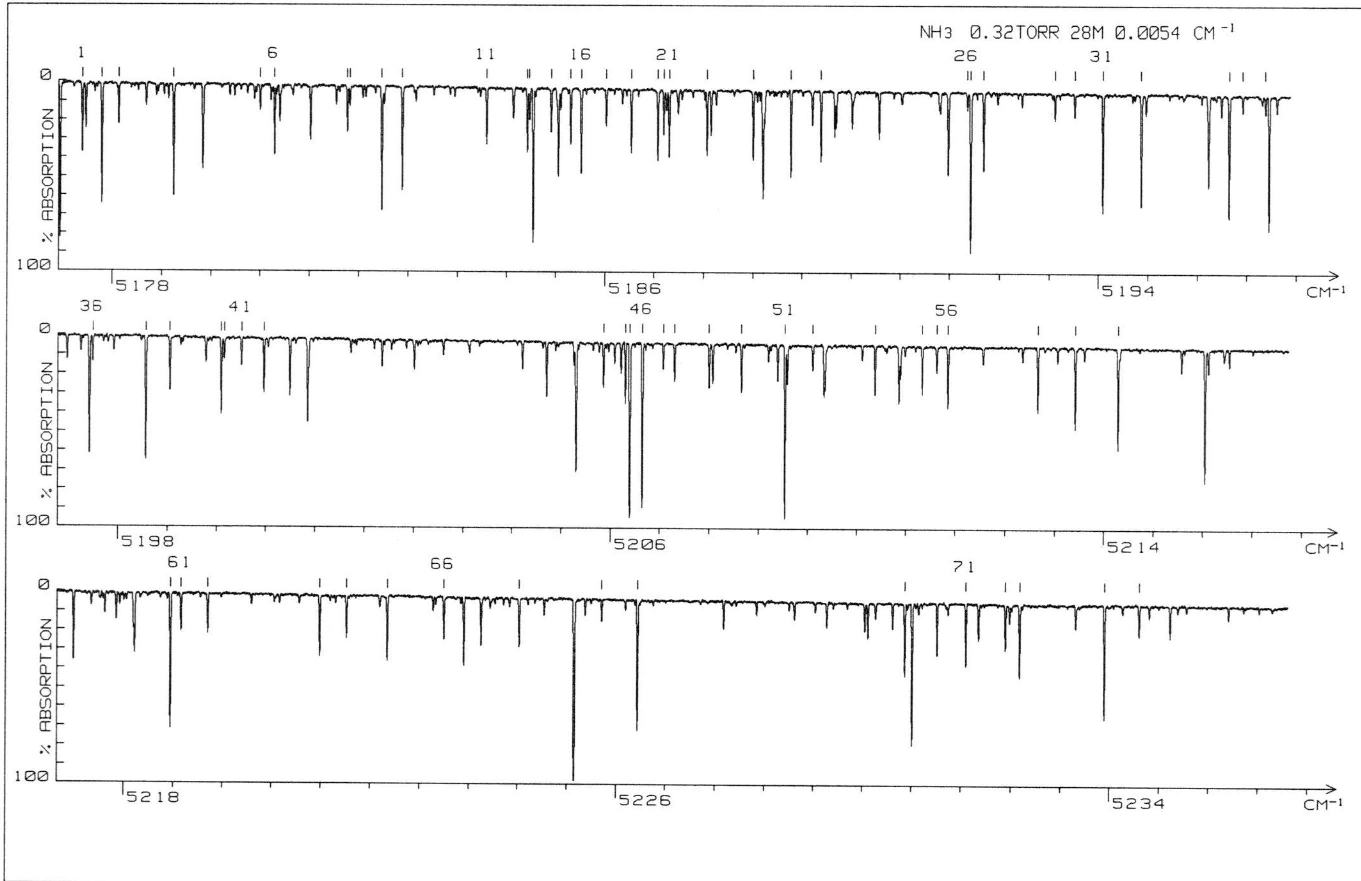
NH3 0.32TORR 28M 0.0054 CM-1
1
6
11
16
21
26
31
36
41
46
51
56
61
66
71
0
100
% ABSORPTION
5178
5186
5194
5198
5206
5214
5218
5226
5234
CM-1

LINE NO	WAVENUMBER CM^{-1}, VAC	LINE NO	WAVENUMBER CM^{-1}, VAC	LINE NO	WAVENUMBER CM^{-1}, VAC
1	5177.50306(38)	31	5194.06544(28)	61	5218.93050(70)
2	5177.82845(28)	32	5194.68447(29)	62	5219.36550(68)
3	5178.09936(55)	33	5196.10799(28)	63	5221.18795(47)
4	5178.98660(29)	34	5196.32697(115)	64	5221.62333(62)
5	5180.39675(87)	35	5196.69150(118)	65	5222.28723(42)
6	5180.62993(36)	36	5197.58795(93)	66	5223.19917(63)
7	5181.81137(53)	37	5198.45074(29)	67	5224.43309(57)
8	5181.85214(121)	38	5198.83906(49)	68	5225.76838(107)
9	5182.37285(28)	39	5199.67313(37)	69	5226.35184(29)
10	5182.70046(29)	40	5199.72284(115)	70	5230.69629(41)
11	5184.06642(44)	41	5200.00480(89)	71	5231.68724(43)
12	5184.73566(41)	42	5200.37172(49)	72	5232.31726(58)
13	5184.77237(76)	43	5205.88288(57)	73	5232.54406(40)
14	5185.12088(53)	44	5206.23927(42)	74	5233.92243(30)
15	5185.43808(45)	45	5206.31195(27)	75	5234.48266(79)
16	5185.60901(34)	46	5206.51592(26)		
17	5186.01197(63)	47	5206.85959(95)		
18	5186.42135(40)	48	5207.04178(67)		
19	5186.85698(38)	49	5207.59506(59)		
20	5186.95190(53)	50	5208.11774(50)		
21	5187.04013(39)	51	5208.82854(27)		
22	5187.64760(39)	52	5209.28143(102)		
23	5188.39335(38)	53	5210.29786(52)		
24	5189.01047(32)	54	5211.06654(54)		
25	5189.49904(37)	55	5211.30509(96)		
26	5191.88008(118)	56	5211.48588(44)		
27	5191.93333(26)	57	5212.92262(43)		
28	5192.14096(35)	58	5213.53666(36)		
29	5193.28347(83)	59	5214.23012(32)		
30	5193.60929(98)	60	5218.75923(28)		

- "LINE NO.": serial number identifying the spectral line in the map on the opposite page.
- "WAVENUMBER": experimental value.
- Uncertainty in the last digit of the wavenumber is given in parentheses.

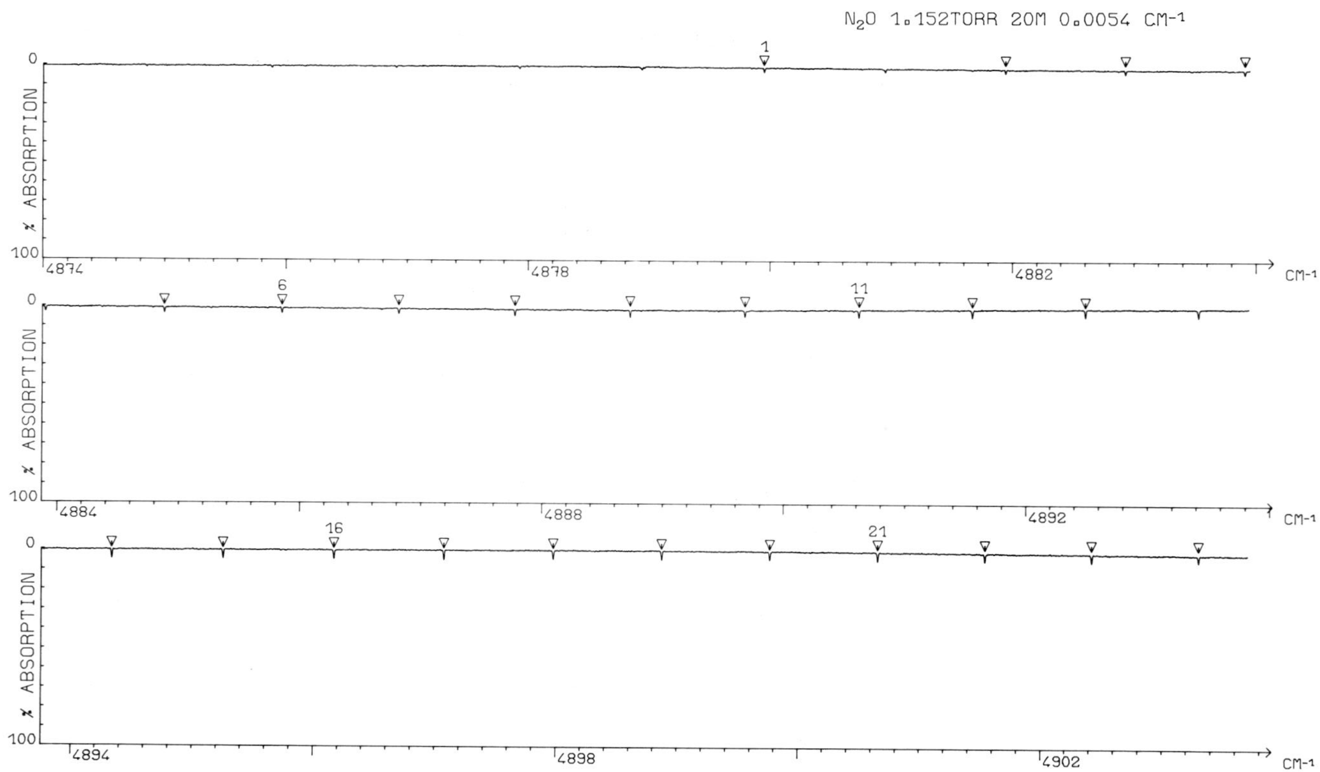

ISOTOPIC IDENTIFICATION – '▽' indicates $^{14}N^{14}N^{16}O$; '◻': $^{15}N^{14}N^{16}O$; '◇': $^{14}N^{15}N^{16}O$; 'Y': $^{14}N^{14}N^{18}O$;
'✧' indicates $^{14}N^{14}N^{17}O$; '⍖': $^{15}N^{15}N^{16}O$; '△': $^{15}N^{14}N^{18}O$; '⍏': $^{14}N^{15}N^{18}O$.

LINE NO.	WAVENUMBER CM^{-1}, VAC.	ROVIBRATIONAL ASSIGNMENT	ISOTOPIC SPECIES
1	4879.94247(52)	P34 2400E-0000E	14N14N160
2	4881.93453(64)	P32 2400E-0000E	14N14N160
3	4882.92219(82)	P31 2400E-0000E	14N14N160
4	4883.90324(50)	P30 2400E-0000E	14N14N160
5	4884.87915(42)	P29 2400E-0000E	14N14N160
6	4885.84899(50)	P28 2400E-0000E	14N14N160
7	4886.81300(42)	P27 2400E-0000E	14N14N160
8	4887.77190(39)	P26 2400E-0000E	14N14N160
9	4888.72518(35)	P25 2400E-0000E	14N14N160
10	4889.67290(34)	P24 2400E-0000E	14N14N160
11	4890.61555(35)	P23 2400E-0000E	14N14N160
12	4891.55299(34)	P22 2400E-0000E	14N14N160
13	4892.48532(32)	P21 2400E-0000E	14N14N160
14	4894.33506(31)	P19 2400E-0000E	14N14N160
15	4895.25260(30)	P18 2400E-0000E	14N14N160
16	4896.16510(32)	P17 2400E-0000E	14N14N160
17	4897.07311(30)	P16 2400E-0000E	14N14N160
18	4897.97646(30)	P15 2400E-0000E	14N14N160
19	4898.87519(30)	P14 2400E-0000E	14N14N160
20	4899.76933(50)	P13 2400E-0000E	14N14N160
21	4900.65896(31)	P12 2400E-0000E	14N14N160
22	4901.54410(38)	P11 2400E-0000E	14N14N160
23	4902.42476(32)	P10 2400E-0000E	14N14N160
24	4903.30121(35)	P 9 2400E-0000E	14N14N160

- "LINE NO.": serial number identifying the spectral line in the map on the opposite page.
- "WAVENUMBER": experimental value.
- Uncertainty in the last digit of the wavenumber is given in parentheses.

N_2O 1.152TORR 20M 0.0054 CM-1

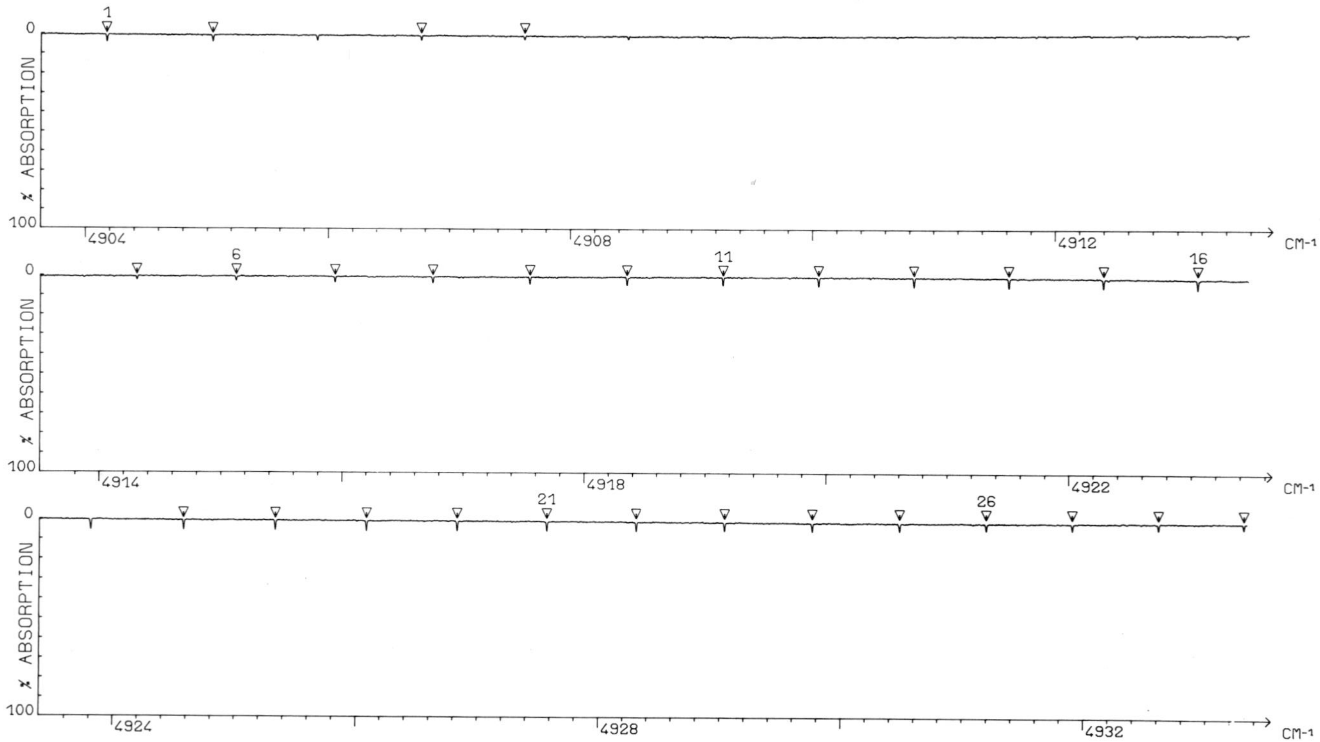

ISOTOPIC IDENTIFICATION – '▽' indicates $^{14}N^{14}N^{16}O$; '◫': $^{15}N^{14}N^{16}O$; '◇': $^{14}N^{15}N^{16}O$; 'Y': $^{14}N^{14}N^{18}O$;
'✧' indicates $^{14}N^{14}N^{17}O$; '⊥': $^{15}N^{15}N^{16}O$; 'Δ': $^{15}N^{14}N^{18}O$; '♠': $^{14}N^{15}N^{18}O$.

LINE NO.	WAVENUMBER CM^{-1} ,VAC.	ROVIBRATIONAL ASSIGNMENT	ISOTOPIC SPECIES
1	4904.17348(44)	P 8 2400E-0000E	14N14N160
2	4905.04098(40)	P 7 2400E-0000E	14N14N160
3	4906.76399(61)	P 5 2400E-0000E	14N14N160
4	4907.61838(66)	P 4 2400E-0000E	14N14N160
5	4914.30544(67)	R 3 2400E-0000E	14N14N160
6	4915.12260(59)	R 4 2400E-0000E	14N14N160
7	4915.93479(43)	R 5 2400E-0000E	14N14N160
8	4916.74294(38)	R 6 2400E-0000E	14N14N160
9	4917.54670(34)	R 7 2400E-0000E	14N14N160
10	4918.34577(35)	R 8 2400E-0000E	14N14N160
11	4919.14089(32)	R 9 2400E-0000E	14N14N160
12	4919.93118(31)	R10 2400E-0000E	14N14N160
13	4920.71708(30)	R11 2400E-0000E	14N14N160
14	4921.49821(29)	R12 2400E-0000E	14N14N160
15	4922.27495(30)	R13 2400E-0000E	14N14N160
16	4923.04666(34)	R14 2400E-0000E	14N14N160
17	4924.57558(29)	R16 2400E-0000E	14N14N160
18	4925.33293(29)	R17 2400E-0000E	14N14N160
19	4926.08493(29)	R18 2400E-0000E	14N14N160
20	4926.83206(30)	R19 2400E-0000E	14N14N160
21	4927.57393(30)	R20 2400E-0000E	14N14N160
22	4928.31046(30)	R21 2400E-0000E	14N14N160
23	4929.04160(34)	R22 2400E-0000E	14N14N160
24	4929.76739(31)	R23 2400E-0000E	14N14N160
25	4930.48773(41)	R24 2400E-0000E	14N14N160
26	4931.20212(37)	R25 2400E-0000E	14N14N160
27	4931.91073(37)	R26 2400E-0000E	14N14N160
28	4932.61368(61)	R27 2400E-0000E	14N14N160
29	4933.31077(69)	R28 2400E-0000E	14N14N160

- "LINE NO.": serial number identifying the spectral line in the map on the opposite page.
- "WAVENUMBER": experimental value.
- Uncertainty in the last digit of the wavenumber is given in parentheses.

N_2O 1.152TORR 20M 0.0054 CM-1

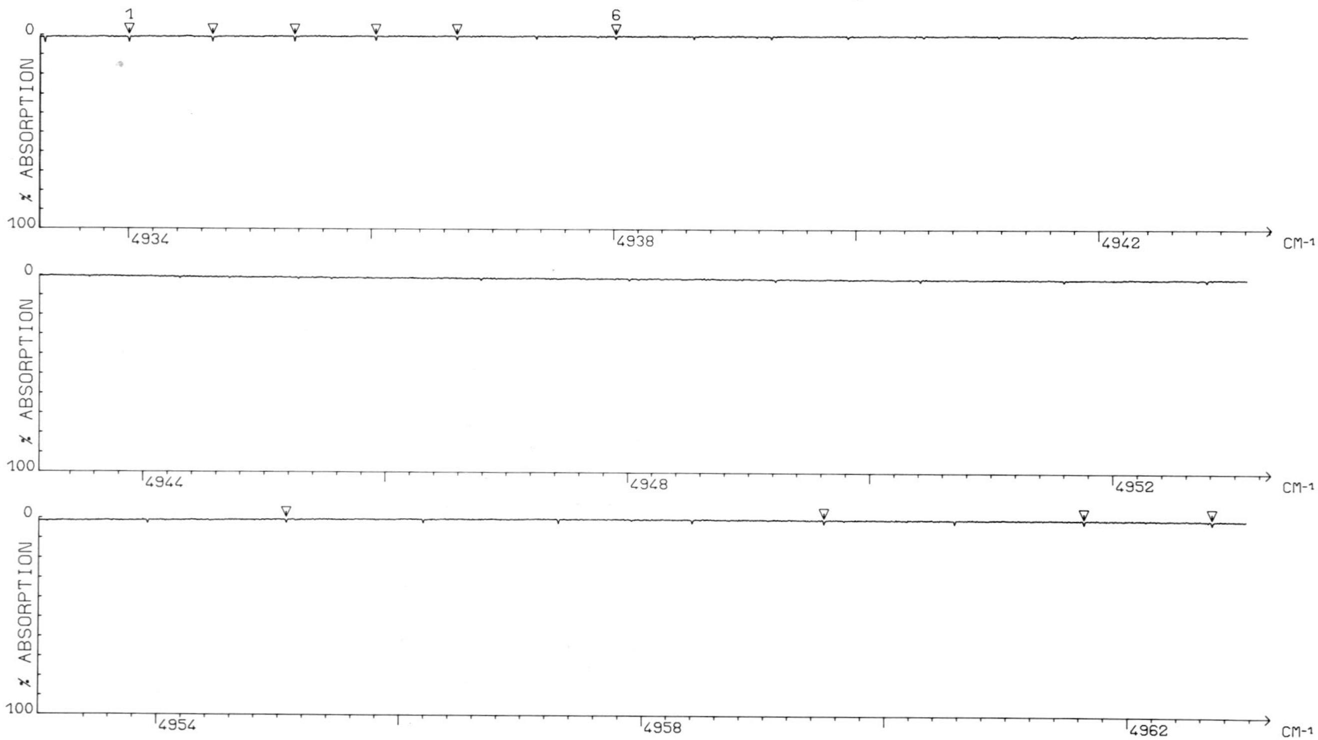

ISOTOPIC IDENTIFICATION – '▽' indicates $^{14}N^{14}N^{16}O$; 'ⓞ': $^{15}N^{14}N^{16}O$; '⌽': $^{14}N^{15}N^{16}O$; 'Y': $^{14}N^{14}N^{18}O$;
'✧' indicates $^{14}N^{14}N^{17}O$; '⏚': $^{15}N^{15}N^{16}O$; '△': $^{15}N^{14}N^{18}O$; '✢': $^{14}N^{15}N^{18}O$.

LINE NO.	WAVENUMBER CM^{-1}, VAC.	ROVIBRATIONAL ASSIGNMENT	ISOTOPIC SPECIES
1	4934.00137(51)	R29 2400E-0000E	14N14N160
2	4934.68606(47)	R30 2400E-0000E	14N14N160
3	4935.36449(48)	R31 2400E-0000E	14N14N160
4	4936.03641(54)	R32 2400E-0000E	14N14N160
5	4936.70214(69)	R33 2400E-0000E	14N14N160
6	4938.01342(85)	R35 2400E-0000E	14N14N160
7	4955.06481(79)	P23 0121E-0000E	14N14N160
8	4959.50432(83)	P19 0121E-0000E	14N14N160
9	4961.64518(71)	P17 0121E-0000E	14N14N160
10	4962.69562(69)	P16 0121E-0000E	14N14N160

- "LINE NO.": serial number identifying the spectral line in the map on the opposite page.
- "WAVENUMBER": experimental value.
- Uncertainty in the last digit of the wavenumber is given in parentheses.

N_2O 1.152TORR 20M 0.0054 CM-1

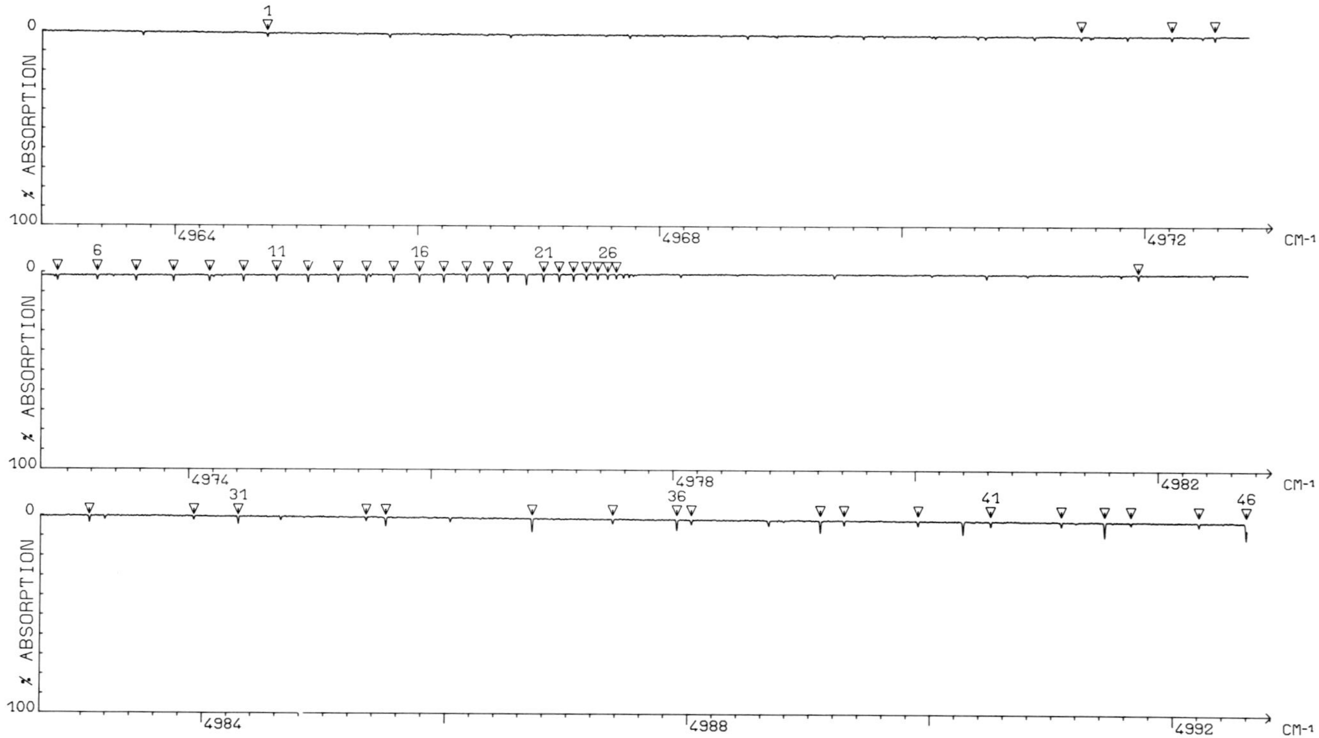

ISOTOPIC IDENTIFICATION – '▽' indicates $^{14}N^{14}N^{16}O$; '◫': $^{15}N^{14}N^{16}O$; '◇': $^{14}N^{15}N^{16}O$; 'Y': $^{14}N^{14}N^{18}O$;
'✧' indicates $^{14}N^{14}N^{17}O$; '⊥': $^{15}N^{15}N^{16}O$; '△': $^{15}N^{14}N^{18}O$; '⊕': $^{14}N^{15}N^{18}O$.

LINE NO.	WAVENUMBER CM^{-1}, VAC.	ROVIBRATIONAL ASSIGNMENT	ISOTOPIC SPECIES
1	4964.75664(67)	P14 0121E-0000E	14N14N160
2	4971.47512(89)	Q32 0121F-0000E	14N14N160
3	4972.21743(70)	Q30 0121F-0000E	14N14N160
4	4972.57120(60)	Q29 0121F-0000E	14N14N160
5	4972.91280(48)	Q28 0121F-0000E	14N14N160
6	4973.24324(60)	Q27 0121F-0000E	14N14N160
7	4973.56170(54)	Q26 0121F-0000E	14N14N160
8	4973.86809(50)	Q25 0121F-0000E	14N14N160
9	4974.16252(45)	Q24 0121F-0000E	14N14N160
10	4974.44548(39)	Q23 0121F-0000E	14N14N160
11	4974.71670(37)	Q22 0121F-0000E	14N14N160
12	4974.97581(45)	Q21 0121F-0000E	14N14N160
13	4975.22330(36)	Q20 0121F-0000E	14N14N160
14	4975.45916(34)	Q19 0121F-0000E	14N14N160
15	4975.68313(34)	Q18 0121F-0000E	14N14N160
16	4975.89531(34)	Q17 0121F-0000E	14N14N160
17	4976.09562(33)	Q16 0121F-0000E	14N14N160
18	4976.28406(35)	Q15 0121F-0000E	14N14N160
19	4976.46078(32)	Q14 0121F-0000E	14N14N160
20	4976.62596(35)	Q13 0121F-0000E	14N14N160
21	4976.92053(35)	Q11 0121F-0000E	14N14N160
22	4977.05022(35)	Q10 0121F-0000E	14N14N160
23	4977.16778(38)	Q 9 0121F-0000E	14N14N160
24	4977.27386(41)	Q 8 0121F-0000E	14N14N160
25	4977.36812(46)	Q 7 0121F-0000E	14N14N160
26	4977.45076(60)	Q 6 0121F-0000E	14N14N160
27	4977.52128(55)	Q 5 0121F-0000E	14N14N160
28	4981.82824(46)	P43 3200E-0000E	14N14N160
29	4983.06448(47)	P42 3200E-0000E	14N14N160
30	4983.92305(82)	R 7 0121E-0000E	14N14N160

LINE NO.	WAVENUMBER CM^{-1}, VAC.	ROVIBRATIONAL ASSIGNMENT	ISOTOPIC SPECIES
31	4984.29141(43)	P41 3200E-0000E	14N14N160
32	4985.34551(73)	R 9 0121E-0000E	14N14N160
33	4985.50779(34)	P40 3200E-0000E	14N14N160
34	4986.71465(26)	R11 0121E-0000E	14N14N160
34		P39 3200E-0000E	14N14N160
35	4987.37968(97)	R12 0121E-0000E	14N14N160
36	4987.91092(28)	P38 3200E-0000E	14N14N160
37	4988.03114(53)	R13 0121E-0000E	14N14N160
38	4989.09788(27)	P37 3200E-0000E	14N14N160
39	4989.29317(86)	R15 0121E-0000E	14N14N160
39		P30 4000E-0000E	15N15N160
40	4989.90477(51)	R16 0121E-0000E	14N14N160
41	4990.50247(51)	R17 0121E-0000E	14N14N160
42	4991.08696(52)	R18 0121E-0000E	14N14N160
43	4991.44184(24)	P35 3200E-0000E	14N14N160
44	4991.65729(71)	R19 0121E-0000E	14N14N160
45	4992.21528(71)	R20 0121E-0000E	14N14N160
46	4992.59903(23)	P34 3200E-0000E	14N14N160

- "LINE NO.": serial number identifying the spectral line in the map on the opposite page.
- "WAVENUMBER": experimental value.
- Uncertainty in the last digit of the wavenumber is given in parentheses.

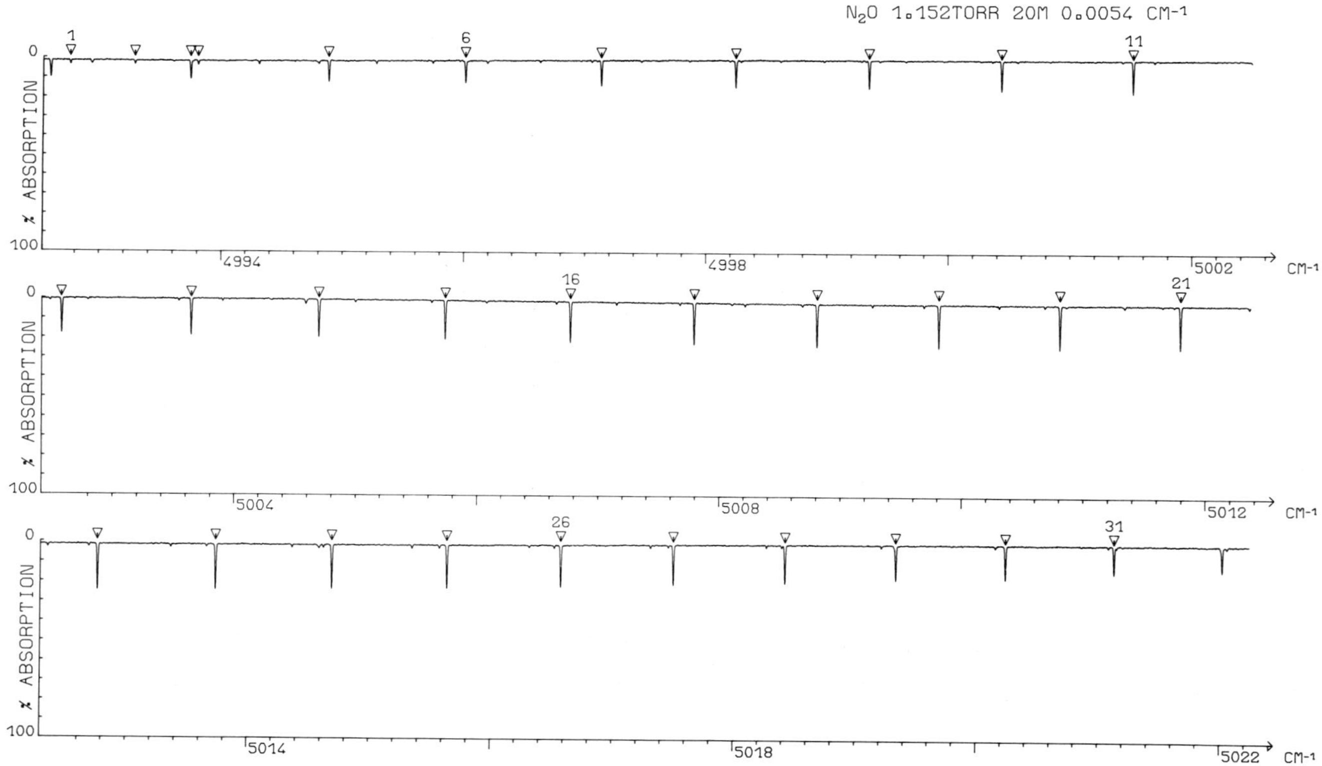

ISOTOPIC IDENTIFICATION – '▽' Indicates $^{14}N^{14}N^{16}O$; '◫': $^{15}N^{14}N^{16}O$; '◇': $^{14}N^{15}N^{16}O$; 'Y': $^{14}N^{14}N^{18}O$;
'✧' Indicates $^{14}N^{14}N^{17}O$; '⏚': $^{15}N^{15}N^{16}O$; '△': $^{15}N^{14}N^{18}O$; '♠': $^{14}N^{15}N^{18}O$.

LINE NO.	WAVENUMBER CM^{-1}, VAC.	ROVIBRATIONAL ASSIGNMENT	ISOTOPIC SPECIES
1	4992.75964(73)	R21 0121E-0000E	14N14N16O
2	4993.29074(75)	R22 0121E-0000E	14N14N16O
3	4993.74661(23)	P33 3200E-0000E	14N14N16O
4	4993.80750(82)	R23 0121E-0000E	14N14N16O
5	4994.88442(22)	P32 3200E-0000E	14N14N16O
6	4996.01244(22)	P31 3200E-0000E	14N14N16O
7	4997.13088(21)	P30 3200E-0000E	14N14N16O
8	4998.23983(21)	P29 3200E-0000E	14N14N16O
9	4999.33883(21)	P28 3200E-0000E	14N14N16O
10	5000.42846(20)	P27 3200E-0000E	14N14N16O
11	5001.50847(20)	P26 3200E-0000E	14N14N16O
12	5002.57900(20)	P25 3200E-0000E	14N14N16O
13	5003.64002(20)	P24 3200E-0000E	14N14N16O
14	5004.69154(20)	P23 3200E-0000E	14N14N16O
15	5005.73359(20)	P22 3200E-0000E	14N14N16O
16	5006.76624(19)	P21 3200E-0000E	14N14N16O
17	5007.78946(19)	P20 3200E-0000E	14N14N16O
18	5008.80331(19)	P19 3200E-0000E	14N14N16O
19	5009.80782(19)	P18 3200E-0000E	14N14N16O
20	5010.80299(19)	P17 3200E-0000E	14N14N16O
21	5011.78876(19)	P16 3200E-0000E	14N14N16O
22	5012.76531(19)	P15 3200E-0000E	14N14N16O
23	5013.73254(19)	P14 3200E-0000E	14N14N16O
24	5014.69043(19)	P13 3200E-0000E	14N14N16O
25	5015.63911(19)	P12 3200E-0000E	14N14N16O
26	5016.57848(20)	P11 3200E-0000E	14N14N16O
27	5017.50867(20)	P10 3200E-0000E	14N14N16O
28	5018.42957(20)	P 9 3200E-0000E	14N14N16O
29	5019.34125(20)	P 8 3200E-0000E	14N14N16O
30	5020.24396(20)	P 7 3200E-0000E	14N14N16O
31	5021.13707(21)	P 6 3200E-0000E	14N14N16O

- "LINE NO.": serial number identifying the spectral line in the map on the opposite page.
- "WAVENUMBER": experimental value.
- Uncertainty in the last digit of the wavenumber is given in parentheses.

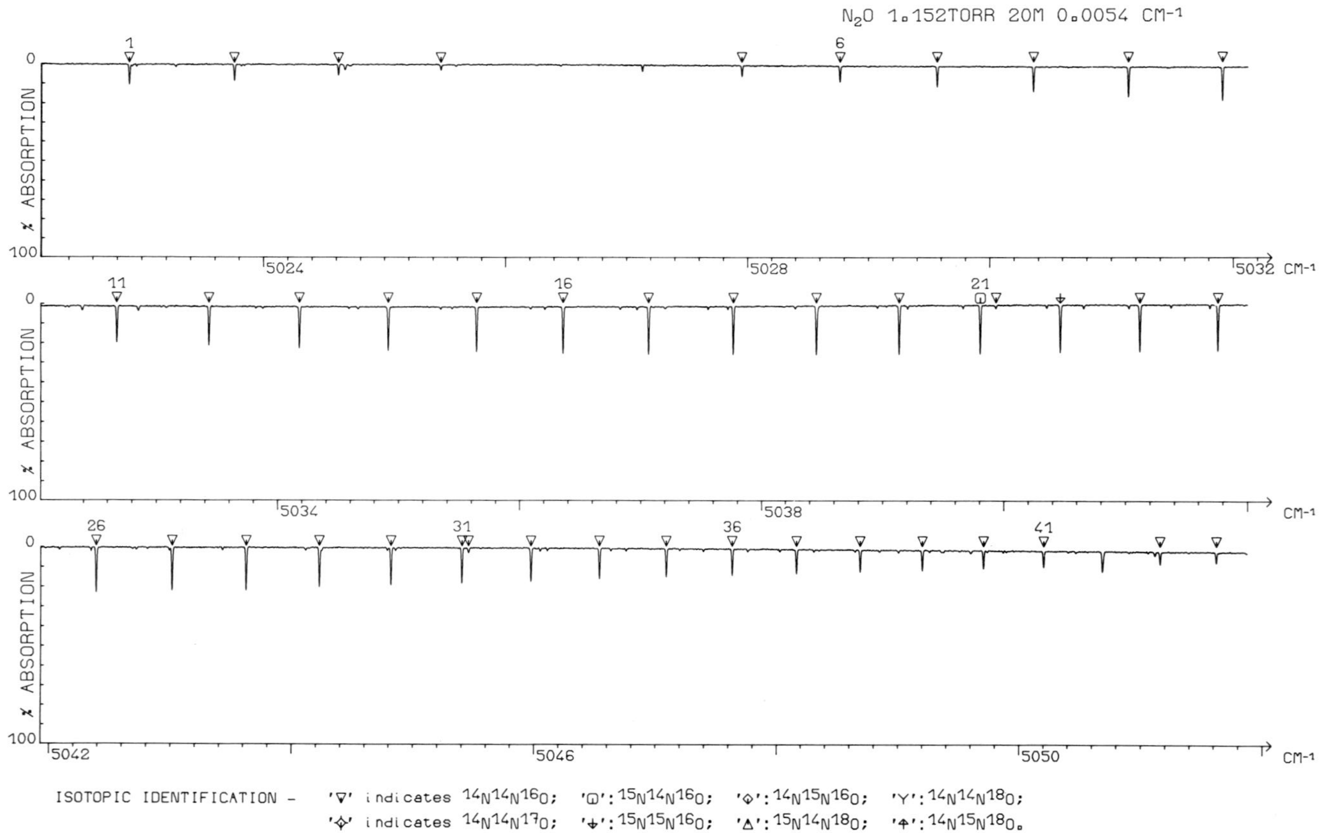
N2O 1.152TORR 20M 0.0054 CM-1
% ABSORPTION
0
100
1
6
11
16
21
26
31
36
41
5024
5028
5032 CM-1
5034
5038
CM-1
5042
5046
5050
CM-1
ISOTOPIC IDENTIFICATION – '▽' indicates 14N14N16O; '⊡': 15N14N16O; '◇': 14N15N16O; 'Y': 14N14N18O;
'◇' indicates 14N14N17O; '⊥': 15N15N16O; '△': 15N14N18O; '+': 14N15N18O.

LINE NO.	WAVENUMBER CM^{-1}, VAC.	ROVIBRATIONAL ASSIGNMENT	ISOTOPIC SPECIES
1	5022.89595(22)	P 4 3200E-0000E	14N14N160
2	5023.76162(24)	P 3 3200E-0000E	14N14N160
3	5024.61793(29)	P 2 3200E-0000E	14N14N160
4	5025.46533(46)	P 1 3200E-0000E	14N14N160
5	5027.95152(30)	R 1 3200E-0000E	14N14N160
6	5028.76207(24)	R 2 3200E-0000E	14N14N160
7	5029.56321(22)	R 3 3200E-0000E	14N14N160
8	5030.35508(21)	R 4 3200E-0000E	14N14N160
9	5031.13765(21)	R 5 3200E-0000E	14N14N160
10	5031.91100(20)	R 6 3200E-0000E	14N14N160
11	5032.67506(20)	R 7 3200E-0000E	14N14N160
12	5033.42985(20)	R 8 3200E-0000E	14N14N160
13	5034.17534(20)	R 9 3200E-0000E	14N14N160
14	5034.91141(20)	R10 3200E-0000E	14N14N160
15	5035.63820(19)	R11 3200E-0000E	14N14N160
16	5036.35562(19)	R12 3200E-0000E	14N14N160
17	5037.06357(19)	R13 3200E-0000E	14N14N160
18	5037.76220(19)	R14 3200E-0000E	14N14N160
19	5038.45138(19)	R15 3200E-0000E	14N14N160
20	5039.13099(19)	R16 3200E-0000E	14N14N160
21	5039.80116(19)	P17 4000E-0000E	15N14N160
21		R17 3200E-0000E	14N14N160
22	5039.93035(88)	R13 3301F-0101F	14N14N160
23	5040.46187(19)	R37 4000E-0000E	15N15N160
23		R18 3200E-0000E	14N14N160
24	5041.11302(19)	R19 3200E-0000E	14N14N160
25	5041.75453(20)	R20 3200E-0000E	14N14N160
26	5042.38644(20)	R21 3200E-0000E	14N14N160
27	5043.00867(20)	R22 3200E-0000E	14N14N160
28	5043.62119(20)	R23 3200E-0000E	14N14N160
29	5044.22424(20)	R24 3200E-0000E	14N14N160
30	5044.81735(20)	R25 3200E-0000E	14N14N160
31	5045.40073(20)	R26 3200E-0000E	14N14N160
32	5045.45521(107)	R21 3301F-0101F	14N14N160
32		R22 3301E-0101E	14N14N160
32		P11 4000E-0000E	15N14N160
33	5045.97435(20)	R27 3200E-0000E	14N14N160
34	5046.53801(21)	R28 3200E-0000E	14N14N160
35	5047.09184(21)	R29 3200E-0000E	14N14N160
36	5047.63575(21)	R30 3200E-0000E	14N14N160
37	5048.16964(21)	R31 3200E-0000E	14N14N160
38	5048.69351(22)	R32 3200E-0000E	14N14N160
39	5049.20744(22)	R33 3200E-0000E	14N14N160
40	5049.71109(23)	R34 3200E-0000E	14N14N160
41	5050.20483(24)	R35 3200E-0000E	14N14N160
41		P23 4000E-0000E	14N15N160
42	5051.16152(27)	R37 3200E-0000E	14N14N160
43	5051.62458(28)	R38 3200E-0000E	14N14N160

- "LINE NO.": serial number identifying the spectral line in the map on the opposite page.
- "WAVENUMBER": experimental value.
- Uncertainty in the last digit of the wavenumber is given in parentheses.

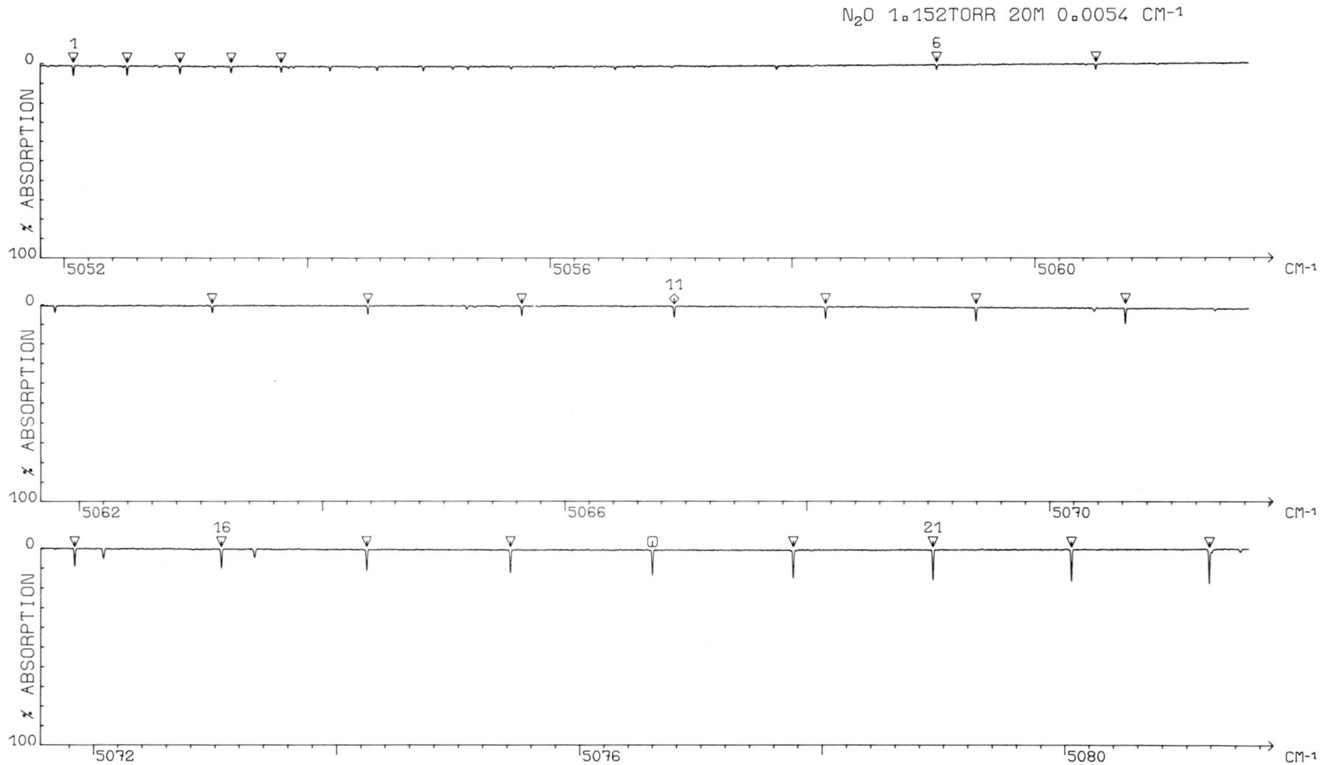

ISOTOPIC IDENTIFICATION – '▽' indicates $^{14}N^{14}N^{16}O$; '⊡': $^{15}N^{14}N^{16}O$; '◇': $^{14}N^{15}N^{16}O$; 'Y': $^{14}N^{14}N^{18}O$; '✧' indicates $^{14}N^{14}N^{17}O$; '⇣': $^{15}N^{15}N^{16}O$; '△': $^{15}N^{14}N^{18}O$; '⇡': $^{14}N^{15}N^{18}O$.

LINE NO.	WAVENUMBER CM^{-1}, VAC.	ROVIBRATIONAL ASSIGNMENT	ISOTOPIC SPECIES
1	5052.07742(30)	R39 3200E-0000E	14N14N16O
2	5052.51952(39)	R40 3200E-0000E	14N14N16O
3	5052.95174(36)	R41 3200E-0000E	14N14N16O
4	5053.37339(39)	R42 3200E-0000E	14N14N16O
5	5053.78450(44)	R43 3200E-0000E	14N14N16O
6	5059.18984(69)	P43 4000E-0000E	14N14N16O
7	5060.50083(60)	P42 4000E-0000E	14N14N16O
8	5063.09274(38)	P40 4000E-0000E	14N14N16O
9	5064.37349(34)	P39 4000E-0000E	14N14N16O
10	5065.64405(32)	P38 4000E-0000E	14N14N16O
11	5066.90424(29)	P 7 4000E-0000E	14N15N16O
11		P37 4000E-0000E	14N14N16O
12	5068.15418(27)	P36 4000E-0000E	14N14N16O
13	5069.39332(26)	P35 4000E-0000E	14N14N16O
14	5070.62183(25)	P34 4000E-0000E	14N14N16O
15	5071.83965(24)	P33 4000E-0000E	14N14N16O
16	5073.04684(23)	P32 4000E-0000E	14N14N16O
17	5074.24318(23)	P31 4000E-0000E	14N14N16O
18	5075.42856(22)	P30 4000E-0000E	14N14N16O
19	5076.60294(22)	R33 4000E-0000E	15N14N16O
19		P29 4000E-0000E	14N14N16O
20	5077.76627(21)	P28 4000E-0000E	14N14N16O
21	5078.91859(21)	P27 4000E-0000E	14N14N16O
22	5080.05967(21)	P26 4000E-0000E	14N14N16O
23	5081.18956(20)	P25 4000E-0000E	14N14N16O

- "LINE NO.": serial number identifying the spectral line in the map on the opposite page.
- "WAVENUMBER": experimental value.
- Uncertainty in the last digit of the wavenumber is given in parentheses.

N_2O 1.152TORR 20M 0.0054 CM-1

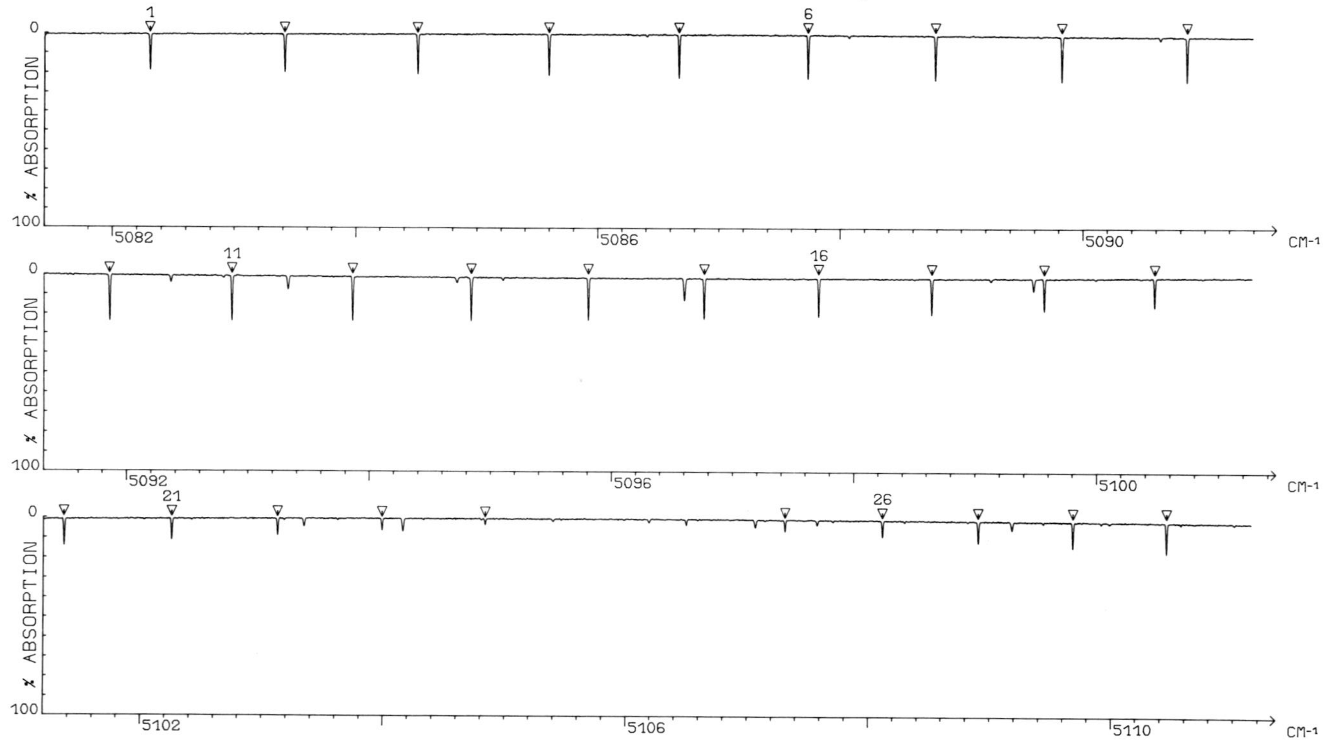

ISOTOPIC IDENTIFICATION – '▽' indicates $^{14}N^{14}N^{16}O$; '◫': $^{15}N^{14}N^{16}O$; '◇': $^{14}N^{15}N^{16}O$; 'Y': $^{14}N^{14}N^{18}O$;
'✧' indicates $^{14}N^{14}N^{17}O$; '⊥': $^{15}N^{15}N^{16}O$; 'Δ': $^{15}N^{14}N^{18}O$; '✦': $^{14}N^{15}N^{18}O$.

LINE NO.	WAVENUMBER CM^{-1}, VAC.	ROVIBRATIONAL ASSIGNMENT	ISOTOPIC SPECIES
1	5082.30819(20)	P24 4000E-0000E	14N14N16O
2	5083.41553(20)	P23 4000E-0000E	14N14N16O
3	5084.51144(20)	P22 4000E-0000E	14N14N16O
4	5085.59594(20)	P21 4000E-0000E	14N14N16O
5	5086.66910(20)	P20 4000E-0000E	14N14N16O
6	5087.73066(20)	P19 4000E-0000E	14N14N16O
7	5088.78078(20)	P18 4000E-0000E	14N14N16O
8	5089.81925(20)	P17 4000E-0000E	14N14N16O
9	5090.84619(20)	P16 4000E-0000E	14N14N16O
10	5091.86148(20)	P15 4000E-0000E	14N14N16O
11	5092.86499(20)	P14 4000E-0000E	14N14N16O
12	5093.85691(20)	P13 4000E-0000E	14N14N16O
13	5094.83713(20)	P12 4000E-0000E	14N14N16O
14	5095.80555(20)	P11 4000E-0000E	14N14N16O
15	5096.76219(20)	P10 4000E-0000E	14N14N16O
16	5097.70694(20)	P 9 4000E-0000E	14N14N16O
17	5098.64010(20)	P 8 4000E-0000E	14N14N16O
18	5099.56128(21)	P 7 4000E-0000E	14N14N16O
19	5100.47061(21)	P 6 4000E-0000E	14N14N16O
20	5101.36808(22)	P 5 4000E-0000E	14N14N16O
21	5102.25371(23)	P 4 4000E-0000E	14N14N16O
22	5103.12752(29)	P 3 4000E-0000E	14N14N16O
23	5103.98940(30)	P 2 4000E-0000E	14N14N16O
24	5104.83927(47)	P 1 4000E-0000E	14N14N16O
25	5107.31745(30)	R 1 4000E-0000E	14N14N16O
26	5108.11987(25)	R 2 4000E-0000E	14N14N16O
27	5108.91024(23)	R 3 4000E-0000E	14N14N16O
28	5109.68854(22)	R 4 4000E-0000E	14N14N16O
29	5110.45506(21)	R 5 4000E-0000E	14N14N16O

- "LINE NO.": serial number identifying the spectral line in the map on the opposite page.
- "WAVENUMBER": experimental value.
- Uncertainty in the last digit of the wavenumber is given in parentheses.

N_2O 1.152TORR 20M 0.0054 CM-1

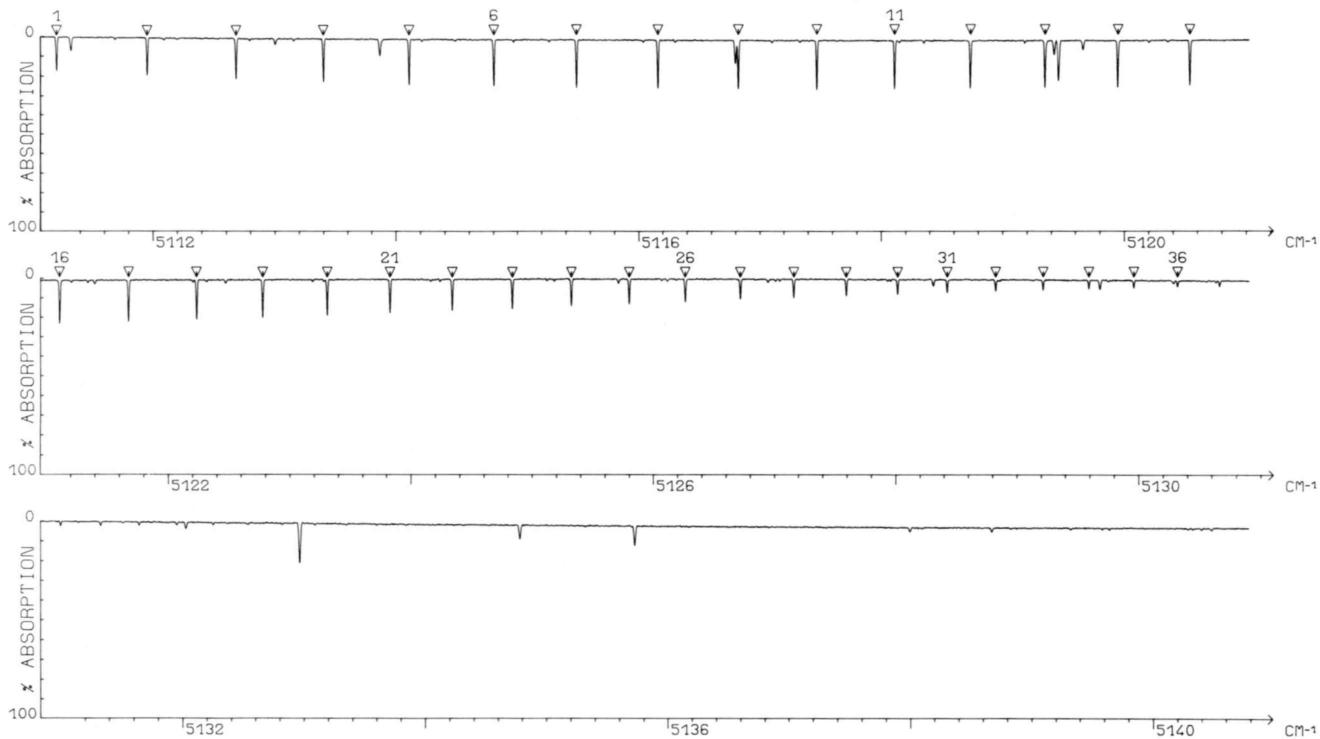

ISOTOPIC IDENTIFICATION – '▽' indicates $^{14}N^{14}N^{16}O$; 'Ⓞ': $^{15}N^{14}N^{16}O$; '◈': $^{14}N^{15}N^{16}O$; 'Y': $^{14}N^{14}N^{18}O$;
'✧' indicates $^{14}N^{14}N^{17}O$; '⏚': $^{15}N^{15}N^{16}O$; '△': $^{15}N^{14}N^{18}O$; '⍏': $^{14}N^{15}N^{18}O$.

LINE NO.	WAVENUMBER CM^{-1}, VAC.	ROVIBRATIONAL ASSIGNMENT	ISOTOPIC SPECIES
1	5111.20973(21)	R 6 4000E-0000E	14N14N160
2	5111.95249(20)	R 7 4000E-0000E	14N14N160
3	5112.68338(20)	R 8 4000E-0000E	14N14N160
4	5113.40233(20)	R 9 4000E-0000E	14N14N160
5	5114.10949(20)	R10 4000E-0000E	14N14N160
6	5114.80469(20)	R11 4000E-0000E	14N14N160
7	5115.48813(20)	R12 4000E-0000E	14N14N160
8	5116.15979(20)	R13 4000E-0000E	14N14N160
9	5116.81960(20)	R14 4000E-0000E	14N14N160
10	5117.46768(20)	R15 4000E-0000E	14N14N160
11	5118.10400(20)	R16 4000E-0000E	14N14N160
12	5118.72860(20)	R17 4000E-0000E	14N14N160
13	5119.34150(20)	R18 4000E-0000E	14N14N160
14	5119.94274(20)	R19 4000E-0000E	14N14N160
15	5120.53239(20)	R20 4000E-0000E	14N14N160
16	5121.11044(20)	R21 4000E-0000E	14N14N160
17	5121.67692(20)	R22 4000E-0000E	14N14N160
18	5122.23194(20)	R23 4000E-0000E	14N14N160
19	5122.77541(20)	R24 4000E-0000E	14N14N160
20	5123.30752(21)	R25 4000E-0000E	14N14N160
21	5123.82810(21)	R26 4000E-0000E	14N14N160
22	5124.33752(21)	R27 4000E-0000E	14N14N160
23	5124.83559(21)	R28 4000E-0000E	14N14N160
24	5125.32244(22)	R29 4000E-0000E	14N14N160
25	5125.79816(22)	R30 4000E-0000E	14N14N160
26	5126.26275(23)	R31 4000E-0000E	14N14N160
27	5126.71626(24)	R32 4000E-0000E	14N14N160
28	5127.15887(24)	R33 4000E-0000E	14N14N160
29	5127.59047(25)	R34 4000E-0000E	14N14N160
30	5128.01125(27)	R35 4000E-0000E	14N14N160
31	5128.42139(28)	R36 4000E-0000E	14N14N160
32	5128.82081(30)	R37 4000E-0000E	14N14N160
33	5129.20937(32)	R38 4000E-0000E	14N14N160
34	5129.58773(36)	R39 4000E-0000E	14N14N160
35	5129.95547(46)	R40 4000E-0000E	14N14N160
36	5130.31304(48)	R41 4000E-0000E	14N14N160

- "LINE NO.": serial number identifying the spectral line in the map on the opposite page.
- "WAVENUMBER": experimental value.
- Uncertainty in the last digit of the wavenumber is given in parentheses.

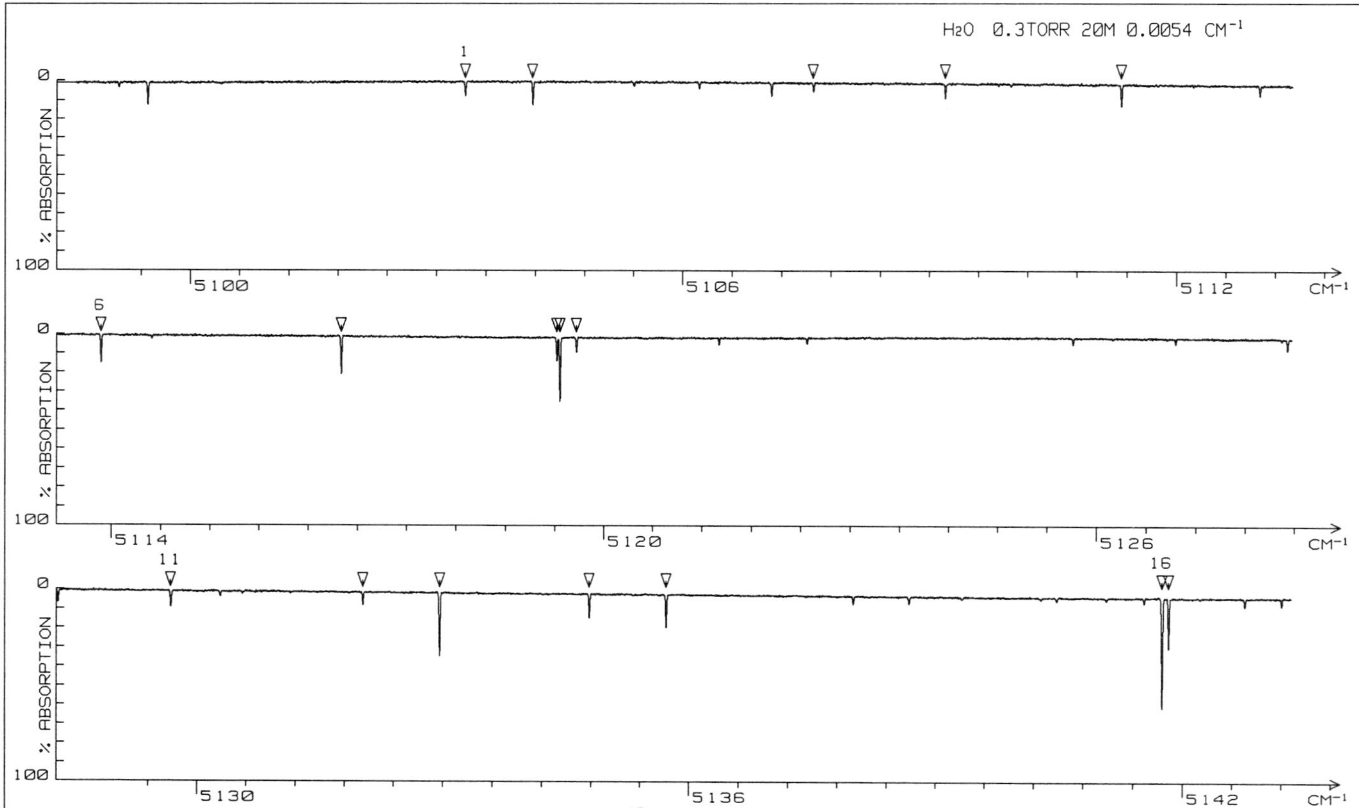

ISOTOPIC IDENTIFICATION- On this page: '∇' indicates $H_2{}^{16}O$.

On the opposite page in column 'IDENT', under 'IS': '161' indicates $H_2{}^{16}O$.

LINE NO.	WAVENUMBER CM^{-1}, VAC.	J'	K_A'	K_C'	J"	K_A"	K_C"	IDENT. BD.	IS.
1	5103.34390(65)	2	1	2	3	2	1	1	161
2	5104.16085(41)	4	0	4	5	2	3	2	161
3	5107.58209(97)	5	0	5	6	1	6	1	161
4	5109.18520(61)	3	1	2	4	2	3	1	161
5	5111.32529(45)	8	3	5	9	3	6	2	161
6	5113.86989(36)	9	2	8	10	2	9	2	161
7	5116.79554(30)	8	2	6	9	2	7	2	161
8	5119.41648(42)	9	0	9	10	0	10	2	161
9	5119.45554(24)	9	1	9	10	1	10	2	161
10	5119.65738(63)	8	4	4	9	4	5	2	161
11	5129.67844(59)	7	3	4	8	3	5	2	161
12	5132.02197(75)	4	1	4	5	0	5	1	161
13	5132.96015(24)	8	1	7	9	1	8	2	161
14	5134.77750(42)	8	2	7	9	2	8	2	161
15	5135.72260(34)	7	2	5	8	2	6	2	161
16	5141.75113(21)	8	0	8	9	0	9	2	161
17	5141.83042(27)	8	1	8	9	1	9	2	161

- "LINE NO.": serial number identifying the spectral line in the map on the opposite page.
- Uncertainty in the last digit of the wavenumber is given in parentheses.
- In column "IDENT." under "BD" 1 represents the band 110–000; 2: 011–000.

For isotopic assignment given under "IS" see comments on the opposite page.

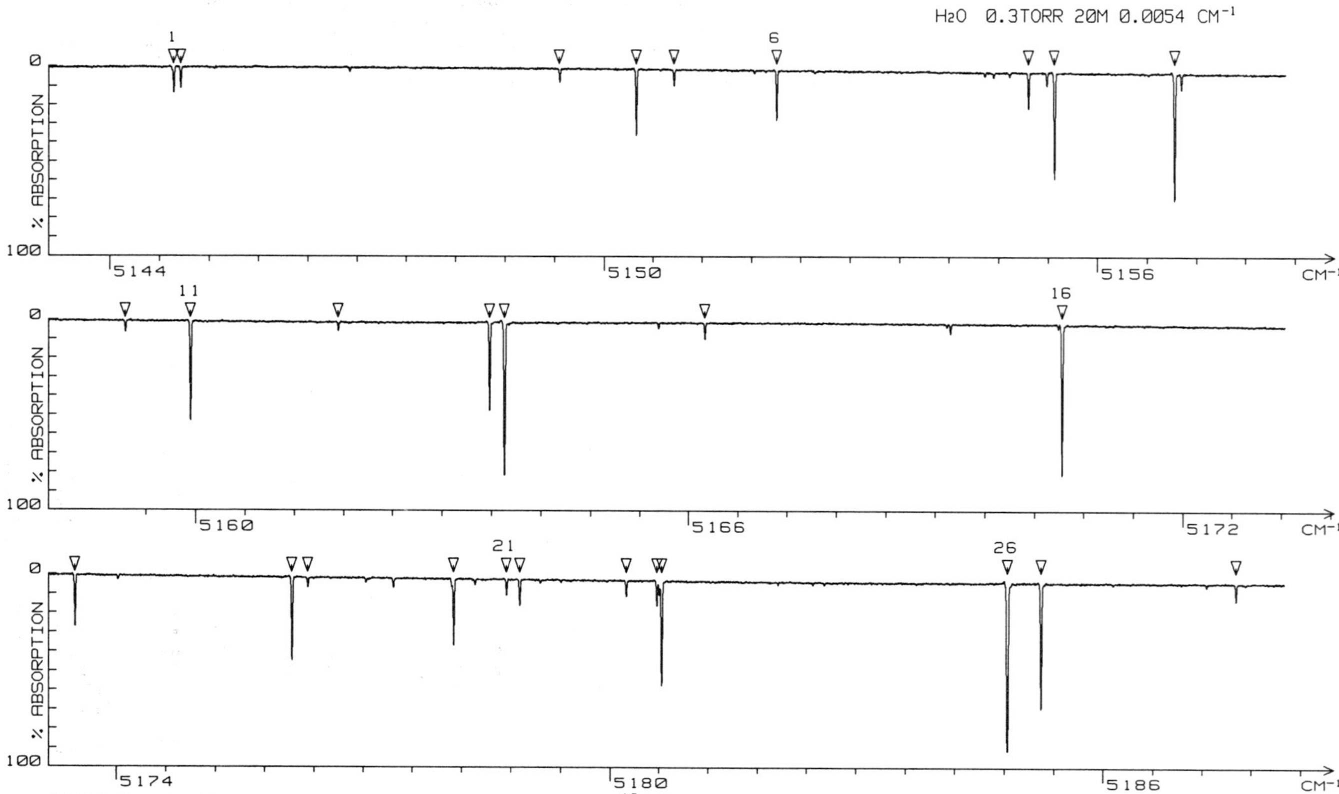

ISOTOPIC IDENTIFICATION- On this page: '∇' indicates $H_2{}^{16}O$.

On the opposite page in column 'IDENT', under 'IS': '161' indicates $H_2{}^{16}O$.

LINE NO.	WAVENUMBER CM^{-1}, VAC.	J'	K_A'	K_C'	J"	K_A"	K_C"	IDENT. BD.	IS.
1	5144.76968(44)	1	1	0	2	2	1	1	161
2	5144.85382(51)	3	0	3	4	1	4	1	161
3	5149.45819(68)	7	4	3	8	4	4	2	161
4	5150.38684(24)	7	3	5	8	3	6	2	161
5	5150.85458(65)	3	0	3	4	2	2	2	161
6	5152.09934(27)	7	1	6	8	1	7	2	161
7	5155.15471(33)	7	4	4	8	4	5	2	161
8	5155.47130(21)	7	2	6	8	2	7	2	161
9	5156.92670(21)	6	2	4	7	2	5	2	161
10	5159.14704(91)	4	2	3	4	3	2	1	161
11	5159.93439(22)	6	3	3	7	3	4	2	161
12	5161.72816(102)	2	0	2	3	1	3	1	161
13	5163.57279(22)	7	0	7	8	0	8	2	161
14	5163.75318(21)	7	1	7	8	1	8	2	161
15	5166.19353(63)	7	5	3	8	5	4	2	161
16	5170.53270(21)	6	1	5	7	1	6	2	161
17	5173.48588(28)	6	3	4	7	3	5	2	161
18	5176.12751(23)	6	2	5	7	2	6	2	161
19	5176.31901(107)	8	0	8	8	2	7	2	161
20	5178.09931(25)	6	4	2	7	4	3	2	161
21	5178.73851(67)	2	1	2	3	0	3	1	161
22	5178.90328(45)	1	0	1	2	1	2	1	161
23	5180.18975(67)	6	4	3	7	3	4	1	161
24	5180.57042(44)	6	4	3	7	4	4	2	161
25	5180.62987(22)	5	2	3	6	2	4	2	161
26	5184.83092(21)	6	0	6	7	0	7	2	161
27	5185.23745(21)	6	1	6	7	1	7	2	161
28	5187.61079(61)	5	1	4	5	2	3	1	161

- "LINE NO.": serial number identifying the spectral line in the map on the opposite page.
- Uncertainty in the last digit of the wavenumber is given in parentheses.
- In column "IDENT." under "BD" 1 represents the band 110-000; 2: 011-000.

For isotopic assignment given under "IS" see comments on the opposite page.

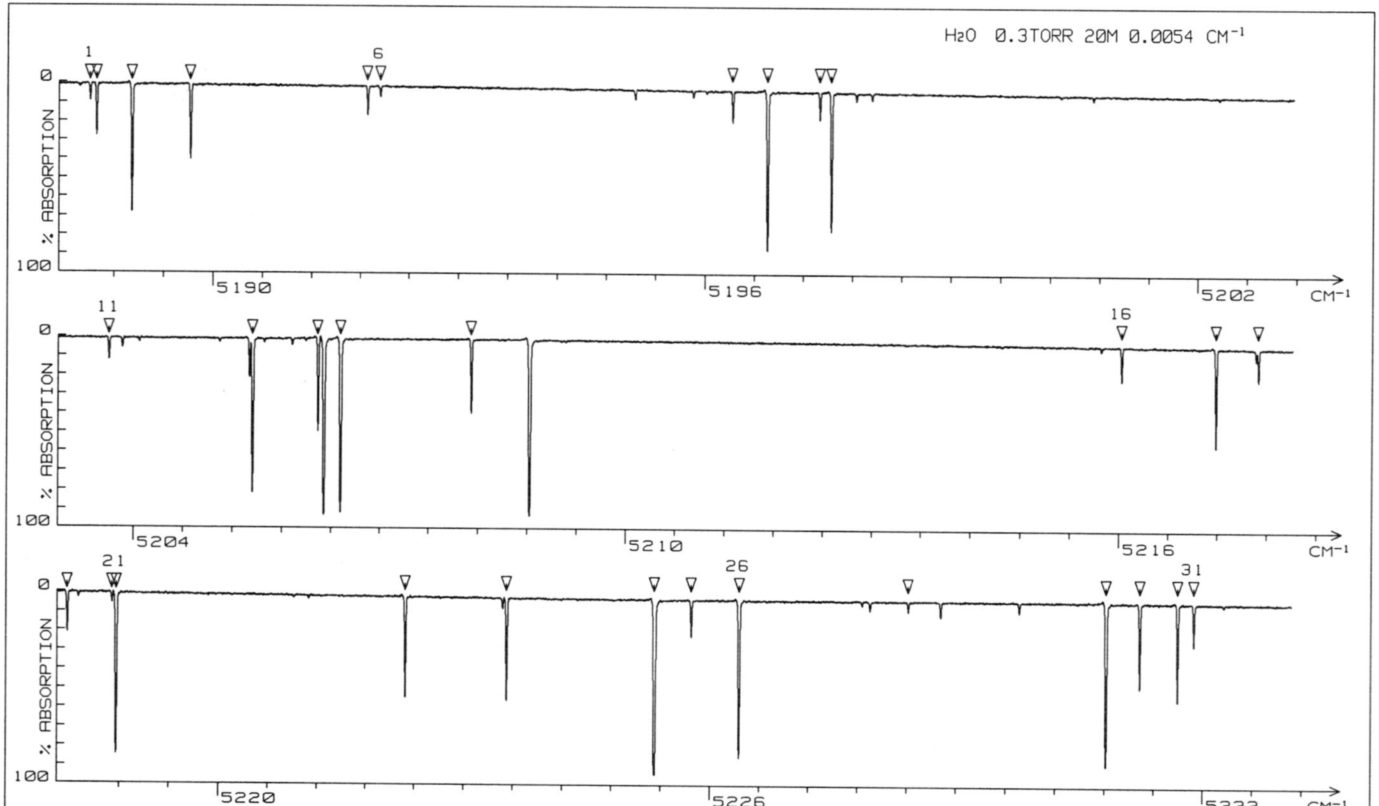

ISOTOPIC IDENTIFICATION- On this page: '∇' indicates $H_2{}^{16}O$.

On the opposite page in column 'IDENT', under 'IS': '161' indicates $H_2{}^{16}O$.

LINE NO.	WAVENUMBER CM^{-1}, VAC.	J'	K_A'	K_C'	J"	K_A"	K_C"	IDENT. BD.	IS.
1	5188.50379(69)	4	3	2	5	2	3	1	161
2	5188.58077(29)	2	0	2	3	2	1	2	161
3	5189.01041(21)	5	1	4	6	1	5	2	161
4	5189.71997(24)	5	3	2	6	3	3	2	161
5	5191.87928(41)	6	5	1	7	5	2	2	161
6	5192.03521(106)	6	5	2	7	5	3	2	161
7	5196.32649(39)	3	0	3	3	1	2	1	161
8	5196.75154(21)	5	2	4	6	2	5	2	161
9	5197.39132(42)	3	1	2	3	2	1	1	161
10	5197.53322(21)	5	3	3	6	3	4	2	161
11	5203.70167(56)	7	1	7	7	1	6	2	161
12	5205.44351(21)	5	0	5	6	0	6	2	161
13	5206.23919(23)	5	4	2	6	4	3	2	161
14	5206.51571(22)	4	2	2	5	2	3	2	161
15	5208.11774(25)	5	3	2	6	2	5	1	161
16	5216.02793(38)	1	0	1	1	1	0	1	161
17	5217.18320(23)	4	3	1	5	2	4	1	161
18	5217.69853(40)	5	5	1	6	5	2	2	161
19	5218.16310(35)	6	0	6	6	2	5	2	161
20	5218.70675(107)	1	0	1	2	2	0	2	161
21	5218.75888(21)	4	3	1	5	3	2	2	161
22	5222.28704(23)	4	3	2	5	3	3	2	161
23	5223.52212(23)	4	2	3	5	2	4	2	161
24	5225.32390(32)	4	0	4	5	0	5	2	161
25	5225.76836(38)	6	1	5	6	3	4	2	161
26	5226.35169(22)	4	1	4	5	1	5	2	161
27	5228.41458(106)	5	4	1	6	3	4	1	161
28	5230.81459(22)	3	1	2	4	1	3	2	161
29	5231.22630(24)	4	2	2	5	1	5	1	161
30	5231.68700(23)	4	4	0	5	4	1	2	161
31	5231.88793(34)	4	4	1	5	4	2	2	161

- "LINE NO.": serial number identifying the spectral line in the map on the opposite page.
- Uncertainty in the last digit of the wavenumber is given in parentheses.
- In column "IDENT." under "BD" 1 represents the band 110-000; 2: 011-000.

For isotopic assignment given under "IS" see comments on the opposite page.

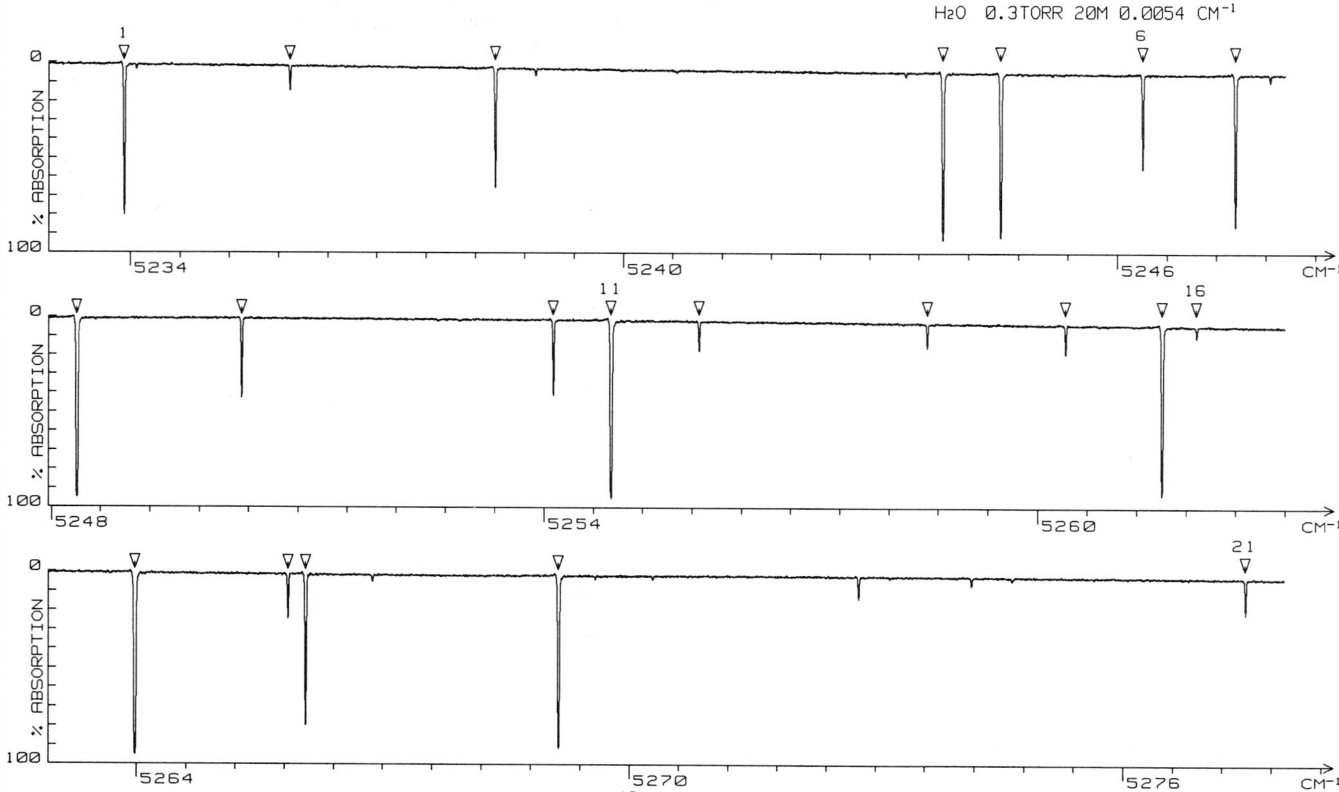

ISOTOPIC IDENTIFICATION- On this page: '∇' indicates $H_2{}^{16}O$.

On the opposite page in column 'IDENT', under 'IS': '161' indicates $H_2{}^{16}O$.

LINE NO.	WAVENUMBER CM^{-1}, VAC.	J'	K_A'	K_C'	J"	K_A"	K_C"	IDENT. BD.	IS.
1	5233.92233(22)	3	2	1	4	2	2	2	161
2	5235.93176(54)	5	0	5	5	2	4	2	161
3	5238.44013(22)	3	3	0	4	2	3	1	161
4	5243.88673(35)	3	2	2	4	2	3	2	161
5	5244.58165(22)	3	0	3	4	0	4	2	161
6	5246.30024(24)	3	3	0	4	3	1	2	161
7	5247.42649(22)	3	3	1	4	3	2	2	161
8	5248.30655(33)	3	1	3	4	1	4	2	161
9	5250.30932(26)	4	0	4	4	2	3	2	161
10	5254.10661(27)	5	1	5	5	1	4	2	161
11	5254.80657(33)	2	1	1	3	1	2	2	161
12	5255.87719(46)	1	1	0	1	0	1	1	161
13	5258.66170(57)	7	2	6	7	2	5	2	161
14	5260.33299(48)	3	0	3	3	2	2	2	161
15	5261.50029(35)	2	2	0	3	2	1	2	161
16	5261.92060(107)	2	1	1	2	0	2	1	161
17	5263.97563(33)	2	0	2	3	0	3	2	161
18	5265.83520(37)	2	0	2	2	2	1	2	161
19	5266.04969(22)	2	2	1	3	2	2	2	161
20	5269.13295(23)	2	1	2	3	1	3	2	161
21	5277.48022(43)	4	1	4	4	1	3	2	161

- "LINE NO.": serial number identifying the spectral line in the map on the opposite page.
- Uncertainty in the last digit of the wavenumber is given in parentheses.
- In column "IDENT." under "BD" 1 represents the band 110-000; 2: 011-000.

For isotopic assignment given under "IS" see comments on the opposite page.

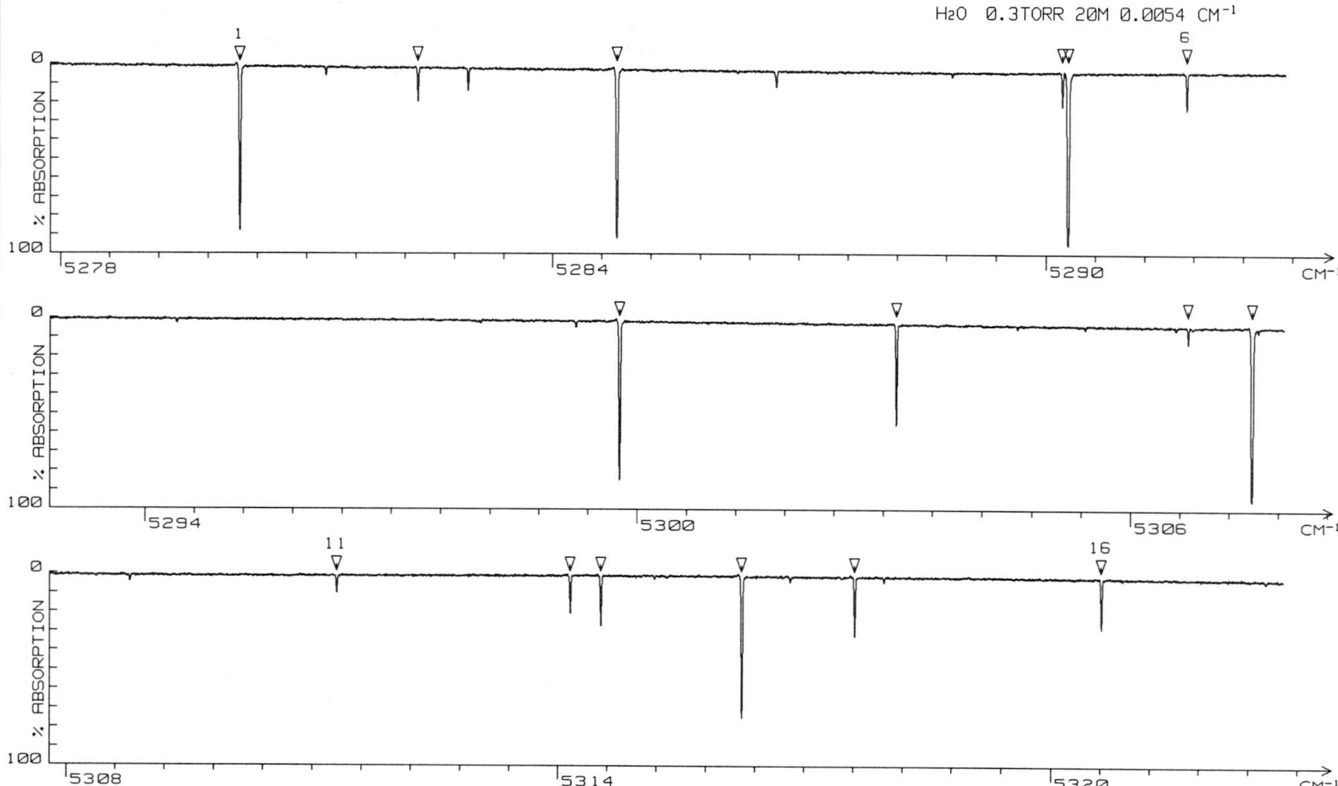

ISOTOPIC IDENTIFICATION- On this page: '∇' indicates $H_2{}^{16}O$.

On the opposite page in column 'IDENT', under 'IS': '161' indicates $H_2{}^{16}O$.

LINE NO.	WAVENUMBER CM^{-1}, VAC.	J'	K_A'	K_C'	J"	K_A"	K_C"	IDENT. BD.	IS.
1	5280.18778(23)	1	1	0	2	1	1	2	161
2	5282.35458(45)	4	2	2	4	1	3	1	161
3	5284.78039(23)	1	0	1	2	0	2	2	161
4	5290.19588(45)	3	0	3	2	1	2	1	161
5	5290.26618(34)	1	1	1	2	1	2	2	161
6	5291.70613(43)	2	1	2	1	0	1	1	161
7	5299.77904(23)	3	1	3	3	1	2	2	161
8	5303.15227(25)	5	2	4	5	2	3	2	161
9	5306.69424(84)	3	1	3	2	0	2	1	161
10	5307.47340(35)	0	0	0	1	0	1	2	161
11	5311.29218(94)	4	0	4	3	1	3	1	161
12	5314.14626(45)	7	3	5	7	3	4	2	161
13	5314.51839(37)	5	3	2	5	2	3	1	161
14	5316.23548(24)	2	1	2	2	1	1	2	161
15	5317.61264(33)	4	3	1	4	2	2	1	161
16	5320.60835(38)	4	1	4	3	0	3	1	161

- "LINE NO.": serial number identifying the spectral line in the map on the opposite page.
- Uncertainty in the last digit of the wavenumber is given in parentheses.
- In column "IDENT." under "BD" 1 represents the band 110–000; 2: 011–000.

For isotopic assignment given under "IS" see comments on the opposite page.

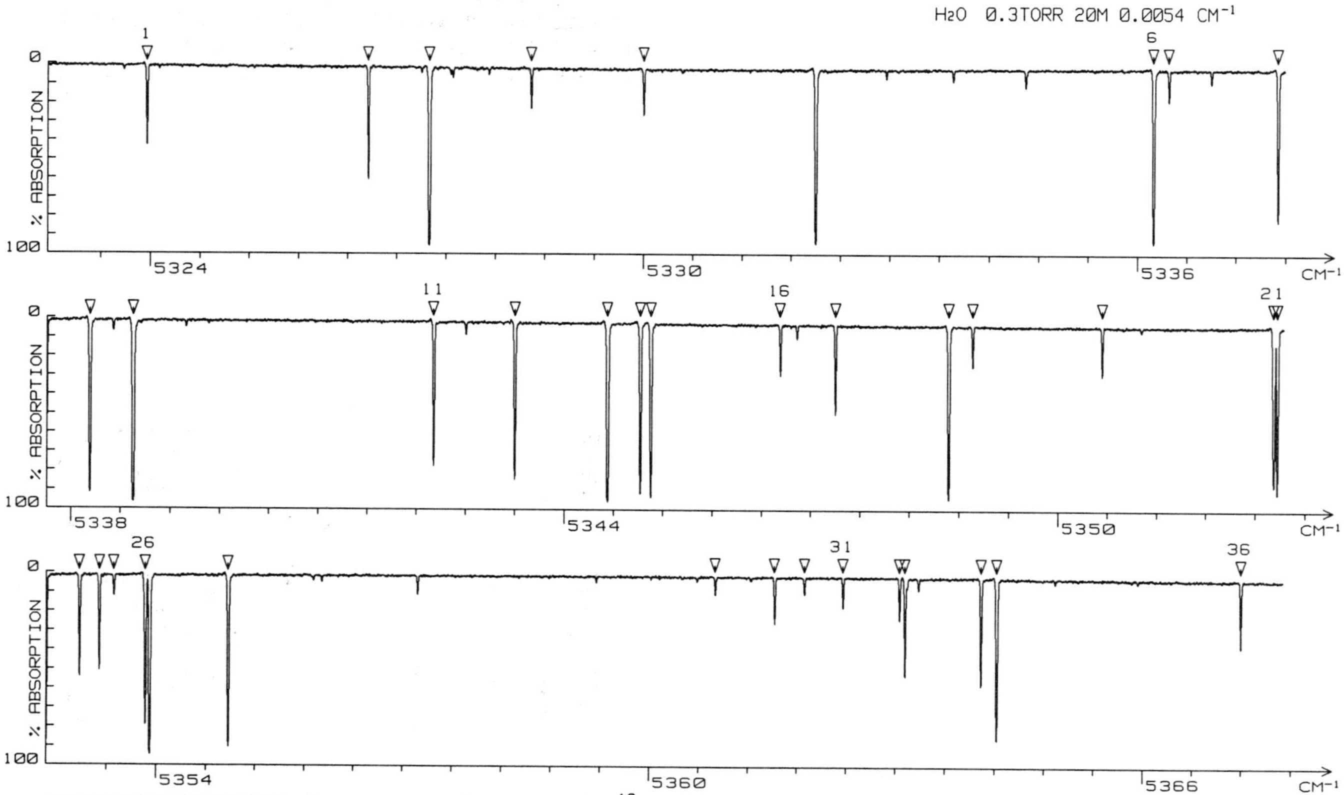

ISOTOPIC IDENTIFICATION– On this page: '∇' indicates $H_2{}^{16}O$.

On the opposite page in column 'IDENT', under 'IS': '161' indicates $H_2{}^{16}O$.

LINE NO.	WAVENUMBER CM^{-1}, VAC.	J'	K_A'	K_C'	J"	K_A"	K_C"	IDENT. BD.	IS.
1	5323.95128(29)	4	2	3	4	2	2	2	161
2	5326.64595(25)	3	3	0	3	2	1	1	161
3	5327.39115(34)	1	1	1	1	1	0	2	161
4	5328.63100(45)	6	3	4	6	3	3	2	161
5	5329.99608(39)	5	0	5	4	1	4	1	161
6	5336.18762(39)	2	2	1	2	2	0	2	161
7	5336.37303(54)	2	2	1	1	1	0	1	161
8	5337.69993(24)	5	3	3	5	3	2	2	161
9	5338.22667(24)	1	1	0	1	1	1	2	161
10	5338.75508(35)	2	2	0	2	2	1	2	161
11	5342.41266(24)	4	3	2	4	3	1	2	161
12	5343.40046(24)	3	2	1	3	2	2	2	161
13	5344.52516(35)	3	3	1	3	3	0	2	161
14	5344.92344(24)	3	3	0	3	3	1	2	161
15	5345.05429(38)	4	3	1	4	3	2	2	161
16	5346.62612(37)	7	4	4	7	4	3	2	161
17	5347.30045(27)	5	3	2	5	3	3	2	161
18	5348.67582(39)	2	1	1	2	1	2	2	161
19	5348.96159(44)	6	1	6	5	0	5	1	161
20	5350.53418(38)	6	4	3	6	4	2	2	161
21	5352.61970(29)	5	4	2	5	4	1	2	161
22	5352.66388(24)	4	2	2	4	2	3	2	161
23	5353.06983(26)	5	4	1	5	4	2	2	161
24	5353.31234(27)	6	3	3	6	3	4	2	161
25	5353.49000(82)	7	4	3	7	4	4	2	161
26	5353.86831(63)	4	4	1	4	4	0	2	161
27	5354.87112(24)	1	0	1	0	0	0	2	161
28	5360.80276(84)	8	5	3	8	5	4	2	161
29	5361.52471(40)	7	5	3	7	5	2	2	161
30	5361.89288(97)	7	5	2	7	5	3	2	161

LINE NO.	WAVENUMBER CM^{-1}, VAC.	J'	K_A'	K_C'	J"	K_A"	K_C"	IDENT. BD.	IS.
31	5362.36171(58)	7	0	7	6	1	6	1	161
32	5363.04963(42)	6	5	2	6	5	1	2	161
33	5363.11604(27)	6	5	1	6	5	2	2	161
34	5364.03300(26)	3	1	2	3	1	3	2	161
35	5364.22225(25)	5	5	1	5	5	0	2	161
36	5367.19464(32)	5	2	3	5	2	4	2	161

- "LINE NO.": serial number identifying the spectral line in the map on the opposite page.
- Uncertainty in the last digit of the wavenumber is given in parentheses.
- In column "IDENT." under "BD" 1 represents the band 110–000; 2: 011–000.

For isotopic assignment given under "IS" see comments on the opposite page.

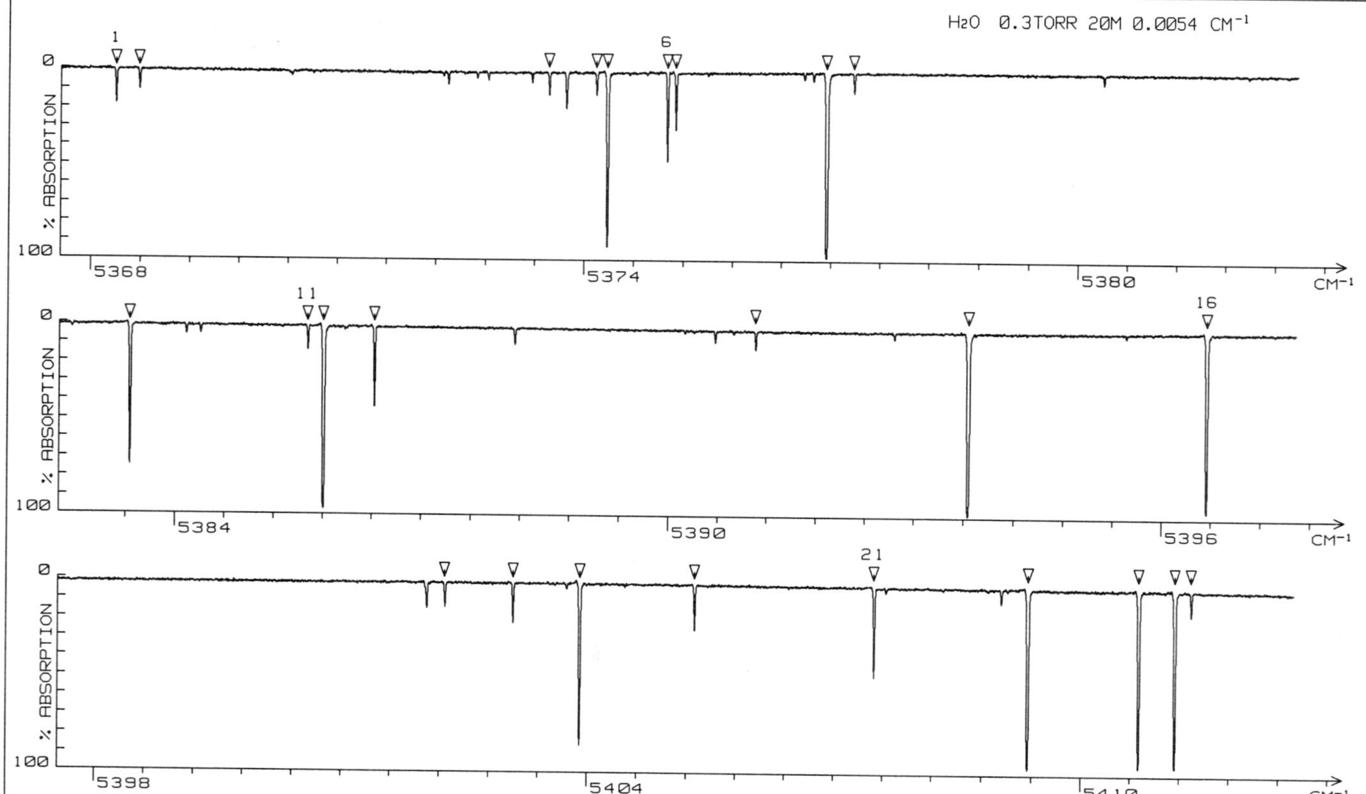

ISOTOPIC IDENTIFICATION- On this page: '∇' indicates $H_2{}^{16}O$.

On the opposite page in column 'IDENT', under 'IS': '161' indicates $H_2{}^{16}O$.

LINE NO.	WAVENUMBER CM^{-1}, VAC.	J'	K_A'	K_C'	J"	K_A"	K_C"	IDENT. BD.	IS.
1	5368.29791(54)	4	2	3	3	1	2	1	161
2	5368.58104(88)	5	4	1	5	3	2	1	161
3	5373.56748(75)	6	4	3	6	3	4	1	161
4	5374.14159(77)	4	4	1	4	3	2	1	161
5	5374.27415(24)	2	1	2	1	1	1	2	161
6	5374.99878(28)	6	6	1	6	6	0	2	161
7	5375.09988(36)	3	2	1	2	1	2	1	161
8	5376.94330(37)	2	0	2	1	0	1	2	161
9	5377.27566(88)	8	1	8	7	0	7	1	161
10	5383.44747(24)	4	1	3	4	1	4	2	161
11	5385.60668(68)	7	7	1	7	7	0	2	161
11		7	7	0	7	7	1	2	161
12	5385.80095(37)	2	1	1	1	1	0	2	161
13	5386.42525(30)	6	2	4	6	2	5	2	161
14	5391.05342(91)	6	2	5	5	1	4	1	161
15	5393.64904(38)	3	1	3	2	1	2	2	161
16	5396.54403(38)	3	0	3	2	0	2	2	161
17	5402.26065(74)	2	2	1	2	0	2	2	161
18	5403.08841(48)	3	3	1	2	2	0	1	161
19	5403.90084(24)	3	3	0	2	2	1	1	161
20	5405.29074(42)	5	1	4	5	1	5	2	161
21	5407.48704(29)	3	2	2	3	0	3	2	161
22	5409.34768(40)	3	2	2	2	2	1	2	161
23	5410.69894(43)	4	1	4	3	1	3	2	161
24	5411.13607(41)	3	1	2	2	1	1	2	161
25	5411.33924(68)	4	3	1	4	0	4	1	161

- "LINE NO.": serial number identifying the spectral line in the map on the opposite page.
- Uncertainty in the last digit of the wavenumber is given in parentheses.
- In column "IDENT." under "BD" 1 represents the band 110-000; 2: 011-000.

For isotopic assignment given under "IS" see comments on the opposite page.

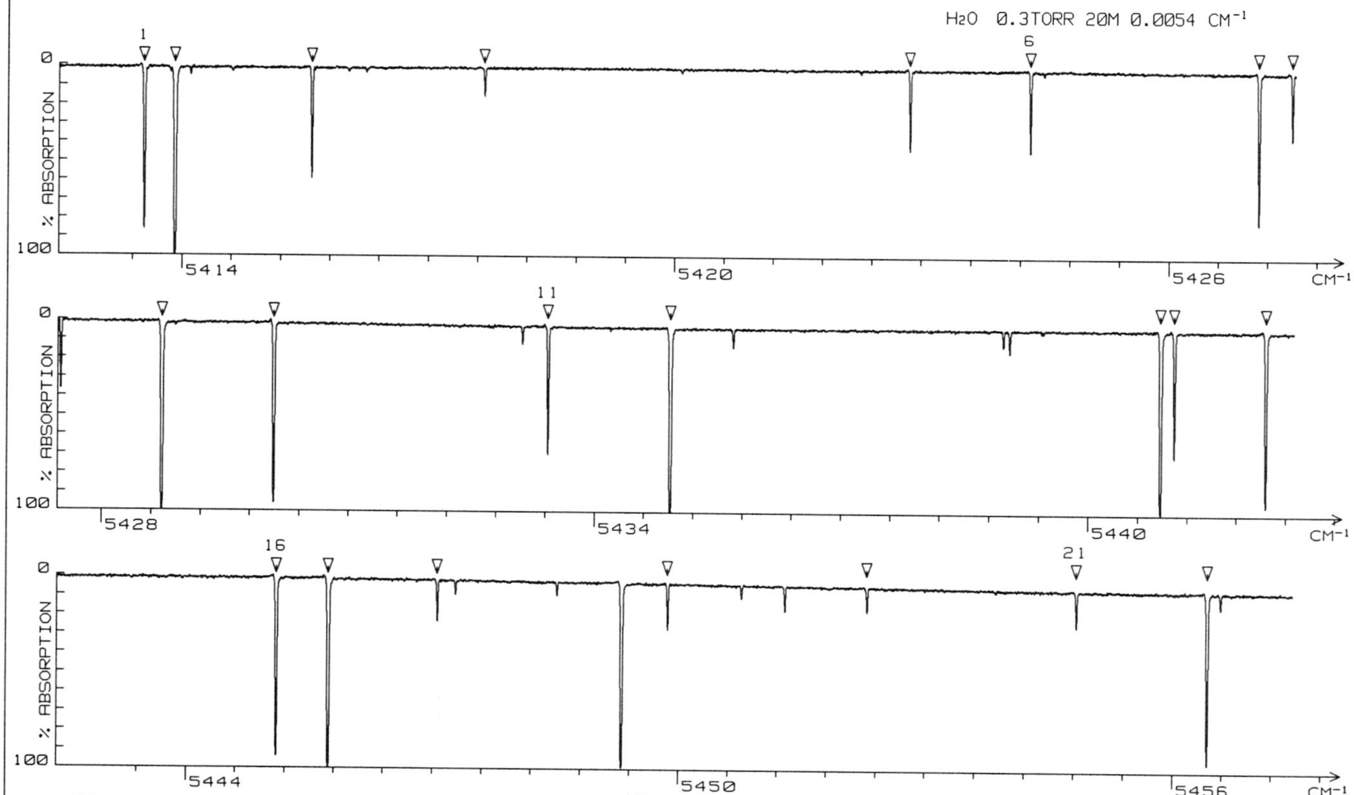

ISOTOPIC IDENTIFICATION- On this page: '∇' indicates $H_2{}^{16}O$.

On the opposite page in column 'IDENT', under 'IS': '161' indicates $H_2{}^{16}O$.

LINE NO.	WAVENUMBER CM^{-1}, VAC.	J'	K_A'	K_C'	J"	K_A"	K_C"	IDENT. BD.	IS.
1	5413.53776(24)	3	2	1	2	2	0	2	161
2	5413.91048(38)	4	0	4	3	0	3	2	161
3	5415.57312(27)	4	2	2	3	1	3	1	161
4	5417.67806(64)	4	2	3	4	0	4	2	161
5	5422.85809(31)	4	3	2	3	2	1	1	161
6	5424.31510(30)	5	2	4	5	0	5	2	161
7	5427.09068(25)	4	3	1	3	2	2	1	161
8	5427.49431(34)	6	1	5	6	1	6	2	161
9	5428.72600(39)	5	1	5	4	1	4	2	161
10	5430.08736(25)	5	0	5	4	0	4	2	161
11	5433.42945(26)	4	2	3	3	2	2	2	161
12	5434.92039(39)	4	1	3	3	1	2	2	161
13	5440.87034(40)	4	2	2	3	2	1	2	161
14	5441.03573(26)	4	3	2	3	3	1	2	161
15	5442.15246(25)	4	3	1	3	3	0	2	161
16	5445.09133(25)	6	1	6	5	1	5	2	161
17	5445.72704(38)	6	0	6	5	0	5	2	161
18	5447.05463(50)	5	3	3	5	1	4	2	161
19	5449.86210(46)	2	2	0	1	0	1	2	161
20	5452.28873(77)	7	3	5	7	1	6	2	161
21	5454.82820(55)	7	2	6	7	0	7	2	161
22	5456.41904(26)	5	1	4	4	1	3	2	161

- "LINE NO.": serial number identifying the spectral line in the map on the opposite page.
- Uncertainty in the last digit of the wavenumber is given in parentheses.
- In column "IDENT." under "BD" 1 represents the band 110-000; 2: 011-000.

For isotopic assignment given under "IS" see comments on the opposite page.

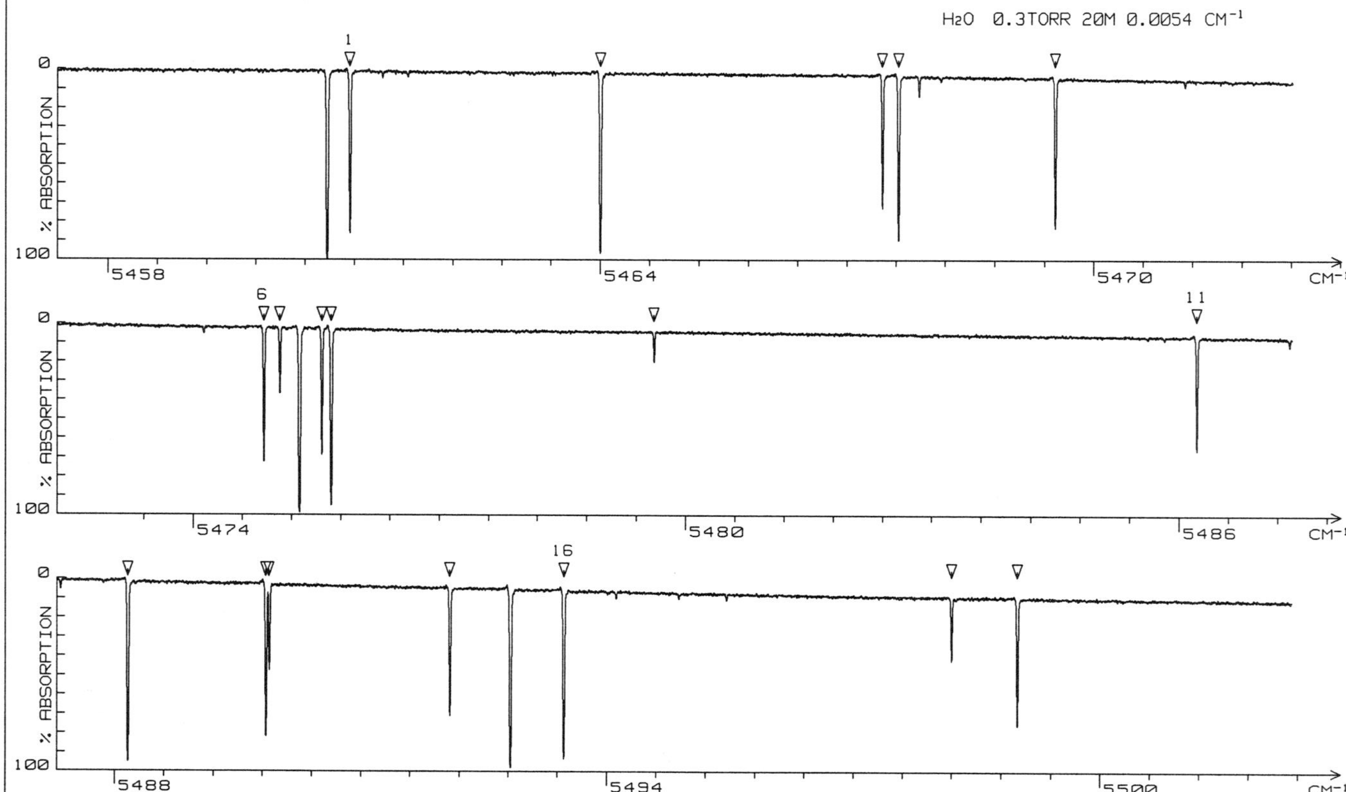

ISOTOPIC IDENTIFICATION- On this page: '∇' indicates $H_2{}^{16}O$.

On the opposite page in column 'IDENT', under 'IS': '161' indicates $H_2{}^{16}O$.

LINE NO.	WAVENUMBER CM^{-1}, VAC.	J'	K_A'	K_C'	J"	K_A"	K_C"	IDENT. BD.	IS.
1	5460.93989(26)	7	0	7	6	0	6	2	161
2	5463.99533(26)	5	3	3	4	3	2	2	161
3	5467.42656(27)	5	3	2	4	3	1	2	161
4	5467.62410(26)	5	2	3	4	2	2	2	161
5	5469.52723(26)	6	2	5	5	2	4	2	161
6	5474.85647(28)	5	4	2	4	4	1	2	161
7	5475.05035(40)	5	4	1	4	4	0	2	161
8	5475.56135(28)	8	1	8	7	1	7	2	161
9	5475.67607(27)	8	0	8	7	0	7	2	161
10	5479.61112(72)	3	2	1	2	0	2	2	161
11	5486.21205(30)	6	3	4	5	3	3	2	161
12	5488.16060(27)	7	2	6	6	2	5	2	161
13	5489.84314(28)	9	1	9	8	1	8	2	161
14	5489.88527(36)	9	0	9	8	0	8	2	161
15	5492.08433(29)	7	1	6	6	1	5	2	161
16	5493.47927(28)	6	3	3	5	3	2	2	161
17	5498.20008(43)	6	4	3	5	4	2	2	161
18	5498.99548(30)	6	4	2	5	4	1	2	161

- "LINE NO.": serial number identifying the spectral line in the map on the opposite page.
- Uncertainty in the last digit of the wavenumber is given in parentheses.
- In column "IDENT." under "BD" 1 represents the band 110–000; 2: 011–000.

For isotopic assignment given under "IS" see comments on the opposite page.

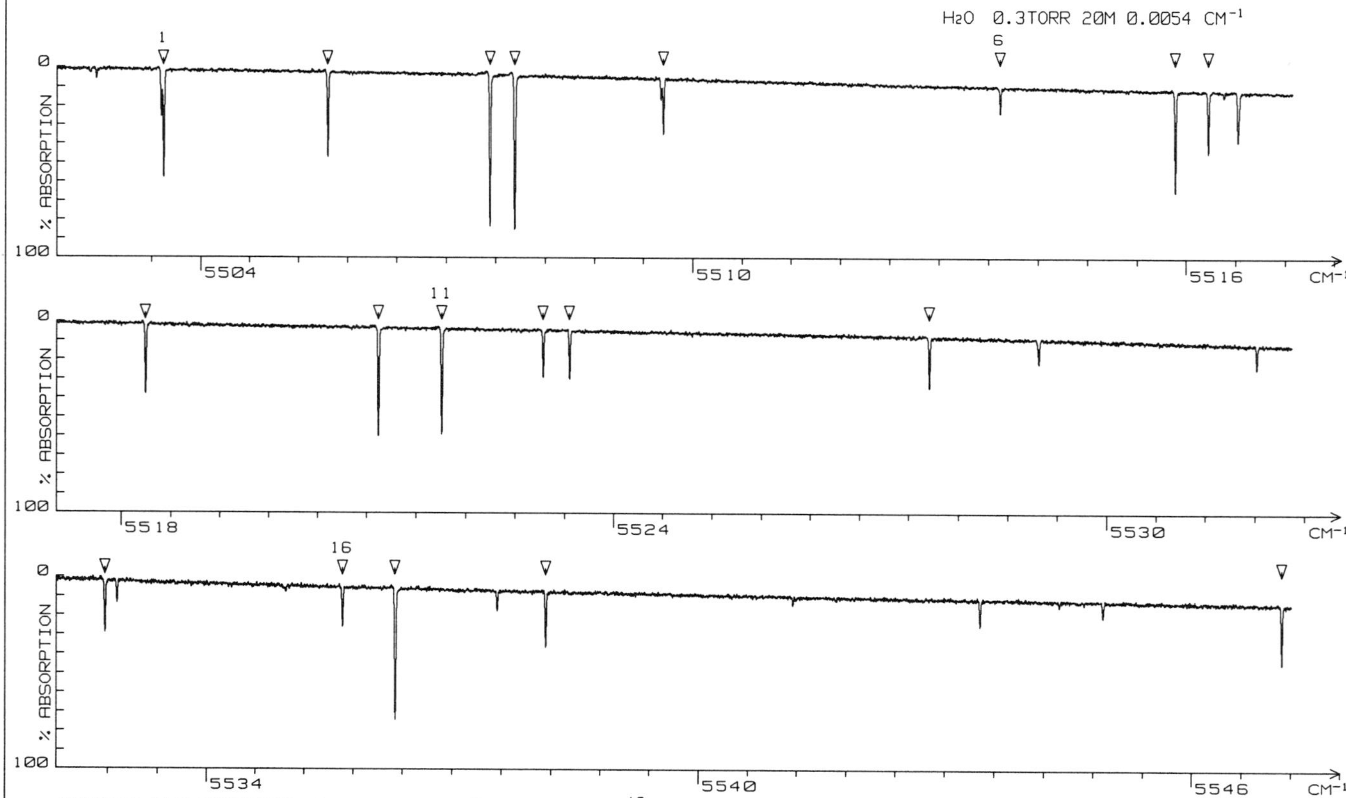

ISOTOPIC IDENTIFICATION- On this page: '∇' indicates $H_2{}^{16}O$.

On the opposite page in column 'IDENT', under 'IS': '161' indicates $H_2{}^{16}O$.

LINE NO.	WAVENUMBER CM^{-1}, VAC.	J'	K_A'	K_C'	J"	K_A"	K_C"	IDENT. BD.	IS.
1	5503.54165(34)	10	0	10	9	0	9	2	161
2	5505.55501(39)	8	2	7	7	2	6	2	161
3	5507.52452(30)	7	3	5	6	3	4	2	161
4	5507.82615(30)	8	1	7	7	1	6	2	161
5	5509.63879(51)	6	5	1	5	5	0	2	161
6	5513.73449(101)	6	4	3	5	3	2	1	161
7	5515.86130(35)	7	2	5	6	2	4	2	161
8	5516.26520(48)	4	2	2	3	0	3	2	161
9	5518.28734(50)	7	3	4	6	3	3	2	161
10	5521.13887(38)	7	4	4	6	4	3	2	161
11	5521.90416(38)	9	2	8	8	2	7	2	161
12	5523.13329(68)	9	1	8	8	1	7	2	161
13	5523.45414(64)	7	4	3	6	4	2	2	161
14	5527.84516(65)	8	3	6	7	3	5	2	161
15	5532.76170(71)	7	5	3	6	5	2	2	161
16	5535.66229(88)	8	4	5	7	3	4	1	161
17	5536.30744(37)	8	2	6	7	2	5	2	161
18	5538.12933(66)	10	1	9	9	1	8	2	161
19	5547.08328(65)	9	3	7	8	3	6	2	161

- "LINE NO.": serial number identifying the spectral line in the map on the opposite page.
- Uncertainty in the last digit of the wavenumber is given in parentheses.
- In column "IDENT." under "BD" 1 represents the band 110–000; 2: 011–000.

For isotopic assignment given under "IS" see comments on the opposite page.

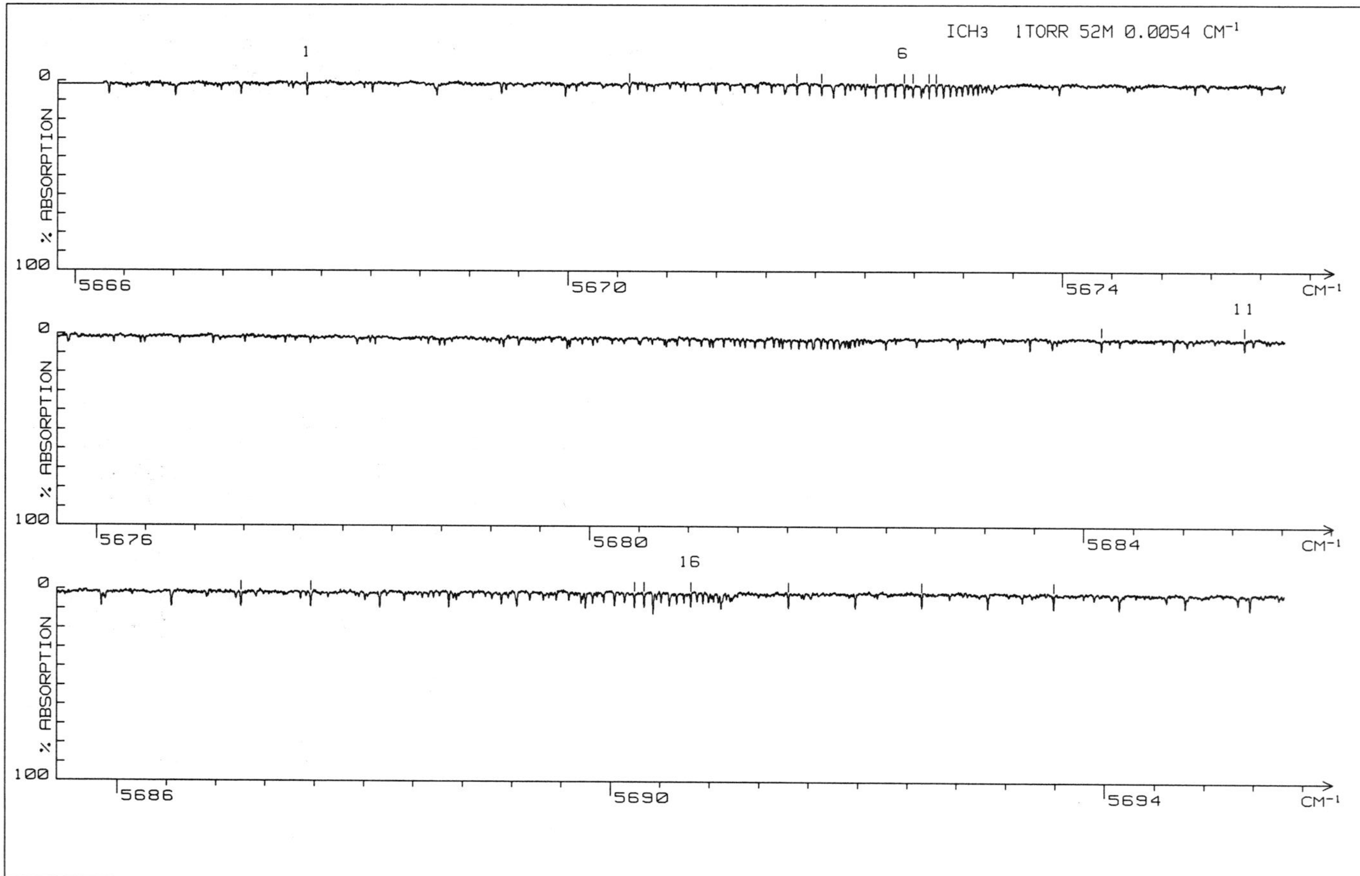
ICH3 1TORR 52M 0.0054 CM⁻¹
1
6
11
16
0
100
% ABSORPTION
5666
5670
5674
5676
5680
5684
5686
5690
5694
CM⁻¹

LINE NO	WAVENUMBER CM^{-1}, VAC
1	5667.88139(76)
2	5670.49253(76)
3	5671.85048(74)
4	5672.04899(79)
5	5672.48525(61)
6	5672.71615(60)
7	5672.78636(69)
8	5672.91734(62)
9	5672.97838(73)
10	5684.14364(70)
11	5685.29573(76)
12	5687.00375(67)
13	5687.56750(61)
14	5690.18996(56)
15	5690.26671(58)
16	5690.64738(58)
17	5691.43527(57)
18	5692.51558(57)
19	5693.58470(56)

- "LINE NO.": serial number identifying the spectral line in the map on the opposite page.
- "WAVENUMBER": experimental value.
- Uncertainty in the last digit of the wavenumber is given in parentheses.

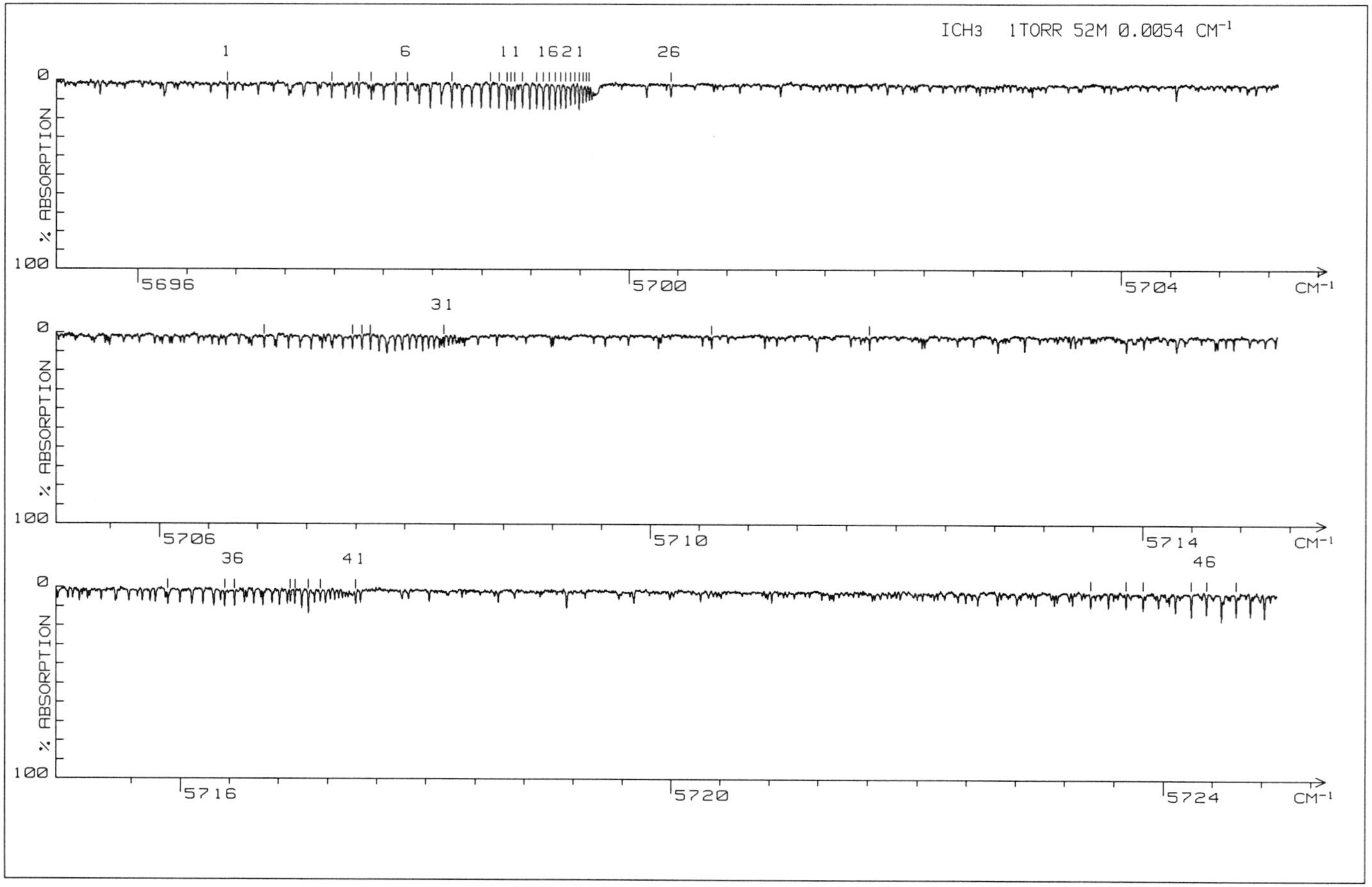
ICH3 1TORR 52M 0.0054 CM⁻¹
1
6
11
16
21
26
31
36
41
46
0
% ABSORPTION
100
5696
5700
5704
5706
5710
5714
5716
5720
5724
CM⁻¹

LINE NO	WAVENUMBER CM^{-1}, VAC	LINE NO	WAVENUMBER CM^{-1}, VAC
1	5696.72491(63)	31	5708.31773(72)
2	5697.57368(59)	32	5710.49862(64)
3	5697.79238(54)	33	5711.78422(64)
4	5697.89722(53)	34	5715.89287(71)
5	5698.09846(42)	35	5716.35662(49)
6	5698.19524(52)	36	5716.43947(53)
7	5698.55211(39)	37	5716.89343(67)
8	5698.86302(38)	38	5716.92934(57)
9	5698.93339(39)	39	5717.03902(40)
10	5699.00078(39)	40	5717.13504(78)
11	5699.03544(66)	41	5717.42118(77)
12	5699.06546(39)	42	5723.40688(62)
13	5699.12726(41)	43	5723.69116(57)
14	5699.24229(38)	44	5723.82914(53)
15	5699.29544(38)	45	5724.22163(41)
16	5699.34586(39)	46	5724.34720(44)
17	5699.39355(38)	47	5724.58897(41)
18	5699.43804(39)		
19	5699.47912(41)		
20	5699.51861(43)		
21	5699.55447(45)		
22	5699.58785(40)		
23	5699.61833(54)		
24	5699.64570(53)		
25	5699.67056(59)		
26	5700.33984(80)		
27	5706.85406(68)		
28	5707.56859(63)		
29	5707.64512(60)		
30	5707.71790(54)		

- "LINE NO.": serial number identifying the spectral line in the map on the opposite page.
- "WAVENUMBER": experimental value.
- Uncertainty in the last digit of the wavenumber is given in parentheses.

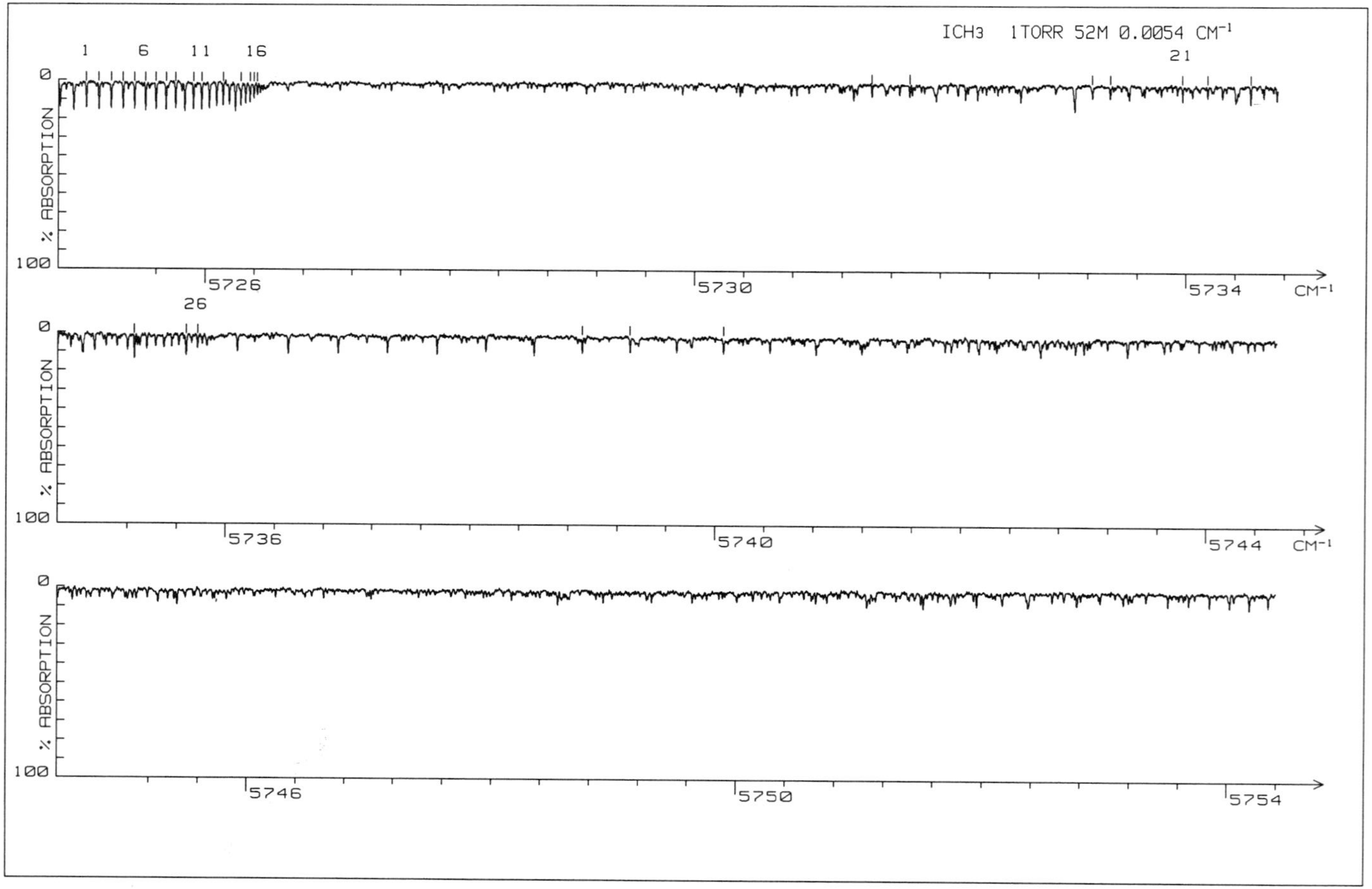
ICH3 1TORR 52M 0.0054 CM-1
1
6
11
16
21
26
0
% ABSORPTION
100
5726
5730
5734
CM-1
5736
5740
5744
5746
5750
5754

LINE NO	WAVENUMBER CM^{-1}, VAC
1	5725.03320(39)
2	5725.13585(38)
3	5725.23495(38)
4	5725.33057(37)
5	5725.42301(37)
6	5725.51160(36)
7	5725.59669(40)
8	5725.67817(37)
9	5725.75606(36)
10	5725.90071(38)
11	5725.96751(38)
12	5726.14484(42)
13	5726.28789(46)
14	5726.36291(49)
15	5726.39493(60)
16	5726.42302(71)
17	5731.43677(72)
18	5731.74524(76)
19	5733.23300(77)
20	5733.37656(79)
21	5733.96692(58)
22	5734.17316(63)
23	5734.52313(48)
24	5735.25328(40)
25	5735.67314(56)
26	5735.76575(69)
27	5738.90989(57)
28	5739.29806(56)
29	5740.06455(62)

- "LINE NO.": serial number identifying the spectral line in the map on the opposite page.
- "WAVENUMBER": experimental value.
- Uncertainty in the last digit of the wavenumber is given in parentheses.

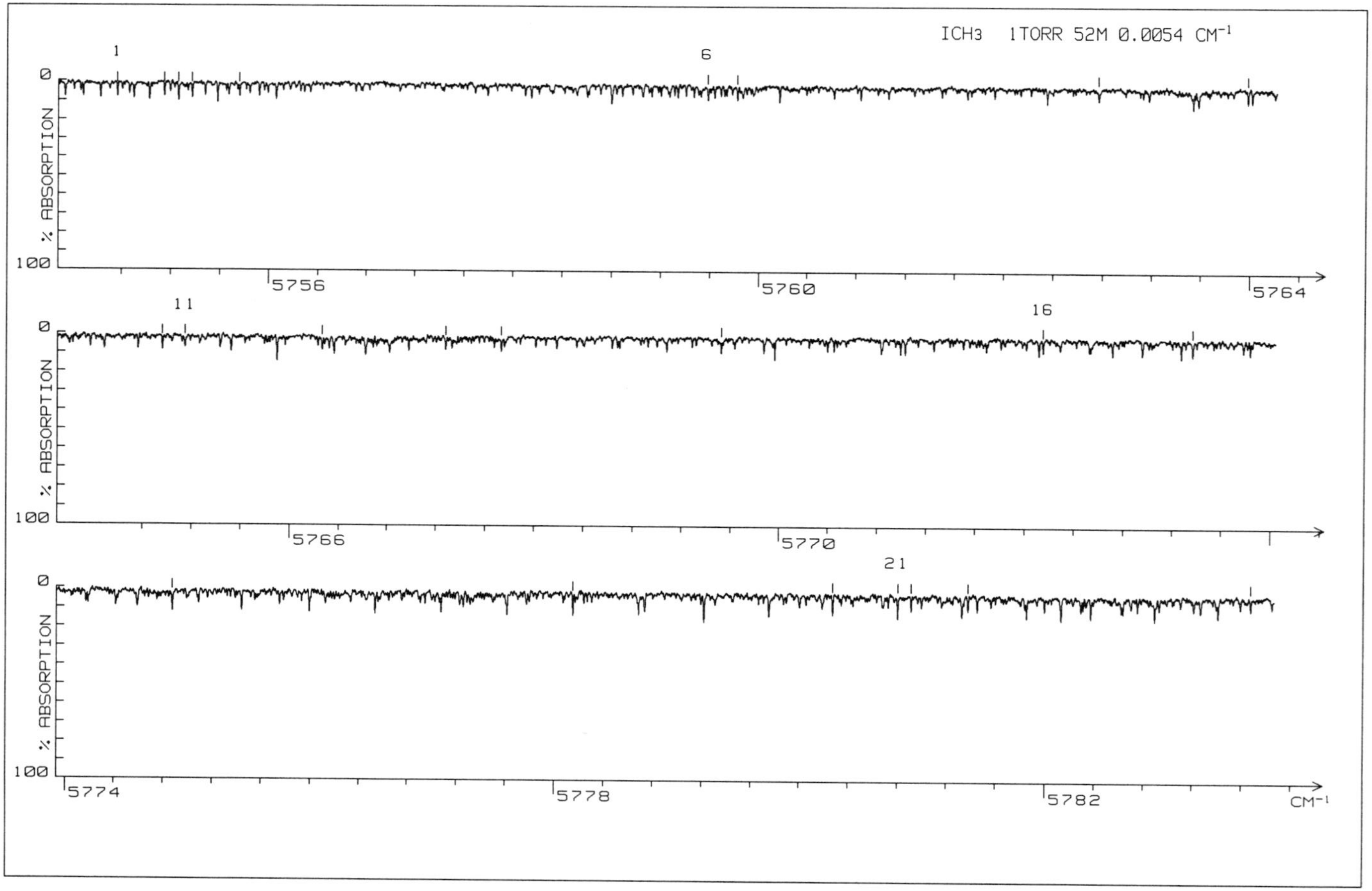
ICH3 1TORR 52M 0.0054 CM-1
1
6
11
16
21
0
% ABSORPTION
100
5756
5760
5764
5766
5770
5774
5778
5782
CM-1

LINE NO	WAVENUMBER CM^{-1}, VAC
1	5754.76433(62)
2	5755.14641(68)
3	5755.26185(58)
4	5755.37331(67)
5	5755.76000(72)
6	5759.58230(77)
7	5759.82096(74)
8	5762.76398(102)
9	5763.98039(65)
10	5764.96252(69)
11	5765.14755(91)
12	5766.26362(95)
13	5767.27970(73)
14	5767.73459(65)
15	5769.52378(62)
16	5772.14861(64)
17	5773.36911(65)
18	5774.87584(54)
19	5778.15031(40)
20	5780.27445(51)
21	5780.80118(42)
22	5780.91037(59)
23	5781.37628(62)
24	5783.67482(73)

- "LINE NO.": serial number identifying the spectral line in the map on the opposite page.
- "WAVENUMBER": experimental value.
- Uncertainty in the last digit of the wavenumber is given in parentheses.

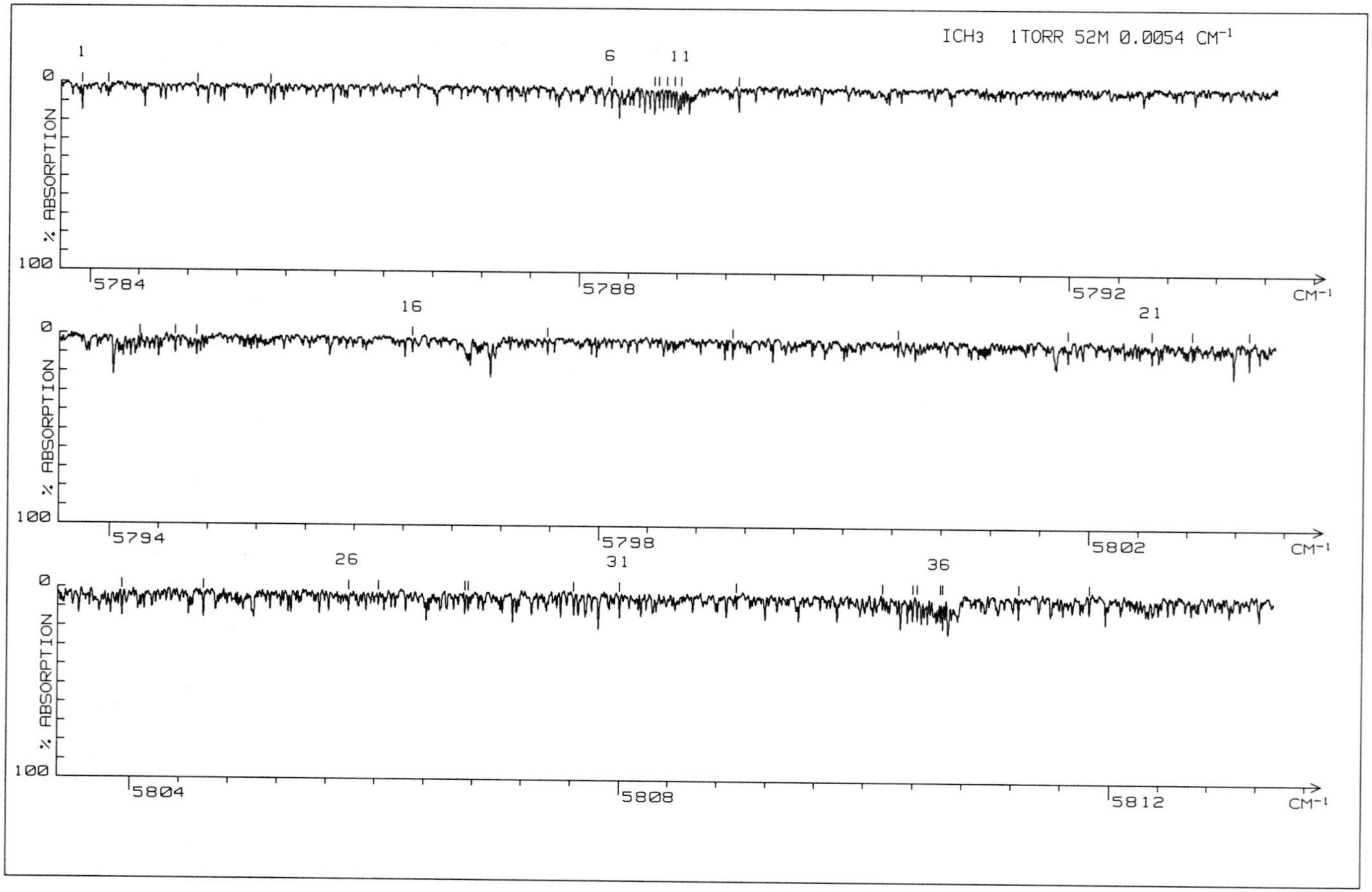
ICH3 1TORR 52M 0.0054 CM-1
1
6
11
0
% ABSORPTION
100
5784
5788
5792
CM-1
16
21
0
% ABSORPTION
100
5794
5798
5802
CM-1
26
31
36
0
% ABSORPTION
100
5804
5808
5812
CM-1

LINE NO	WAVENUMBER CM^{-1}, VAC	LINE NO	WAVENUMBER CM^{-1}, VAC
1	5783.92649(46)	31	5807.99844(46)
2	5784.14117(91)	32	5808.95850(66)
3	5784.87343(81)	33	5810.15112(75)
4	5785.46803(58)	34	5810.39552(51)
5	5786.67070(74)	35	5810.43286(51)
6	5788.25596(52)	36	5810.62354(47)
7	5788.60907(51)	37	5810.64237(44)
8	5788.64722(59)	38	5811.25688(46)
9	5788.71385(56)	39	5811.83180(58)
10	5788.77379(56)		
11	5788.82605(54)		
12	5789.29939(47)		
13	5794.23848(71)		
14	5794.52833(68)		
15	5794.70246(70)		
16	5796.46676(82)		
17	5797.57238(60)		
18	5799.08716(52)		
19	5800.43779(64)		
20	5801.82244(57)		
21	5802.50263(56)		
22	5802.83623(62)		
23	5803.30178(43)		
24	5803.93210(49)		
25	5804.59636(44)		
26	5805.78207(64)		
27	5806.02714(86)		
28	5806.73347(53)		
29	5806.76126(57)		
30	5807.62541(53)		

- "LINE NO.": serial number identifying the spectral line in the map on the opposite page.
- "WAVENUMBER": experimental value.
- Uncertainty in the last digit of the wavenumber is given in parentheses.

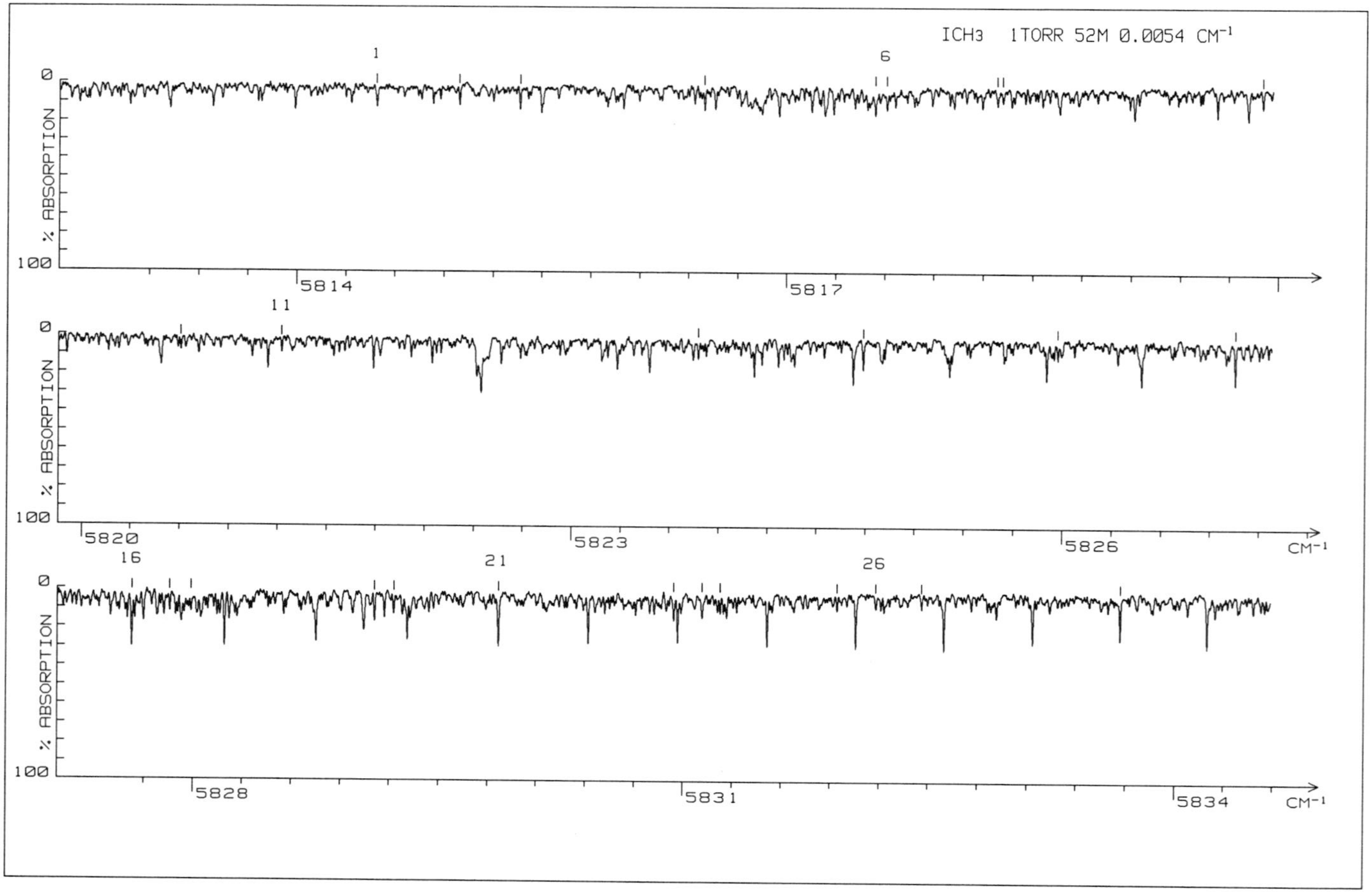
ICH3 1TORR 52M 0.0054 CM-1
1
6
11
16
21
26
0
100
% ABSORPTION
5814
5817
5820
5823
5826
5828
5831
5834
CM-1

LINE NO	WAVENUMBER CM^{-1}, VAC
1	5814.48909(61)
2	5814.99195(50)
3	5815.36585(48)
4	5816.49565(49)
5	5817.54061(50)
6	5817.61123(60)
7	5818.28541(77)
8	5818.31844(91)
9	5819.90416(57)
10	5820.60694(71)
11	5821.22371(63)
12	5823.77534(61)
13	5824.78446(39)
14	5825.96986(71)
15	5827.05415(31)
16	5827.62512(29)
17	5827.85764(56)
18	5827.98705(69)
19	5829.11055(43)
20	5829.22851(61)
21	5829.86617(30)
22	5830.94378(49)
23	5831.11682(61)
24	5831.22729(56)
25	5831.94354(62)
26	5832.18053(88)
27	5832.46112(69)
28	5833.66854(30)

- "LINE NO.": serial number identifying the spectral line in the map on the opposite page.
- "WAVENUMBER": experimental value.
- Uncertainty in the last digit of the wavenumber is given in parentheses.

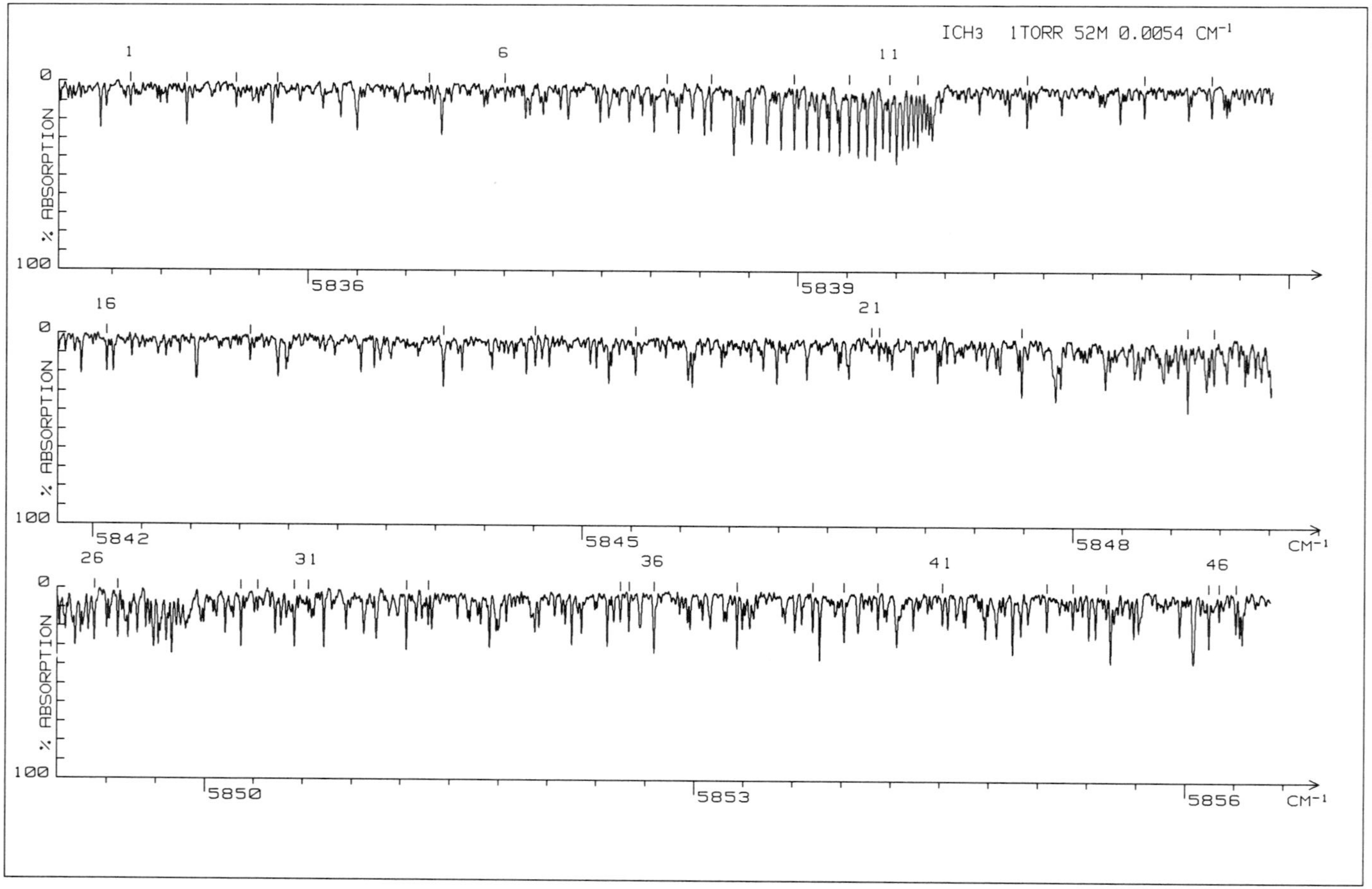
ICH3 1TORR 52M 0.0054 CM-1
1
6
11
16
21
26
31
36
41
46
0
% ABSORPTION
100
5836
5839
5842
5845
5848
5850
5853
5856
CM-1

LINE NO	WAVENUMBER CM^{-1}, VAC	LINE NO	WAVENUMBER CM^{-1}, VAC
1	5834.91161(51)	31	5850.63373(54)
2	5835.25439(32)	32	5851.23609(28)
3	5835.55851(47)	33	5851.36895(39)
4	5835.81253(84)	34	5852.54725(44)
5	5836.74106(62)	35	5852.59900(32)
6	5837.20487(78)	36	5852.75392(30)
7	5838.19908(53)	37	5853.25998(27)
8	5838.46709(30)	38	5853.72326(33)
9	5838.97612(26)	39	5853.91525(34)
10	5839.31177(27)	40	5854.12183(36)
11	5839.55961(27)	41	5854.51706(40)
12	5839.73147(27)	42	5855.15347(33)
13	5840.39674(33)	43	5855.31322(38)
14	5841.11368(37)	44	5855.51651(60)
15	5841.52706(41)	45	5856.14553(30)
16	5842.08390(36)	46	5856.20795(45)
17	5842.96133(45)	47	5856.30901(39)
18	5844.14487(32)		
19	5844.70684(48)		
20	5845.32213(36)		
21	5846.76681(93)		
22	5846.81103(52)		
23	5847.68274(29)		
24	5848.69737(27)		
25	5848.85905(38)		
26	5849.32241(31)		
27	5849.46766(31)		
28	5850.22185(28)		
29	5850.32349(55)		
30	5850.54837(30)		

- "LINE NO.": serial number identifying the spectral line in the map on the opposite page.
- "WAVENUMBER": experimental value.
- Uncertainty in the last digit of the wavenumber is given in parentheses.

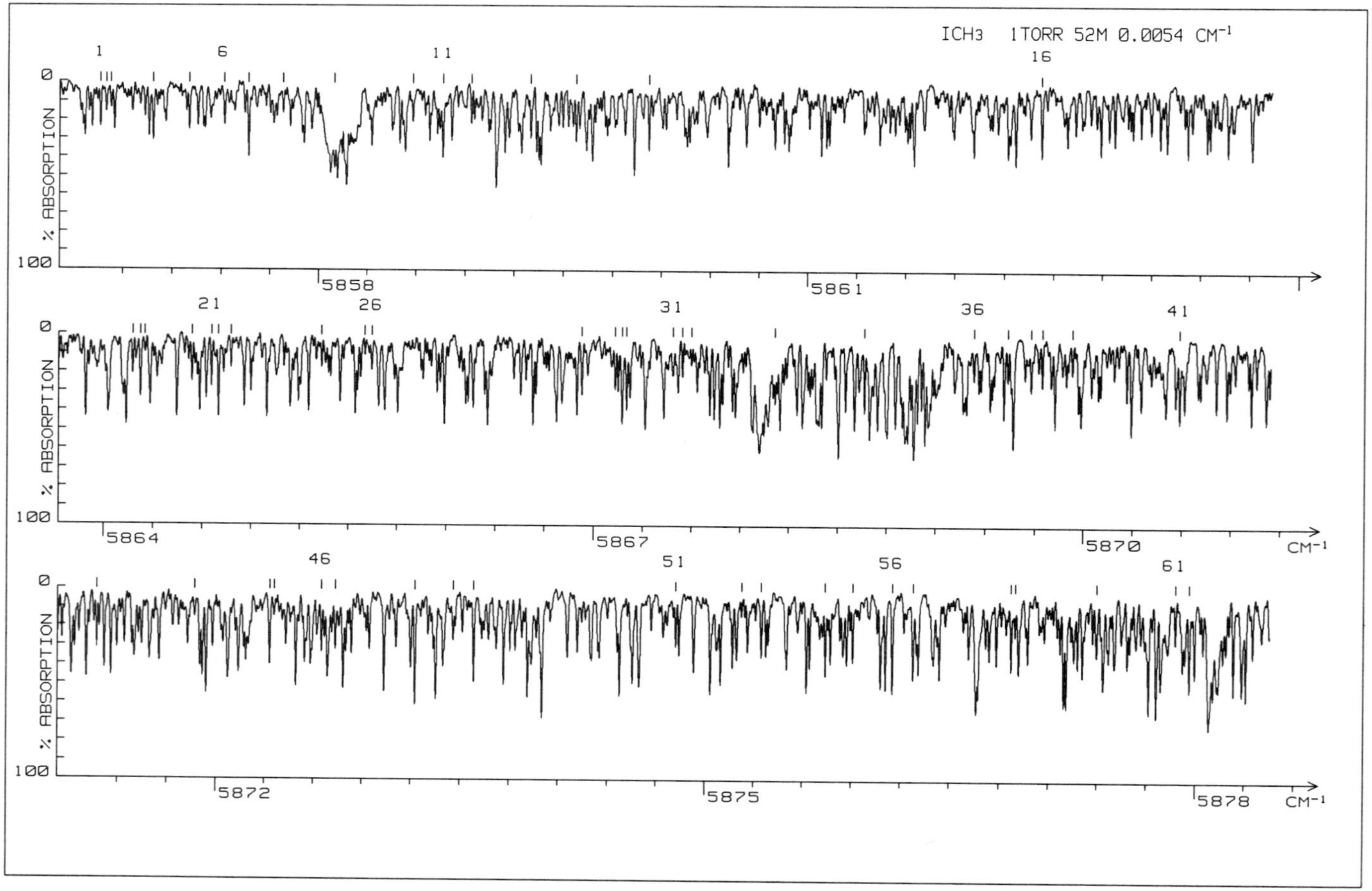
ICH3 1TORR 52M 0.0054 CM-1
% ABSORPTION
0
100
5858
5861
5864
5867
5870
5872
5875
5878
CM-1
1
6
11
16
21
26
31
36
41
46
51
56
61

LINE NO	WAVENUMBER CM^{-1}, VAC	LINE NO	WAVENUMBER CM^{-1}, VAC	LINE NO	WAVENUMBER CM^{-1}, VAC
1	5856.66177(31)	31	5867.48577(61)	61	5877.87988(31)
2	5856.69702(48)	32	5867.54426(34)	62	5877.96495(25)
3	5856.72649(57)	33	5867.59954(54)		
4	5856.98223(30)	34	5868.11409(36)		
5	5857.20146(31)	35	5868.66338(26)		
6	5857.41681(36)	36	5869.33192(41)		
7	5857.56620(26)	37	5869.53600(37)		
8	5857.77862(43)	38	5869.67608(31)		
9	5858.09573(48)	39	5869.74492(30)		
10	5858.57782(38)	40	5869.93257(39)		
11	5858.76055(29)	41	5870.58662(27)		
12	5858.93404(32)	42	5871.27444(28)		
13	5859.29913(31)	43	5871.87076(59)		
14	5859.57412(31)	44	5872.33283(27)		
15	5860.02109(27)	45	5872.36027(64)		
16	5862.42249(27)	46	5872.65191(29)		
17	5864.17518(36)	47	5872.73797(31)		
18	5864.22272(29)	48	5873.22851(24)		
19	5864.25121(81)	49	5873.46017(36)		
20	5864.53593(32)	50	5873.58578(25)		
21	5864.65535(29)	51	5874.82162(29)		
22	5864.69808(26)	52	5875.22615(53)		
23	5864.77597(39)	53	5875.34819(28)		
24	5865.33379(39)	54	5875.74279(26)		
25	5865.60147(55)	55	5875.91093(27)		
26	5865.64540(42)	56	5876.15393(24)		
27	5866.92760(30)	57	5876.28017(27)		
28	5867.12964(32)	58	5876.87500(26)		
29	5867.17273(27)	59	5876.89967(40)		
30	5867.20042(28)	60	5877.40025(35)		

- "LINE NO.": serial number identifying the spectral line in the map on the opposite page.
- "WAVENUMBER": experimental value.
- Uncertainty in the last digit of the wavenumber is given in parentheses.

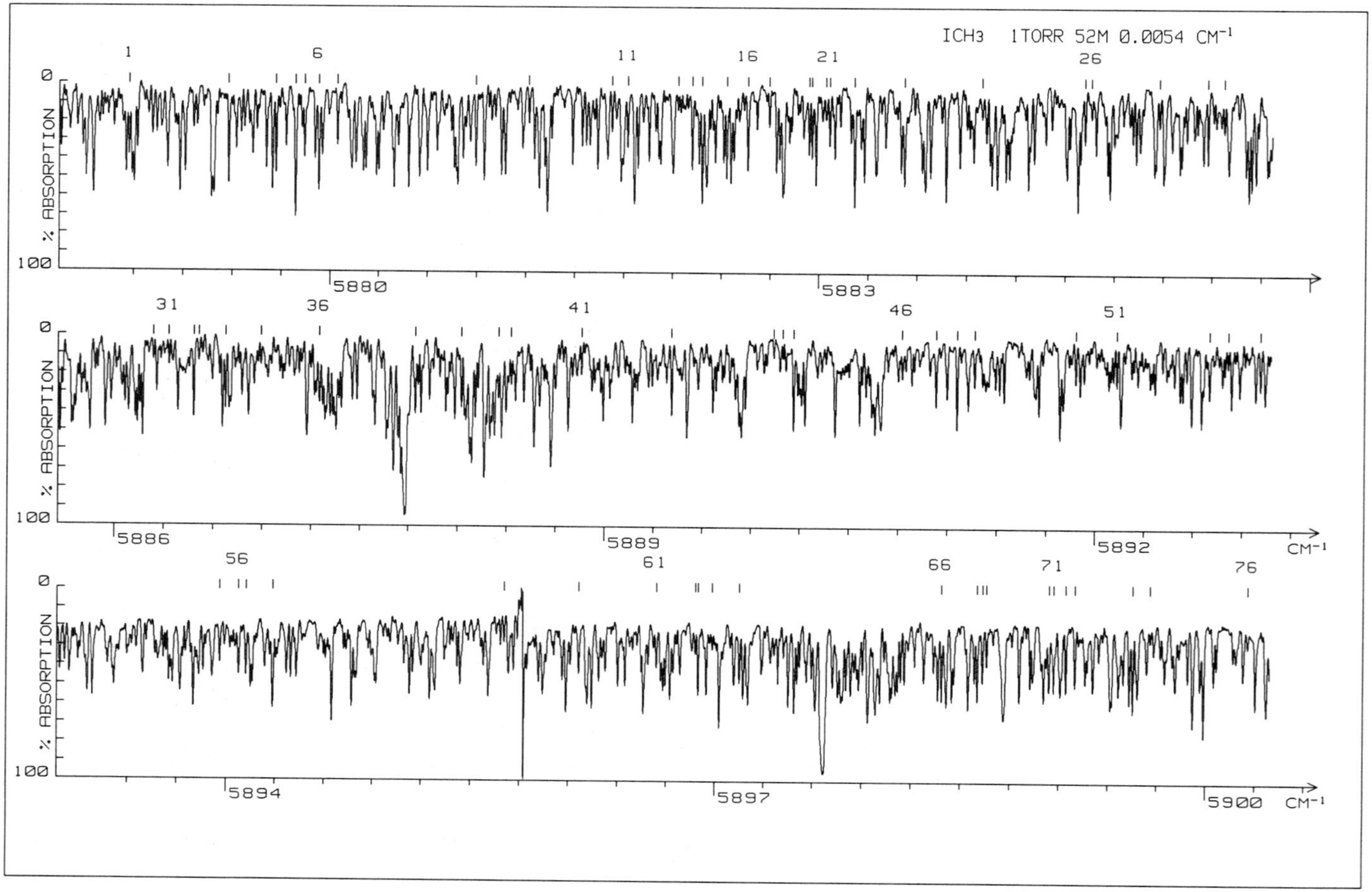
ICH3 1TORR 52M 0.0054 CM⁻¹
1
6
11
16
21
26
31
36
41
46
51
56
61
66
71
76
0
% ABSORPTION
100
5880
5883
5886
5889
5892
5894
5897
5900
CM⁻¹

LINE NO	WAVENUMBER CM^{-1}, VAC	LINE NO	WAVENUMBER CM^{-1}, VAC	LINE NO	WAVENUMBER CM^{-1}, VAC
1	5878.77099(30)	31	5886.33202(44)	61	5896.63788(33)
2	5879.37737(24)	32	5886.48248(25)	62	5896.87782(43)
3	5879.66701(25)	33	5886.51301(56)	63	5896.89503(25)
4	5879.78606(23)	34	5886.67563(29)	64	5896.98275(51)
5	5879.84123(27)	35	5886.89325(50)	65	5897.14847(27)
6	5879.92893(24)	36	5887.25087(36)	66	5898.38604(25)
7	5880.04254(27)	37	5887.84334(29)	67	5898.60283(25)
8	5880.89279(26)	38	5888.12377(28)	68	5898.63390(27)
9	5881.21736(34)	39	5888.34960(39)	69	5898.65962(31)
10	5881.72578(31)	40	5888.42868(44)	70	5899.03757(28)
11	5881.82106(29)	41	5888.85926(36)	71	5899.06767(27)
12	5882.13120(57)	42	5889.41147(26)	72	5899.14264(25)
13	5882.21852(25)	43	5890.04121(61)	73	5899.19799(26)
14	5882.27999(25)	44	5890.09853(57)	74	5899.55216(24)
15	5882.43082(25)	45	5890.16314(25)	75	5899.65926(65)
16	5882.56122(26)	46	5890.82524(43)	76	5900.25539(58)
17	5882.69235(33)	47	5891.03245(28)		
18	5882.93776(27)	48	5891.15888(25)		
19	5882.95654(27)	49	5891.26597(34)		
20	5883.04262(60)	50	5891.88224(34)		
21	5883.06432(29)	51	5892.13324(32)		
22	5883.21721(24)	52	5892.70250(33)		
23	5883.51958(26)	53	5892.81560(52)		
24	5883.99244(54)	54	5893.01313(34)		
25	5884.61925(29)	55	5893.95914(31)		
26	5884.65942(47)	56	5894.07658(32)		
27	5885.07611(39)	57	5894.12034(29)		
28	5885.37345(26)	58	5894.28463(26)		
29	5885.47223(38)	59	5895.70403(40)		
30	5886.23691(50)	60	5896.16191(31)		

- "LINE NO.": serial number identifying the spectral line in the map on the opposite page.
- "WAVENUMBER": experimental value.
- Uncertainty in the last digit of the wavenumber is given in parentheses.

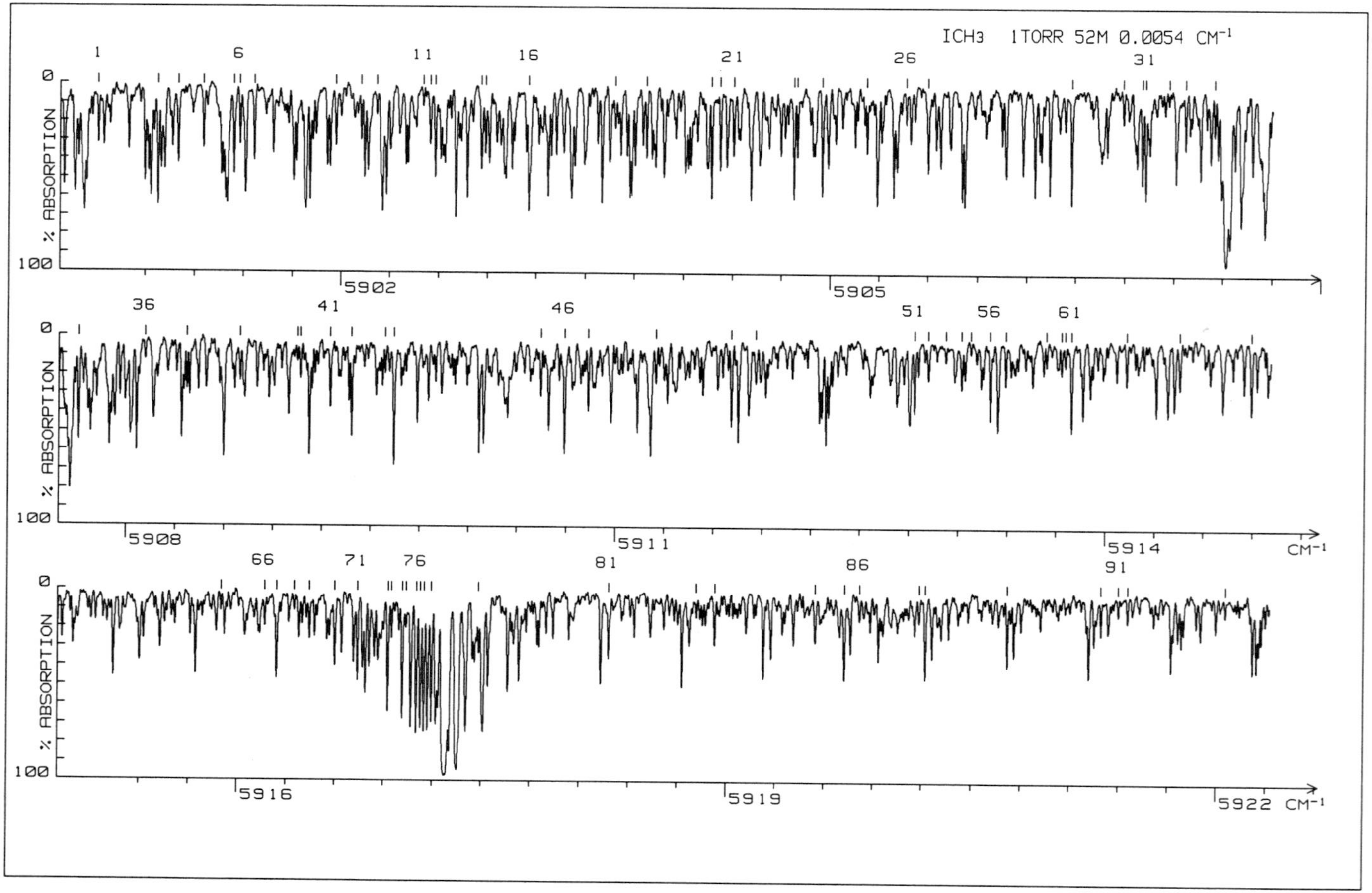

ICH3 1TORR 52M 0.0054 CM-1
% ABSORPTION
0
100
1
6
11
16
21
26
31
36
41
46
51
56
61
66
71
76
81
86
91
5902
5905
5908
5911
5914
5916
5919
5922
CM-1

LINE NO	WAVENUMBER CM^{-1}, VAC	LINE NO	WAVENUMBER CM^{-1}, VAC	LINE NO	WAVENUMBER CM^{-1}, VAC	LINE NO	WAVENUMBER CM^{-1}, VAC
1	5900.51393(28)	31	5906.92628(24)	61	5913.79423(25)	91	5921.39904(52)
2	5900.87671(24)	32	5907.06809(68)	62	5914.13016(34)	92	5921.45492(43)
3	5900.99980(26)	33	5907.17210(28)	63	5914.45485(35)	93	5922.05479(53)
4	5901.15347(28)	34	5907.34917(33)	64	5914.89373(28)		
5	5901.34095(88)	35	5907.70982(25)	65	5915.89765(49)		
6	5901.37779(28)	36	5908.11771(61)	66	5916.16965(39)		
7	5901.46596(26)	37	5908.37095(31)	67	5916.24197(26)		
8	5901.96734(30)	38	5908.69770(35)	68	5916.34900(58)		
9	5902.12405(30)	39	5909.04749(55)	69	5916.44338(33)		
10	5902.21953(45)	40	5909.06759(36)	70	5916.59977(28)		
11	5902.50228(55)	41	5909.25213(28)	71	5916.73884(69)		
12	5902.54565(31)	42	5909.38235(76)	72	5916.92756(24)		
13	5902.57727(25)	43	5909.59134(40)	73	5916.95037(33)		
14	5902.86083(115)	44	5909.64431(24)	74	5917.01501(57)		
15	5902.88525(29)	45	5910.54683(31)	75	5917.03801(56)		
16	5903.15398(102)	46	5910.69043(24)	76	5917.10054(24)		
17	5903.68005(45)	47	5910.83431(26)	77	5917.12587(24)		
18	5903.87348(26)	48	5911.24993(39)	78	5917.14737(24)		
19	5904.27091(24)	49	5911.71089(25)	79	5917.19356(24)		
20	5904.32334(25)	50	5911.86079(45)	80	5917.48369(37)		
21	5904.40468(28)	51	5912.83710(27)	81	5918.27830(28)		
22	5904.77589(24)	52	5912.92156(41)	82	5918.81680(56)		
23	5904.79778(27)	53	5913.02675(65)	83	5918.92782(31)		
24	5904.94986(74)	54	5913.12152(36)	84	5919.54582(37)		
25	5905.22458(28)	55	5913.17735(72)	85	5919.72442(27)		
26	5905.46568(36)	56	5913.29589(26)	86	5919.81813(53)		
27	5905.60006(25)	57	5913.38957(32)	87	5920.18512(51)		
28	5906.47515(63)	58	5913.64181(50)	88	5920.22259(26)		
29	5906.79092(39)	59	5913.73394(40)	89	5920.71965(26)		
30	5906.90495(25)	60	5913.75766(49)	90	5921.29374(34)		

- "LINE NO.": serial number identifying the spectral line in the map on the opposite page.
- "WAVENUMBER": experimental value.
- Uncertainty in the last digit of the wavenumber is given in parentheses.

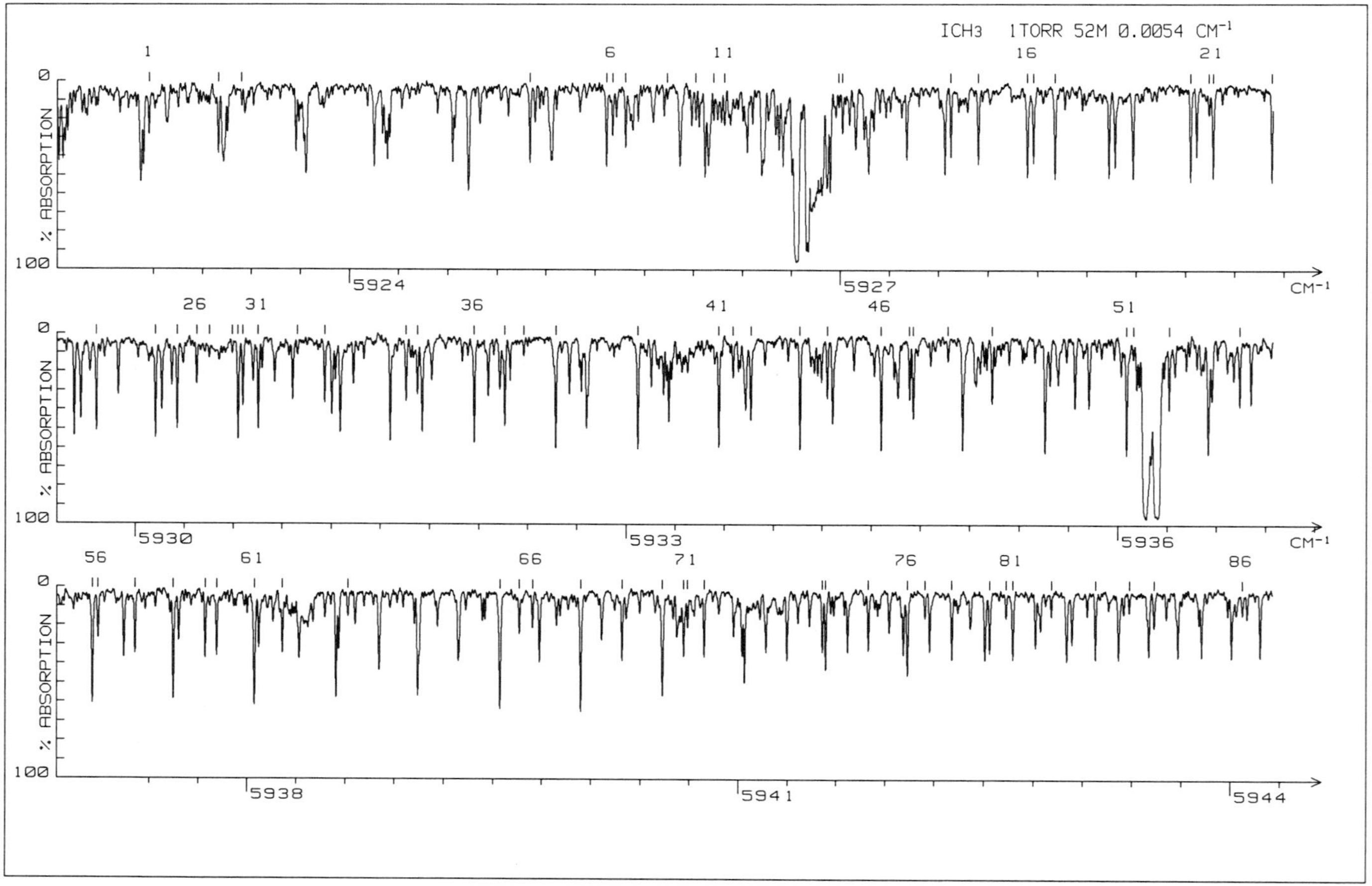
ICH3 1TORR 52M 0.0054 CM-1
1 6 11 16 21
26 31 36 41 46 51
56 61 66 71 76 81 86
0
% ABSORPTION
100
5924
5927
5930
5933
5936
5938
5941
5944
CM-1

LINE NO	WAVENUMBER CM^{-1}, VAC	LINE NO	WAVENUMBER CM^{-1}, VAC	LINE NO	WAVENUMBER CM^{-1}, VAC
1	5922.77219(31)	31	5930.75632(24)	61	5938.04532(24)
2	5923.19633(27)	32	5930.99552(90)	62	5938.21693(28)
3	5923.34179(57)	33	5931.16033(86)	63	5938.61611(53)
4	5925.10302(25)	34	5931.65871(28)	64	5939.54211(24)
5	5925.56938(25)	35	5931.72764(30)	65	5939.66323(33)
6	5925.60574(30)	36	5932.07286(24)	66	5939.74482(35)
7	5925.68421(27)	37	5932.25857(68)	67	5940.03873(24)
8	5925.94097(92)	38	5932.37757(53)	68	5940.28870(26)
9	5926.11468(38)	39	5932.57338(55)	69	5940.53737(25)
10	5926.22438(40)	40	5933.07172(24)	70	5940.66706(28)
11	5926.29038(40)	41	5933.56895(24)	71	5940.69086(39)
12	5926.99170(66)	42	5933.65520(34)	72	5940.79224(73)
13	5927.01392(32)	43	5933.76352(26)	73	5941.51615(28)
14	5927.67493(26)	44	5934.06605(24)	74	5941.53698(25)
15	5927.84030(26)	45	5934.23417(28)	75	5941.79748(83)
16	5928.13767(70)	46	5934.56281(24)	76	5942.03657(25)
17	5928.17345(25)	47	5934.73593(28)	77	5942.14265(45)
18	5928.30345(25)	48	5934.75910(80)	78	5942.30729(27)
19	5929.12975(24)	49	5934.97203(50)	79	5942.53604(28)
20	5929.24383(52)	50	5935.23779(27)	80	5942.63569(43)
21	5929.26912(25)	51	5936.05467(52)	81	5942.67610(27)
22	5929.62622(24)	52	5936.09649(68)	82	5942.90806(44)
23	5929.76169(25)	53	5936.31352(74)	83	5943.17808(26)
24	5930.12549(24)	54	5936.74394(27)	84	5943.38443(50)
25	5930.25824(25)	55	5937.04958(24)	85	5943.53556(39)
26	5930.37713(32)	56	5937.08405(33)	86	5944.07334(58)
27	5930.45368(76)	57	5937.31287(27)		
28	5930.59454(74)	58	5937.54754(24)		
29	5930.63267(24)	59	5937.74168(77)		
30	5930.66205(26)	60	5937.81243(27)		

- "LINE NO.": serial number identifying the spectral line in the map on the opposite page.
- "WAVENUMBER": experimental value.
- Uncertainty in the last digit of the wavenumber is given in parentheses.

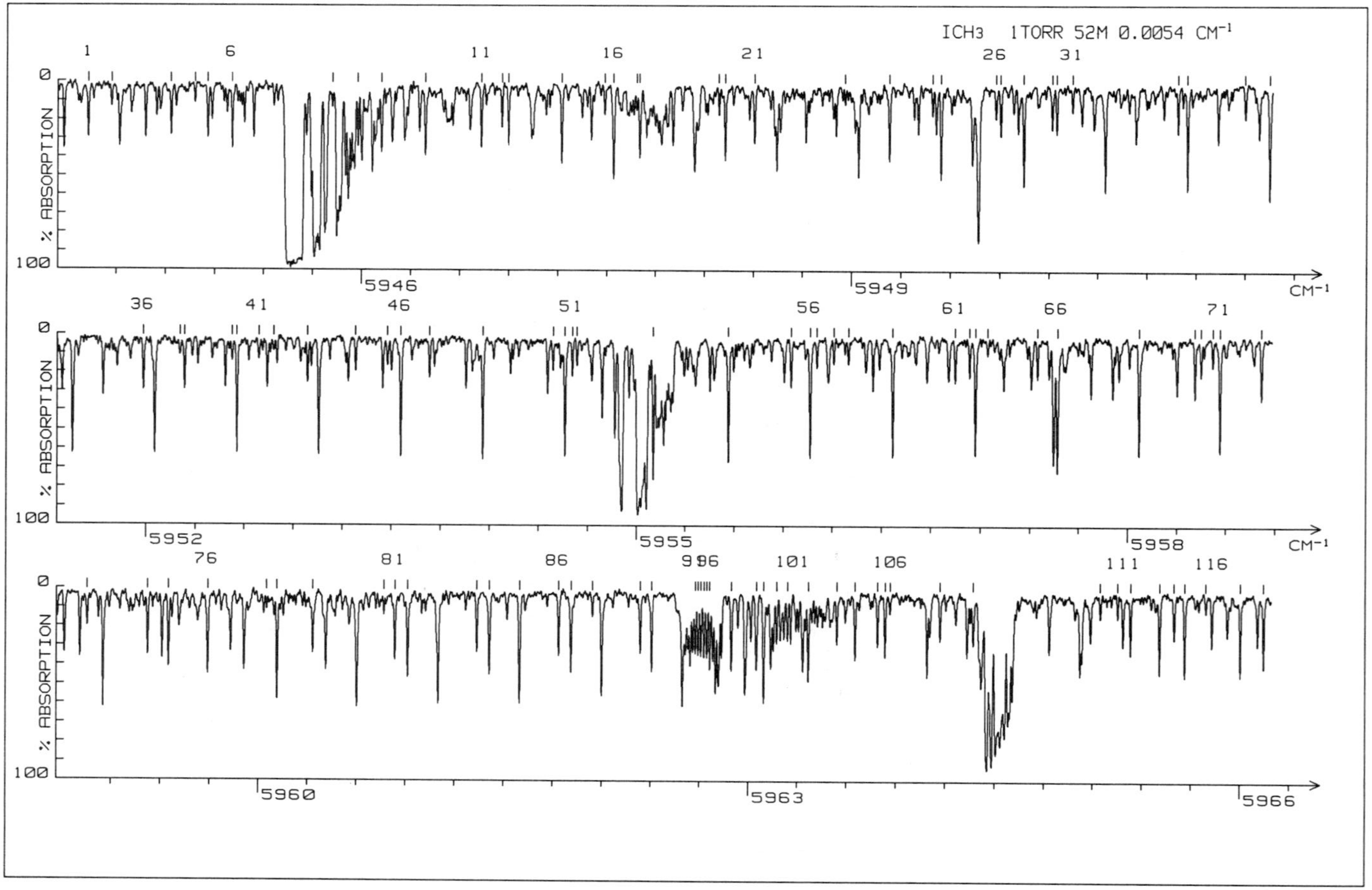
ICH3 1TORR 52M 0.0054 CM⁻¹
1
6
11
16
21
26
31
36
41
46
51
56
61
66
71
76
81
86
96
101
106
111
116
0
% ABSORPTION
100
5946
5949
5952
5955
5958
5960
5963
5966
CM⁻¹

LINE NO	WAVENUMBER CM^{-1}, VAC	LINE NO	WAVENUMBER CM^{-1}, VAC	LINE NO	WAVENUMBER CM^{-1}, VAC	LINE NO	WAVENUMBER CM^{-1}, VAC
1	5944.32942(29)	31	5950.34635(44)	61	5956.94558(33)	91	5962.67625(31)
2	5944.47249(50)	32	5950.98933(30)	62	5957.03245(33)	92	5962.69317(30)
3	5944.83519(29)	33	5951.04807(24)	63	5957.06740(24)	93	5962.70980(29)
4	5944.97988(50)	34	5951.40228(41)	64	5957.14191(66)	94	5962.72884(28)
5	5945.05819(29)	35	5951.55030(52)	65	5957.44512(34)	95	5962.74647(29)
6	5945.20791(27)	36	5951.98277(30)	66	5957.56716(49)	96	5962.76450(27)
7	5945.82250(63)	37	5952.20451(88)	67	5958.06778(44)	97	5962.89586(56)
8	5945.97874(28)	38	5952.23227(29)	68	5958.40917(28)	98	5963.04839(67)
9	5946.12404(28)	39	5952.52219(57)	69	5958.44806(38)	99	5963.09324(24)
10	5946.39019(82)	40	5952.55289(24)	70	5958.52058(42)	100	5963.17277(31)
11	5946.73294(79)	41	5952.68746(67)	71	5958.56444(24)	101	5963.23930(33)
12	5946.85910(32)	42	5952.77846(91)	72	5958.81567(28)	102	5963.36891(25)
13	5946.89983(27)	43	5952.98592(33)	73	5958.95403(39)	103	5963.54280(30)
14	5947.22336(25)	44	5953.28012(38)	74	5959.32414(27)	104	5963.65352(27)
15	5947.48688(42)	45	5953.47590(47)	75	5959.45090(26)	105	5963.78999(29)
16	5947.54077(25)	46	5953.55638(24)	76	5959.68911(57)	106	5963.83518(27)
17	5947.68327(63)	47	5953.73220(35)	77	5960.05140(68)	107	5963.86568(58)
18	5947.70113(74)	48	5954.05813(24)	78	5960.11482(24)	108	5964.17004(31)
19	5948.18872(38)	49	5954.48702(39)	79	5960.33348(28)	109	5964.37050(33)
20	5948.22687(26)	50	5954.55970(24)	80	5960.76988(47)	110	5965.14812(49)
21	5948.40652(29)	51	5954.60397(34)	81	5960.83597(27)	111	5965.25376(42)
22	5948.96131(60)	52	5954.63292(45)	82	5960.91402(25)	112	5965.33174(27)
23	5949.23023(26)	53	5955.10362(24)	83	5961.33701(28)	113	5965.50869(25)
24	5949.49309(40)	54	5955.56310(24)	84	5961.41348(26)	114	5965.59806(34)
25	5949.54394(24)	55	5955.94790(32)	85	5961.59931(24)	115	5965.66292(25)
26	5949.87573(41)	56	5956.06462(24)	86	5961.83710(27)	116	5965.79471(61)
27	5949.90559(30)	57	5956.10584(41)	87	5961.91184(25)	117	5966.00325(25)
28	5950.04522(24)	58	5956.20834(45)	88	5962.04347(45)	118	5966.14472(25)
29	5950.21924(33)	59	5956.29770(48)	89	5962.33656(73)		
30	5950.24729(32)	60	5956.56625(24)	90	5962.40683(25)		

- "LINE NO.": serial number identifying the spectral line in the map on the opposite page.
- "WAVENUMBER": experimental value.
- Uncertainty in the last digit of the wavenumber is given in parentheses.

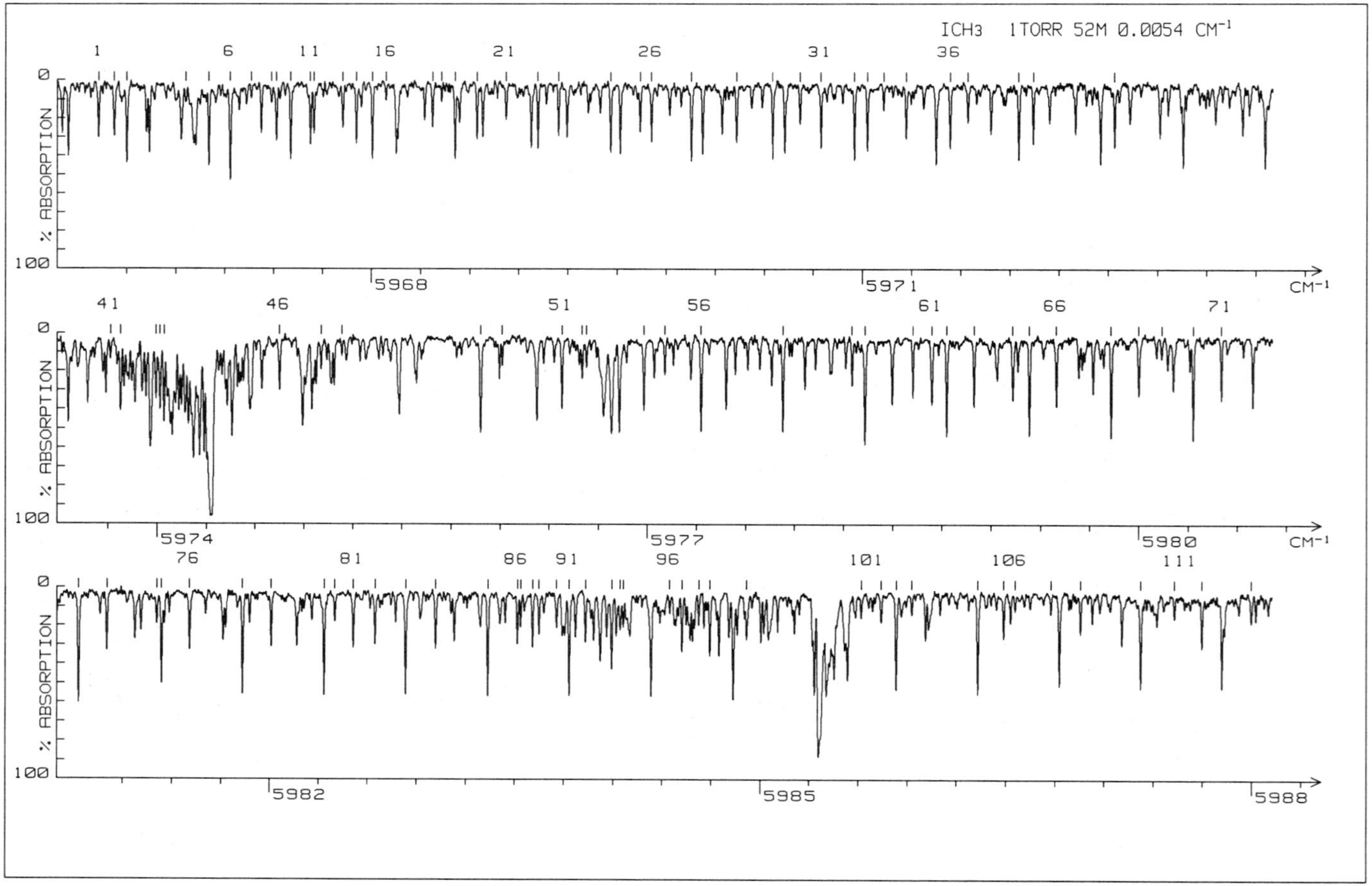
ICH3 1TORR 52M 0.0054 CM-1
% ABSORPTION
0
100
CM-1
5968
5971
5974
5977
5980
5982
5985
5988
1
6
11
16
21
26
31
36
41
46
51
56
61
66
71
76
81
86
91
96
101
106
111

LINE NO	WAVENUMBER CM^{-1}, VAC	LINE NO	WAVENUMBER CM^{-1}, VAC	LINE NO	WAVENUMBER CM^{-1}, VAC	LINE NO	WAVENUMBER CM^{-1}, VAC
1	5966.32997(28)	31	5970.74907(27)	61	5978.73921(30)	91	5983.83451(24)
2	5966.42528(32)	32	5970.95183(25)	62	5978.82957(24)	92	5983.93277(31)
3	5966.50182(25)	33	5971.03069(27)	63	5978.99536(58)	93	5984.09229(26)
4	5966.86089(93)	34	5971.13003(41)	64	5979.23077(29)	94	5984.14405(44)
5	5967.00198(25)	35	5971.26664(74)	65	5979.33029(24)	95	5984.16359(42)
6	5967.13507(25)	36	5971.53707(28)	66	5979.49737(27)	96	5984.44862(55)
7	5967.26453(75)	37	5971.64487(34)	67	5979.83084(24)	97	5984.52347(61)
8	5967.38810(73)	38	5971.95272(26)	68	5980.00000(67)	98	5984.62572(37)
9	5967.41823(30)	39	5972.03976(27)	69	5980.13795(59)	99	5984.69384(62)
10	5967.50476(25)	40	5972.53908(27)	70	5980.33119(24)	100	5984.91734(91)
11	5967.62597(68)	41	5973.71478(54)	71	5980.50277(26)	101	5985.61883(48)
12	5967.64804(30)	42	5973.77558(28)	72	5980.83074(24)	102	5985.74081(47)
13	5967.82508(32)	43	5973.99240(30)	73	5981.00549(28)	103	5985.83292(40)
14	5967.90767(28)	44	5974.01662(28)	74	5981.30956(42)	104	5985.92776(53)
15	5968.00738(25)	45	5974.04121(26)	75	5981.33853(44)	105	5986.33069(40)
16	5968.09022(61)	46	5974.75011(31)	76	5981.50839(28)	106	5986.48751(31)
17	5968.37298(32)	47	5975.00512(40)	77	5981.83527(24)	107	5986.55725(60)
18	5968.42744(48)	48	5975.13234(58)	78	5982.01151(28)	108	5986.77287(47)
19	5968.51063(26)	49	5975.97923(25)	79	5982.33568(24)	109	5986.95685(33)
20	5968.64401(28)	50	5976.11416(43)	80	5982.40139(40)	110	5987.32394(25)
21	5968.82449(37)	51	5976.47800(26)	81	5982.51416(28)	111	5987.52770(49)
22	5969.01927(27)	52	5976.60012(36)	82	5982.64923(29)	112	5987.69737(61)
23	5969.14421(28)	53	5976.62647(40)	83	5982.83559(24)	113	5987.99733(47)
24	5969.46066(26)	54	5976.97904(26)	84	5983.01736(66)		
25	5969.64324(31)	55	5977.10586(35)	85	5983.33564(24)		
26	5969.71270(28)	56	5977.32743(49)	86	5983.51892(76)		
27	5969.95496(58)	57	5977.82794(47)	87	5983.53917(35)		
28	5970.23065(30)	58	5978.25064(75)	88	5983.61258(29)		
29	5970.45222(55)	59	5978.32860(24)	89	5983.65150(33)		
30	5970.62080(34)	60	5978.62197(28)	90	5983.75807(40)		

- "LINE NO.": serial number identifying the spectral line in the map on the opposite page.
- "WAVENUMBER": experimental value.
- Uncertainty in the last digit of the wavenumber is given in parentheses.

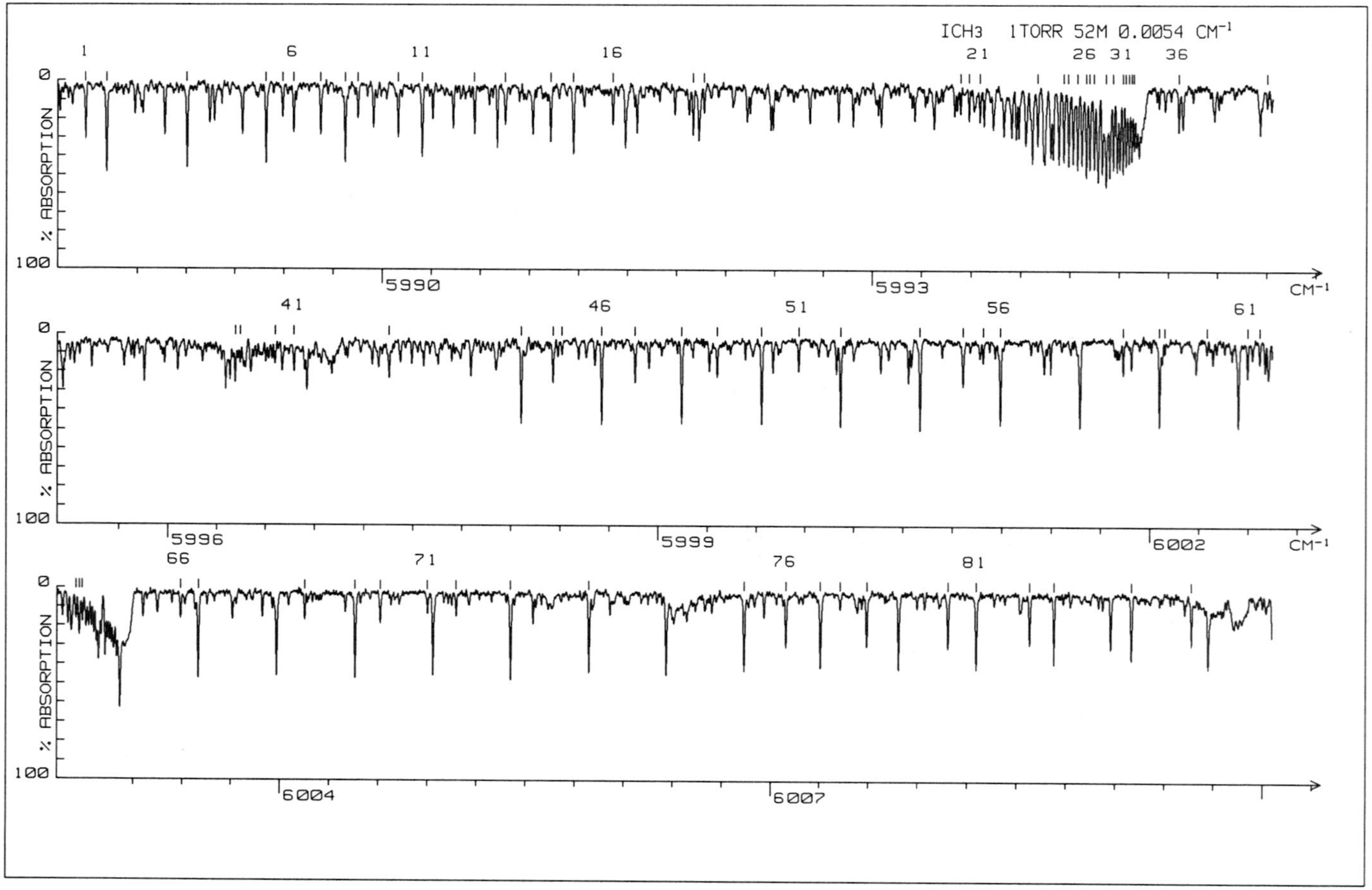
ICH3 1TORR 52M 0.0054 CM⁻¹
% ABSORPTION
0
100
5990
5993
CM⁻¹
5996
5999
6002
6004
6007
1
6
11
16
21
26
31
36
41
46
51
56
61
66
71
76
81

LINE NO	WAVENUMBER CM^{-1}, VAC	LINE NO	WAVENUMBER CM^{-1}, VAC	LINE NO	WAVENUMBER CM^{-1}, VAC
1	5988.18237(28)	31	5994.52090(25)	61	6002.59494(36)
2	5988.31113(25)	32	5994.53989(55)	62	6002.66905(48)
3	5988.80322(25)	33	5994.55681(27)	63	6002.75091(46)
4	5989.28960(45)	34	5994.57352(28)	64	6002.77051(37)
5	5989.39068(38)	35	5994.58912(34)	65	6002.78914(46)
6	5989.45813(31)	36	5994.86019(74)	66	6003.39099(44)
7	5989.62177(66)	37	5995.40080(66)	67	6003.49979(25)
8	5989.77097(25)	38	5996.41483(35)	68	6004.15411(40)
9	5989.84811(37)	39	5996.44540(65)	69	6004.45927(25)
10	5990.09512(29)	40	5996.65824(63)	70	6004.61421(38)
11	5990.24129(27)	41	5996.77423(45)	71	6004.90251(45)
12	5990.56343(30)	42	5997.35403(39)	72	6005.07885(45)
13	5990.75196(37)	43	5998.16282(42)	73	6005.41259(26)
14	5991.02688(29)	44	5998.35540(76)	74	6005.88879(41)
15	5991.16561(57)	45	5998.41090(74)	75	6006.83762(25)
16	5991.40808(34)	46	5998.65345(25)	76	6007.09741(55)
17	5991.89807(76)	47	5998.85731(32)	77	6007.30920(42)
18	5991.96557(43)	48	5999.14294(25)	78	6007.43051(58)
19	5993.53355(40)	49	5999.35965(36)	79	6007.59270(54)
20	5993.58441(33)	50	5999.63117(25)	80	6008.08719(28)
21	5993.65057(36)	51	5999.86146(38)	81	6008.25765(25)
22	5994.00081(58)	52	6000.11852(25)	82	6008.58134(30)
23	5994.16128(26)	53	6000.60462(25)	83	6008.72956(26)
24	5994.18988(25)	54	6000.86298(33)	84	6009.20046(47)
25	5994.24458(25)	55	6000.98544(44)	85	6009.56797(59)
26	5994.29763(25)	56	6001.09002(43)		
27	5994.31906(26)	57	6001.83662(34)		
28	5994.34592(25)	58	6002.05724(41)		
29	5994.41796(27)	59	6002.09026(59)		
30	5994.46202(26)	60	6002.34818(82)		

- "LINE NO.": serial number identifying the spectral line in the map on the opposite page.
- "WAVENUMBER": experimental value.
- Uncertainty in the last digit of the wavenumber is given in parentheses.

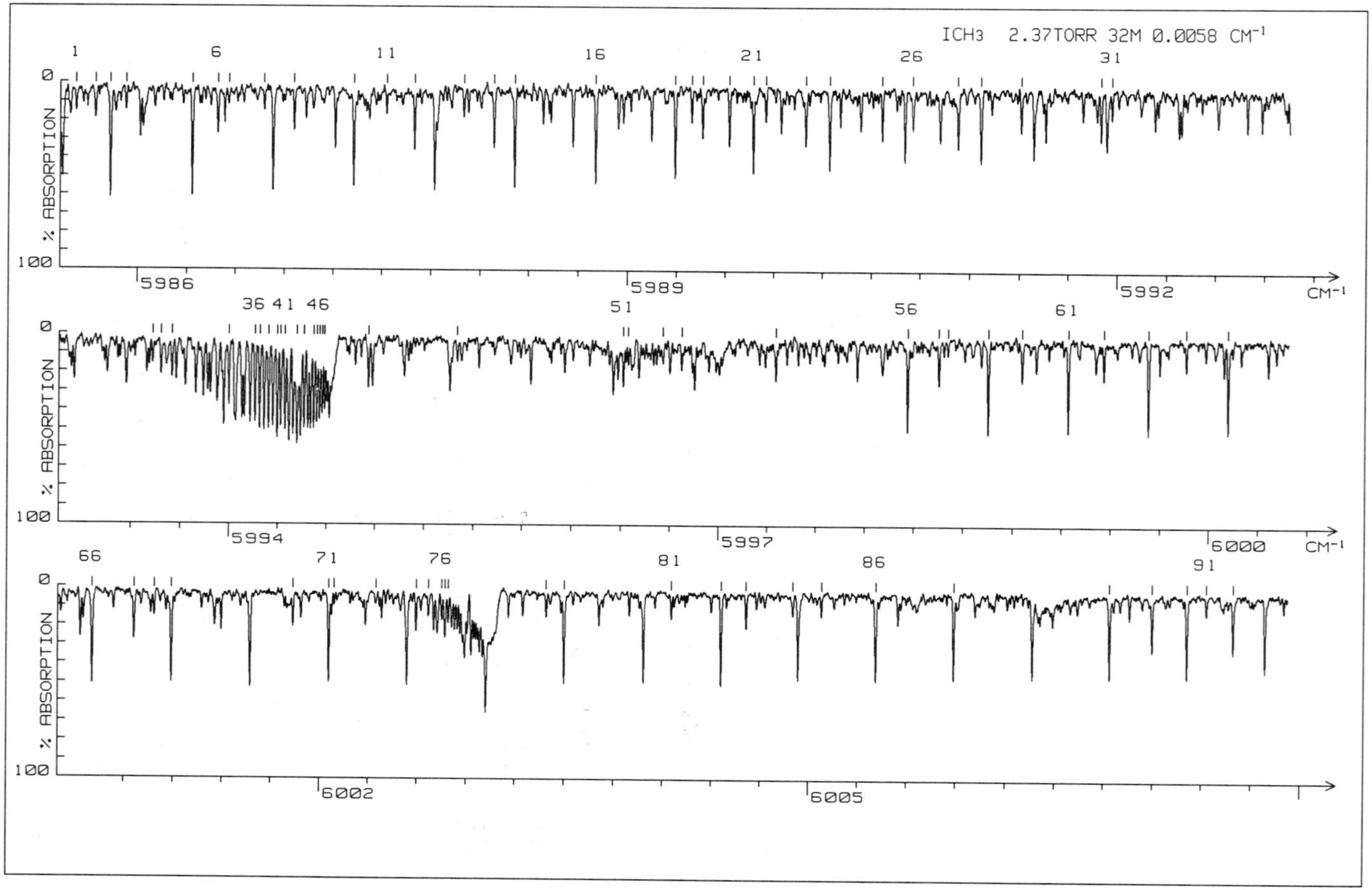
ICH3 2.37TORR 32M 0.0058 CM⁻¹
1
6
11
16
21
26
31
0
% ABSORPTION
100
5986
5989
5992
CM⁻¹
36 41 46
51
56
61
0
% ABSORPTION
100
5994
5997
6000
CM⁻¹
66
71
76
81
86
91
0
% ABSORPTION
100
6002
6005

LINE NO	WAVENUMBER CM^{-1}, VAC	LINE NO	WAVENUMBER CM^{-1}, VAC	LINE NO	WAVENUMBER CM^{-1}, VAC	LINE NO	WAVENUMBER CM^{-1}, VAC
1	5985.61883(48)	31	5991.96557(43)	61	5999.14294(25)	91	6007.43051(58)
2	5985.74081(47)	32	5993.53355(40)	62	5999.35965(36)	92	6007.59270(54)
3	5985.83292(40)	33	5993.58441(33)	63	5999.63117(25)		
4	5985.92776(53)	34	5993.65057(36)	64	5999.86146(38)		
5	5986.33069(40)	35	5994.00081(58)	65	6000.11852(25)		
6	5986.48751(31)	36	5994.16128(26)	66	6000.60462(25)		
7	5986.55725(60)	37	5994.18988(25)	67	6000.86298(33)		
8	5986.77287(47)	38	5994.24458(25)	68	6000.98544(44)		
9	5986.95685(33)	39	5994.29763(25)	69	6001.09002(43)		
10	5987.32394(25)	40	5994.31906(26)	70	6001.83662(34)		
11	5987.52770(49)	41	5994.34592(25)	71	6002.05724(41)		
12	5987.69737(61)	42	5994.41796(27)	72	6002.09026(59)		
13	5987.99733(47)	43	5994.46202(26)	73	6002.34818(82)		
14	5988.18237(28)	44	5994.52090(25)	74	6002.59494(36)		
15	5988.31113(25)	45	5994.53989(55)	75	6002.66905(48)		
16	5988.80322(25)	46	5994.55681(27)	76	6002.75091(46)		
17	5989.28960(45)	47	5994.57352(28)	77	6002.77051(37)		
18	5989.39068(38)	48	5994.58912(34)	78	6002.78914(46)		
19	5989.45813(31)	49	5994.86019(74)	79	6003.39099(44)		
20	5989.62177(66)	50	5995.40080(66)	80	6003.49979(25)		
21	5989.77097(25)	51	5996.41483(35)	81	6004.15411(40)		
22	5989.84811(37)	52	5996.44540(65)	82	6004.45927(25)		
23	5990.09512(29)	53	5996.65824(63)	83	6004.61421(38)		
24	5990.24129(27)	54	5996.77423(45)	84	6004.90251(45)		
25	5990.56343(30)	55	5997.35403(39)	85	6005.07885(45)		
26	5990.75196(37)	56	5998.16282(42)	86	6005.41259(26)		
27	5991.02688(29)	57	5998.35540(76)	87	6005.88879(41)		
28	5991.16561(57)	58	5998.41090(74)	88	6006.83762(25)		
29	5991.40808(34)	59	5998.65345(25)	89	6007.09741(55)		
30	5991.89807(76)	60	5998.85731(32)	90	6007.30920(42)		

- "LINE NO.": serial number identifying the spectral line in the map on the opposite page.
- "WAVENUMBER": experimental value.
- Uncertainty in the last digit of the wavenumber is given in parentheses.

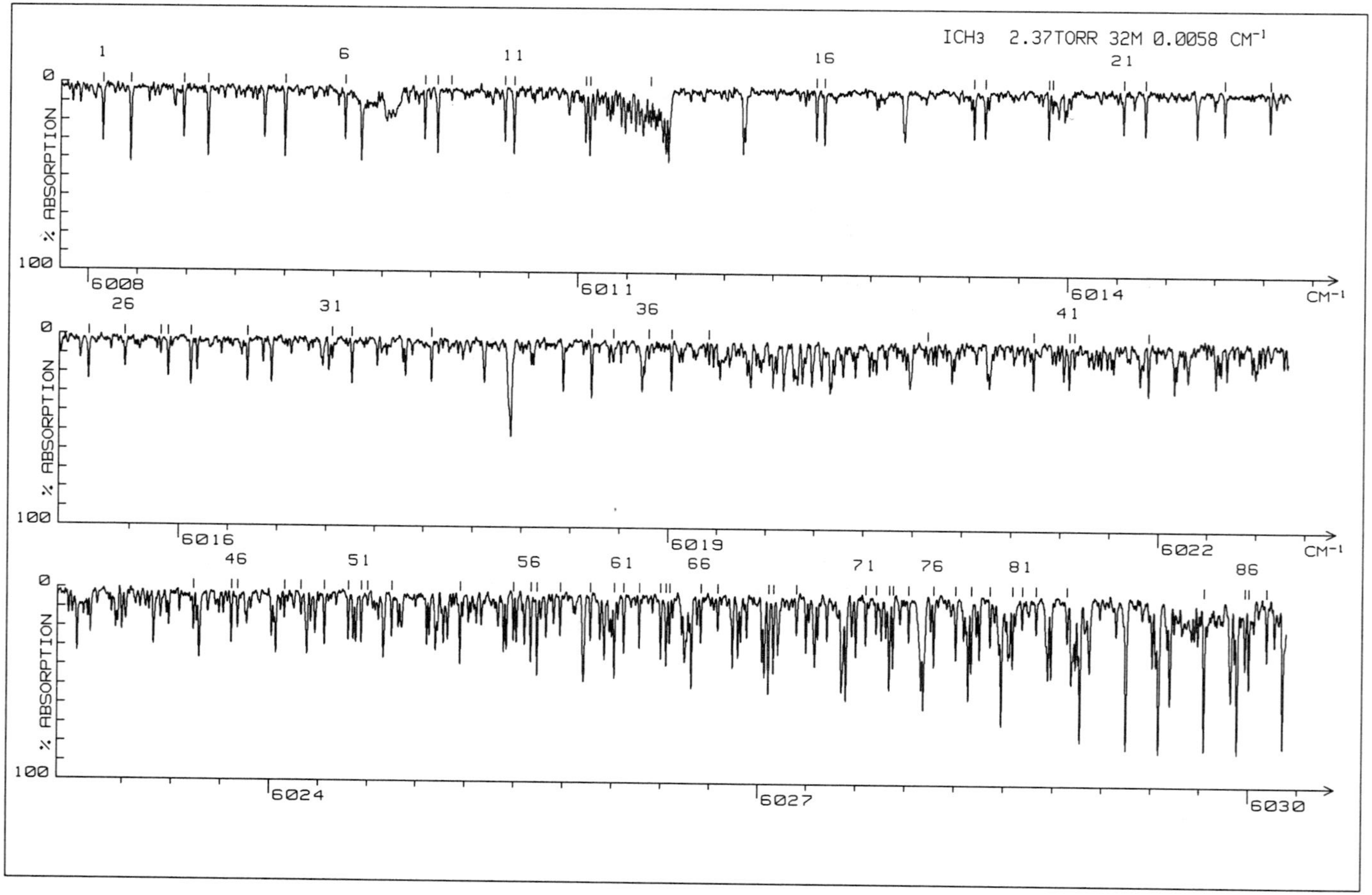

ICH3 2.37TORR 32M 0.0058 CM⁻¹
% ABSORPTION
0
100
6008
6011
6014
6016
6019
6022
6024
6027
6030
CM⁻¹
1
6
11
16
21
26
31
36
41
46
51
56
61
66
71
76
81
86

LINE NO	WAVENUMBER CM^{-1}, VAC	LINE NO	WAVENUMBER CM^{-1}, VAC	LINE NO	WAVENUMBER CM^{-1}, VAC
1	6008.08719(28)	31	6016.93459(69)	61	6026.17059(27)
2	6008.25765(25)	32	6017.05535(66)	62	6026.26505(28)
3	6008.58134(30)	33	6017.54710(31)	63	6026.39737(44)
4	6008.72956(26)	34	6018.52798(29)	64	6026.43060(26)
5	6009.20046(47)	35	6018.66031(54)	65	6026.45145(58)
6	6009.56797(59)	36	6018.87725(61)	66	6026.64477(54)
7	6010.06071(30)	37	6019.01613(30)	67	6026.74979(33)
8	6010.13978(27)	38	6019.24768(65)	68	6027.06097(34)
9	6010.22324(76)	39	6020.59182(69)	69	6027.09158(26)
10	6010.55254(31)	40	6021.23254(32)	70	6027.23630(36)
11	6010.60747(50)	41	6021.45296(35)	71	6027.65871(51)
12	6011.04414(30)	42	6021.48414(68)	72	6027.72218(35)
13	6011.07048(29)	43	6021.93703(32)	73	6027.80211(26)
14	6011.44243(60)	44	6023.53214(31)	74	6027.82420(41)
15	6012.46541(57)	45	6023.76473(28)	75	6027.92013(55)
16	6012.51637(29)	46	6023.80307(36)	76	6028.07338(26)
17	6013.42630(32)	47	6024.08813(42)	77	6028.20730(49)
18	6013.49440(70)	48	6024.19009(48)	78	6028.30196(42)
19	6013.88119(59)	49	6024.33358(29)	79	6028.41645(54)
20	6013.90490(73)	50	6024.47854(31)	80	6028.55223(42)
21	6014.33677(32)	51	6024.55868(55)	81	6028.61108(39)
22	6014.47127(32)	52	6024.59867(68)	82	6028.69660(71)
23	6014.95900(32)	53	6024.74852(60)	83	6028.88670(70)
24	6015.23691(71)	54	6025.16823(27)	84	6029.72727(23)
25	6015.44555(74)	55	6025.49492(52)	85	6029.97667(29)
26	6015.66818(44)	56	6025.60164(26)	86	6030.00026(25)
27	6015.88745(71)	57	6025.64024(25)	87	6030.11007(27)
28	6015.93191(80)	58	6025.78152(31)		
29	6016.07031(31)	59	6025.96621(56)		
30	6016.41710(31)	60	6026.11328(25)		

- "LINE NO.": serial number identifying the spectral line in the map on the opposite page.
- "WAVENUMBER": experimental value.
- Uncertainty in the last digit of the wavenumber is given in parentheses.

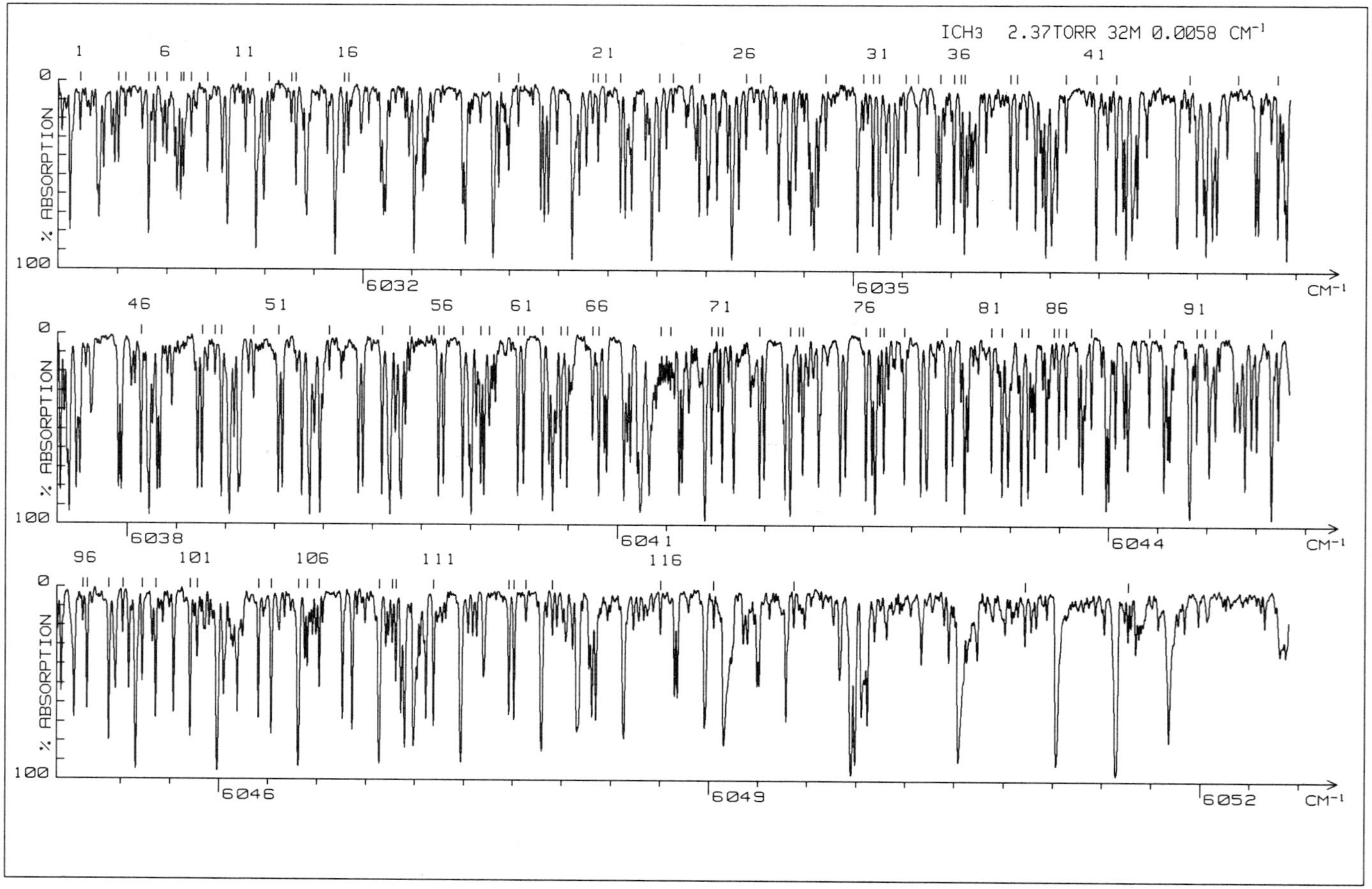
ICH3 2.37TORR 32M 0.0058 CM⁻¹
% ABSORPTION
0
100
CM⁻¹
6032
6035
6038
6041
6044
6046
6049
6052

LINE NO	WAVENUMBER CM^{-1}, VAC	LINE NO	WAVENUMBER CM^{-1}, VAC	LINE NO	WAVENUMBER CM^{-1}, VAC	LINE NO	WAVENUMBER CM^{-1}, VAC
1	6030.26956(31)	31	6035.15823(23)	61	6040.42837(23)	91	6044.53618(24)
2	6030.50529(25)	32	6035.31949(27)	62	6040.54393(23)	92	6044.58604(70)
3	6030.54630(45)	33	6035.39908(25)	63	6040.65291(23)	93	6044.65027(25)
4	6030.68886(30)	34	6035.53408(24)	64	6040.69161(24)	94	6044.99136(29)
5	6030.72627(25)	35	6035.61610(23)	65	6040.84652(25)	95	6045.16445(39)
6	6030.79828(47)	36	6035.65673(31)	66	6040.88431(23)	96	6045.19054(24)
7	6030.88472(24)	37	6035.68008(24)	67	6041.26193(38)	97	6045.32635(23)
8	6030.90173(28)	38	6035.95671(24)	68	6041.32154(34)	98	6045.41041(32)
9	6030.94828(29)	39	6035.99820(23)	69	6041.57345(24)	99	6045.52871(25)
10	6031.04736(25)	40	6036.29851(29)	70	6041.61449(38)	100	6045.61084(24)
11	6031.28120(42)	41	6036.48376(23)	71	6041.63894(28)	101	6045.82216(23)
12	6031.42515(51)	42	6036.60384(23)	72	6041.86784(28)	102	6045.86445(29)
13	6031.55997(37)	43	6037.05018(32)	73	6042.05744(30)	103	6046.24038(24)
14	6031.58912(24)	44	6037.34693(58)	74	6042.10857(26)	104	6046.31861(28)
15	6031.88294(26)	45	6037.58908(28)	75	6042.13273(24)	105	6046.48420(31)
16	6031.91078(29)	46	6038.08173(24)	76	6042.51778(23)	106	6046.53905(38)
17	6032.83200(24)	47	6038.45546(23)	77	6042.60247(27)	107	6046.61312(24)
18	6032.95450(57)	48	6038.53412(39)	78	6042.62717(24)	108	6046.98191(32)
19	6033.40629(45)	49	6038.57372(23)	79	6042.75394(28)	109	6047.06081(31)
20	6033.43927(25)	50	6038.77066(28)	80	6043.01196(23)	110	6047.08370(25)
21	6033.48404(70)	51	6038.92211(29)	81	6043.28374(23)	111	6047.31231(28)
22	6033.57520(24)	52	6039.23398(68)	82	6043.34729(23)	112	6047.77578(24)
23	6033.81329(30)	53	6039.55856(23)	83	6043.46532(24)	113	6047.80973(24)
24	6033.89713(51)	54	6039.72561(35)	84	6043.50668(23)	114	6047.88229(48)
25	6034.05843(24)	55	6039.90244(23)	85	6043.66392(53)	115	6048.04313(54)
26	6034.34348(30)	56	6039.93312(29)	86	6043.69398(25)	116	6048.70118(31)
27	6034.42981(29)	57	6040.05097(28)	87	6043.73847(32)	117	6049.02632(36)
28	6034.83147(27)	58	6040.16084(24)	88	6043.89374(25)	118	6049.51810(37)
29	6035.05971(32)	59	6040.21446(29)	89	6044.24583(25)	119	6050.92691(51)
30	6035.11982(23)	60	6040.39297(23)	90	6044.33616(23)	120	6051.55703(55)

- "LINE NO.": serial number identifying the spectral line in the map on the opposite page.
- "WAVENUMBER": experimental value.
- Uncertainty in the last digit of the wavenumber is given in parentheses.

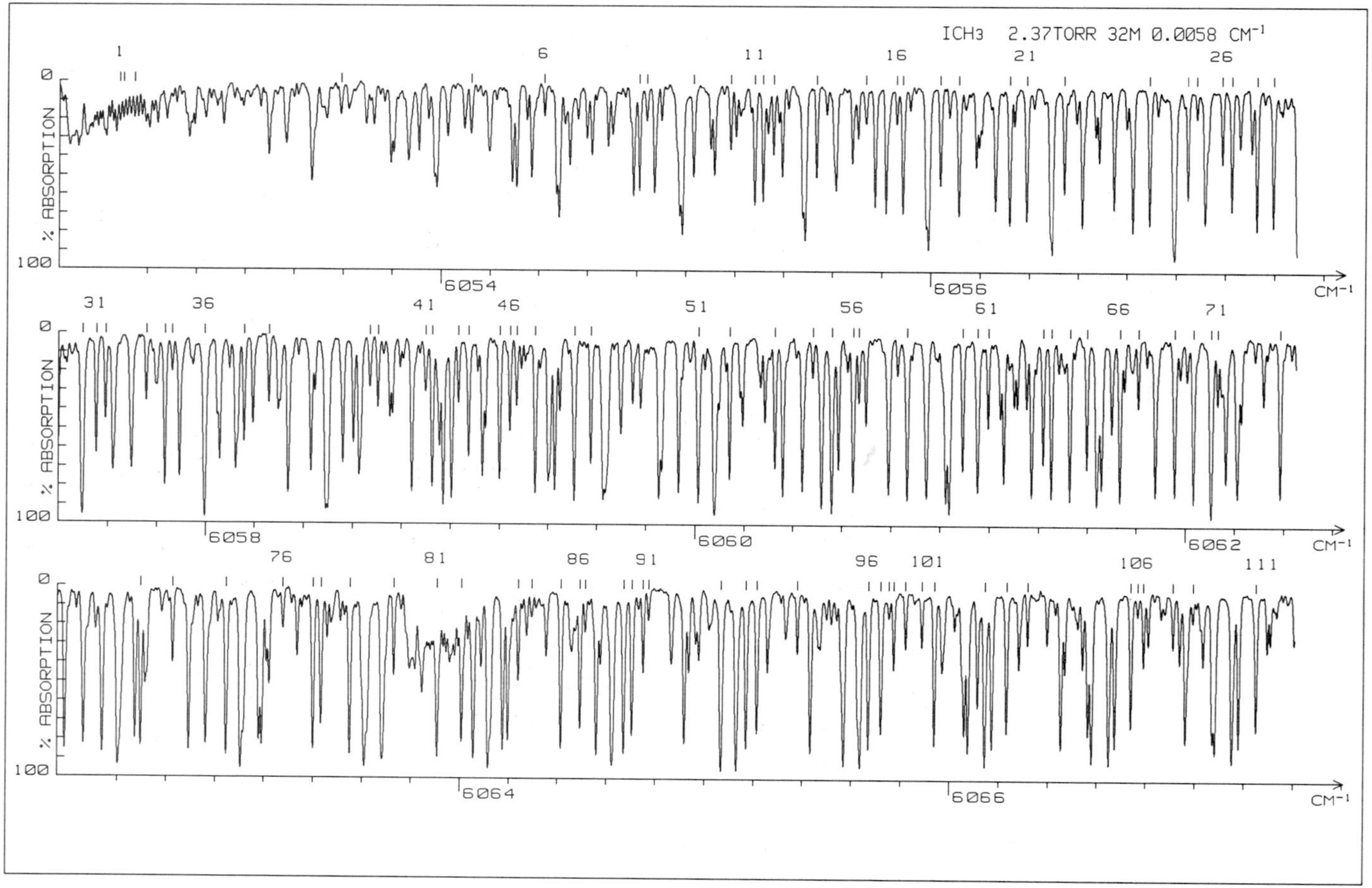
ICH3 2.37TORR 32M 0.0058 CM⁻¹
% ABSORPTION
0
100
6054
6056
6058
6060
6062
6064
6066
CM⁻¹

LINE NO	WAVENUMBER CM^{-1}, VAC	LINE NO	WAVENUMBER CM^{-1}, VAC	LINE NO	WAVENUMBER CM^{-1}, VAC	LINE NO	WAVENUMBER CM^{-1}, VAC
1	6052.68842(44)	31	6057.55253(24)	61	6061.19473(25)	91	6064.77247(67)
2	6052.70276(47)	32	6057.59072(25)	62	6061.41754(24)	92	6065.06647(27)
3	6052.74776(37)	33	6057.75705(26)	63	6061.45108(23)	93	6065.16920(23)
4	6053.59109(39)	34	6057.83327(23)	64	6061.52827(23)	94	6065.21290(23)
5	6054.12230(36)	35	6057.86327(64)	65	6061.59743(23)	95	6065.37970(26)
6	6054.42428(61)	36	6057.99588(28)	66	6061.73343(23)	96	6065.66970(23)
7	6054.81230(24)	37	6058.15768(24)	67	6061.80864(26)	97	6065.72040(24)
8	6054.84293(68)	38	6058.25947(26)	68	6061.95625(23)	98	6065.75258(61)
9	6055.03219(25)	39	6058.67178(30)	69	6062.03335(26)	99	6065.77211(25)
10	6055.18257(26)	40	6058.70423(26)	70	6062.10495(27)	100	6065.81999(38)
11	6055.28128(29)	41	6058.89837(31)	71	6062.13144(41)	101	6065.88747(28)
12	6055.31526(29)	42	6058.92493(23)	72	6062.38495(23)	102	6065.93885(23)
13	6055.35816(26)	43	6059.03229(26)	73	6062.69569(24)	103	6066.14312(24)
14	6055.53602(24)	44	6059.07397(24)	74	6062.82624(34)	104	6066.23185(23)
15	6055.73637(36)	45	6059.19974(24)	75	6063.04422(23)	105	6066.31639(27)
16	6055.86180(34)	46	6059.24354(24)	76	6063.27776(32)	106	6066.74022(23)
17	6055.88641(24)	47	6059.27071(36)	77	6063.39959(23)	107	6066.76729(69)
18	6056.03985(30)	48	6059.34765(27)	78	6063.43222(24)	108	6066.79117(27)
19	6056.11458(24)	49	6059.50843(23)	79	6063.54956(23)	109	6066.91092(28)
20	6056.32110(24)	50	6059.57572(23)	80	6063.73063(25)	110	6066.99157(44)
21	6056.38939(23)	51	6060.01307(27)	81	6063.90689(24)	111	6067.24904(24)
22	6056.54390(24)	52	6060.14004(27)	82	6064.00657(28)		
23	6056.89280(23)	53	6060.32529(24)	83	6064.23898(25)		
24	6057.04809(24)	54	6060.48180(33)	84	6064.29476(53)		
25	6057.08610(64)	55	6060.55955(27)	85	6064.41487(23)		
26	6057.18876(25)	56	6060.64675(24)	86	6064.49349(23)		
27	6057.22683(24)	57	6060.66978(27)	87	6064.51453(76)		
28	6057.32888(23)	58	6060.86499(23)	88	6064.67043(23)		
29	6057.39563(23)	59	6061.09187(24)	89	6064.70396(23)		
30	6057.49475(29)	60	6061.15200(28)	90	6064.74931(25)		

- "LINE NO.": serial number identifying the spectral line in the map on the opposite page.
- "WAVENUMBER": experimental value.
- Uncertainty in the last digit of the wavenumber is given in parentheses.

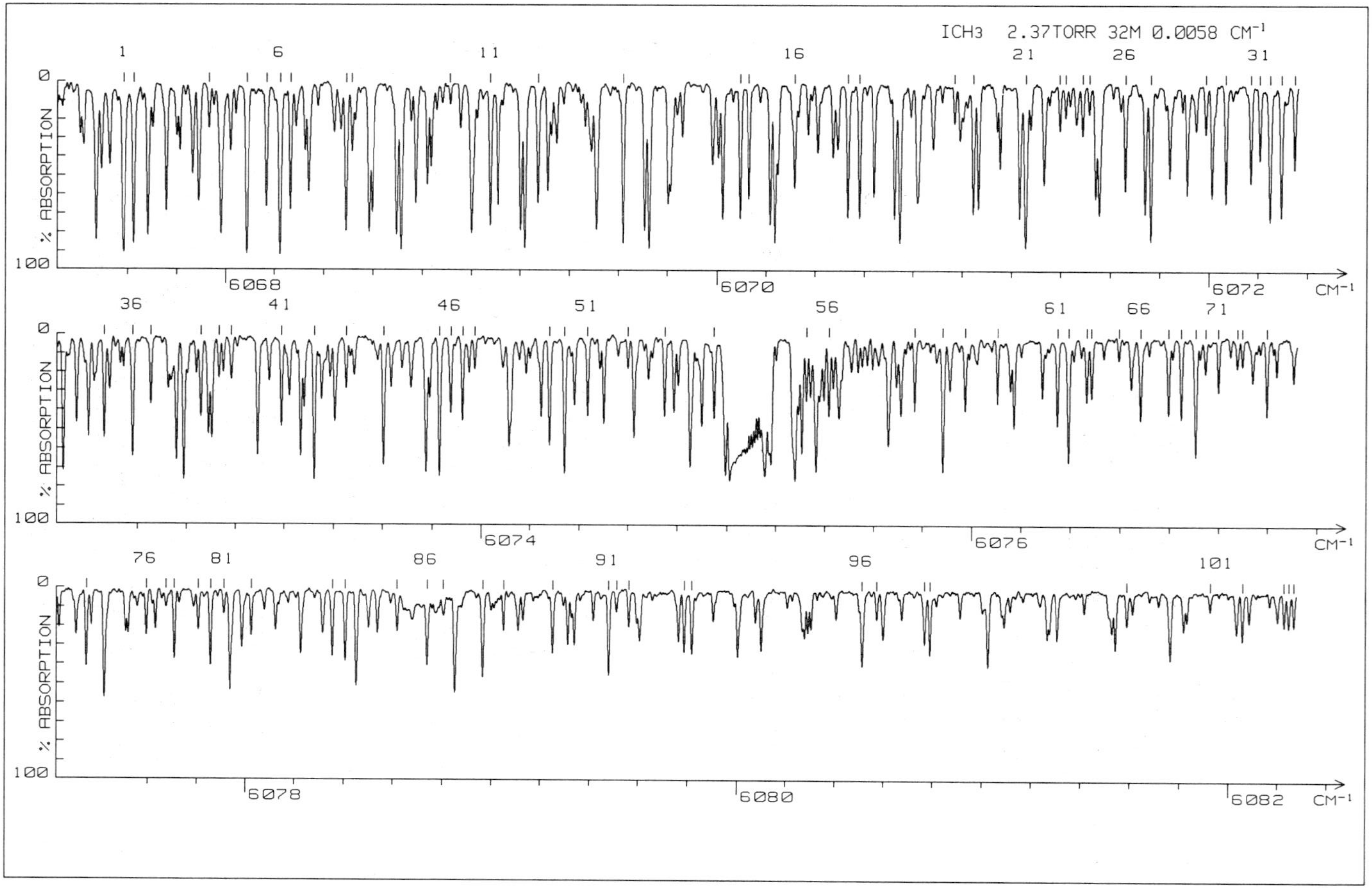
ICH3 2.37TORR 32M 0.0058 CM-1
1
6
11
16
21
26
31
0
% ABSORPTION
100
6068
6070
6072
CM-1
36
41
46
51
56
61
66
71
0
% ABSORPTION
100
6074
6076
CM-1
76
81
86
91
96
101
0
% ABSORPTION
100
6078
6080
6082
CM-1

LINE NO	WAVENUMBER CM^{-1}, VAC	LINE NO	WAVENUMBER CM^{-1}, VAC	LINE NO	WAVENUMBER CM^{-1}, VAC	LINE NO	WAVENUMBER CM^{-1}, VAC
1	6067.58293(28)	31	6072.20834(25)	61	6076.34789(25)	91	6079.47935(25)
2	6067.62480(26)	32	6072.24964(24)	62	6076.39265(24)	92	6079.51142(52)
3	6067.93198(43)	33	6072.29538(23)	63	6076.46572(26)	93	6079.56294(46)
4	6068.08733(26)	34	6072.34899(25)	64	6076.48536(28)	94	6079.78701(34)
5	6068.16857(24)	35	6072.46263(28)	65	6076.59470(45)	95	6079.81765(36)
6	6068.22409(26)	36	6072.58013(24)	66	6076.68415(31)	96	6080.51113(29)
7	6068.26637(23)	37	6072.65335(25)	67	6076.79567(25)	97	6080.57246(38)
8	6068.49078(23)	38	6072.85764(32)	68	6076.84684(25)	98	6080.76816(39)
9	6068.51493(27)	39	6072.93105(46)	69	6076.90652(27)	99	6080.79003(27)
10	6068.91274(45)	40	6072.98297(33)	70	6076.94598(33)	100	6081.58816(54)
11	6069.07711(23)	41	6073.18888(29)	71	6076.99862(27)	101	6081.92515(54)
12	6069.27361(24)	42	6073.32150(23)	72	6077.07575(45)	102	6082.05587(29)
13	6069.62018(23)	43	6073.45034(28)	73	6077.09588(60)	103	6082.22650(35)
14	6070.09357(23)	44	6073.60392(26)	74	6077.19690(25)	104	6082.24492(51)
15	6070.13011(24)	45	6073.82918(23)	75	6077.34711(25)	105	6082.26588(33)
16	6070.31818(25)	46	6073.87456(27)	76	6077.59399(29)		
17	6070.53389(23)	47	6073.92318(25)	77	6077.67419(52)		
18	6070.58118(23)	48	6073.97308(34)	78	6077.70863(31)		
19	6070.97008(30)	49	6074.27984(24)	79	6077.80589(46)		
20	6071.04519(26)	50	6074.34069(23)	80	6077.85563(25)		
21	6071.25791(26)	51	6074.43413(25)	81	6077.91018(73)		
22	6071.39723(29)	52	6074.60009(55)	82	6078.02519(40)		
23	6071.42032(33)	53	6074.74983(25)	83	6078.35355(26)		
24	6071.48879(28)	54	6074.94758(26)	84	6078.40561(26)		
25	6071.51634(38)	55	6075.32553(28)	85	6078.61765(49)		
26	6071.66209(28)	56	6075.41701(33)	86	6078.73824(34)		
27	6071.76504(23)	57	6075.76778(26)	87	6078.80442(75)		
28	6071.98655(42)	58	6075.88308(23)	88	6078.96466(25)		
29	6072.06835(23)	59	6075.97263(33)	89	6079.05134(45)		
30	6072.17120(29)	60	6076.10346(26)	90	6079.25131(27)		

- "LINE NO.": serial number identifying the spectral line in the map on the opposite page.
- "WAVENUMBER": experimental value.
- Uncertainty in the last digit of the wavenumber is given in parentheses.

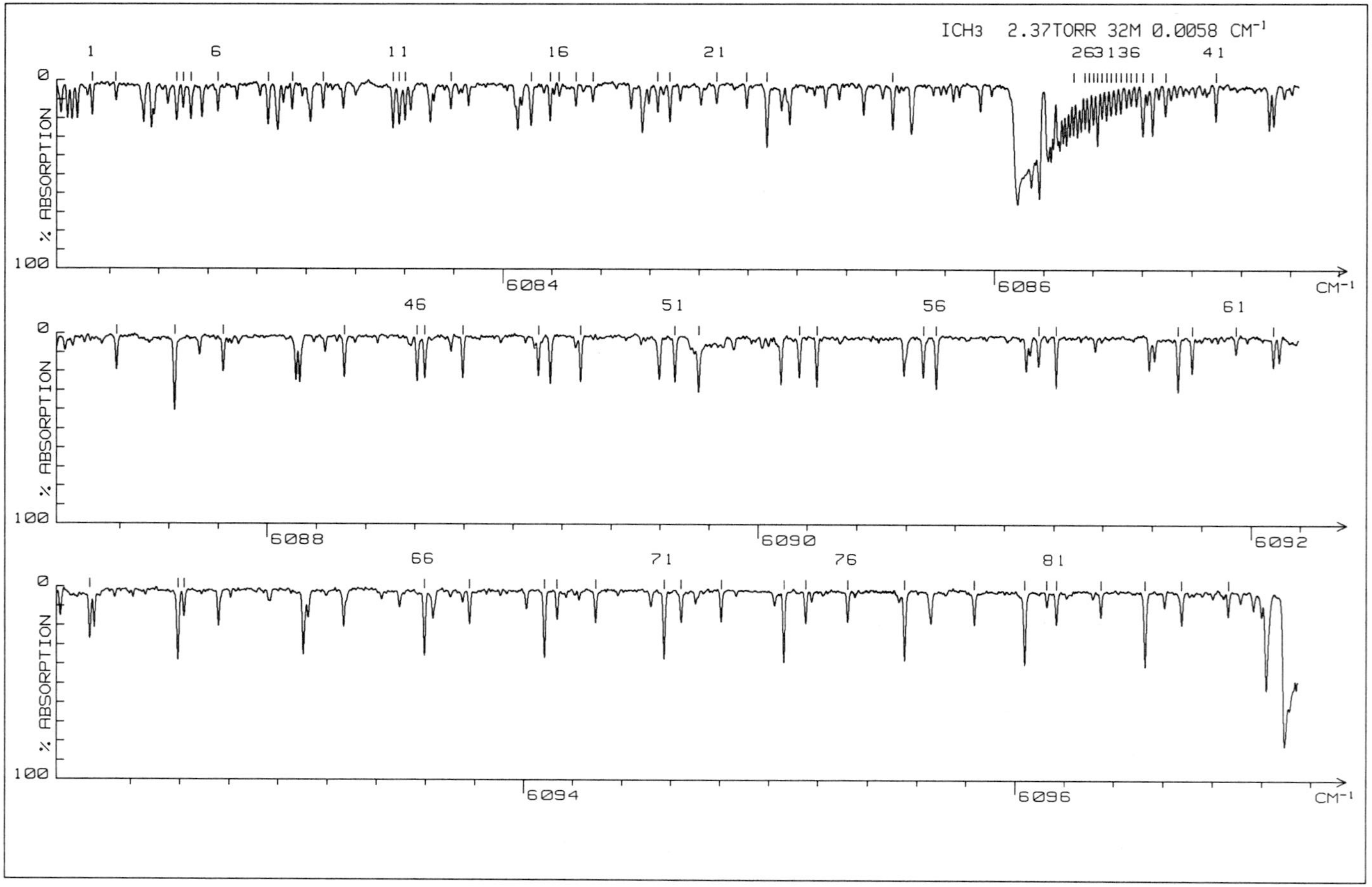
ICH3 2.37TORR 32M 0.0058 CM⁻¹
1
6
11
16
21
26 31 36
41
46
51
56
61
66
71
76
81
0
% ABSORPTION
100
6084
6086
6088
6090
6092
6094
6096
CM⁻¹

LINE NO	WAVENUMBER CM^{-1}, VAC	LINE NO	WAVENUMBER CM^{-1}, VAC	LINE NO	WAVENUMBER CM^{-1}, VAC
1	6082.32887(33)	31	6086.45488(33)	61	6091.93750(52)
2	6082.42545(55)	32	6086.47336(57)	62	6092.08842(33)
3	6082.67378(37)	33	6086.49318(54)	63	6092.23251(30)
4	6082.69962(66)	34	6086.51273(56)	64	6092.59339(32)
5	6082.73097(35)	35	6086.53648(40)	65	6092.61802(60)
6	6082.83984(47)	36	6086.55412(42)	66	6093.59671(26)
7	6083.04607(34)	37	6086.57617(45)	67	6093.77938(32)
8	6083.14381(47)	38	6086.60243(31)	68	6094.08572(25)
9	6083.26879(36)	39	6086.64096(37)	69	6094.13742(54)
10	6083.55325(43)	40	6086.69264(40)	70	6094.29350(49)
11	6083.57844(49)	41	6086.89719(31)	71	6094.57399(25)
12	6083.60183(47)	42	6087.38574(32)	72	6094.64395(33)
13	6083.78841(46)	43	6087.62290(25)	73	6094.80639(49)
14	6084.11557(33)	44	6087.81950(32)	74	6095.06226(25)
15	6084.19274(30)	45	6088.31286(41)	75	6095.15159(49)
16	6084.23027(63)	46	6088.60910(40)	76	6095.32084(33)
17	6084.29849(64)	47	6088.64030(32)	77	6095.55053(30)
18	6084.36663(56)	48	6088.79453(29)	78	6095.83470(47)
19	6084.63126(44)	49	6089.10322(33)	79	6096.03908(25)
20	6084.68115(31)	50	6089.27539(38)	80	6096.12841(66)
21	6084.87150(86)	51	6089.66045(41)	81	6096.16726(49)
22	6084.99462(65)	52	6089.75683(28)	82	6096.34932(36)
23	6085.07668(26)	53	6090.16813(30)	83	6096.52844(25)
24	6085.58549(41)	54	6090.23899(28)	84	6096.67511(38)
25	6086.32147(34)	55	6090.66946(43)	85	6096.86419(37)
26	6086.36804(32)	56	6090.72266(27)		
27	6086.38446(28)	57	6091.13685(55)		
28	6086.40175(34)	58	6091.20886(35)		
29	6086.41876(26)	59	6091.70235(27)		
30	6086.43647(35)	60	6091.75905(32)		

- "LINE NO.": serial number identifying the spectral line in the map on the opposite page.
- "WAVENUMBER": experimental value.
- Uncertainty in the last digit of the wavenumber is given in parentheses.

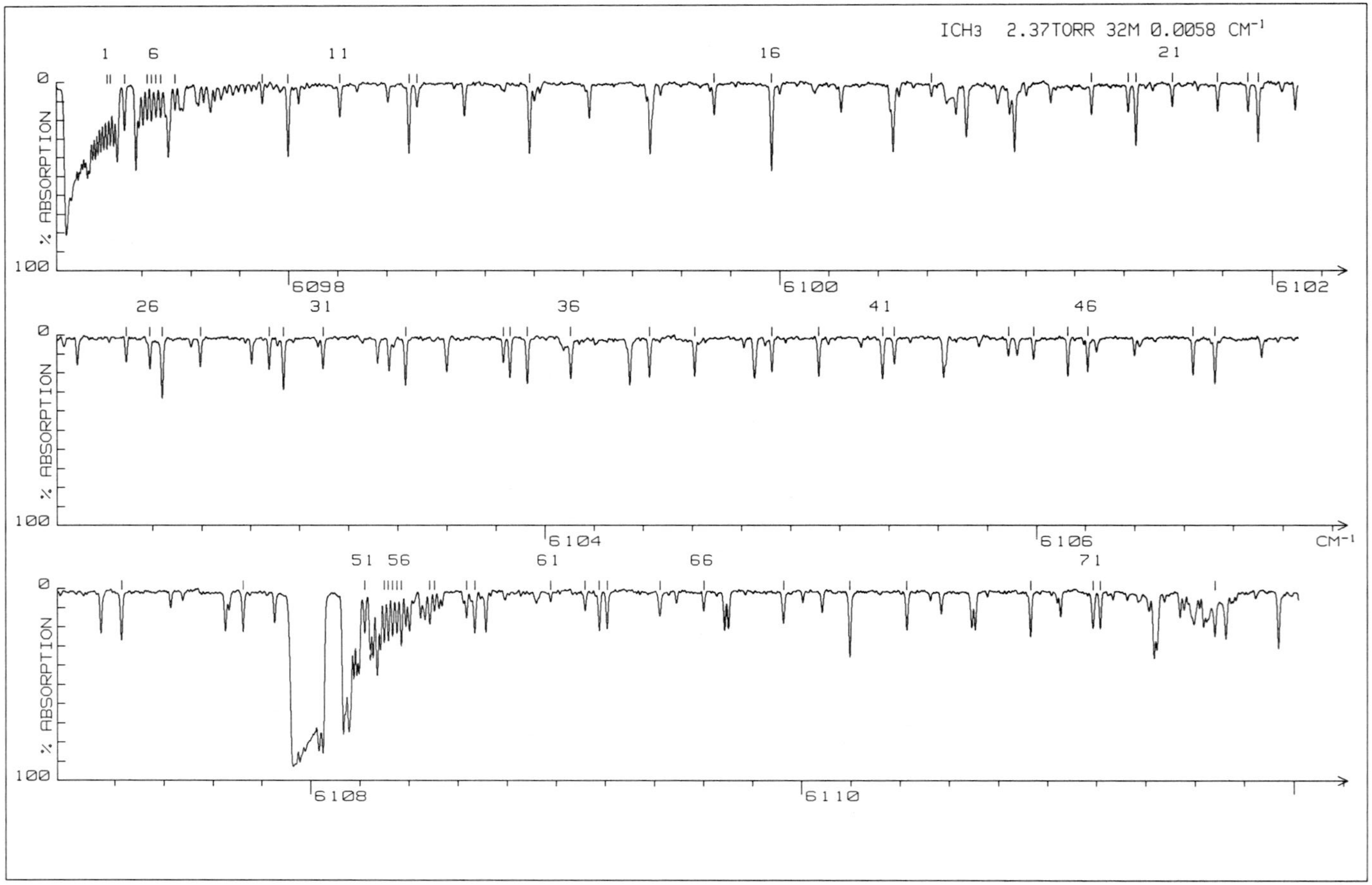
ICH3 2.37TORR 32M 0.0058 CM⁻¹
1
6
11
16
21
26
31
36
41
46
51
56
61
66
71
0
% ABSORPTION
100
6098
6100
6102
6104
6106
6108
6110
CM⁻¹

LINE NO	WAVENUMBER CM^{-1}, VAC	LINE NO	WAVENUMBER CM^{-1}, VAC	LINE NO	WAVENUMBER CM^{-1}, VAC
1	6097.25985(33)	31	6103.09685(38)	61	6108.97757(57)
2	6097.27261(35)	32	6103.43149(28)	62	6109.11870(69)
3	6097.33041(29)	33	6103.83006(55)	63	6109.17696(30)
4	6097.42214(30)	34	6103.85643(29)	64	6109.20841(41)
5	6097.44008(33)	35	6103.92870(28)	65	6109.42460(53)
6	6097.45854(51)	36	6104.10454(43)	66	6109.60153(41)
7	6097.47742(51)	37	6104.42587(41)	67	6109.92612(52)
8	6097.53583(58)	38	6104.60982(31)	68	6110.19664(26)
9	6097.89317(66)	39	6104.92368(33)	69	6110.42818(41)
10	6097.99869(25)	40	6105.11561(31)	70	6110.93099(28)
11	6098.20929(34)	41	6105.37378(33)	71	6111.18683(47)
12	6098.49025(25)	42	6105.42185(38)	72	6111.21618(30)
13	6098.52243(50)	43	6105.88497(53)	73	6111.68178(36)
14	6098.98176(25)	44	6105.98740(67)		
15	6099.73398(50)	45	6106.12848(40)		
16	6099.96713(24)	46	6106.20945(31)		
17	6100.61812(62)	47	6106.63531(41)		
18	6101.26930(35)	48	6106.72598(28)		
19	6101.41742(53)	49	6107.22837(36)		
20	6101.44862(26)	50	6107.72470(30)		
21	6101.59510(42)	51	6108.22043(31)		
22	6101.78036(36)	52	6108.30044(35)		
23	6101.90275(35)	53	6108.31697(28)		
24	6101.94357(26)	54	6108.33420(29)		
25	6102.29244(33)	55	6108.35161(31)		
26	6102.38873(41)	56	6108.36936(27)		
27	6102.43880(26)	57	6108.48587(36)		
28	6102.59369(50)	58	6108.50486(71)		
29	6102.87647(33)	59	6108.63538(47)		
30	6102.93506(34)	60	6108.66865(35)		

- "LINE NO.": serial number identifying the spectral line in the map on the opposite page.
- "WAVENUMBER": experimental value.
- Uncertainty in the last digit of the wavenumber is given in parentheses.

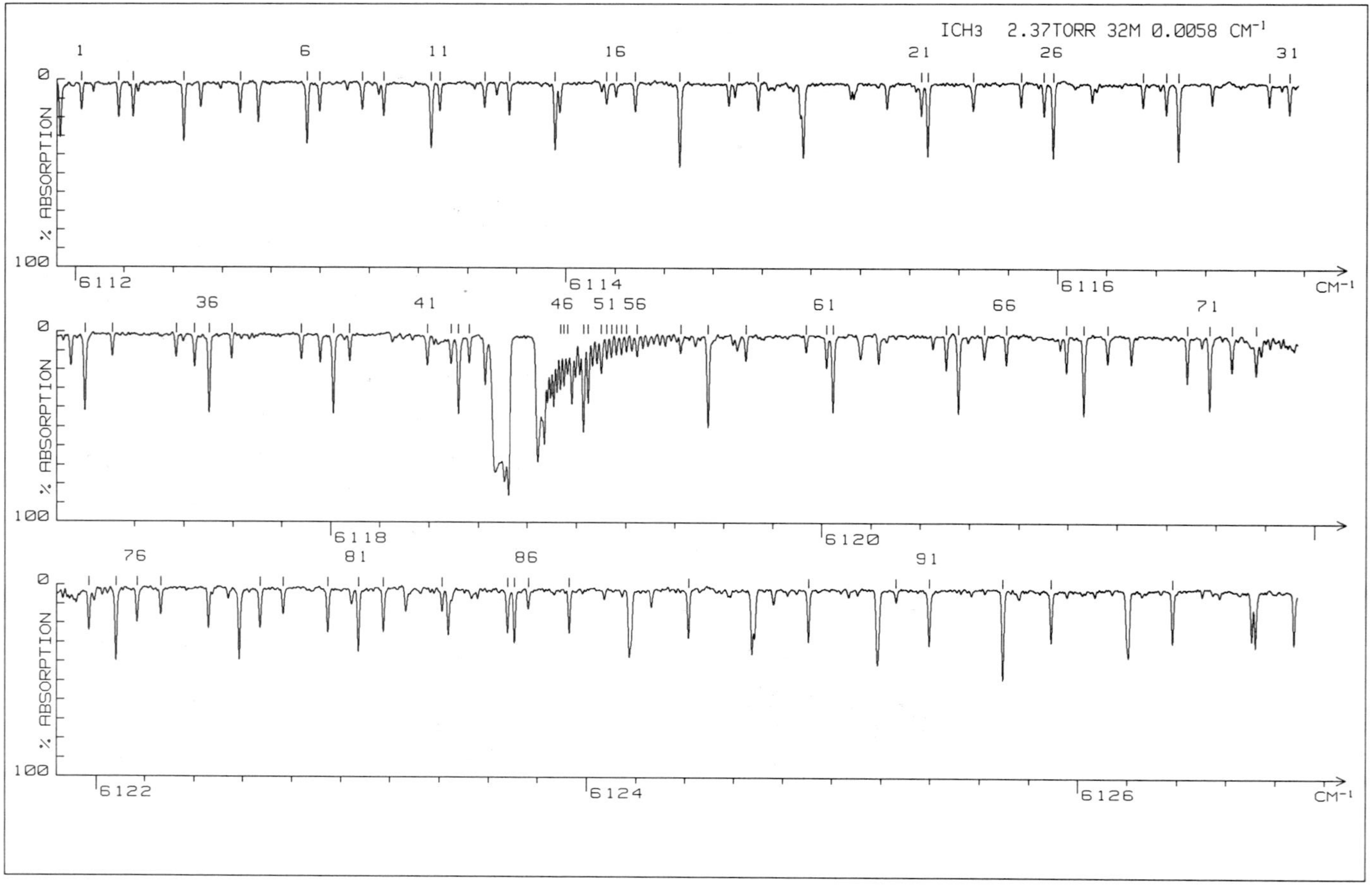
ICH3 2.37TORR 32M 0.0058 CM-1
% ABSORPTION
0
100
6112
6114
6116
6118
6120
6122
6124
6126
CM-1
1
6
11
16
21
26
31
36
41
46
51
56
61
66
71
76
81
86
91

LINE NO	WAVENUMBER CM^{-1}, VAC	LINE NO	WAVENUMBER CM^{-1}, VAC	LINE NO	WAVENUMBER CM^{-1}, VAC	LINE NO	WAVENUMBER CM^{-1}, VAC
1	6112.02410(51)	31	6116.93882(33)	61	6120.02089(42)	91	6125.39747(26)
2	6112.17791(31)	32	6116.99561(26)	62	6120.04684(25)	92	6125.69399(26)
3	6112.23708(43)	33	6117.10955(56)	63	6120.50743(40)	93	6125.88971(31)
4	6112.44181(31)	34	6117.36868(38)	64	6120.55624(27)	94	6126.38245(31)
5	6112.67401(33)	35	6117.44384(32)	65	6120.66083(59)		
6	6112.94606(25)	36	6117.50315(24)	66	6120.74944(34)		
7	6112.99749(51)	37	6117.59541(36)	67	6120.99428(29)		
8	6113.17018(37)	38	6117.88139(38)	68	6121.06550(24)		
9	6113.25879(32)	39	6118.01120(24)	69	6121.16251(37)		
10	6113.45090(30)	40	6118.07751(36)	70	6121.48208(27)		
11	6113.48594(38)	41	6118.39410(35)	71	6121.57433(28)		
12	6113.66732(57)	42	6118.48949(54)	72	6121.66427(42)		
13	6113.76977(43)	43	6118.51974(24)	73	6121.76243(43)		
14	6113.95611(29)	44	6118.56306(49)	74	6121.96937(37)		
15	6114.16482(55)	45	6118.93579(36)	75	6122.08207(25)		
16	6114.20306(66)	46	6118.95057(31)	76	6122.16860(36)		
17	6114.28002(48)	47	6118.96435(50)	77	6122.26403(35)		
18	6114.46254(24)	48	6119.02904(24)	78	6122.67079(38)		
19	6114.66334(62)	49	6119.04837(30)	79	6122.76462(51)		
20	6114.78280(34)	50	6119.10137(32)	80	6122.94660(28)		
21	6115.44836(44)	51	6119.12358(52)	81	6123.07092(25)		
22	6115.47399(25)	52	6119.14208(57)	82	6123.17387(28)		
23	6115.65686(39)	53	6119.16211(53)	83	6123.41152(54)		
24	6115.84963(40)	54	6119.18307(62)	84	6123.67728(36)		
25	6115.94265(32)	55	6119.20365(48)	85	6123.70684(26)		
26	6115.98074(25)	56	6119.24555(63)	86	6123.76341(45)		
27	6116.34598(53)	57	6119.42443(77)	87	6123.92955(34)		
28	6116.43918(41)	58	6119.53695(24)	88	6124.41489(32)		
29	6116.48797(24)	59	6119.69158(36)	89	6124.90617(27)		
30	6116.85631(55)	60	6119.93830(50)	90	6125.26050(91)		

- "LINE NO.": serial number identifying the spectral line in the map on the opposite page.
- "WAVENUMBER": experimental value.
- Uncertainty in the last digit of the wavenumber is given in parentheses.

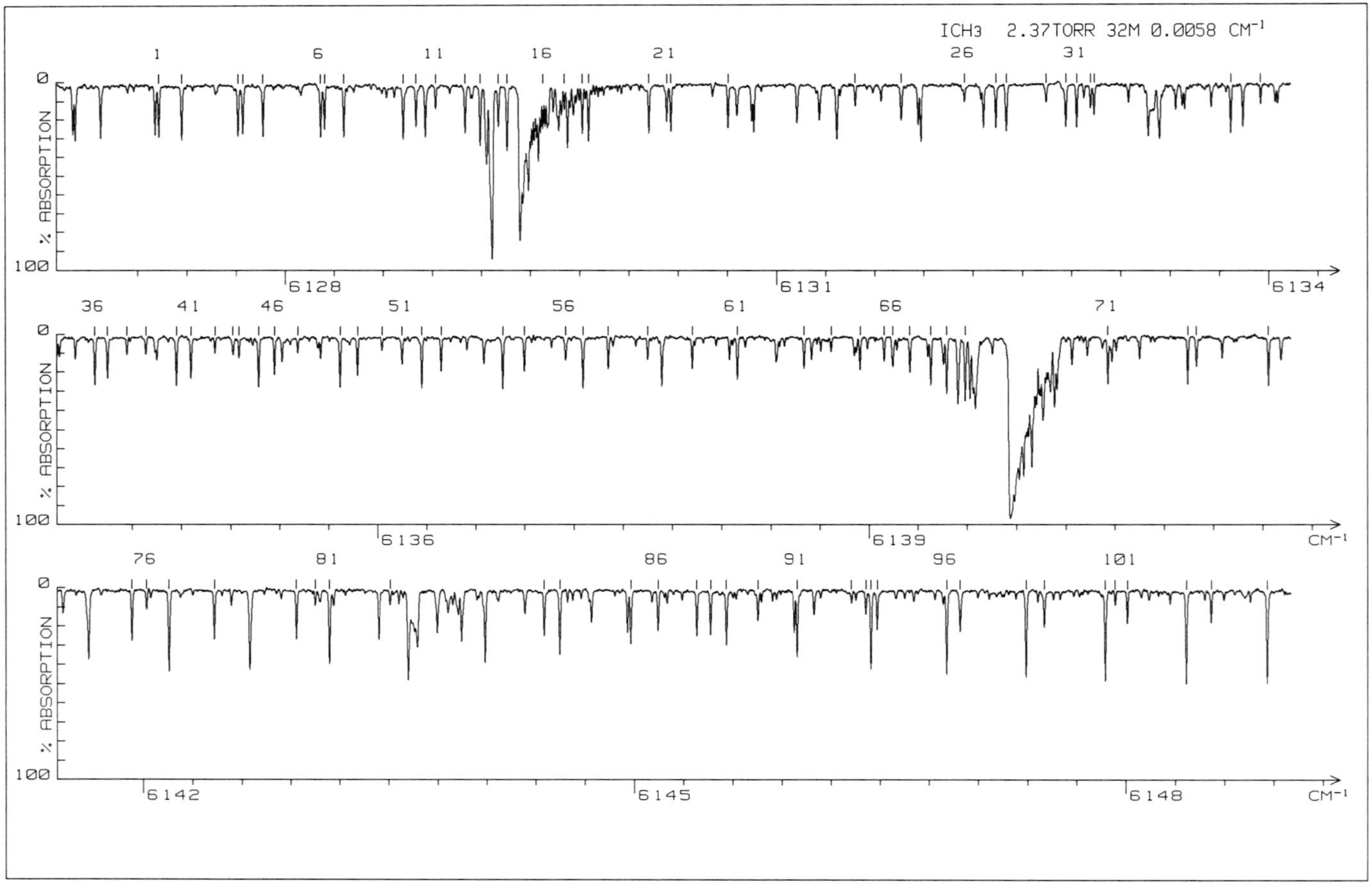
ICH3 2.37TORR 32M 0.0058 CM⁻¹
1
6
11
16
21
26
31
0
% ABSORPTION
100
6128
6131
6134
36
41
46
51
56
61
66
71
0
% ABSORPTION
100
6136
6139
CM⁻¹
76
81
86
91
96
101
0
% ABSORPTION
100
6142
6145
6148
CM⁻¹

LINE NO	WAVENUMBER CM^{-1}, VAC	LINE NO	WAVENUMBER CM^{-1}, VAC	LINE NO	WAVENUMBER CM^{-1}, VAC	LINE NO	WAVENUMBER CM^{-1}, VAC
1	6127.23149(31)	31	6132.83230(28)	61	6138.19579(28)	91	6145.99429(28)
2	6127.37001(30)	32	6132.91657(52)	62	6138.60266(32)	92	6146.32513(58)
3	6127.71343(27)	33	6132.93870(33)	63	6138.76778(54)	93	6146.41262(52)
4	6127.74287(26)	34	6133.76905(27)	64	6138.94414(40)	94	6146.44458(25)
5	6127.86489(26)	35	6133.95174(54)	65	6139.09250(37)	95	6146.48291(34)
6	6128.21802(26)	36	6134.27023(27)	66	6139.14573(44)	96	6146.91024(26)
7	6128.24249(34)	37	6134.35027(29)	67	6139.25010(31)	97	6146.99197(31)
8	6128.36095(26)	38	6134.46901(47)	68	6139.37943(27)	98	6147.39051(24)
9	6128.72341(29)	39	6134.58490(75)	69	6139.47629(26)	99	6147.50080(37)
10	6128.80121(28)	40	6134.77100(27)	70	6139.58770(26)	100	6147.87562(24)
11	6128.92017(53)	41	6134.85849(28)	71	6140.45711(33)	101	6147.93574(74)
12	6129.10091(26)	42	6135.00428(68)	72	6140.94090(27)	102	6148.01016(38)
13	6129.19351(26)	43	6135.11386(78)	73	6140.99362(33)	103	6148.36633(24)
14	6129.30425(29)	44	6135.15193(58)	74	6141.43384(26)	104	6148.51955(40)
15	6129.35721(30)	45	6135.27139(32)	75	6141.93169(26)	105	6148.86118(24)
16	6129.57684(36)	46	6135.36767(29)	76	6142.02138(45)		
17	6129.70729(65)	47	6135.51122(73)	77	6142.15893(25)		
18	6129.81747(27)	48	6135.77091(27)	78	6142.43277(26)		
19	6129.85593(26)	49	6135.87758(29)	79	6142.93571(31)		
20	6130.22414(27)	50	6136.02882(87)	80	6143.05054(72)		
21	6130.33265(38)	51	6136.15008(50)	81	6143.13813(25)		
22	6130.35800(27)	52	6136.26900(31)	82	6143.50938(87)		
23	6130.70547(28)	53	6136.38783(30)	83	6144.45198(27)		
24	6131.48181(58)	54	6136.76431(26)	84	6144.54619(25)		
25	6131.76365(39)	55	6136.89819(39)	85	6144.97931(31)		
26	6132.14942(50)	56	6137.14902(58)	86	6145.14647(29)		
27	6132.33950(28)	57	6137.25500(26)	87	6145.38210(28)		
28	6132.40256(34)	58	6137.40939(33)	88	6145.46609(28)		
29	6132.64379(66)	59	6137.64849(38)	89	6145.56235(30)		
30	6132.76596(31)	60	6137.92089(40)	90	6145.75355(42)		

- "LINE NO.": serial number identifying the spectral line in the map on the opposite page.
- "WAVENUMBER": experimental value.
- Uncertainty in the last digit of the wavenumber is given in parentheses.

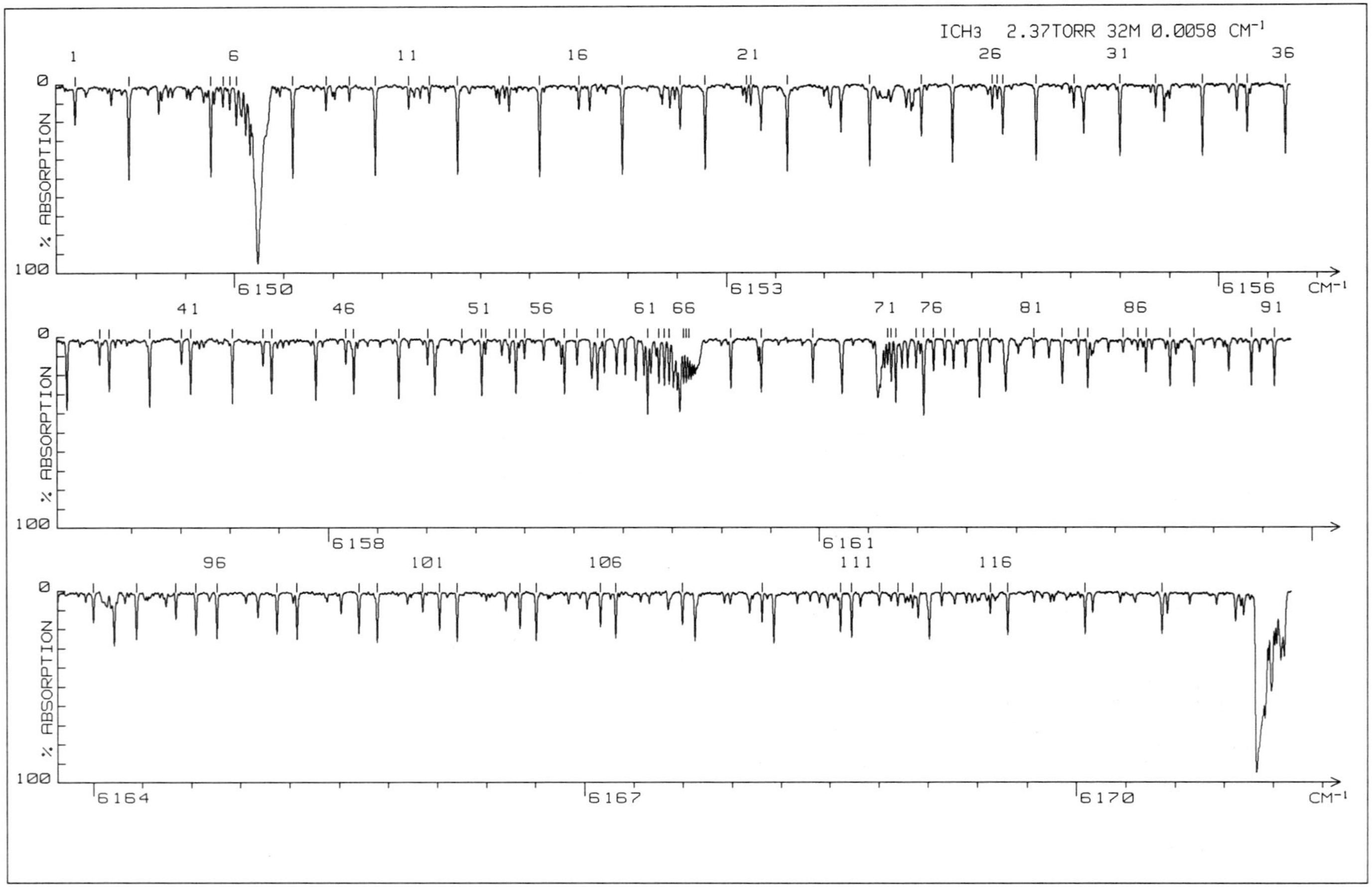
ICH3 2.37TORR 32M 0.0058 CM-1
% ABSORPTION
CM-1
6150
6153
6156
6158
6161
6164
6167
6170

LINE NO	WAVENUMBER CM^{-1}, VAC	LINE NO	WAVENUMBER CM^{-1}, VAC	LINE NO	WAVENUMBER CM^{-1}, VAC	LINE NO	WAVENUMBER CM^{-1}, VAC
1	6149.02855(31)	31	6155.40027(25)	61	6159.95356(25)	91	6163.77409(27)
2	6149.35803(25)	32	6155.61706(46)	62	6160.02235(33)	92	6164.00063(33)
3	6149.85692(24)	33	6155.90266(27)	63	6160.05607(32)	93	6164.26565(27)
4	6149.93137(41)	34	6156.11306(37)	64	6160.08498(27)	94	6164.50467(42)
5	6149.97466(51)	35	6156.17589(27)	65	6160.17235(31)	95	6164.62774(32)
6	6150.01521(33)	36	6156.40828(25)	66	6160.18915(31)	96	6164.75576(27)
7	6150.35699(24)	37	6156.61053(38)	67	6160.20481(35)	97	6165.12407(33)
8	6150.55825(46)	38	6156.66910(26)	68	6160.45930(27)	98	6165.24527(31)
9	6150.70054(59)	39	6156.91433(27)	69	6160.64489(29)	99	6165.61970(29)
10	6150.85842(24)	40	6157.10872(46)	70	6160.96539(32)	100	6165.73356(31)
11	6151.06325(48)	41	6157.16410(29)	71	6161.42087(36)	101	6166.01166(53)
12	6151.18909(43)	42	6157.42046(25)	72	6161.44310(33)	102	6166.11436(34)
13	6151.36056(24)	43	6157.60827(37)	73	6161.47065(25)	103	6166.22058(27)
14	6151.67857(40)	44	6157.66016(26)	74	6161.59432(34)	104	6166.60755(35)
15	6151.86329(24)	45	6157.92739(25)	75	6161.64279(27)	105	6166.70629(27)
16	6152.10262(56)	46	6158.10880(45)	76	6161.70300(38)	106	6167.09770(36)
17	6152.36693(24)	47	6158.15655(30)	77	6161.77077(44)	107	6167.19045(32)
18	6152.71998(28)	48	6158.43341(28)	78	6161.82337(42)	108	6167.59752(38)
19	6152.87082(24)	49	6158.61011(46)	79	6161.98083(28)	109	6168.08215(39)
20	6153.12310(68)	50	6158.81681(81)	80	6162.04407(46)	110	6168.56635(28)
21	6153.14942(53)	51	6158.93967(25)	81	6162.30786(54)	111	6168.63449(27)
22	6153.37525(24)	52	6158.96383(71)	82	6162.48276(33)	112	6168.80321(78)
23	6153.87996(26)	53	6159.11299(48)	83	6162.58103(60)	113	6168.91565(81)
24	6154.19699(27)	54	6159.15108(26)	84	6162.63784(26)	114	6169.00633(74)
25	6154.38477(24)	55	6159.20343(53)	85	6162.85214(83)	115	6169.18515(76)
26	6154.62695(46)	56	6159.32059(51)	86	6162.94028(74)	116	6169.47923(37)
27	6154.65719(72)	57	6159.44625(29)	87	6162.99011(36)	117	6169.58432(28)
28	6154.68923(31)	58	6159.52310(44)	88	6163.13579(31)	118	6170.05465(27)
29	6154.88979(26)	59	6159.64795(31)	89	6163.28184(31)	119	6170.52054(35)
30	6155.12157(38)	60	6159.68948(30)	90	6163.63343(27)		

- "LINE NO.": serial number identifying the spectral line in the map on the opposite page.
- "WAVENUMBER": experimental value.
- Uncertainty in the last digit of the wavenumber is given in parentheses.

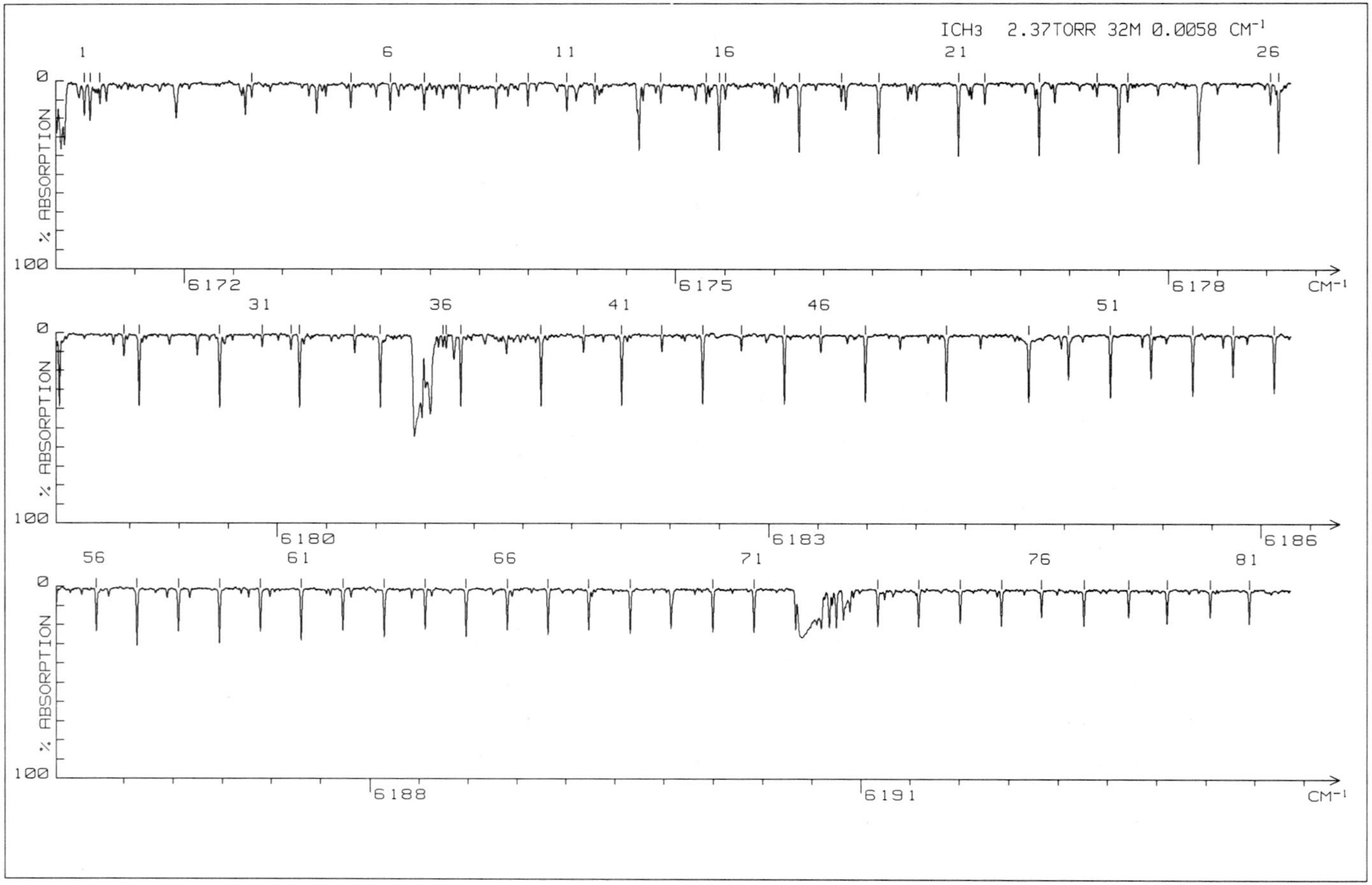
ICH3 2.37TORR 32M 0.0058 CM-1
1
6
11
16
21
26
0
% ABSORPTION
100
6172
6175
6178
CM-1
31
36
41
46
51
0
% ABSORPTION
100
6180
6183
6186
56
61
66
71
76
81
0
% ABSORPTION
100
6188
6191
CM-1

LINE NO	WAVENUMBER CM^{-1}, VAC	LINE NO	WAVENUMBER CM^{-1}, VAC	LINE NO	WAVENUMBER CM^{-1}, VAC
1	6171.38947(39)	31	6179.91037(74)	61	6187.58280(26)
2	6171.42505(31)	32	6180.08671(64)	62	6187.83524(31)
3	6171.48716(54)	33	6180.13861(24)	63	6188.08751(26)
4	6172.41246(68)	34	6180.47212(38)	64	6188.33700(28)
5	6173.01419(33)	35	6180.62898(24)	65	6188.58776(26)
6	6173.25456(41)	36	6181.01096(54)	66	6188.83888(28)
7	6173.46070(37)	37	6181.03115(67)	67	6189.09036(27)
8	6173.67832(46)	38	6181.12044(24)	68	6189.34130(28)
9	6173.90170(36)	39	6181.61268(25)	69	6189.59348(30)
10	6174.09741(45)	40	6181.87405(54)	70	6190.09702(27)
11	6174.33605(36)	41	6182.10580(25)	71	6190.34624(31)
12	6174.50823(50)	42	6182.35251(56)	72	6191.10376(29)
13	6174.91022(53)	43	6182.59995(25)	73	6191.35197(29)
14	6175.18582(36)	44	6182.83329(58)	74	6191.60651(30)
15	6175.26420(25)	45	6183.09490(26)	75	6191.85512(38)
16	6175.30173(64)	46	6183.31799(57)	76	6192.09745(32)
17	6175.60019(40)	47	6183.59057(25)	77	6192.35847(29)
18	6175.75131(25)	48	6184.08718(25)	78	6192.62943(38)
19	6176.00742(58)	49	6184.58457(25)	79	6192.86181(29)
20	6176.23787(25)	50	6184.82806(27)	80	6193.12601(38)
21	6176.72438(26)	51	6185.08256(25)	81	6193.36543(30)
22	6176.88274(48)	52	6185.32880(28)		
23	6177.21096(25)	53	6185.58134(27)		
24	6177.56481(76)	54	6185.82975(31)		
25	6177.75159(53)	55	6186.08119(27)		
26	6178.62207(39)	56	6186.33087(31)		
27	6178.67179(25)	57	6186.58098(25)		
28	6179.06746(48)	58	6186.83199(27)		
29	6179.16000(26)	59	6187.08157(26)		
30	6179.64887(26)	60	6187.33359(28)		

- "LINE NO.": serial number identifying the spectral line in the map on the opposite page.
- "WAVENUMBER": experimental value.
- Uncertainty in the last digit of the wavenumber is given in parentheses.

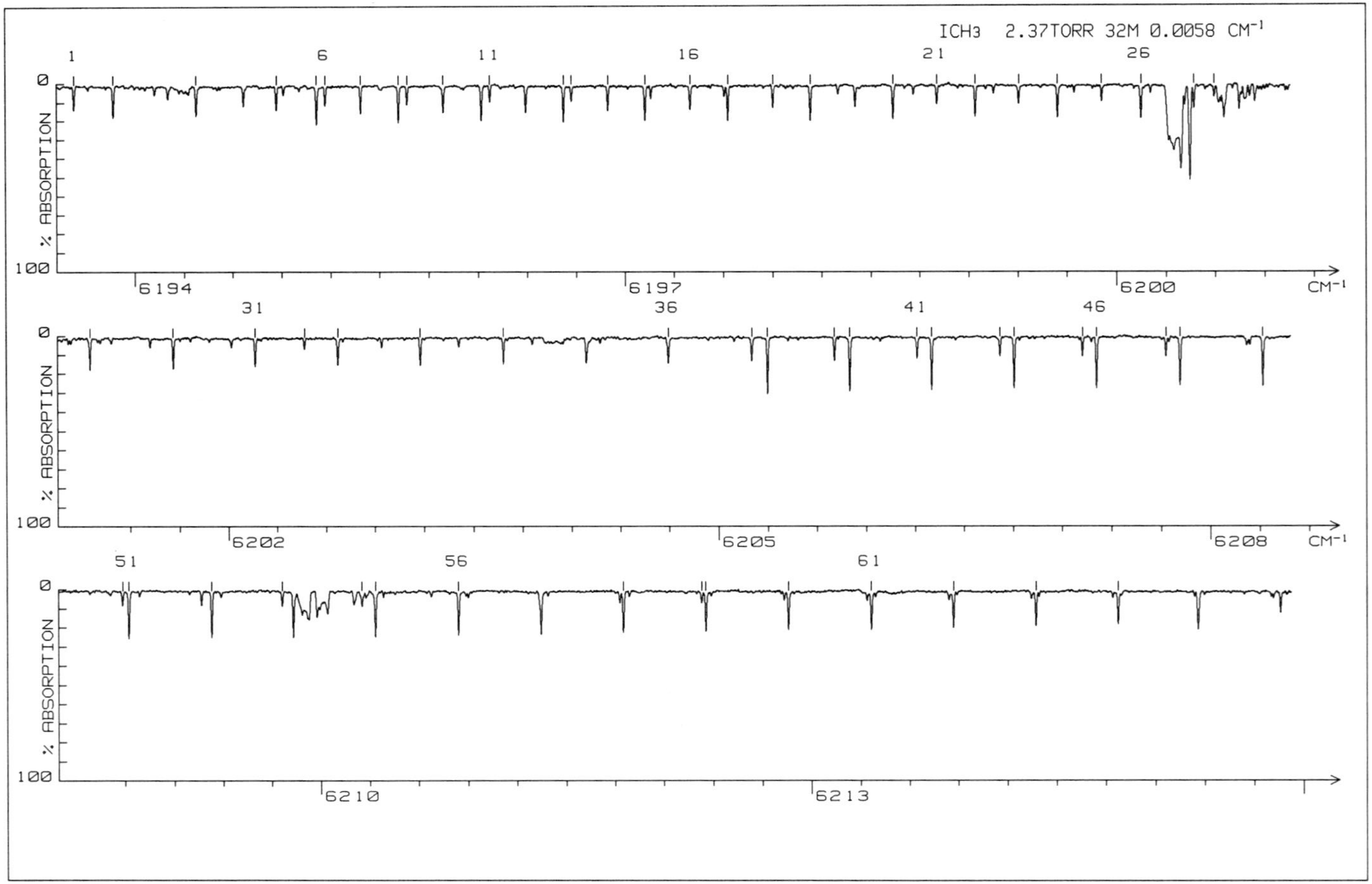

ICH3 2.37TORR 32M 0.0058 CM⁻¹
1
6
11
16
21
26
31
36
41
46
51
56
61
0
% ABSORPTION
100
6194
6197
6200
6202
6205
6208
6210
6213
CM⁻¹

LINE NO	WAVENUMBER CM^{-1}, VAC	LINE NO	WAVENUMBER CM^{-1}, VAC	LINE NO	WAVENUMBER CM^{-1}, VAC
1	6193.62799(35)	31	6202.16727(38)	61	6213.36367(29)
2	6193.86904(35)	32	6202.46888(79)	62	6213.86787(33)
3	6194.37250(31)	33	6202.67244(32)	63	6214.37143(34)
4	6194.86701(43)	34	6203.17829(38)	64	6214.87255(35)
5	6195.11292(28)	35	6203.68409(33)		
6	6195.16511(39)	36	6204.69619(48)		
7	6195.38278(39)	37	6205.20302(44)		
8	6195.61547(29)	38	6205.29841(25)		
9	6195.66764(50)	39	6205.70813(42)		
10	6195.88646(40)	40	6205.80116(27)		
11	6196.16992(55)	41	6206.21433(46)		
12	6196.62077(29)	42	6206.30402(26)		
13	6196.66936(44)	43	6206.72035(49)		
14	6196.89428(33)	44	6206.80696(28)		
15	6197.12395(34)	45	6207.22571(49)		
16	6197.39765(35)	46	6207.31053(28)		
17	6197.62715(28)	47	6207.72938(50)		
18	6197.90071(48)	48	6207.81375(26)		
19	6198.13085(34)	49	6208.31862(26)		
20	6198.63467(34)	50	6208.78428(64)		
21	6198.90487(49)	51	6208.82236(29)		
22	6199.13843(29)	52	6209.32637(29)		
23	6199.40365(52)	53	6209.76186(62)		
24	6199.64298(35)	54	6210.25040(72)		
25	6199.90972(54)	55	6210.33476(27)		
26	6200.14685(31)	56	6210.83981(27)		
27	6200.47006(47)	57	6211.84932(28)		
28	6200.59134(71)	58	6212.32835(86)		
29	6201.15655(31)	59	6212.35408(28)		
30	6201.66199(31)	60	6212.85894(32)		

- "LINE NO.": serial number identifying the spectral line in the map on the opposite page.
- "WAVENUMBER": experimental value.
- Uncertainty in the last digit of the wavenumber is given in parentheses.

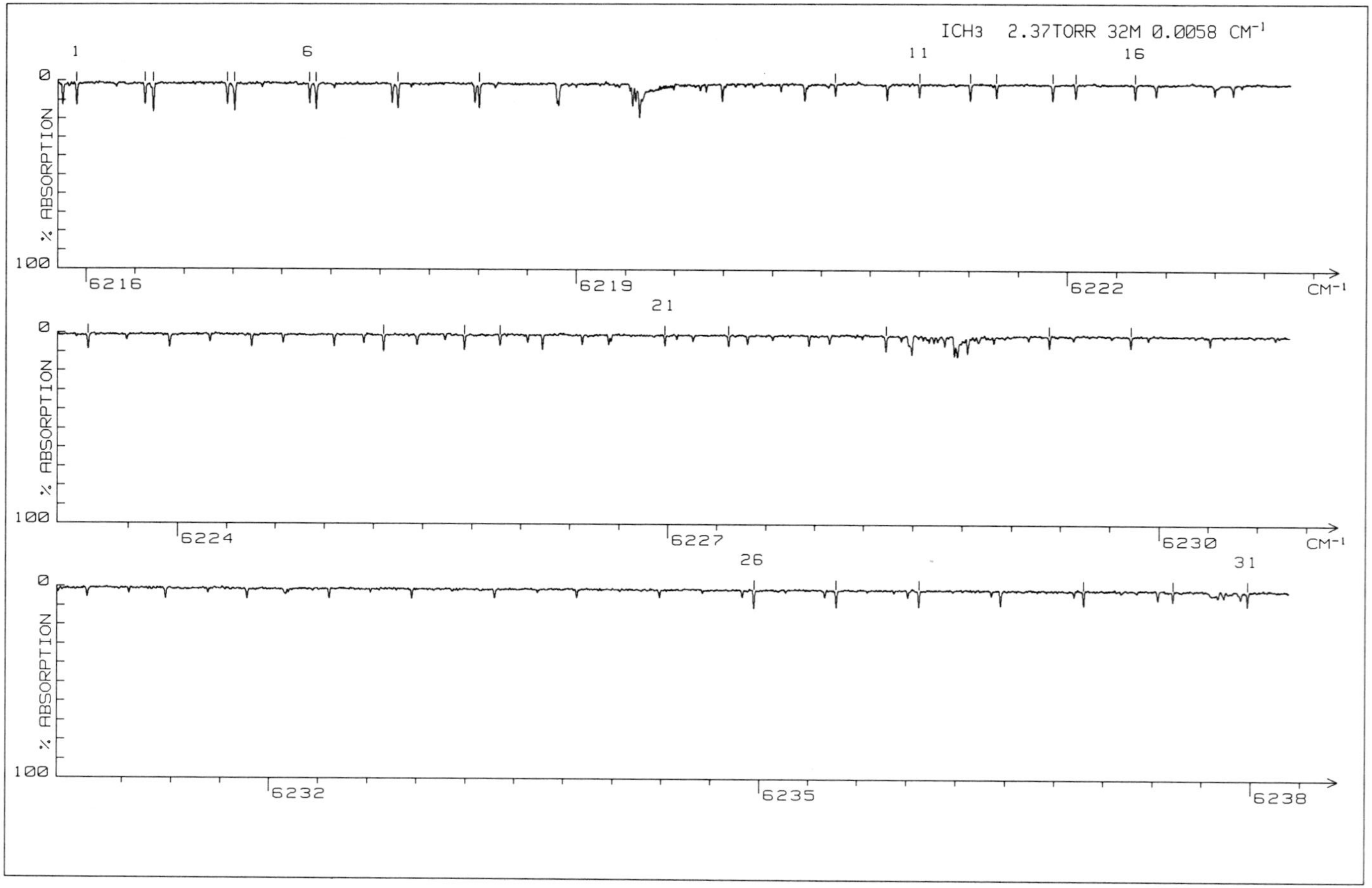

ICH3 2.37TORR 32M 0.0058 CM-1
1
6
11
16
21
26
31
0
% ABSORPTION
100
6216
6219
6222
6224
6227
6230
6232
6235
6238
CM-1

LINE NO	WAVENUMBER CM^{-1}, VAC	LINE NO	WAVENUMBER CM^{-1}, VAC
1	6215.94226(34)	31	6237.97731(57)
2	6216.36161(48)		
3	6216.41168(38)		
4	6216.86514(47)		
5	6216.90921(42)		
6	6217.36923(48)		
7	6217.40953(41)		
8	6217.90860(42)		
9	6218.40433(41)		
10	6220.58602(69)		
11	6221.09998(65)		
12	6221.40755(58)		
13	6221.56587(47)		
14	6221.91097(45)		
15	6222.05072(63)		
16	6222.41122(57)		
17	6223.45085(62)		
18	6225.26045(43)		
19	6225.75300(58)		
20	6225.96957(82)		
21	6226.98099(72)		
22	6227.37024(72)		
23	6228.33504(56)		
24	6229.32800(51)		
25	6229.82514(74)		
26	6234.96382(67)		
27	6235.47015(40)		
28	6235.97405(43)		
29	6236.97936(48)		
30	6237.52067(77)		

- "LINE NO.": serial number identifying the spectral line in the map on the opposite page.
- "WAVENUMBER": experimental value.
- Uncertainty in the last digit of the wavenumber is given in parentheses.

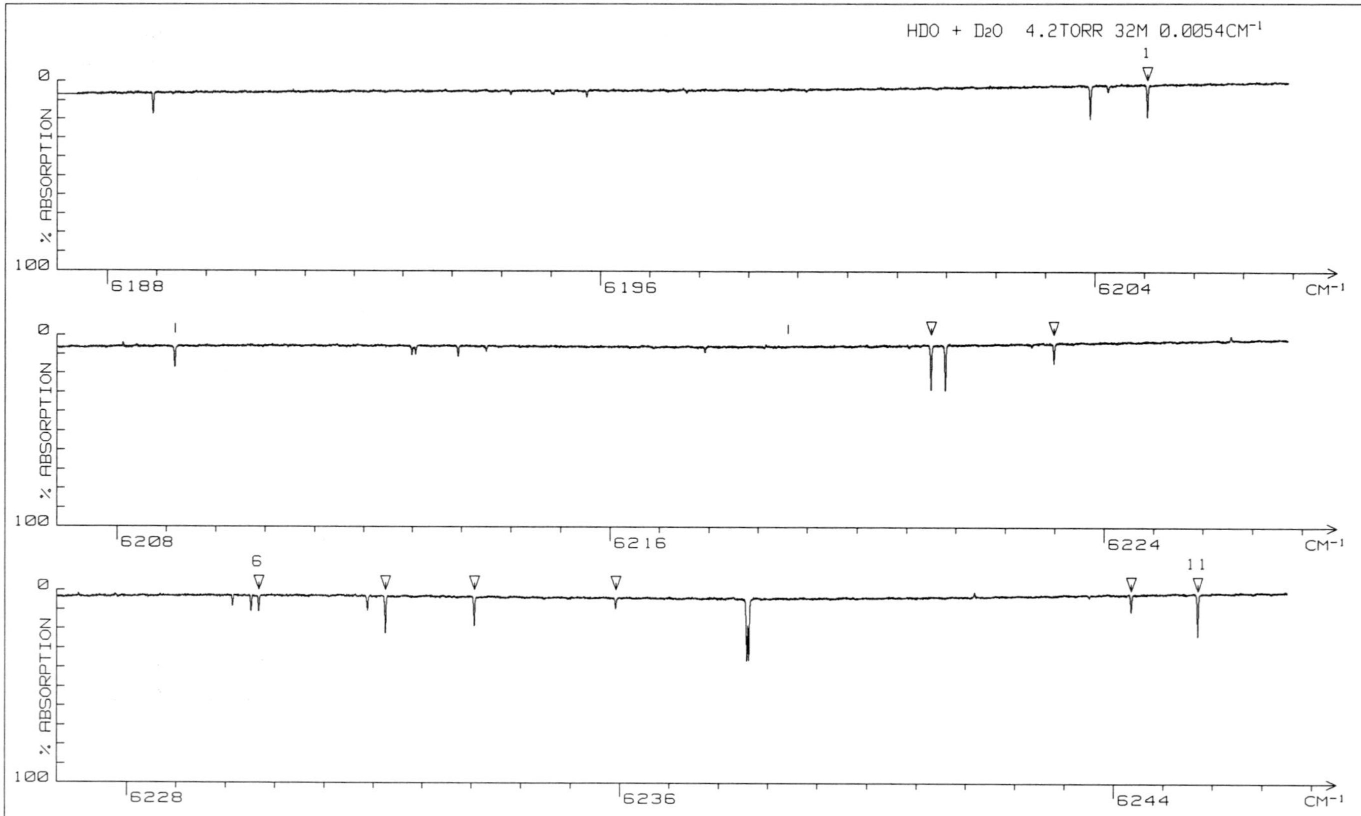

ISOTOPIC IDENTIFICATION – '∇' indicates HDO; 'Ⓘ': D2O; 'ı': unidentified.

LINE NO.	WAVENUMBER CM^{-1}, VAC.	J'	K_A'	K_C'	J''	K_A''	K_C''	IDENT. BD. IS.
1	6204.84443(41)							HDO
2	6208.93924(56)							
3	6218.88857(62)							
4	6221.19703(33)							HDO
5	6223.18459(59)							HDO
6	6230.14161(87)							HDO
7	6232.20467(45)							HDO
8	6233.65112(43)							HDO
9	6235.93445(135)							HDO
10	6244.28533(84)							HDO
11	6245.35776(34)							HDO

LINE NO.	WAVENUMBER CM^{-1}, VAC.	J'	K_A'	K_C'	J''	K_A''	K_C''	IDENT. BD. IS.

- "LINE NO.": serial number identifying the spectral line in the map on the opposite page.
- Uncertainty in the last digit of the wavenumber is given in parentheses.
- In column "IDENT." under "BD" 1 represents the band 101-000; 2: 111-000; 3: 021-000; 4: 210-000; 5: 031-000.

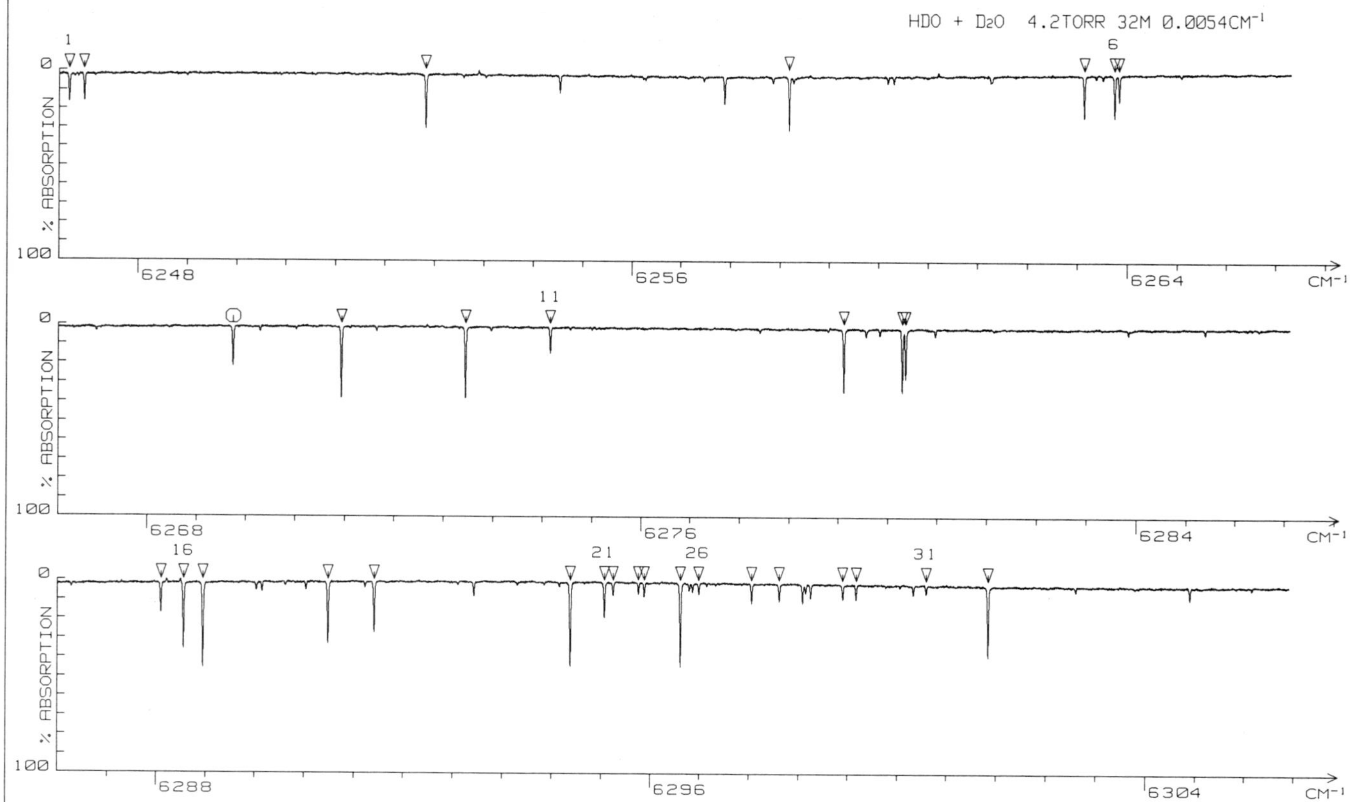

ISOTOPIC IDENTIFICATION – '∇' indicates HDO; 'Ⓘ': D_2O; 'ı': unidentified.

LINE NO.	WAVENUMBER CM^{-1}, VAC.	J'	K_A'	K_C'	J"	$K_A"$	$K_C"$	IDENT. BD.	IS.
1	6246.87128(44)								HDO
2	6247.11665(59)								HDO
3	6252.66010(31)								HDO
4	6258.54633(31)								HDO
5	6263.31377(33)								HDO
6	6263.79803(33)								HDO
7	6263.87159(45)								HDO
8	6269.38896(44)	3	0	3	2	0	2	5	D2O
8									HDO
9	6271.14506(28)								HDO
10	6273.15447(28)								HDO
11	6274.52920(46)								HDO
12	6279.28949(29)								HDO
13	6280.23314(29)								HDO
14	6280.28820(32)								HDO
15	6288.07944(54)								HDO
16	6288.44675(29)								HDO
17	6288.75843(27)								HDO
18	6290.78515(29)								HDO
19	6291.53230(32)								HDO
20	6294.70490(27)								HDO
21	6295.25611(48)								HDO
22	6295.39641(84)								D2O
23	6295.80975(119)								HDO
24	6295.90445(114)								HDO
25	6296.50157(27)								HDO
26	6296.79771(116)								HDO
27	6297.64935(57)								HDO
28	6298.09455(63)								HDO
29	6299.12673(107)								HDO
30	6299.34125(71)								HDO
31	6300.47446(124)								HDO
32	6301.47845(32)								HDO

- "LINE NO.": serial number identifying the spectral line in the map on the opposite page.
- Uncertainty in the last digit of the wavenumber is given in parentheses.
- In column "IDENT." under "BD" 1 represents the band 101-000; 2: 111-000; 3: 021-000; 4: 210-000; 5: 031-000.

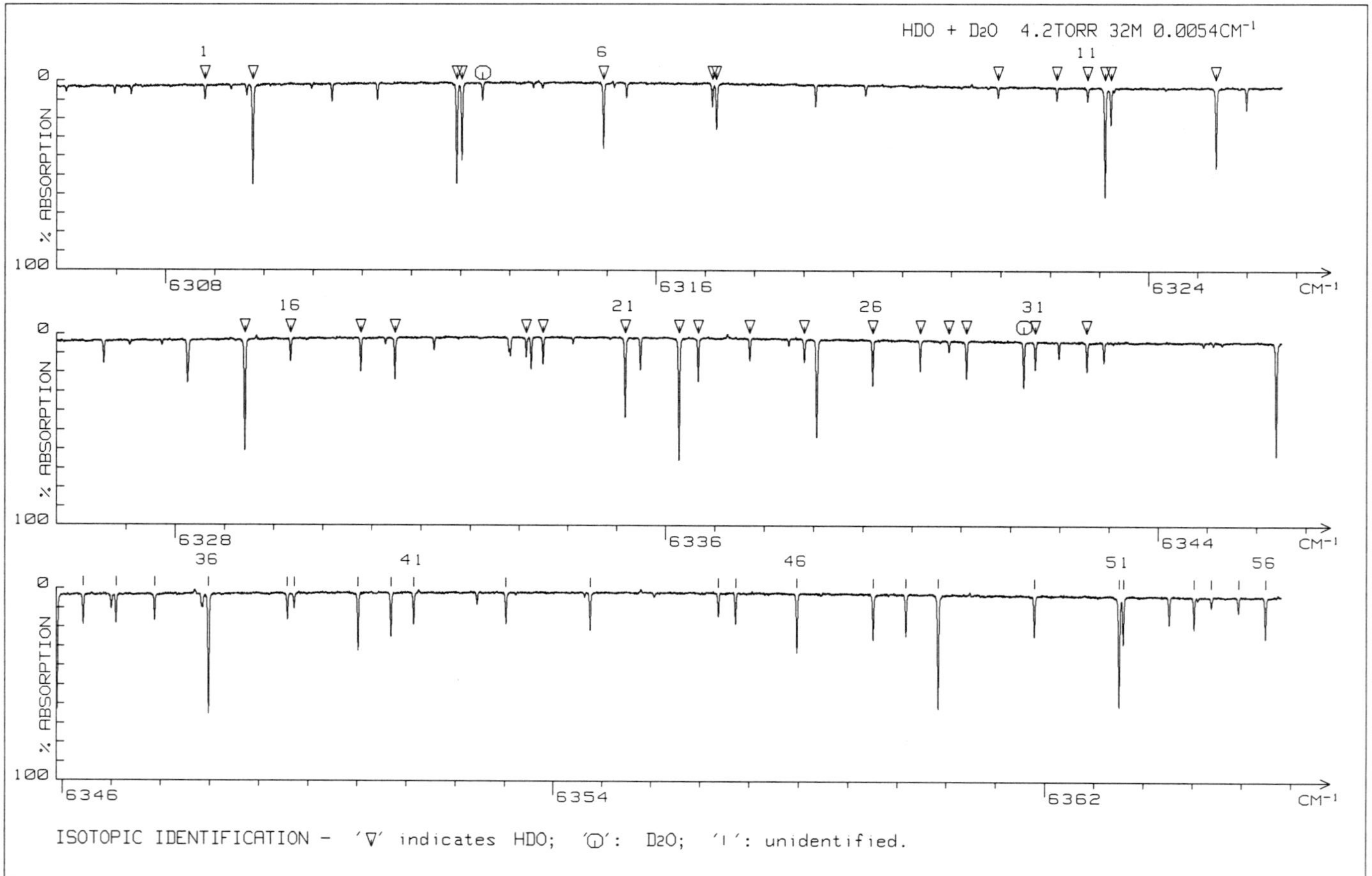
HDO + D2O 4.2TORR 32M 0.0054CM⁻¹
% ABSORPTION
0
100
6308
6316
6324
6328
6336
6344
6346
6354
6362
CM⁻¹
1
6
11
16
21
26
31
36
41
46
51
56
ISOTOPIC IDENTIFICATION - '∇' indicates HDO; 'Ⓘ': D2O; 'I': unidentified.

LINE NO.	WAVENUMBER CM^{-1}, VAC.	J'	K_A'	K_C'	J"	K_A"	K_C"	IDENT. BD.	IS.
1	6308.64396(82)								HDO
2	6309.42431(26)								HDO
3	6312.73910(26)								HDO
4	6312.82377(27)								HDO
5	6313.15326(78)	7	5	2	7	5	3	5	D_2O
6	6315.14170(32)								HDO
7	6316.91500(66)								HDO
8	6316.98528(33)								HDO
9	6321.55700(137)								HDO
10	6322.51007(108)								HDO
11	6323.00697(94)								HDO
12	6323.29047(26)								HDO
13	6323.38559(37)								HDO
14	6325.10091(30)								HDO
15	6329.13917(25)								HDO
16	6329.87854(65)								HDO
17	6331.02502(43)								HDO
18	6331.58306(35)								HDO
19	6333.71781(72)								HDO
20	6333.99433(57)								HDO
21	6335.34265(30)								HDO
22	6336.22353(25)								HDO
23	6336.53449(41)								HDO
24	6337.37364(65)								HDO
25	6338.25505(70)								HDO
26	6339.36069(40)								HDO
27	6340.14215(52)								HDO
28	6340.60974(123)								HDO
29	6340.89234(45)								HDO
30	6341.81740(32)	6	3	3	5	3	2	5	D_2O
30									HDO

LINE NO.	WAVENUMBER CM^{-1}, VAC.	J'	K_A'	K_C'	J"	K_A"	K_C"	IDENT. BD.	IS.
31	6342.00186(57)								HDO
32	6342.84299(54)								HDO
33	6346.34373(50)								
34	6346.88228(44)								
35	6347.51270(45)								
36	6348.38890(25)								
37	6349.67062(45)								
38	6349.78044(94)								
39	6350.82500(30)								
40	6351.36291(33)								
41	6351.72841(52)								
42	6353.22606(40)								
43	6354.61507(45)								
44	6356.70740(68)								
45	6356.98675(41)								
46	6357.97980(34)								
47	6359.21259(39)								
48	6359.74838(42)								
49	6360.27905(26)								
50	6361.83384(42)								
51	6363.21258(26)								
52	6363.27989(33)								
53	6364.43133(55)								
54	6364.71827(145)								
55	6365.15582(91)								
56	6365.59281(34)								

- "LINE NO.": serial number identifying the spectral line in the map on the opposite page.
- Uncertainty in the last digit of the wavenumber is given in parentheses.
- In column "IDENT." under "BD" 1 represents the band 101-000; 2: 111-000; 3: 021-000; 4: 210-000; 5: 031-000.

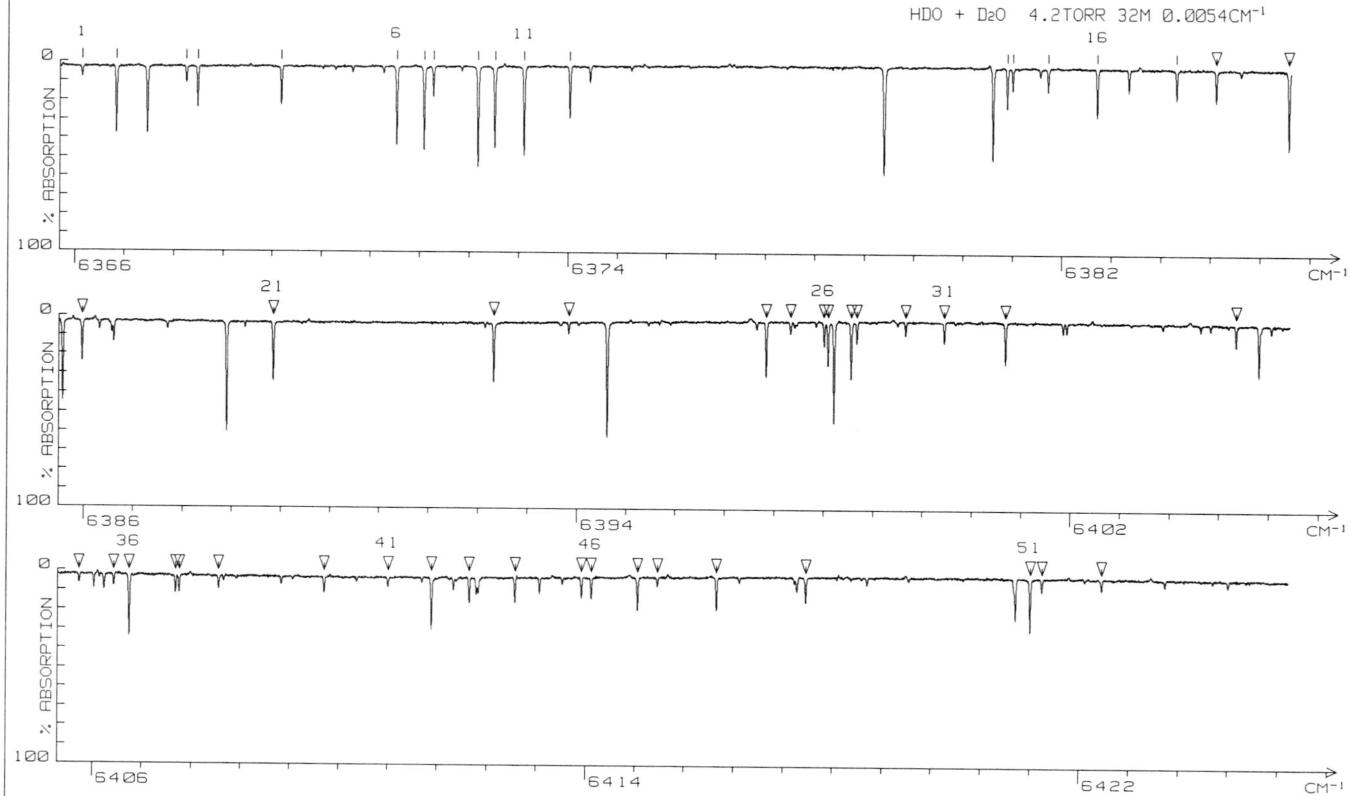

ISOTOPIC IDENTIFICATION – '∇' indicates HDO; 'Ⓘ': D_2O; 'ı': unidentified.

LINE NO.	WAVENUMBER CM^{-1}, VAC.	J'	K_A'	K_C'	J"	K_A"	K_C"	IDENT. BD.	IS.
1	6366.10822(153)								
2	6366.66995(29)								
3	6367.79345(94)								
4	6367.97495(35)								
5	6369.33958(44)								
6	6371.22191(28)								
7	6371.66574(27)								
8	6371.81504(55)								
9	6372.53215(26)								
10	6372.80054(30)								
11	6373.27871(27)								
12	6374.01753(37)								
13	6381.10467(35)								
14	6381.19299(67)								
15	6381.76326(72)								
16	6382.56647(38)								
17	6383.83925(41)								
18	6384.48577(40)	1	0	1	2	0	2	1	HDO
19	6385.65278(30)	0	0	0	1	1	1	1	HDO
20	6385.96712(44)	4	2	3	5	2	4	3	HDO
21	6389.07216(30)	3	1	2	4	1	3	3	HDO
22	6392.64515(34)	3	0	3	4	0	4	3	HDO
23	6393.85864(123)	5	3	2	6	2	5	1	HDO
24	6397.06315(31)	3	1	3	4	1	4	3	HDO
25	6397.46148(123)	4	3	2	5	2	3	1	HDO
26	6398.00061(65)	4	3	1	5	3	2	3	HDO
27	6398.06556(35)	3	2	1	4	2	2	3	HDO
28	6398.43860(36)	1	1	1	2	0	2	1	HDO
29	6398.53062(70)	4	3	2	5	3	3	3	HDO
30	6399.31942(100)	4	4	0	4	4	1	1	HDO
30		4	4	1	4	4	0	1	HDO
31	6399.95313(58)	0	0	0	1	0	1	1	HDO
32	6400.93992(35)	3	2	2	4	2	3	3	HDO
33	6404.66947(79)	2	1	1	3	1	2	3	HDO
34	6405.77881(180)	4	3	2	4	3	1	1	HDO
35	6406.34708(131)	4	2	2	5	1	5	1	HDO
36	6406.59672(38)	2	0	2	3	0	3	3	HDO
37	6407.33963(90)	3	3	1	3	3	0	1	HDO
38	6407.39848(70)	3	3	0	3	3	1	1	HDO
39	6408.03661(108)	4	3	1	5	2	4	1	HDO
39		8	3	5	8	3	6	1	HDO
40	6409.75804(91)	3	2	2	3	2	1	1	HDO
41	6410.80245(149)	4	1	4	3	2	1	1	HDO
42	6411.50970(32)	2	1	2	3	1	3	3	HDO
43	6412.11570(51)	1	1	1	1	1	0	1	HDO
44	6412.85517(49)	2	2	0	2	2	1	1	HDO
45	6413.92713(57)	3	3	0	4	3	1	3	HDO
46	6414.08614(54)	3	3	1	4	3	2	3	HDO
47	6414.84092(52)	2	2	0	3	2	1	3	HDO
48	6415.16961(168)	4	2	2	4	2	3	1	HDO
49	6416.12413(52)	2	2	1	3	2	2	3	HDO
50	6417.56907(51)	1	1	0	1	1	1	1	HDO
51	6421.20191(41)	1	0	1	2	0	2	3	HDO
52	6421.39030(80)	2	2	0	3	1	3	1	HDO
53	6422.36385(153)	2	1	1	2	1	2	1	HDO

- "LINE NO.": serial number identifying the spectral line in the map on the opposite page.
- Uncertainty in the last digit of the wavenumber is given in parentheses.
- In column "IDENT." under "BD" 1 represents the band 101-000; 2: 111-000; 3: 021-000; 4: 210-000; 5: 031-000.

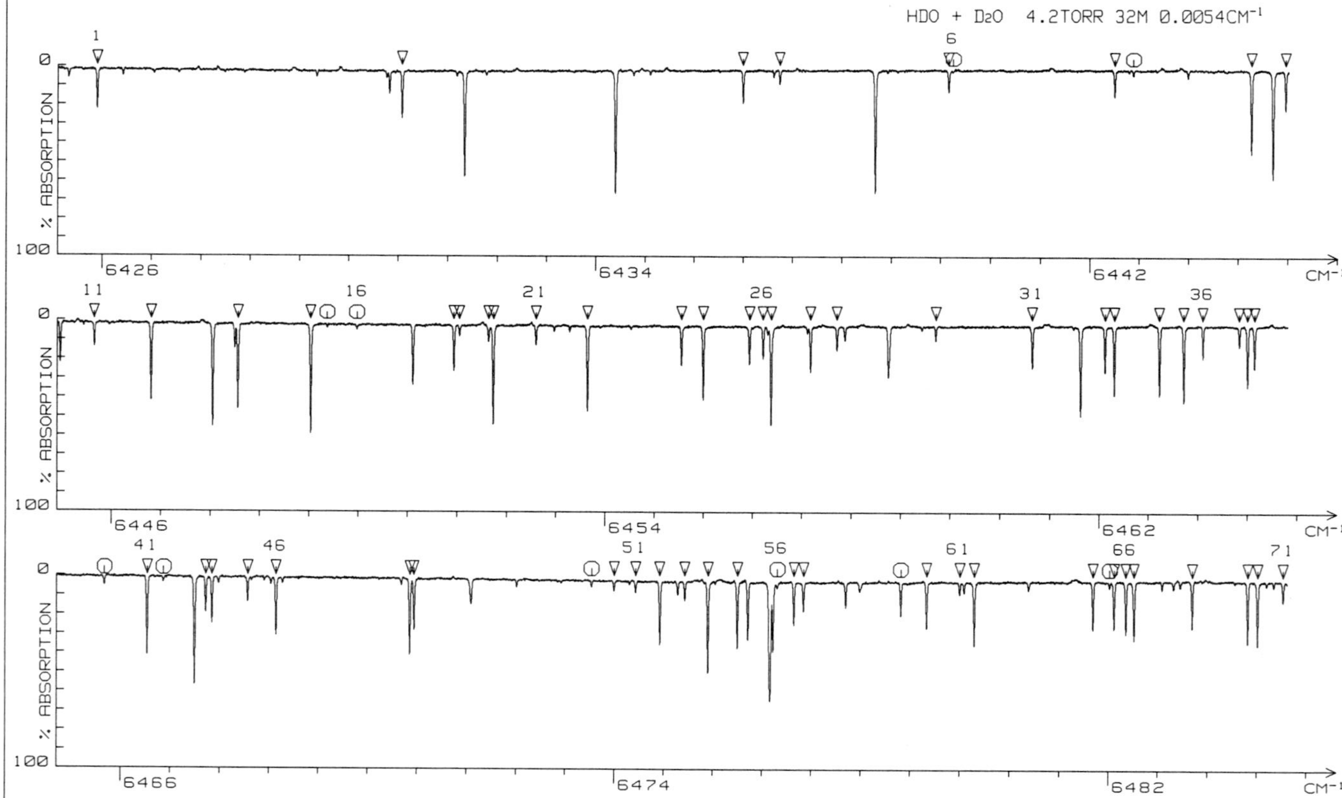

ISOTOPIC IDENTIFICATION - '▽' indicates HDO; 'Ⓘ': D_2O; 'ı': unidentified.

LINE NO.	WAVENUMBER CM^{-1}, VAC.	J'	K_A'	K_C'	J"	K_A"	K_C"	IDENT. BD.	IS.
1	6425.91730(37)	1	1	1	2	1	2	3	HDO
2	6430.86134(35)	2	0	2	1	1	1	1	HDO
3	6436.39194(51)	0	0	0	1	0	1	3	HDO
4	6436.98460(114)	3	1	3	3	1	2	3	HDO
5	6439.72574(76)	4	1	3	3	2	2	1	HDO
6	6439.79399(174)	6	2	4	7	2	5	2	D_2O
7	6442.39769(61)	2	1	2	1	1	1	1	HDO
8	6442.70195(73)	7	0	7	8	0	8	2	D_2O
9	6444.61156(31)	1	1	1	0	0	0	1	HDO
10	6445.16186(37)	2	0	2	1	0	1	1	HDO
11	6445.71580(86)	2	1	2	2	1	1	3	HDO
12	6446.64437(28)	3	0	3	2	1	2	1	HDO
13	6448.04905(28)								HDO
14	6449.22879(26)	4	2	2	4	1	3	1	HDO
15	6449.49491(146)	6	3	4	7	3	5	2	D_2O
16	6449.97811(116)	6	2	5	7	2	6	2	D_2O
17	6451.54746(35)	1	1	1	1	1	0	3	HDO
18	6451.63729(136)	10	3	7	10	2	8	1	HDO
19	6452.10097(91)	5	2	4	5	2	3	3	HDO
20	6452.18044(27)	3	2	1	3	1	2	1	HDO
21	6452.87766(81)	9	3	6	9	2	7	1	HDO
22	6453.71467(28)	5	1	4	5	0	5	1	HDO
23	6455.24213(48)	3	1	3	2	1	2	1	HDO
24	6455.59764(29)	2	2	0	2	1	1	1	HDO
25	6456.35030(38)	8	2	6	8	1	7	1	HDO
26	6456.56925(54)	8	3	5	8	2	6	1	HDO
27	6456.69796(29)	2	1	2	1	0	1	1	HDO
28	6457.34184(35)	1	1	0	1	1	1	3	HDO
29	6457.76782(68)	4	2	3	4	2	2	3	HDO
30	6459.37249(106)	3	2	1	2	2	0	1	HDO
31	6460.91677(37)	3	2	2	3	2	1	3	HDO
32	6462.09288(34)	7	3	4	7	2	5	1	HDO
32		6	1	5	6	1	6	1	HDO
33	6462.24408(29)	2	2	1	2	2	0	3	HDO
34	6462.97982(29)	2	2	0	2	2	1	3	HDO
34		2	1	1	2	1	2	3	HDO
35	6463.37107(28)	2	2	1	2	1	2	1	HDO
36	6463.68109(56)	3	1	2	2	1	1	1	HDO
37	6464.26758(85)	9	2	7	9	1	8	1	HDO
38	6464.40090(30)	6	1	5	6	0	6	1	HDO
39	6464.51296(36)	3	2	1	3	2	2	3	HDO
40	6465.73514(60)	5	5	1	6	5	2	2	D_2O
40		5	5	0	6	5	1	2	D_2O
40		5	0	5	6	0	6	2	D_2O
41	6466.43175(32)	3	2	2	3	1	3	1	HDO
42	6466.69562(111)	5	1	5	6	1	6	2	D_2O
43	6467.37544(51)	1	0	1	0	0	0	3	HDO
44	6467.47586(35)	4	1	4	3	1	3	1	HDO
45	6468.05791(69)	4	2	2	4	2	3	3	HDO
46	6468.51570(31)	6	3	3	6	2	4	1	HDO
47	6470.68832(33)	4	2	3	4	1	4	1	HDO
48	6470.75806(33)	4	0	4	3	0	3	1	HDO
49	6473.63026(65)	4	2	3	5	2	4	2	D_2O
50	6473.99621(147)	10	2	8	10	1	9	1	HDO
51	6474.34428(135)	5	2	3	5	2	4	3	HDO
52	6474.74034(30)	5	3	2	5	2	3	1	HDO
53	6475.14736(63)	4	2	2	3	2	1	1	HDO
54	6475.52416(30)	5	0	5	4	1	4	1	HDO
55	6476.00322(34)	5	2	4	5	1	5	1	HDO
56	6476.64865(107)	4	0	4	5	0	5	2	D_2O
57	6476.92341(48)	4	2	3	4	0	4	1	HDO
57		4	3	1	4	3	2	3	HDO

- "LINE NO.": serial number identifying the spectral line in the map on the opposite page.
- Uncertainty in the last digit of the wavenumber is given in parentheses.
- In column "IDENT." under "BD" 1 represents the band 101-000; 2: 111-000; 3: 021-000; 4: 210-000; 5: 031-000.

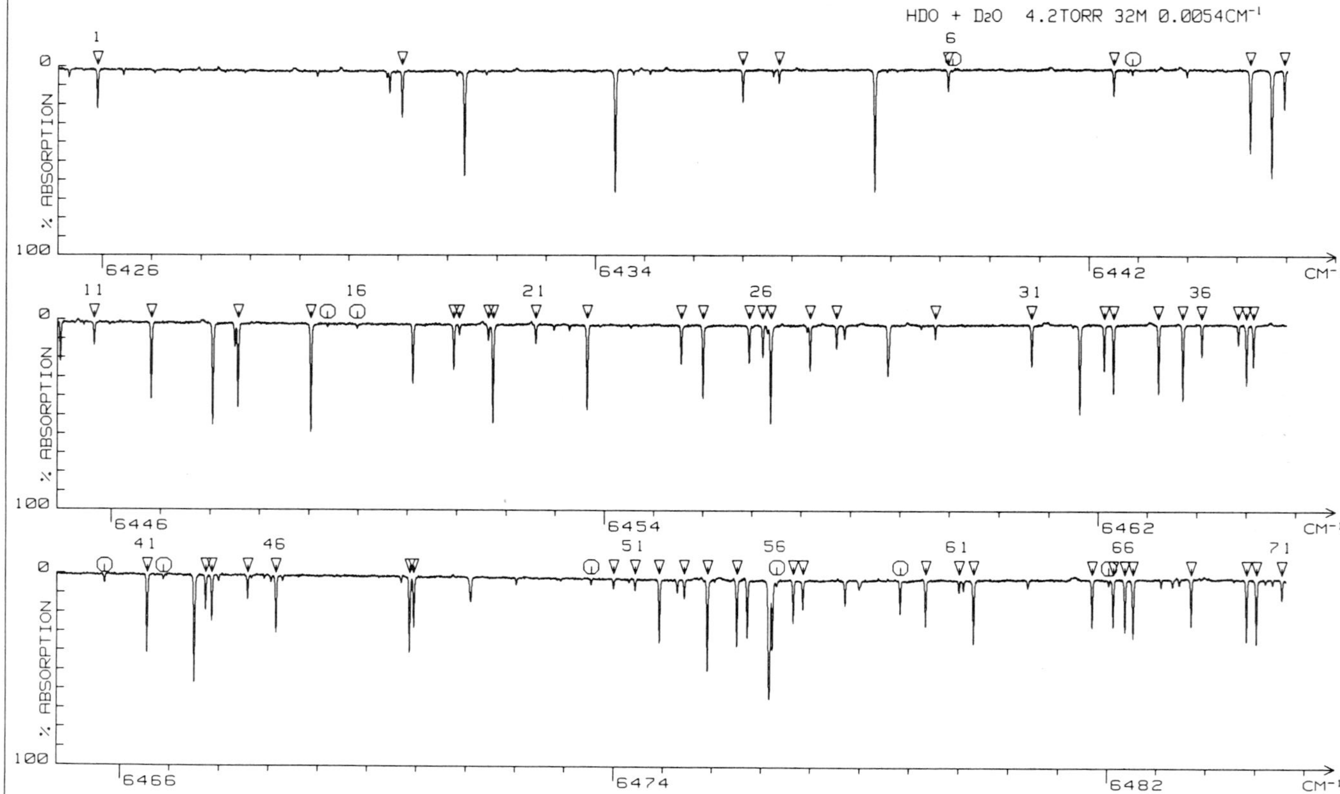

ISOTOPIC IDENTIFICATION – '▽' indicates HDO; '(I)': D_2O; 'I': unidentified.

LINE NO.	WAVENUMBER CM^{-1}, VAC.	J'	K_A'	K_C'	J"	K_A"	K_C"	IDENT. BD.	IS.
58	6477.07901(46)	6	1	5	5	2	4	1	HDO
59	6478.65365(42)	8	2	7	8	2	6	2	D_2O
59		4	1	3	3	1	2	1	HDO
60	6479.07171(34)	5	1	5	4	1	4	1	HDO
61	6479.60318(137)	6	3	3	6	3	4	3	HDO
62	6479.83663(30)	4	3	1	4	2	2	1	HDO
63	6481.75008(41)	5	0	5	4	0	4	1	HDO
64	6482.01633(87)	3	2	1	4	2	2	2	D_2O
65	6482.09201(34)	2	1	2	1	1	1	3	HDO
66	6482.28544(32)	6	2	5	6	1	6	1	HDO
67	6482.41895(42)	2	0	2	1	0	1	3	HDO
68	6483.35691(34)	3	3	0	3	2	1	1	HDO
69	6484.25544(30)	5	3	3	5	2	4	1	HDO
70	6484.41415(30)	4	3	2	4	2	3	1	HDO
71	6484.82618(79)	5	2	4	4	2	3	1	HDO

LINE NO.	WAVENUMBER CM^{-1}, VAC.	J'	K_A'	K_C'	J"	K_A"	K_C"	IDENT. BD.	IS.

- "LINE NO.": serial number identifying the spectral line in the map on the opposite page.
- Uncertainty in the last digit of the wavenumber is given in parentheses.
- In column "IDENT." under "BD" 1 represents the band 101–000; 2: 111–000; 3: 021–000; 4: 210–000; 5: 031–000.

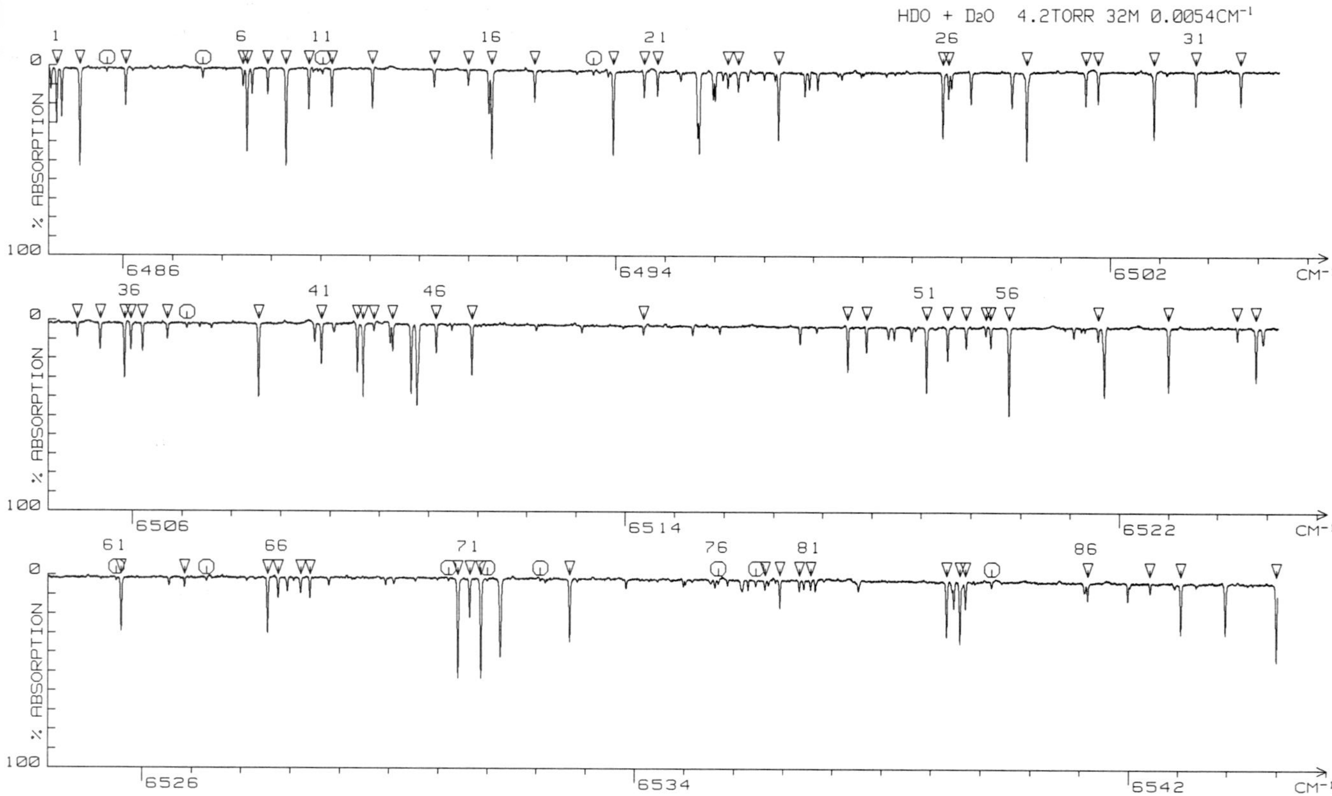

ISOTOPIC IDENTIFICATION – '∇' indicates HDO; 'Ⓘ': D_2O; 'I': unidentified.

LINE NO.	WAVENUMBER CM^{-1}, VAC.	J'	K_A'	K_C'	J"	K_A"	K_C"	IDENT. BD.	IDENT. IS.
1	6484.91927(32)	6	3	4	6	2	5	1	HDO
2	6485.29758(29)	5	1	5	4	0	4	1	HDO
3	6485.73947(155)	3	2	2	4	2	3	2	D_2O
4	6486.04129(48)	7	3	5	7	2	6	1	HDO
5	6487.28090(143)	3	0	3	4	0	4	2	D_2O
6	6487.93934(84)	8	4	4	8	3	5	1	HDO
7	6488.00534(28)	6	0	6	5	1	5	1	HDO
8	6488.33812(68)	8	3	6	8	2	7	1	HDO
9	6488.63447(27)	2	1	1	1	1	0	3	HDO
10	6489.00164(38)	2	2	1	1	1	0	1	HDO
11	6489.22422(105)	3	1	3	4	1	4	2	D_2O
12	6489.37394(38)	7	2	6	7	1	7	1	HDO
13	6490.03536(36)	6	1	6	5	1	5	1	HDO
14	6491.03648(66)	5	2	3	4	2	2	1	HDO
15	6491.58858(73)	9	3	7	9	2	8	1	HDO
16	6491.97373(28)	2	2	0	1	1	1	1	HDO
17	6492.67689(54)	5	1	4	4	1	3	1	HDO
18	6493.62790(129)	2	1	1	3	1	2	2	D_2O
19	6493.95443(31)	6	1	6	5	0	5	1	HDO
20	6494.46077(62)	7	1	6	6	2	5	1	HDO
21	6494.67397(65)	7	4	3	7	3	4	1	HDO
22	6495.80438(71)	6	4	3	6	4	2	3	HDO
23	6495.97120(97)	6	4	2	6	4	3	3	HDO
23		7	4	4	7	4	3	3	HDO
24	6496.62875(30)	3	0	3	2	0	2	3	HDO
25	6499.29640(30)	7	0	7	6	1	6	1	HDO
26	6499.38523(65)	7	4	4	7	3	5	1	HDO
27	6500.63810(31)	3	2	2	2	1	1	1	HDO
28	6501.59722(51)	7	0	7	6	0	6	1	HDO
29	6501.79699(43)	6	4	3	6	3	4	1	HDO
30	6502.70267(30)	7	1	7	6	0	6	1	HDO
31	6503.39291(40)	5	4	1	5	3	2	1	HDO
32	6504.11663(41)	5	4	2	5	3	3	1	HDO
33	6505.10462(112)	9	2	8	9	1	9	1	HDO
34	6505.47995(67)	6	1	5	5	1	4	1	HDO
35	6505.87079(32)	3	1	2	2	1	1	3	HDO
36	6505.97294(66)	4	4	0	4	3	1	1	HDO
37	6506.16107(59)	4	4	1	4	3	2	1	HDO
38	6506.56094(101)	6	2	4	5	2	3	1	HDO
39	6506.87504(94)	1	1	0	2	1	1	2	D_2O
40	6508.04052(29)	4	1	4	3	1	3	3	HDO
41	6509.05556(46)	3	2	2	2	2	1	3	HDO
42	6509.63499(40)	8	0	8	7	1	7	1	HDO
43	6509.73287(33)	4	0	4	3	0	3	3	HDO
44	6509.91074(160)	10	1	9	10	0	10	1	HDO
45	6510.21630(66)	8	1	8	7	1	7	1	HDO
46	6510.91914(44)	8	0	8	7	0	7	1	HDO
47	6511.49972(33)	8	1	8	7	0	7	1	HDO
48	6514.29992(168)	8	5	3	8	4	4	1	HDO
49	6517.60280(42)	5	5	1	5	5	0	3	HDO
49		5	5	0	5	5	1	3	HDO
50	6517.90653(70)	6	5	2	6	5	1	3	HDO
50		6	5	1	6	5	2	3	HDO
51	6518.88462(30)	5	2	4	4	1	3	1	HDO
52	6519.22355(42)	9	0	9	8	1	8	1	HDO
53	6519.52075(76)	9	1	9	8	1	8	1	HDO
54	6519.83703(173)	8	2	7	7	2	6	1	HDO
55	6519.91470(75)	9	0	9	8	0	8	1	HDO
56	6520.20979(28)	5	1	5	4	1	4	3	HDO
56		9	1	9	8	0	8	1	HDO
57	6521.64884(108)	6	5	1	6	4	2	1	HDO
58	6522.79255(35)	4	1	3	3	1	2	3	HDO
59	6523.91485(111)	9	1	8	8	2	7	1	HDO
60	6524.21461(32)	4	2	3	3	2	2	3	HDO

- "LINE NO.": serial number identifying the spectral line in the map on the opposite page.
- Uncertainty in the last digit of the wavenumber is given in parentheses.
- In column "IDENT." under "BD" 1 represents the band 101-000; 2: 111-000; 3: 021-000; 4: 210-000; 5: 031-000.

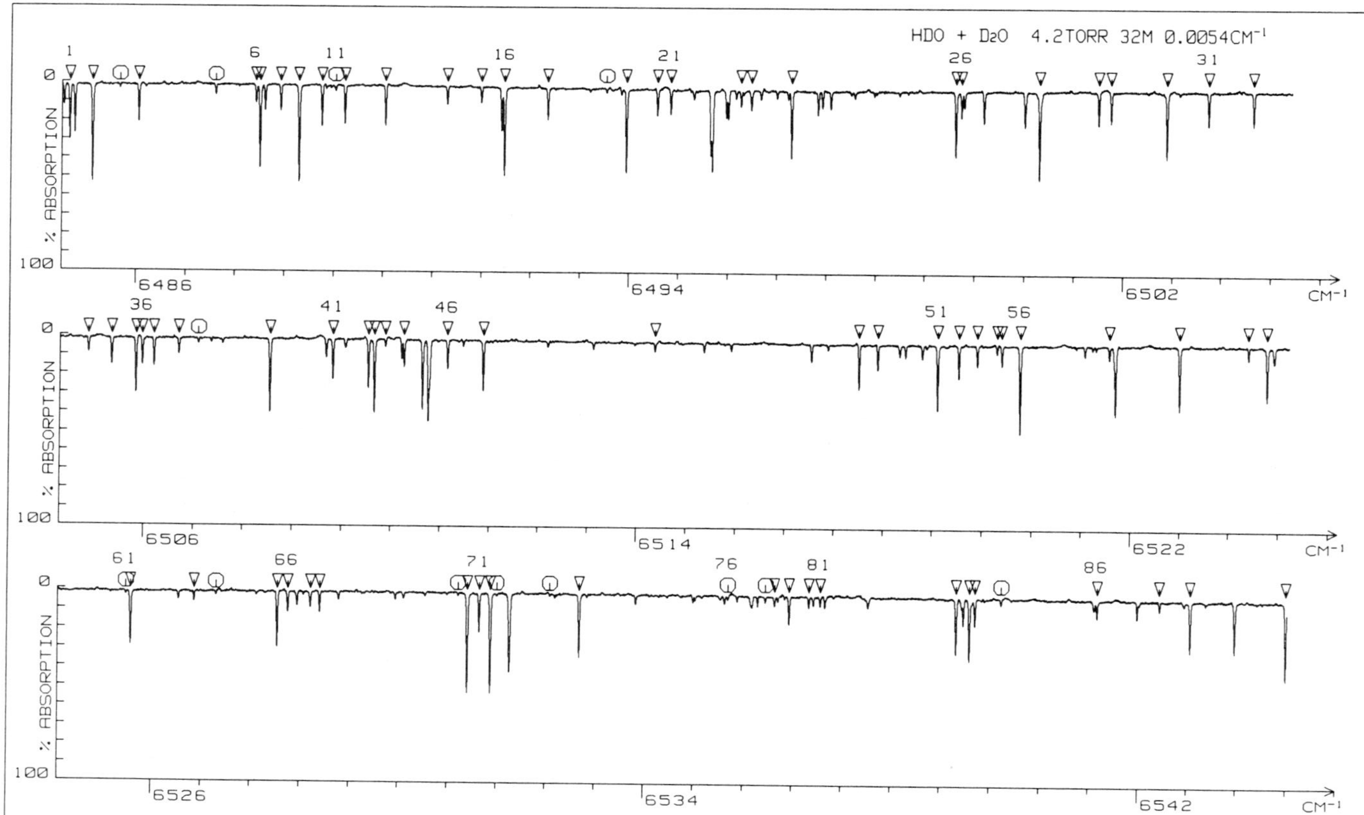

ISOTOPIC IDENTIFICATION - '∇' indicates HDO; 'Ⓘ': D2O; 'ı': unidentified.

LINE NO.	WAVENUMBER CM^{-1}, VAC.	J'	K_A'	K_C'	J"	K_A"	K_C"	IDENT. BD.	IDENT. IS.
61	6525.58529(158)	2	1	2	2	1	1	2	D_2O
62	6525.66483(33)	6	2	5	5	1	4	1	HDO
63	6526.68384(99)	8	1	7	7	1	6	1	HDO
64	6527.03745(151)	4	2	3	4	2	2	2	D_2O
65	6528.03511(32)	4	2	2	3	2	1	3	HDO
66	6528.20471(82)	10	0	10	9	1	9	1	HDO
67	6528.56603(131)	10	0	10	9	0	9	1	HDO
68	6528.71705(56)	10	1	10	9	0	9	1	HDO
69	6530.96827(148)	1	1	1	1	1	0	2	D_2O
70	6531.12141(30)	3	3	1	2	2	0	1	HDO
71	6531.31253(39)	7	2	6	6	1	5	1	HDO
72	6531.49573(27)	3	3	0	2	2	1	1	HDO
73	6531.59782(161)	3	2	2	3	2	1	2	D_2O
74	6532.46426(168)	6	4	3	6	4	2	2	D_2O
75	6532.94407(31)	6	0	6	5	0	5	3	HDO
76	6535.34897(109)	5	4	1	5	4	2	2	D_2O
77	6535.95604(107)	1	1	0	1	1	1	2	D_2O
78	6536.10287(170)	7	6	1	7	5	2	1	HDO
78		7	6	2	7	5	3	1	HDO
79	6536.34649(49)	8	2	7	7	1	6	1	HDO
80	6536.66545(159)	11	0	11	10	1	10	1	HDO
81	6536.85095(147)	11	0	11	10	0	10	1	HDO
82	6539.05750(32)	5	2	4	4	2	3	3	HDO
83	6539.27198(31)	5	1	4	4	1	3	3	HDO
84	6539.35921(49)	4	3	1	3	3	0	3	HDO
85	6539.77979(64)	5	5	1	5	5	0	2	D_2O
85		5	5	0	5	5	1	2	D_2O
86	6541.33101(95)	9	2	8	8	1	7	1	HDO
87	6542.34183(107)	7	6	1	7	6	2	3	HDO
87		7	6	2	7	6	1	3	HDO
88	6542.84661(41)	7	1	7	6	1	6	3	HDO
89	6544.39123(29)	4	3	2	3	2	1	1	HDO

LINE NO.	WAVENUMBER CM^{-1}, VAC.	J'	K_A'	K_C'	J"	K_A"	K_C"	IDENT. BD. IS.

- "LINE NO.": serial number identifying the spectral line in the map on the opposite page.
- Uncertainty in the last digit of the wavenumber is given in parentheses.
- In column "IDENT." under "BD" 1 represents the band 101-000; 2: 111-000; 3: 021-000; 4: 210-000; 5: 031-000.

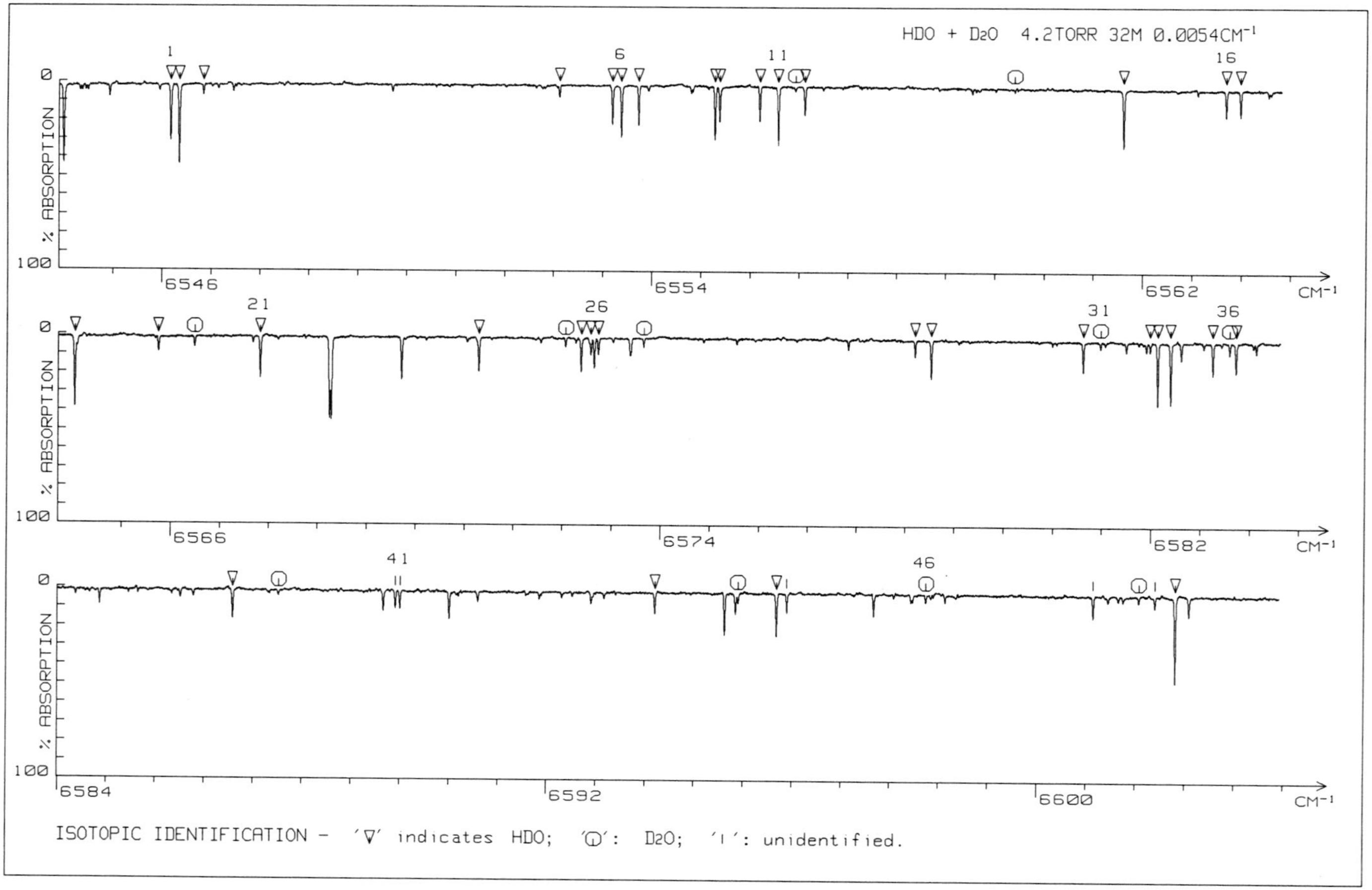
HDO + D2O 4.2TORR 32M 0.0054CM-1
0
% ABSORPTION
100
1
6
11
16
6546
6554
6562
CM-1
21
26
31
36
6566
6574
6582
41
46
6584
6592
6600
ISOTOPIC IDENTIFICATION - '▽' indicates HDO; 'Ⓘ': D2O; 'ı': unidentified.

LINE NO.	WAVENUMBER CM^{-1}, VAC.	J'	K_A'	K_C'	J"	K_A"	K_C"	IDENT. BD.	IDENT. IS.
1	6546.14425(33)	5	2	3	4	2	2	3	HDO
2	6546.28401(29)	4	3	1	3	2	2	1	HDO
3	6546.67425(152)	10	2	9	9	1	8	1	HDO
4	6552.49486(136)	11	2	10	10	1	9	1	HDO
4		8	0	8	7	1	7	3	HDO
5	6553.35220(39)	8	1	8	7	1	7	3	HDO
6	6553.49793(40)	6	2	5	5	2	4	3	HDO
7	6553.77675(38)	8	0	8	7	0	7	3	HDO
8	6555.02461(33)	6	1	5	5	1	4	3	HDO
9	6555.10066(54)	5	3	3	4	3	2	3	HDO
10	6555.74938(42)	5	3	2	4	3	1	3	HDO
11	6556.05573(32)	5	3	3	4	2	2	1	HDO
12	6556.33767(129)	2	0	2	1	0	1	2	D_2O
13	6556.49045(49)	5	2	3	4	1	4	1	HDO
14	6559.91529(158)	2	1	1	1	1	0	2	D_2O
15	6561.69328(32)	5	3	2	4	2	3	1	HDO
16	6563.35173(69)	9	1	9	8	1	8	3	HDO
17	6563.58463(48)	9	0	9	8	0	8	3	HDO
18	6564.42433(35)	6	2	4	5	2	3	3	HDO
19	6565.80015(117)	6	3	4	5	2	3	1	HDO
20	6566.39018(62)	3	0	3	2	0	2	2	D_2O
21	6567.46218(46)	7	2	6	6	2	5	3	HDO
22	6571.03174(57)	6	3	4	5	3	3	3	HDO
23	6572.45577(67)	3	1	2	2	1	1	2	D_2O
24	6572.70786(43)	6	3	3	5	3	2	3	HDO
25	6572.86386(103)	10	1	10	9	1	9	3	HDO
26	6572.98588(75)	10	0	10	9	0	9	3	HDO
27	6573.72236(169)	4	1	4	3	1	3	2	D_2O
28	6578.16126(68)	8	3	6	7	2	5	1	HDO
29	6578.42841(50)	6	3	3	5	2	4	1	HDO
30	6580.89585(46)	8	2	7	7	2	6	3	HDO

LINE NO.	WAVENUMBER CM^{-1}, VAC.	J'	K_A'	K_C'	J"	K_A"	K_C"	IDENT. BD.	IDENT. IS.
31	6581.18191(74)	4	2	3	3	2	2	2	D_2O
32	6581.98275(141)	11	0	11	10	0	10	3	HDO
33	6582.10547(31)	5	4	2	4	3	1	1	HDO
33		11	1	11	10	0	10	3	HDO
34	6582.31532(31)	5	4	1	4	3	2	1	HDO
35	6582.99942(61)	7	2	5	6	2	4	3	HDO
36	6583.27152(53)	5	0	5	4	0	4	2	D_2O
37	6583.37483(62)	8	1	7	7	1	6	3	HDO
38	6586.86376(48)	7	3	5	6	3	4	3	HDO
39	6587.61965(103)	6	2	4	5	1	5	4	D_2O
40	6589.53231(75)								
41	6589.60649(106)								
42	6593.76772(86)	9	2	8	8	2	7	3	HDO
43	6595.12044(60)	5	1	4	4	1	3	2	D_2O
44	6595.74295(38)	6	4	2	5	3	3	1	HDO
45	6595.91532(59)								
46	6598.19934(63)	7	0	7	6	0	6	2	D_2O
47	6600.92019(55)								
48	6601.65941(81)	6	2	5	5	2	4	2	D_2O
49	6601.92059(130)								
50	6602.24876(28)								HDO

- "LINE NO.": serial number identifying the spectral line in the map on the opposite page.
- Uncertainty in the last digit of the wavenumber is given in parentheses.
- In column "IDENT." under "BD" 1 represents the band 101-000; 2: 111-000; 3: 021-000; 4: 210-000; 5: 031-000.

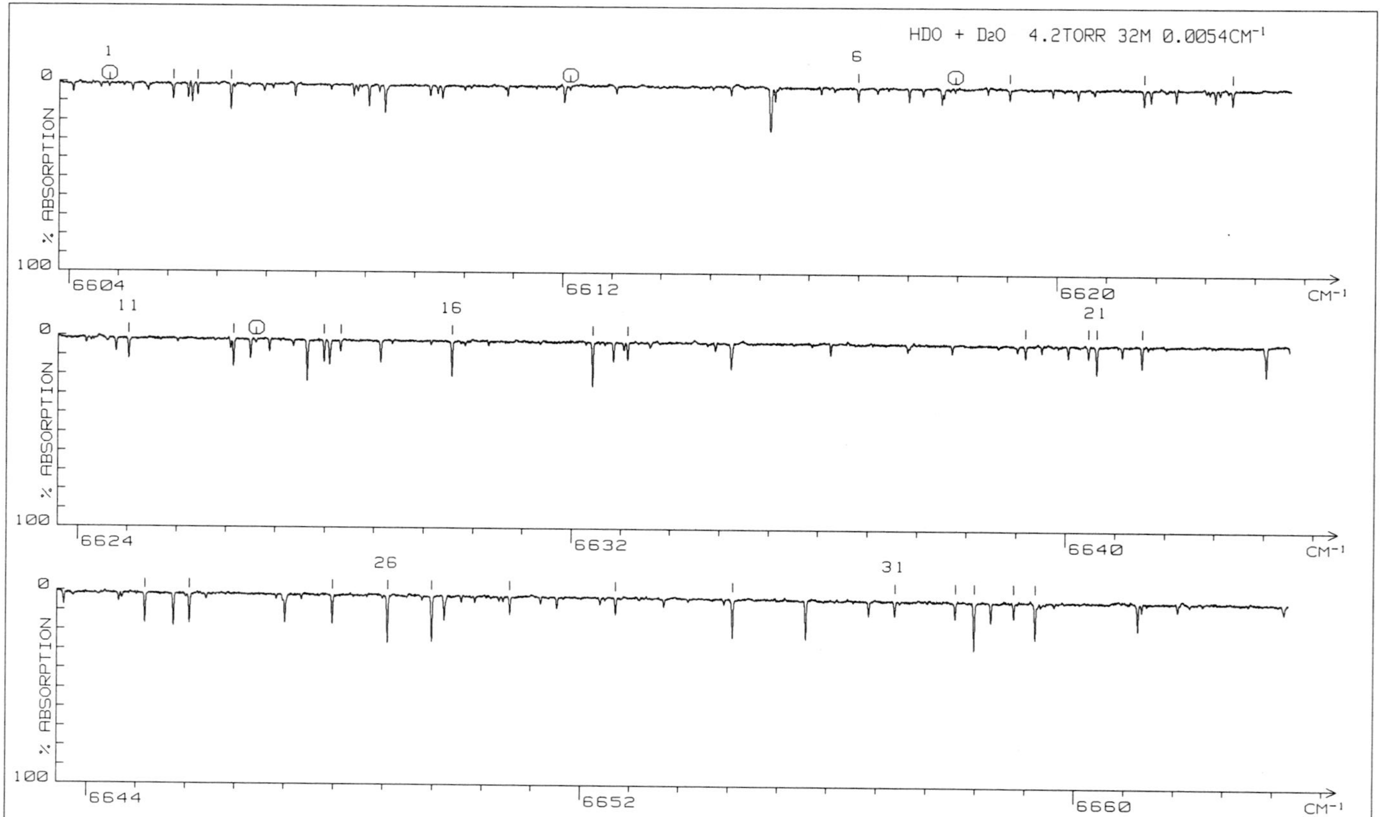

ISOTOPIC IDENTIFICATION – '∇' indicates HDO; '⦶': D_2O; '|': unidentified.

LINE NO.	WAVENUMBER CM^{-1}, VAC.	J'	K_A'	K_C'	J"	K_A"	K_C"	IDENT. BD.	IDENT. IS.
1	6604.64827(118)	6	1	5	5	1	4	2	D_2O
2	6605.68236(117)								
3	6606.07644(99)								
4	6606.62588(73)								
5	6612.11019(110)	9	0	9	8	0	8	2	D_2O
6	6616.78809(101)								
7	6618.35510(111)	10	1	10	9	1	9	2	D_2O
8	6619.23093(171)								
9	6621.39789(115)								
10	6622.82995(126)								
11	6624.82472(89)								
12	6626.52825(69)								
13	6626.89173(134)	9	1	8	8	1	7	2	D_2O
14	6627.98843(82)								
15	6628.25634(156)								
16	6630.05498(54)								
17	6632.34060(37)								
18	6632.91206(104)								
19	6639.34763(133)								
20	6640.36845(131)								
21	6640.50075(48)								
22	6641.22556(82)								
23	6644.94557(50)								
24	6645.66707(68)								
25	6647.98371(68)								
26	6648.87919(47)								
27	6649.58799(47)								
28	6650.85811(108)								
29	6652.57443(112)								
30	6654.47290(150)								
31	6657.10154(147)								
32	6658.08047(148)								
33	6658.38196(45)								
34	6659.02038(109)								
35	6659.36393(55)								

- "LINE NO.": serial number identifying the spectral line in the map on the opposite page.
- Uncertainty in the last digit of the wavenumber is given in parentheses.
- In column "IDENT." under "BD" 1 represents the band 101-000; 2: 111-000; 3: 021-000; 4: 210-000; 5: 031-000.

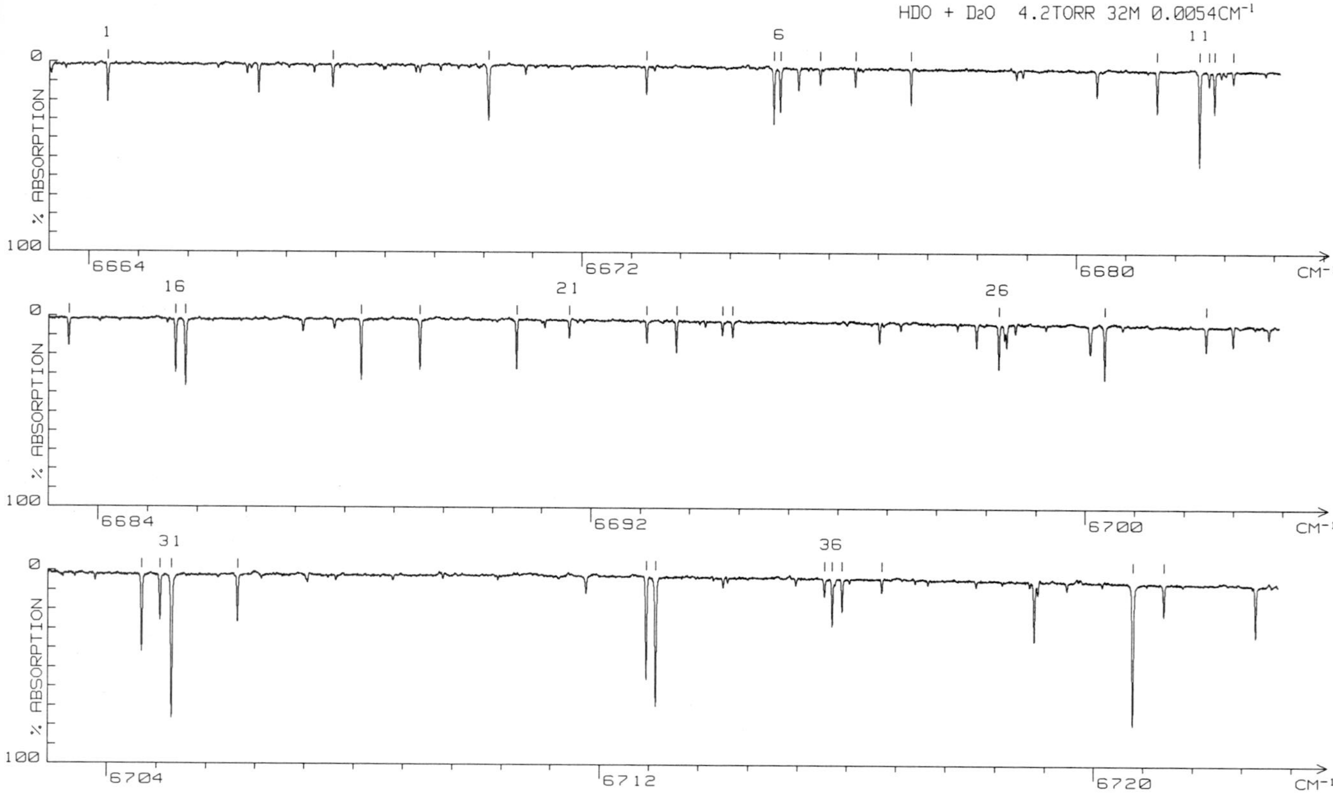

ISOTOPIC IDENTIFICATION – '∇' indicates HDO; 'Ф': D_2O; 'ı': unidentified.

LINE NO.	WAVENUMBER CM^{-1}, VAC.	J'	K_A'	K_C'	J"	$K_A"$	$K_C"$	IDENT. BD. IS.
1	6664.30502(55)							
2	6667.95089(91)							
3	6670.47731(50)							
4	6673.04536(70)							
5	6675.10426(41)							
6	6675.20997(39)							
7	6675.85515(80)							
8	6676.42760(107)							
9	6677.33061(56)							
10	6681.30558(50)							
11	6681.99617(31)							
12	6682.14762(122)							
13	6682.23519(48)							
14	6682.53974(157)							
15	6683.52151(54)							
16	6685.25138(44)							
17	6685.41236(38)							
18	6688.26470(41)							
19	6689.21718(110)							
20	6690.79232(38)							
21	6691.64965(104)							
22	6692.90621(88)							
23	6693.38650(66)							
24	6694.12642(131)							
25	6694.29235(119)							
26	6698.59800(50)							
27	6700.31410(36)							
28	6701.95715(86)							
29	6704.55792(36)							
30	6704.85179(50)							

LINE NO.	WAVENUMBER CM^{-1}, VAC.	J'	K_A'	K_C'	J"	$K_A"$	$K_C"$	IDENT. BD. IS.
31	6705.03939(29)							
32	6706.10922(51)							
33	6712.75094(32)							
34	6712.90529(30)							
35	6715.63044(122)							
36	6715.75971(55)							
37	6715.91923(51)							
38	6716.56440(142)							
39	6720.62742(42)							
40	6721.12515(66)							

- "LINE NO.": serial number identifying the spectral line in the map on the opposite page.
- Uncertainty in the last digit of the wavenumber is given in parentheses.
- In column "IDENT." under "BD" 1 represents the band 101–000; 2: 111–000; 3: 021–000; 4: 210–000; 5: 031–000.

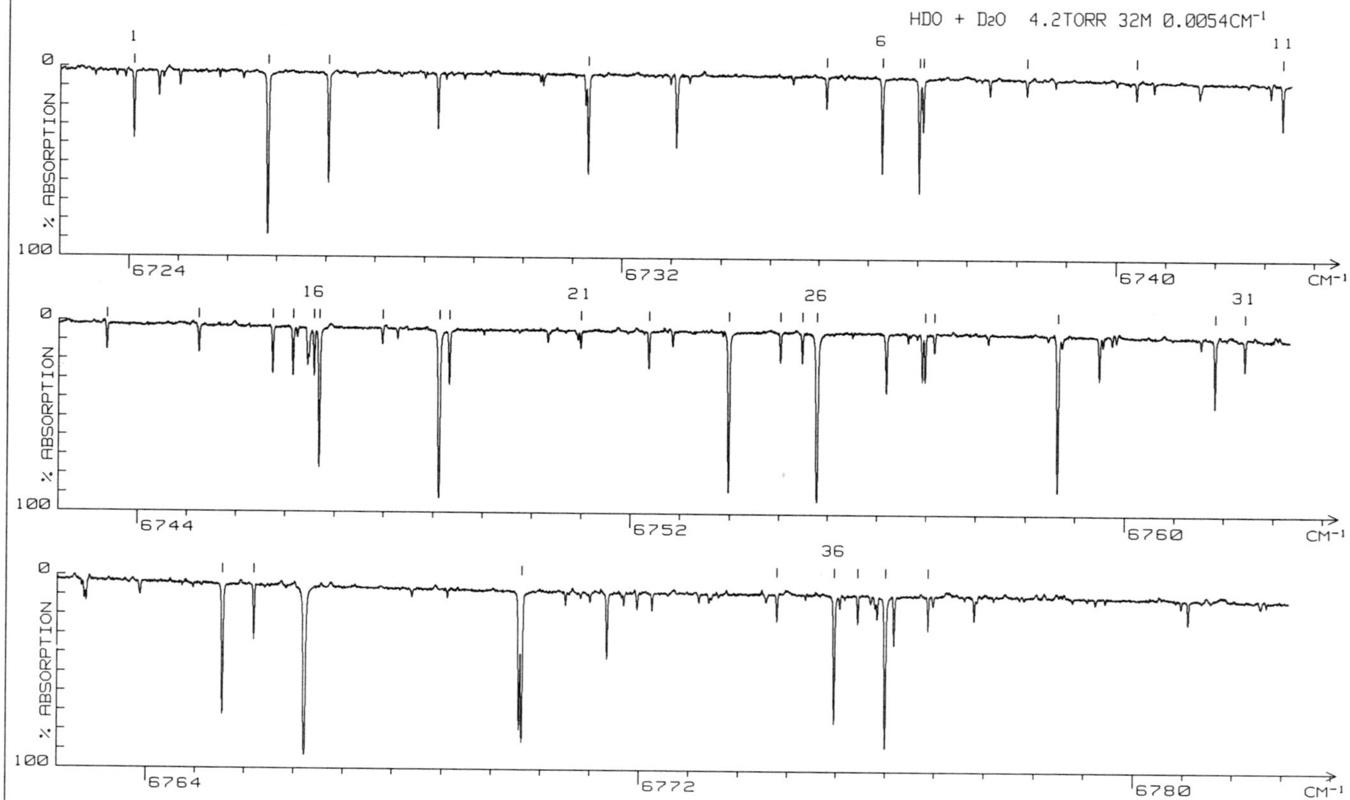

ISOTOPIC IDENTIFICATION – '∇' indicates HDO; '⏀': D_2O; 'ı': unidentified.

LINE NO.	WAVENUMBER CM^{-1}, VAC.	J'	K_A'	K_C'	J"	K_A"	K_C"	IDENT. BD. IS.
1	6724.07004(39)							
2	6726.24999(29)							
3	6727.23288(32)							
4	6731.44587(33)							
5	6735.31096(72)							
6	6736.21551(35)							
7	6736.81944(31)							
8	6736.88500(49)							
9	6738.54778(129)							
10	6740.31523(88)							
11	6742.67780(56)							
12	6743.49852(85)							
13	6744.98772(83)							
14	6746.19232(54)							
15	6746.51876(54)							
16	6746.86224(43)							
17	6746.94857(31)							
18	6747.97559(123)							
19	6748.89099(30)							
20	6749.05537(54)							
21	6751.18496(100)							
22	6752.29535(52)							
23	6753.57873(30)							
24	6754.41593(57)							
25	6754.77839(60)							
26	6755.02204(34)							
27	6756.78066(44)							
28	6756.93262(87)							
29	6758.90993(30)							
30	6761.45260(100)							
31	6761.93268(59)							
32	6765.23730(34)							
33	6765.75098(41)							
34	6770.09758(43)							
35	6774.22351(87)							
36	6775.16587(34)							
37	6775.54797(85)							
38	6775.99496(34)							
39	6776.69158(74)							

- "LINE NO.": serial number identifying the spectral line in the map on the opposite page.
- Uncertainty in the last digit of the wavenumber is given in parentheses.
- In column "IDENT." under "BD" 1 represents the band 101-000; 2: 111-000; 3: 021-000; 4: 210-000; 5: 031-000.

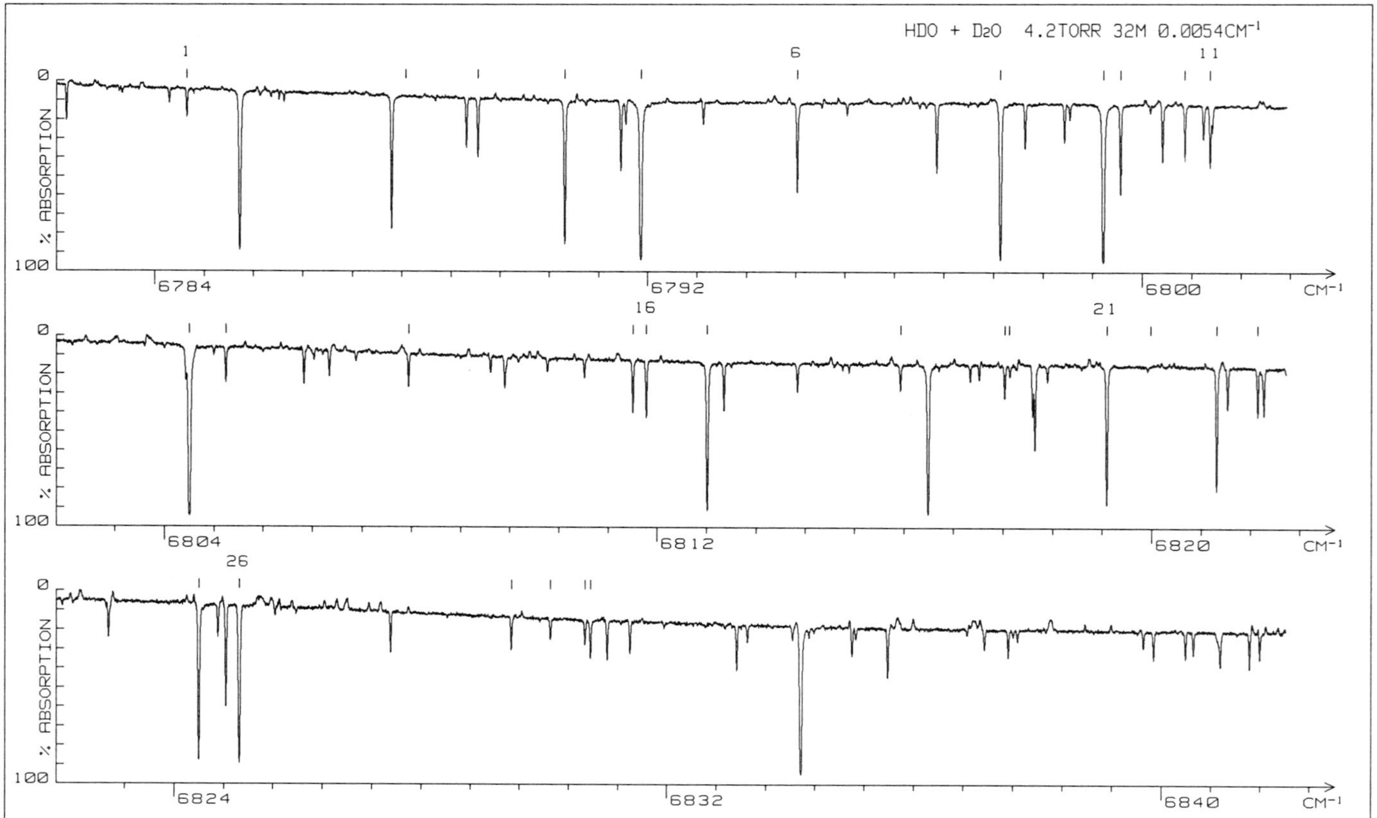

ISOTOPIC IDENTIFICATION – '∇' indicates HDO; '◯': D_2O; '|': unidentified.

LINE NO.	WAVENUMBER CM^{-1}, VAC.	J'	K_A'	K_C'	J"	$K_A"$	$K_C"$	IDENT. BD. IS.	LINE NO.	WAVENUMBER CM^{-1}, VAC.	J'	K_A'	K_C'	J"	$K_A"$	$K_C"$	IDENT. BD. IS.
1	6784.51728(92)																
2	6788.06068(108)																
3	6789.23493(53)																
4	6790.65008(33)																
5	6791.88908(40)																
6	6794.42365(106)																
7	6797.70854(38)																
8	6799.37466(42)																
9	6799.64986(41)																
10	6800.68405(54)																
11	6801.09361(57)																
12	6804.40283(44)																
13	6804.99510(84)																
14	6807.95184(97)																
15	6811.59763(65)																
16	6811.82415(45)																
17	6812.81158(40)																
18	6815.94129(117)																
19	6817.62333(96)																
20	6817.70060(148)																
21	6819.27709(38)																
22	6819.97675(106)																
23	6821.04965(70)																
24	6821.71661(67)																
25	6824.39675(36)																
26	6825.05709(54)																
27	6829.47016(110)																
28	6830.10783(195)																
29	6830.66564(155)																
30	6830.75492(95)																

- "LINE NO.": serial number identifying the spectral line in the map on the opposite page.
- Uncertainty in the last digit of the wavenumber is given in parentheses.
- In column "IDENT." under "BD" 1 represents the band 101-000; 2: 111-000; 3: 021-000; 4: 210-000; 5: 031-000.

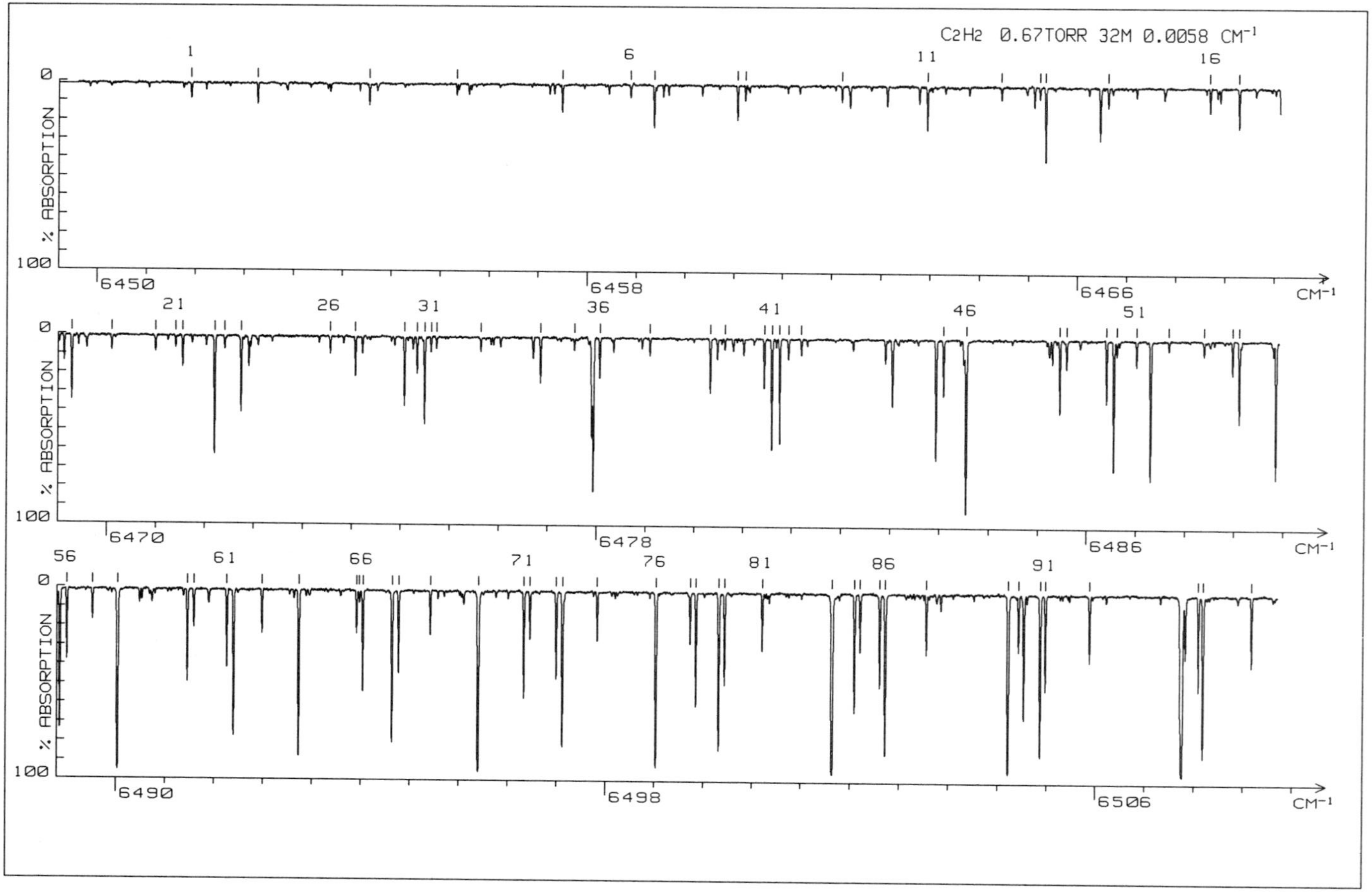

C2H2 0.67TORR 32M 0.0058 CM-1
% ABSORPTION
0
100
6450
6458
6466
6470
6478
6486
6490
6498
6506
CM-1

LINE NO	WAVENUMBER CM^{-1}, VAC	LINE NO	WAVENUMBER CM^{-1}, VAC	LINE NO	WAVENUMBER CM^{-1}, VAC	LINE NO	WAVENUMBER CM^{-1}, VAC
1	6451.53437 (106)	31	6475.29318 (103)	61	6491.80427 (32)	91	6505.17027 (30)
2	6452.61840 (77)	32	6475.38283 (123)	62	6492.37931 (44)	92	6505.89866 (34)
3	6454.44786 (79)	33	6476.10539 (101)	63	6492.97904 (27)	93	6507.66905 (30)
4	6455.86912 (135)	34	6477.08454 (42)	64	6493.92494 (44)	94	6507.74717 (27)
5	6457.59251 (62)	35	6477.63182 (117)	65	6493.97141 (130)	95	6508.53111 (33)
6	6458.70643 (130)	36	6478.04856 (47)	66	6494.02762 (29)		
7	6459.09440 (44)	37	6478.87150 (90)	67	6494.50940 (27)		
8	6460.45731 (50)	38	6479.85625 (38)	68	6494.61513 (32)		
9	6460.58599 (101)	39	6480.09544 (141)	69	6495.13061 (43)		
10	6462.16269 (91)	40	6480.74245 (40)	70	6495.91252 (29)		
11	6463.55597 (43)	41	6480.86619 (28)	71	6496.66428 (29)		
12	6464.75386 (118)	42	6480.99901 (29)	72	6496.76360 (41)		
13	6465.37865 (120)	43	6481.14345 (88)	73	6497.18986 (31)		
14	6465.47383 (32)	44	6481.35376 (97)	74	6497.29117 (27)		
15	6466.50151 (76)	45	6483.66483 (37)	75	6497.85788 (41)		
16	6468.15829 (65)	46	6484.03079 (27)	76	6498.82132 (27)		
17	6468.62762 (44)	47	6485.55780 (32)	77	6499.37885 (40)		
18	6469.42213 (34)	48	6485.66866 (61)	78	6499.47435 (29)		
19	6470.07845 (108)	49	6486.32169 (34)	79	6499.84645 (27)		
20	6470.79028 (100)	50	6486.48876 (101)	80	6499.94222 (30)		
21	6471.11999 (129)	51	6486.80750 (61)	81	6500.56141 (37)		
22	6471.23683 (55)	52	6487.33546 (140)	82	6501.70526 (30)		
23	6471.75707 (28)	53	6487.91036 (104)	83	6502.06854 (28)		
24	6471.92192 (132)	54	6488.37156 (50)	84	6502.15991 (37)		
25	6472.19399 (32)	55	6488.48024 (31)	85	6502.47821 (30)		
26	6473.64476 (88)	56	6489.18898 (33)	86	6502.56888 (27)		
27	6474.05633 (46)	57	6489.60527 (56)	87	6503.24179 (35)		
28	6474.86201 (33)	58	6490.02095 (28)	88	6504.56451 (28)		
29	6475.06725 (48)	59	6491.16054 (30)	89	6504.73344 (37)		
30	6475.18909 (31)	60	6491.26657 (48)	90	6505.08608 (27)		

- "LINE NO.": serial number identifying the spectral line in the map on the opposite page.
- "WAVENUMBER": experimental value.
- Uncertainty in the last digit of the wavenumber is given in parentheses.

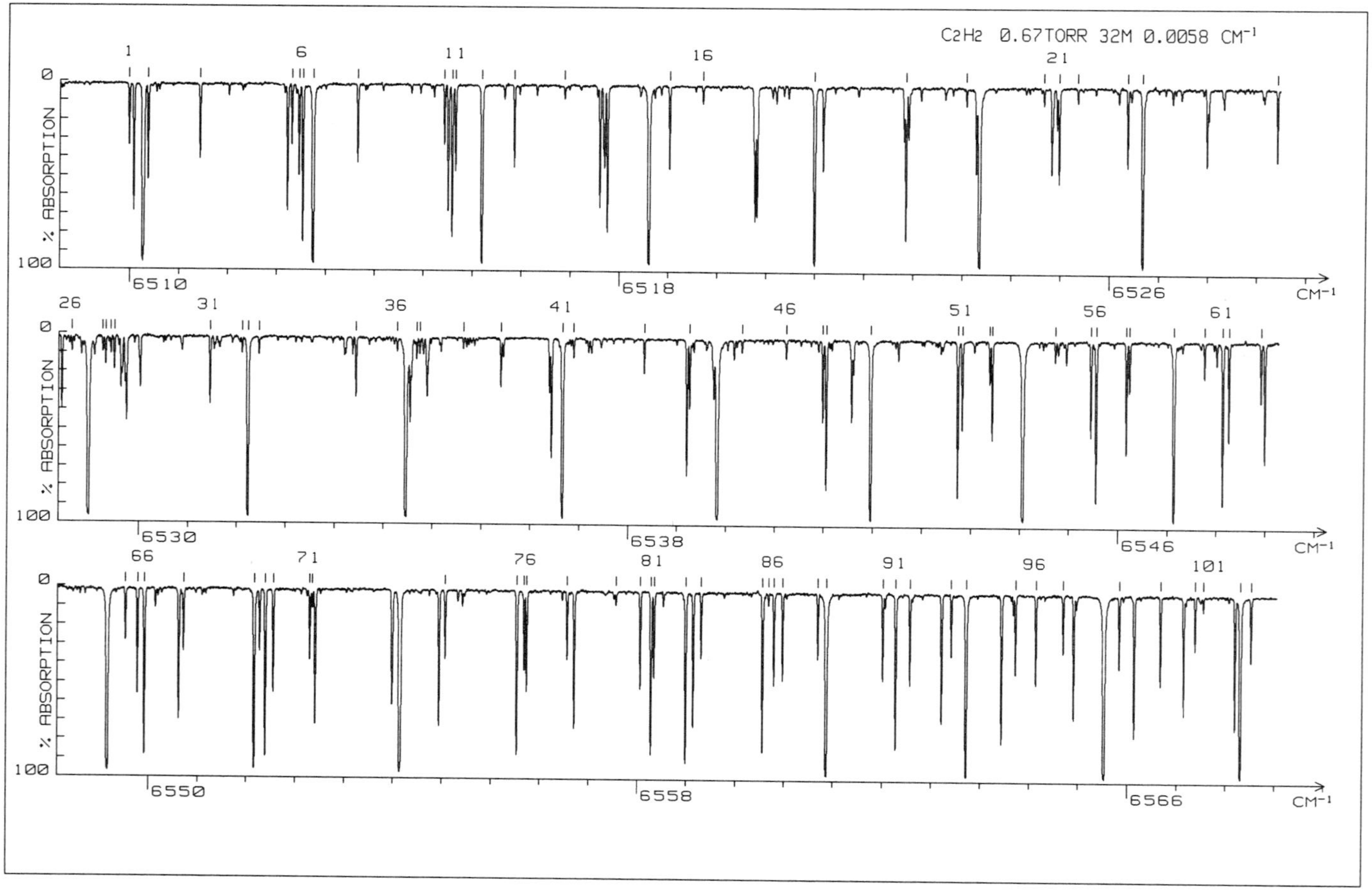

C2H2 0.67TORR 32M 0.0058 CM-1
1
6
11
16
21
26
31
36
41
46
51
56
61
66
71
76
81
86
91
96
101
0
% ABSORPTION
100
6510
6518
6526
6530
6538
6546
6550
6558
6566
CM-1

LINE NO	WAVENUMBER CM^{-1}, VAC	LINE NO	WAVENUMBER CM^{-1}, VAC	LINE NO	WAVENUMBER CM^{-1}, VAC	LINE NO	WAVENUMBER CM^{-1}, VAC
1	6509.98761(36)	31	6531.15302(34)	61	6547.69996(28)	91	6562.21940(28)
2	6510.29879(30)	32	6531.68482(106)	62	6547.79960(30)	92	6562.45521(31)
3	6511.14016(32)	33	6531.78120(30)	63	6548.32256(38)	93	6563.12730(37)
4	6512.64758(36)	34	6531.96046(101)	64	6549.62421(42)	94	6563.37093(30)
5	6512.76173(30)	35	6533.54653(37)	65	6549.82135(30)	95	6564.17768(33)
6	6512.82567(27)	36	6534.22101(126)	66	6549.93542(28)	96	6564.50754(31)
7	6512.99281(32)	37	6534.53809(85)	67	6550.57047(37)	97	6564.95739(38)
8	6513.72541(32)	38	6534.58917(110)	68	6551.73340(29)	98	6565.86702(34)
9	6515.14189(36)	39	6535.30127(135)	69	6551.91721(27)	99	6566.53552(31)
10	6515.27129(27)	40	6535.91589(42)	70	6552.04593(30)	100	6567.10581(40)
11	6515.32718(31)	41	6536.92193(31)	71	6552.63864(35)	101	6567.23879(104)
12	6515.75220(30)	42	6537.10347(85)	72	6552.68157(115)	102	6567.84515(31)
13	6516.28677(31)	43	6538.26134(54)	73	6554.85946(35)	103	6568.01811(36)
14	6517.10608(138)	44	6539.00462(35)	74	6556.03169(28)		
15	6518.82422(31)	45	6539.85257(139)	75	6556.14995(33)		
16	6519.37037(96)	46	6540.58246(89)	76	6556.19131(30)		
17	6521.19588(30)	47	6541.18223(32)	77	6556.84994(35)		
18	6522.68307(28)	48	6541.24118(28)	78	6557.64587(136)		
19	6523.66592(88)	49	6541.96111(31)	79	6558.05023(31)		
20	6524.93575(99)	50	6543.38041(27)	80	6558.22613(28)		
21	6525.18349(30)	51	6543.45236(32)	81	6558.27604(32)		
22	6525.48764(106)	52	6543.90242(46)	82	6558.79313(28)		
23	6526.29332(32)	53	6543.94388(30)	83	6559.04530(35)		
24	6526.53898(30)	54	6544.97287(85)	84	6560.04320(28)		
25	6528.73519(33)	55	6545.55292(30)	85	6560.14173(141)		
26	6528.89805(107)	56	6545.63871(28)	86	6560.23539(31)		
27	6529.40617(108)	57	6546.12551(29)	87	6560.37792(31)		
28	6529.45188(65)	58	6546.17840(42)	88	6560.95633(36)		
29	6529.53825(95)	59	6546.89846(30)	89	6561.09524(32)		
30	6529.59386(55)	60	6547.40027(52)	90	6562.01016(32)		

- "LINE NO.": serial number identifying the spectral line in the map on the opposite page.
- "WAVENUMBER": experimental value.
- Uncertainty in the last digit of the wavenumber is given in parentheses.

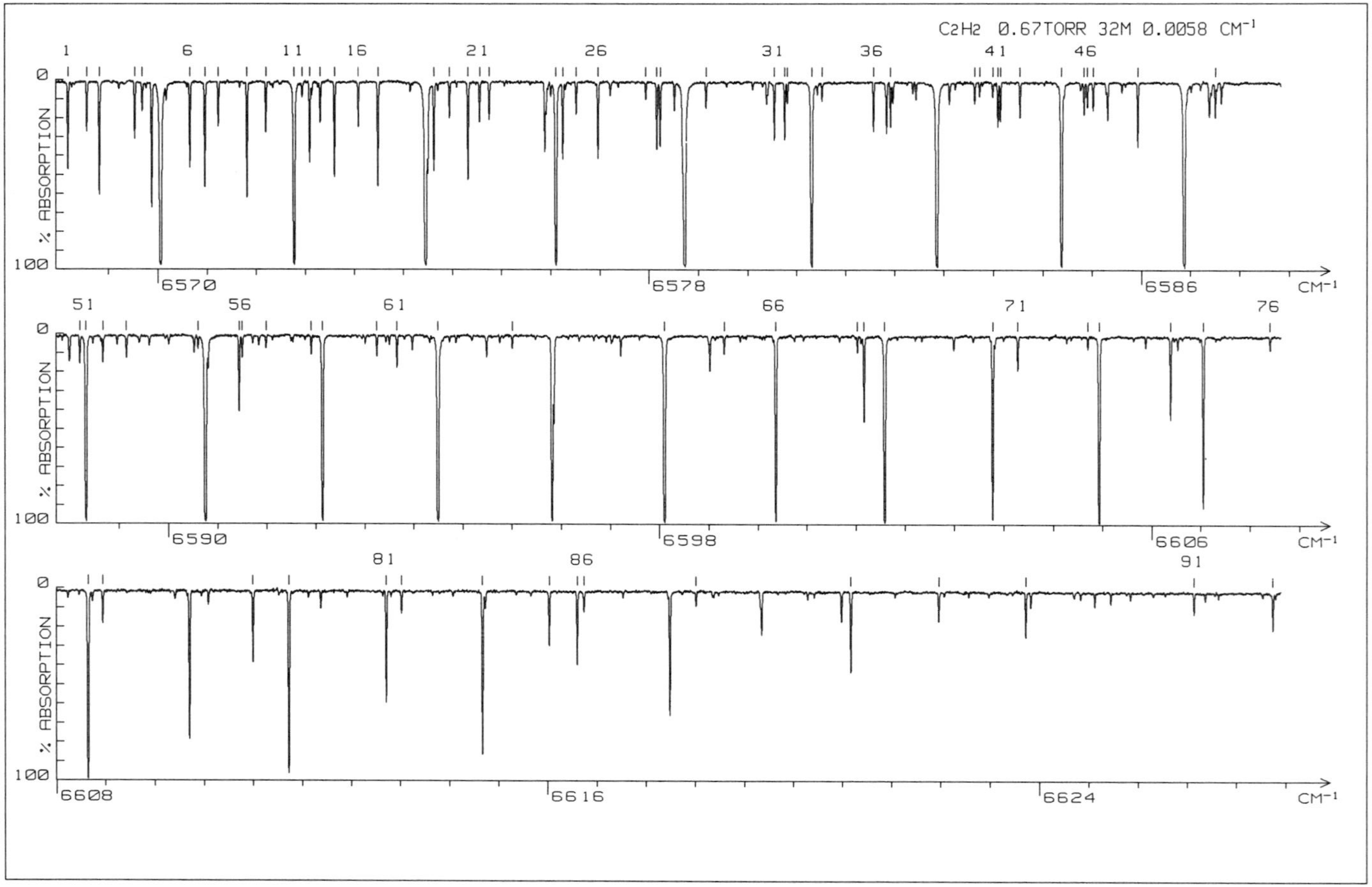
C2H2 0.67TORR 32M 0.0058 CM-1
% ABSORPTION
CM-1
6570
6578
6586
6590
6598
6606
6608
6616
6624

LINE NO	WAVENUMBER CM^{-1}, VAC	LINE NO	WAVENUMBER CM^{-1}, VAC	LINE NO	WAVENUMBER CM^{-1}, VAC	LINE NO	WAVENUMBER CM^{-1}, VAC
1	6568.53873(32)	31	6580.03786(38)	61	6593.72297(59)	91	6626.50924(88)
2	6568.85271(43)	32	6580.20674(39)	62	6594.38815(33)	92	6627.79152(57)
3	6569.05565(29)	33	6580.25161(89)	63	6595.60349(129)		
4	6569.62650(39)	34	6580.64479(32)	64	6598.07804(32)		
5	6569.74404(64)	35	6580.81729(94)	65	6599.04243(98)		
6	6570.51714(32)	36	6581.65257(43)	66	6599.88350(29)		
7	6570.76100(30)	37	6581.93299(46)	67	6601.21716(100)		
8	6570.97945(46)	38	6583.29784(87)	68	6601.32496(33)		
9	6571.45247(29)	39	6583.37507(117)	69	6601.66240(31)		
10	6571.75667(42)	40	6583.58841(126)	70	6603.41488(28)		
11	6572.21577(31)	41	6583.67185(45)	71	6603.82060(58)		
12	6572.34680(114)	42	6583.71344(50)	72	6604.95694(129)		
13	6572.47093(32)	43	6584.02984(54)	73	6605.14082(30)		
14	6572.64258(49)	44	6584.70268(32)	74	6606.29731(33)		
15	6572.87745(31)	45	6585.07028(61)	75	6606.84029(28)		
16	6573.26555(46)	46	6585.12325(72)	76	6607.91971(120)		
17	6573.58794(30)	47	6585.21733(67)	77	6608.51323(29)		
18	6574.49791(32)	48	6585.93686(37)	78	6608.75617(62)		
19	6574.74876(51)	49	6587.20582(55)	79	6611.19689(36)		
20	6575.05194(31)	50	6588.55390(68)	80	6611.77924(28)		
21	6575.24011(49)	51	6588.65562(31)	81	6613.37241(30)		
22	6575.39578(51)	52	6588.93562(69)	82	6613.61988(88)		
23	6576.48238(31)	53	6589.31609(84)	83	6614.93883(28)		
24	6576.59448(34)	54	6590.47592(128)	84	6616.02543(42)		
25	6576.81132(57)	55	6591.14880(34)	85	6616.47869(35)		
26	6577.16493(34)	56	6591.20021(86)	86	6616.58710(90)		
27	6577.95046(100)	57	6591.58408(144)	87	6618.41312(133)		
28	6578.12850(36)	58	6592.32002(91)	88	6620.93791(34)		
29	6578.18324(36)	59	6592.50363(30)	89	6622.37182(69)		
30	6578.92280(69)	60	6593.39595(88)	90	6623.77715(47)		

- "LINE NO.": serial number identifying the spectral line in the map on the opposite page.
- "WAVENUMBER": experimental value.
- Uncertainty in the last digit of the wavenumber is given in parentheses.

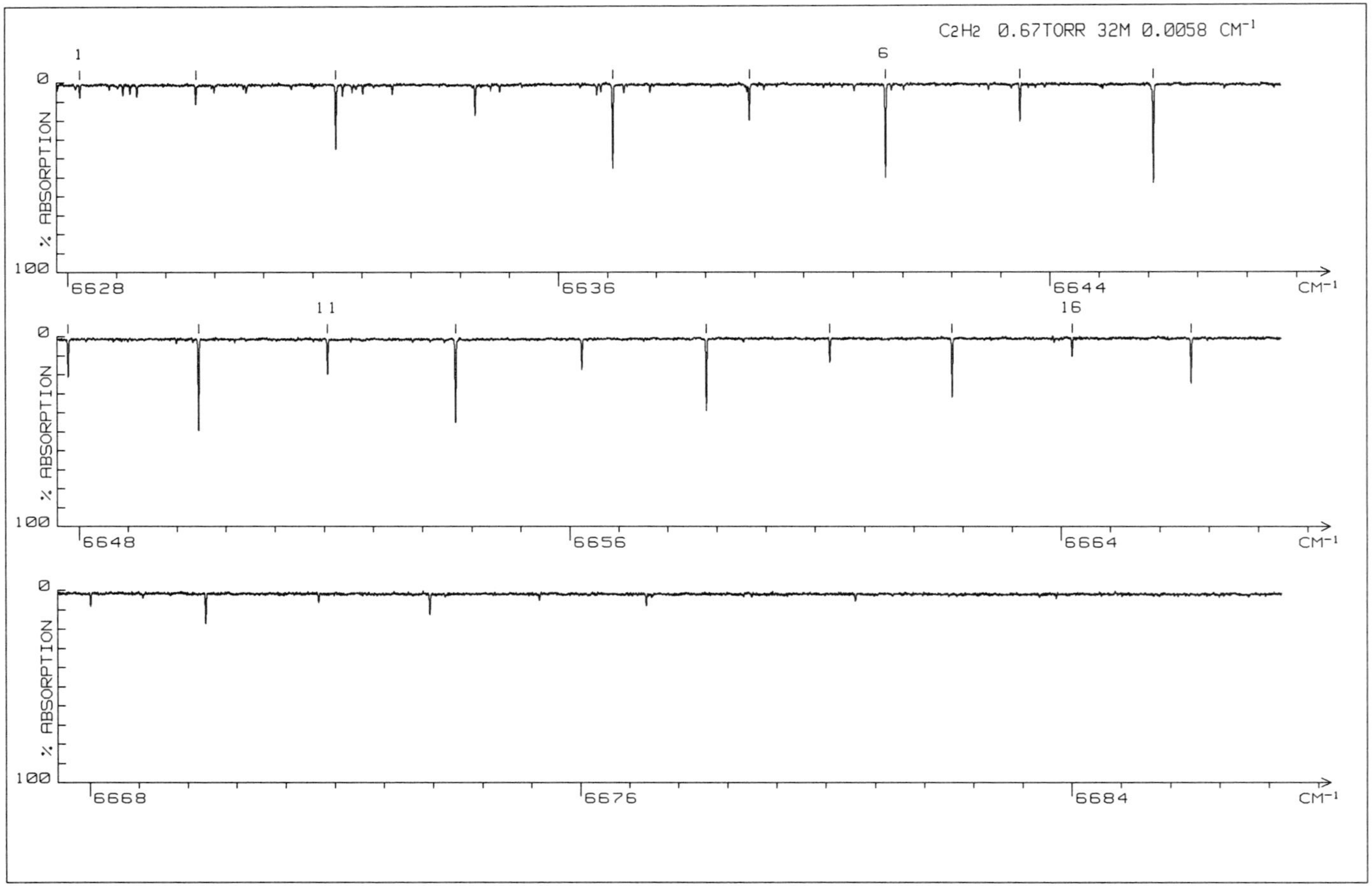

C2H2 0.67TORR 32M 0.0058 CM-1
1
6
11
16
% ABSORPTION
0
100
6628
6636
6644
6648
6656
6664
6668
6676
6684
CM-1

LINE NO	WAVENUMBER CM^{-1}, VAC
1	6628.20589 (134)
2	6630.09189 (102)
3	6632.37519 (39)
4	6636.88674 (34)
5	6639.11453 (58)
6	6641.32306 (33)
7	6643.51169 (57)
8	6645.68082 (32)
9	6647.82877 (55)
10	6649.95562 (33)
11	6652.06056 (60)
12	6654.14244 (35)
13	6658.23579 (37)
14	6660.24565 (82)
15	6662.22948 (43)
16	6664.18664 (113)
17	6666.11691 (52)

- "LINE NO.": serial number identifying the spectral line in the map on the opposite page.
- "WAVENUMBER": experimental value.
- Uncertainty in the last digit of the wavenumber is given in parentheses.

SECTION II

HETERODYNE FREQUENCY MEASUREMENTS

It is well known that frequencies determined by the heterodyne techniques have high accuracies. To supplement the frequency information given in the previous handbook,[1] we have collected most [23] of the recent measurements and included them in pages 574–616. A summary of these data appears on the opposite page (573).

[23]For instance, the "Heterodyne Frequency Measurements of $^{12}C^{16}O$ Laser Transitions" by M. Schneider, K. M. Evenson, M. D. Vanek, D. A. Jennings, J. S. Wells, A. Stahn, and W. Urban, *J. Mol. Spectrosc.* 135, 197–206 (1989)" are not included here because of their specialized nature.

Heterodyne frequency measurements are given between pages 574 and 610 as summarized below. In converting frequencies to wavenumbers in cm^{-1}, the speed of light $c = 2.99792458 \times 10^{10}$ cm/sec has been used.

Table No.	Molecular Species	Spectral Range (cm^{-1}	Page
B.1	CO	3.8 - 144.8	574
B.2	$H^{37}Cl$	20.8 - 206.4	575
B.3	$H^{35}Cl$	20.8 - 206.8	576
B.4	OCS	26.3 - 37.6	577–578
	N_2O	50.1 - 50.3	578
B.5	HF	41.1 - 204.6	579
		956.2 - 1086.0	579
B.6	$^{12}C^{16}O_2$	888.5 - 960.0	580–583
B.7	N_2O	896.9 - 989.7	584
B.8	N_2O	1037.1 - 1084.6	585
B.9	N_2O	1257.3 - 1335.1	586
B.10	N_2O	1257.3 - 1339.9	587
B.11	SO_2	1349.5 - 1394.2	588–590
B.12	OCS	1363.6 - 1397.5	591
B.13	CS_2	1462.9 - 1549.1	592
B.14	N_2O	1591.3 - 1672.8	593
B.15	OCS	1650.7 - 1850.8	594–595
B.16	NO	1750.0 - 1931.8	596–597
B.17	OCS	1850.7 - 1914.7	598–600
B.18	OCS	1991.1 - 2085.4	601–602
B.19	CO	1995.1 - 2068.9	603
B.20	CO	2008.5 - 2081.3	604
B.21	CO	2082.0 - 2115.7	605
B.22	OCS	2898.6 - 2912.8	606
B.23	HF	3693.4 - 3877.8	607
B.24	CO_2	3697.9 - 3733.5	608
B.25	C_2H_2, HCN, NH_3, D_2O and HDO	6593.3 - 6564.6	609–610

TABLE B.1 Far Infrared Pure Rotational Frequency Measurements[1] on CO in Its Ground Electronic State

J''	*Frequency* [a] *(MHz)*	*Wavenumber* [a] *(cm^{-1})*
0	115 271.204(5)	3.845 033 49(17)
1	230 538.000(1)	7.689 919 94(33)
2	345 795.991(1)	11.534 512 69(33)
3	461 040.769(2)	15.378 664 70(67)
4	576 267.922(32)	19.222 228 80(107)
5	691 473.090(12)	23.065 059 56(40)
6	806 651.806(10)	26.907 007 98(33)
7	921 799.711(10)	30.747 928 66(33)
8	1 036 912.403(10)	34.587 674 75(33)
9	1 151 985.454(11)	38.426 098 56(37)
10	1 264 014.504(10)	42.162 985 43(33)
11	1 381 995.110(13)	46.098 394 84(43)
12	1 496 922.915(12)	49.931 973 77(40)
13	1 611 793.517(11)	53.763 644 61(37)
15	1 841 345.512(11)	61.420 674 97(37)
16	1 956 018.142(11)	65.245 742 17(37)
17	2 070 615.998(14)	69.068 315 19(47)
18	2 185 134.685(13)	72.888 247 41(43)
19	2 299 569.847(10)	76.705 393 54(33)
20	2 413 917.118(11)	80.519 607 93(37)
21	2 528 172.060(11)	84.330 742 57(37)
23	2 756 387.590(17)	91.943 193 25(57)
24	2 870 339.414(13)	95.744 216 95(43)
25	2 984 181.461(14)	99.541 578 89(47)
26	3 097 909.366(17)	103.335 133 47(57)
29	3 438 364.618(10)	114.691 498 28(33)
30	3 551 592.365(10)	118.468 369 37(33)
32	3 777 635.731(16)	126.008 364 46(53)
33	3 890 442.724(13)	129.771 200 72(43)
35	4 115 605.592(22)	137.281 825 55(73)
37	4 340 138.124(43)	144.771 424 64 (140)

[1] T. D. Varberg and K. M. Evenson, *Astrophys. J.* **385**, 763-765 (1992).
[a] The estimated uncertainty in the last digits is given in parentheses.

TABLE B.2 Pure Rotational Frequency Measurements[1] for $H^{37}Cl$

J''	*Frequency* [a] *(MHz)*	*Wavenumber* [a] *(cm^{-1})*
0	624 975.04(10)	20.846 923(3)
1	1 249 571.39(20)	41.681 215(7)
2	1 873 410.07(20)	62.490 233(7)
3	2 496 115.33(20)	83.261 445(7)
4	3 117 308.72(20)	103.982 226(7)
5	3 736 615.72(25)	124.640 084(8)
6	4 353 662.88(25)	145.222 562(8)
7	4 968 078.97(30)	165.717 277(10)
8	5 579 495.42(30)	186.111 934(10)
9	6 187 546.55(40)	206.394 337(13)

[1] I. G. Nolt, J. V. Radostitz, G. DiLonardo, K. M. Evenson, D. A. Jennings, K. R. Leopold, M. D. Vanek, L. R. Zink, A. Hinz, and K. V. Chance, *J. Mol. Spectrosc.* **125**, 274-287 (1987).

[a] The estimated uncertainty in the last digits is given in parentheses.

TABLE B.3 Pure Rotational Frequency Measurements[1] for $H^{35}Cl$

J''	*Frequency*[a] *(MHz)*	*Wavenumber*[a] *(cm^{-1})*
0	625 915.20(10)	20.878 284(3)
1	1 251 450.48(20)	41.743 895(7)
2	1 876 226.60(20)	62.584 183(7)
3	2 499 864.42(20)	83.386 501(7)
4	3 121 986.49(20)	104.138 260(7)
5	3 742 216.58(25)	124.826 909(8)
6	4 360 179.99(25)	145.439 949(8)
7	4 975 504.67(30)	165.964 971(10)
8	5 587 820.10(40)	186.389 616(13)
9	6 196 759.71(35)	206.701 655(12)

[1] I. G. Nolt, J. V. Radostitz, G. DiLonardo, K. M. Evenson, D. A. Jennings, K. R. Leopold, M. D. Vanek, L. R. Zink, A. Hinz, and K. V. Chance, *J. Mol. Spectrosc.* **125**, 274-287 (1987).

[a] The estimated uncertainty in the last digits is given in parentheses.

TABLE B.4.I Rotational Transitions Measured[1] for OCS and N_2O

Rot. Trans. $J' \leftarrow J''$	*Vib. State*	*Frequency* [a] *(MHz)*	*Wavenumber* [a] *(cm^{-1})*
		$^{16}O^{12}C^{32}S$	
65-64	$00^0\ 0$	78 916 3.801(200)	26.323 67(1)
65-64	$02^{2f}\ 1$	78 981 4.147(200)	26.345 36(1)
65-64	$01^{1e}\ 0$	79 010 5.845(200)	26.355 09(1)
66-65	$00^0\ 1$	79 883 4.732(200)	26.646 26(1)
66-65	$02^{2e}\ 1$	80 158 4.243(200)	26.737 97(1)
66-65	$02^0\ 1$	80 190 1.657(200)	26.748 56(1)
66-65	$01^{1f}\ 0$	80 305 0.609(200)	26.786 88(1)
66-65	$02^{2e}\ 0$	80 372 4.281(200)	26.809 36(1)
66-65	$02^0\ 0$	80 396 9.157(200)	26.817 52(1)
66-65	$02^{2f}\ 0$	80 399 3.512(200)	26.818 34(1)
66-65	$03^{1e}\ 0$	80 411 5.917(200)	26.822 42(1)
66-65	$03^{3e}\ 0$	80 521 4.913(200)	26.859 08(1)
66-65	$03^{3f}\ 0$	80 524 4.142(200)	26.869 05(1)
66-65	$03^{1f}\ 0$	80 571 8.448(200)	26.875 87(1)
67-66	$00^0\ 2$	80 837 8.994(200)	26.964 62(1)
67-66	$01^{1f}\ 1$	81 293 2.683(200)	27.116 52(1)
91-90	$00^0\ 0$	110 290 5.314(200)	36.788 96(1)
91-90	$02^{2e}\ 0$	110 595 1.322(200)	36.890 57(1)
91-90	$02^0\ 0$	110 689 6.674(200)	36.922 10(1)
92-91	$01^{1e}\ 0$	111 624 1.324(200)	37.233 80(1)
92-91	$01^{1f}\ 0$	111 739 9.237(200)	37.272 43(1)
92-91	$02^{2f}\ 0$	111 868 4.910(200)	37.315 31(1)
93-92	$00^0\ 1$	112 350 8.720(200)	37.476 22(1)
93-92	$00^0\ 0$	112 696 6.568(200)	37.591 56(1)

[1] M. D. Vanek, D. A. Jennings, J. S. Wells, and A. G. Maki, *J. Mol. Spectrosc.* **138**, 79-83 (1989).

[a] The estimated uncertainty in the last digits is given in parentheses.

Rot. Trans. $J' \leftarrow J''$	*Vib. State*	*Frequency* [a] *(MHz)*	*Wavenumber* [a] *(cm^{-1})*
		$^{16}O^{12}C^{34}S$	
67-66	00^0 1	79 110 7.050(200)	26.388 49(1)
68-67	00^0 0	80 530 2.973(200)	26.862 02(1)
68-67	02^{2e} 0	80 776 3.668(200)	26.944 10(1)
68-67	02^{2f} 0	80 804 9.576(200)	26.953 63(1)
68-67	02^0 0	80 806 2.907(200)	26.954 08(1)
69-68	01^{1e} 0	81 808 1.708(200)	27.288 27(1)
69-68	01^{1f} 0	81 891 4.201(200)	26.982 47(1)
		$^{14}N^{14}N^{16}O$	
60-59	00^0 0	150 283 2.222(200)	50.129 09(1)
60-59	01^{1e} 0	150 337 4.990(200)	50.147 19(1)
60-59	01^{1f} 0	150 619 8.421(200)	50.241 37(1)

[a] The estimated uncertainty in the last digits is given in parentheses.

TABLE B.5 Pure Rotational Frequency Measurements[1, 2] for HF

J''	Frequency[a] (MHz)	Wavenumber[a] (cm^{-1})
0	1 232 476.27(14)	41.110 983 2(47)
1	2 463 428.14(21)	82.171 117 9(70)
2	3 691 334.65(28)	123.129 670 2(93)
3	4 914 682.57(62)	163.936 164 (21)
4	6 131 968.40(130)	204.540 449 (43)
...	...	...
26	2 866 669 99(8)	956.228 16 (27)
32	3 255 713 2(8)	1 085.989 03 (27)

[1] I. G. Nolt, J. V. Radostitz, G. DiLonardo, K. M. Evenson, D. A. Jennings, K. R. Leopold, M. D. Vanek, L. R. Zink, A. Hinz, and K. V. Chance, *J. Mol. Spectrosc.* **125**, 274-287 (1987) for J" = 0 - 4.

[2] D. A. Jennings and J. S. Wells, *J. Mol. Spectrosc.* **130**, 267-268 (1988) for J" = 26 and 32.

[a] The estimated uncertainty in the last digits is given in parentheses.

TABLE B.6.I Measured $^{12}C^{16}O_2$ Hot Band 01^11-$[11^10,03^10]$ Frequencies[1]

J	Doppler Limited Measurements (NRC)[a,b] *Measured Freq. (MHz)*	*Wavenumber (cm^{-1})*
P-branch		
43	26 636 675.00(60)	888.503 840(20)
42	26 647 450.85(60)	888.863 283(20)
41	26 697 509.31(60)	890.533 054(20)
40	26 709 734.71(60)	890.940 849(20)
39	26 757 659.53(60)	892.539 449(20)
38	26 771 244.65(60)	892.992 600(20)
37	26 817 128.96(60)	894.523 136(20)
36	26 831 984.34(60)	895.018 658(20)
35	26 875 918.37(60)	896.484 139(20)
34	26 891 955.39(60)	897.019 077(20)
33	26 934 029.43(60)	898.422 516(20)
9	27 578 702.70(60)	919.926 501(20)

[1] F. R. Peterson, J. S. Wells, K. J. Siemsen, A. M. Robinson, and A. G. Maki, *J. Mol. Spectrosc.* **105**, 324-330 (1984).

[a] The estimated uncertainty in the last digits is listed in parentheses.

[b] P(43) to P(33) and P(9) are new measurements made at NRC on lines observed in laser emission.

Saturated Absorption Measurements (NBS)[a, b]

J	*Measured Freq.* (MHz)	*Wavenumber* (cm^{-1})
P-branch		
35	26 875 918.5 346(150)	896.484 144 861(50)
32	26 951 160.8 098(150)	898.993 957 006(50)
29	27 048 219.3 655(150)	902.231 481 937(50)
26	27 124 201.4 833(150)	904.765 972 574(50)
23	27 214 437.5 437(150)	907.775 923 559(50)
20	27 290 413.4 286(150)	910.310 206 289(50)
18	27 344 305.9 459(150)	912.107 867 166(50)
15	27 426 627.6 987(150)	914.853 825 265(50)
12	27 501 461.2 833(150)	917.350 004 959(50)

[a] The estimated uncertainty in the last digits is listed in parentheses.

[b] New measurements made at NBS. Each laser frequency was actively stabilized to the saturated absorption line with a room temperature absorption cell filled to 5.3 Pa pressure with CO_2.

Saturated Absorption Measurements (NRC)[a,b]

J	*Measured Freq. (MHz)*	*Wavenumber (cm^{-1})*
P-branch		
29	27 048 219.29(20)	902.231 479(7)
24	27 180 362.18(20)	906.639 292(7)
23	27 214 437.86(20)	907.775 934(7)
22	27 235 765.80(20)	908.487 358(7)
20	27 290 413.26(20)	910.310 201(7)
19	27 321 878.92(20)	911.359 782(7)
17	27 374 589.83(20)	913.118 029(7)
16	27 397 444.06(20)	913.880 364(7)
15	27 426 627.60(20)	914.853 822(7)
14	27 449 828.96(20)	915.627 736(7)
12	27 501 461.11(20)	917.349 999(7)
11	27 528 684.40(20)	918.258 070(7)
Q-branch		
11	27 786 661.90(20)	926.863 274(7)
R-branch		
23	28 306 272.85(20)	944.195 629(7)

[a] Measurements made earlier at NRC, but corrected to reflect new values for the CO_2 reference lasers.

[b] The estimated uncertainty in the last digits is listed in parentheses.

	Doppler Limited Measurements (NBS)[a, b]	
J	*Measured Freq.* (*MHz*)	*Wavenumber* (cm^{-1})
R-branch		
4	27 909 939.0(40)	930.975 355(13)
12	28 083 279.4(40)	936.757 368(13)
26	28 357 183.7(100)	945.893 832(33)
32	28 462 935.1(50)	949.421 319(17)
47	28 717 240.6(50)	957.904 038(17)
52	28 763 525.2(50)	959.447 926(17)

[a] The Doppler limited R-branch lines were measured by tunable diode laser heterodyne techniques.
[b] The estimated uncertainty in the last digits is listed in parentheses.

TABLE B.7 Heterodyne Frequency Measurements[1] on N_2O Near 930 cm^{-1}

Transition [a]	*Frequency* [b] *(MHz)*	*Wavenumber* [b] *(cm^{-1})*
P(24)F	26 889 735.4(60)	896.945 03(20)
P(46)A	26 891 673.1(50)	897.009 66(17)
P(24)E	26 893 253.4(50)	897.062 37(17)
P(34)A	27 239 172.6(50)	908.601 00(17)
P(11)F	27 239 823.9(80)	908.622 72(27)
P(11)E	27 240 857.4(50)	908.657 20(17)
R(15)F	27 908 130.1(30)	930.915 02(10)
R(15)E	27 908 336.9(30)	930.921 91(10)
R(22)F	28 068 189.7(30)	936.254 03(10)
R(22)E	28 069 142.1(30)	936.285 80(10)
P(3)A	28 070 758.0(30)	936.339 70(10)
R(30)F	28 244 287.3(40)	942.128 01(13)
R(3)A	28 245 169.6(40)	942.157 44(13)
R(30)E	28 246 618.2(40)	942.205 76(13)
R(38)F	28 412 973.5(70)	947.754 78(23)
R(10)A	28 414 571.2(70)	947.808 07(23)
R(48)F	28 613 129.8(90)	954.431 28(30)
R(71)A	29 669 511.3(100)	989.668 37(33)

[1] A. G. Maki, J. S. Wells, and M. D. Vanek, *J. Mol. Spectrosc.* **138**, 84-88 (1989).

[a] The band definitions are as follows: $A = 10^00 - 00^01$, $E = 11^{1e}0 - 01^{1e}1$, and $F = 11^{1f}0 - 01^{1f}1$.

[b] The estimated uncertainty in the last digits is given in parentheses.

TABLE B.8 Heterodyne Frequency Measurements[1] for the N_2O Band at 9.5μm

Trans.	*Frequency* [a] *(MHz)*	*Wavenumber* [a] *(cm^{-1})*
P(20)	31 094 136.7(40)	1037.188 76(13)
P(14)	31 270 692.8(40)	1043.078 04(13)
P(12)	31 327 521.1(40)	1044.973 62(13)
P(6)	31 491 840.5(80)	1050.454 73(27)
P(4)	31 544 541.7(40)	1052.212 65(13)
R(21)	32 134 816.8(40)	1071.902 11(13)
R(30)	32 298 843.1(40)	1077.373 44(13)
R(37)	32 412 768.3(40)	1081.173 57(13)
R(44)	32 515 225.9(70)	1084.591 19(23)

[1] L. R. Zink, J. S. Wells, and A. G. Maki, *J. Mol. Spectrosc.* **123**, 426-433 (1987).
[a] The estimated uncertainty in the last digits is given in parentheses.

TABLE B.9 Heterodyne Frequency Measurements[1] on Various Bands of N_2O Near 1280 cm^{-1}

Trans. Rot.[c]	Band[a]	Measured Freq.[b] (MHz)	Wavenumber[b] (cm^{-1})
P(26)	I	37 695 138.6(20)	1257.37 448(7)
P(15)	J	37 696 318.9(20)	1257.41 385 (7)
P(24)	U	37 699 186.0(20)	1257.50 948(7)
P(8)	J	37 873 541.4(20)	1263.32 535(7)
P(34)	S	37 874 751.3(20)	1263.36 571(7)
P(4)	U	38 225 885.4(20)	1275.07 828(7)
P(25)[e]	T[c]	38 226 482.9(30)	1275.09 821(10)
P(25)[f]	T	38 226 668.5(30)	1275.10 440(10)
P(15)	S	38 400 350.6(20)	1280.89 782(7)
R(2)	U	38 401 010.9(20)	1280.91 984(7)
P(23)	W	38 404 371.3(20)	1281.03 193(7)
R(0)	I	38 409 060.2(40)[d]	1281.18 834(13)
P(15)	T	38 496 873.0(30)	1284.11 746(10)
R(6)	U	38 498 898.3(30)	1284.18 501(10)
P(8)	S	38 584 684.3(20)	1287.04 653(7)
R(33)	J	38 838 294.4(40)	1295.50 605(13)
R(1)	S	38 839 105.2(40)	1295.53 309(13)
R(18)	I	38 840 091.7(20)	1295.56 600(7)
R(21)	U	38 851 769.4(20)	1295.95 553(7)
R(29)	U	39 030 777.7(60)	1301.92 660(20)
R(5)	T	39 033 831.1(60)	1302.02 845(20)
R(40)	U	39 266 556.4(90)[d]	1309.79 133(30)
R(19)	S	39 270 101.8(60)	1309.90 959(20)
R(47)	I	39 456 943.8(80)	1316.14 197(27)
R(69)	U	39 833 272.8(80)	1328.69 496(27)
R(40)[e]	T[c]	39 834 054.1(70)	1328.72 102(23)
R(40)[f]	T	39 834 244.5(70)	1328.72 737(23)
R(50)	S	39 929 466.5(100)	1331.90 363(33)
R(49)[f]	V[c]	40 017 155.4(100)	1334.82 862(33)
R(49)[e]	T	40 020 283.8(60)	1334.93 297(20)
R(49)[f]	T	40 020 875.4(60)	1334.95 270(20)
R(43)	W	40 022 207.0(60)	1334.99 712(20)
R(50)[e]	V	40 022 468.7(105)[d]	1335.00 585(35)

[1] A. Hinz, J. S. Wells, and A. G. Maki, *Z. Phys. D, Atoms, Molecules and Clusters* **5**, 351-358 (1987).

[a] The notation for the $^{14}N_2{}^{16}O$ bands is as follows: I = 10^00-00^00, J = 10^00-00^00, S = 02^01-02^00, T = 02^21-02^20, U = 00^02-00^01, V = 03^11-03^10, W = 03^31-03^30.

[b] The estimated uncertainty in the last digits is given in parentheses.

[c] The superscripts e and f given after some rotational quantum numbers refer to the parity of the doublet levels.

[d] Average uncertainties.

TABLE B.10 Heterodyne Frequency Measurements[1] of the $01^{1e,f}1$ - $01^{1e,f}0$ and $00^{0}1$ - $00^{0}0$ Bands of N_2O Near 7.7 μm

Trans. Rot; Band[a]	*Measured Freq.*[b] *(MHz)*	*Wavenumber*[b] *(cm^{-1})*
P(31)A	37 693 378.1(30)	1257.315 76(10)
P(1)A	38 495 307.9(20)	1284.065 25(7)
R(40)A	39 458 714.5(50)	1316.201 04(17)
R(64)A	39 924 104.2(20)	1331.724 77(7)
R(78)A	40 167 482.0(80)	1339.842 98(27)
P(5)B	38 591 443.6(40)	1287.272 00(13)
R(8)B	38 939 693.7(30)	1298.888 37(10)
R(26)B	39 357 539.0(40)	1312.826 19(13)
R(44)B	39 741 356.7(50)	1325.628 97(17)
R(58)B	40 016 257.3(100)	1334.798 67(33)
P(5)C	38 591 275.8(40)	1287.266 40(13)
R(8)C	38 940 435.5(30)	1298.913 11(10)
R(26)C	39 361 529.2(40)	1312.959 29(13)
R(65)C	40 167 112.7(40)	1339.830 66(13)

[1] J. S. Wells, A. Hinz, and A. G. Maki, *J. Mol. Spectrosc.* **114**, 84-96 (1985).

[a] The band notation is as follows: A = $00^{0}1$ - $00^{0}0$, B = $01^{1e}1$ - $01^{1e}0$, and C = $01^{1f}1$ - $01^{1f}0$.

[b] The estimated uncertainty in the last digits is given in parentheses.

TABLE B.11.i Heterodyne Frequency Measurements[1] on the υ_3 Band of SO_2

Transition						Obs. Frequency[a]	Wavenumber[a]
J'	K_a'	K_c'	J''	K_a''	K_c''	(MHz)	(cm^{-1})
17	6	11	18	6	12	40 457 019.8(50)	1349.50 092(17)
19	0	19	20	0	20	40 458 986.4(50)	1349.56 651(17)
15	10	5	16	10	6	40 460 775.5(90)	1349.62 619(30)
16	8	9	17	8	10	40 461 362.9(50)	1349.64 579(17)
18	2	17	19	2	18	40 462 193.6(50)	1349.67 349(17)
17	5	12	18	5	13	40 463 046.4(50)	1349.70 194(17)
17	2	15	18	2	16	40 464 934.6(50)	1349.76 492(17)
21	12	9	21	12	10	40 734 536.6(40)	1358.75 788(13)
4	2	3	5	2	4	40 734 999.1(40)	1358.77 331(13)
20	12	9	20	12	8	40 735 926.4(40)	1358.80 424(13)
19	12	7	19	12	8	40 737 249.7(40)	1358.84 838(13)
18	12	7	18	12	6	40 738 508.3(40)	1358.89 036(13)
4	1	4	5	1	5	40 740 637.5(40)	1358.96 138(13)
16	12	5	16	12	4	40 740 829.0(40)	1358.96 777(13)

[1] M. D. Vanek, J. S. Wells, A. G. Maki, and James B. Burkholder, *J. Mol. Spectrosc.* **141,** 346-347 (1990).

[a] The estimated uncertainty in the last digits is given in parentheses.

Transition J'	K_a'	K_c'	J''	K_a''	K_c''	*Obs. Frequency* [a] *(MHz)*	*Wavenumber* [a] *(cm^{-1})*
24	11	14	25	11	15	40 741 496.8(40)	1358.99 005(13)
15	12	3	15	12	4	40 741 887.7(40)	1359.00 309(13)
13	12	1	13	12	2	40 743 809.8(60)	1359.06 720(20)
3	0	3	2	0	2	40 890 362.9(60)	1363.95 569(20)
8	3	6	7	3	5	40 979 127.8(200)	1366.91 656(67)
9	6	3	8	6	2	40 981 712.6(100)	1367.00 278(33)
10	8	3	9	8	2	40 983 895.9(90)	1367.07 561(30)
9	5	4	8	5	3	40 988 165.8(150)	1367.21 804(50)
15	8	7	14	8	6	41 075 376.3(30)	1370.12 707(10)
17	11	6	16	11	5	41 078 597.4(60)	1370.23 451(20)
14	1	14	13	1	13	41 079 955.2(50)	1370.27 980(17)
14	5	10	13	5	9	41 080 193.8(30)	1370.28 776(10)
27	10	17	26	10	16	41 267 578.4(50)	1376.53 824(17)
25	6	19	24	6	18	41 271 076.8(40)	1376.65 493(13)
26	8	19	25	8	18	41 271 318.4(40)	1376.66 299(13)
28	11	18	27	11	17	41 272 779.1(40)	1376.71 172(13)

[a] The estimated uncertainty in the last digits is given in parentheses.

Transition J'	K_a'	K_c'	J''	K_a''	K_c''	*Obs. Frequency*[a] *(MHz)*	*Wavenumber*[a] *(cm^{-1})*
26	1	26	25	1	25	41 275 291.4(40)	1376.79 552(13)
39	8	31	38	8	30	41 494 155.7(60)	1384.09 605(20)
42	12	31	41	12	30	41 495 464.1(80)	1384.13 969(27)
38	6	33	37	6	32	41 496 666.6(40)	1384.17 980(13)
38	4	35	37	4	34	41 499 389.5(40)	1384.27 063(13)
37	3	34	36	3	33	41 500 253.4(40)	1384.29 944(13)
38	5	34	37	5	33	41 502 232.2(40)	1384.36 545(13)
41	0	41	40	0	40	41 504 180.9(60)	1384.43 045(20)
39	7	32	38	7	31	41 504 742.1(50)	1384.44 917(17)
41	10	31	40	10	30	41 504 915.0(60)	1384.45 494(20)
39	2	37	38	2	36	41 505 369.2(40)	1384.47 009(13)
57	8	49	56	8	48	41 792 579.5(40)	1394.05 039(13)
59	10	49	58	10	48	41 793 377.8(50)	1394.07 702(17)
60	11	50	59	11	49	41 794 759.0(40)	1394.12 309(13)
59	3	56	58	3	55	41 796 023.8(60)	1394.16 528(20)
61	1	60	60	1	59	41 796 222.6(50)	1394.17 191(17)
58	5	54	57	5	53	41 796 860.0(50)	1394.19 317(17)

[a] The estimated uncertainty in the last digits is given in parentheses.

TABLE B.12 Heterodyne Measurements[1] on OCS Near 1372 cm^{-1}

OCS Transition[a]	*Frequency*[b] *(MHz)*	*Wavenumber*[b] *(cm^{-1})*
P(20)B	40 880 607.3(200)	1363.63 027(67)
P(21)A	40 885 958.5(150)	1363.80 877(50)
P(13)A	40 985 706.8(150)	1367.13 602(50)
R(9)A	41 265 881.6(50)	1376.48 164(17)
R(11)B	41 271 214.9(350)	1376.65 954(117)
R(17)A	41 360 969.2(80)	1379.65 342(27)
R(19)B	41 367 444.2(400)	1379.86 940(133)
R(30)B	41 898 893.7(100)	1397.59 665(33)
R(29)A	41 501 244.6(50)	1384.33 251(17)
R(46)A	41 694 907.5(50)	1390.79 240(17)
R(47)B	41 700 480.9(200)	1390.97 831(67)
R(55)A	41 794 911.1(50)	1394.12 817(17)
R(64)A	41 893 079.4(100)	1397.40 271(33)

[1] M. Schneider, A. G. Maki, M. D. Vanek, and J. S. Wells, *J Mol. Spectrosc.* **134**, 349-353 (1989).

[a] A denotes the 01^10 - 00^00 and B denotes the $02^0 1$ - 01^10 band.

[b] The estimated uncertainty in the last digits is given in parentheses.

TABLE B.13 Heterodyne Frequency Measurements[1] on CS_2

Transition	*Frequency*[a] *(MHz)*	*Wavenumber*[a] *(cm^{-1})*
$^{13}CS_2$		
P(82)	43 856 762.7(30)	1462.904 14(10)
P(72)	43 953 404.7(30)	1799.691 86(10)
P(22)	44 375 671.6(30)	1480.213 07(10)
P(8)	44 475 613.6(20)	1483.546 78(7)
R(24)	44 679 304.2(30)	1490.341 17(10)
R(44)	44 780 863.1(30)	1493.728 81(10)
R(68)	44 880 808.8(50)	1497.062 64(17)
R(102)	44 981 253.5(30)	1500.413 11(10)
$^{12}CS_2$		
P(116)	44 986 965.3(20)	1500.603 64(7)
P(106)	45 099 103.2(50)	1504.344 16(17)
P(76)	45 410 368.2(30)	1514.726 84(10)
P(30)	45 813 975.8(40)	1528.189 74(13)
P(14)	45 933 333.0(20)	1532.171 07(7)
R(0)	46 035 322.3(30)	1535.573 06(10)
R(88)	46 438 997.8(70)	1549.038 23(23)

[1] J. S. Wells, M. Schneider, and A. G. Maki, *J. Mol. Spectrosc.* **132**, 422-428 (1988).
[a] The estimated uncertainty in the last digits is given in parentheses.

TABLE B.14 Heterodyne Measurements[1] on the $10^0 0$ - $01^1 0$ Band of N_2O

Transition	*Frequency (MHz)*	*Wavenumber (cm^{-1})*
P(44)	47 706 760.3(100)	1591.32 623(33)
P(28	48 230 503.8(100)	1608.79 643(33)
P(10)	48 754 661.7(100)	1626.28 046(33)
Q(18)	48 970 527.3(50)	1633.48 096(17)
Q(17)	48 975 279.7(50)	1633.63 948(17)
R(6)	49 185 589.0(70)	1640.65 464(23)
R(20)	49 493 233.9(50)	1650.91 657(17)
R(61)	50 146 501.0(100)	1672.70 722(33)

[1] M. D. Vanek, M. Schneider, J. S. Wells, and A. G. Maki, *J. Mol. Spectrosc.* **134**, 154-158 (1989).

"The estimated uncertainty in the last digits is given in parentheses. This is one-tenth of the TDL jitter linewidth plus one half of the Doppler width divided by the signal-to-noise ratio of the first derivative lock signal."

TABLE B.15.I Heterodyne Measurements[1] on OCS Near 1700 cm^{-1}

Transition[a]	*Frequency*[b] *(MHz)*	*Wavenumber*[b] *(cm^{-1})*
P(78)E	49 488 266.7(200)	1650.750 89(71)
P(65)E	49 708 361.8(50)	1658.092 47(17)
P(76)B	49 773 465.4(300)	1660.264 10(100)
P(80)F	49 819 241.4(200)	1661.791 02(67)
P(73)B	49 825 200.8(50)	1661.989 80(17)
P(67)C	49 929 960.4(50)	1665.484 21(17)
P(51)E	49 931 791.0(50)	1665.545 27(17)
P(85)A	49 998 816.2(100)	1667.780 99(33)
P(61)D	49 999 475.5(90)	1667.802 98(30)
P(39)D	50 347 398.4(50)	1679.408 44(17)
P(24)D	50 563 908.0(200)	1686.630 42(67)
P(52)A	50 564 525.8(30)	1686.651 03(10)
P(34)F	50 565 605.5(100)	1686.687 04(33)
P(5)E	50 567 619.9(100)	1686.754 24(33)
P(25)B	50 568 164.9(50)	1686.772 42(17)
P(25)C	50 568 423.5(50)	1686.781 04(17)
P(17)B, C	50 676 729.5(150)	1690.393 74(50)
P(26)F	50 679 663.3(80)	1690.491 60(27)
P(12)B, C	50 742 338.6(60)	1692.582 23 (20)
P(21)F	50 748 608.2(80)	1692.791 36(27)
P(10)D	50 750 887.1(30)	1692.867 37(10)
R(44)E	51 087 031.1(50)	1704.079 93(17)

[1] J. S. Wells, M. D. Vanek, and A. G. Maki, *J. Mol. Spectrosc.* **135**, 84-88 (1989).

[a] The band definitions are as follows: A = 00^02 - 00^00, B = $01^{1e}2$ - $01^{1e}0$, C = $01^{1f}2$ - $01^{1f}0$, D = 00^03 - 00^01, all for the normal isotope. E and F are the 00^02 - 00^00 transitions for $OC^{34}S$ and $O^{13}CS$, respectively.

[b] The estimated uncertainty in the last digits is given in parentheses.

Transition[a]	*Frequency*[b] *(MHz)*	*Wavenumber*[b] *(cm^{-1})*
R(16)B	51 089 542.7(40)	1704.163 71(13)
R(16)C	51 090 060.7(50)	1704.180 99(17)
P(16)A	51 090 402.1(120)	1704.192 37(40)
R(18)D	51 091 675.4(150)	1704.234 85(50)
R(26)C	51 196 971.4(60)	1707.747 14(20)
R(58)E	51 199 695.8(60)	1707.838 02(20)
R(15)F	51 202 495.5(60)	1707.931 41(20)
R(29)D	51 204 381.2(30)	1707.994 31(10)
R(27)B	51 206 151.0(80)	1708.053 34(27)
R(38)B	51 314 334.1(60)	1711.661 94(20)
R(38)C	51 316 312.6(60)	1711.727 94(20)
R(41)D	51 316 789.1(60)	1711.743 83(20)
R(44)B	51 369 747.4(40)	1713.510 33(13)
R(44)C	51 372 281.2(40)	1713.594 85(13)
R(58)F	51 603 367.7(90)	1721.303 07(30)
R(27)A	51 604 457.8(60)	1721.339 43(20)
R(73)C	51 606 832.2(60)	1721.418 63(20)
R(74)B	51 607 782.9(80)	1725.185 49(27)
R(39)A	51 719 760.0(60)	1728.941 86(20)
R(52)A	51 823 373.1(60)	1732.969 66(20)
R(68)A	51 953 123.5(90)	1736.698 03(30)
R(86)A	52 064 897.1(90)	1850.779 77(30)

[a] The band definitions are as follows: A = $00^02 - 00^00$, B = $01^{1e}2 - 01^{1e}0$, C = $01^{1f}2 - 01^{1f}0$, D = $00^03 - 00^01$, all for the normal isotope. E and F are the $00^02 - 00^00$ transitions for $OC^{34}S$ and $O^{13}CS$, respectively.

[b] The estimated uncertainty in the last digits is given in parentheses.

TABLE B.16.i Frequency Measurements[1] on NO with Respect to CO_2 Laser Standard Frequencies via a CO Laser Transfer Oscillator

a) Transitions for the $^2\Pi_{1/2}$ - $^2\Pi_{1/2}$ States

Transition	*Frequency*[a] *(MHz)*	*Wavenumber*[a] *(cm^{-1})*
P(32.5)	52 465 656.4(100)	1750.065 92(33)
P(30.5)	52 728 472.6(40)	1758.832 53(13)
P(28.5)	52 987 432.4(80)	1767.470 49(27)
P(23.5)	53 617 709.7(80)	1788.494 28(27)
	53 617 877.6(80)	1788.499 88(27)
P(9.5)	55 249 808.1(60)	1842.935 22(20)
	55 250 126.7(60)	1842.945 85(20)
R(0.5)	56 392 005.4(60)	1881.034 83(20)
	56 392 368.3(60)	1881.046 93(20)
R(1.5)	56 489 691.6(50)	1884.293 29(17)
	56 490 049.9(50)	1884.305 24(17)
R(8.5)	57 144 652.8(50)	1906.140 44(17)
	57 144 967.9(50)	1906.150 95(17)
R(11.5)	57 409 807.5(50)	1914.985 05(17)
	57 410 093.7(50)	1914.994 60(17)
R(17.5)	57 911 619.8(80)	1931.723 71(27)
	57 911 862.5(80)	1931.731 80(27)

[1] A. Hinz, J. S. Wells, and A. G. Maki, *J. Mol. Spectrosc.* **119**, 120-125 (1986).
[a] The estimated uncertainty in the last digits is given in parentheses.

b) Transitions for the $^2\Pi_{3/2}$ - $^2\Pi_{3/2}$ States

Transition	*Frequency*[a] *(MHz)*	*Wavenumber*[a] *(cm^{-1})*
P(28.5)	52 901 080.8(40)	1764.590 12(13)
	52 901 353.1(40)	1764.599 20(13)
P(24.5)	53 416 752.9(90)	1781.791 08(30)
	53 416 967.5(90)	1781.798 24(30)
P(4.5)	55 764 932.9(80)	1860.117 94(27)
Q(16.5)	56 081 098.5(50)	1870.664 09(17)
	56 082 391.2(50)	1870.707 21(17)
Q(13.5)	56 131 466.8(50)	1872.344 19(17)
	56 132 213.7(50)	1872.369 11(17)
R(1.5)	56 490 591.7(50)	1884.323 31(17)
R(15.5)	57 778 254.0(90)	1927.275 10(30)

[a] The estimated uncertainty in the last digits is given in parentheses.

TABLE B.17.I Heterodyne Frequency Measurements[1] for OCS in the 5.3 μm Region

Band [a]	*Rot.*	*Measured Frequency* [b] *(MHz)*	*Wavenumber* *(cm^{-1})*
D	P(55)	55 484 981.5(100)[b]	1850.77 976(33)
D	P(34)	55 727 828.1(100)	1858.88 025(33)
E	P(51)	55 754 952.8(60)	1859.78 503(20)
E	P(40)	55 881 343.1(90)	1864.00 096(30)
D	P(21)	55 881 589.4(70)	1864.00 918(23)
B	P(58)	56 000 991.4(100)	1867.99 200(33)
D	P(11)	56 001 251.1(100)	1868.00 066(33)
F	P(54)	56 007 893.8(80)	1868.22 224(27)
E	P(29)	56 009 092.1(150)	1868.26 221(50)
E	P(19)	56 126 183.5(100)	1872.16 796(33)
C	P(49)	56 126 694.0(80)	1872.18 499(27)
F	P(44)	56 127 605.0(50)	1872.21 537(17)
H	P(46)	56 128 346.7(60)	1872.24 011(20)
F	P(34)	56 248 736.7(60)	1876.25 589(20)
H	P(36)	56 249 754.1(60)	1876.28 983(20)
B	P(28)	56 369 492.0(200)	1880.28 385(67)
F	P(24)	56 370 671.3(150)	1880.32 319 (50)
G	P(26)	56 371 205.8(300)	1880.34 102(100)
H	P(26)	56 371 436.8(150)	1880.34 873(50)
C	P(28)	56 374 232.6(80)	1880.44 198(27)
D	R(28)	56 490 095.7(80)	1884.30 676(27)
B	P(18)	56 492 005.5(100)	1884.37 047(33)
F	P(14)	56 492 912.0(100)	1884.40 070(33)

[1] A. G. Maki, W. B. Olson, J. S. Wells, and M. D. Vanek, *J Mol. Spectrosc.* **130**, 69-80 (1988).

[a] The band designation (for the normal isotopes, unless otherwise indicated) is as follows: B = $03^{1e}1 - 01^{1e}0$, C = $03^{1f}1 - 01^{1f}0$, D = $02^{0}2 - 00^{0}1$, E = $02^{0}1 - 00^{0}0$ (for $^{16}O^{12}C^{34}S$), F = $04^{0}1 - 02^{0}0$, G = $04^{2e}1 - 02^{2e}0$, H = $04^{2f}1 - 02^{2f}0$.

[b] The estimated uncertainty in the last digits is given in parentheses for the heterodyne measurements; for example, ...2912.0(100) means ...2912.0 ± 10.0.

Band [a]	*Rot.*	*Measured Frequency (MHz)*	*Wavenumber (cm^{-1})*
G,H	P(16)	56 493 721.0(300)	1884.42 769(100)
C	P(18)	56 493 861.2(100)	1884.43 236(33)
B	P(16)	56 516 461.2(80)	1885.18 622(27)
G,H	P(14)	56 517 654.1(80)	1885.22 601(27)
C	P(16)	56 517 891.6(80)	1885.23 393(27)
E	R(21)	56 613 092.8(120)	1888.40 950(40)
B	P(8)	56 614 218.2(120)	1888.44 704(40)
C	P(8)	56 614 487.1(120)	1888.45 601(40)
G,H	P(6)	56 615 207.0(120)	1888.48 002(40)
E	R(23)	56 637 097.0(80)	1889.21 020(27)
B	P(6)	56 638 641.6(300)	1889.26 172(100)
C	P(6)	56 638 744.5(200)	1889.26 515(67)
D	R(40)	56 640 158.6(120)	1889.31 232 (40)
F	R(7)	56 761 434.0(200)	1893.35 763(67)
D	R(50)	56 766 766.2(150)	1893.53 550(50)
B	R(4)	56 772 578.9 (60)	1893.72 939(20)
C	R(4)	56 772 847.1(60)	1893.73 833(20)
F	R(8)	56 773 603.5(80)	1893.76 356(27)
G,H	R(6)	56 773 918.2(80)	1893.77 406(27)
G,H	R(15)	56 883 782.4(100)	1897.43 874(33)
B	R(14)	56 893 945.0(70)	1897.77 772(23)
D	R(60)	56 895 093.3(400)	1897.81 603(133)
F	R(18)	56 895 292.1(200)	1897.82 266(67)
C	R(14)	56 895 757.3(200)	1897.83 818(67)
G,H	R(16)	56 895 977.5(400)	1897.84 552(133)

[a] The band designation (for the normal isotopes, unless otherwise indicated) is as follows: B = $03^{1e}1 - 01^{1e}0$, C = $03^{1f}1 - 01^{1f}0$, D = $02^{0}2 - 00^{0}1$, E = $02^{0}1 - 00^{0}0$ (for $^{16}O^{12}C^{34}S$), F = $04^{0}1 - 02^{0}0$, G = $04^{2e}1 - 02^{2e}0$, H = $04^{2f}1 - 02^{2f}0$.

[b] The estimated uncertainty in the last digits is given in parentheses for the heterodyne measurements; for example, ...5977.5(400) means ...5977.5 ± 40.0.

Band [a]	*Rot.*	*Measured Frequency* [b] *(MHz)*	*Wavenumber* [b] *(cm^{-1})*
B	R(24)	57 014 898.5(150)	1901.81 230(50)
F	R(28)	57 017 035.1(100)	1901.88 357(33)
C	R(33)	57 131 856.3(100)	1905.71 359(33)
E	R(64)	57 137 040.8(120)	1905.88 653(40)
B	R(35)	57 147 493.1(100)	1906.23 518(33)
F	R(39)	57 151 473.8(200)	1906.36 796(67)
F	R(49)	57 274 716.8(60)	1910.47 890(20)
B	R(46)	57 279 619.6(70)	1910.64 244(23)
C	R(54)	57 396 711.5(100)	1914.54 821(33)
B	R(56)	57 399 428.5(100)	1914.63 884(33)

[a] The band designation (for the normal isotopes, unless otherwise indicated) is as follows: B = $03^{1e}1 - 01^{1e}0$, C = $03^{1f}1 - 01^{1f}0$, D = $02^{0}2 - 00^{0}1$, E = $02^{0}1 - 00^{0}0$ (for $^{16}O^{12}C^{34}S$), F = $04^{0}1 - 02^{0}0$.

[b] The estimated uncertainty in the last digits is given in parentheses for the heterodyne measurements; for example, ...9428.5(100) means ...9428.5 ± 10.0.

TABLE B.18.i Heterodyne Frequency Measurements [1] on OCS at 4.9 μm

Transition [a] *P(J")*	*Frequency* [b] *(MHz)*	*Wavenumber* [b] *(cm⁻¹)*
^{3}P(26)C	59 693 603.1(60)	1991.16 427(20)
^{3}P(26)B	59 693 840.1(60)	1991.17 217(20)
^{3}P(32)A	59 813 307.6(100)	1995.15 718(33)
^{3}P(2)A	60 210 827.3(50)	2008.41 701(17)
^{3}R(15)B	60 216 271.2(70)	2008.59 860(23)
^{3}R(15)C	60 216 519.6(70)	2008.60 688(23)
^{3}R(8)A	60 341 171.7(40)	2012.76 483(13)
^{3}R(27)B	60 343 564.4(40)	2012.84 464(13)
^{3}R(27)C	60 344 056.2(40)	2012.86 105(13)
P(82)A	60 586 208.3(50)	2020.93 837(17)
^{3}R(31)A	60 586 692.9(40)	2020.95 453(13)
P(75)A	60 710 590.4(70)	2025.08 731(23)
P(62)C	60 713 866.0(50)	2025.19 657(17)
P(62)B	60 714 374.4(50)	2025.21 353(17)
P(61)A	60 948 914.5(30)	2033.03 695(10)
P(47)C	60 954 112.4(50)	2033.21 033(17)
P(47)B	60 954 526.3(50)	2033.22 414(17)
^{4}P(53)A	61 074 639.0(40)	2037.23 066(13)
P(39)C	61 075 763.7(40)	2037.26 818(13)
P(39)B	61 076 131.6(50)	2037.28 045(17)

[1] J. S. Wells, M. Schneider, and A. G. Maki, *J Mol. Spectrosc.* **140,** 170-176 (1990).

[a] A denotes the 10^00-00^00 band, B and C refer to the 11^10-01^10 band. The left superscript 3 on the transition (P or R) denotes the $O^{13}CS$ isotope; a 4 indicates the $OC^{34}S$. Otherwise, the most abundant isotope is indicated.

[b] The estimated uncertainty in the last digits is given in parentheses.

Transition[a] *P(J")*	*Frequency*[b] *(MHz)*	*Wavenumber*[b] *(cm^{-1})*
P(36)C	61 120 218.3(50)	2038.75 103(17)
P(36)B	61 120 561.9(50)	2038.76 249(17)
^{4}P(42)A	61 241 514.1(80)	2042.79 702(27)
P(34)A	61 368 989.2(100)	2047.04 913(33)
P(9)B,C	61 491 676.5(150)	2051.14 154(50)
^{4}P(15)A	61 615 169.5(100)	2055.26 082(33)
P(7)A	61 736 557.5(50)	2059.30 989(17)
^{4}R(7)A	61 892 945.4(70)	2064.52 643(23)
R(25)B	61 895 273.9(80)	2064.60 410(27)
R(25)C	61 895 672.3(80)	2064.61 739(27)
R(38)B	62 022 563.9(80)	2068.85 004(27)
^{4}R(19)A	62 022 962.6(120)	2068.86 334(40)
R(38)C	62 023 197.7(80)	2068.87 118(27)
^{4}R(31)A	62 142 677.4(100)	2072.85 659(33)
R(52)B	62 145 828.5(80)	2072.96 170(27)
R(41)A	62 267 829.9(80)	2077.03 123(27)
^{4}R(45)A	62 269 217.9(100)	2077.07 753(33)
R(56)A	62 394 930.8(60)	2081.27 086(20)
^{4}R(61)A	62 396 400.3(100)	2081.31 988(33)
R(73)A	62 518 558.6(80)	2085.39 464(27)

[a] The A denotes the $10^00 - 00^00$ band, B and C refer to the $11^10 - 01^10$ band. The left superscript 4 on the transition (P or R) denotes the $OC^{34}S$ isotope. Otherwise, the most abundant isotope is indicated.

[b] The estimated uncertainty in the last digits is given in parentheses.

TABLE B.19 Heterodyne Frequency Measurements[1] on $^{12}C^{16}O$ with $\Delta v = 1$

Trans. $P_{v''}(J'')$	*Obs. Frequency* *(MHz)*	*Wavenumber* *(cm^{-1})*
$P_4(11)$	59 811 747.50	1995.105 14
$P_4(10)$	59 933 071.30	1999.152 07
$P_4(\ 9)$	60 053 387.80	2003.165 40
$P_3(14)$	60 214 872.90	2008.551 96
$P_4(\ 7)$	60 290 990.10	2011.090 96
$P_3(13)$	60 340 233.21	2012.733 53
$P_3(12)$	60 464 604.60	2016.882 11
$P_3(11)$	60 587 978.00	2020.997 41
$P_3(10)$	60 710 350.10	2025.079 30
$P_3(\ 9)$	60 831 718.20	2029.127 70
$P_3(\ 8)$	60 952 077.10	2033.142 44
$P_3(\ 7)$	61 071 422.90	2037.123 39
$P_2(13)$	61 116 439.50	2038.624 98
$P_2(12)$	61 241 858.90	2042.808 53
$P_2(11)$	61 366 283.80	2046.958 89
$P_2(10)$	61 489 706.37	2051.075 83
$P_2(\ 9)$	61 612 123.87	2055.159 24
$P_2(\ 8)$	61 733 533.40	2059.209 02
$P_1(13)$	61 894 667.40	2064.583 87
$P_1(12)$	62 021 140.60	2068.802 56
$P_1(11)$	62 146 613.40	2072.987 89
$P_1(10)$	62 271 086.60	2077.139 87
$P_1(\ 9)$	62 394 555.00	2081.258 33
$P_0(31)$	60 214 073.80	2008.525 30
$P_0(18)$	62 022 472.10	2068.846 98

[1] M. Schneider, J. S. Wells, and A. G. Maki, *J. Mol. Spectrosc.* **139**, 432-438 (1990); *erratum* **141**, 351 (1990).

TABLE B.20 Heterodyne Frequency Measurements[1] of CO

Transition $P_{v''}(J'')$	*Frequency*[a] *(MHz)*	*Wavenumber*[a] *(cm^{-1})*
$^{12}C^{16}O$		
$P_0(31)$	60 214 084.9(100)	2008.525 67(33)
$P_0(31)$[b]	60 214 073.8(50)	2008.525 30(17)
$P_0(18)$[b]	62 022 472.1(40)	2068.846 98(13)
$R_0(\ 2)$[b]	64 593 150.8(30)	2154.595 59(10)
$P_1(13)$[b]	61 894 667.4(30)	2064.583 87(10)
$P_1(12)$[b]	62 021 140.6(30)	2068.802 56(10)
$P_1(11)$[b]	62 146 613.4(30)	2072.987 89(10)
$P_1(10)$[b]	62 271 086.6(30)	2077.139 87(10)
$P_1(\ 9)$[b]	62 394 555.0(30)	2081.258 33(10)
$^{13}C^{16}O$		
$P_0(18)$	60 708 713.5(40)	2025.024 7 (13)
$P_0(\ 7)$	62 046 719.4(100)	2069.655 78(33)
$^{12}C^{18}O$		
$P_0(21)$	60 216 141.4(150)	2008.594 27(50)
$P_0(\ 6)$	62 047 042.9(150)	2069.666 57(50)
$^{13}C^{18}O$		
$R_0(6)$	61 975 673.9(150)	2067.285 96(50)

[1] A. G. Maki, J. S. Wells, and D. A. Jennings, *J. Mol. Spectrosc.* **144**, 224-229 (1990).

[a] The estimated uncertainty in the last digits is given in parentheses.

[b] Measurements reported by M. Schneider, J. S. Wells, and A. G. Maki, *J. Mol. Spectrosc.* **139**, 432-438 (1990); *erratum* **141**, 351 (1990).

TABLE B.21 Frequency Measurements[1] for Some of the 1-0 Band Transitions of CO

Transition	*Frequency*[a] *(MHz)*	*Wavenumber*[a] *(cm^{-1})*
P(15)	62 416 858.3(4)	2 082.002 29(1)
P(14)	62 546 359.4(4)	2 086.321 98(1)
P(13)	62 674 872.6(4)	2 090.608 72(1)
P(12)	62 802 393.7(4)	2 094.862 36(1)
P(11)	62 928 918.1(4)	2 099.082 76(1)
P(10)	63 054 441.5(5)	2 103.269 77(2)
P(9)	63 178 959.6(5)	2 107.423 25(2)
P(8)	63 302 467.8(5)	2 111.543 04(2)
P(7)	63 424 961.8(5)	2 115.629 00(2)

[1] W. Wu, T. George, M. Schneider, W. Urban, and B. Nelles, *Appl. Phys.* **B52**, 1-6 (1991).

[a] The estimated uncertainty in the last digits is given in parentheses.

TABLE B.22 Heterodyne Frequency Measurements[1] for the 11^10 - 01^10 Band of OCS

Transition	*Frequency*[a] *(MHz)*	*Wavenumber*[a] *(cm^{-1})*
P(12)B, C[b]	86 898 356.3(90)	2 898.617 16(30)
P(3)B, C[b]	87 014 402.9(50)	2 902.488 06(17)
R(24)B	87 322 335.8(50)	2 912.759 59(17)
R(24)C	87 323 138.4(50)	2 912.786 37(17)

[1] A. Dax, M. Mürtz, J. S. Wells, M. Schneider, E. Bachem, W. Urban, and A. G. Maki, *J. Mol. Spectrosc.* **155**, 98-103 (1992).

[a] The estimated uncertainty in the last digits is given in parentheses.

[b] The frequency value here is an average of the two (equally intense) unresolved lines. The $11^{1e}1$ - $01^{1e}0$ band is denoted by B, and the $11^{1f}1$ - $01^{1f}0$ band by C.

TABLE B.23 Heterodyne Frequency Measurements[1] for the 1-0 Band of HF

v'	J'	v''	J''	*Frequency (MHz)*	*Wavenumber*[a] *(cm^{-1})*
1	5	0	6	110 725 718.65(91)	3693.412 421(30)
1	4	0	5	112 166 130.10(67)	3741.459 370(22)
1	3	0	4	113 568 197.58(61)	3788.227 308(20)
1	2	0	3	114 930 273.44(61)	3833.661 266(20)
1	1	0	2	116 250 733.33(87)	3877.707 067(29)

[1] D. Goddon, A. Groh, H. J. Hanses, M. Schneider, and W. Urban, *J. Mol. Spectrosc.* **147**, 392-397 (1991).

[a] The estimated uncertainty in the last digits is given in parentheses.

TABLE B.24 Sub-Doppler Heterodyne Frequency Measurements[1] for the 10°11 - 00°01 band of CO_2

Transition	*Frequency*[a] *(MHz)*	*Wavenumber*[a] *(cm^{-1})*
P(20)	110 862 678.84(55)	3 697.980 916 (18)
P(18)	110 916 409.81(40)	3 699.773 188 (13)
P(12)	111 073 143.81(61)	3 705.001 272 (20)
P(6)	111 223 141.91(55)	3 710.004 670 (18)
R(8)	111 568 408.56(57)	3 721.521 525 (19)
R(14)	111 694 584.29(110)	3 725.730 295 (37)
R(16)	111 735 129.97(59)	3 727.082 753 (20)
R(22)	111 852 244.29(40)	3 730.989 266 (13)
R(26)	111 926 565.89(69)	3 733.468 368 (23)

[1] A. Groh, D. Goddon, M. Schneider, W. Zimmermann, and W. Urban, *J. Mol. Spectrosc.* **146**, 161-168 (1991).

[a] The estimated uncertainty in the last digits is given in parentheses.

TABLE B.25.i Semiconductor-Laser Heterodyne Frequency Measurements[1] on C_2H_2, HCN, NH_3, D_2O and HDO at 1.52 μm

Molecule	*Transition*		*Measured Frequency*[a] *(GHz)*	*Wavenumber*[a] *(cm^{-1})*
	Band Assignment	*Rotational Assignment*		
$^{12}C_2H_2$	$\nu_1+\nu_3$	R(2)	196 764.891(26)	6563.370 28(87)
$^{12}C_2H_2$	$\nu_1+\nu_3+\nu_4-\nu_4$	R(15)	196 782.302(26)	6563.951 05(87)
$^{12}C_2H_2$	$\nu_1+\nu_2+(\nu_4+\nu_5)^0$	P(23)	196 787.987(28)	6564.140 68(93)
$^{12}C_2H_2$	$\nu_1+\nu_3+\nu_4-\nu_4$	R(15)	196 789.081(27)	6564.177 17(90)
$^{12}C^{13}CH_2$	$\nu_1+\nu_3$	R(9)	196 798.970(29)	6564.507 04(97)
$^{13}C_2H_2$	$2\nu_2+(\nu_4+3\nu_5)^2$	R(18)	196 771.148(33)	6563.578 99(110)
$^{13}C_2H_2$	$\nu_1+\nu_3$	R(20)	196 776.026(27)	6563.741 71(90)
$^{13}C_2H_2$	$\nu_1+\nu_2+(\nu_4+\nu_5)^0$	R(13)	196 798.150(33)	6564.479 68(110)
$H^{12}C^{14}N$	$2\nu_1$	R(16)	196 764.158(26)	6563.345 83(87)
$^{14}NH_3$	?	?	196 763.841(39)	6563.335 26(130)
$^{14}NH_3$	?	?	196 768.394(26)	6563.487 13(87)
$^{14}NH_3$	?	?	196 783.417(48)	6563.988 24(160)
$^{14}NH_3$	?	?	196 786.923(28)	6564.105 19(93)
$^{14}NH_3$	?	?	196 790.140(31)	6564.212 50(103)
$^{14}NH_3$	?	?	196 791.840(47)	6564.269 21(157)
$^{14}NH_3$	?	?	196 793.043(29)	6564.309 33(97)
$^{14}NH_3$	?	?	196 794.851(30)	6564.369 64(110)

[1] H. Sasada, S. Takeuchi, M. Iritani, and K. Nakatani, *J. Opt. Soc. Am. B* **8**, 713-718 (1991).

[a] The estimated uncertainty in the last digits is given in parentheses.

Molecule	*Transition*		*Measured Frequency* [a] *(GHz)*	*Wavenumber* [a] *(cm^{-1})*
	Band Assignment	*Rotational Assignment*		
$^{15}NH_3$	?	?	196 761.303(39)	6563.250 60(130)
$^{15}NH_3$	?	?	196 763.429(48)	6563.321 52(160)
$^{15}NH_3$	?	?	196 771.182(31)	6563.580 13(103)
$^{15}NH_3$	?	?	196 772.459(65)	6563.622 72(217)
$^{15}NH_3$	?	?	196 777.232(32)	6563.781 93(107)
$^{15}NH_3$	?	?	196 778.790(33)	6563.833 90(110)
$^{15}NH_3$	?	?	196 780.603(32)	6563.894 38(107)
$^{15}NH_3$	?	?	196 787.821(26)	6564.135 15(87)
$^{15}NH_3$	?	?	196 792.123(28)	6564.278 65(93)
D_2O	$\nu_1+\nu_2+\nu_3$	$3_{13}\leftarrow 2_{12}$	196 797.754(31)	6564.466 48(103)
HDO	$2\nu_1+\nu_3$	$9_{19}\leftarrow 8_{18}$	196 764.299(35)	6563.350 54(117)
HDO	$2\nu_1+\nu_3$	$9_{09}\leftarrow 8_{08}$	196 771.284(31)	6563.583 53(103)
HDO	$2\nu_1+\nu_3$	$6_{24}\leftarrow 5_{23}$	196 796.446(30)	6564.422 84(110)

[a] The estimated uncertainty in the last digits is given in parentheses.

SECTION III

SATURATION ABSORPTION SPECTROSCOPY; UPDATE OF OsO_4 MEASUREMENTS

The latest high accuracy standards for some of the OsO_4 lines are presented in Tables Cl (page 612) and C2 (pages 613–616). They have been determined by using saturation absorption resonances of OsO_4 and CO_2.

TABLE C.1 Absolute Frequencies[1] of OsO_4 Lines Used as Secondary Frequency Standards in the Vicinity of CO_2 Laser Lines

Nearby $^{12}C^{16}O_2$ *Laser Line*	*Absolute Frequency (kHz) of* OsO_4 *Reference Lines*		*Wavenumber vac.* (cm^{-1})	
P(24)	28 196 916 356.59	(0.68)	940.547 889 186	(23)
P(22)	28 251 929 514.77	(0.68)	942.382 930 620	(23)
P(20)	28 306 234 121.32	(0.14)	944.194 337 314 4	(48)
P(18)	28 359 769 956.74	(0.83)	945.980 100 565	(28)
P(16)	28 412 623 110.76	(0.82)	947.743 092 015	(27)
P(14)	**28 464 676 938.787**	**(0.095)**	**949.479 420 819 4**	**(32)**
P(12)	28 516 051 989.12	(0.88)	951.193 108 038	(29)
P(10)	28 566 630 371.3	(1.0)	952.880 221 265	(35)
P(8)	28 616 553 629.62	(0.95)	954.545 481 915	(32)
P(6)	28 665 682 602.40	(0.95)	956.184 248 051	(32)
P(4)	28 714 112 424.32	(0.64)	957.799 693 024	(21)
R(4)	28 923 030 675.2	(1.7)	964.768 455 755	(58)
R(6)	28 967 423 192.8	(1.4)	966.249 230 752	(47)
R(8)	29 011 137 371.4	(1.0)	967.707 378 795	(33)
R(10)	**29 054 057 446.660**	**(0.050)**	**969.139 038 403 0**	**(17)**
R(12)	29 096 274 952.343	(0.058)	970.547 262 811 5	(19)
R(14)	29 137 747 033.57	(0.91)	971.930 622 537	(31)
R(16)	29 178 468 914.7	(1.2)	973.288 958 280	(39)
R(18)	29 218 450 029.5	(1.2)	974.622 584 719	(41)
R(20)	29 257 633 822.8	(1.2)	975.929 615 373	(39)
R(22)	29 296 146 358.18	(0.92)	977.214 255 276	(31)
R(24)	29 333 876 263.32	(0.92)	978.472 789 443	(31)
R(26)	29 370 814 078.409	(0.068)	979.704 902 329 8	(23)

[1] Ch. Chardonnet and Ch. J. Bordé, (Private Communication). This is an updated version of Table 1 of the previous handbook by Guelachvili and Rao. New measurements performed at LPTF (Paris) have been taken into account. [For R(10), see A. Clairon, B. Dahmani, O. Acef, M. Granvaud, Y.U.S. Domnin, S.B. Pouchkine, V. M. Tatarenkov, and R. Felder, *Metrologia* **25**, 9-16 (1988): The frequency for R(10) of CO_2 given here is 29 054 072 700.965(0.050) kHz for which the wavenumber is 969.139 547 231 9(17)cm^{-1}.

TABLE C.2.i Accurate Measurements[1] (with Assignments)[2] of Several OsO_4 Lines at 10.4 μm

Nearby CO_2 Laser Line	*Os Isotope*	*Line Assignment*	*Absolute Frequency*[a] *(kHz)*	*Wavenumber*[a] *(cm^{-1})*
P(24)	192	P(81) $A_1^0(-)$	28 197 056 189.19 (68)	940.552 553 500 (23)
P(22)	192	P(74) $A_1^5(+)$	28 251 929 514.77 (68)	942.382 930 620 (23)
	192	P(67) (?)	28 306 127 573.28 (19)	944.190 783 254 4(65)
	192	P(67) (?)	28 306 127 719.96 (19)	944.190 788 147 1(65)
P(20)	190	P(69) $A_1^1(+)$	28 306 359 584.29 (9)	944.198 522 308 6(63)
	189	P(70) $A_1^4(-)$	28 306 374 158.53 (5)	944.199 008 453 (12)
	192	P(60) $A_1^4(-)$	28 359 569 366.96 (83)	945.973 409 610 (28)
	192	P(60) (?)	28 359 583 609.87 (83)	945.973 884 702 (28)
	192	P(60) $A_1^{4-3}(-)$	28 359 591 916.49 (83)	945.974 161 781 (28)
	190	P(62) $A_1^{3-2}(-)$	28 359 617 962.60 (83)	945.975 030 586 (28)
	190	P(62) (?)	28 359 662 702.36 (83)	945.976 522 944 (28)
	190	P(62) (?)	28 359 666 115.20 (83)	945.976 636 784 (28)
	187	P(65) $A_1^{2-3}(+)$	28 359 758 554.74 (83)	945.979 720 235 (28)
P(18)	188	P(64) $A_1^2(-)$(?)	28 359 828 967.11 (83)	945.982 068 939 (28)
	188	P(64) $A_1^2(-)$(?)	28 359 835 411.54 (83)	945.982 283 902 (28)

[1] Ch. Chardonnet and Ch. J. Bordé, (Private Communication), In the case of ^{189}Os and ^{187}Os, the hyperfine structure has been fully resolved and the frequencies correspond to the center of gravity of the structure.

[2] B. Bobin, Ch. J. Bordé, Ch. Chardonnet, L. Henry and A. Valentin. Proceedings of the Xth Conference on high resolution molecular spectroscopy, Dijon, France. The notation of Landau-Berger has been adopted. The assignments with (?) are still uncertain.

[a] The estimated uncertainty in the last digits is given in parentheses.

TABLE C.2.ii *(Continued)*

Nearby CO_2 Laser Line	*Os Isotope*	*Line Assignment*	*Absolute Frequency* [a] *(kHz)*	*Wavenumber* [a] *(cm^{-1})*
	189	P(63) $A_1^1(-)$	28 359 837 050.86 (83)	945.982 338 584 (28)
	190	P(62) (?)	28 359 878 907.64 (83)	945.983 734 776 (28)
	190	P(62) (?)	28 359 883 229.68 (83)	945.983 878 943 (28)
	190	P(62) (?)	28 359 884 565.28 (83)	945.983 923 494 (28)
	190	P(62) $A_1^3(-)$	28 359 908 183.22 (83)	945.984 711 304 (28)
	190	P(62) (?)	28 359 913 265.66 (83)	945.984 880 836 (28)
	189	P(56) $A_1^2(+)$	28 412 365 663.17 (82)	947.734 504 488 (28)
P(16)	190	P(55) (?)	28 412 648 819.30 (82)	947.743 949 559 (27)
	187	P(58) $A_1^1(+)$	28 412 728 129.78 (82)	947.746 595 072 (27)
	189	P(49) $A_1^3(-)$	28 464 739 395.37 (11)	949.481 504 146 7(36)
	189	P(49) $A_1^3(+)$	28 464 742 138.75 (11)	949.481 595 656 0(37)
	190	P(48) $A_1^1(+)$ (?)	28 464 746 562.03 (10)	949.481 743 200 7(34)
P(14)	190	P(48) $A_1^1(+)$ (?)	28 464 755 528.33 (12)	949.482 042 284 3(39)
	192	P(46) $A_1^2(-)$	28 464 788 929.31 (10)	949.483 156 421 1(34)
	190	(?)	28 464 883 627.69 (11)	949.486 315 219 0(36)
	190	P(48) $A_1^2(+)$	28 464 888 460.64 (11)	949.486 476 428 9(36)
P(12)	192	P(39) $A_1^3(-)$	28 516 051 989.12 (88)	951.193 108 038 (29)
	192	P(39) $A_1^2(+)$	28 516 052 446.33 (88)	951.193 123 288 (29)
P(10)	187	P(38) $A_1^2(-)$	28 566 884 676.9 (10)	952.888 703 989 (35)
P(8)	188	P(30) $A_1^1(+)$	28 616 371 073.35 (95)	954.539 392 494 (32)

TABLE C.2.iii *(Continued)*

Nearby CO_2 Laser Line	Os Isotope	Line Assignment	Absolute Frequency[a] (kHz)	Wavenumber[a] (cm^{-1})
P(8)	188	P(30) $A_1^1(-)$	28 616 553 629.62 (95)	954.545 481 915 (32)
R(6)	192	R(23) $A_1^1(-)$	28 967 504 893.9 (14)	966.251 956 007 (47)
	192	R(23) $A_1^0(+)$	28 967 602 799.8 (14)	966.255 221 796 (47)
R(8)	189	R(26) $A_1^0(+)$	29 011 181 441.9 (10)	967.708 848 830 (33)
	189	R(26) $A_1^0(-)$	29 011 266 499.1 (10)	967.711 686 032 (34)
R(10)	187	R(30) $A_1^1(+)$	29 054 016 716.063 (54)	969.137 679 776 5(18)
	187	R(30) $A_1^1(-)$	29 054 169 284.803 (71)	969.142 768 921 9(24)
	192	R(36) $A_1^0(+)$	29 054 231 144.19 (15)	969.144 832 329 1(50)
	192	R(36) $A_1^0(-)$	29 054 231 673.41 (15)	969.144 849 981 9(50)
R(12)	192	R(42) $A_1^3(-)$	29 096 042 436.04 (16)	970.539 506 902 6(54)
	190	R(40) $A_1^0(+)$	29 096 080 710.893 (77)	970.540 783 614 1(26)
	190	R(40) $A_1^0(-)$	29 096 085 508.633 (71)	970.540 943 649 5(24)
	190	R(40) $A_1^1(+)$	29 096 404 957.283 (71)	970.551 599 309 5(24)
R(14)	190	R(46) $A_1^3(-)$	29 137 848 760.40 (91)	971.934 015 778 (31)
	189	R(45) $A_1^0(-)$	29 137 927 780.59 (92)	971.936 651 608 (31)
R(16)	187	R(49) $A_1^2(-)$	29 178 325 373.3 (12)	973.284 170 254 (39)
	187	R(49) $A_1^2(+)$	29 178 468 914.7 (12)	973.288 958 280 (39)
	188	R(50) $A_1^2(-)$(?)	29 178 529 536.1 (12)	973.290 980 393 (39)
	192	R(55) $A_1^{2-3}(-)$	29 178 571 254.4 (12)	973.292 371 964 (39)
R(18)	189	R(57) $A_1^0(+)$ (?)	29 218 273 695.2 (12)	974.616 702 838 (41)

TABLE C.2.iv *(Continued)* — OsO_4 Frequencies

Nearby CO_2 Laser Line	*Os Isotope*	*Line Assignment*	*Absolute Frequency [a] (kHz)*	*Wavenumber [a] (cm^{-1})*
R(18)	189	R(57) A_1^0 (+) (?)	29 218 382 714.4 (12)	974.620 339 329 (41)
	189	(?)	29 218 407 574.2 (12)	974.621 168 562 (41)
	187	R(55) A_1^0 (+)	29 218 542 755.8 (12)	974.625 677 735 (41)
	190	R(59) A_1^3 (−)	29 218 631 278.7 (12)	974.628 630 542 (41)
R(20)	189	R(64) A_1^1 (+)	29 257 467 463.6 (12)	975.924 066 227 (40)
	189	R(64) A_1^1 (−)	29 257 512 517.1 (12)	975.925 569 052 (39)
	192	R(67) (?)	29 257 633 822.8 (12)	975.929 615 373 (39)
	190	R(65) A_1^1 (−)	29 257 709 157.3 (12)	975.932 128 261 (39)
	189	R(64) A_1^2 (+)	29 257 866 486.6 (12)	975.937 376 205 (39)
R(22)	192	R(74) A_1^0 (+)	29 295 949 757.72 (92)	977.207 697 390 (31)
	192	R(74) A_1^0 (−)	29 295 950 006.41 (92)	977.207 705 686 (31)
	192	R(73) A_1^0 (+)	29 296 129 586.46 (92)	977.213 695 832 (31)
	187	R(67) A_1^0 (−)	29 296 138 221.23 (92)	977.213 983 857 (31)
	190	R(71) A_1^1 (−)	29 296 239 199.55 (92)	977.217 352 131 (31)
R(24)	189	R(76) A_1^4 (+)	29 333 619 383.26 (92)	978.464 220 846 (31)
	192	R(80) A_1^2 (−)	29 333 674 950.78 (92)	978.466 074 379 (31)
	187	R(74) A_1^0 (+)	29 334 078 417.86 (93)	978.479 532 592 (31)
	187	R(74) A_1^0 (−)	29 334 078 673.01 (93)	978.479 541 103 (31)
R(26)	187	R(80) A_1^1 (+)	29 370 835 588.20 (13)	979.705 619 819 1(43)
	187	R(80) A_1^1 (−)	29 370 880 770.30 (13)	979.707 126 931 8(43)
	192	R(85) A_1^0 (+)	29 370 885 011.729 (74)	979.707 268 410 6(25)

SECTION IV

METHYL CHLORIDE SPECTRAL MAPS AND WAVENUMBER DATA: 6.2 μm–7.7 μm

For high resolution research in the regions of strong atmospheric absorptions due to water vapor and carbon dioxide, there is need to have reference spectra for molecular species other than these two molecules, H_2O and CO_2. It has therefore been considered useful to present the spectra of CH_3Cl, which have been recorded by the same FTS facility as was used in generating the information given in Section I of this handbook. These spectra observed between 1309 cm^{-1} (7.7 μm) and 1594 cm^{-1} (6.2 μm) have been included in the following pages (618–715) by adopting the format of a spectral map on the even-numbered pages and wavenumber data on the opposite odd-numbered pages. If one examines the summary map attached to the inside of the back cover page of this book, one notices that for this spectral region, mostly water vapor standards have been listed in the previous handbook.[1] Only between 1305–1345 cm^{-1} N_2O standards are also available. Incidentally, when looking at the same spectral map, we realize that there has been a scarcity of wavenumber standards in the far infrared. For the spectral region 25–350 cm^{-1}, pure rotational water vapor data were furnished. This deficiency of standards in the infrared has been partly rectified by FTS measurements on CH_3OH in the region 100–350 cm^{-1}. The results have been reported in extensive publications,[24,25] each of which is 60 to 70 pages in length and includes spectral positions as well as spectral maps.

[24]G. Moruzzi, P. Riminucci, F. Strumia, B. Carli, M. Carlotti, R. M. Lees, I. Mukhopadhyay, J. W. C. Johns, B. P. Winnewisser, and M. Winnewisser, *J. of Mol. Spectrosc.* 144, 139–200, (1990).

[25]G. Moruzzi, F. Strumia, J. C. S. Moraes, R. M. Lees, I. Mukhopadhyay, J. W. C. Johns, B. P. Winnewisser and M. Winnewisser, *J. of Mol. Spectrosc.* 153, 511–577 (1992).

CH$_3$ Cl 0.047TORR 28.17M 0.0054CM-1

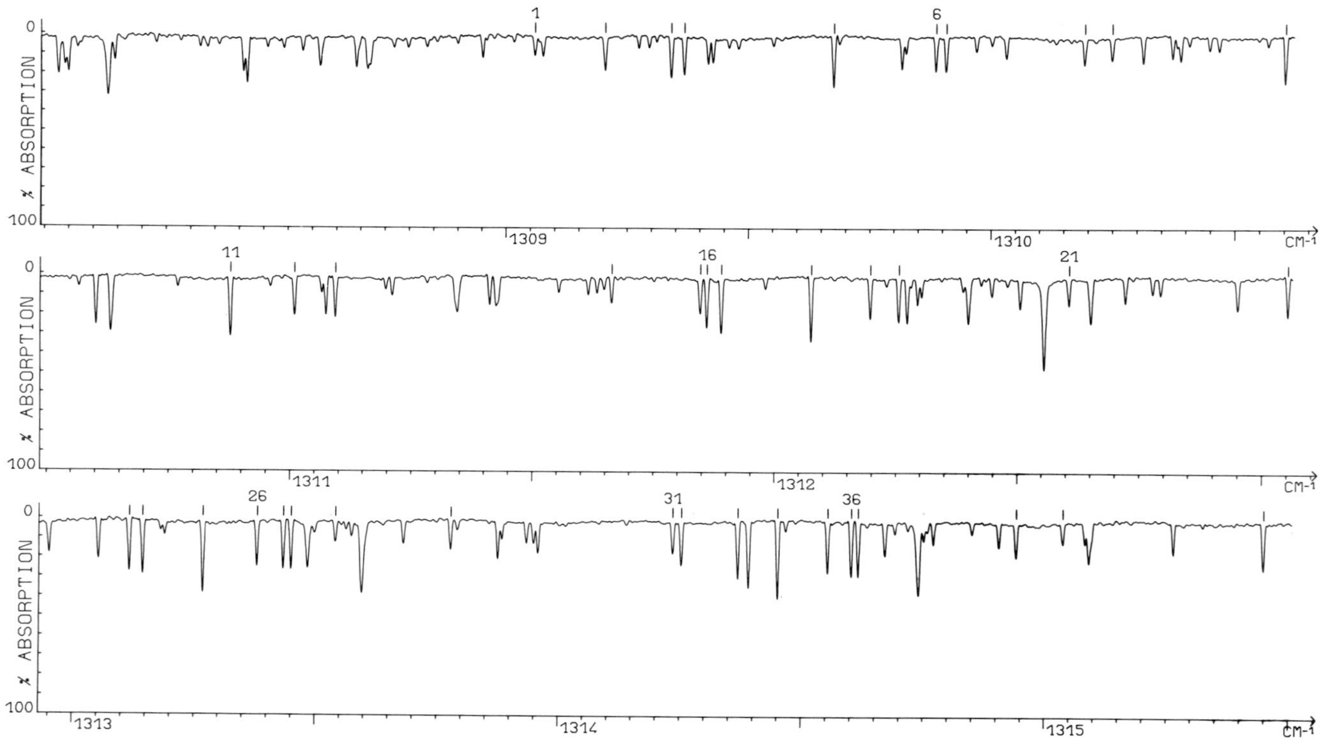

ISOTOPIC IDENTIFICATION – '∇' indicates $CH_3{}^{35}Cl$; 'I': Unidentified.

LINE NO.	WAVENUMBER CM^{-1} ;VAC.		ROTATIONAL ASSIGNMENT	BAND	ISOTOPIC SPECIES	LINE NO.	WAVENUMBER CM^{-1} ;VAC.		ROTATIONAL ASSIGNMENT	BAND	ISOTOPIC SPECIES
1	1309.05928	*				31	1314.23665	*			
2	1309.20385	*				32	1314.25439	*			
3	1309.34065	*				33	1314.37062	*			
4	1309.36735	*				34	1314.45319	*			
5	1309.67446	*				35	1314.55591	*			
6	1309.88495	*				36	1314.60441	*			
7	1309.90592	*				37	1314.61859	*			
8	1310.19086	*				38	1314.94362	*			
9	1310.24717	*				39	1315.04022	*			
10	1310.60164	*				40	1315.44945	*			
11	1310.87768	*									
12	1311.00888	*									
13	1311.09262	*									
14	1311.66328	*									
15	1311.84671	*									
16	1311.86030	*									
17	1311.88995	*									
18	1312.07527	*									
19	1312.19692	*									
20	1312.25615	*									
21	1312.60663	*									
22	1313.05340	*									
23	1313.11655	*									
24	1313.14376	*									
25	1313.26708	*									
26	1313.37934	*									
27	1313.43312	*									
28	1313.44905	*									
29	1313.53994	*									
30	1313.77802	*									

- "LINE NO.": serial number identifying the spectral line in the map on the opposite page.
- "WAVENUMBER": observed value.
- * : all given lines are selected according to intensity, halfwidth and symmetry criteria.

Lines with an asterisk (*) are selected but not identified.

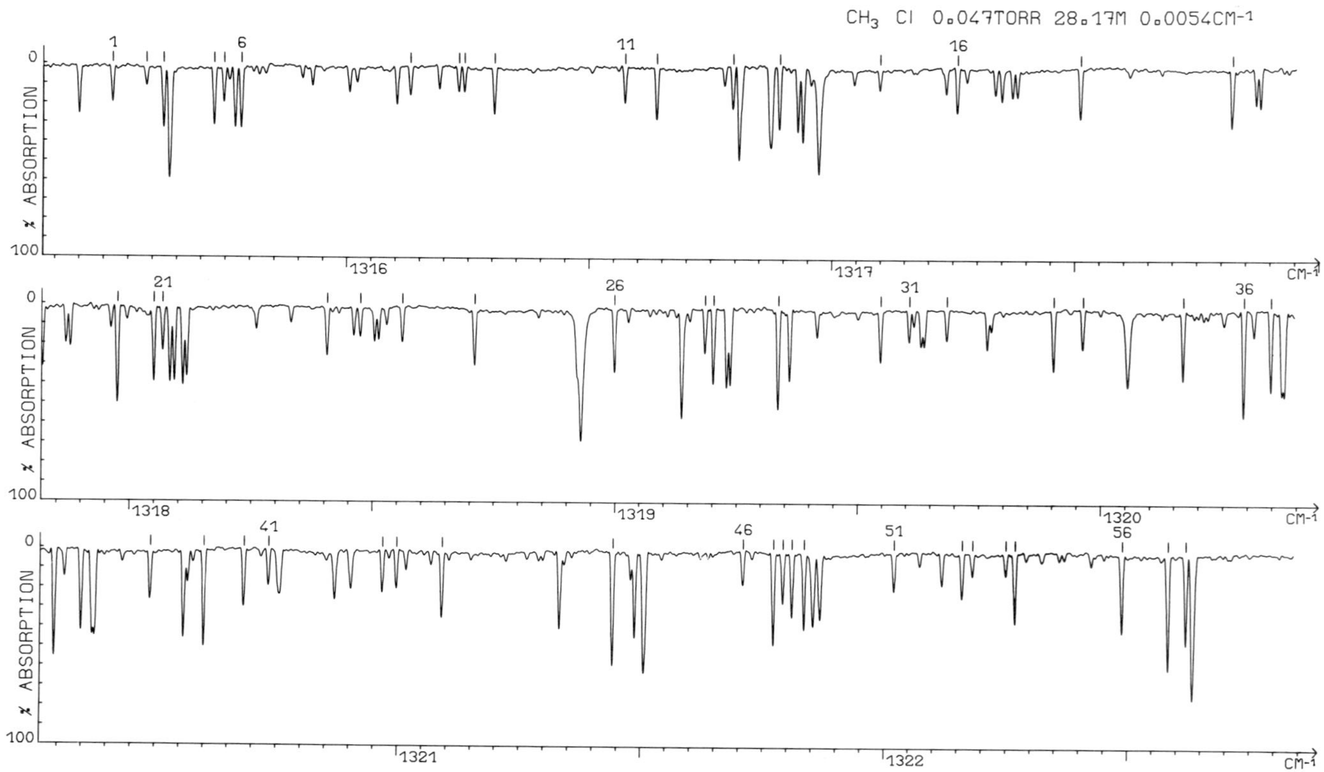

ISOTOPIC IDENTIFICATION – '∇' indicates $CH_3{}^{35}Cl$; 'I': Unidentified.

LINE NO.	WAVENUMBER CM^{-1} ;VAC.	ROTATIONAL ASSIGNMENT	BAND	ISOTOPIC SPECIES
1	1315.51771 *			
2	1315.58702 *			
3	1315.62192 *			
4	1315.72654 *			
5	1315.74662 *			
6	1315.78229 *			
7	1316.13106 *			
8	1316.23063 *			
9	1316.24239 *			
10	1316.30477 *			
11	1316.57331 *			
12	1316.63849 *			
13	1316.79474 *			
14	1316.89113 *			
15	1317.09728 *			
16	1317.25722 *			
17	1317.51031 *			
18	1317.82148 *			
19	1317.97546 *			
20	1318.04959 *			
21	1318.06770 *			
22	1318.40507 *			
23	1318.47400 *			
24	1318.56053 *			
25	1318.70975 *			
26	1318.99829 *			
27	1319.18486 *			
28	1319.20207 *			
29	1319.33613 *			
30	1319.54674 *			

LINE NO.	WAVENUMBER CM^{-1} ;VAC.	ROTATIONAL ASSIGNMENT	BAND	ISOTOPIC SPECIES
31	1319.60643 *			
32	1319.68377 *			
33	1319.90309 *			
34	1319.96292 *			
35	1320.16862 *			
36	1320.29265 *			
37	1320.34812 *			
38	1320.48935 *			
39	1320.59956 *			
40	1320.68221 *			
41	1320.73278 *			
42	1320.96744 *			
43	1320.99639 *			
44	1321.08995 *			
45	1321.44169 *			
46	1321.70992 *			
47	1321.77347 *			
48	1321.79287 *			
49	1321.81154 *			
50	1321.83652 *			
51	1322.02143 *			
52	1322.16190 *			
53	1322.18329 *			
54	1322.25151 *			
55	1322.27020 *			
56	1322.48984 *			
57	1322.58411 *			
58	1322.62104 *			

- "LINE NO.": serial number identifying the spectral line in the map on the opposite page.
- "WAVENUMBER": observed value.
- * : all given lines are selected according to intensity, halfwidth and symmetry criteria.

Lines with an asterisk (*) are selected but not identified.

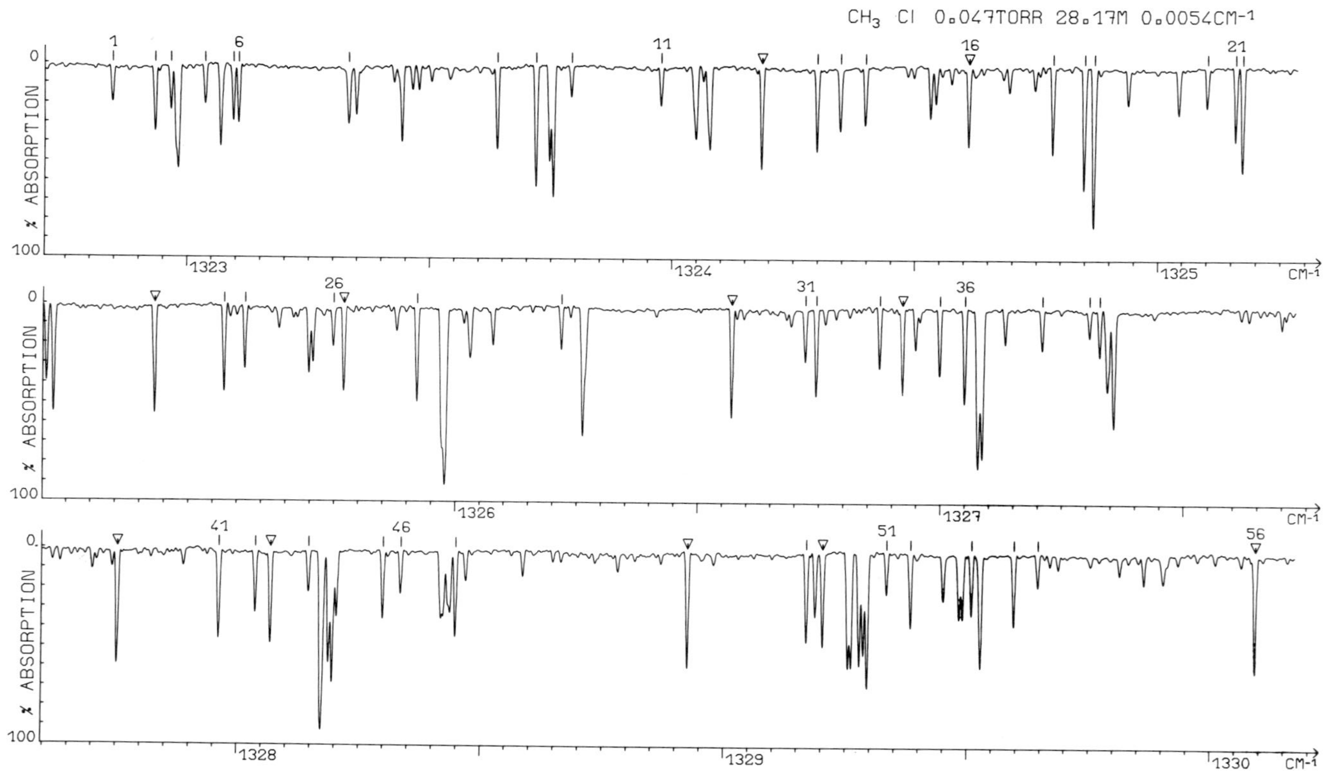

ISOTOPIC IDENTIFICATION – '∇' indicates $CH_3{}^{35}Cl$; 'I': Unidentified.

LINE NO.	WAVENUMBER CM^{-1} ;VAC.	ROTATIONAL ASSIGNMENT	BAND	ISOTOPIC SPECIES
1	1322.84717 *			
2	1322.93410 *			
3	1322.96655 *			
4	1323.03786 *			
5	1323.09569 *			
6	1323.10656 *			
7	1323.33405 *			
8	1323.64035 *			
9	1323.71990 *			
10	1323.79250 *			
11	1323.97783 *			
12	1324.18545	QP 6(30)	NU2	CH3 35C1
13	1324.29890 *			
14	1324.34653 *			
15	1324.39758 *			
16	1324.61045	QP 5(30)	NU2	CH3 35C1
17	1324.78407 *			
18	1324.84897 *			
19	1324.86896 *			
20	1325.10157 *			
21	1325.15980 *			
22	1325.17421 *			
23	1325.38136	QP 6(29)	NU2	CH3 35C1
24	1325.52409 *			
25	1325.56679 *			
26	1325.74838 *			
27	1325.77001	QP 5(29)	NU2	CH3 35C1
28	1325.92085 *			
29	1326.21856 *			
30	1326.57034	QP 6(28)	NU2	CH3 35C1

LINE NO.	WAVENUMBER CM^{-1} ;VAC.	ROTATIONAL ASSIGNMENT	BAND	ISOTOPIC SPECIES
31	1326.72236 *			
32	1326.74487 *			
33	1326.87487 *			
34	1326.92239	QP 5(28)	NU2	CH3 35C1
35	1326.99895 *			
36	1327.05032 *			
37	1327.21025 *			
38	1327.30769 *			
39	1327.32852 *			
40	1327.75203	QP 6(27)	NU2	CH3 35C1
41	1327.96080 *			
42	1328.03589 *			
43	1328.06728	QP 5(27)	NU2	CH3 35C1
44	1328.14525 *			
45	1328.29856 *			
46	1328.33526 *			
47	1328.44779 *			
48	1328.92621	QP 6(26)	NU2	CH3 35C1
49	1329.17149 *			
50	1329.20468	QP 5(26)	NU2	CH3 35C1
51	1329.33592 *			
52	1329.38574 *			
53	1329.51082 *			
54	1329.59753 *			
55	1329.64675 *			
56	1330.09270	QP 6(25)	NU2	CH3 35C1

- "LINE NO.": serial number identifying the spectral line in the map on the opposite page.
- "WAVENUMBER": observed value.
- * : all given lines are selected according to intensity, halfwidth and symmetry criteria.

Lines with an asterisk (*) are selected but not identified.

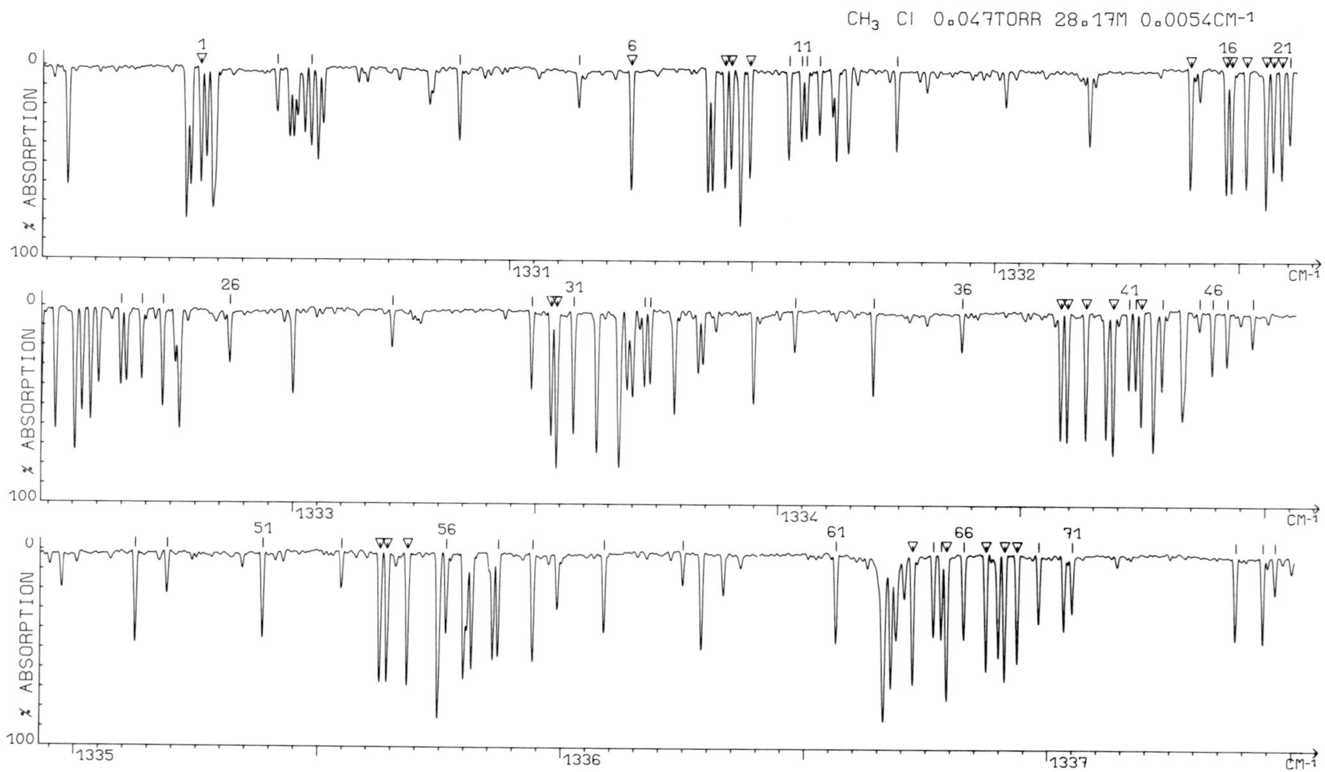

ISOTOPIC IDENTIFICATION – '▽' indicates $CH_3{}^{35}Cl$; 'I': Unidentified.

LINE NO.	WAVENUMBER CM^{-1} ;VAC.	ROTATIONAL ASSIGNMENT	BAND	ISOTOPIC SPECIES
1	1330.36505	PQ 2(25)	NU2	CH3 35C1
2	1330.52080 *			
3	1330.59076 *			
4	1330.89603 *			
5	1331.14236 *			
6	1331.25105	PQ 6(24)	NU2	CH3 35C1
7	1331.44326	PQ 2(24)	NU2	CH3 35C1
8	1331.45587	PQ 5(24)	NU2	CH3 35C1
9	1331.49482	PQ 4(24)	NU2	CH3 35C1
10	1331.57484 *			
11	1331.60072 *			
12	1331.61062 *			
13	1331.63762 *			
14	1331.79725 *			
15	1332.40117	PQ 6(23)	NU2	CH3 35C1
16	1332.47368	PQ 0(23)	NU2	CH3 35C1
17	1332.48449	PQ 1(23)	NU2	CH3 35C1
18	1332.51436	PQ 2(23)	NU2	CH3 35C1
19	1332.55474	PQ 3(23)	NU2	CH3 35C1
20	1332.56936	PQ 5(23)	NU2	CH3 35C1
21	1332.58683	PQ 4(23)	NU2	CH3 35C1
22	1332.60314 *			
23	1332.64819 *			
24	1332.69032 *			
25	1332.73313 *			
26	1332.87030 *			
27	1333.20366 *			
28	1333.49181 *			
29	1333.53197	PQ 0(22)	NU2	CH3 35C1
30	1333.54337	PQ 1(22)	NU2	CH3 35C1
30		QP 6(22)	NU2	CH3 35C1

LINE NO.	WAVENUMBER CM^{-1} ;VAC.	ROTATIONAL ASSIGNMENT	BAND	ISOTOPIC SPECIES
31	1333.57803 *			
32	1333.72432 *			
33	1333.73606 *			
34	1334.03423 *			
35	1334.19613 *			
36	1334.37914 *			
37	1334.58298	PQ 0(21)	NU2	CH3 35C1
38	1334.59640	PQ 1(21)	NU2	CH3 35C1
39	1334.63436	PQ 2(21)	NU2	CH3 35C1
40	1334.68988	PQ 3(21)	NU2	CH3 35C1
41	1334.72182 *			
42	1334.73553 *			
43	1334.74716	PQ 4(21)	NU2	CH3 35C1
44	1334.78958 *			
45	1334.86596 *			
46	1334.89243 *			
47	1334.92232 *			
48	1334.97358 *			
49	1335.12414 *			
50	1335.18891 *			
51	1335.38513 *			
52	1335.54715 *			
53	1335.62670	PQ 0(20)	NU2	CH3 35C1
54	1335.64131	PQ 1(20)	NU2	CH3 35C1
55	1335.68301	PQ 2(20)	NU2	CH3 35C1
56	1335.76281 *			
57	1335.86992 *			
58	1335.94129 *			
59	1336.08754 *			
60	1336.25019 *			

- "LINE NO.": serial number identifying the spectral line in the map on the opposite page.
- "WAVENUMBER": observed value.
- * : all given lines are selected according to intensity, halfwidth and symmetry criteria.

Lines with an asterisk (*) are selected but not identified.

CH$_3$ Cl 0.047TORR 28.17M 0.0054CM-1

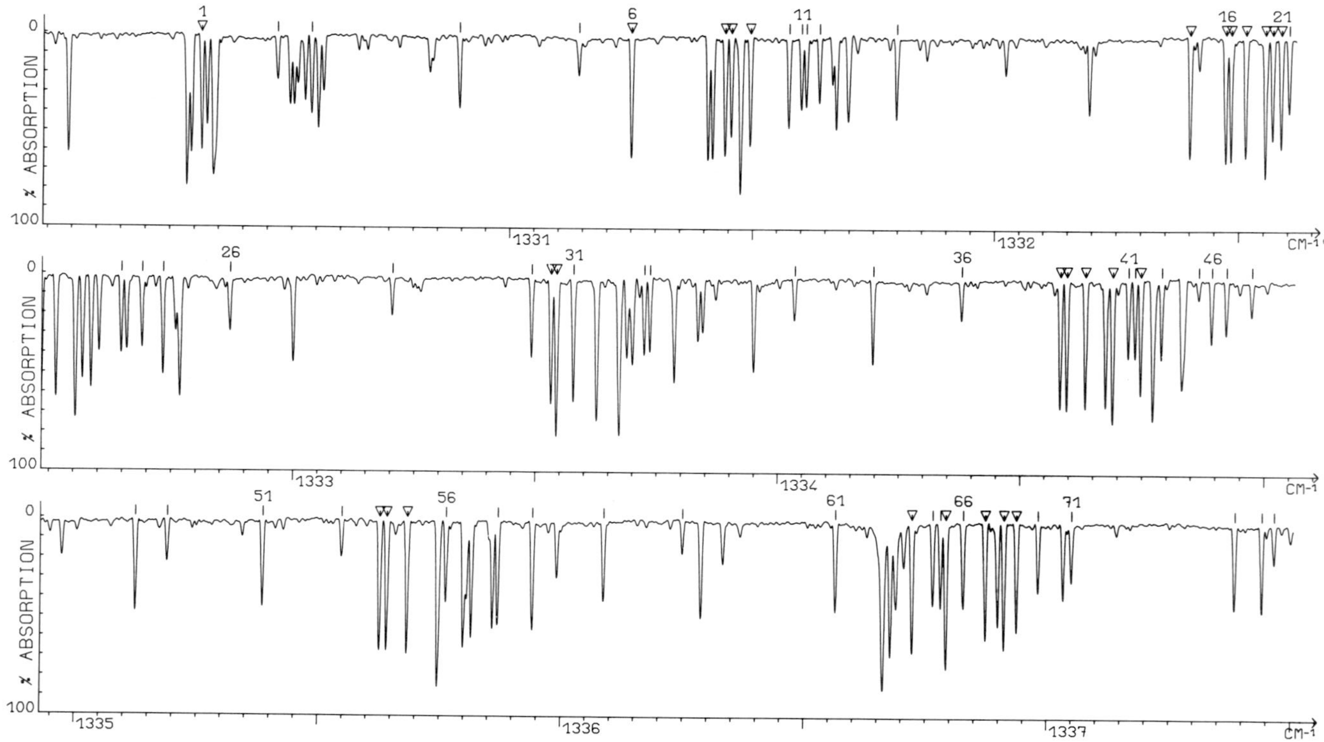

ISOTOPIC IDENTIFICATION – '▽' indicates CH$_3{}^{35}$Cl ; 'I': Unidentified.

LINE NO.	WAVENUMBER CM^{-1} ;VAC.		ROTATIONAL ASSIGNMENT	BAND	ISOTOPIC SPECIES
61	1336.56582	*			
62	1336.72428		PQ 2(19)	NU2	CH3 35C1
63	1336.76697	*			
64	1336.78300	*			
65	1336.79406		PQ 3(19)	NU2	CH3 35C1
66	1336.82947	*			
67	1336.87504		PQ 4(19)	NU2	CH3 35C1
68	1336.91292		PQ 6(19)	NU2	CH3 35C1
69	1336.93884		PQ 5(19)	NU2	CH3 35C1
70	1336.98383	*			
71	1337.05217	*			
72	1337.38591	*			
73	1337.44173	*			
74	1337.46672	*			

- "LINE NO.": serial number identifying the spectral line in the map on the opposite page.
- "WAVENUMBER": observed value.
- * : all given lines are selected according to intensity, halfwidth and symmetry criteria.

Lines with an asterisk (*) are selected but not identified.

CH$_3$ Cl 0.047TORR 28.17M 0.0054CM-1

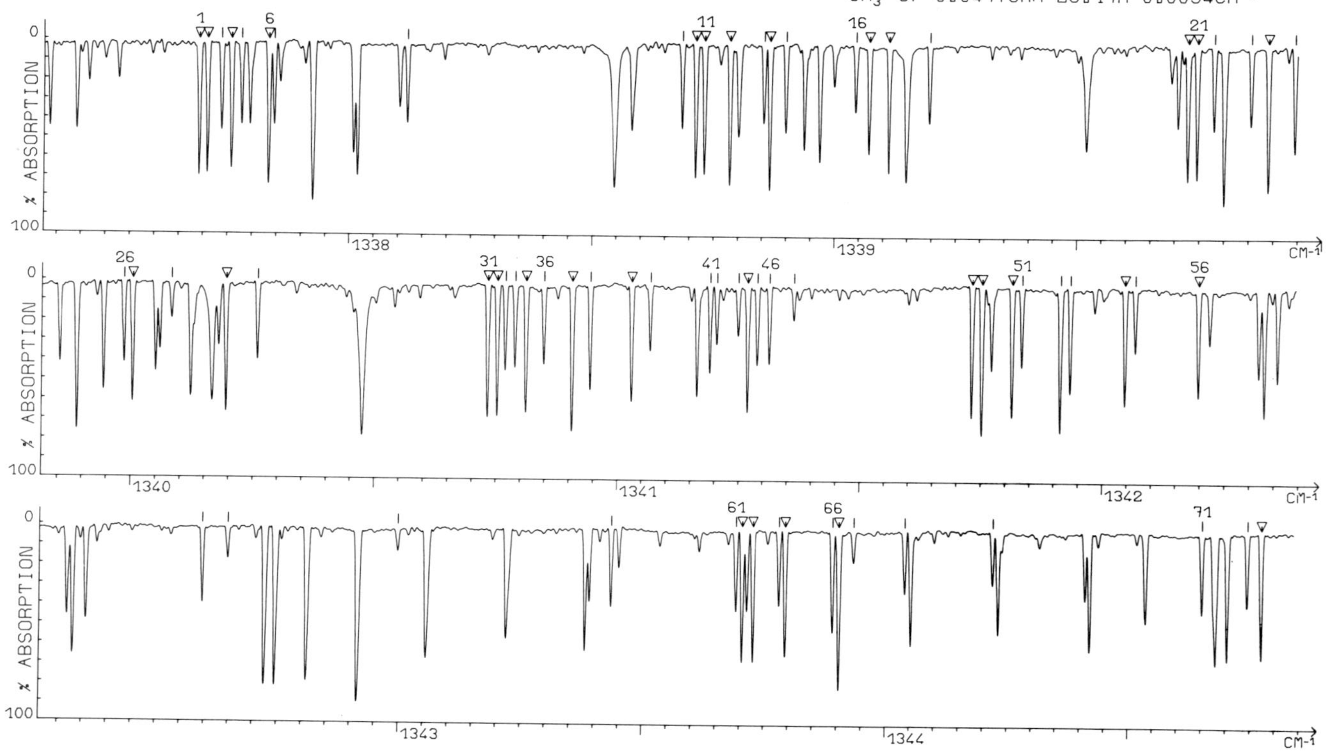

ISOTOPIC IDENTIFICATION – '▽' indicates CH$_3$^{35}Cl ; 'I': Unidentified.

LINE NO.	WAVENUMBER CM^{-1} ;VAC.	ROTATIONAL ASSIGNMENT	BAND	ISOTOPIC SPECIES
1	1337.69186	PQ 0(18)	NU2	CH3 35Cl
2	1337.70883	PQ 1(18)	NU2	CH3 35Cl
3	1337.73751 *			
4	1337.75798	PQ 2(18)	NU2	CH3 35Cl
5	1337.77860 *			
6	1337.83430	PQ 3(18)	NU2	CH3 35Cl
7	1337.84534 *			
8	1338.11990 *			
9	1338.68537 *			
10	1338.71342	PQ 0(17)	NU2	CH3 35Cl
11	1338.73133	PQ 1(17)	NU2	CH3 35Cl
12	1338.78362	PQ 2(17)	NU2	CH3 35Cl
13	1338.85420 *			
14	1338.86657	PQ 3(17)	NU2	CH3 35Cl
15	1338.89942 *			
16	1339.04268 *			
17	1339.07039	PQ 5(17)	NU2	CH3 35Cl
18	1339.11155	PQ 6(17)	NU2	CH3 35Cl
19	1339.19495 *			
20	1339.72732	PQ 0(16)	NU2	CH3 35Cl
21	1339.74615	PQ 1(16)	NU2	CH3 35Cl
22	1339.78003 *			
23	1339.85529 *			
24	1339.89069	PQ 3(16)	NU2	CH3 35Cl
25	1339.94484 *			
26	1339.98630 *			
27	1340.00394	PQ 4(16)	NU2	CH3 35Cl
28	1340.08386 *			
29	1340.19550	PQ 6(16)	NU2	CH3 35Cl
30	1340.25985 *			

LINE NO.	WAVENUMBER CM^{-1} ;VAC.	ROTATIONAL ASSIGNMENT	BAND	ISOTOPIC SPECIES
31	1340.73349	PQ 0(15)	NU2	CH3 35Cl
32	1340.75337	PQ 1(15)	NU2	CH3 35Cl
33	1340.76958 *			
34	1340.78960 *			
35	1340.81224	PQ 2(15)	NU2	CH3 35Cl
36	1340.84872 *			
37	1340.90678	PQ 3(15)	NU2	CH3 35Cl
38	1340.94388 *			
39	1341.02973	PQ 4(15)	NU2	CH3 35Cl
40	1341.06802 *			
41	1341.19123 *			
42	1341.20527 *			
43	1341.24923 *			
44	1341.26886	PQ 6(15)	NU2	CH3 35Cl
45	1341.28939 *			
46	1341.31433 *			
47	1341.36467 *			
48	1341.73206	PQ 0(14)	NU2	CH3 35Cl
49	1341.75253	PQ 1(14)	NU2	CH3 35Cl
50	1341.81460	PQ 2(14)	NU2	CH3 35Cl
51	1341.83444 *			
52	1341.91457 *			
53	1341.93488 *			
54	1342.04672	PQ 4(14)	NU2	CH3 35Cl
55	1342.06790 *			
56	1342.19754	PQ 5(14)	NU2	CH3 35Cl
57	1342.59585 *			
58	1342.64812 *			
59	1342.99700 *			
60	1343.43538 *			

- "LINE NO.": serial number identifying the spectral line in the map on the opposite page.
- "WAVENUMBER": observed value.
- * : all given lines are selected according to intensity, halfwidth and symmetry criteria.

Lines with an asterisk (*) are selected but not identified.

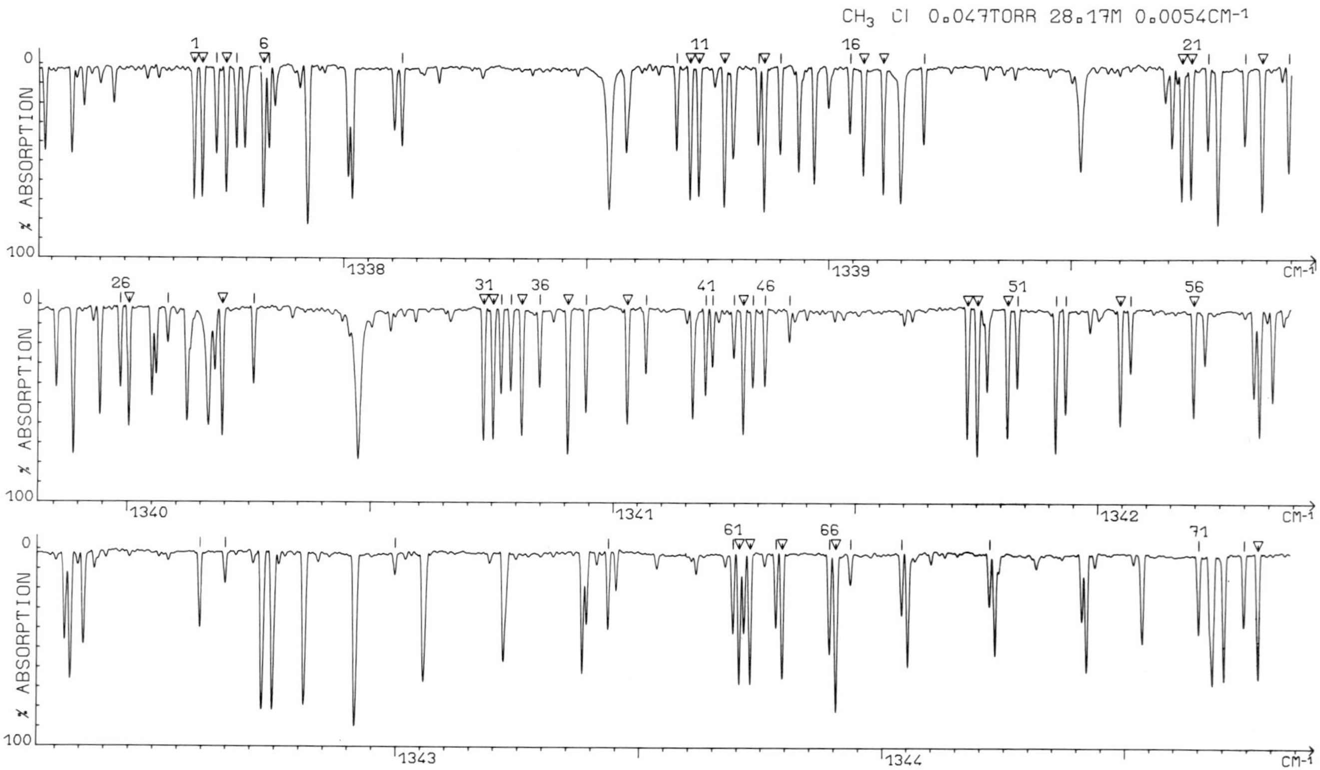

ISOTOPIC IDENTIFICATION – '▽' indicates $CH_3{}^{35}Cl$; 'I': Unidentified.

LINE NO.	WAVENUMBER CM^{-1} ;VAC.		ROTATIONAL ASSIGNMENT	BAND	ISOTOPIC SPECIES
61	1343.69309	*			
62	1343.70594		PQ 0(12)	NU2	CH3 35C1
63	1343.72846		PQ 1(12)	NU2	CH3 35C1
64	1343.78258	*			
65	1343.79538		PQ 2(12)	NU2	CH3 35C1
66	1343.89251	*			
67	1343.90559		PQ 3(12)	NU2	CH3 35C1
68	1343.93571	*			
69	1344.04167	*			
70	1344.22244	*			
71	1344.65253	*			
72	1344.74488	*			
73	1344.77367		PQ 2(11)	NU2	CH3 35C1

- "LINE NO.": serial number identifying the spectral line in the map on the opposite page.
- "WAVENUMBER": observed value.
- * : all given lines are selected according to intensity, halfwidth and symmetry criteria.

Lines with an asterisk (*) are selected but not identified.

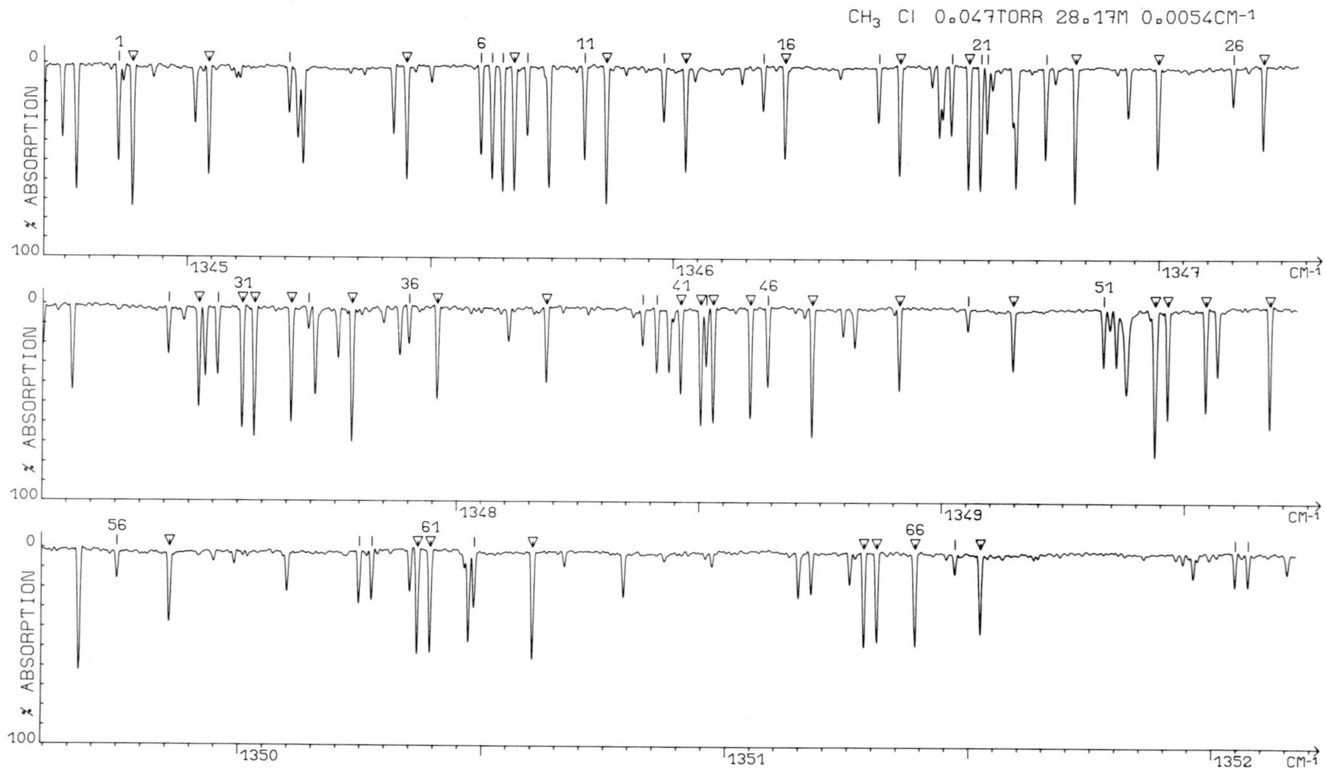

ISOTOPIC IDENTIFICATION – '▽' indicates $CH_3{}^{35}Cl$; 'I': Unidentified.

LINE NO.	WAVENUMBER CM^{-1} ;VAC.	ROTATIONAL ASSIGNMENT	BAND	ISOTOPIC SPECIES	LINE NO.	WAVENUMBER CM^{-1} ;VAC.	ROTATIONAL ASSIGNMENT	BAND	ISOTOPIC SPECIES
1	1344.85902 *				31	1347.55964	PQ 0(8)	NU2	CH3 35C1
2	1344.88782	PQ 3(11)	NU2	CH3 35C1	32	1347.58447	PQ 1(8)	NU2	CH3 35C1
3	1345.04422	PQ 4(11)	NU2	CH3 35C1	33	1347.65960	PQ 2(8)	NU2	CH3 35C1
4	1345.20860 *				34	1347.69475 *			
5	1345.45044	PQ 6(11)	NU2	CH3 35C1	35	1347.78466	PQ 3(8)	NU2	CH3 35C1
6	1345.60395 *				36	1347.90146 *			
7	1345.62695 *				37	1347.95964	PQ 4(8)	NU2	CH3 35C1
8	1345.64857 *				38	1348.18498	PQ 5(8)	NU2	CH3 35C1
9	1345.67246	PQ 1(10)	NU2	CH3 35C1	39	1348.38370 *			
10	1345.69927 *				40	1348.41275 *			
11	1345.81723 *				41	1348.46198	PQ 6(8)	NU2	CH3 35C1
12	1345.86198	PQ 3(10)	NU2	CH3 35C1	42	1348.50316	PQ 0(7)	NU2	CH3 35C1
13	1345.98032 *				43	1348.51429 *			
14	1346.02527	PQ 4(10)	NU2	CH3 35C1	44	1348.52863	PQ 1(7)	NU2	CH3 35C1
15	1346.18505 *				45	1348.60510	PQ 2(7)	NU2	CH3 35C1
16	1346.23002	PQ 5(10)	NU2	CH3 35C1	46	1348.64168 *			
17	1346.42209 *				47	1348.73294	PQ 3(7)	NU2	CH3 35C1
18	1346.46664	PQ 6(10)	NU2	CH3 35C1	48	1348.91286	PQ 4(7)	NU2	CH3 35C1
19	1346.57252 *				49	1349.05379 *			
20	1346.60811	PQ 0(9)	NU2	CH3 35C1	50	1349.14683	PQ 5(7)	NU2	CH3 35C1
21	1346.63259 *				51	1349.33332 *			
22	1346.64574 *				52	1349.43871	PQ 0(6)	NU2	CH3 35C1
23	1346.76708 *				53	1349.46455	PQ 1(6)	NU2	CH3 35C1
24	1346.82759	PQ 3(9)	NU2	CH3 35C1	54	1349.54237	PQ 2(6)	NU2	CH3 35C1
25	1346.99708	PQ 4(9)	NU2	CH3 35C1	55	1349.67258	PQ 3(6)	NU2	CH3 35C1
26	1347.15140 *				56	1349.74947 *			
27	1347.21270	PQ 5(9)	NU2	CH3 35C1	57	1349.85671	PQ 4(6)	NU2	CH3 35C1
28	1347.40890 *				58	1350.24613 *			
29	1347.47042	PQ 6(9)	NU2	CH3 35C1	59	1350.27221 *			
30	1347.50934 *				60	1350.36614	PQ 0(5)	NU2	CH3 35C1

- "LINE NO.": serial number identifying the spectral line in the map on the opposite page.
- "WAVENUMBER": observed value.
- * : all given lines are selected according to intensity, halfwidth and symmetry criteria.

Lines with an asterisk (*) are selected but not identified.

CH$_3$ Cl 0.047TORR 28.17M 0.0054CM-1

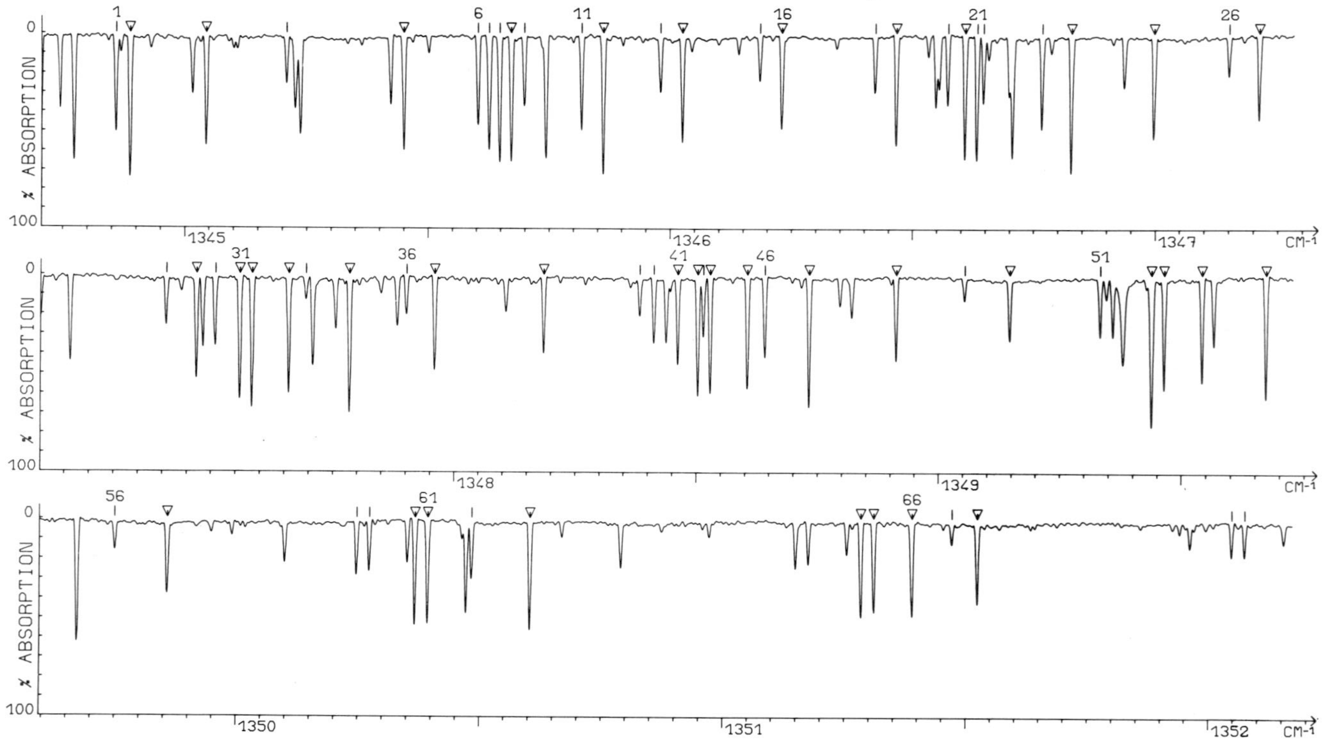

ISOTOPIC IDENTIFICATION – '∇' indicates $CH_3{}^{35}Cl$; 'I': Unidentified.

LINE NO.	WAVENUMBER CM^{-1} ;VAC.		ROTATIONAL ASSIGNMENT	BAND	ISOTOPIC SPECIES
61	1350.39232		PQ 1(5)	NU2	CH3 35Cl
62	1350.48226	*			
63	1350.60349		PQ 3(5)	NU2	CH3 35Cl
64	1351.28551		PQ 0(4)	NU2	CH3 35Cl
65	1351.31198		PQ 1(4)	NU2	CH3 35Cl
66	1351.39120		PQ 2(4)	NU2	CH3 35Cl
67	1351.47328	*			
68	1351.52559		PQ 3(4)	NU2	CH3 35Cl
69	1352.04798	*			
70	1352.07443	*			

- "LINE NO.": serial number identifying the spectral line in the map on the opposite page.
- "WAVENUMBER": observed value.
- * : all given lines are selected according to intensity, halfwidth and symmetry criteria.

Lines with an asterisk (*) are selected but not identified.

CH3 Cl 0.047TORR 28.17M 0.0054CM-1

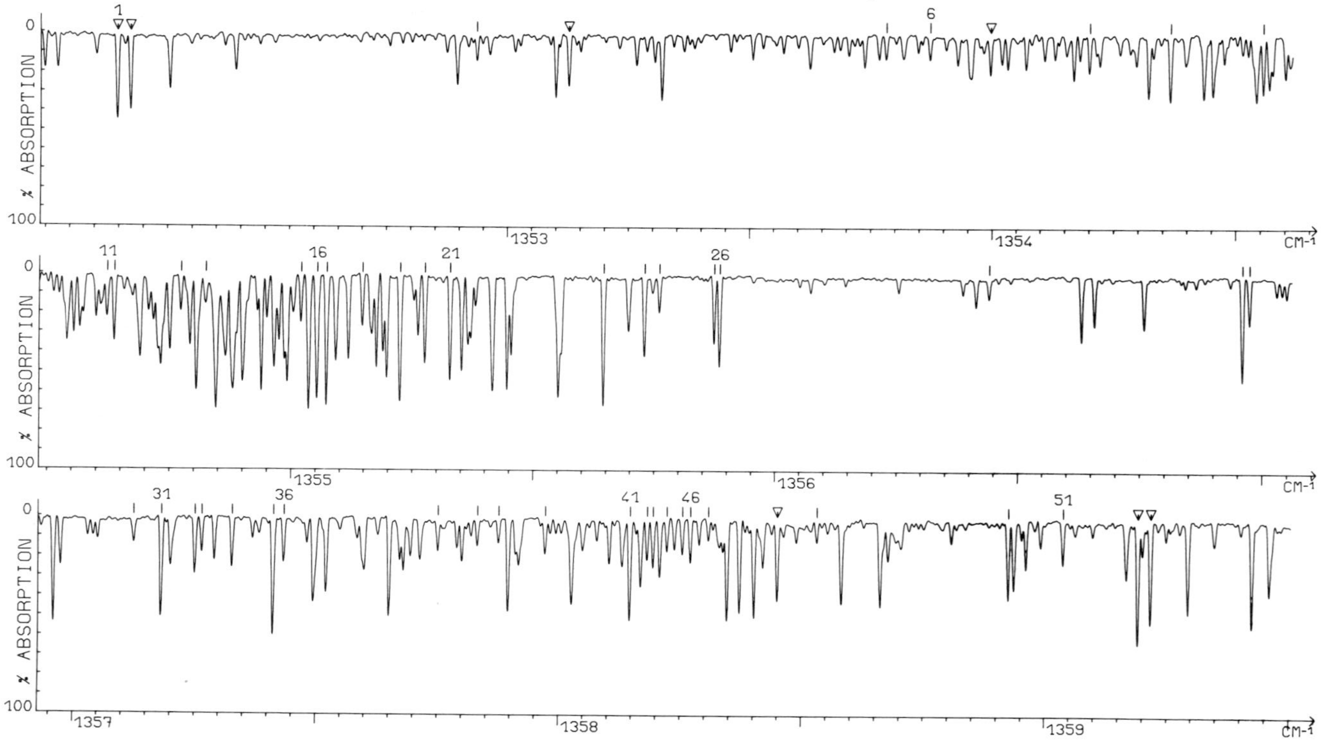

ISOTOPIC IDENTIFICATION – '▽' indicates $CH_3{}^{35}Cl$; 'I': Unidentified.

LINE NO.	WAVENUMBER CM^{-1} ;VAC.	ROTATIONAL ASSIGNMENT	BAND	ISOTOPIC SPECIES
1	1352.19685	PQ O(3)	NU2	CH3 35Cl
2	1352.22343	PQ 1(3)	NU2	CH3 35Cl
3	1352.93717 *			
4	1353.12670	PQ 1(2)	NU2	CH3 35Cl
5	1353.78055 *			
6	1353.87108 *			
7	1353.99486	PQ O(1)	NU2	CH3 35Cl
8	1354.19891 *			
9	1354.36632 *			
10	1354.55561 *			
11	1354.62333 *			
12	1354.63809 *			
13	1354.77366 *			
14	1354.82481 *			
15	1355.01974 *			
16	1355.05287 *			
17	1355.07222 *			
18	1355.14661 *			
19	1355.22440 *			
20	1355.27551 *			
21	1355.32745 *			
22	1355.64430 *			
23	1355.72968 *			
24	1355.76035 *			
25	1355.87341 *			
26	1355.88469 *			
27	1356.43971 *			
28	1356.95912 *			
29	1356.97387 *			
30	1357.12372 *			
31	1357.18018 *			
32	1357.24866 *			
33	1357.26284 *			
34	1357.32558 *			
35	1357.41049 *			
36	1357.43198 *			
37	1357.74974 *			
38	1357.83100 *			
39	1357.87549 *			
40	1357.97153 *			
41	1358.14700 *			
42	1358.18250 *			
43	1358.19473 *			
44	1358.22336 *			
45	1358.25500 *			
46	1358.27167 *			
47	1358.30880 *			
48	1358.45150	RQ 2(3)	NU2	CH3 35Cl
49	1358.53320 *			
50	1358.92657 *			
51	1359.03979 *			
52	1359.19264	RQ O(4)	NU2	CH3 35Cl
53	1359.21843	RQ 1(4)	NU2	CH3 35Cl

- "LINE NO.": serial number identifying the spectral line in the map on the opposite page.
- "WAVENUMBER": observed value.
- * : all given lines are selected according to intensity, halfwidth and symmetry criteria.

Lines with an asterisk (*) are selected but not identified.

CH_3 Cl 0.047TORR 28.17M 0.0054CM-1

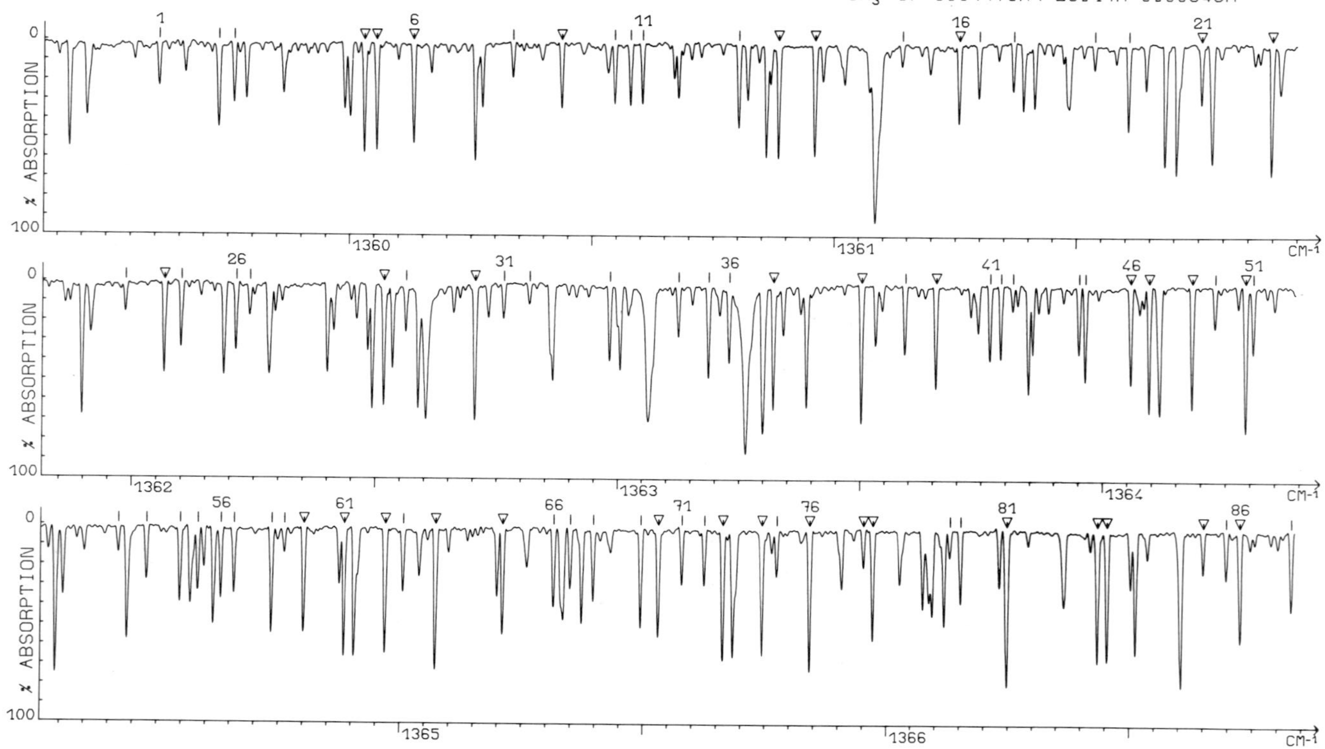

ISOTOPIC IDENTIFICATION – '▽' indicates $CH_3{}^{35}Cl$; 'I': Unidentified.

LINE NO.	WAVENUMBER CM^{-1} ;VAC.		ROTATIONAL ASSIGNMENT	BAND	ISOTOPIC SPECIES	LINE NO.	WAVENUMBER CM^{-1} ;VAC.		ROTATIONAL ASSIGNMENT	BAND	ISOTOPIC SPECIES
1	1359.60838	*				31	1362.76453	*			
2	1359.73075	*				32	1362.81745	*			
3	1359.76233	*				33	1362.98311	*			
4	1360.03026		RQ 0(5)	NU2	CH3 35C1	34	1363.12544	*			
5	1360.05556		RQ 1(5)	NU2	CH3 35C1	35	1363.18734	*			
6	1360.13154		RQ 2(5)	NU2	CH3 35C1	36	1363.22972	*			
7	1360.33688	*				37	1363.32144		RQ 1(9)	NU2	CH3 35C1
8	1360.43725		RQ 4(5)	NU2	CH3 35C1	38	1363.50263		RQ 3(9)	NU2	CH3 35C1
9	1360.54695	*				39	1363.59317	*			
10	1360.57956	*				40	1363.65704		RQ 4(9)	NU2	CH3 35C1
11	1360.60431	*				41	1363.76844	*			
12	1360.80200	*				42	1363.79059	*			
13	1360.88446		RQ 1(6)	NU2	CH3 35C1	43	1363.81507	*			
14	1360.95882		RQ 2(6)	NU2	CH3 35C1	44	1363.95203	*			
15	1361.13883	*				45	1363.96510	*			
16	1361.25648		RQ 4(6)	NU2	CH3 35C1	46	1364.05776		RQ 6(9)	NU2	CH3 35C1
17	1361.29723	*				47	1364.09514		RQ 0(10)	NU2	CH3 35C1
18	1361.36785	*				48	1364.18331		RQ 2(10)	NU2	CH3 35C1
19	1361.53557	*				49	1364.22951	*			
20	1361.60579	*				50	1364.29143		RQ 3(10)	NU2	CH3 35C1
21	1361.75505		RQ 6(6)	NU2	CH3 35C1	51	1364.30665	*			
22	1361.89830		RQ 3(7)	NU2	CH3 35C1	52	1364.42006	*			
23	1361.98646	*				53	1364.47754	*			
24	1362.06611		RQ 4(7)	NU2	CH3 35C1	54	1364.54576	*			
25	1362.10056	*				55	1364.58254	*			
26	1362.21355	*				56	1364.63065	*			
27	1362.24160	*				57	1364.65749	*			
28	1362.51736		RQ 1(8)	NU2	CH3 35C1	58	1364.73455	*			
29	1362.56220	*				59	1364.76073	*			
30	1362.70484		RQ 3(8)	NU2	CH3 35C1	60	1364.80056		RQ 6(10)	NU2	CH3 35C1

- "LINE NO.": serial number identifying the spectral line in the map on the opposite page.
- "WAVENUMBER": observed value.
- * : all given lines are selected according to intensity, halfwidth and symmetry criteria.

Lines with an asterisk (*) are selected but not identified.

CH_3 Cl 0.047TORR 28.17M 0.0054CM-1

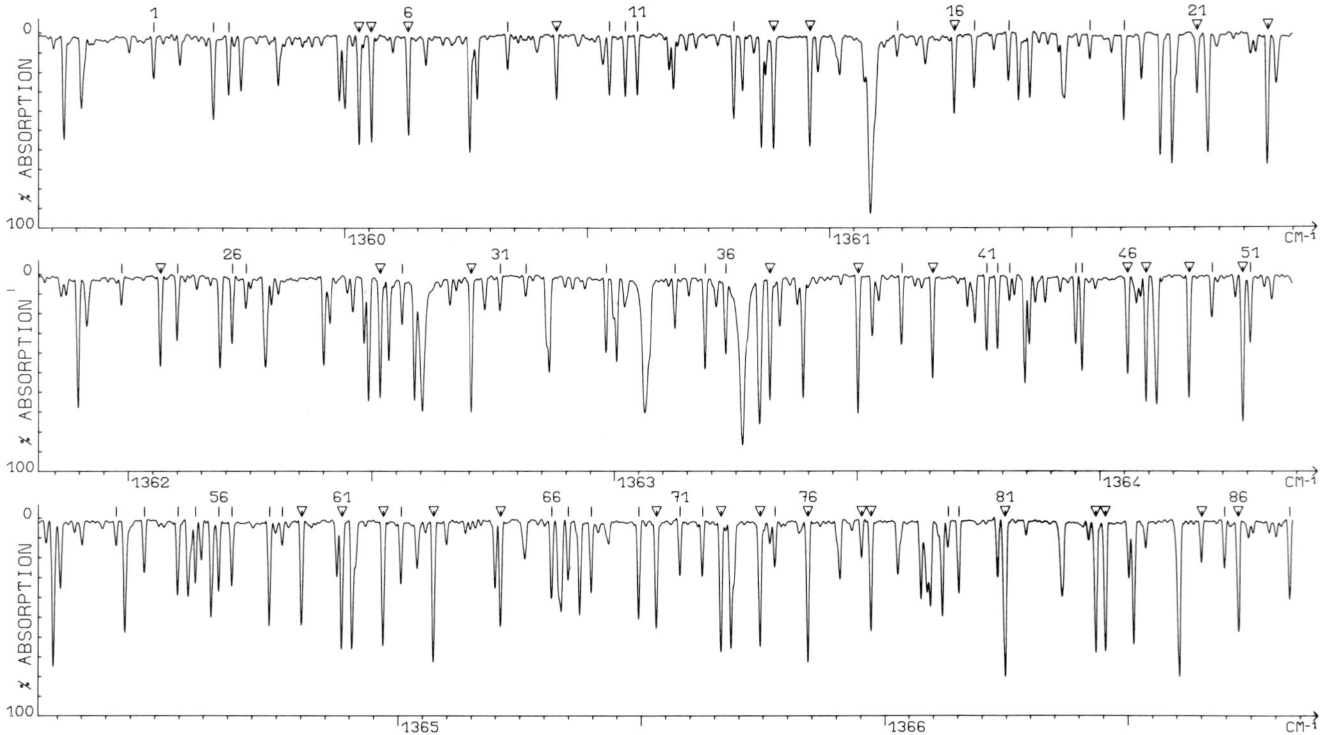

ISOTOPIC IDENTIFICATION – '▽' indicates $CH_3{}^{35}Cl$; 'I': Unidentified.

LINE NO.	WAVENUMBER CM^{-1} ;VAC.		ROTATIONAL ASSIGNMENT	BAND	ISOTOPIC SPECIES
61	1364.88356		RQ 0(11)	NU2	CH3 35Cl
62	1364.96840		RQ 2(11)	NU2	CH3 35Cl
63	1365.00487	*			
64	1365.07172		RQ 3(11)	NU2	CH3 35Cl
65	1365.21023		RQ 4(11)	NU2	CH3 35Cl
66	1365.31505	*			
67	1365.34915	*			
68	1365.39658	*			
69	1365.49516	*			
70	1365.53133		RQ 6(11)	NU2	CH3 35Cl
71	1365.57972	*			
72	1365.62580	*			
73	1365.66391		RQ 0(12)	NU2	CH3 35Cl
74	1365.74490		RQ 2(12)	NU2	CH3 35Cl
75	1365.77534	*			
76	1365.84317		RQ 3(12)	NU2	CH3 35Cl
77	1365.95372		PP 6(25)	NU5	CH3 35Cl
78	1365.97283		RQ 4(12)	NU2	CH3 35Cl
79	1366.13200	*			
80	1366.15416	*			
81	1366.24897		RQ 6(12)	NU2	CH3 35Cl
82	1366.43596		RQ 0(13)	NU2	CH3 35Cl
83	1366.45554		RQ 1(13)	NU2	CH3 35Cl
84	1366.65132		PP 6(24)	NU5	CH3 35Cl
85	1366.69823	*			
86	1366.72616		RQ 4(13)	NU2	CH3 35Cl
87	1366.83006	*			

- "LINE NO.": serial number identifying the spectral line in the map on the opposite page.
- "WAVENUMBER": observed value.
- * : all given lines are selected according to intensity, halfwidth and symmetry criteria.

Lines with an asterisk (*) are selected but not identified.

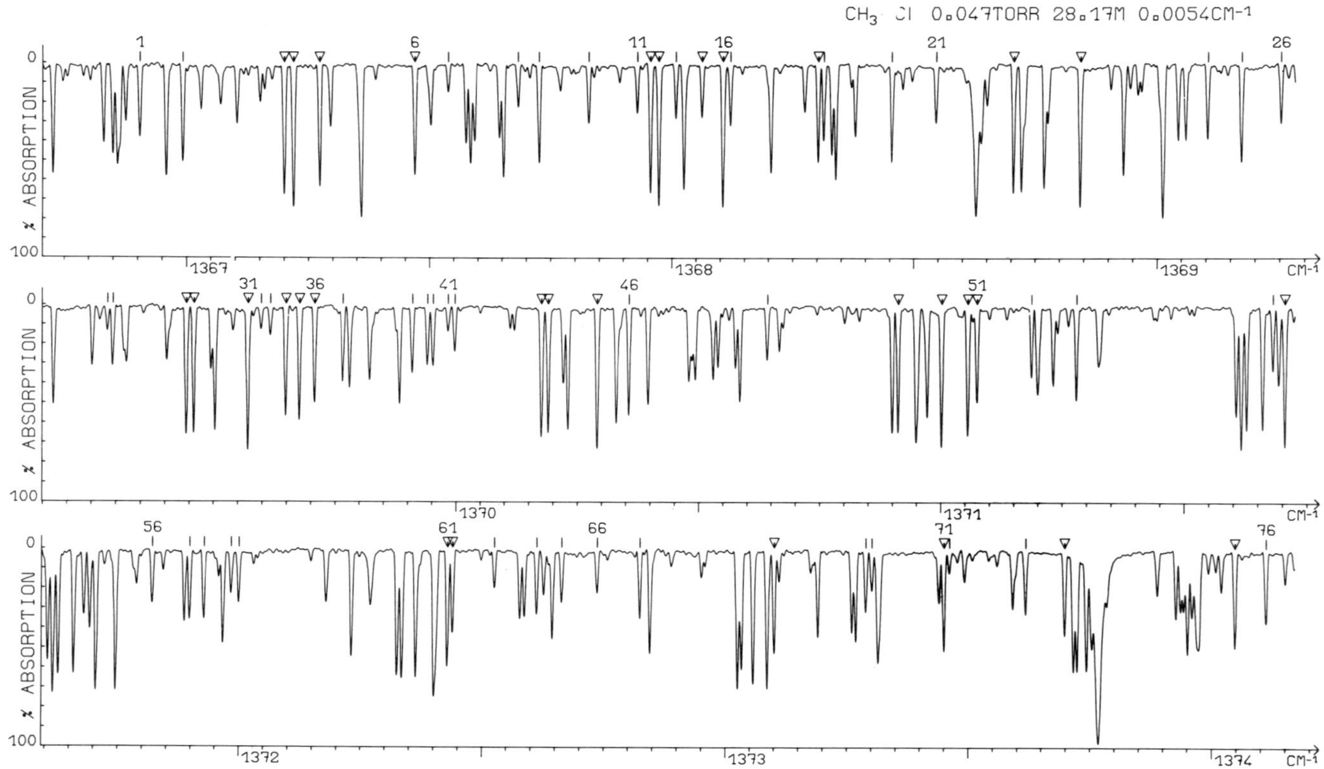

ISOTOPIC IDENTIFICATION – '▽' indicates $CH_3{}^{35}Cl$; 'I': Unidentified.

LINE NO.	WAVENUMBER CM^{-1} ;VAC.	ROTATIONAL ASSIGNMENT	BAND	ISOTOPIC SPECIES
1	1366.90373 *			
2	1366.99132 *			
3	1367.20013	RQ 0(14)	NU2	CH3 35C1
4	1367.21913	RQ 1(14)	NU2	CH3 35C1
5	1367.27325	RQ 2(14)	NU2	CH3 35C1
6	1367.47034	RQ 4(14)	NU2	CH3 35C1
7	1367.53941 *			
8	1367.68390 *			
9	1367.72682 *			
10	1367.82863 *			
11	1367.92905 *			
12	1367.95624	RQ 0(15)	NU2	CH3 35C1
13	1367.97343	RQ 1(15)	NU2	CH3 35C1
14	1368.00834 *			
15	1368.06214	PP 6(22)	NU5	CH3 35C1
16	1368.10548	RQ 3(15)	NU2	CH3 35C1
17	1368.12091 *			
18	1368.30229	RQ 5(15)	NU2	CH3 35C1
19	1368.31364 *			
20	1368.45394 *			
21	1368.54513 *			
22	1368.70423	RQ 0(16)	NU2	CH3 35C1
23	1368.84239	RQ 3(16)	NU2	CH3 35C1
24	1369.10394 *			
25	1369.17280 *			
26	1369.25289 *			
27	1369.28343 *			
28	1369.29463 *			
29	1369.44422	RQ 0(17)	NU2	CH3 35C1
30	1369.45946	RQ 1(17)	NU2	CH3 35C1

LINE NO.	WAVENUMBER CM^{-1} ;VAC.	ROTATIONAL ASSIGNMENT	BAND	ISOTOPIC SPECIES
31	1369.57084	RQ 3(17)	NU2	CH3 35C1
32	1369.59732 *			
33	1369.61644 *			
34	1369.64832	RQ 4(17)	NU2	CH3 35C1
35	1369.67642	RQ 6(17)	NU2	CH3 35C1
36	1369.70773	RQ 5(17)	NU2	CH3 35C1
37	1369.76591 *			
38	1369.90930 *			
39	1369.94059 *			
40	1369.95212 *			
41	1369.98329 *			
42	1369.99705 *			
43	1370.17625	RQ 0(18)	NU2	CH3 35C1
44	1370.19028	RQ 1(18)	NU2	CH3 35C1
45	1370.29074	RQ 3(18)	NU2	CH3 35C1
46	1370.35647 *			
47	1370.64261 *			
48	1370.91313	RQ 1(19)	NU2	CH3 35C1
49	1371.00235	RQ 3(19)	NU2	CH3 35C1
50	1371.05620	RQ 4(19)	NU2	CH3 35C1
51	1371.07508	RQ 5(19)	NU2	CH3 35C1
52	1371.18726 *			
53	1371.27994 *			
54	1371.68106 *			
55	1371.70557	RQ 3(20)	NU2	CH3 35C1
56	1371.82035 *			
57	1371.89699 *			
58	1371.92624 *			
59	1371.98246 *			
60	1371.99824 *			

- "LINE NO.": serial number identifying the spectral line in the map on the opposite page.
- "WAVENUMBER": observed value.
- * : all given lines are selected according to intensity, halfwidth and symmetry criteria.
Lines with an asterisk (*) are selected but not identified.

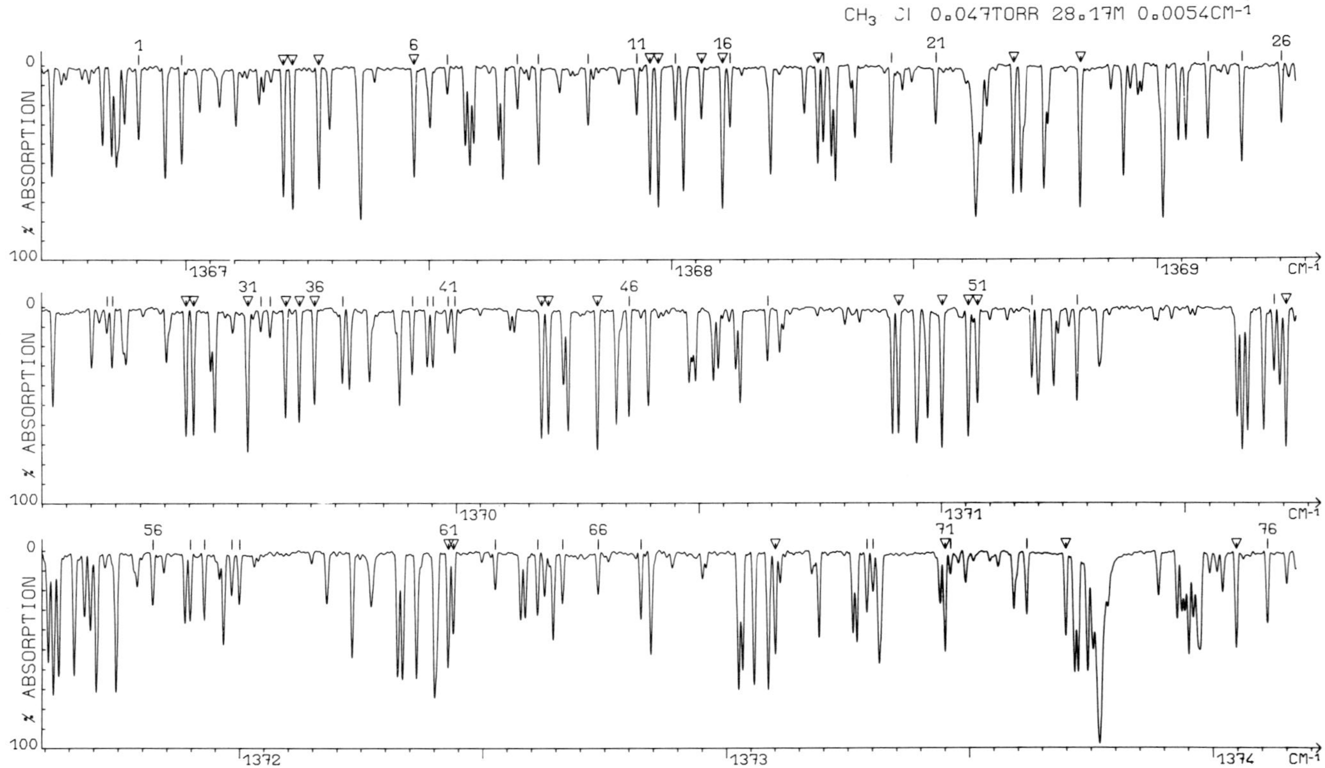

ISOTOPIC IDENTIFICATION – '▽' indicates $CH_3{}^{35}Cl$; 'I': Unidentified.

LINE NO.	WAVENUMBER CM^{-1} ;VAC.		ROTATIONAL ASSIGNMENT	BAND	ISOTOPIC SPECIES
61	1372.42848		RQ 4(21)	NU2	CH3 35C1
62	1372.43880		PP 6(16)	NU5	CH3 35C1
63	1372.52557	*			
64	1372.61224	*			
65	1372.66348	*			
66	1372.73636	*			
67	1372.82506	*			
68	1373.10196		RQ 4(22)	NU2	CH3 35C1
69	1373.29052	*			
70	1373.30301	*			
71	1373.45168		RQ 6(23)	NU2	CH3 35C1
72	1373.46316	*			
73	1373.61910	*			
74	1373.70034		RQ 5(23)	NU2	CH3 35C1
75	1374.04886		RQ 6(24)	NU2	CH3 35C1
76	1374.11189	*			

- "LINE NO.": serial number identifying the spectral line in the map on the opposite page.
- "WAVENUMBER": observed value.
- * : all given lines are selected according to intensity, halfwidth and symmetry criteria.

Lines with an asterisk (*) are selected but not identified.

CH_3 Cl 0.047TORR 28.17M 0.0054CM-1

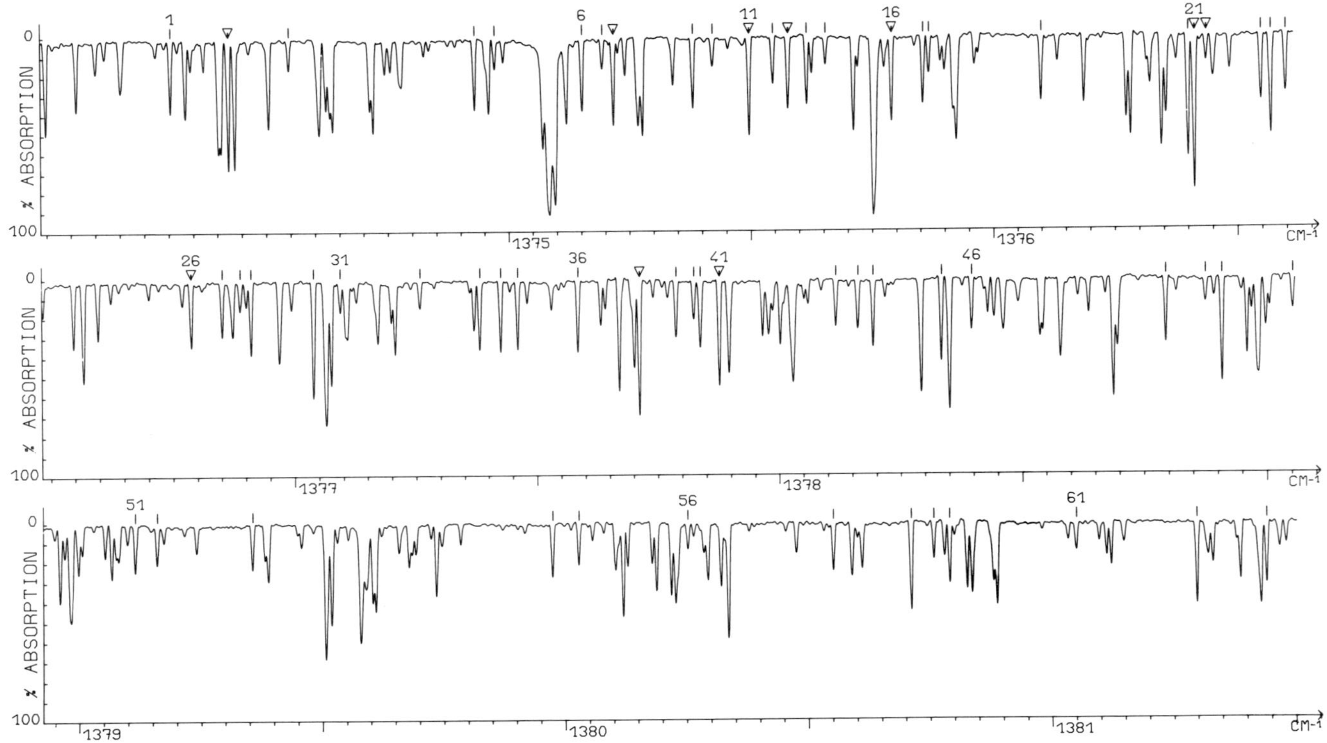

ISOTOPIC IDENTIFICATION – '∇' indicates $CH_3{}^{35}Cl$; 'I': Unidentified.

LINE NO.	WAVENUMBER CM^{-1} ;VAC.		ROTATIONAL ASSIGNMENT	BAND	ISOTOPIC SPECIES
1	1374.30444	*			
2	1374.42371		RQ 2(24)	NU2	CH3 35C1
2			QR 4(24)	NU2	CH3 35C1
3	1374.54791	*			
4	1374.93063	*			
5	1374.97332	*			
6	1375.15335	*			
7	1375.19515	*			
8	1375.21792		RQ 6(26)	NU2	CH3 35C1
9	1375.38215	*			
10	1375.42253	*			
11	1375.49774		PP 6(12)	NU5	CH3 35C1
12	1375.54708	*			
13	1375.57794		RQ 5(26)	NU2	CH3 35C1
14	1375.61632	*			
15	1375.65484	*			
16	1375.79021		RQ 6(27)	NU2	CH3 35C1
17	1375.85546	*			
18	1375.86757	*			
19	1376.09916	*			
20	1376.40037	*			
21	1376.41315		RQ O(27)	NU2	CH3 35C1
22	1376.43677		PP 5(27)	NU5	CH3 35C1
23	1376.54828	*			
24	1376.56825	*			
25	1376.59811	*			
26	1376.78830		RQ 5(28)	NU2	CH3 35C1
27	1376.85271	*			
28	1376.88957	*			
29	1376.91162	*			
30	1377.03935	*			

LINE NO.	WAVENUMBER CM^{-1} ;VAC.		ROTATIONAL ASSIGNMENT	BAND	ISOTOPIC SPECIES
31	1377.09409	*			
32	1377.25833	*			
33	1377.38196	*			
34	1377.42595	*			
35	1377.46112	*			
36	1377.58539	*			
37	1377.71285		RQ O(29)	NU2	CH3 35C1
38	1377.78842	*			
39	1377.82493	*			
40	1377.83841	*			
41	1377.87750		PP 6(9)	NU5	CH3 35C1
42	1378.11750	*			
43	1378.16272	*			
44	1378.19411	*			
45	1378.33528	*			
46	1378.39736	*			
47	1378.79518	*			
48	1378.87693	*			
49	1378.91029	*			
50	1379.05372	*			
51	1379.11509	*			
52	1379.15943	*			
53	1379.35607	*			
54	1379.97466	*			
55	1380.02858	*			
56	1380.25386	*			
57	1380.55327	*			
58	1380.71380	*			
59	1380.76034	*			
60	1380.79276	*			

- "LINE NO.": serial number identifying the spectral line in the map on the opposite page.
- "WAVENUMBER": observed value.
- * : all given lines are selected according to intensity, halfwidth and symmetry criteria.

Lines with an asterisk (*) are selected but not identified.

CH_3 Cl 0.047TORR 28.17M 0.0054CM-1

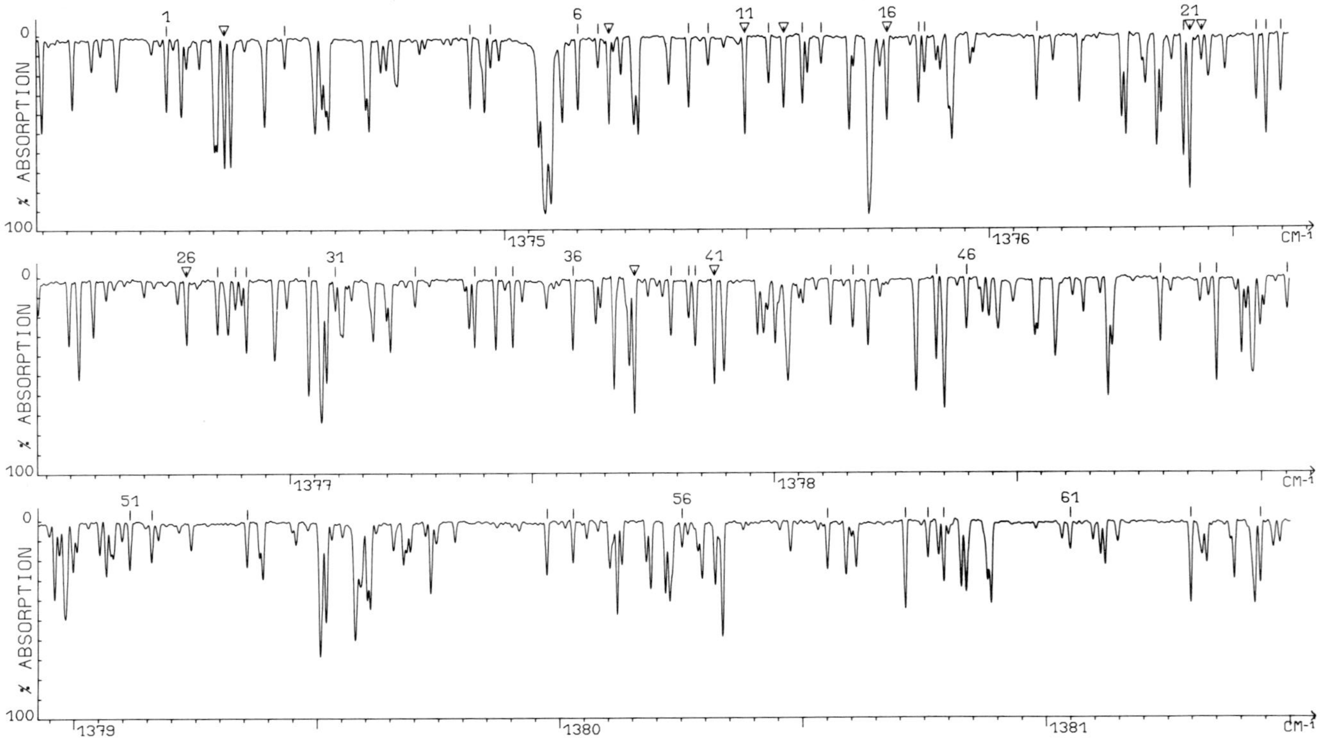

ISOTOPIC IDENTIFICATION – '▽' indicates $CH_3{}^{35}Cl$; '|': Unidentified.

LINE NO.	WAVENUMBER CM^{-1} ;VAC.		ROTATIONAL ASSIGNMENT	BAND	ISOTOPIC SPECIES
61	1381.05344	*			
62	1381.30034	*			
63	1381.44153	*			

- "LINE NO.": serial number identifying the spectral line in the map on the opposite page.
- "WAVENUMBER": observed value.
- * : all given lines are selected according to intensity, halfwidth and symmetry criteria. Lines with an asterisk (*) are selected but not identified.

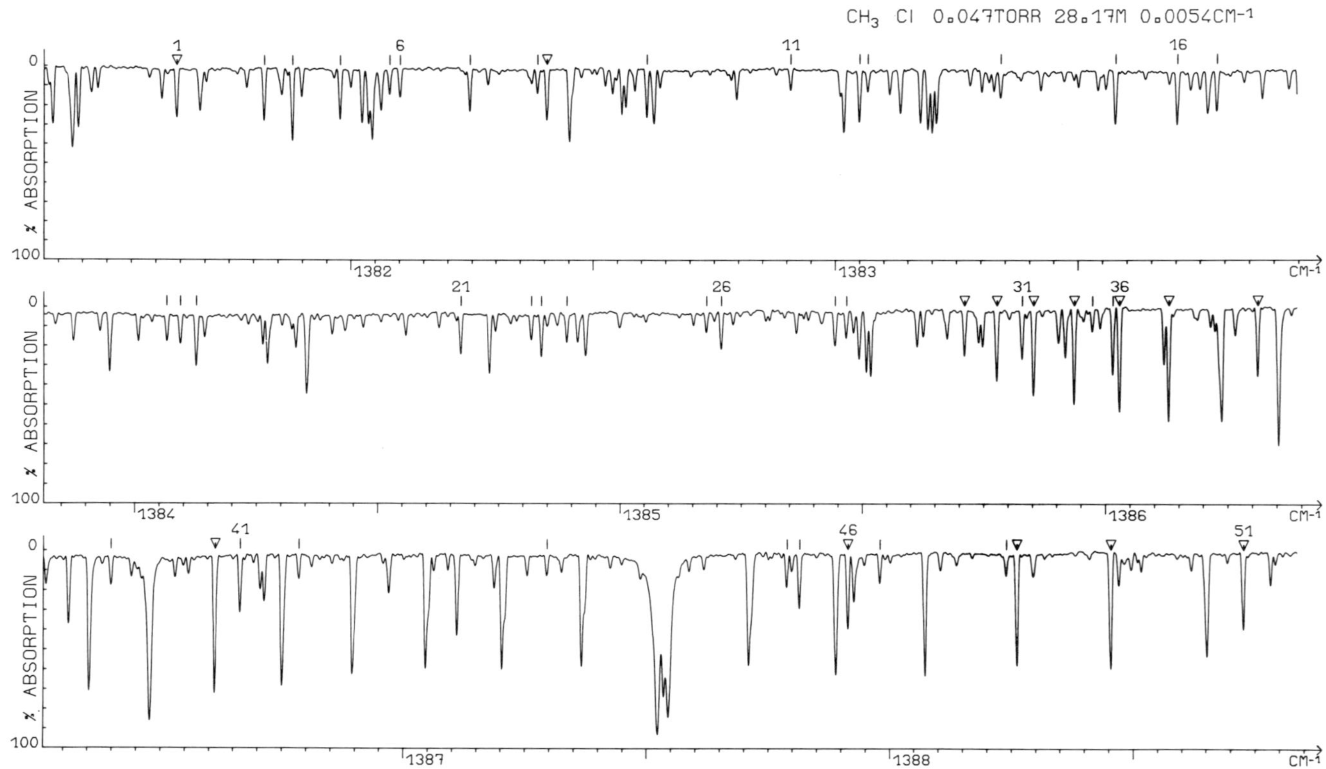

ISOTOPIC IDENTIFICATION – '∇' indicates $CH_3{}^{35}Cl$; 'I': Unidentified.

LINE NO.	WAVENUMBER CM^{-1} ;VAC.	ROTATIONAL ASSIGNMENT	BAND	ISOTOPIC SPECIES
1	1381.64187	PP 5(20)	NU5	CH3 35C1
2	1381.82177 *			
3	1381.87973 *			
4	1381.97790 *			
5	1382.08095 *			
6	1382.10247 *			
7	1382.24633 *			
8	1382.38654 *			
9	1382.40583	PP 5(19)	NU5	CH3 35C1
10	1382.61135 *			
11	1382.90770 *			
12	1383.04893 *			
13	1383.06615 *			
14	1383.33957 *			
15	1383.57548 *			
16	1383.70293 *			
17	1383.78362 *			
18	1384.06680 *			
19	1384.09416 *			
20	1384.12669 *			
21	1384.67153 *			
22	1384.81771 *			
23	1384.83846 *			
24	1384.89079 *			
25	1385.17891 *			
26	1385.20933 *			
27	1385.44440 *			
28	1385.46749 *			
29	1385.71135	QP 6(6)	NU5	CH3 35C1
30	1385.77756	QP 6(7)	NU5	CH3 35C1
31	1385.82908 *			
32	1385.85276	QP 6(8)	NU5	CH3 35C1
33	1385.93652	QP 6(9)	NU5	CH3 35C1
34	1385.97368 *			
35	1386.01525 *			
36	1386.02886	QP 6(10)	NU5	CH3 35C1
37	1386.12956	QP 6(11)	NU5	CH3 35C1
38	1386.31105	PP 5(14)	NU5	CH3 35C1
39	1386.39785 *			
40	1386.61085	QP 6(15)	NU5	CH3 35C1
41	1386.66276 *			
42	1386.78322 *			
43	1387.29501 *			
44	1387.78978 *			
45	1387.81584 *			
46	1387.91555	PP 5(12)	NU5	CH3 35C1
47	1387.98120 *			
48	1388.24131 *			
49	1388.26319	QP 6(25)	NU5	CH3 35C1
50	1388.45586	QP 6(26)	NU5	CH3 35C1
51	1388.72724	PP 5(11)	NU5	CH3 35C1

- "LINE NO.": serial number identifying the spectral line in the map on the opposite page.
- "WAVENUMBER": observed value.
- * : all given lines are selected according to intensity, halfwidth and symmetry criteria.

Lines with an asterisk (*) are selected but not identified.

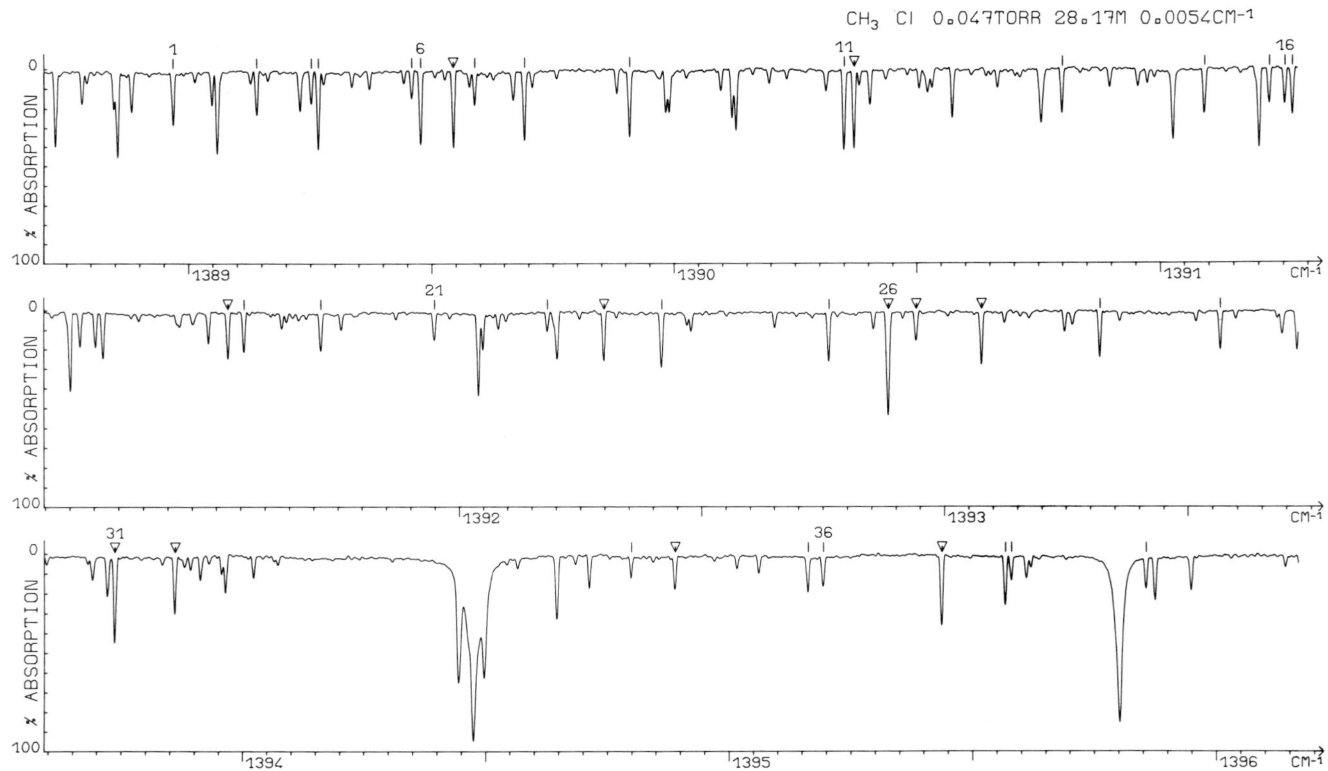

ISOTOPIC IDENTIFICATION – '▽' indicates $CH_3{}^{35}Cl$; 'I': Unidentified.

LINE NO.	WAVENUMBER CM^{-1} ;VAC.		ROTATIONAL ASSIGNMENT	BAND	ISOTOPIC SPECIES
1	1388.96811	*			
2	1389.14045	*			
3	1389.25275	*			
4	1389.26712	*			
5	1389.46003	*			
6	1389.47815	*			
7	1389.54546		PP 5(10)	NU5	CH3 35Cl
8	1389.58923	*			
9	1389.69205	*			
10	1389.90880	*			
11	1390.34964	*			
12	1390.37022		PP 5(9)	NU5	CH3 35Cl
13	1390.79806	*			
14	1391.09021	*			
15	1391.22262	*			
16	1391.25406	*			
17	1391.26953	*			
18	1391.52483		PP 4(23)	NU5	CH3 35Cl
19	1391.55744	*			
20	1391.71519	*			
21	1391.94888	*			
22	1392.18216	*			
23	1392.29856		PP 4(22)	NU5	CH3 35Cl
24	1392.41802	*			
25	1392.76249	*			
26	1392.88439		PP 5(6)	NU5	CH3 35Cl
27	1392.94243		RP 6(7)	NU5	CH3 35Cl
28	1393.07669		PP 4(21)	NU5	CH3 35Cl
29	1393.32019	*			
30	1393.56623	*			

LINE NO.	WAVENUMBER CM^{-1} ;VAC.		ROTATIONAL ASSIGNMENT	BAND	ISOTOPIC SPECIES
31	1393.73671		PP 5(5)	NU5	CH3 35Cl
32	1393.85924		PP 4(20)	NU5	CH3 35Cl
33	1394.79912	*			
34	1394.88967		RP 6(9)	NU5	CH3 35Cl
35	1395.16363	*			
36	1395.19472	*			
37	1395.43866		PP 4(18)	NU5	CH3 35Cl
38	1395.56892	*			
39	1395.58140	*			
40	1395.85774	*			

- "LINE NO.": serial number identifying the spectral line in the map on the opposite page.
- "WAVENUMBER": observed value.
- * : all given lines are selected according to intensity, halfwidth and symmetry criteria.

Lines with an asterisk (*) are selected but not identified.

CH_3 Cl 0.047TORR 28.17M 0.0054CM-1

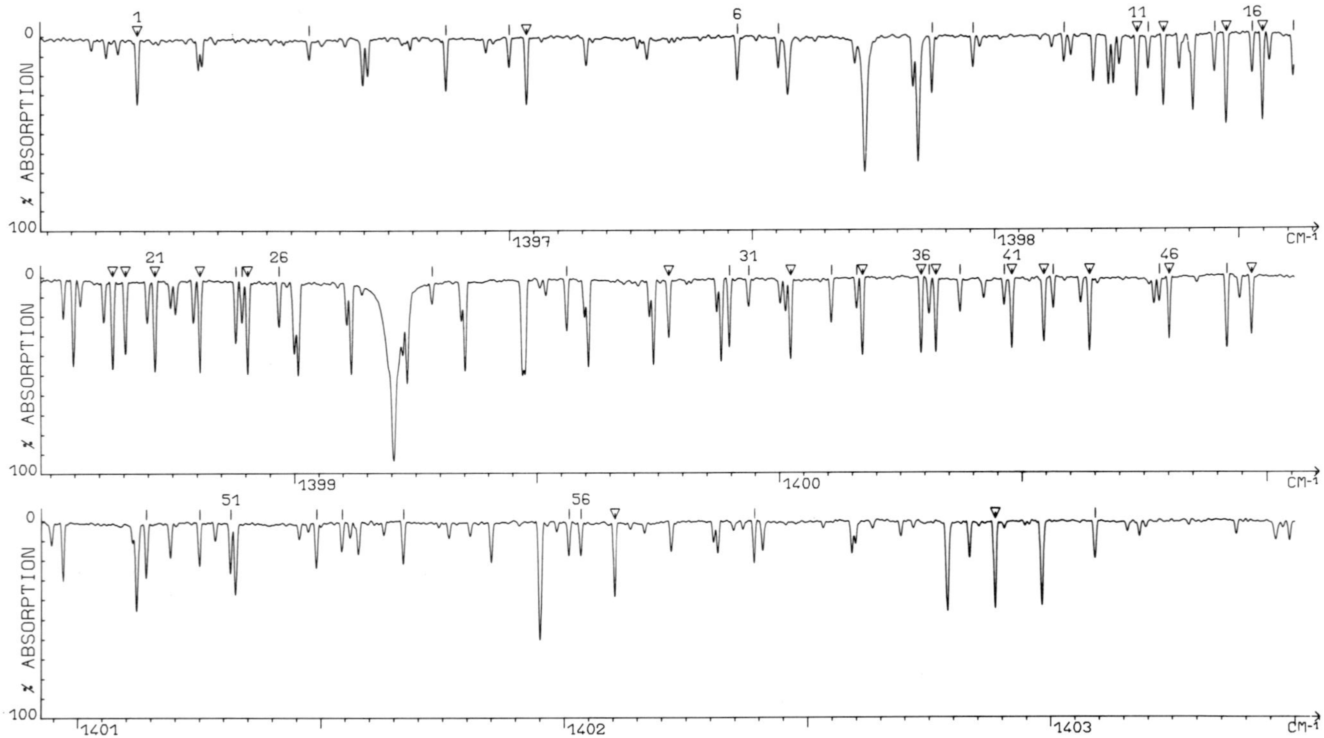

ISOTOPIC IDENTIFICATION - '▽' indicates $CH_3{}^{35}Cl$; 'I': Unidentified.

LINE NO.	WAVENUMBER CM^{-1} ;VAC.	ROTATIONAL ASSIGNMENT	BAND	ISOTOPIC SPECIES	LINE NO.	WAVENUMBER CM^{-1} ;VAC.	ROTATIONAL ASSIGNMENT	BAND	ISOTOPIC SPECIES
1	1396.23523	PP 4(17)	NU5	CH3 35C1	31	1399.93669 *			
2	1396.58822 *				32	1400.02406	QP 5(24)	NU5	CH3 35C1
3	1396.87013 *				33	1400.10785 *			
4	1397.00120 *				34	1400.16012 *			
5	1397.03685	PP 4(16)	NU5	CH3 35C1	35	1400.17197	QP 5(25)	NU5	CH3 35C1
6	1397.47022 *				36	1400.29338	PP 4(12)	NU5	CH3 35C1
7	1397.55507 *				37	1400.30987 *			
8	1397.87172 *				38	1400.32398	QP 5(26)	NU5	CH3 35C1
9	1397.95585 *				39	1400.37360 *			
10	1398.14334 *				40	1400.46424 *			
11	1398.29222	QP 5(7)	NU5	CH3 35C1	41	1400.48000	QP 5(27)	NU5	CH3 35C1
12	1398.31517 *				42	1400.54642	PP 3(27)	NU5	CH3 35C1
13	1398.34672	QP 5(8)	NU5	CH3 35C1	43	1400.56565 *			
14	1398.45095 *				44	1400.63978	QP 5(28)	NU5	CH3 35C1
15	1398.47505	QP 5(10)	NU5	CH3 35C1	45	1400.78212 *			
16	1398.52800 *				46	1400.80309	QP 5(29)	NU5	CH3 35C1
17	1398.54900	QP 5(11)	NU5	CH3 35C1	47	1400.91995 *			
18	1398.61090 *				48	1400.96993	QP 5(30)	NU5	CH3 35C1
19	1398.62899	QP 5(12)	NU5	CH3 35C1	49	1401.14011 *			
20	1398.65495	PP 4(14)	NU5	CH3 35C1	50	1401.24903 *			
21	1398.71507	QP 5(13)	NU5	CH3 35C1	51	1401.31336 *			
22	1398.80716	QP 5(14)	NU5	CH3 35C1	52	1401.48960 *			
23	1398.88077 *				53	1401.54255 *			
24	1398.89330 *				54	1401.66865 *			
25	1398.90496	QP 5(15)	NU5	CH3 35C1	55	1402.01025 *			
26	1398.96898 *				56	1402.03473 *			
27	1399.28371 *				57	1402.10451	PP 3(25)	NU5	CH3 35C1
28	1399.56207 *				58	1402.39208 *			
29	1399.77253	PP 3(28)	NU5	CH3 35C1	59	1402.88864	PP 3(24)	NU5	CH3 35C1
30	1399.89702 *				60	1403.09423 *			

- "LINE NO.": serial number identifying the spectral line in the map on the opposite page.
- "WAVENUMBER": observed value.
- * : all given lines are selected according to intensity, halfwidth and symmetry criteria.
Lines with an asterisk (*) are selected but not identified.

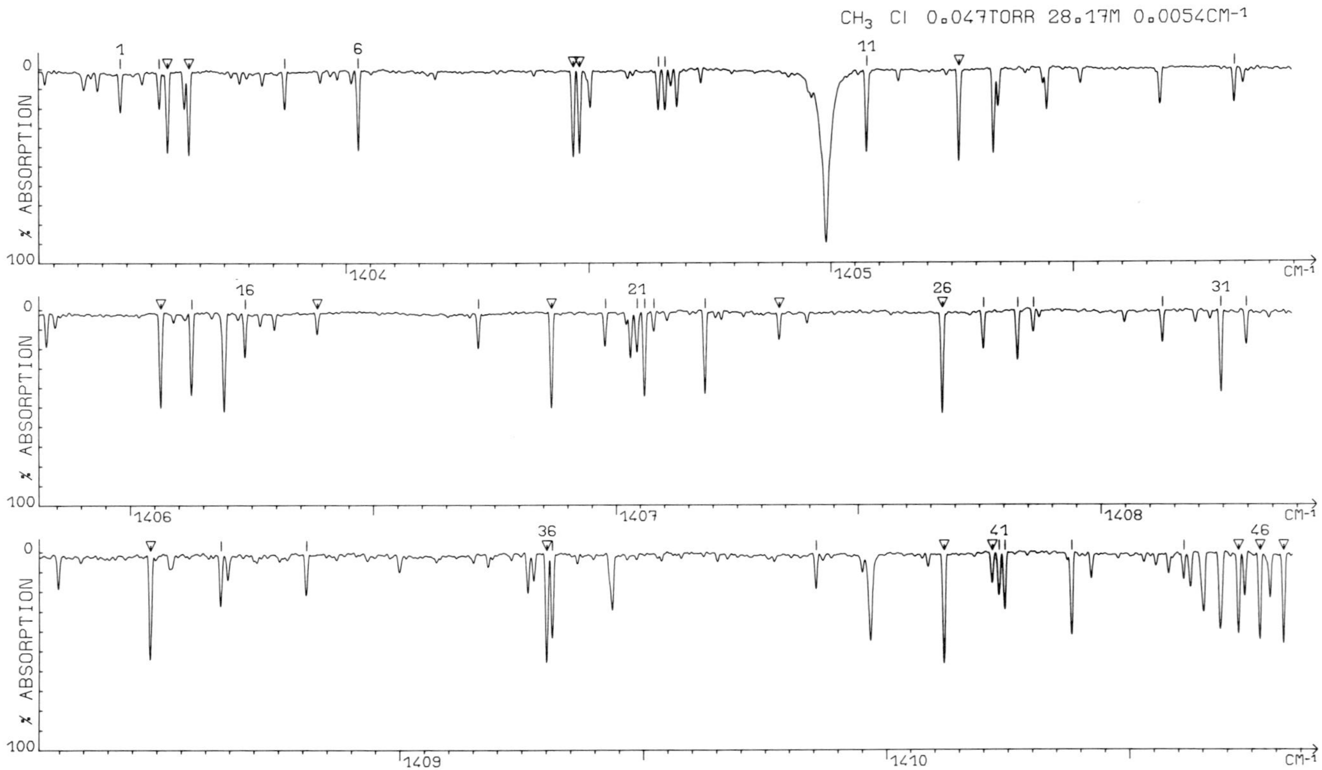

ISOTOPIC IDENTIFICATION – '∇' indicates $CH_3{}^{35}Cl$; 'I': Unidentified.

LINE NO.	WAVENUMBER CM^{-1} ;VAC.	ROTATIONAL ASSIGNMENT	BAND	ISOTOPIC SPECIES
1	1403.53673 *			
2	1403.61616 *			
3	1403.63313	PP 4(8)	NU5	CH3 35C1
4	1403.67683	PP 3(23)	NU5	CH3 35C1
5	1403.87441 *			
6	1404.02660 *			
7	1404.46853	PP 3(22)	NU5	CH3 35C1
8	1404.48150	PP 4(7)	NU5	CH3 35C1
9	1404.64428 *			
10	1404.65815 *			
11	1405.07373 *			
12	1405.26408	PP 3(21)	NU5	CH3 35C1
13	1405.82964 *			
14	1406.06349	PP 3(20)	NU5	CH3 35C1
15	1406.12624 *			
16	1406.23616 *			
17	1406.38425	RP 5(8)	NU5	CH3 35C1
18	1406.71572 *			
19	1406.86683	PP 3(19)	NU5	CH3 35C1
20	1406.97755 *			
21	1407.04382 *			
22	1407.05916 *			
23	1407.07837 *			
24	1407.18374 *			
25	1407.33754	RP 5(9)	NU5	CH3 35C1
26	1407.67424	PP 3(18)	NU5	CH3 35C1
27	1407.75858 *			
28	1407.82862 *			
29	1407.86109 *			
30	1408.12603 *			
31	1408.24601 *			
32	1408.29705 *			
33	1408.48569	PP 3(17)	NU5	CH3 35C1
34	1408.63045 *			
35	1408.80601 *			
36	1409.30129	PP 3(16)	NU5	CH3 35C1
37	1409.31268 *			
38	1409.85699 *			
39	1410.12109	PP 3(15)	NU5	CH3 35C1
40	1410.22010	PP 2(30)	NU5	CH3 35C1
41	1410.23398 *			
42	1410.24586 *			
43	1410.38366 *			
44	1410.61330 *			
45	1410.72468	QP 4(7)	NU5	CH3 35C1
46	1410.76780	QP 4(8)	NU5	CH3 35C1
47	1410.81618	QP 4(9)	NU5	CH3 35C1

- "LINE NO.": serial number identifying the spectral line in the map on the opposite page.
- "WAVENUMBER": observed value.
- * : all given lines are selected according to intensity, halfwidth and symmetry criteria.

Lines with an asterisk (*) are selected but not identified.

CH3 Cl 0.047TORR 28.17M 0.0054CM-1

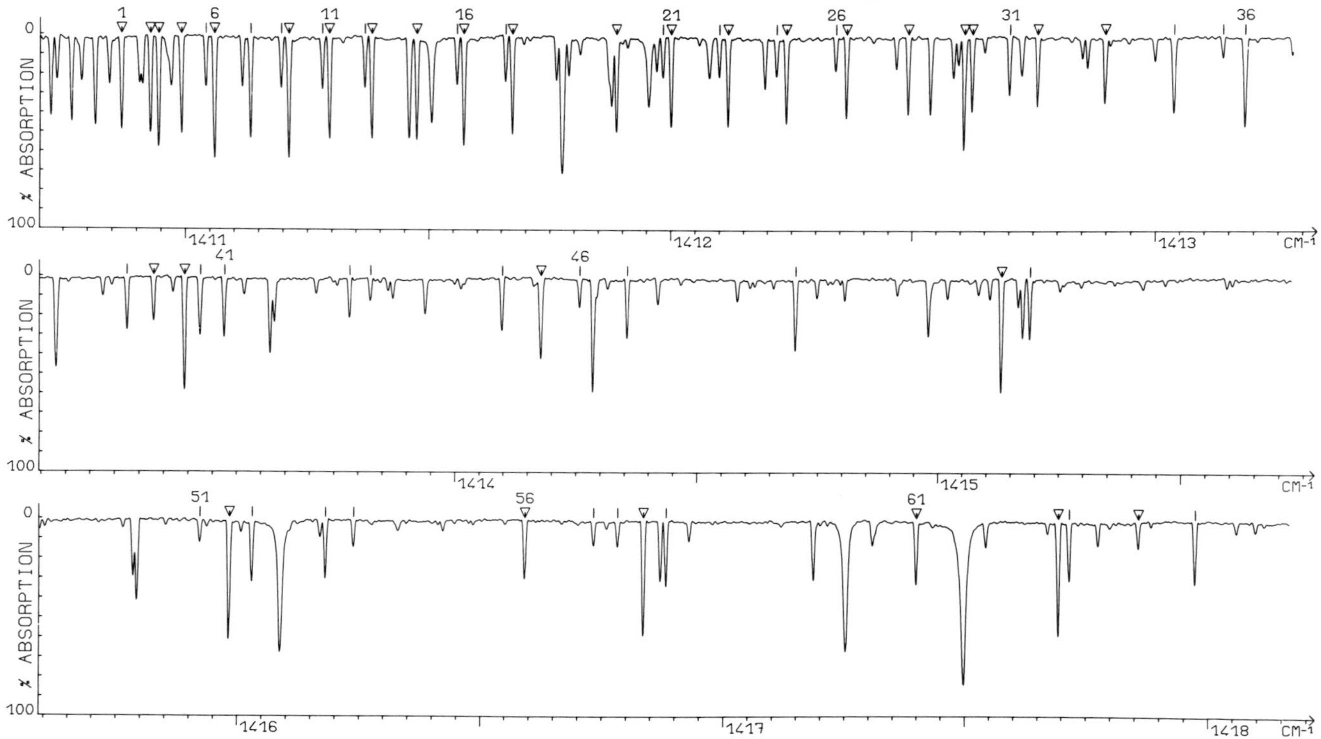

ISOTOPIC IDENTIFICATION – '∇' indicates $CH_3{}^{35}Cl$; 'I': Unidentified.

LINE NO.	WAVENUMBER CM^{-1} ;VAC.	ROTATIONAL ASSIGNMENT	BAND	ISOTOPIC SPECIES	LINE NO.	WAVENUMBER CM^{-1} ;VAC.	ROTATIONAL ASSIGNMENT	BAND	ISOTOPIC SPECIES
1	1410.86968	QP 4(10)	NU5	CH3 35C1	31	1412.69897 *			
2	1410.92834	QP 4(11)	NU5	CH3 35C1	32	1412.75706	QP 4(29)	NU5	CH3 35C1
3	1410.94515	PP 3(14)	NU5	CH3 35C1	33	1412.89495	QP 4(30)	NU5	CH3 35C1
4	1410.99201	QP 4(12)	NU5	CH3 35C1	34	1413.03511 *			
5	1411.04209 *				35	1413.13536 *			
6	1411.06037	QP 4(13)	NU5	CH3 35C1	36	1413.18097 *			
7	1411.13413 *				37	1413.32631 *			
8	1411.19666 *				38	1413.38041	PP 2(26)	NU5	CH3 35C1
9	1411.21223	QP 4(15)	NU5	CH3 35C1	39	1413.44326	PP 3(11)	NU5	CH3 35C1
10	1411.28074 *				40	1413.47502 *			
11	1411.29547	QP 4(16)	NU5	CH3 35C1	41	1413.52493 *			
12	1411.36892 *				42	1413.78145 *			
13	1411.38309	QP 4(17)	NU5	CH3 35C1	43	1413.82493 *			
14	1411.47527	QP 4(18)	NU5	CH3 35C1	44	1414.09697 *			
15	1411.55802 *				45	1414.17728	PP 2(25)	NU5	CH3 35C1
16	1411.57205	QP 4(19)	NU5	CH3 35C1	46	1414.25681 *			
17	1411.65875 *				47	1414.35506 *			
18	1411.67273	QP 4(20)	NU5	CH3 35C1	48	1414.70385 *			
19	1411.88711	QP 4(22)	NU5	CH3 35C1	49	1415.13067	PP 3(9)	NU5	CH3 35C1
20	1411.98354 *				50	1415.18936 *			
21	1412.00033	QP 4(23)	NU5	CH3 35C1	51	1415.92043 *			
22	1412.09911 *				52	1415.98119	PP 3(8)	NU5	CH3 35C1
23	1412.11742	QP 4(24)	NU5	CH3 35C1	53	1416.02791 *			
24	1412.21805 *				54	1416.17849 *			
25	1412.23834	QP 4(25)	NU5	CH3 35C1	55	1416.23631 *			
26	1412.34029 *				56	1416.59080	PP 2(22)	NU5	CH3 35C1
27	1412.36279	QP 4(26)	NU5	CH3 35C1	57	1416.73257 *			
28	1412.49091	QP 4(27)	NU5	CH3 35C1	58	1416.78190 *			
29	1412.60613	PP 3(12)	NU5	CH3 35C1	59	1416.83599	PP 3(7)	NU5	CH3 35C1
30	1412.62237	QP 4(28)	NU5	CH3 35C1	60	1416.88235 *			

- "LINE NO.": serial number identifying the spectral line in the map on the opposite page.
- "WAVENUMBER": observed value.
- * : all given lines are selected according to intensity, halfwidth and symmetry criteria.
Lines with an asterisk (*) are selected but not identified.

CH_3 Cl 0.047TORR 28.17M 0.0054CM-1

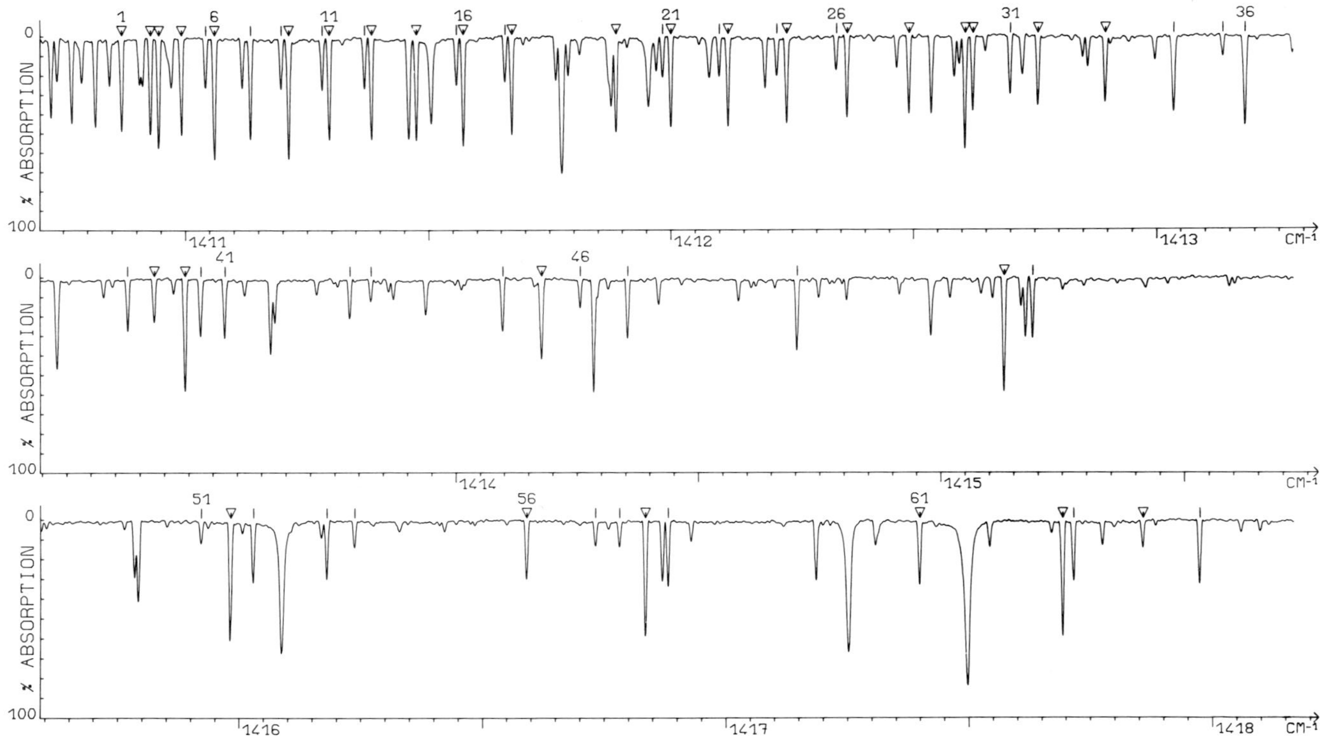

ISOTOPIC IDENTIFICATION – '▽' indicates $CH_3{}^{35}Cl$; 'I': Unidentified.

LINE NO.	WAVENUMBER CM^{-1} ;VAC.		ROTATIONAL ASSIGNMENT	BAND	ISOTOPIC SPECIES
61	1417.40074		PP 2(21)	NU5	CH3 35Cl
62	1417.69539		PP 3(6)	NU5	CH3 35Cl
63	1417.71785	*			
64	1417.85940		RP 4(7)	NU5	CH3 35Cl
65	1417.97541	*			

- "LINE NO.": serial number identifying the spectral line in the map on the opposite page.
- "WAVENUMBER": observed value.
- * : all given lines are selected according to intensity, halfwidth and symmetry criteria.

Lines with an asterisk (*) are selected but not identified.

CH$_3$ Cl 0.047TORR 28.17M 0.0054CM-1

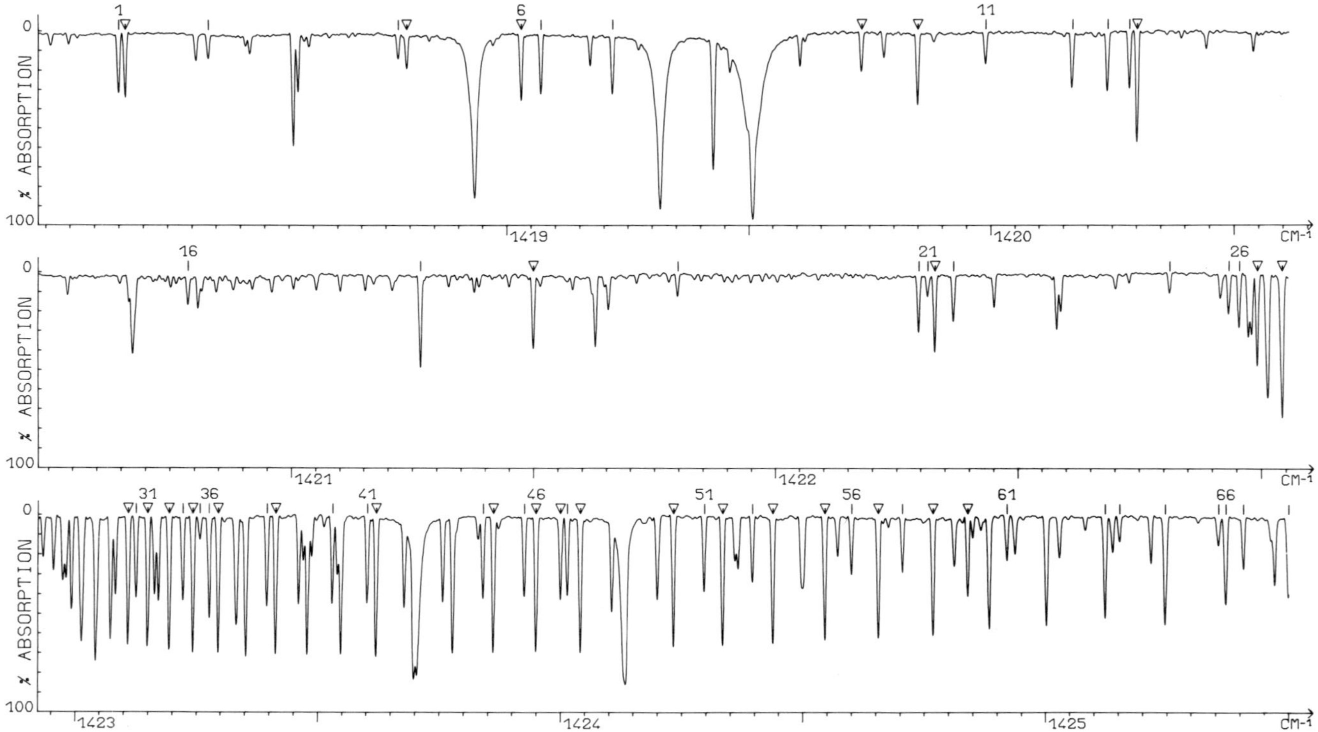

ISOTOPIC IDENTIFICATION – '∇' indicates CH$_3$^{35}Cl ; 'I': Unidentified.

LINE NO.	WAVENUMBER CM^{-1} ;VAC.	ROTATIONAL ASSIGNMENT	BAND	ISOTOPIC SPECIES	LINE NO.	WAVENUMBER CM^{-1} ;VAC.	ROTATIONAL ASSIGNMENT	BAND	ISOTOPIC SPECIES
1	1418.20028 *				31	1423.14918	QP 3(9)	NU5	CH3 35Cl
2	1418.21377	PP 2(20)	NU5	CH3 35Cl	32	1423.19359	QP 3(10)	NU5	CH3 35Cl
3	1418.38404 *				33	1423.22150 *			
4	1418.77615 *				34	1423.24225	QP 3(11)	NU5	CH3 35Cl
5	1418.79380	RP 4(8)	NU5	CH3 35Cl	35	1423.25689 *			
6	1419.03010	PP 2(19)	NU5	CH3 35Cl	36	1423.27597 *			
7	1419.07067 *				37	1423.29510	QP 3(12)	NU5	CH3 35Cl
8	1419.21807 *				38	1423.39519 *			
9	1419.73320	RP 4(9)	NU5	CH3 35Cl	39	1423.41324	QP 3(14)	NU5	CH3 35Cl
10	1419.84973	PP 2(18)	NU5	CH3 35Cl	40	1423.52956 *			
11	1419.98880 *				41	1423.60208 *			
12	1420.16797 *				42	1423.62049	QP 3(17)	NU5	CH3 35Cl
13	1420.23997 *				43	1423.84000 *			
14	1420.28513 *				44	1423.86213	QP 3(20)	NU5	CH3 35Cl
15	1420.30103	PP 3(3)	NU5	CH3 35Cl	45	1423.92566 *			
16	1420.78657 *				46	1423.94993	QP 3(21)	NU5	CH3 35Cl
17	1421.26648 *				47	1424.00043	PP 2(13)	NU5	CH3 35Cl
18	1421.49942	PP 2(16)	NU5	CH3 35Cl	48	1424.01431 *			
19	1421.79773 *				49	1424.04122	QP 3(22)	NU5	CH3 35Cl
20	1422.29610 *				50	1424.23392	QP 3(24)	NU5	CH3 35Cl
21	1422.31467 *				51	1424.29713 *			
22	1422.32946	PP 2(15)	NU5	CH3 35Cl	52	1424.33510	QP 3(25)	NU5	CH3 35Cl
23	1422.36769 *				53	1424.39646 *			
24	1422.81239 *				54	1424.43938	QP 3(26)	NU5	CH3 35Cl
25	1422.93419 *				55	1424.54667	QP 3(27)	NU5	CH3 35Cl
26	1422.95556 *				56	1424.60156 *			
27	1422.99258	QP 3(4)	NU5	CH3 35Cl	57	1424.65690	QP 3(28)	NU5	CH3 35Cl
28	1423.04297	QP 3(6)	NU5	CH3 35Cl	58	1424.70689 *			
29	1423.10906	QP 3(8)	NU5	CH3 35Cl	59	1424.76988	QP 3(29)	NU5	CH3 35Cl
30	1423.12517 *				60	1424.84138	PP 2(12)	NU5	CH3 35Cl

- "LINE NO.": serial number identifying the spectral line in the map on the opposite page.
- "WAVENUMBER": observed value.
- * : all given lines are selected according to intensity, halfwidth and symmetry criteria.
Lines with an asterisk (*) are selected but not identified.

CH_3 Cl 0.047TORR 28.17M 0.0054CM-1

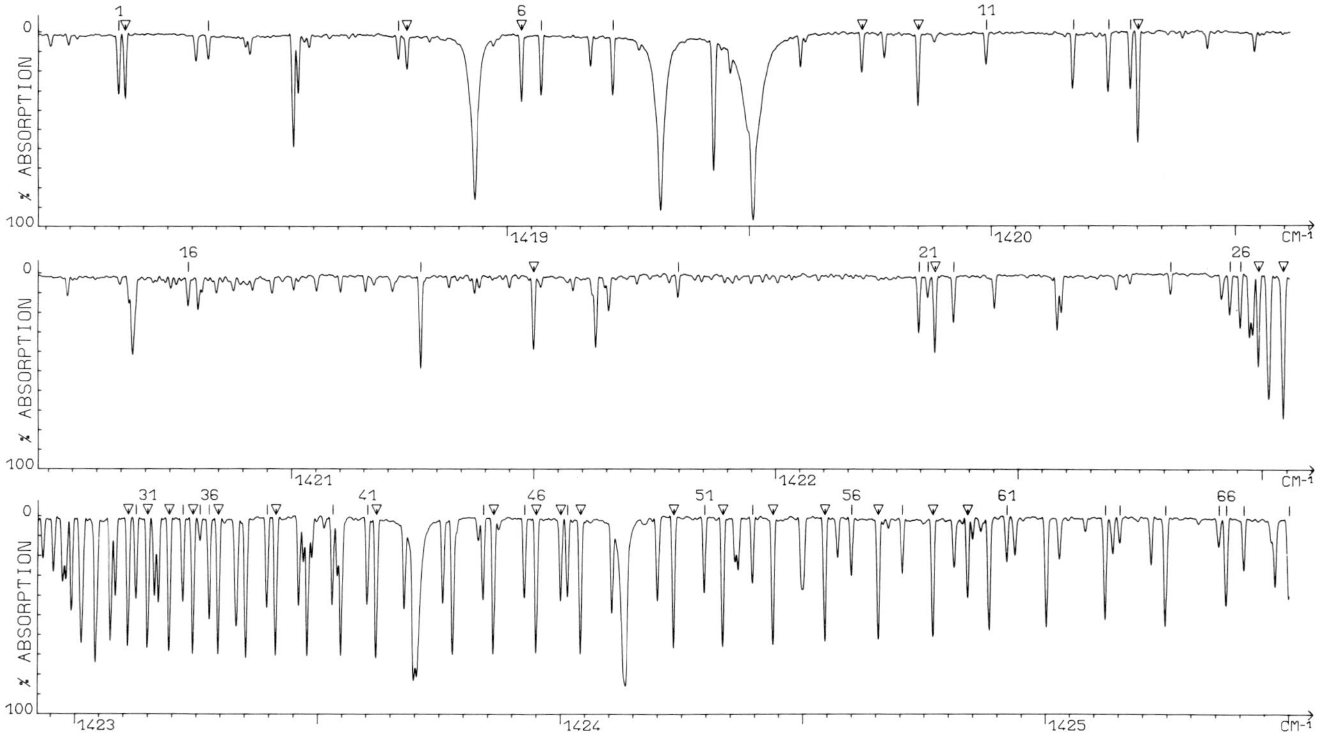

ISOTOPIC IDENTIFICATION – '▽' indicates $CH_3{}^{35}Cl$; '|': Unidentified.

LINE NO.	WAVENUMBER CM^{-1} ;VAC.	ROTATIONAL ASSIGNMENT	BAND	ISOTOPIC SPECIES
61	1424.92170 *			
62	1425.12424 *			
63	1425.15424 *			
64	1425.24746 *			
65	1425.35714 *			
66	1425.37204 *			
67	1425.40797 *			
68	1425.49905 *			

- "LINE NO.": serial number identifying the spectral line in the map on the opposite page.
- "WAVENUMBER": observed value.
- * : all given lines are selected according to intensity, halfwidth and symmetry criteria.

Lines with an asterisk (*) are selected but not identified.

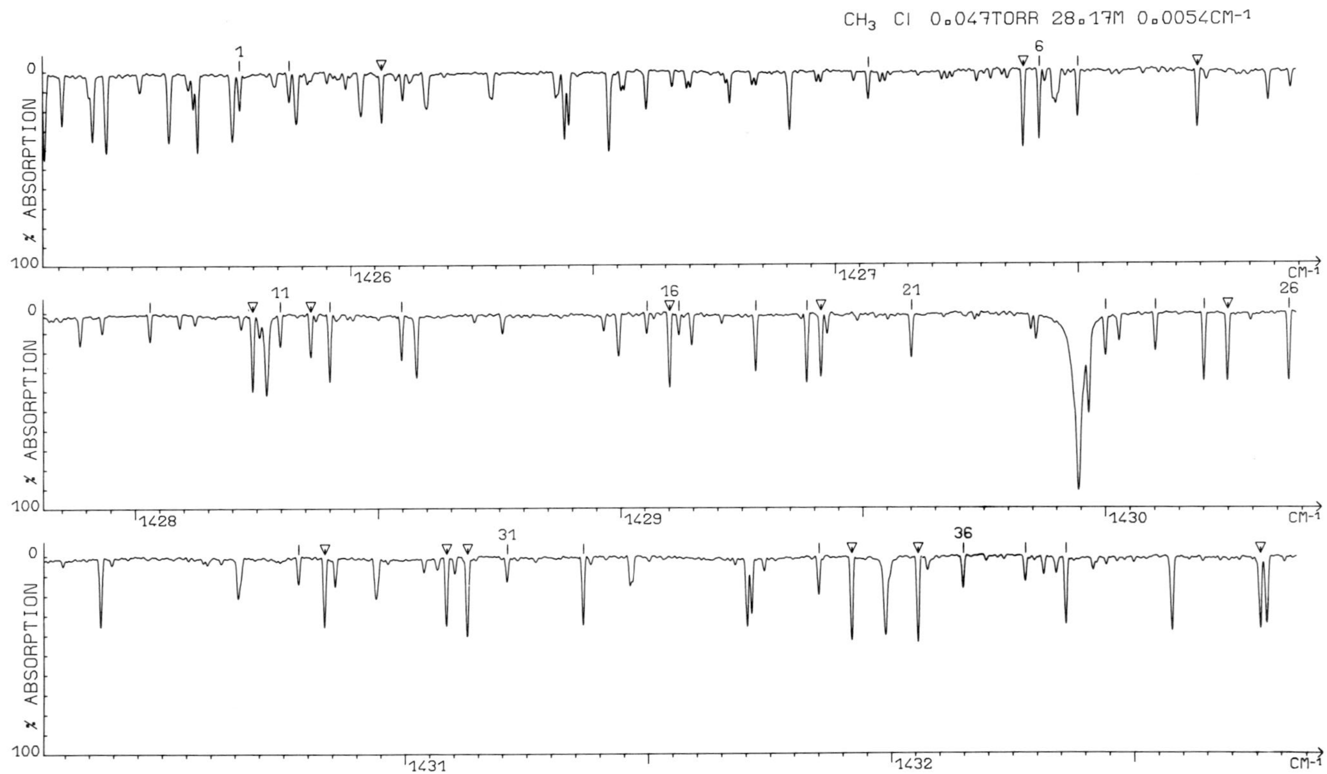

ISOTOPIC IDENTIFICATION – '▽' indicates $CH_3{}^{35}Cl$; 'I': Unidentified.

LINE NO.	WAVENUMBER CM^{-1} ;VAC.	ROTATIONAL ASSIGNMENT	BAND	ISOTOPIC SPECIES
1	1425.77366 *			
2	1425.87530 *			
3	1426.06589	PP 1(25)	NU5	CH3 35Cl
4	1427.06826 *			
5	1427.38655	PP 2(9)	NU5	CH3 35Cl
6	1427.41990 *			
7	1427.49898 *			
8	1427.74462	PP 1(23)	NU5	CH3 35Cl
9	1428.03136 *			
10	1428.24255	PP 2(8)	NU5	CH3 35Cl
11	1428.29917 *			
12	1428.36160	RP 3(5)	NU5	CH3 35Cl
13	1428.40057 *			
14	1428.54884 *			
15	1429.05559 *			
16	1429.10235	PP 2(7)	NU5	CH3 35Cl
17	1429.12137 *			
18	1429.27931 *			
19	1429.38529 *			
20	1429.41452	PP 1(21)	NU5	CH3 35Cl
21	1429.60137 *			
22	1430.00023 *			
23	1430.10169 *			
24	1430.20134 *			
25	1430.24920	PP 1(20)	NU5	CH3 35Cl
26	1430.37394 *			
27	1430.77982 *			
28	1430.83343	PP 2(5)	NU5	CH3 35Cl
29	1431.08461	PP 1(19)	NU5	CH3 35Cl
30	1431.12760	RP 3(8)	NU5	CH3 35Cl
31	1431.20986 *			
32	1431.36643 *			
33	1431.85311 *			
34	1431.92110	PP 1(18)	NU5	CH3 35Cl
35	1432.05805	RP 3(9)	NU5	CH3 35Cl
36	1432.15092 *			
37	1432.27848 *			
38	1432.36259 *			
39	1432.75938	PP 1(17)	NU5	CH3 35Cl

- "LINE NO.": serial number identifying the spectral line in the map on the opposite page.
- "WAVENUMBER": observed value.
- * : all given lines are selected according to intensity, halfwidth and symmetry criteria.

Lines with an asterisk (*) are selected but not identified.

CH$_3$ Cl 0.047TORR 28.17M 0.0054CM-1

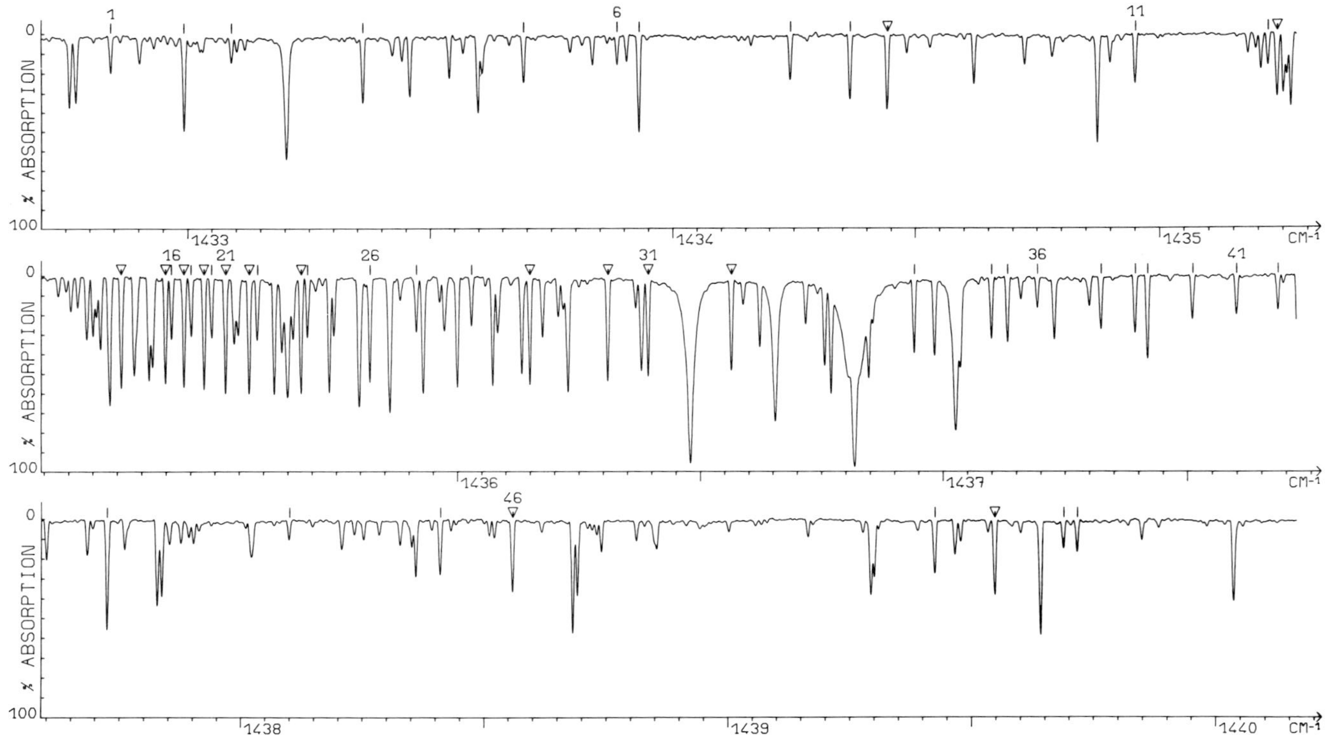

ISOTOPIC IDENTIFICATION – '▽' indicates CH$_3{}^{35}$Cl ; 'I': Unidentified.

LINE NO.	WAVENUMBER CM^{-1} ;VAC.	ROTATIONAL ASSIGNMENT	BAND	ISOTOPIC SPECIES
1	1432.84342 *			
2	1432.99255 *			
3	1433.09079 *			
4	1433.36231 *			
5	1433.69388 *			
6	1433.88592 *			
7	1433.93114 *			
8	1434.24280 *			
9	1434.36545 *			
10	1434.44181	PP 1(15)	NU5	CH3 35C1
11	1434.94978 *			
12	1435.22053 *			
13	1435.23904	QP 2(2)	NU5	CH3 35C1
14	1435.30920	QP 2(6)	NU5	CH3 35C1
15	1435.39952	QP 2(9)	NU5	CH3 35C1
16	1435.41167 *			
17	1435.43704	QP 2(10)	NU5	CH3 35C1
18	1435.45159 *			
19	1435.47821	QP 2(11)	NU5	CH3 35C1
20	1435.49436 *			
21	1435.52294	QP 2(12)	NU5	CH3 35C1
22	1435.57113	QP 2(13)	NU5	CH3 35C1
23	1435.58792 *			
24	1435.67759	QP 2(15)	NU5	CH3 35C1
25	1435.69089 *			
26	1435.82026 *			
27	1435.91568 *			
28	1436.02920 *			
29	1436.14983	QP 2(22)	NU5	CH3 35C1
30	1436.30997	QP 2(24)	NU5	CH3 35C1
31	1436.39349	QP 2(25)	NU5	CH3 35C1
32	1436.56488	QP 2(27)	NU5	CH3 35C1
33	1436.94193 *			
34	1437.10013 *			
35	1437.13288 *			
36	1437.19418 *			
37	1437.32462 *			
38	1437.39393 *			
39	1437.41930 *			
40	1437.51141 *			
41	1437.60026 *			
42	1437.68415 *			
43	1437.72461 *			
44	1438.09830 *			
45	1438.40930 *			
46	1438.55881	PR O(26)	NU5	CH3 35C1
47	1439.42742 *			
48	1439.55017	PP 1(9)	NU5	CH3 35C1
49	1439.69139 *			
50	1439.71938 *			

- "LINE NO.": serial number identifying the spectral line in the map on the opposite page.
- "WAVENUMBER": observed value.
- * : all given lines are selected according to intensity, halfwidth and symmetry criteria.
Lines with an asterisk (*) are selected but not identified.

CH3 Cl 0.047TORR 28.17M 0.0054CM-1

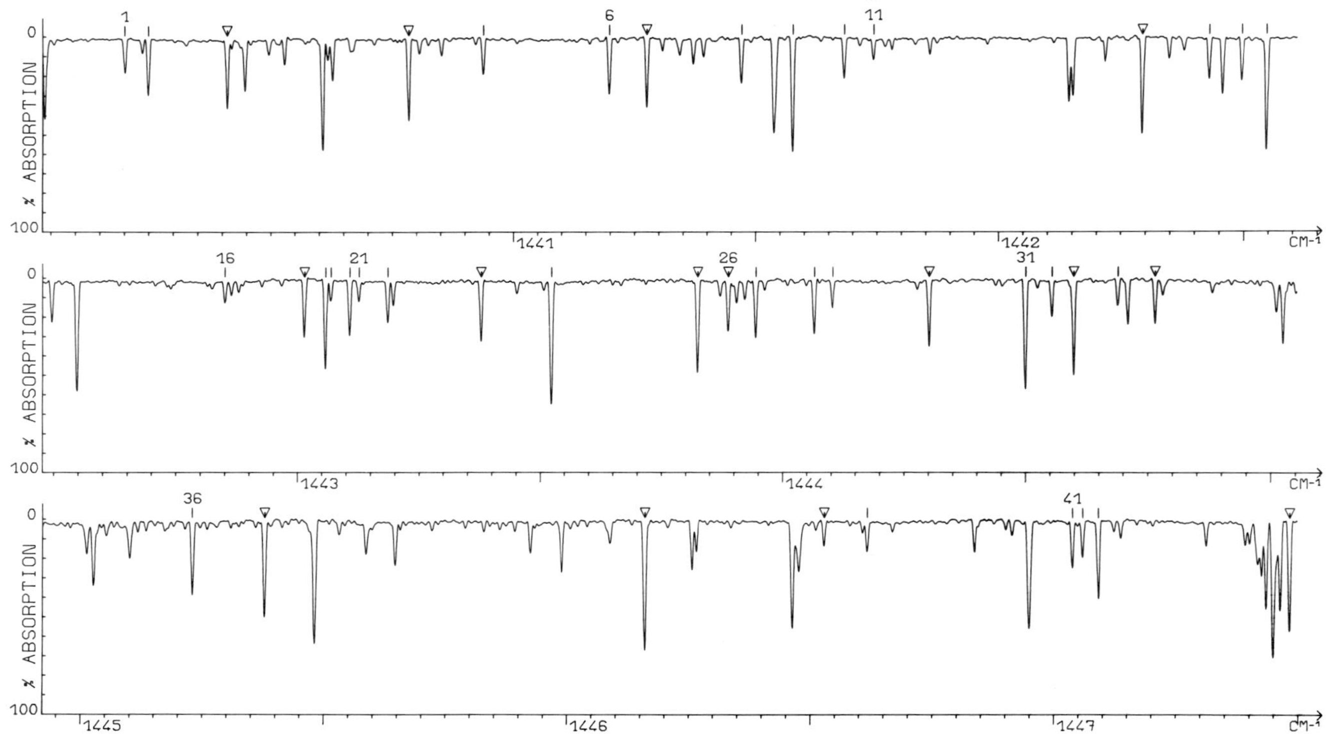

ISOTOPIC IDENTIFICATION – '∇' indicates $CH_3{}^{35}Cl$; 'I': Unidentified.

LINE NO.	WAVENUMBER CM^{-1} ;VAC.	ROTATIONAL ASSIGNMENT	BAND	ISOTOPIC SPECIES
1	1440.20137 *			
2	1440.24960 *			
3	1440.41169	PP 1(8)	NU5	CH3 35Cl
4	1440.78507	PR O(23)	NU5	CH3 35Cl
5	1440.93925 *			
6	1441.19931 *			
7	1441.27630	PP 1(7)	NU5	CH3 35Cl
8	1441.47150 *			
9	1441.57654 *			
10	1441.68267 *			
11	1441.74258 *			
12	1442.29479	PR O(21)	NU5	CH3 35Cl
13	1442.43134 *			
14	1442.49706 *			
15	1442.54801 *			
16	1442.85165 *			
17	1443.01479	PP 1(5)	NU5	CH3 35Cl
18	1443.05778 *			
19	1443.06880 *			
20	1443.10746 *			
21	1443.12624 *			
22	1443.18558 *			
23	1443.37853	RP 2(8)	NU5	CH3 35Cl
24	1443.52299 *			
25	1443.82580	PR O(19)	NU5	CH3 35Cl
26	1443.88880	PP 1(4)	NU5	CH3 35Cl
27	1443.94545 *			
28	1444.06561 *			
29	1444.10331 *			
30	1444.30215	RP 2(9)	NU5	CH3 35Cl

LINE NO.	WAVENUMBER CM^{-1} ;VAC.	ROTATIONAL ASSIGNMENT	BAND	ISOTOPIC SPECIES
31	1444.50029 *			
32	1444.55449 *			
33	1444.59902	PR O(18)	NU5	CH3 35Cl
34	1444.68984 *			
35	1444.76597	PP 1(3)	NU5	CH3 35Cl
36	1445.22937 *			
37	1445.37752	PR O(17)	NU5	CH3 35Cl
38	1446.16064	PR O(16)	NU5	CH3 35Cl
39	1446.53006	PP 1(1)	NU5	CH3 35Cl
40	1446.61939 *			
41	1447.04041 *			
42	1447.06122 *			
43	1447.09351 *			
44	1447.48319	QP 1(6)	NU5	CH3 35Cl

- "LINE NO.": serial number identifying the spectral line in the map on the opposite page.
- "WAVENUMBER": observed value.
- * : all given lines are selected according to intensity, halfwidth and symmetry criteria.

Lines with an asterisk (*) are selected but not identified.

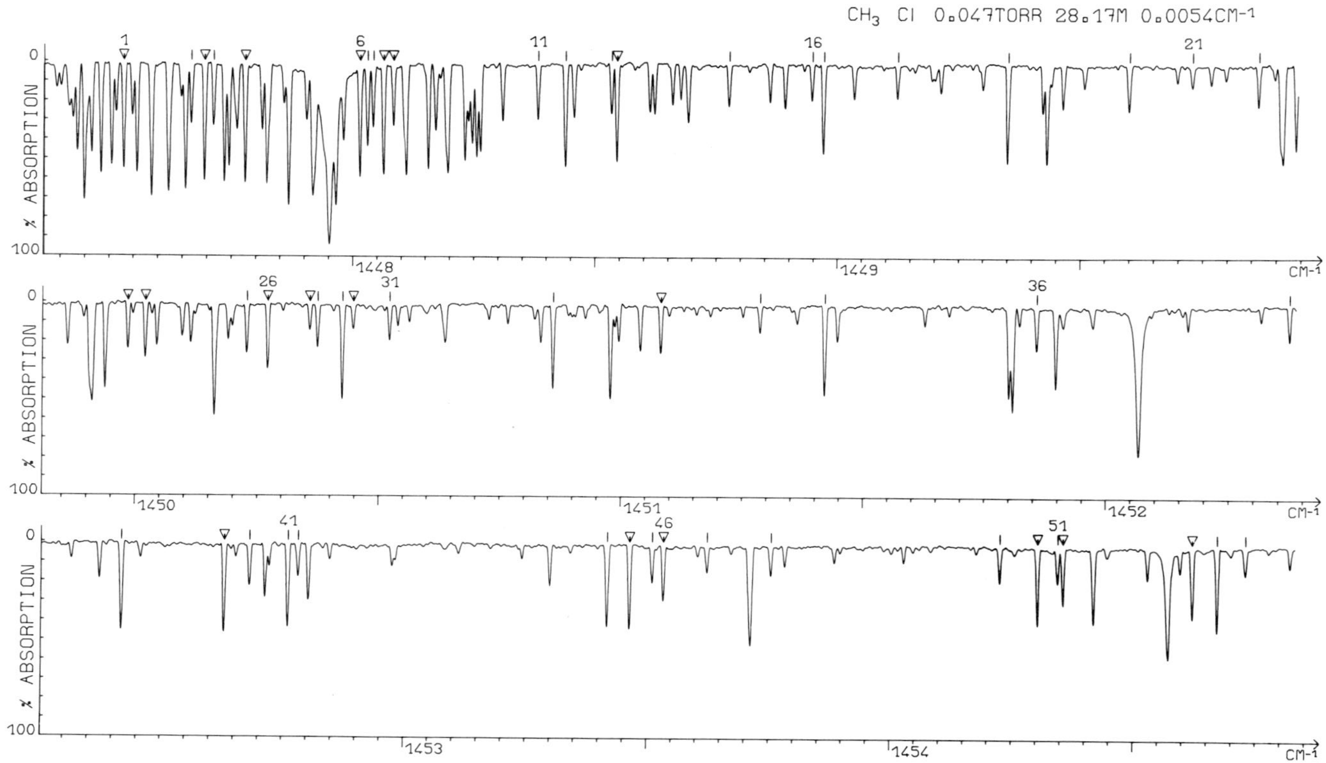

ISOTOPIC IDENTIFICATION – '▽' indicates $CH_3{}^{35}Cl$; '|': Unidentified.

LINE NO.	WAVENUMBER CM^{-1} ;VAC.		ROTATIONAL ASSIGNMENT	BAND	ISOTOPIC SPECIES
1	1447.52952		QP 1(8)	NU5	CH3 35C1
2	1447.66773	*			
3	1447.69496		QP 1(13)	NU5	CH3 35C1
4	1447.71346	*			
5	1447.77833		QP 1(15)	NU5	CH3 35C1
6	1448.01487		QP 1(20)	NU5	CH3 35C1
7	1448.03057	*			
8	1448.04234	*			
9	1448.06367		QP 1(21)	NU5	CH3 35C1
10	1448.08418		QP 1(30)	NU5	CH3 35C1
11	1448.38254	*			
12	1448.43970	*			
13	1448.53401	*			
14	1448.54500		PR O(13)	NU5	CH3 35C1
15	1448.77599	*			
16	1448.94632	*			
17	1448.97086	*			
18	1449.12292	*			
19	1449.35051	*			
20	1449.59899	*			
21	1449.72805	*			
22	1449.86234	*			
23	1449.98544		QP 1(37)	NU5	CH3 35C1
24	1450.02147		QP 1(32)	NU5	CH3 35C1
25	1450.22812	*			
26	1450.27133		PR 1(25)	NU5	CH3 35C1
27	1450.35794		QP 1(42)	NU5	CH3 35C1
28	1450.37399	*			
29	1450.42484	*			
30	1450.44713		QP 1(43)	NU5	CH3 35C1

LINE NO.	WAVENUMBER CM^{-1} ;VAC.		ROTATIONAL ASSIGNMENT	BAND	ISOTOPIC SPECIES
31	1450.52203	*			
32	1450.86023	*			
33	1451.08189		PR 1(24)	NU5	CH3 35C1
34	1451.28691	*			
35	1451.42085	*			
36	1451.85763	*			
37	1452.37396	*			
38	1452.41900	*			
39	1452.62984		PR O(8)	NU5	CH3 35C1
40	1452.68163	*			
41	1452.76126	*			
42	1452.78195	*			
43	1453.41904	*			
44	1453.46501		PR O(7)	NU5	CH3 35C1
45	1453.51161	*			
46	1453.53462		PR 1(21)	NU5	CH3 35C1
47	1453.62530	*			
48	1453.75638	*			
49	1454.22904	*			
50	1454.30663		PR O(6)	NU5	CH3 35C1
51	1454.34769	*			
52	1454.35872		PR 1(20)	NU5	CH3 35C1
53	1454.62272		RP 1(7)	NU5	CH3 35C1
54	1454.67302	*			
55	1454.73076	*			

- "LINE NO.": serial number identifying the spectral line in the map on the opposite page.
- "WAVENUMBER": observed value.
- * : all given lines are selected according to intensity, halfwidth and symmetry criteria.

Lines with an asterisk (*) are selected but not identified.

CH_3 Cl 0.047TORR 28.17M 0.0054CM-1

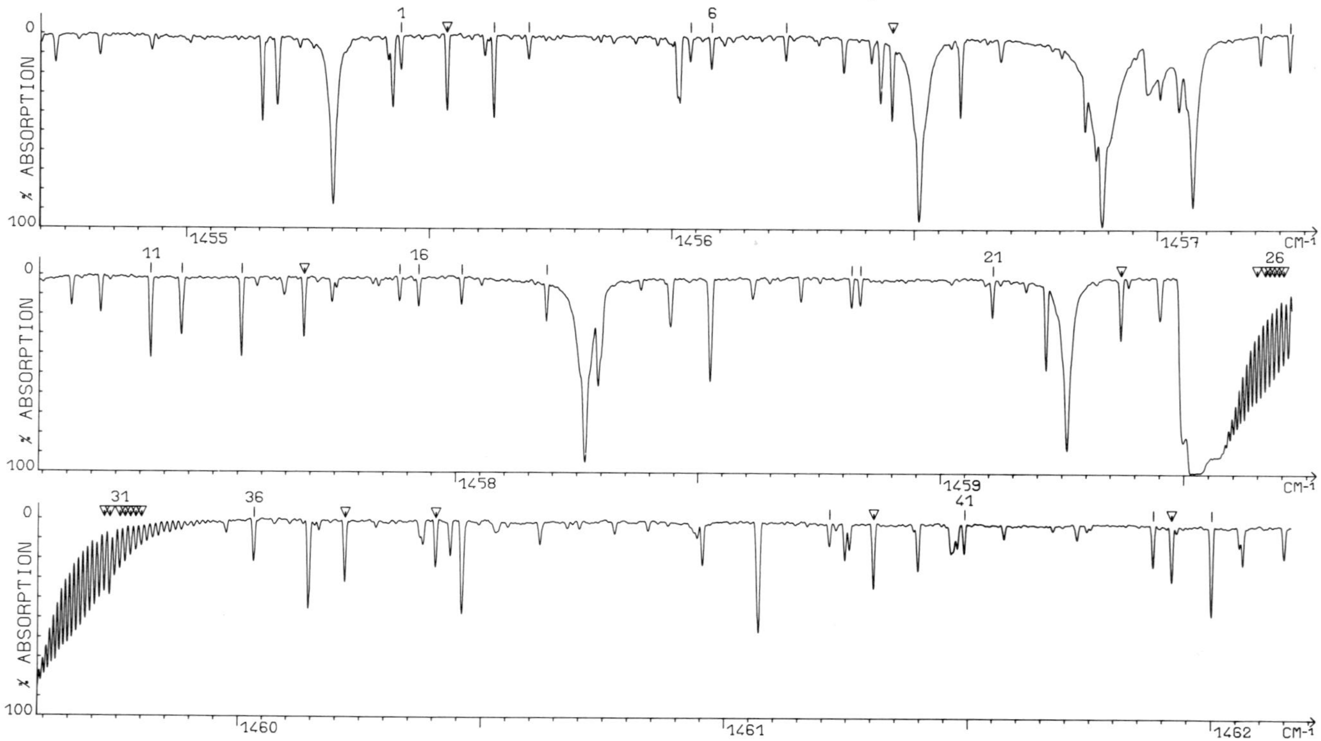

ISOTOPIC IDENTIFICATION – '∇' indicates $CH_3{}^{35}Cl$; 'I': Unidentified.

LINE NO.	WAVENUMBER CM^{-1} ;VAC.	ROTATIONAL ASSIGNMENT	BAND	ISOTOPIC SPECIES
1	1455.44142 *			
2	1455.53664	RP 1(8)	NU5	CH3 35C1
3	1455.63269 *			
4	1455.70453 *			
5	1456.03834 *			
6	1456.08196 *			
7	1456.23452 *			
8	1456.45315	RP 1(9)	NU5	CH3 35C1
9	1457.20961 *			
10	1457.26915 *			
11	1457.37237 *			
12	1457.43578 *			
13	1457.55865 *			
14	1457.68786	PR 1(16)	NU5	CH3 35C1
15	1457.88330 *			
16	1457.92334 *			
17	1458.01164 *			
18	1458.18703 *			
19	1458.81689 *			
20	1458.83472 *			
21	1459.10747 *			
22	1459.37167	PR 1(14)	NU5	CH3 35C1
23	1459.64848	QR O(33)	NU5	CH3 35C1
24	1459.66587	QR O(35)	NU5	CH3 35C1
25	1459.67488	QR O(36)	NU5	CH3 35C1
26	1459.68416	QR O(37)	NU5	CH3 35C1
27	1459.69359	QR O(38)	NU5	CH3 35C1
28	1459.70328	QR O(39)	NU5	CH3 35C1
29	1459.72310	QR O(41)	NU5	CH3 35C1
30	1459.73383	QR O(42)	NU5	CH3 35C1
31	1459.75436	QR O(44)	NU5	CH3 35C1
32	1459.76506	QR O(45)	NU5	CH3 35C1
33	1459.77607	QR O(46)	NU5	CH3 35C1
34	1459.78730	QR O(47)	NU5	CH3 35C1
35	1459.79873	QR O(48)	NU5	CH3 35C1
36	1460.03033 *			
37	1460.21833	PR 1(13)	NU5	CH3 35C1
38	1460.40546	RR O(0)	NU5	CH3 35C1
39	1461.21753 *			
40	1461.30812	RR O(1)	NU5	CH3 35C1
41	1461.49471 *			
42	1461.88326 *			
43	1461.92119	PR 1(11)	NU5	CH3 35C1
44	1462.00180 *			

- "LINE NO.": serial number identifying the spectral line in the map on the opposite page.
- "WAVENUMBER": observed value.
- * : all given lines are selected according to intensity, halfwidth and symmetry criteria.

Lines with an asterisk (*) are selected but not identified.

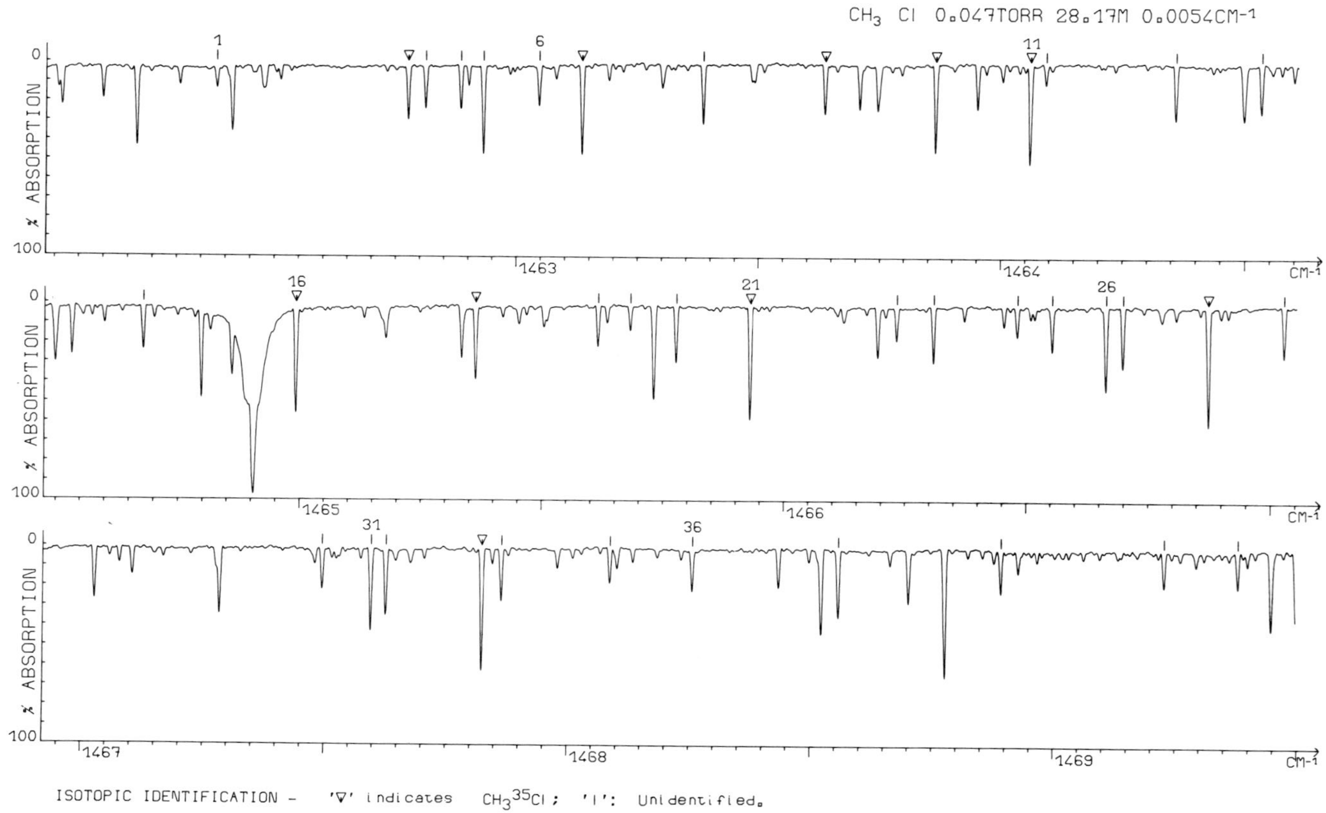
CH3 Cl 0.047TORR 28.17M 0.0054CM-1
1
6
11
16
21
26
31
36
0
% ABSORPTION
100
1463
1464
1465
1466
1467
1468
1469
CM-1
ISOTOPIC IDENTIFICATION – '▽' indicates CH3 35Cl ; '|': Unidentified.

LINE NO.	WAVENUMBER CM^{-1} ;VAC.		ROTATIONAL ASSIGNMENT	BAND	ISOTOPIC SPECIES
1	1462.38276	*			
2	1462.77730		PR 1(10)	NU5	CH3 35C1
3	1462.81311	*			
4	1462.88579	*			
5	1462.93298	*			
6	1463.04776	*			
7	1463.13599		RR O(3)	NU5	CH3 35C1
8	1463.38546	*			
9	1463.63655		PR 1(9)	NU5	CH3 35C1
10	1463.86527		RP 1(17)	NU5	CH3 35C1
11	1464.06067		RR O(4)	NU5	CH3 35C1
12	1464.09259	*			
13	1464.35928	*			
14	1464.53359	*			
15	1464.67929	*			
16	1464.99206		RR O(5)	NU5	CH3 35C1
17	1465.36306		PR 1(7)	NU5	CH3 35C1
18	1465.61553	*			
19	1465.68256	*			
20	1465.77682	*			
21	1465.92987		RR O(6)	NU5	CH3 35C1
22	1466.23288	*			
23	1466.30885	*			
24	1466.48133	*			
25	1466.55377	*			
26	1466.66395	*			
27	1466.69832	*			
28	1466.87376		RR O(7)	NU5	CH3 35C1
29	1467.02722	*			
30	1467.49396	*			
31	1467.59514	*			
32	1467.62570	*			
33	1467.82348		RR O(8)	NU5	CH3 35C1
34	1467.86408	*			
35	1468.08724	*			
36	1468.25730	*			
37	1468.55880	*			
38	1468.89425	*			
39	1469.22939	*			
40	1469.38002	*			

- "LINE NO.": serial number identifying the spectral line in the map on the opposite page.
- "WAVENUMBER": observed value.
- * : all given lines are selected according to intensity, halfwidth and symmetry criteria.

Lines with an asterisk (*) are selected but not identified.

CH_3 Cl 0.047TORR 28.17M 0.0054CM-1

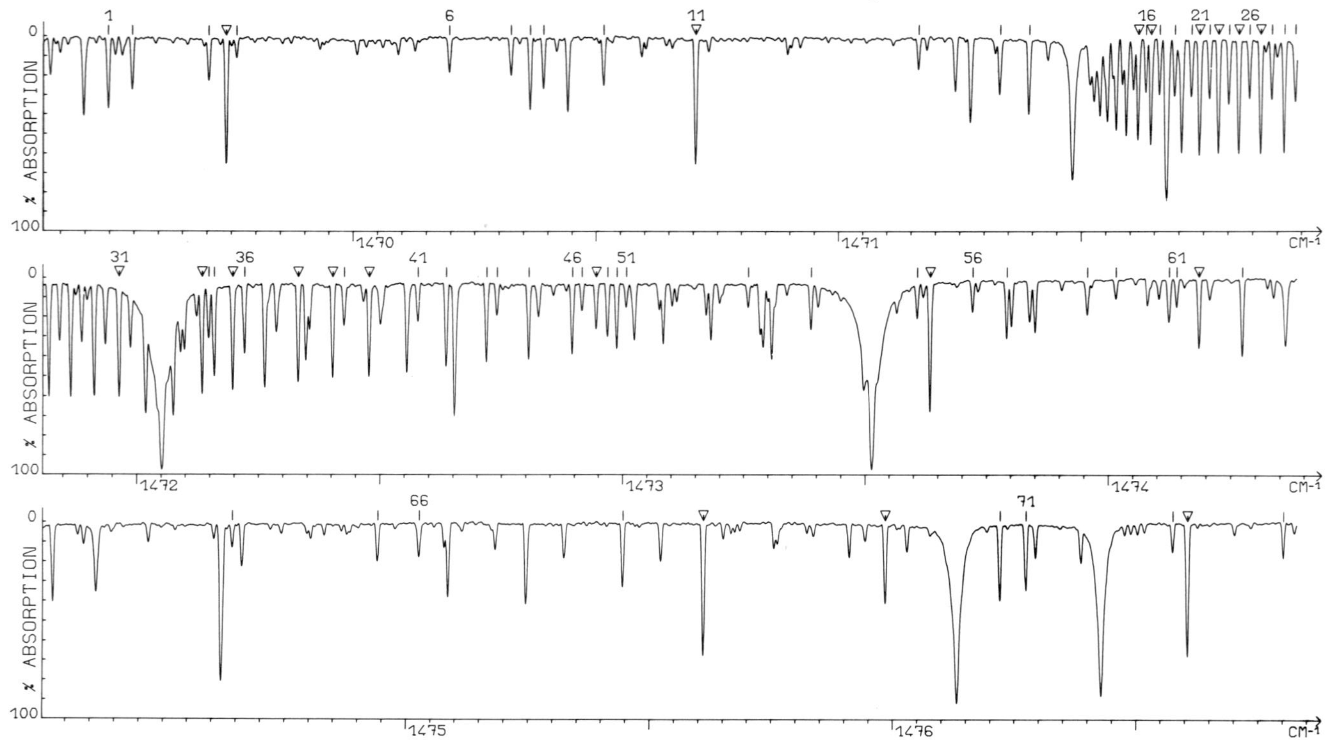

ISOTOPIC IDENTIFICATION – '∇' indicates $CH_3{}^{35}Cl$; 'I': Unidentified.

LINE NO.	WAVENUMBER CM^{-1} ;VAC.		ROTATIONAL ASSIGNMENT	BAND	ISOTOPIC SPECIES
1	1469.49766	*			
2	1469.54655	*			
3	1469.70379	*			
4	1469.73960		RR O(10)	NU5	CH3 35Cl
5	1469.76097	*			
6	1470.19861	*			
7	1470.32551	*			
8	1470.36454	*			
9	1470.39204	*			
10	1470.51625	*			
11	1470.70565		RR O(11)	NU5	CH3 35Cl
12	1471.16416	*			
13	1471.33136	*			
14	1471.39193	*			
15	1471.61588		QR 1(8)	NU5	CH3 35Cl
16	1471.63175	*			
17	1471.64280		QR 1(9)	NU5	CH3 35Cl
18	1471.66000	*			
19	1471.69101	*			
20	1471.72499	*			
21	1471.74149		QR 1(12)	NU5	CH3 35Cl
22	1471.76176	*			
23	1471.78014		QR 1(13)	NU5	CH3 35Cl
24	1471.80137	*			
25	1471.82190		QR 1(14)	NU5	CH3 35Cl
26	1471.84382	*			
27	1471.86664		QR 1(15)	NU5	CH3 35Cl
28	1471.88915	*			
29	1471.91437	*			
30	1471.93731	*			
31	1471.96505		QR 1(17)	NU5	CH3 35Cl
32	1472.13483		QR 1(20)	NU5	CH3 35Cl
33	1472.14856	*			
34	1472.15957	*			
35	1472.19747		QR 1(21)	NU5	CH3 35Cl
36	1472.22151	*			
37	1472.33156		QR 1(23)	NU5	CH3 35Cl
38	1472.40308		QR 1(24)	NU5	CH3 35Cl
39	1472.42644	*			
40	1472.47765		QR 1(25)	NU5	CH3 35Cl
41	1472.57767	*			
42	1472.63570	*			
43	1472.71921	*			
44	1472.74068	*			
45	1472.80580	*			
46	1472.89525	*			
47	1472.91539	*			
48	1472.94532		PR 2(12)	NU5	CH3 35Cl
49	1472.96876	*			
50	1472.98794	*			
51	1473.00725	*			
52	1473.25932	*			
53	1473.38871	*			
54	1473.60742	*			
55	1473.63427		RR O(14)	NU5	CH3 35Cl
56	1473.72131	*			
57	1473.79162	*			
58	1473.95743	*			
59	1474.01593	*			
60	1474.12518	*			

- "LINE NO.": serial number identifying the spectral line in the map on the opposite page.
- "WAVENUMBER": observed value.
- * : all given lines are selected according to intensity, halfwidth and symmetry criteria.

Lines with an asterisk (*) are selected but not identified.

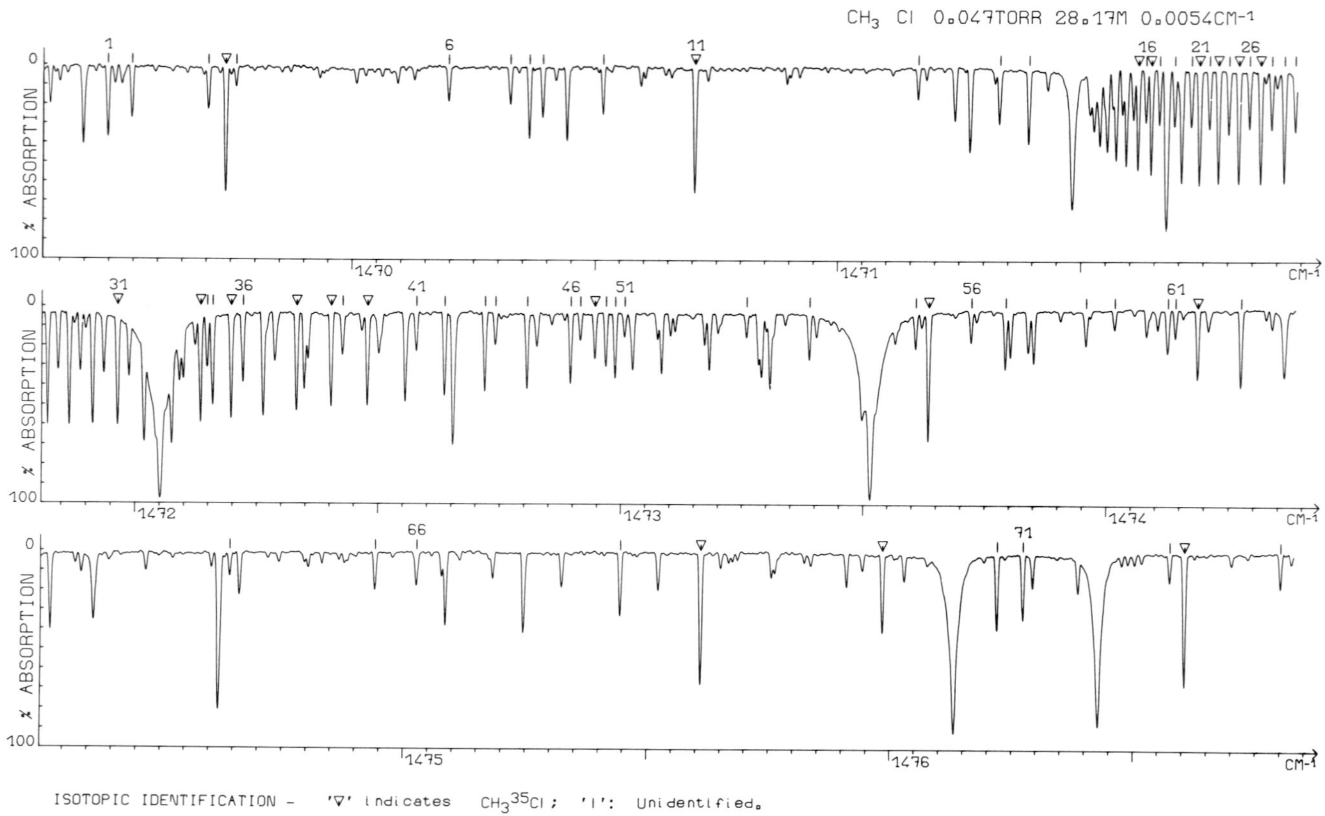
CH3 Cl 0.047TORR 28.17M 0.0054CM-1
% ABSORPTION
0
100
1470
1471
1472
1473
1474
1475
1476
CM-1
ISOTOPIC IDENTIFICATION - '▽' Indicates CH3 35Cl ; 'I': Unidentified.

LINE NO.	WAVENUMBER CM^{-1} ;VAC.		ROTATIONAL ASSIGNMENT	BAND	ISOTOPIC SPECIES
61	1474.14094	*			
62	1474.18651		RR 1(2)	NU5	CH3 35C1
63	1474.27430	*			
64	1474.64178	*			
65	1474.94046	*			
66	1475.02557	*			
67	1475.44532	*			
68	1475.61122		RR O(16)	NU5	CH3 35C1
69	1475.98677		RR 1(4)	NU5	CH3 35C1
70	1476.22205	*			
71	1476.27615	*			
72	1476.57699	*			
73	1476.60696		RR O(17)	NU5	CH3 35C1
74	1476.80232	*			

- "LINE NO.": serial number identifying the spectral line in the map on the opposite page.
- "WAVENUMBER": observed value.
- * : all given lines are selected according to intensity, halfwidth and symmetry criteria.

Lines with an asterisk (*) are selected but not identified.

CH_3 Cl 0.047TORR 28.17M 0.0054CM-1

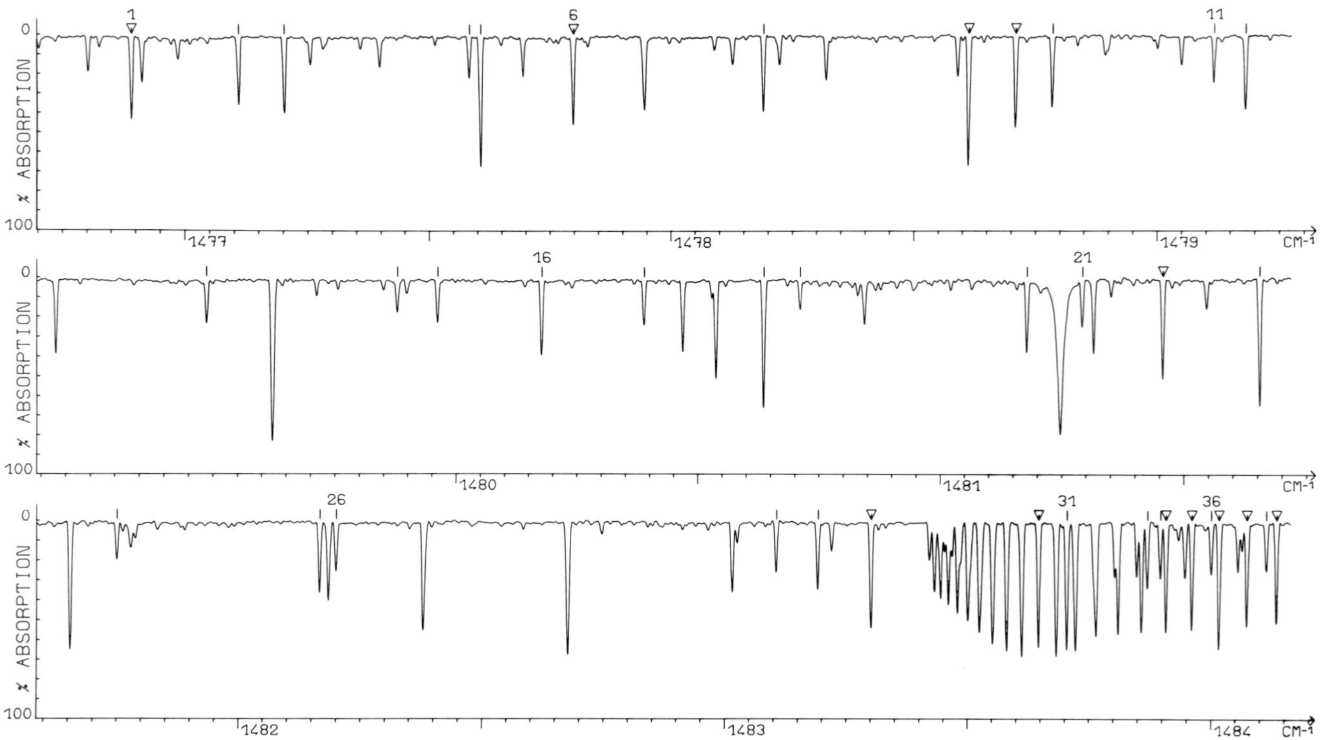

ISOTOPIC IDENTIFICATION – '∇' indicates $CH_3{}^{35}Cl$; '|': Unidentified.

LINE NO.	WAVENUMBER CM^{-1} ;VAC.		ROTATIONAL ASSIGNMENT	BAND	ISOTOPIC SPECIES
1	1476.89131		RR 1(5)	NU5	CH3 35Cl
2	1477.10959	*			
3	1477.20354	*			
4	1477.58371	*			
5	1477.60729	*			
6	1477.79870		RR 1(6)	NU5	CH3 35Cl
7	1478.19000	*			
8	1478.61230		RR 0(19)	NU5	CH3 35Cl
9	1478.70901		RR 1(7)	NU5	CH3 35Cl
10	1478.78427	*			
11	1479.11640	*			
12	1479.18138	*			
13	1479.48825	*			
14	1479.87928	*			
15	1479.96230	*			
16	1480.17757	*			
17	1480.38899	*			
18	1480.63622	*			
19	1480.71146	*			
20	1481.17841	*			
21	1481.29247	*			
22	1481.45704		RR 1(10)	NU5	CH3 35Cl
23	1481.65495	*			
24	1481.75041	*			
25	1482.16513	*			
26	1482.19858	*			
27	1483.10732	*			
28	1483.19380	*			
29	1483.30317		RR 1(12)	NU5	CH3 35Cl
30	1483.64804		QR 2(13)	NU5	CH3 35Cl
31	1483.70584	*			
32	1483.87100	*			
33	1483.89740	*			
34	1483.90893		QR 2(19)	NU5	CH3 35Cl
35	1483.96145		QR 2(20)	NU5	CH3 35Cl
36	1484.00066	*			
37	1484.01716		QR 2(21)	NU5	CH3 35Cl
38	1484.07424		QR 2(22)	NU5	CH3 35Cl
39	1484.11365	*			
40	1484.13442		QR 2(23)	NU5	CH3 35Cl

- "LINE NO.": serial number identifying the spectral line in the map on the opposite page.
- "WAVENUMBER": observed value.
- * : all given lines are selected according to intensity, halfwidth and symmetry criteria.

Lines with an asterisk (*) are selected but not identified.

CH_3 Cl 0.047TORR 28.17M 0.0054CM-1

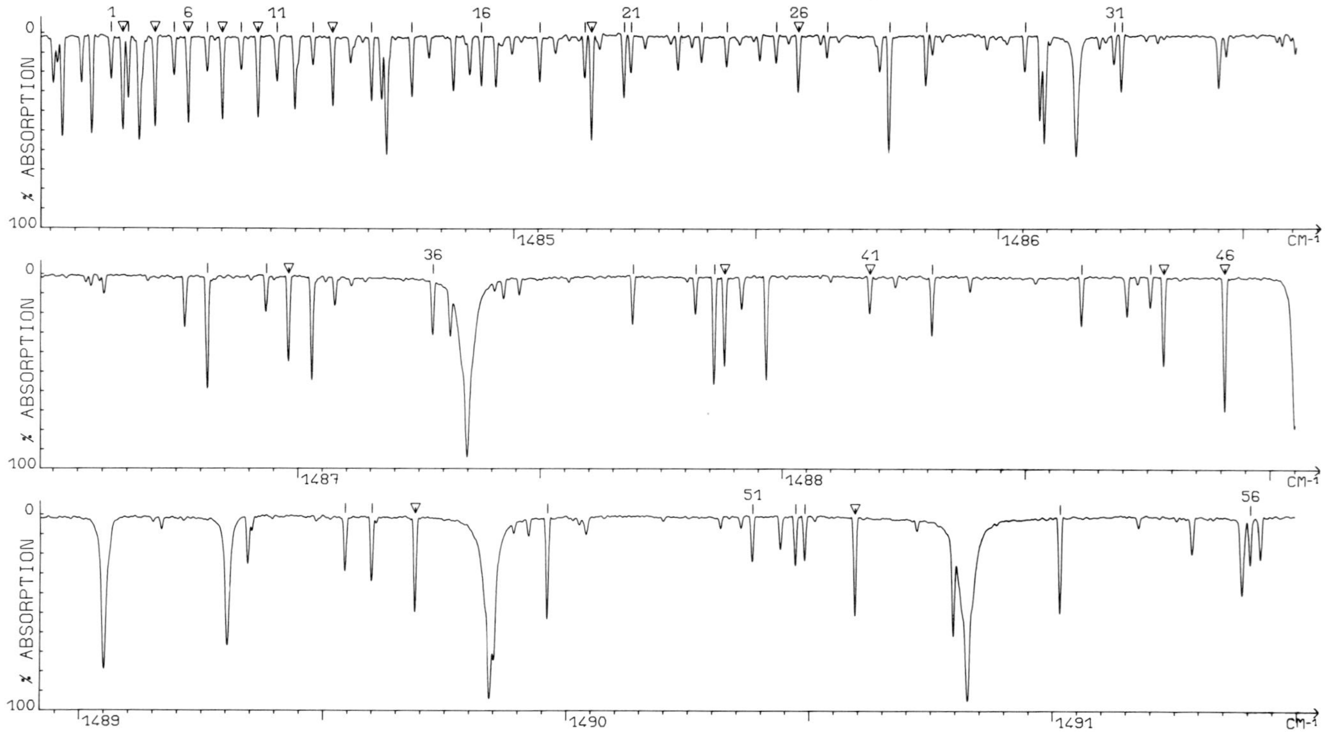

ISOTOPIC IDENTIFICATION – '▽' indicates $CH_3{}^{35}Cl$; 'I': Unidentified.

LINE NO.	WAVENUMBER CM^{-1} ;VAC.	ROTATIONAL ASSIGNMENT	BAND	ISOTOPIC SPECIES	LINE NO.	WAVENUMBER CM^{-1} ;VAC.	ROTATIONAL ASSIGNMENT	BAND	ISOTOPIC SPECIES
1	1484.17391 *				31	1486.23464 *			
2	1484.19725	QR 2(24)	NU5	CH3 35C1	32	1486.25014 *			
3	1484.20824 *				33	1486.81437 *			
4	1484.26254	QR 2(25)	NU5	CH3 35C1	34	1486.93493 *			
5	1484.30165 *				35	1486.98073	RR 2(3)	NU5	CH3 35C1
6	1484.33039	QR 2(26)	NU5	CH3 35C1	36	1487.27742 *			
7	1484.36931 *				37	1487.69111 *			
8	1484.40078	QR 2(27)	NU5	CH3 35C1	38	1487.82050 *			
9	1484.43940 *				39	1487.85899 *			
10	1484.47366	QR 2(28)	NU5	CH3 35C1	40	1487.88060	RR 2(4)	NU5	CH3 35C1
11	1484.51266 *				41	1488.18056	PR 3(8)	NU5	CH3 35C1
12	1484.58666 *				42	1488.30905 *			
13	1484.62702	QR 2(30)	NU5	CH3 35C1	43	1488.61579 *			
14	1484.70739 *				44	1488.75604 *			
15	1484.79028 *				45	1488.78304	RR 2(5)	NU5	CH3 35C1
16	1484.93298 *				46	1488.90775	RR 1(18)	NU5	CH3 35C1
17	1485.05369 *				47	1489.54319 *			
18	1485.14639 *				48	1489.59873 *			
19	1485.16037	RR 1(14)	NU5	CH3 35C1	49	1489.68804	RR 2(6)	NU5	CH3 35C1
20	1485.22705 *				50	1489.96036 *			
21	1485.24153 *				51	1490.38335 *			
22	1485.33905 *				52	1490.47291 *			
23	1485.38751 *				53	1490.49207 *			
24	1485.43907 *				54	1490.59550	RR 2(7)	NU5	CH3 35C1
25	1485.54145 *				55	1491.01702 *			
26	1485.58727	PR 3(11)	NU5	CH3 35C1	56	1491.40537 *			
27	1485.64605 *								
28	1485.77394 *								
29	1485.84972 *								
30	1486.05176 *								

- "LINE NO.": serial number identifying the spectral line in the map on the opposite page.
- "WAVENUMBER": observed value.
- * : all given lines are selected according to intensity, halfwidth and symmetry criteria.

Lines with an asterisk (*) are selected but not identified.

CH_3 Cl 0.047TORR 28.17M 0.0054CM-1

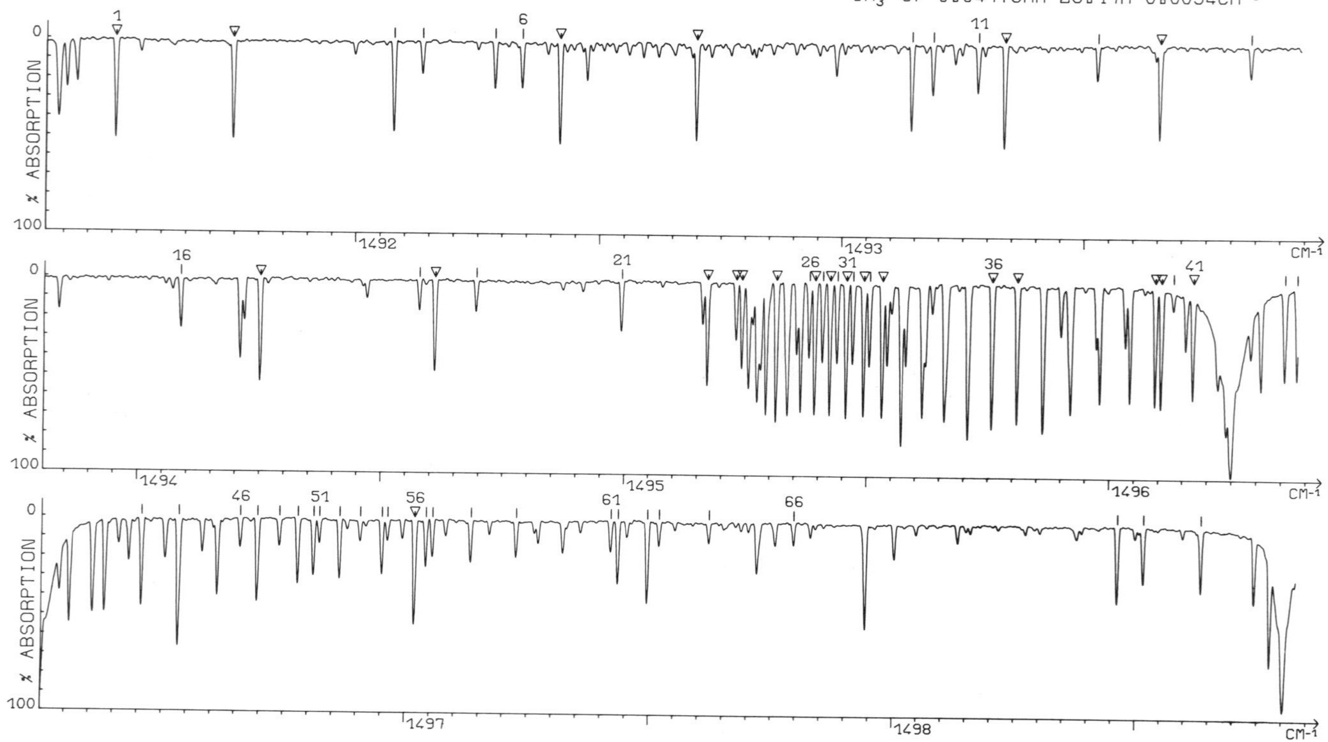

ISOTOPIC IDENTIFICATION – '∇' indicates $CH_3{}^{35}Cl$; 'I': Unidentified.

LINE NO.	WAVENUMBER CM^{-1} ;VAC.	ROTATIONAL ASSIGNMENT	BAND	ISOTOPIC SPECIES
1	1491.50562	RR 2(8)	NU5	CH3 35C1
2	1491.74674	RR 1(21)	NU5	CH3 35C1
3	1492.07753 *			
4	1492.13611 *			
5	1492.28521 *			
6	1492.34042 *			
7	1492.41815	RR 2(9)	NU5	CH3 35C1
8	1492.69846	RR 1(22)	NU5	CH3 35C1
9	1493.14183 *			
10	1493.18530 *			
11	1493.27799 *			
12	1493.33324	RR 2(10)	NU5	CH3 35C1
13	1493.52365 *			
14	1493.65271	RR 1(23)	NU5	CH3 35C1
15	1493.83878 *			
16	1494.08788 *			
17	1494.25072	RR 2(11)	NU5	CH3 35C1
18	1494.57764 *			
19	1494.60969	RR 1(24)	NU5	CH3 35C1
20	1494.69373 *			
21	1494.99276 *			
22	1495.17068	RR 2(12)	NU5	CH3 35C1
23	1495.22967	QR 3(4)	NU5	CH3 35C1
24	1495.24163	QR 3(5)	NU5	CH3 35C1
25	1495.31319	QR 3(9)	NU5	CH3 35C1
26	1495.38087 *			
27	1495.39268	QR 3(12)	NU5	CH3 35C1
28	1495.40847 *			
29	1495.42377	QR 3(13)	NU5	CH3 35C1
30	1495.43831 *			

LINE NO.	WAVENUMBER CM^{-1} ;VAC.	ROTATIONAL ASSIGNMENT	BAND	ISOTOPIC SPECIES
31	1495.45725	QR 3(14)	NU5	CH3 35C1
32	1495.47045 *			
33	1495.49312	QR 3(15)	NU5	CH3 35C1
34	1495.50489 *			
35	1495.53135	QR 3(16)	NU5	CH3 35C1
36	1495.75786	QR 3(21)	NU5	CH3 35C1
37	1495.80954	QR 3(22)	NU5	CH3 35C1
38	1496.09299	RR 2(13)	NU5	CH3 35C1
39	1496.10534	QR 3(27)	NU5	CH3 35C1
40	1496.12958 *			
41	1496.17105	QR 3(28)	NU5	CH3 35C1
42	1496.35746 *			
43	1496.38206 *			
44	1496.45684 *			
45	1496.53261 *			
46	1496.65937 *			
47	1496.69457 *			
48	1496.73963 *			
49	1496.77769 *			
50	1496.80973 *			
51	1496.82187 *			
52	1496.86315 *			
53	1496.90623 *			
54	1496.95084 *			
55	1496.96241 *			
56	1497.01784	RR 2(14)	NU5	CH3 35C1
57	1497.04060 *			
58	1497.05409 *			
59	1497.13249 *			
60	1497.22633 *			

- "LINE NO.": serial number identifying the spectral line in the map on the opposite page.
- "WAVENUMBER": observed value.
- * : all given lines are selected according to intensity, halfwidth and symmetry criteria.

Lines with an asterisk (*) are selected but not identified.

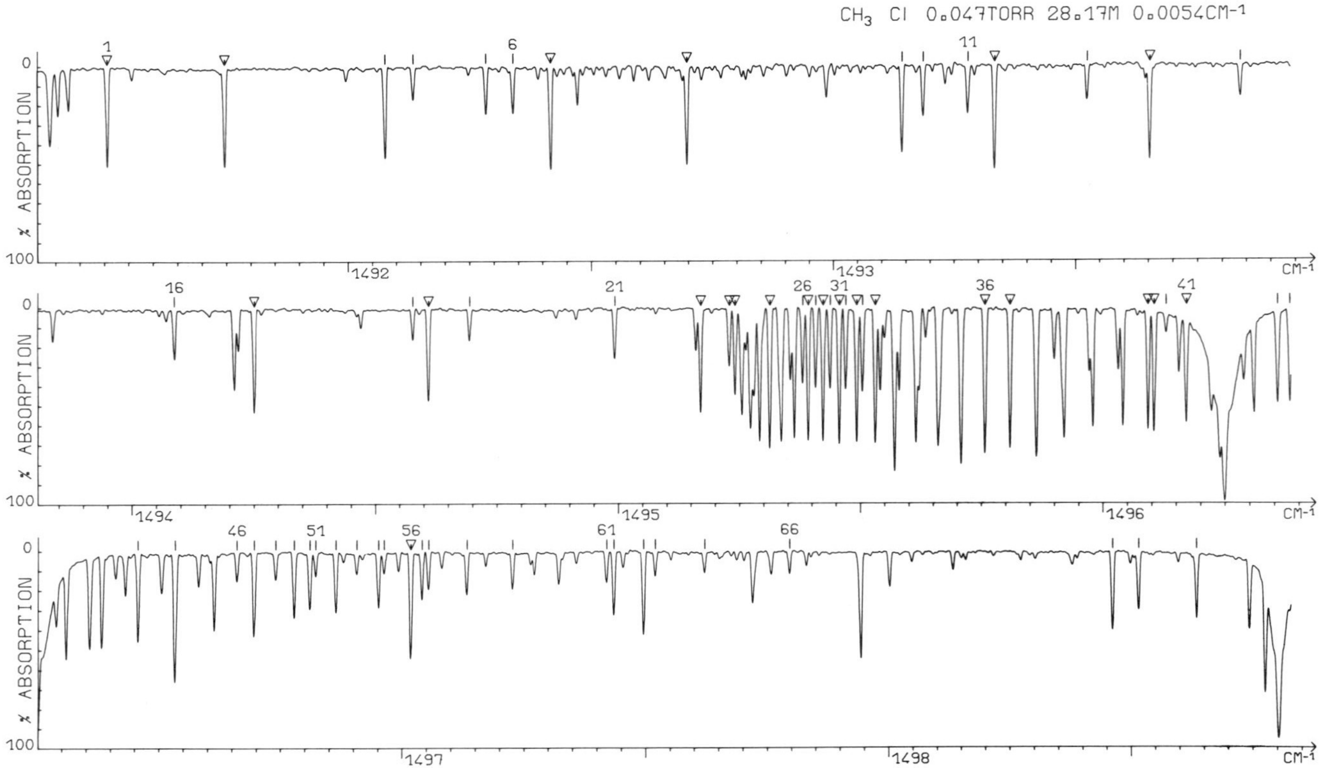

ISOTOPIC IDENTIFICATION - '▽' indicates $CH_3{}^{35}Cl$; 'I': Unidentified.

LINE NO.	WAVENUMBER CM^{-1} ;VAC.		ROTATIONAL ASSIGNMENT	BAND	ISOTOPIC SPECIES
61	1497.42029	*			
62	1497.43519	*			
63	1497.49636	*			
64	1497.52036	*			
65	1497.62250	*			
66	1497.79770	*			
67	1498.46380	*			
68	1498.51721	*			
69	1498.63518	*			

- "LINE NO.": serial number identifying the spectral line in the map on the opposite page.
- "WAVENUMBER": observed value.
- * : all given lines are selected according to intensity, halfwidth and symmetry criteria.

Lines with an asterisk (*) are selected but not identified.

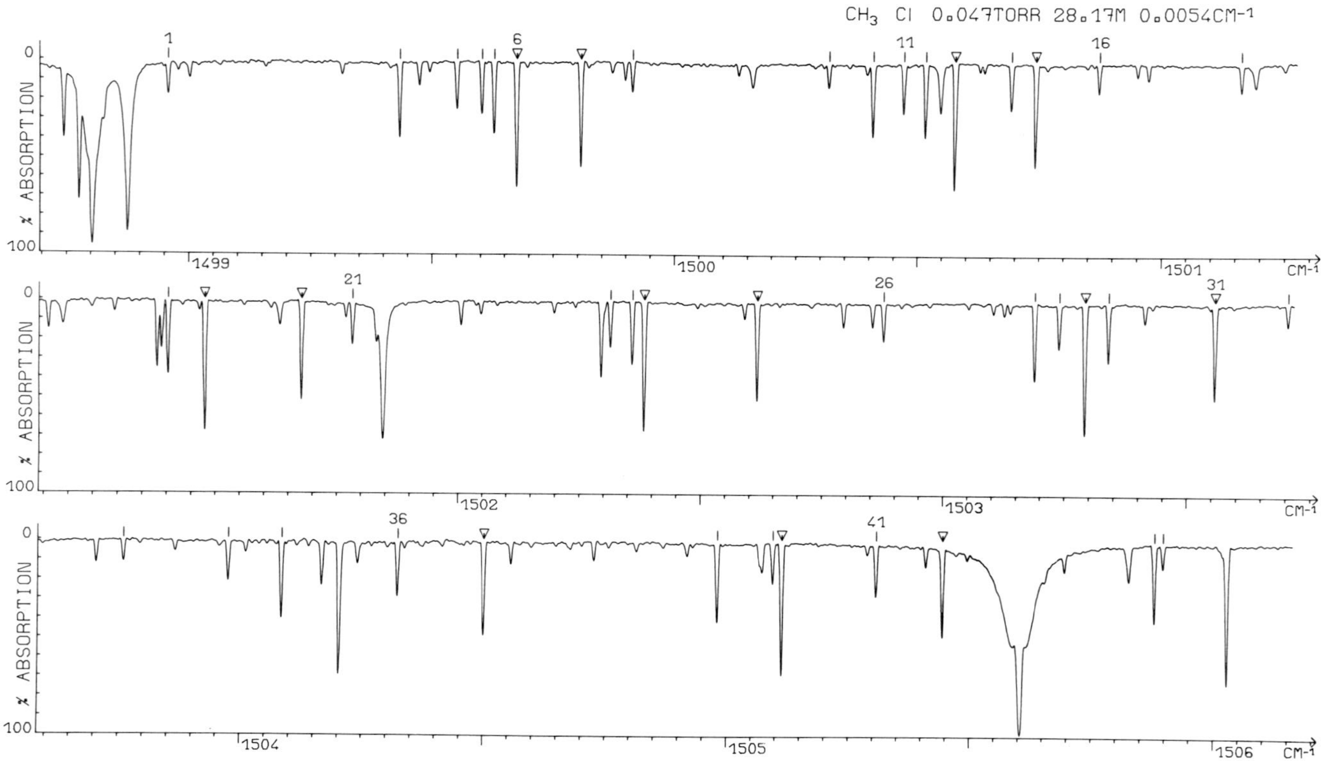

ISOTOPIC IDENTIFICATION – '▽' indicates $CH_3{}^{35}Cl$; 'I': Unidentified.

LINE NO.	WAVENUMBER CM^{-1} ;VAC.	ROTATIONAL ASSIGNMENT	BAND	ISOTOPIC SPECIES
1	1498.95734 *			
2	1499.43385 *			
3	1499.55161 *			
4	1499.60240 *			
5	1499.62842 *			
6	1499.67477	RR 3(4)	NU5	CH3 35Cl
7	1499.80642	RR 2(17)	NU5	CH3 35Cl
8	1499.91286 *			
9	1500.31592 *			
10	1500.40654 *			
11	1500.47014 *			
12	1500.51505 *			
13	1500.57564	RR 3(5)	NU5	CH3 35Cl
14	1500.69071 *			
15	1500.74067	RR 2(18)	NU5	CH3 35Cl
16	1500.87074 *			
17	1501.15992 *			
18	1501.40391 *			
19	1501.47882	RR 3(6)	NU5	CH3 35Cl
20	1501.67718	RR 2(19)	NU5	CH3 35Cl
21	1501.78224 *			
22	1502.31384 *			
23	1502.35957 *			
24	1502.38430	RR 3(7)	NU5	CH3 35Cl
25	1502.61600	RR 2(20)	NU5	CH3 35Cl
26	1502.87663 *			
27	1503.18817 *			
28	1503.23908 *			
29	1503.29202	RR 3(8)	NU5	CH3 35Cl
30	1503.33990 *			

LINE NO.	WAVENUMBER CM^{-1} ;VAC.	ROTATIONAL ASSIGNMENT	BAND	ISOTOPIC SPECIES
31	1503.55708	RR 2(21)	NU5	CH3 35Cl
32	1503.70440 *			
33	1503.75952 *			
34	1503.97425 *			
35	1504.08360 *			
36	1504.32285 *			
37	1504.50044	RR 2(22)	NU5	CH3 35Cl
38	1504.98110 *			
39	1505.09582 *			
40	1505.11428	RR 3(10)	NU5	CH3 35Cl
41	1505.30838 *			
42	1505.44588	RR 2(23)	NU5	CH3 35Cl
43	1505.88081 *			
44	1505.89837 *			

- "LINE NO.": serial number identifying the spectral line in the map on the opposite page.
- "WAVENUMBER": observed value.
- * : all given lines are selected according to intensity, halfwidth and symmetry criteria.

Lines with an asterisk (*) are selected but not identified.

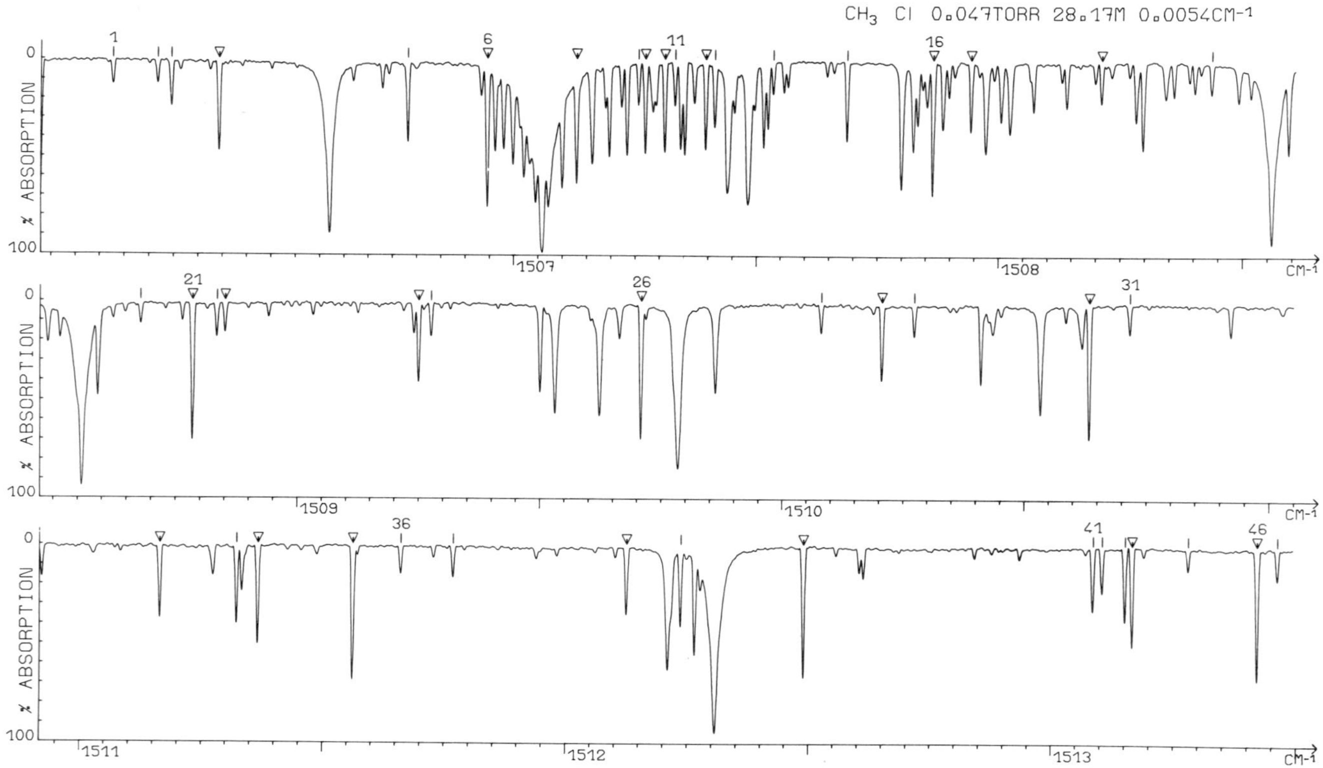

ISOTOPIC IDENTIFICATION – '∇' indicates $CH_3{}^{35}Cl$; 'I': Unidentified.

LINE NO.	WAVENUMBER CM^{-1} ;VAC.	ROTATIONAL ASSIGNMENT	BAND	ISOTOPIC SPECIES
1	1506.17701 *			
2	1506.26854 *			
3	1506.29627 *			
4	1506.39363	RR 2(24)	NU5	CH3 35C1
5	1506.78261 *			
6	1506.94560	QR 4(6)	NU5	CH3 35C1
7	1507.12906	QR 4(14)	NU5	CH3 35C1
8	1507.25641 *			
9	1507.27077	QR 4(18)	NU5	CH3 35C1
10	1507.31069	QR 4(19)	NU5	CH3 35C1
11	1507.33163 *			
12	1507.39475	QR 4(21)	NU5	CH3 35C1
13	1507.41351 *			
14	1507.53387 *			
15	1507.68648 *			
16	1507.86427	RR 3(13)	NU5	CH3 35C1
17	1507.94199	QR 4(30)	NU5	CH3 35C1
18	1508.20958	QR 4(34)	NU5	CH3 35C1
19	1508.43550 *			
20	1508.67968 *			
21	1508.78519	RR 3(14)	NU5	CH3 35C1
22	1508.83456 *			
23	1508.85226	PR 5(11)	NU5	CH3 35C1
24	1509.24973	RR 2(27)	NU5	CH3 35C1
25	1509.27530 *			
26	1509.70841	RR 3(15)	NU5	CH3 35C1
27	1510.08037 *			
28	1510.20599	RR 2(28)	NU5	CH3 35C1
29	1510.27329 *			
30	1510.63371	RR 3(16)	NU5	CH3 35C1

LINE NO.	WAVENUMBER CM^{-1} ;VAC.	ROTATIONAL ASSIGNMENT	BAND	ISOTOPIC SPECIES
31	1510.71620 *			
32	1511.16431	RR 2(29)	NU5	CH3 35C1
33	1511.32259 *			
34	1511.36613	RR 4(4)	NU5	CH3 35C1
35	1511.56105	RR 3(17)	NU5	CH3 35C1
36	1511.66002 *			
37	1511.76749 *			
38	1512.12471	RR 2(30)	NU5	CH3 35C1
39	1512.23671 *			
40	1512.49049	RR 3(18)	NU5	CH3 35C1
41	1513.08715 *			
42	1513.10612 *			
43	1513.15274 *			
44	1513.16731	RR 4(6)	NU5	CH3 35C1
45	1513.28194 *			
46	1513.42203	RR 3(19)	NU5	CH3 35C1
47	1513.46293 *			

- "LINE NO.": serial number identifying the spectral line in the map on the opposite page.
- "WAVENUMBER": observed value.
- * : all given lines are selected according to intensity, halfwidth and symmetry criteria.
Lines with an asterisk (*) are selected but not identified.

CH$_3$ Cl 0.047TORR 28.17M 0.0054CM-1

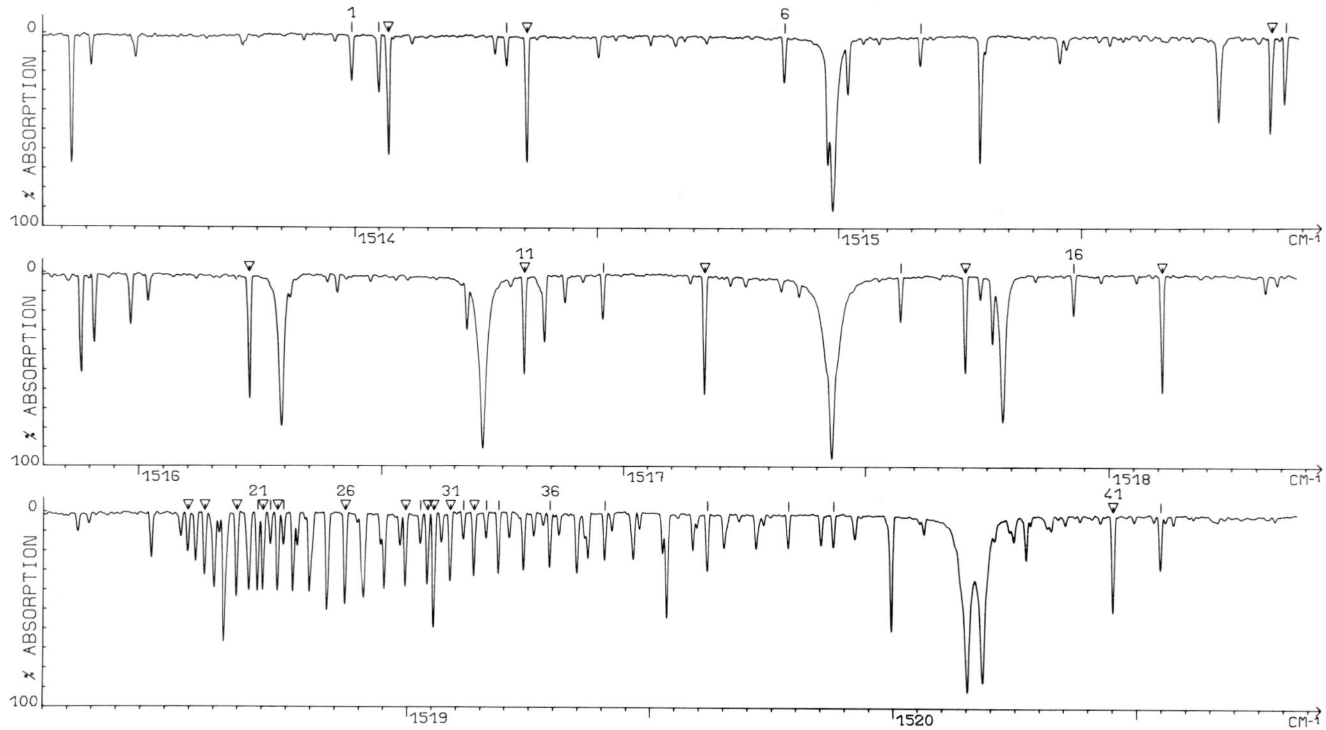

ISOTOPIC IDENTIFICATION – '▽' indicates CH$_3{}^{35}$Cl ; '|': Unidentified.

LINE NO.	WAVENUMBER CM^{-1} ;VAC.	ROTATIONAL ASSIGNMENT	BAND	ISOTOPIC SPECIES
1	1513.99551 *			
2	1514.05164 *			
3	1514.07091	RR 4(7)	NU5	CH3 35C1
4	1514.31382 *			
5	1514.35557	RR 3(20)	NU5	CH3 35C1
6	1514.88692 *			
7	1515.16667 *			
8	1515.88453	RR 4(9)	NU5	CH3 35C1
9	1515.91257 *			
10	1516.22866	RR 3(22)	NU5	CH3 35C1
11	1516.79431	RR 4(10)	NU5	CH3 35C1
12	1516.95672 *			
13	1517.16812	RR 3(23)	NU5	CH3 35C1
14	1517.57245 *			
15	1517.70603	RR 4(11)	NU5	CH3 35C1
16	1517.92893 *			
17	1518.10953	RR 3(24)	NU5	CH3 35C1
18	1518.54603	QR 5(7)	NU5	CH3 35C1
19	1518.58030	QR 5(9)	NU5	CH3 35C1
20	1518.64656	QR 5(12)	NU5	CH3 35C1
21	1518.68971 *			
22	1518.70070	QR 5(14)	NU5	CH3 35C1
23	1518.71573 *			
24	1518.73072	QR 5(15)	NU5	CH3 35C1
25	1518.74270 *			
26	1518.86973	QR 5(19)	NU5	CH3 35C1
27	1518.99436	QR 5(22)	NU5	CH3 35C1
28	1519.02494 *			
29	1519.03955	QR 5(23)	NU5	CH3 35C1
30	1519.05291	RR 3(25)	NU5	CH3 35C1
31	1519.08668	QR 5(24)	NU5	CH3 35C1
32	1519.11375 *			
33	1519.13551	QR 5(25)	NU5	CH3 35C1
34	1519.16064 *			
35	1519.18623 *			
36	1519.29291 *			
37	1519.40642 *			
38	1519.61857 *			
39	1519.78604 *			
40	1519.87901 *			
41	1520.45271	RR 4(14)	NU5	CH3 35C1
42	1520.54952 *			

- "LINE NO.": serial number identifying the spectral line in the map on the opposite page.
- "WAVENUMBER": observed value.
- * : all given lines are selected according to intensity, halfwidth and symmetry criteria.

Lines with an asterisk (*) are selected but not identified.

CH_3 Cl 0.047TORR 28.17M 0.0054CM-1

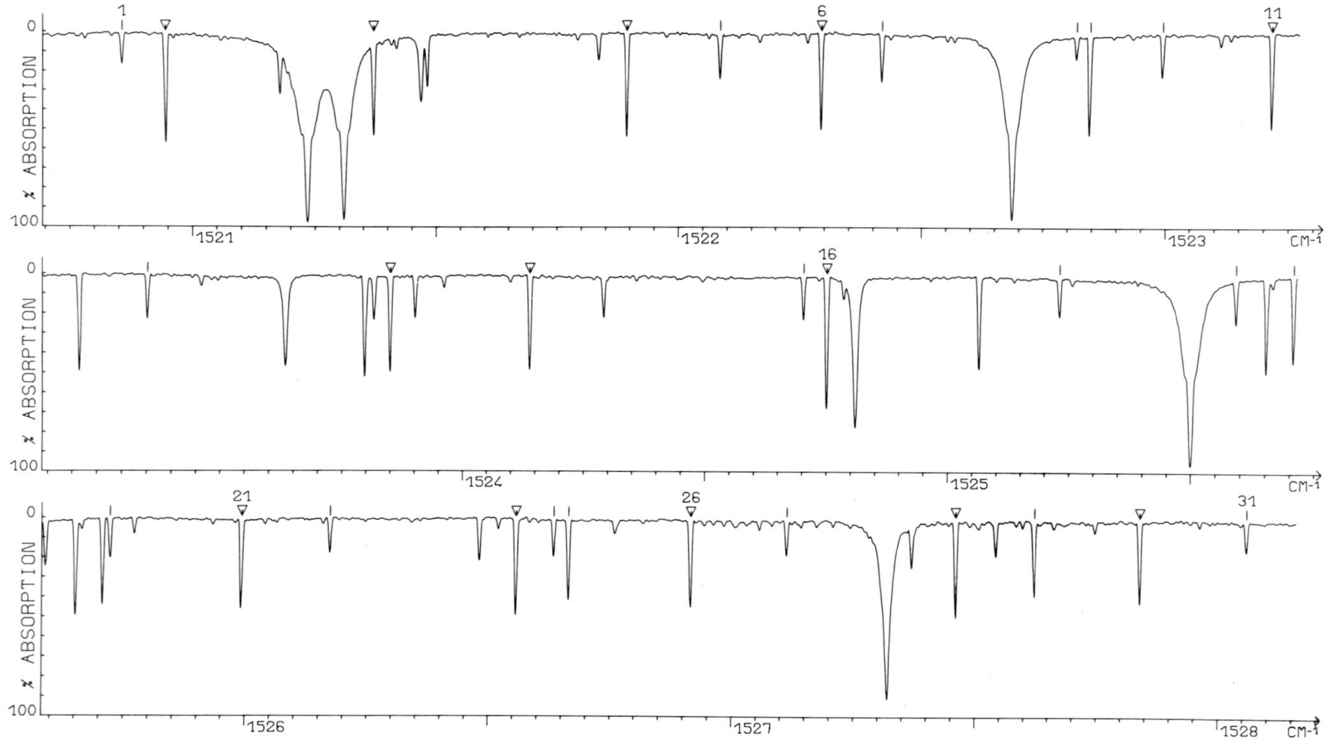

ISOTOPIC IDENTIFICATION – '▽' indicates $CH_3{}^{35}Cl$; 'I': Unidentified.

LINE NO.	WAVENUMBER CM^{-1} ;VAC.		ROTATIONAL ASSIGNMENT	BAND	ISOTOPIC SPECIES
1	1520.85675	*			
2	1520.94522		RR 3(27)	NU5	CH3 35Cl
3	1521.37188		RR 4(15)	NU5	CH3 35Cl
4	1521.89409		RR 3(28)	NU5	CH3 35Cl
5	1522.08492	*			
6	1522.29282		RR 4(16)	NU5	CH3 35Cl
7	1522.41695	*			
8	1522.81764	*			
9	1522.84484	*			
10	1522.99252	*			
11	1523.21534		RR 4(17)	NU5	CH3 35Cl
12	1523.35322	*			
13	1523.85029		RR 5(5)	NU5	CH3 35Cl
14	1524.13940		RR 4(18)	NU5	CH3 35Cl
15	1524.70315	*			
16	1524.75110		RR 5(6)	NU5	CH3 35Cl
17	1525.23076	*			
18	1525.59104	*			
19	1525.70759	*			
20	1525.72304	*			
21	1525.99085		RR 4(20)	NU5	CH3 35Cl
22	1526.17212	*			
23	1526.55688		RR 5(8)	NU5	CH3 35Cl
24	1526.63460	*			
25	1526.66521	*			
26	1526.91703		RR 4(21)	NU5	CH3 35Cl
27	1527.11484	*			
28	1527.46273		RR 5(9)	NU5	CH3 35Cl
29	1527.62457	*			
30	1527.84164		RR 4(22)	NU5	CH3 35Cl
31	1528.05930	*			

- "LINE NO.": serial number identifying the spectral line in the map on the opposite page.
- "WAVENUMBER": observed value.
- * : all given lines are selected according to intensity, halfwidth and symmetry criteria.

Lines with an asterisk (*) are selected but not identified.

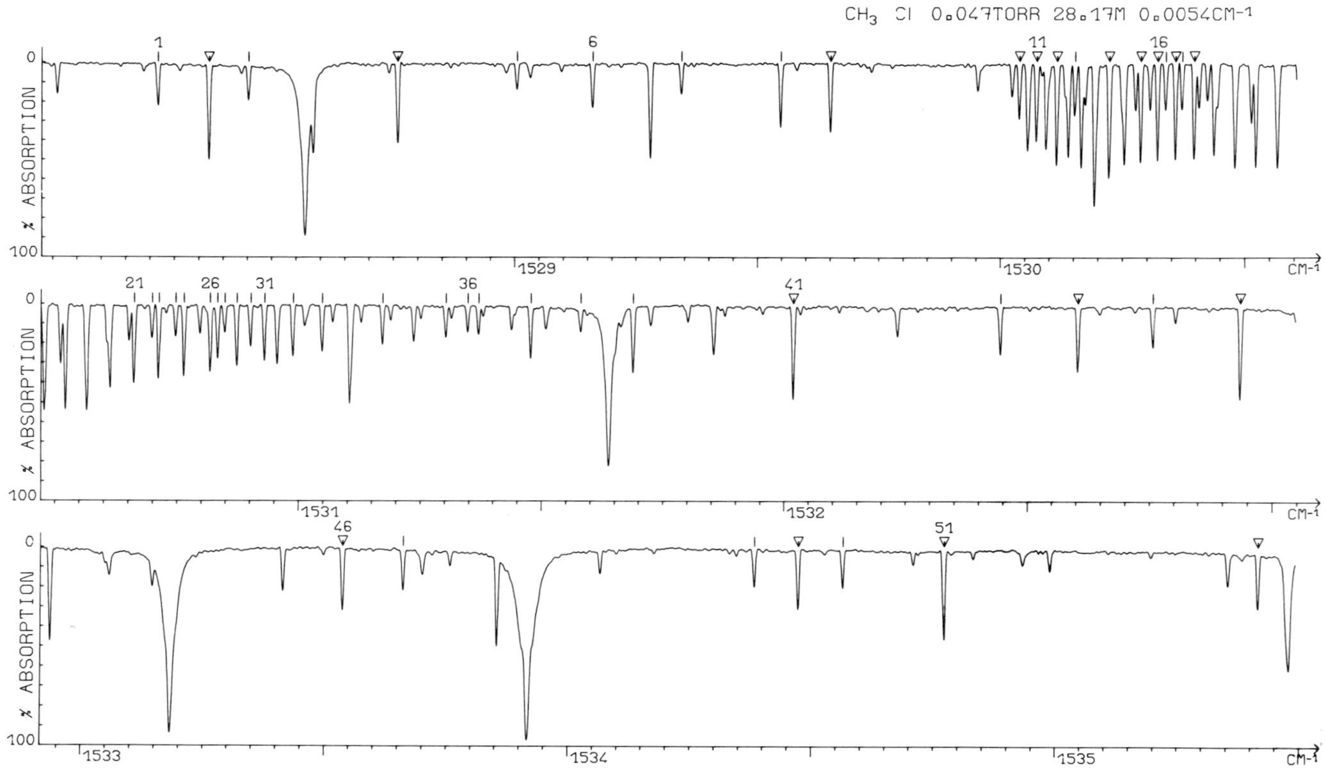

ISOTOPIC IDENTIFICATION – '▽' Indicates $CH_3{}^{35}Cl$; 'I': Unidentified.

LINE NO.	WAVENUMBER CM^{-1} ;VAC.	ROTATIONAL ASSIGNMENT	BAND	ISOTOPIC SPECIES
1	1528.26557 *			
2	1528.37053	RR 5(10)	NU5	CH3 35C1
3	1528.45217 *			
4	1528.75951	RR 4(23)	NU5	CH3 35C1
5	1529.00538 *			
6	1529.16062 *			
7	1529.34368 *			
8	1529.54817 *			
9	1529.65026	RR 4(24)	NU5	CH3 35C1
10	1530.03736	QR 6(8)	NU5	CH3 35C1
11	1530.07266	QR 6(10)	NU5	CH3 35C1
12	1530.11523	QR 6(12)	NU5	CH3 35C1
13	1530.15200 *			
14	1530.22212	QR 6(16)	NU5	CH3 35C1
15	1530.28668	QR 6(18)	NU5	CH3 35C1
16	1530.32151	QR 6(19)	NU5	CH3 35C1
17	1530.33789 *			
18	1530.35807	QR 6(20)	NU5	CH3 35C1
19	1530.37144 *			
20	1530.39648	QR 6(21)	NU5	CH3 35C1
21	1530.66264 *			
22	1530.69917 *			
23	1530.71288 *			
24	1530.74748 *			
25	1530.76472 *			
26	1530.81829 *			
27	1530.83426 *			
28	1530.84888 *			
29	1530.87347 *			
30	1530.90189 *			

LINE NO.	WAVENUMBER CM^{-1} ;VAC.	ROTATIONAL ASSIGNMENT	BAND	ISOTOPIC SPECIES
31	1530.93019 *			
32	1530.98847 *			
33	1531.04834 *			
34	1531.17268 *			
35	1531.30286 *			
36	1531.34770 *			
37	1531.37005 *			
38	1531.47795 *			
39	1531.58054 *			
40	1531.68930 *			
41	1532.01964	RR 5(14)	NU5	CH3 35C1
42	1532.44511 *			
43	1532.60469	RR 4(27)	NU5	CH3 35C1
44	1532.75780 *			
45	1532.93627	RR 5(15)	NU5	CH3 35C1
46	1533.53636	RR 4(28)	NU5	CH3 35C1
47	1533.66115 *			
48	1534.38407 *			
49	1534.47468	RR 4(29)	NU5	CH3 35C1
50	1534.56622 *			
51	1534.77468	RR 5(17)	NU5	CH3 35C1
52	1535.41708	RR 4(30)	NU5	CH3 35C1

- "LINE NO.": serial number identifying the spectral line in the map on the opposite page.
- "WAVENUMBER": observed value.
- * : all given lines are selected according to intensity, halfwidth and symmetry criteria.

Lines with an asterisk (*) are selected but not identified.

CH_3 Cl 0.047TORR 28.17M 0.0054CM-1

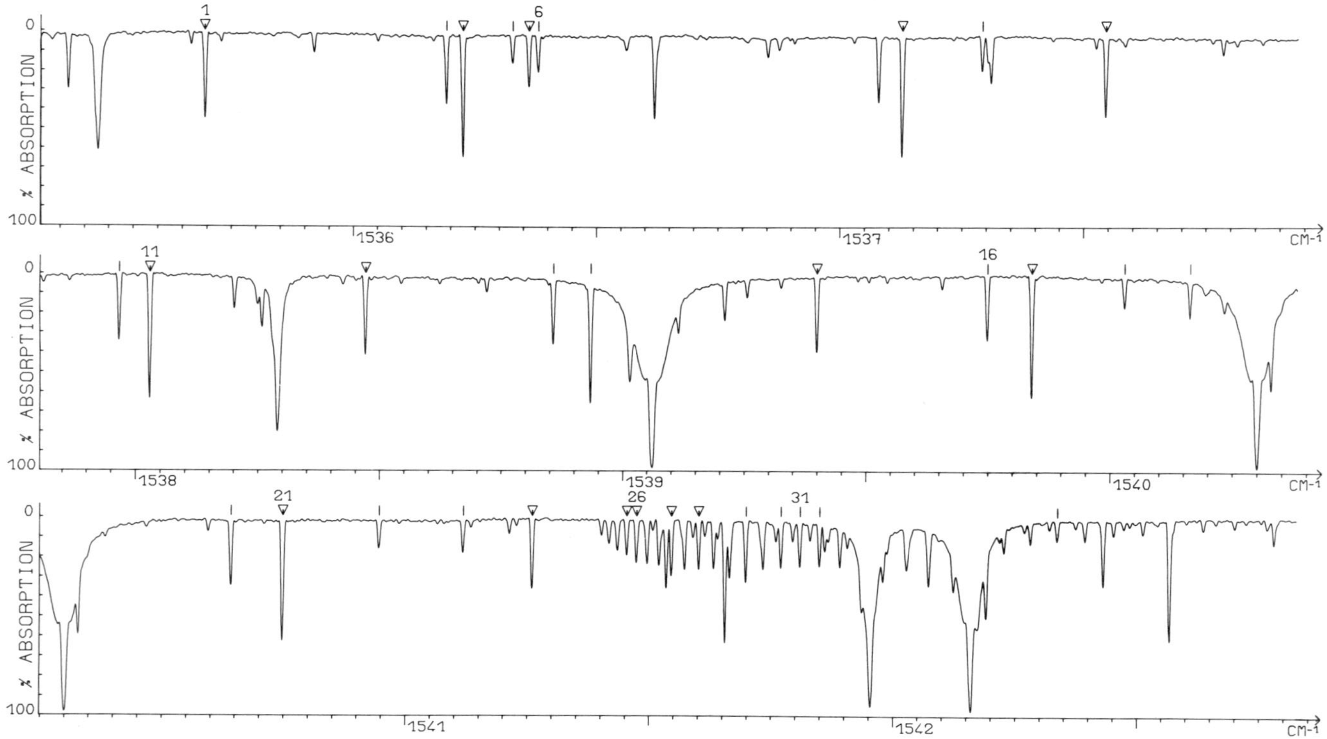

ISOTOPIC IDENTIFICATION – '▽' indicates $CH_3{}^{35}Cl$; 'I': Unidentified.

LINE NO.	WAVENUMBER CM^{-1} ;VAC.	ROTATIONAL ASSIGNMENT	BAND	ISOTOPIC SPECIES
1	1535.69637	RR 5(18)	NU5	CH3 35C1
2	1536.19221 *			
3	1536.22599	RR 6(6)	NU5	CH3 35C1
4	1536.32869 *			
5	1536.36214	RR 4(31)	NU5	CH3 35C1
6	1536.38118 *			
7	1537.12679	RR 6(7)	NU5	CH3 35C1
8	1537.29099 *			
9	1537.54463	RR 5(20)	NU5	CH3 35C1
10	1537.96698 *			
11	1538.02931	RR 6(8)	NU5	CH3 35C1
12	1538.47108	RR 5(21)	NU5	CH3 35C1
13	1538.85688 *			
14	1538.93357 *			
15	1539.39913	RR 5(22)	NU5	CH3 35C1
16	1539.74833 *			
17	1539.83947	RR 6(10)	NU5	CH3 35C1
18	1540.02911 *			
19	1540.16278 *			
20	1540.64141 *			
21	1540.74707	RR 6(11)	NU5	CH3 35C1
22	1540.94470 *			
23	1541.11703 *			
24	1541.25968	RR 5(24)	NU5	CH3 35C1
25	1541.45396	QR 7(11)	NU5	CH3 35C1
26	1541.47462	QR 7(12)	NU5	CH3 35C1
27	1541.54630	QR 7(15)	NU5	CH3 35C1
28	1541.60247	QR 7(17)	NU5	CH3 35C1
29	1541.69904 *			
30	1541.77179 *			
31	1541.81051 *			
32	1541.85078 *			
33	1542.33823 *			

- "LINE NO.": serial number identifying the spectral line in the map on the opposite page.
- "WAVENUMBER": observed value.
- * : all given lines are selected according to intensity, halfwidth and symmetry criteria.

Lines with an asterisk (*) are selected but not identified.

CH$_3$ Cl 0.047TORR 28.17M 0.0054CM-1

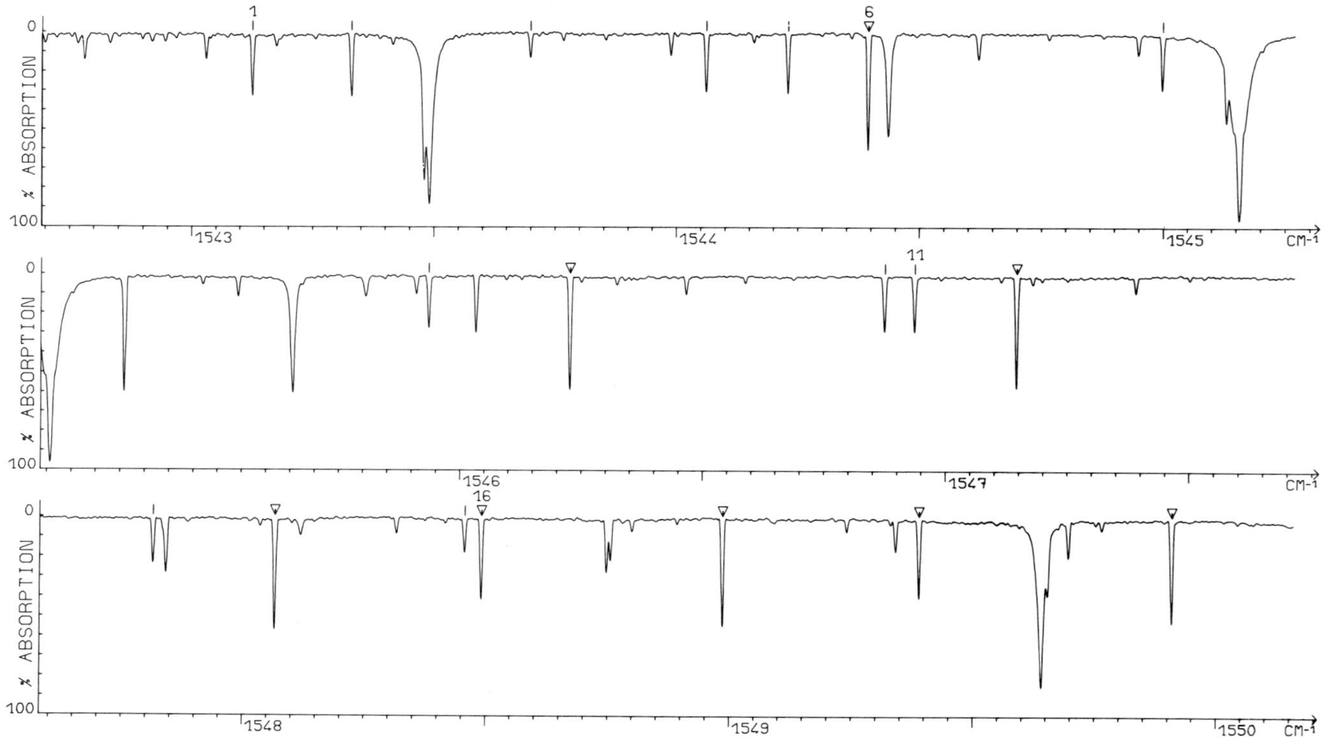

ISOTOPIC IDENTIFICATION – '▽' indicates CH$_3$^{35}Cl ; '|': Unidentified.

LINE NO.	WAVENUMBER CM^{-1} ;VAC.		ROTATIONAL ASSIGNMENT	BAND	ISOTOPIC SPECIES
1	1543.12599	*			
2	1543.33024	*			
3	1543.69966	*			
4	1544.06116	*			
5	1544.22947	*			
6	1544.39373		RR 6(15)	NU5	CH3 35Cl
7	1544.99774	*			
8	1545.93551	*			
9	1546.22652		RR 6(17)	NU5	CH3 35Cl
10	1546.87408	*			
11	1546.93628	*			
12	1547.14519		RR 6(18)	NU5	CH3 35Cl
13	1547.81481	*			
14	1548.06538		RR 6(19)	NU5	CH3 35Cl
15	1548.45598	*			
16	1548.49045		RR 7(7)	NU5	CH3 35Cl
17	1548.98700		RR 6(20)	NU5	CH3 35Cl
18	1549.39152		RR 7(8)	NU5	CH3 35Cl
19	1549.91005		RR 6(21)	NU5	CH3 35Cl

- "LINE NO.": serial number identifying the spectral line in the map on the opposite page.
- "WAVENUMBER": observed value.
- * : all given lines are selected according to intensity, halfwidth and symmetry criteria. Lines with an asterisk (*) are selected but not identified.

CH_3 Cl 0.047TORR 28.17M 0.0054CM-1

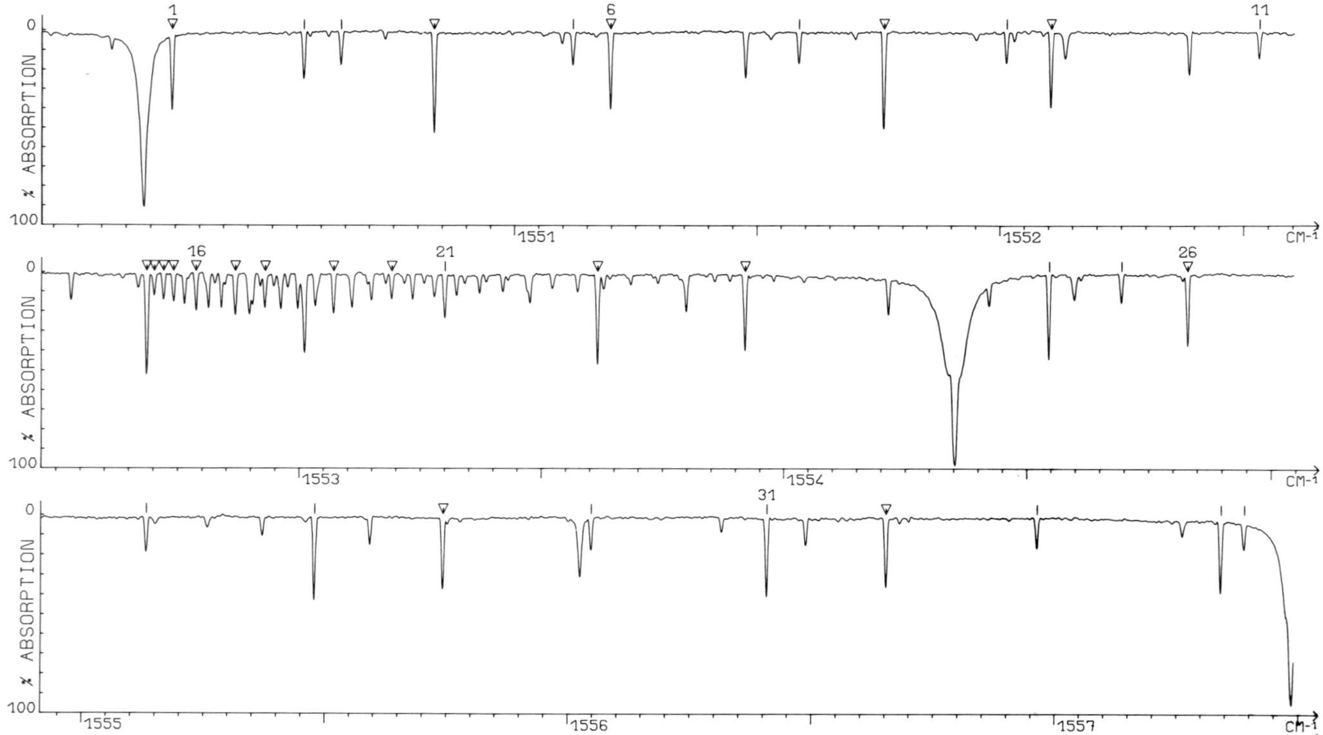

ISOTOPIC IDENTIFICATION – '∇' Indicates $CH_3{}^{35}Cl$; 'I': Unidentified.

LINE NO.	WAVENUMBER CM^{-1} ;VAC.	ROTATIONAL ASSIGNMENT	BAND	ISOTOPIC SPECIES
1	1550.29423	RR 7(9)	NU5	CH3 35C1
2	1550.56533 *			
3	1550.64185 *			
4	1550.83454	RR 6(22)	NU5	CH3 35C1
5	1551.12111 *			
6	1551.19841	RR 7(10)	NU5	CH3 35C1
7	1551.58609 *			
8	1551.76040	RR 6(23)	NU5	CH3 35C1
9	1552.01251 *			
10	1552.10426	RR 7(11)	NU5	CH3 35C1
11	1552.53127 *			
12	1552.68735	RR 6(24)	NU5	CH3 35C1
13	1552.70258	QR 8(11)	NU5	CH3 35C1
14	1552.72167	QR 8(12)	NU5	CH3 35C1
15	1552.74228	QR 8(13)	NU5	CH3 35C1
16	1552.78839	QR 8(15)	NU5	CH3 35C1
17	1552.86925	QR 8(18)	NU5	CH3 35C1
18	1552.93004	QR 8(20)	NU5	CH3 35C1
19	1553.07082	QR 8(24)	NU5	CH3 35C1
20	1553.19136	QR 8(27)	NU5	CH3 35C1
21	1553.30076 *			
22	1553.61614	RR 6(25)	NU5	CH3 35C1
23	1553.92021	RR 7(13)	NU5	CH3 35C1
24	1554.54597 *			
25	1554.69507 *			
26	1554.83054	RR 7(14)	NU5	CH3 35C1
27	1555.13123 *			
28	1555.47712 *			
29	1555.74221	RR 7(15)	NU5	CH3 35C1
30	1556.04810 *			
31	1556.40942 *			
32	1556.65535	RR 7(16)	NU5	CH3 35C1
33	1556.96604 *			
34	1557.34297 *			
35	1557.39028 *			

- "LINE NO.": serial number identifying the spectral line in the map on the opposite page.
- "WAVENUMBER": observed value.
- * : all given lines are selected according to intensity, halfwidth and symmetry criteria.
Lines with an asterisk (*) are selected but not identified.

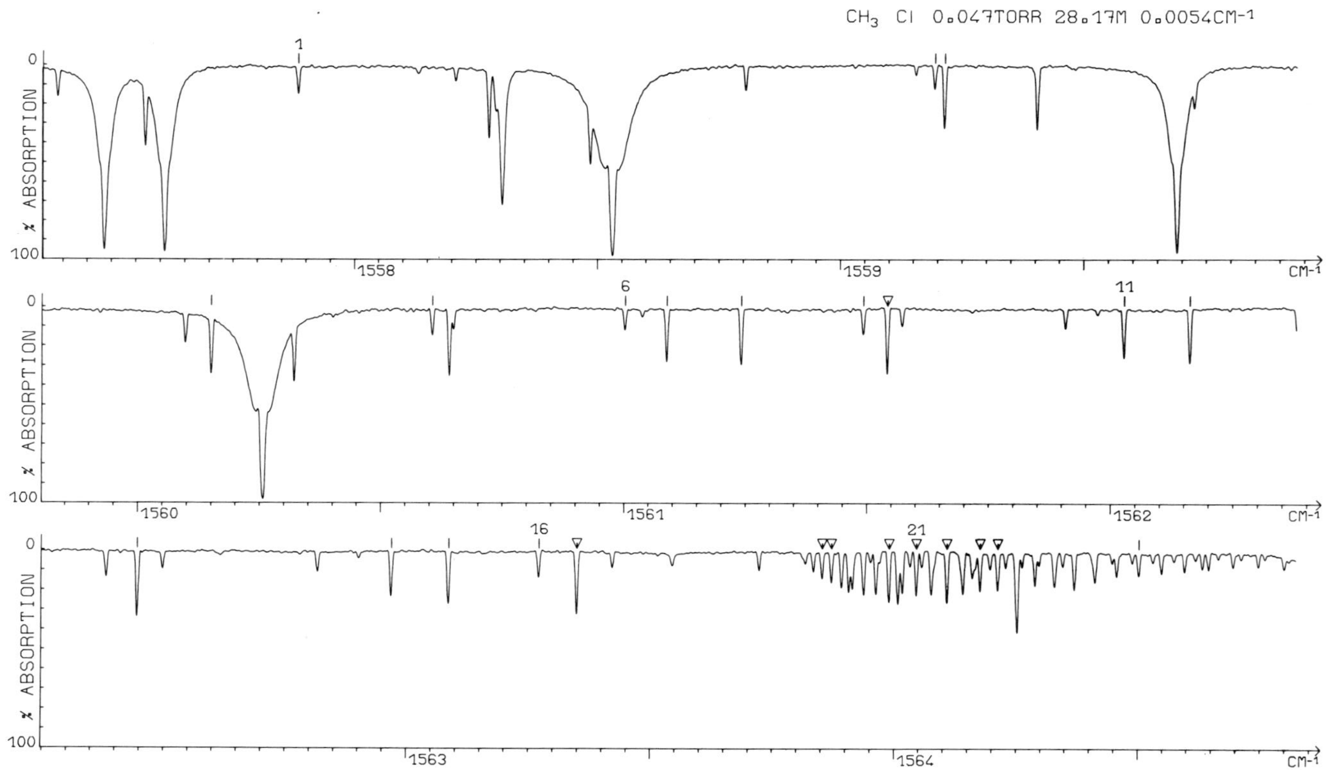

ISOTOPIC IDENTIFICATION – '∇' indicates $CH_3{}^{35}Cl$; 'I': Unidentified.

LINE NO.	WAVENUMBER CM^{-1} ;VAC.		ROTATIONAL ASSIGNMENT	BAND	ISOTOPIC SPECIES
1	1557.88529	*			
2	1559.19361	*			
3	1559.21358	*			
4	1560.15062	*			
5	1560.60605	*			
6	1561.00215	*			
7	1561.08857	*			
8	1561.24160	*			
9	1561.49298	*			
10	1561.54227		RR 8(9)	NU5	CH3 35Cl
11	1562.02762	*			
12	1562.16274	*			
13	1562.44489	*			
14	1562.96776	*			
15	1563.08518	*			
16	1563.27102	*			
17	1563.34902		RR 8(11)	NU5	CH3 35Cl
18	1563.85428		RQ 9(12)	NU5	CH3 35Cl
19	1563.87340		QR 9(13)	NU5	CH3 35Cl
20	1563.99107		QR 9(18)	NU5	CH3 35Cl
21	1564.04789		QR 9(20)	NU5	CH3 35Cl
22	1564.11041		QR 9(22)	NU5	CH3 35Cl
23	1564.17827		QR 9(24)	NU5	CH3 35Cl
24	1564.21431		RQ 9(25)	NU5	CH3 35Cl
25	1564.50339	*			

- "LINE NO.": serial number identifying the spectral line in the map on the opposite page.
- "WAVENUMBER": observed value.
- * : all given lines are selected according to intensity, halfwidth and symmetry criteria.

Lines with an asterisk (*) are selected but not identified.

CH_3 Cl 0.047TORR 28.17M 0.0054CM-1

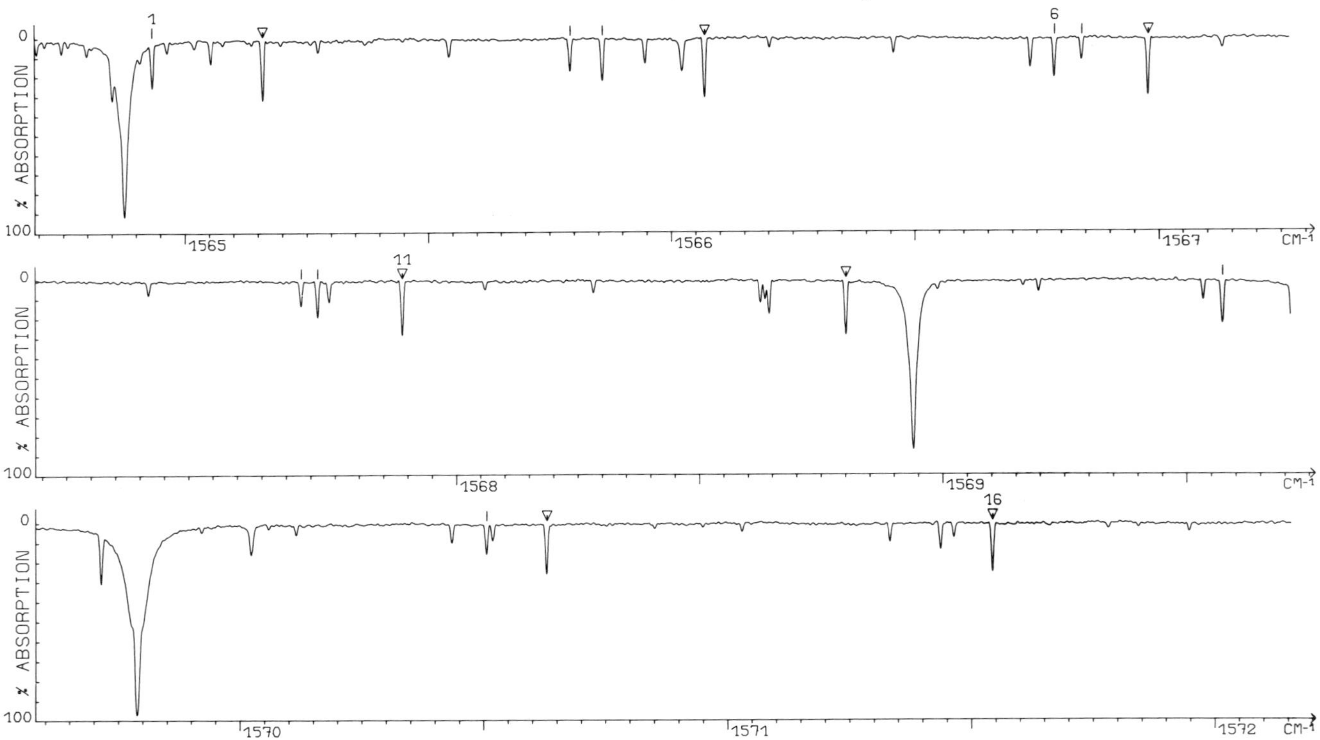

ISOTOPIC IDENTIFICATION – '∇' indicates $CH_3{}^{35}Cl$; 'I': Unidentified.

LINE NO.	WAVENUMBER CM^{-1} ;VAC.		ROTATIONAL ASSIGNMENT	BAND	ISOTOPIC SPECIES
1	1564.93376	*			
2	1565.16128		RR 8(13)	NU5	CH3 35C1
3	1565.79360	*			
4	1565.85979	*			
5	1566.06947		RR 8(14)	NU5	CH3 35C1
6	1566.78704	*			
7	1566.84283	*			
8	1566.97900		RR 8(15)	NU5	CH3 35C1
9	1567.68169	*			
10	1567.71531	*			
11	1567.88988		RR 8(16)	NU5	CH3 35C1
12	1568.80193		RR 8(17)	NU5	CH3 35C1
13	1569.57449	*			
14	1570.50673	*			
15	1570.62991		RR 8(19)	NU5	CH3 35C1
16	1571.54569		RR 8(20)	NU5	CH3 35C1

- "LINE NO.": serial number identifying the spectral line in the map on the opposite page.
- "WAVENUMBER": observed value.
- * : all given lines are selected according to intensity, halfwidth and symmetry criteria.

Lines with an asterisk (*) are selected but not identified.

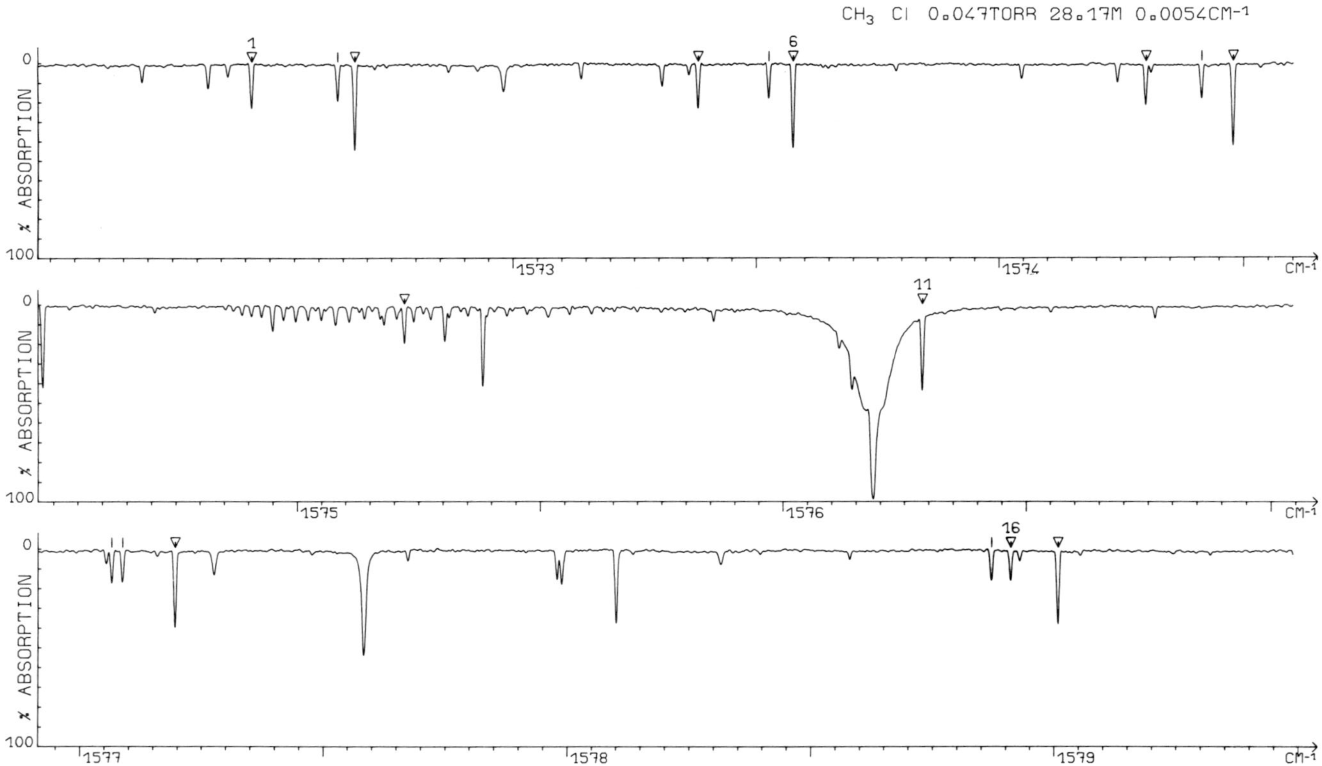

ISOTOPIC IDENTIFICATION – '∇' indicates $CH_3{}^{35}Cl$; 'I': Unidentified.

LINE NO.	WAVENUMBER CM^{-1} ;VAC.		ROTATIONAL ASSIGNMENT	BAND	ISOTOPIC SPECIES
1	1572.46271		RR 8(21)	NU5	CH3 35Cl
2	1572.63954	*			
3	1572.67520		RR 9(9)	NU5	CH3 35Cl
4	1573.38089		RR 8(22)	NU5	CH3 35Cl
5	1573.52647	*			
6	1573.57645		RR 9(10)	NU5	CH3 35Cl
7	1574.30022		RR 8(23)	NU5	CH3 35Cl
8	1574.41463	*			
9	1574.47893		RR 9(11)	NU5	CH3 35Cl
10	1575.22063		RR 8(24)	NU5	CH3 35Cl
11	1576.28777		RR 9(13)	NU5	CH3 35Cl
12	1577.06472	*			
13	1577.08639	*			
14	1577.19407		RR 9(14)	NU5	CH3 35Cl
15	1578.87348	*			
16	1578.91283		RR 8(28)	NU5	CH3 35Cl
17	1579.01026		RR 9(16)	NU5	CH3 35Cl

- "LINE NO.": serial number identifying the spectral line in the map on the opposite page.
- "WAVENUMBER": observed value.
- * : all given lines are selected according to intensity, halfwidth and symmetry criteria.

Lines with an asterisk (*) are selected but not identified.

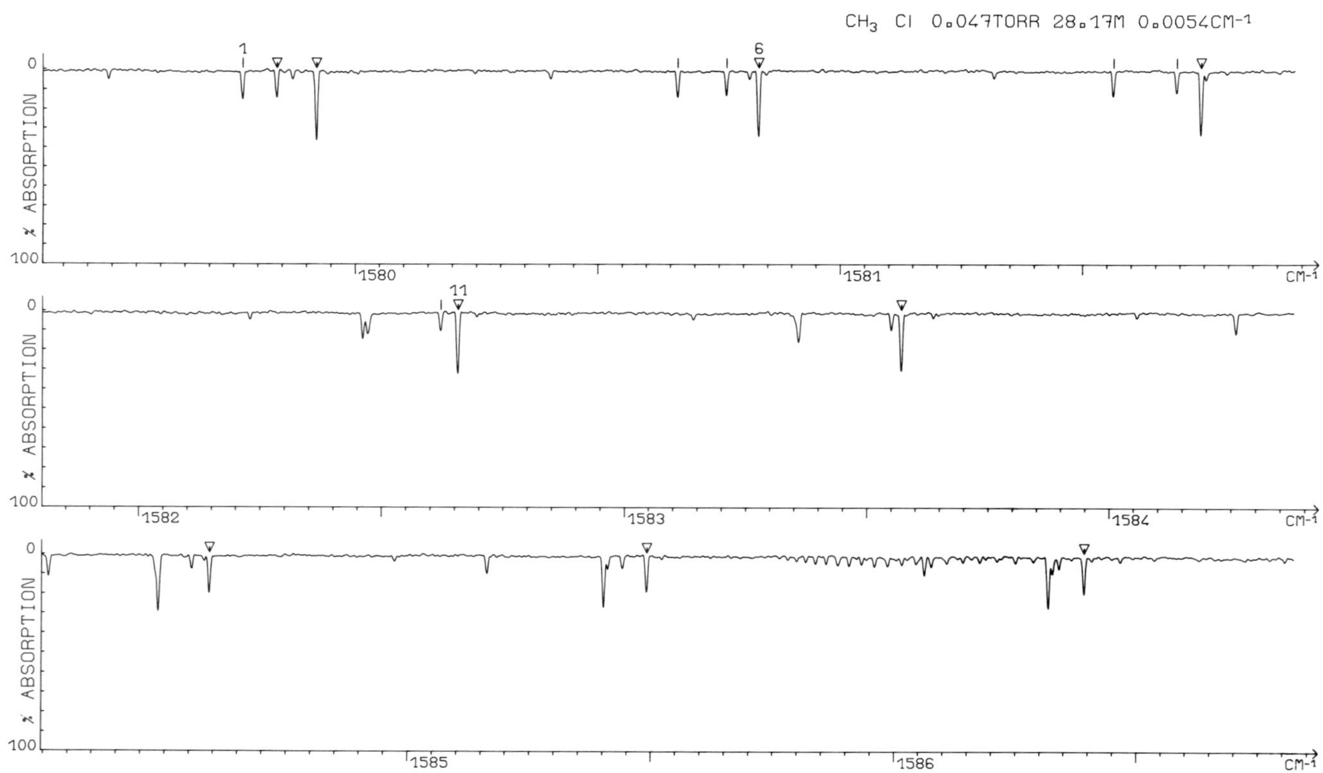

ISOTOPIC IDENTIFICATION – '▽' indicates $CH_3{}^{35}Cl$; 'I': Unidentified.

LINE NO.	WAVENUMBER CM^{-1} ;VAC.	ROTATIONAL ASSIGNMENT	BAND	ISOTOPIC SPECIES
1	1579.76862 *			
2	1579.83858	RR 8(29)	NU5	CH3 35C1
3	1579.92004	RR 9(17)	NU5	CH3 35C1
4	1580.66494 *			
5	1580.76511 *			
6	1580.83108	RR 9(18)	NU5	CH3 35C1
7	1581.56236 *			
8	1581.69244 *			
9	1581.74327	RR 9(19)	NU5	CH3 35C1
10	1582.62104 *			
11	1582.65653	RR 9(20)	NU5	CH3 35C1
12	1583.57075	RR 9(21)	NU5	CH3 35C1
13	1584.59061	RR10(10)	NU5	CH3 35C1
14	1585.49167	RR10(11)	NU5	CH3 35C1
15	1586.39380	RR10(12)	NU5	CH3 35C1

- "LINE NO.": serial number identifying the spectral line in the map on the opposite page.
- "WAVENUMBER": observed value.
- * : all given lines are selected according to intensity, halfwidth and symmetry criteria.

Lines with an asterisk (*) are selected but not identified.

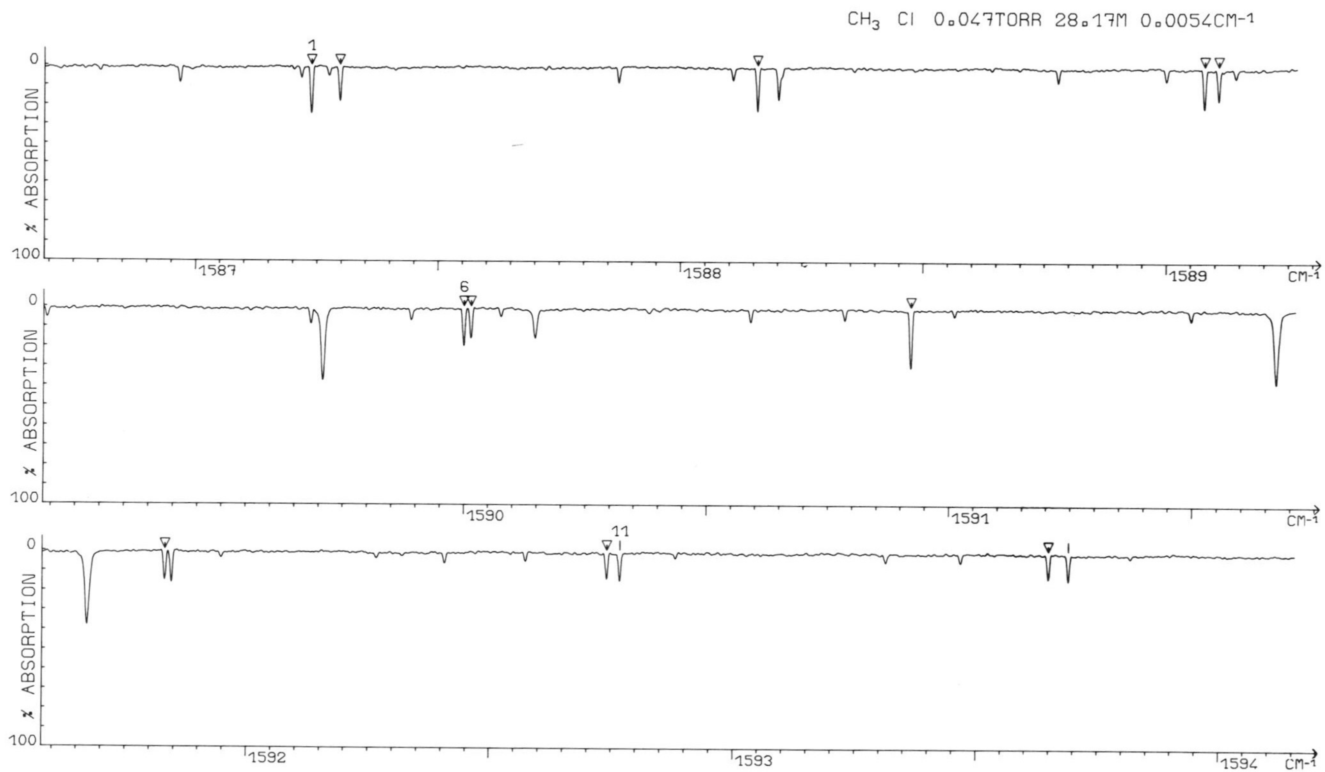

ISOTOPIC IDENTIFICATION – '▽' indicates $CH_3{}^{35}Cl$; 'I': Unidentified.

LINE NO.	WAVENUMBER CM^{-1} ;VAC.	ROTATIONAL ASSIGNMENT	BAND	ISOTOPIC SPECIES
1	1587.23835	RR 9(25)	NU5	CH3 35C1
2	1587.29711	RR10(13)	NU5	CH3 35C1
3	1588.15779	RR 9(26)	NU5	CH3 35C1
4	1589.07808	RR 9(27)	NU5	CH3 35C1
5	1589.10727	RR10(15)	NU5	CH3 35C1
6	1589.99925	RR 9(28)	NU5	CH3 35C1
7	1590.01397	RR10(16)	NU5	CH3 35C1
8	1590.92146	RR10(17)	NU5	CH3 35C1
9	1591.83050	RR10(18)	NU5	CH3 35C1
10	1592.74037	RR10(19)	NU5	CH3 35C1
11	1592.76791 *			
12	1593.65125	RR10(20)	NU5	CH3 35C1
13	1593.69241 *			

- "LINE NO.": serial number identifying the spectral line in the map on the opposite page.
- "WAVENUMBER": observed value.
- * : all given lines are selected according to intensity, halfwidth and symmetry criteria.

Lines with an asterisk (*) are selected but not identified.

ISBN 0-12-305362-5

90038

9 780123 053626